Neem and Environment

Volume 1

Editors
R.P. Singh • M.S. Chari
A.K. Raheja • W. Kraus

SCIENCE PUBLISHERS, INC.

SCIENCE PUBLISHERS, INC.
10 Water Street, # 310
Lebanon, New Hampshire 03766
United States of America

ISBN Set 1-886106-32-0
ISBN Vol. 1 1-886106-33-9
ISBN Vol. 2 1-886106-34-7

1-L6-SP.2

Foreword

For centuries, neem has been held in high esteem in India because of its medicinal and insecticidal properties. Neem tree, also known as margosa or the Indian lilac, grows extensively in many Asian and African countries. This multipurpose tree lives for 200 years and can grow well even under semi-arid and subhumid conditions up to 700 m above sea-level. It is tolerant to soil salinity, alkalinity and acidity and thrives well on all types of soils, besides establishing well even in the drought-prone areas. It is known to arrest desertification and soil erosion. It is used as a pesticide, nitrification inhibitor and is an effective component in the pharmaceuticals, and cosmetics. Research on neem is an excellent example of blending traditional wisdom with scientific knowledge.

The World Neem Conference was held for the first time in India at Bangalore from 24 to 28 February, 1993. I understand that this was the largest gathering on this subject, which comprised over 400 delegates from 30 countries who deliberated on the various aspects of neem. Research work done around the world in the last two decades has confirmed the immense value of neem. This tree will not only usher in a new era in pest management and medicine, but is also likely to play a significant role in preventing environmental degradation by reducing soil erosion and deforestation. Thus, the theme of the conference, 'Neem and Environment' had been in keeping with the spirit of the Earth Summit held at Rio de Janeiro in 1992.

It has been a matter of great privilege and pleasure for me to be associated with the hosting of the World Neem Conference in India. I congratulate the editors for their painstaking efforts in reviewing the papers presented at the Conference and bringing out the Proceedings. I am sure, the information contained in these Proceedings will be of immense value to all those concerned about the conservation and improvement of our environment. Indeed, the information is a valuable compilation of the available literature and knowledge on neem — the wonder tree of the World.

R.S. Paroda
Director General, ICAR and
Secretary, Department of Agricultural Research
and Education, Government of India

Preface

All efforts to increase the productivity of crops go waste if plants are not protected. Pests and diseases are important components of the adverse environmental factors. Putting of tonnes of toxic material in the environment to manage them is certainly not a right approach of solving pest problems. Cost development of resistance etc. are the other drawbacks associated with the use of synthetic pesticides. Neem (*Azadirachta indica* A. Juss.) has emerged as a single most important safe and cheap alternative to synthetic pesticides. We sincerely extend our compliments to the Indian Council of Agricultural Research and Indian Tobacco Company for sponsoring, and Indian Society of Tobacco Sciėnce for organising the World Neem Conference in the homeland of the tree, in 1993 at Bangalore.

The Conference received overwhelming response from all those who are directly or indirectly associated with neem. In the 5-day-deliberation, about 400 delegates from 30 countries exchanged their views on different aspects of neem. It was highly satisfying that almost all eminent researchers such as Prof. H. Schmutterer, Dr. R.C. Saxena, Prof. H. Rembold and Dr. Martin Rice participated in the conference and contributed papers.

Because of large number of papers presented in the Conference, it was decided to take out proceedings in two volumes. In preparing the articles, the editors have fully exercised their privilege of altering and reducing some of the manuscripts. It is likely that in doing so certain parts in the manuscript may have been altered unintentionally. For that editors tender their sincere apologies.

We express our deep sense of gratitude to Dr. Balram Jakhar, then Hon'ble Minister of Agriculture, Govt. of India, New Delhi for his wholehearted support and blessing for the Conference.

We express our sincere thanks to Mr. Verrapa Moily then Chief Minister of Karnataka for kindly inaugurating the Conference and Dr. V.L. Chopra, then Director-General, Indian Council of Agricultural Research, New Delhi and Secretary, DARE, Govt. of India, New Delhi for presiding over the Conference. We are also equally grateful to Shri G.K.P. Reddi, Director, ITL Ltd. for his keen interest and encouragement.

We sincerely thank Dr. R.S. Paroda, Chairman of the Organizing Committee for taking initiative in organizing the Conference and providing constant supervision and guidance.

The editors are grateful to Dr. T.S. Subramaniam, Senior Scientist, ITC Ltd. who played a very important role in conducting the conference seminar and providing all the necessary facilities to delegates.

The editor (Dr. R.P. Singh) is grateful to Prof. R.B. Singh, Director, Indian Agricultural Research Institute for providing facilities and permission for editing the proceedings.

We express thanks to the eminent persons of the Steering Committee and Members of the Executive Committee for providing guidance and help in organizing the Conference.

Finally we would like to express our great appreciation to all those who were directly or indirectly involved in making the Conference a success and delegates who responded to our invitation and participated in the Conference.

The Editors

WORLD NEEM CONFERENCE, FEB. 24–28, 1993, BANGALORE

Recommendations

I. *Sustainable Agriculture*

- While we need to produce more the environment is often harmed for immediate economic gains, sustainability of agro-ecosystem should, therefore, be the guiding policy for future research activity.
- Increased awareness of the potential of neem in agricultural sustainability — mass media, farming community and group meetings.
- Create favourable policy environment — at national, regional and international levels for careful and effective utilization of neem as sustainable substitute to synthetic agro-chemicals.

II. *Management of Pests and Diseases of Crops*

- For wider adoptibility and effective management, the pest control efficiency of neem-based pesticides should be evaluated for diverse biological effects using innovative bioassays and application methods against major insect pests and diseases of crops through on-farm trials.
- Relative safety to pests' natural enemies and other non-target organisms especially invertebrates needs to be determined.
- Evaluate the compatibility with synthetic pesticides especially plant derived pesticides and biocontrol agents.
- Encourage the traditional use of neem at farm level. Educate farmers about the advantages of using neem.
- Data on effect on seed germination, vegetative vigour phytotoxicity data etc. should be generated.
- Regulatory agencies should re-examine the existing pesticide registration requirements to encourage the use of neem formulations.
- Investigate systemic activity of bioactive principles and its effect on plant.
- Establish standard method for residue analysis of end product.
- Evaluate suppressive role of neem constituents on aflatoxins.

III. *Botany, Silviculture and Production*

- Genetic variability needs to be assessed keeping in view different utilization objectives especially bioactivity against pests.
- Understand neem biology and tree requirements to determine the adaptability of neem ecotypes/genotypes under diverse ecosystems, soil nutrients, etc.
- Develop techniques for seed collection preservation to increase the viability and deterioration of the active components.
- Investigate neem's nursery management including protection against pest and diseases.
- Investigate the value of monoculture versus mixed plantation with other economic trees including its role as wind breaker and in soil erosion.
- Promote neem cultivation especially in degraded land and investigate the impact of large-scale plantation on environment.

IV. *Neem Chemistry*

- Identify bioactive principles including non-terpenoids content from leaves, seed coat kernel, bark and flowers.
- Evaluate the biological effects of neem volatiles especially against stored grain pests.
- Investigate effect of temperature and photo-degradation on active compounds especially azadirachtin.
- Standardization of bio-assays to determine pest control effectiveness of neem-based products.
- To match synthetic pesticides develop as many neem formulations.
- Increase the shelf-life of neem formulations.
- Study structure-activity relationships of neem compounds.
- * Identify metabolic pathways to production of bioactive principles.

V. *Neem Industry; Commercialization*

- Standardization of neem products.
- Improved shelf-life and stability.
- Cost effectiveness.
- Manufacturers to submit pertinent data as per requirement of the government. Compatibility with other pesticides should be declared.
- Aflatoxin content to be established.
- Azadirachtin should be declared.

VI. *Neem in Human and Veterinary Medicines and Public Health*

- Evaluate antibacterial, antiviral and fungicidal activities of neem.
- Evaluate neem products against human, poultry livestock diseases.
- Evaluate neem products against vectors and parasites affecting human, livestock and poultry.
- Evaluate immunostimulating properties of neem.
- Evaluate neem for its possible application in population control.
- Test neem for its efficacy against human disease (malaria, tuberculosis, leprosy, etc.).

Contents

Foreword iii
Preface v

VOLUME 1

1. INVITED LECTURES AND GENERAL PAPERS

1. Neem and Its General Development for Pest Control 3
Heinz Rembold

2. Interspecific Differences in the Susceptibility of Noctuid Larvae to Behavioural and Physiological Effects of Azadirachtin 11
Murray B. Isman

3. Repellent Action of Neem Oil on Insects of Public Health Importance 19
V.P. Sharma

4. Neem and Transfer of Technology 27
M.S. Chari

5. The Use of Neem by Farmers in the Dominican Republic 39
Carsten Hellpap

6. Neem in Holland 47
Arnold A. Ruiter

7. Neem as a Safe Trend for Pest Control in the Future 51
Nadia Z. Dimetry

8. Neem Pest Control Research in West Africa 61
Joanny O. Zongo

9. Improving the Productivity of Neem Trees 69
Narayan G. Hegde

10. Neem for Vegetable IPM in Nicaragua: Experiments and Small Farmers' Perspective 81
D. Gómez, F. Miranda, B. Guerrero, I. Rivas, J. Mercado and F. Guharay

11. Potential for Neem (*Azadirachta indica* A. Juss.) in Armyworm Control in Africa 91
Paul B. Tanzubil

12. Strategies in Management of Insect Pests with Neem (*Azadirachta indica* A. Juss.) 103
R.P. Singh and A.K. Raheja

2. BOTANY

13. An Assay of Genetic Variability through Phenological Studies on Neem (*Azadirachta indica* A. Juss.) Plantation 123
M.G. Gogate and D.R. Gujar

14. Engineering Properties of Neem Nut 129
P.T. Palanisamy, R. Vishwanathan, S.S. Sivakumar, L. Gothandapani and K.R. Swaminathan

15. Location of Neem Triterpenoids and Other Secretory Structures 141
P. Dayanandan, A. Stephen and B. Muruganandam

16. Studies on the Resin Canals and Ducts of *Azadirachta indica* A. Juss. (Meliaceae) 151
K. Arumugasamy, K. Udaiyan, S. Manian, C. Balasubramanian and P.S. Mohan

17. Neem Seedlings — Growth Potentials of Different Organs 159
A. Stephen, B. Muruganandam, P. Ravichandran and P. Dayanandan

18. Variation Studies and Heritable Components of Seed Parameters in Neem (*Azadirachta indica* A. Juss.) 167
K. Kumaran, C. Surendran and R.S.V. Rai

19. *De novo* Synthesis of Secondary Metabolites in Cultured Cells of Neem 173
K. Ramesh and M.A. Padhya

3. NEEM CHEMISTRY

20. Chemistry of Neem (*Azadirachta indica*), a Sustainable Source of Natural Pesticides 187
Bina Shaheen Siddiqui, Shaheen Fazi, Ghiasuddin and Salimuzzaman Siddiqui

21. Photo-oxidiation of Teranortriterpenoids 199
S.R. Rojatkar, S.P. Joshi and B.A. Nagasampagi

22. New Tetra and Pentanortriterpenoids from *Azadirachta indica* A. Juss. 203
S.R. Rojatkar, D.D. Sawaikar and B.A. Nagasampagi

23. Azadirachtin Content and Bioactivity of Some Neem Ecotypes of India 207
Seeni Rengasamy, Nutan Kaushik, Jitendra Kumar, Opender Koul and Balraj Singh Parmar

4. PHYSIOLOGICAL AND BEHAVIOURAL EFFECTS

24. The Effects of Various Neem Formulations on Mortality Rate and Morphogenetic Defects Upon *Schistocerca gregaria* (Forskal) Larvae 221
H. Wilps, O. Nasseh and S. Krall

25. Antifeedant and Insecticidal Activity of Some Neem Fractions 237
M. Narayana Rao, K. Nagaiah and G. Srimannarayana

26. Antifeedant and Growth-inhibiting Effects of Azadirachtin-rich Neem Fractions on Rice White-backed Planthopper *Sogatella furcifera* Horvath, *Spodoptera litura* Fb. and *Helicoverpa armigera* Hb. 243
S.J. Nelson, P.C. Sundarababu, D.S. Rajvel, G. Sreemannarayana and Y. Geethanjali

27. Effects of Extracts from Neem on Aphids (Homoptera: Aphididae) and Their Natural Enemies 253
D. Thomas Lowery and Murray B. Isman

28. Differential Sensitivity of *Spodoptera litura* (Fabricius) (Lepidoptera: Noctuidae) Larvae to Multifarious Activity of Azadirachtin and Possible Implications in Integrated Pest Management 265
S.N. Mukherjee and R.N. Sharma

29. Effect of Neem-based Insecticides on Growth and Development of White-backed Planthopper, *Sogatella furcifera* (Horvath) 281
S.S. Sandhu, G.S. Dhaliwal and V.K. Dilawari

30. Toxic, Developmental and Reproductive Effects of AZT-VR-K and Some Neem Products against Mustard Aphid, *Lipaphis erysimi* (Kaltenbach) 287
Surinder S. Bhathal and Darshan Singh

31. Effects of Azadirachtin on Neural Regulation of Midgut Peristalsis in *Locusta migratoria* 297
A. Dorn and P. Trumm

32. Fate of Azadirachtin in the Variegated Cutworm, *Peridroma saucia* 307
Opender Koul

33. Effect of Neem Oil on Trehalase Activity in Accessory Reproductive Glands of the Male *Odontopus varicornis* (Dist.) Hemiptera: Pyrrhocoridae 315
N. Selvi Sabhanayakam, K. Ramalingam, Senthamil Selvi and R. Krishnakumari

34. Effect of Neem Kernel Extract and Neem Oil on Nutritive and Reproductive Physiology of *Heliothis armigera* Hub. 321
K. Murugan, P. Jahanmohini and R. Babu

35. Insecticide, Antfieeding and Juvenilizing Effects of Neem (*Azadirachta indica* A. Juss.) Oil against *Corcyra cephalonica* Staint. and *Epilachna vigintioctopunctata* F. 335
Y.K. Mathur and Suman Nigam

36. Olfactory, Oviposition and Developmental Response of Four Species of Pulse Beetles to Four Neem Oil Treated Legumes 343
T.D. Yadav

37. Effectivity of Certain Plant Extracts against Larvae of *Mythimna separata* Wlk. and Adults of *Callosobruchus chinensis* Linn. 349
T.G.M. Singh and T.B. Gour

38. Neem Effects on *Spodoptera litura* (Fabr.): A Holistic Study 357
P.J. Rao, Shashi Gupta, D. Mohan Raj and K.R. Kranthi

39. Effect of Neem Extracts on Brown Planthopper, *Nilaparvata lugens* and Its Symbionts 375
S. Raguraman, S. Jayaraj and R.C. Saxena

40. Identification of Effective and Inexpensive Neem (*Azadirachta indica* A. Juss.) Seed Kernel Extract for Management of *Helicoverpa armigera* (Hubner) 381
B.R. Jhansi and R.P. Singh

5. FIELD EVALUATION

41. Effects of Neem Oil on the Grasshopper *Zonocerus variegatus* L. in the Field 405
Martin Baumgart

42. Environmental Impact and Bioefficacy of Neem Products against Insect Pests in Rice 413
N.V. Krishnaiah and M.B. Kalode

43. Potential of Neem in Insect Pest Management in Rice 425
G.S. Dhaliwal, Jaswant Singh and V.K. Dilawari

44. Evaluation of Neem Products and Standard Chemicals for the Management of Bean Fly *Ophiomyia phaseoli* (Tryon) in French Beans 433
P.N. Krishna Moorthy and K. Srinivasan

45. Neem in Integrated Management of *Helicoverpa armigera* Hubn. Infesting Chickpea 439
S.N. Sinha

46. Evaluation of Neem Products and Other Standard Chemicals for the Management of Major Pest Complex on Cabbage: Comparison between Standard Spray Regime and I.P.M. Involving Mustard as a Trap Crop 447
K. Srinivasan and P.N. Krishna Moorthy

47. Neem in *Helicoverpa* [*Heliothis*] *armigera* Management: Pulses Ecosystem 459
J.N. Sachan and Gururaj Katti

48. Use of Neem in Management of Pest Complex in Cotton 469
G.P. Gupta

49. Field Evaluation of Some Botanical Insecticides Alone and in Combination with Other Insecticides for Management of Bollworm Complex on Cotton 485
Ashok K. Dhawan and G.S. Simwat

50. Evaluation of Botanical Insecticides Alone and in Combination with Carbaryl against Insect Pest Complex of Okra 493
S. Venkateswara Rao and B. Rosaiah

51. Evaluation of Neem Products against Insect Pests of Rice and Pulse Crops 497
B.C. Shukla, U.K. Kaushik and R. Gupta

52. Efficacy of Neem and *Melia* in Overcoming Tolerance of *Plutella xylostella* to Some Commonly Used Synthetic Insecticides 505
S. Facknath

53. Bioefficacy of Neem Derivatives against the Insect Pest Complex of Rice 517
U.K. Nanda, B. Parija, N.C. Pradhan, B. Nanda and D.D. Dash

54. Evaluation of Neem Products for the Control of Vectors of Japanese Encephalitis Virus in Rice Fields — A Five-year Study (1987–1992) 529
D. Raghunatha Rao, R. Reuben and C. Hellpap

55. The Performance of an IPM Package Including Neem for the Control of Potato Pests in Farmers Fields 541
S.A. Siddiq and A. Khalafalla

56. Potential of Neem for Controlling Leaf-eating Caterpillar Complex on Cabbage 547
K.A.N.P. Bandara and C. Kudagamage

57. Neem Oil as Possible Biorational Insecticide for the Management of Spotted Pod Borer (*Maruca testulalis* Geyer) in Early Pigeon Pea 559
R.K. Akhauri, M.M. Sinha and R.P. Yadav

58. Effects of the Extracted Substances from Thai Neem Seed, *Azadirachta indica* var. *siamensis*, on the Insect Pests of Chinese Kale 565
Kwanchai Sombatsiri

59. Exploratory Evaluation of Neem Extract Sprays on Maize for Control of Stalk Borers in Coastal Kenya 571
M. Kiarie and R.C. Saxena

60. Repelin — An Effective Botanical Insecticide against Ber Fruitfly *Carpomyia vesuviana* Costa 575
V.V.S. Reddy, T. Ramesh Babu and B. Narasimha Rao

61. Evaluation of Enriched Neem (*Azadirachta indica* A. Juss.) Seed Extracts against Mango Hoppers, *Idioscopus nitidulus* Walker and *Amritodus atkinsoni* Leth. 581
R.P. Srivastava, R.P. Singh, M. Haseeb and S.R. Abbas

62. Evaluation of Neem Derivatives and *Lantana camara* L. against Cardamom Pests 589
B. Gopakumar, D. Kumaresen, S.S. Chandrasekar and S. Varadarasan

63. Bioefficacy of Some Neem Formulations against *Spodoptera litura* F. 595
M.S. Chari, G. Ramaprasad, S. Sitaramaiah and P.S.N. Murthy

64. Bioefficacy of Neem Extracts against *S. litura* on Tobacco 613
P.S.N. Murthy, G. Ramaprasad, S. Sitaramaiah, C.V. Narasimha Rao, B.V. Kameswara Rao, S.R. Prabhu and S. Nageswara Rao

VOLUME 2

6. EVALUATIONS AGAINST PHYTONEMATODES

65. Use of Neem Cake in Control of Lesion Nematode, Cyst Nematode and Root Knot Nematode 625
P. Gupta, S. Sebastian, L.R. Sobita and S.M. Mishra

66. Efficacy of Neem Cake against Soil Nematodes and Fungi on Chickpea 627
Sartaj A. Tiyagi and M. Mashkoor Alam

67. Efficacy of Cakes in the Management of Root Knot Nematode (*M. arenaria*) in Groundnut 643
M.U. Vaishnav, I.U. Dhruj and R.P. Lakhataria

68. Integrated Management of the Citrus Nematode, *Tylenchulus semipenetrans*, Using Neem Cake and the Parasitic Fungus *Paecilomyces lilacinus* 647
P. Parvatha Reddy, R.M. Khan and M.S. Rao

69. Effective Use of Wellgro and RD-9 Repelin for the Management of Root Knot Nematode on Tomato 651
M.S. Rao and P. Parvatha Reddy

70. Nematicidal Efficacy of Decomposed Neem Products on *Meloidogyne incognita* Infecting Mungbean and Their Residual Effect in Chickpea 657
S.D. Mishra and Vijayalakshmi Mojumder

71. Effect of Neem Cake, Neem Oil, Repelin and Carbofuran on Control of Soybean Nematodes 663
S.C. Vyas

72. Evaluation of Nematicidal Potential in Neem Allelochemicals 669
Mansoor A. Siddiqui and M. Mashkoor Alam

73. Management of Root Knot Nematode in Betelvine by Neem Cake 679
K. Sitaramaiah, A. Acharya, E.I. Jonathan, A.R. Wasnikar, R.V. Nakat, D. Subba Rao and Satyabrata Maiti

74. Nematode-controlling Efficiency of *Paecilomyces lilacinus* in Relation to Neem Products 685
Anjum Ahmad and M. Mashkoor Alam

75. Management of *Meloidogyne incognita* Infesting Pulse Crops through Seed Treatments with Neem Products 699
Vijayalakshmi Mojumder and S.D. Mishra

76. Neem Cake for Management of Root Knot Disease in Green Gram and Tomato Nursery 705
D.J. Patel, S.K. Patel and H.V. Patel

7. PLANT DISEASE MANAGEMENT

77. Neem and Control of Aflatoxin Contamination 713
Hampden J. Zeringue, Jr. and Deepak Bhatnagar

78. Management of Pest and Diseases of Cotton and Soybean Using Neem Products 729
F. Vidya, K. Sethuraman, V. Mariappan, K.R. Karunakaran and P. Vidyasekaran

79. Efficacy of Plant Derivatives on Rice Blast Pathogen *Pyricularia oryzae* cav., Blast Infection and Enzymes of Rice Plants 735
E. Rajeswari and V. Mariappan

80. Evaluation of *Azadirachta indica* and *Lawsonia inermis* for Antifungal Activity and Studies on Neem Products for Control of Damping-off in Tobacco Nurseries 745
M.M. Shenoi, S.M. Abdul Wajid and S.S. Sreenivas

81. Management of Pests and Diseases of Rice Crop with Neem Products 755
V. Mariappan, K. Sethuraman, F. Vidya, K.R. Karunakaran and P. Vidhyasekaran

82. Neem for the Control of Aphid and Whitefly Vectors and Virus Diseases of Plants 763
R. Roychoudhury and R.K. Jain

83. Effect of Neem and Other Plant Extracts on *Xanthomonas capestris citri*, the Incitant of Citrus Canker 771
B. Chendramohan and G.J. Moses

84. Effect of Plant Derivatives on the Transmission of Potato Virus Y Infecting Chilli and Its Aphid Vectors 777
L. Duleep Kumar Samuel and V. Mariappan

85. Efficacy of Neem Products in the Control of Rice Bacterial Blight 783
S. Eswaramurthy, V. Mariappan, M. Muthusamy, M.N. Alagianagalingam and K.S. Subramaniam

86. Effect of Non-edible Seed Oils in Aphid Transmitted Chilli Mosaic Virus in Tamil Nadu 787
V. Mariappan and L. Duleep Kumar Samuel

87. Use of Neem in Integrated Management of Rot Diseases of Small Cardamom (*Elettaria cardamomum* Maton.) 793
K. Dhanapal, Thomas Joseph and R. Naidu

88. Antimycotic Activity of a Methanolic Extract from *Azadirachta indica* A. Juss. 799
B. Steinhauer

89. Management of Chilli (*Capsicum annuum* L.) Diseases by Neem-based Preparations 809
R.P. Maharishi

8. ROLE IN INCREASING FERTILIZER NITROGEN EFFICIENCY

90. Potential Use of Urea Coated with Neem Cake to Improve Nutrient Uptake and N Recoveries and to Reduce N Losses in a Perennial Aromatic Grass, Java citronella (*Cymbopogon winterianus* Jowitt.) 815
E.V.S. Prakasa Rao

91. Relative Agronomic Efficiency of Neem Extract-coated Urea versus Prilled Urea for Rice 821
A.L. Bhandari, K.N. Sharma, R.S. Brar and D.S. Rana

92. Development of Neem-coated Urea for Increasing Nitrogen Use Efficiency and Crop Yields 827
B.N. Vyas, N.B. Godrej and K.B. Mistry

93. Use of Different Oil Cakes as Fertilizers: Comparison of Processes of Immobilization, Mineralization and Nitrification, Inhibition in Soil and Nitrogen Uptake 835
B.F. Hulagur

94. Comparison of Nimin-coated Urea with Other Sources at Graded Level of Nitrogen in Lowland Rice 843
M.K. Porwal, G.S. Bhatnagar and P.C. Chaplot

95. Use of Neem as Nimin to Increase N-use Efficiency in Transplanted Rice 847
S.S. Tomar and P.K. Gupta

96. Influence of Neem Oil Coating on Prilled Urea on Volatilization Loss from Swell-shrink Soil Amended with Organic Matter under Different Moisture Regimes 851
Muneshwar Singh and P.N. Takkar

9. COMMERCIAL NEEM INSECTICIDES

97. RD-9 Repelin: An Effective Botanical for Integrated Pest Management 857
T.S. Subramaniam, B. Sivaprasad and V.V.L.N. Prasad

98. Development of a Neem Formulation and Its Evaluation for Control of Crop Pests 865
K. Raman, S. Ganesan, B.N. Vyas, N.B. Godrej and K.B. Mistry

99. The NeemAzal Conception: Future Possibilities of the Use of Neem in Biological and Integrated Pest Management 875
Hubertus Kleeberg

100. The Use of Margosan-O® in the Greenhouse Environment 883
J.F. Walter and J.F. Knauss

101. Efficacy Profile of a Commercial Neem Insecticide 897
John Immaraju, Timothy Wood, Reese Nelson, William Ruggero and Samuel Wells

10. TECHNOLOGICAL DEVELOPMENT

102. Machineries for Neem Processing 909
S.S. Sivakumar, P.T. Palanisamy, N. Varadharaju, L. Gothandapani and K.R. Swaminathan

103. Technology for Production of Insecticides of Plant Origin at Rural Level 921
A. Dakshinamurthy

104. Cold Processing of Neem Seed 931
G. Ramakrishna, N.B.L. Prasad and G. Azeemoddin

11. PROPAGATION

105. Site Suitability for Neem in Queensland 941
Axel Bosselmann

106. Somatic Embryogenesis and Organogenesis in Neem (*Azadirachta indica* A. Juss.) 955
R. Islam, N. Joarder, O.I. Joarder, M. Hossain, S.M. Rahman and A.T.M. Naderuzzaman

107. Micropropagation of Neem through Axillary Bud Culture 961
O.I. Joarder, A.T.M. Naderuzzaman, R. Islam, M. Hossain, N. Joarder and B.K. Biswas

108. Potential Application of *in vitro* Methods for Propagation of Neem (*Azadirachta indica* A. Juss.) 967
M.S. Joshi and S.R. Thengane

109. High Frequency Regeneration of Neem *in vitro* (*Azadirachta indica* A. Juss.) from Immature Embryo Explants 975
M. Thiagarajan and P.M. Murali

110. Tissue Culture Studies in Neem (*Azadirachta indica* A Juss.) 981
A. Nirmalakumari, N.M. Ramaswamy and S.R. Sree Rangaswamy

111. Investigations on Neem and *Melia* for Sustainable Production 993
M.V.R. Prasad, A.R.G. Ranganatha, M. Sujatha, Mukta Dhingra M. Ramachandram and G. Nagaraj

112. Clonal Propagation of Neem by Tissue Culture 999
R.A. Drew

113. Twig Blight — A New Disease of Neem Seedlings 1007
J.C. Kaushik, S.S. Bisla and D.P. Singh

114. Vegetative Propagation of *Azadirachta indica* A. Juss. 1013
K.P. Srivasuki, N.S. Philomina and R. Vijaya Kumar

12. OTHER INSECTICIDAL PLANTS AND MISCELLANEOUS

115. Juvenomimetic Activity of Benzene Extracts of Twenty Plants from Kerala on the Red Cotton Bug *Dysdercus cingulatus* (Fb.) 1019
K. Saradamma, D. Dale and N.M. Das

116. Biological Effects of Certain Oils and Terpene Fractions of Plant Origin of Insects 1029
B. Subrahmanyam and P.J. Rao

117. Antifeedant Effects of *Jatropha curcas* L. Seed Oil and Extracts on the Variegated Grasshopper *Zonocerus variegatus* and the Eggplant Defoliator, *Selepa docilis* 1037
J.R. Cobbinah

118. Phytophagous Insects Living on *Ajuga* Species *A. bracteosa, A. chamaepitys, A. genevensis, A. pyramidalis. A. reptans* var. *reptans* and *A. reptans* var. *atropupurea* 1043
Béla Darvas, László Alfonz Polgár, Abdel-Monem Mokthar, Piroska Szabo, Mária Torma-Gazdag, Zoltán Ilovai, Ede Petro, Chih-Hua Tsou, Yaw-Huei Lin and Arild Anderson

119. Biological Activity of the Constituents of *Dalbergia sissoo* and Their Derivatives 1057
P. Venkat Kumar, D.B. Saxena and G.T. Gujar

120. Effects of *Acorus calamus* Oil and Its Main Compound β-Asarone on the Larger Corn Borer, *Prostephanus truncatus* (Horn) the Main Pest of Corn in Central Africa 1069
Gerhard H. Schmidt and Martin Streloke

121. Efficacy of Botanical Pesticides in the Control of Citrus Butterfly *Papilio demoleus* Linn. (Papilionidae: Lepidoptera) 1081
P. Kameshwara Rao and Ch. Balagangadhar Tilak

122. Efficacy of *Ajuga* (*A. chamaepitys, A. reptans,* var. *reptans,* and var. *atropurpurea*) Extracts on a Wide Variety of Non-adapted Insect Species 1085
Béla Darvas, Lászlő Alfonz Polgár, Ahmad Saber Bream, Imre Csatlos, Ali Ibrahim Farag, Mária Torma-Gazdag, Zoltán Ilovai, Maria Pia Calcagno and Josep Coll Toledano

123. Utility of Botanicals and Biocides as Additives to Insecticides in *Helicoverpa armigera* Hübner Management on Cotton 1101
B. Rosaiah and A.S. Reddy

124. Performance of Some Plant Products as Additives to Insecticides against *Helicoverpa armigera* Hübner (Noctuidae: Lepidoptera) 1109
N. Venugopal Rao, P. Raja Sekhar, M. Venkataiah and A.S. Reddy

125. Effect of Botanicals on Certain Parasitoids 1117
K. Srinivasa Babu, M.S.N. Murthy and T. Ramesh Babu

126. Effectiveness of Neem-based Products (Azadirachtin) against *Tetranychus macfarlanei* (Acarina: Tetranychidae) on Okra and Their Safety to Predatory Mites 1127
A.B. Rai, G.R. Bhanderi and C.B. Patel

127. Use of Botanical Formulations as Potential Repellents in Bird Pest Management 1135
P. Syamsunder Rao, K. Seshachalam Babu, M. Anand Rao and B.P. Vardhani

13. THERAPEUTIC POTENTIAL AND OTHER MEDICAL APPLICATION

128. *Azadirachta indica* A. Juss. Stem Bark as an Anti-Leprosy Source 1143
K.S. Subramanian and K.K. Lakshmanan

129. Identification and Characterization of the Immunomodulatory Fraction from Neem Seed Extract Responsible for Long-term Antifertility Activity 1151
Sanjay Garg, G.P. Talwar and S.N. Upadhyay

130. Clinical Studies with Praneem Polyherbal Cream in Chlamydial Cervicitis 1155
Sanjay Garg, A. Mittal, S.K. Kapur, S. Gupta, S.N. Upadhyay and G.P. Talwar

131. Rat Toxicity Studies with Neem Oil 1159
P. Badri Sriman Narayanan, R. Sampathraj and G. Vanitha Kumari

132. Alteration in Testicular Histoarchitecture Following Neem Oil Administration in Albino Rats 1165
M.P. Manoranjitham, A. Prem Anandhi, R. Sampathraj and C. Vanithakumari

133. Effect of Neem Oil: Structural and Functional Changes in the Epidymis of Rats 1173
R. Sampathraj, P. Badri Sriman Narayanan and G. Vanithakumari

134. An Experimental Evaluation of *Azadirachta indica* (A. Juss.) for Adaptogenic Potential 1183
P. Sen, P. Mediratta and A. Ray

135. Thyroid Hormones and Intermediary Metabolism in Fish: Influence of Neem Kernel Extract 1189
M.C. Subash Peter

Appendix: List of Contributors and Keynote Speakers 1199

1

INVITED LECTURES AND GENERAL PAPERS

1

Neem and Its General Development for Pest Control

HEINZ REMBOLD

Max-Planck-Institute for Biochemistry, Insect Biochemistry Unit
D-8033, Martinsried, Germany

ABSTRACT

The neem tree, *Azadirachta indica* A. Juss., is one of the most promising carriers of botanical insect growth inhibitors. The primary bio-active products have been identified as a group of azadirachtins, the main representatives of which are azadirachtins A and B. Three aspects of how to make use of the insect growth inhibiting activity of neem kernels are discussed: home-made crude extracts, crude azadirachtin, and synthetics with azadirachtin activity. Prerequisite for the development of a neem-based agrochemical is the guarantee of a constant seed source and for constant quality of the crude azadirachtin. The price of the latter should be near US$ 2 per gram to be competitive with the synthetic insecticides.

In the course of plant and insect coevolution which took place over millions of years, development of tolerance and resistance against herbivorous insects was reflected, *inter alia*, by a plurality of secondary compounds in the higher plants. It is primarily these plant chemicals which medicate and influence the plant-herbivore interaction, much more than other plant-based cues like physical stimuli. This long-standing interaction between insects and plants took place under fairly constant environmental conditions and ended in stable wild plants with plentiful survival strategies. Then the green revolution of agriculture initiated its own rules. The present problems in protecting our crops from their often new pests are caused by the strategy of breeding for high-yielding varieties. With our modern agriculture, therefore, a completely new type of coevolution has started between insects and the cultured plants which were and still are being bred for high yields much more than for resistance against their pests. We want plants to produce materials which meet our demand, and this can only work at the expense of the wild ancestors' resistance factors which to quite an extent are identical with their secondary plant compounds. The high-yielding plant genotypes consequently need to be grown under the umbrella of agrochemicals, whereas their traditional herbivores still benefit from a long coevolution with the wild ancestors of our crop plants. The insect pests, therefore, profit from the lack of chemical resistance factors of their host plants, which in addition are grown in monoculture, in an ideal breeding area for the fast-growing pests. With this view in mind we can improve tolerance of our crops against their pests either by increasing their resistance factors—which still have to be identified in most cases and can be increased only at the expense of their primary metabolites—or by identifying botanical pesticides in organisms which possess a high pest resistance. One promising candidate for such natural pesticides is the neem tree, *Azadirachta indica* A. Juss.

It is a fascinating challenge for the biologist and for the natural products chemist to identify resistance factors, the genetic pathways of which have been masked and suppressed during the breeding process of a crop plant and under practical aspects apparently lost. One aspect we should not miss when thinking of new strategies in plant protection is the interconnectedness of biosynthetic pathways which may explain the high susceptibility to agricultural pests of some promising new genetic varieties when exposed to different agroclimates. Breeding for resistance or tolerance of crop plants against diseases and pests has a long history which, however, has often been unsatisfactory, in most cases because of our ignorance about the factors which are finally responsible for resistance against pests. One of the many reasons is that some of the chemical factors which for example are resistance factors in the wild ancestors of our vegetables are not edible or are even toxic also to humans, like phenolics or alkaloids, and therefore had to be eliminated during plant breeding. Another reason for the empiricism in this field is the fact that for most of our crop plants we know neither the chemistry nor the metabolic pathways of signals which attract or repel the herbivore. A phytophagous insect, however, identifies its host plant as a unity which is composed of many signals out of which the semiochemicals are only one important component. This brings us to another important aspect in the field of plant protection—to find and develop a strategy of minimal risk to mammals, plants, and other nontargets and consequently directed a specific target. One of them could be the insect-specific control of their growth and development. A high rate of reproduction is one of the insects' survival strategies. One female of the pod borer, *Helicoverpa armigera*, is able to lay up to 3000 eggs out of which, under favourable conditions, 1 per cent may reach the reproductive stage. This means that in one generation each female represents a power to multiply by a factor 15! This fantastic reproductive power also provides the pest insects with a rapidly increasing tolerance against every type of insecticide.

The great number of insect species, their tremendous capability to reproduce, and their long evolutionary history—the oldest insect records go back to fossils from the middle Devonian period some 350 million years ago—enable them to occupy almost every environmental niche. By virtue of this adaptive power some of them have become vectors of infectious diseases while others are important agricultural pests. Most of them can still be controlled by toxic compounds like organochlorines, organophosphates, pyrethroids, or carbamates. However, a rapidly increasing resistance against every type of insecticide along with an acute off-target neurotoxicity makes further insecticide application an impending threat to ecology and human health. For this reason, more target-specific methods of insect control are the goal of worldwide efforts. The search for safer and better compounds has also focussed on phytochemicals. Mainly in the last three decades many plant products have been identified which could be of interest as potential protective allelochemicals against insects and other pests (Rosenthal and Bernebaum, 1991). Insect feeding deterrents have, because they are easily detected, been identified in huge numbers (Morgan and Mandava, 1990). Also, other naturally occurring pest bioregulators (Hedin, 1991) which affect insect behaviour and development or phytoalexins and phototoxins which are responsible for plant defense reactions have attracted scientists interested in new plant protection strategies.

Practical application of plant-based pesticides has been limited up to now for several reasons. The most important drawback of use of the crude plant extracts is a restrictive registration policy in the industrialized countries. As it is impossible to give a complete toxicity for each of at least 10 main compounds of such an extract, no preparation which is based on botanical pesticides could up to now enter the big markets of edible plants, forest, or drinking water protection. This limitation can only be overcome by purification and subsequent characterization of the active ingredients, as in the case of pyrethrum or abamectin (Putter et al., 1981). Another limitation is the guarantee for quality and availability of the resources. An agricultural product is always unpredictable, as exemplified by the many problems of our highly specialized agricultural practices. And even if all these problems can be overcome by application of well-designed technologies, it is finally vital for each promising botanical pesticide to end up as a semisynthetic and finally synthetic product which can chemically be modified whenever necessary for overcoming an increasing tolerance from the target's side. This, in principle, is the pyrethroid story with all its consequences in terms of the time which had to be invested in R&D and the time which passed till the first tolerant and finally resistant pest insects appeared. How much of this story already parallels that of neem? Let us consider this question under the three different aspects of crude extracts, pure compounds, and synthetics. Each aspect has its own significance for a general development of neem for pest control.

What is the potential of neem as a home-made insecticide in the form of seed powder or a crude water extract in sustainable agriculture? Three international conferences have been organized during the last decade which discussed this and other topics (Schmutterer et al., 1982; Schmutterer and Ascher, 1984, 1987). Various parts of the neem tree have traditionally been used for pest control. The dried leaves and powdered seeds were applied to prevent insects from feeding and oviposition and water extract from the seed powder to protect crops from their insect pests. Extracts of the ground dried seed kernels are the most potent neem preparations (Schmutterer and Hellpap, 1988). Being a new pesticide, crude extract from neem seed could even break resistance against commercial insecticides. In recent years, and as a consequence of these international conferences, the tree has become popular especially among agricultural scientists who are engaged in development. An increasing number of farmers applied neem for its potential use as a natural insecticide.

There are good reasons for their optimism. Neem can easily be propagated by seed and reproduces after five years. Under favourable conditions fresh fruit yield per fully grown tree is about 50 kg per year, which gives about 6.5 kg of kernel from which about 3 kg of neem oil and 3.5 kg of cake can be obtained. The oil as well as the cake can be used for plant protection directly.

Since neem trees are also grown in plantations for wood, their combined use as a very potent natural pesticide and for timber and firewood makes them an attractive component in agroforestry systems for establishing a sustainable land-use system. One model for the economic potential of such an agroforestry system with neem has recently been described by Marz (1992). He studied an alley-cropping system with neem and found that the change in cropping pattern could become crucial due to reallocation of farm labour when neem fruits are to be harvested. The introduction of this agroforestry system may therefore have strong impacts on the organization and economic performance of a farm. However, income and liquidity of small farms are increased significantly. The profitability of alley-cropping systems with neem was mostly determined by the costs for processing neem fruits into an insecticidal powder and the value of this powder on the market. Processing costs are mainly labour costs and the market value of the neem powder is determined by its efficiency against insect pests and the costs of commercial insecticides. The replacement of widely used insecticides by neem seed powder should be studied therefore through more on-farm experiments also in view of economic criteria, and under the aspect of a rapidly increasing market for high quality neem kernels.

What are the chances of obtaining a marketable, partly purified neem material which is effective, safe, chemically stable, degradable, available in commercial quantities, and economical? In principle this means how can a crude azadirachtin preparation be made which can be offered at a moderate and therefore competitive price? There are, it is estimated, 15–25 million neem trees in India. Under the optimistic assumption of an azadirachtin content of 5 g per kg of dried kernels, this would represent a reservoir of about 150 tonnes of azadirachtin per year but also a yield of less than 50 g per tree.

The criteria which apply for a commercial product and have been mentioned above are also met by azadirachtin. It is highly effective against most insects, with the restriction, however, that it can be selective. In his latest review on organisms which are affected by neem, Warthen (1989) lists for different insect species a species-specific activity of azadirachtin as well as against chemoreceptors as against the endocrine system. Therefore, the different authors describe azadirachtin as antifeedant, control agent, repellent, feeding inhibitor, and growth regulator. Whereas some insects, like the desert locust, *Schistocerca gregaria*, do not even touch a plant sprayed with trace amounts of azadirachtin, the Mexican bean beetle, *Epilachna varivestis*, is only sensitive against its growth inhibiting effect, at doses which are lower than the concentration for antifeedant activity by more than one order of magnitude (Rembold, 1988).

With reference to safety, azadirachtin by itself seems to be directed against the neuroendocrine system of insects (Rembold, 1988) and therefore has been found to be of low risk against the non-target organisms.

The compound is chemically very unstable and degrades quickly when exposed to air and light, especially if contained in crude extracts. It is mandatory for this reason to prepare the crude azadirachtin as a dry powder. The dry azadirachtin can then be stored safely under the usual precautions. Nevertheless, azadirachtin formulations such as Margosan-O® and Azatin® are commercially available and registered for non-crop and ornamental uses in the United States and registration is pending in Europe. Margosan-O is a formulated neem seed extract containing a high and consistent quantity of azadirachtin which is stabilized by adjusting the pH and adding a sunscreen agent (Larson, 1988).

If azadirachtin is to be produced in bulk quantities, 100 kg at least, the fruits must be dried and processed under carefully controlled conditions in order to avoid fungal infection which may cause aflatoxin pollution. Neem cake seems to be the most economic starting material for production of crude azadirachtin. The seed kernels must be already dry, therefore, in order to avoid extraction of the water-soluble azadirachtin into the oil. First of all, however, an efficient cultivation of high-yielding neem genotypes would be necessary, as is the case for pyrethrum production. Harvest, drying, extraction, and precipitation of the powdered crude azadirachtin in centrally located plants will be important if the compound is to be produced in commercial quantities at reasonable costs. One can assume that a standardized crude azadirachtin should be available at around US$ 2 per gram. Only then can it compete with new synthetic insecticides. Industrial interest in commercial application of azadirachtin has been low up to now, because of the limited and unexploited resources for crude azadirachtin and the high costs of obtaining the toxicological and environmental fate data which are required for registration. In sum, there exists a great potential for use of neem in a major agricultural market.

The most potent neem product in controlling insect growth, development, and reproduction is the group of azadirachtins, out of which 10 have been isolated and 9 were fully elucidated in their chemical structure. They all under standardized conditions interfere with growth and development of last-instar larvae of *Epilachna varivestis* (Rembold et al., 1980). They all have the same limonoid (tetranortriterpenoid) nucleus with minute differences in the ligands. This fact is highly suggestive of a minimum structural element which is responsible for their activity (Rembold, 1988).

Azadirachtins A and B are the major components in neem seeds; all the other azadirachtins are present in minor quantities only. There are still other azadirachtins which, due to their natural occurrence in minute quantities, have not been isolated.

The mode of azadirachtin, on a molecular level, is still unknown. By use of 22,23-tritiated and ^{125}I labelled dihydroazadirachtin we showed in *Locusta migratoria* that the neurohaemal organ, the corpus cardiacum, is the real target of azadirachtin (Rembold et al., 1989). A consequence of its interference with neuroendocrine control is a high disturbance in ecdysone and juvenile hormone titers (Rembold et al., 1987). This is a completely new target for an insecticide, i.e., the interference with the insect's specific neuroendocrine control system, which is so different from the mammalian system that no toxic side effects have been found even after application of gram amounts of azadirachtin per kilogram body weight. Quite some effort is therefore being invested in the total synthesis of the azadirachtin molecule. The strategy followed by Ley and his co-workers is to synthesize the two moieties of the azadirachtin molecule, the decalin and the hydroxytetrahydrofuran fragment, separately and to combine them at the end only (Kolb et al., 1992). Besides its interest as a synthetic proof for the complicated azadirachtin structure, this effort may yield some interesting mimics which may finally be used for plant protection. However, due to the highly competitive insecticide market, it is hardly to be expected that an even drastically reduced structure could be produced at a low price, which is a prerequisite for every agrochemical.

Our own strategy for obtaining a reduced bio-active structure, which also means elucidating an economically feasible molecule which still induces similar effects, is to make use of the highly sensitive and specific *Epilachna* bioassay, as it is fairly insensitive to feeding deterrents. Consequently, a growth disrupting effect is not masked by the starvation of the insect as is the case with locusts or lepidopterous insects. Starvation can induce hormonal disturbances and therefore misguide the search for natural products with antihormonal effects. Azadirachtins A and B have, with 1.66 and 1.30 ppm respectively, almost the same biological effects in the *Epilachna* bioassay, although they have characteristic structural differences in ring A substitution and at position 11, which is reduced in azadirachtin B. The other members of the azadirachtin group have an activity which is around 1 ± 0.5 ppm with the exception of azadirachtin G, which lacks the epoxide ring (Rembold, 1988).

Chemical modification of the azadirachtin molecule could be one routine method in overcoming an increasing tolerance of the target insects. We therefore modified some of the substituents in order to come to an increased biological activity. The azadirachtin A and B molecules have such a high biological activity, however, that its increase, also for the purpose of overcoming resistance of a potential pest, was only found for the 22,23-hydrogenated 1-detigloyl-3-deacetyl azadirachtin, which, in the *Epilachna* bioassay, exceeds the activity of azadirachtin A by more than one order of magnitude (Rembold, 1988). This chemical modification is of interest under commercial aspects as the stability of the molecule increases concomitantly.

For a better understanding of the essential structural elements or the anchors which interact with some unknown receptor molecule, it is also of interest to find derivatives with reduced biological activity. There are indeed some critical substitutions which reduce the growth disrupting activity of the original azadirachtins in the bioassay. Methylation of the hydroxyl groups at C-11 and C-20 reduces the azadirachtin A activity to 15 per cent, acetylation of the C-11 group reduces it to 3 per cent, and permethylation of all the three free hydroxyl groups even abolishes the biological activity. Also, cyclization to the azadirachtinin skeleton (Bilton et al., 1988) by ring opening of the epoxide with subsequent reaction of the hydroxyl group at position 7 yields a completely *Epilachna*-inactive compound. From these results we have proposed, as a working hypothesis, that most of the polar substituents of the azadirachtin structure are interacting with the hypothetical receptor molecule (Rembold and Puhlmann, 1993).

A final answer about the essential anchor points in the azadirachtin molecule and their function in the blockade of a hypothetic receptor protein in the corpus cardiacum is still speculative. However, after all our present results there is a good chance of obtaining a simpler chemical structure which does a similar job to that of the complex azadirachtin molecule by interfering with the insects' endocrine control. Such a strategy can only be successful by also using screening methods which look into the molecular basis of azadirachtin's mode of action.

The new technique of high-resolution two-dimensional electrophoresis allows screening of individual organs from the treated and untreated insects. A study of azadirachtin-treated *S. gregaria* revealed a dramatic effect of the compound on the polypeptide pattern of all the endocrine organs studied. The number of individual polypeptides was specifically reduced in the treated insects and this effect persisted through the 45 days examined and was most distinct in the brain. Some polypeptides had also been induced in the treated locust, indicating that the mode of azadirachtin action is not a simple interference with protein synthesis (Annadurai and Rembold, 1993).

An even more sensitive technique makes use of an insect cell line (Rembold and Annadurai, 1993). In similar concentrations as *in vivo*, azadirachtin inhibits in ppm quantity proliferation and monolayer formation of the Sf 9 cells in monolayer culture. The growth inhibiting effect is specific for azadirachtin and was not found with three other botanicals which also are limonoids. We believe that this combination of cell culture with high-resolution two-dimensional electrophoresis will provide another tool for designing chemically new types of insecticides, one of which is the fascinating and complex molecule of azadirachtin.

REFERENCES

Annadurai, R.S. and H. Rembold. 1993. Azadirachtin A modulates the tissue-specific 2D polypeptide patterns of the desert locust, *Schistocerca gregaria*. *Naturwissenschaften* 80: 127–130.

Bilton, J.N., P.S. Jones, S.V. Ley, N.G. Robinson and R.N. Sheppard. 1988. Chemistry of insect antifeedants from *Azadirachta indica* (Part I): Conversion from the azadirachtin to the azadirachtinin skeletons. *Tetrahedron Lett.* 29: 1849–1852.

Hedin, P.A. (ed.). 1991. *Naturally Occurring Pest Bioregulators* ACS Symp. 449. American Chemical Society, Washington, DC.

Kolb, H.C., S.V. Ley, A.M.Z. Slawin and D.J. Williams. 1992. Chemistry of insect antifeedants from *Azadirachta indica* (Part 12): Use of silicon as a control element in the synthesis of a highly functionalized decalin fragment of azadirachtin. *Chem. Soc. Perkin Trans.*, pp. 2735–2762.

Larson, R.O. 1988. The commercialization of neem. In: *Focus on Phytochemical Pesticides*, Vol. 1, *The Neem Tree*. M. Jacobson (ed.). CRC Press, Boca Raton, pp. 165–168.

Marz, U. 1992. The economic potential of an agroforestry system with neem (*Azadirachta indica* A. Juss) for small farms in the Sudano-Sahelian zone of Burkina Faso. *Der Tropenlandwirt* 93: 23–38.

Morgan E.D. and N.B. Mandava (eds.). 1990. *Handbook of Natural Pesticides* Vol. VI, *Insect Attractants and Repellents,* CRC Press, Boca Raton.

Putter, I., J.G. MacConnell, F.A. Preiser, A.A. Haidri, S.S. Ristich and R.A. Dybas. 1981. Avermectins: Novel insecticides, acaricides and nematicides from a soil microorganism. *Experientia* 37: 963–964.

Rembold, H. 1988. Isomeric azadirachtins and their mode of action. In: *Focus on Phytochemical Pesticides,* Vol. I, *The Neem Tree*, M. Jacobson (ed.). CRC Press, Boca Raton, pp. 47–67.

Rembold, H. and R.S. Annadurai. 1993. Azadirachtin inhibits proliferation of Sf9 cells in monolayer culture. *Z. Naturforsch.* 48c: 495–499.

Rembold, H. and I. Puhlmann. 1993. Phytochemistry and biological activity of metabolites from tropical Meliaceae. In: *Phytochemical Potential of Tropical Plants* K.R. Downum et al., (eds.), Plenum Press, New York, pp. 153–165.

Rembold, H., G.K. Sharma, Ch. Czoppelt and H. Schmutterer. 1980. Evidence of growth disruption in insects without feeding inhibition by neem seed fractions. *Z. Pflanzenkr. Pflanzenschutz* 87: 290–297.

Rembold, H., M. Uhl and T. Muller. 1987. Effect of azadirachtin A on hormone titers during the gonadotrophic cycle of *Locusta migratoria*. In: *Natural Pesticides from the Neem Tree and other Tropical Plants*, Proc. 3rd. Int. Neem Conf., Nairobi, (eds.), 1986, H. Schmutterer and K.R.S. Ascher GTZ Schriftenreihe, Eschborn, pp. 289–298.

Rembold, H., B. Subrahmanyam and T. Muller. 1989. Corpus cardiacum—A target for azadirachtin. *Experientia* 45: 361–363.

Rosenthal G.A. and M.R. Bernebaum (eds.). 1991. *Herbivores, Their Interaction with Secondary Plant Metabolites,* 2nd ed., Vol. 1, *The Chemical Participants*, Academic Press, San Diego.

Schmutterer, H. and C. Hellpap. 1988. Effects of neem on pests of vegetables and fruit trees. In: *Focus on Phytochemical Pesticides*, Vol. 1, *The Neem Tree* (ed.) M. Jacobson, CRC Press, Boca Raton, pp. 69–86.

Schmutterer, H., K.R.S. Ascher and H. Rembold. (eds.) 1982. *Natural Pesticides from the Neem Tree,* Proc. 1st Int. Neem Conf., Rottach-Egern, 1980, GTZ Schriftenreihe, Eschborn.

Schmutterer, H. and K.R.S. Ascher (eds.). 1984. *Natural Pesticides from the Neem Tree and Other Tropical Plants,* Proc. 2nd Int. Neem Conf., Rauischholzhausen, GTZ Schriftenreihe, Eschborn.

Schmutterer, H. and K.R.S. Ascher (eds.). 1987. *Natural Pesticides from the Neem Tree and Other Tropical Plants*, Proc. 3rd Int. Neem Conf., Nairobi, GTZ Schriftenreihe, Eschborn.

Warthen, J.D. 1989. Neem (*Azadirachta indica* A. Juss.): Organisms affected and reference list update. *Proc. Entomol. Soc. Wash.* 91: 367–388.

2

Interspecific Differences in the Susceptibility of Noctuid Larvae to Behavioural and Physiological Effects of Azadirachtin

MURRAY B. ISMAN

Department of Plant Science, University of British Columbia
Vancouver, B.C., Canada V6T 1Z4

ABSTRACT

Comparative efficacy of azadirachtin, as a larval growth inhibitor and feeding deterrent, was determined for six species of noctuids of economic importance: the black army cutworm, *Actebia fennica* (Tausch.), the bertha armyworm, *Mamestra configurata* Walker, the variegated cutworm, *Peridroma saucia* Hubner, the zebra caterpillar, *Melanchra picta* (Harr.), the tobacco cutworm, *Spodoptera litura* (Fab.), and the cabbage looper, *Trichoplusia ni* Hubner. Azadirachtin added to an artificial diet inhibited neonate larval growth of all species in a dose-dependent manner. EC_{50} values (dietary concentration reducing larval growth by 50 per cent relative to controls, following 10 days of feeding) ranged from 0.12 to 0.24 ppm, but without significant differences between species. However, when second-instar larvae were offered a choice, only larvae of *P. saucia* and *S. litura* discriminated between a control diet and diet containing azadirachtin at concentrations up to 0.4 ppm. In a more sensitive behavioural bioassay with fifth-instar larvae, *S. litura* was the most sensitive to the antifeedant effects of azadirachtin (EC_{50} = 1.25 ng/cm), whereas *A. fennica* was the least (EC_{50} = 40.7 ng/cm). Topical treatment of fourth-instar larvae with 50 or 100 ng of azadirachtin resulted in significant inhibition of subsequent growth, diet consumption, and dietary utilization in all six species. However, *A. fennica* and *S. litura* are less inhibited than the other four species. This study indicates that behaviour-based bioassays do not fully characterize the total effect of azadirachtin on noctuid larvae.

INTRODUCTION

The efficacy of neem extracts for the control of certain pest noctuids has been extensively investigated (Schmutterer, 1990) as has the mode-of-action of azadirachtin in some noctuids (Barnby and Klocke, 1987, 1990; Simmonds et al., 1990; Koul and

Isman, 1991). Although the bioactivity of neem or azadirachtin has been reported for more than two dozen species of pest noctuids, particularly within the genera *Heliothis* and *Spodoptera* (Warthen, 1989), a number of noctuid species which are of economic importance have yet to be tested. Also, there are few studies which permit direct comparison of the efficacy of azadirachtin between noctuid species. Simmonds and Blaney (1984) examined neonate larval growth and mortality of six species of pest noctuids (*Spodoptera frugiperda* [J.E. Smith], *S. littoralis* [Bolsd.], *S. exempta* Walker, *Helicoverpa zea* [Boddie], *H. virescens* [Fab.] and *Trichoplusia ni* Hubner) fed an artificial diet containing azadirachtin, and Kubo and Klocke (1982) used a similar method to compare the effects of azadirachtin against *H. zea, H. virescens,* and *S. frugiperda*.

In the present paper I assess the effects of azadirachtin on six species of noctuids, the black army cutworm, *Actebia fennica* (Tausch.), the bertha armyworm, *Mamestra configurata* Walker, the variegated cutworm, *Peridroma saucia* Hubner, the cabbage looper, *Trichoplusia ni* Hubner, the zebra caterpillar, *Melanchra picta* (Harr.), and the tobacco cutworm, *Spodoptera litura* (Fab.). All six of these species are highly polyphagous insects. In examining interspecific differences in sensitivity to azadirachtin, I compare results of two behaviour-based bioassays (diet choice, leaf disc choice) with those obtained from a chronic growth bioassay and following topical application of azadirachtin.

MATERIALS AND METHODS

All larvae except those of *M. picta* were from long-established laboratory colonies (> 50 generations). Larvae of *M. picta* used were one or two generations removed from larvae collected at Langley, British Columbia, Canada on potato and blueberry crops. Larvae of all six species were reared on an artificial diet (No. 9795, BioServ. Inc., Frenchtown, NJ) supplemented with finely ground alfalfa meal to improve its acceptability to neonate larvae.

Azadirachtin (> 95 per cent purity based on FAB-mass spectroscopy) was a gift from J.T. Arnason, University of Ottawa.

Chronic Growth Bioassay

Azadirachtin was dissolved in methanol and added to the dry ingredients of the diet prior to diet preparation as previously described (Isman and Rodriguez, 1983). For each species, five dietary concentrations were tested along with a solvent-treated control. Concentrations ranged from 0.05 to 0.4 ppm (based on diet fresh weight) for *M. configurata*, *M. picta, S. litura*, and *P. saucia*, 0.025 to 0.3 ppm for *T. ni*, and 0.1 to 0.5 ppm for *A. fennica*. For each species and concentration, two neonate larvae were placed in individual compartments of a plastic tray and provided with about 1 g of treated or control diet. Twenty replicates (each with two larvae) were prepared for each treatment. The experiment was conducted in a growth chamber at 25 $\pm$ 1°C and 16L:8D photoperiod. After 10 days, all larvae were individually weighed and mean weight for each treatment expressed as a percentage of controls. EC_{50} (effective concentration to inhibit growth by 50 per cent relative to controls) and 95 per cent confidence limits were calculated by a computer program (Anon., 1991).

Behavioural Boiassays

Diet choice test

The ability of second instar larvae to discriminate between azadirachtin-treated and untreated diet was assessed in a diet choice test, as previously described (Xie and Isman, 1992). All species were tested with diets containing from 0.05–0.4 ppm azadirachtin. Five replicates were prepared for each concentration with twenty larvae in each replicate (dish). For this and the leaf disc choice test (described below), deterrency index (per cent) for each treatment was calculated as (C-T)/(C+T) × 100, where C is the number of larvae on control diets (or control leaf area consumed) and T is the number of larvae on treated diets (or treated leaf area consumed). (Xie and Isman, 1992 or Isman et al., 1990).

Leaf disc choice test

Antifeedant effects of azadirachtin were further assessed using fifth instar larvae with a leaf disc choice test as previously described (Isman et al., 1990). Cabbage leaf discs (3.5 cm^2) were treated on each side with 15 µl of azadirachtin solutions emulsified with 0.1 per cent Trition X-100. Four concentrations were tested for each species, ranging from 4.4 to 35.2 ng/cm, except for *A. fennica*, where the concentrations ranged from 8.8 to 70.5 ng/cm. The duration of the bioassay was 10 hs. EC_{50} values and 95 per cent confidence intervals were calculated as indicated above. Interspecific differences in deterrence were subjected to analysis of variance (ANOVA) with means separated by the LSD test (Anon., 1991).

Topical Application

Effects of azadirachtin on larval growth, consumption, and dietary utilization independent of behavioural effects were assessed via topical application to fourth-instar larvae (12 $\pm$ 2 mg). Twenty larvae were treated with 100 ng of azadirachtin in 1 µl acetone, or the carrier alone, then allowed to feed for three days on untreated diet before larvae and remaining diet were weighed after drying to constant weight at 60°C. Initial fresh weights of diets and larvae were recorded at the outset of the experiment; dry weights were calculated using a wet weight/dry weight conversion established empirically using a separate cohort of larvae and diet cubes. The following parameters were determined: (1) relative growth rate per unit weight of insect at the outset (RGRi); (2) relative consumption rate per unit weight of insect at the outset (RCRi); and (3) efficiency of conversion of ingested food (= gross dietary utilization), per cent ECI = 100 × G/C where G = dry weight gain of insect and C = dry weight of food consumed (Farrar et al., 1989). ANOVA was performed and differences between treatment means (for each species) were determined by the LSD test (Anon., 1991).

RESULTS

Chronic Growth Bioassay

When added to artificial diet, azadirachtin inhibited larval growth of all six species in a dose-dependent manner, with EC_{50} values after 10 days of feeding ranging from 0.12

to 0.24 ppm (Table 1). However, differences in EC_{50} values between species were not significant, based on overlap of 95 per cent confidence intervals. Mortality did not exceed 2.5 per cent in any of the treatments with the lone exception of *T. ni* at the highest concentration (0.4 ppm), where 23 per cent mortality was observed.

Table 1. EC_{50} values for azadirachtin to six species of noctuids: dietary exposure and antifeedant effects*

Species	Chronic growth (First instar)** (ppm)	Diet Choice (Second instar)** (ppm)	Leaf Disc Choice (Fourth instar)*** (ng/cm^2)
A. fennica	**0.23** (0.07–0.39)	>> **0.40**	**40.7** (35.0–46.5)
M. configurata	**0.18** (0.02–0.34)	>> **0.40**	**17.6** (12.8–22.4)
M. picta	**0.17** (0.02–0.31)	>> **0.40**	**21.2** (6.0–36.4)
P. saucia	**0.19** (0.05–0.34)	**0.18** (0.09–0.26)	**13.0** (9.2–16.7)
S. litura	**0.24** (0.12–0.36)	**0.15** (0.09–0.22)	**1.25** (0.86–1.64)
T. ni	**0.12** (0.07–0.37)	>> **0.40**	**16.2** (12.8–19.6)

* EC_{50} = concentration resulting in 50% inhibition of growth or deterrency, relative to controls. Numbers in parentheses are the 95% confidence interval for each EC_{50} value.

** dietary concentration in ppm

*** concentration in ng/cm^2

Behavioural Bioassays

In the diet choice test, the deterrent EC_{50} values for second-instar *P. saucia* and *S. litura* larvae were 0.18 and 0.15 ppm, respectively (Table 1). None of the other four species were significantly deterred by azadirachtin in diets at the highest concentration tested, 0.4 ppm.

Feeding deterrence of fifth-instar larvae was, however, dose-dependent for all six species, although in this bioassay, the response is curvilinear. Based on this bioassay, *S. litura* is clearly the most sensitive to the direct antifeedant effects of azadirachtin, whereas *A. fennica* is the least (Table 1).

Topical Application

The RGRi, RCRi, and ECI, were significantly reduced following topical application of 100 ng of azadirachtin in all six species (Table 2). Although mean reductions were not statistically compared between species, it appears that larval growth of *S. litura* and *A. fennica* is considerably less impaired than that of the other four species following topical administration of azadirachtin. In that this trend is not reflected in reductions in dietary efficiency among the species (ECI, Table 2), interspecific reductions in RGRi are more likely attributable to interspecific differences in post-treatment consumption rates (RCRi, Table 2).

DISCUSSION

There are few studies which permit direct comparison of the efficacy of azadirachtin against noctuid species. Simmonds and Blaney (1984) examined larval growth and

Table 2. Relative growth rate (RGRi), consumption rate (RCRi), and dietary utilization (ECI) in six species of noctuids following topical administration of azadirachtin (100 ng)

Species	RGRi*	RCRi*	ECI*
A. fennica	55.3	75.2	73.6
M. configurata	30.9	46.3	66.9
M. picta	36.3	47.5	79.8
P. saucia	41.0	62.3	67.8
S. litura	72.1	77.1	69.4
T. ni	41.2	60.2	73.1

* Values shown are percentages relative to respective control larvae. All values are significantly different from respective controls ($p < 0.05$, LSD test)

mortality of six species of noctuids feeding on an artificial diet containing azadirachtin. However, the lowest dietary concentration they tested was 10^{-5} M, equivalent to 7.2 ppm—more than an order of magnitude higher than the highest concentration used in the present report. In their study, larval mortality in the various species ranged from 51 to 77 per cent, with larval growth severely inhibited in only one species (*T. ni*). It is unlikely that their data for growth inhibition accurately portrays the sublethal bioactivity of azadirachtin against noctuids, as significant growth inhibition has been reported for some of the species they used or closely related ones at far lower concentrations (as in the present report).

Lee et al. (1991) compared the efficacy of azadirachtin via dietary incorporation against *H. virescens* and *S. frugiperda* using a dose-response method similar to that used herein, and reported EC_{50} values of 0.07 and 0.08 ppm, respectively, with LC_{50} values (50 per cent mortality) of 0.80 and 1.00 ppm, respectively. These latter results lend support for the mortality data of Simmonds and Blaney (1984), but suggest that the dietary concentrations used in the former study were too great to evaluate larval growth inhibition. The present results for *P. saucia* are similar to those obtained earlier for this species. Champagne et al. (1989) reported an EC_{50} value (seven-day) of 0.36 ppm and an LC_{50} value of 2.7 ppm for azadirachtin; Isman et al. (1990) reported an EC_{50} value (nine-day) of 0.26 ppm.

One interesting finding of the present study is the interspecific feeding responses of second-instar larvae to azadirachtin, given the lack of differences in inhibition of larval growth (Table 1). *Peridroma saucia* and *S. litura* were the only species deterred from feeding on azadirachtin-containing diet. These two were also the most deterred from feeding on azadirachtin-treated leaf discs, although *S. litura* is significantly more sensitive than *P. saucia* in this bioassay. The larval chronic growth bioassay suffers from its inability to discriminate between behavioural (antifeedant) and physiological effects, but for *P. saucia* and *S. litura*, it seems likely that feeding deterrence is an important factor in growth inhibition, as previously reported (Koul and Isman, 1991). However, for the other four species tested, it appears that post-ingestive physiological effects are the likely cause of growth inhibition, in that they were not deterred from feeding on diets containing azadirachtin at concentrations exceeding those which significantly impaired growth in the chronic bioassay.

From the present data, *S. litura* represents somewhat of an enigma. It is one of only two species sensitive to the antifeedant effects of azadirachtin in the early instars, and is clearly the most sensitive of the six species in the leaf disc choice test using fifth-instar larvae (Table 1). On the other hand, growth of early-instar larvae is not unduly inhibited (relative to the other species) when larvae are reared on diet containing azadirachtin, and the species is the least susceptible to the inhibitory effects of azadirachtin applied topically to fourth-instar larvae. Previous studies with grasshoppers (Bernays, 1990, 1991) argue against a strong link between feeding deterrency and internal toxicity in insects, and the experimental design used in the present report demonstrates how these bioactivities can be uncoupled in the laboratory.

Taken together, my results suggest that behavioural bioassays may not provide an adequate representation of the total effect of azadirachtin on noctuid species, and therefore caution needs to be exercised in predicting the susceptibility of a particular insect to neem insecticides in the field on the basis of behavioural bioassay data alone.

Nonetheless, the present results, along with earlier findings, suggest that neem should be efficacious for the control of most noctuid species, assuming that control is aimed at the early instars. However, the ultimate value of neem-based insecticides for the control of noctuid species can only be determined through actual field trials.

ACKNOWLEDGEMENTS

I thank Yanfen Zheng for expert technical assistance and Yongshou Xie for statistical analyses. Azadirachtin was kindly provided by Dr. J.T. Arnason, University of Ottawa; eggs of noctuids were provided by the following: Peter Mason, Agriculture Canada, Saskatoon (*Mamestra*); Maynard Milks, Plant Science, University of British Columbia (*Trichoplusia*); Shiyou Li, E.S. Cropconsult Ltd., Vancouver (*Melanchra*); Pierre Escoubas, Research Development Corporation of Japan, Eniwa-Shi, Japan (*Spodoptera*); and the Insect Production Unit, Forestry Canada, Sault Ste. Marie (*Actebia*). Supported by grants from the Natural Sciences and Engineering Research Council of Canada.

REFERENCES

Anonymous. 1991. *Statistix (v3.5) User's Manual*, Analytical Software, St. Paul, Minnesota, 308 pp.

Barnby, M.A. and J.A. Klocke. 1987. Effects of azadirachtin on the nutrition and development of the tobacco budworm, *Heliothis virescens* (Fabr.) (Noctuidae). *J. Insect Physiol.* 33: 69–75.

Barnby, M.A. and J.A. Klocke. 1990. Effects of azadirachtin on levels of ecdysteroids and prothoracicotropic hormone-like activity in *Heliothis virescens* (Fabr.) larvae. *J. Insect Physiol.* 36: 125–131.

Bernays, E.A. 1990. Plant secondary compounds deterrent but not toxic to the grass specialist acridid *Locusta migratoria*: Implications for the evolution of graminivory. *Entomol. Exp. Appl.* 54: 53–56.

Bernays, E.A. 1991. Relationship between deterrence and toxicity of plant secondary compounds for the grasshopper *Schistocerca americana*. *J. Chem. Ecol.* 17: 2519–2526.

Champagne, D.E., M.B. Isman and G.H.N. Towers. 1989. Insecticidal activity of phytochemicals and extracts of the Meliaceae. In: *Insecticides of Plant Origin*. J.T. Arnason, B.J.R. Philogene and P. Morand, (eds.), American Chemical Society, Symposium Series 387, pp. 95–109.

Farrar, R.R., Jr., J.D. Barbour and G.G. Kennedy. 1989. Quantifying food consumption and growth in insects. *Ann. Entomol. Soc. Am.* 82: 593–598.

Isman, M.B. and E. Rodriguez. 1983. Larval growth inhibitors from species of *Parthenium* (Asteraceae). *Phytochemistry* 22: 2709-2713.

Isman, M.B., O. Koul, A. Luczynski and J. Kaminski. 1990. Insecticidal and antifeedant bioactivities of neem oils and their relationship to azadirachtin content. *J. Agric. Food Chem.* 38: 1406–1411.

Koul, O. and M.B. Isman. 1991. Effects of azadirachtin on the dietary utilization and development of the variegated cutworm (*Peridroma saucia*.) *J. Insect Physiol.* 37: 591–598.

Kubo, I. and J.A. Klocke. 1982. Azadirachtin, insect ecdysis inhibitor. *Agric. Biol. Chem.* 46: 1951–1953.

Lee, S.M., J.A. Klocke, M.A. Barnby, R.B. Yamasaki and M.F. Balandrin. 1991. Insecticidal constituents of *Azadirachta indica* and *Melia azedarach* (Meliaceae). In: *Naturally Occurring Pest Bioregulators*, P.A. Hedin (ed.), American Chemical Society Symposium Series 449, pp. 293–304.

Schmutterer, H. 1990. Properties and potential of natural pesticides from the neem tree, *Azadirachta indica*. *Ann. Rev. Entomol.* 35: 271-297.

Simmonds, M.S.J. and W.M. Blaney. 1984. Some neurophysiological effects of azadirachtin on lepidopterous larvae and their feeding response. In: *Natural Pesticides from the Neem Tree and Other Tropical Plants*. H. Schmutterer and K.R.S. Ascher (eds.), *Schriftenreihe der GTZ* 161, GTZ, Eschborn, pp. 163–180.

Simmonds, M.S.J., W.M. Blaney, S.V. Ley, J.C. Anderson and P.L. Toogood 1990. Azadirachtin: Structural requirements for reducing growth and increasing mortality in lepidopterous larvae. *Entomol. Exp. Appl.* 55: 169–181.

Warthen, J.D., Jr. 1989. Neem (*Azadirachta indica* A. Juss). Organisms affected and reference list update. *Proc. Entomol. Soc. Washington* 91: 367–388.

Xie, Y.S. and M.B. Isman. 1992. Antifeedant and growth inhibitory effects of tall oil and derivatives against the variegated cutworm, *Peridroma saucia* Hubner (Lepidoptera: Noctuidae). *Can. Ent.* 124: 861–869.

Lee, S.M., [illegible] and M.[illegible] 1991. [illegible] environmental [illegible] and [illegible] [illegible]. Washington, D.C., American Chemical Society [illegible] pp. [illegible]

Sandermann, H. 1992. [illegible] Rev. [illegible] 31: [illegible]–[illegible]

Simmonds, M.S.J. and W.M. Blaney [illegible] larvae and their [illegible] In: [illegible] M. [illegible] and [illegible] Academic [illegible] pp. [illegible]

Simmonds, M.S.J., W.M. Blaney, S.V. Ley, [illegible] structural requirements [illegible] J. [illegible] 16: [illegible]–151

Wright, J.D. [illegible] 1990. [illegible] Entomol. [illegible] 31: [illegible]

Xie, Y.S. and M.B. Isman 1992. [illegible] varies [illegible] Entomol. [illegible]

3

Repellent Action of Neem Oil on Insects of Public Health Importance

V.P. SHARMA

Malaria Research Centre, 22 Sham Nath Marg, Delhi 110 054, India

ABSTRACT

Neem oil extracted from the seeds of *Azadirachta indica* A. Juss. (Family Meliaceae) was tested for its repellent action on mosquitoes and sandflies. Results of application of low concentrations of neem oil applied on the exposed body parts in coconut oil or allowed to vaporize in mats heated with low voltage current or burnt in kerosene lamp repelled mosquitoes and sandflies. Protection provided from the bites of *Anopheles* was 100 per cent, from *Culex* about 80 per cent, and from *Phlebotomus* 100 per cent for 12 hrs. Cost analysis revealed that neem oil is the cheapest and most effective and safe method of personal protection.

INTRODUCTION

A variety of mats and coils are marketed in the Asian countries to protect against mosquito bites. However, literature on similar methods of protection against sandflies, the vectors of leishmaniasis and phlebotomus fever, is scanty. The currently marketed devices generally contain synthetic pyrethroids such as allethrin, bioallethrin, and related compounds. These chemicals are imported, and their use on a regular basis may be harmful. Discussions with clinicians in many urban areas revealed allergic reactions in the users to the vapours of these repellents. At the Malaria Research Centre we are working on the insect repellents, and work done on the repellent action of neem oil on mosquitoes and sand flies is reviewed below.

METHODOLOGY

Neem oil is extracted from the seeds of *Azadirachta indica* A. Juss. (Family Meliaceae) using an expeller or a chemical process. Neem oil used in the study was marketed by Unjha Ayurvedic Pharmacy, Gujarat. Three methods were used in experiments on the repellent action of neem oil. In the first method, neem oil was diluted in coconut oil at 0.01, 0.05, 0.5, 1 and 2 per cent strength. About 3–5 ml diluted neem oil was applied on the exposed body parts of volunteers, and they were allowed to lie on cots from 1800 hrs to 0600 hrs. Mosquitoes landing on the body were collected by an insect collector with the help of a suction tube. Similarly, for *Phlebotomus argentipes*, collections were made all night in an endemic village in Bihar. For *Phlebotomus papatasi* 2 per cent neem oil was applied on the arms and a known number of sandflies were released in cages. Sandflies were allowed to feed on the treated and control arms inside the cage. Biting was allowed for 40 min. at each hour and the experiment continued from 1800 hrs to 0600 hrs. Sandflies biting on the arm were collected and counted.

In the second method of treatment neem oil was mixed in kerosene at 0.01, 0.05, 0.5, 1, 2, and 4 per cent. Tin lamps (100 ml capacity) containing neem oil in kerosene and a wick were allowed to burn all night. Mosquitoes that entered the room were collected from 1800 hrs to 0600 hrs. In addition, volunteers were allowed to lie on cots with exposed body and landing mosquitoes were collected.

In the third method of treatment neem oil was diluted in acetone at 5 and 10 per cent strength and 1 ml diluted neem oil was applied on the mats. Mats were inserted in an electrical device marketed under the trade name and connected with the electric current for heating the mats. Concurrent controls were held with (4 per cent w/w) allethrin-treated mats purchased locally and without mat. Mosquitoes landing on volunteers were collected from 1800 hrs to 0600 hrs. At 0615 hrs all resting mosquitoes inside the room were also collected. In all the studies mosquitoes were collected by insect collectors using a suction tube and a flash light.

The cost of protection from mosquito or sandfly bites with neem oil was computed and compared with the currently used methods. Cost calculations were made to provide protection in a room or per person in Indian Rupees for one year at 1993 prices.

RESULTS

A recent review of neem in malaria control revealed that it is a good larvicide, but its effect on *Plasmodium* spp. was not very pronounced and its action as anti-relapse drug or sporontocidal or gametocytocidal action was not examined. For brevity we have tabulated results of various studies and selected the optimum dose of neem oil to achieve maximum repellent action. The results of these studies are given in Table 1.

In the first experiment when 2 per cent neem oil was applied on the exposed body parts, no *Anopheles* mosquito was collected biting the human volunteers. There was 100 per cent protection for 12 hrs (Sharma et al., 1993).

The application of 2 per cent neem oil also repelled all *P. argentipes*, the vector of kala azar, from biting the human volunteers in an endemic village of Bihar near Patna. In the laboratory tests with 2 per cent neem oil, when *P. papatasi*, the vector of phlebotomus fever, were released in the cages for biting on the arms of volunteers,

Table 1. Protection from bites of mosquitoes and sandfly using neem oil

Treatment	Insect vectors and diseases they transmit	No. of insects entered/landed on baits		Protection (duration in hrs)	Remarks
		Experi-mental	Control		
2% neem oil in coconut oil	*A. culicifacies* Malaria	0	48	100%	Field study was carried out in two villages situated 30 km east of Delhi. Mosquito collections were made for 10 nights from 1800 hrs to 0600 hrs in Oct. 1992. (Sharma et al., 1993 a).
	All anophelines	0	144	100% (12 hrs)	
2% neem oil in coconut oil	*P. argentipes* Kala azar	0	167	100% (12 hrs)	Five-night collection in an endemic village, Govindpur, near Patna, Bihar, in Nov. 1992 (Sharma and Dhiman, 1993).
2% neem oil in coconut	*P. papatasi* Phlebotomus fever				Six-night observation in Malaria Research Centre Lab. in Delhi in Oct. 1992 (Sharma and Dhiman, 1993).
	Nos. released (79)	11		100% (6 hr) 86% from 6 to 12 hrs	
	Nos. released (24)		24	0	
1% neem oil in kerosene and lamp burnt	*A. culicifacies* in living room Malaria	10	1013	99% (12 hrs)	Study villages are located 30 km east of Delhi. Two collections were made, one hourly collection of mosquitoes in the room and another man-biting collection. A total of 10 night collections were made from March to Sept. 1992 between 1800 and 0600 hrs (Sharma and Ansari, 1993).
	A. culicifacies on human bait Malaria	0	41	100% (12 hrs)	
	total Anophelines in living room	106	1838	94.2% (12 hrs)	
	Total Anophelines on human bait	0	108	100% (12 hrs)	
	C. quinquefasciatus in living room	896	4711	80.9% (12 hrs)	
	on human baits Filaria	569	2729	79.1% (12 hrs)	

contd.

Table 1 contd.

Treatment	Insect vectors and diseases they transmit	No. of insects entered/landed on baits		Protection (duration in hrs)	Remarks
		Experi-mental	Control		
	Total mosquitoes in living room	1002	6549	84.6% (12 hrs)	
	on human baits	569	2837	79.9% (12 hrs)	
5% neem oil diluted in acetone and applied on mats	*C. quinquefasciatus* on human baits Filaria	1290	7290	82.3% (12 hrs)	Two collections were made: biting on human volunteers and resting mosquitoes in the room.
	in living room *Anopheles*	2	95	98%	Mosquitoes were collected for 10 nights in Rithala village situated 30 km from the centre in West Delhi. Mosquito collections were made on human baits at hourly intervals from 1800 to 0600 hrs and at 0615 hrs. resting mosquitoes were collected in the room (Sharma et al., 1993 b).
	Culex	78	212	63%	
	Total	95	307	69%	

there was 100 per cent protection for 6 hrs and 86 per cent protection was observed from 6 to 12 hrs. In the control 100 per cent sandflies were collected biting the arms held inside the cage. In the laboratory experiments sandflies were forced to feed on the arm, but even under such stringent conditions 6 hr protection was remarkable.

Burning of lamp containing 1 per cent neem oil in kerosene prevented 99 per cent *Anopheles culicifacies*, the vector of malaria from entering the room and there was 100 per cent protection to human baits sleeping in the same room. Burning of neem oil lamp provided 94.2 per cent protection from *Anopheles* mosquitoes and no *Anopheles* landed on human bait. The protection from *Culex quinquefasciatus*, the vector of filaria, was 80.9 per cent and on the human baits in the same room it was 79.1 per cent. When all mosquitoes were pooled together there was 84.6 per cent protection from the entry of mosquitoes and 79.9 per cent protection from biting on human host. This level of protection lasted for 12 hrs in each case (Sharma and Ansari, 1993).

The application of neem oil was also tested on the mats. The application of 1 ml 5 per cent neem oil diluted in acetone (Sharma et al., 1993) on the mats provided 82.3 per cent protection from *C. quinquefasciatus*. Collection of mosquitoes in the rooms in the morning revealed 98 per cent less *Anopheles* and 63 per cent less *Culex* than in control (Sharma et al., 1993 b).

The above experiments clearly demonstrated the action of low concentrations of neem oil in repelling *A. culicifacies, C. quinquefasciatus, P. argentipes*, and *P. papatasi*. Protection provided lasted for at least 12 hrs, and such an extended protection was not provided by any other repellent marketed in India.

DISCUSSION

This development is to be seen in the background of the load of vector-borne diseases in India. There are at least 2 million parasite-positive cases in the country every year, and 38 per cent of them are due to *Plasmodium falciparum*. *P. falciparum* gives malignant infection and causes almost all deaths due to malaria. This parasite has become resistant to chloroquine. Several studies have reported that malaria incidence in India is grossly underreported (Sharma, 1991 a). The other important mosquito-borne disease is filariasis, transmitted by *C. quinquefasciatus*. In 1981 it was estimated that 308 million people were at risk of filariasis, 24 million people had circulating microfilaria, and 16 million people had disease manifestation. Delimitation surveys are still in progress. Studies have also revealed that areas at one time free from the disease are coming under active transmission. Limited control efforts are directed in some urban areas. There is also a small focus of *Brugia malayi* filariasis transmitted by Mansonia mosquitoes. Dengue, dengue haemorrhagic fever, and dengue shock syndrome transmitted by *Aedes aegypti* are spreading in urban areas and epidemics have been reported from Tamil Nadu, West Bengal, Uttar Pradesh, Maharashtra, and Gujarat. Besides the transmission of diseases, the mosquito nuisance is enormous throughout the country. In Bihar, kala azar has resurged in a big way and in 1992 there were at least 72,000 cases. The disease is spreading to Uttar Pradesh and West Bengal and to southern Bihar from its focus in the north (Sharma, 1991 b).

Against this background personal protection methods are most important. They can prevent human-vector contact, thus eliminating the possibility of transmission of diseases and the nuisance they cause. These methods are very useful for people going into endemic areas and for those who can apply them on a routine basis in endemic areas.

In recent years an important development has been the use of insecticide-impregnated bed nets. Cotton or nylon mosquito nets are treated with synthetic pyrethroids such as deltamethrin, cyfluthrin, or lambdacyhalothrin at 25 mg/sq. m or permethrin at higher dose of 200 mg/sq. m. Treated bed nets provide almost complete protection from mosquito bites, and one treatment lasts for six to eight months.

Field trials in Assam and Orissa have shown that bed nets successfully interrupt malaria transmission. However, in Madhya Pradesh impregnated bed nets were not very successful. Studies on the biting rhythm of mosquitoes revealed that *A. minimus*, the vector of malaria in Sonapur, and *A. fluviatilis*, the vector of malaria in Orissa, start biting only after 2100 hrs and by that time most people start using the bed nets. But in Madhya Pradesh the vector is *A. culicifacies*, a complex of four sibling species A, B, C, and D (Subbaroa et al., 1988). Species C and D start biting after 1900 hrs and by 2100 hrs at least 30 per cent of the bites are over. In case of species A about 5–10 per cent bites are taken by 2100 hrs (Bhatt et al., 1991).

In India *A. culicifacies* is the vector of almost all rural malaria and peri-urban malaria. It is responsible for about 70 per cent of malaria cases every year. *Anopheles culicifacies* has become resistant to DDT and BHC almost throughout the country and to malathion in Gujarat and Maharashtra. Most efforts and the budget of the National Malaria Eradication Programme (NMEP) go in the control of *A. culicifacies*, and there are no hopes of any major improvement in the field operations.

Anopheles culicifacies is invading new areas in the country. In Uttar Pradesh Terai and Orissa it was not a vector during the pre-eradication period in the 1930s and 1940s, but now it is a well-established vector. It is also spreading into new areas along with canal irrigation and thus increasing the receptivity of rural areas to species A, and as a result areas at one time free from the disease have come under active transmission (Sharma, 1991 b).

In areas where insecticide-impregnated bed nets are recommended as a method of malaria control, there is need of a method that can be used either alone or in combination with the nets in providing protection during early evening hours before people start using the nets (Bermejo and Veckan, 1992). Personal protection methods available in the market are not suitable for this purpose as the protection provided is generally 60–70 per cent (Ansari et al., 1990). These methods are also expensive, and some require the use of electricity. Besides, their prolonged use may be harmful (Liu et al., 1987). Use of neem oil in repelling mosquitoes and sandflies is the safest method of personal protection. Neem oil applied on the exposed body parts in cream, vaseline, coconut oil, or mustard oil or burned in kerosene or applied on mats provides complete protection from the mosquito and sandfly bites that transmit malaria and kala azar. Protection provided against *C. quinquefasciatus* is 80 per cent; such protection is not provided by any other repellent. Besides, filaria transmission is weak and reduction of bites by 80 per cent in most areas may be helpful in curtailing transmission.

Table 2 gives the comparative cost of personal protection methods for one year. Space spraying by Finit was the most expensive method and may cost Rs. 780. This was followed by mats (Rs. 494) and mosquito coils (Rs. 475). Ropes developed by the Malaria Research Centre (Sharma et al., 1989) were also expensive and one year's use required Rs. 230. Preparations of neem oil were the cheapest. Neem oil is marketed in the country at Rs. 30–50/kg. and low concentrations of neem oil used for one year may cost Rs. 1 or so. The cost is basically of the oils used as carrier. In comparing impregnated bed nets with neem oil first-year cost is comparable but in subsequent years bed nets would be cheaper as they may last for five to seven years, and re-impregnation may cost only Rs. 14 to 20 per year.

CONCLUSION

Neem oil is indigenous, cheap, and widely available throughout India. Its application does not require any training or technology and provides better protection than any other method used to repel mosquitoes (Chadwick, 1985; Curtis et al., 1987; Coene et al., 1989; Mosha et al., 1992). Neem leaves are eaten as vegetable, and twigs are used as toothbrushes. Neem is used in the treatment of various skin diseases and it has antibiotic properties (Sharma, 1993). Toxicological studies have shown that neem alkaloids are safe for humans and neem oil is edible (Bhide et al., 1958; Gaitonde and Seth, 1958; Rukmini, 1987) after processing. Widespread use of neem oil in the country can strengthen our fight against malaria and other vector-borne diseases.

Table 2. Comparative cost of personal protection methods in India at 1993 prices

Protection method	Product	Cost calculation	Cost of protection per room per head for one year (Rs.)
Space spraying	Finit	100 ml/day at Rs. 20/l sprayer 2 year Rs. 50	780
Mosquito coil	Tortoise	Rs. 6.50 for 10 coils, 2 coils/night. Cost per night Rs. 1.30	475
Mat	Good Night	Heating unit = Rs. 35 35 units electricity = Rs. 15 Cost of impregnated mat Rs. 37 for 30 for one year	494
Mat (neem oil)	MRC	Heating unit = Rs. 35 Electricity = Rs. 15 Neem oil = Rs. 5	55
Rope (jute)	MRC	Rs. 0.6/pm per night = Rs. 219 Insecticide = Rs. 11	230
Lamp (kerosene)	MRC	Kerosene 30 l = Rs. 143 Tin lamp = Rs. 1 Wick = Rs. 2 Neem oil = Rs. 5	151
Neem oil in coconut oil	MRC	Coconut oil 3 ml/day at Rs. 80/kg = Rs. 80 Neem oil = Rs. 4	81
Neem oil in mustard oil	MRC	Mustard oil 3 ml/day at Rs. 32/kg = Rs. 32 Neem oil = Rs. 1	33
Neem oil in vaseline	MRC	Vaseline/gin/day Rs. 200/kg = Rs. 73 Neem oil = Rs. 1	74
Neem oil in cream	MRC	Cream 1 g/day Rs. 400/kg = Rs. 146 Neem oil = Rs. 1	147
Insecticide-impregnated bed nets	MRC		
Single cotton		Bed net Rs. 50 2 treatments Rs. 14	64
Single nylon		Bed net Rs. 70 2 treatments Rs. 14	84
Double cotton		Bed net Rs. 80 2 treatments Rs. 20	100
Double nylon		Bed net Rs. 135 2 treatments Rs. 20	155

REFERENCES

Ansari, M.A., V.P. Sharma, R.K. Razdan and P.K. Mittal 1990. Evaluation of certain mosquito repellents marketed in India. *Indian. J. Malariol.* 27: 57–64.

Bermejo, A. and H. Veckan 1992. Insecticide impregnated bed nets for malaria control: A review of the field trials. *Bull. World Hlth. Org.* 70(3): 293–296.

Bhatt, R.M., R.C. Sharma, V.K. Kohli, A.S. Gautam and D.K. Gupta 1991. Biting rhythms of malaria vector *Anopheles culicifacies* in Kheda district, Gujarat. *Indian. J. Malariol.* 28:91–99.

Bhide, N.K., D.J. Mehta, W.W. Alterkar and R.A. Lewis 1958. Toxicity of sodium nimbidinate. *Ind. J. Med. Res.* 12: 146–149.

Chadwick, P.R. 1985. Mosquito coils protect against bites. *Parasitol. Today* 1(3): 91–92.

Coene, J., N.P. Ngimbi, M.P. Mulumba and M. Wery 1989. Ineffectiveness of mosquito coils in Kinshasa, Zaire. *Trans. R. Soc. Trop. Med. Hyg.* 83: 560–569.

Curtis, C.F., J.D. Lines, J. Ijumba, A. Callaghan, W. Hill and M.A. Karimzad 1987. The relative efficacy of repellents against mosquito vectors of disease. *Med. Vet. Entomol.* 1: 109–119.

Gaitonde, B.B. and U.K. Seth 1958. Pharmacological studies of sodium nimbidinate. *Ind. J. Med. Res.* 13: 156–161.

Liu, W.K., M.H. Wong and Y.L. Mui 1987. Toxic effects of mosquito coils (a mosquito repellent) smoke on rats. I. Properties of the mosquito coil and its smoke. *Toxicol. Letters* 39: 223–230.

Mosha, F.W., R.A.J. Njau and J. Alfred 1992. Efficacy of esbiothrin mosquito coils at community level in northern Tanzania. *Med. Vet. Entomol.* 6: 44–46.

Rukmini, C. 1987. Chemical and nutritional evaluation of neem oil. *Food Chem.* 26: 119–124.

Sharma, V.P. 1991a. Presidential address: Malaria: trends and approaches for its control. Diamond Jubilee Session. The National Academy of Sciences, Allahabad. Malhotra Publishing House, New Delhi.

Sharma, V.P. 1991 b. The changing scenario of disease vector control in India. In: *Glimpses of Science in India.* Diamond Jubilee Commemoration Volume. U.S. Srivastava (ed.), The National Academy of Sciences, Allahabad, Malhotra Publishing House, New Delhi, pp. 69–83.

Sharma, V.P. 1993. Malaria control. In: *Neem Research and Development.* N.S. Randhawa and B.S. Parmar (eds.), Publication No. 3, Society of Pesticide Science, India, pp. 235–241.

Sharma, V.P. and M.A. Ansari 1993. Protection from mosquitoes by burning neem oil in kerosene. *J. Med. Entomol.* (Submitted).

Sharma, V.P., M.A. Ansari, P.K. Mittal and R.K. Razdan 1989. Insecticide impregnated ropes as mosquito repellent. *Indian J. Malariol.* 26: 179–185.

Sharma, V.P., M.A. Ansari and R.K. Razdan 1993 a. Mosquito repellent action of neem (*Azadirachta indica*) oil. *J. Amer. Mosq. Control Assoc.* (Accepted).

Sharma, V.P. and R.C. Dhiman 1993. Neem oil as sandfly repellent. *J. Amer. Mosq. Control Assoc.* (Submitted).

Sharma, V.P., B.N. Nagpal and Aruna Srivastava 1993 b. Effectiveness of neem oil mats in repelling mosquitoes. *Trans. R. Soc. Trop. Med. Hyg.* (Accepted).

Subbarao, S.K., K. Vasantha and V.P. Sharma. 1988. Cytotaxonomy of certain malaria vectors in India. In: *Biosystematics of Haematophagous Insects*, M.W. Service. (ed.), Clarendon Press, Oxford, pp. 25–37.

4

Neem and Transfer of Technology

M.S. CHARI

Director, Central Tobacco Research Institute, Rajahmundry 533 105, India

ABSTRACT

Pest control properties of Neem were known to India in ancient times. Farmers in India protected the crops and food grains insect pests using leaves and seeds. Since the discovery of the antifeedant properties of neem against the insect pests, its use as pest control agent has been exploited. Use of neem kernel suspension in tobacco nurseries to protect the seedlings against the damage of leaf eating caterpillar was first demonstrated on large-scale in the commercial tobacco nurseries. The practical utility of neem products like neem seed kernel, neem oil for pest control on other crops like cotton, paddy, pulses and castor, neem oil as a suckericide on tobacco, neem cake as a nematicide and as an organic manure has been discussed. The importance of transfer of technology through National Demonstration Project, Operational Research Project, Lab-to-Land Programme and Krishi Vigyan Kendra is highlighted.

The major efforts of the Indian Council of Agricultural Research (ICAR) in terms of information and communication apply to the transfer of technology programmes. These programmes include farm advisory units, vocational training units, operational research and on-farm trials, and national demonstrations. In general, the front-line extension activities of the ICAR are carried out to provide a backup to the state and central Departments of Agriculture for suitably demonstrating the farm-worthiness of technologies developed by ICAR.

The first-line extension education effort of ICAR aims at (1) promptly disseminating the latest agricultural technologies to farmers as well as extension agencies; (2) testing and verifying the latest technologies in the socioeconomic conditions of the farmers; (3) analysing the constraints of adoption and giving first-hand scientific feedback to farm scientists; (4) evolving models of transfer of technology as well as farming practices appropriate to local conditions; and (5) promoting research in transfer of technology. Since 1965, the ICAR extension system has been a strong catalytic mechanism for accelerating agricultural production.

The ICAR extension system deals with the first-line extension projects and activities described below.

National Demonstration Project: The Government of India's Ministry of Agriculture initiated a country-wide programme in 1965–66 in which demonstrations were conducted in farmers' fields. One of the important objectives of the national demonstration is to maximize the production as well as income per unit or per unit time. Taking this into account, output/input analysis has been conducted for various crops grown in the demonstrations. The cost-benefit ratios for various crops have clearly indicated that the new agricultural technology is not only production-oriented but also economically viable.

Operational Research Project: The Operational Research Project is primarily devoted to demonstrating the impact of new technologies on a large scale involving a whole village or cluster of villages at a time. Operational Research Projects are concerned with solving community-based problems through appropriate technology and identifying the constraints preventing adoption.

Lab-to-Land Programme: The Lab-to-Land Programme was launched to adopt small-scale, marginal and landless agricultural labourers to educate them about the proven technologies. Critical inputs were given to them.

Krishi Vigyan Kendras: The Krishi Vigyan Kendra (KVK) project is sponsored by the ICAR and is implemented by ICAR institutes, state agricultural universities, reputed voluntary organizations, and state Departments of Agriculture. It has three functions:

1) farm advisory services (will incorporate functions of national demonstrations and Lab-to-Land)
2) vocational training of the farmers (original function of the KVK)
3) operational research, on-farm trials, and adaptive trials

So far, 105 KVKs have been established in India.

Neem (*Azadirachta indica* A. Juss.) is a large, evergreen glabrous tree native to the Indian sub-continent and grows in dry forest regions, even in nutrient-deficient soils. It is commonly grown throughout Southeast Asia, including Pakistan, Sri Lanka, Thailand, Malaysia, and Indonesia. It was introduced into African countries in the last century and is now found in more than 30 other countries including Fiji, the Philippines, the Carribean, Mauritius, Latin America, and the central and southern United States. It was recently introduced into Saudi Arabia and 50,000 neem trees have been planted in Mecca to protect Haj pilgrims from the fiery desert sun (Ahmed, 1988).

The neem tree can survive high temperatures, altitudes between 50 and 1000 m, as little rainfall as 130 mm per year, and prolonged drought. It grows well on moist, dry, stony, clayey, or shallow soils. Although neem grows well at pH 5, it brings surface soil to neutral by its leaf litter. It propagates easily by seed, without pretreatment. Nine- to twelve-month-old seedlings are good for transplanting. Fruiting begins after five years. In India, neem flowers from the beginning of January to May and the fruit matures from May to August. The fruit yield per tree ranges from 37 to 55 kg.

The fresh fruit, on an average, gives 60 per cent dry fruit. The kernel yields 45 per cent fixed oil (Ketkar, 1982).

The uses of neem are very well known in India and are documented in the earliest Sanskrit medical writings (Watt, 1972). For centuries the tree has been held in esteem for its medicinal and pesticidal properties. In fact, the tree has served as a village dispensary in rural areas and served as a place in the Indian Pharmacopoeia. Neem is an effective component in pharmaceuticals, cosmetics, toothpaste, and soap. Neem leaves are used as fodder, neem husk as fuel, and neem pulp for biogas generation. Neem leaves possess excellent antiseptic and antifungal characters. Neem oil is reported to be effective in fighting the dreaded AIDS and also as a promising contraceptive.

USE OF NEEM IN PEST CONTROL

The pest control properties of neem were known to Indians in ancient times. Farmers in India protected crops with natural repellents found in neem seeds and leaves (Pruthi and Singh, 1944). Dried neem leaves are mixed with stored grains or placed with woollen clothes to ward off insects. The insect repellent property of neem seems to have been first reported by Mann and Burns (1927), who observed that neem leaves were not eaten by locust *Schistocerca gregaria* during the locust cycle in 1926–27. However, the antifeeding property of neem was first demonstrated by Pradhan et al. (1962) against desert locust *S. gregaria* and migratory locust *Locusta migratoria* when they observed that as low as 0.1 per cent neem kernel suspension was able to provide complete protection to the treated foliage. This discovery attracted the attention of biologists and chemists all over the world on the use of neem as a pest control agent. Research on chemical constituents of neem led to the isolation and identification of several limonoid compounds which act as antifeedants and growth regulators in insects. Today more than 80 chemical compounds are known to have been isolated and identified from neem (Kulkarni, 1987; Rojatkar, 1989). Of these compounds, azadirachtin is the most important, having strong insect antifeedant and growth disruptant properties. It was first isolated by Butterworth and Morgan (1968). Notable among others having antifeedant effect are meliantriol (Lavie et al., 1967) and salannin (Henderson et al., 1964). Another compound, vepaol, having both insect antifeedant and growth disrupting property has been recently isolated from neem by Sankaram et al. (1987). Extensive investigations have been carried out by scientists all over the world on the use of these compounds and different products of neem such as oil, cake, aqueous or organic solvent extracts of neem seed kernel, and enriched extracts such as neemrich, and valuable information has been obtained.

TRANSFER OF NEEM TECHNOLOGY TO THE FARMERS IN INDIA

Neem research in India has been thoroughly reviewed and documented in the proceedings of the second and third International Neem Conferences (Jotwani and Srivastava, 1984; Parmar, 1987) held in Germany and Kenya respectively. Subsequently, the research on neem has been intensified and several workers have reported the usefulness of neem products in agriculture especially for the control of various pests on crops. Very little effort is made, however, in the vital field of transfer of technology.

AS PEST CONTROL AGENT

The practical utility of neem was first demonstrated by Pradhan et al. (1962) when locust swarms invaded Delhi during July-August 1962. The standing crops in the farm of the Indian Agricultural Research Institute were sprayed with 0.1 per cent neem kernel suspension at 300–600 l/ha depending on the crop and its stage of growth. No feeding was observed on the sprayed crop, although the adults settled on them. The locusts collected from the visiting swarms did not feed on the sprayed foliage in the laboratory, but there was voracious feeding by the same locusts on unsprayed leaves (Anon., 1983).

Patel et al. (1968) demonstrated in field trials in Gujarat that 0.1 per cent and 1.0 per cent seed paste suspension effectively protected sunnhemp (*Crotolaria juncea*) crop from hairy caterpillar, *Amsacta moorei* But. The treated plants were not consumed at all, while the untreated (control) plants were completely defoliated.

On Tobacco

At the Central Tobacco Research Institute (CTRI), Rajahmundry, systematic work on transfer of neem technology to the farmers was taken up by Joshi and his team (1978a), who have established the antifeedant (Joshi and Ramaprasad, 1975; Joshi et al., 1978b) and ovipositional repellent properties of neem (Joshi and Sitaramaiah, 1979) for protection of tobacco nurseries from *Spodoptera litura* damage.

On-Farm Trial

Based on phytotoxicity studies (Ramaprasad, et al., 1979) a spray schedule of neem seed kernel suspension (NSKS)—first spray with 1 per cent NSKS on three-week-old seedlings and subsequently three 2 per cent NSKS sprays at 10-day intervals—was experimented with in a 120 sq m tobacco nursery in 1976. The spray schedule gave excellent protection to tobacco seedlings from *S. litura* damage until the seedlings were ready for transplanting and no phytotoxicity was observed. Only 4.6 per cent seedlings were damaged in treated beds as against 83.5 per cent in untreated control beds (Joshi et al., 1978a).

Demonstration Trials

Six location trials were conducted in farmers' nurseries. In each location, an area of 400 sq m was given the NSKS spray schedule and a corresponding area, where approved pesticides were used, was taken as control. At all locations the schedule of NSKS spray gave significantly superior protection to the transplants over pesticide schedule (Table 1). This superior performance of NSKS can be explained by its double action: it repels the *S. litura* moths, thereby reducing the egg laying, and has immediate effect as an antifeedant. The NSKS behaved similarly at all locations (Joshi et al., 1978a).

Table 1. Per cent seedlings damaged by *Spodoptera litura* at different locations following treatment with (NSKS) vs pesticides

Location no.	Treatment		
	NSKS spray schedule		Control (pesticide spray schedule)
1	10.0 (3.0)		22.6 (14.8)
2	9.6 (2.8)		20.2 (11.9)
3	8.4 (2.2)		13.8 (5.7)
4	11.6 (4.0)		19.3 (10.9)
5	13.8 (5.7)		21.4 (13.3)
6	7.8 (1.8)		12.4 (4.6)
Mean	10.2 (3.1)		18.3 (9.8)
SEM for locations		1.67	
CV%		74.4	
CD at 5%		4.6	
SEM for NSKS vs pesticides		1.0	
CD at 5%		2.7	

Figures in parentheses are retransformed values.

Lab-to-Land Programme

From 1979 to 1980, the usefulness of NSKS technology was also demonstrated to 25 farmers in East Godavari district and 17 farmers in West Godavari district of Andhra Pradesh under the Lab-to-Land programme (Wilson and Naidu, 1981). The majority of the farmers felt that NSKS effectively protected the seedlings from tobacco caterpillar damage (Table 2).

The chi-square value was significant, which indicates that the effectiveness of NSKS and the locations are independent (Table 2); NSKS was equally effective at different locations. The majority of the farmers expressed that the seedlings protected by NSKS exhibited vigorous growth which might be due to good protection from caterpillar damage and additional supply of nitrogen (Table 3). These demonstrations helped in successful transfer of NSKS technology to the farmers and in convincing them of the usefulness of NSKS in protecting the tobacco nurseries from *S. litura* damage. Neem is becoming popular with tobacco nursery growers because it is safe, indigenous and economical (Joshi, 1987).

Table 2. Farmers' opinion on effectiveness of NSKS

District	Level of effectiveness				Total farmers
	Very effective	Effective	Moderately effective	Not effective	
East Godavari	15 (60%)	5 (20%)	4 (16%)	1 (4%)	25
West Godavari	6 (35.3%)	9 (53.9%)	0 (0)	2 (11.7%)	17

Chi-square (X_1^2) value = 3.841 at 5 per cent.

Taking its cue from the neem technology developed by CTRI, the ILTD Division of Indian Tobacco Company Limited developed the neem-based formulation Wellgro, which is being marketed and distributed to the nursery growers.

Table 3. Farmers' observations on health of seedlings protected by NSKS

District	Level of healthiness			Total farmers
	Very healthy	Healthy	Moderate	
East Godavari	15 (65.2%)	8 (34.8%)	0	23
West Godavari	7 (14.2%)	9 (53.0%)	1	17

Observed chi-square (X_1^2) value = 2.85 at 5 per cent.

NSKS IN INTEGRATED PEST MANAGEMENT

An integrated schedule involving spraying of antifeedant neem seed endosperm suspension, nuclear polyhedrosis virus, and chitin inhibitor diflubenzuron with castor as trap crop was also developed for effective management of *S. litura* in bidi tobacco nurseries in Gujarat (Chari et al., 1985). Similarly, integrated management for *S. litura* in flue-cured tobacco nurseries and planted flue-cured Virginia tobacco of Andhra Pradesh and Karnataka involving NSKS spray as one of the components was developed (Chari and Ramaprasad, 1987; Krishnaprasad et al., 1990; Shivayogeswara et al., 1991).

Use of NSKS as one of the components in integrated management of *S. litura* has been recently demonstrated in three farmers' fields in East and West Godavari districts of Andhra Pradesh. The results showed that an additional income of Rs. 21,875 (17.5 per cent) can be obtained per hectare of tobacco nursery by adopting the technology (Anon., 1991).

On Cotton

The efficacy of Repelin, a botanical pesticide formulated by the Research Department of ITC Limited containing products of neem (*A. indica*), mahua (*Madhuca indica*), and sesame (*Sesamum indicum*), was demonstrated in large-scale field trials (12 ha plots) for control of the bollworm *Heliothis armigera* and white-fly on cotton by the Andhra Pradesh Agricultural University in two locations of Guntur district during 1986–1988. Repelin was applied eight times in Pedanandipadu village and four times in Parchuru village. Its effects was more pronounced in Pedanandipadu village. Repelin-treated plots recorded only 1 per cent damage by bollworm as compared to 27 per cent damage in plots treated with insecticides. The mean white-fly adult population was 11.4 in Repelin-treated plots as against 34.1 in control plots. The cotton yield was 2625 kg/ha in Repelin-treated plots as against 820 kg/ha in control plot (Table 4). In Parchuru village also the cotton yield was higher in Repelin-treated plot than in control (Subramaniam, 1988).

ILTD Division of ITC Limited also laid out 115 demonstration plots of 1 acre each in Guntur district where Repelin was sprayed three times at the peak of *H. armigera* outbreak. Repelin was alternated with conventional insecticides like monocrotophos, phosalone, endosulfan, quinalphos, acephate, echion and triazophos. Except in a few places, yields were considerably higher in Repelin-treated plots (Table 5).

Table 4. Incidence of key pests of cotton in Repelin-treated plots

Pest	Pedanandipadu (MCU 5)		Parchuru (LRA 5166)	
	Sprayed	Control	Sprayed	Control
White-fly				
Adults (no.)/leaf*	11.4	34.1	16.9	23.2
Heliothis				
Eggs (no.)/plant*	0.08	8.4	2.3	15.2
Larvae (no.)/plant*	0.12	13.2	5.4	19.2
Boll damage (%)[1]	0–4.7 (1.14)	7–74 (27.1)	3–28 (9.1)	7–84 (37.7)
Yield (q/ha)[2]	26.25	8.2	16.25	5.0

* Indicates mean

[1] Indicates range and mean (within parentheses).

[2] Quintal (q) = 100 kg

Table 5. Effect of Repelin application on cotton yields in plots infested with *Heliothis armigera*

Demonstration[1] plots (no.)	Spray applications (no.)		Yield (kg/ha)	
	Repelin + insecticide	Insecticide	Repelin + insecticide	Insecticide
87	3 + 12	19	1411	871
16	3 + 13	21	1206	1051
12	3 + 14	20	1008	1094

[1] Trials conducted on 115 demonstration plots, each measuring about 0.4 ha.

On Paddy

Field demonstration trials conducted by the State Department of Agriculture with Repelin in farmers' paddy fields of East Godavari district in Andhra Pradesh revealed that white-backed planthopper, *Sogatella furcifera*, and the brown planthopper, *Nilaparvata lugens*, were effectively controlled. In Repelin-treated plots, yield increased by 400 to 600 kg/ha over controls (Subramaniam, 1988). Field trials carried out by Krishnaiah et al. (1990) showed that ready-to-use neem formulations like Repelin and Neemta 2100 in combination with half the recommended dose of recommended insecticide monocrotophos effectively controlled the stem borer and leaf folder. Samalo et al. (1990) conducted field trials with neem oil for control of major pests of wet season rice which revealed that 3 per cent neem oil sprayed at 10, 25, 40, and 55 days after transplanting significantly reduced the gall midge infestation.

On Pulses

The efficacy of 3 and 5 per cent extracts of neem kernel and neem oil was determined against *H. armigera* Hubn. in multi-location trials with chick-pea carried out under the All India Coordinated Pulses Improvement Project (AICPIP) during 1979–80 (Saxena, 1979–80). Sachan and Lal (1990) also evaluated the performance of neem seed kernel extract (5 per cent), neem leaf extract (5 per cent), and neem oil along with recommended insecticides against *H. armigera* on chick-pea and pigeon-pea under AICPIP during 1982–83 and concluded that neem seed kernel extract and leaf extract had good scope for control of *H. armigera* on chick-pea.

On Castor

Chari and Muralidharan (1985) showed that neem acts as a feeding deterrent of castor semilooper, *Achoea janata* Linn., on castor in large-scale trials.

Use of Neem Cake as Nematicide

Mishra and Gaur (1984) showed in field trials that neem oil cake at 17 q/ha provided protection to mung crop against *Tylenchorhynchus* sp., *Helicotylenchus* sp., *Rotylenchus reniformis*, and *Meloidogyne incognita* and increased grain yield by 42 per cent as compared with 63 per cent by fensulfothion seed treatment and 8.9 per cent by aldicarb granules. Ravindran Nair et al. (1984) observed that 15 g neem cake per plant reduced the population buildup of the rootknot (*M. incognita*) nematode of brinjal and increased the crop yield. Furthermore, Gaur and Mishra (1984) reported that a combination of neem cake and phorate seed treatment gave better control of nematodes viz., *Tylenchorhynchus* sp., *Helicotylenchus* sp., *H. indicus*., *Rotylenchus reniformis* and *Hemicriconemoides* sp., on lentils and also gave better yields of lentils and subsequent mung crop.

Ahmed (1988) reported that about 88 per cent of cardamom growers in Kerala relied on neem cake for nematode control. They used at 100–250 kg/ha, incorporating it into the soil around each cardamom plant once a year. They expressed that no synthetic insecticide came even near to neem cake's performance against nematodes. Very recently Hussaini et al. (1992) conducted nursery trials on fcv tobacco with neem cake at 100, 200, and 400 g/m^2 alone and in combination with soil solarization (50G low density polyethelene, LDPE, LDPE sheet used as cover for beds) for four weeks after application of neem cake powder for their efficacy against (*M. incognita*). The trials showed neem cake alone at all the doses significantly reduced the root knot infection on seedlings up to 80 days from sowing. Furthermore, the efficacy of neemcake was enhanced by soil solarization.

Use of Neem Oil as a Suckericide on Tobacco

When the apical dominance is removed to improve the quality of tobacco, auxiliary shoots—known as suckers—develop. It was shown for the first time at the CTRI that neem oil emulsion at 10 per cent is an effective suckericide for tobacco (Bangarayya et al., 1982). It was also superior to manual desuckering, as were the other oil emulsions. Cured leaf yield and bright grade yield were improved when neem oil emulsion was used as a suckericide (Bangarayya et al., 1982). The chemical quality attributes were not affected by neem oil emulsion application (Table 6). The 10 per cent neem oil emulsion is more economical than other oil emulsions and can be recommended as a suckericide (Table 7).

Table 6. Effect of neem oil emulsion (NOE) on sucker control, yield and chemical quality of fcv tobacco

Treatment	Dry weight of suckers	Cured leaf yield (kg/ha)	Bright grades leaf(kg/ha)	TBLE* (kg/ha)	Reducing sugars (%)	Nicotine (%)
NOE 20%	108.1	960.5	286.7	514.6	16.6	1.74
NOE 10%	213.3	1029.2	323.1	573.1	15.9	1.66
NOE 5%	343.9	918.3	361.4	564.8	18.5	1.54
Topped and desuckered	536.0	899.1	298.0	511.1	16.8	1.83
Topped and unsuckered	1051.6	827.4	232.5	462.8	15.4	1.50
SEM	50.2	56.8	35.8	40.1	0.9	0.14
CV %	40.0	18.8	35.6	23.1	16.0	26.93
CD at 5%	139.1	157.4	99.3	111.2	n.s.	n.s.

n.s. = non-significant.
*TBLE = bright grade + (0.465 x medium grade).

Table 7. Economics of promising suckericide treatments

Treatment	Cost of emulsion (Rs.)	Cost of application (Rs.)	Total (Rs.)
Three manual desuckerings	—	195	195
Neem oil emulsion 10%	95	47	142
Mahua oil emulsion 20%	218	47	265
Peanut oil emulsion 25%	255	47	302

Neem Cake as Organic Manure

Neem cake is the by-product after oil has been extracted from neem seed. Compounds in neem cake called 'triterpenes' retard the activity of the nitrifying bacteria, regulating nitrogen supply to plants.

Bains et al. (1971) were the first to show under field conditions that treatment of urea with an acetone extract of cake (which includes most fat and lipid associates) compares well with proven nitrification inhibitors and was superior to sulphur-coated urea under actual field conditions. Data generated from experiments under actual field conditions revealed that neem-coated urea was as effective as N-Serve when applied in wheat and paddy fields. Further trials at the Indian Agricultural Research Institute, New Delhi, showed that when 160 kg of urea and 40 kg of neem cake are applied on 1 ha of paddy, the yield is 0.6 tonnes higher than when 200 kg of urea alone is applied. Though urea neem cake fertilizer is more effective, it contains less nitrogen for the same weight and accordingly the price will be lower, which is a disincentive to manufacturers. However, Godrej Agrovet is marketing a new product called 'Nimin', which contains neem-based triterpenes to be mixed with urea (Anjani Khanna and Gopalakrishna Warrier, 1992). The Konkan Krishi Vidyapeeth has recommended Nimin-coated urea. Similarly, research experiments have been conducted at agricultural universities in Rajasthan, Punjab, and Tamil Nadu and at various research institutes of ICAR which have confirmed the efficacy of Nimin-coated urea. Sales of Nimin have been steadily increasing (Fig. 1).

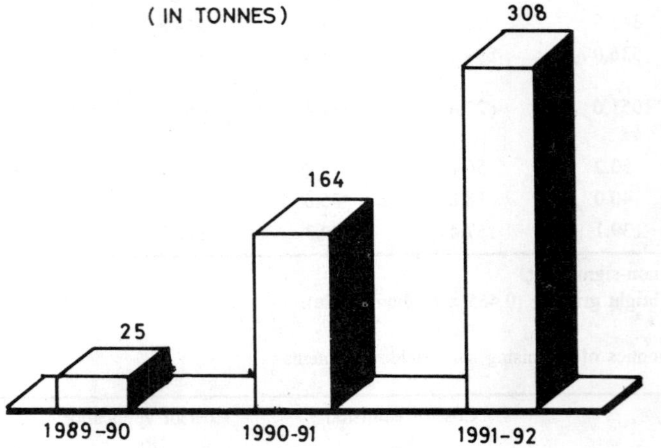

Fig. 1. Sales of Nimin, a product to reduce nitrogen loss of urea (Godrej Agrovet)

It can be inferred from the information given above that the efforts made so far on the transfer of neem technology are not commensurate with the voluminous data collected on use of neem leaves, seed kernel, oil, cake, enriched and aqueous formulations of seed kernel and cake, and ready-to-use neem formulations as insecticides, antifeedants repellents, growth disruptors, ovipositional repellents, and chemosterilants for control of several insect pests on crops and also as nematicides and nitrogen regulators. There is an urgent need to strengthen and gear up the extension wings of state agriculture departments and universities, Krishi Vigyan Kendras, and research institutes of ICAR in collaboration with the manufacturers of neem products to disseminate the information on neem to farmers.

REFERENCES

Ahmed, S. 1988. Potential of using the neem tree (*Azadirachta indica*) for pest control and rural development. Seminar held at IARI, New Delhi, May 27, 1988. *Neem Newsl.* 5(4): 49–55.

Anjani Khanna and S. Gopalakrishna Warrier. 1992. Tapping commercial potential. *Down to Earth. Sci. Environ.* (Soc. Environ. Comm., New Delhi) 1(10): 8–9.

Anonymous. 1983. Neem in insect control. In: *Neem in Agriculture. Res. Bull.* 40: 18–40.

Anonymous. 1991. Annual Report of Central Tobacco Research Institute, Rajahmundry, 1990–91.

Bains, S.S., R. Prasad and P.C. Bhatia. 1971. *Fertil. News* 16(3): 30–32.

Bangarayya, M., C.B. Sarma, Y.Ch. Narasimhamurty and D. Prabhakara Babu. 1982. Suppression of suckers with edible oils in FCV tobacco. *Tob. Res.* 8(1): 25–29.

Butterworth, J.H. and E.D. Morgan. 1968. Isolation of a substance that suppresses feeding in locusts. *J. Chem. Soc. Chem. Comm.* 1: 23–34.

Chari, M.S. and C.M. Muralidharan. 1985. Neem (*Azadirachta indica*) as feeding deterrent of castor semilooper (*Achoea janta* Linn.). *J. Entomol. Res.* 9(2): 243–245.

Chari, M.S. and G. Ramaprasad. 1987. Integrated control of pests on tobacco. *Proc. Nat. Symp. on Integrated Pest Control—Progress and Perspectives*. Trivandrum, Nov. 1988, pp. 42–52.

Chari, M.S., T.M. Bharpoda and S.N. Patel. 1985. Studies on integrated management of *Spodoptera litura* Fb. in tobacco nursery. *Tob. Res.* 11(2): 93–98.

Gaur, H.S. and S.D. Mishra. 1984. Integrated control of nematodes in lentil with aldicarb, neem cake and seed treatment with thimet and its residual effect on subsequent mung crop. Souvenir and Abstracts, *Nat. Sym. on Soil Pest and Soil Organisms* (Banaras Hindu Univ., Varanasi), Oct. 29–31, Abstr. No. 108, p. 73.

Henderson, R., R. McCrindle and K.H. Overton. 1964. Salannin. *Tetrahedron Lett.* 24: 1517–1523.

Hussaini, S.S., K.S. Bharata and R.V.V. Prasada Rao. 1992. Combined efficacy of soil solarization and neem-cake application against root-knot (personal communication).

Joshi, B.G. 1987. Use of neem products in tobacco in India. In: *Natural pesticides from the Neem Tree and Other Tropical Plants*. Proc. 3d Int. Neem Conf., Nairobi, 1986, H. Schmutterer and K.R.S. Ascher (eds.), GTZ, Eschborn, pp. 479–494.

Joshi, B.G. and G. Ramaprasad. 1975. Neem kernel as an antifeedant against the tobacco caterpillar (*Spodoptera litura* F.). *Phytoparasitica* 3(1): 59–61.

Joshi, B.G., G. Ramaprasad and S. Sitaramaiah. 1978a. Neem kernel suspension protects tobacco nurseries. *Indian Fmg.* 28(9): 17–18.

Joshi, B.G., G. Ramaprasad and S.V.V. Satyanarayana. 1978b. Relative efficacy of neem kernel, fentin acetate and fentin hydroxide as antifeedants against tobacco caterpillar, *Spodoptera litura* Fabricius, in the nursery. *Indian J. Agric. Sci.* 48(1): 19–22.

Joshi, B.G. and S. Sitaramaiah. 1979. Neem kernel as an ovipositional repellent for *Spodoptera litura* (F.). *Phytoparasitica* 7(3): 199–202.

Jotwani, M.G and K.P. Srivastava. 1984. A review of neem research in India in relation to insects. In: *Natural Pesticides from the Neem Tree and Other Tropical Plants*. Proc. 2nd Int. Neem Conf., Rauischholzhausen, 1983. H. Schmutterr and K.R.S. Ascher (eds.), GTZ Eschborn, pp. 43–56.

Ketkar, C.M. 1982. Properties and uses of neem (*Azadirachta indica* A. Juss.): Its products and by-products. In: *Cultivation and Utilisation of Medicinal Plants*, RRL, Jammu Tawi, CSIR, New Delhi.

Krishnaprasad, N.K., B. Shivayogeswara, H. Mallikarjunaiah and K.R. Vasanth. 1990. Management of *Spodoptera litura* F. (Noctuidae: Lepidoptera) in FCV tobacco nursery by integration of different approaches including botanical pesticides. *Proc. Nat. Symp. on Problems and Prospects of Botanical Pesticies in Integrated Pest Management*, Rajahmundry, Jan. 21–22 (in press).

Krishnaiah, N.V., M.B. Kalode and I.C. Pasalu. 1990. New approaches in utilisation of botanicals in rice insect pest control. *Proc. Nat. Symp. on Problems and Prospects of Botanical Pesticides in Integrated Pest Management*, Rajahmundry, Jan. 21–22 (in press).

Kulkarni, M.M. 1987. *Chemistry and biological activity of natural products*. Ph.D. thesis, Univ. of Poona, Pune.

Lavie, D., M.K. Jain and S.R. Shpan-Gabrielith. 1967. A locust phagorepellent from two *Melia* species. *J. Chem. Soc. Comm.* 910–911.

Mishra, S.D. and H.S. Gaur. 1984. Control of nematodes infesting mung with nematicidal seed treatment and field applications of dasanit, aldicarb and neem cake. Souvenir and Abstracts. *Nat. Symp. on Soil Pest and Soil Organisms* (Banaras Hindu Univ., Varanasi), Oct. 29–31, Abstr. No. 185, p.71.

Mann, N.H. and W. Burns. 1927. A locust of 1927 in Kathiawar, Sind and Gujarat. *Agric. J. India* (Calcutta) 22(5): 325–332.

Parmar, B.S. 1987. An overview of neem research and use in India during the years 1983–1986. In: *Natural Pesticides from the Neem Tree and Other Tropical Plants. Proc. 3d Int. Neem Conf.*, Nairobi, 1986. H. Schmutterer and K.R.S. Ascher (eds.), GTZ, Eschborn, pp. 55–80.

Patel, H.K., V.C. Patel, M.S. Chari, J.C. Patel. and J.R. Patel. 1968. Neem seed paste suspension—A sure deterrent to hairy caterpillar, *Amsacta moorei* But. *Madras Agric. J.* 55(11): 509–510.

Pradhan, S., M.G. Jotwani and B.K. Rai. 1962. The neem seed deterrent to locusts. *Indian Fmg.* 12(8): 7–11.

Pruthi, H.S., and M. Singh. 1944. Stored grain pests and their control. *Imperial Counc. Agric. Res. Misc. Bull.* No. 57.

Rojatkar, R.S. 1989. *Chemical investigation of some Indian patents*. M.Sc. thesis, Univ. of Poona, Pune.

Ramaprasad, G., S.V.V. Satyanarayana and S. Sitaramaiah. 1979. Studies on phytotoxic effect of neem kernel suspension on tobacco seedlings. *Indian J. Plant Prot.* 7: 100–101.

Ravindran Nair, K.K., Kamalakshmi Amma and K.J. Kurian. 1984. Use of the oil cakes for the control of root-knot nematodes in brinjal. Souvenir and Abstracts, *Nat. Symp. on Soil Pest and Soil Organisms* (Banaras Hindu Univ., Varanasi), Oct. 29–31, Abstr. 107, p. 72.

Sachan, J.N. and S.S. Lal. 1990. Role of botanical insecticides in *Helicoverpa armigera* management in pulses. *Proc. Sym. on Problems and Prospects of Botanical Pesticides in Integrated Pest Management*, Rajahmundry, Jan. 21–22 (in press).

Samalo, A.P., B. Senapati, C.R. Satpathy and T.J. Jacob. 1990. Effect of neem derivatives on incidence of some major insect pests in wet season rice. *Proc. Nat. Symp. on Problems and Prospects of Botanical Pesticides in Integrated Pest Management*, Rajahmundry, Jan. 21–22 (in press).

Sankaram, A.V.B., M. Marthanda Murthy, K. Bhaskaraiah, M. Subramanyam, N. Sultana, H.C. Sharma, K. Leuschner, G. Ramaprasad, S. Sitaramaiah, C. Rukmini and P. Udayasekhara Rao. 1987. Chemistry, biological activity and utilisation aspects of some promising neem extractives. In: *Natural Pesticides from the Neem Tree and Other Tropical Plants*. Proc. 3d Int. Neem Conf., Nairobi, 1986, H. Schmutterer and K.R.S. Ascher. (eds.), GTZ, Eschborn, pp. 127–148.

Shivayogeswara, B., H. Mallikarjunaiah, N.K. Krishnaprasad and M.V.N. Shetty. 1991. Integrated management of *Spodoptera litura* Fabricius (Noctuidae: Lepidoptera) in FCV tobacco crop. *Tob. Res.* 17(2): 59–61.

Saxena, H.P. (ed.) 1979–80. *Report of Research Work of Rabi Pulses*. All India Coordinated Pulse Improvement Project, ICAR, New Delhi, 65 pp.

Subramaniam, T.S. 1988. RD-9 Repelin—A botanical insecticide for controlling certain crop pests. *Final Workshop of IRRI-ADB-EWC Project on Botanical Pest Control in Rice-based Cropping Systems*. International Rice Research Institute, Los Banos, Laguna, Philippines, Dec. 12–16.

Watt, G. 1972. *A Dictionary of the Economic Products of India* Vol. 15. Periodical Experts, New Delhi, p. 676 (reprinted from 1891 edition).

Wilson, M.J. and S.K. Naidu. 1981. Impact of Lab-to-Land programme on the production of healthy vigorous seedlings of flue-cured Virginia tobacco. *Tob. News* 5(2): 12–18.

5

The Use of Neem by Farmers in the Dominican Republic

CARSTEN HELLPAP

GTZ / Institut fur Phytopathologie und Angewandte Zoologie,
Ludwigstr. 23, 6300 Giessen, Dominican Republic

ABSTRACT

In 1987 the Deutsche Gesellschaft für Technische Zusammenarbeit (GTZ) and the Instituto Politécnico Loyola started a joint project to promote and disseminate the use of local, home-made insecticides from neem by small-scale vegetable farmers in the Dominican Republic. Since then more than 400,000 neem trees have been planted in the country. HPLC analysis of the first harvests shows average contents of azadirachtin in the seeds of 4–5 mg/g.

In more than 60 field trials simple aqueous seed extracts proved to be effective against the most important pests of beans, cabbage, cucumbers, eggplant, maize, melon, okra, and tomato. Some results will be presented here. Of special interest was the good controlling effect of neem against the white-fly *Bemisia tabaci*, as this pest has become resistant against a broad range of synthetic insecticides.

The project is cooperating closely with a group of 60 pilot farmers, who visited once a week and whose experiences with neem are regularly evaluated. They use home-made aqueous neem extracts prepared in barrels with a specially designed wire mesh filter. The method is easy and is accepted by the farmers.

Socioeconomic data demonstrate that the use of neem extracts is profitable for the farmers.

INTRODUCTION

Due to encouraging research findings in the 1970's and early 1980's on the insecticidal properties of neem, the project Natural Pesticides from Tropical and Subtropical Plants of the Deutsche Gesellschaft für Technische Zusammenarbeit (GTZ) and the Institute of Phytopathology and Applied Zoology of the University of Giessen put main emphasis on promoting and disseminating the use of local, home-made insecticides from neem by small-scale farmers. For this purpose joint programmes with organizations in several countries were begun which included planting neem trees, supplying farmers with raw material, and elaborating and distributing brochures, video tapes, slides and other informative material to farmers in connection with intensive extension work.

This paper describes some of the main experiences of the neem programme in the Dominican Republic which was carried out in cooperation with Instituto Politécnico Loyola, San Cristóbal.

PLANTING OF NEEM

When the programme started in 1987 the Dominican Republic had very few neem trees, most of them less than five years old. In contrast, the number of trees in neighbouring Haiti was rather high due to reforestation programmes in the 1960's where neem was planted specially in dry and eroded areas.

With seeds from Togo and local trees of the island Hispaniola which probably originate from Nigeria, the project started to plant thousands of neem trees in different regions of the Dominican Republic. About 900,000 seedlings were produced and about 400,000 seedlings distributed to farmers or planted along roadsides and irrigation canals or on small plots of fallow land. The concept was not to promote large plantations, but to motivate farmers or farmer groups to plant neem on small plots of their own or municipal land, as the programme was promoting a decentralized use of neem by farmers and not a central processing by an extraction plant.

However, when neem became more and more popular several institutions started to establish neem plantations with thousands of trees. Sugarcane companies in particular planted large numbers on eroded and saline fields. Even under these unfavourable conditions neem proved to develop relatively well.

Both planting patterns, the dispersed planting by farmers as well as the propagation of plants, offer advantages.

The advantages of dispersed planting are:

1) Farmers feel responsible for trees.
2) Farmers can supply themselves.
3) Farmers develop their own harvesting and drying techniques.
4) Farmers can get additional income.
5) Low risk of pest attack on neem and destruction by fire.
6) Low costs for planting and maintaining.

The advantages of neem plantations are:

1) High, uniform quality of seeds.
2) Harvest of a high percentage of the fruit/seed potential.
3) Quick and proper processing of fruits.
4) Possibility for mechanization of harvest and processing.

According to our experience the best growth and fruit production are obtained under dry climatic conditions and high groundwater level. Thus, two-year-old neem trees along irrigation canals in dry areas were full of fruits and gave a yield of several kilograms of seeds.

MARKETING AND QUALITY OF NEEM SEEDS

In many countries including the Dominican Republic neem trees are planted first of all in dry areas. However, the demand for seeds is low there, whereas it is much higher in the humid and fertile regions, where vegetables are grown. Due to this gap between supply and demand the use of home-made neem insecticides depends largely on an efficient marketing of seeds between the regions. As a first step in this direction the project provided vegetable growers with seeds purchased in the dry areas. Additional amounts of raw material were imported from Haiti. The quality of the neem seeds was at first not satisfying because of improper harvesting. Seeds were picked sporadically, and not systematically in short time intervals. Consequently, a considerable part of them lay on the ground for several days and became mouldy, thus reducing the azadirachtin content. Seeds from Haiti had a slightly poorer quality than those from the Dominican Republic (Fig. 1). The project is now trying to use a higher proportion of the Dominican neem potential, which is estimated at 30–50 tonnes of seeds per year. This amount would be sufficient to cover the actual demand.

To improve the quality of the raw material the project has put more emphasis on promoting a proper harvest, which is organized by village traders. These are often women, who buy seeds from pickers living in the village and sell them to farmers or organizations. The traders are now trained in the special collecting and drying requirements of neem. Furthermore, prices for seeds will be graded according to quality parameters such as humidity, which will be measured directly at the village with a mobile moisture meter. It is estimated that due to these measures the average azadirachtin content in the kernels can be increased to a level above 0.45 per cent.

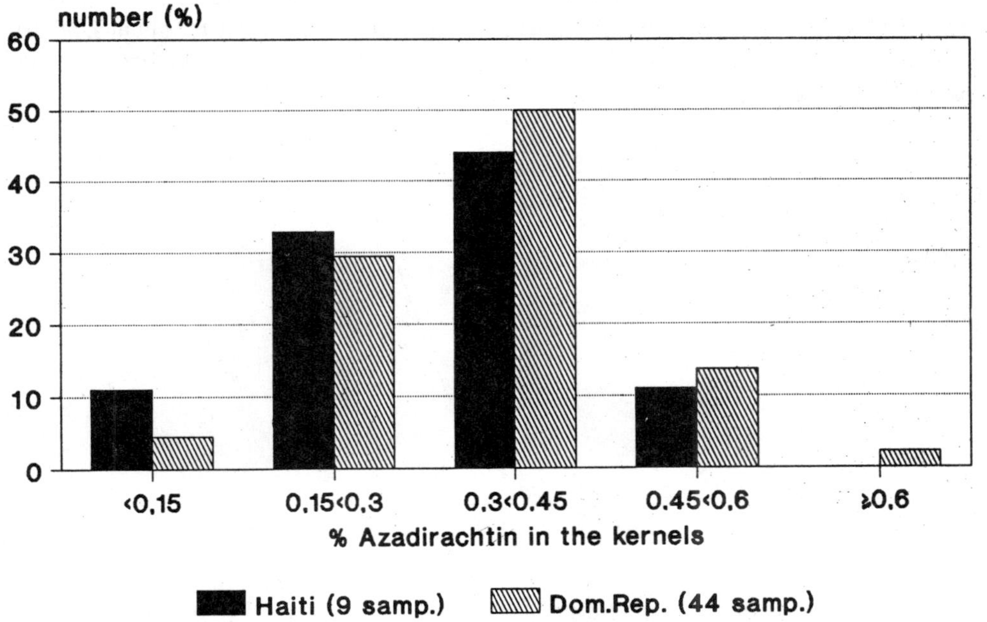

Fig.1. Classification of neem kernels according to the azadirachtin content

Additional elements in the marketing structure are regional distribution centres equipped with threshers and mills. They shall provide farmers with already processed raw material, such as ground neem kernels, to make the preparation of aqueous extracts easier.

PILOT FARMERS PROGRAMME

Whenever innovations are introduced to agricultural practice they are accepted not at once but in several steps (Fig. 2). At the beginning some few pioneer farmers, who are more open to new ideas, apply and test the new technology. If the results are convincing neighbouring farmers follow the example. If the introduction fails in this stage the dissemination of the technology can be blocked for a long time. Otherwise, the innovation spreads more and more and becomes part of the common agricultural practice.

In the Dominican Republic the use of neem insecticides was completely new so that the extension work had to start from point zero. One of the first steps was to establish a close cooperation with so-called pilot farmers in distinct regions. They are chosen representatives of farmer associations which showed high interest in neem. The pilot farmers were trained repeatedly on quality of raw material, preparation of simple extracts, application methods, dose determination, and range of target pests and received periodical advice about proper use of neem.

An important result of the cooperation with the pilot farmers was the development of a new method for preparing aqueous extracts appropriate to their conditions. The original recommendation to use buckets for the seed-water mixture, which is then poured through a coarse cloth or gauze, was not accepted as it was too labour-intensive and difficult when larger amounts were needed. The new method uses a 220 l barrel and a cylindrical wire mesh which farmers can buy from the project for about US$ 10. The wire mesh is packed in a bag of fine cloth or gauze. One day before spraying,

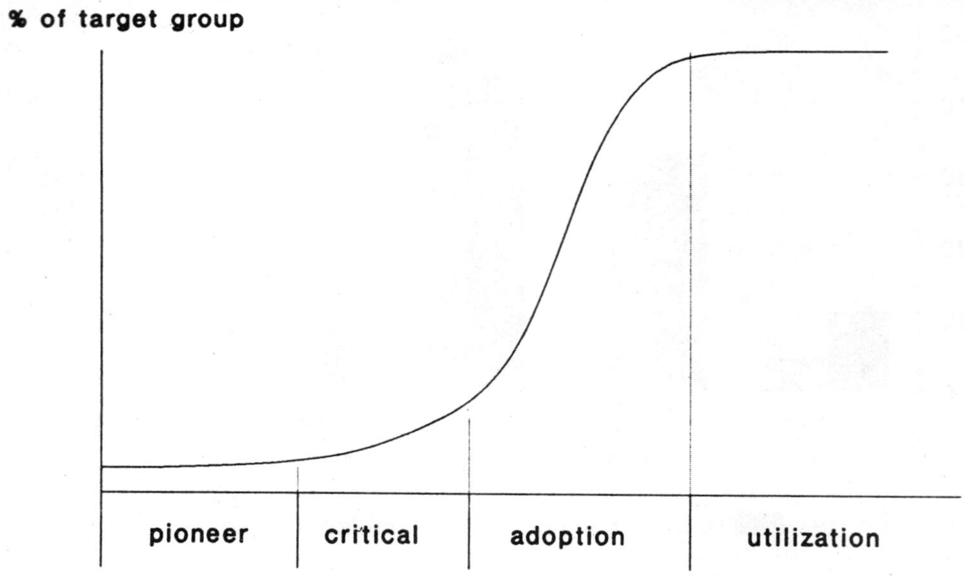

Fig. 2. Diffusion process of innovations

ground seeds are mixed with water and left to stand overnight. The next day the bag with the wire mesh is put into the barrel with the open side to the top, so that it slowly sinks down to the bottom. In this way the seed-water-suspension is filtered as the extraction liquid flows into the bag through the mesh, keeping particles between the cloth and the wall of the barrel. The technique has been proved for several seasons and is now widely accepted among the pilot farmers, who discussed their experiences with neem in regular meetings.

Due to the great interest in the vegetable growing area the number of pilot farmers has steadily increased to 60 at the present time. It is planned to extend the programme to about 100 within the next two years, which would be the limit of the extension capacity of the project, although more farmers would like to join it. We think that the 100 pilot farmers are a good base for the dissemination of the technology to neighbours and others interested vegetable growers during a critical phase.

A second key element of the extension work has been demonstrative trials on farm level, so that farmers could see the excellent effects of neem against many pests. Tables 1 and 3 (Dreyer and Hellpap, 1991) and Table 2 (Serra, 1992) give some examples of the more than 60 field trials conducted.

Table 1. Results of a field trial in cabbage in the Dominican Republic

	Index of leaf damage (4 count)		Yield (kg/plot)
	P. xylostella 0–3*	*L. erysimi* 0–3**	
Untreated	2.2 b	2.3 b	0
Thuricide 2 g/l + molasses 10 g/l	0.5 a	2.3 b	0
Neem seed water extract 60 g/l	0.4 a	0 a	52
Probability of error	p = 5%		

Source: Dreyer and Hellpap, 1991

* Damage of four outer leaves:; 0 = without damage; 1 = up to 20%; 2 = 20–60%; 3 = > 60%.

** Damage of four outer leaves: 0 = without damage; 1 = few chlorotic spots; 2 = chlorotic leaves with necrotic edges; 3 = > 50% necrotic surface of the leaves.

Additional data about the field trial in cabbage in the Dominican Republic:

Variety: Marion Market

Date of sowing in the nursery bed: 25.11.87

Date of transplanting: 5.1.88

Date of application:	15.1.88	
	22.1.88	
	30.1.88	
	5.2.88	
	12.2.88	
	19.2.88	
	26.2.88	
Date of evaluation:	20.1.88	(1. count)
	3.2.88	(2. count)
	18.2.88	(3. count)
	3.3.88	(4. count)

Consequently 4. (4th) count means data recording after 7 applications or 47 days after transplanting.

Table 2. Results of a trial against *Bemisia tabaci* on tomato (1992)

	No. of *B. tabaci*/cm² 3 counts		
	Eggs	Nymphs	Pupae
Comb. synth. insecticides	25.7	56.3	0.80
Mineral oil 0.75% + comb. synth. insecticides	19.2	23.9	0.36
Neem oil 1.5%	5.0	5.7	0.03
NSWE 50 g/l	8.7	7.7	0.10

Source: Serra, 1992

Additional data about the field trial in tomato in the Dominican Republic:

Variety: Napoli-VF
Date of direct sowing: 14.4.89
Date of application and type of treatment:
Type of treatment

	type of treatment			
Date of appl.	Comb. synth. insecticides	Comb. synth. insecticides + mineral oil 0.75%	Neem oil	Neem seed water extract
10.5.89	Dimethoate 0.26%	Deltamethrin 0.08%	1%	50g/1
16.5.89	Monocrotophos 0.3%	Endosulfan 0.25%	1%	50g/1
23.5.89	Methamidophos 0.18%	Malathion 0.25%	1.5%	50g/1
30.5.89	Dimethoate 0.26%	Oxamyl 1.0%	1.5%	50g/1
6.6.89	Monocrotophos 0.3%	Permethrin 0.6%	1.5%	50g/1
13.6.89	Methamidophos 0.18%	Endosulfan 0.25%	1.5%	50g/1
22.6.89	Dimethoate 0.26%	Malathion 0.25%	1.5%	50g/1

Date of evaluation: 23.5.89 (1. count)
6.6.89 (2. count)
22.6.89 (3. count)

Consequently 3. (3rd) count means data recording after 6 applications or 69 days after sowing.

Table 3. Results of a trial in cucumber (1991)

	No. of pests (4. counts)				Yield
	*Diaphania hyalinata**		*Aphis gossypii***		
	L1–2/L3–5/pup.	%loss (leaf)	Inf. Index 0–7		Pou/plot
Untreated	30.3/20.5/1.5	21.0a	3.5a		11.4
NSWE 50g/l	0.5/8.8/0.3	3.8b	0.8b		44.8
Ambush (0.5 ml/l)	6.3/2.7/0.3	1.2b	1.4b		46.5
Probability of error		P = 1%			

Source: Dreyer and Hellpap, 1991

* No. of different instars and % loss of leaf area (estimated)

** No. of aphids/leaf: 0 = no aphids, 7 = > 50 aphids, pou = pound, inf. = infestation

Farmers were especially interested in the efficacy of neem against pest species which are difficult to control, such as *Plutella xylostella* and *Bemisia tabaci*. Both pest species are resistant to many synthetic pesticides and can cause complete loss of crops. Although neem can be successfully applied against both pests the programme recommends that it be applied not alone but in combination with other environmentally sound control methods. The frequent and exclusive use of neem insecticides may provoke resistance in the long run and can have negative effects on the yield of some crops.

SOCIOECONOMICS OF NEEM

Pilot farmers receive neem seeds from the project regularly (up to once a week), sold as ground powder in packages of 5 kg for 50 US cents per kg. However, this price cannot be kept in the long run and will increase to about 80 US cents. The use of 15–25 kg seeds per ha, which corresponds to an azadirachtin amount of approximately 30 to 50 g/ha, is recommended. Thus, farmers would have to spend US$ 12 to 20/ha for each spraying. These costs are comparable with those of many synthetic products. Home-made neem products have the disadvantage that the farmer has to handle larger amounts of raw material to prepare his spraying. On the other hand, there is no risk that he will intoxicate himself or his family when using the natural insecticides.

Commercial neem insecticides, such as Azatin or other products, which are more easy to handle, are at present considerably more expensive. Thus, they will preferably be applied for special problems such as high resistance of pests against synthetic products or ecological farming. In the long run they may win a significant market share only when the price of 1 g azadirachtin in the product drops below US$ 1.50.

REFERENCES

Dreyer, M. and C. Hellpap. 1991. Neem—A promising natural insecticide for small scale vegetable production in tropical and subtropical countries. *Plant Dis. Prot.* 98: 428–437.

Serra, C.A. 1992. Umweltschonender Einsatz von Niemextrakten zur integrierten Bekampfung von Schadlingen im kleinbauerlichen Tomatenanbau in der Dominikanischen Republik. Dissertation FB Angewandte Biologie und Umweltsicherung, Justus-Liebig-Universitat, Giessen.

6

Neem in Holland

ARNOLD A. RUITER

Rasun Natuurprodukten, Postbus 740, 9400 As Assen, Holland

ABSTRACT

Trials with neem extracts in orchards and commercial products in green house showed promising results against lepidopterous larvae, white fly, leafminer, mealy bugs nematode etc. In some cases the results were very good. One of the major problem is to get plant-derived approved. Dutch government plans to cut down the use of chemicals by at least 50 per cent by the year 2000. It is suggested that policy makers should look into the problems and act accordingly.

INTEREST AND RESEARCH

Interest in neem as an insecticide and nematicide was initially evinced by the Foundation for Eco-Development Alternatives. This foundation looks at the neem tree as the most promising source of all biological active compounds. The Foundation collects knowledge on neem and disseminates it to scientists, policy makers, and the pest control operators and made them aware of the advantages of neem.

The World Neem conferences contributed to the knowledge of neem in Holland (Putters, 1985; Aerden and Broek, 1987; Brock and Aerden, 1988; Gerrits and van Latum, 1988). The Dutch Agricultural University at Wageningen has done some field, and desk research.

The trials with neem extracts conducted in the orchards showed promising results on Lepidopterous larvae (Aerden and Broek, 1987; Broek and Aerden, 1988).

It is considered that neem extract is very important in biological, insect pest management and IRM systems. In the Dutch integrated pest management system only eight insecticides have been cleared selectively for 40 insect pests (Gerrits and van Latum, 1988).

The Dutch Research Stations for floriculture, glasshouse cultures and fruitculture have taken a great interest in neem (*Azadirachta indica*) as an insecticide and nematicide.

Presently the activities of the official research stations are slowed down by the position of the herbal pesticides.

Due to the pesticide legislation, especially in regard to plant-derived pesticides, one of the major problems is to get the products approved. This is complicated by the fact that in Holland, for example, there are 5000 pests and diseases and 600 different crops. Dutch agriculture is vulnerable to diseases and pests, and is yet heavily dependent on pesticides because of highly specialized farming systems, which result in limited crop rotations, and because of climatic conditions.

Inside the EEC, the disparities among the member states in climate, production and environmental requirements, have lead to differences in pesticide approval and use. Because of the needed research and the high cost of approval the Dutch agriculture is more or less dependent on the progress in main crops in the big countries.

In floriculture soil-borne diseases are among the major problems. Biological and integrated controls are encouraged. Rasun has started in collaboration with Grent, Consultancy Biological Products, to practice the use of neem in floriculture on an experimental basis.

Suneem G, a granular product

Suneem G is a 15 per cent Indian-made granular product, used to regulate the growth of nematodes, soil-borne insects in roses, bouvardia and chrysanthemum in greenhouses. For potplants Suneem (15g) is mixed at the rate of 0.2 kg per cu m of soil. Alternatively 3 gm of Suneem G is strewn into plantholes per sq m, or in perennial crops it is added between the plants and mixed into the soil (Grent, 1992, 1993).

Suneem liquid

Suneem liquid is an oil-based watermissible concentrate and has been used as a foliar spray and as a soil application on greenhouse roses and chrysanthemums at a dosage of 3 ml/litre of water. The main insect targets are the white fly, leafminers, mealy bug, spider mites, thrips tabacci and centipeds. The foliar spray was used three times with an interval of 15 days. The results are encouraging.

A foliar spray with 0.3 per cent Suneem liquid used three times with a 14-day interval gives very good results on Cocoidae in Asplenium under greenhouse conditions. No chemical pesticide gives that good a result (Grent, 1992, 1993). So far trials have been conducted but no statistical data is available. However the visual results are promising.

Grent Consultancy, Rasun and Binab AB, Sweden, will jointly undertake work on Suneem and trichoderma for the soil-related problems, nematodes, soil-borne insects and diseases including *Fusarium, Pythium, Rhizoctonia, Sclerothinia*. The lab tests are very promising (Grent 1992, 1993).

POLICY OF THE GOVERNMENT

For the ninetees, the Dutch government plans to make agriculture safe, competitive and sustainable. The use of chemical pesticides has to be cut down by at least 50 per cent by the year 2000. At this moment the use of chemical pesticides is estimated at 21 million kg of active compounds a year. The government has planned to stop the use of several pesticides presently being used (The Dutch Multi-year Crop Protection Plan, 1991).

The implementation of this "Multi-Year Crop Protection Plan" makes it necessary for the government to put in additional resources. This can be achieved provided there are sufficient alternatives available. Herbal pesticides such as neem can be one of the safe alternatives.

The Netherlands are part of the EEC. Only if the policy of banning several chemical products and the registration policy of new (herbal) products is harmonized then the policy can be successful. In the Netherlands, 2000 chemical pesticide products have been registered and only 12 biological products. From the biological products only one, Pyrethrum, is a plant-derived pesticide.

INTERNATIONAL COOPERATION

It is time that policy-makers of the EEC followed the directions of registration of the USA and India to register biorational pesticides. Unlike the USA, none of the EEC-Member States have provisions in their pesticides legislations for plant-derived pesticides or pesticides to be used on low-acreage crops. The recommendation from the Council of Europe (August 1991) gives no special status for plant-derived pesticides nor is there any made for small market volumes (Journal of the ECCD, 1991).

Because of the policy of cutting down the amount of chemical pesticides, the time becomes right to push for the introduction of plant-derived pesticides. It is necessary to mobilize the existing interest in establishing a situation in which some plant-derived pesticides can be used legally. It will be necessary that interested companies, environmental organizations, development-aid organizations and agriculture organizations as the International Federation of Organic Agriculture Movements (IFOAM) should form a working committee and work out a plan of action.

That plan of action has to include the following:

1) The demand of potential user groups.
2) Requirements for an official application with preliminary available data.
3) A decision on starting an application procedure.
4) Ways to fill research gaps and raise sufficient funds to perform tests.

The European governments are now making major efforts to cut down the use of chemical pesticides so as to achieve a healthy and chemically save environment. They should give the developing countries support to develop basic knowledge for registration, the culture of neem trees, the production structure, and use of the end-products.

REFERENCES

Aerden, R.S.P.M. and R.C.F.M. Broek van den. 1987. De werking van neem aftreksels tegen vruchtbandroller, fruitmot en groene appelwants in de Nederlandse fruitteelt. L.U. Wageningen.

Broek, R.C.F.M van den and R.S.P.M. Aerden. 1988. De gewasbeschermende werking van neem-extracten onder de Nederlandse omstandigheden, L.U. Wageningen.

Gerrits, R. and E. van Latum. 1991. Plant-derived Pesticides in Developing Countries: Possibilities and Research Needs. Foundation for Ecodevelopment Hoofddorp.

Grent. T.N.H. 1992, 1993. Counsultancy Biological Products, personal communications.

Official Journal of the European Communities, Council Directive of 15 July 1991, concerning the placing of plant protection products on the market, L230, Luxembourg Aug 1991.

Putters, Ton. 1985. The Neem Tree, Toekomstige leverancier van pesticiden? Deventer, R.H.L.S.

The Dutch Multi-year Crop Protection Plan. 1991. 's–Gravenhage Sdu.

7

Neem as a Safe Trend for Pest Control in the Future

NADIA Z. DIMETRY

Pests and Plant Protection Department,
National Research Centre, Dokki, Cairo, Egypt

ABSTRACT

The use of biological active substances from neem for the control of agricultural pests, has been undertaken in different parts of the world. The studies done in Egypt indicate the possibility of using the neem-seed kernel extracts with insecticidal properties, for the control of different pests. The repellent, antifeedant and growth-regulator properties of the toxic principles from the seeds of Meliaceae, combined with low cost, local availability, safety to the environment, and compatibility with the agroecosystem, enhance their potential in insect pest management systems.

INTRODUCTION

One of the greatest challenges facing the world is to produce adequate food for the growing population, which is expected to reach about seven billion by the year 2000. The situation is especially difficult in developing countries where food needs are likely to double over the next decade. An increase in agricultural production will continue to be of vital importance in third world countries in the future.

Approximately one third of the global food production is estimated to be destroyed annually by over 20,000 species of insects, diseases, weeds, mites, nematodes, rodents and other field and storage pests (Agrios, 1978; McEwen, 1978). The highest losses occur in the developing countries of Asia and Africa reducing total agricultural production there by about 42 per cent annually (Table 1).

Developed countries also have their share of losses, in the United States for example, losses amount to about 29 per cent of the potential annual production (McEwen, 1978).

Table 1. Estimated crop losses due to pests in various regions (represented from Agrios, 1978)

Region	1980 Crop production (mm mt)	Estimated loss % of potential production
Asia	910	43
Africa	286	42
South America	403	33
USSR and China	939	30
North and Central America	690	29
Oceania	57	29
Europe	633	25

Crop data from FAO, 1980.

Synthetically produced pesticides are the principal means of crop protection in developed countries. They are valued for their uniform and rapid effectiveness against target pests, their ease of shipment, storage, and use, and their relatively long shelf-life under proper storage conditions. In 1980, approximately one million tonnes of pesticides were consumed globally, and over one-third of this was in the United States alone (FAO, 1980).

Pesticide use, however, is not without problems, unforeseen side effects, such as toxicity to non target organisms, development of resistance in pests to the materials used and environmental contamination with the potential to affect the entire food chain are some major drawbacks. In developing countries, additional problems are a lack of understanding of their proper use, non availability of suitable application equipment, inadequate storage conditions and high prices.

Scientists are thus coming under increasing pressure to investigate alternative plant protection technologies to help meet the food needs of the 21st century. A 1981 policy statement by the US Department of Agriculture (USDA) specifically recommends a general reorientation of research and extension to develop new technologies to reduce costs, increase efficiency and facilitate the economic viability of small to medium sized farms. In this context great demands will be made on plant and post harvest protection. Considerable attention must be focussed on biological pest control. Extraction and production of natural insecticides assumes an important function in this regard. In this connections neem products are being adapted to the needs of developing countries. During the last decade intensive studies have been carried out by several research groups under laboratory and field conditions aiming to throw light on the potency of neem (*Azadirachta indica* A. Juss.) components as safe substitutes for chemical pesticides against different agricultural, medical, and veterinary pests. The three International Neem Conferences that were held for neem in 1980, 1983, and 1986 (Schmutterer et al., 1981; Schmutterer and Ascher, 1984, 1987) and the first workshop on practice-oriented results on use and production of neem ingredients in 1992 greatly advanced research activities on the ingredients of "*A. indica*" in many countries, and new and interesting results have been obtained. The side effects against natural enemies, investigations into ecotypes of neem tree, and large field experiments were tackled by scientists. More recently, the potential offered by the wider use of specialized plants in combination with major crops to control weeds and insect pests has been emphasized (National Research Council, 1983). The systemic and in-depth investigation of plant materials for pest control purposes therefore merits consideration. Jacobson et al., (1984) indicated that *A. indica* has come to be considered possibly the world's most fantastically effective natural insect antifeedant known today. This reputation is based on solid foundation of evidence in the scientific literature.

GEOGRAPHICAL DISTRIBUTION OF NEEM TREE

Neem is the common name give to different species of large evergreen glabrous shade trees up to 50 ft high grown throughout India. These trees are also found in Bangladesh, Pakistan, Sri Lanka, Burma, Malaysia, Indonesia, and tropical regions of America, Australia, and Africa (Fig.1). In India, the common neem tree is also popularly known as margosa. Other common species are gora neem or Persian lilac (*Melia azedarach* L.), a native of West Asia, and Malabar neem (*Melia composita* Willd.)

According to the Directorate of the Non-Edible Oils and Soap Industry, Khadi and Village Industries Commission, there are 13,899,067 neem trees in India with a potential of yielding 418,633 tonnes of seed per year (Ketkar, 1976), but only 25 per cent of the seed is actually collected. The number of trees increased to over 14 million trees (Saxena, 1983). The neem flowers from January to May and fruits mature from May to August. One tree produces 30–50 kg fruit per year and 30 kg of seed yields 6 kg of oil and 24 kg of cake. Neem cake, the residue remaining after oil expulsion, can be mixed with urea and applied to the soil of rice crop. Every part of the plant is used: the root bark, the stem bark, the blossoms, the young fruits, the seeds, the oil, the leaves, and the gum.

BIOACTIVITY OF NEEM COMPONENTS IN EGYPT

In Egypt, experiments were carried out on the efficacy of extracts of different plant species against some agricultural pests, the most important of which are the cotton leaf worm, *Spodoptera littoralis* (Boisd.) (Ahmed, 1983; Mohamed, 1983; Dimetry and Abdalla 1988–1989a), the cowpea aphid, *Aphis craccivora* Koch. (Dimetry and Abdalla 1988–1989b) and the two-spotted spider mite *Tetranychus urticae* Kach. (Barakat et al., 1984–1985; Amer et al., 1989 and Dimetry et al., 1990).

In the last decade, much attention on NSKE fractions and formulations were carried out by Egyptian scientists. The work of El-Sayed (1982 and 1982–1983a and b), Kelany et al., (1991); Salem (1991a & b); Salem and Matter (1991); Dimetry and Schmidt (1992) and Salem (1992) are worthy of mention.

Methodology

Bioassay of the antifeedant and repellent effects were generally done in the laboratory by the leaf disc method in case of *S. littoralis*. After treatments, the area of the leaves consumed was determined and compared with the untreated leaves.

In case of *T. urticae*, the deterrent effect was examined in an arena where one half of each disc was treated with different concentrations of the extract or formulation and the other half was treated with the solvent and served as control. Orientations of the female *T. urticae* on treated or control half were recorded after different periods for 24

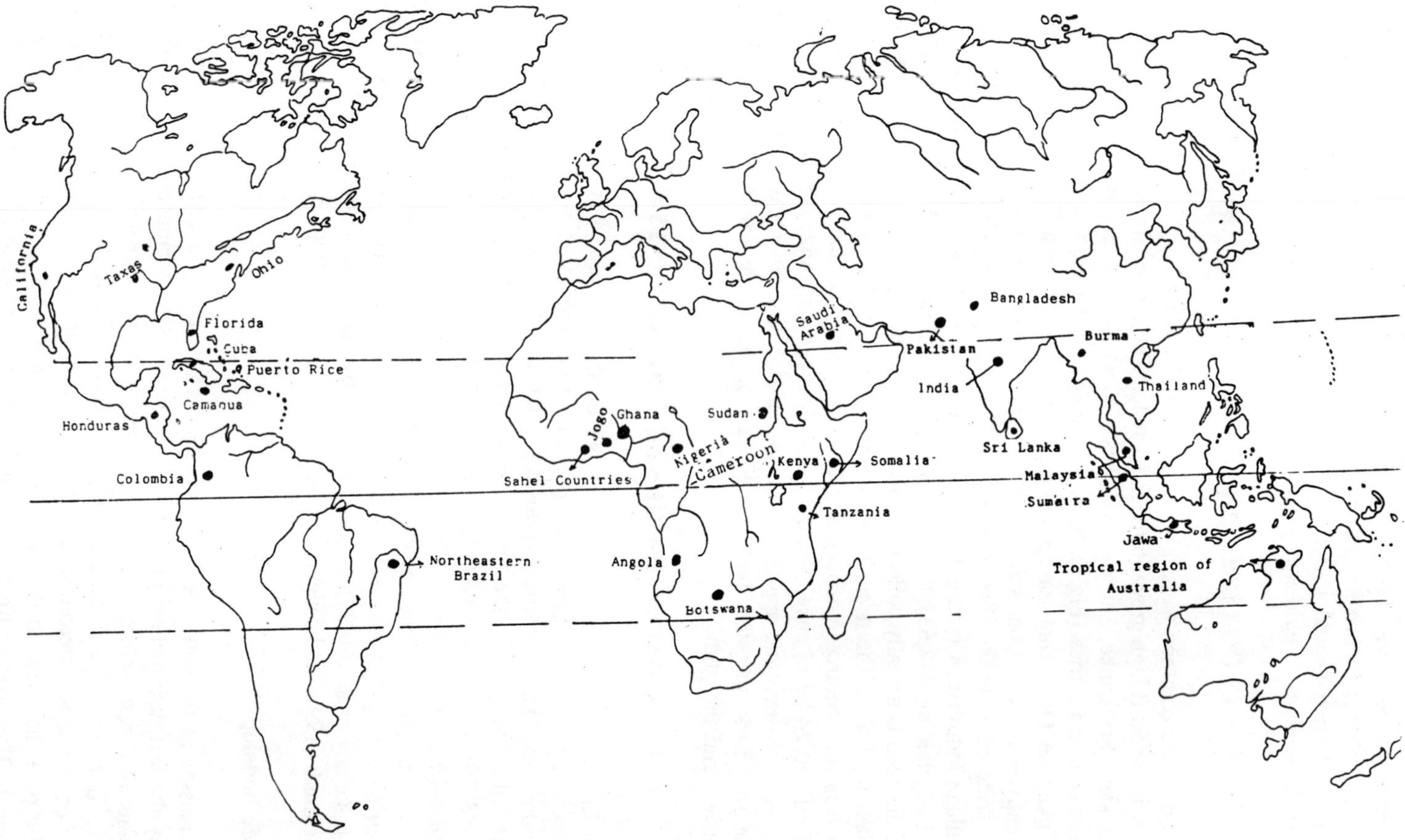

Fig. 1. Geographical distribution of neem trees allover the world

hrs from the beginning of the experiment. Also, the eggs laid on control versus the treated half were counted after 24 hrs.

For the bean aphid and the cowpea aphid, orientation of virginoparous females was carried out in a big arena; the treated seedlings were arranged versus the control in alternation, the aphids were released in the middle of the arena and orientation was recorded as in case of *T. urticae.*

For other tests, the food plants were either sprayed with or dipped into the plant extract.

For evaluating growth disrupting effects, topical applications in the dorsal thorax region with a calibrated micropipetter or incorporation of the plant material in the artificial diet was generally used in case of *S. littoralis* and *Tropinota squalida*. Treated larvae were observed for malformation, mortality, and duration of larval and pupal periods.

In case of agromyzid, white-flies, and tomato bugs, semi-field tests were carried out and the population density of agromyzid, white-fly or tomato bug adults and immature stages was recorded on hour before and after spraying, at one, three, five and seven days, and compared with the control.

Results

The data obtained by Moustafa (1993) showed that in case of *S. littoralis*, both polar and non polar components of the plant materials seem to be responsible for the antifeedant properties. Petroleum ether extract of both fruits and leaves of neem caused significant elongation of larval and pupal periods in comparison to the control, while chloroform extract prevented the larvae from reaching the pupal stage, and all the larvae died in the fourth instar. Also, the percentage of malformation in the resulting pupae reached its maximum when the larvae were treated with petroleum ether extract of neem leaves (20 per cent). The percentage of adult emergence greatly decreased for all the extracts tested. Also, the different extracts were tested in a similar manner with the third instar larvae of *S. littoralis* which were followed up till the adult stage. The fecundity and the fertility of the resulting eggs were recorded. In case of petroleum ether extract of neem fruits, the adults were completely sterile but those adults whose larvae were offered leaves treated with castor oil with the other extracts laid very small egg batches with the exception of ethanol extract of neem leaves, the adults laid on the average five egg batches (one medium and four small).

Regarding the percentage of hatching, all the resulting egg masses failed to hatch with the exception of the egg masses laid by females whose larvae were offered leaves treated with petroleum ether extract of neem leaves where the hatching reached 25.8 per cent.

Further information on the activity of pure neem seed oil compared to chemical insecticides (recommended for application) were tested against the cotton bollworms *Pectinophora gossypiella* and *Earias insulana* (Salem, 1991). The highest concentration of neem oil used (150 ppm) caused reduction in the percentage of infestation very near to that of the chemical insecticides applied. The yield obtained at the end of the season was 8.5 $\pm$ 0.61 and 7.7 $\pm$ 0.63 kentar in 1989 for areas sprayed with chemical insecticides and neem oil (150 ppm) respectively. The percentage of infestation decreased as the concentration of neem oil increased.

The work of neem seed kernel extract (NSKE) in Egypt was extended also to the piercing sucking pests like aphids, white flies, leaf miners and the two-spotted spider mite (Dimetry and Schmidt, 1992 and Dimetry, et al., 1993). The results obtained illustrate the efficacy of NSKE formulations against the tested pests. Again, Margosan-0 and Neem azal-S were tested for their adverse effect against some predacious mites and neem seed oils against different predators and it was found that both formulations of NSKE and neem oil were found to be safe in varied manner against the tested predators (Table 2).

FUTURE CONSIDERATION OF NEEM AND CONCLUSIONS

An enlightened attitude toward pest control would be helpful in promoting the use of botanical pest control agents. Natural products from various plants in Egypt are a rich source of bioactive substances which have been exploited to a limited extent as insecticides, insect growth regulators, and phagodeterrents. The use of neem seed derivatives offers a novel approach in the management of insect pests especially if it is integrated with other methods of pest control. Neem derivatives are relatively safe to natural enemies, humans, domestic animals, and other beneficial organisms.

The neem tree is widespread in several African and Asian countries. It can withstand variations in climate and geography and thrives even on poor soils. In spite of the fact that this tree was cultivated in the botanic garden in Aswan Governerate many years ago, a trial to cultivate it in Egypt on a large scale was started only few months ago in Giza Governerate in the hope that it will spread all over the country. Enough seed oil and cake would then become available to the farmers at low cost for crop protection. Productivity can be increased and stabilized through selection of superior genotypes.

The use of simple, crude botanicals such as neem oil and neem cake is ideally suited for crop protection by resource limited farmers in developing countries. Neem seed can also be processed at village level and crude fractions can be extracted using simple methods. Efficacy of neem derivatives can be enhanced by using additives and synergists especially under our climatic conditions of high temperature and ultraviolet rays during summer in Egypt. The chances of insect pests developing resistance to neem seed derivatives are remote (Saxena, 1983). The national extension services should be encouraged to conduct demonstration trials with neem products to convince the farmers to use the new safer and cheaper method of pest control.

Table 2. Bioactivity of neem products against the most important pests in Egypt

Test product	Test animal	Activity observed	Reference
Neem seed suspension + insecticides	*Spodoptera littoralis* (Boisd.)	varying levels of antagonism	El-Sayed (1982)
Neem seed suspension	*S. littoralis* (Boisd.)	insecticidal	El-Sayed (1982–1983a)
Neem seed suspension	*S. littoralis* (Boisd.)	antifeedant + oviposition repellent	El-Sayed (1982–1983b)
Aqueous neem kernel extract (ANSKE)	*Musca domestica* L.	delayment of pupation, reduced pupal weight + malformation	Kelany et al. (1991)
Neem seed kernel pure oil	*Earias insulana* Boisd. and *Pectinophora gossypiella* Saund.	oviposition deterrency	Salem (1991a)
Neem seed kernel pure oil	*Spodoptera littoralis* (Boisd.)	acute reduction in in number of insects + antifeedant	Salem and Matter (1991)
Neem seed kernel pure oil	*Paederus alfierii* (Koch.) (predator)	No detrimental effect	Salem and Matter (1991)
Neem seed kernel pure oil	*Chrysopa carnea* Steph. (predator)	Sharp drop in population (92.5%)	Salem and Matter (1991)
Neem seed kernel pure oil	*Coccinella undecimpunctata* L. (predator)	initial reduction in No. of predator	Salem and Matter (1991)
Neem seed kernel pure oil	*Orius* species	slight decrease in early count	Salem and Matter (1991)
Neem seed kernel pure oil	*Phthorimaea operculella* Zell.	insecticidal + deformity in adults + decreased fecundity	Salem (1991b)
Neem seed kernel pure oil	*Ceratitis capitata* Wied.	decrease percent infestation + deterrent + insecticidal	Salem (1992)
Neem extracts (crude)			
Petroleum ether extract (fruit)	*Spodoptera littoralis* (Boisd.)	insecticidal + developmental retardation + complete sterility	Moustafa Ph.D thesis (1993)
Petroleum ether extract (leaves)	*S. littoralis* (Boisd.)	insecticidal + malformation + partial sterility	
Chloroform extract (fruits) + leaves	*S. littoralis* (Boisd.)	insecticidal	
Ethanol extract (fruits)	*S. littoralis* (Boisd.)	developmental retardation + insecticidal + sterility	
Ethanol extract (leaves)	*S. littoralis* (Boisd.)	slight effect (33% mortalities)	
Water extract	*S. littoralis* (Boisd.)	developmental retardation + complete sterility	

(contd.)

(Table 2 contd.)

Test product	Test animal	Activity observed	Reference
Commercial preparation of NSKE			
Margosan-O	*Aphis fabae* Scop.	deterrent + reducing fecundity	Dimetry and Schmidt (1992)
Neem Azal-s	*Aphis fabae* Scop.	deterrent + growth disruption + insecticidal + reducing fecundity	Dimetry and Schmidt (1992)
Neem Azal-F	*Aphis craccivora* Koch.	larviposition detterent + developmental retardation + antifeedant + insecticidal	Dimetry and El-Hawary (1993)
Margosan-O and Neem Azal-S	*Liriomyza trifolii* Burgess	insecticidal + oviposition deterrency	Ezzat M.Sc. thesis (1993)
Margosan-O	*Cyrtopeltius tenuis* Reut.	insecticidal	Hussein M.Sc. thesis (1993)
	Bemisia tabaci Genn.	insecticidal	
Neem Azal-S	*Cyrtopeltius tenuis* Reut.	insecticidal + developmental retardation	
	Bemisia tabaci Genn.	insecticidal + developmental retardation	
Margosan-O and Neem Azal-S	*Tropinota squalida* Scop.	insecticidal + developmental retardation	Dimetry and Husseain (Unpublished data)
Margosan-O and Neem Azal-S	*Tetranychus urticae* Koch.	deterrent + insecticidal + reducing fecundity	Dimetry et al., (1993a)
Margosan-O and Neem Azal-S	*Amblyseius barkeri* (Hughes) and *Typhlodromus richteri* Karg (predators)	reducing food intake	Dimetry et al., (1993b)

REFERENCES

Agrios, G.N. 1978. *Plant Pathology*. Academic Press, New York.

Ahmed, M.A. 1983. *Studies of some insecticidal effects of seven plant extracts against S. littoralis* Boisd. & *Tribolium confusum* (Duv.), M.Sc. Thesis, Fac. Agric., Cairo University.

Amer, S.A.A., A.S. Reda and N.Z. Dimetry. 1989. Activity of *Abrus precatorius* L. extracts against the two-spotted spider mite *Tetranychus urticae* Koch. (Acari.: Tetranychidae) *Acarologia* XXX (3): 209–215.

Ascher, K.R.S., Miriam Eliahu, Nadia, E. Nemny and J. Meisner. 1984. Neem seed kernel extract as an inhibitor of growth and fecundity in *Spodoptera littoralis*. In: *Proc. 2nd Int. Neem Conf.*, Rauischholzhausen, 1983, pp. 331–344.

Barakat, A.A., G.M. Shereef, S.A. Abdallah and S.A.A. Amer. 1984–1985. Toxic action of some plant extracts against *Tetranychus urticae* Koch. *Bull. Ent. Soc. Egypt*, Econ. Ser., 14: 233–242.

Dimetry, N.Z. and E.F. Abdalla. 1988–1989a. A feeding deterrent effect of various extracts of *Abrus precatorius* seeds towards the cotton leafworm *Spodoptera littoralis* Boisd. *Bull. Entomol. Soc. Egypt*, Econ. Ser., 17: 85–92.

Dimetry, N.Z. and E.F. Abdalla. 1988–1989b. Bioactivity of *Abrus precatorius* seeds extracts against the cowpea aphid, *Aphis craccivora* Koch. *Bull. Entomol. Soc. Egypt*, Econ. Ser., 17: 159–167.

Dimetry, N.Z. and F.M.A. El-Hawary. 1993. Neem Azal-F as an inhibitor of growth and reproduction in the cowpea aphid *Aphis craccivora*. *J. Appl. Entomol.* (in press).

Dimetry, N.Z., S. El-Gengaihi, A.S. Reda and S.A.A. Amer. 1990. Toxicity of some compounds isolated from *Abrus precatorius* L. seeds towards the two-spotted spider mite *Tetranychus urticae* Koch. *Acarologia* XXXI (4): 361–366.

Dimetry, N.Z. and G.H. Schmidt. 1992. Efficacy of Neem-Azal-S and Margosan-O against the bean aphid *Aphis fabae* Scop. *Anz. Schadlingskde, Pflanzenschutz, Umweltschutz*, 65: 75–79.

Dimetry, N.Z., S.A.A. Amer and A.S. Reda. 1993a. Biological activity of two neem seed kernel extracts against the two-spotted spider mite, *Tetranychus urticae*. *J. Appl. Entomol.* (in press).

Dimetry, N.Z., S.A.A. Amer and F. Moemen. 1993b. Laboratory assessments for testing the adverse effects of two neem products on the predatory mites *Amblyseius barkeri* and *Typhlodromus richteri*. *Bull. Zool. Agr.* (in press).

El-Sayed, E.I. 1982. Effect of different insecticides in combination with neem (*Azadirachta indica* A. Juss.) seed suspension on the Egyptian cotton leafworm *Spodoptera littoralis* (Boisd.). *Zag. J. Agric. Res.*, 9(2): 457–469.

El-Sayed, E.I. 1982–1983a. Evaluation of the insecticidal properties of the common Indian neem, *Azadirachta indica* A. Juss. seeds against the Egyptian cotton leafworm *Spodoptera littoralis* (Boisd.). *Bull. Entomol. Soc. Egypt*, Econ. Ser., pp. 39–47.

El-Sayed, E.I. 1982–1983b. Neem (*Azadirachta indica* A. Juss.) seeds as antifeedant and ovipositional repellent for the Egyptian cotton leafworm *Spodoptera littoralis* (Boisd.). *Bull. Entomol. Soc. Egypt*, Econ. Ser., pp. 49–58.

Ezzat, A. 1993. Ecological and toxicological studies on some pests infesting faba beans "*Vicia faba*" in Egypt. M.Sc. Thesis, Fac. Agric., Cairo Univ., Egypt.

FAO. 1980. *Food Production Year Book*, Vol. 34. FAO, Rome.

Feuerhake, K.J. 1984. Effectiveness and selectivity of technical solvents for the extraction of neem seed components with insecticidal activity. In: *Proc. 2nd Int. Neem Conf.*, Rauischholzhausen 1983, pp. 103–114.

Hussein, A.S. 1993. Ecological studies on some important tomato insect pests and the susceptibility of different varieties of tomato to pest infestation. M.Sc. Thesis, Fac. Agric., Ain Shams Univ. Egypt.

Jacobson, M., J.B. Stokes, J.D. Warthen Jr., R.E. Redfern, D.K. Reed, R.E. Webb and L. Telek. 1984. Neem research in the U.S. Department of Agriculture: An update *Proc. 2nd Int. Neem Conf.*, Rauischholzhausen, 1983, pp. 31–42.

Kelany, I.M., M.A. El-Deeb and A.A. El-Fishawi. 1991. The effect of aqueous neem *Azadirachta indica* A. Juss. seed kernel extract on the house fly, *Musca domestica* L. *Egypt J. Biol. Pest Control* 1(1): 121–128.

Ketkar, C.M. 1976. Utilization of Neem (*Azadirachta indica* A. Juss.) and Its By-products. Directorate of Non-edible Oils and Soap Industry. Khadi and Village Industries Commission, Bombay.

Kleeberg, H. 1992. Practice oriented results on use and production of neem ingredients, *Proc. 1st Workshop* 127 pp.

Kraus, W. and M. Bokel. 1981. Neue Tetranor-triterpenoide aus *Melia azedarach* Linn. (Meliaceae) *Chem. Ber.* 114: 267–275.

Kraus, W. and R. Cramer. 1981. Pentanortriterpenoide aus *Azadirachta indica* A. Juss. (Meliaceae) *Chem. Ber.*, 114: 2375–2381.

Kraus, W., R. Cramer and G. Sawitzki. 1981. New tetranor-triterpenoids from seeds of *Azadirachta indica* (Neem tree). *Phytochemistry*, 117–120.

Mansour, F.A. and K.R.S. Ascher. 1984. Effects of neem (*Azadirachta indica*) seed kernel extracts from different solvents on the carmine spider mite, *Tetranychus cinnabarinus*. *Proc. 2nd Int. Neem Conf.*, Rauischholzhausen, 1983, pp. 461–470.

McEwen, F.L. 1978. Food production—The Challenge of pesticides. *Bioscience* 28: 773.

Mohamed, N.M. 1983. Studies on the efficacy of some plant extracts against certain insect pests. M.Sc. Thesis, Fac. Agric., Cairo Univ.

Morgan, E.D. and M.D. Thornton. 1973. Azadirachtin in the fruit of *Melia azedarach* (Meliaceae), *Phytochemistry* 12: 391–392.

Moustafa, S.M. 1993. Biochemical studies on some medicinal plants having insecticidal activity. Ph.D. Thesis, Fac. Agric., Cairo Univ.

National Research Council. 1983. *Bostid Developments*, Vol. 1, No 3, Washington, DC.

Salem, S.A. 1991a. Response of cotton bollworms *Pectinophora gossypiella* Saund. and *Earias insulana* Boisd. to seed neem kernel pure oil. *Annals Agric. Sc. Moshtohor* 29(1): 597–607.

Salem, S.A. 1991b. Evaluation of neem seed oil as tuber protectant against *Phthorimaea operculella* Zell. (Lepidoptera, Gelechiidae). *Annals of Agric. Sc., Moshtohor* 29(1): 589–596.

Salem, S.A. 1992. Relative efficacy of dimethoate insecticide and neem seed kernel oil suspension against med fly, *Ceratitis capitata* Wied. African *J. Agric. Soc.* Vol. 19 (In press).

Salem, S.A. and M.M. Matter, 1991. Relative effects of neem seed oil and Deenate on the cotton leafworm, *Spodoptera littoralis* Boisd. and the most prevalent predators in cotton fields at Menoufyia Governorate. *Bull. Fac. of Agric.*, Univ. of Cairo, 42: (3) 941–952.

Saxena, R.C. 1983. Naturally occurring pesticides and their potential. In: *Conf. on Chemistry Applied to World Needs* L.E. Shemilt (ed.), (Manila, Philippines, Dec., 1982) pp. 143–161.

Schmutterer, H., K.R.S. Ascher and H. Rembold (eds.). 1981. Natural pesticides from the neem tree (*Azadirachta indica* A. Juss). *Proc. 1st Int. Neem Conf.*, Rottach-Egern, FRG, 1980, 297 pp.

Schmutter, H. and K.R.S. Ascher (eds.). 1984. Natural pesticides from the neem tree and other tropical Plants. *Proc. 2nd. Int. Neem Conf.*, Rauischholzhausen, FRG, 1983, 587 pp.

Schmutterer, H. and K.R.S. Ascher (eds.). 1987. Natural pesticides from the neem tree and other tropical plants. *Proc. 3d Int. Neem Conf.*, Nairobi, 1986, 703 pp.

U.S. Dep. Agric. 1981. Time to Choose. Summary Report on the structure of Agriculture. Government Printing Office Washington, D.C.

Warthen, J.D., Jr. 1979. *Azadirachta indica*: A source of insect feeding inhibitors and growth regulators. US Dept. Agric. Rev. Man., ARM-NE-4.

8

Neem Pest Control Research in West Africa

JOANNY O. ZONGO

Laboratoire de Recherches Protection des Végétaux, BP 403 Bobo-Dioulasso, Burkina Faso, West Africa

ABSTRACT

Since the three International Neem Conferences, much attention has been paid to neem tree products as a pest control tactic in West Africa. This review paper deals with neem pest control research in West Africa with special reference to Burkina Faso, Niger and Togo. Field and laboratory studies have showed that neem tree products could be a technically acceptable input of integrated pest management. Original unpublished data will also be discussed, including the traditional use of neem as a pest control strategy by farmers.

Insect pests are an important component of the adverse environmental factors in food production particularly in developing countries. They greatly reduce crop yields in the field, destroy stored food, and transmit the causal agents of many dangerous human diseases. To control these pests, most developing countries depend heavily on imported pesticides, which are generally expensive. This dependence could be reduced by using natural products readily available in the tropics and subtropics. Neem (*Azadirachta indica* A. Juss., Meliaceae) products are known to have strong pesticidal properties (Schmutterer et al., 1981, Schmutterer and Ascher, 1984, 1987; Jacobson, 1986) and could provide alternative pest control strategies for the farmers of Sahelian countries (Zongo et al., 1992). The tree grows well in the Sahel and produces fruit, leaves, and wood, which are used for a variety of purposes by farmers. This paper will examine neem pest control research in West Africa with special reference to Burkina Faso, Niger, and Togo. Results from Nigeria and Ghana will also be discussed.

Neem research has received great attention from many scientists from India, Pakistan, Israel, Japan, the United Kingdom, Federal Republic of Germany, Canada, and the United States. Although neem products are effective against many insect species, few papers have been published in West Africa. Since the three International Neem Conferences, much attention has been paid to neem pest control research in Sahelian countries. Zongo (1990a), in a review paper discussed neem's pesticidal properties and future considerations. This paper helped local scientists to get more information on neem and to consider its potential as a pest control strategy. Current research has demonstrated the great potential of neem for pest control. New and interesting results have been obtained from Burkina Faso (Zongo et al., 1992), Niger (Radcliffe et al., 1991; Wilps et al., 1992), and Nigeria (Okorie et al., 1990). These results give rise to the hope that in Sahelian countries, where sufficient neem trees are available, the use of imported pesticides could be replaced partly by the use of locally produced neem. Neem pest control research in West Africa demonstrated that about 19 insect pests could be controlled, the main groups being lepidopterous, polyphagous homopterous, and particularly stored grain insect pests (Table 1).

In Burkina Faso, Zongo (unpublished data) reported after a survey done in 1986 and 1987 that neem leaves were traditionally used in warehouses to control stored grain pests. Farmers used alternating layers of neem leaves and grains in the warehouses. This methods is still practised by some farmers in Diébougou. Neem products are used in various medicinal applications. Neem leaves and bark are extensively used in treatment of fevers. In cases of fever due to malaria, a decoction is prepared by boiling neem leaves or the inner layer of the bark. When warm, the decotion is taken as a bath, and when cold it is administered twice daily. Traoré (1989) found that 4 ml of neem seed oil per kg of cowpea seeds protected cowpea seeds against *Callosobruchus maculatus* F. (Coleoptera: Bruchidae) for three months. Zongo (1990b) noted that neem seed kernels were seriously damaged by *Oryzaephilus surinamensis* L. (Coleoptera: Cucujidae) and recommended that general control measures against stored grain pests be applied to protect stored neem seed kernels. Lynch (1990) and Batchomo (1991) reported inconclusive results in controlling groundnut arthropod pests (termites and millipedes) with neem leaf, seed, or kernel extracts in the central part of Burkina Faso (Gampela, Ouagadougou). Studying the effects of neem seed kernel extracts on egg and larval survival of the sorghum shoot fly, *Atherigona soccata* Rondani (Diptera: Muscidae), Zongo et al. (1992) found that unhatched eggs were completely decomposed 24 hrs after the treatment with neem aqueous extracts while they collapsed with neem oil. They found that neem seed kernel extracts (commercial neem oil containing 0.63 per cent azadirachtin, local neem oil, 40 g seed kernels/l water, 40 g seed kernels/l water + 5 ml adhesol, 60 g seed kernels/l water) caused 81.5 to 91.7 per cent egg mortality and 50 to 55 per cent larval mortality of *A. soccata* in laboratory conditions. In field conditions, they found significant differences among treatments in the number of eggs laid and the percentage of dead hearts. In both 1990 and 1991, they observed fewer eggs laid in plots treated with neem extracts (20 kg/ha neem seeds/500 l water, 20 kg/ha neem seeds/500 l water + 2.5 l/ha adhesol, 30 kg/ha wild neem seeds/500 l water + 2.5 l/ha adhesol) than in control plots. Variations in the number of eggs laid, the percentage of eggs hatching, and dead hearts have been observed between 1990 and 1991. They explained that these variations were due to the qualitative difference recorded in azadirachtin content. Need seed kernels and seed oil originated from Koudougou (12° 43'N, 4° 40'WA), a city located in the central western part of Burkina Faso, and contained 447 ppm and 93 ppm of azadirachtin in solution respectively in 1990 whereas

Table 1. Insect pests controlled by neem tree extract products in West Africa

Insect pest	Plant or product	Country	Reference
Atherigona soccata Rond. (sorghum shoot fly)	Sorghum	Burkina Faso	Zongo et al. (1992)
Bemisia tabaci Genn. (tobacco whitefly)	Courgette	Togo	Dreyer (1984)
Callosobruchus maculatus F. (cowpea beetle)	Cowpea	Burkina Faso Togo Ghana Nigeria	Traoré (1989) Zehrer (1984) Tanzubil (1987) Ivbijaro (1987)
Dermestes maculatus DeGeer (dermestid beetle)	Dried fish	Nigeria	Okorie et al. (1990)
Hellula undalis F. (cabbage webwork)	Cabbage	Togo	Dreyer (1987)
Henosepilachna elaterii Rossi (spotted cucumber beetle)	Courgette	Togo	Dreyer (1984)
Jacobiella facialis Jac. (leaf sucker)	Eggplant	Togo	Dreyer (1987)
Kraussaria angulifera Krauss (Sahelian grasshopper)	Millet and sorghum	Niger	Radcliffe et al. (1991)
Phytica melongenae Aina (leaf roller)	Eggplant	Togo	Dreyer (1987)
Plutella xylostella L. (diamondback moth)	Cabbage	Togo	Adhikary (1981) Dreyer (1987)
Schistocerca gregaria Forsk. (desert locust)	Multiple crops	Niger	Wilps et al. (1992)
Scrobipalpa ergasima Mayr (leafminer)	Gboma (local vegetable)	Togo	Dreyer (1987)
Selepa docilis Butl. (noctuid moth)	Eggplant	Togo	Dreyer (1987)
Sesamia calamistis Hmps. (corn stem borer)	Corn	Togo	Dreyer (1987)
Sitophilus oryzae L. (rice weevil)	Rice	Nigeria	Ivbijaro (1987)
Sitophilus zeamais Motsch (corn weevil)	Corn	Togo Nigeria	Akou-Edi (1984) Ivbijaro (1987)
Tribolium castaneum Herbst (flour beetle)	Sorghum	Nigeria	Ivbijaro (1987)
Tribolium confusum J. du Val. (flour beetle)	Corn	Togo	Akou-Edi (1984)
Zonocerus variegatus L. (variegated grasshopper)	Cowpea	Nigeria	Olaifa and Akingbohungbe (1987)

they contained 508 ppm and 100 ppm respectively in 1991. Experiments are currently being undertaken on the effects of neem extracts on cowpea (*Vigna unguiculata L.* Walp) pod borers and thrips (Dabiré, personal communication).

In Niger, Radcliffe et al. (1991) found in field experiments that a water extract of locally grown neem kernels (150 g neem seed/3 l of water) significantly prevented feeding on sorghum seedlings by the Sahelian grasshopper, *Kraussaria angulifera* Krauss (Orthoptera: Acrididae). Neem extracts were more effective against nymphs than adults of *K. angulifera*. Wilps et al. (1992) found that neem seed oil (0.80, 0.5 per cent active ingredient) at a rate of 1 l/ha caused 8 to 22 per cent adult mortality of *Schistocerca gregaria* Forsk (Orthoptera: Acrididae) on the ground. On flying insects, they reported 70–85 per cent adult mortality and concluded that insects treated with neem products were incapable of flying a long distance. Practical handouts, excellent photographs, and posters on how to protect cowpea seeds and how to extract neem seed oil have been provided to extension service by Niger Plant Protection Direction. These materials are appropriate tools to ensure good training for farmers.

A survey of the literature showed that Togo seems to be one of the first West African countries to initiate neem pest control research. In 1979, a project on the use of plants as a source for obtaining natural insecticides was created by the German Agency for Technical Co-operation (GTZ). Working in this project, Adhikary (1981) published during the First International Neem Conference the first results which indicated that methanol extract of neem seed kernels (4 per cent at a rate of 225 l/ha) was effective in controlling *Plutella xylostella* L. (Lepidoptera: Plutellidae) on cabbage. He noted that methanol extract of neem seed kernels gave significantly lower damaged cabbage heads (9 per cent) than mevinphos (19 per cent). Another interesting result was that no adverse effect of neem on the taste or smell has been observed on treated cabbage. Siebeneicher (1982) studied the effects of fresh neem leaves, ground leaf powder, and ground neem seeds on root nematodes (*Meloidogyne incognita* Kofoid & White) of vegetables. He found that fresh leaves and leaf power applied to potted lettuce increased yields 19.4 times over the controls. Dreyer (1984) found that simple aqueous extracts of neem seed powder (50 g and 100 g/l of water) and neem oil (10 l/ha and 20 l/ha) reduced significantly damage done by *Bemisia tabaci* Genn. (Homoptera: Aleyrodidae) on courgettes, *Cucurbita pepo* L. He observed that feeding by larvae of the spotted beetle, *Henosepilachna elaterii* (*Epilachna chrysomelinia* F.), was prevented almost completely, and that their development was negatively affected. Akou-Edi (1984), in laboratory experiments, concluded that neem seed powder (20–80 g/kg of red corn) and neem oil (1–8 ml/kg of red corn) repelled *Tribolium confusum* J. du Val (Coleoptera: Tenebrionidae) and *Sitophilus zeamais* Motsch (Coleoptera: Curculionidae). To protect the seeds of cowpea against the bruchids during storage, Zehrer (1984) used neem oil (0.5 per cent v/v) and found that cowpeas lost only 5 per cent of their initial weight after six months. The oil has to be thoroughly mixed with the seeds. The neem oil treatment of cowpea seeds is effective for six months or longer (Stoll, 1986), making neem oil more successful than all synthetic pesticides used in storage pest control, including synthetic pyrethroids (Schmutterer, no year). One problem of using neem oil is the bitter taste of oil that remains after treatment. To remove this bitter taste from grains before eating, they should be covered with hot water for a few minutes and drained after. Research at the International Rice Research Institute demonstrated that the bitters (limonoids) may be extracted from neem seed as a crystalline brownish powder which could be water soluble, non-phytotoxic, and photostable (Saxena, 1989). Dreyer (1987), studying the effects of weekly applications of aqueous extracts of neem kernel powder (25 or 50 g/l of water) and neem oil (5–10 l/ha) on several insect pests of vegetables and field crops, obtained good results in controlling *P. xylostella* and *Hellula undalis* F. (Lepidoptera: Pyralidae) on cabbage, *Jacobiella facialis* Jac. (Homoptera: Cicadellidae) on eggplant, and *Scrobipalpa ergasima* Mayr (Lepidoptera: Gelechiidae) on gboma (*Solanum aethiopicum* L.).

In Nigeria, Olaifa and Akingbohungbe (1987), in choice and non-choice experiments, concluded that 3.3 g of mixed neem fruits and leaves (1:1) per ml of water was effective as an antifeedant against *Zonocerus variegatus* L. (Orthoptera: Pyrgomorphidae). They obtained 100 per cent antifeedant effect using neem and *Petiveria allaiacea* L. mixture. Ivbijaro (1987) discussed the prospects for neem in the Nigerian agroforestry system and concluded that neem research has been focussed on the preservation of grains which represent over 47 per cent of the total food produced in Nigeria. He mentioned that the maize weevil, *Sitophilus zeamais,* and the rice weevil, *S. oryzae* L. (Coleoptera: Curculionidae), were effectively controlled using 1–2.5 g of ground neem seed per 20 g of corn grains. Other stored grain pests such as *T. castaneum* and *C. maculatus* were also negatively affected by neem tree products. Okorie et al. (1990) studied the effects of neem seed powder on *Dermestes maculatus* DeGeer (Coleoptera: Dermestidae) feeding on dried *Tilapia* fish. They found that 0.25 to 2 g of neem seed powder per 25 g of *Tilapia* fish significantly inhibited oviposition and killed the adult *D. maculatus.* A larval mortality of 93 per cent has been observed after 30 days of exposure.

In Ghana, tests on *C. maculatus* have been done by Tanzubil (1987), who concluded that neem seed kernel oil at a rate of 5 ml/kg of cowpea seeds, and neem fruit dust at 10 per cent by weight of cowpea seeds, gave good results in protecting cowpea against weevil damage for four months with no adverse effect on cowpea seed viability or quality.

CONCLUSIONS

During the First International Neem Conference, Radwanski (1981) noted that neem components have been studied by many scientists over the world, but these efforts have been both isolated and sporadic. This suggests that measures should be taken on a national, regional, or international basis to coordinate and promote results in a practical context.

To promote neem tree products as a pest control strategy, important problems should be solved at the farmer level, particularly in developing countries. It is well known that formulation greatly influences the effectiveness of the active ingredient. The residual effects of neem products last mostly for five to seven days (Schmutterer, 1990). Consequently, intensive research leading to appropriate formulation and extraction of neem compounds which will last longer than those currently used should be pursued. Another problem to be considered is how to protect stored neem seed kernels against stored grain pests. Specific protection should be provided against both stored neem seed kernel pests and all insect pests attacking neem leaves and flowers.

Neem pest control technology in West Africa can be improved by application of the classic system of research, demonstration, and training. This requires a well-organized and decentralized national plant protection service to provide widespread assistance to farmers. Much attention must be given to sustained training of peasant farmers, technicians, and researchers. This training has three benefits: secure knowledge base, facilities to establish demonstrations, and access to appropriate technologies.

ACKNOWLEDGEMENTS

I wish to express my great gratitude to the International Development Research Centre (IDRC), Ottawa, Canada, for the financial support in attending this conference. Thanks are extended to the Canadian International Development Agency (CIDA) Plant Protection Project (960325) managed by Agriculture Canada, Research Station at Saint-Jean-sur-Richelieu, Quebec, Canada, for establishing communication with the secretariat of the conference.

REFERENCES

Adhikary, S. 1981. The Togo experience in moving from neem research to its practical application for plant protection. In: *Natural Pesticides from the Neem Tree* Azadirachta indica *A. Juss., Proc. 1st Int. Neem Conf.*, Rottachegrern, 16–18 June 1980. H. Schmutterer, K.R.S. Ascher and H. Rembold (eds.), GTZ, Eschborn, pp. 215–222.

Akou-Edi, D. 1984. Effects of neem seed powder and oil on *Tribolium confusum* and *Sitophilus zeamais*. In: *Proc. 2nd Int. Neem Conf.,* Rauischholzhausen, 1983. H. Schmutterer and K.R.S. Ascher (eds.), GTZ, Eschborn, pp. 445–452.

Batchomo, H. 1991. Evaluation de trois composantes de lutte intégrée pour le contrôle des populations et dégâts des arthropodes nuisibles de l'arachide *Arachis hypogaea* L. Mémoire de fin d'études, option Agronomie, Université de Ouagadougou, Institut des Sciences de la Nature, IDR. 83 pp.

Dreyer, M. 1984. Effects of aqueous neem extracts and neem oil on the main pests of *Cucurbita pepo* in Togo. In: *Proc. 2nd Int. Neem Conf.,* Rauischholzhausen, 1983. H. Schmutterer and K.R.S. Ascher (eds.), GTZ, Eschbron, pp. 435–444.

Dreyer, M. 1987. Field and laboratory trials with simple neem products as protectants against pests of vegetable and field crops in Togo. In: *Proc. 3rd Int. Neem Conf.* Nairobi, 10–15 July 1986. H. Schmutterer and K.R.S. Ascher (eds.), GTZ, Eschborn, pp. 431–447.

Ivbijaro, M.F. 1987. Prospects for neem in Nigerian agriculture. In: *Proc. 3rd Int. Neem Conf.*, Nairobi, 1986. H. Schmutterer and K.R.S. Ascher (eds.), GTZ, Eschborn, pp. 525–533.

Jacobson, M. 1986. The neem tree: Natural resistance par excellence. In: *Natural Resistance of Plants to Pests. Roles of Allechemicals*. M.B. Green and P.A. Hedin (eds.), American Chemical Society Symposium Series No. 296. Washington, D.C., pp. 220–232.

Lynch, R.E. 1990. *Groundnut Pest Management in Burkina Faso*. Entomology, University of Georgia, Coastal Plain Exp. Station, USA.

Okorie, T.G., O.O. Siyanbola and V.O. Ebochuo. 1990. Neem seed powder, as a protectant for dried *Tilapia* fish against *Dermestes maculatus* DeGeer infestation. *Insect Sci. Appl.*, 11: 153–157.

Olaifa, J.I. and A.E. Akingbohungbe. 1987. Antifeedant and insecticidal effects of extracts of *Azadirachta indica, Petiveria alliacea* and *Piper quineense* on the variegated grasshopper, *Zonocerus variegatus*. In: *Proc. 3rd Int. Neem Conf.* Nairboi, 1986. H. Schmutterer and K.R.S. Ascher (eds.), GTZ, Eschborn, pp. 405–418.

Radcliffe, E.B., F.V. Dunkel, P.P. Strzok and S. Adam 1991. Antifeedant effect of neem, *Azadirachta indica* A. Juss. kernel extracts on *Kraussaria angulifera* Krauss (Orthoptera: Acrididae), a Sahelian grasshopper. *Trop. Agric.* (Trinidad), 68: 95–101.

Radwanski, S.A. 1981. Multiple land utilization in the tropics: An integrated approach with proposals for an international neem tree research and development program. In: *Proc. 1st Int. Neem Conf.* Rottachegrern, 16–18 June 1980. H. Schmutterer, K.R.S. Ascher and H. Rembold (eds.), GTZ, Eschborn, pp. 267–277.

Saxena, R.C. 1989. Insecticides from neem. In: *Insecticides of Plant origin.* J.T. Arnason, B.J.R. Philogène and P. Morand (eds.), American Chemical Society, Washington, D.C., pp. 110–135.

Schmutterer, H. 1990. Properties and potential of natural pesticides from the neem tree, *Azadirachta indica*. *Annu. Rev. Entomol.* 35: 271–297.

Schmutterer, H. no year. Some simple methods of application of neem (*Azadirachta indica*) products for pest control by African farmers. Institute of Phytopathology and Applied Zoology, Univ. of Giessen, FRG. 11 pp.

Schmutterer, H. and K.R.S. Ascher. 1984. *Natural Pesticides from the Neem Trees* Azadirachta indica *A. Juss. and Other Tropical Plants. Proc. 2nd Int. Neem Conf.*, Rauischholzhausen, 25–28 May 1983. GTZ, Eschborn, 587 pp.

Schmutterer H. and K.R.S. Ascher. 1987. *Natural Pesticides from the Neem Tree* Azadirachta indica *A. Juss. and Other Tropical Plants. Proc. 3rd Int. Neem Conf.* Nairobi, 10–15 July 1986. GTZ, Eschborn, 703 pp.

Schmutterer, H., K.R.S. Ascher and H. Rembold. 1981. *Natural Pesticides from the Neem Tree* Azadirachta indica *A. Juss., Proc. 1st Int. Neem Conf.,* Rottachegrern, 16–18 June 1980. GTZ, Eschborn, 297 pp.

Siebeneicher, F. 1982. Untersuchungen über die Eignung von Neem (*Azadirachta indica* A. Juss.) zur Bekämpfung von gallbildenden Nematoden (*Meloidogyne* spp.) in Togo. Dipl. thesis, Univ. of Giessen, FRG.

Stoll, G. 1986. *Natural Crop Protection, Based on Local Resources in the Tropic and Subtropics.* Josef Margraf, AIchtal, Germany, 186 pp.

Tanzubil, P.B. 1987. The use of neem products in controlling the cowpea weevil, *Callosobruchus maculatus*. In: *Proc. 3rd Int. Neem Conf.,* Nairobi, Kenya, 10–15 July 1986. H. Schmutterer and K.R.S. Ascher. (eds.), GTZ, Eschborn, pp. 517–523.

Traoré, A. 1989. Evaluation du pouvoir protectant de quelques huiles végétales du séchoir solaire et de quelques insecticides de synthèse contre la bruche du niébé *Callosobruchus maculatus. Sci. et Technique* 19: 48–60.

Wilps, H., O. Nasseh, S. Krall and O. Kabo 1992. Lutte contre les *Schistocerca gregaria* adultes au moyen de nouvelles préparations et méthodes. SAHEL PV INFO Bulletin d'Information en Protection des Végétaux de I'UCTR/PV (Bamako, Mali), 44: 7–14.

Zehrer, W. 1984. The effect of the traditional preservatives used in Northern Togo and of neem oil for control of storage pests. In: *Proc. 2nd Int. Neem Conf.*, Rauischholzhausen, 1983. H. Schmutterer and K.R.S. Ascher (eds.), GTZ, Eschborn, pp. 453–460.

Zongo, J.O. 1990a. Le neem *Azadirachta indica* A. Juss.: Revue sur ses Propriétés Pesticides. SAHEL PV INFO Bulletin d'Information en Protection des Végétaux de I'UCTR/PV (Bamako, Mali), 20: 12–20.

Zongo, J.O. 1990b. Les insectes ravageurs des amandes et des feuilles du neem. SAHEL PV INFO Bulletin d'Information en Protection des Végétaux de I'UCTR/PV (Bamako. Mali), 27: 19–20.

Zongo, J.O., C. Vincent and R.K. Stewart. 1992. Effects of neem seed kernel extracts on egg and larval survival of the sorghum shoot fly, *Atherigona soccata* Rondani (Diptera: Muscidae). *J. Appl. Entomol.* (in press).

9

Improving the Productivity of Neem Trees

NARAYAN G. HEGDE

BAIF Development Research Foundation, Pune, India

ABSTRACT

Neem is a versatile tree of the tropics with immense potential to protect the environment while developing sustainable agriculture. Apart from its tolerance to drought and saline conditions, neem can contribute to the income of the rural people through production of seeds. However, neem cultivation is not popular because of lack of awareness about the benefits and economic advantages.

To improve the productivity of neem and to popularise the cultivation, it is necessary to formulate a strategy which will include the standardisation of propagation techniques, selection of superior germplasm, development of low cost silvicultural techniques, orientation of farmers to proper collection and processing of seeds and establishment of a strong extension network to highlight the benefits of neem in improving the economy and protecting the environment.

INTRODUCTION

Neem tree is gaining importance through out the world, because it provides solution to several environmental problems such as rehabilitation to degraded ecosystems, reduction in the use of agrochemicals and generation of income for small farmers. Although neem is well known for its medicinal properties, very little attention was paid till recently to popularise the cultivation of this tree.

DISTRIBUTION

Neem tree (*Azadirachta indica*, A. Juss. formerly known as *Melia azadirachta*) belonging to the family Meliaceae is known for thousands of years in the Indian sub-continent by different names such as Indian lilac and margosa tree. Neem is a native species of the Indo-Burma region, found in open shrub forests and dry zones of both India and Myanmar. In India it is found naturally occuring on the Shivalik hills of Uttar Pradesh and hillocks of Deccan, in Karnataka and adjoining states. In Myanmar neem grows in the natural forests of the central dry zone. Excepting these locations, neem is generally found growing around the backyards, along roads and field bunds, in and around the villages, than in the forests. This is an important reason why it was neglected by the foresters in the past.

In India, it is estimated in 1959 that about 14 million trees were growing all over the country with about 55.7 per cent trees in Uttar Pradesh state alone (Ketkar, 1976). The current population of neem is estimated at 16–18 million trees. It is believed that neem was introduced from India to other countries by the Indian emigrants in the earlier part of this century. Neem is a popular tree in drier regions of Pakistan. In Sri Lanka, neem grows abundantly in the drier parts of southern and northern provenances, mostly in home gardens as a medicinal plant. In Nepal, neem grows wild as well as in homesteads in Tarai and inner Tarai regions upto an altitude of 900 m. However, the trees have established well in the drier parts of the western region, compared to the wet eastern region. In Indonesia neem trees are found in the islands of Java, Bali, Madura, West Nusa Tenggara and East Nusa Tenggara from sea coast upto an altitude of 400 m. In Bangladesh, neem is a household tree, widely distributed in north-west part of the country, particularly in Rajashahi, Sirajganj, Bogra and Dinajpur districts.

In India, only one species of the genera *Azadirachta* has been reported so far. However, in Thailand, there are two species of *Azadirachta* namely *Azadirachta indica* and *Azadirachta excelsa*. *Azadirachta excelsa* with large size leaflets and less bitterness, is native to Thailand. *Azadirachta indica* var. *siamensis* is believed to be the native plant which resembles neem in most respects but for its erect growth habit and larger leaves and fruits. This variety is less bitter and found extensively on farmlands throughout Thailand, particularly in south eastern region and western coast. The local people use the tender leaves and flowers of this variety for salad and other food preparations. The other variety is the bitter neem, *A. indica* introduced from India some time in 1920s for its medicinal uses. However, this variety is confined to a small region in Southern Thailand.

Neem was introduced to Malaysia in 1940s by the migrating Indians as a shade tree with medicinal uses in home gardens in northern states of peninsular region.

Neem was introduced from India to Africa in early 1920s by the Indian migrants. It reached Nigeria around 1928 and to Senegal in 1944. In Senegal neem was the most popular tree species adopted by the foresters and farmers for large scale planting as avenue tree and wind break. More than 50 per cent of the seedling raised during the years from 1966 to 1979 in this country was neem. This tree has also been introduced to other tropical regions of Africa, Middle East, West Indies and Australia.

AGROCLIMATIC ADAPTABILITY

Neem has been held in great esteem in India from time immemorial and references on the diverse uses of different parts of this plant are found in ancient religious literature. People in India have been using various parts of neem in herbal medicines and to control storage pests. The neem oil was extracted from seeds for lightening lamps and cake was used a manure. However, the use of neem was confined to a shade tree and indigenous medicines in other parts of Asia except in Myanmar and Thailand, where tender leaves and flowers are used as food.

Neem is a large evergreen tree 15–20 m in height, with semi-straight to straight trunk, 30–80 cm in diameter and spreading branches, forming a broad crown. Being a hardy species, neem tree performs better than many fast growing species in arid regions. It is a popular tree species planted on road sides in India with a long life of over 100 years.

Neem has adapted to a wide range of climate. It thrives well in hot weather, where the maximum shade temperature is as high as 49°C and tolerates cold up to 0°C, on altitudes upto 1500 m.

The tree grows on almost all types of soils including clayey, saline and alkaline soils, with pH upto 8.5, but does well on black cotton soils and deep well-drained soil with good sub-soil water. Unlike most other multipurpose tree species (MPTS), it thrives well on dry, stony, shallow soils and even on soils having hard calcareous or clay pan, at a shallow depth. The tree improves the soil fertility and water-holding capacity as it has a unique property of calcium mining, which changes the acidic soils into neutral. Neem grows naturally in areas where the rainfall is in the range from 450 to 1200 mm. However, it has been introduced successfully even in areas where the rainfall is as low as 250 mm. It cannot withstand water-logged areas and poorly drained soils.

Establishment and Growth

Neem is established by direct seeding or seedlings raised in nurseries. Most of the trees growing in India are established naturally through bird distributed seeds. New methods have also been developed to establish neem plants through propagules raised by tissue culture and stem cuttings.

Neem grows slow during the first year of planting. Young neem plants cannot tolerate intensive shade, frost or excessive cold conditions. After establishment, the saplings start growing better and compare well with other multipurpose tree species. Under a multi-locational silvicultural study carried out in Maharashtra (at Urulikanchan, on latitude 18.3° North, altitude 560 m, annual rainfall 450 mm, with temperature ranging between 8 and 42°C, with a thin layer of gravel to sandy clay soil on basalt rocks, with pH 7.8), Karnataka (at Lakkihalli, latitude 13.12° North, altitude 238.2 m, annual rainfall 615 mm, temperature ranging between 15 and 38°C, with deep loamy sand having pH 6.4) and Gujarat (at Nanodara, latitude 23° North, altitude 44.5 m, annual rainfall 860 mm, temperature in the range between 6 and 44°C, with saline sandy loam, having calcareous pan at 1.0–1.2 m depth, with pH 8.7), the data indicated that neem can adapt very well to high pH, saline soils as well to other shallow and sandy soils. From Table 1, it can be seen that during the first five years of the trial, neem ranked third in Gujarat next to *Eucalyptus* and *Leucaena*, both in height and DBH, which can be attributed to high calcium content and adequate moisture supply. Its performance in low rainfall region of Maharashtra on shallow soils was medium, while the growth was poor in dry sandy tract of Karnataka. Such poor performance in Karnataka despite higher rainfall is probably because of low soil fertility. This indicates that fertility of the soil is more critical than moisture supply for the growth of neem tree.

Table 1. Performance of selected MPTS which have performed better on different locations

Species	Urulikanchan HT Cm	Urulikanchan BD Cm	Urulikanchan DBH Cm	Urulikanchan Canopy Cm	Lakkihalli HT Cm	Lakkihalli BD Cm	Lakkihalli DBH Cm	Lakkihalli Canopy Cm	Nanodara HT Cm	Nanodara BD Cm	Nanodara DBH Cm	Nanodara Canopy Cm
	Average Annual Growth Rates (Av. of five years)											
1. *Eucalyptus tereticornis*	194	2.34	1.71	57	145	1.83	1.46	50	260	3.65	3.05	50
2. *Eucalyptus camaldulensis*	160	1.89	1.29	60	162	2.2	1.55	57	—	—	—	—
3. *Leucaena leucocephala*	150	2.09	1.66	75	84	1.26	0.73	35	211	3.35	2.83	89
4. *Prosopis juliflora*	81	1.52	0.89	102	52	0.62	0.38	61	125	2.58	1.95	142
5. *Acacia auriculiforsis*	107	1.86	0.79	62	127	1.96	1.12	75	188	3.42	2.31	83
6. *Gliricidia sepius*	130	1.47	1.91	83	83	0.96	0.50	48	—	—	—	—
7. *Melia azedarach*	131	2.16	1.53	73	76	1.51	1.15	25	—	—	—	—
8. *Pithecellobius dulce*	110	1.85	1.19	89	51	0.73	0.48	51	37	0.40	0.13	10
9. *Albizia lebbeck*	102	1.44	0.94	51	96	2.07	1.29	84	172	3.27	2.52	148
10. *Acacia nilotica* var. *telia*	63	1.32	2.17	58	40	0.89	0.39	31	191	2.81	2.85	103
11. *Azadirachta indica*	99	2.00	1.44	68	61	1.23	0.66	34	192	3.28	2.73	114

Source: BAIF, 1991. MPTS (India) Project, final report submitted to IDRC.
HT: Height, BD: Basel Diameter, DBH: Diameter at Breast Height

Neem is a hardy tree adapted to severe drought and saline conditions. It also has an ability to withstand salty and non-salty industrial effluent, polluted gases like sulfur-di-oxide and carbon-di-oxide and attract dust and carbon particulates. Hence it has a special use in establishing industrial green-belts and avenue plantations.

The annual biomass increment of neem plantation has been reported at 3 to 10 m^3/ha, which is considered as medium fast growth, next to fast growing tree species such as *Eucalyptus*, *Casuarina*, *Leucaena* and *Acacia* of humid region. The wood makes good fuel, although it emits more smoke compared to *Acacia* and *Casuarina* wood (Hegde, 1991).

PRODUCTION AND USES

Neem tree has several economic advantages over other MPTS grown in the tropics. The tree can be harvested for timber, after 35–40 years of planting. The sapwood of this tree is grayish white and the heartwood is red to reddish brown.

The wood is aromatic, moderately heavy (specific gravity 0.72 to 0.83), narrowly interlocked, with uneven grains, and medium to coarse texture, durable and not easily attacked by insects. The timber is medium refractory and seasons well even when sawn wet. It is easy to work with this wood by hand and on machines, but does not take polish well. The wood properties presented in Table 2, clearly indicate that neem timber can be closely compared with teak in many respects.

The wood is used for building houses as posts, beams, door and window frames, furnitures, carts, axles, yokes, naves and fallows, ship and boat building, helms and oars, oil mills, cigar boxes, carved images, toys, drums and agricultural implements (CSIR, 1985; Troup, 1986). However, as it's special qualities have not been fully explored and hence wood is sold at Rs. 7000 per cubic m as against Rs. 24000 for teak wood.

Table 2. Quality index of neem timber in comparison to teak

Properties	Teak	Neem
Weight	100	124
Strength (as beam)	100	87
Stiffness (as beam)	100	81
Suitability (as post)	100	82
Shock resistance	100	105
Shape resistance	100	77
Shear	100	129
Surface hardness	100	131
Nail holding ability	100	144

Source: CSIR (1985)

Non-wood Products

Apart from its favourable characteristics as a timer tree, neem can make a significant contribution to the income of rural people, through seed production every year. Neem seeds are not only rich in oil but also contain different compounds which have medicinal value and pesticidal properties. Neem leaves, which are known for bitterness also contain useful compounds like nimbin, nimbinene, nimbandiol, nimbolide, 6-desacetylnimbinene and quercetin, which are useful for ayurvedic medicines, pest and disease control. The leaves are also fed to livestock mixed with other fodder. In certain parts of Andhra Pradesh, farmers specially feed neem leaves to cattle and goats immediately after parturition for increasing milk secretion (CSIR, 1985).

The leaves contain 12.40 to 18.27 per cent crude protein, 11.40 to 23.08 per cent crude fibre, 43.32 to 66.60 per cent N free extract, 2.27 to 6.24 per cent ether extract, 7.73 to 18.37 per cent total ash, 0.89 to 3.96 per cent calcium and 0.10 to 0.30 per cent phosphorus. These chemical compositions vary with the season and locality. The foliage of neem is a good fodder, although it is not possible to lop the trees regularly. In general, the farmers in North India lop the trees in winter while in South India the trees located in agricultural fields are lopped at the time of sowing *kharif* or *rabi* crops, to reduce shade. Neem flowers and tender leaves are used as human food in Myanmar and Thailand.

Neem starts producing seeds after 5–6 years and a grown up tree yields on an average about 25 kg seeds every year. With an average price of Rs. 2 per kg seed, a tree generates a gross income of Rs. 50 per year and with about 150–200 trees per hectare, the annual income is Rs. 10,000 (US $ 350), which is higher than the returns from many food crops.

Although, neem seeds contain azadirachtin and many other compounds which are useful as medicines, pesticides, bactericides and fungicides, most of it is crushed for oil and about 95 per cent of it is presently used in soap production. This is in fact an inferior use of the produce, because the same material in different form can be used as pesticide of very high value. Processing into value added products can fetch higher price for seeds.

Neem cake is used as livestock feed, manure, insecticide and nematicide. It not only provides organic nitrogen, but also inhibits the nitrification process, when mixed with urea, before applying in the field. Such use of neem coated urea (90 : 10) can save about 30 per cent of the total chemical nitrogen requirement of crops, which otherwise would go waste (Hegde, 1991; Ketkar, 1976) and help in reducing the cost of agricultural production.

Environmental Safety

The neem based pesticides are safe, non-toxic and do not keep any residual effects on agricultural produce. As the pesticidal preparations are very simple, farmers can make direct use of these products locally without any cost.

Use of neem products for plant protection can help in reducing the atmospheric pollution and preventing food poisoning. It can also reduce the demand for chemical pesticides, which will subsequently reduce industrial production of agrochemicals.

In spite of several benefits from neem, there is no significant increase in area under neem cultivation. Neem products are yet to be popular among Indian farmers. Therefore, it is necessary to solve the problems related with productivity of neem trees and develop suitable strategy to popularise its cultivation particularly on wastelands. Similar efforts will also have to be made in other neem growing countries.

PROBLEMS OF NEEM PRODUCTION AND UTILIZATION

Neem production is presently neither attractive nor remunerative because of poor establishment, stunted growth, poor yield, deterioration of seeds and inadequate marketing infrastructure. Thus efforts should be made to identify superior germplasm and to introduce ideal silvicultural practices for improving the growth and yield of neem. It is also necessary to promote its use in agriculture to increase the demand for neem products.

Propagation of Neem

Neem is propagated mainly through seed, which has a short viability of 3–4 weeks. The seed contains 35 to 45 per cent oil and about 30 per cent moisture and needs proper handling, not only to extend its viability but also to prevent rancidity. Although the normal practice is to collect fresh seeds naturally dropped on the ground for sowing, new techniques have now been developed to extend the seed viability by improved handling and storage practices. Roederer and Bellefontaine (1989) have reported 42% germination of neem seeds which were stored for five years at 4°C. According to Vandenbeldt and Bhumibhamon (1992), collection of just ripe fruits, depulping, quick washing, shade drying and storage of seeds in cloth bag, at low temperature upto 4°C can improve the seed viability. Surendran et al. (1992) have reported that the storage of depulped and shade dried seeds in earthen pots containing wet sand (30 per cent moisture) can retain the viability upto 62 per cent, at the end of three months.

As the seeds contain certain chemicals which inhibit germination, it has been recommended to soak the seeds in warm water at 65–70°C for 30 minutes for improving the germination.

As the flowering pattern of neem is influenced by the altitude and temperature, fruiting is delayed in areas where winter is severe or prolonged. The fruits start ripening from early April in Tamil Nadu and Karnataka, while peak fruiting season in central and northern India is during June–August. Hence the nursery activity to raise neem seedlings in most parts of the country can start only in June-July, although the recommended practice is to establish the nursery in February-March, to plant seedlings of 3–4 months age at the onset of rains, in June-July. Such practice will ensure better establishment and the saplings can withstand the dry spell of summer. Any delay in raising the nursery will result in either using small seedlings or delayed planting and in both cases, the survival will be poor. The other option of direct sowing of seeds is also not ideal because of slow growth of neem seedlings, particularly during the first year. Hence for successful promotion of neem cultivation, it is essential to develop low cost technology for extending the seed viability.

Another option for successful establishment of neem plantations is through stump planting, which is considered to be better than direct seeding or planting seedlings. Stumps are generally prepared from two year old seedlings, raised in seedbeds, by trimming the plant to retain 22 cm root and 5 cm shoot portions. The stumps are prepared under shade and transported in wet gunny bags (Singh, 1982).

Although, the most and economic method of propagation is through seeds, propagules can also be produced through root suckers and stem cuttings. It has been reported by Surendran et al. (1992) that application of indole butyric acid (IBA) and indole acetic acid (IAA) (1000 ppm) induced rooting in neem stem cuttings. Air-laying can also be produced with the help of IBA or naphthalen acetic acid (NAA). As vegetative propagation can induce early fruiting, without losing any economic traits of the mother trees, perfection of this technique for large scale production is necessary to popularise neem cultivation in a big way.

Variation in Germplasm

Neem being an open pollinated species, a wide variation in the germplasm has been reported from neem growing regions. The neem seeds produced in Niger and other African countries are rich in azadirachtin, although neem was introduced in these countries from India. A variety of neem *Azadirachta indica* var. *siamensis* in Thailand grows fast with clear bole, larger leaves, dark pink to light brown coloured bark and a long drupe, almost twice the size of the fruits found in India, but with low azadirachtin content.

In spite of its pesticidal properties, neem is susceptible to several diseases and pests which have been listed below:

Organism	Nature of damages
Nursery Diseases	
Fusarium oxysporium	damping off, due to excessive moisture in the nursery
Rhyzoctonia solani	leaf web blight, causes gray brown spots on foliage
Colletotrichium-gloeosporides	leaf spot and blight, cover a large areas of foliage
Alternaria alternata	damage foliage in nursery
Pseudocercospora subsessilis	

Organism	Nature of damages
Foliage Diseases	
Odium azadirachtae	powdery mildew of neem foliage
Pseudomonas viticola	cause leaf spot diseases
P. azadirachtae	
Xanthomonas azadirachtii	
Root Diseases	
Ganoderma lucidum	root-rot
Stem Diseases	
Corticium salmonicolor	pink disease of stem resulting in dead bark
Phomopsis sp.	twig blight, causes branch dieback
Insects	
Solenopsis sp.	fire ant causing severe defoliation
Acromyrex sp.	leaf cutting ant
Ascotis selenaria	defoliator
Boarmia variegata	defoliates in nursery
Eucema sp.	defoliator
Apate monachus	powder post beetle bones stem and wood
Helopeltis antonii	tea mosquito attacks terminal shoots, causes drying of foliage and terminal shoots
Aonidiella orientalis	oriental yellow scale or cochineal attack young stems and foliage and feed heavily
Pulvinaria maxima	feeds on sap of tender shoots and leaves
Pinnaspis strachani	lesser snow scale, attack shoots and foliage of neem
Calepitrimerus-azadirachtae	infesting tender shoots and foliage

There is a serious disorder of neem with the symptoms of yellowing of older leaves, loss of foliage with leaves concentrated at the branch tips have been noticed on neem trees in Nigeria since last few years. The typical symptoms also include distorted branches, dieback of branches, stunted and dry foliage, red discoloration of the cambium layer and exudation of sap from the branch tip. Earlier it was thought of fungal disease reported from Burkina Faso, Cameroon, Chad, Mali and Nigeria. But it has been now reported that the cause of neem decline is due to low soil moisture, competition and biotic pressure. However further studies are being carried to investigate in association of plant virus or micoplasma that are responsible for causing such disorders.

There is good scope for improving the productivity of neem through selection of superior germplasm for higher yield, quality and resistance to many of the above pests and diseases.

Silvicultural Practices to Improve Productivity

Although neem has an edge over many other MPTS in arid regions and saline soils, it is necessary to develop ideal silvicultural practices to improve its productivity.

Establishment of neem

Suitable site preparation practices help in better establishment of neem seedlings. Soil and water conservation measures such as contour bunding and trenching, plugging of gullies, formation of basins are also helpful in ensuring better establishment and growth.

The neem seedlings may fail to establish due in water-logging areas. In such areas, it is necessary to establish plantation on raised mounds or on ridges. Nutritional deficiency in infertile soils may affect the growth. On such sites, application of organic matter and chemical fertilizers can help to boost the growth.

Mulching of the basin is helpful to conserve moisture. It has been observed that even young trees are affected by moisture stress, resulting in wilting of terminal buds. In extreme cases of drought, the plants shed the foliage and enter the phase of dormancy till the rainy season. In the absence of terminal buds, the auxiliary buds sprout after the rains and the plants remain bushy, stunted and fruiting is delayed (Hegde, 1991). Such moisture stress can be avoided by mulching the plant basins during the initial 2–3 years.

For improving the productivity of the plantation, depending on the agroclimatic conditions of the site, a spacing of 6 × 6 m or 8 × 8 m may be provided. As it will take atleast 8–10 years for the trees to attain good size, mixed plantation of neem can be established with leguminous trees of short rotation.

Fruiting

The agro-climatic conditions of the mite also affect the yield and quality of neem seeds. A good rainy season helps the trees to maintain a good crown to synthesize good amount of carbohydrates needed for seed development. A few good showers after the fruit sets will also influence the fruit size, which in turn increases the yield (Hegde, 1991). However, systematic studies are needed to understand the flowering behaviour and tendency of alternate bearing, if any.

Seed Collection and Processing

The traditional practice of neem seed collection is to hand pick or broom the fruits fallen on the ground. The seed coated with sweet pulp in a thin skinny epicarp needs to be separated, sun dried and stored till crushed for oil extraction. Any delay in separating the seed and drying, will affect the quality both with respect to oil and azadirachtin content. Properly dried seeds can be stored upto one year, although it has been recommended to store for a minimum of three months after collection for maximum oil recovery (CSIR, 1985). The seed contains about 44.7 per cent kernel and about 55.3 per cent shell. The kernels constitute only about 35 per cent of the fresh fruits. The seeds are decorticated before extracting the oil. Decorticated seeds can also be stored for 8–12 months without any deterioration in the quality. In India, the commercial oil extraction process has been well developed, but the real problem lies in collection and drying of seeds at the village level.

In spite of good demand and an attractive price, only about 25 to 30 per cent of the neem seed is collected in India and the value of the uncollected seeds is estimated to be over Rs. 10 crores (US$ 3.5 million), every year.

The most important reasons for lack of interest among farmers is lack of knowledge about the handling of seeds and inadequate marketing facilities. The price realised by the seed collectors is in the range of Rs. 500 to 2500 per tonne seed, while mill owners are willing to pay upto Rs. 4000 per tonne. An important reason for poor price realisation is poor quality seeds. Some farmers have the tendency to keep the fruits after collection for a few days without depulping. In the absence of bright sunlight, the drying is slow and the seeds having very high content of moisture and oil get rancid and turn black. Due to poor recovery of oil, the price paid for such seeds is low. Thus the seed collectors need to be oriented for clean handling and drying of seeds.

The other problem is collection of clean seeds. Many collectors, broom the ground under the tree to collect the seeds. In this process, gravel and dirt get mixed with the seed, which fetch a low price. This can be avoided by cleaning the ground before fruits start ripening. Temporary flooring and spreading gunny or plastic sheets under the tree can also help in collecting clean seeds, without any wastage of labour.

Organising seed collectors' cooperatives for marketing and oil extraction can help to realise better price.

PROMOTION OF NEEM PRODUCTION

Considering the present scope for improving the productivity of neem trees and involvement of small farmers to adopt neem cultivation, the following activities can be initiated at the grass-root level in India and other developing countries.

1. Packaging of Silvicultural Practices

It is necessary to develop a package of silvicultural practices which should include superior provenances, seed storage, economic production of propagules, techniques of soil and water conservation and establishment of plantation. These silvicultural practices should ensure better growth, early fruiting, higher yield and better recovery of oil and other components.

2. Popularising the Use of Neem Products

Neem products, particularly the seed is still not considered as a dependable source for plant protection. Hence, use of neem products should be popularised among farmers through field demonstrations. The villagers should also be educated about the preparation of pesticides on their own, using neem seeds, instead of depending on commercial neem products. This requires a strong extension network of both government and non-government organisations (NGOs) and information back up about the technology and cost benefit analysis of neem cultivation.

3. Marketing of the Produce

It is necessary to develop grass-root level organisations by the farmers or NGOs to motivate the villagers to collect, clean, dry and sell the seeds without any deterioration in the quality. Small scale industrial units for oil extraction and pesticidal preparation can also be organised through such organisations. They can also take up the promotion of neem products, which will have to compete with well organised multinational pesticide firms.

REFERENCES

BAIF. 1991. MPTS (India) Project—final report submitted to IDRC.

CSIR. 1985. The Wealth of India—I: Raw materials, Government of India, New Delhi: 504–511.

Hegde, N.G. 1991. Neem: Silviculture production and uses. Paper presented at the Neem Symposium at the Pacific Science Congress held in Honolulu, Hawaii: 13 pp. May 1991.

Ketkar, C.M. 1976. Utilization of neem (*Azadirachta indica* A. Juss) and its by products. KVIC, Bombay. 234 pp.

Roederer, Y. and R. Bellefontaine. 1989. Can neem seeds be expected to keep their germination capacity for several years after collection? *Forest Genetic Resources Information* 17: 30–33.

Singh, R.V. 1982. *Fodder Trees of India*. Oxford and IBH Publishing Co. Pvt. Ltd., New Delhi: 77–84.

Surendran, C., R.S.V. Rai, K. Sivagnanam, G. Kumaravelu, M. Shanmugam, A. Ragupathy, S. Preetha, P. Srimathi and K. Kumaran. 1992. Paper presented at the National Workshop on Non-wood Forest Produce in Pune on March 8–11, BAIF, Pune.

Troup, R.S. 1986. Indian woods and their uses. Soni Reprints Agency, New Delhi: 187–188.

Vandenbeldt, S. and Bhumibhamon. 1992. The Need for Cooperative Research on Neem. Paper presented at the National Workshop on Non-wood Forest Produce in Pune on March 8–11, BAIF, Pune.

10

Neem for Vegetable IPM in Nicaragua: Experiments and Small Farmers' Perspective

D. GÓMEZ[1], F. MIRANDA[2], B. GUERRERO[3], I. RIVAS[4], J. MERCADO[4] AND F. GUHARAY[1]

[1]Proyecto Catie-Mag/Mip (Nicaragua)
[2]Universidad Nacional Agraria
[3]Estación Experimental del Valle de Sébaco
[4]Centro Nacional de Protección Vegetal, Mag
Apartado Postal P-116, Managua, Nicaragua

ABSTRACT

Use of insecticides produced from neem is becoming a real alternative for the management of pests in vegetable crops grown by small farmers in Nicaragua. Three factors for the use of neem by farmers are accessibility, effectiveness, and low cost. Informal surveys among small farmers of tomato show that they consider neem products accessible through planting neem trees on their own land or by purchasing formulated products. Secondly, experimental data collected from 1989 to 1992 show that application of neem insecticides reduces the population and damage caused by important pests like *Plutella xylostella* in cabbage and *Bemisia tabaci* and *Helicoverpa zea* in tomato. For management of *P. xylostella*, water extract of ground seed kernel is effective, while formulations based on neem oil reduce the incidence of *B. tabaci* in tomato. Partial budget analysis shows that management of these pests with neem insecticides costs less than use of chemical insecticides. These factors lead to a low-risk pest management strategy preferred by the small farmers of Nicaragua.

INTRODUCTION

In the vegetable-growing regions of Nicaragua, a small farmer rarely goes to his field alone. He is always accompanied by his knapsack sprayer. Under the conventional pest control strategy, farmers use repeated applications of synthetic pesticides. For *Plutella xylostella* control in cabbage, applications may range from 10 to 20 per cycle (Miranda, 1989; Barahona, 1992). In recent years the principal pests of tomato have been geminivirus transmitted by adults of sweet potato white-fly (*Bemisia tabaci*) and fruit borers (*Helicoverpa* sp. and *Spodoptera* sp.). The 10–15 applications during a tomato cycle account for approximately 50 per cent of the cost of external inputs (Gómez et al., 1990). Unfortunately, even with such elevated insecticide use, yields are low and of poor quality, mostly due to the ineffective conventional pest control strategy.

Recent research on critical periods and economic damage levels of these pests has provided the basis for more rational insecticide use. For example, chemical control of *P. xylostella* is recommended only after 20 days from transplanting if more than 0.5 larvae per plant are found on 50 plants per ha (Miranda, 1992). More than four eggs of *Helicoverpa* sp. in 30 tomato leaves is the control threshold (CATIE, 1990) and for management of white-flies as vector, farmers have recently started using 40 adults in 50 leaves as threshold for insecticide application (Guharay and Monterrey, 1993, unpublished data).

To further increase the efficiency of pest management systems via direct and indirect cost reduction and an increase in natural biocontrol, Nicaragua has been evaluating a long list of biorational inputs that can substitute conventional pesticides. Neem is one of the most important.

In this paper we will present results and reflections about the use of neem by Nicaraguan small vegetable farmers for the management of cabbage and tomato pests. Accessibility of the technology, effectiveness, and low cost are proving to be three fundamental requirements for the use of new rational alternatives of pest management. Widespread use of neem for experimentation and further commercial use for the vegetable pest management in Nicaragua is a result of the fact that these requirements are being met from the small farmers' perspective.

ACCESSIBILITY OF NEEM TO SMALL FARMERS

Small farmers have access to neem in two ways: they can plant neem trees on their own land to produce seeds, or they can purchase neem insecticides.

For planting neem trees, good quality seed is available to the farmers at a reasonable cost from the Botanical Insecticide Project of Nicaragua (PIBN) and Neem Cooperative (COPINIM). In the month of July or August a farmer can buy a kilogram of neem seed for US$ 4, which is sufficient for raising 3500 seedlings good for sowing 12 ha. Farmers can also purchase ready-to-plant seedlings from commercial nurseries. During 1992 about 150,000 seedlings were produced, enough to plant 535 ha. Each seedling costs US$ 0.20, bringing the cost of planting material to US$ 60/ha (personal communication, K. Gruber, PIBN).

Training and information is also available on planting, maintenance, and use of neem trees. In 1992 PIBN provided about 30 such training sessions to Nicaraguan farmers (personal communication, K. Gruber, PIBN).

As a part of the insecticide fabrication programme, COPINIM purchases neem seeds at a good price for quality seeds (US$ 1.60/kg), making seed production an attractive proposal for Nicaraguan farmers.

The neem technology is also available to the small farmers through the purchase of neem insecticides fabricated by COPINIM. There are 125,000 neem trees planted in Nicaragua, of which 25,000 will produce seeds in 1993. COPINIM purchases much of the production and transforms it into three pesticide formulations. These products are sold to the farmers at a reasonable price, making it easier for them to conduct experiments with these biorational pesticides. For example, during 1992 COPINIM procured and processed 45 tonnes of matured fruits, transforming them into 1.78 tonnes of powdered neem seed kernel (Nim 20), 326 l of formulated neem oil (Nim 80 EC), and 300 kg of powdered neem seed cake (Nim 25). The current local prices of these products are US$ 4/kg, 5/l and 3/kg respectively, although a differential price rate is applied for foreign buyers (personal communication, K. Gruber, PIBN). Currently COPINIM has licence to sell these products for experimental purpose but shortly they will be registered for commercial use in vegetable and other crops (personal communication, Ing. E. Mora, Dept. of Registration of Agricultural Chemicals, Director of Plant Protection, Ministry of Agriculture).

EFFECTIVENESS OF NEEM PRODUCTS FOR DAMAGE REDUCTION OF KEY PESTS

Since 1988 experiments have been conducted in experiment stations as well as in the farmers' fields to evaluate the efficacy of neem products for the reduction of key pests and their damage in vegetable crops like cabbage and tomato.

A research group at the School of Plant Protection of the National Agricultural University has conducted experiments in two different regions of the country during dry as well as rainy seasons from 1989 to 1992 which demonstrate that rational applications of water extract of powdered neem seed kernel (40 g/l) maintain the population of *P. xylostella* larvae at acceptable levels permitting the harvest of better quality, higher-priced heads (Barahona, 1990; Gómez and Miranda, 1992). A four-season study conducted by Miranda (1992) also demonstrated that application of water extract of neem seeds can be as effective as the commercially used biological insecticide based on *Bacillus thuringiensis* (DIPEL: Abbott Corp.) for the population and damage reduction of *P. xylostella* (Fig. 1). In some of the experiments it was possible to obtain even better quality, higher-priced heads in plots treated with neem seed extract than in Bt-treated plots (Table 1) (Miranda, 1992).

Zeledon (1988) demonstrated that neem oil treatment of bean leaves reduces the oviposition of white-flies in a small cage study indicating the possibility of a repellent action of neem oil. Based on such notions neem oil formulation is being evaluated in experiment station as well as farmers' fields for the management of white-fly-transmitted virus in tomato. Guerrero (1992) reported that two applications per week of 1 per cent of formulated neem oil (Nim 80 EC: COPINIM) reduces significantly the population of white-flies in 70 days after sowing (Fig 2.). In the plots treated with formulated neem oil a higher commercial harvest was obtained possibly due to production of bigger and heavier fruits. The production of non-commercial, small, and unevenly ripe fruits was significantly reduced in the plots treated with neem oil as compared to the control (Table 2).

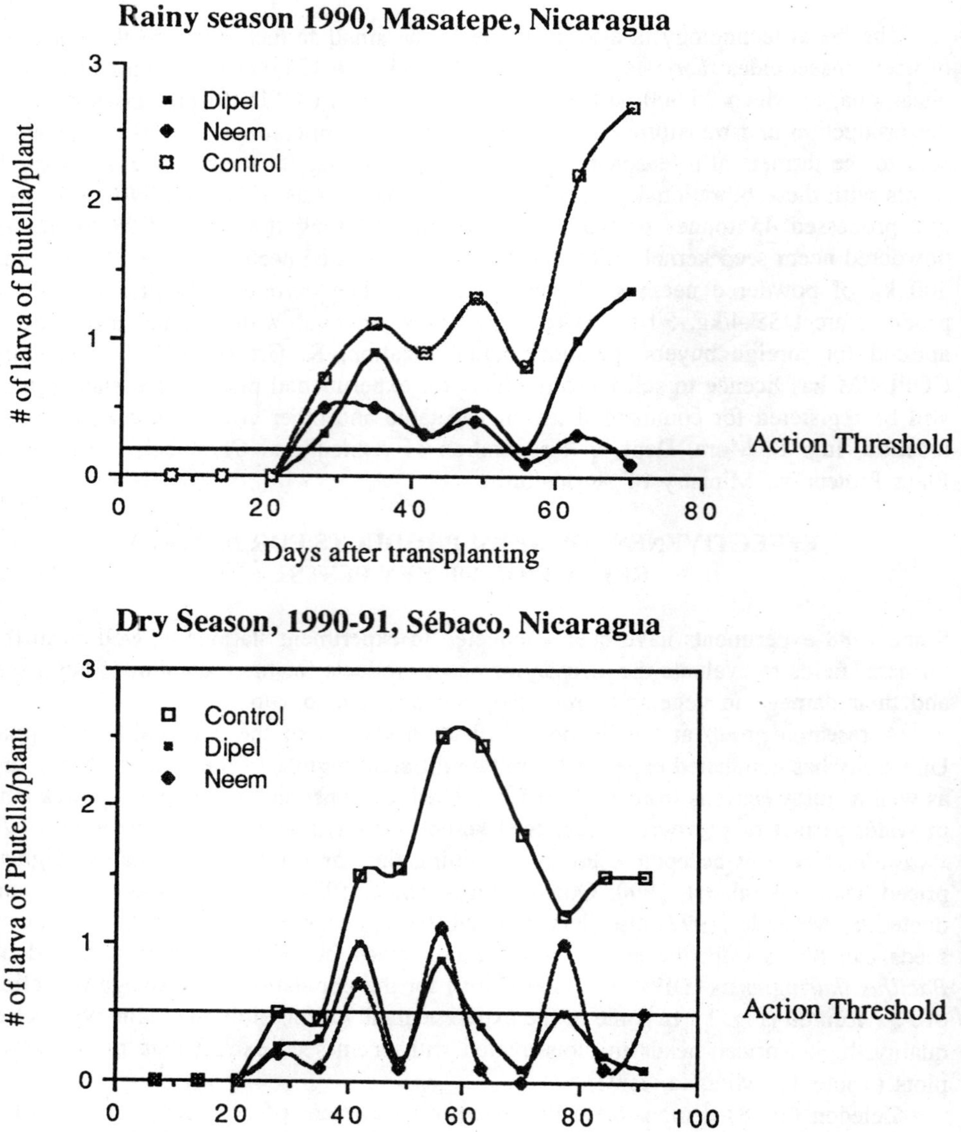

Fig. 1. Effect of applications of water extract of powdered neem seed kernel (40g/l) and DIPEL (Bt. 16,000 units/g; 1000 g/ha) on the population of larvae of *P. xylostella* in cabbage. In the rainy season experiment an action threshold of 0.2 larva/plant was employed; in the dry season it was modified to 0.5 larva/plant.

Experiments conducted in the farmers' fields in the same area during the dry season 1992–93 show good reduction of migrant white-fly population by application of 0.4 per cent of neem oil formulation. The replacement of the insecticides with formulated neem oil when the white-fly adult population is relatively low (20–40 adults/50

Table 1. Effect of applications of biorational insecticides on the damage caused by *P. xylostella* larvae in cabbage (var. Superette)

Year	Treatments	Head weight (kg)	Head quality*	Price (US$)
Rainy season 1990	Neem seed extract	1.60	1.7	0.19
	DIPEL (Bt)	1.58	2.8	0.11
	Control	1.60	3.4	0.12
Dry season 1990–91	Neem seed extract	1.43	3.0	0.77
	DIPEL (Bt)	1.46	3.0	0.98
	Control	1.32	4.6	0.67
Rainy season 1991	Neem seed extract	1.37	1.8	0.70
	DIPEL (Bt)	1.19	2.7	0.54
	Control	1.37	2.8	0.50

* Head quality was evaluated by using a scale proposed by Chalfant and Brett (1965) of 1 to 6 depending on the increasing level of damage.

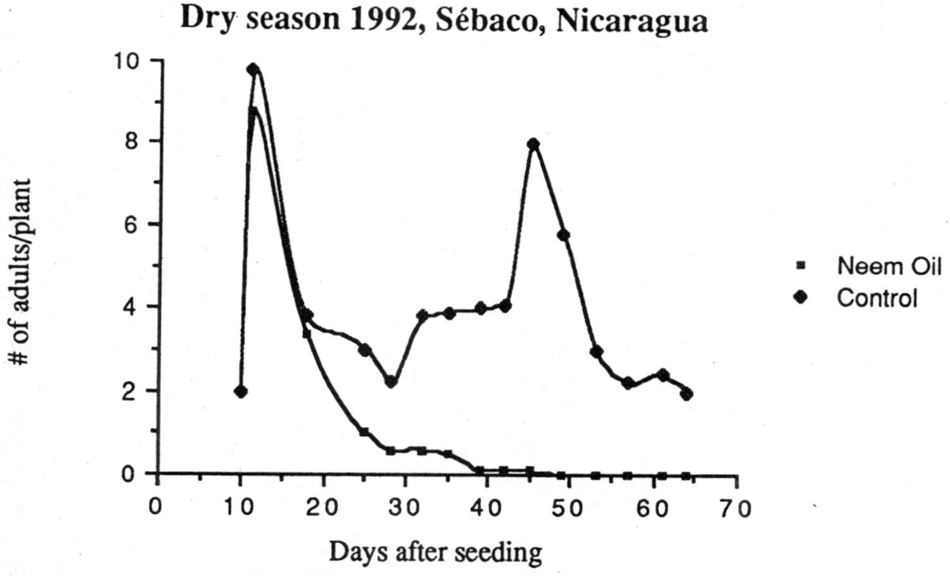

Fig. 2. Effect of two applications per week of 1 per cent of formulated neem oil (Nim 80 EC: COPINIM) on the adult white-fly population in tomato. (Dry Season 1992, Sébaco, Nicaragua).

leaves sampled) can produce the same amount of harvest at much lower cost (unpublished data, Guharay and Monterrey, 1993).

Few studies have reported neem use for the reduction of fruit borers (*Helicoverpa zea* and *Spodoptera* spp.) and their damage in tomato. Rivas and Méndez (1991) reported that seven applications of water extract of powdered neem seed kernel (15 g/l) 18–66 days after transplanting reduced the population of *Helicoverpa* larvae. Mercado and Guerrero (1992) demonstrated that three applications of 100 ppm of ethanolic extract of neem seed kernel reduced damage caused by the fruit borers compared to control, but damage levels were not significantly different from three applications of

Table 2. Effect of applications of 1 per cent of formulated neem oil on the yield and quality of tomato, dry season 1992

Treatments	Fruit wt (g)	Commercial harvest (t/ha)	Non-commercial harvest (t/ha)
1% formulated neem oil (Nim 80 EC)	70	9.0	1.4
Control	30	3.0	3.0

insecticide (TAMBO: Cipermethrin and Profenofos) (Fig. 3). However, the number of fruits damaged by the fruit borers was much less in the plots treated with neem than in plots treated with insecticides (Table 3).

LOW COST OF PEST MANAGEMENT WITH NEEM INSECTICIDES

Economic analysis of the experiment conducted in the farmers' fields or experiment station can give us an estimate of the profitability of the rational alternatives. Although we can not ignore the scaling effect of converting costs and returns from areas as small as 50 sq to 1 ha, the estimates permit at least an approximate comparison.

Miranda (1992) provides detailed information on the cost of different rational pest management strategies in relation to the net benefits, including data on water extract of neem seed kernel and DIPEL (Bt, Table 4). In all the estimates net benefit produced by the application of neem is superior to that DIPEL. In two of the cases neem even presents less variable cost than DIPEL. Depending on the fluctuation of the price of cabbage a farmer can obtain a marginal return of 1057 to 15,702 per cent for the investment incurred by employing water extract of neem seed kernel. This makes neem use for the management of *P. xylostella* in cabbage extremely profitable.

Gómez et al. (1991) performed partial budget analysis of costs and returns obtained in experiments conducted with the participation of farmers in the Sébaco Valley. In these trials the integrated pest management strategies included use of neem oil for the management of white-flies and a mixture of JAVELIN (Bt: 32,000 units/g) and Nudrin (Methomil) in combination with scouting for the management of fruit borers. This was compared with a treatment where the farmers used the conventional insecticides for the management of these pests. In almost all the fields but one, the use of integrated pest management strategies incurred less cost and produce higher net benefits (Table 5). In the remaining field the application of insecticides produced higher net benefit but at a higher cost.

CONCLUSION

The use of biorational insecticides such as neem for the management of insect pests of vegetable crops is being experimented with and used by the small vegetable farmers for management of problem vegetable pests. Its use may even spread to some export crops like honeydew melon. The easy accessibility of the technology, effectiveness of the products for damage reduction of target pests, and lower costs are very important

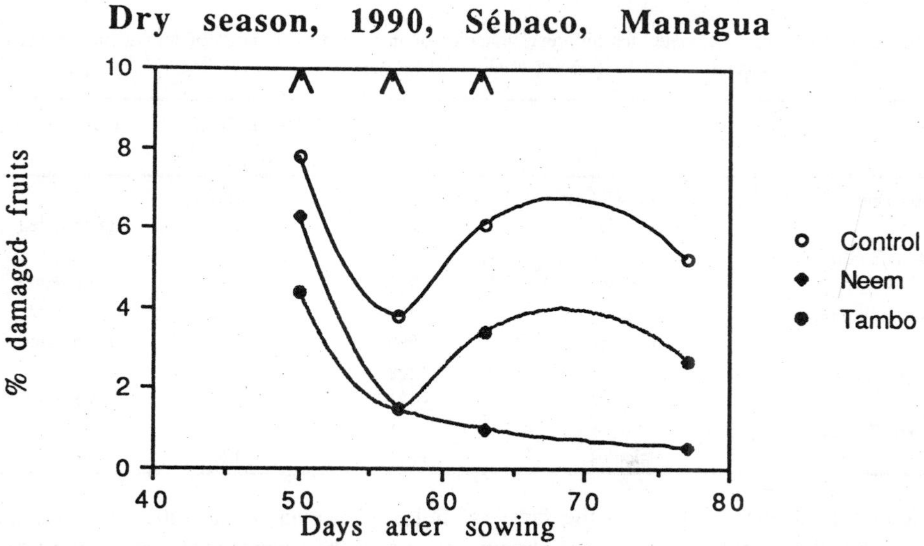

Fig. 3. Effect of applications of 100 ppm of ethanolic extract of neem seed kernel and TAMBO (Cypermethrin and Profenofos) on the incidence of fruits damaged by *H. zea* in tomato (Dry season, 1990, Sébaco, Nicaragua).

Table 3. Effect of applications of 100 ppm of ethanolic extract of neem seed kernel and Tambo (Cipermethrin Profenofos) on the yield of tomato (Mercado and Guerrero, 1992), dry season 1990

Treatments	No. healthy fruits/ 10 plants	No. damaged fruits/ 10 plants	Commercial harvest, kg/10 plants
Ethanolic extract 100 ppm	232	16.5	8.24
TAMBO Cipemethrin + Profenofos	137	39.2	8.51
Control	117	32.0	6.38

Table 4. Partial budget estimates for different strategies of management of *P. xylostella* (diamondback moth) larvae in cabbage (Mirinda, 1992)

Year	Treatments	Net benefit (US$/ha)	Variable costs (US$/ha)	Marginal rate of return (%)
Rainy	Neem seed extract	7,069	305	1057
season	DIPEL (Bt)	3,423	279	Dominated
1990	Control	3,844	0	
Dry	Neem seed extract	23,804	115	10,541
season	DIPEL (Bt)	22,237	120	Dominated
1990–91	Control	11,681	0	
Rainy	Neem seed extract	17,217	38	15,207
season	DIPEL (Bt)	12,070	80	Dominated
1991	Control	11,250	0	

Table 5. Partial budget estimates for different strategies of management of pests of tomato in the Sébaco Valley, dry season 1990–91 (Gómez et al., 1991)

Farm	Treatment	Net benefit (US$/ha)	Variable costs (US$/ha)	Observation
Coop Daniel Teller	Neem oil + Bt	2644	106	
	Insecticides	2192	443	Dominated
Ludin Gurdian	Neem oil + Bt	2455	122	
	insecticides	1891	204	Dominated
José León Torres	Neem oil + Bt	1080	104	
	insecticides	690	217	Dominated
Coop Rigoberto L.	Neem oil + Bt	1228	99	
	insecticides	1421	487	RMR = 49.9%
Alvaro Torres	Neem oil + Bt	225	372	
	insecticides	–124	421	Dominated

factors for the use of neem in the farmers' fields. However, even with these requirements fulfilled, technologies do not always get to find their appropriate use in the small farmers' fields.

In Nicaragua, most of the institutions working in the generation and technology transfer of integrated pest management in vegetable crops are intending to work in a participative mode, where farmers have a chance to select the technologies that they can experiment with in their fields. A good part of the quick adoption of use of neem products by the small farmers in Nicaragua probably can be attributed to that participative process.

REFERENCES

Barahona, L. 1990. Effects of botanical and biological insecticides on the entomofauna present in cabbage (in Spanish). Thesis Ing. Agr. ISCA, Managua, Nicaragua.

Barahona, L. 1992. Preliminary diagnostics about management of cabbage in the zone of Pacaya (Masaya, IV region, Nicaragua) (in Spanish) *Revista de la Escuela de Sanidad Vegetal* 2(3): 62.

CATIE. 1990. Manual for the integrated management of tomato pests. Tech. Rep. No. 153. Turrailba, Costa Rica.

Chalfant, R. and C. Brett. 1965. Cabbage looper and imported cabbage worms; feeding damage and control on cabbage in Western North Carolina. *J. Econ. Entomol.* 58: 28–33.

Gomez, C. and F. Mirnada. 1992. Effect of biological and botanical insecticides on the control of *Plutella xylostella* in cabbage during the rainy season (in Spanish). *Revista de la Esccuela de Sanidad Vegetal* 2(3): 66.

Gómez, D., J. Simán and K. Nelson. 1991. Report for the producers about costs, yields, soil and nematodes. Junio 1991 (in Spanish). Internal Tech. Rep. Project CATIE-MAG/MIP, Nicaragua (mimeo).

Gómez, D., J. Simán, I. Rivas, P. Rosset, R. Dávila and J.B. Guerrero. 1990. Diagnostics of the factors that influence the yield of tomato in the sixth region (in Spanish). Memoria Resumen, 4to Congreso Nacional MIP, 3er Congreso Internacional "Humberto Tapia Barquero" (in Memoriam), Managua, Nicaragua, 65 pp.

Guerrero, B. 1992. Evaluation of different formulations and doses of neem (*azadirachtina*) for the control of white-flies *Bemisia tabaci* Genn (Homoptera, Aleyrodidae) in tomato *Lycopersicum esculentum*. Mill in the valley of Sébaco, Nicaragua (in Spanish). Memoría Jornada Científica-técnica sobre el manejo de mosca blanca en el cultivo de tomate. Grupo Interinstitutional de Tomate, Managua 8 pp. (mimeo).

Mercado, J. and B. Guerrero. 1992. Different concentrations of formulated azadirachtin for the control of *Heliothis* spp. in tomato (in Spanish). *Revista de la Escuela de Sanidad Vegetal* 2(3): 64.

Miranda, F. 1989. Estimation of economic damage level of diamondback moth larvae (*Plutella xylostella*) in cabbage (*Brassica oleracea* L.) var. Superette (in Spanish). Thesis Ing. Agr. ISCA, Managua, Nicaragua.

Miranda, F. 1992. Assessment of the effect of biological and botanical insecticides against *Plutella xylostella* in cabbage (*Brassica oleracea*) in two regions of Nicaragua. VAXTSKYDDSRAPPORTER. JORDBRUK 59. Swedish University of Agricultural Sciences, Research Information Center, Uppsala, Sweden.

Rivas, I. and M. Méndez. 1991. Effectivity of water extract and preformulated neem products for the control of Lepidopteran pests in tomato, 1989–1990 (in Spanish). Annual Tech. Rep. Centro Nacional Proteccion Vegetal, Managua, Nicaragua.

Zeledon, B. 1988. Effect of neem oil on the oviposition of white-fly (*Bemisia tabaci*) in bean seedlings (in Spanish). Thesis Ing. ISCA, Managua, Nicaragua.

11

Potential for Neem (*Azadirachta indica* A. Juss.) in Armyworm Control in Africa

PAUL B. TANZUBIL

Nyankpala Agricultural Experiment Station, P.O. Box 483,
Tamale, Ghana

ABSTRACT

The African armyworm (*Spodoptera exempta*, Walker) is a devastating pest of cereal crops and grassland in Africa. Current control is achieved via spraying infested fields with synthetic insecticides that are not only costly but can also produce grave sociological as well as ecological consequences. Traditional pest control agents such as the neem tree (*Azadirachta indica* A. Juss.) are widely available in Africa, possess marked insect control potential, and could thus be used as substitutes or even complements to synthetic insecticides.

Azadirachtin, a tetranortriterpenoid obtained from neem seeds, reduces food intake and utilization efficiency in larvae of *S. exempta*. At 1 μg/larva, azadirachtin also curtailed development of penultimate instars into adults. Lower doses (0.1 μg/larva and below) produced larval-pupal mosaics or abnormal pupae which failed to develop further or gave rise to malformed adults. Such adults had impaired ovarian development and vitellogenesis, resulting in up to 43 per cent reduction in fecundity. Although azadirachtin did not decrease fertility, progeny from treated individuals or eggs were less viable, fewer than 40 per cent reaching the fourth instar. Prospects for using these results for practical control of this and other insects are discussed.

INTRODUCTION

The African armyworm (*Spodoptera exempta*, Walker) is one of the most devastating insect pest in Africa. It attacks and damages cereal crops and pasture in many parts of the continent annually. The destructive stage is the larva, which is usually referred to as 'armyworm'. This is probably due to the fact that during outbreaks, hordes of larvae can be seen marching like an army, devouring any foliage in their way. It is estimated that two larvae can destroy a 10-day-old maize plant, a single larva being capable of consuming up to 200 mg dry weight of maize foliage in the course of the sixth (final) instar alone (Brown and Odiyo, 1968). In outbreak years, therefore, armyworm attack can usually lead to total crop loss, requiring replanting. This is particularly the case when the invasions occur when plants are still very young (Brown, 1972).

Spodoptera exempta larvae usually show a density-dependent phase variation (Faure, 1943) in which the passive phase larvae, characteristic of low populations, go unnoticed while the active forms, characteristic of high population densities, are recognized as armyworms (Rose, 1975). These forms have been referred to as *solitaria* and *gregaria* respectively, the latter changing from green to black in colour as they develop while the former remain pale green up to pupation (Faure, 1943). The larval period in Africa lasts usually for two to three weeks, during which time substantial damage is inflicted on staple food crops such as maize, rice, sorghum, and millet. Females can lay up to 1000 eggs each in their lifetime (Gunn and Gatehouse, 1985) and this prolificity coupled with a relatively short generation period of about four weeks has served to increase the frequency and intensity of outbreaks over the last decade.

Chemical control remains the most effective measure to date and most African governments have contingency plans to deal with outbreaks. Logistic problems coupled with the suddenness of attacks have, however, often militated against effective execution of such plans in many parts of the continent. Help for the farmer often arrives too late from the urban centres to prevent crop damage. Any material that can protect the growing crop or reduce feeding damage at least until synthetic insecticides arrive would therefore be a big relief to the farmer. The environmental and sociological hazards associated with extensive reliance on synthetic insecticides needs also to be emphasized, especially in African countries where facilities for residue monitoring and effective health care are quite rudimentary.

Botanical pesticides have of late attracted worldwide attention as possible substitutes or even complements to synthetic insecticides for pest control. Of these, the neem tree (*Azadirachta indica* A. Juss.) probably has the greatest promise as extracts from its seeds and leaves have shown marked insect control potential (Schmutterer and Ascher, 1987). Studies on the effects of neem extracts and azadirachtin, the principal active ingredient in neem seeds, on the African armyworm have been conducted in laboratory and semi-field conditions in Ghana. This paper summarizes results from these studies and examines the potential role of neem extracts in armyworm control in Africa.

EFFECTS OF NEEM ON THE AFRICAN ARMYWORM

Antifeedant Action

Through a series of experiments conducted at both Reading (UK) and Nyankpala (Ghana), feeding inhibition has been established as a major biological effect of neem extracts on the African armyworm. A strong antifeedant action was observed when wheat or sorghum foliage treated with azadirachtin and aqueous neem seed extract (ANSE) were offered to penultimate (fifth) instar larvae. This was also the case when azadirachtin was applied to the larvae rather than to their food (Table 1). At higher doses, both routes of application produced about 95 per cent feeding inhibition and 100 per cent larval mortality within five days. Topical treatment of larvae with azadirachtin

Table 1. Effects of azadirachtin on food consumption and utilization in fifth-instar *S. exempta* larvae (topical application)

Treatment	FI/Larva	AD	ECI	ECD	RGR	HAT
Control	22.96	60.07	14.98	25.46	0.37	24
Acetone	22.72	60.22	11.97	16.86	0.33	24
0.1 μg/larva	22.34	60.05	12.00	20.44	0.29	24
1 μg/larva	22.32	63.23	7.53	13.20	0.26	24
10 μg/larva	19.71	59.17	8.77	16.80	0.27	24
Control	56.01a	79.73c	13.44a	16.86a	0.46a	48
Acetone	43.37a	82.25bc	14.14a	15.33a	0.43a	48
0.1 μg/larva	24.29a	87.21b	7.53b	8.62b	0.17b	48
1 μg/ha	27.07b	87.22b	6.69b	7.93b	0.14b	48
10 μg/larva	19.24b	96.25a	–6.88c	9.90c	0.16c	48
Control	67.79a	52.74c	16.03a	29.82a	0.38a	72
Acetone	65.04a	49.12c	15.63a	35.20a	0.33a	72
0.1 μg/larva	53.32a	59.85bc	15.83a	25.69a	0.35a	72
1 μg/larva	34.83b	73.66ab	12.10a	17.49a	0.21b	72
10 μg/larva	13.54c	80.83a	–0.70a	–1.91b	–.01c	72

In a column, means followed by the same letter are not significantly different for that time after treatment (HAT) at p = 0.05

FI/Larva = Food Ingested/Larva
AD = Approximate Digestibility
ECI = Efficiency of Conversion of Ingested food
ECD = Efficiency of Conversion of Digested food
RGR = Relative Growth Rate
HAT = Hours after treatment

(1 μg/larva) also inhibited feeding by up to 50 per cent. Treated insects also had reduced food conversion and utilization efficiencies resulting in decreased growth rates (Table 2). Two types of antifeedant action in the African armyworm can thus be induced by neem. The primary effect which is observed when neem is applied to the food of the insect presumably results from sensory detection and avoidance by the insect, as has been shown by Schoonhoven and Jermy (1977) and Simmonds and Blaney (1984). A secondary antifeedant effect results when neem extracts are applied directly to the insect rather than to its food. This action cannot be attributed to sensory detection as applied products presumably bypass the taste receptors responsible for this. It is reasonable to associate this effect with interferences or adverse effects of the neem extracts on centres in the insect that control and regulate feeding and metabolism (Tanzubil, 1989). The decreases in food conversion and utilization efficiency observed in neem-treated larvae may be interpreted as representing a diversion of metabolic energy away from biomass production into detoxifying or otherwise dealing with azadirachtin and other active ingredients in the extracts.

The above have important implications for field control of the insect. Applications of neem extracts in the field could protect the growing crop in two ways:

1. The crop would be instantly protected via the primary, gustatory repellent action of extracts on the surface of the foliage.

Table 2. Effects of ANSE nutrition on *S. exempta* larvae

Treatment F1:	wt. µg/larva	% feeding inhibition	days to mortality 50%	days to mortality 100%	max. larval (mg fresh)
Control	185.0a	0.00a			271.1a
1% dipped	107.1b	42.10b	6		219.9a
1% fed	25.7c	86.10a	3	9	96.7b
10% dipped	8.9d	95.20a	4	5	72.4b
10% fed	8.0d	95.70a	3	5	68.4b

In a column, means followed by the same letter are not significantly different at $p = 0.01$

2. Larvae picking up residues during spray application or through foraging can suffer feeding inhibition as a result of decreased activity and metabolic inhibitions (Tanzubil, 1989).

The net result would be high larval mortality and decreased crop damage. Thus, although neem application alone may not result in ultimate and comprehensive control of *S. exempta*, it could guarantee a reasonable level of protection to the growing crop at least until insecticides arrive.

Growth Regulation Effects

Both azadirachtin and ANSE induce growth-disruptive effects when applied to African armyworm larvae (Tables 3 and 4). High doses of azadirachtin of up to 10 µg/larva resulted in 100 per cent mortality, while lower doses inhibited and/or disrupted moulting, resulting in the formation of larval-pupal intermediates or abnormal pupae. The few pupae obtained from larvae treated with the lower doses (0.01 and 0.1 µg/larva) failed to develop further, produced adults that died in a pharate condition during eclosion, or had frizzled, curled wings, incapable of embarking on the mandatory migratory flights typical of this insect (Tanzubil and McCaffery, 1990a).

The inability of treated larvae to develop normally could be a consequence of interferences of neem with hormonal events in the insect. Formation of larvae-pupal intermediates can result from interference of neem with ecdysteriod titres and/or function (Seiber and Rembold, 1983; Garcia and Rembold, 1984), while an interference with juvenile hormone titre and activity could produce larval-pupal mosaics (Wigglesworth, 1972). Another hormonal effect observed during the studies was that on eclosion. Seiber and Rembold (1983) reported that azadirachtin directly interfered with the synthesis and release of this hormone in *Locusta migratoria*. The high mortality of African armyworm moths observed during eclosion in the current studies could thus be attributed to such an interference.

The overall effect of applying neem products for armyworm control could be a decrease in the sizes of subsequent generations of the insect and a significant reduction in the frequency and intensity of outbreaks.

Table 3. Effect of azadirachtin on growth and development of *S. exempta* (Tanzubil and McCaffery, 1990a)

Treatment	Pupal RGR	Max. larval wt (mg.)	% Surviving to			
			Pre-pupa	Pupa	Adult	Wt (mg)
Control	0.46[a]	236.1[a]	95.24[a]	90.48[a]	86.09[a]	115.8[a]
Acetone	0.48[a]	218.9[ab]	80.95[a]	80.95[a]	74.60[a]	103.7[a]
0.01 μg	0.42[a]	200.9[ab]	90.48[a]	76.19[a]	33.33[b]	100.3[a]
0.1 μg	0.33[b]	170.0[b]	80.95[a]	76.19[a]	9.53[c]	105.2[a]
1μg	0.31[b]	174.3[b]	61.91[b]	33.33[b]	0.00[c]	87.8
10 μg	0.16[c]	92.3[c]	0.00[c]	0.00[c]	0.00[c]	

Figures within any column followed by the same letter are not significantly different from each other at p=0.05.
RGR = Relative growth rate (Calculated after Waldbauer, 1968).

Table 4. Effect of ANSE on development of *S. exempta* (Tanzubil and Mccaffery, 1990a)

Treatment	Max. larval wt (mg)	% surviving to			Mortality %		
		Pre-pupa	Pupa	Adult	2 DAT	5DAT	8DAT
Control	269.9a	100a	100a	100a	0.0c	0.0c	0.0c
1% dip	245.0a	71.4a	42.9b	0.0b	0.0c	0.0c	14.3c
1% food	100.3b	0.0b	0.0b	9.0b	28.6a	57.1b	71.4a
10% dip	73.0c	0.0b	0.0b	0.0b	14.3b	100.0a	100.0a
10% food	69.2	0.0b	0.0b	0.0b	28.6a	100.0a	100.0a

Figures within any column followed by the same letter are not significantly different at p = 0.05.
DAT = days after treatment.

Anti-reproductive Effects

Treating penultimate or final-instar larvae African armyworm with azadirachtin or ANSE produces a range of anti-reproductive effects that could be of value in practical control of the insect. Azadirachtin treatment significantly reduced growth and development of oocytes in subsequent adults, thus delaying the attainment of reproductive maturity. Fecundity in moths was also decreased by up to 45 per cent. Azadirachtin treatment did not significantly affect the proportion of eggs that hatched or the rate at which they did so. Emerging larvae were, however, less viable with high mortality occurring during the first three instars. Topical treatment of eggs with azadirachtin produced similar results (Table 5).

Physiological studies revealed that azadirachtin adversely affected development and function of fat body as lower proteins levels were extracted from fat bodies of treated than of untreated insects. In the same studies, it was found that ovarian proteins were lower in treated insects, implying poor uptake of yolk proteins (vitellogenins) by developing oocytes in such insects.

It is evident, therefore, that azadirachtin decreases reproductive potential in *S. exempta* principally by interfering with oogenesis through its action on vitellogenesis by the fat body and its uptake by developing oocytes. Similar results have been reported in *Aedes aegypti* by Ludlum and Seiber (1988).

Table 5. Hatchability of *S. exempta* eggs topically treated with azadirachtin or ANSE (Tanzubil and McCaffery, 1990b)

Treatment	% Hatch	% Surviving to	
		3rd instar	4th instar
Control	79.88a	74.53a	72.96a
Acetone	76.65a	72.04a	64.61a
0.01 μ azadirachtin	69.97a	37.12b	30.65b
1% ANSE	71.84a	49.72b	43.52b

Figures within any column followed by the same letter are not significantly different at p = 0.01.

These effects could be relevant for practical field control. Delays in attaining reproductive maturation could influence the migratory behaviour of moths. Since moths terminate migratory flights only upon attainment of reproductive maturity, the current results imply that treated moths could spend a large proportion of their adult life and nutrient reserves in flight at the expense of reproduction. This together with the decreases in fecundity and the high incidence of mortality in progeny of contaminated moths and eggs could contribute directly to reducing population buildups and crop damage.

DESIRABLE ATTRIBUTES OF NEEM AS A SOURCE OF PEST CONTROL MATERIALS

Available evidence from the studies described above and from elsewhere point to a great potential for neem-based pesticides for armyworm control in Africa. Some of the desirable attributes are described below.

Safety and Selectivity

Neem products are reportedly safe to non-target organisms and exhibit a reasonable degree of selectivity. This means that they can be used with fewer ecological side effects than current insecticides. In fact, concentrations of up to 8500 mg/kg body weight of neem extract were not toxic to rats when applied orally (Schmutterer, 1984). Also, Sardeshpande (1976) noted no carcinogenic effects in rats fed with 25 per cent neem cake protein for nine months. Neem cake is also fed to animals with no side effects in some parts of the world. All these support the contention that neem pesticides would be far safer to animals than currently used synthetic insecticides.

Neem extracts are further reputed to be relatively safe to predators and parasites as well as pollinators (Srivastava and Parmar, 1985; Schmutterer and Holst, 1987). This enhances their potential as practical pest control agents since their use could help augment natural enemy populations and thus enhance natural control.

Availability and Economy

Neem trees grow widely in Africa. This means that development of pest management strategies based on neem extracts could be cheaper and more convenient, especially in the rural communities where farmers lack the necessary funds to purchase synthetic insecticides. Such farmers only need to gather neem seeds and employ relatively simple procedures to obtain home-made pesticides.

In West Africa (Nigeria), an average fruit yield of 20.5 kg/tree has been obtained (Schmutterer, 1990), but Saxena (1982) estimates that each tree can produce 30–50 kg seed annually, enough to protect one hectare of rice against insect damage. The several other uses of neem by-products make (Radwanski, 1977) it an economically important tree that deserves greater attention.

Ease of Cultivation and Use

The neem tree is hardy and easy to grow, requiring little attention from the farmer after establishment. It can withstand variations in climate and thrives well even in the most degraded soils. In fact it is widely used in agroforestry and reforestation programmes in many parts of the continent. The active principles are also obtainable via simple extraction procedures and applicable in different forms. For instance, in the current studies simple aqueous extracts were effective and these could be prepared by any farmer. Additionally, neem seed dust has proved effective against stored product pests (Ivbijaro, 1983; Tanzubil, 1991) and shoot pests of sorghum (Tanzubil and Dekuku, 1991). Neem extracts can also be applied using simple garden equipment such as the watering can (Dreyer, 1983). With some improvements, therefore, neem-based pesticides could be useful in insect control in general and control of African armyworm in particular. In fact, it should be possible for resource-poor farmers in Africa to grow their own neem trees, produce simple home-made insecticides, and apply these to protect their crops against insect infestation and damage.

Systemic Action

There have been some reports of systemic action of neem against insects by Gill (1972), Gill and Lewis (1977), and Saxena (1982), among others. In Ghana, neem seed powder applied to the soil protected young sorghum seedlings against shootfly and stemborer attack (Tanzubil and Dekuku, 1991). This implies systemic action of the powder as both insects reside in the stem upon entry. Systemic action, if proved and exploited, could raise new hopes for employing neem-based pesticides in pest management programmes. This would be a result of the added selectivity and improvement in residual activity of neem extracts associated usually with systemic action.

Compatibility with Other Control Options

Perhaps the most promising use of neem would be in integrated pest management. Its relative safety to predators and parasites makes it compatible with biocontrol. As with other antifeedants, neem extracts would tend to affect only insects which actually attempt to feed on the crop and not their natural enemies. Neem-treated *S. exempta* larvae slow down in activity and become weak. Such larvae can become more vulnerable to predation and natural enemy attack.

In fact, work at Nyankpala (Tanzubil, 1990) has shown that treating cowpeas with neem extracts resulted in increased exposure of *Maruca testulalis* larvae on the pods. Apart from enhancing biocontrol, this could increase the window of activity and hence efficacy of insecticides applied. This might mean lower doses being employed and hence reduced cost and ecological degradation.

Neem extracts also exhibit synergistic action with some insecticides such as pyrethrins (Ahmed et al., 1984), malathion, quinalphos, and monocrotophos (Parmar and Dutta, 1986). Synergistic action of neem extracts with *Bacillus thuringiensis* has also been reported by Hellpap (1984), while Joshi et al. (1984) successfully integrated neem seed kernel suspension with two natural enemies (*Telenomus remus* and *Chrysopa scelestes*) in the control of *Spodoptera litura* in tobacco.

All these findings point to the possible application of neem extracts with suboptimal doses of insecticides for effective pest control. The reductions in the quantity of insecticides used under these conditions would mean fewer socioecological hazards.

Diverse Modes of Action

As shown for this and other susceptible insects, neem extracts usually produce a range of adverse effects on larvae. In principle, it can be argued that it would be unlikely or at least more difficult for insects to develop resistance to neem than to many currently used synthetic insecticides. Saxena (1983) described the chances of insects developing resistance to neem seed extracts as remote. Working with *Plutella xylostella*, Vollinger (1987) in fact found no signs of resistance in moths treated with neem seed kernel extract after 42 generations, while deltamethrin treatment resulted in resistance levels 35 times that of controls. This aspect of resistance, however, still needs to be pursued in greater details and neem extracts must be used judiciously as resistance to other insect growth regulators has been documented (Cerf and Georghiou, 1972; Brown and Brown, 1974).

PRACTICAL PROBLEMS IN THE USE OF NEEM FOR FIELD CONTROL

Although the potential of neem extracts for insect control has been established over the last decade by many workers, their commercial use in the field is still very limited. This is due in part to some practical problems that need to be addressed to transform neem from a laboratory insecticide to one available to and widely used by farmers for field pest control. Apart from competitive inhibition by established chemical companies that produce synthetic insecticide, the following need emphasis.

Ecotypes

The concentration of active ingredients (e.g., azadirachtin) in neem seeds varies with location, temperature, relative humidity, and genetic makeup (Schmutterer, 1990). Ermel et al. (1984) extracted 4.8 per cent and 3.3 per cent azadirachtin from seed samples originating from Indonesia and Togo respectively, while those from Niger gave the lowest yield of 1.5 per cent. Similarly, Singh (1987) obtained higher extract yield and antifeedant potency from neem trees growing in arid areas near the desert than from those growing in the coastal areas of India and refers to these as being different 'neem ecotypes'. This variability in levels of active ingredients can pose problems when it comes to making recommendations for on-farm formulation of neem insecticides. One would be very uncertain about the quality of seed (in terms of percentage of active ingredient) being dealt with so that recommendations made may fail to produce the desired results and hence lead to a loss of confidence on the part of the farmer in neem extracts. There is a need for simpler methods of estimating active ingredient concentrations in neem seeds so as to overcome this problem.

Poor Residual Activity

Botanical pesticides generally have short residual life on treated foliage in the field. Temperature, ultraviolet light, pH on treated plant parts, rainfall, and other environmental factors may exert negative effects on the active principles (Schmutterer, 1990). The residual effect of neem products does not usually exceed five to seven days under tropical field conditions (Schmutterer, 1990; Tanzubil, 1991). This calls for repeated treatment of the growing crop at frequent intervals. This may be regarded as a disadvantage, but one should remember that most currently recommended insecticides, especially the pyrethroids, have about the same residual life and in fact high persistence in the environment is increasingly becoming objectionable as far as chemical control of pests is concerned. Under field conditions in Ghana, the residual activity of neem seed extract applied to control flower bud thrips (*Megalurothrips sjostedti*) in cowpea was as effective and protected the crop for the same length of time (seven days) as decis (deltamethrin) (Tanzubil, 1991). For armyworm control, a residual activity period of seven days would probably be adequate as this is a long enough time for insecticides to be mobilized and used as back-up treatments.

Delayed Action

Neem products do not exert acute toxic action against the target insect. Rather, they tend to incapacitate it through feeding inhibition and growth disruption. Thus, although neem products may curtail or limit feeding by the target insect, the farmer may still find live insects on his crop hours after treatment. This is likely to irritate and discourage the farmer who has been used to the rapid knockdown action characteristic of most synthetic insecticides.

It is essential to educate farmers on this special mode of action of neem products so as to reduce their expectations when using them. Farmers need to be adequately convinced that although insects may still abound in a neem-treated crop, their ability to cause damage is greatly curtailed and yield losses may not necessarily follow.

High Cost of Exploitation

Some have argued that developing pesticides from neem would entail a high capital outlay. This is probably true if the objective is to produce pesticides based on the purified active ingredients of neem if large-scale isolation and synthesis for commercial use is contemplated. When it comes to the resource-poor farmers, however, it could still be worth developing biologically active crudes such as alcohol or water extracts and improving their field activity and residual life via the use of additives.

This is the aspect with the greatest potential and therefore deserves greater emphasis in Africa.

REFERENCES

Ahmed, S., M. Grainage, J.W. Hylin, W.C. Micthel and J.A. Litsingir. 1984. Some promising plant species for use as pest control agents under traditional farming system. In: *Proc. 2nd Int. Neem Conf.* Eds. H. Schmutterer and K.R.S. Ascher. GTZ, Eschborn, Germany. pp. 565–587.

Brown, E.S. 1972. Armyworm control. *PANS*, 18(2): 197–204.

Brown, E.S. and P. Odiyo. 1968. The genus *Spodoptera* in Africa and the near east. *Bull. Entomol. Res.* 65: 221--262.

Brown, T.M. and A.W.A. Brown. 1974. Experimental induction of resistance to a juvenile hormone mimic. *J. Econ. Entomol.* 67: 799–801.

Cerf, D.C. and G.P. Georghiou. 1972. Evidence of cross-resistance to a juvenile hormone analogue in some insecticide-resistant houseflies. *Nature* 239: 401–402.

Dreyer, M. 1984. Effects of aqueous neem seed extract and neem oil on the main pests of *Curcurbita pepo in Togo*. In: *Natural Pesticides from Neem Tree and Other Tropical Plants*. Eds. H. Schmutterer and K.R.S. Ascher, GTZ, Eschborn, Germany, pp. 435–444.

Ermel, K., E. Pahlich and H. Schmutterer. 1984. Comparison of azadirachtin content of neem seeds from ecotypes of Asian and African origin. In: *Proc. 2nd Int. Neem Conf.* Eds. H. Schmutterer and K.R.S. Ascher, GTZ, Eschborn, Germany, pp. 91–94.

Faure, J.C. 1943. Phase variation in the African armyworm (cited by Brown, 1972).

Garcia, E. and H. Rembold. 1984. Effect of azadirachtin on ecdysis of *Rhodnius prolixus*. *J. Insect Physiol.* 30: 939–941.

Gill, J.S. 1972. Studies on insect feeding deterrents with special reference to fruit extracts of the neem tree. Ph.D. thesis, University of London.

Gill, J.S. and C.T. Lewis. 1977. Systemic action of an insect feeding deterrent. *Nature* 232: 402–403.

Gunn, A. and A.G. Gatehouse. (1985). Effects of availability of food and water on reproduction in the African armyworm. *Physiol. Entomol.* 10: 53–63.

Haggis, M.J. 1979. African armyworm (*Spodoptera exempta*) and wind convergence in the Kenyan Rift valley, May 1970. *East African Agric .& Forestry Journal* 44(4): 332–346.

Hellpap, C. 1984. Effects of neem kernel extract on the fall armyworm, *Spodoptera frugiperda*. In: *Proc. 2nd Int. Neem Conf.* Eds. H. Schmutterer and K.R.S. Ascher, GTZ, Eschborn, Germany, pp. 353–363.

Ivbijaro, M.F. 1983. Toxicity of neem seeds to *Sitophilus oryzae* in stored maize. *Prot. Ecol.* 5(4): 353–357.

Joshi, B.G., S. Sitaramaiah, G. Ramaprasad and R.S.N. Rao. 1984. Efficacy of integrated use of two natural enemies, *Telenomus remus* and *Chrysopa scelestes* and neem seed kernel suspension for management of *Spodoptera litura* in tobacco. *5th Nat. Symp. on Tobacco*, Rajahmundhry, Feb. 15–17.

Ludlum, C.T. and K.P. Seiber. 1988. Effect of azadirachtin on oogenesis in *Aedes aegypti*. *Physiol. Entomol.* 13: 177–184.

Parmar, B.S. and S. Dutta. 1986. Neem oil as a synergist for insecticides. *Neem Newsletter* 3: 3–5.

Randwanski, S.A. 1977. Neem Tree. *World Crops & Livestock* 29: 61–66 & 111–113.

Rose, D.J.W. 1975. Field development and quality changes in successive generation of *Spodoptera exempta*, the African armyworm. *J. Appl. Ecol.* 12: 727–739.

Sadre, N.L., V.Y. Deshpande, K.N. Mendulkar and D.H. Nandal. 1984. Male antifertility activity of *Azadirachta indica* in different species. *Proc. 2nd Int. Neem Conf.* 473–482.

Saxena, R.C. 1982. Note on the use of neem kernel for protection of dewgram against *Amsacta morei*. *Indian J. Agric. Sci.* 1: 51–52.

Saxena, R.C. 1983. Naturally occurring pesticides and their potential. In:*Chemistry and World Food Supplies: the frontiers*, Ed. E. Shelmit Cherurawn II, Pergamon, New york. pp. 143–161.

Schmutterer, H. 1984. Neem research in the FRG since the 1st neem conference. In: *Proc. 2nd Int. Neem Conf.* pp. 21–30.

Schmutterer, H. 1990. Properties and potential of natural pesticides from the neem tree. *Annu. Reps. Entomol.* 35: 77–297.

Schmutterer, H. and K.R.S. Ascher, eds. 1987. *Natural Pesticides from the Neem Tree and Other Tropical Plants*. GTZ, Eschborn, Germany pp. 703.

Schmutterer, H. and H. Holst. 1987. Cited by Schmutterer, 1990.

Schoonhoven, L.M. and T. Jermy. 1977. A behavioural and electrophysiological analysis of insect feeding deterrents. In: *Crop Protection Agents: Their Biological Evaluation* Ed. McFarlane. pp. 133–145.

Seiber, K.P. and H. Rembold. 1983. The effects of azadirachtin on endocrine control of moulting in *Locusta migratoria, J. Insect Physiol*. 29: 523–527.

Singh, R.P. 1987. Comparison of antifeedant efficacy and extract yields from different parts and ecotypes of neem trees. In: *Natural Pesticides from Neem Tree and Other Tropical Plants*. Eds. H. Schmutterer and K.R.S. Ascher, GTZ, Eschborn, FRG. pp. 185–194.

Srivastava, K.P. and B.S. Parmar. 1985. Evaluation of neem oil emulsifiable concentrate against sorghum aphids. *Neem Newsletter* 2(1): 7.

Tanzubil, P.B. 1989. Effects of neem on the Africa armyworm and their implications for field control. M.Phil. thesis, Univ. of Reading, UK.

Tanzubil, P.B. 1991. Control of some insect pest of cowpeas with neem in Northern Ghana. *Trop. Pest Mgmt*. 37(3): 216–217.

Tanzubil, P.B. and R.C. Dekuku. 1991. Insect pests of cereal crops in the northern savanna of Ghana. Paper presented at the 9th Symp. of AAIS, Sept. 1991, Legon, Accra, Ghana.

Tanzubil, P.B. and A.M. McCaffery. 1990a. Effects of azadirachtin and aqueous neem seed extract on survival, growth and development of the African armyworm *Spodoptera exempta*. *Crop Prot*. 9: 383–386.

Tanzubil, P.B. and A. M. McCaffery. 1990. Effect of neem in reproduction in the African armyworm. *Entomol. Exp. Appl*. 57: 115–121.

Vollinger, M. 1987. The possible development of resistance against neem seed kernel extract and deltamethrin in *Plutella xylostella*. In: *Proc. 3rd Int. Neem Conf.* Eds. H. Schmutterer and K.R.S. Ascher, GTZ, Eschborn, Germany, pp. 543–554.

Waldbauer, G.P. 1968. The consumption and utilization of food by insects. *Adv. in Insect Physiol* 5: 229–288.

Wigglesworth, V.B. 1972. *The Principles of Insect Physiology* 7th ed. Chapman and Hall, London pp. 827.

12

Strategies in Management of Insect Pests with Neem (*Azadirachta indica* A. Juss.)

R.P. SINGH[1] AND A.K. RAHEJA[2]

[1]Division of Entomology, Indian Agricultural Research Institute
New Delhi 110012, India
[2]Indian Council of Agricultural Research, Krishi Bhavan
New Delhi 110 001, India

ABSTRACT

Despite demonstrated insecticidal efficacy of neem against about 350 species of Arthropods, it is still not used as substitute to synthetic insecticides. One of the reasons is the inconsistancy in the results of field evaluation even with the standard neem products. This is mainly because unlike synthetic insecticides, neem has diverse biological effects on insects and the insects sensitivity to different biological effects vary greatly irrespective of genera or species of insects. It thus, become imperative to know which of these effect(s) predominate and adversely affect the insects and consequently provide protection to the crop. Hence a laboratory screening to know the sensitivity of insects prior to field evaluation is suggested for effective control.

Insect-growth-regulatory activity of neem is the most important biological effect. Its effect is, however, limited to the developmental stages, the adults generally remain unaffected (except in cases where they get sterilized) and continue to be a potential threat. Strategies for the management of development stages as well as adults of five economically important orders, Orthoptera, Hemiptera, Lepidoptera, Diptera and Coleoptera are given as follows:

Simple aqueous extract of seed kernel may be as effective or even better than synthetic insecticides against freely feeding larvae of Lepidoptera, Coleoptera and nymphs of Orthoptera because they are subjected to both contact and ingestion effect of neem. Nymphs of Hemiptera are exposed only to contact effect and hence difficult to manage them with aqueous or even ethanolic extracts unless high concentration is used. Azadirachtin-rich oil-based formulation has been suggested for the management of borers belonging to Diptera and Lepidoptera and nymphs and adults of Hemiptera.

While it may be easy to manage freely feeding larvae and nymphs, it is quite difficult to manage adults as they quickly leave the plant after finding the food unattractive. Need for more studies on the effect of neem on oviposition, sterilization and mating behaviour is stressed.

INTRODUCTION

Neem tree has unequivocally emerged as the single most important source of insecticide. There is also now no place of scepticism regarding its potential as an effective substitute to synthetic insecticides. A recent compilation by Prof. H. Schmutterer, Institute of Phytopathology and Applied Zoology, University of Giessen, FRG (Personal communication) revealed that neem has been evaluated against about 350 arthropods. In India alone, 105 insect pests have been evaluated with neem (Singh and Kataria, 1991) and almost all insects have been found susceptible. The lists contain all the key pests including those which have developed resistance to practically all insecticides. In spite of these established facts, neem is still not being used as a substitute to synthetic insecticides. One of the important reasons is the inconsistency in the results of field trials not only with crude extracts but also with the standardized commercial products. In the present paper attempts have been made to analyse the reasons for inconsistency/ ineffectiveness of neem preparations as also to suggest ways and means of achieving greater success with neem products.

MECHANISM OF ACTION

There is no plant or synthetic chemical known which has such a diverse biological effects on insects as neem. Also, there is no plant which is endowed with such an array of unique chemicals whose major role is defensive, especially against insects. The tree's ability to grow in diverse climatic conditions is indicative of the fact that it must have been subjected to diverse biotic factors of environment in the past which enabled it to develop large defensive chemicals, especially against insects, because the overall pressure by insects on the plant is more than any other organism. It seems that the tree has elaborated these chemicals in a very systematic manner to combat insect pests. It is seen in Fig. 1 that all steps involved in the host selection by insects are affected by neem. Insects escaping one effect (repellent) have to face other behavioural effects (antifeedant, oviposition deterent). If however, they cross these barriers and feed they succumb to physiological effects.

NEED FOR LABORATORY SCREENING

The sensitivity of insects to different behavioural and physiological effects varies greatly irrespective of genera or species of insects. For example, the desert locust, *Schistocerca gregaria* is about 50 times more susceptible than the migratory locust, *Locusta migratoria* to feeding deterent effect of aqueous suspension of seed kernel (Pradhan and Jotwani, 1968). Ethanolic extract of petroleum ether extract of seed kernel at 2.5 per cent deterred the oviposition of *Dacus cucurbitae*. However, a concentration of 20 per cent was necessary to obtain the same effect with *D. dorsalis* (Singh and Srivastava, 1983). Insect growth-regulatory (IGR) activity of neem is the most important biological effect and rarely have insects not succumbed to its effect. Significant to highly significant differences in sensitivity of developing stages are reported to occur when insects are fed or subjected to topical application. ED_{50} of azadirachtin when injected into hemocoel of final instar larvae of *Achoea janata* and *Spodoptera litura* for adult emergence was 4.1

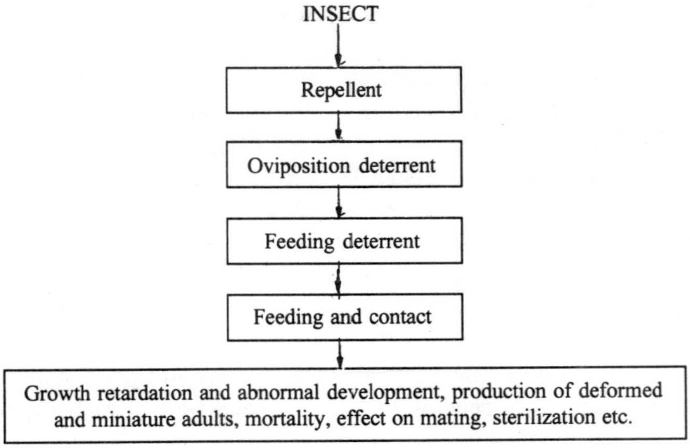

Fig. 1. Biological effects of neem on different steps of host plant selection by insects

and 1.1 mg/g body weight respectively (Rao and Subramaniam, 1987). Under actual field or storage conditions contact and feeding of neem material by developing stages determines the IGR efficacy of neem. The sensitivity of insects to both the effects too vary greatly. Insect-growth-regulatory effect due to feeding is more pronounced and quicker than by contact. Larvae which are less sensitive to feeding are likely to succumb faster due to the ingestion of larger quantity of the extract than those which are more sensitive. Sensitivity to feeding deterrent effect, thus, plays important role in determining the efficacy of the neem products. The sensitivity of insects to feeding on neem treated plants, may be more than a 1000 times. Only 0.003 per cent ethanolic extract of seed kernel was required to inhibit completely the feeding by desert locust, *Schistocerca gregaria* (Singh, 1987) whereas the same extract at 4 per cent failed to prevent feeding by *Heliothis armigera* (Jhansi and Singh, 1993).

In view of diverse biological effects and large variations in the sensitivity to these effects, it becomes necessary to know which of these effects will predominate and adversely affect insects. To know this, it is important that the target pests should first be screened in the laboratory to know its sensitivity to various effects before contemplating field trials. Though it is a known fact that even excellent results in the laboratory do not automatically mean that the same has to be expected under actual field conditions yet in view of above detailed laboratory study seems to be prerequisite for large scale field trials.

Orthoptera

Orthopterans are external feeders so not much problem in their management with neem. Desert locust, *Schistocerca gregaria* and migratory locust, *Locusta migratoria* were the first to be evaluated with neem and fortunately both responded favourably to the feeding deterrent effect of neem as only 0.001 and 0.05 per cent aqueous suspension respectively was needed for absolute antifeedant effect (Pradhan et al., 1962; Pradhan and Jotwani, 1968). Ever since this finding neem has been tested against numerous species of locusts and grasshoppers. The results show that there exists large variations in their sensitivity.

Butterworth and Morgan (1971) reported that azadirachtin at incredible amount of 1 ng/cm^2 filter paper inhibited the feeding of *S. gregaria*. Desert locust thus seems to be the most sensitive insect to feeding deterrent effect of neem. On the other hand Mulkern and Mongolkiti (1975) screened azadirachtin against several North American grasshoppers, *Encoptolophus sordidus, Melanoplus bivittatus, M. femurrubrum, M. foedus, M. keeleri, M. sanguinipes, Phoetaliotes nebrascensis, Schistocerca lineata, M. differentialis* and *S. vaga* and reported that azadirachtin at 1 ′ 10^{-5} g/g of carrier had no effect on feeding. Surprisingly there were also no observed effects on the survival and development of nymphs following the ingestion of azadirachtin. Barring the above mentioned North American grasshoppers all other locusts and grasshoppers tested so far have shown varying degree of sensitivity and are managable even with simple aqueous extracts.

Neem oil and aqueous extract of seed kernel was reported to be superior than lindane, cypermethrin and monocrotophos against *Zonocerus variegatus* (Olaifa and Adenuga, 1989). Rice grasshopper, *Hieroglyphus banian* could attack only a few plants in rice fields treated with 5 per cent aqueous neem seed kernel extract. The treatment was at par with 0.1 per cent monocrotophos (Mohan et al., 1991). Aqueous extract of kernel was as effective as ethanolic extract against Senegalese grasshopper, *Kraussaria angulifera* on caged sorghum seedlings. Nymphs were more susceptible than adults (Radcliffe et al., 1991). Egg plants sprayed with aqueous extract at 25–50 g pulverized kernel/1 or 5–10 1 neem oil (ULV)/h protected the damage by millet grasshopper *Acanthacris ruficornis* in Togo (Dreyer, 1987). Methanolic extract of seed kernel at 10^{-4} per cent applied on bean leaves caused 100 per cent protection against 4th instar nymphs of *Eyprepocnemis plorans*. The extract was more effective than commercial preparation Neemark' (Ascher et al., 1989). Corn leaf (2 × 2 cm) treated with 20 litres of 10 to 50 per cent of a commercial product Margosan-O caused grasshopper, *Dissosteira carolina* (Adler and Uebel, 1984). Absolute feeding deterrency at 2.5 per cent of kernel suspension against *Acrida exaltata* was reported by Mane (1968). Aqueous extract of neem cake, a bye-product of neem oil industry, at 0.006 per cent was found to deter completely the feeding by *S. gregaria* adults on cabbage leaves (Jhansi and Singh, unpublished). They also reported that only 0.003 per cent of ethanolic extract of cake was needed to get the same effect. Singh (1987) tested different parts of neem tree and reported that even seed coat and fallen leaves at 0.012 and 1.6 per cent respectively can give complete protection against *S. gregaria*. Passerini and Hill (1993) evaluated aqueous extract of neem seed kernel under laboratory and field conditions against 1st and 3rd instar grasshopper, *Kraussaria angulifera*. The treatment caused growth inhibition. Nasseh et al. (1992) found neem seed kernel extract as effective as synthetic growth regulator. Alsysten (triflumuron) and triflubenzuron against immature stages of *S. gregaria*. The treatment caused 100 per cent mortality. Thus it is evident from above that simple aqueous extract of seed kernel can be used very effectively for the management of locusts and grasshopper.

Neem oil obtained by solvent extraction has poor feeding deterrent and growth-regulatory activity. An experiment conducted by Attri (1975) revealed that antifeeding activity of oil was 40 times less active than aqueous extract of seed kernel against

S. gregaria. The best fraction of oil at 0.25 could afford only 92.9 per cent protection of cabbage leaves against the same insect (Narayanan et al., 1980). On the other hand, oil obtained by cold pressing of seed kernel in several developing countries showed marked feeding deterrent and growth-regulatory effect due to the enrichment of oil with azadirachtin (AZA) during pressing

Schmutterer (1993) in a field trial compared the efficacy of following four neem products against red locust, *Nomadaris septumfaciata*:

1. Neem oil ULV formulation (25 g AZA/1, 2 1/ha applied by ULV A8 sprayer.
2. Neem seed kernel powder (Ca.5g AZA/kg, 5 kg/ha)
3. Neem seed kernel water extract (50 g kernel/1, Ca 2.2 mg AZA/g kernel, applied with a broom consisting small branches 600 1/ha.
4. Neem seed kernel water extract (50g kernel/1 ca. 2.2 mg AZA/g kernel applied with ULA A8 sprayer, 200 L/ha.

All neem products exhibited almost the same results. This shows that neem oil with 50 g AZA has only as much activity as does the aqueous extract with only half (22 g) AZA.

Neem oil containing 0.09 per cent AZA when sprayed at the rate of 2.5–10 1/ha, nymphs of *S. gregaria* not only suffered from morphogenetic defects but also showed tendency of solitarization (Nicol and Schmutterer, 1991). Oil enriched with 0.05 per cent AZA applied in green house and 0.2–0.4 per cent in field test against 1st and 2nd instar nymphs of *S. gregaria* produced adults that failed to aggregate (Langewald and Schmutterer, 1992).

While it is easy to manage nymphs with aqueous extracts or AZA rich oil, it may be quite difficult to control adult locusts and grasshoppers as they are unaffected by growth regulatory activity and quickly leave the field after finding the food unattractive. Very little work has been done on the effect of neem on adult locust. Wilps et al. (1993) found neem products ineffective on sitting *S. gregaria* but when sprayed on flying locust, these products caused 70–90 per cent mortality. Wilps et al. (1992) earlier reported that neem oil reduced flight performance up to 70 per cent. High metabolic rate during flight must have absorbed a higher quantity of the neem products in a shorter time, but how the oil affects flight needs further probing.

Drenching of the breeding ground with aqueous extracts may provide an effective control of locusts and grasshoppers by deterring oviposition/affecting emergence of nymphs. Bhanotar and Srivastava (1984) reported that sand treated with 0.125 per cent aqueous extract had only 22.6 per cent emergence of nymphs and none in 0.25, 0.5 and 1.0 per cent whereas 60 nymphs emerged in control. Five egg pods were laid in 0.125 per cent concentration and one each in other treatments. This shows that neem extract has not only deterred oviposition of *S. gregaria* but has also affected the emergence of nymphs. Singh and Singh (1987a) found reduced egg hatching in sand treated with 0.1, 0.5 or 0.1 per cent aqueous extract. They also reported (1987b) that at the above concentration significantly fewer eggs were laid by *S. gregaria*.

The following conclusions may be drawn from foregoing results:

1. Barring a few North American grasshoppers which are unaffected by azadirachtin, all other locust and grasshoppers evaluated so far could be managed with simple aqueous extract of seed kernel.

2. While it is easy to manage nymphs with aqueous extract, it may be difficult to control adults as they fly away on finding the food unattractive. To prevent migration of both nymphs and adults a medium doses may be more effective than using high concentration.
3. Drenching of breeding grounds with aqueous extract may nip in the bud the building up of population. This may take care of other soil-borne pests and even may be translocated and protect the crop against other pests.
4. More studies should be conducted on feeding and contact effect of neem on adult Orthopterans.

Lepidoptera

Lepidopterans are economically the most important order of class Insecta as they cause severe damage to almost all important crops. Depending on the nature of damage, they can be broadly divided as external and internal feeders.

There are several examples which clearly suggest that there is not much problem in managing external feeders. Lepidopterans are generally less sensitive to feeding deterrent effect of neem. Initially they show reluctancy but start feeding if no choice is given. This is more so against lethargic larvae which cannot quickly search alternative food. Two per cent aqueous extract is currently being used as an effective substitute to synthetic insecticide for the management of tobacco caterpillar. *Spodoptera litura* in South India (Joshi et al., 1978). Aqueous extract of kernel at 4–5 per cent significantly reduced the damage to cabbage by diamond back moth. *Plutella xylostella* (Srinivasan and Krishana Moorthy, 1993). Aqueous extract of unshelled neem seeds at the rate 25 and 50 g powder per litre of water prevented feeding damage by *P. xylostella* and oriental cabbage worm, *Hellula undales* very well. Dipel, a microbial biological product based on *Bacillus thuringiensis* proved to be considerably less active than the neem extracts (Dreyer, 1987). This pest has developed resistance to practically all insecticides Gypsymoth, *Lymantria dispar* was found highly susceptible to even low concentration of 0.5 per cent aqueous seed kernel extract. The larvae of this polyphagous pest were completely eliminated within 15 days (Skatulla and Meisner, 1975). Ethanolic extract of seed kernel at 2 per cent sprayed weekly on cabbage reduced 80 per cent larvae (Reed and Reed, 1986). Under laboratory condition more than 2 per cent concentration was needed to inhibit the feeding of *Euproctis lunata*, *Utethesia pulchella* and *Laphygma exigua* in Laboratory (Anonymous, 1983).

.Polyphagous fast-moving larvae with strong locomotary orgnas may desert the sprayed field and cause damage to adjacent crops. Patel et al., (1968) reported that all the hairy caterpillars (*Amsacta moorei*) infesting sun hemp (*Crotalaria juncea*) migrated to the unsprayed field following application of neem seed kernel suspension. Further, the presence of hairs considerably reduces the contact of spray with the body. Unless thoroughly sprayed, such larvae will be affected only partly by neem products. Margosan-O failed to control beet armyworm, *Spodoptera exigua* on chrysanthemum (*Dendranthema grandiflora*) (Harris et al., 1987). A medium dose may be more effective as larvae will then remain in the field and consume treated plants.

Lepidopterous leafminers are easy to manage. Simple aqueous extract does the job very well as it can penetrate thin leaf epidermis under which the larvae live. Aqueous extract of kernel prepared with 12.5, 25 and 50 g/1 of water reduced the number of leafminer, *Scrobipalpa ergasima* by more than 80 per cent on gboma (*Solanum aethiopicum*) an important vegetable in western Africa (Dreyer, 1986).

Internal feeders (borers) have always been difficult to manage even with powerful synthetic insecticides. This is due to the fact that larvae remain outside for only a brief period and if the spray does not coincide with hatching they quickly bore inside and go out of control. Thus timing of the application as well as strength of the neem product are very important. Aqueous extract has low contact toxicity and hence not suitable for use. Since little feeding is involved or larvae even may avoid feeding, it would be better to use organic solvent-based azadirachtin-rich formulation. Oil enriched with azadirachtin would perhaps be the best option. A slow release formulation may be ideal wherever the possibility of their placement exists. *Chilo partellus* a serious pest of maize and sorghum in India was effectively controlled by neem granules prepared with neem oil and seed kernel extract. The granules were placed in whorls. The efficacy was at par with quinalphos (Anonymous, 1983).

Limited screening done in the past show that lepidopterans are susceptible to oviposition deterrent effect of neem. Jhansi and Singh (1993) tested several extracts and reported 60–90 per cent reduction in egg laying by *Heliothis armigera* under choice condition. Azadirachtin failed to deter oviposition of *H. armigera*. Neem oil at 0.05 per cent almost completely inhibited oviposition of potato tuber moth, *Phthorimaea operculella* (Shelke, et al., 1985). In a choice test egg laying by *Crocidolomia binotalis* was considerably reduced when cabbage leaves treated with methanolic extract of neem leaves were offered as oviposition sites (Fagoonee, 1980).

Other moths which have been found sensitive to the oviposition deterrent effect of neem extracts are: *Spodoptera littoralis* to 2 per cent aqueous suspension (El-Sayed, 1985), *S. furgiperda* to aqueous kernel extract (Hellpap and Mercado, 1986). *S. litura* to 2 per cent aqueous extract (Joshi and Sitaramaiah, 1979), *Cnaphalocrocis medinalis* to neem oil (Saxena et al., 1981), and *Sitotroga cerealilla* to neem oil (1 mg/kg) (Verma et al., 1985).

There are also reports of ovicidal activity of neem products. Neem oil at 12 per cent reduced egg hatching of *Mythimina separata* and *Spodoptera mauritia* (Saxena and Barrion, 1987). Shelke et al. (1987) reported that neem oil at 5 and 10 per cent was an effective ovicide when sprayed on stored potato. Jhansi and Singh (1993) tested several extract for ovicidal effect on eggs of *H. armigera*. All extracts (aqueous, ethanolic, hexane, chloroform extract of deoiled kernel and ethanolic extract of hexane extract) effected egg hatching. Chloroform extract caused maximum effect as only 24 per cent eggs hatched in treatment as against 60 per cent in control.

Following strategies may be helpful in management lepidopterous pests:

1. External feeders and leaf miners succumb easily to feeding and contact effects of aqueous extract of seed kernel. For fast moving larvae medium doses to retain them in the treated field for slow poisoning may be more effective than high concentration.
2. For the management of borers, the timing of application as well as strength of the neem products are important. Aqueous extract may not be suitable for use against borers. Oil enriched with azadirachtin is the best option.

3. Since oviposition deterrent effect is an important biological effect, the spraying of crops should be taken up immediately after spotting moths in the field. A few applications at quick intervals may provide effective control. A single spray of ethanolic extract at 50 per cent flowering of pigeon pea (*Cajanus cajan*) gives substantial control of *H. armigera* (Singh et al., 1985).

Diptera

Maggots cause damage either as leaf miner, pod or stem borer. As the aqueous extract can move from leaf surface to inside the leaf, there is no problem in managing leaf miners. Webb et al. (1983) found that aqueous extract of neem seed was effective against Agromyzid, *Liriomyza trifolii* and *L. sativae*. The latter experienced 95 per cent mortality if leaves were treated with 1 per cent neem seed extract when leaf miners were egg, or at first, second or third instar. The results indicated that the extract disrupted development of internally feeding larvae. Soil drenching of chrysanthemum with 0.4 per cent crude extract (9.2 ppm AZA) seed killed older larvae and pupae in greenhouse test but did not inhibit female feeding and oviposition on treated plants. It was as effective as cyromazine, an efficacious insecticide (Larew et al., 1985). Foliage sprays (0.4 to 0.5 per cent) neem seed extract were also effective and caused quicker mortality of the larvae than did drenches. The speed of kill of *L. trifolii* was concentration dependent (Knodel-Montz and Larew, 1985).

Agromyzid, *Melanagromyza obtusa* which lays egg inside the pod was managed effectively with 5–8 per cent neem kernel extract on pigeonpea (*Cajanus cajan*). The results were comparable with synthetic insecticide fenvalerate and quinalphos (Srivastava et al., 1984). Since the eggs are laid inside the pod and larvae are out of reach and aqueous extract cannot kill the adult, the extract must have repelled or deterred the oviposition especially when it used at 5–8 per cent concentration.

Gall midge, *Orseolia oryzae* (Cecidomyiidae) a serious pest of rice next only to stem borer, has been reported to be sensitive to oviposition deterrent when released on 0.5 per cent emulsified oil (Chiu, 1985; Chiu et al., 1985). Midge lays eggs on ligule hair or basal part of leaves and the larvae creep down between leaf sheaths. It is during this period, that it should be managed. Neem cake or seed kernel extract are less likely to give significant control of the midge. The pest infestation is more in alkaline soil. Cake as basal application may be of little value because azadirachtin is highly sensitive to alkalinity. Four sprays of 3 per cent neem oil gave effective control of the midge and was at par with monocrotophos (Samalo et al., 1993). Application of urea and crude neem oil to root zone of rice seedlings was found as effective as three broadcast application of carbofuran against this pest and also whorl maggot, *Hydrellia*. Neem oil at 1 per cent was effective against chilli gall midge. The damage reduction was 56.9 per cent (David et al., 1990a).

Neem oil at 2 per cent gave effective control of *Contarinia* sp. on jasmine and was better than endosulfan and chlorpyrifos (David et al., 1990). Sorghum midge, *Contarinia sorghicola* was, however not affected by neem fraction (Sharma et al., 1984). It seems that spray could not reach larvae as they are well protected. It seems neem oil had a two pronged effect toxic and oviposition deterrent on tender Dipterans.

Tephritids also seems to be sensitive to oviposition deterrent effect. Methanolic extract of neem seed kernel at 1 per cent when applied to banana fruits in the laboratory, completely deterred the oviposition of *Dacus dorsalis* for two days. The number of eggs thereafter increased rapidly (Sombatsiri and Tigvattanont, 1984). Mediterranean fruitfly, *Ceratitis capitata* under field condition on citrus fruits was not sufficiently deterred (Steffens and Schmutterer, 1982). Singh and Srivastava (1983) reported that ethanolic extract of petroleum ether extract of seed kernel was an effective deterrent at 2.5 per cent against melonfly *Dacus cucurbitae* on bittergourd. However, a much higher concentration of 20 per cent was required to get the same effect against fruit fly *D. dorsalis* on guava.

Repelin, a commercial neem-based product at 2 per cent, gave very effective control of *Carpomyia vesuviana* on ber (*Zizyphus* sp.). The control was better than dichlorvos, carbaryl, phosalone and monocrotophos under actual field conditions (Reddy et al., 1993). Aqueous extract of unshelled seeds (50 and 100 g/1) and neem oil (10 and 20 1/ha, ULV) were sprayed in Togo against fruit fly *D. ciliatus* on zucchini (*Cururbita pepo*) in the field. The neem treatment could not reduce the attack (Dreyer, 1984).

In a detailed laboratory evaluation of aqueous extract at 2 per cent exhibited strong feeding deterrent effect on *D. cucurbitae*. Flies preferred death rather than drinking treated water. At lower concentration (0.25 per cent) the extract caused adverse physiological effect leading to death. Larvae when allowed to pupate in sand treated with 0.5 and 0.25 per cent aqueous extract failed to reach adult stage. Low concentration of aqueous extract (0.5, 0.2 and 0.1) when offered as water source to female adults inhibited ovarian development (Shivendra Singh, Sr Research Fellow, Division of Entomology, IARI, Personal communication).

Sorghum shoot fly *Atherigona soccata* was only partially controlled with aqueous extract (0.5 and 1.0 per cent) in India (Kareem, et al., 1974) and methanolic and leaf extract in Togo (Adhikary, 1981). Azadirachtin-rich oil may provide effective control. From the above and some other reports it is evident that:

1) There is not much problem in the management of Dipterous leaf miners with simple aqueous extract.
2) Gall midge seems to be sensitive to oviposition deterrent as well as toxic effect of oil. Thus timing of application as well as placement of spray are important.
3) Irrigation of field with low concentration of aqueous extract may be beneficial against those Depterans which pupate in soil. Water and food source are very important for the survival and egg laying of Tephritids. Contamination of the crop with aqueous extract thus, may provide effective control.

Hemiptera

The vast majority of the members though small (eg., aphids jassids, white flies, etc.) cause enormous damage to the crops. The most characteristic feature of the order is the possession of piercing and sucking mouth part. Many have extraordinary rapid rate of reproduction. In aphids reproduction can be parthenogenetic and birth is given to young ones (viviparity). In scale insect all stages from oviparity through oviparity to viviparity are found. White flies can often breed parthenogenetically.

Due to a piercing mouth part there is no effect of neem on feeding until it is absorbed and translocated in the plant system. Aqueous extract, due to contact effect alone can affect only nymphal development, the adults remain mostly unaffected and quickly produce the next generation. Thus frequent application of aqueous extract would be necessary to contain proliferation. Neem oil on the other hand has mild toxicity (Singh et al., 1988) and kill both nymphs and adults. Thus oil containing little quantity of azadirachtin may offer effective control of both nymphs and adults of not only aphids but also other small Hemipterans. Such formulation when prepared and sprayed against mango hoppers *Amritodus atkinsoni* gave excellent results in India (Srivastava et al., 1993). However, acqueous and ethanolic extracts may be more useful for those Hemipetrans which have long life cycle with fewer generation in a season rather than small Hemipetrans with rapid rate of reproduction.

All leaf hoppers (*Nephotettix virescens*) died four days after spraying rice plant with 5per cent neem oil (Mariappan et al., 1988) but extracts of neem seed cake or leaves were less effective (Narasimhan and Mariappan, 1988). Neem oil proved significantly better than monocrotophos in controlling *Bemisia tabaci* on cotton as only 14.3 and 13per cent nymphs could reach the adult stage following treatment with 0.5 and 1per cent oil respectively as against 84.3per cent in plot sprayed with monocrotophos (Natarajan and Sundaramurthy, 1990). Patel et al. (1993) reported that aqueous extract of neem seed kernel at 5 per cent was less effective than 0.5 per cent seed oil against whitefly, *Bemisia tabaci* on cotton. One per cent neem oil was found better than commercial products (Neemark, Replen, Biosol) against *Aphis gossypii* on cotton (Bhatnagar and Kandasamy, 1993). In a field trial ethanolic extract of seed kernel at 2 per cent and neem oil at 1 per cent were as effective as pyrethrum against *Myzus persicae* and *A. gossypii* on sweet pepper (Lowery et al., 1993).

Saxena and Boncodin (1988) compared the survival of *N. virescens* on two-week-old rice seedling whose roots were immersed for 24 hours 2,500 ppm solution in seed bitters or plant sprayed with at the same concentration. Newly emerged female adults were released for three days on these plants. The survival of the hoppers on systematically treated plant decreased significantly after two days but this did not occur on plants receiving foliar spray. Bitters when applied systematically to 21-day-old seedling by soaking in 500 ppm solution for 12 hours or as foliar spray with 10,000 ppm solution gave pest mortality *N. virescens* equal to that observed with monocrotophos at 0.75 kg a.i./h (Saxena et al., 1989).

Sharma et al. (1984) obtained less spectacular results with neem seed fraction against earhead bug *Calocoris angustatus* a serious pest of sorghum. High fecundity of this pests may be a reason for extracts ineffectiveness as each female lays about 150–200 egg, the population of the pest increases by 75 to 100 times within about a week since the crop begin to attract adults from surrounding uncultivated grasses.

There is also report on effect of neem on, oviposition, settling response and mating behaviour. Root dipping of paddy in aqueous extract reduced the oviposition by *N. virescens* (Kareem et al., 1988). The treated plant received 7.6 eggs as against 50 in control. In laboratory and field studies treated rice plants with neem seed bitters and leaf bitters had significantly less number of eggs. Neem oil sprayed rice plants was found to deter the oviposition by *Nilaparvata lugens* (Saxena et al., 1981). Hexane extract of neem of neem kernel was reported to repel California red scale *Aonidiella aurantii* and yellow scale *A. citrina* on citrus (Jacobson et al., 1978).

Haque and Isman (1988) reported that methanolic extracts of seeds of neem and China berry had strong repellent and antifeedent effect on nymphs of *N. nigropictus*. Some other insects which are reported to be repelled are: *Aphis gossypii* to neem oil (Matter et al., 1993); *Sogetella furcifera* to neem oil (Sandhu et al., 1993); *Dialeyrodes pallida* to neem leaf extract (Das and Pande, 1991).

Neem oil was reported to disrupt normal courtship signal production and mating behaviour of female *N. lugens* (Saxena et al., 1989a). *Azadirachtin* reduced the probing of *Rophalosiphum padi* (West and Mordue, 1992).

The following strategies may be useful in managing Hemipterous insects.

1. Repeated application of aqueous extract may be necessary for the management of small Hemipterans with high fecundity. Neem oil enriched with azadirachtin will provide much faster and effective control than the former.
2. For mediumsized insects with low fecundity aqueous extract may be useful. However, higher doses than generally used for managing externally feeding lepidopterous larvae or nymphs of Orthoptera will be needed for effective control.
3. Since large number Hemipterans especially rice pests are reportedly sensitive of oviposition deterrent effect of neem, spraying of crop at early stage of infestation may provide better control.

Coleoptera

Both grubs and adults cause damage to crops. Grubs inflict injury by feeding on roots or boring into stem or feeding externally on different parts of the plant. Adults of all the three categories are also pests and cause damage by feeding mainly on leaves and flowers.

Except for stem borers, all other grubs whether they live in soil or on the surface of the plant including those which burrow into the leaf can be managed with simple aqueous extract of seed kernel. They succumb easily to feeding deterrent and growth regulatory effect because they are sluggish and hence cannot search an unsprayed portion within the plant for feeding and thus prolonging their survival. Adults too have generally been found sensitive to feeding deterrent effect of neem as is evident from below.

Aqueous seed extract, neem oil and ethanolic extract reduced the flea beetle, *Phyllotreta downsei* population on young radish plants. The best result was obtained with aqueous extract giving 60–80 per cent protection (Kareem, 1981).

In Gambia weekly application of aqueous extract of neem fruits (berries at 112.5 g/1) efficiently controlled Africa melon lady bird, *Henosepilachna elaterri* (=*Epilachna chrysomelina*). The number of cucumber per plant was higher than malathion treatment (Redknap, 1981). Application of 4 per cent aqueous neem seed kernel extract was very effective against *Ootheca bennigseni* on bean plant in Tanzania (Ladd, 1981). Tortoise beetle, *Chirida bipucntata* fed least on sweet potato leaves sprayed with 1 per cent acetone extract of seed kernel. The treatment was superior than synthetic antifeedants (Kareem, 1981). In green house and field studies Karel (1987) found 1 per cent aqueous extract of neem leaf and kernel highly effective against *Ootheca bennigseni* in green house. However, 2 and 4 per cent kernel extract had deterrent effect but failed to give complete protection under field conditions. Neem seed extract at low concentration of 0.4 per cent had strong inhibitory effect on feeding of flea beetle *Phyllotreta striolata*, the extract was more effective than methanol and hexane extract (Meisner and Mitchell, 1982).

Japanese beetle, *Popillia japonica* was found highly sensitive to neem seed kernel extract as only 0.5 to 2.0 per cent were required to almost completely inhibit the feeding of adults on sassafras (*Sassafras albidum*) and soybean (*Glycine max*) (Ladd et al., 1978; Anon., 1979). The spray remained effective for 12 days. Interestingly the beetles were not deterred on rose blossom and grape foliage (Ladd, 1981).

Under no choice conditions aqueous and ethanolic extract of seed kernel at 2.5 and 1.0 per cent respectively inhibited feeding by chafer beetle *Apogonia blanchardi* infesting roses (Doharey and Singh, 1989).

Alcoholic extract of seed kernel and enriched fractions caused high mortality of rice hispa, *Dicladispa armigera*. These products caused strong influence on the fecundity of the beetle (Islam, 1987). Egg plants sprayed with 2 per cent ethanolic extract offered very effective control of flea beetle, *Epitrix fuscula* and the Colorado potato beetle, *Leptinotarsa decemlineata*. There was no significant difference between the treatment of ethanolic extract and the synthetic insecticide cypermethrin (Reed and Reed, 1986).

In laboratory and greenhouse experiments Schmutterer (1987a, b) demonstrated that neem seed kernel extract greatly reduced the fecundity of the Colorado potato beetle. Females laid only few eggs, some were completely sterile when they consumed leaves treated with enriched formulated extract or aqueous extract. There was 86 per cent reduction in oviposition and 92 per cent in egg hatching.

Egg plant leaves treated with 0.025, 0.05 per cent neem oil when given to *Epilachna sparsa* in laboratory, the pre-oviposition period was 21 per cent longer than the insects feeding on untreated leaves. Also females had shorter oviposition period, laid few eggs and eggs were smaller in size (Mishra et al., 1990). Feeding on potato leaves treated with neem extracts or oil caused morphogenetic defects in F1 progeny of *L. decemlineata*. Female laid few egg masses; some eggs had chorion defects and dried in a short time (Kaethner, 1992). *Epilachna varivestis* fed for five days on bean leaves treated with azadirachtin (50 ppm) and then transferred to fresh untreated leaves laid only 58 eggs/female as against 730 in control but treatment of male alone did not affect fecundity (Steets and Schmutterer, 1975).

Jacobson et al. (1978) noted reduced egg laying (44 eggs) by dried fruit beetle, *Carpophilus hemipterus* on twigs treated with, ether extract of hexane extracted neem seed kernel in comparison to 556 eggs on control twigs.

Neem oil at 2 per cent was at par with 0.07 per cent endosulfan against flea beetle, *Phyllotreta crucifera* on crucifers and bean (Rajan and Nambisan, 1993). Neem oil was reported to affect egg hatching of rice hispa, *Dicladispa armigera* (Krishnaiah and Kalode, 1991). Soxhlet extracted petroleum ether (40–60 °C) extract of neem leaves at 1 per cent gave 94.7 per cent control of grubs of *Henosepilachna vigintioctopunctata* on brinjal crop (Venkataramireddy et al., 1993).

Kaethner (1991) reported that neem oil at 3 per cent or kernel extract (500–1000 ppm) reduced damage by cockchafers, *Melolontha hippocastani* and *M. melolontha* on several crop through repellent action. In both laboratory and field experiments, fecundity of beetles was significantly lower and food intake was lesser than those fed with untreated plants. Male beetles treated with kernel extract lost their ability to fly thus making their mating impossible in the fields.

Interestingly while seed kernel extracts deter feeding of several cockchafers, neem leaves are powerful attractant for *Holotrichia consanguinea*, *H. serrata* and *H. insularis*. This was confirmed in the laboratory and utilised in capturing 1,00,000 adults of *H. serrata* in Merrut, U.P., India in 1971. The technique consisted of planting fresh branches of neem twigs (1.25 – 2.0 m long) at the rate of two twigs/ha daily before sun set in the infested area. The adults congregated on branches for copulation and feeding were captured and destroyed (Gupta, 1973).

From the above and also other available literatures on management of pests belonging to Coleoptera with neem, it seems that:

1. Most adults and larvae which feed externally or mine the leaves could be controlled with aqueous seed kernel extract.
2. Though little work has been done, it appears quite likely that soil drenching with aqueous extract may nip in the bud soil inhibiting larvae (Scarabaeidae, Meloidae and others). Highly acidic or alkaline soil, however, may adversely affect azadirachtin content.

Curculionids, however may be difficult to manage as vast majorly are internal or subterranean feeder. Since these adults also feed on exposed parts, aqueous extracts can affect them through feeding as there is increasing evidence of sterlizing effect of neem on Coleoptera. The strategies suggested for management of Lepidopterous borers may be used to kill the grubs before boring.

CONCLUSION/SUGGESTIONS

It is seen in the preceding pages that aqueous extract has offered very effective control against variety of pests affecting crops. The advantages of using aqueous extract of seed kernel are many folds:

1. It is the cheapest of all the extracts.
2. Can be prepared easily by even illiterate farmers in developing countries.
3. Neem seed kernel is nature's readymade formulation as it contains natural emulsifier and with little efforts seeds can be stored for a period of one year without loss of activity.
4. Its preparation does not involve any hazard.
5. It is relatively safer to parasitoids than other formulations.
6. Unlike oil it can be translocated in the plant system.

Very little data are available on the residual activity of neem products both on the plant surface and in the soil. For effective utilization, therefore, it is essential to know the persistence of neem products. This is more so when it is used as feeding or oviposition deterrent.

Neem in several cases has shown better activity than synthetic insecticides. To make it more versatile it should have as many formulations too including synergised formulations.

Pretreatment of crops may provide better and effective control as large number of insects have been found sensitive to behavioural effects of neem.

More studies should be done on the feeding and contact effects of neem on adult insects.

Desert locust, *Schistocerca gregaria* is perhaps the most sensitive insect to feeding deterrent effect of neem. On the other hand a few Noth American grasshoppers which are not pests of neem seems to be immune to the feeding and growth-regulatory effects of neem. Physiological studies should be undertaken to unravel the reason for such large variation within a family of insects. This would be of help in developing more effective neem formulations.

Toxicological data generated so far shows that neem is safe but we still do not know whether small amount as residue on the plant or foodgrain if eaten daily is good or bad for us. Hence acceptability daily intake (ADI) and tolerance limit should be established.

Development of resistance is a natural phenomenon. Insects are gifted with extraordinary power to detoxify chemical shortest possible time. Neem would, thus, not be an exception. If *Plutella xylostella* can develop resistance to practically all the synthetic insecticides, why cannot it develop resistance to various neem compounds. Hence it would be wise not to over use neem, the best course would be to incorporate it as strong component in IPM programmes.

ACKNOWLEDGEMENT

The authors express thanks to Shri Shivendra Singh, Senior Research Fellow, DBT Project on Neem, Division of Entomology, Indian Agricultural Research Institute, New Delhi 110 012 for his help in preparation of the manuscript.

REFERENCES

Adhikary, S. 1981. The Togo experience in moving from neem research to its practical application for plant protection. In: *Natural Pesticides from Neem Tree (Azadirachta indica* A.Juss.). H. Schmutter, K.R.S. Ascher and H. Rembold, (eds.), GTZ, Eschborn, West Germany, pp. 215–222.

Adler, V.F. and E.C. Uebel. 1984. Antifeedant and bioassays of neem extracts against the Carolina grasshopper, walking stick and field cricket. *J. Environ. Sci. and Health*: A Environmental Science and Engineering 19: 393–403.

Anonymous. 1979. *Neem tree*: Seed extracts repel Japanese beetle. *Agric. Res.* March: 8–10.

Anonymous. 1983. Neem in Agriculture, Indian Agricultural Research Institute, New Delhi, Research Bulletin No. 40, 62 p.

Ascher, K.R.S., M. Streloke, G.H. Schimat and J.D. Warthen Jr. 1989. The antifeedant effect of neem seed kernel extract and azadirachtin on nymphs of *Eyprepocnemis plorans*. *Phytoparasitica* 17(13): 167–174.

Attri, B.S. 1975. Utility of neem oil extractive as feeding deterrants to locust. *Indian J. Ent.* 37: 417–419.

Bhanotar, R.K. and Y.N. Srivastava. 1984. Effect of neem kernel suspension on the development of eggs of desert locust, *Schistocerca gregaria* F. *Neem Newsl.* 1(3): 30.

Bhatnagar, P. and C. Kandasamy. 1993. Evaluation of a neem-based fromulation against insect pests of cotton. *Pestology* 17(1): 13–15.

Butterworth, J.H. and E.D. Morgan. 1971. Investigation of the locust feeding inhibition of the seeds of neem tree, *Azadirachta indica*. *J. Insect Physiology* 17: 969–977.

Chiu, S.F. 1985. Recent research findings on Meliaceae and other promising botanical insecticides in China. *Zeitschrift, fur pflanzen kreikheiten und Pflanzen Schutz* 92(3): 310–319.

Chiu, S.F., B.Q. Huans and M.Y. Hu. 1985. Deterrent effect of seed oil and extracts of some meliaceous plants on rice gall midge (GM) oviposition. *Int. Rice Res. Newslelt.* 10(3): 25.

Das, M.L. and Y.D. Pande. 1991. Evaluation of insecticides and leaf extract of *Azadirachta indica* against *Dialeurodes pallida* Uttar Pradesh *J. Zool.* 11(1): 25–30.

David, P.M.M., A.M. Hanifa and S. Natarajan. 1990. Biology and control of blossom midge, *Contarinia* sp. (Diptera : Cecidomyiidae) on *Jasminum sambac* in Tamil Nadu. *Entomon.* 15 (3/4): 193–196.

David, P.M.M., M. Jayasekhar and S. Natarajan. 1990a. Evaluation of insecticide and botanicals for control of flower gall midge on chillies. *Madras Agric. J.* 77: 249–252.

Doharey, K.L. and R.P. Singh. 1989. Evaluation of neem (*Azadirachta indica* A. Juss.) seed kernel extracts against chafer beetles. *Indian J. Ent.* 51: 217–220.

Dreyer, M. 1984. Effect of aqueous neem extracts and neem oil on the main pests of *Cucurbita pepo* in Togo. In: *Natural Pesticides from Neem tree and other Tropical Plants*, H. Schmutterer and K.R.S. Ascher, (eds.), GTZ, Eschborn, West Germany, pp. 435.

Dreyer, M. 1986. Investigations on the effect of Aqueous extracts and other products from neem seeds against pests of vegetable and field crops in Togo, Ph.D. Thesis, University of Giessen, West Germany.

Dreyer M. 1987. Results of field experiments on the effect of neem (*Azadirachta indica*) aqueous extracts and neem oil on the pests of egg plant (*Solanum melongene*) in Togo. *Mitteilungen der Deutschen Gesellschaft fur Allgemeine und Angewandte. Entomologie* 50: 144–151.

Dreyer, M. 1987. Field and laboratory trials with simple neem products as protectants against pests of vegetables and field crops in Togo. In: *Natural Pesticides from Neem tree* (*Azadirachta indica* A. Juss.) H. Schmutterer and K.R.S. Ascher, (eds.), GTZ, Eschborn, West Germany, 431–447.

El-Sayed. 1985. Neem (*Azadirachta indica* A. Juss) seeds as antifeedant and ovipositional repellent for the Egyptian cotton leaf worm, *Spodoptera littoralis* (Boisd). *Bull. Ent. Sco. Egypt*, Economic series, 13: 49–58.

Fagoonee, I. 1980. The potential of natural products in crop protection in Mauritius. *Proc. Neth. Agric. Prod. Cont.*, 1979, Reduit, Mauritius, 201.

Haque, M.A. and B.J. Islam, 1988. Effect of methanolic neem and China berry seed extracts on rice green leafhopper (*Nephotettix nigropictus* Stal. *Bangladesh J. Agric.* 13(1): 53–57.

Gupta, K.M. 1973. Neem leaves attract white grub beetles. *Indian J. Ent.* 35: 276–278.

Harris, M., D. Warkentin and J. Begley. 1987. Margosan-O contact and residual efficacy against beet army worm larvae. *Insecticide and Acaricide Tests* 12: 331.

Hellpap, C. and J.C. Mercado. 1986. Effect of neem on the oviposition behaviour of the fall armyworm *Spodoptera furgiperda. J. Appl. Entomol.* 101: 463–467.

Islam, B.N. 1987. Use of some extracts from Meliaceae and Annonaceae for control of rice hispa, *Dicladispa armigera* and the pulse beetle, *Callosobruchus chinensis*. In: *Natural Pesticides from Neem Tree and Other Tropical Plants*, H. Schmutterer and K.R.S. Ascher, (eds.), GTZ Eschborn, West Germany, pp. 217–242.

Jacobson, M., D.K. Reed, M.M. Crystal, D.S. Moreno and E.I. Soderstorm. 1978. Chemistry and biological activity of insect feeding deterrent from certain weed and crop plants. *Entomologia Exp. Appl.* 24: 448–457,

Jhansi, B.R. and R.P. Singh. 1993. Identification of effective and inexpensive neem (*Azadirachta indica* A. Juss.) seed kernel extract for management of *Helicoverpa armigera* Hubner. Abstract World Neem Conf. 24–28, February, Bangalore, India, p.7.

Joshi, B.G., G. Ramaprasad and S. Sitaramaiah. 1978. Neem seed kernel suspension protects tobacco nurseries. *Indian Fmg.* 28(9): 17–18.

Joshi, B.G. and S. Sitaramaiah. 1979. Neem kernel as an ovipositional repellent for *Spodoptera litura* (F.) moths. *Phytoparasitica* 7: 199–202.

Kaethner, M. 1991. Potential of neem seed kernel products for the control of the cockchafers, *Melolontha hippocastani* F. and *M. melolontha* L. (Coleoptera: Scarabbacidae). *J. Appl. Entomol.* 112: 345–352.

Kaethner, M. 1992. Fitness reduction and mortality effects of neem-based pesticides on the Colorado potato beetles, *Leptinotarsa decemlineata* Say (Coleoptera: Chrysomelidae). *J. Applied Ent.* 112: 345–352.

Kareem, A. 1981. Neem as an antifeedant for certain phytophagous insects and a bruchid on pulses. In: *Natural Pesticides from the Neem Tree* (*Azadirachta indica* A. Juss.), H. Schmutterer, K.R.S. Ascher and H. Rembold, (eds.), GTZ. Eschborn, West Germany, pp. 223.

Kareem, A., R.C. Saxena, E.L. Palanginam, E.M. Boncodin and M.T. Malayba. 1988. Effects of neem derivative on insect pests of rice and certain other crops in the Philippines (Lab and Field evaluation, 1986-88). *Int. Rice Res. Inst.*, Manila, pp. 87.

Kareem, A., S. Sadakathulla, M.S. Venugopal and T.R. Subramaniam. 1974. Efficacy of two organotin compounds and neem extracts against sorghum shoot fly. *Phytoparasitica* 2: 127–129.

Karel, A.K. 1987. Response of *Ootheca bennigseni* (Coleoptera: Chrysomelidae) to neem extracts. In: *Natural Pesticides from the Neem Tree* (*Azadirachta indica* A. Juss) *and Other Tropical Plants*, H. Schmutterer and K.R.S. Ascher, (eds.), GTZ, Eschborn, West Germany, pp. 393–403.

Knodel-Montz, J.J. and H.G. Larew. 1985. Control of leaf miner using neem (*Azadirachta indica* A. Juss.) products and other insecticides as foliar sprays. *Insecticide and Acaricide Tests* 10: 282.

Krishnaiah, N.V. and M.B. Kalode. 1991. Efficacy of neem oil against rice insect pests under greenhouse and field conditions. *Indian J. Plant Protec.* 19: 11–16.

Ladd, T.L. 1981. Neem seed extracts as feeding deterrents for the Japanese beetle, *Popillia japonica*. In: *Natural Pesticides from the Neem Tree* (*Azadirachta indica* A. Juss), H. Schmutterer and K.R.S. Ascher, (eds.), GTZ, Eschborn, West Germany, pp. 149–156.

Ladd, T.L., M. Jacobson and C.R. Burff. 1978. Japanese beetle : Extracts from neem tree seeds as feeding deterrents. *J. Econ. Entomol.* 71: 810.

Langewald, J. and H. Schmutterer. 1992. Effects of neem oil treatment on the phase status of desert locust, *Schistocerca gregaria*. Proceedings of Workshop on Biological Control of Locusts and Grasshoppers. CAB International, pp. 142–154.

Larew, H.G., J.J. Knodel-Montz, R.E. Webb and J.D. (Jr.) Warthen. 1985. *Liriomyza trifolii* (Burgess) (Diptera: Agromyzidae): Control on chrysanthemum by neem seed extract applied to soil. *J. Econ. Entomol.* 78: 80–84.

Lowery, D.T., M.B. Isman and N.L. Brard. 1993. Laboratory and field evaluation of neem for the control of aphids (Homoptera: Aphididae). *J. Econ. Entomol.* 86: 864–870.

Mane, S.D. 1968. Neem spray as repellent against some foliage feeding insects. M.Sc. Thesis, Indian Agricultural Research Institute, New Delhi.

Mariappan, V., S. Jayaraj and R.C. Saxena. 1988. Effect of non-edible oils on survival of *Nephotettix virescenes* (Homoptera: Cicadellidae) and on transmission of rice tungro virus. *J. Econ. Entomol.* 81: 1369–1372.

Matter, M.M., S.S. Marei, S.M. Moawad and S. El-Gengaihi. 1993. The reaction of *Aphis gossypii* and v/s predator, *Coccinella unidecimpunctata* to some plant extract. *Bull. Faculty of Agriculture*, University of Cairo 44(2): 417–432.

Meisner, J. and E.K. Mitchell. 1982. Phagodeterrent effect of neem extracts and azadirachtin on the flea beetles, *Phyllotreta striolata* (F.) Zeitschrift fur pflanzen krankheiten und, pflanzensachutz, 8–9: 463–467.

Mishra, B.K., P.R. Mishra, N.C. Patnaik and H. Mahapatra. 1990. Effect of neem oil on the oviposition behaviour of epilachna beetle. *Orissa J. Agric. Res.* 2(1): 70–71.

Mohan, K., M. Gopalan and G. Balasubramaniam. 1991. Studies on the effects of neem products and monocrotophos against major pests of rice and their safety to natural enemies. *Indian J. Pl. Prot.* 19: 23–30.

Mulkern, G.B. and S. Mongolikti. 1975. Desert locust feeding deterrent ineffective against North American Grass-hoppers (Orthoptera: Acrididae). *Acrida* 4(2): 95–103.

Narasimhan. V. and V. Mariappan. 1988. Effect of plant derivatives on green leafhopper (GLH) and rice tungro (RTV) transmission. *Int. Rice Res. Newsl.* 13(1): 28–29.

Narayanan, C.R., R.P. Singh and D.D. Sawalker. 1980. Phogodeterrency of various fractions of neem oil against *Schistocerca gregaria* F. *Indian J. Ent.* 48: 462–472.

Nasseh, O., S. Krall, H. Wilps and G.B. Salisson. 1992. The effects of growth inhibitors and of biotic insecticides on larvae, *Schistocerca gregaria* Sahd PV info No. 45, 5–19.

Natarajan, K. and V.T. Sundaramurthy. 1990. Effect of neem oil on cotton whitefly (*Bemisia tabaci*). *Indian J. Agri. Sci.* 60: 290–291.

Nicol, C.M.Y. and H. Schmutterer. 1991. Contact effects of seed oil from the neem tree, *Azadirachta indica* on nymphs of gregarious phase of the desert locust *Schistocerca gregaria* (Forsk.). *J. Appl. Ent.* 111(2): 197–205.

Olaifa, J.I. and A.O. Adenuga. 1989. Neem product for protecting field cassava from grasshopper damage. *Insect Science and Its Application* 9(2): 267–270.

Passerini, J. and S.B. Hill. 1993. Field and laboratory trials using a locally produced neem insecticide against Sahelian grasshopper, *Kraussaria augulioera* (Orthoptera: Acrididae) on millet in Mali. *Bull. Entomol. Res.* 83(1): 121–126.

Patel, B.V., B.S. Nandihalli, Parameswar Hugar, B.R. Thimmanagouda and S. Lingappa. 1993. Efficacy of some plant and animal products against cotton pest complex. In: *Botanical Pesticides in Integrated Pest Management*, M.S. Chari and G. Ramaprasad, (eds.), Indian Society of Tobacco Science, Rajamundry, India, pp. 183–189.

Patel, H.K., V.C. Patel, M.S. Chari, J.J. Patel and J.R. Patel. 1968. Neem seed suspension—a sure deterrent to hairy caterpillar (*Amsacta moorei* Butl.). *Madras Agric. J.* 55(11): 509–510.

Pradhan, S. and M.G. Jotwani. 1968. Neem as an insect deterrent. *Chemical Age India* 19(9): 756–760.

Pradhan, S., M.G. Jotwani and B.K. Rai. 1962. The neem seed deterrent of locust. *Indian Fmg.* 12: 7–11.

Radcliffe, E.B., F.V. Dunkel, P.P. Strzok and S. Adam. 1991. Antifeedant effect of neem. *Azadirachta indica*, kernel extracts on *Kraussaria angulifera*, a Sahelian grasshopper. *Tropical Agriculture* 68: 95–101.

Rajan, S.K. and K.M.P. Nambisan. 1993. Effectiveness of neem against five temperate crop pests in the upper palani hills in Abstracts: World Neem Conf. 24–28 Feb., Bangalore, India, p. 28.

Rao, P.J. and B. Subramanyam. 1987. Effect of azadirachtin on *Achoea janata* Linn and *Spodoptera litura* (F.). *J. Entomol. Res.* 11(1): 166–169.

Reddy, V.V.S., T. Ramesh Babu and B. Narasimha Rao. 1993. Repelin an effective botanical insecticide against *Carpomyia vesuviana* costs. *Abstract World Neem Conference* 24–28th February, Bangalore, India, pp. 24.

Redknap, R.S. 1981. The use of crushed neem berries in the control of some insect pests in Gambia. In: *Natural Pesticides from the Neem Tree (Azadirachta indica* A. Juss.), H. Schmutterer and K.R.S. Ascher, (eds.), GTZ, Eschborn, West Germany, pp. 205–214

Reed, D.K. and G.L. Reed. 1986. Control of vegetable insects with neem seed extracts. *Proc. Indian Acad. Sci.* 94: 335.

Samalo, A.P., B. Senapati, C.R. Satpathy and T.J. Jacob. 1993. Effect of neem derivatives on incidence of some major insect pests in wet season rice. In: *Botanical Pesticides in Integrated Pest Management*, M.S. Chari and G. Ramaprasad, (eds.), Indian Society of Tobacco Science. Rajahmundry, Indian, pp. 197–202.

Sandhu, S.S., G.S. Dhaliwal and V.K. Dilwari. 1993. Effect of neem-based insecticide on growth and development of white backed planthopper. Abstract: World Neem Conf. Bangalore, India, 24–28 Fed., p. 46.

Saxena, R.C. and M.E.M. Boncodin. 1988. Effect of neem seed bitters (NSB) on green leafhopper (GLH). *Int. Rice Res. Newsl.*, 13(1): 27.

Saxena, R.C. and A.A. Barrion. 1987. Cytogenetic effects of neem seed "bitters" (NSB) on green leaf hopper (GLH) males. *Int. Rice Res. Newsl.* 12(5): 24–25.

Saxena, R.C., J.N. Liquido and H.D. Justo Jr. 1981. Neem seed oil a potential antifeedant for control of the rice brown plant hopper, *Nilaparvata lugens*. In: *Natural Pesticides from Neem tree* (*Azadirachta indica* A. Juss), H. Schmutterer, K.R.S. Ascher and Rembold (eds.), GTZ, Exchborn, West Germany, pp. 171–188,

Saxena, R.C., A. Kareem, E.L. Palanginan and M.T. Malayba. 1989a. Systemic and foliar application of neem seed bitters (NBS) to control green leafhopper (GLH) and rice tungro virus (RTV) diseases. *Int. Rice Res. Newsl.*, 14(1): 31

Saxena, R.C., Z.T. Zhang and M.E.M. Boncodin. 1989. Effect of neem oil on courtship signals and mating behaviour of brown plant hopper (BHP) female. *Int. Rice Res. Newsl.* 14(6): 28–29.

Schmutterer, H. 1987a. Fecundity reducing and sterilizing effects of neem seed kernel extracts in the Colorado potato beetle *Leptinotarsa decemlineata*. In: *Natural Pesticides from Neem Tree and other Tropical Plants*. H. Schmutterer and K.R.S. Ascher, (eds.), GTZ Eschborn, West Germany, pp. 351–360.

Schmutterer, H. 1987b. Experiments on reduction in fecundity and sterilization of the Colarado potato beetle. *Leptinotarsa decemlineata* by application of neem seed extract. *Mitteilunger der Deutschen Gesellschaft fur Allgemcine und Argewaandte Entomoligies*, 5: 152, 161.

Schmutterer, H. 1993. Some effects of neem (*Azadirachta indica*) products on locusts and grasshoppers. Souvenir World Neem Conf. Indian Society of Tobacco Science, CTRI, Rajahmundry, India, pp. 44–52.

Sharma, H.C., K. Leuschner, A.V.B. Sankaran, D. Gunasekhar, M. Marthandamorti, K. Bhaskariah, M. Subramanyam and N. Sultana. 1984. Insect antifeedants and growth inhibitors from *Azadirachta indica* and *Plumbago zeylanica*. In: *Natural Pesticides from the Neem Tree and Other Tropical Plants*, H. Schmutterer, K.R.S. Ascher, (eds.), GTZ Eschborn, West Germany, pp. 291–320.

Shelke, S.S., L.O. Jadhav and G.N. Salunkhe. 1985. Ovipositional and adult repellent action of some vegetable oils/extracts against potato tuber moth. *J. Maharashtra Agric. Univ.* 10: 284–286.

Shelke, S.S. L.D. Jadhav and G.N. Salunkhe. 1987. Ovicidal action of some vegetable oils and extracts in the storage of potato, *Phthorimaea operculella* Zell. *Biovigyanam* 13(1): 40–41.

Singh B. and A.P. Singh. 1987a. Neem suspension as an oviposition deterrent to the desert locust (*Schistocèrca gregaria* Forsk.). *Plant Prot. Bull.* 39 (1–2): 9–16.

Singh B. and A.P. Singh. 1987b. Effect of neem kernel suspension on the hatchability of eggs of the desert locust *Schistocerca gregaria* Forskall. *J. Adv. Zool.* 8(1): 52–54.

Singh, R.P. 1987. Comparison of antifeedant efficacy and extract yield from different parts and ecotypes of neem (*Azadirachta indica* A. Juss) trees. In: *Natural Pesticides from the Neem Tree and other Tropical Plants*. H. Schumutterer and K.R.S. Ascher (eds). GTZ Eschbron, W. Germany, pp. 185–194.

Singh, R.P. and P.K. Kataria. 1991. Insects nematodes and fungi evaluated with neem (*Azadirachta indica* A. Juss.). *Neem Newsl.* 8(1): 3–10.

Singh, R.P. and B.G. Srivastava. 1983. Alcohol extract of neem (*Azadirachta indica* A. Juss.) seed oil as oviposition deterrent for *Dacus cucurbitae* (Coq.). *Indian J. Ent.* 45: 497–498.

Singh, R.P., Y. Singh and S.P. Singh. 1985. Field evaluation of neem (*Azadirachta indica* A. Juss.) seed kernel extracts against the pod borers of pigeon pea. *Cajanus cajan* (L.) Mill sp. *Indian J. Ent.* 47: 111–112.

Singh, R.P., C. Devakumar and S. Dhingra. 1988. Activity of neem seed kernel extract against the mustard aphid. *Lipaphis erysimi. Phytoparasitica* 16: 225–229.

Skatulla, U. and J. Meisner. 1975. Laboratory experiments using neem seed extract for control of gypsy moth, *Lymantria dispar* L. *Anz. Schaedlingskd. Pflanz. Umweltschutz* 48: 38.

Sombatsiri, K. and S. Tigvattanont. 1984. Effects of neem extracts on some insects pests of economics importance in Thailand. In: *Natural Pesticides from the Neem Tree and Other Tropical Plants*. H. Schmutterer and K.R.S. Ascher (eds.), G.T.Z. Eschborn, W. Germany, pp. 95–100.

Srinivasan, K. and P.N. Krishna Moorthy. 1993. Evaluation of neem products and other standard cnemicals for the management of major pest complex on cabbage : Comparison between standard spray-regime and IPM involving mustard as trap crop. *Abstract World Neem Conf.*, Bangalore, India, 24–28 February, p. 19.

Srivastava, K.P., N.P. Agnihotri and H.K. Jain. 1984. Relative efficacy of fenvalerate, quinalphos and neem kernel extracts for the control of pod fly, *Melanagromyza obtusa* and pod borer, *Heliothis armigera* infesting red gram, *Cajanus cajan* together with their residues. *J. Ento. Res.* 1–4.

Srivastava, R.P., R.P. Singh and Masarrat Haseeb. 1993. Evaluation of formulated neem (*Azadirachta indica* A. Juss.) kernel extract against mangohoppers and mealybug. Abstracts, *World Neem Conf.*, Bangalore, India, 24–28, February, p. 18.

Steets, R. and H. Schmutterer. 1975. The effect of azadirachtin on the life span and reproduction capacity of *Epilachna varivestis* Muls. (Coleoptera: Coccinellidae). *Zeitschrift Pflanzenkran* 82: 176–179.

Steffens, R. and H. Schmutterer. 1982. The effect of a crude methanolic neem (*Azadirachta indica*) seed kernel extract on metamorphosis and quality of adults of the Mediterranean fruit fly, *Ceratitis capitata* (Diptera: Tephritidae). *Z. Angew. Ent.* 14: 98–103.

Venkataramireddy, B., K.C. Chitra and P. Kameswara Rao. 1993. Efficacy of the plant extracts in the control of brinjal spotted leaf beetle, *Henosepilachna vigintioctopunctata* F. In: *Bolanical Pesticides in Integrated Pest Management*, M.C. Chari and G. Ramaprasad. (eds.), Indian Soc. of Tobacco Science, Rajamundry, India, pp 225–227.

Verma, S.P., B. Singh and Y.P. Singh. 1985. Studies on comparative efficacy of certain grain protectants against *Sitotroga cerealela* Oliver. *Bull. grain Tech.* 24: 37–42.

West, A.J. and A.J. Mordue. 1992. The influence of azadirachtin on the feeding behaviour of cereal aphids and slugs. *Ent. Exp. Appl.* 62(1): 75–79.

Wilps, H., O. Nasseh and S. Krall. 1992 The effect of various neem products on the survival and the flight activity of adult *Schistocerca gregaria. Proc. of Workshop on Biological Control of locusts an Grasshoppers*, CAB International, pp. 337–346.

Wilps, H., O. Nasseh and S. Krall. 1993. Application to flying locusts a powerful method for combating *Schistocerca gregaria* (Forskal) adults. *Adstract, World Neem Conf.* 24–28 Fed., Bangalore, India, p. 8.

Webb, R.E., M.A. Hinebaugh, R.K. Lindquist and M. Jacobson. 1983. Evaluation of aqueous solution of neem seed extract against *Liriomyza sativae* and *L. trifolii* (Diptera : Agromyzidae). *J. Econ. Entomol.* 76: 357.

2

BOTANY

13

An Assay of Genetic Variability through Phenological Studies on Neem (*Azadirachta indica* A. Juss.) Plantation

M.G. GOGATE[1] AND D.R. GUJAR[2]

[1] Conservator of Forests (Research), Maharashtra State PMT Building, Pune, India
[2] Asst. Silviculturist, Ravinagar Nursery, Nagpur 440 001, Maharashtra, India

ABSTRACT

Information has been collected on crown shape and size and fruit production capabilities of neem trees in a 50- to 60-year-old plantation in Vidarbha region of Maharashtra State. It reveals that the shape of the crown may be mushroom, conical, or inverted conical and the occurrence of these shapes in the population is 39, 38, and 23 per cent respectively. Crown volume varies from 471.08 to 2095.64 m^3 and there is correlation between crown size and bole diameter. Flowering behaviour is not uniform; out of 331 trees under study, a total of 292 trees flowered and fruited in the 1992 season. Fruiting starts in late April and fruit fall is from mid-May to mid-July. Fruit production varies from 7.5 to 17.325 kg (dry weight) per tree, the average being 12.795 kg. Fruit production is in consonance with crown size and bole diameter. Selections for specified traits (as indicated) and concerted efforts at international and national levels are recommended for genetic improvement.

INTRODUCTION

The genetic potential of neem (*Azadirachta indica* A. Juss.), a valuable multipurpose species, is an untapped natural resource (Dogra and Thapliyal, 1993). Neem is a hardy evergreen tree commonly found in South Asia and parts of Africa. It thrives under humid to semiarid conditions (Chaturvedi, 1993). In India alone, over 14 million trees exist according to the Directorate of the Non-edible Oils and Soap Industries, Khadi and Village Industries Commission, with a potential of yielding of 418,000 tonnes of seed per year (Ketkar, 1976). It is highly efficient in restoring soil productivity and simultaneously provides fodder, fuelwood, and other products used in medicine, pesticides, fertilizer, food, soap, lubricants, gums, toothpaste etc.(Anon., 1968; Chaturvedi, 1985; Arora, 1993; Gill and Deb Roy, 1993; Riar, 1993). Although neem tree is widely recognized as a forest crop with multiple uses, most published research has dealt with the isolation, identification, formulation, and field testing as pest control agents derived from the seed kernels and leaves (Ahuja et al., 1976; Morgan, 1982; Anon., 1985; Siddiqui et al., 1986; Rembold et al., 1987; Ara et al., 1990; Tewari, 1992). Studies related to selection of superior trees, provenances, and other genetic improvements are lacking. The wide distribution of neem trees implies genetic variability (Percy, 1972; Zanno et al., 1975; Morgan, 1981; Schmutterer and Zebitz, 1984) and this offers wide scope for selection of the required traits for its improvement.

MATERIALS AND METHODS

A neem plantation was raised around 1935 as part of the Forest Timber Depot established in Amravati District of Maharashtra State. The study area is part of 5A/C3 Southern dry mixed deciduous forests (Champion and Seth, 1968). The temperature ranges from 11° to 47°C, rainfall from 750 to 1200 mm, and relative humidity from 7 to 90 per cent depending upon the season. Overall, the weather during summer is hot and during winter it is cold.

The plantation site has deep black cotton soil. The trees are spaced at 10 × 10 m or so. At present there are 331 trees surviving.

All the trees were considered for the phenological studies. Average diameter at breast height for each tree was calculated by taking two diameters at right angle to each other with callipers. The trees were then grouped in 10 cm diameter classes.

The crown height of each tree was measured by Abney's level and crown diameter was measured by Ravi Pentaprism. For crown diameter also, two readings at right angles to each other were taken for each tree. Shape of crown of each tree was recorded by ocular estimation for all the trees.

The volume of crown was calculated by the following formulae:

$$\text{Volume of mushroom shaped crown} = \frac{1}{2} \times \frac{4}{3} \pi r^3$$

$$\text{Volume of conical and inverted conical crown} = \frac{1}{3} \pi r^2 h$$

where r = crown radius and h = crown height.

The period of flowering, fruiting, and fruit fall was recorded. For fruit production study, five well-grown trees from each diameter class were selected and fallen fruits were collected daily till all fruits were shed. Fresh weight for fruits of each trees was recorded immediately after collection and then allowed to air dry in shade and weighed after every 15 days, till three consecutive weights remained the same.

OBSERVATIONS AND DISCUSSIONS

It was observed that the neem trees were evergreen throughout the year, although in February-March the leaf fall was heavy. However, Tewari (1992) reported that in dry localities the tree becomes nearly leafless for a short period during February-March.

The neem trees under study showed variation in size and shape; and the growth is not uniform though the trees are generally free-standing without any congestion.

Observations on bole diameter were recorded at breast height and were seen to range from 31 to 91 cm. The percentage of trees in each diameter class is shown in Table 1 and it follows a definite pattern typical for any plantation. The crown height and crown diameter also varied from tree to tree and ranged from 11.70 to 17.05 m and 11.77 to 20.05 m for crown height and crown diameter, respectively. The rise in crown height and crown diameter is in consonance with increase in bole diameter of

Table 1. Phenological characters of neem (*Azadirachta indica*) plantation at Paratwada

Diam. class (cm)	Tree %	Av. crown ht (m)	Av. crown Diam.	Crown (Vol.m^3)	Fruit prod. (kg)	
					Fresh wt.	Dry wt.
31–40	21.10	11.70	11.77	471.08	10.000	7.500
41–50	39.44	13.36	11.86	489.71	13.250	11.250
51–60	23.85	15.25	13.75	754.16	18.000	11.950
61–70	11.00	17.61	14.10	813.10	22.000	15.950
71–80	3.66	17.20	16.12	1098.12	25.700	17.325
91–100	0.95	17.05	20.05	2095.64		

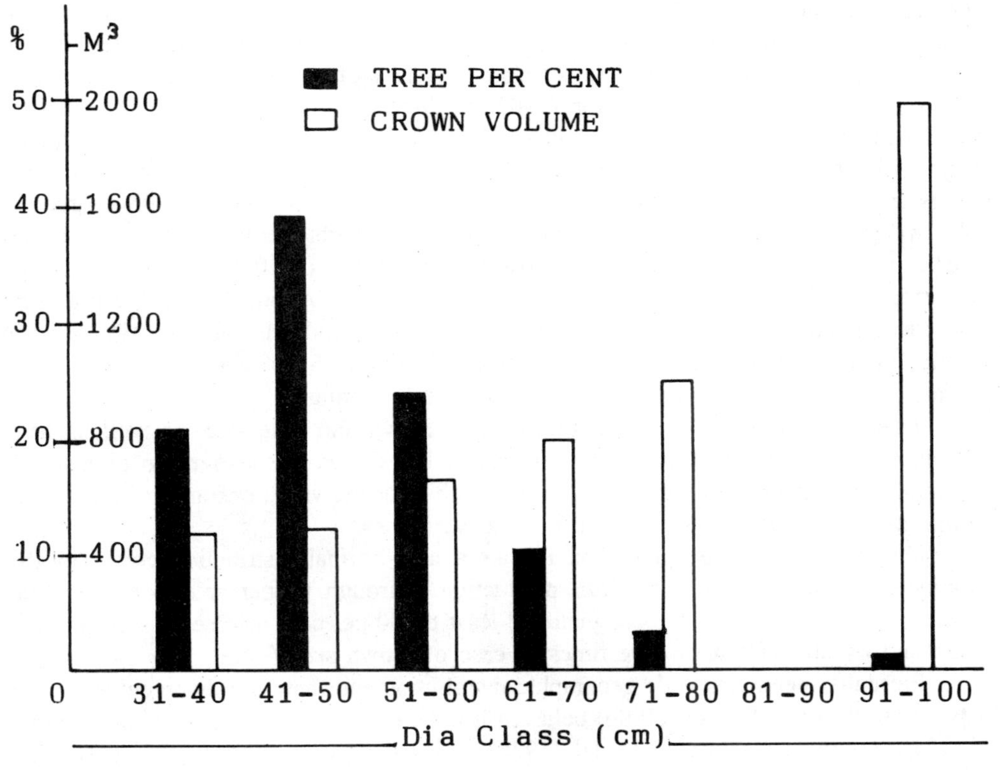

Fig. 1. Histogram showing relationship among diameter class, percentage of trees and crown volume of neem

the tree. Similarly, the volume of crown also showed an upward increasing trend from 471.08 to 2095.64 m^3 and there is a definite correlation between crown size and bole diameter (Table 1, Fig. 1).

Studies on shape of crown of the neem trees revealed mushroom-shaped (39 per cent), conical (38 per cent), and inverted conical crown (23 per cent).

Studies on flowering behaviour revealed that the panicles of small, honey-scented white flowers started appearing in mid-March and flowering ended by late April. Out of 331 trees under study, a total of 292 trees flowered and subsequently fruited in 1992 season. It has been reported that flowering appear from January–March in southern parts of India and later towards the north (Anon., 1981). The flowering starts in the beginning of January in Kerala, in February-March in Karnataka, Tamil Nadu, and Andhra Pradesh, in the first week of April in Central India, getting progressively delayed in northern India, and going up to the first week of May in the sub-Himalayan area (Anon., 1981; Tewari, 1992). Mitra (1963) reported the restricted bloom of neem in August-September in the plains of northern India. Guhabakshi (1984) observed the flowering and fruiting of neem in Murshidabad district of West Bengal occasionally throughout the year.

The fruiting also follows a similar pattern and fruits ripen from June to August (Anon., 1981). In the present study, fruiting started in late April and fruit fall was between mid-May and mid-July. Fruit production varied from tree to tree, depending upon its size and shape of crown. The details of fruit production on average basis for different crown volume and bole diameter class are given in Table 1. It revealed that fruit production in 31–40 cm diameter class was 10 kg/tree on fresh weight basis and 7.5 kg/tree on dry weight basis; for 71–80 cm diameter class the fruit production was 25.7 kg/tree on fresh weight basis and 17.325 kg/tree on dry weight basis. Thus, there occurred a definite correlationship between the fruit production and bole diameter *vis-a-vis* crown volume of the tree. The overall average fruit production 12.795 kg/tree on dry weight basis. It has been reported that the neem starts fruiting at the age of five years but economic yield of fruit is obtained at the age of 10–12 years and, on an average, a medium-sized tree produces 37–55 kg of fruits (Anon., 1978). The low fruit production rates in the present study may be due to a bad seed year in 1992, and a definite conclusion is possible only after data collection of 5–10 years in series, as there is a possibility of cyclic changes in flowering and fruiting.

There are variations in size and shape of crown and bole size, as well as fruit production. As the stand is of an even age and has been raised over an almost flat terrain, local factors alone may not be responsible for the variations and genetic variations must have contributed significantly to variations.

An evaluation of the plantation reveals a near normal distribution curve for the diameter, crown volume, and fruit production. Through proper selection it would, therefore, be possible to increase yields at least by 50 per cent in case of seed production and by almost four to five times in case of crown size.

Presently, neem has wide geographic distribution and there are divergent views on its origin. Whereas Brandis (1906) believed it to be endemic to upper Myanmar, Duthie (1903), Troup (1921), and Kanjilal (1928) believed it to occur naturally in Siwalik forests. The present wide range of distribution is not a recent phenomenon and therefore, through natural processes, appropriate selection must have taken place for perpetuation of the species under very diverse agroclimatic conditions. One should, therefore, expect geographic variability leading to distinct provenances. In some species there is 4 : 1 difference in growth rate and correspondingly large differences in other traits between different races of the same species (Wright, 1976). Considerable work has already been done on provenance trials for species having wide range of distribution and substantial gains are possible through selection of appropriate provenance, for example, in *Eucalyptus comendulensis*, *E. territicornis*, and *Acacia nilotica*. No provenance work has been systematically carried out in neem (Tewari, 1992), except identification of an isolated natural provenance of *A. indica* from Dio Tao demarcated from Chiang Mai province of Thailand for seed collection (Lauridsen et al., 1991). Against this background, it is necessary to examine the extent of provenance or geographic races that could be distinguished. Critical study of the entire range of present neem occurrence is called for. A coordinated study at the international or national level is recommended.

Neem being a multiple-use species, selection of 'superior' types will have to be made on different criteria, for example, production of foliage, fruit production, percentage of oil and active ingredients etc.

Germplasm collection, provenance research, and *ex situ* conservation have been hampered by short seed viability. Clonal propagation through tissue culture or macro propagation will be inevitable. However, before undertaking such ventures, it is necessary to understand the broad spectrum of natural variations through phenological studies and make selections on different criteria, with the ultimate use in view.

ACKNOWLEDGEMENTS

The authors are thankful to Dr. A.G. Raddi, IFS, Chief Conservator of Forests (Coordination Cell), Pune, for his constant encouragement and interest. The help rendered by Mr. G.H. Tathod, Range Forest Officer (Research), Paratwada, and his staff in collecting the field data is very much appreciated.

REFERENCES

Agrawal, I.L. and S.B. Mall. 1968. Studies on the insecticidal and antifeedant activity of some plant extracts on Bihar Hairy Caterpillars. *J. Appl. Entomol.* 105(5): 529–532.

Ahuja, M.M., R.R. Gupta, K.N. Agrawal and A.C. Gupta. 1976. Miscella refining of neem seed and mahua oils. *Indian J. Technol.* 14(5): 257–259.

Anonymous 1968. Report on two consignments of neem (*Azadirachta indica*) from the Republic of Sudan. Cons. 1307 and 1374. Reports on Overseas Timber. Princess Risbourgh Lab. No. 11 U.R.

Anonymous 1978. *Tree-borne oilseeds*. Directorate, Non-edible Oils and Soap Industry. Khadi and Village Industries Commission, Bombay.

Anonymous 1981. Troup's *The Silviculture of Indian Trees*, Vol. III. Editorial Board, Forest Research Institute and College, Dehra Dun, pp. 147–154.

Anonymous 1985. *Wealth of India. Raw Materials*, Vol. 1A. Publication and Information Directorate, CSIR, New Delhi, pp. 504–511.

Ara, L., B.S. Siddiqui, S. Faizi and S. Siddiqui. 1990. Three new diterpenoids from the stem bark of *Azadirachta indica*. *J. Nat. Products*, 53(5): 816–820.

Arora, P.K. 1993. Genetic diversity and ethnobotany. In: *Neem Research and Development*. N.S. Randhava and B.S. Parmar (eds.), Publ. No. 3 (1993). Society of Pesticide Science, India, pp. 33–37.

Brandis, D. 1906. *Indian Trees*. London.

Champion, H.G. and S.K. Seth. 1968. A revised study of the forest types of India.

Chaturvedi, A.N. 1985. Firewood farming on degraded lands. U.P. Forest Bulletin No. 50.

Chaturvedi, A.N. 1993. Silviculture. In: *Neem Research and Development*. N.S. Randhava and B.S. Parmar (eds.), Publ. No. 3. Society of Pesticide Science, India, pp. 38–48.

Dogra, P.D. and R.C. Thapliyal. 1993. Gene resource and breeding potential. In: *Neem Research and Development*. N.S. Randhava and B.S. Parmar (eds.), Publ. No. 3. Society of Pesticide Science, India, pp. 27–32.

Duthie, J.F. 1903. *Flora of Upper Gangetic Plain and Adjacent Siwalik and Sub-Himalayan Tracts*, Vol. I. Calcutta.

Gill, A.S. and R. Deb Roy. 1993. Agroforestry uses. In: *Neem Research and Development*. N.S. Randhava and B.S. Parmar, (eds.), Publ. No. 3. Society of Pesticide Science, India. pp. 49–98.

Guhabakshi, D.N. 1984. *Flora of Murshidabad District, West Bengal*. India Scientific Publishers, Jodhpur.

Kanjilal, U.N. 1928. *Forest Flora of Chakrata*. Dehra Dun and Saharanpur Forest Division, U.P. Manager of Publications, Delhi.

Ketkar, C.M. 1976. Final technical report—Utilization of neem (*Azadirachta indica*, A. Juss.) and its byproducts. Nana Dengle Sadhana Press, Pune.

Lauridsen, E.B., C. Kanchanaburagura and S. Boonsermsuk. 1991. Neem (*A. indica* A. Juss.) in Thailand. *Forest Genetics Resources Info. FAO*, Rome, 19: 25–33.

Mitra, C.R. 1963. *Neem*. The Indian Central Oil Seeds Commission, Hyderabad.

Morgan, E.D. 1981. Strategy in the isolation of insect control substances from plants. *Proc. 1st Int. Neem Conf.*, Rottach-Egern, 1980. GTZ, Eschborn, pp. 43–52.

Percy, G.R. 1972. Studies on the structure of the insect feeding inhibitor, azadirachtin. Ph.D. thesis, University of Keele, UK.

Rembold, H., H. Foster and C.H. Czoppeit. 1987. Structure and biological activity of azadirachtin A and B: *Natural Pesticides from the Neem Tree* (*Azadirachta indica* A. Juss.) *and Other Tropical Plants*. Proc. 3d Int. Neem Conf., Nairobi, 1986. GTZ, Eschborn, pp. 149–169.

Riar, S.S. 1993. Antifertility and other medicinal applications. In: *Neem Research and Development*. N.S. Randhava and B.S. Parmar (eds.), Publ. No. 3. Society of Pesticide Sciences, India. pp. 220–226.

Schmutterer, H. and C.P.W. Zebitz. 1984. Effect of methanolic extract from seeds of single neem tree of African and Asian origin on *Epilachna varivestis* and *Aedes aegypti*. Proc. 2nd Int. Neem Conf. Rauischholzhausen, 1984. GTZ, Eschborn, pp. 83–90.

Siddiqui, S., B.S. Siddiqui, S. Faizi and T. Mahamood. (1986). Studies on the chemical constituents of *Azadirachta indica* A. Juss. (Meliaceae). Part VI. *J. Chem. Soc. Pakistan*, 8(3): 341–347.

Tewari, D.N. 1992. *Monograph on Neem*. International Book Distributors, Dehra Dun.

Troup, R.S. 1921. *Silviculture of Indian Trees*, Vols 3. 1–3. Clarendon Press, Oxford.

Wright, J.W. 1976. *Introduction to Forest Genetics*. Academic Press, New York, San Francisco, London.

Zanno, P.R., E. Miura, K. Nakanishi and D.L. Elder. 1975. Structure of the insect phagorepellent azadirachtin. Application of PRFT/CWD Carbon-13, Nuclear Magnetic Resonance. *J. Amer. Chem. Soc.* 97: 1975–1977.

14

Engineering Properties of Neem Nut

P.T. PALANISAMY, R. VISWANATHAN, S.S. SIVAKUMAR, L. GOTHANDAPANI AND K.R. SWAMINATHAN

Department of Agriculture Processing, College of Agricultural Engineering, Tamil Nadu Agricultural University, Coimbatore 641003, India

ABSTRACT

Neem oil and neem powder are being considered as effective substitutes for chemical pesticides. Research is going on in India and several other countries on use of neem products as plant protection agents. The properties of the neem nut—kernel-shell ratio, length, diameter, weight of 1,000 nuts, angle of repose, porosity, bulk density, particle density, coefficient of friction—were determined for various moisture levels. They may be useful in design of neem processing equipment and other applications.

The kernel-shell ratio decreased with increased moisture content. The length and diameter at stem end of the nut and the weight of 1,000 nuts increased with moisture content. The angle of repose increased linearly and the hardness along the length and diameter at stem decreased non-linearly with moisture content. Porosity, bulk density, and particle density decreased linearly with moisture content. The coefficient of friction increased with moisture content against all surfaces studied. The maximum friction was offered by plywood, followed by mild steel, galvanized iron, and glass.

INTRODUCTION

Neem (*Azadirachta indica*) is an evergreen tree found in Asia and tropical regions of Australia and Africa. The timber from neem is durable and not usually attacked by insects because of the oil content of the wood (Mitra, 1963). However, this tree is grown more often for shade and ornament than for its timber. The leaves and nuts are extensively used for medicinal purpose. The neem nut or seed is obtained by rubbing the ovoid-shaped, yellow to brown, bluntly pointed, smooth-druped neem fruit, which possesses a very scanty pulp, against a coarse surface. When the nut dries it is hard and brittle and can be decorticated or shelled. The greenish-brown kernels are obtained from the nut by shelling, and constitute, about 45 per cent of the nut (Mitra, 1963).

In the process of extracting neem oil the nuts undergo a series of unit operations. A knowledge of physical properties and their dependence on the moisture content is useful for the design development of methods and equipment. The studies done so far on the physical properties of neem nut are limited. This study was carried out to determine the moisture-dependent physical properties of neem nut in the moisture range of 7.67 to 20.89 per cent (wet basis).

REVIEW OF LITERATURE

The major moisture-dependent physical properties of biological materials are shape and size, hardness, bulk density, angle of repose, porosity, mass of the seeds, and friction against various surfaces. (Richey, 1961; Mohsenin, 1980). These properties have been studied for rough rice (Wratten et al., 1975), canola meal (Kulkelko et al., 1988), oilseeds (Muir and Sinha, 1988), gram (Datta et al., 1988), pigeon-pea (Shephard and Bhardwaj, 1986), popcorn (Hangh et al., 1976), faba bean (Frasser et al., 1987), and wheat (Nelson, 1980). The bulk density and particle density of these crops decreased with moisture content (Brown, 1962; Brusewitz, 1975; Day, 1964; Gustafson and Hall, 1972; Nelson, 1980). Day (1964) developed a device for measuring the voids in biological materials. He found that porosity decreased linearly with moisture content.

The earlier researchers also found that angle of repose of the biological materials increased with increasing moisture content. Plywood, wood, and galvanized steel sheet were the commonly used surfaces in friction studies for canola meal (Kulkelko et al., 1988), gram (Datta et al., 1988), and pigeon-pea (Shephard and Bhardwaj, 1986), and the coefficient of friction depended on the moisture content. The static coefficient of friction against all surfaces increased with moisture content. The smooth surface offered less friction than the rough surface.

MATERIALS AND METHODS

Apparatus and Procedure

Dimensions and Hardness

The length and the diameter at the stem-end of the neem nut were measured by a travelling microscope with 0.01 mm graduations. Five nuts were randomly selected and the mean value was recorded.

Hardness was measured using a hardness tester with a flat loading pointer. The nuts were loaded along the length and diameter at the stem-end at various moisture contents. The compressive force corresponding to the failure of the nut was taken as the hardness of the nut and expressed in Newton.

Nut Weight and Kernel-shell Mass Ratio

One thousand neem nuts were picked at random and weighed by electronic balance with 0.01 g graduations. Manual peeling was carried out to obtain 50 g of nuts and from the weights of the kernel and shell, their ratios were calculated for different samples at selected moisture contents.

Angle of Repose

The filling angle of repose of the nuts was determined from the diameter and height of a heap on a circular plate. In the apparatus, the nuts were allowed to fall on a circular plate of 20 cm diameter from a height of 15 cm and a natural heap was formed. With the help of the pointer and scale provided, the height of the heap was noted directly from the scale with 1 mm accuracy. The angle of repose was calculated from the diameter and height of the heap.

Porosity

An experimental apparatus similar to the one developed by Day (1964) was used to determine the porosity of neem nut. The apparatus consisted of two glass bottles of equal volume (2,000 ml), water-filled U-tube manometer, and glass tubes with valves. With the neem nut in grain tank filled gently without compaction, air was supplied to air tank keeping two valves closed and one open. After achieving appropriate pressure of air in the air tank, the last valve was closed. This was noted as the initial pressure (p_1). One closed valve was opened allowing the air to enter the grain tank and fill the pore spaces and the new pressure (p_2) was noted. The per cent porosity (P_t) was calculated using the formula

$$P_t = \frac{p_1 - p_2}{p_2} \times 100 \tag{1}$$

Bulk Density and Particle Density

Bulk density was determined by filling a circular container of known volume (124.68×10^{-6} m^3) with neem nuts and weighing the content. The bulk density (*b*) was calculated from the mass and volume.

The particle density (true density) was calculated from the experimental values of bulk density and porosity at corresponding moisture contents with the following relation:

$$\rho_p = \frac{\rho_b}{1 - P_t} \tag{2}$$

Friction

The experimental apparatus used in the friction studies consisted of a frictionless pulley fitted on a frame, a bottomless cylindrical grain container, loading pan, and test surfaces. The grain container placed on the test surface was filled with a known quantity of neem nut and weights were added to the loading pan until the container began to slide. The mass of the nut and the added weights comprise the normal force (N_f) and frictional force (F) respectively expressed in N. The coefficient of static friction (μ) was calculated as the following ratio:

$$\mu = \frac{F}{N_f} \tag{3}$$

The experiment was performed at different moisture contents of neem nut on various test surfaces such as plywood, galvanized iron sheet, mild steel, and glass.

Preparation of Samples

Dried neem nuts obtained from the local market at 11.6 per cent (w.b.) moisture content were used in the experiments. The experiments were conducted in the moisture range of 7.67 to 20.89 per cent (w.b.). The samples of lower moisture contents were obtained by drying in a mechanically ventilated oven at 45°C with a 15 mm nut layer. The desired moisture content was obtained from the weight of the sample calculated as below (Bisht, 1986).

$$B = \frac{A(100 - a)}{100 - b} \quad (4)$$

The samples of higher moisture contents were obtained by adding distilled water as calculated from the following relation (Bisht, 1986):

$$Q = \frac{A(b - a)}{100 - b} \quad (5)$$

where A is initial weight of the sample, *Kg*; B is final weight of the sample after drying, *Kg*; a is initial moisture content of sample, % (w.b.); b is final (desired) moisture content of sample, % (w.b.); and Q is amount of water to be added, *Kg*.

The samples of lower and higher moisture content were stored in airtight container and polyethylene bags at 5°C in a refrigerator respectively and used in the experiments. The moisture content of the samples was determined (AOCC, 1976) by drying the sample at 130 ± 1°C for 1 hr.

RESULTS AND DISCUSSION

Dimensions and Hardness

The shape of the neem nut can be regarded as ovate, that is, egg shape, broader at the stem end. The length and diameter at the stem end measured at the moisture content range of 7 to 20 per cent (w.b.) are 14.83 mm and 7.12 mm respectively (Table 1).

The effect of moisture content on hardness of the neem nut is given in Fig. 1. The hardness of the neem nut along the length and diameter at stem end decreased linearly for the moisture content range of 7.67 to 20.89 per cent (w.b.). The decrease in the

Table 1. Physical dimensions and weight of neem nut at various moisture contents

Moisture content % (w.b.)	Length (mm)	Diameter at stem end (mm)	Weight 1000 nuts (g)
7.67	12.87	6.86	207.50
11.60	15.09	7.24	219.83
16.12	15.10	7.24	233.00
20.89	16.20	8.52	239.91

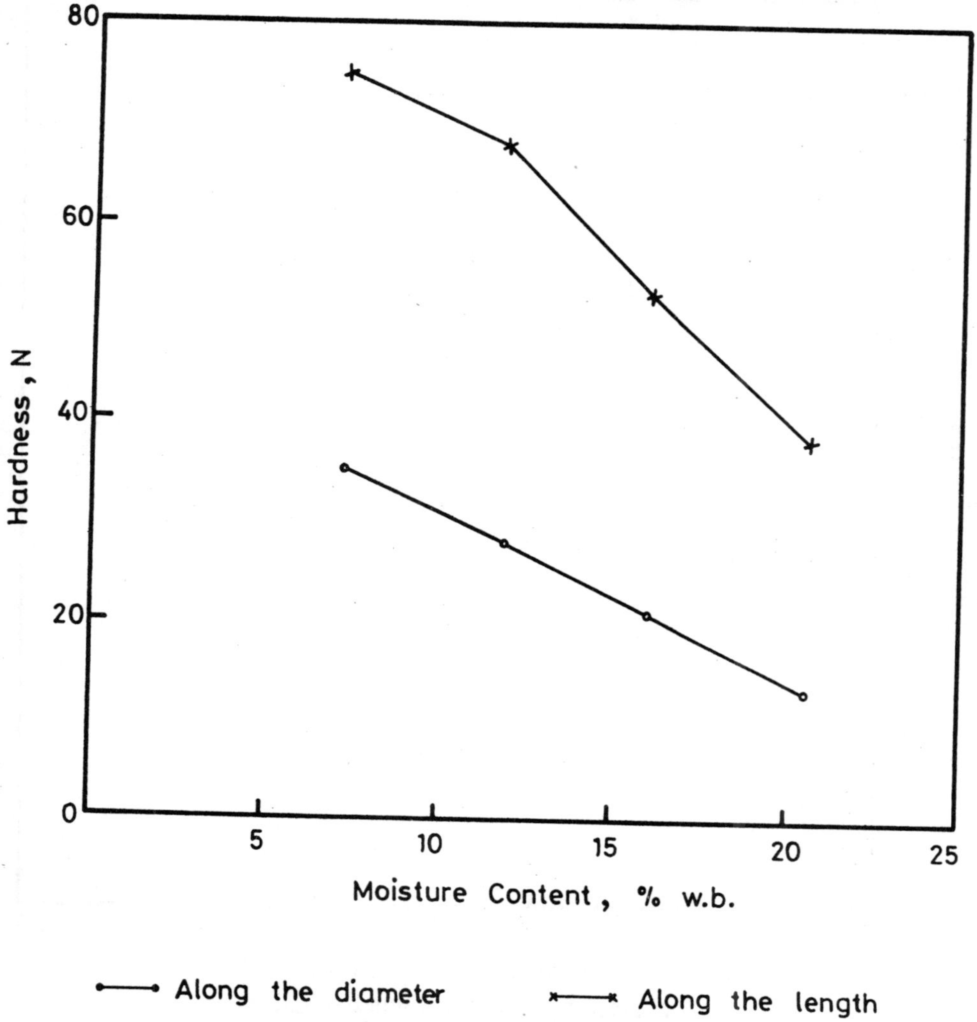

Fig. 1. Effect of moisture content of neem nut on hardness

hardness may be due to the nut's becoming softer at higher moisture contents. The regression equations relating these properties and moisture content are

$$\text{hardness (along length)}, f_l = 50 - 1.75M_n \ (r = 0.98) \quad (6)$$

$$\text{hardness (at stem end)}, f_2 = 97 - 2.72\ M_n \ (r = 0.99) \quad (7)$$

where M_n is the moisture content of the neem nut, per cent, w.b. (Table 2).

Nut Weight and Kernel-Shell Mass Ratio

The mass of 1,000 nuts determined at different moisture contents is given in Fig. 2. It shows a linear increase with moisture content. An increase in the 1,000 nuts mass was

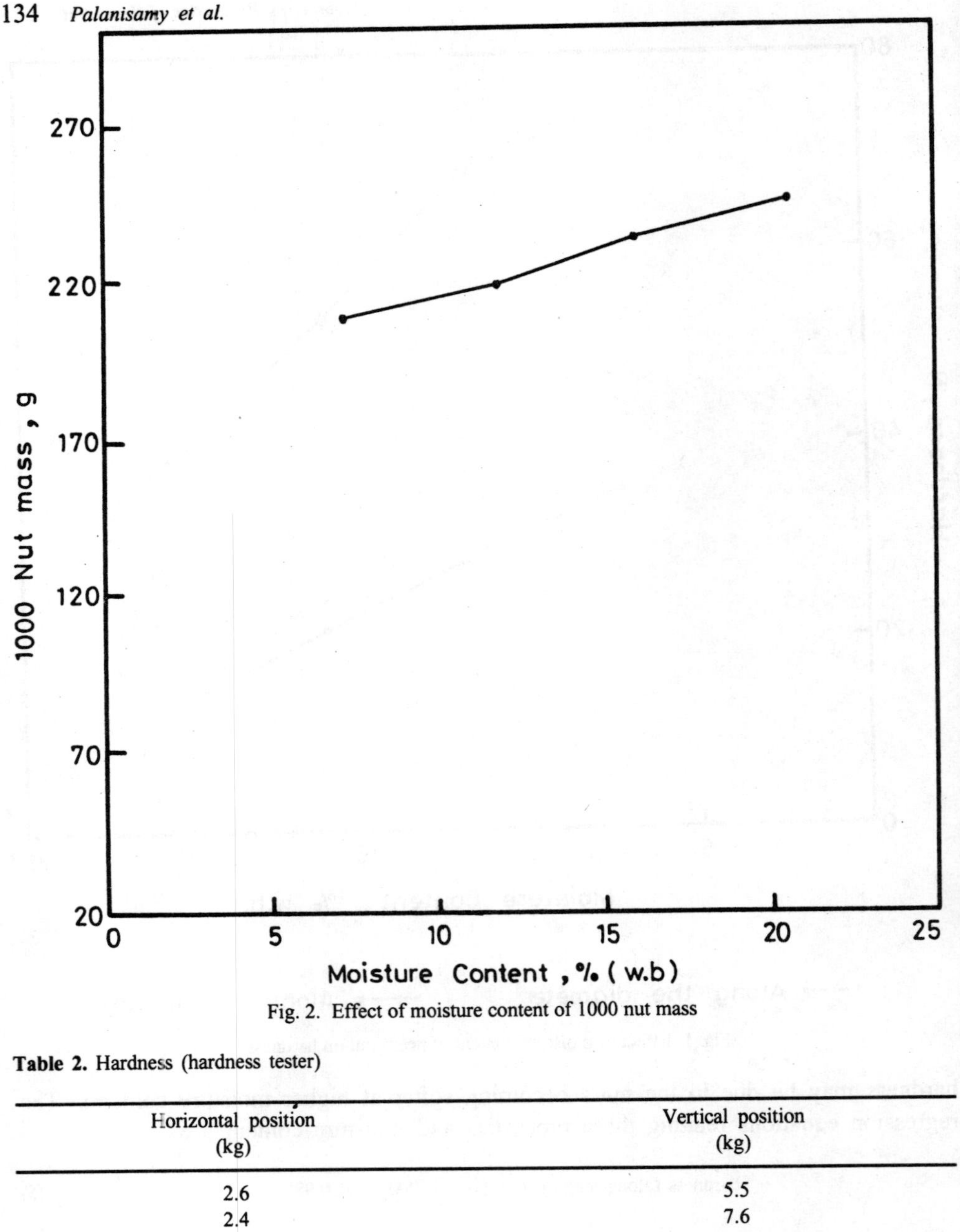

Fig. 2. Effect of moisture content of 1000 nut mass

Table 2. Hardness (hardness tester)

Horizontal position (kg)	Vertical position (kg)
2.6	5.5
2.4	7.6
3.2	3.4
4.2	2.7
3.1	5.8
1.8	7.8

Average: Horizontal Position = 2.871 kg; vertical position = 5.56 kg.

found over that of pigeon-pea and gram with increased moisture content (Datta et al., 1988; Hangh et al., 1976; Shephard and Bhardwaj, 1986). The relationship to quantify the mass of 1,000 nuts (w) at any moisture is as follows:

$$w = 184 + 2.9\ M_n \qquad (r = 0.99) \tag{8}$$

The kernel-shell mass ratio and the moisture content of the neem kernel and shell at various moisture contents of neem nut are given in Table 3. For a particular moisture content of neem nut, the moisture contents of kernel and shell are different. The moisture content of both kernel and shell increased with the moisture content of the nut and followed the following regression equation:

$$M_n = 0.82\ M_n + 0.66 \quad (r = 0.998) \tag{9}$$

$$M_s = 0.84\ M_n + 1.45 \quad (r = 0.994) \tag{10}$$

where M_n, M_k, and M_s are the moisture contents (% w.b.) of neem nut, kernel, and shell respectively.

The content of the kernel in the neem nut increased from 55.83 to 60.5 per cent with the increase in moisture content from 7.17 to 20.89 per cent w.b. and the shell content of the nut decreased from 44.17 to 39.5 per cent. The following regression equations are proposed from the values presented in Table 3:

$$k = 0.33\ M_n + 54 \qquad (r = 0.996) \tag{11}$$

$$s = -\ 0.33\ M_n + 46 \qquad (r = 0.996) \tag{12}$$

where k and s are the per cent content of kernel and shell respectively in the neem nut.

Table 3. Kernel-shell ratio and moisture content of kernel and shell at various moisture content of neem nut

Moisture content % (w.b.)			Kernel-shell ratio
Nut	Kernel	Shell	
7.67	7.16	8.06	1:0.791
11.60	9.87	10.95	1:0.786
16.42	13.75	15.13	1:0.703
20.89	17.88	19.06	1:0.658

The values reported are average of replications.

Angle of Repose and Porosity

The effect of moisture content on angle of repose of neem nut is presented in Fig. 3. The filling angle of repose is found to increase with moisture content linearly from 30.75° to 37.65° for the moisture range of 7.67 to 20.89 per cent (w.b.) and all the biological materials exhibit an increase in angle of repose with moisture content

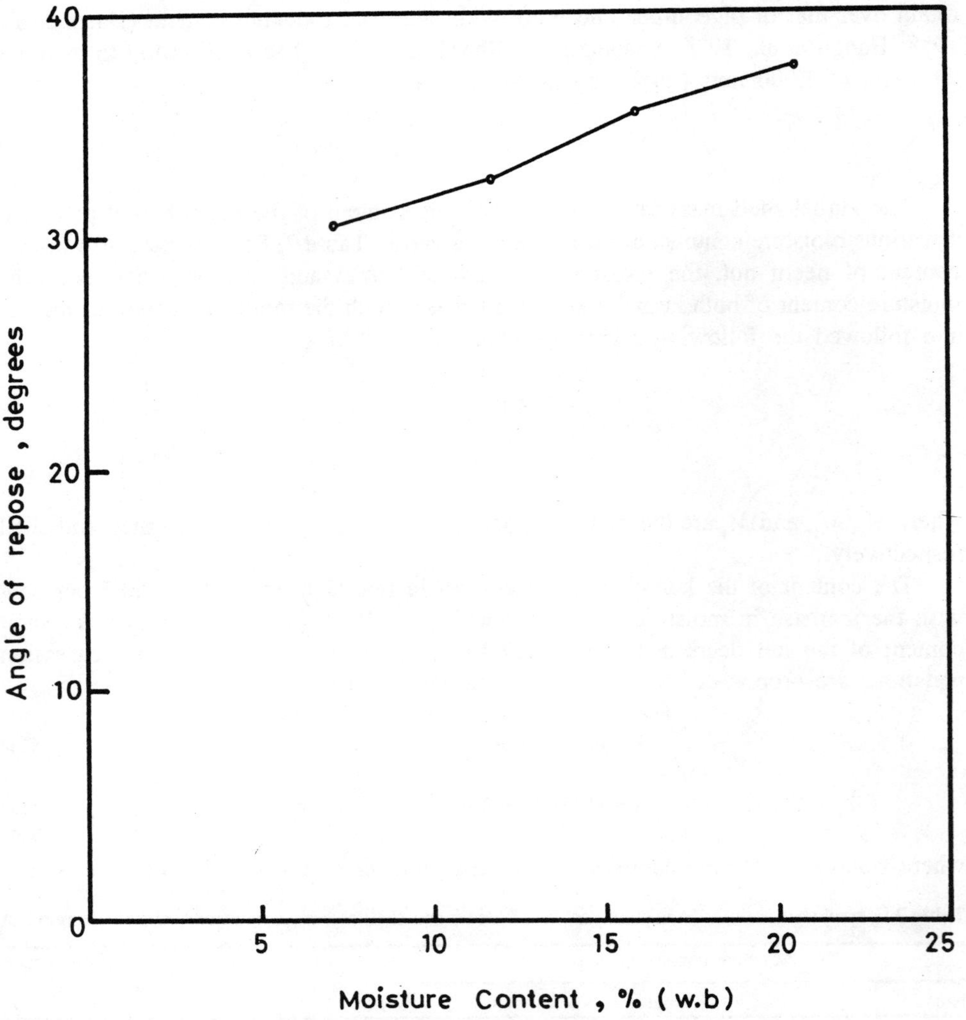

Fig. 3. Effect of moisture content of neem nut on angle of repose

Table 4. Angle of Repose

Diam. (d) (cm)	height (h) (cm)	L = $\tan^{-1} 2h/d$
20	7.4	36.50
15	5	33.70
10	3.4	34.2
		Average = 34.8

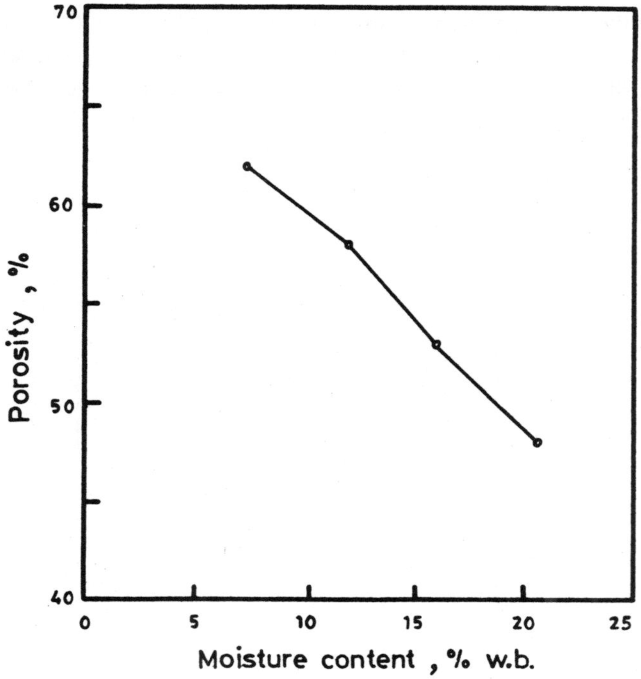

Fig. 4. Effect of moisture content of neem nut on porosity

(Mohsenin, 1980). The regression equation relating the angle of repose and moisture content of neem nut is

$$\theta = 0.47\, M_n + 27 \quad (r = 0.963) \tag{13}$$

where θ and M_n are the angle of repose (degrees) and moisture content per cent (w.b.) respectively (Table 4).

The porosity of the neem nut and its relation with moisture content is given in Fig. 4. Porosity of the neem nut also decreased with the moisture content, as in the other biological materials. Porosity decreased linearly for the moisture range of 7.67 to 20.89 per cent (w.b.) The regression equation relating the moisture content and the porosity of the neem nut is

$$P_t = 69 - M_n \qquad (r = 0.997) \tag{14}$$

where P_t is the porosity in percentage.

Bulk Density and Particle Density

Figure 5 shows the effect of moisture content on bulk density and particle density of neem nut. The bulk density and particle density decreased linearly for the moisture range of 7.67 to 20.89 per cent (w.b.). This decrease in bulk density and particle density of neem nut with increase in moisture content is similar to that found for other biological materials (Brown, 1962; Brusewitz, 1975; Day, 1964; Gustafson and Hall, 1972; Nelson, 1980). The regression equations proposed to quantify these properties are as follows (Table 5):

$$\rho_b = 569 - 2\,M_n \qquad (r = 0.999) \tag{15}$$

$$\rho_p = 1705 - 33\,M_n \qquad (r = 0.997) \tag{16}$$

where ρb and ρp are the bulk density and particle density of neem nut respectively.

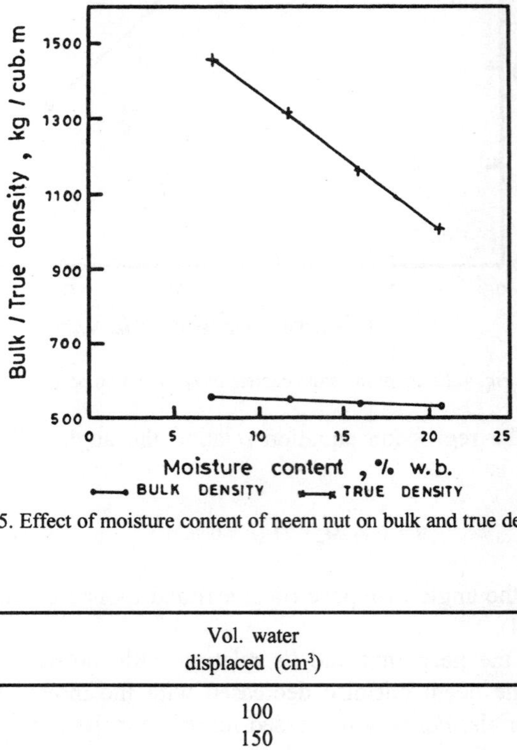

Fig. 5. Effect of moisture content of neem nut on bulk and true density

Table 5. Bulk Density

Weight of nut (g)	Vol. water displaced (cm^3)	Bulk density (kg/m^3)
75	100	0.75
100	150	0.66
90	120	0.75

Average = 0.72

Coefficient of Friction

The effect of moisture content of neem nut on friction against various test surfaces—glass, galvanized iron, mild steel, and plywood—is given in Fig. 6. The friction coefficient increased linearly with moisture content for all contact surfaces. The maximum friction was offered by plywood at all moisture contents, followed by mild steel and galvanized iron. Glass offered the least friction. Plywood also offered the maximum friction for rapeseed, pigeon-pea, and gram and the friction coefficient increased with

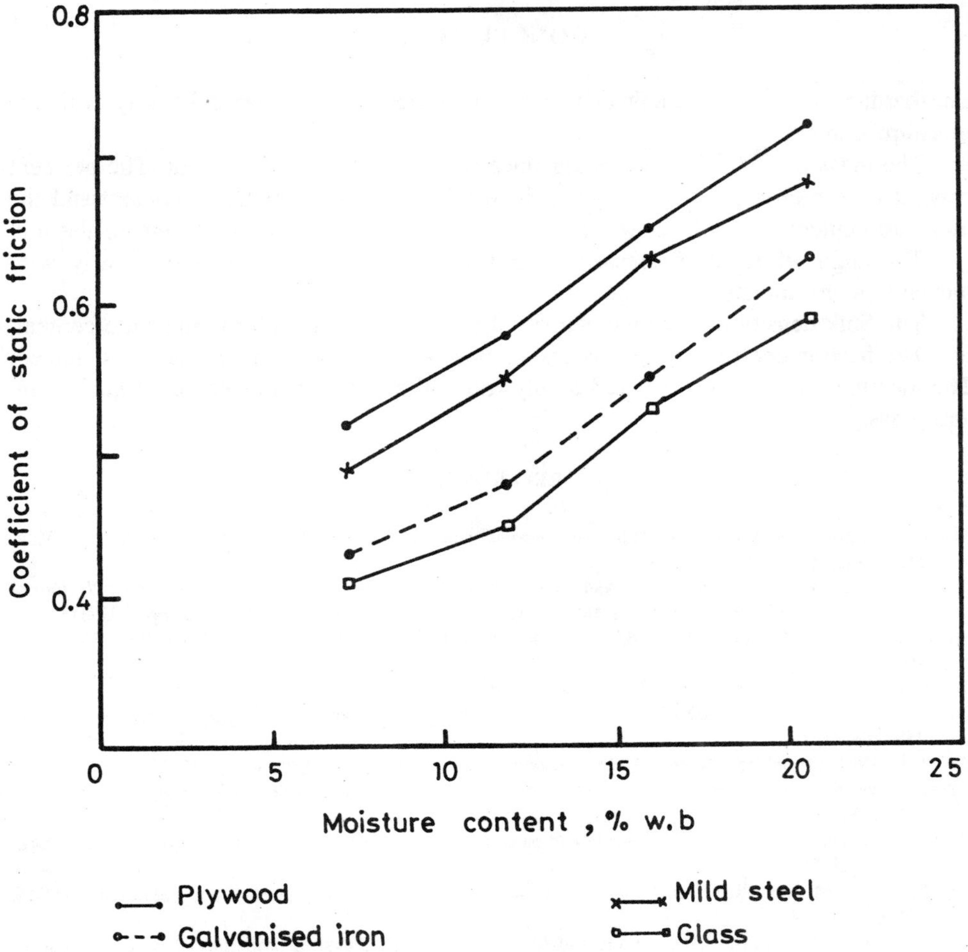

Fig. 6. Effect of moisture content of neem nut on friction against various surface

the moisture content (Datta et al., 1988; Kulkelko et al., 1988; Shephard and Bhardwaj, 1986).

The increased friction coefficient at higher moisture content may be due to the water present in the nut offering a cohesive force on the surface of contact. To quantify the friction coefficient of the nut at various moisture content on various surfaces, the following regression equations are proposed:

Glass, $\mu_g = 0.012\ M_n + 0.33$ $(r = 0.994)$ (17)

Galvanized iron, $\mu_{gi} = 0.015\ M_n + 0.32$ $(r = 0.990)$ (18)

Mild steel, $\mu_{ms} = 0.014\ M_n + 0.39$ $(r = 0.998)$ (19)

Plywood, $\mu_{pw} = 0.016\ M_n + 0.38$ $(r = 0.999)$ (20)

CONCLUSIONS

The hardness along the length and diameter at stem-end decreased linearly with the moisture content.

The mass of 1,000 nuts increased linearly with the moisture content. The per cent content of kernel in the neem nut increased with increase in moisture content and the moisture contents of both kernel and shell increased with moisture content of the nut.

The angle of repose increased linearly and the porosity decreased linearly with increase in the moisture content.

The bulk density and particle density decreased linearly with the moisture content.

The friction coefficient increased with moisture content against all surfaces studied. The maximum friction was offered by plywood followed by mild steel, galvanized iron, and glass.

REFERENCES

AOCC. *Approved Methods.* (1976). *Moisture Determination.* Association of Cereal Chemists, Inc., St. Paul, Minnesota, pp. 44–45A.

Bisht, B.S. 1986. Grain moisture meters and their calibration. *Proc. Course on Instrumentation for Testing of Agricultural Machinery.* Central Institute of Agricultural Engineering, Bhopal, India. pp. 295–311.

Brown, D.A. 1962. Variation of bulk density of cereals with moisture content. *Agric. Engg. Res.*, 7(4): 288–290.

Brusewitz, G.H. 1975. Density of rewetted high moisture grains. *Trans. ASAE*, 18(8): 935–938.

Datta, S.K., V.K. Nema and R.K. Bhardwaj. 1988. Physical properties of gram. *J. Agric. Engg. Res.*, 39(4): 259–268.

Day, C.L. 1964. Device for measuring voids in porous materials. *Agric. Engg.*, 45(1): 36–37.

Frasser, B.M., S.S. Verma and W.E. Muir. 1987. Some physical properties of faba bean. *J. Agric. Engg. Res.*, 23(3):160–163.

Gustafson, R.J. and G.E. Hall. 1972. Density and porosity changes of shelled corn during drying. *Trans. ASAE*, 15(4): 523–525.

Hangh, C.G., R.M. Lien, R.E. Hares and R.B. Ashman. 1976. Physical properties of pop corn. *Trans. ASAE*, 19(4): 620–624.

Kulkelko, D.A., D.S. Jayas, N.D.G. White and M.G. Britton. 1988. Physical properties of canola (rapeseed) meal. *Can. Agric. Engg.*, 3(1): 61–64.

Mitra, C.R. 1963. *Neem.* The Indian Central Oil Seeds Committee, Bombay.

Mohsenin, N.N. 1980 *Physical Properties of Plant and Animal Materials.* Gordon and Breach Science Publishers, New York.

Muir, W.E. and R.N. Sinha. 1988. Physical properties of cereal and oilseed cultivars grown in Western Canada. *Can. Agric. Engg.*, 30(1): 51–55.

Nelson, S.O. 1980. Moisture-dependent kernel and bulk density relationships for wheat and corn. *Trans. ASAE* 23(1): 139–143.

Richey, C.B. (ed.) 1961. *Agricultural Engineers' Handbook.* McGraw-Hill, New York.

Shephard, H. and R.K. Bhardwaj. 1986. Moisture-dependent physical properties of pigeon pea. *Agric. Engg. Res.*, 35(4): 227–234.

Wratten, F.T., W.D. Poole, J.L. Chesness, S. Bal and V. Rama Rao. 1975. Physical and thermal properties of rough rice. *Trans. ASAE*, 18(2): 370–375.

15

Location of Neem Triterpenoids and Other Secretory Structures

P. DAYANANDAN, A. STEPHEN AND B. MURUGANANDAM

Department of Botany, Madras Christian College, Tambaram
Madras 600 059, India

ABSTRACT

Neem triterpenoids have been isolated from a variety of tissues but the cell types that accumulate the limonoids are not known. Identification of such secretory cells would help in rapid screening of clones rich in limonoids and formulating treatments that enhance yield. Based on microscopic and histochemical studies we have established the following secretory structures in neem plant: Secretory cells (SC) that accumulate triterpenoids; glandular trichomes; gum ducts; laticifers and extrafloral nectaries. Triterpenoids containing SC were identified on the basis of strong positive reaction in free-hand sections with the following reagents employed in chromatographic studies: iodine vapour, Carr-Price reagent (antimony III chloride + chloroform), and Liebermann-Burchard reagent (sulphuric acid + acetic anhydride). The SCs are found in the cortex and pith regions of stems, petioles, and pedicels, in the cortex of roots, between palisade and spongy tissues in leaves, and abundantly in cotyledons. They are oval and contain many pale orange-brown vesicles whose size appears to be dependent upon age as they become larger in older cotyledons. The contents of SCs are readily extracted with alcohol but not with non-polar solvents such as benzene and hexane. The density of SC distribution is related to intensity of bitterness of the organs. With the identification of the SC it should be possible to study the dynamics of synthesis, storage, interconversions, and transport when specific histochemical reagents for different triterpenoids become available. Laticifers develop only in the pericarp. They are articulate, contain milky white latex, and persist with opaque contents even in ripe fruits. Laticifers may be modified SCs and hence may contain some triterpenoids. Some uniseriate hairs secrete an unknown substance. Multicellular trichomes secrete resin that protects young apices. Secretions of extrafloral nectaries attract ants and gum ducts develop due to injury and stress.

INTRODUCTION

The neem tree (*Azadirachta indica* A. Juss.) is known to contain several classes of useful chemicals such as carbohydrates, proteins, sulphurous compounds, phenolics, aromatic esters, steroids, and terpenoids (Van der Nat et al., 1991). Terpenoids present in neem include diterpenes and tetra-, penta-, and hexanortriterpenoids (Govindachari, 1992). Although triterpenoids have been isolated from a variety of organs such as green and dry leaves, stem and root bark, wood, flowers, fruits, and seeds (Van der Nat et al., 1991; Tewari, 1992), the precise location of these chemicals is not known. We report here the distribution of secretory cells (SCs) in different parts of the plants which are presumably the sites of storage of nearly 100 known triterpenoids. In addition we describe the presence of other secretory structures in neem: glandular trichomes, extrafloral nectaries, laticifers, and gum ducts.

MATERIALS AND METHODS

Seeds, seedlings of different ages, and mature plants collected or grown in the department were surveyed for the presence of external and internal secretory structures. Various stains used for staining free-hand and microtome cut sections were safranin, toluidine blue O, ruthenium red, Sudan III, phloroglucianol, and I_2KI for routine examination. No specific histochemical reaction is known for the localization of any of the triterpenoids. However, terpenoids in general can be detected using iodine vapour, Carr-Price, or Liebermann-Burchard reagents employed in thin-layer chromatography of terpenoids (Harborne, 1993; Jork et al., 1990). Free-hand sections attached to cover-slips were exposed for 10 min. to iodine vapour in a closed vial. I_2KI, in addition to staining starch, also stains the contents of secretory cells. Carr-Price reagent was prepared fresh by dissolving 2 g antimony (III) chloride in 50 ml of chloroform. Liebermann-Burchard reagent was prepared fresh by adding 1 ml of sulphuric acid to 9 ml of acetic anhydride. Sections obtained from fresh materials were mounted between a slide and cover-slip in a few drops of either one of the reagents and examined

---

Figs. 1–10. Diagrammatic representation of secretory structures in neem. Fig. 1. Longsection of a multicellular glandular trichome from abaxial surface of a young leaflet. The gland secretes resins. Scale = 12 μm. Fig. 2. Three-celled and single-celled glandular trichomes from the adaxial surface of cotyledonary petiole compared with non-glandular trichomes present on the shoot system. Scale = 50 μm. Fig. 3 Young leaflet with three extrafloral nectaries over the midrib on adaxial surface and numerous resin-secreting multicellular glands on the abaxial surface. Scale = 4 mm. Fig. 4. Two large and several smaller extrafloral nectaries on the petiole. Scale = 5 mm. Fig. 5. Sectional view of an extrafloral nectary. Scale = 30 μm. Fig. 6. Transverse section of a cotyledon from a 10-day-old seedling. Secretory cells are distributed throughout the cotyledon. They are more isodiametric in the peripheral regions and elongated between the adaxial surface and the vascular bundles. Scale = 1 mm. Fig. 7. A single SC from the cotyledon seen in unstained sections. Numerous vesicles are present in the secretory cell that is surrounded by epithelium-like adjacent cells. Scale = 50 μm. Fig. 7a. Periclinal cell wall formation in cells adjoining the SC in 10-day-old excised hypocotyl segments. The SC is distorted and its contents are not vesicular. Scale = 50 μm. Fig. 8. Longitudinal section of a hypocotyl with SCs arranged in a series and surrounded by epithelium-like cells. Contents are vesicular. Scale = 30 μm. Fig. 9. Transverse section of a leaflet with three SCs located at the interface between palisade and spongy parenchyma cells. Scale = 60 μm. Fig. 10. Transverse section of an ovary soon after fertilization. Laticifers occur in a ring in the middle of the ovary wall. Scale = 400 μm.

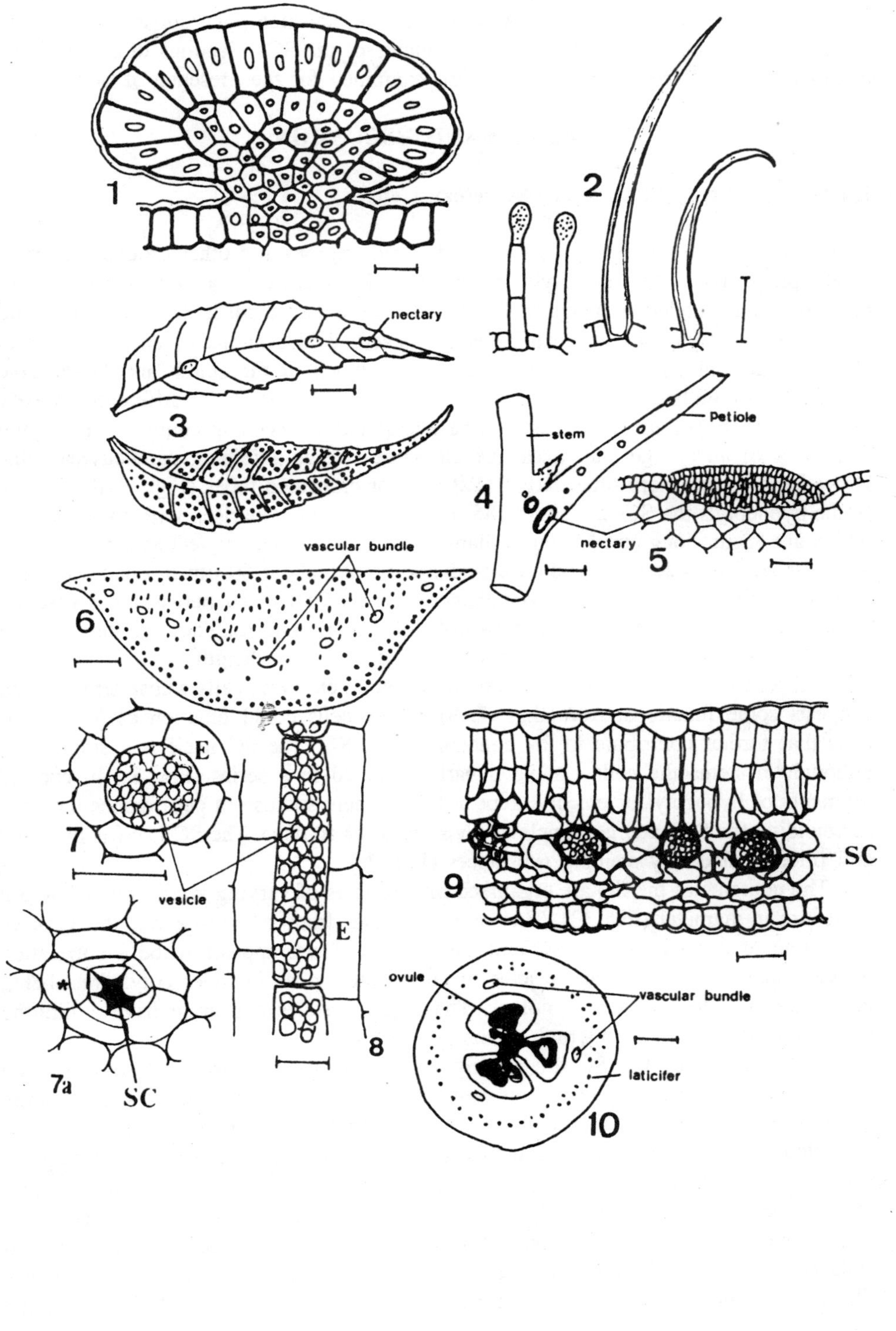
1
2
nectary
3
stem
Petiole
4
nectary
5
vascular bundle
6
E
7
vesicle
E
8
7a
SC
9
SC
E
ovule
vascular bundle
laticifer
10

immediately. Sections were also incubated in ethanol for 5 to 30 min. to extract the triterpenoids before being stained. A series of polar and non-polar solvents were used to observe their capacity to dissolve the contents of the SCs. Sections of callus tissue obtained from different explants were also examined for the presence of SCs.

RESULTS AND DISCUSSION

Internal Secretory Structures—Secretory Cells

Secretory cells are present in the cortical and pith regions and other ground tissues of shoot apices, hypocotyls, epicotyls, young and old stems, roots, petioles, leaves, staminal tubes, and cotyledons (Figs. 6–8, 11–13). The SCs are particularly abundant and prominent in cotyledons and can be easily observed in sections of cotyledons from about 10-day-old seedlings. In these cotyledons much of the starch and lipids have already been mobilized, affording a clearer view of the SC. Each SC differentiates early in the ontogeny of different tissues and accumulates vesicular contents of varying intensities of an orange hue. It can be the same size as adjoining parenchyma cells. However, SCs are more often larger oval cells or elongated tubes (Figs. 7–9; 12, 13). In the hypocotyls the SCs are about 185 μm × 30 μm in comparison with adjacent cells which are 85 μm long and 35 μm in diameter (Fig. 8). In the cotyledons the SCs close to the abaxial hypodermal region are isodiametric and about 60 μm in diameter. Non-secretory cells are about 40 μm in diameter. The SCs located between the vascular bundles and the abaxial epidermis tend to be elongated (Fig. 6). They are three to five times longer than the adjacent parenchyma cells of 50 μm length.

The SCs are often surrounded by six or seven parenchyma cells whose arrangement suggests some functional role (Figs. 7, 9). These cells could function as 'epithelial' cells that further contribute to the contents of the SC. The SC itself is a single cell, although in stems the cells can be linearly arranged in a series (Fig. 8). In excised segments of hypocotyl or epicotyl kept in 0.1 per cent sucrose in petri dishes the cells surrounding the SC undergo periclinal divisions after 10 days. The SC itself is distorted and filled with darkly staining substances (Fig. 7a).

The contents of the SC are organized into vesicles of varying sizes from a few μm to 50 μm in diameter (Fig. 13). Their number varies from a few to about 100 vesicles depending on the size of the SC and the vesicles. There appears to be a correlation between age and size of the vesicles in cotyledons, those in older cotyledons being larger and a deeper orange in colour. The larger vesicles appear to have a limiting membrane with granular internal contents.

The contents of the SC are readily stained by iodine and the Carr-Price and Liebermann-Burchard reagents. In iodine-treated cells the contents lose vesicular nature and turn deep orange and then dark brown. Varying shades of pink, green, and brown or yellow to brown are obtained with Carr-Price and Liebermann-Burchard reagents (Fig. 12). Pretreatment of the sections with the following reagents results in the complete extraction of the contents of the SC and negative reaction with the above reagents: ethanol, methanol, butanol, 2-propanol, tertiary butyl alcohol, and acetone. The contents are not extracted by hexane, petroleum ether, chloroform, xylene, and benzene. The vesicles remain intact in water but are slowly extracted in 1 N HCl, NaOH, and

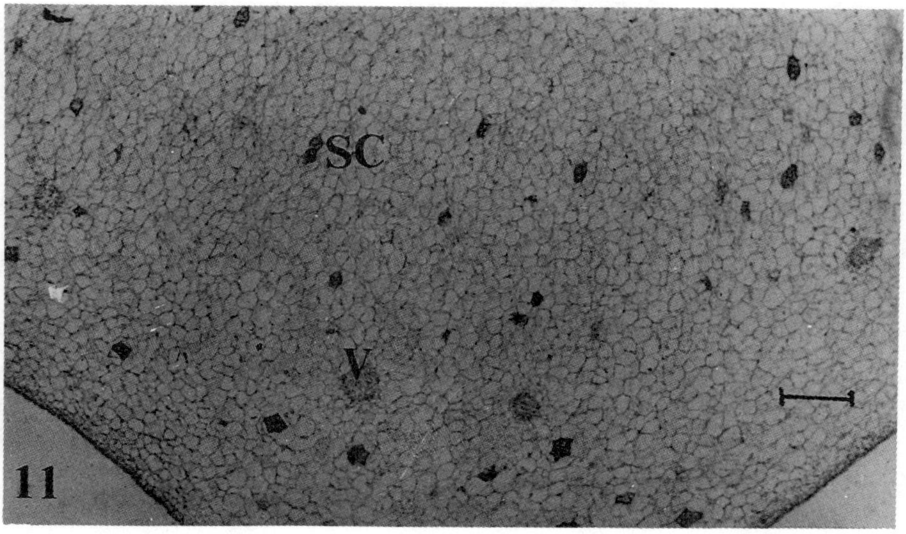

Fig. 11. Transverse section of a portion of cotyledon from an unripe fruit. Secretory cells (SC) are just differentiating, Vascular traces (v) are also seen. Scale = 100 μm

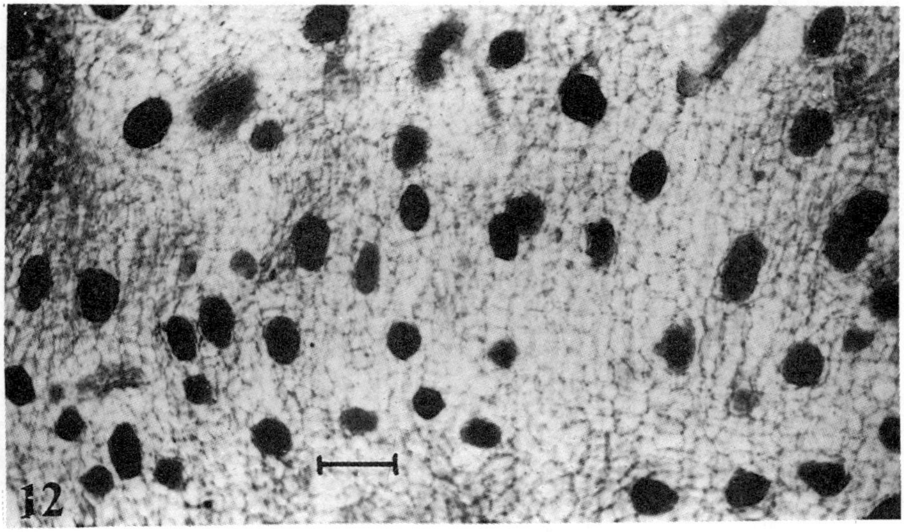

Fig. 12. Transverse section of a portion of a cotyledon from a 10-day-old seedling stained with Carr-Price reagent. Darkly stained spots indicate the contents of SCs. Scale = 100 μm

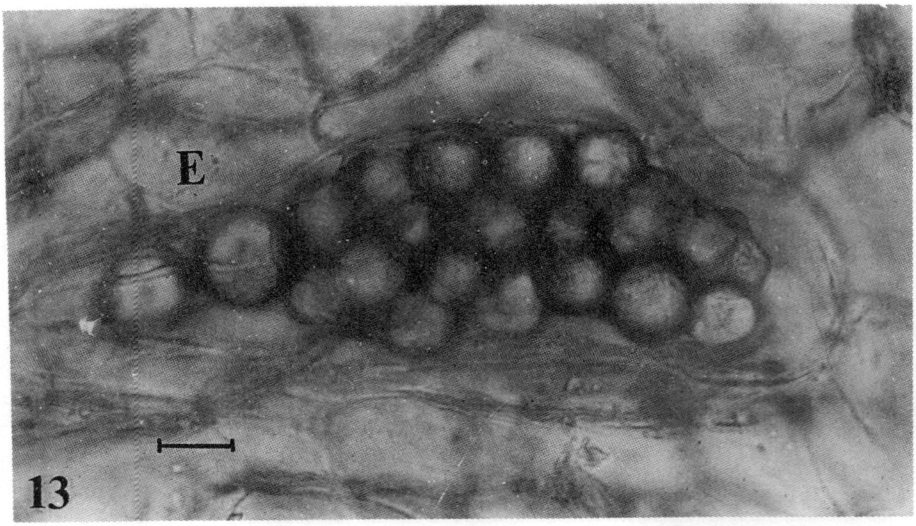

Fig. 13. A single SC from a cotyledon without any staining. Contents appear as vesicles. The SC is surrounded by epithelium-like cells (E). Scale = 20 µm.

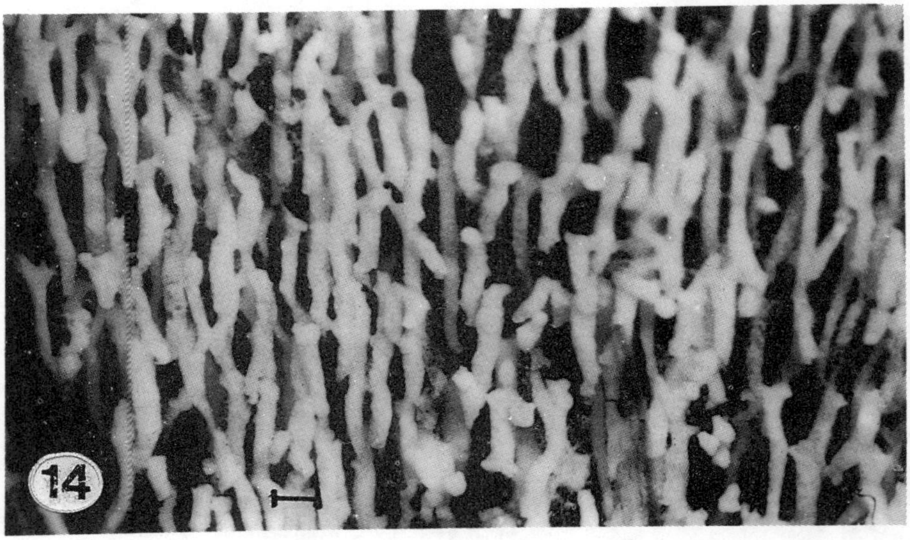

Fig. 14. Articulated laticifers seen in surface view in the pericarp of a ripe fruit. A segment of a pericarp was treated with 1 per cent NaOH for 3 hrs. This resulted in the clearing of cellular materials, except the contents of the laticifers. Scale = 200 µm

acetic acid. Secretory cells are difficult to locate in microtome sections since the chemicals used in fixation and tissue processing readily dissolve their contents. It is interesting to note that ethanol is the preferred solvent for the extraction of triterpenoids from neem and that hexane, petroleum, and others are used to remove the lipids during such extraction procedures.

It is highly likely that the SCs are the major sites of storage of the nearly 100 known triterpenoids. The sites of synthesis may include the SC as well as the adjacent epithelium-like cells. The contents of these cells are stained with reagents that are known to react with terpenoids and the contents are readily extracted with solvents known to dissolve them. The distribution density of SCs is also correlated with bitterness of the organs and yields which have been reported in the literature. Thus, seed kernels, i.e., the cotyledons, are known to yield as much as 4.8 per cent triterpenoids and the nimbidin and azadirachtin contents can be 2 per cent and 0.2 to 0.9 per cent respectively (Tewari, 1992; Vietmeyer, 1992). Interestingly, the cotyledons possess 5–6 per cent SCs.

Seed oil contains up to 2 per cent triterpenoids (Anon., 1985) and this may represent the substances derived from the SC during the crushing procedure. Such correlations suggest the potential of the simple anatomical technique of determining the number of SCs in an organ to gather data on elite clones that may contain higher concentrations of triterpenoids. Seeds from different countries are known to contain differing amounts of azadirachtin (Tewari, 1992; Morgan, 1991) and it will be rewarding to determine the density and size distribution of SCs in seeds from different sources. There are conflicting reports as to the age of the seeds, whether fresh or older, when the seeds yield higher amounts of triterpenoids (Morgan, 1991; Nagasampagi et al., 1991). The morphological changes observed in the contents of the SC during aging could be further investigated to determine the relationship between the SC characteristics and yield. The biogenesis of important neem chemicals such as azadirachtin involves progressive rearrangements and oxidation of a tetracyclic triterpene, tirucallol (Govindachari, 1992). The SC might be expected to contain such starting material as well as other oxidation products with side chain such as meliantriol, and the tetranortriterpenoids including limonoids such as azadirone and nimbin and the unique azadirachtin and its cogeners. Ultrastructural and fractionation studies of the SC could reveal the compartmentalization and timing of biogenesis. Specific histochemical, fluorescence, and immunochemical reactions developed for different triterpenoids could help localize the tissues in which they are present and the time at which they occur in maximum concentrations and mobilization to other sites.

Laticifers

Laticifers in neem are confined to the pericarp of fruits (Figs. 10, 14). Laticifers differentiate as single cells early in the development of the ovary. Each cell is long and cylindrical and may be branched. The end walls of adjacent laticifer cells are in close contact with each other. Thus, the neem laticifers are of the articulated type (Esau, 1965). In unripe fruits the latex is milky white and is under pressure within the laticifers. The latex contains amyloplasts and lipid droplets and several unknown substances. Preliminary analysis (personal communication, Dr. G. Suresh, SPIC Science

Foundation) reveals an absence of triterpenoids. However, the latex is very bitter and might contain a trace amount of triterpenoids. Siddiqui et al. (1978) have reported the isolation of 17-0 hydroxyazadiradione from fresh fruit pulp.

The functions of laticifers in fruits are not known. During fruit ripening the contents of the laticifers remain opaque within the cells. The outer pericarp separates from the seed due to degradation of parenchyma cells of the pericarp around the laticifers. Based on the similarities between the SC and laticifers during the early ontogeny it may be surmised that both cell types share a common origin but differ in function based on their specific locations.

Gum Ducts

Gum ducts develop in mature shoots in response to stress and injury. Groom (1926) has described the development of gum ducts in Meliaceae. Neem gum, known as East India gum, is of commercial importance. It does not appear to contain triterpenoids but is rich in proteins, amino acids, and sugars (Anon., 1985; Tewari, 1992). Older neem trunks also exude a sap rich in amino acids and sugars (Tewari, 1992).

External Secretory Structures—Glandular Trichomes

Younger regions of the shoot and floral organs of neem possess trichomes of three basic types (Figs. 1, 2). The more numerous unicellular hairs do not appear to be secretory in function. Secretory trichomes with a single cell or a row of three cells are present on the adaxial surface of the short petiole of the cotyledons (Fig. 2). The tips of these hairs are enlarged and contain unknown substances.

Multicellular glandular trichomes are found in large numbers at the shoot apex, on young developing petioles, leaflets, and stems (Fig. 3). These glands are confined to the abaxial surface of the leaflets and never occur on the adaxial surface. The glandular trichomes secrete a sticky and tasteless resinous substance. The glands in the shoot apices may secrete such copious amounts of resins that even the glands get submerged in the secretions. Sudan III stains this region a deep red. Glandular trichomes do not develop in older regions. Those in the young regions soon become non-functional, turn colour, and may wither, leaving tiny scars of attachment.

The resin glands appear to offer protection against desiccation and injury by insects. Resin secretions are more abundant during the dry season, and form a thick covering over the dormant buds. Several larvae and nymphs can be seen caught in the sticky resins in shoot apices. Plants growing in the shade develop very few glandular trichomes.

Extrafloral Nectaries

Extrafloral nectaries are present at the base of a petiole where it joins the stem, along the petiole, and in one or a few places over or close to the midrib on the adaxial surface of leaflets (Figs. 3, 4). The nectaries are of the 'flack nectarian' type and secrete droplets of sweet nectar. Ants gather at the nectaries and parade up to the young regions of the shoot The extrafloral nectaries appear to be involved in some association between herbivore and protective ants. A detailed investigation of this association will reveal evolutionary adaptations of neem and its pests.

This survey has revealed the presence of diverse secretory structures involved in the biology of the neem plant during the different stages of growth. Secretory cells containing triterpenoids occur in tissues of young and old plant organs while glandular trichomes are confined to young growing regions. The protective roles of triterpenoids are supplemented by resin glands and extrafloral nectaries. Laticifers are confined to developing fruits. The functional evolutionary and ecological significance of each of these structures deserves further investigation. Recently we have detected the presence of SC in callus tissues derived from the explants of cotyledon, hypocotyls, and leaf discs. This potential of the callus cells to differentiate SCs could be useful in the production of triterpenoids in cultured cells.

ACKNOWLEDGEMENTS

The authors thank the Indian Council of Agricultural Research for financial support through a grant and Dr. S. Chinnamani and Dr. G. Suresh for helpful discussions and assistance.

REFERENCES

Anonymous 1985. *Azadirachta indica* A. Juss. Meliaceae. In: *The Wealth of India*, Raw Materials, Vol. 1A. Publication and Information Directorate, Council of Scientific and Industrial Research (CSIR), Hill Side Road, New Delhi, pp. 504–509.

Esau, K. 1965. *Plant Anatomy*, 2nd. ed. Wiley, New York.

Govindachari, T.R. 1992. Chemical and biochemical investigations on *Azadirachta indica* (the neem tree). *Curr. Sci.*, 63: 117–122.

Groom, P. 1926. Excretory system in the secondary xylem of Meliaceae. *Ann. Bot.*, 40: 631–649.

Harborne, J.B. 1973. *Phytochemical Methods*. Chapman and Hall, London.

Jork, H., W. Funk, W. Fischer and H. Wimmer. 1990. *Thin-layer Chromatography—Reagents and Detection Methods*, Vol. 1A. VCH Verlagsgesellschaft, FRG.

Morgan, E.D. 1991. Strategy in the isolation of insect control substances from plants. In: *Natural Pesticides from the Neem Tree (Azadirachta indica* A. Juss.). *Proc. Ist. Int. Neem Conference*, 16–18 June, 1980, Rottach-Egern, F.R.G., H. Schmutterer, K.S.R. Ascher and H. Rembold, (eds.), pp. 43–52.

Nagasampagi, B.A., S.P. Joshi and S.R. Rojatkar. 1991. Development of pest control agents of botanical origin. *Proc. Nat. Semi. on Sustainable Agricultural Development*. Ed. Sanjoy Sengupta. Voluntary Health Association of India. New Delhi, pp. 63–69.

Siddiqui, S., S. Fuchs, J. Luke and W. Voelter. 1978. Structure eines neuen Naturstoffes aus *Melia azadirachta* Linn.: 17-hydroxyazdiradion. *Tetrahedron Letters*, 611-612.

Tewari, D.N. 1992. Monograph on Neem (*Azadirachta indica* A. Juss.). International Book Distributors, Dehra Dun.

Van der Nat, J.M., W.G. Van der Sluis, K.T.D. de Silva and R.P. Labadie. 1991. Ethnopharmacognostical survey of *Azadirachta indica* A. Juss. (Meliaceae). *J. Ethnopharmacol.*, 35: 1–24.

Vietmeyer, N.D. 1992. *Neem: A Tree for Solving Global Problems*. National Academy of Sciences, Washington, DC.

16

Studies on the Resin Canals and Ducts of *Azadirachta indica* A. Juss. (Meliaceae)

K. ARUMUGASAMY [1], K. UDAIYAN [1], S. MANIAN [1],
C. BALASUBRAMANIAN [2] AND P.S. MOHAN [2]

[1] Department of Botany, Bharathiar University, Coimbatore 641 046, India
[2] Department of Chemistry, Bharathiar University
Coimbatore 641 046, India

ABSTRACT

Resin canals and ducts in *Azadirachta indica* A. Juss are embryonic in origin. Their initials are mostly confined to the peripheral regions of the embryo and endosperm. They are distributed naturally in the fruit walls, shoot apices, cortex of young stem, and midrib regions. These resin canals are unbranched, elongated, tube-like structures without an accompanying well-organized epithelium of secretory cells. At a few places, the canals exactly resemble resin ducts due to their discontinuity. The resin canals contain many dense, globular bodies inside the lumen and their resinous nature is revealed by fluorescence and electromicroscopic studies. However, these resin canals differ from the resin glands and nectaries of Meliaceae both in structure and in nature of the secretory products.

INTRODUCTION

Neem is being extensively investigated by chemists and biologists for its potential uses. A number of organic compounds have been isolated from the various parts of neem (Tewari, 1992). The presence of resin glands, gum canals, induction of traumatic gum cavities, and extrafloral nectaries have already been reported by many workers (Spiekerkoetter, 1924; Inamdar et al., 1986; Ghosh et al., 1963; Nair et al., 1983; Lersten and Pohl, 1985). In the present studies, we have investigated the structure and development of naturally occurring resin canals/ducts, and a very few of the chemical constituents of resin glands.

MATERIALS AND METHODS

For the present study, the young embryos, stems of young seedlings, and fruits of *Azadirachta indica* were fixed in Formalin-acetic alcohol. Transverse and longitudinal sections 10μ thick were cut on a rotary microtome and stained with haematoxylin. Young stems were pre-fixed in cold glutaraldehyde in 0.1 μ phosphate buffer (pH 7.8) and post-fixed in 2 per cent osmium tetraoxide (OSO_4) in the same buffer dehydrated through a graded acetone series and embedded in a low-viscosity resin (Spurr, 1969). There sections (pale gold to silver) were cut on Reichert OMU_3 ultra microtome using glass knives. Sections were collected on formvar coated grids stained with uranyl acetate followed by lead citrate (Reynolds, 1963), and examined with a Philips 400 electron microscope. Thick sections of 1–2 μm were stained with toluidine blue 'O'(O'Brien et al., 1964).

The presence of resinous materials inside the lumen of the canals was observed in free-hand sections stained positively with (1) oil red 'O' differentiating with 4 mg of dansyl chloride in 10 ml isopropyl alcohol, (2) OSO_4, and (3) neutral red 0.25 per cent in distilled water. The sections were thoroughly washed in distilled water and mounted in 50 per cent glycerine (Analar). Photomicrographs were taken by Carl-Zeiss epifluorescence microscope using ultraviolet range filters. The detailed observations are given in Table 1.

OBSERVATION

In the embryo of *A. indica* 5–10 resin canal and duct initials confined mostly to the peripheral regions were recorded. In the cotyledon the resin canals and ducts were associated with procambial stands and later extended to all other parts. Very few ducts were found in the endosperm. In a mature embryo they were densely stained with a prominent nucleus giving the appearance of a duct and filled with resinous materials and many vacuoles (Fig. 1 A, B, C, D).

Table. 1. Histochemical Tests to localize Resin

Stains used for bright field and epifluorescence microscopy	Reference	Duration of staining, min.	Staining reaction with resin
Oil red 'O' (B and F)	Pearse, 1968	1–6	Red colour (B) and faint red fluorescence (F)
OSO_4 (B)	Jensen, 1962	10–15	Black colour
Neutral red (F)	Kirk, 1970	1–2	Greenish fluorescence
Sudan black B	O' Brien and McCully, 1981	2–5	Blue-black
Sudan IV	Johansen,1940	2–5	Orange red

B = bright field, F = fluorescence.

DISTRIBUTION OF RESIN CANALS AND DUCTS

Resin canals and ducts occur naturally in the embryo, endosperm, fruit walls, shoot apices, cortex regions of young stem, petiole, and midrib regions.

STRUCTURE OF RESIN CANALS AND DUCTS

The longitudinal sections of young stem show the occurrence of many resin canals and ducts in the pericycle, cortex, pith, and petiolar regions. The resin canals are elongated tube-like structures without any specialized epithelial cells, but surrounded by normal parenchyma cells only (Fig. 1 E, F, G). At a few places these exactly resemble resin ducts due to their discontinuity. They run parallel to the long main axis of the stem. The resinous materials accumulate in the lumen of the cavity. The resinous secretion of the ill-defined epithelial cells accumulates in the duct lumen and stains positively with sudan black 'B', sudan IV, OSO_4, oil red 'O', and neutral red. In longitudinal and transverse section, the resinous material in the lumen showed strong white autofluorescence under ultraviolet light (Fig. 2 K, L). The resin gave yellow and greenish-white fluorescence with auromine 'O' and neutral red respectively. The fresh hand-cut sections stained with oil red 'O' and sections deresinated by acetone also indicated positively. Moreover, the secretory products of canals and ducts were readily dissolved in preheated acetone and alcohol-xylene, which are resin mixtures (Fig. 2 M). The electron microscopic observations revealed the presence of many dense, globular bodies inside the lumen with well-defined nucleus, many vacuoles, and undulated cell walls (Fig. 1H, I, J).

DISCUSSION

Azadirachta indica has naturally occurring resin canals and ducts apart from resin glands and the induction of traumatic gum and resin ducts in the wood has already been reported (Inamdar et al., 1986; Nair et al., 1983). In the present study the origin, structure, and development of resin canals and ducts are described for the first time. The resin canals and ducts which originate in the mature embryo were associated with procambial strands. The resin canals are tube-like structures with discontinuity in many places resembling large ducts.

In the primary plant body, resin canals and ducts are a normal feature. In the secondary plant body their occurrence is more liable to be influenced by external factors. In some plants they appear as a normal feature in the secondary body, while in others they are produced as a result of injury, and are known as traumatic ducts (Bannan, 1933a, b, 1936, 1941). The embryological reinvestigations by Nair and Kanta (1959) did not reveal anything on the presence of resin canals and ducts in the embryo. Lersten and Curtis' (1989) hypothesis states that though the development of cavities and ducts in the same plant is similar, they differ both morphologically and anatomically. Nair et al. (1983) have also reported that the schizogenous development of ducts in *A. indica* differs from the development of resin canals in embryo. Though these canals differ developmentally and structurally from the traumatic ducts of the same species reported by Nair et al. (1983, 1985), they resemble kino veins of *Eucalyptus*

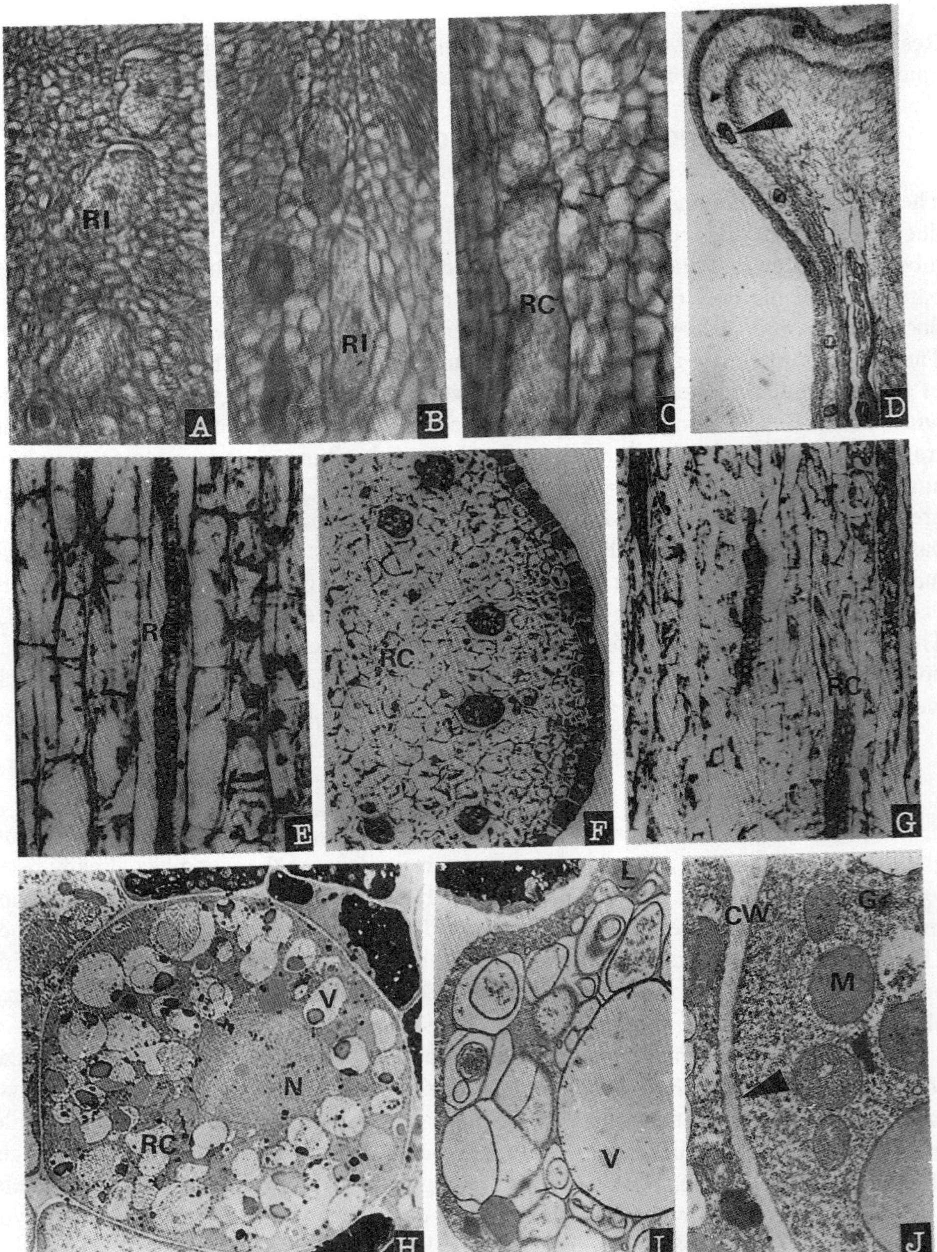

Fig. 1. (**A, B, C**) Developmental stages of resin canals in mature embryo × 450; × 450; × 350. (**D**) Resin ducts in endosperm (arrow) × 450. (**E, F**) Longitudinal and transverse sections of young stem showing resin canals × 300; × 350. (**G**) Golgi body (**H, I**) EM photomicrographs showing resin canals with well-defined nucleus × 3000. (**J**) EM photomicrographs showing undulated cell-wall × 5000. (RI—resin canal initials, RC— resin canal, N—nucleus, V—vacuole, L—lipid body, CW—cell wall, G—golgi body, M—mitochondria.)

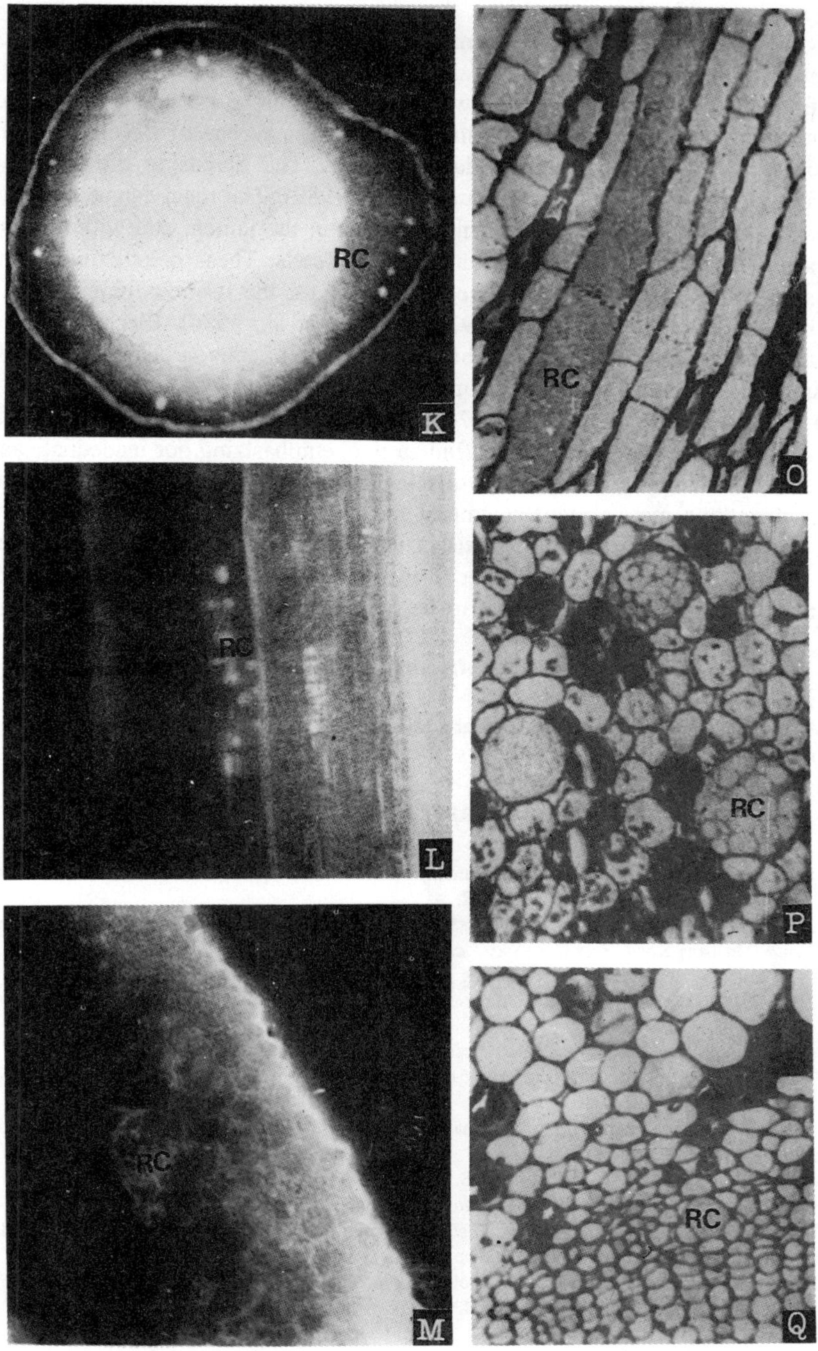

Fig. 2. (**K, L**) Epifluorescence photomicrographs showing resin canals × 250; × 250. (**O, P, Q**) L.S. and T.S. of resin canals in the cortex and pith regions × 550; × 500; × 450. (RC-resin canals.)

(Hillis, 1962) in their longitudinal orientation and isolated distribution but not in origin. Fahn (1979) also has advocated that the glands with similar structures may secrete entirely different compounds, while similar compounds may be secreted by glands with different structures. Subramaniam et al. (1990) reported that two types of glands develop from a group of initials and show similarities of tissue systems and secretion, but the nature of secretory substance is different. The occurrence of developmentally and structurally dissimilar resin canals, ducts, cavities, and glands in the same plant is contradictory to the view of Lersten and Crutis (1989). The resin canals and ducts are filled with resinous contents, with many vacuoles in the lumen, and without any septa and resemble non-articulated unbranched laticifer cells. These cells exhibit a number of dense globular structures and stained positively for the resinous materials as stated by many authors (Fahn and Evert, 1974; Inamdar et al., 1986). Electron microscopic study showed that the middle lamella between a group of cells swells and exhibits many electron-dense globular structures. Similar contents also have been reported in *Rhus glabra* Author by Fahn and Evert (1974). This paper describes a new configuration for resin canals and ducts in the neem, an Indian tree, emphasizing our inadequate anatomical knowledge about secretory reservoirs in Meliaceae. This suggests that some anatomical variations may also exist among the secretory structures. The difference in origin and development of resin canals in neem can be due to its inherent characteristics. It indicates the need for further ultrastructural studies to resolve the phylogeny of secretory glands, since it appears to be independent as suggested by Fahn (1979).

ACKNOWLEDGEMENTS

Dr. K.M. Marimuth, Vice Chancellor, Bharathiar University is thanked for encouragement. K. Arumugasamy and C. Balasubramanian are thankful to CSIR New Delhi for financial assistance.

REFERENCES

Bannan, M.W. 1933a. Vigour of growth and traumatic resin cyst development in the hemlock *Tsuga canadensis* L. Carr. *Trans. R. Soc. Can.*, 27: 197–202.

Bannan, M.W. 1993b. Factors influencing the distribution of vertical resin canals in the wood of the larch *Larix laricina* (Du Roi) Koch. *Trans. R. Soc. Can.*, 27: 203–218.

Bannan, M.W. 1936. Vertical resin ducts in the secondary wood of the Abicineae. *New Phytol.*, 35: 11–16.

Bannan, M.W. 1941. Variability in wood structure roots of native ontario connifers. *Bull. Torrey Bot. Club*, 68: 173–193.

Fahn, A. 1979. *Secretory Tissues in Plants*. Academic Press, London.

Fahn, A. and R.F. Evert. 1974. Ultrastructure of the secretory ducts of *Rhus glabra* L. *Amer. J. Bot.* 61(1): 1–14.

Ghosh, S.S., S.K. Purkayastha and S.K. Krishnalal. 1963. Meliaceae. In: *Indian Woods*, Vol. 2. Manager of Publications, New Delhi.

Govindachari, T.R., Gopalan Sandhya and S.P. Ganesh Raj. 1992. Azadirachtins H and I: Two new tetranortriterpenoids from *Azadirachta indica. J. Natur. Prod.* 55(5): 596–601.

Hillis, W.E. 1962. In: *Wood Extractives and Their Significance to Pulp and Paper Industry*. W.E. Hillis. (ed.), New York, Academic Press, New York pp. 59–131.

Inamdar, J.A., R.B. Subramanian and J.S.S. Mohan. 1986. Studies on the resin glands of *Azadirachta indica* A. Juss (Meliaceae). *Ann. Bot.*, 58: 425–429.

Jensen, W.A. 1962. *Botanical Histochemistry: Principles and Practice*. W.H. Freman and Company, San Francisco. pp. 226.

Johansen, D.A. 1940. *Plant Microtechnique*. Tata McGraw-Hill Publishing Co. Ltd., Bombay.

Kirk, P.W. (1970). Neutral red as lipid fluorochrome. *Stain Technol.*, 45: 1–4.

Lersten, N.R. and R.W. Pohl. 1985. Extrafloral nectaries in Cipadessa (Meliaceae). *Ann. Bot.*, 56: 363–366.

Lersten, N.R. and J.D. Curtis. 1989. Secretory reservoirs (ducts) of two kinds in giant ragweed (*Ambrosia trifida*; Asteraceae). *Amer. J. Bot.*, 75(9): 1313–1323.

Nair, N.C. and K. Kanta. 1959. Studies on Meliaceae. IV. Morphology and embroyology of *Azadirachta indica* A. Juss., a reinvestigation. *J. Indian Bot. Soc.*, 40: 382–396.

Nair, M.N.B., J.J. Shah and S.V. Subramaniyam. 1983. Ultrastructure and histochemistry of traumatic gum ducts in the wood of *Azadirachta indica* A. Juss. *IAWA Bull* n.s. 4(2, 3): 103–112.

Nair, M.N.B., J.R. Bhatt and J.J. Shah. 1985. Induction of traumatic gum cavities in sapwood of the neem (*Azadirachta indica* A. Juss.) by ethephon and paraquat. *Indian J. Exp. Bio.*, 23: 60–62.

O'Brien, T.P. and M.E. McCully. 1981. *The Study of Plant Structure: Principles and Selected Methods*. Termacarphi Pty., Melbourne.

O'Brien, T.P., N. Feder and M.E. McCully. 1964. Polychromatic staining of plant cell walls by toluidine blue O. *Protoplasma*, 59: 368–373.

Pearse, A.G.E. 1968. *Histochemistry, Theoretical and Applied*, Vol I. Churchill Livingstone, Edinburgh. pp. 415–697.

Reynolds, E.S. 1963. The use of lead citrate at high pH as an electron opaque stain in electron microscopy. *J. Cell Bio.*, 17: 208–212.

Spiekerkoetter, H. 1924. Utersuchungen zur Anatomie und systematik Ostafrikanischer Meliaceen, Burseraceeu und Simarubaceeu. *Botanisches Archiv.*, 7: 274–320.

Spurr, A.R. 1969. A low viscosity epoxy resin embedding medium for electromicroscopy. *J. Ultrastructural Res.*, 26: 31–34.

Subramaniam, R.B., K. Arumugasamy and J.A. Inamdar. 1990. Studies in the secretory glands of *Hiptage sericeae* (Malpighiaceae). *Nord. J. Bot.*, 10: 57–62.

Tewari, D.N. 1992. *Monograph on Neem* (Azadirachta indica *A. Juss*). International Book Distributors, Dehra Dun, India.

17

Neem Seedlings — Growth Potentials of Different Organs

A. STEPHEN, B. MURUGANANDAM, P. RAVICHANDRAN AND P. DAYANANDAN

Department of Botany, Madras Christian College, Tambaram
Madras 600 059, India

ABSTRACT

We have determined the growth pattern of pot-grown neem seedlings in light and dark and the response of these seedlings to spray application of brassinolide (BR), gibberellic acid A3 (GA), benzyl aminopurine (BAP), ethylene as chloroethylphosphonic acid (ET), and cycocel (CCC) either alone or in combination. One centimetre hypocotyl and epicotyl segments were excised and treated with the above plant growth regulators in petri dishes to observe elongation and developmental changes. Segments were also treated with indole butyric acid (IBA) to promote rooting in vermiculite. Both in dark and in light seedling growth was not continuous but sigmoidal with a plateau phase setting in after 10 days. The elongation response was primarily due to the growth of the hypocotyl. Also, hypocotyls elongated twice as much in dark as in light.

Brassinolide induced dramatic changes including promotion of elongation, retention of cotyledons, extension of cotyledonary petioles, and expansion of leaf blades and basal regions of major veins and petioles. Hypocotyl segments responded to IAA by elongating 150 per cent more than control segments in 48 hrs. Brassinolide-treated segments increased in diameter. This swelling response was due to expansion of cells in the region of the statenchyma near the endodermis. Cut segments produced callus on both the cut ends. Shoot buds developed from the distal cut ends directly or after callus production. The outer surfaces of segments treated with IAA or ET produced about 25 lenticels per segment. Shoot buds with or without roots developed from the lenticels. Segments 2–4 cm long with cotyledons and epicotyl produced profuse roots when treated with IBA. Segments of cotyledons on MS medium supplemented with 2, 4-D, 2, 4, 5-T, or trichloropropionic acid produced profuse callus while explants grown on 2-isopentenyl adenine produced callus as well as shoot buds. Segments of hypocotyls, epicotyls, and seedling shoot apices on MS medium with zeatin, kinetin, or BAP produced callus and numerous shoot buds. Seedling organs hold great potential for conventional and tissue culture methods of propagation.

INTRODUCTION

Neem is conventionally propagated through raising seedlings (Joshi, 1983), although root and shoot cuttings also can be used to establish plants (Tewari, 1992). Propagation through seeds is limited by the seasonal availability of seeds and the rapid loss of viability (Ezumah, 1986; Ponnuswamy et al., 1990). An understanding of the morphogenetic potentials of different regions of neem seedling is necessary in order to make efficient use of seedlings as a source of propagating material. In this investigation we have studied (1) time course of growth of seedlings in light and dark, (2) response of seedlings to selected plant growth regulators (PGRs), (3) growth and morphogenetic potential of excised segments of hypocotyls, epicotyls, and cotyledons, and (4) growth of seedling organs in tissue culture.

MATERIALS AND METHODS

Experiments were carried out with freshly collected neem (*Azadirachta indica*) seeds in June, July, and August. Seeds without the endocarp were sterilized with 0.5 per cent mercuric chloride for a few minutes and germinated in petri dishes lined with filter paper or on perlite-vermiculite (50:50) in trays. Seedlings were germinated under fluorescent light or in complete darkness. In one set of experiments seedlings in the dark were irradiated every day with red light (660 nm) for 30 min. Observations were made on the percentage of germination, pattern of rupture of seed coat, emergence of radicle, and growth of whole seedlings as well as hypocotyls and epicotyls. Seven-day-old light-grown seedlings were sprayed with the following PGRs: gibberellic acid A3 (GA), brassin (BR), benzyl aminopurine (BAP), ethylene (ET) as chloroethylphosphonic acid, and cycocel (CCC). Seedlings were also sprayed with BR + BAP or BR + GA. The concentrations of PGRs were 10^{-5} M for all treatments. Control plants were sprayed with distilled water. Tween-80 was added to all solutions to facilitate wetting of the plant surfaces. Morphological changes in seedling growth were analysed and epicotyl and hypocotyl lengths were measured on the 15th and 25th days.

Ten-to twelve-day-old seedlings were cut at different regions to determine rooting efficiency. Cut ends of hypocotyls and epicotyls were treated with distilled water or 10^{-3} M indole butyric acid (IBA), planted in vermiculite, and grown in a mist chamber. The response of excised segments of seedlings to PGRs was investigated by placing 1 cm segments in petri dishes containing GA, BR, IAA, or ET. Sucrose 0.1 per cent was supplied along with the PGRs.

For tissue culture studies 10- to 15-day-old seedlings were washed under running water, washed in the detergent Teepol, and again rinsed with water. Explants of hypocotyls, epicotyls, cotyledons, and shoot tips were surface-sterilized with mercuric chloride (0.05) per cent and washed several times with sterile, distilled water. They were cultured in MS medium supplemented with kinetin, zeatin, zeatin riboside (ZR), 2-isopentenyl adenine (2iP) BAP, 2, 4-dichlorophenoxyacetic acid (2,4-D), 2, 4, 5-trichlorophenoxyacetic acid (2, 4, 5-T), trichloropropionic acid (TCPA), and naphthalene acetic acid (NAA). The cultures were incubated in the dark or under white fluorescent light (1000 lux) for 16/8 hrs light/dark at 26°C.

RESULTS AND DISCUSSION

The seeds were 100 per cent viable. Emergence of radicle can be observed within 24 hrs of incubation. The thin seed coat can delay the emergence of radicle by about 24 hrs if the initial rupture does not occur at the radicle end of the seed. The time course of seedling growth and the elongation of hypocotyl and epicotyl are shown in Fig. 1. The pattern of seedling growth is similar in light and dark up to the sixth day. Thereafter, seedlings in light exhibit a decrease in growth rate. Both light-grown and dark-grown seedlings show a sigmoidal growth curve indicating the presence of factors that slow down growth. This growth pattern is also observed in seedlings growing in nature and it may be an adaptation to strengthen the seedlings before they commence growth at a later period.

Analysis of the response of hypocotyl and epicotyl reveals interesting differences in their response to dark, light, and red light regimes (Fig. 1). Growth of hypocotyl is clearly inhibited by white light as well as red light. However, inhibition by white light is greater and commences earlier. Epicotyl growth, on the other hand, is promoted by light and dramatically so by red light. The lengths of dark-grown and light-grown epicotyls are similar only after 10 days. It is likely that the differences between seedlings grown in red and white light are due to the inhibitory effect of far red light present in the white light.

Figure 2 summarizes the response of seedlings sprayed with PGRs. These measurements were taken 15 days after spraying of seven-day-old seedlings. Seedling growth is promoted by BR and GA. Benzyl aminopurine and ET do not influence elongation while CCC suppresses growth. Growth induced by BR and GA is primarily due to their effect on elongation of the hypocotyl. Surprisingly, simultaneous application of BR and

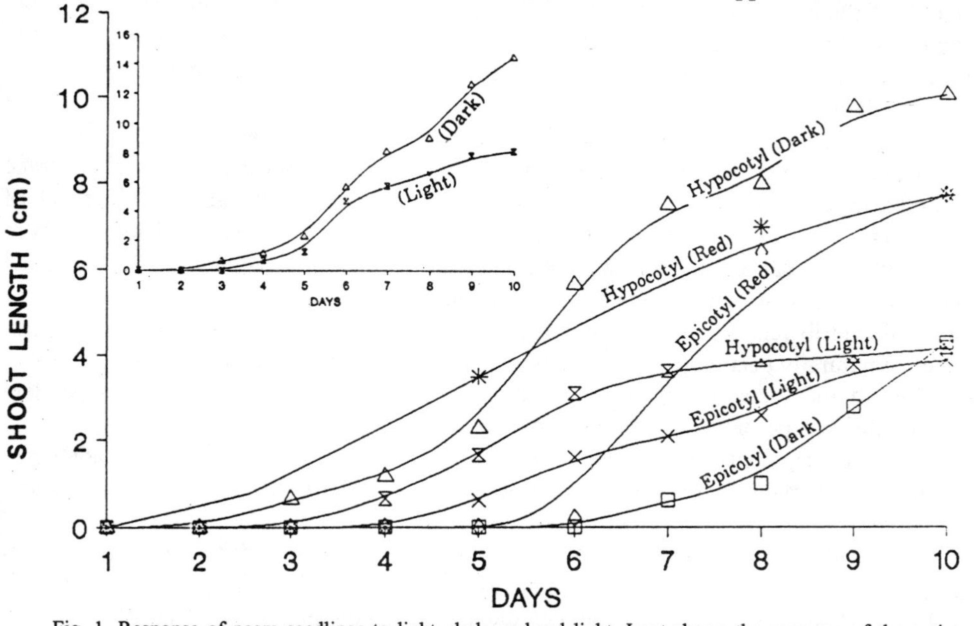

Fig. 1. Response of neem seedlings to light, dark, and red light. Inset shows the response of the entire seedling

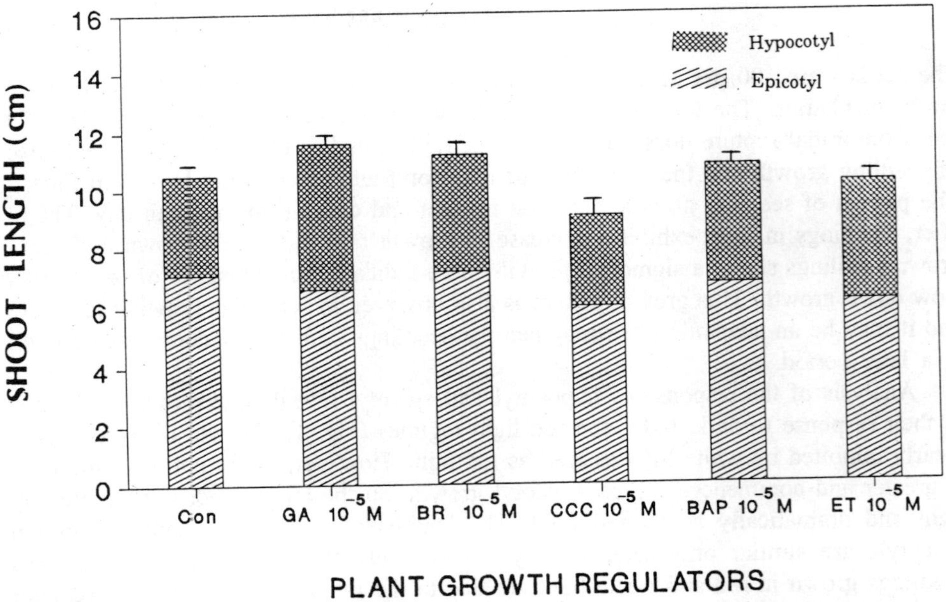

Fig. 2. Response of seedlings to spray application of PGRs. Bars indicate the total height of the seedlings. The separate responses of hypocotyls and epicotyls are indicated within each bar

GA tends to inhibit growth. Seedlings treated with BR show several morphological changes, besides elongation. The cotyledons expand (Fig. 3(1)) and remain attached to the axis for a longer period. The petioles of cotyledons elongate. Leaf blades and the bases of major veins as well as bases of leaflets also expand. Seedling leaflets show epinasty. These changes are similar to the effects of BR already reported in other seedlings and pulvinus and leaf sheath-blade joints of grasses (Dayanandan and Kaufman, 1984; Mandava, 1988).

Seedlings of neem possess morphogenetic potential that is age dependent. Freshly collected seeds that are 100 per cent viable produce normal seedlings. After about three months much variability is observed in viability and the length and development of seedlings and seedling organs. In some cases the two cotyledons are displaced from each other so that an apparent internodal region of 1–4 cm separates them. Where the cotyledons remain close to the vermiculite medium they tend to produce roots at the adaxial midrib regions and cut ends (Fig. 3(2)). These observations indicate the influence of naturally occurring growth regulating substances elaborated during aging. Further studies should focus on identifying these substances and exploiting the age-dependent morphological potential of the seedling axes and cotyledons for propagation.

Excised segments of epicotyl and hypocotyl readily root at the proximal ends when planted in vermiculite. Dipping the cut ends in 10^{-3} M IBA for about 5 min. enhances the rooting capacity considerably.

Excised 1 cm segments can be kept relatively undamaged for as long as 60 days in petri dishes with H_2O, 0.1 per cent sucrose, or various PGRs (Fig. 3(3–10)). Significant changes observed include development of a shoot on the distal end (Fig. 3(5)), production of callus on the distal or proximal ends (Fig. 3(3, 5, 9)), and development of lenticels (Fig. 3(4, 7, 9, 10)). Lenticels are more numerous on the hypocotyls than on the epicotyls. About 25 lenticels develop on a 1 cm segment incubated for 8–12 days in sucrose medium. Figure 3(4–10) shows the response of segments to IAA and ET. Treatment with IAA induces lenticels that can be about 5 mm long. Shoot buds develop from these lenticels. Along with shoot buds root development has also been observed. Shoot buds and roots develop superficially from local proliferations of lenticels and the buds do not have direct vascular connection with the central vascular cylinder.

Shoot buds are poorly developed on epicotyl segments. However, in segments that include the cotyledonary nodal regions of the seedling, callus is produce and shoot regeneration occurs even on the proximal end.

Thus, excised segments of seedlings can develop shoot buds directly on cut ends, produce callus on both ends, develop shoot buds from the callus, proliferate lenticels, and produce shoot buds and roots from the lenticels. Seedlings of neem possess morphogenetic potential that is age dependent. Excised segments of epicotyls and hypocotyls respond to external application of IAA and BR. After 48 hrs in dark, hypocotyls elongate 150 per cent over control in 10^{-5} M IAA, whereas epicotyls elongate only 100 per cent control at this concentration. Epicotyls also curve in response to IAA. At 10^{-5} M, IAA does not affect epicotyl elongation, while it promotes hypocotyl growth by 66 per cent.

Thus, IAA is able to induce growth in the otherwise slow or non-growing, more mature region of a seedling, namely hypocotyl. Brassin induces swelling of the hypocotyl segments. Anatomical studies reveal that the swelling is due to enlargement of cells near the statolith-containing endodermal region. Similar response to BR has been reported for grass pulvinus (Dayanandan and Kaufman, 1984). The hormonal and growth conditions necessary for root production and enhanced production of lenticels and shoot buds with roots are under investigation.

We have explored the potential for micropropagation through explants of cotyledon, epicotyl, hypocotyl, and seedling shoot tips. Cotyledonary explants on MS medium with 2, 4-D, 2, 4, 5-T, and TCPA produced profuse callus. In light, callusing occurred at the cut ends after 25 days. In dark, the cut ends as well the entire surface of the segments developed callus tissue 15 days after inoculation. In cotyledonary explants on 2iP, cut ends produced callus after 15 days and numerous shoot buds developed after 20 days on the same medium. Cotyledonary explants on NAA-supplemented medium produced callus as well as roots after one month in light.

Hypocotyls and epicotyls were cultured on MS medium containing zeatin, BAP, or kinetin of 1×10^{-5} to 4×10^{-5} M and 3 per cent sucrose. Explants begin to enlarge three days after inoculation. Visible callus production was observed on the fifth day. By the 15th day callus was present on all surfaces of the explants.

In some of the explants on zeatin or kinetin shoot buds developed on the distal cut ends (Fig. 3(11)). Shoot buds proliferated into plantlets on the cut ends of all surfaces.

Shoot tips from seedlings were cultured on MS medium with 4×10^{-5} M zeatin. Callus developed from the cut ends after 15 days followed by differentiation of multiple shoots on the same medium after 20 to 50 days of inoculation.

Micropropagation of neem was previously attempted by Rangaswamy and Promila ((1972), Sanyal et al. (1981), Schulz (1983), Narayan and Jaiswal (1985), Rao et al. (1988), and Ramesh and Padhya (1990). These authors used explants of young leaf

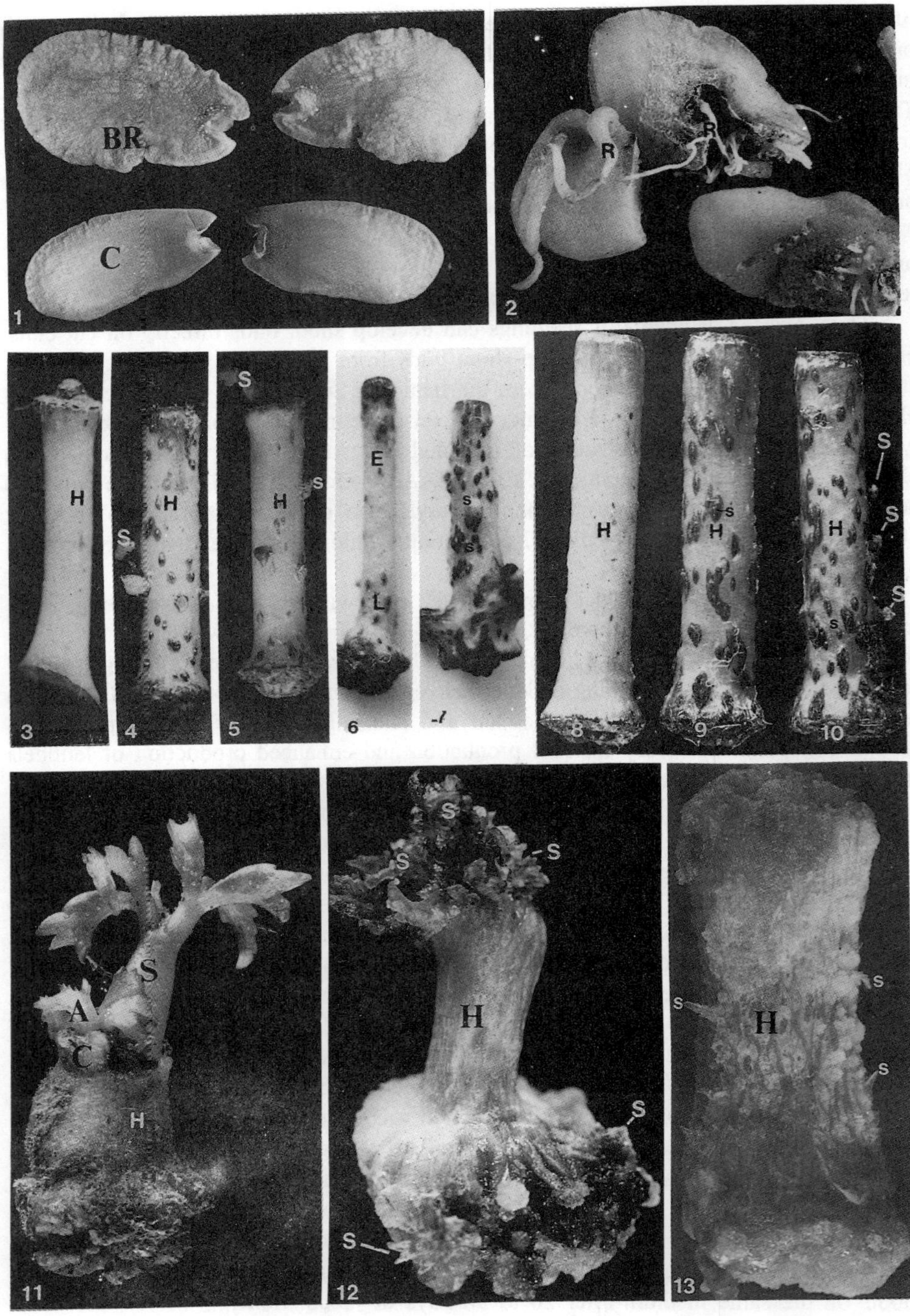
1
BR
C
2
R
R
3
H
4
H
S
5
S
H
S
6
E
L
s
s
8
H
9
H
s
10
H
S
S
S
s
11
S
A
C
H
12
S
S
S
S
S
H
13
H
s
s
s

discs and petioles. Schulz (1983) observed shoot bud production on MS medium supplemented with IAA. However, rooting was not observed by him. Ramesh and Padhya (1990) were able to induce the production of plantlets with shoot and root on Wood and Braun medium supplemented with kinetin or BAP.

The present investigation indicates that segments of hypocotyl, epicotyl, cotyledon, and seedling shoot apex could all be used for the induction of shoot buds. These explants respond favourably to cytokinins and relatively more solid medium of 3 per cent agar. In the cut ends of seedling axis where profuse callusing occurs, many secretory cells have been observed. Dayanandan et al. (1993) consider these secretory cells to be the site of storage of triterpenoids. Thus, the production of triterpenoids in cultured cells is possible when conditions for enhanced production of secretory cells are determined.

ACKNOWLEDGEMENTS

The authors thank the Indian Council of Agricultural Research for financial support through a grant and Dr. S. Chinnamani, Mr. V.R. Subbiah and Mr. N.P. Samson for their help and suggestions.

REFERENCES

Dayanandan, P. and P.B. Kaufman. 1984. Analysis and significance of gravity-induced asymmetric growth in the grass leaf-sheeth pulvinus. *Ann. Bot.*, 53: 29–44.

Dayanandan, P., A. Stephen and B. Muruganandam. 1993. Location of neem triterpenoids and other secretory structures. *Proc. 4th Int. Neem Conf.*, Bangalore.

Ezuma, R.S. 1986. Germination and storage of neem (*Azadirachta indica* A. Juss.) seeds. *Seed Sci. Technol.*, 14: 593–600.

Joshi, H.B. 1983. *The Silviculture of Indian Trees,* Vol. III. Government of India, Controller of Publication, Delhi, pp. 147–154.

Mandava, N.B. 1988. Plant growth-promoting brassinosteriods. *Annu. Rev. Plant Physiol. Plant Mol. Biol.*, 39: 23–53.

Narayan, Pratap, and V.S. Jaiswal. 1985. Plantlet regeneration from leaflet callus of *Azadirachta indica* A. Juss. *Tree Sci.*, 4(2): 65–68.

Ponnuswamy, R.S., Vinaya Rai, C. Surendran and T.V. Karivaratharaju. 1990. Studies on maintaining seed longevity and the effect of fruit grades in neem (*Azadirachta indica*). *J. Trop. For. Sci.*, 3: 258–290.

Ramesh, K. and M. Padhya. 1990. *In vitro* propagation of neem, *Azadirachta indica* (A. Juss.) from leaf discs. *Indian J. Exp. Biol.*, 28: 932–935.

Rangaswamy, N.S. and Promila. 1972. Morphogenesis of the adult embryo of *Azdirachta indica* A. Juss. *Z. Pflphzsiol.*, 67: 377–379.

Fig. 3(1-13). Response of seedlings organs of neem. (1) Comparison of 15-day-old cotyledons treated with brassin (BR) with controls (C). (2) Development of roots from margins and adaxial surface of cotyledons. Roots (R) develop more readily in damaged cotyledons. (3) Hypocotyl segments (H) of 1 cm after 15 days' incubation in 0.1 per cent sucrose. Callus has developed at the distal end but no lenticels are produced. (4, 5) Similar to (3) but treated with 10^5 M IAA. Note the development of lenticels and shoot buds (S). (6, 7) Epicotyl segments (E) treated with ethylene showing lenticels and shoot buds (S). (8–10) Hypocotyls (H) 1 cm long, treated with sucrose alone (8) or with ethylene (9, 10). Ethylene treatment has resulted in the production of many lenticels and shoot buds (S). (11) Hypocotyl segment (H) grown on MS medium supplemented with 2×10^5 M zeatin. The proximal end has produced callus and the distal end has produced a shoot from the cut surface. One of the axillary buds (A) can be seen in the axil of the remaining base of cotyledonary petiole (C). (12) Hypocotyl segment (H) on MS medium with 10^5 M BAP. Callus has developed on both ends and after 15 days shoot buds (S) have proliferated. (13) Hypocotyl explants (H) grown on MS medium with 4×10^5 M showing warty growth on the surface and shoot buds (S)

Rao, M.V.S., Y.Y. Rao, Y.S. Rao and V. Manga. 1988. Induction and growth of callus in *Azadirachta indica* A. Juss. *Crop Improvement*, 15: 203–205.
Sanyal, M., A. Das, M. Banerjee and P.C. Datta. 1981. *In vitro* hormone induced chemical and histological differentiation in stem callus of neem, *Azadirachta indica* A. Juss. *Indian J. Exp. Biol.*, 19: 1067–1068.
Schulz, F.A. 1983. Tissue culture of *Azadirachta indica*. In: *Proc. 2nd Int. Neem Conf.* Rauischholzhausen. GTZ, Eschborn, pp. 539–542.
Tewari, D.N. 1992. *Neem*. International Book Distributors, Dehradun.

18

Variation Studies and Heritable Components of Seed Parameters in Neem (*Azadirachta indica* A. Juss.)

K. KUMARAN [1], C. SURENDRAN [2] AND R.S.V. RAI [2]

[1] Forest College and Research Institute, Tamil Nadu
[2] Agricultural University, Mettupalayam 641 301, Tamil Nadu, India

ABSTRACT

Seeds were collected from 28 one-parent families of neem (*Azadirachta indica* A. Juss.) in seven agroclimatic zones of Tamil Nadu, India and studied for the variation and heritability of seed length, seed breadth, seed length-breadth ratio, 100 seed weight, and oil content. Results showed significant differences among the one-parent families for all the seed parameters studied. Hundred seed weight recorded moderate genotypic (12.9 per cent) and phenotypic (15.8 per cent) coefficients of variance. Heritability was the highest for oil content (99.8 per cent). Seed length showed high genetic gain. From these results, it is concluded that hundred seed weight may serve as a reliable selection index by virtue of its high genotypic coefficient of variance, high genetic advance, and moderate heritability.

INTRODUCTION

Native to the Indo-Pakistan subcontinent (Chaney and Knudson, 1988), neem is a versatile tree with a variety of uses in industry, medicine and energy and has aesthetic and timber values. Neem cake, the residue left after extraction of oil from the seed, is an excellent fertilizer and soil conditioner (Ahmed and Grainge, 1985; Manavu, 1984) and is fed to animals in a few districts in Andhra Pradesh (Christopher, 1970). Among the uses that have most interested scientists, especially entomologists, are the pesticidal proclivities of neem fruits (Atwal and Panji, 1964), kernels (Girish and Jain, 1978), and oil (Attri and Ravi Prashad, 1980; Ali et al., 1983). The need for developing countries to minimize their dependance on imported pesticides by using indigenous resources has been emphasized by many workers (Fagoonee, 1980; Jacobson, 1981, 1986). There is thus a need for the improvement in need to increase the fruiting potential of this species. All tree improvement programmes must have seed production at some stage of their development if continued gains are to be achieved. This is true even for programmes using vegetative propagules (Zobel and Talbert, 1984). Despite seed being such an important variable, studies on genetic parameters in respect of seed are scant. An attempt was therefore made to study the heritable component of seeds of neem.

MATERIALS AND METHODS

Seeds were collected from the selected one-parent families of neem (*Azadirachta indica* A. Juss.) representing 15 districts of Tamil Nadu: Dindigul Anna, Pasumpon Thevar Thirumagan, Pudukottai, Kamarajar, Nellai Kattabomman, Trichy, Ramnad, Chengai M.G.R., South Arcot, North Arcot, Salem, Periyar, Coimbatore, Dharmapuri, and Thanjavur. The details of the 28 one-parent families are given in Table 1 and the geographic sources are shown in Fig. 1. The seeds were then measured for length, breadth, length-breadth ratio, and 100 seed weight. The oil content was analysed using the Sochlete apparatus.

The data were subjected to analysis of variance (Panse and Sukhatme, 1961). Phenotypic and genotypic coefficients of variance (PCV and GCV) were computed following Burton (1952).

$$\text{PCV \%} = \text{Phenotypic variance}^{1/2} \times 100$$

$$\text{GCV \%} = \text{Genotypic variance}^{1/2} \times 100$$

Heritability in the broad sense was calculated according to Lush (1940):

$$h^2 \text{ (Broad sense)} = \frac{\sigma^2 g}{\sigma^2 p}$$

where σ^2 g is the genotypic variance and p is the phenotypic variance.

Genetic advance (GA) was worked out after Johnson et al. (1955):

$$\text{GA} = \frac{\text{Genotypic variance}^{1/2}}{\text{Phenotypic variance}^{1/2}} \times K$$

where K = 2.06, a selection differential at 5 per cent selection intensity.

$$\text{GA as \% of mean} = \frac{\text{Genetic Advance}}{\text{Grand Mean}} \times 100$$

Among the seed parameters, for describing genetic parameters the values recorded in the study were classified as detailed below.

Genetic Parameter	Low	Moderate	High
GCV and PVC	0–10%	10–20%	20% and above
Heritability	0–30%	30–60%	60% and above
GA as % of Mean	0–10%	10–20%	20% and above

Table 1. Parental tree characters of *Azadirachta indica*

Map No.	Location	Age (Yr)	Height (m)	Girth (cm)
1.	Vedasandur	16	11	87
2.	Oddanchatram	15	14	87
3.	Pudupatty	12	10	80
4.	N-Pudur	16	13	95
5.	Paiyur	15	8.29	95
6.	Prungolur	10	13.3	114
7.	Vamban-5	10	10	80
8.	Vamban-1	14	12	78
9.	Kulithalai	9	10	69
10.	Paramakudi	10	6.5	85
11.	Muduranthagam	18	15	92
12.	Kanchipuram	14	12	86
13.	Virudhachalam	13	10	126
14.	Cuddalore	20	12	148
15.	Panruti	17	15	138
16.	Salem-1	15	13	124
17.	Salem-2	18	15.5	121
18.	Gobi	12	13	96
19.	Bannari	15	13	92
20.	Sathyamangalam	16	14	108
21.	Periyanaiakanpalayam	15	12	99
22.	Annur	18	16	120
23.	Sirumugai	20	15	116
24.	Dharmapuri 3	12	9	90
25.	Dharmapuri 2	13	11	160
26.	Dharmapuri 1	15	13	162
27.	Thanjavur	22	17	142
28.	Nagapattinam	15	14	98

RESULTS AND DISCUSSION

Variation Studies

Significant differences at 1 per cent level were found among the 28 one-parent families for all the seed parameters studied (seed length, seed breadth, seed length-breadth ratio, 100 seed weight, and oil content) (Table 2). Regarding oil content, the highest value of 52.66 per cent was recorded by the one-parent family 23 (Kulithalai, Trichy). Four one-parent families (16, 2, 22 and 1) showed high values for seed length, seed breadth, seed length-breadth ratio, and 100 seed weight respectively (Table 2). Similar results of variability studies were reported in *Santalum album* (Bagchi and Sharma, 1989) and variability in 100 seed weight was reported for slash pine (USFS, 1948; Wakeley, 1954; Squillace, 1966).

Heritable Components

Among seed parameters, 100 seed weight recorded moderate GCV (12.90 per cent) and PCV (15.81 per cent) followed by seed length, seed length-breadth ratio, seed breadth,

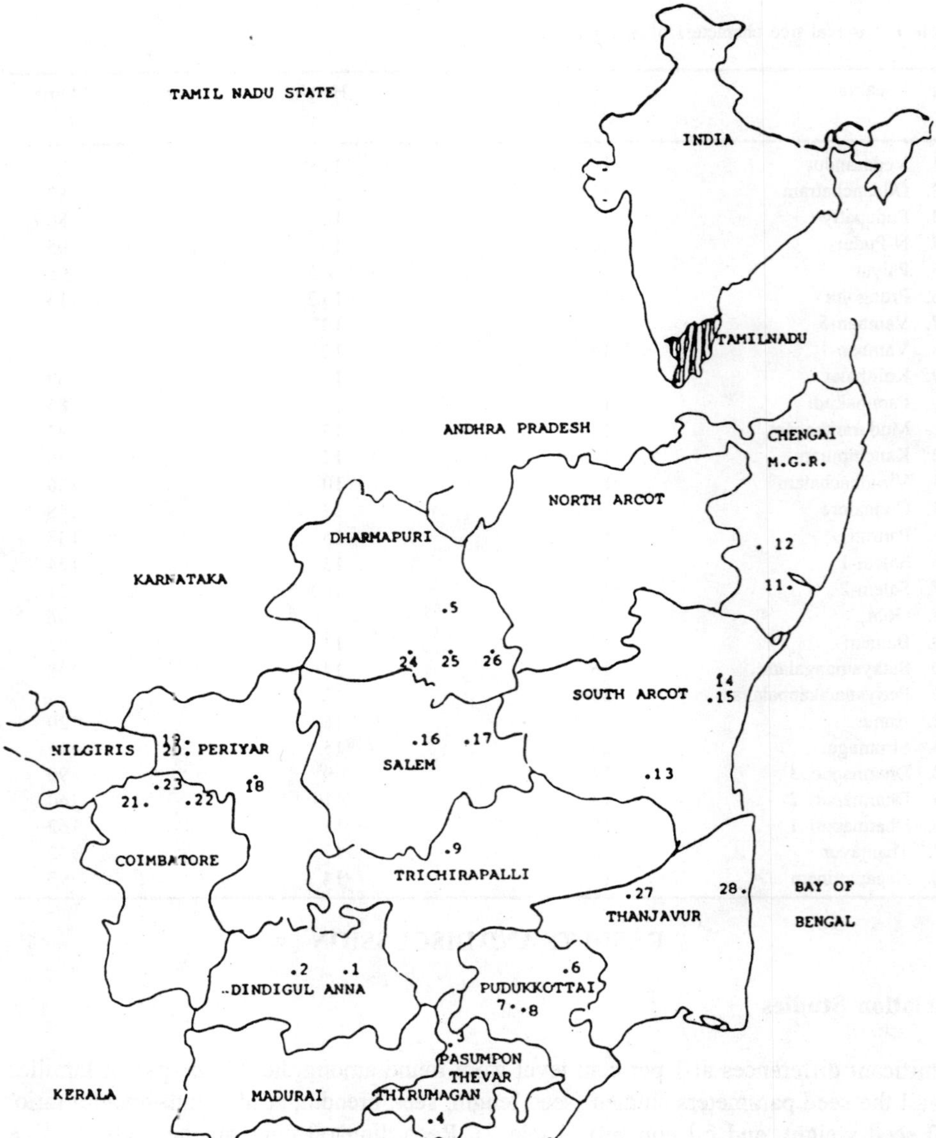

Fig. 1. Distribution of 28 one-parent families of *Azadirachta indica*

and oil content. Regarding heritability, oil content recorded the highest value of 99.83 per cent followed by seed length, which recorded 98.20 per cent. Seed breadth and seed length-breadth ratio recorded fairly high heritability, while 100 seed weight recorded a low value. Seed length recorded a high value for GA as per cent of mean (23.88 per cent), followed by 100 seed weight (21.99 per cent) and seed length-breadth ratio (20.69 per cent). Seed breadth recorded a moderate value of GA as per cent of mean, while oil content recorded a low value (Table 3). Among the seed parameters studied 100 seed weight recorded high GCV, moderate heritability, and genetic advance. According to Burton (1952), GCV in conjunction with heritability estimate gave a clear picture of extent of advance to be expected by selection. Heritability, however, indicate the effectiveness with which the selection of the genotype is based on the phenotype but fails to convey the genetic progress (Johnson et al., 1955). High heritability therefore does not necessarily imply greater gain, which was emphasized by Ramanujam and Jhirumalachar (1967). According to them heritability estimates in a broad sense will be reliable if accompanied by high genetic advance as well. High GCV (17.72 per cent) for 100 seed weight and identical values (6.7 per cent) for seed length and seed breadth were reported for *Santalum album* by Bagchi and Sharma (1989).

Evaluated by this perspective, 100 seed weight came into prominence with high value for heritability and genetic advance, and it can be used as the selection index among the seed parameters.

Table 2. Variability studies (seed parameters)

Character	Mean squares	
	Error DF=27	One-parent family DF=27
Seed length	0.0005	0.059**
Seed breadth	0.0009	0.010**
Seed length-breadth	0.001	0.122**
Oil content	0.016	18.949**
100 seed weight	3.976	20.458**

* Significant at 5 per cent level
** Significant at 1 per cent level

Table 3. Genetic parameters for seed characters

Characters	GCV (%)	PCV (%)	h^2 (broad)	GA as % of mean
Seed length	11.69	11.89	98.20	23.88
Seed breadth	9.94	10.90	83.25	18.71
Seed length-breadth ratio	10.95	11.94	84.03	20.69
Oil content	6.50	6.50	99.83	13.38
100 seed weight	12.99	15.81	67.46	21.99

REFERENCES

Ahmed, S.A. and M. Grainge. 1985. Use of indigenous plant resources in rural development potential of the neem tree. *Int. J. Dev. Tech.*, 3: 23–130.

Ahmed, S.A. and M. Grainge. 1986. Potential of the neem tree (*Azadirachta indica*) for pest control and rural development. *Econ. Bot.*, 40: 201–209.

Ali, S.I., O.P. Singh and V.S. Misra. 1983. Effectiveness of plant oils against pulses beetle (*Collosobruchus chinensis* Linn.). *Indian J. Entomol.*, 45: 6–9.

Attri, B.S. and G. Ravi Prashad. 1980. Studies on the pesticidal values of neem oil by-products. *Pestology*, 4: 16–20

Atwal, A.S. and H.R. Panji. 1964. Preliminary studies on the insecticidal properties of drupes of *Melia azaderach* against caterpillars of *Pieris brassicaes* (L.). *Indian J. Entomol.*, 26: 221–227.

Bagchi, S.K. and V.P. Sharma. 1989. Biometrical studies on seed characters of *Saritalum album* L. *Silvae Genet.*, 38: 152–153.

Burton, G.W. 1952. Quantitative inheritance in grasses. *Proc. 6th Int. Grassland Cong.*, 7: 277–283.

Chaney, W.R. and D.M. Knudson. 1988. Germination of seeds of *Azadirachta indica* enhanced by endocarp removal. *Int. Tree Crops J.*, 5: 153–161.

Christopher, J. 1970. Neem seed cake is also good for cattle. *Indian Fmg.*, 20: 38.

Fagoonee, I. 1980. The potential of natural products in crop protection in Mauritius. In: *National Agricultural Production Conference*, Univ. Mauritius, 1979. A. Peerully, I. Fagoonee and E.H. Unmole (eds.), pp. 201–217.

Girish, G.K. and S.K. Jain. 1974. Studies on the efficacy of neem seed kernel powder against stored grain pests. *Bull. Grain. Tech.*, 12: 225–228.

Jacobson, M. 1981. Neem research in the US Department of Agriculture, Chemical, biological and cultural aspects: *Proc. 1st Int. Neem Conf.*, Rottach-Egern, West Germany, 1980. H. Schmutterer, K.R.S. Asche. and H. Rembold (eds.) pp. 33–42.

Jacobson, M. 1986. The neem tree natural resistance for excellence. American Chemical Society Symposium Series No. 296. M.B. Green and P.A. Green (eds.), American Chemical Society, Washington pp. 220–232.

Johnson, H.W., H.F. Robinson and R.E. Comestock. 1955. Estimates of genetic and environmental variability in soybeans. *Agron. J.*, 314–318.

Lush, K.L. 1940. Intra-sire correlation and regression offspring on dams as a method of establishing heritability of characters. *Proc. Amer. Soc. Animal Production*, 33: 293–301.

Manavu, R.M. 1984. Non-conventional vegetable oils for fuel in Kenya. In: *Bioenergy* 84, Vol. III, *Biomass Resources. Proc. Conf.*, Gotebrogy, Sweden. H. Egneus and A. Ellegard (eds.), Elsevier Applied Science Publishers, London. pp. 166–169.

Panse, V.G. and P.V. Sukhatme. 1961. *Statistical methods for Agricultural Workers*, 2nd ed. ICAR, New Delhi, 361 pp.

Ramanujam, S. and D.K. Thirumalachar. 1967. Genetic variability of certain characters in red piper (*Capsicum annum*). Mysore. *J. Agric. Sci.*, 1: 30–36.

Squillace, A.N. 1966. Geographic variation in slash pine. *For. Sci. Mon.*, 10.

USFS. 1948. *Woody Plant Seed Manual*. USDA Misc. Pub., 654 pp.

Wakeley, P.C. 1954. *Planting the Southern Pines. USDA Monoq.*, 18. 233 pp.

Zobel, B. and J. Talbert. 1984. *Applied Forest Tree Improvement.* John Wiely and sons, New York, 505 pp.

19

De novo Synthesis of Secondary Metabolites in Cultured Cells of Neem

K. RAMESH AND M.A. PADHYA

Tissue Culture Laboratory, Department of Botany
M.S. University, Baroda 390 002, India

ABSTRACT

Flavonoids are the natural protective compounds which form an array of characteristic secondary metabolites. This study describes the biosynthesis of these compounds in neem cell and tissue cultures which can be enhanced by altering the culture conditions. The product accumulation/synthesis was correlated with the biomass yield where the synthesis occurred when cells attain the stationary phase. Flavonoids, in particular, kaempferol and quercetin were found maximum in flowers (1.93 per cent and 1.43 per cent) and their cell cultures (0.84 per cent and 0.68 per cent). Sucrose level at 3 per cent enhanced their quantities *in vitro*. Low levels of inorganic nitrogen (30 µM/1) in WB (1961) medium favoured the biomass yield and flavonoid synthesis, while MS (1962) medium rich in nitrogen source (60 µM/1) found inhibitory. L-phenylalanine (flavonoid precursor) at 20 µM/1 is the optimum concentration for biomass and product yield and further rise in the precursor was found to be toxic to the cell culture. Cell lines resistant to paraflourophenylalanine (the precursor analog) at 50 µM/l have shown 3-fold increase in the quantities of kaempferol (2.51 per cent) and quercetin (1.01 per cent) over the control. There was no significant change in the flavonoid quantities even after subsequent passages of subcultures or cell generations. The stability and the frequency of PFP resistance indicate that it is due to mutation.

INTRODUCTION

Investigations in the area of agrochemicals, medicine and pharmaceuticals identified neem as a renewable resource of various biologically active compounds, which form a great variety of secondary metabolites (Govindachari, 1992). It also offers an excellent environmentally safe pest control agent within the framework of IPM. Neem has a narrow genetic base which shows a wide variation in germplasm (i.e., environment vs genotype, seed yield and composition of bioactive principles). Propagation, conventionally through seeds exhibits considerable heterogeneity which further hastens genetic erosion, leading to the loss in the potential properties and its superior architecture. Application of *in vitro* culture techniques provides a promising method for uniform propagation of 'elite' genotypes. Little initiatives were made in early 70's in neem with encouraging reports of regeneration through tissue culture (Rangaswamy and Promilla, 1972). Further efforts were made to standardise the protocol for a rapid multiplication of neem using shoot tips, cotyledons (Ramesh and Padhya, unpublished) and leaflets (Ramesh and Padhya, 1990). Plant cell cultures have long been regarded as a source for commercially important compounds, as they sustain uniformity at a desired state (Zenk et al., 1977a). Increase of production efficiency per cell is an essential factor for commercial exploitation of tissue cultures, which need to take a fresh look over the biochemical control of metabolic flux to manipulate the activity of secondary pathways and to select the mutants with practical value (Chaleff, 1983).

Flavonoids are widely used in the treatment of capillary fragility, retinal haemorrhage and in radiation diseases (Tripathi and Rastogi, 1981). Flavonoids in particular, kaempferol and quercetin in their methyl derivatives exhibit pronounced antiviral activity against herpes simplex (Vlietinck et al., 1988). Biosynthesis of flavonoids has been studied in *Petroselenium hortense* cell cultures with respect to enzymes involved in the biosynthesis and its controlling factors (Hahlbrock, 1976). In our study, the ability of neem cell cultures to synthesize/accumulate kaempferol and quercetin has been described. Also, an attempt was made to identify the mutants to produce highly productive and stable lines for a commercial range application.

MATERIALS AND METHODS

Cell Suspension Culture

Callus was initiated from stem, leaf and flower explants of an elite plant and grown on WB standard medium (Ramesh and Padhya, 1987). Healthy growing callus was transferred to liquid standard medium and the culture flasks were continuously agitated at 100 rpm on rotary shaker kept under 1000 lux illumination for 16/8 photoperiod at 25 ± 2°C. After 14 days of incubation when cells were separated from callus tissue, they were filtered through 0.22μ pore into the liquid medium to obtain homogenous cell suspension cultures.

Measurement of Growth

Cells from suspension culture were harvested at regular intervals by suction filtration and measured their fresh weights. The tissue was dried at 60°C to take the dry weights. Cell number was determined according to the procedures followed by Mac Carthy et al., (1980) using haemocytometer. The viability of cells was tested by staining them with Evan's blue solution.

Selection of Cell Lines

The minimum inhibitory concentration of parafluorophenylalanine (PFP) was determined by dose response curve with respect to the growth inhibition of cells in suspension cultures (Gonzales et al., 1984). Depending on the sensitivity of the cell line, the resistant cells to PFP concentration of 10–70 μM/1 were obtained through cell plating technique, with an initial inoculum of 1.5 × 10^5 cells/ml/plate.

Extraction, Isolation and Estimation of Flavonoids

Each powdered sample was extracted with aqueous methanol in reflux. The methanolic extract was sequentially separated with petroleum ether, chloroform and ethyl acetate (Wahlen and Mabry, 1979). Ethyl acetate extract was hydrolysed with 2N HCl to remove sugar moieties. Further separation of flavonoids from chloroform and ethyl acetate fractions was done on sephadex LH20 column chromatography. The purity of isolated compounds was confirmed by UV and IR spectra. The quantitative estimation was made spectrophotometrically according to the method described by Block et al. (1958).

RESULTS AND DISCUSSION

The highest amounts of flavonoids, kaempferol (1.93 per cent) and quercetin (1.43 per cent) were obtained from flowers which is in accordance with the results obtained by us from different parts of the plant (Ramesh, 1992). The results agree with the differences observed in different callus strains and showed that, flower originated cultures possess a greater capacity for kaempferol (0.8 per cent) and quercetin (0.68 per cent) synthesis. This reveals the fact that secondary product formation is organ specific where the phase dependent enzymes associated with membranes of ER might demonstrate the biosynthesis and secretion of flavonoids (del Rio et al., 1992). Detailed studies on cell cultures of flower callus showed that the maximum flavonoids accumulation was obtained at the end of three week culture period where they show a decline in the growth rate (Fig. 1). The inhibition of secondary product synthesis at initial phase of rapidly growing cultures could be the result of competition for a common substrate between secondary and primary pathways and, a change in the utilization of intermediates might occur only at stationary phase. This pattern of product accumulation during stationary phase of cell cultures lead to the introduction of 2-phase nutrient concept an initial growth medium which stimulates cell proliferation, and the production medium which inhibits growth and stimulate the secondary product biosynthesis (Zenk et al., 1977a) for the establishment of productive cell culture processes. The nutrient growth media used for cell cultures consist of carbon source, inorganic nutrients, vitamins, aminoacids and growth regulators. Several growth media used to obtain optimal biomass production showed that, Wood and Braun medium (1961) containing low levels of inorganic nutrients, B5 vitamins and aminoacids supplemented with sucrose (2 per cent), Kn (2μM/1) and NAA 4μM/1) were suitable for supporting the optimal biomass yield and also for the biosynthesis of products.

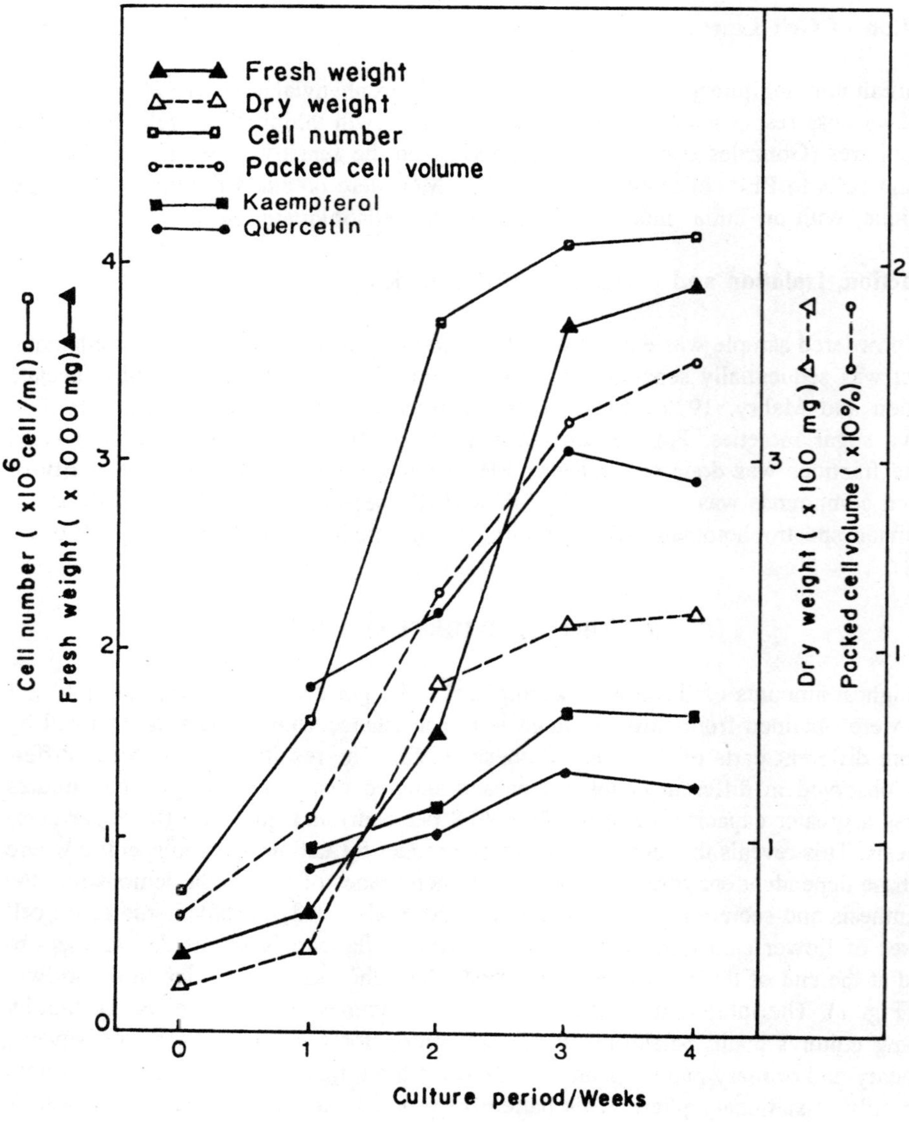

Fig. 1. Growth kinetics and flavonoid accumulation in cell suspension cultures of petal callus

Effect of Carbon Source

Sucrose is the preferential carbon source in cell cultures. When it is replaced by alternative sources of soluble carbohydrates, the yield of flavonoids as well as tissue biomass was declined.

Even sucrose concentration greatly influences the production of metabolites of the phenylpropanoid pathway (Ibrahim, 1987). The concentration of sucrose has been varied from 0–6 per cent in the medium showed a linear increase in fresh and dry weights and also in the amounts of flavonoids upto 3 per cent sucrose at its optimal concentration, beyond which a sharp decline was observed in these values (Fig. 2). The best production of rosmarinic acid was achieved in *Anchusa officinalis* with 3 per cent sucrose (De Eknamkul and Ellis, 1984) although varied optimal concentrations of sucrose were tried in *Coleus blumei* cell cultures (Ulbrich et al., 1985; Zenk et al., 1977b).

Effect of Nitrogen

Nitrogen in the culture medium is supplied as nitrate and ammonia in inorganic form and also as aminoacids. The relationship between nitrogen supply, growth and product accumulation, as shown in Fig. 3a indicates that lower levels of inorganic nitrogen inhibited the growth of cell culture and flavonoid synthesis. Also, when the nitrogen concentration was increased to 60 μM/l equivalent to that of MS medium, there was a slight reduction in the biomass yield as well as in flavonoid contents. The results showed that the inorganic nitrogen present in WB medium (30 μM/l) equally supported the growth of cell suspension and the accumulation of flavonoids. Hahlbrock (1977) reported a significant increase in the production of phenolic compounds and the activity of some enzymes of phenylpropanoid biosynthesis at decreased nitrate concentrations. This attributes to the fact that nitrogen level of WB medium supported better than those levels in MS medium for flavonoid accumulation.

Nitrogen as organic source in the medium enhanced the biomass production when supplemented with casein hydrolysate (400 ppm). However, no significant increase in flavonoid content was observed (Fig. 3b). The increased fresh and dry weights with CH can be related to the increased proportions of starch accumulation.

Effect of Plant Growth Regulators

Growth regulators are effective triggers for the synthesis of secondary products as well as primary metabolism (Mantell and Smith, 1983). In our study, highest yield of kaempferol and quercetin was obtained at reduced levels of cytokinins and auxins. Cytokinins were found effective only when added along with auxin, since the growth and product accumulation reduced in cell suspension when Kn or BAP alone was present in the medium. Kinetin at 2 μM/l was superior over BAP for flavonoid synthesis. Further increase in Kn showed inhibition of product synthesis as well as growth in cell suspension cultures as shown in Fig. 4. Auxins were effective at 4 μM/l for product synthesis and biomass yield when supplemented in medium containing 2 μM/l kinetin. A linear increase in fresh and dry weights of tissue and in flavonoid contents was observed with the increased concentrations of either IAA or NAA upto 4 μM/l. Further increase in auxin concentration (8 μM/l) reduced both the values. Biomass production and flavonoid synthesis were comparatively more in NAA (4 μM/l) supplemented medium than in the medium containing IAA. Similarly in *Morinda citrifolia* NAA stimulated authraquinone production, while 2, 4-D totally suppressed the growth and product synthesis (Zenk et al., 1975), as was noticed in this system.

Effect of Precursor Feeding

Phenylalanine is the precursor for flavonoids (Hahlbrock, 1981). When added to the standard medium at different concentrations (10, 20, . . . 100 μM/l) it was observed

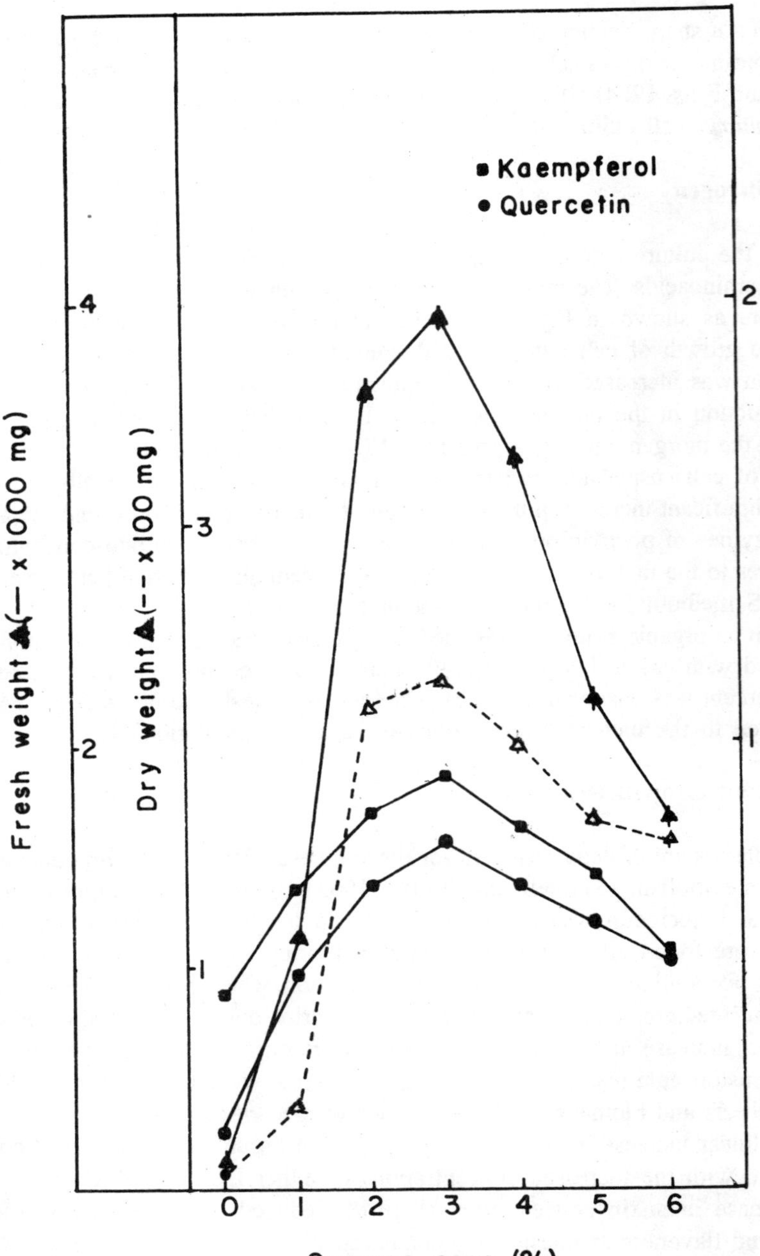

Fig. 2. Effect of sucrose levels on growth and flavonoid accumulation in cell suspension

that the dry matter decreased from control values with the increase in aminoacid concentration. The highest values of kaempferol (1.42 per cent) and quercetin (1.19 per cent) were obtained at 20 μM/l of phenylalanine (Table 1), which are very insignificant. It is reported that, the effects of phenylalanine differ as per the medium used. On

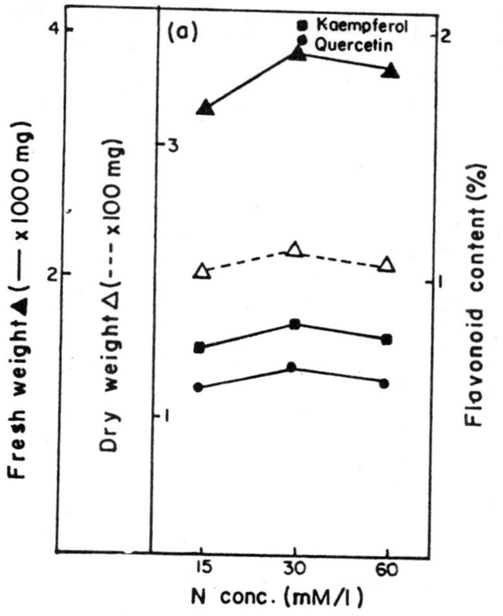

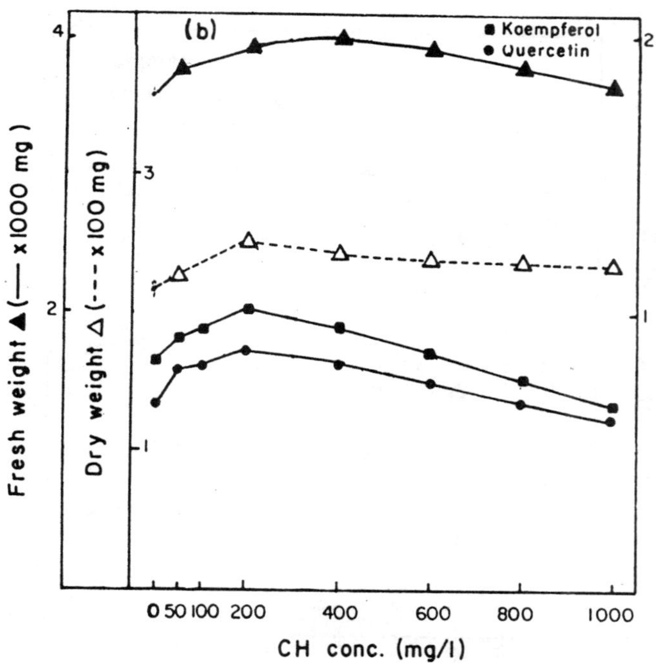

Fig. 3. Effect of (a) nitrogen levels and (b) Casein hydrolysate on growth and flavonoid synthesis in cell suspension

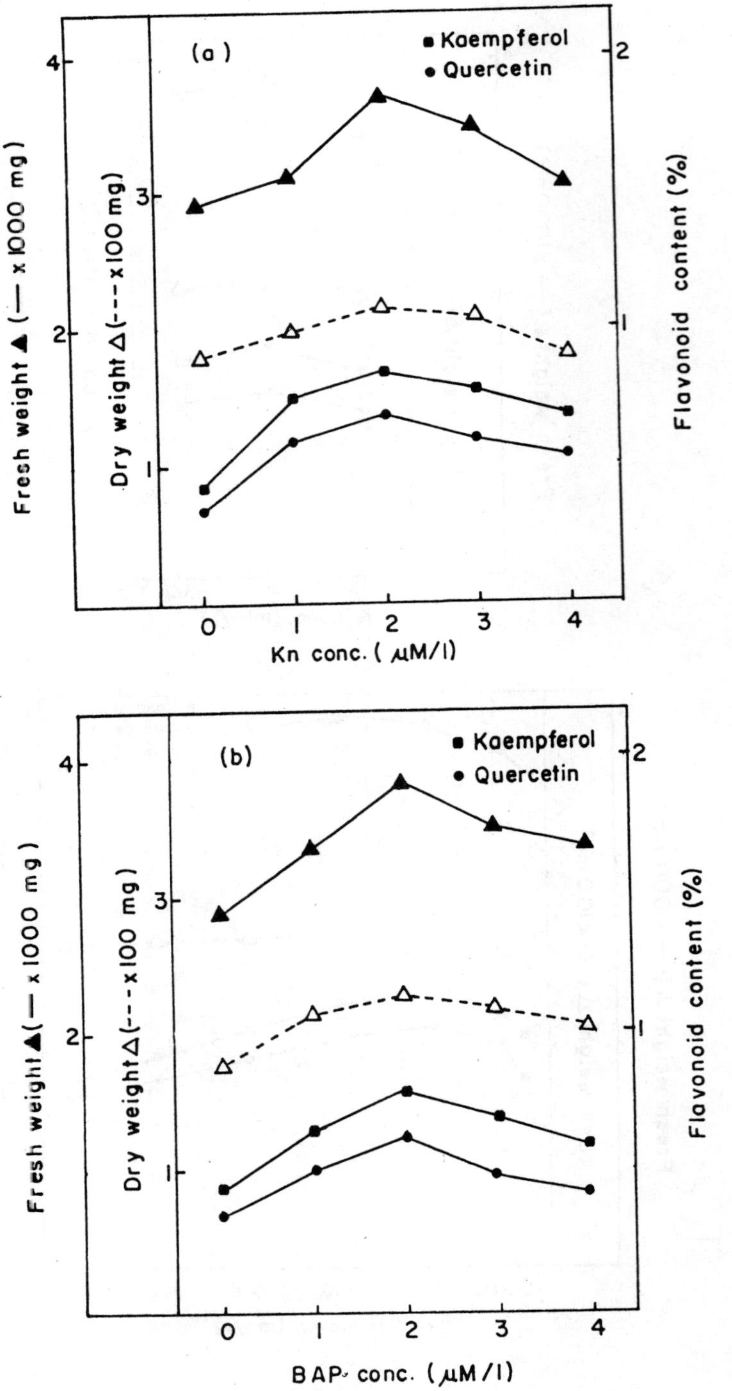

Fig. 4. Effect of phytohormones on growth and flavonoid accumulation in cell suspension

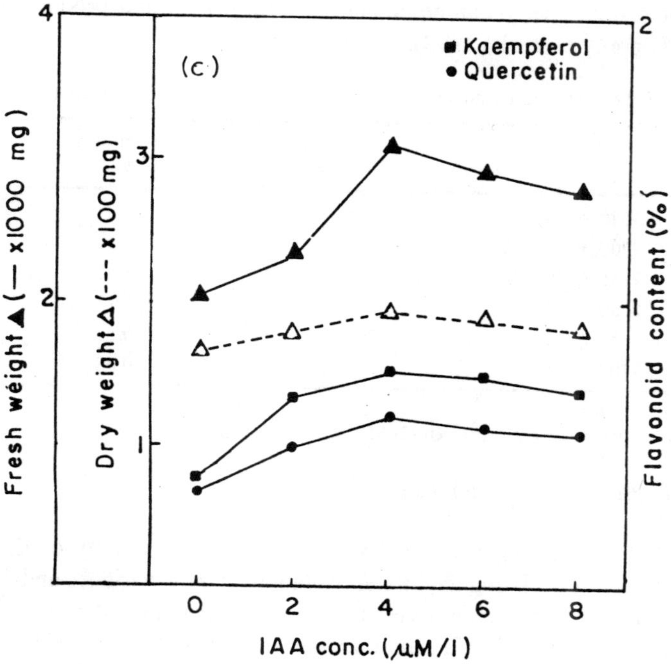
(c)
■ Kaempferol
● Quercetin
Fresh weight ▲ (— x1000 mg)
Dry weight △ (--- x100 mg)
Flavonoid content (%)
4
3
2
1
2
1
0
2
4
6
8
IAA conc. (µM/l)

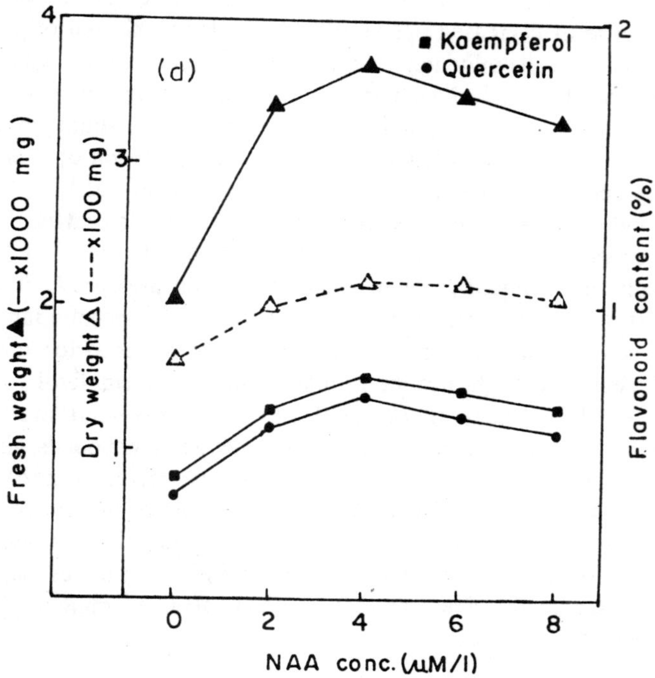
(d)
■ Kaempferol
● Quercetin
Fresh weight ▲ (— x1000 mg)
Dry weight △ (--- x100 mg)
Flavonoid content (%)
4
3
2
1
2
1
0
2
4
6
8
NAA conc. (µM/l)

Fig. 4. (contd . . . .)

contrary to our results phenylalanine enhanced both growth and productivity in *Salvia* cell cultures (Hippolyte et al., 1992).

Table 1. Effect of l-phenylalanine on growth and flavonoid synthesis in cell suspension culture

L-phenylalanine (μM/l)	Fresh wt. mg	Dry wt. mg	Kaempferol %	Quercetin %
0	3680 ± 55	215 ± 15	0.84	0.68
10	2700 ± 50	220 ± 15	1.34	1.00
20	2385 ± 45	215 ± 15	1.42	1.19
30	2250 ± 45	210 ± 15	1.29	0.96
40	1850 ± 20	208 ± 15	0.85	0.59
50	1700 ± 15	205 ± 15	0.68	0.53

Mean of 6 replicates with S.D.
Medium: WB basal, sucrose (2 per cent), Kn (2μM/l) and NAA (4μM/l).

Selection of High Yeilding Cell Lines

Suspension culture is heterogenous in terms of biosynthetic activity of its cells. Selection of 'clones' via cell plating facilitates isolation of rare variants and high yielding cell lines as a distinct independent cell colonies (Conner and Meredith, 1989).

Parafluorophenylalanine—PFP, an analogue of l-phenylalanine is toxic as a result of being incorporated into protein synthesis and producing altered protein molecules. These cells become resistant either due to over-synthesis of corresponding natural aminoacid or because of decreased uptake of PFP. Cells (from petal originated callus) were plated (1.5×10^5 cells/ml/plate) on the medium supplemented with 10, 20, . . . , 120 μM/l PFP and incubated under normal culture conditions for 4 weeks. The cell colonies were obtained in control and PFP supplemented medium up to 50 μM/l. Chemical analysis of these cell colonies showed the accumulation of kaempferol—2.5 per cent and quercetin—2.01 per cent, being maximum at 50 μM/l PFP. The plating efficiency is found to be 9 per cent when compared to 32 per cent in the control. A low profile of secondary compounds was recorded beyond 50 μM/l PFP. Isolated cell colonies were cultured in liquid medium without selection pressure (PFP). They were once again plated on medium containing corresponding PFP concentration as described above. The cell colonies obtained from these plates were screened for kaempferol and quercetin which showed a slight decline in their quantities, kaempferol—2.42 per cent and quercetin—1.83 per cent at 50 μM/l PFP. PFP resistant mutants excrete phenylpyruvate, hydroxyphenylpyruvate and other phenolic compounds due to the loss of feedback control (Barker and Lewis, 1974). Similarly, increased phenolic contents were found in the stable PFP resistant cell lines of tobacco and carrot cell cultures where the PFP resistance resulted due to increased production of phenylalanine (Palmer and Widholm, 1975) and the decreased uptake of PFP contributes partially to the overall resistance due to the oversynthesis of phenylalanine ammonialyase, the key enzyme of flavonoids (Berlin et al., 1981).

ACKNOWLEDGEMENT

The authors wish to thank scientists at Natural Products Chemistry Division of Regional Research Laboratory (CSIR) Jammu, for their help in chemical analysis.

REFERENCES

Barker, C. and D. Lewis. 1974. Impaired regulation of aromatic amino acid synthesis in a mutant resistant to p-fluorophenylalanine. *J. Gen. Microbiol.*, 82: 337–343.

Berlin, J., K.G. Ku ɔschoe and K.H. Knolboch. 1981. Selection of tobacco cell lines with high yields of cinnamoyl putresiences. Planta Med., 42: 162–172.

Block, R.J., E.L. Durraum and G. Zweig. 1958. A mannual of paper chromatography and electrophoresis. Academic press, Inc. Publishers, NewYork., pp. 329–336.

Chaleff, R.S. 1983. Isolation of agronomically useful mutants from plant cell cultures. *Science* 219: 676–682.

Conner, A.J. and C.P. Meredith. 1989. Genetic manipulation of plant cells. The biochemistry of plants, vol. 15 (A Marcus, ed.) Academi press, Inc. California. pp. 653–688.

De Eknamkul, W. and B.E. Ellis. 1984. Rosmarinic acid production and growth characterization of *Anchusa officinalis* cell suspension cultures. *Planta Med.*, 50: 346–350.

del Rio, J.A., A. Ortuno, F.R. Marin, D. Garcia Puig and F. Sabater. 1992. Bioproduction of neohesperidin and naringin in callus cultures of *Citrus aurantium. Plant Cell Rep.*, 11: 592–596.

Gonzales, R.A., P.K. Das and J.M. Widholm. 1984. Characterization of cultured tobacco cell lines resistant to ethionine, a methionine analog. *Plant Physiol.*, 74: 640–644.

Govindachari, T.R. 1992. Chemical and biological investigations on *Azadirachta indica* (the neem tree). *Current Sci.*, 63: 117–122.

Hahlbrock, K. 1976. Regulation of phenylalanine ammonia-lyase activity in cell suspension cultures of *Petroselinum hortense*. Apparent rates of enzyme synthesis and degradation. *Eur. J. Biochem.*, 63: 137–145.

Hahlbrock, K. 1977. Regulatory aspects of phenylpropanoid biosynthesis in cell cultures. In: *Plant tissue culture and its Biotechnological Application*. W. Barz, E. Reinhard, M.H. Zenk (eds.), Springer-Verlag, Berlin. pp. 95–11.

Hahlbrock, K. 1981. Flavonoids. In: *The biochemistry of plants*. Vol. 7 E.E. Conn, (ed.), Academic Press, New York. pp. 425–429.

Hippolyte, I., B. Marin, J.C. Baccou and R. Jonard. 1992. Growth and rosmarinic acid production in cell suspension cultures of *Salvia officinalis*. *Plant Cell Rep.*, 11: 109–112.

Ibrahim, R.K. 1987. In: *Cell culture and somatic cell genetics of plants*. Vol. 1 F. Constable and I. Vasil (eds.), Academic Press, Inc. London. pp. 77–96.

Mac Carthy, J.J., D. Ratcliffe and H.E. Street. 1980. The effect of nutrient medium composition on the growth cycle of *Catharanthus roseus* G. Don. cells grown in batch culture. *J. Exptl. Bot.*, 31: 1315–1325.

Mantell, S.H. and H. Smith. 1983. Cultural factors that influence secondary metabolite accumulations in plant cell and tissue cultures. In: *Plant Biotechnology* S.H. Mantell and H. Smith (eds.), Cambridge University Press, Cambridge. pp. 75–110.

Palmer, J.E. and J.M. Widholm. 1975. Characterization of carrot and tobacco cell cultures resistant to P-fluoro phenylalanine. *Plant Physiol.*, 56: 233–238.

Ramesh, K. 1992. Tissue culture studies of medicinally important plants. Ph.D. thesis. M.S. University Baroda.

Ramesh, K. and M.A. Padhya. 1987. Prospects of tissue culture studies on neem. Neem Newsletter 4: 25–27.

Ramesh, K. and M.A. Padhya. 1990. *In vitro* propagation of neem, *Azadirachta indica* (A Juss.) from leaf discs. *Ind. J. Expt. Biol.*, 28: 932–935.

Rangaswamy, N.S. and M.A. Promilla.1972. Morphogenesis of the adult embryo of *Azadirachta indica* A Juss. *P Flanzenphysiol.*, 67: 377–379.

Tripathi, V.D. and R.P. Rastogi. 1981. Flavonoids in biology and medicine. *J. Sci. Ind. Res.*, 40: 116–124.

Ulbrich, B., W. Wiesner and H. Arens. 1985. In: *Primary and secondary metabolism of plant cell cultures*. K.H. Neumann, W. Barz and E. Reinhard (eds.), Springer-Verlag, Berlin. pp. 293–303.

Vlietinck, A.J., D.K. Vanden Berghe and A. Haemers. 1988. Present status and prospects of flavonoids as anti viral agent. In: *Plant flavonoids in biology and medicine*. II *Biochemical cellular and medicinal properties*. pp. 283–299.

Wahlen, M.D. and T.J. Mabry. 1979. New 8-hydroxy flavonoids from *Solanum* section *Androceras*. *Phytochem.*, 18: 263–265.

Wood, H.N. and A.C. Braun. 1961. Studies on the regulation of certain essential biosynthetic systems in normal and crown gall tumor cells. *Proc. Natl. Acad. Sci.* (USA) 47: 1907–1913.

Zenk, M.H., H. El Shagi and U. Schulte. 1975. Anthraquinone production by cell suspension cultures of *Morinda citrifolia*. *Planta Med. Suppl.* 79.

Zenk, M.H., H. El Shagi, H. Arens, J. Streckigt, E.W. Weiler and B. Dues. 1977a. Formation of the indole alkaloids serpentin and ajmalicine in cell suspension cultures of *Catharanthus roseus* In: *Plant tissue culture and its biotechnological application.* W. Barz, E. Reinhard and M.H. Zenk (eds.). Springer-Verlag, Berlin. pp. 27–43.

Zenk, M.H., H. El Shagi and B. Ulbrich. 1977b. Production of rosmaric acids by cell suspension cultures of *Coleus blumei*. *Naturwissenschaften.* 64: 585–586.

3

NEEM CHEMISTRY

20

Chemistry of Neem (*Azadirachta indica*), a Sustainable Source of Natural Pesticides

BINA SHAHEEN SIDDIQUI, SHAHEEN FAIZI, GHIASUDDIN AND SALIMUZZAMAN SIDDIQUI

H.E.J. Research Institute of Chemistry, University of Karachi
Karachi-75270, Pakistan

ABSTRACT

Considerable amounts of synthetic pesticides are being used through out the world to control heavy loss in food and other crops by various pest species during growth, harvest and storage. Many of the synthetic pesticides previously used for insect control have been banned or their use seriously curtailed due to their negative impact on health and environment. The discovery and use of safe botanical pesticides has thus become an essential requisite. In view of these facts, studies were undertaken by this group on the various parts of *Azadirachta indica*, a tree heading the list of the plants bearing effectiveness as insect feeding deterrents, repellents and disruptants of insect growth and development. As a result, several effective formulations have been prepared and their major constituents isolated in a pure state. The chemistry of these constituents is based on classical and modern sophisticated spectroscopic methods. More than 100 new compounds have been isolated and their structures determined. An account of their chemistry with reference to the biological activity will be presented.

INTRODUCTION

Azadirachta indica A. Juss. (Neem), an indigenous tree of Indo-Pakistan subcontinent belongs to the family Meliaceae and is widely distributed in Asia, Africa and other tropical parts of the world. Almost every part of the tree has found applications in indigenous medicine for the treatment of a variety of human ailments. It has been an age-old practice in the Indo-Pakistan subcontinent to mix dried neem leaves with stored grains and to place leaves in folds of clothes and woollen blankets to keep insects away (Chopra et al., 1956; Dymock et al., 1980; Manjunath, 1948; Nadkarni, 1976).

On account of the enormous curative properties of neem, efforts were made by various groups of workers since about the beginning of the current century, to isolate its active constituents, particularly the bitter principles to which the therapeutic virtues of its various parts were attributed. Initially these studies were mainly concerned with the fatty acid components and amorphous bitter substances isolated from neem oil which is about 45 per cent of the total weight of seeds.

Siddiqui (1942) undertook a reinvestigation of the oil, employing an analytical procedure based on its partitioning between hexane and slightly diluted alcohol, avoiding its prior saponification. As a result of this study, he reported the isolation and characterization of the first two crystalline bitter constituents nimbin and nimbinin along with an amorphous bitter principle nimbidin from dilute alcoholic extracts of neem oil.

Subsequently, a host of compounds has been isolated from various parts of the tree including meliantriol (Lavie et al., 1967), salannin (Henderson et al., 1968) and azadirachtin (Butterworth and Morgan, 1968) which are active insect feeding deterrents, toxicants and/or disruptants of growth and development against a large variety of insect species and nematodes. Continuing research into the various parts of neem is still producing compounds of considerable chemical and biological interest. More recently,

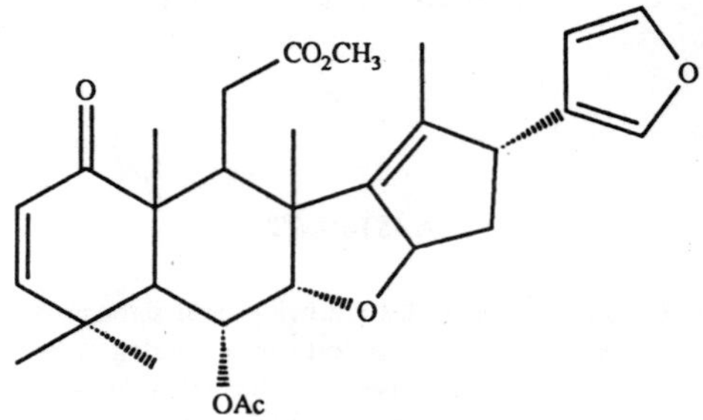

Nimbin

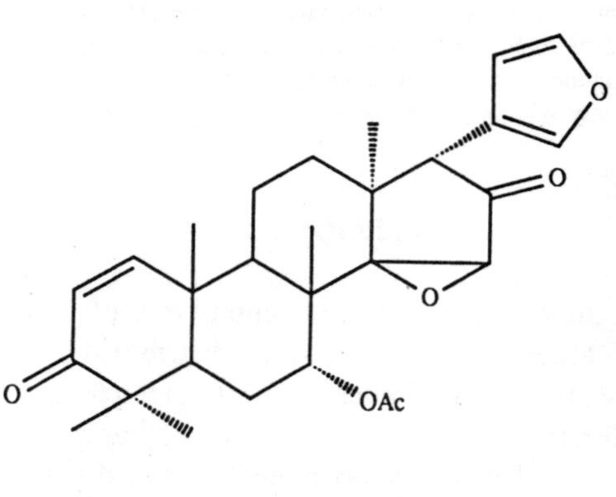

Nimbinin
(Epoxyazadiradione)
(Nimolin)

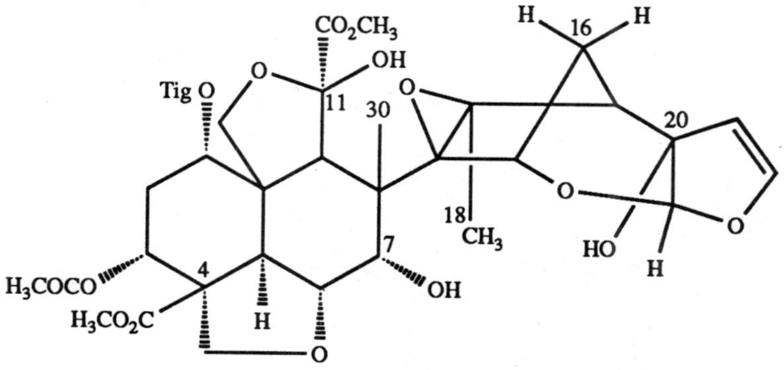

Azadirachtin

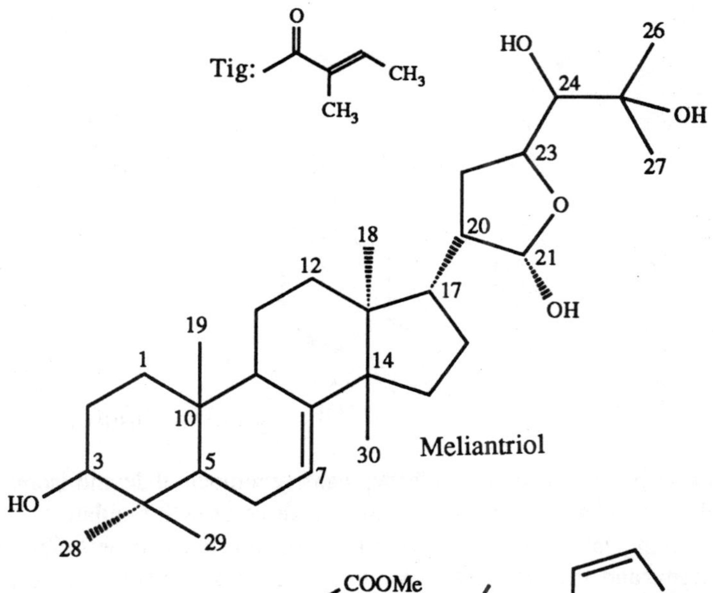

Meliantriol

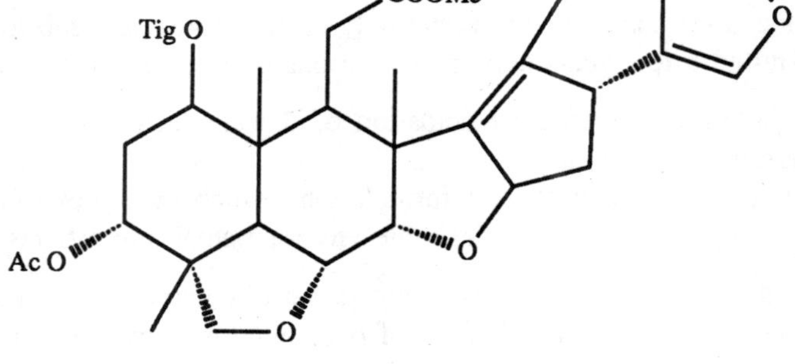

Salannin

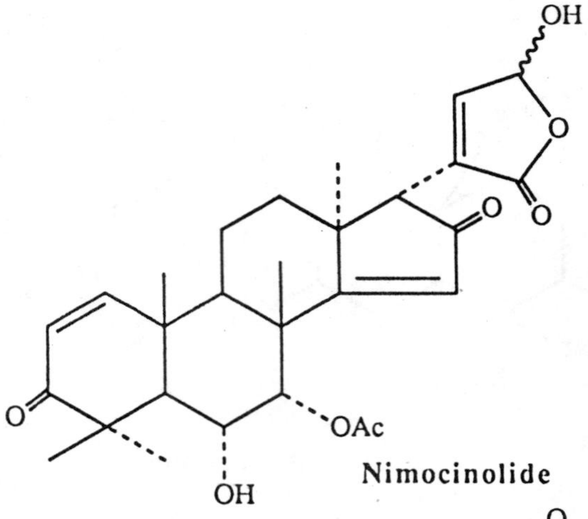

Nimocinolide

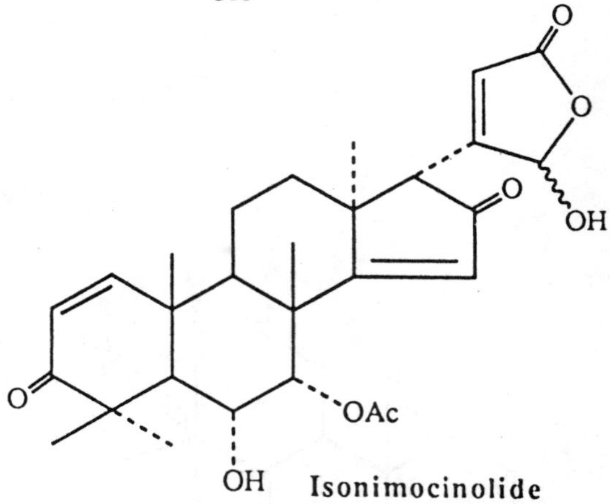

Isonimocinolide

two triterpenoids from neem leaves namely nimocinolide and isonimocinolide (Siddiqui et al., 1986) have been reported to behave as growth regulators against *Aedes aegypti.*

Taking into account the enormous therapeutic and pesticidal properties attributed to neem, and the fact that heavy loss in food and other crops caused by various pest species is a serious problem, particularly in the tropical and sub-tropical regions, studies were started a few years ago on the various parts of the neem tree with the object of:

a) isolation and structure elucidation of the chemical constituents from the various parts of the tree, and
b) obtaining active pesticidal formulations which may prove effective in controlling the target pests with undesirable environmental side effects.

Studies undertaken by this group in various parts of the tree have resulted in the isolation and structure elucidation of over hundred new compounds on the one hand, and preparation of some simple non-hazardous effective formulations on the other. These formulations can be safely applied against domestic and agricultural pests. The insects against which the activity has been tested are shown in Table 1 and some interesting and promising results of the biological activity of various fractions and pure

Table 1.

Fruit Constituents	**Insects**
	1. *Musca domestica*
	2 *Callasobruchus analis*
	3. *Aleurobus barodensis*
	4. *Oxyearenus lugubris*
	5. *Scintherrista notabilis*
	6. *Sitophilus oryzae*
	7. *Amritodus atkinsoni*
	8. *Culex fatigans*
	9. *Tribolium castaneum*
Leaves Constitutents	**Insects**
	1. *Aedes aegypti*
	2. *Musca domestica*
	3. *Blattela germanica*
	4. *Calasobruchus analis*
	5. *Sitophilus oryzae*
	6. *Tribolium castaneum*

compounds obtained from the leaves, and various parts of the fruits are depicted in Table 2.

Some of these results need particular attention. For instance, NFD, a fraction of neem leaves and its two pure compounds nimocinolide (NP1) and isonimocinolide (NP1') have LC_{50}, 0.58, 0.625 and 0.74 ppm against *Aedes aegypti* whereas margosan-O, a neem-based pesticide of U.S. origin has LC_{50} 148 ppm. Similarly, NFB, a fraction of neem leaves has LC_{50} 0.15 per cent against *Musca domestica* which is almost of the same order as that of dimilin (LC_{50} 0.14 per cent) and is about three times more effective than malathion (LC_{50} 0.33 per cent). A pure compound (NE) from the fruits was as effective (LC_{50} 0.85 μg/kg) as coopex (LC_{50} 0.80 μg/kg).

The fruit fractions NFC, RBA and RB-a are effective against *Callosobruchus analis*, *Aleurobus barodensis* and *Amritodus atkinsoni* respectively.

It would be interesting to note that when the activity of neem fruit coats (RB-b) and neem seeds (RB-a) was studied against *M. domestica*, it was noted that RB-b was 200 times more effective (LC_{50} 1.00 ppm) as compared to RB-a (LC_{50} 200 ppm).

A formulation of the same fraction (RB-b) was prepared with piperonyl butoxide and triton X-100 and applied against *Sitophilus oryzae*. It was found that this formulation has LC_{50} 6.28 μg/cm^2 as against 18.85 μg/cm^2 of that of the standard, methoprene.

These observations formed the basis of directing our studies towards the chemical constituents of the fruit coatings (RB-b) in particular and look into their potential biological activity. It may be noted in this connection that azadirachtin has been isolated from the neem seeds.

Fresh fruits collected from the Karachi region were manually separated into seeds and coatings, each with some part of the fruit pulp. Fruit coatings were repeatedly extracted with ethyl alcohol at room temperature and the thickish syrupy residue obtained on removal of the solvent from the combined extracts was separated further through solvent fractionation followed by chromatographic procedures such as vacuum liquid chromatography (VLC), flash column chromatography (FCC), thick layer chromatography (TLC) and high pressure liquid chromatography (HPLC). As a result three

Table 2.

	Leaves			Fruits							
A. aegypti	NFD	NPI	NPI'	RB-b	RBU-9	Margosan-0					
LC_{50}	0.58 ppm	0.625 ppm	0.74 ppm	575 ppm	455 ppm	148 ppm					
	Leaves			Fruits							
M. domestica	NFA	NFB	NU-1	N-7	RB-a	RB-b	NPE	NE	RBS	RBU	Coopex
LD_{50}	1.4μg per fly						1.4μg per fly	0.85μg per fly		5.6μg per fly	0.8μg per fly
		0.15%	1.06%	3.8%		3.7%			4.8%		0.029%
(Swiss Strain) LD_{50}					200 ppm	1.00 ppm					Malathion 0.33% Dimilin 0.14%
	Leaves	Fruits									
C. analis	NFQ	NFC	NPE	NC	Dimilin						
LD_{50}	0.12 mg/cm²	40 μg/cm²	1.15 mg/cm²	0.18%	20 μg/cm²						
A. barodensis	RBA (Fruits)		Malathion								
LC_{50}	1.0%		3.0%								
O. lugubris	RB-A (Fruits)		Malathion								
LC_{50}	9.2%		0.004%								

Table 2 contd.

S. notabilis	RB-b (Fruits)			
EC_{50}	12.5%			
S. oryzae	LPA (Leaves)	RB-b (Fruits) + PBO + TX – 100	Methoprene	Danitol
LC_{50}	0.50mg/cm^2	6.28μg/cm^2	18.85μg/cm^2	0.39μg/cm^2
A. atkinsoni	RB-a (Fruits)	URN (Unripe fruits)	Dimethoate	Perfekthion
LC_{50}	0.40%	1.40%	0.14%	0.14%
T. castaneum	RB-a (Fruits) + PBO + TX – 100	Malathion		
LC_{50}	0.05%	0.007%		

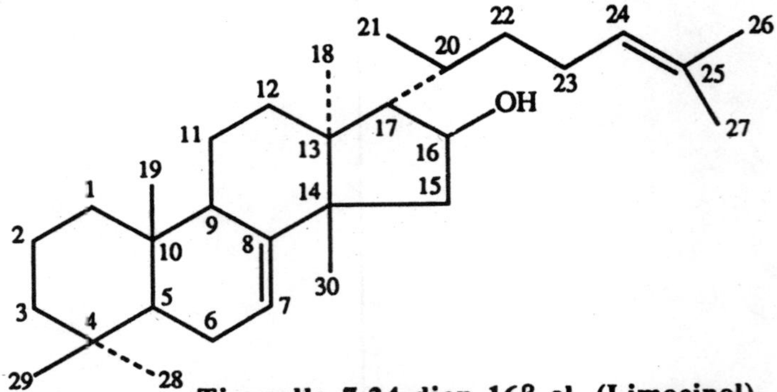

Tirucalla-7,24-dien-16β-ol (Limocinol)

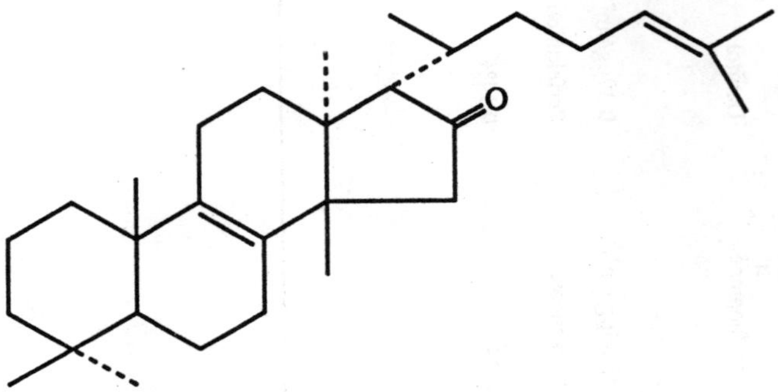

Tirucalla-8,24-dien-16-one (Limocinone)

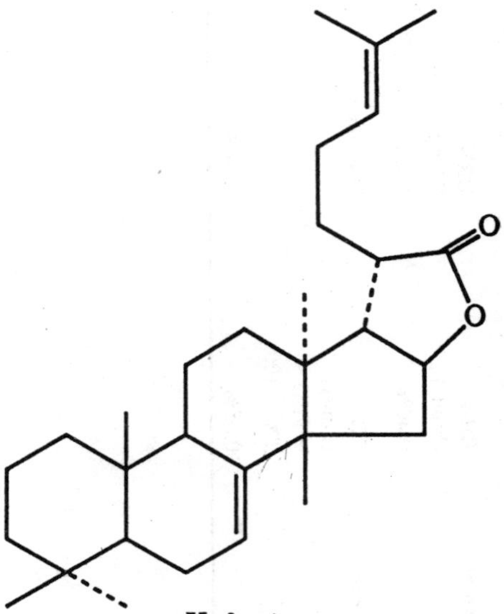

Kulactone

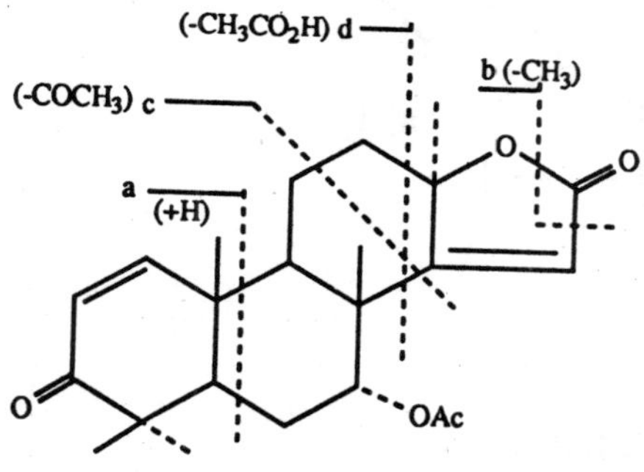

N-191

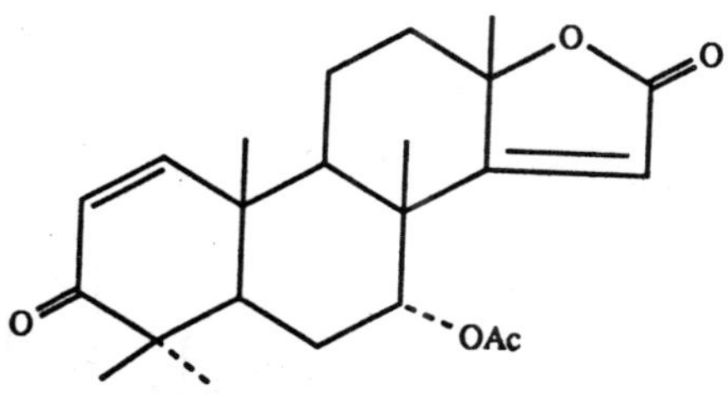

N-192

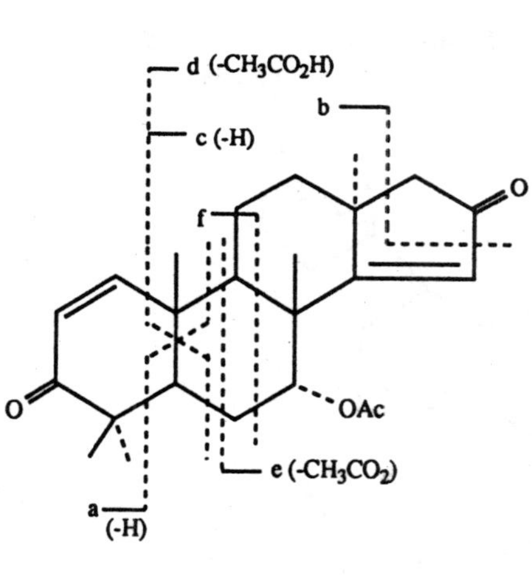

N-193

Fig. 1. Structures of N-191, N-192 and N-193

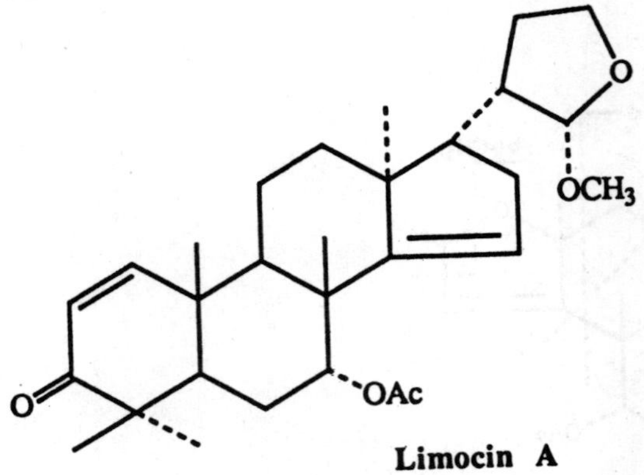

Limocin A

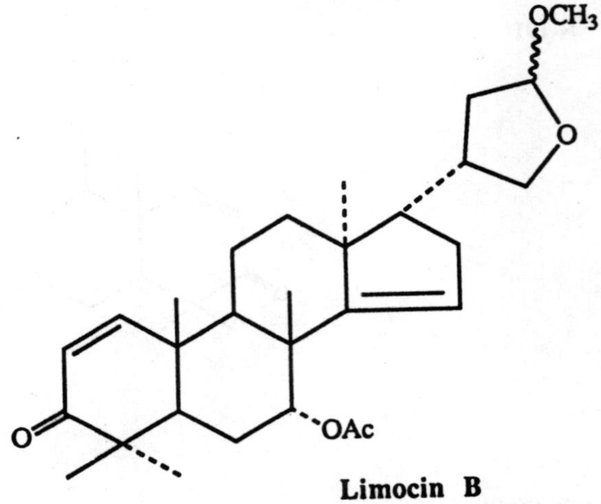

Limocin B

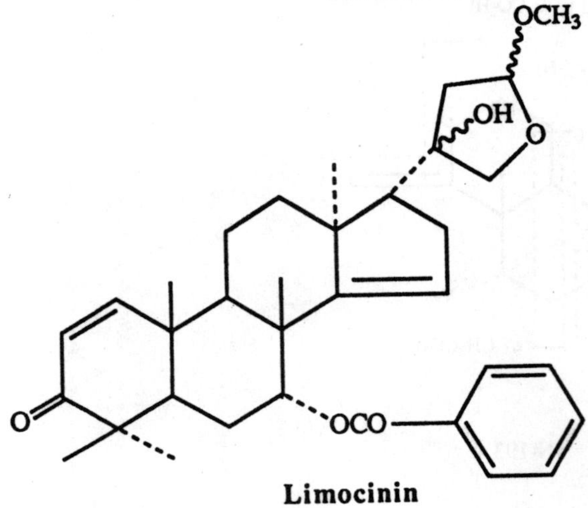

Limocinin

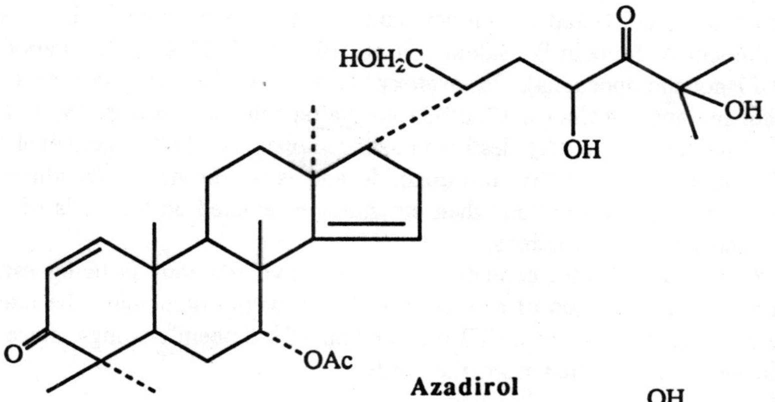

Azadirol

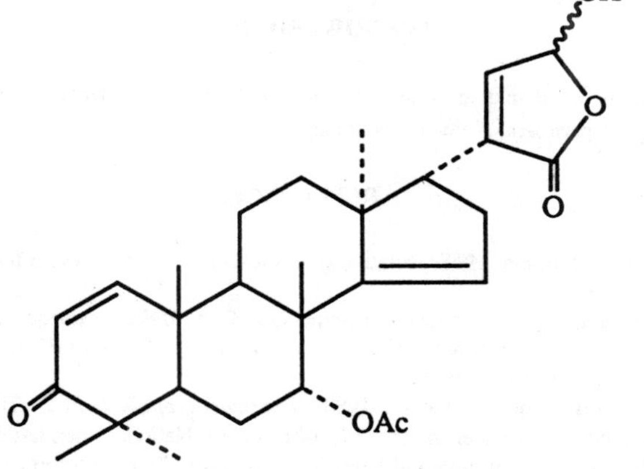

Azadironolide

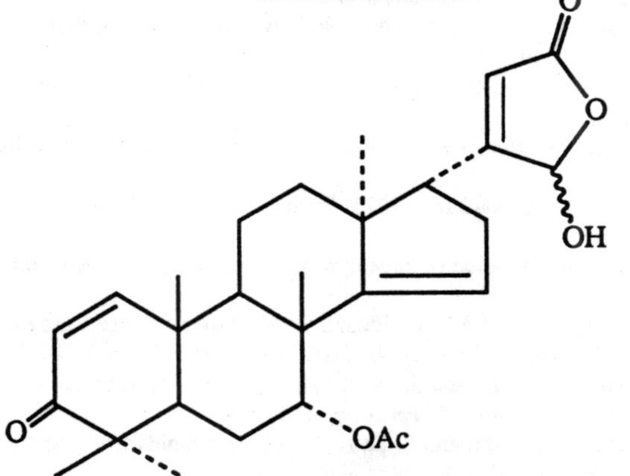

Isoazadironolide

known triterpenoids—epoxyazadiradione, azadiradione (Lavie et al., 1971) and kulactone (Chiang and Chang, 1973) and eleven new compounds namely, limocinol, limocinone, limocinin, limocin A, limocin B (Siddiqui et al., 1991a) 17, 20-27-nonanorapotirucalla (or 17, 20-27-nonanorapoeupha), 7α–acetoxy, 1, 14-dien, 13→16-γ-lactone (N-191), 17, 20-27-nonanorapotirucalla (or 17, 20-27-nonanorapoeupha), 7α-acetoxy, 1, 14-dien, 13-epi, 13→16-γ-lactone (N-192), desfurano-azadiradione (N-193) (Siddiqui et al., 1992), azadirol (Siddiqui et al., 1991b), azadironolide and isoazadironolide (Siddiqui et al., 1990) have so far been isolated and their structures elucidated on the basis of spectroscopic data and chemical reactions.

Presently the azadirachtin content in various active fractions is being estimated, apart from the characterization of new constituents, it would particularly be interesting to note the azadirachtin content in RB-b a fraction of the neem coatings, since, so far azadirachtin has been reported from the seeds.

ACKNOWLEDGEMENT

For the biological results our thanks are due to Dr. S.N.H. Naqvi and coworkers from the Toxicology Laboratory of the Zoology Department, Karachi University.

REFERENCES

Butterworth J.H. and E.D. Morgan. 1968. Isolation of a substance that suppresses feeding in locusts. *J. Chem. Soc. Chem. Commun.*, 23.

Chiang, C. and F.C. Chang. 1973. Tertracyclic triterpenoids from *Melia azedarach* L., *Tetrahedron* 29: 1911.

Chopra, R.N., S.L. Nayar and I.C. Chopra. 1956. *Glossary of Indian Medicinal Plants*, Council of Scientific and Industrial Research, New Delhi, p. 31.

Dymock, W., C.J.H. Warden and D. Hooper. 1980. *'Pharmacographia Indica.'* The Institute of Health and Tibbi Research, republished under the auspices of Hamdard National Foundation Pakistan, Vol. 1, p. 322.

Henderson, R., R. McCrindle, A. Melera and K.H. Overton. 1968. Tetranortriterpenoids-1X, The Constitution and Stereochemistry of Salannin. *Tetrahedron* 24: 1525.

Lavie, D., M.K. Jain and S.R. Shpan-Gabrielith. 1967. A locust phagorepellent from two melia species. *J. Chem. Soc. Chem. Commun.*, 910.

Lavie, D., E.C. Levy and M.K. Jain. 1971. Limonoids of biogenetic interest from *Melia azadirachta*. *Tetrahedron* 27: 3927.

Manjunath B.L. 1948. *'The Wealth of India.'* Council of Scientific and Industrial Research, New Delhi, Vol. 1, p. 140.

Nadkarni, K.M., Revised by A.K. Nadkarni. 1976. *Indian Materia Medica*, Popular Prakashan Ltd., Bombay, Vol. 1, p. 776.

Siddiqui, S. 1942. A note on the isolation of three new bitter principles from neem oil (*Melia azadirachta*) *Curr. Sci.*, 11: 278.

Siddiqui, S., S. Faizi, T. Mahmood and B.S. Siddiqui. 1986. Two new insect growth regulator meliacins from *Azadirachta indica* A. Juss (Meliaceae). *J. Chem. Soc. Perkin Trans* 1: 1021.

Siddiqui, S., B.S. Siddiqui, Ghiasuddin and S. Faizi. 1990. Isolation and structure elucidation of the constituents of *Azadirachta indica*. *Proc. Natl. Chem. Conf.* (II): 275.

Siddiqui, S., B.S. Siddiqui, Ghiasuddin and S. Faizi. 1991a. Terpenoids from the fruit coatings of *Azadirachta indica*. *Phytochemistry* 30: 1615.

Siddiqui, S., B.S. Siddiqui, Ghiasuddin and S. Faizi. 1991b. Tetracyclic triterpenoids of the fruits coats of *Azadirachta indica*. *J. Nat. Prod.*, 54: 408.

Siddiqui, S., B.S. Siddiqui, Ghiasuddin and S. Faizi. 1992. Triterpenoids from the fresh fruit coats of *Azadirachta indica*. *Phytochemistry* 31, 4275 (1992).

21

Photo-oxidation of Tetranortriterpenoids

S.R. ROJATKAR, S.P. JOSHI AND B.A. NAGASAMPAGI

National Chemical Laboratory, Pune 411 008, India

ABSTRACT

Photo-oxidation of salannin [1] and nimbin [4], two biologically active constituents of the seeds of *Azadirachta indica*, in methanol/chloroform have yielded compounds [2] and [3] from salannin and compounds [5] and [6] from nimbin. The structures of the photo-oxidized compounds [2], [3], [5] and [6] have been deduced mainly by ^1H and ^{13}C NMR spectral studies.

INTRODUCTION

An α, β-unsaturated γ-lactone of nimbin was reported from neem oil by Connolly et al. in 1968 but it was considered to be an artefact since it was not detected in fresh neem oil. Recently, a number of α, β-unsaturated γ-hydroxy butenolides have been isolated from different parts of neem (Ara et al., 1988; Garg and Bhakuni, 1984; Siddiqui et al., 1986a, 1986b, 1987a, 1987b).

MATERIALS AND METHODS

Singlet oxidation of two major tetranortriterpenoids, salannin ([1] in Fig. 1) and nimbin [4] isolated from neem seeds yielded four α, β-unsaturated γ-hydroxy butenolides: salannolide [2] and isosalannolide [3] from salannin and nimbinolide [5] and isonimbinolide [6] from nimbin.

A solution of the compound along with rose bengal in methanol/chloroform in a quartz tube was irradiated under a stream of oxygen with medium pressure 200 W Hanovia UV lamp. Reaction was monitored by thin-layer chromatography (TLC). The reaction mixture was separated by chromatographic technics into respective compounds which were identified by their spectroscopic data (UV, (IR), mass, ^1H NMR, and ^{13}C NMR).

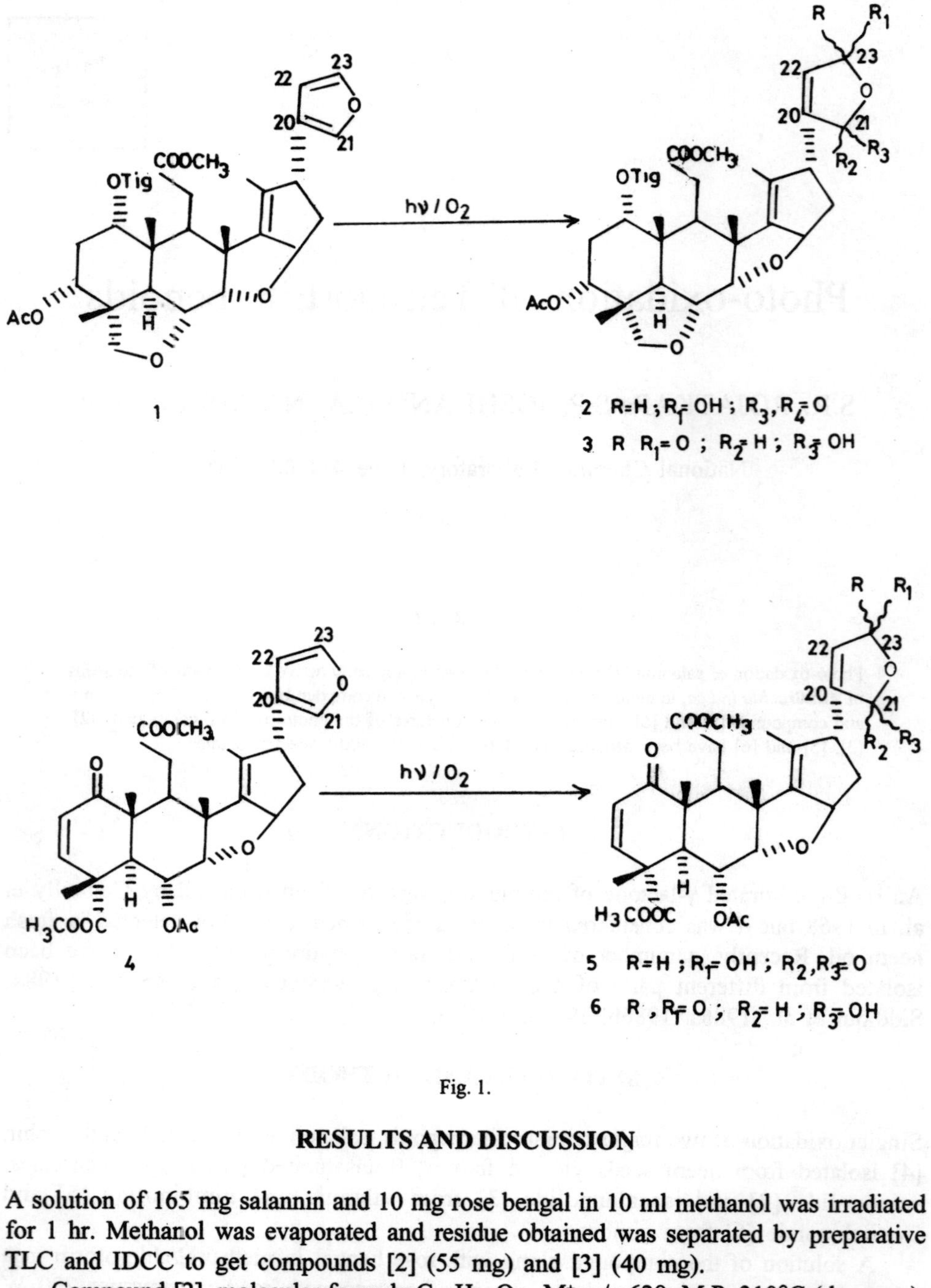

Fig. 1.

RESULTS AND DISCUSSION

A solution of 165 mg salannin and 10 mg rose bengal in 10 ml methanol was irradiated for 1 hr. Methanol was evaporated and residue obtained was separated by preparative TLC and IDCC to get compounds [2] (55 mg) and [3] (40 mg).

Compound [2], molecular formula $C_{34}H_{44}O_{11}$, M^+ m/z 628; M.P. 310°C (decomp.), was characterized as salannolide by comparing its spectral data (IR, UV, mass, ^1H NMR, and ^{13}C NMR) to those reported for salannolide (Garg and Bhakuni, 1983).

The IR, UV, mass, ^1H NMR, and ^{13}C NMR of compound [3] showed similarity to those of salannolide [2]. The presence of a 21-hydroxy-buta-20(22)-ene-21-23-γ-lactone side chain was indicated by the presence of two one-proton signals at δ5.8 (H-21, m) and 5.91 (H-22, m), the former being shifted to δ6.84 (m) upon acetylation in the ^1H NMR spectrum of acetate of [3]. In addition, the presence of signals at δc 165.09 (C-20, s), 97.55 (C-21, d), 119.82 (C-22, d), and 170.19 (C-23, s) in the 13 C NMR of [3] compared to the signals at 137.80 (C-20, s), 171.17 (C-21, s), 141.54 (C-22, d), and 96.71 (C-23, d) in the ^{13}C NMR of salannolide [2] shows that compound [3] can be characterized as isosalannolide.

A solution of 200 mg nimbin and 10 mg rose bengal in 10 ml chloroform was irradiated as in the case of salannin for 10 hrs. The solvent was evaporated and the residue obtained was chromatographed repeatedly by preparative TLC to obtain compound [5] (70 mg) and [6] (80 mg).

Compound [6] was obtained as viscous oil, molecular formula $C_{30}H_{36}O_{11}$, M^+ m/z 572. It was characterized as isonimbinolide by comparing spectral data of [6] (IR, UV, mass, and ^1H NMR) to those reported for isonimbinolide (Ara et al., 1988). ^{13}C NMR of [6] was in full agreement with the assigned structure [201.95, s C-1; 125.69 d, C-2; 147.71, d, C-3; 47.01, s, C-4; 52.97 d, C-5; 68.34, d C-6; 84.50 d, C-7; 47.97, s, C-8; 38.86, d, C-9; 52.27, s, C-10; 34.58, t, C-11; 175.66, s, C-12; 131.21, s, C-13; 147.71, s, C-14; 86.63, d, C-15; 41.53, t, C-16; 47.9, d, C-17; 16.58, q, C-18; 16.34, q, C-19; 168.56, s, C-20; 97.28, d, C-21; 119.91, d, C-22; 170.10, s, C-23; 174.27, s, C-28; 20.79, q, C-29; 13.10, q, C-30; 170.10, s, O<u>C</u>OMe; 16.89, q, OCO<u>Me</u>; 41.8, q, COO<u>Me</u>; 38.9, q, COO<u>Me</u>].

Compound [5], molecular formula $C_{30}H_{36}O_{11}$, M^+ m/z. 572, isolated as a gum, showed spectral characteristics which were very close to those of isonimbinolide. The presence of 23-hydroxy-20(22)-butene-γ-lactone side chain in [5] was shown by chemical shifts in the ^{13}C NMR at δc 137.60 (s, C-20), 170.40 (s-C-21), 141.12 (d, C-22), and 96.88 (d, C-23) compared to that of chemical shifts of isonimbinolide [6] i.e., δc 168.56 (s, C-20), 97.28 (d, C-21), 119.91 (d, C-22), and 170.10 (s, C-23).

REFERENCES

Ara, I., B.S. Siddiqui, S. Faizi and S. Siddiqui. 1988. Terpenoids from the stem bark of *Azadirachta indica*. *Phytochemistry* 27(6): 1801–1804.

Connolly, J.D., K.L. Handa and R. McCrindle. 1968. Further constituents of nim oil: The constitution of meldenin. *Tetrahedron Letters* 437–440.

Garg, H.S. and D.S. Bhakuni. 1984. Salannolide, a meliacin from *Azadirachta indica*. *Phytochemistry* 23(10): 2383–2385.

Siddiqui, S., T. Mahmood, B.S. Siddiqui and S. Faizi. 1986a. Two new tetranortriterpenoids from *Azadirachta indica*. *J. Natural Products* 49(6): 1068–1073.

Siddiqui, S., B.S. Siddiqui, S. Faizi and T. Mahmood. 1986b. Isoazadirolide, a new tetranortriterpenoid from *Azadirachta indica* A. Juss. (Meliacene). *Heterocycles* 24(11): 3163–3167.

Siddiqui, S., T. Mahmood, B.S. Siddiqui and S. Faizi. 1987a. Isonimolide and isolimbolide two tetranortriterpenoids from the twigs of *Azadirachta indica* A. Juss. (Meliacene). *Heterocycles* 26(7): 1823–1827.

Siddiqui, S., T. Mahmood, S. Faizi and B.S. Siddiqui. 1987b. Studies in the chemical constituents of *Azadirachta indica* A. Juss. (Meliaceae) Part 10. Isolation and structure elucidation of isonimolicinolide, the first 17-acetoxy tetranortriterpenoid and nimolicinoic acid, the first hexanortriterpenoid with an Apoeuphane (Apotirucallane) skeleton. *J. Chem. Soc., Perkin Trans* 1: 1429–1432.

22

New Tetra and Pentanortriterpenoids from *Azadirachta indica* A. Juss.

S.R. ROJATKAR, D.D. SAWAIKAR AND B.A. NAGASAMPAGI

National Chemical Laboratory, Pune 411 008, India

ABSTRACT

Two new tetranortriterpenoids and one new pentanortriterpenoid have been isolated along with two known tetranortriterpenoids from the methanol extract of *Azadirachta indica* seeds. Their structures were proposed on the basis of ^{1}H-NMR spectral data.

INTRODUCTION

The Indian neem tree (*Azadirachta indica* A. Juss) has attracted the attention of many chemists and biologists all over the world during the past two decades because of its plethora of biological activities. Azadirachtin (I) (Fig. 1) is one of the most interesting constituents of neem because of its influence on insect feeding behaviour and insect development (Kraus et al., 1985, 1987). A number of derivatives of azadirachtin exhibiting similar activity have been isolated and identified. During our work on the isolation and identification of new constituents of neem seed extract we have isolated one new derivative (II) and (III) each of the azadirachtin and azadirachtinin (IV) respectively along with a new benzoyl derivative (V) of nimbandiol (VI) and two known compounds identified as ochinin acetate (VII) and ochinolide B (VIII).

Compound (V) is a derivative of nimbandiol (VI) (Kraus and Bokel, 1981) in which the hydroxy group at C-4 is replaced by a benzoyl group. This is the first report of the isolation of ochinin acetate (VII) and ochinolide B (VIII) from neem seeds although they are known to occur in the related species *Melia azedarach* (Ochi et al., 1978, Kraus and Bokel, 1981). Compound (VII) has been earlier isolated as acetylated product of its natural precursor. Compound (VIII) is reported to be a good antifeedant against *Spodoptera littoralis.*

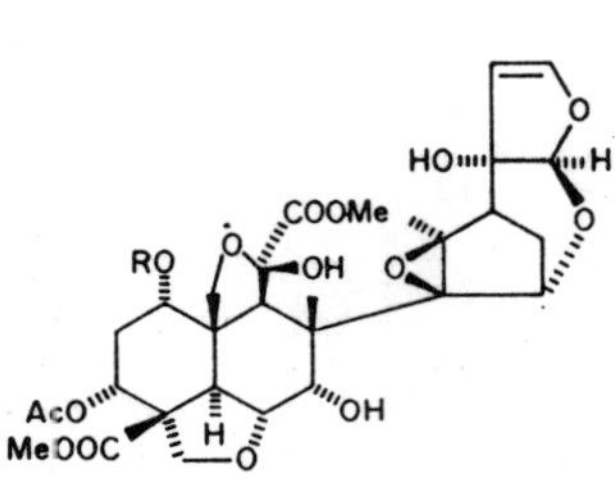

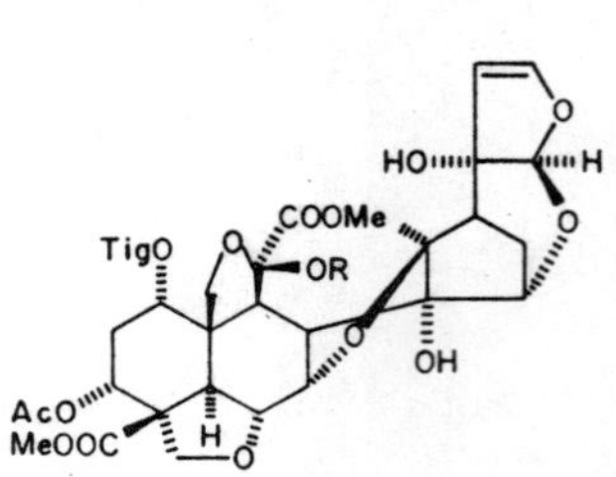

I R = Tig
II R = Bz

III R = H
IV R = Me

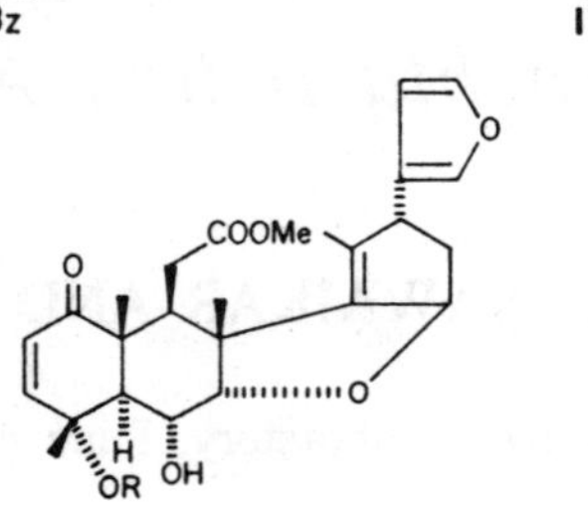

V R = Bz
VI R = H

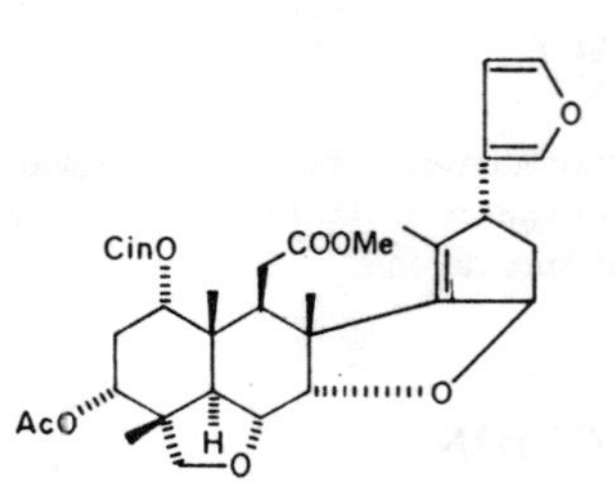

VII

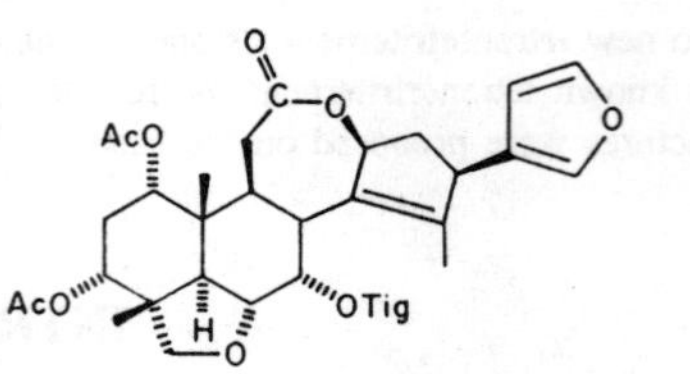

VIII

Fig. 1

RESULTS AND DISCUSSION

1α-Destigloyl-1α-benzoylazadirachtin (II)

The structure of (II) $C_{37}H_{42}O_{16}$ (724) was deduced by comparison of its ^{1}H NMR spectral data with those of (I). The proton signals in the spectra of both (II) and (I) showed striking similarity except for the ester substituent at C-1. The characteristic signals attributable to the tiglate group in (II) were missing and instead a multiplet integrating for five protons at δ 7.32–7.88 attributable to aromatic protons is found. This indicated the presence of a benzoyl group in (II). As the chemical shifts of H-3 in the ^{1}H-NMR spectra of both (I) and (II) were almost the same the benzoyl group could conveniently be placed at C-1 in (II) to assign the structure as 1α-destigloyl-1α-benzoylazadirachtin to it.

11β-Hydroxyazadirachtinin (III)

Compound (III), molecular formula $C_{35}H_{44}O_{16}$, showed in its ^{1}H-NMR spectrum signals assignable to an acetoxy group, a tigloyl group and two carbomethoxy substituents. A critical study of its ^{1}H-NMR spectrum revealed that the compound under discussion belonged to azadirachtinin type and displayed close similarity with the 1H-NMR spectrum of (IV) with respect to the following signals. (1) Two signals at δ 4.88 and 6.36 were assigned to the enol ether protons at 22-H and 23-H and a low field singlet at 8 5.65 to H-21. (2) The four spin system 15-H, 16-Ha, 17-H were assigned the signals at δ 4.18, 2.25, 1.9 and 2.16 respectively. (3) Signals at δ 4.72, 2.28, 2.13 and 5.28 were assignable to another four spin system 1-H, 2-Hα,β and 3-H respectively. (4) An AB system due to 28-Hα, β was attributed to signals at δ 3.69 and 4.04. (5) The signals at δ 3.30, 4.42 and 4.48 were assignable to the three spin system, 5-H, 6-H and 7-H respectively. (6) Two methyl signals assignable to 18-H and 30-H were located at δ 1.55 and 1.68. The above assignments were achieved by ^{1}H-^{1}H homonuclear decoupling. The only difference in the spectrum of [III] was the absence of a signal (singlet assignable to methoxy group present at C-11 in [IV]. Instead it showed the presence of three hydroxy groups as determined by D_2O exchange experiments infering the presence of a hydroxy group at C-11. The small chemical shifts with respect to H-16α, β protons, H-19α, β protons and methyl signals at δ 1.55 to 1.69 in [III] can be attributed to the presence of 11β-hydroxy group in place of a methoxy group as in [IV]. The spectral data compared very well with those of 11(RS)-hydroxyazadirachtinin, a transformation product of [I] (Ley et al., 1989). Based on the above data compound [III] has been assigned to the structure of 11β-hydroxydazadirachtinin.

4α-benzoyl nimbandiol [V]

^{1}H NMR spectrum of [V], $C_{33}H_{36}O_8$, was similar to that of nimbandiol [VI] except for the signals due to hydroxy groups (Kraus and Crammer, 1981). The spectrurn did not show the signal due to 4-hydroxy group at δ 4.29 but displayed a multiplet at δ 7.48–7.82 assignable to benzoyl group indicating the presence of a benzoyl group at C-4. The downfield chemical shift of H-6 at δ 4.27 (ddd, J=10, 7.5, 3.5 Hz) from δ 3.28 (d, J=7.5 Hz) in nimbandiol could be due to the presence of a benzoyl group at C-4. Based on these data the structure 4α-benzoyl nimbandiol has been assigned to [V].

Ochinin Acetate [VII] and Ochinolide B[VIII]

All the ^{1}H NMR spectral features of compound [VII] and [VIII] were exactly identical to those reported in literature (Ochi et al., 1978; Kraus and Bokel, 1981).

REFERENCES

Kraus, W. and M. Bokel. 1981. Neue tetranortriterpenoide aus *Melia azedarach* Linn (Meliaceae). *Chem. Ber.* 114: 267–275.

Kraus, W., M. Bokel, A. Klenk and H. Pohnl. 1985. The structure of azadirachtin and 22,23-dihydro-23β-methoxy azadirachtin. *Tetrahedron Lett.* 26: 6435–6458.

Kraus, W., M. Bokel, A. Bruhn, R. Crammer, I. Klaiber, A. Klenk, G. Nagl, H. Pohnl, H. Sadlo and B. Vogler. 1987. Structure determination by NMR of azadirachtin and related compounds from *Azadirachta indica* A. Juss. *Tetrahedron* 43: 2817–2830.

Kraus, W. and R. Crammer. 1981. Pentanortriterpenoide aus *Azadirachta indica* A. Juss (Meliaceae). *Chem. Ber.* 114: 2375–2381.

Ley, S.V., J.C. Anderson, W.M. Blaney, P.S. Jones, Zev Lidert, E.D. Morgan, N.G. Robinson, D. Santafianos, M.S., J. Simmonds and P.L. Toogood. 1989. Insect antifeedants from *Azadirachta indica* (Part-5): Chemical modification and structure-activity relationships of azadirachtin and some related limonoids. *Tetrahedron* 45: 5175.

Ochi, M., H. Kotsuki, T. Kataoka, T. Tada and T. Tokoroyama. 1978. Limonoids from *Melia azedarach* Linn. Var. *Japonica makino* III. The Structure of Ochinal and Ochinin acetate. *Chem. Letters* 331.

23

Azadirachtin Content and Bioactivity of Some Neem Ecotypes of India

SEENI RENGASAMY, NUTAN KAUSHIK, JITENDRA KUMAR, OPENDER KOUL[1] AND BALRAJ SINGH PARMAR

Division of Agricultural Chemicals, Indian Agricultural Research Institute
New Delhi 110 012, India
[1] Regional Research Laboratory, Jammu 180 001, India

ABSTRACT

Neem ecotypes of India showed varying azadirachtin content depending on the climate, soil type, and altitude. The azadirachtin content in the test samples ranged from 0.14 to 1.66 per cent. Ecotypes growing in regions with moderate climate, red laterite and shallow medium black soils, and altitudes less than 500 m above mean sea level were rich in azadirachtin content. Ecotypes growing in high altitude alluvial soils with extreme hot and cold climates had very low azadirachtin content. Preliminary bioassay against neonate larvae of *Spodoptera litura* indicated that azadirachtin was not the only component responsible for the bioactivity of the neem extractives. Neem concentrate showed mild fungicidal activity against *Alternaria alternata, Helminthosporium oryzae, Pythium aphenidermatum, Rhizoctonia bataticla, and Sclerotium rolfsii.*

INTRODUCTION

With the increasing interest in neem (*Azadirachta indica* A. Juss.) as a source of botanical pesticides, there is a growing need to identify plants with high azadirachtin content. It has already been reported that neem ecotypes vary widely in their azadirachtin content (Ermel et al., 1984, 1987). The oils obtained from different ecotypes have also been reported to vary widely in azadirachtin content (Isman et al., 1990).

India has 21 agro-ecological regions with different climates and soil types (NBSS and LUP, 1990). A survey was undertaken to identify regions where neem trees grow rich in azadirachtin content. The present paper reports the results of 21 samples analysed from eight agroecological regions of the country.

MATERIALS

Seeds of neem grown in eight different agro-ecological regions were collected through Neem Mission, Pune, India (Table 1, Fig. 1). Oil samples were prepared in the laboratory by solvent extraction (Fraction A, Fig. 2). Standard azadirachtin sample was obtained from Trifolio-M GMBH, Germany. Laboratory-grade reagents and solvents were used in extraction and enrichment. Appropriate analytical-grade solvents were used in high performance liquid chromatography (HPLC).

METHODS

Estimation of Azadirachtin Content

Kernels

Dried seeds were cleaned and decorticated manually to obtain kernels. The kernels were powdered using a waring blender and the powder (0.5 g) was extracted by refluxing with ethanol (3 multi 20 ml, 30 minutes each). The contents were cooled, filtered through a G2 septum fitted Buchner funnel and the filtrates collected. The filtrates were pooled, made up to 100 ml and aliquots analysed for azadirachtin content by HPLC.

Oil

The oil (1 g) was dissolved in hexane (10 ml) and partitioned with methanol-water (55:45, 2 multi 5 ml). The combined methanol-water layers were concentrated and the

Table 1. Distribution of neem test samples among the agro-ecological regions of India

Region no.	Region	Climate	Soil	No. of samples
R3	Deccan plateau	Hot arid	Red, black	1
R4	Northern plain, Central highlands	Hot semiarid (cold winter)	Alluvium and its derivatives	2
R5	Central Malwa highlands	Hot semiarid (cold winter)	Medium deep black	2
R6	Deccan plateau	Hot semiarid	Shallow medium black	6
R7	Deccan plateau, Eastern ghats	Hot semiarid	Red, black	2
R8	Eastern ghats (Tamil Nadu uplands), Deccan plateau	Hot semiarid	Red, loamy	5
R19	Eastern Coastal plain	Sub-humid	Alluvium and its derivatives	2
R20	Western ghats, Coastal plains	Hot, humid, per-humid	Red laterite, alluvium	1

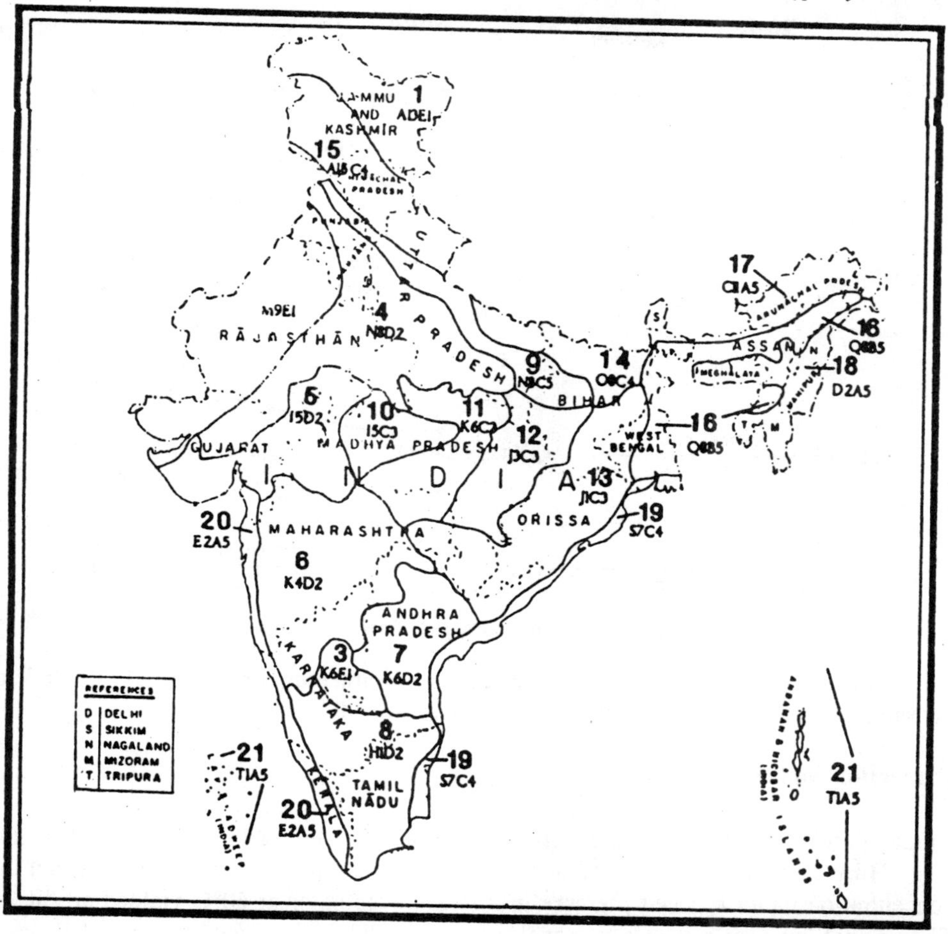

1. Western Himalayas
2. Western Plain and Kutch Peninsula
3. Deccan Plateau
4. Northern Plain and Central Highlands
5. Central (Malwa) Highlands and Kathiawar Peninsula
6. Deccan Plateau
7. Deccan Plateau and Eastern Ghats
8. Eastern Ghats (TN uplands) and Deccan Plateau
9. Northern Plain
10. Central Highlands (Malwa & Bundelkhand)
11. Deccan Plateau and Central Highlands (Bundelkhand)
12. Eastern Plateau (Chhatisgarh)
13. Eastern (Chhotanagpur) Plateau and Eastern Ghats
14. Eastern Plain
15. Western Himalayas
16. Assam and Bengal Plains
17. Eastern Himalayas
18. North-Eastern Hills (Purvachal)
19. Eastern Coastal Plains
20. Western Ghats and Coastal Plains
21. Islands of Andaman-Nicobar and Lakshadweep

Fig. 1. Agro-ecological regions of India

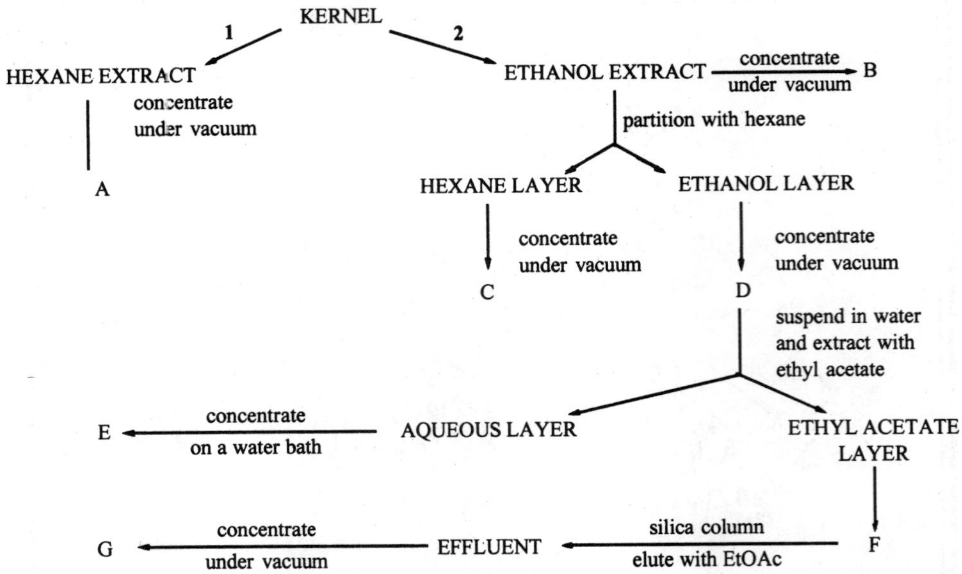

Fig. 2. Extraction and enrichment of neem extractives

volume was made up to 10 ml and analysed for azadirachtin content by HPLC. Spiked oil samples with azadirachtin showed a recovery of 90 per cent.

Extraction and Enrichment

Kernel powder (200 g) was soxhlet-extracted successively with hexane and ethanol (Fig. 2) by a modified method of Schroeder and Nakanishi (1987). Hexane extract on concentration under reduced pressure in a rotary evaporator at 60°C yielded an oil A. The ethanol layer on concentration yielded a semisolid B. The yields of A and B were recorded. B was dissolved in ethanol and analysed for azadirachtin content by HPLC.

The concentrate B was enriched for its azadirachtin content. A portion of B was dissolved in ethanol and partitioned with hexane (6 multi 50 ml) to remove traces of oil. The combined hexane layers and the mother liquor on evaporation yielded C and D respectively. A portion of D was suspended in water and extracted with ethyl acetate. The aqueous layer on evaporation yielded E. The ethyl acetate layer on drying over anhydrous sodium sulphate and evaporation yielded F. F was further enriched by passing through a silica gel column (1.5 cm diameter multi 15 cm, 25 g, 60/120 BSS) eluting with 150 ml of ethyl acetate. The effluent was evaporated to obtain G. The yields of C, D, E, F, and G were recorded.

Physicochemical Properties of Oil

The oil samples were studied for their colour (Munsell, 1975), specific gravity (25°C), refractive index (Toshniwal India Ltd. refractometer, 27°C), iodine value, acid value, and saponification value (AOAC, 1975) and azadirachtin content (described earlier).

Azadirachtin Estimation by HPLC

Shimadzu HPLC system, fitted with LC9A pumps in binary mode, Rheodine 7161 injector with 20 ul loop, Shim Pack, CLC-phenyl stainless steel column (6 mm diameter multi 15 cm), and SPDM6A photodiode array detector was used. The following were the operating conditions: mobile phase, methanol-water (65:35) at 1.0 ml/min; detector wavelengths, 214 and 250 nm; sensitivity, 0.1 AUFS. The data were acquired in a PCS-DG India Ltd. workstation and quantification was done in the post-analysis session. Retention time of azadirachtin was 8.33 minutes. After every four injections the column was cleaned by gradient methanol-water elution. Imperfect cleanings led to erratic column behaviour.

Insect Growth Regulation (IGR)

Preliminary studies on the insect growth inhibition activity of the neem extracts against *Spodoptera litura* were conducted at Regional Research Laboratory, (CSIR), Jammu. Extracts A and B were incorporated in fresh artificial diet at 100 ppm (w/w). Two neonate larvae of *S. litura* were released in 25 ml plastic cups containing 0.5 g diet each in 10 replicates. The larvae were reared for seven days and larval weight (as a measure of growth) was recorded.

Fungicidal Activity

Fungicidal activity was studied by poisoned food technique using potato-dextrose-agar-agar (PDA) medium (Nene and Thapliyal, 1979). In a preliminary study, fractions A to G were incorporated in the medium at 400 ppm and screened against *Sclerotium rolfsii*. Untreated food served as control. Concentrate G was found to be more active and was subsequently tested against *Alternaria alternata, Helminthosporium oryzae, Pythium aphenidermatum, Rhizoctonia bataticola*, and *S. rolfsii*. Inhibitory dose values for 50 per cent inhibition (ID_{50}) were calculated using probit analysis.

RESULTS AND DISCUSSION

Azadirachtin Content of Neem Ecotypes

Kernels

The azadirachtin content of kernels in different ecotypes is given in Table 2. There is wide variation in the azadirachtin content from sample to sample within a region and between the regions. The content ranged from 0.14 to 1.66 per cent among the test samples.

The degree of variation within a region is indicated by the standard deviation (SD), which is high in regions R6 and R8. Such a variation is likely to occur in other regions also where the region is wide and the number of samples analysed is limited (Fig. 1, Table 1).

The sample from region R20 recorded the highest azadirachtin content (average value), followed by samples from regions R8 and R6. These regions are in peninsular

Table 2. Azadirachtin content of neem ecotypes of different regions

Region no.	Sample no.	Azadirachtin content of kernel (%, w/w)	Average azadirachtin content in a region (% w/w)
R3	S-13	0.99	0.99
R4	SMT-2	0.42	0.28 ± 0.20
	SMT-5	0.14	
R5	S-4	0.31	0.30 ± 0.20
	S-18	0.28	
R6	S-1	0.61	0.80 ± 0.38
	S-2	0.65	
	S-3	1.27	
	S-8	1.21	
	S-10	0.27	
	S-14	0.79	
R7	S-17	0.70	0.71 ± 0.01
	S-26	0.72	
R8	S-7	0.49	0.83 ± 0.53
	S-9	0.29	
	S-12	0.46	
	S-15	1.49	
	S-16	1.66	
R19	S-6	0.56	0.65 ± 0.12
	S-20	0.73	
R20	S-11	1.52	1.52

India and are characterized by moderate summer and moderate winter, and hilly uplands with red laterite and shallow medium black soils. Samples from region R4 recorded the lowest (0.28 ± 0.2 per cent), followed by region R5 (0.30 + 0.20 per cent). These regions are characterized by extreme hot summers (maximum 48°C) and extreme cold winters (minimum < 1°C), altitudes 500, 1000 m above mean sea level (MSL) alluvium-derived soils, and medium deep black soil (Terai region of the Himalayan base and Central Malwa Uplands). Region R3, which lies in peninsular India and has hot arid climate, high altitude (500, 1000 m above MSL), and mixed red and black soils, recorded 0.99 per cent. Regions R20 and R19 recorded azadirachtin contents of 1.52 and 0.65 ± 0.12 per cent respectively. Both are coastal with humid climates but differ in soil type.

It appears from the available data that the climate, soil type, and altitude affect the azadirachtin content of the ecotype. Trees grown in red and medium black soils of regions with altitudes up to 500 m above MSL and moderate climate of peninsular India are rich in azadirachtin.

Oil

The oils of different ecotypes showed different azadirachtin content and physicochemical properties (Table 3). Twelve oil samples recorded no azadirachtin content and nine samples recorded very low values ranging from 0.01 to 0.09 per cent (w/w of oil). The colour ranged from pale yellow to olive yellow. Specific gravity (20°C) ranged from 0.83 to 0.98 g/cm^3. Iodine value, which is an index of unsaturation, ranged from 57.38 to 191.92 per cent. Acid value, an index of free acid, ranged from 3.84 to 48.86

per cent. Saponification value, an index of hydrolysable fats and esters, ranged from 164.95 to 240.80 KOH/g of oil.

The region-wise means (Table 4) of saponification value, acid value, and iodine value of the oils were compared with the azadirachtin content of kernels. The following is the decreasing order among the regions:

Azadirachtin content of kernels: R20, R3, R8, R6, R7, R19, R5, R4
Saponification value of oils: R20, R3, R8, R7 = R4, R5, R19, R6
Acid value of oils: R20, R3, R8, R5, R6, R7, R4, R19
Iodine value of oils: R19, R5, R6, R8, R7, R4, R3, R20

The above correlation indicates that the azadirachtin content of the kernel is directly related to the saponification and acid values of the oil inversely related to the iodine value of the oil. It appears that these parameters of the oil are related to the biosynthesis of azadirachtin. Work is in progress to confirm the above inference.

Enrichment

Enrichment by solvent extraction and partition was carried out to prepare neem fractions rich in azadirachtin (Fig. 2). Hexane extraction removed the bulk of oil from the kernel. Ethanol extraction removed the traces of the oil left in the kernel along with azadirachtin, soluble sugars, amino acids, peptides, polypeptides, etc. Washing of the ethanol extract with hexane removed the traces of oil. Suspension of D in water and partitioning with ethyl acetate extracted azadirachtin, leaving sugar, and peptides in water. Passing of F through silica-column removed most of the polyphenols and polar impurities, eluting out azadirachtin-rich G.

The yields of azadirachtin-rich concentrates B, D, F, and G varied widely from sample to sample. This influenced the enrichment factor and hence the azadirachtin content of the enriched concentrates. The azadirachtin content of the ethanolic extract (B) and the enrichment factor for the enrichment of azadirachtin from kernel to ethanol extract are given in Table 5.

The azadirachtin content of the ethanolic extract ranged from 1.14 to 15.69 per cent and the enrichment factor in B ranged from 3.2 to 22.4. The contents and the factors varied from sample to sample within a region and between the regions. No meaningful correlation emerged from the enrichment factor and the corresponding oil content of the kernel. It appears that the chemical composition of the ecotypes varies widely with respect to many other components.

The chemical composition of a bioactive product can influence the behaviour of the bioactive molecules within a phase, between the phases, on a surface, or within a biological system. The variations found in the neem products will have a bearing on their shelf-life and bioefficacy vis-a-vis the quality of the product.

Insect Growth Regulation

The preliminary data on growth inhibition of *S. litura* (Table 6) reveal no correlation between the azadirachtin content of the oil or extract with the inhibition of insect growth.

Table 3. Azadirachtin content and physicochemical properties of oils

Region no.	Sample no.	Azadirachtin (%)	Sp. gravity	Saponification value$	Acid value#	Iodine value@	Colour
R3	S-13	0.09	0.92	229.11	23.27	58.76	Yellow
R4	SMT-2	0.09	0.98	208.44	15.38	73.79	Olive yellow
	SMT-5	ND	0.93	217.76	6.58	59.89	Pale yellow
R5	S-4	ND	0.93	232.41	22.30	86.44	Yellow
	S-18	0.04	0.92	193.42	15.60	91.81	Yellow
R6	S-1	0.03	0.88	164.95	6.58	57.71	Pale yellow
	S-2	ND	0.91	188.01	10.79	65.96	Olive yellow
	S-3	0.03	0.97	216.89	18.05	88.49	Olive yellow
	S-8	ND	0.89	211.00	44.60	102.00	Olive yellow
	S-10	0.01	0.89	196.29	6.58	68.76	Yellow
	S-14	ND	0.83	222.37	14.71	64.32	Yellow
R-7	S-17	ND	0.91	212.46	16.78	73.08	Yellow
	S-26	0.05	0.91	215.39	9.97	66.53	Olive yellow
R8	S-7	ND	0.93	212.03	6.58	60.96	Pale yellow
	S-9	0.02	0.98	240.80	34.88	66.44	Pale yellow
	S-12	0.02	0.88	218.40	17.47	61.01	Pale yellow
	S-15	ND	0.93	217.47	48.86	85.09	Olive yellow
	S-16	ND	0.83	224.54	3.84	79.81	Pale yellow
R19	S-6	0.08	0.98	190.29	7.83	92.58	Yellow
	S-20	ND	0.89	210.85	7.83	191.92	Yellow
R20	S-11	0.01	0.92	233.19	31.33	57.38	Olive yellow

$ - mg KOH/g oil; # - per cent w/w of oil
@ - per cent w/w of oil.

Table 4. Region-wise means of the azadirachtin contents in neem kernel and the physicochemical properties of oils

Region no.	Azadirachtin cont. of kernel	Saponification value of oil$	Acid value of oil#	Iodine value of oil@
R3	0.99	229.11	23.27	58.76
R4	0.28	213.10	10.98	66.84
R5	0.30	212.92	18.95	89.13
R6	0.80	199.92	16.93	74.54
R7	0.71	213.93	13.38	69.81
R8	0.88	222.65	22.33	70.66
R19	0.65	200.57	7.83	144.25
R20	1.52	233.19	31.33	57.38

ND - non-detectable; $-mg KOH/g oil; # - per cent w/w of oil; @ - per cent w/w of oil.

Table 5. Azadirachtin content of ethanolic extract (B) and enrichment factor

Region no.	Sample no.	Azadirachtin Kernel	(%, w/w) B	Enrichment factor
R3	S-13	0.99	4.49	4.5
R4	SMT-2	0.42	4.61	10.9
	SMT-5	0.14	1.97	14.0
R5	S-4	0.31	1.14	3.7
	S-18	0.28	1.87	6.7
R6	S-1	0.61	1.96	3.2
	S-2	0.65	3.38	5.2
	S-3	1.27	5.45	4.3
	S-8	1.21	13.98	11.6
	S-10	0.27	1.61	6.0
	S-14	0.79	5.12	6.5
R7	S-17	0.70	15.69	22.4
	S-26	0.72	6.32	8.8
R8	S-7	0.49	3.03	6.2
	S-9	0.29	1.81	6.2
	S-12	0.46	3.24	7.0
	S-15	1.49	9.71	6.5
	S-16	1.66	12.27	7.4
R19	S-6	0.56	3.50	6.3
	S-20	0.73	6.24	8.5
R20	S-11	1.52	7.56	4.9

Fungicidal Property

Initial screening of extracts and concentrates A to G from sample no. S-9 revealed the following order of per cent inhibition: G, 33.88; F, 13.63; B, 8.34; D, 2.13; A and C, and E, 0. The most active G showed only mild fungicidal activity (Table 7) against *A. alternata, S. rolfsii, H. oryzae, P. aphenidermatum*, and *R. bataticola* in that order.

Table 6. Relative IGR activity of ethanol extracts (B) and oils (extract A) of kernels from different ecotypes against neonate larvae of *S. litura*

Region	Sample No.	Ethanol extract (B)		Oil	
		Azadirachtin (%)	% Inhibition	Azadirachtin (%)	% Inhibition
R3	S-13	4.49	NA	0.09	66.0
R4	SMT-2	4.61	NA	0.09	NA
	SMT-5	1.97	NA	ND	NA
R5	S-4	1.14	NA	ND	NA
	S-18	1.88	35.0	0.04	NA
R6	S-1	1.96	65.9	0.03	NA
	S-2	3.38	86.5	ND	88.5
	S-3	5.45	73.5	0.03	87.8
	S-8	13.98	55.8	ND	NA
	S-10	1.61	24.7	0.01	NA
	S-14	5.12	71.3	ND	NA
R7	S-17	15.69	28.9	ND	NA
	S-26	6.32	100.0	0.05	NA
R8	S-7	3.03	40.5	ND	NA
	S-9	1.86	72.0	0.02	NA
	S-12	9.71	75.3	ND	44.8
	S-15	3.24	77.9	0.02	NA
	S-16	12.27	59.1	ND	NA
R19	S-6	3.49	NA	0.08	95.6
	S-20	6.24	100.0	ND	NA
S20	S-11	7.55	NA	ND	NA

Table 7. Fungicidal spectrum of neem concentrate G

Fungus	ID 50 (ppm)
A. alternata	490
H. oryzae	831
P. aphanidermatum	871
R. bataticola	14,125
S. rolfsii	691

SUMMARY

The neem ecotypes showed varying azadirachtin content depending on the climate, soil type, and altitude. The content ranged from 0.14 to 1.66 per cent. Ecotypes growing in regions with moderate climate, red laterite and shallow medium black soils, and altitudes upto 500 m above MSL showed high azadirachtin content. Ecotypes growing in high altitude alluvial soils with extreme hot and cold climates showed very low azadirachtin content. Saponification value, acid value, and iodine value of the oil may serve as an index of the azadirachtin content in neem kernels. The yield of enriched concentrates and the azadirachtin content of the enriched concentrates varied with the ecotypes. Preliminary bioassay indicated that azadirachtin was not the only component responsible for the bioactivity of the extractives. Neem concentrates showed mild fungicidal activity against *A. alternata, H. oryzae, P. aphenidermatum, R. bataticola,* and *S. rolfsii.*

REFERENCES

AOAC. 1975. *Official Methods of Analysis*, 12th ed. Association of Official Analytical Chemists, Washington, DC.

Ermel, K., E. Pahlich and H. Schmutterer. 1984. Comparison of the azadirachtin content of neem seeds from ecotypes of Asian and African origin. *Proc. 2nd Int. Neem Conf.*, Rauischholzhausen, Germany, pp 91–94.

Ermel, K., E. Pahlich and H. Schmutterer. 1987. Azadirachtin content of neem kernels from different geographical locations and its dependence on temperature, relative humidity and light. *Proc 3rd Int. Neem Conf.*, Nairobi, Kenya, 1986. pp. 185 194.

Isman, M.B., O. Koul, A. Luzynski and J. Kaminski. 1990. Insecticidal and antifeedant bioactivities of neem oils and their relationship to azadirachtin content. *J. Agric. Food Chem.*, 35(6): 1406–1411.

Munsell Color. 1975. *Munsell R Soil Colour Charts*, 2441, North Calvet Station, Baltimore, Maryland, 21218, USA.

NBSS and LUP. 1990. Agroecological regions of India. Bull. 24. National Bureau of Soil Survey and Land Use Planning, Nagpur, India.

Nene, Y.L. and P.N. Thaplial. 1979. *Fungicides in Plant Disease Control*. Oxford & IBH Publishing Co., New Delhi, p. 413.

Schroeder, D.R. and K. Nakanishi. 1987. A simplified isolation procedure for azadirachtin. *J. Nat. Prod.*, 50: 241–244.

4

PHYSIOLOGICAL AND BEHAVIOURAL EFFECTS

24

The Effects of Various Neem Formulations on Mortality Rate and Morphogenetic Defects Upon *Schistocerca gregaria* (Forskal) Larvae*

H. WILPS, O. NASSEH AND S. KRALL

Deutsche Gesellschaft für Technische Zusammenarbeit (GTZ)
GmbH, 6236 Eschborn, Germany

ABSTRACT

The effects of various neem products on nymphs of *Schistocerca gregaria* were tested in field (cage) trials in the southern and western Sahara desert (Republic of Niger). The neem products applied had different sources (Prof. Schmutterer, University of Giessen, and Dr. H. Kleeberg, Trifolio GmbH, Lahnau), and differed with regard to both formulation and concentrations of active compounds. All products were tested either on locusts that hatched from eggs of laboratory cultures or on nymphs caught in the field. The products supplied by Trifolio caused 100 per cent mortality in both test populations (applied quantities: 1 litre per hectare), compared with mortality rates of 60–70 per cent for the products from Giessen (applied quantities 10 litres per hectare). The laboratory-reared nymphs, proved to be more sensitive, i.e., died four to six days earlier and exhibited higher rates of morphogenetic defects after application of sublethal dosages.

INTRODUCTION

Both pure neem oil and neem oil enriched with methanolic neem seed extracts cause morphogenetic defects in nymphs of *Schistocerca gregaria*. Poor physical condition of the nymphs results in increased mortality rates and, in gregarious nymphs, an increased tendency to shift to, the solitary status (Nicol and Schmutterer, 1991; Schmutterer and Freres, 1991). It is also known that neem products cause antifeedant effects in *S.*

* Report on investigations carried out on field and laboratory-reared locusts under natural conditions in the southern Sahara Desert (Republic of Niger).

gregaria nymphs and adults (Pradhan and Jotwani, 1968; Attri and Singh, 1977; Nasseh and Freres, 1990; Nasseh et al., 1991). The precise mode of action in which the active substances contained in neem products—primarily azadirachtin—is not known. It is suggested, however, that azadirachtin directly interacts with the endocrinological processes responsible for the ecdysteroid and juvenile hormone titres (Rembold and Sieber, 1981; Sieber and Rembold, 1983; Schluter et al., 1985; Bidmon et al., 1987; Rembold et al., 1987; Subrahmanyam et al., 1989). It is the relation between the respective concentrations of these hormones that regulates growth and development processes that take place during the insects' metamorphosis.

The field trials conducted in Agadez and Anou Mekkerene (Republic of Niger) were performed with larvae that descended from laboratory-bred lines from Constance, Oldenburg, Saarbrücken and Giessen, and were raised *ex ovo* in Agadez. In addition, larvae were used that had been captured in a *Shcouwia* field in the Air Mountains. In the rest of this report, the former are referred to as laboratory insects or larvae, and the latter as field insects or larvae. The comparative studies carried out were aimed at shedding light on the following questions:

1) Are laboratory-bred insects still capable of developing under their original natural living conditions (compared to the general breeding conditions, with temperatures 10°C higher, relative humidity 10–35 per cent lower, and an utterly different diet)?
2) Can results obtained with laboratory insects and under laboratory conditions be transferred to field populations and conditions?
3) In the studies discussed below, the mortality rates of larvae from both groups were observed following application of different neem products, as well as the further development of the surviving individuals up until attainment of sexual maturity.

MATERIALS AND METHODS

Laboratory Insects

Laboratory insects were raised from egg pods imported from Germany, and were kept in breeding cages (50 × 50 × 50 cm) until they reached the second to fourth larval instars. The breeding cages were placed outdoors at a shady site. Following treatment, the locusts were transferred to field cages measuring 2 × 2 × 2 m. During the trials, which lasted for three months, the days and nights were of equal length (12 hours each), and the daytime and nighttime temperatures were 45° ± 5°C and 25° ± 5°C, respectively. The relative humidity was 10–15 per cent during the daytime, reaching maximum levels of 25–30 per cent at night. The insects were fed each day with fresh *Schouwia thebaica* plants.

Field Insects

Larvae were collected between 6:30 and 11 a.m. in a *Schouwia* field 30 km east of Anou Mekkerene in the foothills of the Air Mountains, and kept in the field cages for about 72 hours prior to the beginning of the trials. They fed on the natural *Schouwia* vegetation growing in the cages. The climatic conditions (temperature, relative humidity) were virtually the same as those described later.

Neem Products

In the studies described here, various neem products from different sources were used in varying concentrations.

a) Enriched neem oil, azadirachtin content: 0.2 per cent (60g azeotropic mixture extract + 439 ml propyl alcohol + 360 ml Span 85 per 3000 ml of neem oil, prepeared as described elsewhere, Feuerhake, 1984).
b) Pure neem oil, origin of seeds: Republic of Niger; azadirachtin content: 0.01 per cent.
c) Azal-F: 80 per cent filler (blank) + 20 per cent neem seed extracts, the latter containing 50 per cent azadirachtin.
d) Pure neem oil: 0.04 per cent azadirachtin and 0.04 per cent 3-tigloyl-azadirachtol + approx. 10 per cent suspended matter.

All of the products and control substances were applied with a hand sprayer (Micro Ulva).

Controls

Two types control were used in all trials:

a) Untreated insects
b) Insects treated with the pure formulation substance and/or additives (vegetable oil, blank, propyl alcohol + Span 85).

Application of the Larvae

The larvae (both field-captured and laboratory-reared) were treated while in the second to fourth instars using ultra low volume techniques. In order to reduce the mobility of the insects, immediately prior to beginning the trials they were kept at a temperature of 4°C for 20 minutes. In each trial, 120 larvae were sprayed between 7:00 and 8:00 a.m. in pens of 16 m^2, area fenced in by plastic sheeting 15 m high and subsequently transferred to field cages in groups of 30 each for further observation.

The mortality rates following application of the various products were determined each day between 7:30 and 11:00 a.m. by counting the dead and surviving insects. For each trial, the observation period lasted up to a maximum of 22 days. The incidence of faulty development and the deformities induced by the various products were also recorded every day. Whenever an individual successfully completed adult moulting, its body weight and any deformities were recorded.

Application of the Vegetation

One hundred and twenty larvae each were placed in cages containing either one treated plant only (application rate of the enriched neem oil: 10 l/ha) or one sprayed and one untreated plant. The mortality rates and deformations occurring were recorded on a daily basis.

RESULTS

Mortality Rates Following Application of Neem Products to Larvae

Two goals were pursued by the studies carried out in the Republic of Niger. One was to determine, under conditions of practical relevance, the extent to which neem products are capable of inducing sufficiently high mortality rates in *S. gregaria* larvae for successful crop protection; the other was to directly compare the impact of neem products on field and laboratory insects. The mortality rates caused by neem products are shown in Figs. 1–4. Figure 1 shows the effect of neem Azal-F made by the Trifolio company on field and laboratory *S. gregaria* larvae of instars 2 through 4.

The curve designated as "blank" represents a control trial. "Blank" is the vehicle used to formulate Azal-F, accounting for 80 per cent of its volume. Since application of this substance does not induce any significant increase in mortality, it was decided to dispense with identical trials on field insects. One liter per hectare of Azal-F (20 per cent of which consists of neem extracts, thereof 50 per cent azadirachtin and azadirachtin isomers) inflicted 100 per cent mortality on both field and laboratory insects. However, in the case of the field insects the last individuals did not succumb until the 13th day after application, compared to the 9th day for the laboratory insects. When laboratory insects were sprayed with this product at a rate of 500 ml/ha, some 80 per cent of the populations died by the 15th day after application (Fig. 1). Up until the 6th day their

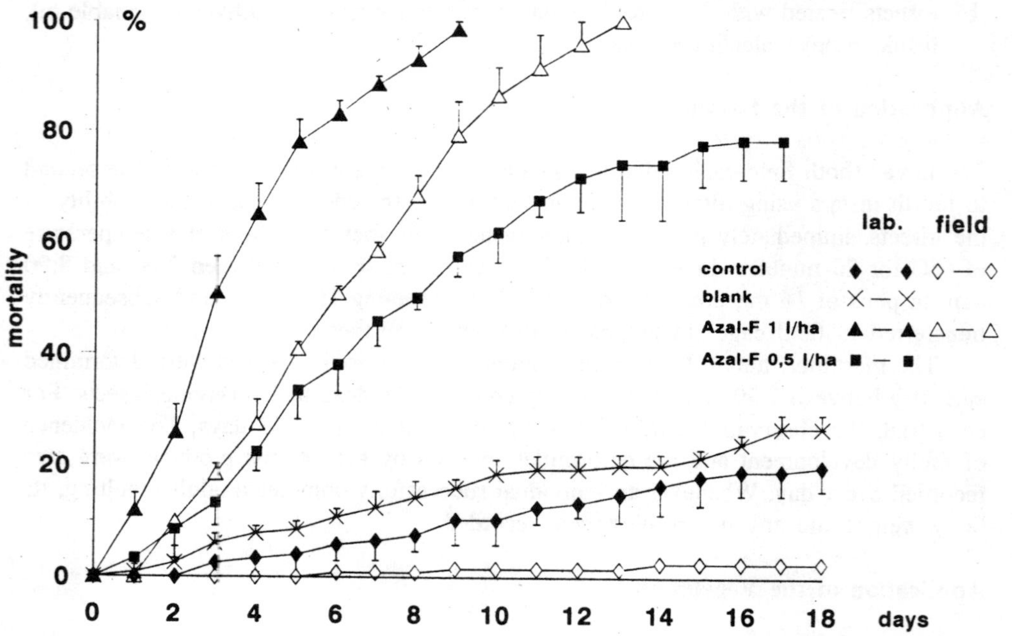

Fig. 1. Mortality rate caused by the application of Neem Azal-F. The individual points represent the mean values ± SD of 4 measurements, and reflect the mortality rates of the various populations on the respective days after treatment (on day 0)

mortality curve runs almost identically to that of the field insects, which had been treated with twice as much Azal-F.

Figure 2 shows the effect of the pure neem oil supplied by the Trifolio company.

This product also induces 100 per cent morality in both field and laboratory insects, with all of the laboratory larvae dying by the 11th day but the field insects not until the 15th day. With increasing incubation times, the incidence of morphogenetic deformation increased (Plates 1–4). The spectrum of observed defects ranged from completely deformed extremities (Plates 1, 2) and antenna deformation (Plate 3) or loss, to the inability to shed the exuviae (Plate 4).

The larval mortality rates caused by application of the neem products that were prepared and formulated at the University of Giessen are shown in Fig. 3.

A control trial was also performed in this group of trials, using the fillers (Span 85 and propyl alcohol) that are contained in this neem product. These compounds by themselves inflict a slight but not significant increase in mortality. With this enriched neem oil, there is no delay in the mortality rates of field insects as compared to the laboratory larvae, as was with the Trifolio products. There were, however, differences in the maximum mortality rates, which after 17 days reached 53 per cent for the field insects and 60 per cent for the laboratory insects. The trials were discontinued on the 17th day, since by that time all of the surviving larvae had undergone adult moulting.

Figure 4 shows the mortality of field and laboratory larvae after treatment with pure neem oil. Supplied by the University of Giessen the same statements regarding the course described by the mortality curves over time and the maximum achieved mor-

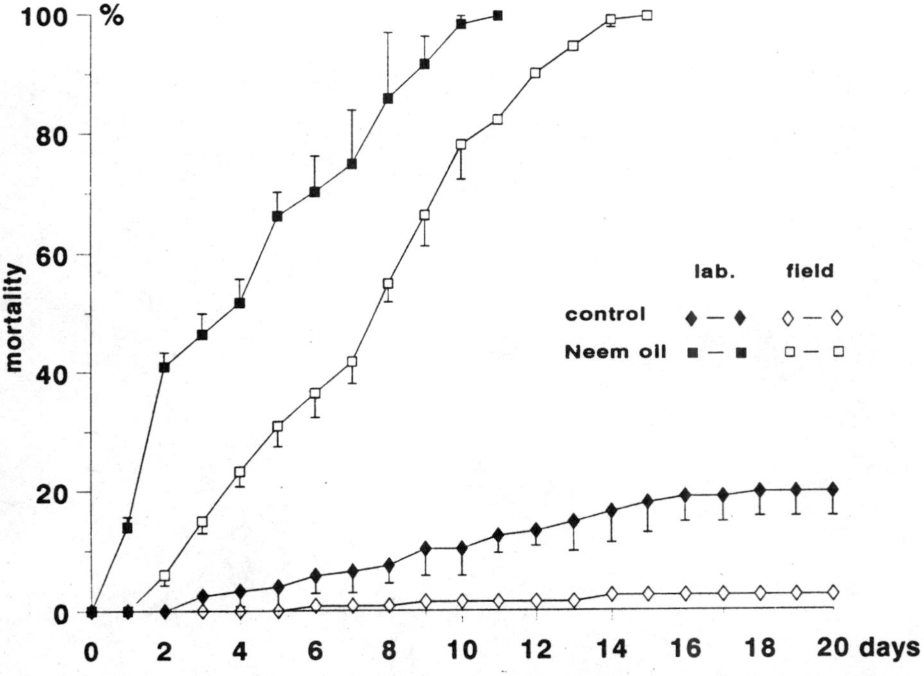

Fig. 2. Mortality rates following application of unfiltered neem oil made by the Trifolio company. The data were obtained in the same way as those for Fig. 1

Plates 1 and 2. Morphogenetic defects in *S. gregaria* larvae after application of Trifolio products

Plates 3 and 4. Morphogenetic defects in *S. gregaria* larvae after application of Trifolio products

Plate 5. Cage with treated and untreated *Schouwia* plants 2 days after introducing the larvae

Plate 6. The same cage after 10 days

Plate 7. Wing deformities in an adult *S. gregaria* that had been treated with 1000 ppm *Melia* extract while still a larva

Plate 8. Wing and leg defects in an adult *S. gregaria* that had been treated with Azal-F (500 ml/ha) while still a larva

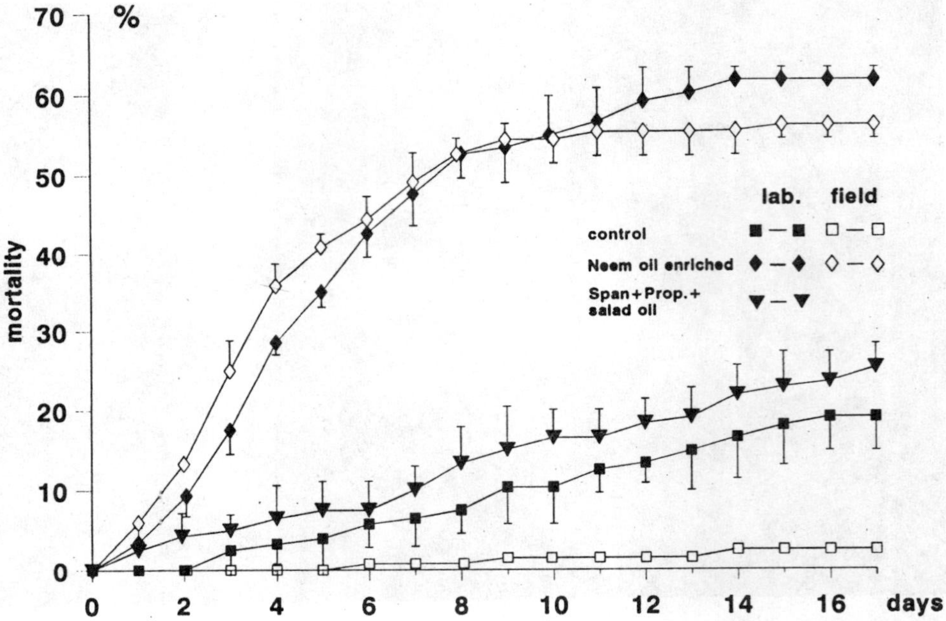

Fig. 3. Mortality rates of laboratory and field larvae after application of enriched neem oil obtained from the University of Giessen (10 1/ha). The data were derived in the same way as described in the caption of Fig. 1

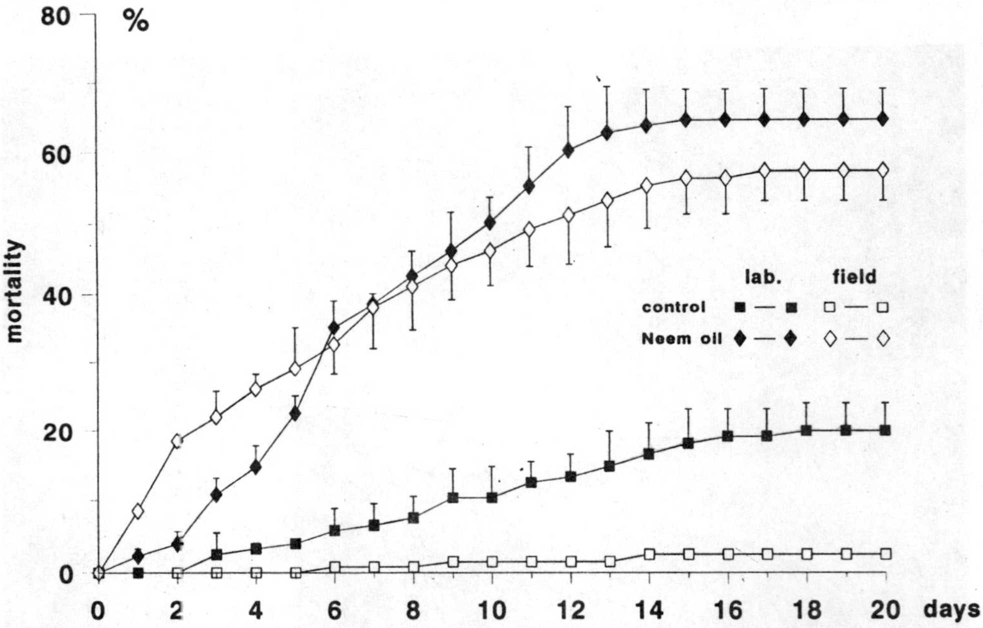

Fig. 4. Mortality rates following application of cold-pressed, pure neem oil obtained from the University of Giessen. The data were derived in the same way as described in the caption of Fig. 1

tality rates apply as have already been made for the previously studied enriched neem oil. However, the mortality rate of the laboratory insects is significantly greater ($p \leq 0.05$) than that of the field larvae, that is by 10 per cent.

Mortality Rates after Treatment of the Plants with Enriched Neem Oil; The Repellent Effect of Neem

The repellent effect of enriched neem oil was investigated in two different trials. In one, larvae were kept in a cage containing only a plant. In the other, a cage held one sprayed and one untreated plant. In both cages, no signs of feeding were observed until the 10th or 11th day after application of neem to the treated plants. Afterwards the repellent effect of the neem product diminished. In the cage with both treated and untreated vegetation, the unsprayed plant was significantly damaged after two days and had been eaten completely bare by the 10th day (Plates 5 and 6).

The mortality rates determined in the two cages are given in Fig. 5.

In both cages the same mortality rate was achieved. This was somewhat surprising, since in the cage with a treated and an untreated plant non contaminated food was obtainable. Surprisingly, the mortality rates were identical in the two cages, i.e., after 14 days only about 40 per cent of the population was still alive in both. In the cage containing treated vegetation only, it is probable that the insects starved to death, although this needs to be reinvestigated. In the cage with an untreated plant, it should have been possible to expect a lower mortality rate, with the insects beginning to die at a later point in time, since sufficient food was available to them until about the sixth

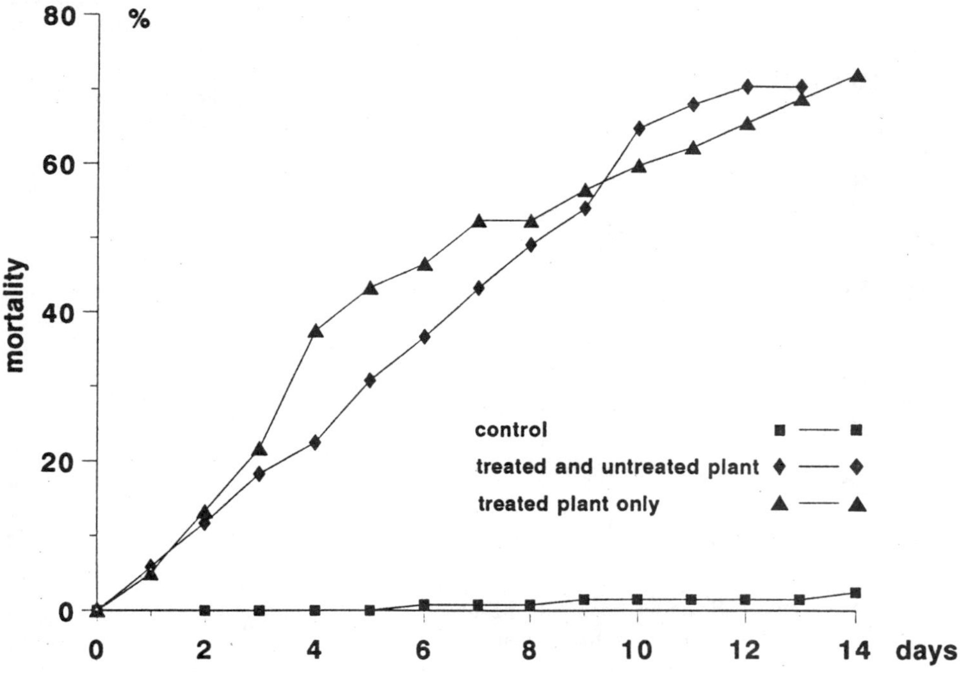

Fig. 5. Mortality rates of field larvae in cages containing either one treated and untreated plant or one treated plant only. The product used was enriched neem oil of Giessen

day following application. It is conceivable that their deaths were caused by contact effects, since the larvae took shelter in the shade of both the treated and the untreated plant. This needs to be studied further, however.

Morphogenetic Defects

The induced mortality rate is not the only criterion for judging the effectiveness of a product. Deformation and concomitant reductions in the fitness of the larvae and the adults that develop from treated larvae can also be of considerable importance. Depending on the severity of this damage, the different types of which are shown in Plates 1–4, the following effects can be assumed: (1) reduction or loss of flight ability; (2) feeding impediments; (3) impairment of communication among individuals of the same species, leading to loss of the ability or readiness to copulate, or disorientation caused by the loss of antennae, deformed "dented" complex eyes. These types of damage were observed in a high percentage of the larvae and the adults that developed from them. Figures 6 and 7 give the percentages of larvae that successfully underwent adult moulting, as well as the percentages of deformed insects. Unfortunately, it was not possible to specify completely the severity and type of these deformities within the scope of this year's trials, so that additional trials must be carried out to elucidate this aspect. For this reason, all of the observed incidences of morphogenetic damage are treated here generically as deformities and lumped together.

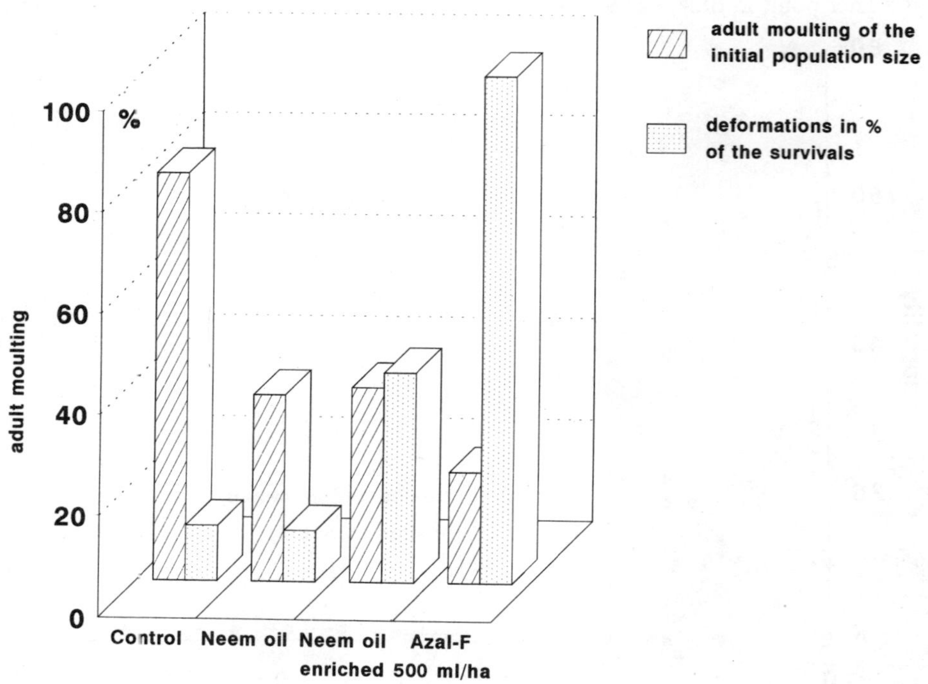

Fig. 6. Successful adult moulting as percentages of control and treated larvae, and the incidence of deformation expressed as percentages of the number of moulted insects

About 80 per cent of the laboratory control insects attained adulthood (Fig. 6). Of these, only roughly 5 per cent exhibited deformation, which in nearly all cases consisted of the loss of a hind led. The survival rate of the insects treated with neem from the University of Giessen was about 37 per cent in both cases, with deformities occurring in approx. 15 per cent (pure neem oil) and 39 per cent (enriched neem oil), respectively. By contrast, all of the insects treated with Azal-F were damaged. The deformities observed included missing legs and a high incidence of antenna and wing defects.

In field insects as well, the incidence of deformities in the control populations was very low and limited to the loss of a hind leg. The survival rate of field larvae treated with neem was slightly greater than that of the corresponding laboratory larvae. The number of adults with deformities was nearly 5 per cent less, however (Fig. 7).

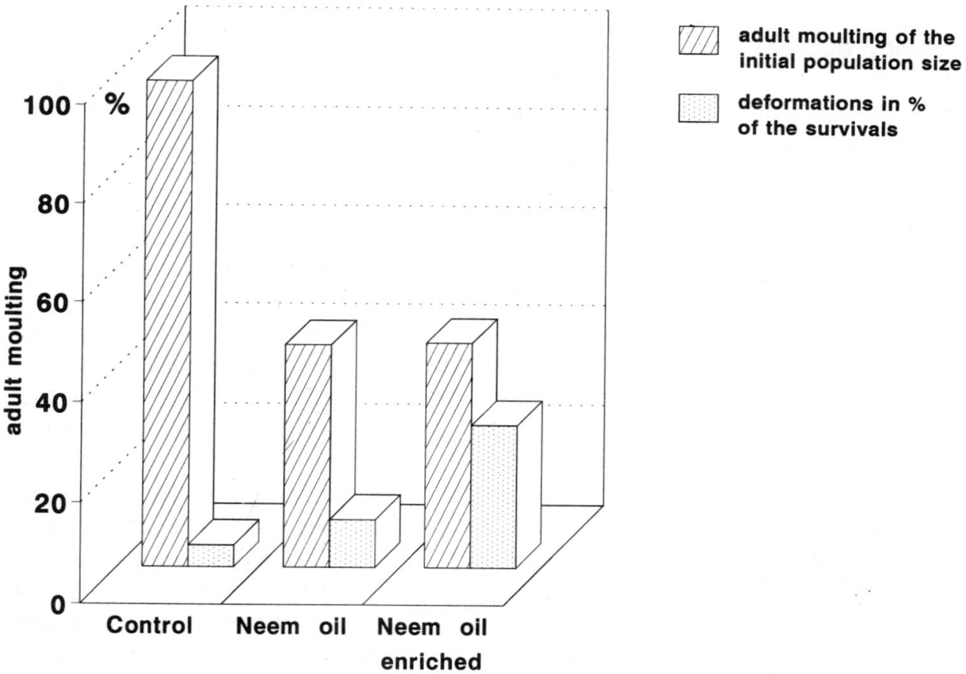

Fig. 7. Adult moulting as percentages of the initial populations and product-related deformities as percentages of the insects that underwent adult moulting (field larvae)

DISCUSSION

The neem products tested on *S. gregaria* nymps in the Tamesna Desert yielded, from the standpoint of effective pest control, satisfactory to excellent mortality. The products of the company Trifolio, with mortality rates of 100 per cent when applied at a concentration of 1 litre per hectare, are clearly superior to those obtained from the University of Giessen (60 per cent mortality, 10 l/ha). This difference could be due to varying contents of active constituents in the pressed neem seeds (Ermel et al., 1984).

But azadirachtin content is not the only important factor; of equal significance can be the amount of azadirachtin isomers, which have possibly the same or even greater effects, depending on the tree species (Rembold, 1989). Nor can additive effects be ruled out. This hypothesis is supported by the finding that disturbances of the lipid metabolism induced by injecting azadirachtin can be overcome by injection of synthetic adipokinetic hormone, but similar disturbances caused by topical application of enriched neem oil (Wilps et al., 1992a; and unpublished data) could not be replaced by hormone injections.

In actual practice, however, 100 per cent mortality is only required in cases of locust outbreaks where the insects have already invaded farmed areas (Wilps et al., 1991a, b; 1992b). For control measures in recession areas, the morphogenetic defects caused by neem products can be just as important as high mortality rates, owing to the fact that the deformed insects (as shown in Plates 7 and 8) are then unable to leave the recession areas. In our field experiments, however, such deformations occurred far less frequently than has been observed in laboratory trials with *S. gregaria* (Nicol and Schmutterer, 1991). These authors found that topical application of 2.5 l/ha caused deformations in nearly 100 per cent of the resulting adults, compared to a deformation rate of only about 45 per cent in our trials using enriched neem oil from Giessen. An explanation for these differences may be the fact that in the experiments of Giessen mostly larvae of stage 1 to 3 were treated whereas in our experiments the nymphal instars ranged between 3 and 5. The application of sublethal doses of Azal-F made by Trifolio yielded high deformation rates. This also suggests that the neem products from Giessen and Trifolio have different contents of biologically active substances.

There are two possible explanations for the differences between our findings and those of Nicol and Schmutterer (1991). For one, in our trials an important role may have been played by the extreme, although natural, abiotic conditions such as low relative humidity (resulting in a high rate of evapotranspiration) and high temperatures with intense UV irradiation (azadirachtin and its isomers are very sensitive to ultraviolet radiation; W. Kraus, personal communication), as well as differences in the test animals used. As our tests showed, laboratory-reared nymphs react more sensitively to neem and other products (Nasseh et al., 1993). Furthermore our studies did not corroborate the tendency of the nymphs to experience a phase shift to solitary status following topical application of neem products that was observed by Schmutterer and Freres (1991) and Nicol and Schmutterer (1991). The principal reasons for this are probably the considerable differences in abiotic conditions between semi-field and laboratory tests, which have already been mentioned, and the different points in time at which the nymphs were treated. Schmutterer and Freres obtained the most prominent solitarization effects when the nymphs were treated during instars one to three. What matters, however, is not just the nymphal stage during which application takes place, but also the exact point in time. The duration of a nymphal stage and phase shifts are possibly greatly influenced by the relation between the concentrations of ecdysteroids and juvenile hormones. But this relation is not a static value, instead being subject to continual changes during the course of a given nymphal instar. However, these hormone titres are directly or indirectly affected by azadirachtin and its isomers (Rembold and Sieber, 1981; Sieber and Rembold, 1983; Mordue et al., 1986; Rembold et al., 1987). When carrying out such studies it is very important to work with nymphs that are not only in the same developmental instar, but also—as far as possible—synchronized within the instar. This is extremely relevant for laboratory studies; under field conditions it is virtually impossible to achieve but less important in any case, since the hopper bands to be controlled are not uniform with respect to either their developmental instar or their instar age. The insect populations used by us were roughly comparable in composition with hopper bands in the wild, i.e. the percentage of individuals highly sensitive to the active constituents of neem products with regard to phase shifts was possibly low, and, in addition their elimination is even probable with respect to the high mortality rates.

The results of the experiment comparing cages with only contaminated vegetation and with treated and untreated plants were unexpected; we had not anticipated that both would exhibit the same mortality. It should have been possible to expect fewer deaths, also beginning later, in the cage containing an untreated plant, because the insects had sufficient food at their disposal until about the sixth day after application. The death of the nymphs in this cage was possibly due to contact effects, seeing as how they took shelter in the shade of both the treated and the untreated plant. But further investigations are needed to illuminate this phenomenon.

ACKNOWLEDGEMENTS

The authors would like to sincerely thank Prof. Schmutterer, Dr. Hellpap, Dr. Ermel and Ms. Nicol for providing us with the neem products, numerous pods of *S. gregaria* eggs, and organizational assistance. We are also indebted to the universities of Saarbrücken, Oldenburg and Constance for having supplied us with *S. gregaria* eggs. Dr. Kleeberg of the company Trifolio also deserves thanks for donating neem products and pure azadirachtin. Our special gratitude is also due to Dr. Pantenius and Dipl. Ing. J. Haag for their help and support in Niamey. We also extend our thanks to the staff of the Centre National Anti-Acridienne in Agadez and our counterparts from the Republic of Niger.

REFERENCES

Attri, B.S. and R.P. Singh. 1977. Effect of time and temperature on the antifeeding activity in neem seed/cake to desert locust *Schistocerca gregaria* Forsk. *Indian J. Entomol.* 39: 383–384.

Bidmon, H.J., G. Kaüser, P. Mobius and J. Koolman. 1987. Action of azadirachtin on blowfly larvae and pupae. *Proc. 3rd Int. Neem Conf.*, Nairobi, 1986. GTZ, Eschborn, 253–272.

Ermel, J., E. Pahlich and H. Schmutterer. 1984. Comparison of the azadirachtin content of neem seeds from extypes of Asian and African origin. *Proc. 2nd Int. Neem Conf.*, Rauischholzhausen, 1983. GTZ, Eschborn, pp. 91–93.

Feuerhake, K.J. 1984. Effectiveness and selectivity of technical solvents for the extraction of neem seed components with insecticidal activity. *Proc. 2nd Int. Neem Conf.*, Rauischholzhausen, 1983. GTZ, Eschborn, pp. 377–390.

Mordue, A.J., K.A. Evans and M. Charlet. 1986. Azadirachtin, ecdysteroids and ecdysis in *Locusta migratoria. Comp. Biochem. Physiol.* 85c: 297–301.

Nasseh, O. and T. Freres. 1990. Käfigfreilandversuche zur Bekämpfung der Larvenstadien und adulten Tieren von *Schistocerca gregaria* (Forskal) mit biologischen und chemischen Präparaten in der Republik Niger (Insenkat/Tamesna). GTZ- Shcrift: Ergebnisse zur Bakämpfung der Wüstenheuschrecke *Schistocerca gregaria* in der Republik Niger, 1990.

Nasseh, O., T. Freres, H. Wilps, E. Kirkilionis and S. Krall. 1991. Field cage trials on the effects of enriched neem oil, insect growth regulators and *Nosema locustae* on desert locusts in the Republic of Niger. *Proc. Workshop on Biological Control of Locusts and Grasshoppers,* Cotonou. pp. 293–302.

Nasseh, O., H. Wilps and S. Krall. 1993. Neem products—Effective biopesticides for combating the desert locust, *Schistocerca gregaria* (Forskal). *Zschr. Pflanzenkr. Pflschutz* (in press).

Nicol, C.M.Y. and H. Schmutterer. 1991. Kontaktwirkungen des Samenöls des Niembaumes *Azadirachta indica* (A. Juss). bei gregären Larven der Wüstenhenschrecke *Schistocerca gregaria* (Forsk). *J. Appl. Entomol.* 111: 197–205.

Pradhan S. and M.G. Jotwani. 1968. Neem as an insect deterrent. *Chem. Age India* 19: 756–760.

Rembold, H. 1989. The azadirachtin—Their potential for insect control. *Econ. Med. Plant Res.* 3: 57–72.

Rembold, H., and K.P. Sieber. 1981. Inhibition of oogenesis and ovarian ecdysteroid synthesis in *Locusta migratoria migratorioides* (R. & F.). *Z. Naturforsch.* 43c: 903–907.

Rembold, H. M. Uhl and T. Muller. 1987. Effect of azadirachtin—A on hormone titres during the gonadotrophic cycle of *Locusta migratoria. Proc. 3d Int. Neem Conf.*, Nairobi, 1986. GTZ, Eschborn, pp. 289–298.

Schlüter, U., H.J. Bidmon and S. Grewe. 1985. Azadirachtin affects growth and endocrine events in larvae of the tobacco hornworm, *Manduca sexta. J. Insect Physiol.* 31: 773–777.

Schmutterer, H. and T. Freres. 1991. Beeinflussung von Metamorphose, Färbung und Verhalten der Wüstenheuschrecke *Schistocerca gregaria* (Forsk.) und der afrikanishchen Wanderheuschrecke *Locusta migratoria migratorioides* (R. & F.) durch Niemsamenöl. *Zschr. Pflkr. Pflschutz* 97(4): 431–438.

Sieber, K.P. and H. Rembold. 1983. The effect of azadirachtin on the endocrine control of moulting in *Locusta migratoria. J. Insect. Physiol.* 29: 523–527.

Subrahmanyam, B., T. Müller and H. Rembold. 1989. Inhibition of turnover of neurosecretion by azadirachtin in *Locusta migratoria. J. Insect Physiol.* 35: 493–500.

Wilps, H., O. Nasseh and S. Krall. 1991a. Neemprodukte—Naturliche Insektizide gegen *Schistocerca gregaria. Mittl. DGaaE* (in press).

Wilps, H., O. Nasseh and S. Krall. 1991b. The effect of various neem products on the fitness of adult *Schistocerca gregaria*. In: *Biological control of Locusts and Grasshoppers. Proc.* Eds. C. Lomer and C. Prior. Cotonou, (Benin), CAB International, pp. 337–347.

Wilps, H., E. Kirkilionis and K. Muschenich. 1992a. The effects of neem oil and azadirachtin on mortality, flight activity, and energy metabolism of *Schistocerca gregaria* Forskal—A comparison between laboratory and field locusts. *Comp. Biochem. Physiol.* 102c(1): 61–71.

Wilps, H., O. Nasseh and S. Krall. 1992b. Lutte contre les *Schistocerca gregaria* adultes au moyen des novelles préparations et méthodes. Sahel PV Info No. 44, pp. 7–14.

25

Antifeedant and Insecticidal Activity of Some Neem Fractions

M. NARAYANA RAO, K. NAGAIAH AND
G. SRIMANNARAYANA

Department of Chemistry, Osmania University, Hyderabad 500 007, India

ABSTRACT

Neem Kernel Powder was first extracted with petroleum ether (40–60), subsequently with ethanol and also directly with ethanol to yield three neem fractions—NK-P, NK-PE and NK-E. Navnem-EC is a formulation containing 5 per cent emulsifiers and the rest mixture of NK-P and NK-PE.

The reactions were tested on laboratory reared, *Spodoptera litura* (IV instar) and *Diaphania indica*. Field studies of NK-P and Navnem-EC was also carried out on groundnut, tobacco nursery and cotton. The results are promising.

In recent years much interest has been evinced in the use of plant products as insecticides to control agricultural pests in view of the environmental and health hazards posed by synthetic organic insecticides. Neem extracts have been studied in detail in the recent past by several scientists for insect repellent, antifeedant, and pesticidal properties. Azadirachtin, the tetracyclic nortriterpenoid compound, isolated from neem seeds by Butterworth and Morgan (1971), drew worldwide attention for its antifeedant activity against desert locusts. Azadirachtin is the most active component in neem (Nakanishi, 1975; Kraus et al., 1981.) The neem extracts have now been found to be highly effective in controlling more than 200 species of insect pests of different crops.

MATERIALS AND METHODS

In the present study neem kernels were powdered and extracted with solvents of different polarity like petroleum ether (40–60°), subsequent ethanol extraction and also direct ethanol extraction of neem kernel powder to yield neem fractions—NK-P, NK-PE and NK-E. Navnem-EC is a formulation containing 5 per cent emulsifiers and the rest mixture of NK-P and NK-PE.

Two methods are often used in the laboratory to study the antifeedant activity of neem on insects. These are choice (Munakata, 1978) and non-choice (Ascher et al., 1980) methods.

Laboratory bioassay of the neem fractions was carried out on *Spodoptera litura* by the non-choice method. A disc of 9.1 cm diameter was cut from a fresh castor leaf. The test compound (1000 ppm in acetone) was sprayed uniformly on either side of the leaf with an atomizer dried by exposing to air for 1 minute and placed on a moist filter paper in a petri dish of 10 cm diameter.

A single fourth-instar larva of *S. litura* (starved for six hours) reared in laboratory was introduced in the petri dish. Two sets of parallel controls were maintained with similar leaf discs, one sprayed with acetone solution and the other with acetone solvent only. The consumed leaf area was measured by graphical method or with a planimeter at regular intervals of 12 hours each till 48 hours. The experiments were triplicated and the average values taken. The percentage leaf protection was calculated by using Singh and Pant's formula (1980):

$$\begin{array}{c}\text{\% leaf protection}\\ \text{or}\\ \text{\% antifeedant activity}\end{array} = \frac{\begin{array}{c}\text{\% leaf protection in}\\ \text{treated discs}\end{array} - \begin{array}{c}\text{\% leaf protection in}\\ \text{control disc}\end{array}}{100 - \text{\% leaf protection in control disc}} \times 100$$

Neem fractions which exhibited 75 per cent and above antifeedant activity are considered to be good antifeedants.

RESULTS AND DISCUSSION

The results of antifeedant activity on laboratory reared fourth-instar *S. litura* is presented in Table 1.

Table 1. Antifeedant activity of NK-P, NK-PE and NK-E against larvae of *Spodoptera litura* (IV instar larva, non-choice method)

Concentrations of neem extracts	% leaf area consumed after (castor leaves)		% leaf area protected after		% of antifeedant activity
	24 hrs	48 hrs	24 hrs	48 hrs	
NK-P					
0.03%	6.9	19.21	93.1	80.79	77.9
0.05%	6.4	13.85	93.6	86.15	84.1
NP-PE					
0.03%	4.8	20.3	95.2	79.3	76.2
0.05%	3.54	15.04	96.46	84.96	82.7
NK-E					
0.03%	2.24	3	97.76	97	96.55*
0.05%	—	—	100	100	100
Control	80.15	86.99	19.85	13.01	Nil

* The best antifeedant

Antifeedant activity of neem extracts on *S. litura* on groundnut plants was also assessed in a field study. Neem fraction NK-P showed promising activity. Values are presented in Table 2.

The effect of Navnem-EC, a formulation of neem, was assessed on *S. litura* in a tobacco nursery. The formulation fared satisfactorily. These are presented in Tables 3 and 4.

However, on the fourth instar larva of *Diaphania indica*, three neem fractions showed only moderate antifeedant activity (66 to 52 per cent). These are shown in Table 5.

Navnem-EC 500 ml/ha and 750 ml/ha fared on par with quinolphos when sprayed on cotton plants in a field study. The percentage of bollworm damage and loculi damage is almost same with Navnem-EC and quinolphos EC (Dhawan et al., 1992). These are given in Table 6.

Further field trials on Navnem-EC is in progress. This formulation is arrived on the basis of the antifeedant activity results of the neem fractions of NK-P, NK-PE and NK-E.

Table 2. Antifeedant activity of neem seed extracts on *Spodoptera litura* in groundnut

Average leaf area fed by IV instar larvae on plant (cm, original leaf area 300 sq cm)			
Time		Control	NK-P
28 hrs		5.4	0.11
48 hrs		16.18	0.57
72 hrs		30.08	1.17
Feeding ratio:			
		24 hrs	72 hrs.
NK-P		2.03	3.88
Leaf area protected			
		24 hrs	72 hrs
NK-P		97.97	96.12
Yield:	Wet wt. of pods from plants	Dry yield per hectare	
Control	3.765 kg	23.5 Qtl	
NK-P	3.915 kg	24.4 Qtl	

Concentrates of 0.1 per cent sprayed @ 1 litre/20 m^2 five times at weekly intervals.

Table 3. Leaf area consumed by the fourth instar larvae of *S. litura* and mean increase/decrease in weight of the larvae 48 hrs after treatment in 1991–92

Sl. No.	Treatment	Leaf area consumed in sq cm: Choice method	Leaf area consumed in sq cm: Non-choice method
1.	Navnem-EC 0.1%	1.76	1.50
2.	Control-water spray	5.45	5.26
3.	Control no spray	6.92	5.19
	S. Em	0.38	0.17
	CV %	28.10	13.79
	CD at 5%	1.12	0.49
	CD at 1%	1.52	0.67

Table 4. Mean number of seedling damaged by *S. litura* in the nursery in 1991–92

Sl. No.	Treatment	Number of seedlings damaged	
		2 days	9 days
1.	Navnem-EC 0.1%	3.10 (9.59)	3.65 (13.30)
2.	Control-water spray	9.52 (20.60)	9.41 (88.56)
3.	Control-No spray	9.70 (94.18)	9.62 (92.65)
	S. Em	0.20	0.21
	CV %	7.66	7.68
	CD at %	0.59	0.62
	CD at 1%	0.80	0.84

Table 5. Antifeedant activity of NK-P, NK-PE, NK-E against larvae of *Diaphania indica* IV instar larva, non-choice method

Concentrations of neem extracts	% leaf area consumed after		% leaf area protected after		% of antifeedant activity
	24 hrs	48 hrs	24 hrs	48 hrs	
NK-P 0.05%	26.58	33.88	73.43	66.13	66
NK-PE 0.05%	50.00	47.95	59.3	52.05	52
NK-E 0.05%	29.34	39.67	70.66	63.33	63
Control	100	—	—	—	—

Table 6. Boll worm damage (Number of bolls %)

Treatment		Boll worm damage %	Loculi damage %
Navnem	500 ml/ha	29.1	11.5
	700 ml/ha	13.9	6.6
Quinolphos		12.1	4.17
Control	75.1	35.1	

ACKNOWLEDGEMENTS

The authors thank ICAR and Union Ministry of Environment and Forests for financial assistance. The authors acknowledge the help rendered by Dr. C. Kanda Swamy, FIPPAT, Madras, Dr. (Ms) Gitanjali, O.U., Dr. Julius Diwakar, CPPTI, Hyderabad, Dr. B.G. Ramaprasad and Dr. M.S. Chari, CTRI, Rajahmundry, Dr. A.K. Dhawan, P.A. U. Ludhiana, Dr. (Ms) Reena, Dr. (Ms) K. Sukumaran, IICT, Hyderabad Dr. V.V.L.N. Prasad, Dr. T.S. Subramaniam, ITC, Hyderabad for discussions and help in arranging the entomological screening of neem fractions.

REFERENCES

Ascher, K.R.S. and G. Rones. 1964. *International Pest Control* (1964). March/April, 6.

Butterworth, J.H. and E.D. Morgan. 1971. Investigation of the locust (*Schistocerca gregaria*) feeding inhibition of the seeds of the neem tree, *Azadirachta indica. J. Insect. Physiol.*, 17: 969.

Dhawan, A.K., G. Srimannarayana, G.S. Simwat and K. Nagaiah. 1992. Management of cotton pests on upland cotton *Gossypium hirsutum* with Navnem-EC. In: *Proc. First Natl. Symp. on Allelopathy in Agroecosystem*, Eds. P. Tauro and S.S. Narwal, Indian Society of Allelopathy, Haryana Agricultural University , Hisar, India, p.156.

Kraus, W., R. Cramer and G. Sawitzki. 1981. Tetranortriterpenoids from the seeds of *Azadirachta indica. Phytochemistry* 20(1): 117.

Munakata, K. 1974. Insect antifeeding substances in plant leaves. In: *Advances in Pesticide Chemistry* IUPAC, 42: 57.

Nakanishi, K. 1975. Structure of the insect antifeedant Azadirachtin. In: *Recent Advances in Phytochemistry*, Ed. V.C. Runeckles, Plenum Press, New York, p. 283.

Singh, R.P. and N.C. Pant. 1980. Lycorine—a resistance factor in the plants sub-family Amarylloidiodeae (Amarylliadeae) against desert locust. *Experientia* 36: 552.

26

Antifeedant and Growth-inhibiting Effects of Azadirachtin-rich Neem Fractions on Rice White-backed Planthopper *Sogatella furcifera* Horvath, *Spodoptera litura* Fb. and *Helicoverpa armigera* Hb.

S.J. NELSON[1], P.C. SUNDARABABU[1], D.S. RAJVEL[1], G. SREEMANNARAYANA[2] AND Y. GEETHANJALI[2]

[1]Tamil Nadu Agricultural University, Agricultural College and Research Institute Madura 625 104, India
[2]Osmania University, Hyderabad 500 001, India

ABSTRACT

Thirteen azadirachtin-rich neem fractions—Neemol, Ne-pet-et, Nemicidin, NS 55/GSN/OU, NS 56/GSN/OU, NS 57/GSN/OU, NS 58/GSN/OU, Vemidin, V-75, NK-72, NK-100, Neem-75, and Nemidin—(all at 1000 ppm) were evaluated for their effect in comparison with neem cake extract (5 per cent) on rice white-backed planthopper (WBPH). Oviposition was highly deterred by the application of NS 58/GSN/OU, in which 89.13 per cent reduction in egg laying was recorded compared to control. Feeding rate measured in terms of honeydew excretion indicated that NK-100 was the best antifeedand (69.34 per cent reduced feeding rate) followed by NS 58/GSN/OU (69.24 per cent), and both the treatments were on par. With reference to nymphal development, only 40.00 per cent nymphs became adults in Vemidin treatment. Moulting disturbances were observed in NS 58/GSN/OU, NK-100, and Nemidin treatments. NS 58/GSN/OU was proved to posses high antifeedant and growth inhibiting effects against WBPH.

Studies with seven azadirachtin-rich frations of neem on *Spodoptera litura* Fb. and *Helicoverpa armigera* Hb. revealed the antifeedant and growth retardant effects of these fractions. In *S. litura* Ne-pet-et showed higher antifeedant effect: 66.68 per cent reduction in leaf area feeding was noticed compared to control. Nemidin inhibited the larvae from reaching the adult stage to the tune of 68.18 per cent and it was on par with Nemidin, Ne-pet-et, Vemidin, and NK-100. In *H. armigera*, 91.56 per cent feeding inhibition was recorded in Nem-75 treatment followed by Vemidin (87.69 per cent) and NK-100 (75.74 per cent). Moulting disturbances were noticed in Ne-pet-et, Vemidin, Nem-75, and Nemidin treatments.

INTRODUCTION

Neem (*Azadirachta indica* A. Juss) trees are widespread in many countries of tropical Asia. Although they are not cultivated commercially, they are grown over fairly large areas in some countries. There are about 14 million neem trees in India and the annual oil and cake production is more than 82,000 and 3,33,000 tonnes respectively (Ketkar, 1987). The increasing cost of synthetic insecticides and the growing awareness of ecological hazards associated with their large-scale use have evoked interest in pest control agents of plant origin. These resulted in the discovery of neem products as effective plant-origin insecticide for the management of crop pests (Joshi et al., 1984; Saxena and Khan, 1985). Neem products possess antifeedant and other effects on insects. These effects are due to a triterpenoid, azadirachtin, present in seeds (Zanno et al., 1975). Although azadirachtin is the most potent principle, its yield is very low and its isolation is a complicated process. Therefore, the attention of chemists and entomologists was directed towards production of azadirachtin-rich neem seed fractions which yielded many fractions such as AZT-VR-NR and MTB/H_2O-NR (Feuerhake, 1983) and Nemol, Vemidin, and Nemidin (Srimannarayana, 1987). This paper reports the effects of azadirachtin-rich fractions developed by Dr. G. Srimannarayana, Natural Products Laboratory, Osmania University, Andhra Pradesh, on rice white-backed plant hopper (WBPH) *Sogatella furcifera* Horvath, *Spodoptera litura* Fb., and *Helicoverpa armigera* Hb.

MATERIALS AND METHODS

The Azadirachtin-rich fractions, namely, Nemol, Ne-pet-et, Nemicidin, NS 55/GSN/OU, NS 56/GSN/OU, NS 57/GSN/OU, NS 58/GSN/OU, Vemidin, V-74, NK-72, NK-100, Nem-75, and Nemidin, were dissolved in acetone separately and made up with water for 1000 ppm concentration. The WBPH was cultured in insectary as suggested by Medrano and Heinrichs (1985). *Spodoptera litura* was reared on castor leaves in insectary. Larvae of *H. armigera* were collected from the field.

White-backed Plant Hopper

Oviposition and hatchability

Three gravid females of WBPH were caged on IR 20 (15 DAT—days after transplanting) plants treated with azadirachtin-rich fractions in three replications. Observations on the nymphs emerged were recorded by removing the nymphs that hatched out from the seventh day onwards for two weeks at three-day intervals. The number of unhatched eggs was determined by dissecting out the tillers under binocular microscope.

Feeding

Feeding of five newly developed female WBPH was estimated by the honeydew excretion measurement method developed by Pathak and Heinrichs (1981).

Development

Ten nymphs of WBPH second/third instar were confined on treated and control IR 20 (15 DAT) plants for a week to study the growth retardant and insecticidal effects of neem fractions. Mortality counts were made on first, third, fifth, and seventh day after treatment. Observations on the malformations and moulting disturbances were also made.

Feeding and Development of *S. litura* Fb.

The feeding of *S. litura* as influenced by spraying of castor leaves and larvae with solution containing fraction was studied in petri dishes in three replicates with five fourth-instar larvae per replicate. Area fed was measured after 24 hrs by graphical method and the faecal pellets were dried in a hot air oven and weighed. Development of the larvae was studied by allowing 10 treated larvae per replication on treated castor leaves. After 24 hrs of feeding, fresh untreated leaves were provided and observations were made on the larval mortality, pupal development, adult emergence, and malformations at larval, pupal, and adult stages.

Feeding and Development of *H. armigera* Hb.

One larva was confined per petri dish on cotton leaves treated with fractions. Five replications were maintained. Feeding area was measured after 24 hrs by graphical method and weight of the faecal pellets was also taken. Observations on malformations in larval, pupal, and adult stages were made.

RESULTS

White-backed Planthopper

Oviposition and Hatchability

Total number of eggs laid, the number of nymphs emerged, and hatchability per cent are furnished in Table 1. NS 58/GSN/OU was found to be the best ovipositional deterrent; only 29.33 eggs were laid, which was 89.13 per cent less than the number of eggs laid in control. The maximum of 324.00 eggs were laid in Nemol, which was more than but on par with control. Nymphs emerged in NS 58/GSN/OU-treated plants numbered only 25.67, which proved to be significantly better than other treatments. It was followed by V-74 and NS 57/GSN/OU, in which 43.00 and 56.67 nymphs emerged. In Nemol-treated plants a maximum of 273.00 nymphs emerged, which was 2.12 per cent more than the control but on par with acetone and control. Hatchability was less (84.25 per cent) in NS 55/GSN/OU followed by Nemol (84.29 per cent) and these two treatments were on par. Maximum (90.33 per cent) hatchability was recorded in NS 57/GSN/OU. There was 2.58 (NS 57/GSN/OU) to 9.14 (NS 55/GSN/OU) per cent reduction in hatchability in different treatments. In control and acetone treatments 92.72 and 90.74 per cent hatchability was observed respectively.

Table 1. Effect of azadirachtin-rich fractions on oviposition and hatchability of WBPH (mean of three replications)

Treatments	Total No. of eggs laid	No. of nymphs emerged	Hatchability %
Nemol	324.00 (18.00) k	273.00 (16.52) j	84.29 (9.18) a
Ne-pet-et	70.00 (8.34) cd	62.00 (7.87) cd	88.50 (9.39) bcd
Nemicidin	111.33 (10.53) dg	100.00 (9.99) fg	89.70 (9.47) cd
NS 55/GSN/OU	85.00 (9.19) def	71.67 (8.46) cde	84.25 (9.18) a
NS 56/GSN/OU	86.00 (9.25) defg	75.00 (8.66) cdef	87.16 (9.34) bc
NS 57/GSN/OU	62.67 (7.80) bc	56.67 (7.49) bc	90.33 (9.50) de
NS 58/GSN/OU	29.33 (5.40) a	25.67 (5.03) a	87.19 (9.34) bc
Vemidin	147.33 (12.11) h	131.67 (11.44) h	89.38 (9.45) bcd
V-74	48.33 (6.95) b	43.00 (6.55) b	89.09 (9.44) bcd
Nk-72	71.00 (8.41) cde	61.67 (7.85) cd	86.89 (9.32) b
Nk-100	92.00 (9.59) defg	81.33 (9.01) defg	88.26 (9.39) bcd
Nem-75	114.67 (10.69) gh	103.00 (10.12) gh	89.86 (9.48) cde
Nemidin	104.00 (10.19) fg	90.67 (9.50) efg	87.19 (9.34) bc
Neem cake extract 5%	96.33 (9.81) efg	85.67 (9.25) efg	88.83 (9.42) bcd
Teepol 0.1%	222.33 (14.90) i	199.33 (14.11) i	89.60 (9.47) cd
Acetone	266.33 (16.31) ij	241.67 (15.55) ij	90.74 (9.53) de
Control	288.33 (16.97) jk	267.33 (16.30) j	92.72 (9.63) e

Nemol to Nemidin: 1000 ppm.
Figures in parentheses are square root values.
In a column, means followed by same letters are not significantly different ($p = 0.05$) by DMRT.

Feeding

Honeydew excreted by five freshly emerged gravid females on treated plants was measured graphically and the data are furnished in Table 2. In NK-100, only 210.33 mm^2 area honeydew was excreted, followed by NS 58/GSN/OU (211.00 mm^2), which was 69.34 and 69.24 per cent less than the control. These two treatments were on par statistically. Maximum area of honeydew was noted in Nem-75 (391.67 mm^2) and it was on par with NS 57/GSN/OU (387.33 mm^2), neem cake extract (380.00 mm^2), and

Table 2. Effect of azadirachtin-rich fractions on feeding and development of WBPH (mean of three replications)

Treatments	Honeydew area, mm^2 *	% decrease from control	Adult developed %**
Nemol	284.00 (2.45) c	58.60	53.33 (46.92) abc
Ne-pet-et	234.67 (2.37) b	65.80	50.00 (45.00) ab
Nemicidin	279.67 (2.45) c	59.23	53.33 (46.92) abc
NS 55/GSN/OU	355.00 (2.55) d	48.25	66.67 (55.08) bc
NS 56/GSN/OU	232.67 (2.37) b	66.08	50.00 (45.00) ab
NS 57/GSN/OU	387.33 (2.59) d	43.54	66.67 (54.78) bc
NS 58/GSN/OU	211.00 (2.32) a	69.24	43.33 (41.15) a
Vemidin	237.00 (2.37) b	65.45	40.00 (39.15) a
V-74	300.67 (2.48) c	56.17	50.00 (45.07) ab
NK-72	285.33 (2.46) c	58.41	70.00 (57.00) c
NK-100	210.33 (2.32) a	69.34	43.33 (41.15) a
Nem-75	391.67 (2.59) d	42.91	70.00 (57.00) c
Nemidin	270.00 (2.43) c	60.64	43.33 (41.15) a
Neem cake extract 5%	380.00 (2.58) d	44.61	53.33 (46.92) abc
Teepol 0.1%	489.00 (2.69) e	28.72	93.33 (77.71) d
Acetone	684.33 (2.84) f	0.24	96.67 (83.85) de
Control	686.00 (2.84) f		100.00 (90.00) e

Neem to Nemidin: 1000 ppm.
* Figures in parentheses are logarithmic values.
** Figures in parentheses are arcsine values.
In a column, means followed by same letters are not significantly different (p = 0.05) by DMRT.

NS 55/GSN/OU (355.00 mm^2). In acetone and control treatments 684.33 mm^2 and 686.00 mm^2 honeydew area was recorded respectively and these were on par.

Development

In Vemidin-treated plants, 40.00 per cent of nymphs became adults followed by NS 58/GSN/OU, NK-100, and Nemidin, in which 43.33 per cent nymphs became adults.

These four treatments were on par statistically. NK-72 and Nem-75 recorded maximum development (70.00 per cent). Moulting disturbances were observed in NS 58/GSN/OU, NK-100, and Nemidin (i.e., exuviae could not come off the body of the insects).

Feeding and Development of *S. litura*

When both the leaves and larvae were treated with azadirachtin-rich fractions, there was a reduction in feeding area and weight of faecal pellets (Table 3). Ne-pet-et was found to be the best antifeedant; 1068.33 mm^2 area was fed and 521.00 mg of faecal pellets was excreted, 66.68 and 46.42 per cent lower than the control respectively. Maximum feeding was noted in Vemidin (2823.00 mm^2) and faecal pellets excreted was also higher (911.67 mg) in Vemidin. Feeding was reduced from 11.95 (Vemidin) to 66.68 per cent (Ne-pet-et) in different treatments.

The pupae and adults developed in different treatments are presented in Table 4. Minimum (53.33 per cent) development of larvae to pupae was observed in Ne-pet-et and NS 58/GSN/OU. Larval death due to exposure of internal contents through the opening formed in dorsal surface and larval-pupal intermediaries were also observed in Ne-pet-et, Nem-75, and Nemidin. When the normal adult emergence was considered, in Ne-pet-et only 23.33 per cent became adults and it was comparable with NS 58/GSN/OU (30.00 per cent), Nemidin (30.00 per cent), NK-100 (33.33 per cent), and Nem-75 (33.33 per cent). In acetone and control 66.67 and 76.67 per cent of larvae became

Table 3. Effect of azadirachtin-rich fractions on feeding of *S. litura* (mean of three replications)

Treatments	Area fed by five fifth-instar larvae, mm^2	% decrease from control	Weight of faecal pellets, mg	% decrease from control
Nemol	2082.33 (3.32) d	35.05	851.00 (2.93) cd	12.48
Ne-pet-et	1068.33 (3.03) a	66.68	521.00 (2.72) a	46.42
NS 58/GSN/OU	1433.67 (3.16) c	55.28	697.00 (2.84) b	28.32
Vemidin	2823.00 (3.45) e	11.95	911.67 (2.96) de	6.24
NK-100	1191.67 (3.08) b	62.83	651.00 (2.81) b	33.05
Nem-75	1951.67 (3.29) d	39.13	800.33 (2.90) c	17.69
Nemidin	2039.00 (3.31) d	36.41	784.00 (2.89) c	19.37
Acetone	3179.67 (3.50) f	0.83	939.00 (2.97) de	3.43
Control	3206.23 (3.51) f		972.33 (2.99) e	

Nemol to Nemidin: 1000 ppm.
Figures in parentheses are logrithimic values.
In a column means followed by same letters are not significantly different (p = 0.05) by DMRT.

Table 4. Effect of azadirachtin-rich fractions on development of *S. litura* (mean of three replications)

Treatments	Larval death, % (including larval-pupal intermediaries)	% increase over control	% pupae developed	% increase from control	Pupal death % (including malformed adults)	% increase over control	Normal adults produced %	% decrease from control
Nemol	36.67 (37.22) a	175.09	63.33 (52.80) a	26.93	23.33 (28.78) ab	133.30	40.00 (39.23) bc	47.83
Ne-pet-et	46.67 (43.08) a	250.11	53.33 (46.95) a	38.47	30.00 (33.00) a	200.00	23.33 (28.78) a	69.57
NS 58/GSN/OU	46.67 (43.08) a	250.11	53.33 (46.95) a	38.47	23.33 (28.78) ab	133.00	30.00 (33.21) ab	60.87
Vemidin	20.00 (26.07) b	50.04	80.00 (63.96) b	7.70	26.67 (30.99) a	166.70	53.33 (46.92) cd	30.44
NK-100	43.33 (41.15) a	225.06	56.67 (48.87) a	34.61	23.33 (28.78) ab	133.00	33.33 (35.22) ab	56.53
Nem-75	43.33 (41.07) a	225.06	56.67 (48.87) a	34.61	23.33 (28.78) ab	133.00	33.33 (34.92) ab	56.53
Nemidin	43.33 (41.15) a	225.06	56.67 (48.87) a	34.61	26.67 (30.99) a	166.70	30.00 (33.21) ab	60.87
Acetone	20.00 (26.56) b	50.04	80.00 (63.47) b	7.70	13.33 (21.15) bc	33.30	66.67 (54.78) de	13.04
Control	13.33 (21.15) b		86.67 (68.89) b		10.00 (18.44) c		76.67 (61.22) e	

Nemol to Nemidin: 1000 ppm.
Figures in parentheses are arcsine values.
In a column means followed by same letters are not significantly different ($p = 0.05$) by DMRT.

adults respectively. In Ne-pet-et treatment one adult could not emerge from the pupae and died. Malformed adults were produced in Ne-pet-et.

Feeding and Development of *H. armigera*

There was a significant reduction in feeding and excretion in azadirachtin-rich fraction treatments compared to control (Table 5). The least area was fed in Nem-75 (84.20 mm^2), which was significantly lower than other treatments and 91.56 per cent lesser than control. In Vemidin, NK-100, and Nemidin, 87.69, 75.74, and 66.61 per cent reduction in feeding respectively was noticed. Weight of faecal pellets was less in Nem-75 (39.00 mg), which was 99.19 per cent lower than the control, followed by Nemicidin (93.20 mg) and NS 58/GSN/OU (89.00 mg). These treatments were on par. Malformed adults emerged in Ne-pet-et, Vemidin, Nem-75 and Nemidin treatments.

DISCUSSION

The results from the laboratory experiments on effect of neem fractions on feeding and development of WBPH, *S. litura*, and *H. armigera* are in conformity with the earlier findings. Heyde et al. (1983) and Shukla et al. (1987) reported the same effects of neem oil on feeding and development of WBPH. Behavioural and physiological effects of

Table 5. Effect of azadirachtin-rich fractions on *H. armigera* (mean of five replications)

Treatments	Area fed by single larvae, mm^2*	% decrease from control	Wt. of faecal pellets, mg**	% decrease from control
Ne-pet-et	519.60 (2.71) e	47.90	74.40 (8.62) d	41.23
Nemicidin	707.60 (2.85) f	29.06	93.20 (9.64) d	26.38
NS 58/GSN/OU	537.20 (2.73) e	46.14	89.00 (9.43) e	29.70
Vemidin	122.80 (2.08) b	87.69	54.80 (7.40) b	56.71
NK-100	242.00 (2.38) c	75.74	61.60 (7.84) bc	51.34
Nem-75	84.20 (1.92) a	91.56	39.00 (6.23)a	69.19
Nemidin	333.00 (2.52) d	66.61	68.20 (8.25) cd	46.13
Acetone	936.60 (2.97) g	6.10	116.00 (10.76) f	8.37
Control	997.40 (2.99) g		126.60 (11.25) f	

* Figures in parentheses are logarithmic values.
** Figures in parentheses are square root values.
Ne-pet-et to Nemidin: 1000 ppm.
In a column, means followed by same letters are not significantly different ($p = 0.05$) by DMRT.

neem seed derivatives, purified fractions, and their formulations on WBPH were clearly demonstrated by Saxena et al. (1985). Unhatchability of eggs of WBPH may be due to systemic action of neem products as reported against *Schistocerca gregaria* F. (Gill and Lewis, 1971), European corn borer *Ostrinia nubilalis* Hb. (Meisner and Ascher, 1986), and leafminer *Liryomyza trifoli* (Burgess) (Hiram, 1988). Neem products served as ovipositional deterrent in *S. litura* (Joshi and Sitaramaiah, 1979) and *Phthorimaea operculella* (Sharma et al., 1983). The results on feeding of *S. litura* corroborate the eariler findings (Joshi et al., 1978, 1984). Azadirachtin had an antifeedant activity against *S. frugiperda* (Klocke and Kubo, 1982; Raffa, 1987) and *S. littoralis* (Meisner et al., 1981). Antifeedant effect of neem products on other insects was also reported by many scientists (Quadri, 1973; Attri and Singh, 1977). Growth retardant activity of neem products against WBPH (Saxena et al., 1983) and *S. litura* (Koul, 1985) was established already. Malformations in larval, pupal, and adult stages may be due to the growth disrupting effects of neem fractions. Growth disrupting effects of azadirachtin-rich fractions confirm the findings of many authors on lepidopterous pests (Kubo and Klocke, 1982; Schlueter et al., 1985; Bidmon, 1986).

This study reveals the ovipositional deterrent, antifeedant, and growth disrupting effects of new fractions of neem and the potential of these new fractions for developing commercial formulations for managing agricultural pests.

REFERENCES

Attri, B.S. and R.P. Singh. 1977. Effect of time and temperature on the antifeeding activity of neem seed cake to desert locust *Schistocerca gregaria* F. *Indian J. Entomol.* 39: 384–385.

Bidmon, H.J. 1986. Ultrastructural changes of the prothorax glands of untreated and with azadirachtin treated *Manduca sexta* larvae (Lepidoptera:Sphingidae). *Entomol. Gen.* 12: 1–17.

Feuerhake, K.J. 1983. Effectiveness and selectivity of technical solvents for the extraction of neem seed components with insecticidal activity. *Proc. 2nd Int. Neem Conf.*, Rauischholzhausen, 1983. German Agency for Technical Co-operation, Eschborn, Germany pp. 103–104.

Gill, J.S., and C.T. Lewis. 1971. Systemic action of insect feeding deterrent. *Nature* 232 (5310): 402–403.

Heyde, J.V.D., R.C. Saxena and H. Schmutterer. 1983. Neem oil and neem extracts as potential insecticides for control of hemipterous rice pest. *Proc. 2nd Int. Neem Conf.*, Rauischholzhausen, 1983. German Agency for Technical co-operation, Eschborn, Germany pp. 377–390.

Hiram, G.L. 1988. Limited occurrence of foliar, root and seed extract toxin in untreated plant parts. *J. Econ. Entomol.* 81: 593–598.

Joshi, B.G. and S. Sitaramaiah. 1979. Neem kernel as an ovipositional repellent for *Spodoptera litura* F. moths. *Phytoparasitica* 7: 199–202.

Joshi, B.G., G. Ramaprasad and S. Nageswararao. 1984. Neem seed kernel suspension as an antifeedant for *Spodoptera litura* in a planted flue-cured virginia tobacco crop. *Phytoparasitica* 7: 199–202.

Joshi, B.G., G. Ramaprasad and S.V.V. Satyanarayana. 1978. Relative efficacy of neem kernel, fentin acetate and fentin hydroxide as antifeedant against tobacco caterpillar *Spodoptera litura* Fb. in the nursery. *Indian J. Agric. Sci.* 48(1): 19–22.

Ketkar, C.M. 1987. Production and availability of neem products in India. Paper presented in the Research Planning Workshop on Botanical Pest Control in Rice-Based Cropping Systems. June 1–6, 1987. Abst. p. 15. Tamil Nadu Agricultural University, Coimbatore.

Klocke, J.A. and I. Kubo. 1982. Citruslimonoid by products as insect control agents. *Entomol. Exp. Appl.* 32: 299–301.

Koul, O. 1985. Azadirachtin interaction with development of *Spodoptera litura* (Fab). *Indian J. Exp. Biol.* 23(3): 160–163.

Kubo, I. and J.A. Klocke. 1982. Azadirachtin insect ecdysis inhibitor. *Agric. Biol. Chem.* 46: 1951–1953.

Medrano, F.G. and E.A. Heinrichs. 1985. *Nilaparvata lugens*. In: *Hand book of Insect Rearing*, Vol. I. Eds. P. Singh and R.F. Moore. Elsevier Publishing Co., New York, pp. 361–372.

Meisner, J. and K.R.S. Ascher. 1986. The influence of neem on the cornborer *Ostrinia nubilalis, Sesamia nonagridoide* and the leaf miner *Liriomyza trifolii. 3rd Int. Neem Conf.*, July 10–15, 1986. Abst. p. 49, ICIPE, Nairobi.

Meisner, J., K.R.S. Ascher, R. Aly and J.D. Warthen. 1981. Response of *Spodoptera littoralis* (Boisd) and *Earias insulana* (Boisd) larvae to azadirachtin and salannin. *Phytoparasitica* 9:27–32.

Pathak, P.K. and E.A. Heinrichs. 1981. Bromocresol green indicator for measuring feeding activity of *Nilaparvata lugens* on rice varieties. *Philip. Ent.* 5: 195–198.

Quadri, S.S.H. 1973. Some new indegenous plant repellents for storage pests. *Pesticides* 7(12): 18–20.

Raffa, K.F. 1987. Influence of host plant on deterrence by azadirachtin of feeding by fall army worm larvae (Lepideptera: Noctuidae). *J. Econ. Entomol.* 80: 384–387.

Saxena, R.C. and Z.R. Khan. 1985. Effect of neem oil on survival of *Nilaparvata lugens* and on grassy stunt virus transmission. *J. Econ. Entomol.* 78: 647–651.

Saxena, R.C., P.B. Epino, Tucheng Wen and B.C. Puma. 1983. Neem, China berry and Custard apple; Antifeedant and insecticidal effects of seed oils on leafhopper and plant hopper pests of rice. *Proc. 2nd Int. Neem Conf.*, Rauischholzhausen, 1983. German Agency for Technical Co-operation Eschborn, Germany pp. 403–412.

Saxena, R.C., S.F. Chiu, V. Mariappan and M.B. Kalode. 1985. Evaluation and utilization of neem *Azadirachta indica* A. Juss. seed derivatives for management of rice insect pests. Forum: *Int. Rice Res. Conf.* 1–5 June, 1985, International Rice Research Institute, Los Banos, Philippines.

Schlueter, U., H.J. Bidmon and S. Grewe. 1985. Azadirachtin affects growth and endocrine events in larvae of the tobacco horn worm *Manduca sexta* V. *J. Insect Physiol.* 31(10): 773–777.

Sharma, B.N., B.A. Nagasampagi, A.S. Bhosale, M.M. Kulkarni and V.B. Tungikar. 1983. Neem-rich: The concept of enriched fractions from neem for behavioural and physiological control of insects. *Proc. 2nd Int. Neem Conf.*, Rauischholzhausen, 1983. German Agency for Technical Co-operation, Eschborn, Germany pp. 115–128.

Shukla, B.C., S.K. Srivastava and R. Gupta. 1987. Evaluation of No against WBPH and gallmidge pests of rice. Paper presented at the Research Planning Workshop on Botanical Pest Control in Rice-Based Cropping Systems. June 1–6, 1987. Abst. p. 14. Tamil Nadu Agricultural University, Coimbatore.

Srimannarayana, G. 1987. Recent advances of insecticidal interest in leguminous plant products and neem. *Proc. Symp. Alternatives to Synthetic Insecticides* March 30-31. Centre for Research in Medical Entomology Madurai, Tamil Nadu, pp. 7–20.

Zanno, P.R., I. Miura, K. Nakanishi and D.L. Elder. 1975. Structure of the insect phagorepellent azadirachtin application of PRFT/CWD carbon-13 nuclear magnetic resonance. *J. Amer. Chem. Soc.* 97: 1975–1977.

27

Effects of Extracts from Neem on Aphids (Homoptera: Aphididae) and Their Natural Enemies

D. THOMAS LOWERY[1] AND MURRAY B. ISMAN

Department of Plant Science, University of British Columbia, Vancouver, B.C., Canada V6T 1Z4

[1]Present address: Agriculture Canada, Research Station, Summerland B.C., Canada V0H 1Z0

ABSTRACT

Laboratory and field trials with formulated neem seed oil and neem seed extract demonstrated that these materials are efficacious as aphicides. Control of aphids results from failure of nymphs to moult and from inhibition of adult reproduction. The effective concentrations of neem or azadirachtin resulting in 50 per cent mortality (EC_{50}) were variable, depending on the aphid species, instar, and host plant. For example, the EC_{50} for second-instar green peach aphids, *Myzus persicae* (Sulzer), on corn was more than 20 times higher than on mustard cabbage, while fourth instars were approximately 20 times less sensitive than first instars. The primary route of entry of azadirachtin appears to be through ingestion of plant fluids, rather than through direct penetration of the aphid's integument. In field trials, populations of aphid natural enemies (predators and parasitoids) were not affected by applications of neem insecticides, suggesting compatibility of neem with biological control.

INTRODUCTION

Aphids are economically important pests that are difficult to control because of their mobility, tremendous reproductive ability, and resistance to many synthetic pesticides. Studies directed against several species of aphids have indicated that neem-based insecticides are potentially effective natural control agents (Schauer, 1984; Patel and

Srivastava, 1989) that may be suitable for inclusion in integrated pest management (IPM) programmes (Saxena, 1989; Hoelmer et al., 1990). Results from studies involving aphids have often been inconclusive or contradictory, however, and some reports have stated that aphids are not good candidates for control with neem-based products (Schmutterer, 1990a; National Research Council, 1992).

Studies of the effects of neem on aphids with respect to their control could help answer several questions pertaining to the potential usefulness of this plant-based pesticide. Research is required to widen the spectrum of neem-sensitive agricultural pests, as certain pest groups have largely been neglected in the past, specifically phloem-feeding insect pests (Schmutterer, 1990b). Moreover, it is well known that good or even excellent control of insects in the laboratory, particularly with insect growth regulators, does not guarantee effective control under field conditions (Schmutterer and Hellpap, 1989).

In this paper we present results from studies of the control of several economically important aphid pests with formulated neem seed oil (NSO) and an ethanolic neem seed extract (NSE) under laboratory and field conditions. Repellent and antifeedant activities of neem, and neem's effect on aphid survival and reproduction, were also evaluated. Finally, because aphids are a food source for numerous predatory and parasitic insects, effects of neem on populations of aphid predators and parasitoids were evaluated to determine the potential usefulness of neem in IPM programmes.

MATERIALS AND METHODS

Azadirachtin Content

The azadirachtin content of the formulated materials, provided by Safer Ltd. (Victoria, Canada), was determined by reverse-phase gradient high performance liquid chromatography (HPLC) (Isman et al., 1990). A comparative standard of purified (> 95 per cent) azadirachtin was supplied by J.T. Arnason (Univ. Ottawa, Ottawa, Canada).

Aphid Control

Four- to six-week-old lattuce, sweet pepper, rutabaga, and strawberry were treated with formulated NSO (0.5, 1.0, or 2.0 per cent) or with emulsifier only (1.06 ml/l, Mazon BSF19, Mazer Chemicals Inc., Gurnee, Illinois) as a control, at a rate of 10 ml/plant. Ten adult green peach aphids, *Myzus persicae* (Sulzer), were transferred to rutabaga and pepper; lettuce aphids, *Nasonovia ribisnigri* (Mosley), and strawberry aphids, *Chaetosiphon fragaefolii* (Cockerell), were transferred to lettuce and strawberry, respectively, 48 hrs before treatment. The individually caged plants were maintained in a growth chamber (20°C, 16L:8D) for one week, after which aphids were counted. Transformed aphid counts were subjected to analysis of variance, and differences between treatments were determined by Tukey's multiple range tests (Wilkinson, 1990).

During 1989 and 1990, plots of lettuce, pepper, cabbage, and strawberry were established at the Plant Science Field Station, U.B.C., Vancouver. Individual plots consisted of two rows of seven plants with aphid counts taken from the centre 10 plants, except for plots of strawberry, which were single rows of 10 plants. Experiments were randomized complete block designs with four replicates per treatment. Two separate trials were conducted on each crop.

Treatments consisted of several rates of neem, pyrethrum, and a mixture of the two materials. For the purpose of this paper, estimated concentrations of NSO or NSE resulting in a 50 per cent reduction in aphid numbers relative to controls from linear regression analyses and inverse predictions are reported. Foliar sprays were applied twice (seven-day interval) using commercial garden sprayers. One week after the second spray, plants were carefully harvested, foliar weights recorded, and aphid numbers assessed. Counts were recorded separately of *M. persicae* and the cotton aphid, *Aphis gossypii* (Glover), on pepper; *C. fragaefolii* and *Fimbriaphis fimbriata* Richards on strawberry; *M. persicae* and the cabbage aphid, *Brevicoryne brassicae* (L.), on cabbage; and *N. ribisnigri* and a complex of aphids consisting mostly of the foxglove aphid, *Aulacorthum solani* (Kalt.), and the potato aphid, *Macrosiphum euphorbiae* (Thomas), on lettuce.

Aphid Natural Enemies

The effect of foliar applications of NSO and NSE on populations of aphid predators and parasitoids was determined during the 1990 field trials. Numbers of larval predators (syrphids, neuropterans, coccinellids, and cecidomiids) and aphid mummies without emergence holes per plant were assessed during the final destructive counts. Data were analysed for natural enemy numbers per plot (10 plants), with the various host plants serving as replicated blocks. Predators and parasitoids may respond numerically to increases in aphid numbers. Therefore, natural enemy numbers per 1,000 aphids were also analysed.

Antifeedant Activity

The antifeedant activity of NSO and azadirachtin was evaluated by leaf disc choice test bioassays. Aphids were placed, two per disc, on four leaf discs (20 mm diameter) cut from leaves with a cork borer and placed in small (9 × 50 mm) petri dishes with 10 small holes in the tight-fitting lids. Dishes were placed upside down on moist Kimwipe™ lining the bottom of clear plastic containers, which were held in a growth chamber (17 ± 2°C). Two leaf discs treated with the test material and two discs treated with emulsifier (Triton X-100) only as a control were allowed to dry and then arranged alternately in each dish with their edges barely touching. Deterrence of NSO and azadirachtin was determined by the proportions of aphids on the treated discs relative to the total number of aphids on treated and control discs. Proportions were transformed by arcsin $\sqrt{x}$ to normalize the variances. Linear regression analyses and inverse predictions (Neter et al., 1985) were used to determine the effective concentrations required to deter 50 per cent of the aphids (i.e., when 25 per cent of the aphids remained on the treated discs).

Aphid Growth and Development

Aphids were reared in petri dishes as outlined for the leaf disc choice bioassays, except that only three discs were placed in each dish. Aphids were reared for three days on treated discs and a further six days on untreated discs, with new discs supplied every three days. The effect of NSO (0.5, 1.0 and 1.5 per cent) on aphid survival was compared with that of azadirachtin (40 and 80 ppm). For each species and instar, aphids were reared two per dish, with six dishes per treatment, and the entire experiment was replicated twice.

The effective concentration of azadirachtin resulting in 50 per cent mortality (EC_{50}) after nine days (initial three days on treated leaf discs) was recorded for second-instar *M. persicae* reared (five per dish) on discs dipped in azadirachtin. Values for various instars on leaf discs of pepper and second instars on several host plants were determined by linear regression analyses and inverse predictions (Neter et al., 1985).

Aphid Reproduction

The effect of NSO (0.5, 1.0 and 1.5 per cent) and azadirachtin (40 and 80 ppm) on aphid reproduction was evaluated by rearing adults or fourth instars (two per dish) for three days on treated leaf discs and a further three (for adult) or six (for fourth instar) days on untreated discs. Aphid survival and numbers of living and dead offspring were recorded daily, and the average reproductive rate (live offspring/aphid/day) calculated. The effective concentration of azadirachtin resulting in a 50 per cent reduction in reproduction was also determined for aphids on leaf discs dipped in azadirachtin at concentrations of 0–1,000 ppm.

RESULTS AND DISCUSSION

Aphid Control

In the laboratory, with the exception of 0.5 per cent NSO to strawberry, foliar applications of NSO significantly reduced aphid numbers on pepper, rutabaga, lettuce, and strawberry at all treatment rates ($p < 0.05$) compared with the control (Table 1). Neem seed oil reduced aphid numbers in a dose-dependent manner, with estimated concentrations resulting in a 50 per cent reduction in aphid numbers (EC_{50}) ranging from 1.4 per cent for *C. fragaefolii* on strawberry to as low as 0.2 per cent for *M. persicae* on pepper. These values correspond to ≈ 28.0 and 4.0 ppm azadirachtin, respectively, based on HPLC analysis.

One per cent NSO, the standard rate used in these trials, reduced *M. persicae* numbers on pepper and rutabaga by 96 and 78 per cent, respectively, compared to their controls, while numbers of *N. ribisnigri* and *C. fragaefolii* were respectively 79 and 58 per cent lower (Table 1). Control of *M. persicae* was significantly ($p < 0.05$) better on pepper than on rutabaga. The variable levels of control for this aphid on pepper as compared with rutabaga indicates that the effectiveness of neem treatment is influenced by the host plant of the aphid.

It has been previously demonstrated that control of white-flies with foliar applications of neem is influenced by the species of host plant (Schmutterer, 1990a), and soil drenches with Margosan-O reduced the number of adult leafhoppers on marigold and chrysanthemum, but not on zinnia (Knodel-Montz et al., 1985). The cause of this difference is currenty unknown, but the systemic activity of neem may vary with the species of plant.

In the field, applications of NSO effectively reduced *C. fragaefolii* and *F. fimbriata* numbers on strawberry (Table 2). The EC_{50} values for *C. fragaefolii* and *F. fimbriata* on strawberry ranged from 0.4 to 0.9 per cent NSO. Although applications of NSO reduced *N. ribisnigri* numbers on lettuce in the laboratory in a dose-dependent relationship (Table 1), neem oil was ineffective for the control of this aphid in the field ($p > 0.05$) (data not shown). The aphid complex on lettuce was also not affected by applications of neem.

Neem sprays might not have controlled aphids on lettuce in the field due to the environmental degradation of azadirachtin, particularly from the action of rain and ultraviolet light (Schmutterer, 1990a; Walter and Knauss, 1990), but changes to the plants cannot be overlookd. Also, temperatures were relatively cool during the lettuce field trials, with average daily temperatures of 11.5°C and 10.6°C during the first and second trials. These cool conditions might have reduced the effectiveness of neem, partly by decreasing aphid feeding (Schmutterer and Hellpap, 1989).

Table 1. Control of aphids on plants in the laboratory with foliar applications of formulated NSO

Treatment	Mean number aphids[a]/plant			
	M. persicae on pepper	*M. persicae* on rutabaga	*N. ribisnigri* on lettuce	*C. fragaefolii* on strawberry
Control	130.2a	150.8a	73.5a	51.3a
NSO 0.5%	16.2b	93.6b	36.7b	37.9a
NSO 1.0%	4.7c	33.1c	15.5c	21.6bc
NSO 1.5%	1.1d	3.1d	6.3c	8.5c
EC_{50} (%NSO)	0.2	0.6	0.5	1.4

Means within a row followed by the same letter are not significantly different ($p > 0.05$, Tukey's multiple range test).

[a] See text for aphid common names.

Table 2. Estimated concentrations of neem (%) required to reduce aphid populations on plants in the field by 50 per cent. Numbers in parentheses are approximate concentrations of azadirachtin in ppm.

Host plant	Aphid species	Trial 1	Trial 2
Strawberry	*C. frgaefolii*	0.6 (12)	0.9 (18)
Strawberry	*F. fimbriata*	0.4 (8)	0.6 (12)
Pepper	*M. persicae*	0.6 (12)	1.7 (34)
Pepper	*A. gossypii*	0.5 (10)	

Note : EC_{50} values based on NSO and NSE treatments for strawberry and pepper, respectively, both containing 2000 ppm AZA.

Except for the second trial involving *A. gossypii*, NSE reduced aphid numbers on pepper in the field in a dose-dependent manner, with estimated EC_{50} values ranging from 0.5 to 1.7 per cent (Table 2). Although slightly better in the laboratory, NSO effectively controlled *M. persicae* on pepper in the laboratory and in the field. NSE was also very effective for the control of both *M. persicae* and *B. brassicae* on cabbage. For *M. persicae*, 1 per cent NSO and 1 per cent NSE reduced populations by 86 to 91 per cent, while the same treatments reduced numbers of *B. brassicae* by 87 to 100 per cent (Lowery et al., 1993).

Control of aphids with foliar applications of pyrethrum (0.5 g/l) was variable and generally poorer than with neem (Lowery et al., 1993). Pyrethrum controlled *B. brassicae, F. fimbriata*, and *N. ribisnigri* to some extent, had no effect on *C. fragaefolii* or *A. gossypii*, and significantly increased numbers of *M. persicae* on pepper. In no instance did the co-application of pyrethrum and neem improve the level of control compared to neem alone.

According to Schmutterer (1990a) aphids are only slightly influenced by oil-free extracts of neem. An unidentified aphid on tobacco was not affected by an enriched NSE, but the addition of neem oil reduced aphid numbers by 80 per cent (Schmutterer, 1985). In the current trials, NSO was superior to NSE for the control of *M. persicae*, but not *A. gossypii*, on pepper; NSE was as effective as NSO for the control of aphids on strawberry and cabbage (Lowery et al., 1993). Also, aqueous neem seed extracts reduced numbers of *A. gossypii* on potato in the Sudan by 26 to 75 per cent (Siddig, 1987).

Aphid Natural Enemies

Foliar applications of 1 per cent NSO and 1 per cent NSE significantly reduced the number of aphid predators compared with controls ($p < 0.05$) (Table 3). In relation to their prey, however, numbers of predators on plants treated with neem were consistently higher, but not significantly so, than in controls. Absolute counts of parasitoids (mummies) did not differ significantly for any treatment ($p > 0.05$), but the number of mummies per 1,000 aphids, which better reflects the parasitism rate, was significantly higher for NSO and NSE treatments than for controls ($p < 0.05$), being, on average, nearly 3.7 times higher (Table 3).

Numbers of predators on plants sprayed with neem may have been lower as a result of lower aphid populations, which would have overestimated the damaging effect of neem. On the other hand, although an increase in host density is typically followed by

Table 3. Effect of NSO and NSE on predator and parasitoid numbers per 280 plants and per 1000 aphids

	Predators		Parasitoids	
	/280 plants	/1,000 aphids	/280 plants	/1,000 aphids
Control	413a	41.3a	85a	8.5a
NSO 1.0%	180b	66.9a	88a	32.7b
NSE 1.0%	255b	73.9a	101a	30.0b

Means within a column followed by the same letter are not significantly different ($p > 0.05$, Tukey's multiple range test).

an increase in numbers of predators and parasitoids, the relationship is often variable and difficult to determine (Stary, 1970). Therefore, numbers per 1000 aphids may not have provided an accurate measure of neem's impact. Neem sprays may have resulted in some mortality to immature predators, but the overall effect on predator populations appears to be minimal.

In greenhouse trials, rates of parasitism of *A. gossypii* and greenhouse white-fly on plants treated up to four times with Margosan-O were comparable to controls and significantly higher than other pesticide treatments (Hoelmer et al., 1990). Applications of NSO to field plots of rice controlled leafhoppers and planthoppers, but did not affect numbers of predatory spiders or mirid bugs (Saxena et al., 1981). Parasitism of rice leaf folder larvae was significantly higher on rice treated with neem, perhaps because treated larvae were unable to fold the leaves properly. Detrimental effects of neem to aphid predators and parasitoids were demonstrated in laboratory trials (Lowery, 1992), and the possible longterm effects of neem on the survival, behaviour, and fitness of insect natural enemies need to be thoroughly investigated (Hoelmer et al., 1990). In spite of these concerns, the use of neem-based products offers an approach to pest control that is less damaging to beneficial insects.

Antifeedant Activity

Based on leaf disc choice tests, NSO was equally deterrent to first-instar, third-instar, and adult *C. fragaefolii* (Fig. 1). Estimated EC_{50} values were almost identical (≈1.2 per

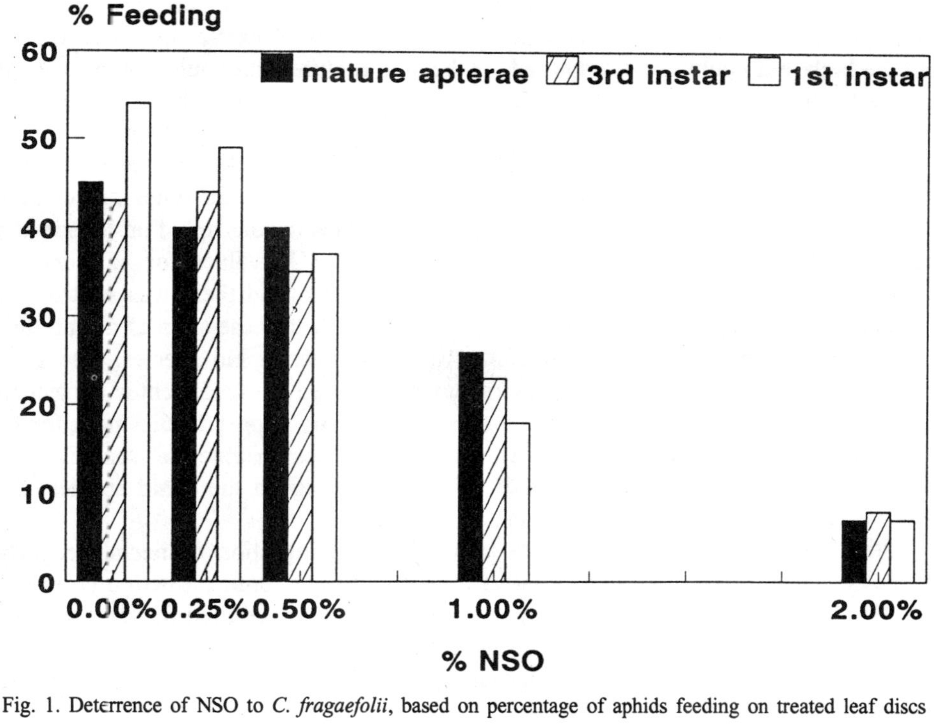

Fig. 1. Deterrence of NSO to *C. fragaefolii*, based on percentage of aphids feeding on treated leaf discs

cent), and slopes of the regression equations were not significantly different ($p > 0.05$). Deterrency of NSO to *C. fragaefolii* was not due solely to the presence of azadirachtin, however (Lowery and Isman, 1993). Deterrence was not related to the azadirachtin content of the oils, and only one-half the aphid species tested responded to NSO over the ranges tested. Most importantly, deterrence of NSO to *C. fragaefolii* disappeared within 12 to 24 hrs following application to strawberry in the greenhouse. According to Wood (1990), the commercial significance of the antifeedant activity of neem may be limited, as there appears to be considerable variation between insect species in their sensitivity to azadirachtin as an antifeedant, high concentrations are required to deter some species, and activity may be short-lived under field conditions. Our results support these concerns (Lowery and Isman, 1993).

Aphid Growth and Development

Representative survival curves for *N. ribisnigri* and *M. persicae* exposed to 40 ppm azadirachtin or 1 per cent NSO containing approximately the same amount of azadirachtin (Fig. 2) demonstrates that azadirachtin likely accounts for much of the observed mortality in these two species. No aphids exposed as second instars to azadirachtin or NSO survived for more than six days, compared to ≈67 per cent for controls. Aphids were observed to swell (apolysis) in preparation for ecdysis, but they either failed to moult, or could not escape the exuviae. Third-instar *N. ribisnigri* and *M. persicae* exposed to 0.5 per cent NSO that successfully moulted to adults were generally smaller than controls, and often possessed deformed wings and tibiae (Lowery, 1992). Studies by Schauer (1984) demonstrated that during moults the exuviae of pea aphids, *Acyrthosiphon pisum* (Harris), on broadbean treated with neem seed kernel extract began to detach from the body but could not be ruptured, or the skin ruptured but could not be cast off completely. In our studies, NSO and azadirachtin were not toxic to adult aphids, and survival did not differ from controls ($p > 0.05$) at any of the tested rates.

Susceptibility of *M. persicae* to azadirachtin is clearly dependent on the host plant. The EC_{50} for second instars on corn (43.3 ppm) was more than 20 times higher than on mustard cabbage (1.8 ppm), and the intermediate values for this aphid on lettuce and broadbean also differed significantly ($p < 0.05$). The EC_{50} values for *M. persicae* increased from first (1.8 ppm) to fourth (37.4 ppm) instar, and there was no toxicity to adults even at the highest rate of 1000 ppm. In addition to these sources of variability, susceptibility of aphids to azadirachtin varied between aphid species. The EC_{50} values after nine days for nine species of aphids exposed to azadirachtin as second instars ranged from as low as 2.4 ppm for *M. persicae* on pepper to 635 ppm for *C. fragaefolii* on strawberry (Lowery, 1992). These results demonstrate that sensitivity to azadirachtin is highly variable for aphids, likely depending on the aphid species, the instar treated, and the aphid's host plant.

Although activity to aphids results primarily from ingestion of neem, properly formulated materials possess some contact toxicity, but not fumigant action (Lowery, 1992). Depending on temperature, seven to nine days may be required for the complete expression of the growth regulating effects. Treated nymphs that successfully moulted to adults often possessed abnormal wings and legs, and these and other fitness-reducing

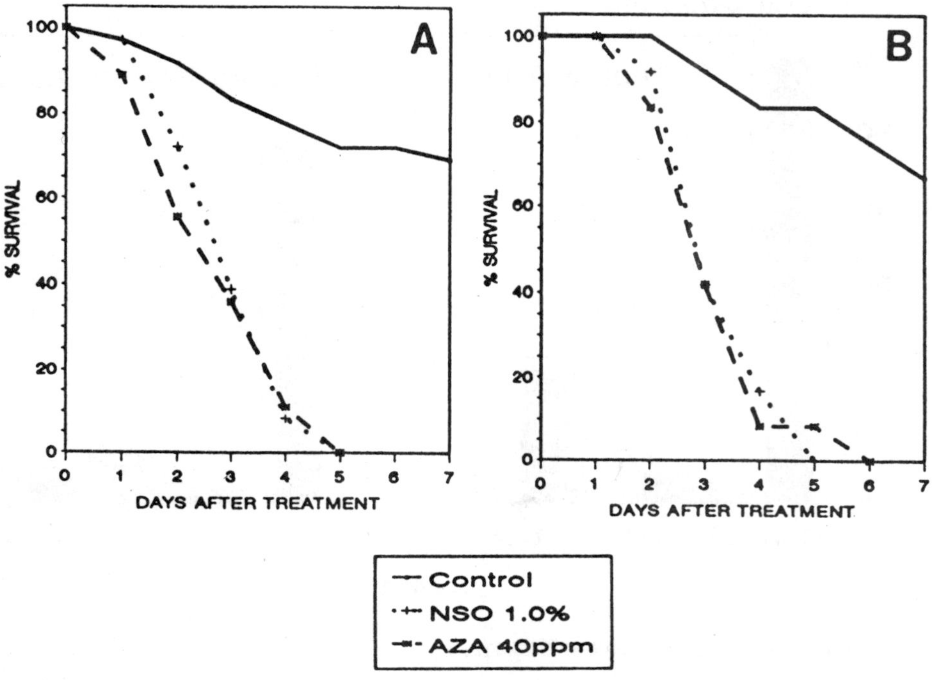

Fig. 2. Survival of second instar (A) *M. persicae* and (B) *N. ribisnigri*, exposed to 40 ppm azadirachtin (AZA) or 1.0 per cent NSO

effects may be important for the control of aphids and other insects over longer periods of time (Lindquist et al., 1990; Schmutterer, 1990a).

Aphid Reproduction

Exposure of adult or fourth-instar *N. ribisnigri* to 40 ppm azadirachtin or 1 per cent NSO resulted in significantly ($p > 0.05$) fewer offspring (Fig. 3). Fecundity of adult *N. ribisnigri* exposed to 1 per cent NSO or 40 ppm azadirachtin was 66 per cent (0.5 live offspring/aphid/day) and 32 per cent (1.01 live offspring/aphid/day) that of controls. Reproduction of adults exposed as fourth instars was suppressed even further, with 87 to 90 per cent fewer offspring relative to controls.

The effective concentrations of azadirachtin required to reduce aphid reproduction by 50 per cent during the first three days following treatment of adults ranged from 14.4 ppm for *N. ribisnigri* on lettuce to 616.4 ppm for *Rhopalosiphum padi* on corn (Lowery, 1992). In addition to a decrease in production of live offspring, aphids treated with azadirachtin produced a large number of dead (embryonic) offspring with darkened cuticle that were clearly nonviable at birth. Dead embryonic offspring accounted for up to 75.5 per cent of the total number of offspring produced.

Our studies have demonstrated that NSO and azadirachtin effectively inhibit aphid reproduction due to both a decrease in the maturation of oocytes and increased mortality of mature embryos. According to Schmutterer (1990a), fecundity of homopterous

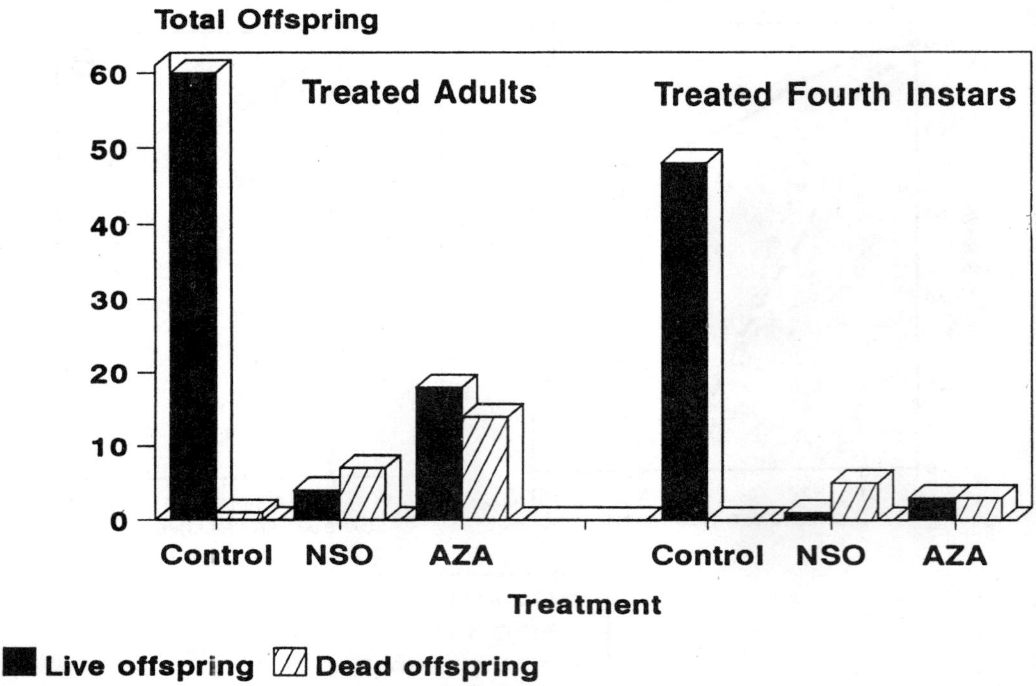

Fig. 3. Total production of live and dead offspring over a three-day period by lettuce aphids, *N. ribisnigri*, exposed as adults or as fourth instars to NSO (1 per cent) or azadirachtin (AZA, 40 ppm)

insects is strongly influenced by treatments of neem or azadirachtin, but the effect on reproduction of viviparous insects has not been thoroughly investigated. In an earlier study, reproduction of adult *A. pisum* placed as first instars on broadbean plants sprayed with 0.002 per cent methanolic neem seed extract produced 1.6 offspring per female per day, compared to 7.1 offspring per aphid per day for untreated controls (Schauer, 1984).

CONCLUSION

Except where virus transmission is a concern, control of aphids is usually required only at relatively high population levels (van Lenteren, 1990). Applications of neem can be used in these situations to maintain numbers below economic injury levels. For the majority of crops, neem will likely prove to be an effective aphicide when used alone, or in conjunction with other control strategies. Because neem is somewhat selective toward phytophagous insects (Saxena, 1989), it may be suitable for inclusion in IPM programmes where aphid control can be enhanced by the activity of natural control agents. Our results support this contention. However, compared to synthetic neurotoxins, neem is slow-acting and rarely results in complete suppression of aphid populations, and is thus unsuitable for certain situations.

Although the deterrent activity of neem appears to be of limited value for the control of aphids, it may contribute to the control of certain species and the viruses they transmit. Undoubtedly, control of aphids results primarily from the combined toxic (growth inhibiting) and sterilizing activities. Exposure to neem or azadirachtin also results in several sub-lethal effects that may be important for the control of aphids over longer periods of time.

ACKNOWLEDGEMENTS

We thank Safer Ltd., Victoria, Canada, for the formulated neem materials, and Dr. J.T. Arnason, Univ. of Ottawa, Ottawa, Canada, for azadirachtin. C.K. Chan provided the initial aphids from colonies maintained at the Agriculture Canada Research Station, Vancouver, Canada. Technical assistance was provided by N.L. Brard. Funding for this research was provided by grants from the Natural Sciences and Engineering Research Council of Canada and the Science Council of British Columbia.

REFERENCES

Hoelmer, K.A., L.S. Osborne and R.K. Yokomi. 1990. Effects of neem extracts on beneficial insects in greenhouse culture. In *Neem's Potential in Pest Management Programs Proc. USDA Neem Workshop,* Eds., J.C. Locke and R.H. Lawson. USDA ARS-86, pp. 100–105.

Isman, M.B., O. Koul, A. Lucynski and J. Kaminsky. 1990. Insecticidal and antifeedant bioactivities of neem oils and their relationship to azadirachtin content. *J. Agric. Food Chem.* 38: 1406–1411.

Knodel-Montz, J.J., H.G. Larew and R.E. Webb. 1985. Efficacy of Margosan-O, a formulation of neem, against *Liriomyza trifolii* (Burgess) on floral crops. *USDA Informal Conf. on Liriomyza leafminers*, Beltsville, Maryland, June 1985, pp. 33–44. As cited by M. Jacobson. Review of neem research in the United States. In: *Neem's Potential in Pest Management Programs, Proc. USDA Neem Workshop.* Eds., J.C. Locke and R.H. Lawson. USDA ARS-86, pp. 4–14, 1990.

Lindquist, R.K., A.J. Adams, F.R. Hall and I.H.H. Adams. 1990. Laboratory and greenhouse evaluations of Margosan-O against bifenthrin-resistant and -susceptible greenhouse whiteflies, *Trialeurodes vaporarium* (Homoptera: Aleyrodidae). In: *Neem's Potential in Pest Management Programs Proc. USDA Neem Workshop,* Eds. J.C. Locke and R.H. Lawson, Proc. USDA ARS-86, pp. 91–99.

Lowery, D.T. 1992. Effects of extracts from neem, *Azadirachta indica* (A. Juss.), on aphids (Homoptera: Aphididae) with respect to their control. Ph.D. thesis, Univ. British Columbia, Vancouver.

Lowery, D.T. and M.B. Isman. 1993. Antifeedant activity of extracts from Neem, *Azadirachta indica* (A. Juss.), to aphids (Homoptera: Aphididae). *J. Chem. Ecol.* (in press).

Lowery, D.T., M.B. Isman and N.L. Brard. 1993. Laboratory and field evaluation of neem for the control of aphids (Homoptera: Aphididae). *J. Econ. Entomol.* (in press).

National Research Council. 1992. *Neem: A Tree for Solving Global Problems.* National Academy Press, Washington, DC, 139 pp.

Neter, J., W. Wasserman and M.H. Kutner. 1985. *Applied Linear Statistical Models: Regression, Analysis of Variance, and Experimental Designs*, 2nd ed. Richard D. Irwin, Inc., Homewood, Illinois, pp. 172–174.

Patel, M.B. and K.P. Srivastava. 1989. Evaluation of neem oil formulation against cowpea aphid (*Aphis craccivora* Koch). *Neem Newsletter* 6(4): 41–44.

Saxena, R.C., N.J. Liquido and H.D. Justo. 1981. Neem seed oil, a potential antifeedant for the control of the rice brown planthopper, *Nilaparvata lugens*. In: *Natural Pesticides from the Neem Tree* (*Azadirachta indica* A. Juss.), *Proc. 1st Int. Neem Conf.*, Rottach-Egern, 1980. Eds., H. Schmutterer, K.R.S. Ascher and H. Rembold, GTZ, Eschborn, pp. 171–188.

Saxena, R.C. 1989. Insecticides from neem. In: *Insecticides of Plant Origin*. Eds., J.T. Arnason, B.J.R. Philogene and P. Morand. ACS Symp. No. 387, Amer. Chem. Soc., Washington, DC, pp. 110–135.

Schauer, M. 1984. Effects of variously formulated neem seed extracts on *Acyrthosiphon pisum* and *Aphis fabae*. In: *Natural Pesticides from the Neem Tree* (*Azadirachta indica* A. Juss.) *and other Topical Plants. Proc. 2nd Int. Neem Conf.*, Rauischholzhausen. Eds., H. Schmutterer and K.R.S. Ascher GTZ, Eschborn, pp. 141–150.

Schmutterer, H. 1985. Which insect pests can be controlled by application of neem seed kernel extracts under field conditions? *Z. Ang. Ent.* 100: 468–475.

Schmutterer, H. 1990a. Properties and potential of natural pesticides from the neem tree, *Azadirachta indica*. *Annu. Rev. Entomol.* 35: 271–297.

Schmutterer, H. 1990b. Future tasks of neem research in relation to agricultural needs worldwide. In: *Neem's Potential in Pest Management Programs, Proc. USDA Neem Workshop*. Eds., J.C. Locke and R.H. Lawson. USDA ARS-86, pp. 15–22.

Schmutterer, H., and C. Hellpap. 1989. Effects of neem on pests of vegetables and fruit trees. In: *Focus on Phytochemical Pesticides*, Vol. 1, *The Neem Tree*, Ed. M. Jacobson CRC Press, Boca Raton, pp. 69–86.

Siddig, S.A. 1987. A proposed pest management program including neem treatments for combating potato pests in the Sudan. In *Natural Pesticides from the Neem Tree* (*Azadirachta indica* A. Juss.) *and other Tropical Plants. Proc. 3rd Int. Neem Conf.*, Nairobi. Ed. H. Schmutterer GTZ, Eschborn, pp. 449–459.

Stary, P. 1970. *Biology of Aphid Parasites* (*Hymenoptera*: *Aphididae*) *with Respect to Integrated Control*. Dr. W. Junk, the Hague.

van Lenteren, J.J. 1990. Biological control in a tritrophic system approach. In: *Aphid-Plant Interactions: Populations to Molecules*. Eds., D.C. Peters, J.A. Webster and C.S. Chlouber USDA/ARS & Oklahoma State University, Stillwater, pp. 3–28.

Walter, J.F. and J.F. Knauss. 1990. Developing a neem-based pest management product. In: *Neem's Potential in Pest Management Programs. Proc. USDA Neem Workshop.* Eds., J.C. Locke and R.H. Lawson USDA ARS-86, pp. 29–31.

Wilkinson, L. 1990. *Systat: The System for Statistics*. Systat Inc., Evanston, Illinios, 677 pp.

Wood, T. 1990. Efficacy of neem extracts and neem derivatives against several agricultural insect pests. In: *Neem's Potential in Pest Management Programs, Proc. USDA Neem Workshop*. Eds., J.C. Locke and R.H. Lawson USDA ARS-86, pp. 76–84.

28

Differential Sensitivity of *Spodoptera litura* (Fabricius) (Lepidoptera: Noctuidae) Larvae to Multifarious Activity of Azadirachtin and Possible Implications in Integrated Pest Management

S.N. MUKHERJEE AND R.N. SHARMA

Entomology Laboratory, National Chemical Laboratory, Pune 411 008, India

ABSTRACT

Effects of azadirachtin, the principal active ingredient from neem, were studied on various chosen biological parameters in different larval stages of the tobacco leaf eating caterpillar, *Spodoptera litura*, at different doses. The parameters included feeding deterrence, esterase activity, growth and development. Neonate larvae were susceptible to azadirachtin in a dose-dependent manner for antifeedant action, moult inhibition, toxicity, growth and development. Antifeedant action was not dose-dependent against older (fourth- and sixth-instar) larvae. In the latter, curiously approximate digestibility of food was not depressed but increased significantly with increase in azadirachtin concentration, as revealed by dietary utilization studies. However, efficiencies of conversion of ingested and digested food were significantly lowered. Effects on growth and development followed a similar pattern in both neonate and older larvae, except that higher doses were required for the latter. Midgut esterase activity increased significantly on feeding sixth-instar larvae at 30 and 50 ppm dose levels. The findings are discussed in light of potential of such azadirachtin activity in enriched neem formulations and its implications in modern pest management strategies.

INTRODUCTION

The hypothesis that plant secondary compounds may serve a protective function against herbivory is well accepted (Feeny, 1976; Rhoades and Cates, 1976), though it may not always be as well documented (Reese, 1979). The wide range of natural toxic products found in present-day plants is the culmination of selective pressures exerted by biotic factors, one of which is selective pressures exerted by biotic factors, one of which is insect predation on plants. In recent years, much effort has been devoted to the search for alternatives to the conventional arsenal of pest control chemicals. Plant-derived extracts and phytochemicals, which once formed the basis of pest control technology, are again being scrutinized for new classes of synthetic insecticides (Hedin, 1982; Balandrin et al., 1985). By its ubiquity, abundance, and historical biomedical importance and use, the Indian neem tree has become the launch pad and sheet anchor of the sustainable low input agriculture movement and integrated pest management. Among all the tetranortriterpenoids the most spectacular is azadirachtin, whose bioactivities include antifeedant and insect growth regulatory action. Azadirachtin is quite patently the single most potent neem constituent with pronounced and multipronged biological activities (Rembold, 1984; Rembold et al., 1984; Jacobson, 1986).

Whereas the antifeedant activity of azadirachtin had raised high hopes when discovered, it has now become apparent that it has limited use as well as a restricted range among target insects (Mukherjee and Ramachandran, 1989; Mukherjee, 1992). On the other hand, the growth inhibitory action of azadirachtin extends to the very vital life processes of several important insect species. Thus, azadirachtin continues to be an important molecule in terms of pest regulatory functions and needs more detailed investigations on target insects.

The present study is a sequel to our work reported earlier (Ramachandran et al., 1989; Mukherjee, 1992), wherein it was shown that there was absence of significant bioactivity of even higher doses or quantities of azadirachtin against the older larvae of *Spodoptera litura*. This prompted an examination of this compound on various larval stages of *S. litura* with respect to their behavioural, physiological, and biochemical responses.

MATERIALS AND METHODS

Experimental *S. litura* was obtained from a laboratory culture maintained on excised castor leaves at 25 ± 2°C, 60 per cent relative humidity. Castor leaves for experiments were obtained from a plantation maintained in the campus.

Azadirachtin used in these experiments was 98 per cent pure and was a gift from Prof. H. Rembold, Max-Planck Institute für Biochemie, Germany.

To obtain freshly hatched first-instar larvae (neonate larvae), eggs laid by the adult mated females were isolated and incubated at 28° ± 2°C and 70 per cent relative humidity.

Neonate sensitivity to azadirachtin was based on the response of the neonate larvae with respect to feeding, moulting, survival, and finally growth and development.

The leaf disc method was used to study the antifeedant activity of azadirachtin against neonate as well as older larvae (fourth and sixth instar) as reported earlier (Ramachandran et al., 1989). Doses used against neonate larvae were 0.1, 0.25, 0.5, 1.0, 2.5, 5.0, and 10.00 ppm, whereas those against the older larvae were 10, 20, 30, and 50 ppm.

Effect of azadirachtin on the moulting of neonate larvae to the second-instar was studied by releasing neonate larvae on castor leaf discs (2.5 cm diameter) treated by dipping for 5 sec in azadirachtin solution of different concentrations (0.1, 0.2, 0.4, 0.8, 1.6, and 3.2 ppm). Controls were treated with solvent alone. One larva was released per leaf disc. Observation on moulting was taken after three days, till all the larvae in the control lots moulted into the second-instar.

Similarly, the effect of azadirachtin on survival of neonate larvae was studied by releasing neonate larvae on castor leaf discs (2.5 cm diameter) treated with various concentrations of azadirachtin solution (0.5, 1.0, 2.0, 4.0 and 8.0 ppm). Controls were treated with solvent alone. This experiment was carried out for seven days and freshly treated and control castor leaves were provided to the larvae every 48 hrs. Mortality counts were taken after the stipulated period of seven days.

To study the effect of azadirachtin on the growth and development of both neonate and fourth-instar larvae, they were fed on castor leaf discs treated with various concentrations of azadirachtin (0.1, 0.25, 0.5, 1.0, 2.5, and 5.0 ppm for neonate larvae and 1, 2.5. 5.0, 10.0 and 20.0 ppm for fourth-instar larvae) for 48 hrs and thereafter the larvae were allowed to feed and grow on untreated castor leaves till pupation or death. Appropriate controls were taken. Observations on larval weight, day on which 100 per cent mortality occurred, and average pupal weight were recorded.

To study the possible effect of azadirachtin on dietary utilization, freshly moulted and weighed fourth- and sixth-instar larvae of *S. litura* were fed on pre-weighed castor leaf discs treated with 10, 20, 30, and 50 ppm of azadirachtin. Control leaf discs were treated with acetone only. The duration of larval feeding was monitored so as to allow the insects to complete a full instar. Both treated and untreated castor leaves were added after weighing as and when required. On the third day following the start of the experiment, the larvae, faeces, and remaining uneaten leaf disc were separated and fresh weights were taken. The larvae, faeces and uneaten leaf were subsequently dried at 60°C for 48 hrs (till constant weights were obtained) and then weighed.

All weighings were carried out in a Sartorious MP 2024 balance with 0.01 mg sensitivity. Nutritional indices of dry weight of food ingested, dry weight gain, approximate digestibility (AD), efficiency of conversion of digested food (ECD), and efficiency of conversion of ingested food (ECI) were calculated as described by Waldbauer (1968) based on dry weights obtained.

To study the effect of azadirachtin on the midgut esterase activity, freshly moulted sixth instar larvae of *S. litura* were fed on castor leaf discs treated with 10, 20, 30 and 50 ppm azadirachtin for 40 hrs and subsequently enzyme activity was determined as reported earlier (Ghumare et al., 1989).

All experiments were replicated five times unless otherwise stated and data obtained were subjected to appropriate statistical tests for their significance.

RESULTS

Response of Neonate *S. litura* Larvae to Azadirachtin (Neonate Sensitivity)

Antifeedant Activity

Azadirachtin generated a linear dose-response curve with respect to percentage protection of leaves against feeding by the neonate larvae and concentration of azadirachtin. With increasing concentration of azadirachtin, there was corresponding increase in percentage protection (Fig. 1).

$$Y = 0.62x + 3.59$$

ED_{50} or the effective dose for 50 per cent protection was 2.27 ppm.

Moult Inhibitory Activity

Moult inhibition in case of neonate larvae to second instar increased with corresponding increase in azadirachtin concentration and thus a linear dose-response relationship was established (Fig. 2).

$$Y = 0.75x + 3.66$$

MI_{50} or the concentration at which 50 per cent of the population could not moult normally was found to be 0.62 ppm.

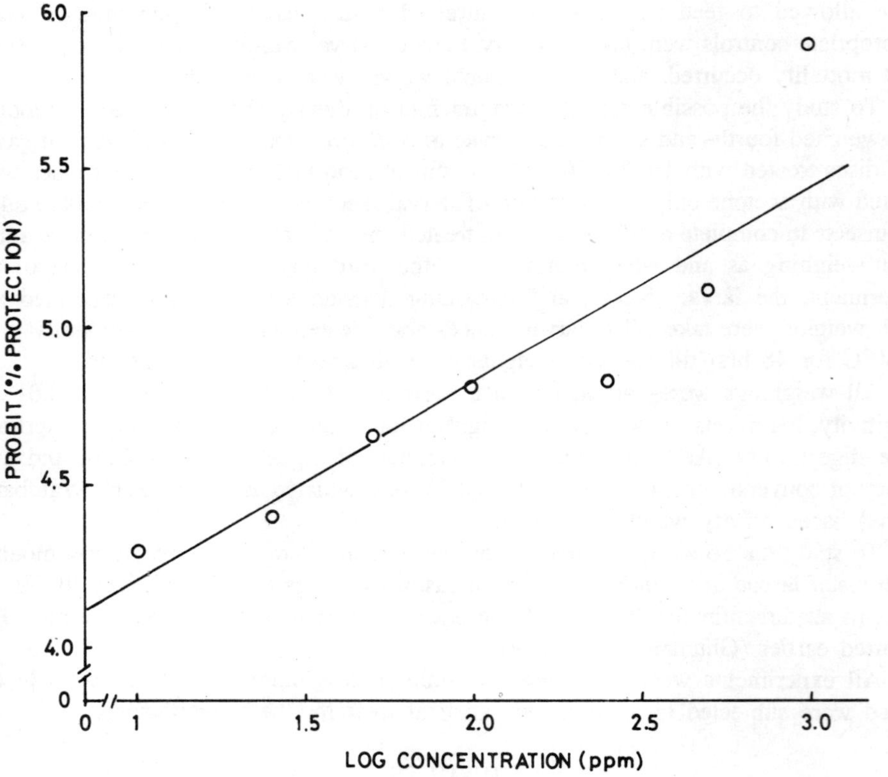

Fig. 1 Antifeedant action of azadirachtin on neonate larvae of *S. litura*

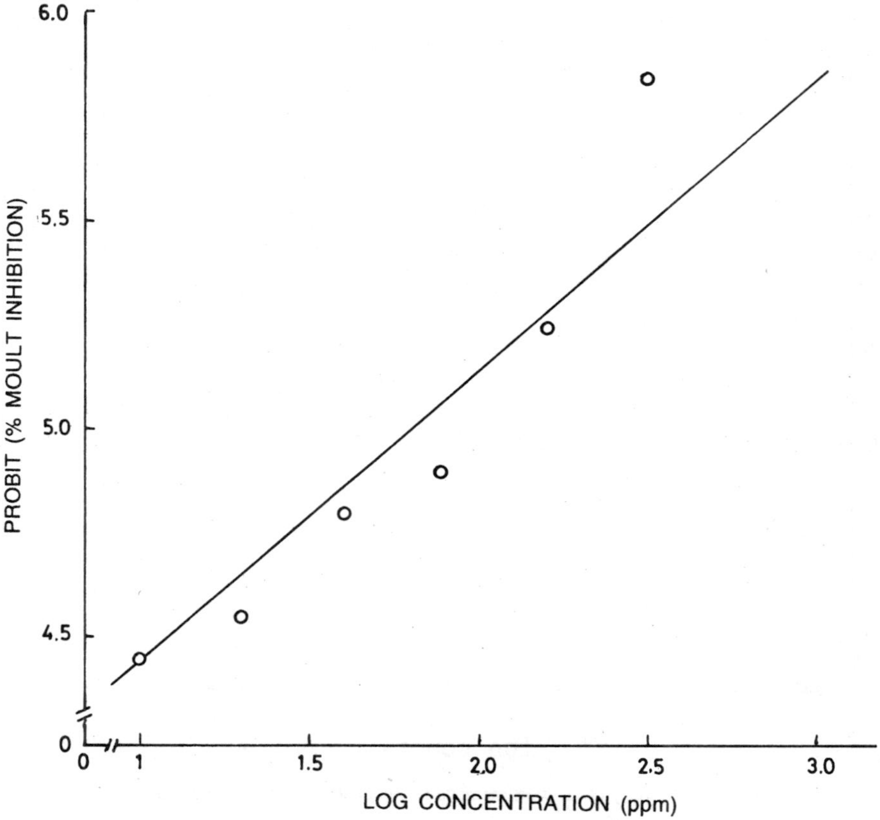

Fig. 2. Inhibition of moulting of neonate larvae of *S. litura* to second-instar on azadirachtin treated castor leaf discs

Survival

Survival of neonate larvae over a seven-day period decreased with corresponding increase in azadirachtin concentration and thus a linear dose response was established (Fig. 3).

$$< Y = 2.48x + 1.53$$

The LC_{50} value was 2.5118 ppm.

Growth and Development

Forty-eight hours of feeding on azadirachtin-treated castor leaves disrupted normal development of neonate larvae of *S. litura* in a more or less dose-dependent manner (Table 1). At the lowest concentration of 0.10 ppm, azadirachtin could not exert influence on the pupal weight, since there was no significant difference in the pupal weight at this concentration compared to the control ($p < 0.05$). However, with subsequent increase in azadirachtin concentration, pupal weights differed significantly

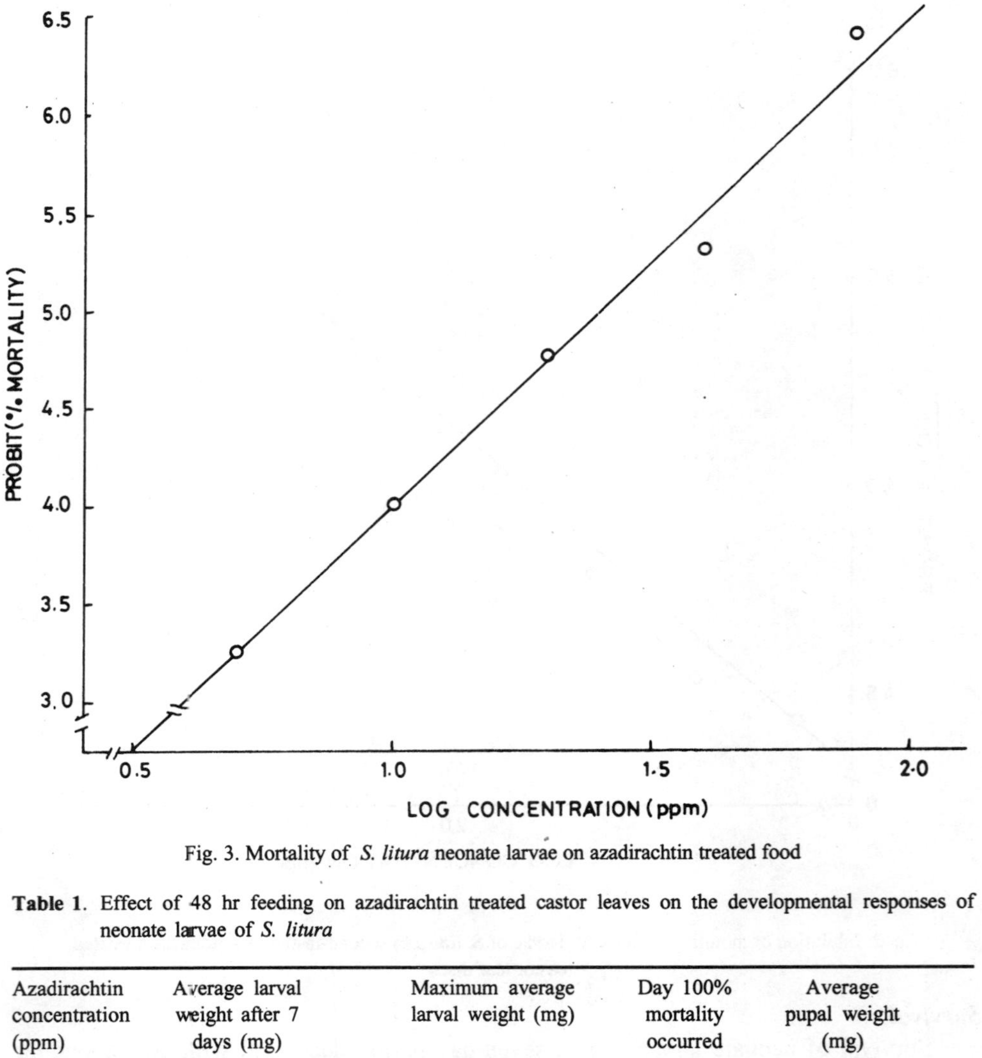

Fig. 3. Mortality of *S. litura* neonate larvae on azadirachtin treated food

Table 1. Effect of 48 hr feeding on azadirachtin treated castor leaves on the developmental responses of neonate larvae of *S. litura*

Azadirachtin concentration (ppm)	Average larval weight after 7 days (mg) X ± SE	Maximum average larval weight (mg) X ± SE	Day 100% mortality occurred	Average pupal weight (mg) X ± SE
CONTROL	78.52 ± 2.61a	506.51 ± 11.50a		208.52 ± 12.12a
0.10	64.46 ± 2.82b	325.63 ± 7.16b		197.47 ± 15.36a
0.25	60.57 ± 2.43b	309.47 ± 15.60b		163.51 ± 18.20b
0.50	52.51 ± 5.01c	270.54 ± 11.76c		124.63 ± 8.56c
1.00	30.27 ± 1.73d	264.32 ± 20.48c	16	
2.50	27.64 ± 1.84d	146.57 ± 14.28d	14	
5.00	14.23 ± 1.43e	70.51 ± 7.36e	12	

Means not followed by the same letters are significantly different ($p < 0.05$; one-way ANOVA).

from that of the control ($p < 0.05$). Concentration-dependent effect of azadirachtin was also reflected in the larval weights taken seven days after the start of the experiment as well as on the maximum average larval weight. At higher doses of 1.0, 2.5, and 5.0 ppm, azadirachtin produced 100 per cent larval mortality on day 16, 14, and 12 respectively.

Response of Older *S. litura* Larvae to Azadirachtin

Antifeedant Studies

Azadirachtin did not deter feeding by fourth- and sixth-instar larvae of *S. litura* on castor leaves in a dose-dependent manner. Maximum protection afforded by azadirachtin against both instars was at a concentration of 20 ppm, where the mean percentage protection against the fourth-instar larvae was 47 $\pm$ 3.5 per cent and against sixth-instar larvae it was 63 $\pm$ 4.5 per cent. Subsequent increase in azadirachtin concentration did not increase the percentage protection in fourth- and sixth-instar larvae. At the highest concentration of azadirachtin (i.e., 50 ppm) mean percentage protection against the fourth- and sixth-instar larvae was 37 $\pm$ 4.0 per cent and 44 $\pm$ 4.5 per cent respectively. However, percentage protection afforded by azadirachtin at identical dose levels was greater against the sixth-instar larvae (Figs. 4, 5, 6).

Growth and Development

Feeding by fourth-instar larvae of *S. litura* on azadirachtin-treated castor leaves for 48 hrs disrupted development in a dose-dependent manner. This disruption of growth was reflected in the maximum mean larval weight (Table 2). Maximum mean larval weight at all concentrations differed significantly ($p < 0.05$) from the control. Similarly, mean pupal weight also differed significantly ($p < 0.05$) from the control at all concentrations tested. However, at doses of 10 and 20 ppm, pupation did not take place since 100 per cent larval death occurred on day 6 from the start of the experiment.

Dietary Utilization

Azadirachtin at all concentrations tested significantly reduced the amount of food ingested and dry weight gained by both fourth-and sixth-instar larvae of *S. litura* ($p < 0.05$) (Tables 3, 4). However, both fourth- and sixth-instar larvae were able to digest azadirachtin-treated castor leaves better than the control leaves; hence, AD was significantly higher ($p < 0.05$) at all concentrations tested compared to the controls. Azadirachtin reduced the ECD and ECI into biomass in both the fourth- and sixth-instar larvae. However, reduction in ECI and ECD occurred at 20 ppm and larger doses in both fourth- and sixth-instar larvae (Tables 3, 4).

Midgut Esterase Activity

Final-instar (sixth-instar) larvae when fed on azadirachtin-treated castor leaves for 48 hrs exhibited increased midgut esterase activity. However, significant increase ($p < 0.05$) compared to the control was exhibited only at 30 and 50 ppm dose levels, where the specific activity was 481 $\pm$ 8.75 and 518 $\pm$ 7.05 nmol/min./mg protein 1-napthol, respectively (Table 5).

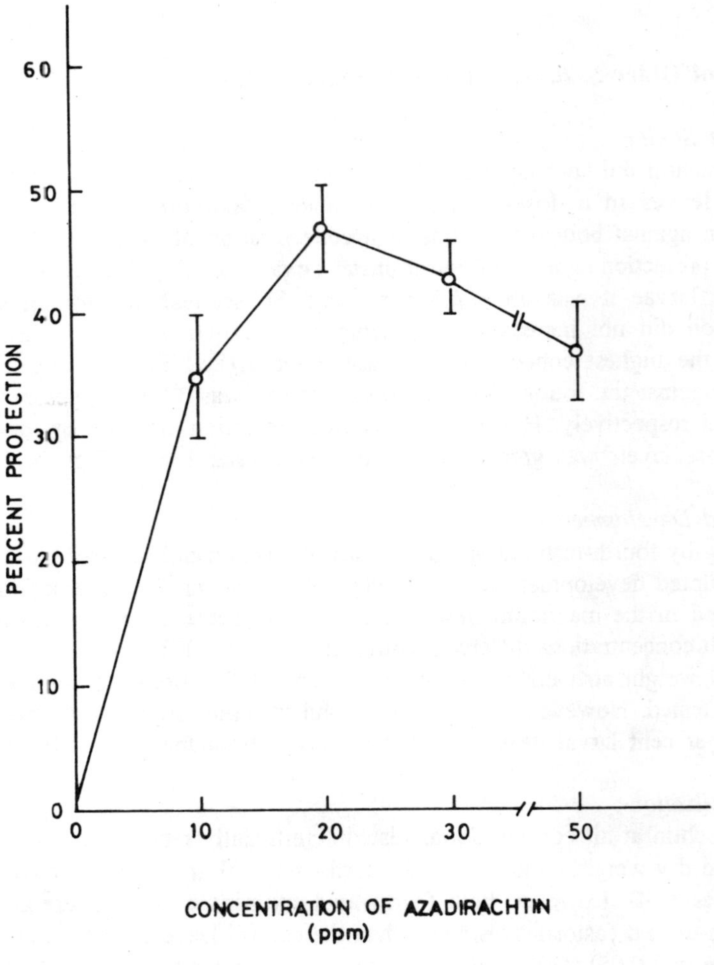

Fig. 4. Antifeedant action of azadirachtin against fourth-instar larvae of *S. litura*

DISCUSSION

The present study reaffirms that azadirachtin not only has strong feeding deterrence against neonate larvae of *S. litura*, but also apparently affects development by inhibiting moulting into the next instar, resulting in eventual mortality of the larvae. Similar action on neonate larvae of this insect has been shown by us to be exerted by a plant glycoside (Ghumare et al., 1989). However, in this case the chemical—rutin—did not show any antifeedant action. Moult inhibitory activity of azadirachtin against neonate larvae of various insect species has been demonstrated earlier (Kubo and Klocke, 1986).

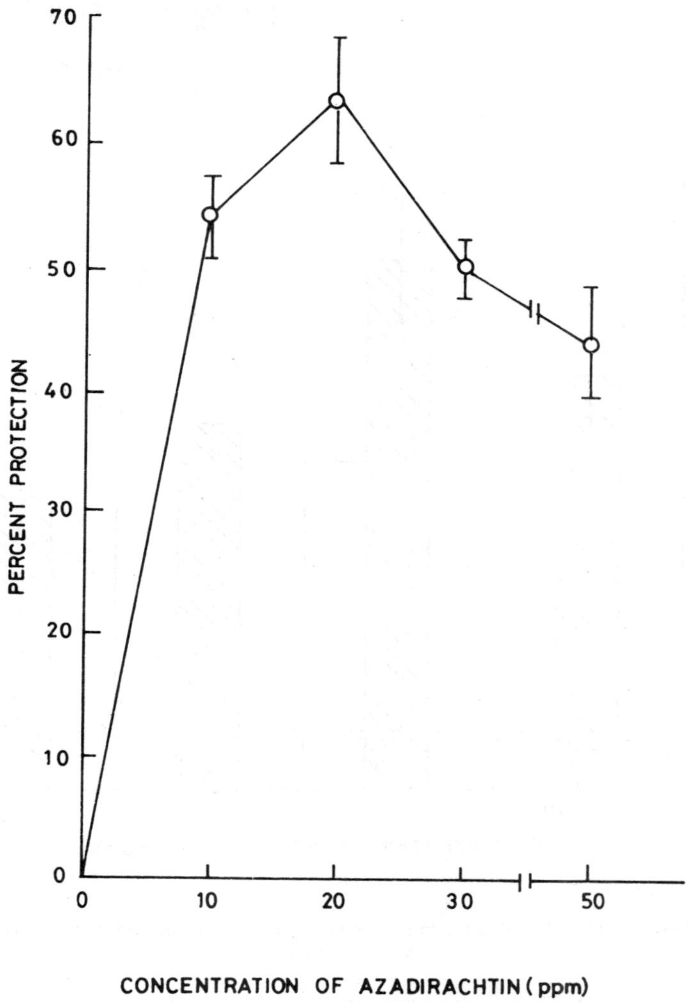

Fig. 5. Antifeedant action of azadirachtin against sixth-instar larvae of *S. litura*

From the present studies it is observed that neonate larvae of *S. litura* are vulnerable to azadirachtin even at 0.25 ppm and larger doses fed to them in treated castor leaves for 48 hrs. Not only was the pupal weight reduced significantly, such larvae were unable to survive beyond 12–16 days at the most. This clearly suggests that the amount of azadirachtin consumed during the exposure period of 48 hrs is sufficient to disrupt growth and development mechanisms. On the other hand, when fourth-instar larvae of this insect were exposed likewise, effects on pupation were obtained only at 10 and 20 ppm doses. At these higher doses the larvae failed to survive beyond six days after the

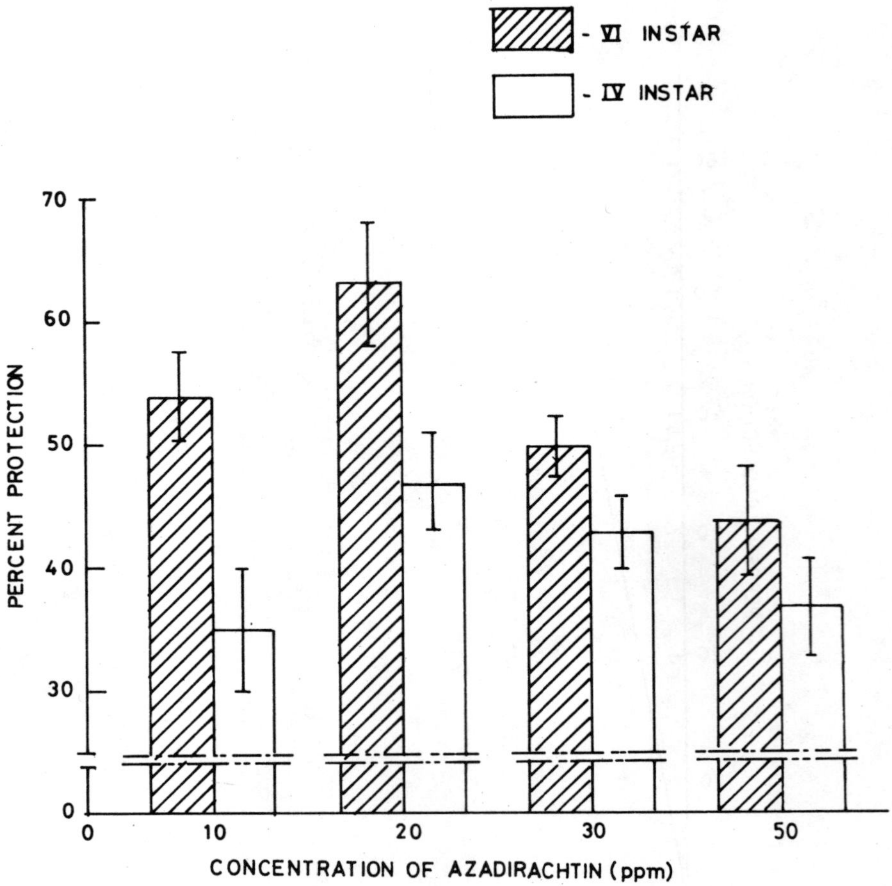

Fig. 6. Comparative efficacy of azadiractin on percent protection afforded against fourth- and sixth-instar larvae of *S. litura*

start of the experiment. There was an apparent correlation between increase in azadirachtin concentration and reduction in average larval and pupal weight. Similar results with azadirachtin have also been reported by Schmutterer et al. (1983) on two lepidopteran rice pests and by Gaaboub and Hayes (1984) on the face fly *Musca autumnalis*. However, Barnby and Klocke (1987) did not observe reduced pupal weights in case of *Heliothis virescens*. The dose-dependent effects of azadirachtin on post-embryonic development of *S. litura* parallel those reported for other insects (Qadri and Narsaiah, 1978; Rembold et al., 1980; 1982; Schmutterer and Rembold, 1980; Redfern et al., 1982; Warthen and Ubel, 1982; Schmutterer et al., 1983; Seiber and Rembold, 1983; Garcia and Rembold, 1983; Jotwani and Srivastava, 1984; Ladd et al., 1984; Schluter et al., 1985; Dorn et al., 1986; Barnby and Klocke, 1987).

In neonate larvae of *S. litura,* the dose required for 50 per cent moult inhibition was 3.66 times less than the dose required to inhibit 50 per cent feeding. Since the

Table 2. Effect of 48-hour feeding on azadirachtin treated castor leaves on the growth and development of fourth-instar larvae of *S. litura*

Azadirachtin concentration (ppm)	Average larval weight at the beginning (mg) X = SE	Maximum average larval weight (mg) X = SE	Day 100% mortality occurred	Average pupal weight (mg) X ± SE
Control	113.51 ± 0.21	568.45 ± 21.4a		209.53 ± 11.9a
1.00	109.46 ± 0.60	468.55 ± 19.6b		188.45 ± 15.4b
2.50	110.57 ± 0.74	428.37 ± 14.3b		178.55 ± 13.8b
5.00	112.35 ± 0.29	220.51 ± 18.5c		162.35 ± 17.5c
10.00	111.53 ± 0.35	145.57 ± 11.3d	6	
20.00	112.57 ± 0.31	138.41 ± 20.4d	6	

Means not followed by the same letters are significantly different ($p < 0.05$; one-way ANOVA).

Table 3. Nutritional indices of *S. litura* fourth-instar larvae on azadirachtin treated and control leaf discs

Azadirachtin concentration (ppm)	Dry weight of food ingested (mg) X ± SE	Dry weight gain (mg) X ± SE	AD X ± SE	ECD X ± SE	ECI X ± SE
CONTROL	330.07 ± 17.92a	40.80 ± 2.40a	49.60 ± 3.41a	23.74 ± 2.44a	12.56 ± 0.91a
10.00	291.49 ± 14.71b	16.59 ± 0.52b	57.54 ± 3.82b	18.04 ± 1.72b	10.42 ± 1.62a
20.00	262.45 ± 11.95c	13.99 ± 0.81b	58.53 ± 5.56b	9.73 ± 0.42c	6.53 ± 0.86b
30.00	259.52 ± 14.35c	13.27 ± 1.42b	61.50 ± 3.91b	9.50 ± 0.61c	6.41 ± 0.93b
50.00	267.61 ± 16.36c	9.51 ± 0.99c	66.46 ± 4.62c	6.88 ± 0.91d	4.77 ± 0.96c

Means not followed by the same letters are significantly different ($p < 0.05$; one-way ANOVA).
AD = approximate digestibility.
ECD = efficiency of conversion of digested food.
ECI = efficiency of conversion of ingested food.

Table 4. Nutritional indices of *S. litura* sixth-instar larvae on azadirachtin-treated and control castor leaf discs

Azadirachtin concentration (ppm)	Dry weight of food ingested (mg) X ± SE	Dry weight gain (mg) X ± SE	AD (mg) X ± SE	ECD X ± SE	ECI X ± SE
Control	312.49 ± 15.35a	32.81 ± 1.91a	42.55 ± 1.92a	24.32 ± 1.11a	10.41 ± 0.55a
10.00	279.63 ± 13.12b	13.45 ± 0.46b	49.36 ± 1.84b	18.48 ± 0.55b	8.64 ± 0.66a
20.00	276.43 ± 14.64b	11.16 ± 0.42b	50.21 ± 0.78b	9.97 ± 0.65c	5.41 ± 0.64b
30.00	284.82 ± 16.24b	10.83 ± 0.61b	52.76 ± 2.01b	9.73 ± 0.78c	5.31 ± 0.58b
50.00	260.85 ± 12.62c	7.55 ± 0.82c	57.02 ± 1.14c	7.05 ± 0.95d	3.96 ± 0.90c

Means not followed by the same letters are significantly different ($p < 0.05$; one-way ANOVA).

magnitude of difference between these two doses is not very large, the antifeedent action may also be responsible for disruption of development. Some insects that are starved and fail to reach a critical weight as larvae cease developing into pupae and/ or adults (Redfern et al., 1982a; Schluter and Schulz, 1984). Alternately, the role of azadirachtin in disrupting post-embryonic development may be due to effects that are

Table 5. Effect of azadirachtin on the midgut esterase activity of sixth-instar larvae of *S. litura*

Azadirachtin concentration (ppm)	Specific activity (nmol/min./mg protein -1- naphthol) X ± SE	Percentage of control
Control	366.3 ± 17.95a	100
10.00	374.7 ± 6.65a	102.3
20.00	395 ± 28.2a	107.8
30.00	481 ± 8.75b	131.3
50.00	518 ± 7.05b	141.4

Means followed by the same letters are not significantly different ($p < 0.05$; one-way ANOVA).

separate from those associated with feeding inhibition and reduced weight gain (Rembold, 1984). Garcia and Rembold (1983) demonstrated that 600 times less azadirachtin was required to disrupt development in *Rhodnius prolixus* than was required to produce antifeedant effect. Also, azadirachtin was still able to inhibit the moulting of *Locusta migratoria* nymphs even though these individuals exceeded the critical weight to become adults (Sieber and Rembold, 1983).

Protection afforded by azadirachtin against feeding by the older larvae (fourth- and sixth-instar) was not at all influenced in a dose-dependent manner, although sixth-instar larvae were more sensitive than the fourth-instar stage (Fig. 6). This may be due to the fact that in this insect, the fourth-instar larvae are voracious feeders, whereas the sixth-instar larvae (the last larval instar stage) have to prepare themselves to enter the pupal stage. It is also possible that perception of antifeedants at the sensory levels may involve different mechanisms at different developmental stages. The existence of changes in receptor sensitivity in relation to age seems to be well established (Blaney et al., 1986). Interestingly, lower protection at higher concentration in fourth as well as sixth instar (Fig. 5) may be a result of reduced sensitivity of the deterrent receptors, as has been shown in *S. exempta* and *S. littoralis*, wherein addition of azadirachtin to an artificial diet for a period of two days reduced sensitivity of their deterrent receptors, whereas sugar-sensitive receptors remain unaltered (Simmonds et al., 1984). However, in *S. litura*, the alteration in sensitivity occurred within a period of 24 hrs unlike in the other two species reported.

Conversion efficiencies (both ECI and ECD) were significantly reduced as a result of feeding on azadirachtin-treated castor leaves in the fourth- and sixth-instar larvae of *S. litura.* Similar findings have been reported in many other insects (Fagoonee, 1984; Arnason et al., 1985; Barnby and Klocke, 1987). Lower utilization efficiencies can possibly be due to diversion of energy from production of biomass to detoxification, especially at the higher doses (Barnby and Klocke, 1987).

Increase in AD with increasing concentrations of azadirachtin in the food (Tables 3, 4) in fourth- and sixth-instar larvae of *S. litura* was similar to that reported for other insects (Fagoonee, 1984; Arnason et al., 1985; Barnby and Klocke, 1987; Champagne et al., 1989) and this may reflect attempts by the insect to compensate for reduced consumption and utilization of treated food in order to maintain growth rates (Slansky and Scriber, 1985). Alternatively, this value may have increased in response to a decreased rate of passage of food through the gut as was shown to occur in *Locusta*

(Mordue et al., 1985). Prolongation of time for passage of food in the gut to facilitate increased exposure to digestive enzymes may increase AD (Slansky and Scriber, 1985).

Inhibition of larval development by azadirachtin may be attributed at least in part to reduced physiological age brought about by reduced food intake and weight gain. On the other hand, such responses may also be explained by disruption of neuroendocrine centre of moulting (Rembold and Sieber, 1981; Redfern et al., 1982; Rembold et al., 1984; Dorn et. al., 1986; Kubo and Klocke, 1986).

Increase in midgut esterase activity in the final-instar larvae after feeding on castor leaf discs treated with higher doses of azadirachtin is being reported for the first time and this finding very well justifies the observation on low utilization efficiencies, since this may be due to diversion of energy for detoxification as suggested by Barnby and Klocke (1987). Esterases cannot be involved in the detoxification of azadirachtin, since the molecule does not have any ester linkage. However, the implications of such increase in midgut esterase activity as a response to allelochemic (azadirachtin) challenge have serious importance in pest control schedules, since increase in esterase activity can result in decreased toxicity to insecticides containing ester linkage. This sort of phenomenon might require serious rethinking before recommending the use of azadirachtin-based formulations in integrated pest management.

Thus, the present study indicates that azadirachtin affects the important pest *S. litura* in a manner similar to that reported for several other insect species. More importantly, these data provide basic information that can enable understanding of effects of such chemicals on hormone levels, leading to, among other phenomena, sensitivity differentials in responses of larvae of different ages. In practical terms, the higher insensitivity of older larvae of *S. litura* to azadirachtin (e.g., in antifeedant action), comapred to neonate larvae in particular, can influence formulation of more realistic and effective application strategies for compounds such as those enriched in azadirachtin (e.g., many neem products). Similar implications also stem from the phenomenon of induction of midgut esterase activity in final-instar larvae of *S. litura* by high doses of azadirachtin, since this will again entail a fair possibility of induction of detoxification mechanisms seriously handicapping the desired effects. Furthermore, knowledge of precise dose-activity relationship of azadirachtin on particulate biological systems can help in formulation of recommendations and prediction of possible results.

REFERENCES

Arnason, J.T., B.J.R. Philogene, N. Donskov, M. Hudon, C. McDougall, G. Fortier, P. Morand, D. Gardner, J. Lambret, C. Morris and C. Nozzolillo. 1985. Antifeedant and insecticidal properties of azadirachtin to the European corn borer, *Ostrinia nubilis. Entomol. Exp. Appl.* 38: 29–34.

Balandrin, M.F., J.A. Klocke, E.S. Wurtele and W.H. Bollinger. 1985. Natural plant chemicals: Sources of industrial and medicinal materials. *Science* 228: 1154–1160.

Barnby, M.A. and J.A. Klocke. 1987. Effects of azadirachtin on the nutrition and development of the tobacco budworm, *Heliothis virescens* (Fabr.) (Noctuidae). *J. Insect Physiol.* 33(2): 69–75.

Blaney, W.M., L.M. Schoonhoven and M.S.J. Simmonds. 1986. Sensitivity variations in insect chemoreceptors; A review. *Experientia* 42: 13–19.

Champagne, D.E., M.B. Isman and G.H. NeilTowers. 1989. Insecticidal activity of phytochemicals and extracts of the Meliaceae. In: *Insecticides of Plant Origin*. Eds., J.T. Arnason, B.J.R. Philogene and P. Morand. ACS Symp. Se. Washington, 387, pp. 95–109.

Dorn, A., J.M. Rademacher and E. Sehn. 1986. Effects of azadirachtin on the moulting cycle, endocrine system, and ovaries in last-instar larvae of the milkweed bug, *Oncopeltus fasciatus. J. Insect Physiol.* 32: 231–238.

Fagoonee, I. 1984. Effect of azadirachtin and of a neem extract on food utilization by *Crocidolomia binotalis*. In: *Natural Pesticides from the Neem Tree* (*Azadirachta indica* A. Juss.) *and Other Tropical Plants. Proc. 2nd Int. Neem Conf.* Eds., H. Schmutterer and K.R.S. Ascher. GTZ, Eschborn, pp. 211–223.

Feeny, P. 1976. Plant apparency and chemical defence. *Rec. Adv. Phytochem.* 10: 1–40.

Gaaboub, I.A. and D.K. Hayes. 1984. Biological activity of azadirachtin, component of the neem tree, inhibiting moulting in the face fly, *Musca autumnalis* De Geer (Diptera: Muscidae). *Envir. Entamol.* 13: 803–812.

Garcia, E.S. and H. Rembold. 1983. Effects of azadirachtin on ecdysis of *Rhodnius prolixus*. *J. Insec Physiol.* 30: 939–941.

Ghumare, S.S., S.N. Mukherjee and R.N. Sharma. 1989. Effect of rutin on the neonate sensitivity, dietary utilization and mid-gut carboxylesterase activity of *Spodoptera litura* (Fabricius) (Lepidoptera: Noctuidae). *Proc. Indian Acad. Sci.* (Anim. Sci.) 98: 399–404.

Hedin, P.A. 1982. New concepts and trends in pesticide chemistry. *J. Agric. Food Chem.*, 30: 201–215.

Jacobson, M. 1986. The neem tree: Natural resistance par excellence. ACS Symp. Ser. 296, pp. 220–232.

Jotwani, M.G. and K.P. Srivastava. 1984. A review of neem research in India in relation to insects. In: *Natural Pesticides from the Neem Tree* (*Azadirachta indica* A. Juss.) *and other Tropical Plants*. Eds., H. Schmutterer and K.R.S. Acher, GTZ, Eschborn, pp. 43–56.

Kubo, I. and J.A. Klocke. 1986. Insect ecdysis inhibitors. In: *Natural Resistance of Plants to Pests*. Eds., M.B. Green and P.A. Hedin. ACS Symp. Ser., Washington, 296, pp. 206–219.

Ladd, T.L. Jr., J.D. Warthen Jr. and M.G. Klein. 1984. Japanese beetle (Coleoptera: Scarabidae): The effects of azadirachtin on the growth and development of the immature forms. *J. Econ. Entomol.* 77: 903–905.

Mordue (Luntz), A.J., P.K. Cottee and D.A. Evans. 1985. Azadirachtin: Its effect on gut motility, growth and moulting in *Locusta. Physiol. Entomol.* 10: 431–437.

Mukherjee, S.N. 1992. Studies on some biological effects of azadirachtin on *Tribolium castaneum* (Herbst) (Coleoptera: Tenebrionidae), *Achaea janata* (Linn.) and *Spodoptera litura* (Fabricius) (Noctuidae: Lepidoptera). Ph.D. thesis, Shivaji University, Kolhapur, India.

Mukherjee, S.N. and R. Ramachandran. 1989. Effects of azadirachtin on the feeding, growth and development of *Tribolium castaneum* (Herbst) (Coleoptera: Tenebrionidae). *J. Appl. Entomol.* 107: 145–149.

Qadri, S.S.H. and J. Narsaiah. 1978. Effect of azadirachtin on the moulting processes of last instar nymphs of *Periplaneta americana* (Linn.). *Indian J. Exp. Biol.* 16: 1141–1143

Ramachandran, R., S.N. Mukherjee and R.N. Sharma. 1989. Effects of food deprivation and concentration of azadirachtin on the performance of *Achoea janata* and *Spodoptera litura* on young and mature leaves of *Ricinus communis. Entomol. Exp. Appl.* 51: 29–35

Redfern, R.E., T.J. Kelly, A.B. Borokovec and D.K. Hayes. 1982a. Ecdysteroid titers and moulting aberrations in last stage *Oncopeltus* nymphs treated with insect growth regulators. *Pestic. Biochem. Physiol.* 18: 351–356.

Redfern, R.E., J.D. Warthen Jr., E.C. Ubel and G.D. Mills Jr. 1982. The antifeedant and growth-disrupting effects of azadirachtin on *Spodoptera frugiperda* and *Oncopeltus fasciatus*. In: *Natural Pesticides from the Neem Tree* (*Azadirachta indica* A. Juss.). *Proc. 1st. Neem Conf.* Eds., H. Schmutterer, K.R.S. Ascher and H. Rembold. GTZ, Eschborn, pp. 129–136.

Reese, J.C. 1979. Interaction of allelochemicals with nutrients in herbivore food. In: *Herbivores: Their Interaction with Secondary Plant Metabolites.* Eds., G.A. Rosanthal and D.A. Zanzen. Academic Press, New York, pp. 309–330.

Rembold, H. 1984. Secondary plant products in insect control, with special reference to the azadirachtins. In: *Advances in Invertebrate Reproductions*, Vol. 3. Ed., W. Engels. Elsevier, Amsterdam, pp. 481–491.

Rembold, H., H. Forster, Ch. Czoppelt, P.J. Rao and K.P. Sieber. 1984. The azadirachtins, a group of insect growth regulators from the neem tree. In: *Natural Pesticides from the Neem Tree* (*Azadirachta indica* A. Juss.) *and Other Tropical Plants. Proc. 2nd Int. Neem Conf.* Eds., H. Schmutterer and K.R.S. Ascher. GTZ, Eschborn, pp. 153–162.

Rembold, H., G.K. Sharma, Ch. Czoppelt and H. Schmutterer. 1980. Evidence of growth disruption in insects without feeding inhibition of neem seed fractions. *Z. Pfl. Krankh Pfl. Scutz.* 87: 290–297.

Rembold, H., G.K. Sharma, Ch. Czoppelt and H. Schmutterer. 1982. Azadirachtin: A potent insect growth regulator of plant origin. *Z. ang. Ent.* 93: 12–17.

Rembold, H. and K.P. Sieber. 1981. Inhibition of oogenesis and ovarian ecdysteroid synthesis by azadirachtin in *Locusta migratoria* (R&F). *Z. Naturforsch.* 36C: 446–469.

Rhoades, D.F. and R.G. Cates. 1976. Towards a general theory of plant antiherbivore chemistry. *Rec. Adv. Phytochem.* 10: 168–213.

Schluter, U, H.J. Bidmon and S. Grewe. 1985. Azadirachtin affects growth and endocrine events in larvae of the tobacco hornworm, *Manduca sexta. J. Insect Physiol.* 31: 773–777.

Schluter, U. and W.D. Schulz. 1984. Structural damages caused by neem in *Epilachna varivesti*: A summary of histological and ultrastructural data I. Tissues affected in larvae. In: *Natural Pesticides from the Neem Tree (Azadirachta indica* A. Juss.) *and Other Tropical Plants. Proc. 2nd Int. Neem Conf.* Eds., H. Schmutterer and K.R.S. Ascher. GTZ, Eschborn, pp. 227–235.

Schmmuterer, H. and Rembold, H. 1980. Zur Wirking elniger Rainfraktionen aus Samen von *Azadirachta indica* auf FrakBaktivitat and Meta morphose von *Epilachna varinestis* (Col: Coccinellidae). *Z. ang. Ent.* 89, 179–188.

Schmutterer, H., R.C. Saxena and J. von der Hyke. 1983. Morphogenetic effects of some partially purified fractions and methanolic extracts of neem seed on *Mythimna separata* (Walker) and *Cnaphalocrocis medinalis* (Guenee). *Z. ang. Ent.* 95: 230–237.

Sieber, K.P. and H. Rembold. 1983. The effects of azadirachtin on the endocrine control of moulting in *Locusta migratoria. J. Insect Physiol.* 29: 523–527.

Simmonds, M.S.J. and W.M. Blaney. 1984. Some effects of azadirachtin on lepidopterous larvae. *Proc. 2nd Int. Neem Conf.* Eds., H. Schmutterer and K.R.S. Ascher. GTZ, Eschborn, pp. 163–180.

Slansky, F. Jr., and J.M. Scriber. 1985. Food consumption and utilization. In: *Comprehensive Insect Physiology, Biochemistry and Pharmacology*, Vol. 4 Eds., G.A. Kerkut and L.I. Gilbert. Pergamon Press, Oxford, pp. 87–163.

Waldbauer, G.P. 1968. The consumption and utilization of food by insects. *Adv. Insect Physiol.* 5: 229–288.

Warthen, J.D. Jr., and E.C. Ubel. 1982. Effect of azadirachtin on house crickets, *Acheta domesticus*. In: *Natural Pesticides from the Neem Tree (Azadirachta indica* A. Juss.). *Proc. 1st Int. Neem Conf.* Eds., H. Schmutterer, K.R.S. Ascher and H. Rembold. GTZ, Eschborn, pp. 137–148.

29

Effect of Neem-based Insecticides on Growth and Development of White-backed Planthopper, *Sogatella furcifera* (Horvath)

S.S. SANDHU, G.S. DHALIWAL AND V.K. DILAWARI

Section of Ecology, Punjab Agricultural University, Ludhiana 141 004, India

ABSTRACT

Five neem-based insecticides—Neemark (0.5, 0.25, 0.125 per cent), RD 9 Repelin (1.0, 0.5, 0.25 per cent), Neemrich I (0.4, 0.2, 0.1 per cent) Neemrich II (0.1, 0.05, 0.025 per cent), and neemoil (5.0, 2.0, 1.0 per cent) were evaluated for the olfactory response and effect on fecundity and growth of white-backed planthopper, *Sogatella furcifera* (Horvath). All the treatments repelled the insect, with maximum repulsion (–0.76) caused by Neemrich I 0.4 per cent followed by Neemark 0.5 per cent (–0.74), RD 9 Répelin 1 per cent (–0.72), Neemrich II 0.1 per cent (–0.68), and neem oil 5 per cent (–0.44). The repellent effect decreased with decrease in concentration. Fecundity of the insect was minimum in case of RD 9 Repelin 1 per cent (40.0 eggs/female) followed by Neemrich II 0.1 per cent (46.0 eggs/female), Neemrich I 0.4 per cent (46.66 eggs/female), RD 9 Repelin 0.5 per cent (52.0 eggs/female), neem oil 1 per cent (52.33 eggs/female), and Neemark 0.5 per cent (55.66 eggs/female) as against 141.0 eggs/female in control. Fecundity increased with decrease in concentration of insecticides. The number of nymphs attaining adult stage was minimum in case of RD 9 Repelin 1 per cent (28.0) followed by Neemrich II 0.1 per cent (30.0), neem oil 1 per cent (32.0), Neemrich I 0.4 per cent (32.0), Neemrich I 0.2 per cent (42.0), and Neemark 0.5 per cent (42.0) as compared to 98.0 in control. Thus, RD 9 Repelin 1 per cent was the most effective in reducing fecundity and growth of *S. furcifera*, while Neemrich I 0.4 per cent caused maximum repulsion.

INTRODUCTION

White-backed planthoppeer (WBPH), *Sogatella furcifera* (Horvath), is one of the important pests of rice in Punjab (Dhaliwal and Singh, 1983). Although the pest can be controlled effectively by synthetic organic insecticides, the environmental hazards and other insect-related problems posed by them discourage their use in integrated pest management programmes. The use of plant-based products to manage insect pest populations is considered safer and suitable. Neem products are being tried against a number of insect pests and have been reported to be quite effective in the rice agroecosystem (Saxena, 1989). It was, therefore, decided to evaluate some of the commercial preparations based on neem against WBPH by studying olfactory response of WBPH and its fecundity and growth as affected by neem products.

MATERIALS AND METHODS

The investigations were carried out at the Entomological Research Farm, Punjab Agricultural University, Ludhiana, using five neem-based products: Neemrich I, Neemrich II, RD 9 Repelin, Neemark, and neem oil. In the experiments, 30-day-old Taichung Native 1 (TN-1) rice plants were used. The stock culture of WBPH was maintained on 30- to 40-day-old TN-1 plants in the screenhouse.

Repellence was studied in a horizontal cylindrical chamber (50 cm × 12.5 cm i.d.) open at the two ends for one hr. Control and the treated TN- 1 plants vertically aligned and with roots immersed in water were placed at the opposite ends of the chamber. From the middle, 20 newly emerged nymphs of WBPH were released into the chamber. The number of insects alighting on the treated (A) and control (B) plants were recorded and orientation was calculated as A – B/A + B as per Dilawari and Atwal (1989). The value in negative indicated repulsion and vice versa.

For determining the effect on fecundity, four pairs of newly emerged WBPH were caged on treated and control plants. Newly emerging nymphs were recorded for 10 days from the start of emergence of first nymph and removed every day. Unhatched eggs were determined by staining the leaf portions where eggs were laid (Gifford and Trahan, 1969) and were counted under the microscope. Fecundity was worked out from fertility and the number of unhatched eggs.

For effect on growth, first-instar WBPH nymphs were caged on 30-day-old treated and control plants of TN-1 at the rate of 10 individuals per cage. Growth was measured in terms of the percentage of nymphs becoming adult, and the period required by nymphs to become adult.

RESULTS AND DISCUSSION

Olfactory response of WBPH as demonstrated by its orientational behaviour to different neem-based insecticides indicated that the insect did not prefer the treated plants at all the concentrations (Table 1). The repellence decreased with decreasing concentrations in all the treatments. Neemrich I (0.4 per cent), Neemrich II (0.1 per cent), Neemark (0.5 per cent), and RD 9 Repelin (1.0 per cent) were found to be equally effective in repelling the insect and were significantly superior over the maximum concentration of neem oil (5 per cent). Repellence of the same order was also reported by Heyde et al. (1984) in case of *Nilaparvata lugens* (Stal) due to 5 per cent neem oil applied on rice plants. Using neem oil, Krishnaiah and Kalode (1984) also observed a similar behavioural response among *N. lugens* and *Nephotettix virescens* (Distant).

Number of eggs laid by WBPH was found to be affected adversely by neem-based insecticides in the screenhouse (Table 2). Although the number of eggs was the mini-

Table 1. Orientation of WBPH when given a choice between treated and untreated TN-1 rice plants

Treatments and concentrations (%)	Arrival on treated plants (%)*	Orientation
Neemrich I		
0.4	12[f]	–0.76
0.2	35[abcd]	–0.30
0.1	46[a]	–0.08
Neemrich II		
0.1	16[f]	–0.68
0.05	38[abcd]	–0.24
0.025	40[abc]	–0.20
Neemark		
0.5	13[f]	–0.74
0.25	32[bcd]	–0.36
0.125	42[ab]	–0.16
RD 9 Repelin		
1.0	14[f]	–0.72
0.5	30[cd]	–0.40
0.25	40[abc]	–0.20
Neem oil		
5	28[d]	–0.44
2	39[abcd]	–0.22
1	39[abcd]	–0.22

*Average of five replications of twenty insects each.
In a column, values followed by the same letter are not significantly different by DMRT (p = 0.05).

mum on plants treated with RD 9 Repelin at the rate of 1.0 per cent (40 eggs/female), it did not differ significantly from that on plants treated with neem oil (1.0, 0.5 per cent), Neemrich I (0.4 per cent), Neemrich II (0.1, 0.05 per cent), and Neemark (0.5, 0.25 per cent). The egg count was significantly lower in all the treatments, irrespective of concentrations over the untreated control. Egg count increased with decrease in the concentrations of different treatments. The decrease in number of eggs has been attributed to the volatile repellent compounds present in neem (Saxena and Rembold, 1984; Rice et al., 1985), to some physiological alteration (Ascher, 1981; Rembold et al., 1984), to male sterilant effects (Dorn et al., 1987), or to failure in production of normal courtship behaviour as in *N. lugens* (Saxena et al., 1989) due to neem products. Neem did not appear to affect hatching of eggs adversely, as is evident from the results.

Neem products were found to affect the growth of WBPH adversely (Table 3). The minimum number of nymphs attained the adult stage in treatment with RD 9 Repelin 1 per cent (28 per cent) followed by Neemrich II 0.1 per cent (30 per cent) and neem oil 1 per cent (32 per cent). The number of nymphs reaching adult stage was lower, but was not significantly different from that on plants treated with Neemark 0.5 per cent, RD 9 Repelin 5 per cent, Neemrich I 0.4 and 0.2 per cent, and Neemrich II 0.5 per cent. The number of nymphs attaining adult stage was, however, significantly lower in all the treatments than in control. Similarly, the growth period was the shortest in control and was significantly different from that on all other treatments. The growth period was the longest on plants treated with Neemrich II 0.1 per cent (12.75 days),

Table 2. Number of eggs laid by WBPH on plants treated with different botanical insecticides

Treatments and concentrations (%)	Mean No. of eggs laid/female*	Hatching (%)
Neemrich I		
0.4	46.66[ab]	73.37
0.2	69.33[bcd]	74.03
0.1	104.33[e]	72.51
Neemrich II		
0.1	46.0[ab]	76.80
0.05	59.66[abc]	73.15
0.025	86.66[de]	73.85
Neemark		
0.5	55.66[ab]	62.27
0.25	59.66[abc]	72.62
0.125	91.0[de]	72.89
RD 9 Repelin		
1.0	40.0[a]	70.82
0.5	52.0[ab]	78.19
0.25	88.66[de]	72.18
Neem oil		
1.0	52.33[ab]	64.32
0.5	62.0[abc]	68.32
0.25	81.33[cde]	76.63
Control	141.0[f]	78.48

*Average of six replications of four pairs in each.
In a column, values followed by the same letter are not significantly different by DMRT ($p = 0.05$).

Table 3. Growth and development of WBPH under the influence of different botanical insecticides

Treatments and concentrations (%)	Nymphs becoming adults* (%)	Growth period (days)
Neemark		
0.5	42[abc]	12.20[bc]
0.25	52[bcde]	12.32[b]
0.125	62[de]	12.12[bc]
RD 9 Repelin		
1.0	28[a]	12.20[bc]
0.5	44[abcd]	12.02[c]
0.25	56[cde]	11.98[c]
Neemrich I		
0.4	38[ab]	12.30[bc]
0.2	42[abc]	12.30[b]
0.1	54[bcde]	12.24[bc]
Neemrich II		
0.1	30[a]	12.75[a]
0.05	40[abc]	12.49[ab]
0.025	58[cde]	12.30[bc]
Neem oil		
1.0	32[a]	12.33[ab]
0.5	58[cde]	12.18[bc]
0.25	66[e]	12.36[ab]
Control	98[f]	11.35[d]

*Average of five replications of ten nymphs each.
In a column, values followed by the same letter are not significantly different by DMRT ($p = 0.05$).

followed by Neemrich II 0.05 per cent (12.49) and neem oil 1 per cent (12.33) and 0.25 per cent (12.36). The differences were, however, non-significant. Growth disrupting effects of various neem products have been reported by several workers (Heyde et al., 1984; Zebitz, 1984; Schmutterer, 1990). Therefore, it is possible that due to repellence, the insect coult not feed on the host sufficiently (Rice et al., 1985), which reduced its vitality and delayed its growth (Schmutterer, 1990).

REFERENCES

Ascher, K.R.S. 1981. Some physical (solubility) prospectives and biological (sterilant for *Epilachna varivestis* females) effects of a dried methanolic neem (*Azadirachta indica*) seed kernel extract. In: *Natural Pesticides from the Neem Tree. Proc. 1st Int. Neem Conf.*, 1980. GTZ Eschborn, pp. 63–74.

Dhaliwal, G.S. and Jaswant Singh. 1983. Outbreaks of whitebacked planthopper in the Punjab, India. *Int. Rice Comm. Newsletter* 32(1): 26–28.

Dilawari, V.K. and Atwal, A.S. 1989. Response of mustard aphid, *Lipaphis erysimi* (Kalt.) to allylisothio cyanate *J. Insect Sci.* 2(2): 103–108.

Dorn, A., S.M. Rademacher and E. Sehn. 1987. Effect of azadirachtin on reproductive organs and fertility in the large milkweed bug, *Oncopeltus fasciatus*. *Proc. 3rd Int. Neem Conf.*, 1986 GTZ, Eschborn, pp. 273–288.

Gifford, J.T. and Trahan, G.B. 1969. Staining technique for eggs of rice water weevil oviposited intracellularly in the tissues of the leaf sheath of rice. *J. econ. Ent.* 62: 740–741.

Heyde, J.V.D., R.C. Saxena and H. Schmutterer. 1984. NO and NE, a potential insecticide for control of hemipterous rice pests. *Proc. 2nd Int. Neem Conf.* GTZ, Eschborn, pp. 377–390.

Krishnaiah, N.V. and M.B. Kalode. 1984. Evaluation of neem oil, neem cake and other non-edible oil cakes against rice pests. *Indian J. Plant Prot.* 12(2): 101–107.

Rembold, H., H. Forster, C. Czoppelt, P.J. Rao and K.P. Sieber. 1984. The azadirachtins, a group of insect growth regulators from the neem tree. In: *Natural Pesticides from the Neem Tree and Other Tropical Plants. Proc. 2nd Int. Neem Conf.*, 1983. GTZ, Eschborn, pp. 153–162.

Rice, M., S. Sexton and A.M. Esmail. 1985. Antifeedant phytochemical blocks oviposition by sheep blowfly. *J. Australian Entomol. Soc.* 24: 16.

Saxena, K.N. and H. Rembold. 1984. Orientation and ovipositional response of *Heliothis armigera* to certain neem constituents. *Proc. 2nd Int. Neem Conf.*, 1983. GTZ, Eschborn, pp. 199–210.

Saxena, R.C. 1989. Insecticides from neem. *Insecticides from Plants*. ACS Symp. 387, Washington, D.C., pp. 110–135.

Saxena, R.C., Z.T. Zang and M.E.M. Boncodin. 1989. Effect of neem oil on courtship signals and mating behaviour of brown planthopper's female. *Int. Rice Res. Newsletter* 14(6): 28–29.

Schmutterer, H. 1990. Properties and potential of natural pesticides from the neem tree, *Azadirachta indica. A. Rev. Entomol.* 35: 271–297.

Zebitz, C.P.W. 1984. Effect of some crude and azadirachtin enriched neem (*Azadirachta indica*) seed kernel extracts on larvae of *Aedes aegypti*. *Entomol. Ext. Appl.* 35: 11–16.

30

Toxic, Developmental and Reproductive Effects of AZT-VR-K and Some Neem Products against Mustard Aphid, *Lipaphis erysimi* (Kaltenbach)

SURINDER S. BHATHAL[1] AND DARSHAN SINGH

Department of Entomology, Punjab Agricultural University
Ludhiana 141 004
[1]District Extension Specialist (Plant Protection), PAU FASS Office
Club Road, Sangrur (Punjab)

ABSTRACT

Two commercial neem products (neemguard and neemark) and two neem extracts (neem oil and AZT-VR-K) were evaluated by topical application for their biological activities against mustard aphid, *Lipaphis erysimi* (Kaltenbach). Commercial products and neem oil proved highly toxic to third- and fourth-instar (presumptive alate) nymphs. On the other hand AZT-VR-K was more toxic to fourth-instar nymphs while in third-instar treated nymphs considerable delayed mortality and abnormality in adults formed was observed. The extent of adultoid formation was 26.7 to 50 per cent when treated with doses of 0.5 to 5 μg/nymph. The reproductive period, fecundity and longevity of adultoids was also greatly reduced as compared to control. Like wise, on treatment of apterous adults with AZT-VR-K besides their mortality, the reproductive period, rate of reproduction and longevity were reduced considerably.

INTRODUCTION

Indiscriminate and injudicious use of synthetic pesticides in agriculture has had many side effects. At present, emphasis is being laid on finding alternative means of pest control. One approach is to search for new bioactive plant products which are environmentally safe and commercially viable. Among the numerous plant ingredients studied so far, the extracts of neem tree, *Azadirachta indica* A. Juss., have attracted the special interest of scientists the world over. Neem is known to contain a diverse array of biologically active principles of which azadirachtin (a tetranortriterpenoid) is the best known derivative (Broughton et al., 1986). It possesses antifeedant, anti-ovipositional, growth-disrupting, and fecundity-reducing properties against insect pests of different orders (Schmutterer, 1990). Recently, some marketable neem-based products having enriched concentration of azadirachtin have been developed and released for commercial use in insect pest management. The present work describes the toxic, developmental, and reproductive effects of AZT-VR-K, Neemguard, Neemark, and neem oil against mustard aphid, *Lipaphis erysimi* (Kaltenbach), a serious pest of *Brassica* oilseeds and other cruciferous crops in India

MATERIALS AND METHODS

The desired dosages of AZT-VR-K (azadirachtin extract of neem seed kernels supplied by Dr. C. Hellpap of Justus-Leibig-University, Giessen, Germany); Neemguard (neem concentrate water miscible formulation of Gharda Chemicals Limited, Bombay); Neemark (water miscible concentrate containing 80 per cent neem biomass from West-Coast Herbochem Pvt. Ltd. Bombay) and Neem oil (Neem seed crude oil purchased from local market) were prepared in acetone, 0.1 μl of each dose was topically applied on to the dorsum of third- and fourth-instar (presumptive alate) nymphs and adults separately, by using Hamilton microsyringe (Ca = 1 μl). In control, only 0.1 μl of acetone was applied to each aphid. There were three such replications with 20 nymphs and 5 adults in each replicate. The treated aphids were transferred on to a fresh untreated radish (*Raphanus sativus* L.) leaf kept on a moist sponge placed in a crystallizing dish (50 × 100 mm.). Each dish was covered with a muslin piece held tightly with a rubber band. The leaves were replaced by a fresh untreated ones at 24 hr intervals. The observations on the aphids were taken daily till adult formation with regard to their mortality, moulting and abnormalities, if any both during nymphal and adult stages.

The daily rate of reproduction, fecundity and longevity of abnormal alate adults (adultoids) formed from third-instar (presumptive alate) nymphs and apterous adults treated with AZT-VR-K extract were also recorded.

To observe the effect of AZT-VR-K on reproduction, five freshly moulted apterous adults were topically treated with 0.1 μl/aphid of each dose. In control, the aphids were treated with 0.1 μl of acetone alone. There were three such replications for each dose. The treated individuals of a replicate were transferred on to a fresh radish leaf placed on a moist sponge in a crystallizing dish. These aphids were observed daily for mortality, fecundity, and longevity. The rate of nymphal production of each surviving group was counted daily till their death while they were provided daily with fresh, untreated leaves.

The fecundity and longevity were also recorded of abnormal alate adults (adultoids) formed from third-instar (presumptive alate) nymphs treated with AZT-VR-K extract at different doses.

RESULTS AND DISCUSSION

On exposure of third- and fourth-instar nymphs to various doses of neem extracts, products both immediate (within 24 hrs) and delayed mortalities were experienced. In addition, a varied type of nymphal and adult deformities was also recorded. The deformities caused during the nymphal development were in the form of incomplete ecdysis in third- and fourth-instar nymphs. The affected individuals were unable to shed their old cuticle during moulting and died in pharate condition. The adult deformities were in the form of reduced, curled, crumpled, and incompletely formed wings (Fig. 1a). Such individuals were called 'adultoids' and were unable to fly. Some of the adultoids failed to shed completely their exuvae, which remained attached to their legs and crumpled wings (Fig. 1b).

On treatment of third-instar nymphs with neem extracts, products both immediate and delayed nymphal mortalities occurred (Table 1). The immediate mortality was significantly the highest in Neemark (80 per cent) and Neemguard (78.3 per cent), both at the dose of 5 μg/nymph. These products registered high mortality (48.3 per cent), even 1 μg/nymph, which was comparable with 5 μg dose of neem oil. AZT-VR-K, on the whole, registered very low immediate mortality at all the doses tried (0.25 to 5 μg). The immediate mortality at 5μg (16.7 per cent) was significantly on par with that obtained with 0.5 μg of Neemark and Neemguard and 1μg of neem oil. Interestingly, in AZT-VR-K at 1–5 μg/nymph, significantly highest delayed mortality effects were manifested during fourth-instar (3.3 to 8.3 per cent) and later in fourth-instar pharate condition (5 to 23.3 per cent), indicating the inability of nymphs to ecdyse normally. In other products, there was no delayed mortality with the exception of 5 per cent in 5 μg of Neemguard. Composite mean values for mortality in third- and fourth-instar

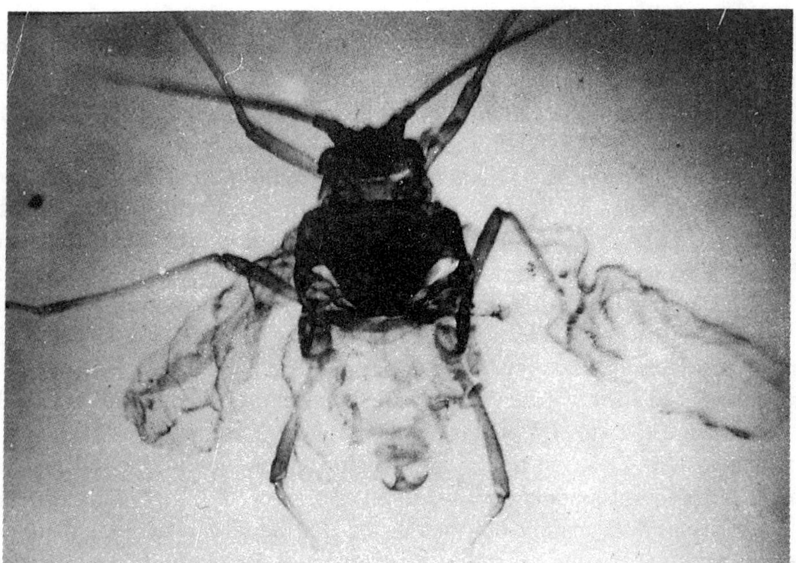

Fig. 1 (a).Adultoid with crumpled and incompletely formed wings

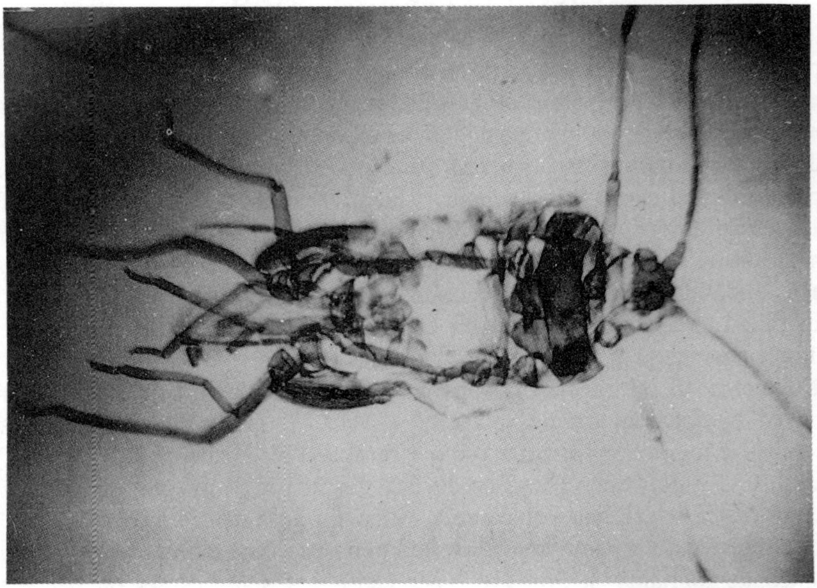

Fig. 1(b). Adultoid with exuvium attached to the posterior region of the body

also revealed Neemguard and Neemark to be highly toxic to third-instar nymphs followed by neem oil and AZT-VR-K. It was worth noticing that out of total mortality of 40 and 26.7 per cent at 5 and 4 μg/nymph of AZT-VR-K, 23.3 and 10 per cent occurred in fourth-instar pharate condition. The data also revealed that almost all the surviving nymphs in Neemguard, Neemark, and neem oil treatments transformed into normal-looking adults with the only exception of Neemguard at 5 μg where 6.7 per cent of the total adults formed were abnormal. In AZT-VR-K, the extent of abnormal adult formation was unusually high at all the doses. Out of 60, 73.3, 83.3, 86.7 and 91.7 per cent total adults formed on treatment with 5, 4, 2, 1 and 0.5 μg, respectively, the extent of adultoids was 28.3, 41.7, 50.0, 36.7 and 26.7 per cent.

On treatment of fourth-instar nymphs with different doses of various neem extracts/products, a considerable level of immediate mortality was registered in all the treatments (Table 2). Even in AZT-VR-K, high immediate mortality was recorded contrary to results obtained on treatment of third-instar nymphs. It varied from 26.7 to 51.7 per cent at doses varying between 1 and 5 μg/nymph. Delayed mortality in fourth-instar pharate condition was also evidenced at all the doses of AZT-VR-K (10 to 26.7 per cent) and 5 μg dose of Neemguard (18.3 per cent) and Neemark (8.3 per cent). The surviving nymphs of different treatments developed into normal adults with only insignificant percentage (3.3 to 6.7 per cent) developing as adultoids.

In the present studies, the rate of reproduction, reproductive period and fecundity was greatly affected on treatment with 4 μg/nymph of AZT-VR-K. The reduced fecundity at lower doses was mainly due to the reduced adult life of treated individuals.

The reproductive period, fecundity, and longevity of adultoids formed from topically treated third-instar nymphs were also greatly reduced (Fig. 2). The adultoids continued to lay young ones for 17.9–24.6 days in different treatments as compared to

Table 1. Effects of some neem extracts and products on mortality and adult formation of *Lipaphis erysimi* (Kaltenbach) following topical treatment to third-instar presumptive alate

Extract/ product	Dose (µg/insect)	*Nymphal mortality (%) during				* Adults formed (%)		
		3rd instar	4th instar	4th pharate stage	Total	Normal	Abnormal	Total
AZT-VR-K	5	16.7 (24.05)[a]	0.0 (0.71)[b]	23.3 (4.88)[b]	40.0 (39.21)[a]	31.7 (34.23)[a]	28.3 (32.01)[a]	60.0 (50.79)[a]
	4	8.3 (16.60)	8.3 (2.94)	10.0 (3.18)	26.7 (31.07)	31.7 (34.23)	41.7 (40.20)	73.3 (58.93)
	2	3.3 (6.57)	8.3 (2.94)	5.0 (2.10)	16.7 (24.05)	33.3 (34.81)	50.0 (44.94)	83.3 (65.95)
	1	1.7 (4.73)	6.7 (2.40)	5.0 (2.10)	13.3 (21.34)	50.0 (45.00)	36.7 (37.22)	86.7 (68.66)
	0.5	3.3 (8.83)	3.3 (1.80)	1.7 (1.26)	8.3 (16.60)	65.0 (54.33)	26.7 (30.29)	91.7 (73.40)
	0.25	0.0 (0.64)	0.0 (0.71)	1.7 (1.26)	1.7 (4.73)	95.0 (79.33)	3.3 (8.83)	98.3 (85.27)
Neemguard	5	78.3 (62.29)	5.0 (2.35)	0.0 (0.71)	83.3 (65.95)	10.0 (15.21)	6.7 (12.12)	16.7 (24.04)
	1	48.3 (44.04)	0.0 (0.71)	0.0 (0.71)	48.3 (44.04)	50.0 (45.00)	1.7 (4.73)	51.7 (45.96)
	0.5	23.3 (28.67)	0.0 (0.71)	0.0 (0.71)	23.3 (28.67)	75.0 (60.08)	1.7 (4.73)	76.7 (61.33)
Neemark	5	80.0 (63.55)	0.0 (0.71)	0.0 (0.71)	80.0 (63.55)	20.0 (26.45)	0.0 (0.64)	20.0 (26.45)
	1	48.3 (44.03)	0.0 (0.71)	0.0 (0.71)	48.3 (44.03)	51.7 (45.97)	0.0 (0.64)	51.7 (45.97)
	0.5	18.3 (25.30)	0.0 (0.71)	0.0 (0.71)	18.3 (25.30)	81.7 (64.70)	0.0 (0.64)	81.7 (64.70)
Neem oil	5	43.3 (41.07)	0.0 (0.71)	3.3 (1.55)	46.7 (43.08)	51.7 (45.97)	1.7 (4.73)	53.3 (46.92)
	1	21.6 (27.52)	0.0 (0.71)	0.0 (0.71)	21.7 (27.52)	76.7 (61.22)	1.7 (4.73)	78.3 (62.48)
	0.5	8.3 (16.60)	0.0 (0.71)	0.0 (0.71)	8.3 (16.60)	91.7 (73.40)	0.0 (0.64)	91.7 (73.40)
Control		0.0 (0.64)	0.0 (0.71)	0.0 (0.71)	0.0 (0.64)	100.0 (89.36)	0.0 (0.64)	100.0 (89.36)
LSD (p = 0.05)		(8.17)	(0.78)	(1.18)	(6.08)	(10.50)	(10.53)	(6.08)

Figures in parentheses are (a) arcsin $\sqrt{\text{percentage}}$ and (b) $\sqrt{n + 0.5}$ transformations,
* Means of three replications.

Table 2. Effects of some neem extracts and products on mortality and adult formation of *Lipaphis erysimi* (Kaltenbach) following topical treatment to fourth-instar presumptive alate

Extract/ product	Dose (μg/insect)	*Nymphal mortality (%) during			*Adults formed (%)		
		4th instar	4th pharate stage	Total	Normal	Abnormal	Total
AZT-VR-K	5	51.7 (45.96)a	10.0 (3.18)b	61.7 (51.81)a	35.0 (36.15)a	3.3	38.3 (38.19)a
	4	33.3 (35.25)	20.0 (4.51)	53.3 (46.92)	40.0 (39.15)	6.7	46.7 (43.08)
	2	40.0 (39.15)	26.7 (5.19)	66.7 (55.08)	30.0 (32.71)	3.3	33.3 (34.92)
	1	26.7 (30.99)	20.0 (4.53)	46.7 (43.08)	53.3 (46.92)	0.0	53.3 (46.92)
Neemguard	5	63.3 (52.78)	18.3 (4.33)	81.7 (65.00)	15.0 (21.90)	3.3	18.3 (25.00)
Neemark	5	66.7 (54.75)	8.3 (2.94)	75.0 (60.08)	25.0 (29.92)	0.0	25.0 (29.92)
Neem oil	5	71.7 (57.99)	0.0 (0.71)	71.7 (57.99)	23.3 (28.78)	5.0	28.3 (32.01)
Control		0.0 (0.64)	0.0 (0.71)	0.0 (0.64)	100.0 (89.36)	0.0	100.0 (89.36)
LSD ($p = 0.05$)		(5.97)	(0.79)	(8.68)	(10.37)	NS	(8.68)

Figures in parentheses are (a) arcsin $\sqrt{\text{percentage}}$ and (b) $\sqrt{n + 0.5}$ transformations.
*Means of three replications.

Table 3. Mortality, reproduction and longevity of freshly moulted apterous adults of *Lipaphis erysimi* (Kaltenbach) treated topically with AZT-VR-K extract

Dose (µg/aphid)	*Mortality (%) within 24 hrs	*Reproductive period (days)	*Total progeny per female	*Daily rate of reproduction	*Longevity (days)
4	73.3 (59.22)	8.0	13.0	0.7	15.0
2	20.0 (22.36)	13.2	39.7	2.5	16.3
1	13.3 (13.93)	13.1	48.3	2.9	16.0
Control	0.0 (1.28)	26.0	91.3	3.2	28.9
LSD (p = 0.05)	(28.35)	7.9	23.3	1.1	6.0

Figures in parentheses are arcsin $\sqrt{\text{percentage}}$ transformations.

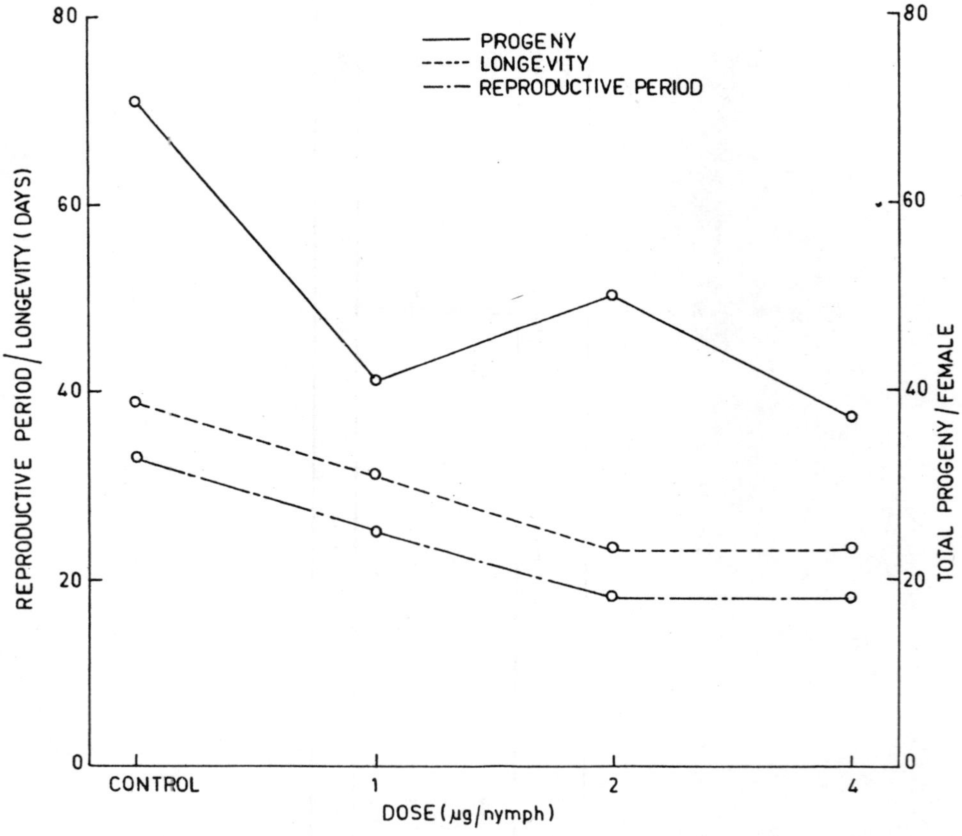

Fig. 2. Reproductive period, fecundity, and longevity of adultoids formed from third-instar presumptive alate treated topically with AZT-VR-K

the reproductive period of 33.2 days in check. The daily rate of reproduction was not much affected, but total progeny per female was greatly reduced. It was 37.3 to 50.5 young ones/female in different treatments as compared to 71.3 young ones/female in control. The adultoids also lived for a shorter period (22.9–30.6 days) in AZT-VR-K treatments as compared to average life of 38.6 days of control adults.

The results thus revealed that Neemguard, Neemark and neem extracts possessed both toxic and growth disruptive effects. However, the growth disruption was much more pronounced in nymphs and apterous adults treated with AZT-VR-K. In literature the detailed studies are available regarding the toxicity of neem extracts to aphids. Schauer (1984) obtained toxic effects of neem seed extracts against *Acyrthosiphon pisium* Harris and *Aphis fabae* Scopoli. The crude extract (1.5 per cent) of neem provided 86.7 per cent kill of *L. erysimi* after 72 hrs of treatment (Pandey et al., 1987). Likewise, a hexane extract obtained from neem seed kernels exhibited higher toxicity (LC_{50} = 0.67 per cent) against *L. erysimi* (Singh et al., 1988). The other aphids known to be similarly affected by neem derivatives are citrus aphid, *Aphis citricola* Van der Groot (Chiu and Zeng, 1986); cotton aphid, *Aphis gossypii* Glover (Siddig, 1987); safflower aphid, *Dactynotus carthami* H.R.L. (Devakumar et al., 1986); sugarcane aphid, *Melanaphis sacchari* (Zehnt.), and corn leaf aphid, *Rhopalosiphum maidis* (Fitch) (Srivastava and Parmar, 1985); green peach aphid, *Myzus persicae* (Sulzer) (Griffiths et al., 1978); and water lily aphid, *Rhopalosiphum nymphaceae* (L.) (Goyal et al., 1971).

The information on the growth regulatory effects of neem against aphids is scanty, but some reports are available against other homopterous insect pests. The foliar application of neem oil and AZT-VR-K extract resulted in disturbances of the moulting process, extension of the nymphal period, and a dose-dependent mortality in brown rice planthopper, *Nilaparvata lugens* (Stal), the whitebacked rice planthopper, *Sogatella furcifera* (Horvath), and the green rice leafhopper, *Nephotettix virescens* (Distant) (Heyde et al., 1984). In the present studies, AZT-VR-K extract also resulted in reduced fecundity of mustard aphid. Similar results had also been reported by other workers. It was reported that no young ones were born to *L. erysimi* adults released on the leaf discs treated with *A. indica* extract at 1 and 1.5 per cent concentrations, 72 hrs after treatment (Pandey et al., 1987). Likewise, pea aphid, *A. pisum*, kept from the first nymphal instar onwards on *Vicia faba* L. plants treated with 20 ppm of a purified methanolic neem seed kernel extract, produced only about one-twelfth the number of young ones of control insects (Schauer, 1985).

ACKNOWLEDGEMENTS

The authors wish to thank Dr. C. Hellpap (Institute fur Phytopathologie, V., Angewandte Zoologie, Justus Liebig Universitat, Giessen, Germany), Dr. V.S. Khandal (Director, West-Coast Herbochem Pvt. Ltd., Bombay), and Gharda Chemicals Limited, Bombay, for gratis supply of AZT-VR-K, Neemguard, and Neemark, respectively. The senior author also gratefully acknowledges financial assistance from ICAR and CSIR in the form of a Senior Research Fellowship.

REFERENCES

Broughton, H.B., S.V. Ley, A.M.Z. Slawin, D.J. Williams and E.D. Morgan. 1986. X-ray crystallographic structure determination of detigloyldihydroazadirachtin and reassignment of the structure of the limonoid insect antifeedant azadirachtin. *J. Chem. Soc. Chem. Commun.* pp. 391–401.

Chiu, S.F. and X.N. Zeng. 1986. Antifeedant activity of some materials from meliaceae against the citrus aphid, *Aphis citricola* Van der Groot. *Neem Newsletter* 3: 48.

Devakumar, C., V.S. Saxena and S.K. Mukerjee. 1986. Evaluation of neem *(Azadirachta indica* A. Juss.) limonoids and azadirachtin against safflower aphid (*Dactynotus carthami* H.R.L.). *Indian J. Ent.* 48: 467–470.

Goyal, R.S., K.C. Gulati, P. Sarup, M.A. Kidwai and D.S. Singh. 1971. Biological activity of various alcohol extracts and isolates of neem (*Azadirachta indica*) seed cake against *Rhopalosiphum nympheae* (Linn.) and *Schistocerca gregaria* Forsk. *Indian J. Ent.* 33: 67–71.

Griffiths, D.C., A.R. Greenway and S.L. Lloyd. 1978. The influence of repellent materials and aphid extracts on settling behaviour and larviposition of *Myzus persicae* (Sulzer) (Hemiptera: Aphididae). *Bull. Ent. Res.*, 68: 613–619.

Heyde, V.D.J., R.C. Saxena and H. Schmutterer. 1984. Neem oil and neem extracts as potential insecticides for control of hemipterous rice pests. In *Natural Pesticides from the Neem Tree and Other Tropical Plants Proc. 2nd Int. Neem Conf.*, Rauischholzhausen. 1983. GTZ, Eschborn, pp. 377–390.

Pandey, N.D., L. Singh, V.P. Singh and R.A. Tripathi. 1987. Effect of certain plant extracts against *Lipaphis erysimi* Kalt. under laboratory conditions. *Indian J. Ent.* 49: 238–242.

Schauer, M. 1984. Effect of variously formulated neem seed extracts on *Acyrthosiphon pisum* and *Aphis fabae*. In *Natural Pesticides from the Neem Tree and Other Tropical Plants, Proc. 2nd Int. Neem Conf.*, Rauischholzhausen, 1983. GTZ, Eschborn, pp. 141–150.

Schauer, M. 1985. Die Wirkung von Nieminnaltsstoffen auf Blatilause und die Rubenblattwanze. Ph.D. thesis, Univ. Giessen, FRG.

Schmutterer, H. 1990. Properties and potential of natural pesticides from the neem tree, *Azadirachta indica*. *A. Rev. Ent.* 35: 271–297.

Siddig, S.A. 1987. A proposed pest management programme including neem treatments for combating potato pests in Sudan. In *Natural Pesticides from the Neem Tree and Other Tropical Plants, Proc. 3rd Int. Neem Conf.*, Nairobi, 1986. GTZ, Eschborn, pp. 449–459.

Singh, R.P., C. Devakumar and S. Dhingra. 1988. Activity of neem (*Azadirachta indica* A. Juss.) seed kernel extracts against the mustard aphid, *Lipaphis erysimi*. *Phytoparasitica* 16: 225–230.

Srivastava, K.P. and B.S. Parmar. 1985. Evaluation of neem oil emulsifiable concentrate against sorghum aphid. *Neem Newsletter* 2: 7.

31

Effects of Azadirachtin on Neural Regulation of Midgut Peristalsis in *Locusta migratoria*

A. DORN AND P. TRUMM

Institute of Zoology, University of Mainz, Saarstr. 21, D-6500 Mainz, Germany

ABSTRACT

A single injection of azadirachtin into larvae and adult locusts results in a decrease of food uptake which is correlated with a gradual inhibition of midgut peristalsis. Gut movement is under the control of the stomatogastric nervous system, especially the ingluvial ganglion, which is shown in an *in vitro* system. Serotonin is the most potent peristalsis-stimulating drug *in vitro* and probably also in the intact system, as is indicated by immunohistochemical studies. Concurrent with the cessation of midgut peristalsis after azadirachtin injection, a depletion of the serotoninergic cell bodies in the frontal ganglion can be observed. When midgut preparations are challenged with azadirachtin *in vitro*, peristalsis is suppressed in a dose-dependent manner. If serotonin is supplied thereafter, only a slight stimulation of midgut peristalsis is achieved at the lower azadirachtin doses. We propose that azadirachtin interferes with the serotoninergic system of the stomatogastric ganglia and thus blocks midgut peristalsis.

INTRODUCTION

Azadirachtin, the most active ingredient in neem preparations, may act as repellent, antifeedant, or growth inhibitor (see Schmutterer et al., 1981). The last effect is best understood. Azadirachtin apparently blocks release of neurohormones—in particular PTTH—from the corpora cardiaca. In consequence, activation of the prothoracic glands is stalled and nymphs do not moult (Sieber and Rembold, 1983; Dorn et al., 1987). Much less is known about how repellence and feeding deterrence are triggered. Olfactory and/or gustatory receptors of the mouth parts may be involved (Schoonhoven, 1981; Simmonds and Blaney, 1984). Mordue et al. (1985) pointed out that azadirachtin also has an inhibitory effect upon gut peristalsis which may contribute to feeding deterrence. In the present investigation, we could demonstrate that midgut peristalsis is indeed affected by azadirachtin *in vivo* and *in vitro*. Since the agent was injected, interactions with chemosensory receptors at the mouth part were excluded. Immunohistological studies on the stomatogastric nervous system and pharmacological studies on the isolated midgut provided information on the neural regulation of midgut peristalsis and its disturbance by azadirachtin.

The studies were carried out within the project on biological-integrated locust control of the Deutsche Gesellschaft für Technische Zusammenarbeit (GTZ) Gmbh.

MATERIALS AND METHODS

A colony of *Locusta migratoria* has been reared in our laboratory for many years under standardized conditions in respect to food supply, photoperiod, temperature, and humidity. The specimens used for the present studies were all reared under crowded conditions (for details see Trumm, 1991). Azadirachtin (ROTH, Karlsruhe, Germany) was dissolved in ethanol (10 per cent) and either injected (1 μl/animal) into the abdomen of the test animal or applied in Ringer solution to the isolated gut. Serotonin (5-HT) and other pharmacological substances were tested for their stimulatory, inhibitory, or modulatory activity on the peristalsis of the isolated midgut. A special apparatus was constructed to monitor spontaneous and experimentally manipulated peristalsis (for details see Trumm, 1991). The presence of putative neurotransmitters in the stomatogastric nervous system and nerve terminals at the midgut muscularis was examined by immunohistochemistry (for details see Trumm, 1991).

RESULTS AND DISCUSSION

Final (fifth) instar larvae and adults received single injections of 3μg azadirachtin per animal at day 1 after previous moult. Whereas controls (injected with 1 μl of the solvent—ethanol 10 per cent showed normal growth (Fig. 1A), treated animals gained only a little weight (Fig. 1B). A comparison in food uptake and faeces production between test animals and controls demonstrates that both procceses are severely reduced in treated larvae and adults (Fig. 2A and B). Calculations concerning food utilization (according to the method reviewed in Waldbauer, 1968) indicated that there is no significant difference between test and control animals. However, the food passage through the gut is considerably slower in azadirachtin-treated larvae and adults than in controls. Food bowls, the passage of which were followed, remained much longer in the crop before they entered the midgut via proventriculus. Also, the midgut passage and formation of faeces was considerably retarded.

These and previous observations from Mordue et al. (1985) led to the hypothesis that inhibition of gut peristalsis may play a major role in the antifeedant activity of azadirachtin—apart from possible effects on chemoreceptors at the mouth parts. Direct proof of peristalsis inhibition by azadirachtin came from *in vitro* experiments.

The isolated midgut exhibited spontaneous contractions when excised with the ingluvial ganglia which are attached to the crop. Removal of the ganglia resulted in the cessation of the peristalsis (Fig. 3). All experiments with azadirachtin were performed

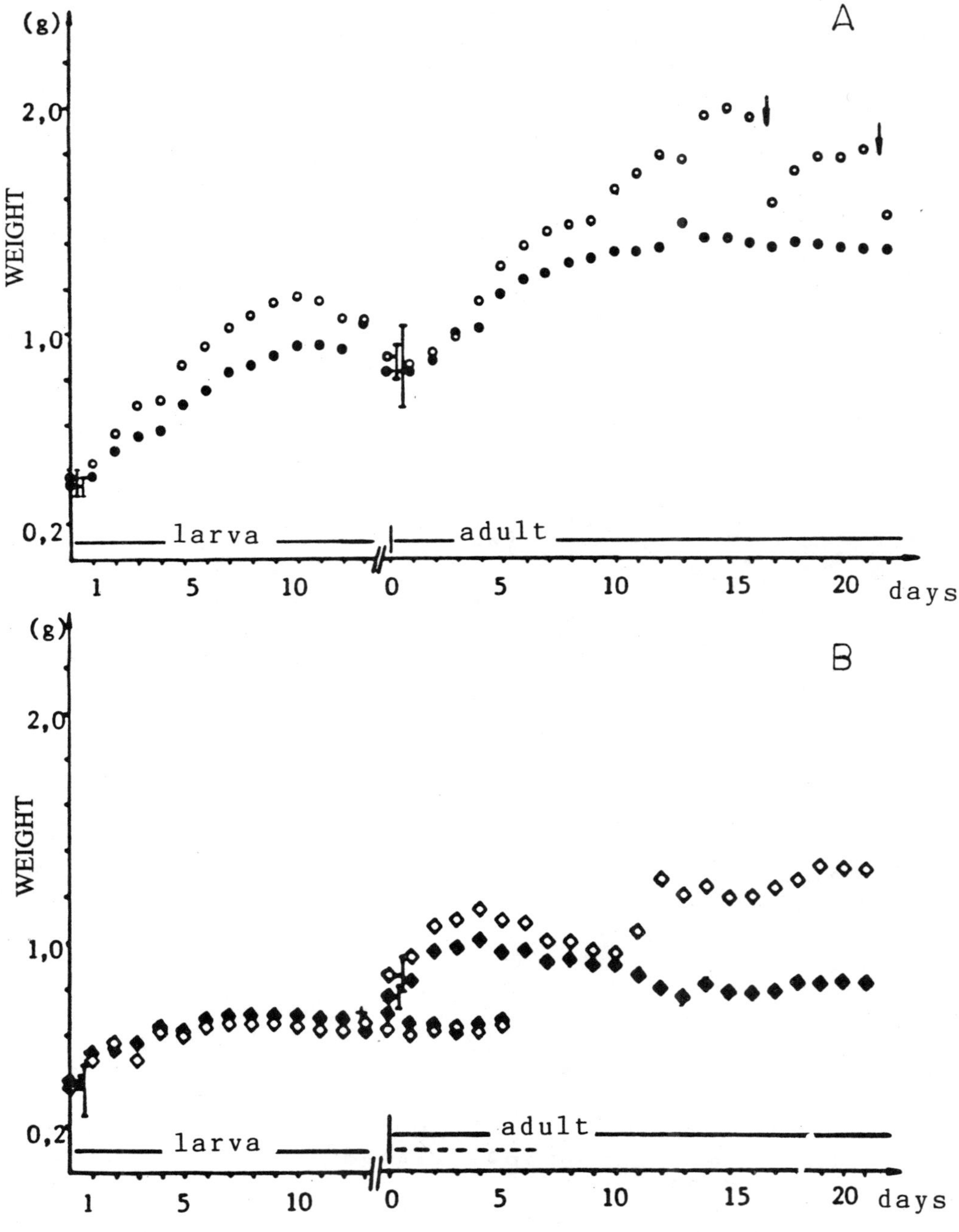

Fig. 1. Weight increase of fifth-instar larvae and adults: (A) Control locusts; females open circles, males solid circles; arrows indicate first and second oviposition. (B) Locusts treated with 3 μg azadirachtin per g body weight; larvae and adults received single injections at day 1 after previous moult; females open squares, males solid squares; treated larvae do not moult any more and become overaged.

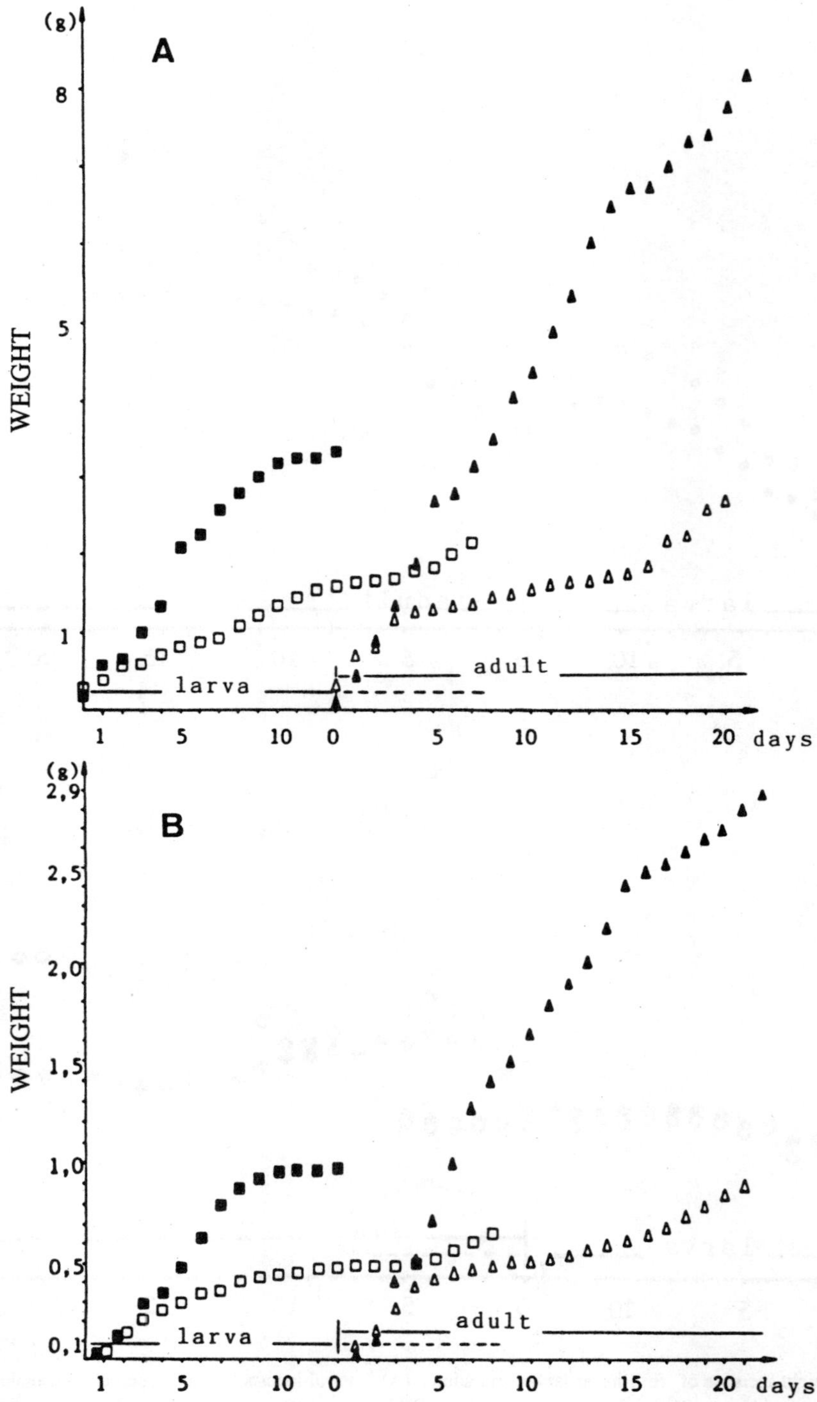

Fig. 2. Food consumption and faeces production in control and azadirachtin treated (3µg/g body weight) larvae and adults; treated larvae do not moult any more and become overaged; (A) Food consumption of control larvae solid squares, of treated larvae open squares, of control adults solid triangles, of treated adults open triangles. (B) Faeces production of control larvae solid squares, of treated larvae open squares, of control adults solid triangles, of treated adults open triangles.

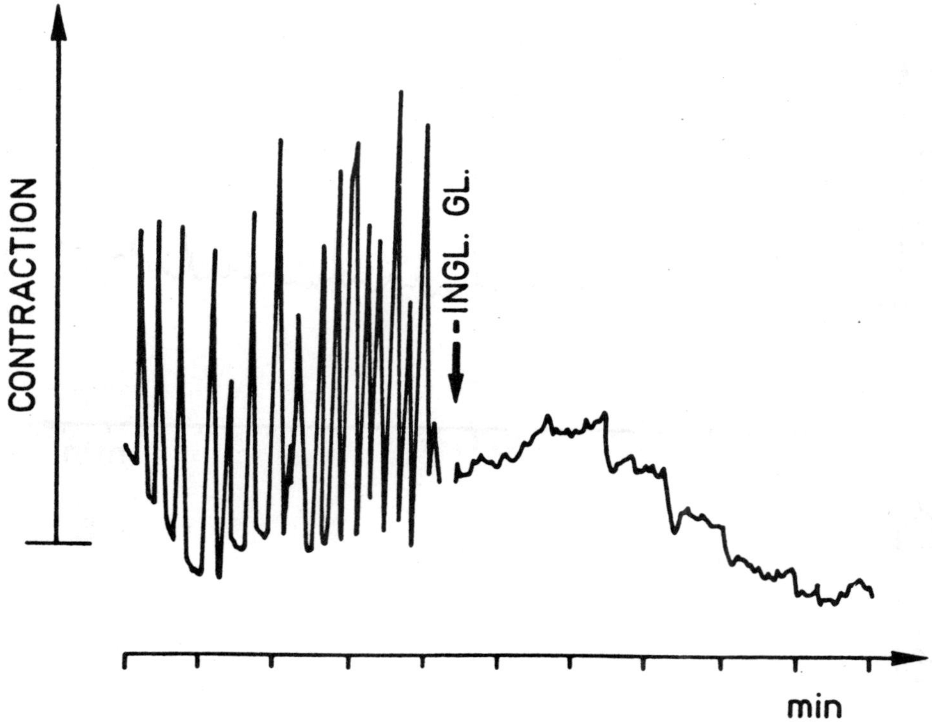

Fig. 3. Peristalsis of the innervated isolated midgut of a five-week-old adult male; the midgut shows spontaneous contractions (to the left) when isolated with the ingluvial ganglia in place (= innervated); the contractions cease when the ganglia are removed (= denervated, -INGL. GL.) (to the right)

on 'innervated' midgut preparations, i.e., with the ingluvial ganglia and ingluvial nerves to the gut muscularis left in place.

In a first set of experiments, adults were injected at day 1 with azadirachtin and sacrificed two and four weeks thereafter. The innervated midgut was isolated and monitored for peristalsis. Contractions were almost absent after two weeks and no longer demonstrable after four weeks of azadirachtin treatment (Fig. 4A and B).

Since Trumm (1991) has shown that serotonin has a strong stimulatory effect on midgut peristalsis by enhancing the frequency of the contractions and their amplitude, we applied serotonin to such preparations. Figure 5 demonstrates that after azadirachtin exposure over two weeks, serotonin could induce normal peristalsis which, however, faded again after repeated rinsing. If azadirachtin exposure lasted for four weeks, even high serotonin doses could hardly stimulate midgut peristalsis (Fig. 6).

In a second set of experiments, we exposed the isolated and innervated midgut of untreated locusts to varying doses of azadirachtin. Figure 7 shows that midgut peristalsis is inhibited in a dose-dependent manner; frequency as well as amplitude decreased. If azadirachtin and serotonin were applied simultaneously, the serotoninergic (= stimulatory) effects prevailed.

Immunological studies on the stomatogastric nervous system revealed that the system contains many serotoninergic elements. But serotoninergic perikarya were only

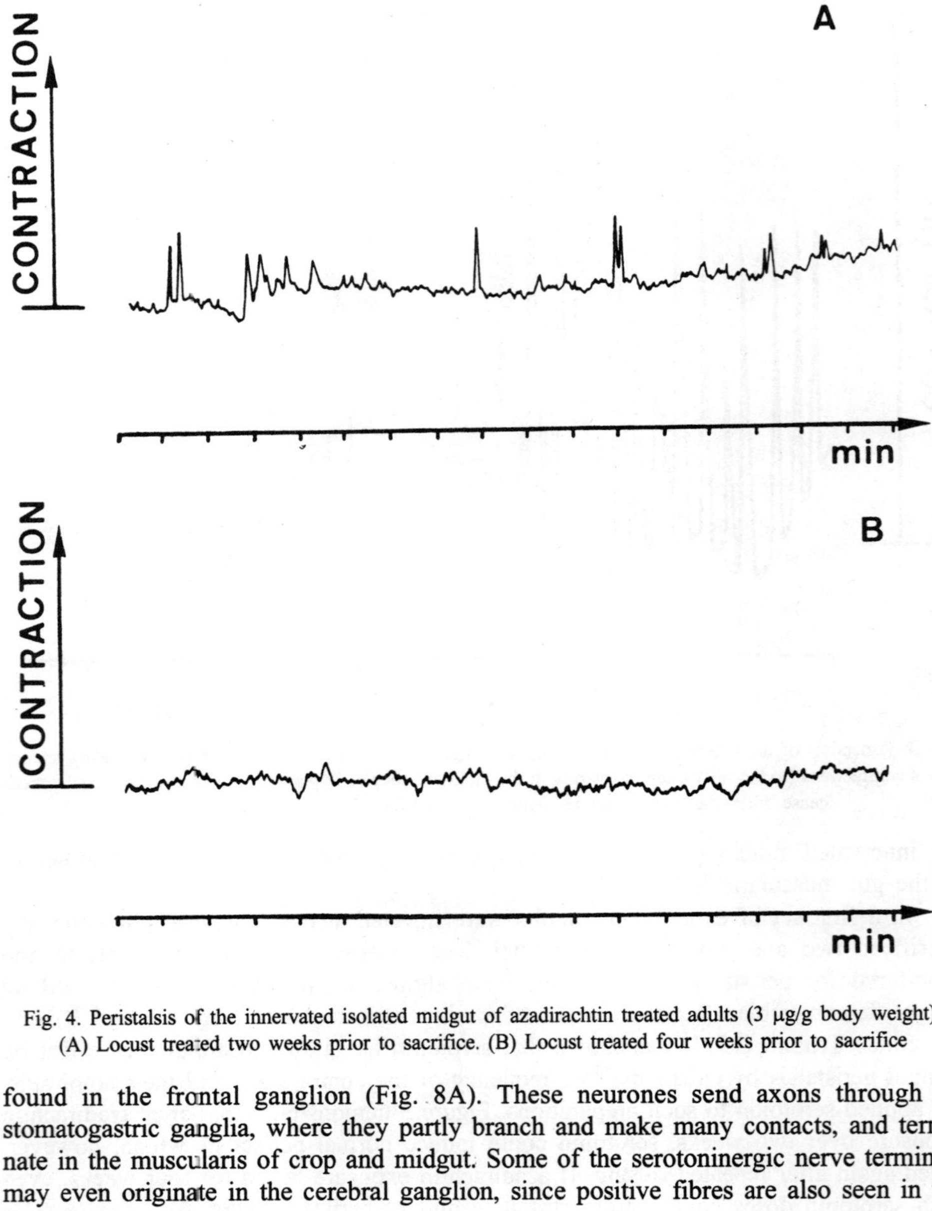

Fig. 4. Peristalsis of the innervated isolated midgut of azadirachtin treated adults (3 μg/g body weight); (A) Locust treated two weeks prior to sacrifice. (B) Locust treated four weeks prior to sacrifice

found in the frontal ganglion (Fig. 8A). These neurones send axons through all stomatogastric ganglia, where they partly branch and make many contacts, and terminate in the muscularis of crop and midgut. Some of the serotoninergic nerve terminals may even originate in the cerebral ganglion, since positive fibres are also seen in the frontal connectives.

Serotonin obviously plays a decisive role in the regulation of midgut peristalsis, and our experiments indicated that azadirachtin might interfere with this neurotransmitter. We therefore examined the immunological stainability of the serotoninergic perikarya in the frontal ganglion in relation to azadirachtin treatment. In treated larvae and adults, the stainability decreased—and in larvae perished completely—with the duration of the azadirachtin exposure after a single injection. Figures 8A and B give a comparison of

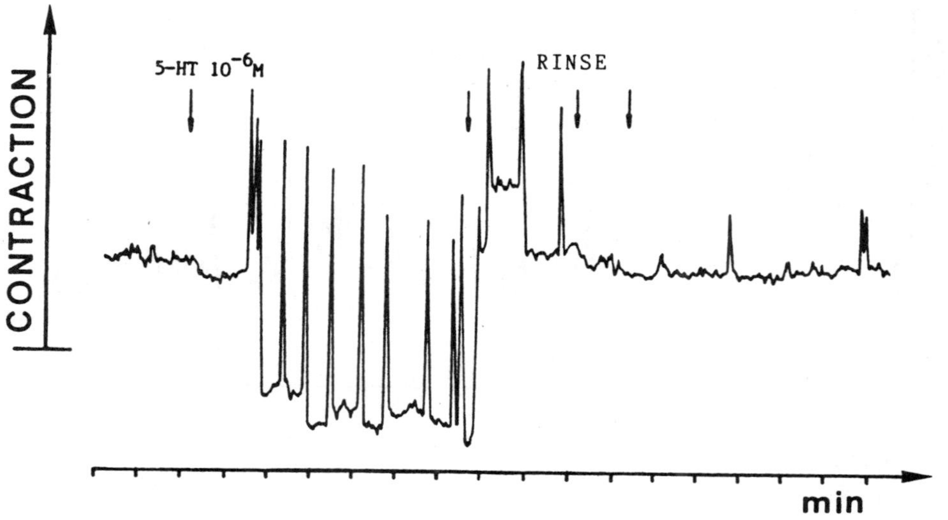

Fig. 5. Response to serotonin (5-HT 10^{-6}M) of the innervated isolated midgut from an adult treated with azadirachtin (3 μg/g body weight) two weeks prior to sacrifice; note the almost quiescent midgut responses with contractions typical for controls; after three rinses contractions almost cease again

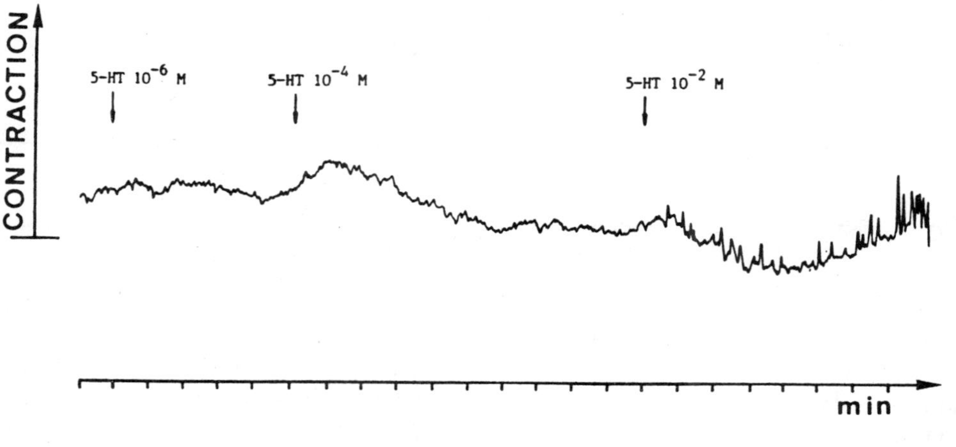

Fig. 6. Response to increasing doses of serotonin of an innervated isolated midgut from an adult treated with azadirachtin (3 μg/g body weight) four weeks prior to sacrifice; note that only an extremely high dose (10^{-2} M) is able to stimulate the quiescent midgut, but very slightly

serotoninergic perikarya in fifth-instar untreated larva and larva injected 12 days prior to preparation at day 13.

CONCLUSION

There is strong evidence that azadirachtin acts directly via haemolymph on the peristalsis of the alimentary canal. The inhibition is dose-dependent and increases with time, even after a single treatment. Azadirachtin may exert its effects via the neural

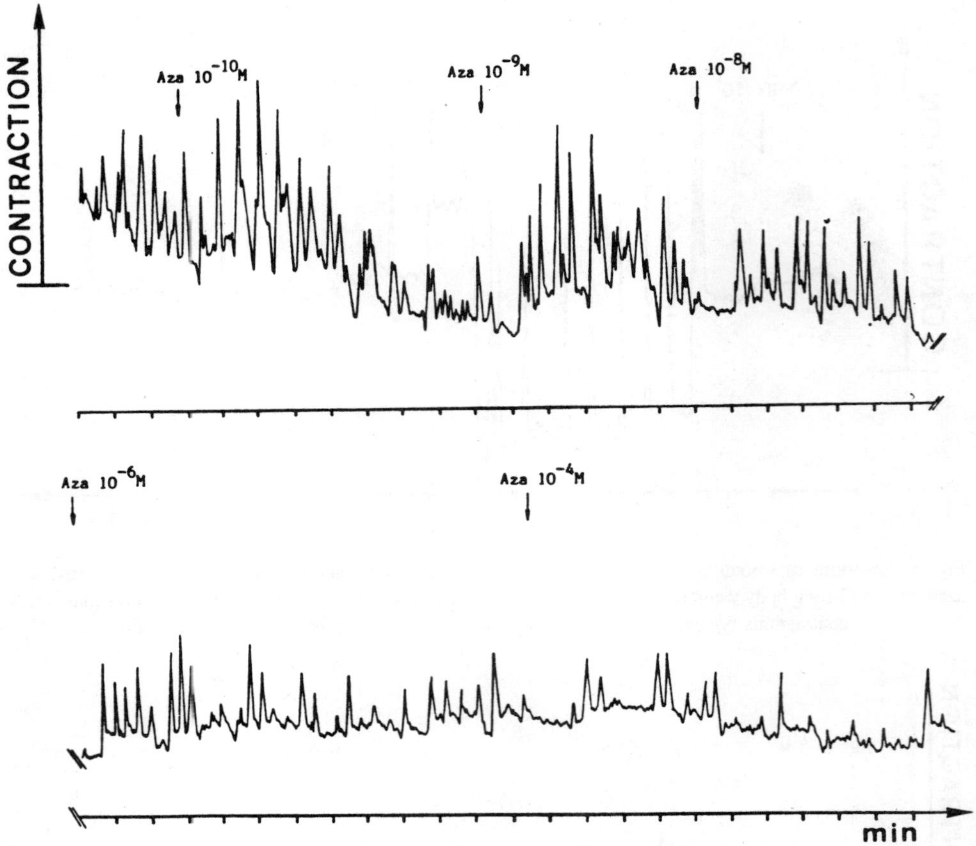

Fig. 7. Effect of azadirachtin administered *in vitro* to innervated midgut preparation from normal, untreated adults; note that increasing doses suppress the contractions in a dose-dependent manner

regulation of the gut peristalsis by the stomatogastric nervous system and probably also directly on the isolated innervated gut. The mode of action on the cellular and molecular level is not yet understood. However, there is clearly an impact on the serotoninergic system of the stomatogastric nervous system leading to the depletion of the serotoninergic perikarya in the frontal ganglion. Effects of azadirachtin on the serotonin pools in the neuroendocrine systems of locusts have been reported recently (Banerjee and Rembold, 1992). Since serotonin is involved in many neurally and hormonally regulated processes, the manifold effects of azadirachtin could be caused by interference with this system and probably also other transmitter hormone-like mediators (cf Bidmon et al., 1987).

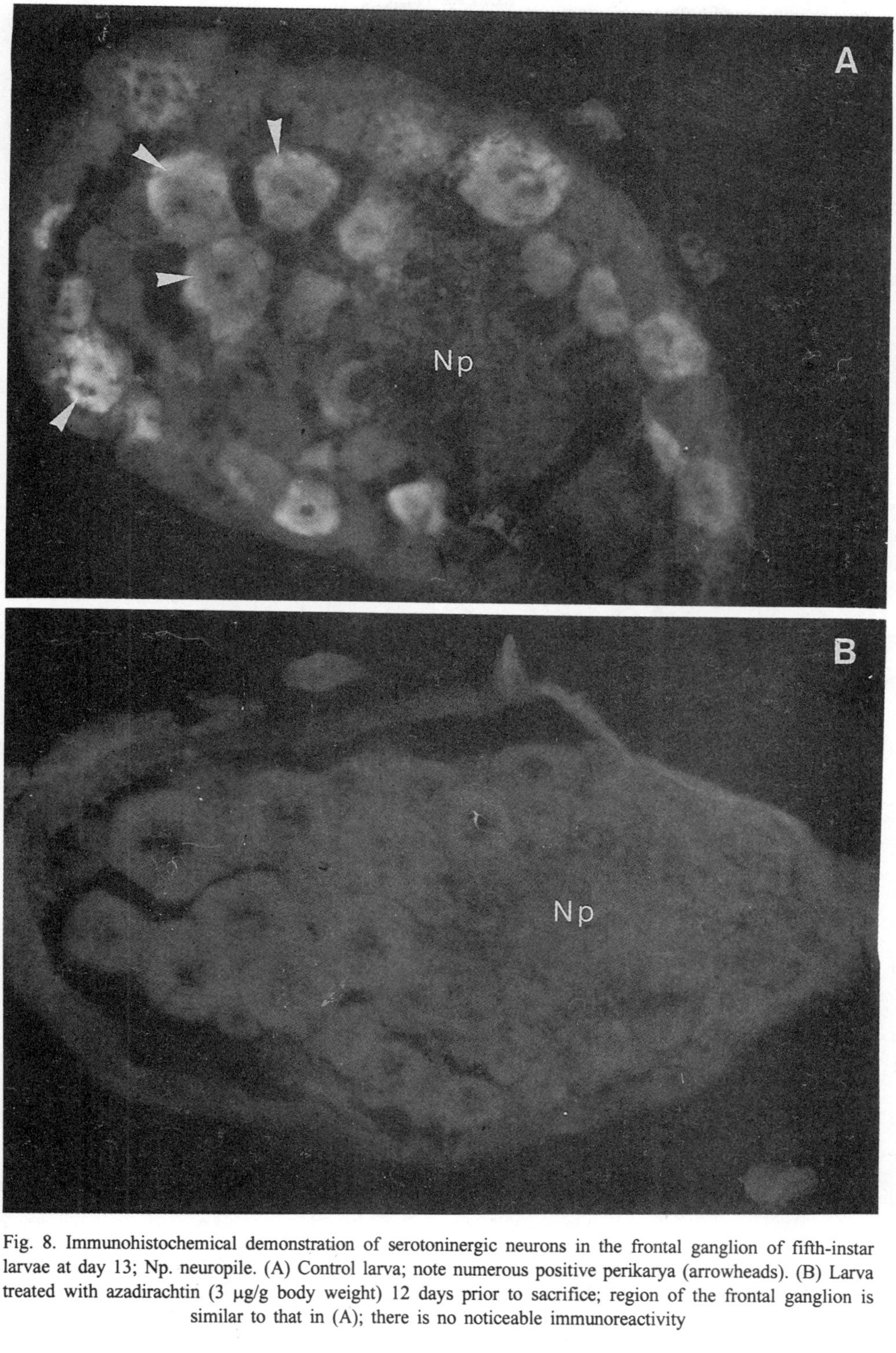

Fig. 8. Immunohistochemical demonstration of serotoninergic neurons in the frontal ganglion of fifth-instar larvae at day 13; Np. neuropile. (A) Control larva; note numerous positive perikarya (arrowheads). (B) Larva treated with azadirachtin (3 μg/g body weight) 12 days prior to sacrifice; region of the frontal ganglion is similar to that in (A); there is no noticeable immunoreactivity

REFERENCES

Banerjee, S. and H. Rembold. 1992. Azadirachtin–A interferes with control of serotoninergic pools in the neuroendocrine system of locusts. *Naturwissenschaften* 79: 81–84.

Bidmon, H.J., G. Käuser, P. Möbus and J. Koolman. 1987. Effect of azadirachtin on blowfly larvae and pupae. *Proc. 3rd Int. Neem Conf.*, Nairobi, 1986. GTZ, Eschborn, pp. 253–271.

Dorn. A., J.M. Rademacher and E. Sehn. 1987. Effects of azadirachtin on reproductive organs and fertility in the large milkweed bug, *Oncopeltus fasciatus. Proc. 3rd Int. Neem Conf.*, Nairobi, 1986. GTZ, Eschborn, pp. 273–288.

Mordue, A.J., P.K. Cottee and K.A. Evans. 1985. Azadirachtin: Its effects on the gut motility, growth and moulting in locusts. *Physiol. Entomol.* 10: 431–437.

Schmutterer, H., K.R.S. Ascher and H. Rembold eds. 1981. *Natural Pesticides from the Neem Tree. Proc. 1st Int. Neem Conf.* Rottach-Egern, 1980. GTZ, Eschborn.

Schoonhoven, L.M. 1981. Perception of azadirachtin by some lepidopterous larvae. *Proc. 1st Int. Neem Conf.*, Rottach-Egern, 1980. GTZ, Eschborn, pp. 105–108.

Sieber, K.P. and H. Rembold. 1983. The effects of azadirachtin on the endocrine control of moulting in *Locusta migratoria. J. Insect Physiol.* 29: 523–527.

Simmonds, M.S.J. and W.M. Blaney. 1984. Some neurophysiological effects of azadirachtin on lepidopterous larvae and their feeding response. *Proc. 2nd Int. Neem Conf.*, Rauischholzhausen, 1983. GTZ, Eschborn, pp. 163–180.

Trumm. P. 1991. Der Einfluss von Azadirachtin auf die nervöse Steuerung der Darmaktivität bei *Locusta migratoria*. Thesis, University of Mainz, Mainz, Germany.

Waldbauer, G.P. 1968. The consumption and utilisation of food by insects. *Adv. Insect Physiol.* 5: 229–282.

32

Fate of Azadirachtin in the Variegated Cutworm, *Peridroma saucia*

OPENDER KOUL

Regional Research Laboratory, Canal Road, Jammu Tawi, 180 001, India

ABSTRACT

This paper describes the fate and metabolism of azadirachtin in *Peridroma saucia* larvae, using tritiated 22, 23-dihydroazadirachtin (DAZ). The physiological and behavioural effects of DAZ and azadirachtin were quantitatively similar except that the former was slightly more active against *P. saucia* larvae.

After oral administration the candidate compound was gradually eliminated from the haemolymph within 72 hrs, whereas elimination was rapid following topical application. In the two treatments, 45.8 and 54.1 per cent respectively of labelled material was retained in the body with the gut and integument identified as major sites of label accumulation respectively. The variation in tissue distribution with respect to mode of application is discussed.

INTRODUCTION

The pest control efficacy and the biological properties of azadirachtin from the Indian neem tree, *Azadirachta indica* A. Juss, have been the subject of numerous studies worldwide and a substantial part of them have been compiled in recent volumes and reviews (Schmutterer and Ascher, 1987; Jacobson, 1988; Warthen, 1989; Schmutterer, 1990; Locke and Lawson, 1990; Koul et al., 1990; Koul, 1992).

The overall knowledge concerning the mechanism of azadirachtin metabolism and its fate in various tissues in insects is far from complete. Most of the studies have used *Locusta migratoria* and *Rhodnius prolixus* as test insects (Rembold et al., 1988; Subramanyam and Rembold, 1989; Garcia et al., 1989). Some preliminary study in a Lepidopteran, *Heliothis virescens* (Barnby et al., 1989), has also been done.

A deeper understanding of azadirachtin action on insects, therefore, requires more information on its distribution, clearance, and metabolism by the insects. The present study demonstrates the fate of the candidate compound in *Peridroma saucia* larvae in order to determine the incorporation of this compound into visceral tissues, as well as its degradation and excretion when administered via different routes.

APPLICATION OF [22, 23-3H_2] - DIHYDROAZADIRACHTIN

In the few insect species studied so far labelled DAZ has been either added to the diet (*R. prolixus*) or mostly administered via injections (*L. migratoria, H. virescens*). In the present study sixth-instar variegated cutworms were treated with 1.2 µg of DAZ either topically in acetone or as a single meal on cabbage leaf discs or via injection at the rate of 1.2 µg/insect using methods described earlier (Isman et al., 1990; Koul and Isman, 1991).

Larvae in each group were subjected to radioactivity determination at different intervals of time. Seventy-two hours after treatment insects were dissected and selected tissues chemically digested prior to liquid scintillation counting. Preparation of haemolymph samples, tissues, and excreta for measuring radioactivity along with metabolite separation methods have been described elsewhere (Koul et al., 1993).

FATE OF INGESTED DAZ

Azadirachtin and DAZ are quantitatively comparable in their physiological effects against *P. saucia* larvae (Fig. 1) with an EC_{50} of 0.27 ppm and 0.22 ppm respectively for neonates (first-instar larvae). This similarity in physiological effects has been shown for *Rhodnius* as well (Rembold et al., 1984; Garcia et al., 1989).

Ingested DAZ is absorbed by *P. saucia* larvae and passes into the haemolymph, wherefrom it is gradually eliminated for 72 hrs (Fig. 2). A large amount of label is recovered in the frass (51.6 per cent) and 47.1 per cent is retained by the same set of insects. A significant accumulation occurs in the digestive tract (24.0 per cent) and nearly 12.2 per cent is retained by the integument. Other tissues like head, gonads, and malpighian tubules also absorb the radioactive material (Fig. 3). These findings are, however, in contrast to *Rhodnius* (Garcia et al., 1989), where even six days after blood meal the haemolymph concentration of DAZ remains the same as at 24 hrs after oral treatment.

Storage of DAZ in various visceral tissues in the adult females of *Rhodnius* has also been observed, showing maximum retention in crop + intestine (39.7 per cent) and integument (18.4 per cent) after oral administration (calculated from Gàrcia et al., 1989). In this insect overall retention amounts to about 50.0 per cent which is comparable to the amount retained by *P. saucia* larvae.

FATE OF TOPICALLY APPLIED DAZ

After topical application of 1200 ng of DAZ to sixth-instar *P. saucia* larvae, the effects on ingestion or growth parameters were not significant at this level of treatment. However, delay in the development and pupation time was a characteristic feature.

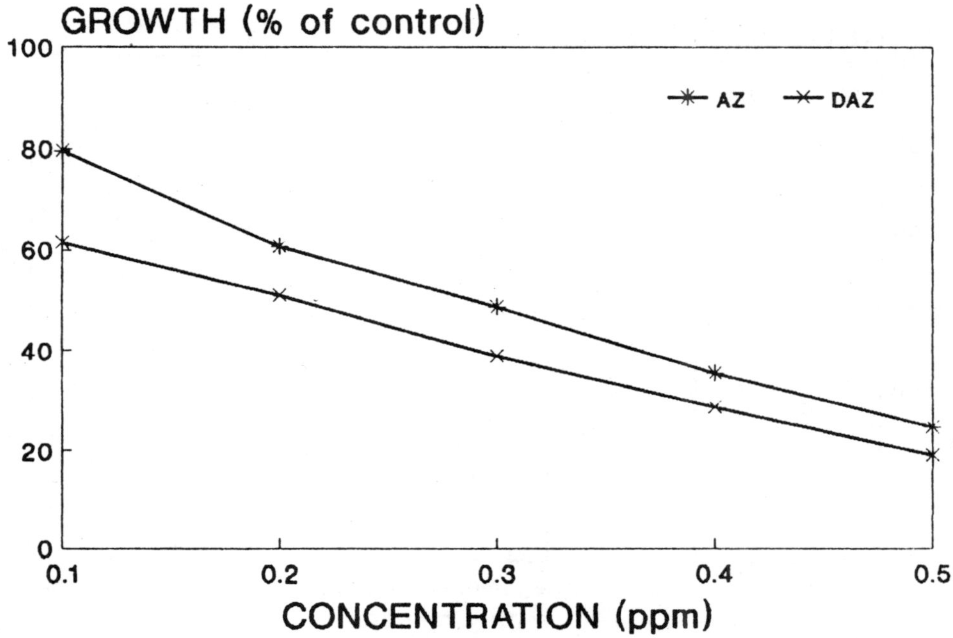

Fig.1. Effect on growth of *P. saucia* neonate larvae after azadirachtin and dihydroazadirachtin treatment in artificial diets

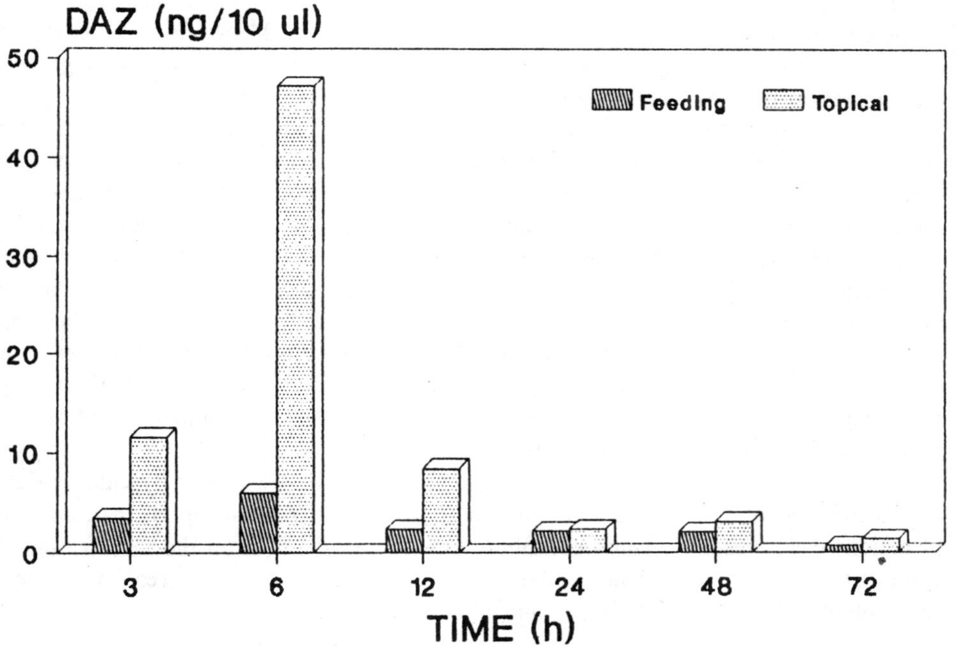

Fig. 2. Recovery of labelled dihydroazadirachtin in the haemolymph of sixth-instar larvae through 72 hrs after treatment after oral feeding and topical application

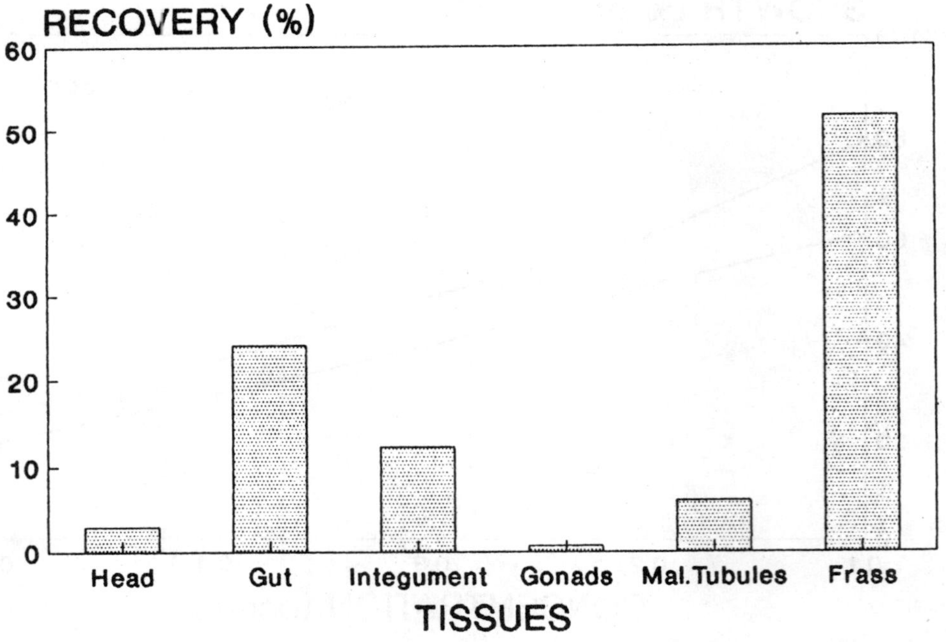

Fig. 3. Recovery of DAZ from various tissues and frass of sixth-instar larvae after oral administration (after 72 hrs post treatment; N = 10)

Recovery of radioactivity in the haemolymph showed a rapid increase up to 6 hrs, followed by a sharp decline and ultimate levelling off after 24 hrs (Fig. 2). This rapid elimination after topical application of DAZ is similar to the one shown in *L. migratoria* after injection of DAZ in the haemocoel of these insects (Rembold, 1984).

The labelled DAZ was detected in frass of *P. saucia* larvae with an increasing trend up to 48 hrs, whereafter recovery in frass and the retention in the body tissues was constant (particularly after 72 hrs). In these experiments, 41.8 per cent label was recovered from frass and 55.4 per cent was retained in the body tissues. The major depot of radiolable retention was the integument followed by gut, and collectively these two tissues accounted for 89.0 per cent of the retained radioactivity. As both tissues are extensively labelled whether the isotope is administered orally or topically, the compound must be widely distributed by haemolymph following treatment. However, much of the applied material is bound in the first tissue it encounters, namely the gut following oral treatment and the integument following topical treatment. Additionally, a considerable amount is absorbed by malpighian tubules (Figs. 3 and 4).

The present findings thus give a reasonable support to our earlier hypothesis (Koul et al., 1987, Koul and Isman, 1991) that it is likely that there are more than one sites of action of azadirachtin which contribute to the overall effect on the insects, and the action obviously is dependent on the route of administration that fixes its course of retention in different target tissues and elimination from the body.

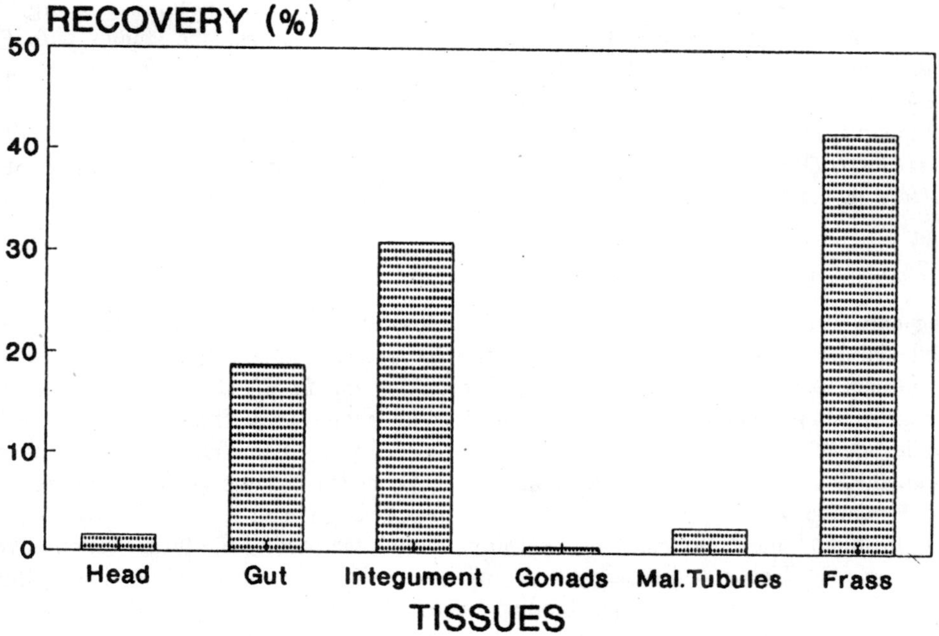

Fig. 4. Recovery of DAZ from various tissues and frass of sixth-instar larvae after topical application (after 72 hrs post treatment; N = 10)

FATE OF DAZ ADMINISTERED VIA INJECTIONS

When 1.2 μg of DAZ was injected into the haemocoel of sixth-instar *P. saucia* larvae, a significant growth inhibition (41.0 per cent) was observed against only 31.1 per cent after oral feeding of DAZ. This indicates that overall distribution of DAZ in injected insects is rapid and reaches the target sites faster than other modes of application. This is further evidenced by the fact that among injected insects more than 50.0 per cent mortality was observed (only 5.0 per cent in controls, N = 20) and it was difficult to ascertain the retention and elimination of DAZ from various tissues due to highly significant variations.

In *Locusta* (Rembold et al., 1988) and *Heliothis* (Barnby et al., 1989) also a rapid elimination has been observed after DAZ injections. Surprisingly, 75.0 per cent radioactivity has been detected in the faeces 1 hr after injection and only 25.0 per cent is retained in the body tissues. In *Locusta*, on the other hand, malpighian tubules have been shown to account for 73.7 per cent of the amount of DAZ which is retained in the whole body after injections (Rembold et al., 1988).

METABOLISM

So far, studies with DAZ have shown that pure urine of *R. prolixus* contained the unchanged labelled compound demonstrating the excretion of unmetabolized DAZ through the malpighian tubules (Garcia et al., 1989). In *Locusta* studies it has been inferred that DAZ acts in its unchanged form through a high-affinity binding to organ-specific membrane receptors (Rembold et al., 1988). In contrast, the studies with *P. saucia* larvae using reverse phase HPLC fractionation of frass extracts have shown the presence of single polar metabolite of DAZ (peak Rt = 2.13 min) against the parent compound (peak Rt = 8.62 min) A polar metabolite has also been indicated by TLC in *Heliothis* (Barnby et al., 1989). However, the metabolic degradation in *P. saucia* and its role in the bioactivity of azadirachtin deserve further investigation.

CONCLUSIONS

The data available today on the fate and metabolism of azadirachtin have provided evidence that DAZ at a physiologically effective dose is absorbed, stored, and excreted by insects like *P. saucia, H. virescens, R. prolixus,* and *L. migratoria* irrespective of the mode of application, i.e., via oral, topical, or injection routes. Though some of these studies show that labelled compound is retained or excreted in chemically unchanged form, there are indications of metabolism of DAZ at least in the gut of Lepidoptera.

Retention of DAZ in malpighian tubules is a characteristic feature of most of the insects studied and *P. saucia* is no exception, though the modes of administration have been different. These studies suggest the existence of high-affinity binding sites in this organ, but the influence of such binding on the excretion physiology of the tubules deserves further investigation (Garcia et al., 1989). Further investigation is also required in terms of the pharmacokinetics of the candidate compound in different insect orders, which will throw more light on its fate and the mechanism of metabolism in insects.

REFERENCES

Barnby, M.A., J.A. Klocke, M.V. Darlington and R.B. Yamasaki. 1989. Uptake, metabolism and excretion of injected tritiated 22, 23-dihydroazadirachtin in last instar larvae of *Heliothis virescens. Entomol. Exp. Appl.* 52: 1–6.

Garcia, E.S., B. Subrahmanyam, J.H. Buller and H. Rembold. 1989. Absorption, storage, organ distribution and excretion of dietary [22, 23-$^3H^2$]-dihydroazadirachtin—A in the blood feeding of *Rhodnius prolixus. J. Insect Physiol.* 35: 743–749.

Isman, M.B., O. Koul, A. Luczynski and J. Kaminski. 1990. Insecticidal and antifeedant bioactivities of neem oils and their relationship to azadirachtin content. *J. Agric. Food Chem.* 38: 1406–1411.

Jacobson, M. 1988. *Phytochemical Pesticides*, Vol I, *The Neem Tree.* CRC Press, Boca Raton, Florida.

Koul, O. 1992. Neem allelochemicals and insect control. In: *Allelopathy: Basic and Applied Aspects.* Eds. S.J.H. Rizvi and V. Rizvi. Chapman and Hall, London, pp. 389–413.

Koul, O., K. Amanai and T. Ohtaki. 1987. Effect of azadirachtin on the endocrine events of *Bombyx mori. J. Insect Physiol.* 33: 103–108.

Koul, O., M.B. Isman and C.M. Ketkar. 1990. Properties and uses of neem *Azadirachta indica* A. Juss. *Can. J. Bot.* 68: 1–11.

Koul, O. and M.B. Isman. 1991. Effect of azadirachtin on the dietary utilization and development of the variegated cutworm, *Peridroma saucia. J. Insect Physiol.* 37: 591–598.

Koul, O., M.B. Isman and J.T. Arnason. 1993. Toxicokinetics of [^{3}H]-dihydroazadirachtin in the variegated cutworm, *Peridroma saucia. Arch. Insect Biochem, Physiol.*

Koul, O., M.B. Isman and C.M. Ketkar. 1990. Properties and uses of neem *Azadirachta indica* A. Juss. *Can. J. Bot.* 68: 1–11.

Locke, J.C. and R.H. Lawson. 1990. Neem's potential in pest management programs. *Proc. USDA Neem Workshop*, ARS 86, USDA.

Rembold, H. 1984. Secondary plant products in insect control with special reference to azadiachtins. In: *Advances in Invertebrate Reproduction*, Vol. 3. Ed. W.E. Engels. Elsevier, Amsterdam, pp. 481–491.

Rembold, H., H. Forster, C. Czoppelt, P.J. Rao and K.P. Sieber. 1984. The azadirachtins, a group of insect growth regulators from the neem tree and other tropical plants. In: *Natural Pesticides from the Neem Tree and Other Tropical Plant,* Eds. H. Schmutterer and K.R.S. Ascher. GTZ, Eschborn, pp. 289–298.

Rembold, H., T. Muller and B. Subrahmanyam. 1988. Tissue specific incorporation of azadirachtin in the Malpighian tubules of *Locusta migratoria. Z. Naturforsch.* 43C: 903–907.

Schmutterer, H. 1990. Properties and potential of natural pesticides from the neem tree, *Azadirachta indica. Ann. Rev. Entomol.* 35: 271–297.

Schmutterer, H. and K.R.S. Ascher. 1987. *Natural Pesticides from the Neem Tree and Other Tropical Plants, Proc. 3rd Int. Neem Conf.*, Nairobi. GTZ, Eschborn.

Subrahmanyam, B. and H. Rembold. 1989. Effect of azadirachtin–A on neuroendocrine activity in *Locusta migratoria. Cell Tissue Res.* 256: 513–517.

Warthen, J.D., Jr. 1989. Neem (*Azadirachta indica* A. Juss.) organisms affected and reference list update. *Proc. Entomol. Soc. Wash.* 91: 367–388.

33

Effect of Neem Oil on Trehalase Activity in Accessory Reproductive Glands of the Male *Odontopus varicornis* (Dist.) Hemiptera: Pyrrhocoridae

N. SELVI SABHANAYAKAM, K. RAMALINGAM, SENTHAMIL SELVI AND R. KRISHNAKUMARI

Department of Zoology, Annamalai University
Annamalai Nagar 608 002, Tamil Nadu, India

ABSTRACT

Male adults of *Odontopus varicornis*, transfer sperms to the female via seminal fluid. This fluid contains the secretion of the accessory reproductive glands (MARGs) of the male bug. The trehalase activity of MARGS of the insect was estimated during different mating periods such as 0, 2, 4, 6 and 7 days after mating making use of the method of Sumida and Yamashita (1983). The total trehalase activity of this gland during 0, 2, 4 and 6 days before mating is found to be respectively 4, 100, 200 and 275 nmol/min, there by indicating an increasing trend of its activity. Its activity is estimated to be 100 nmol/min in insects after mating, suggesting its possible role in mating. The trehalase activity in MARGs during these periods of mating is then estimated in insects each injected with neem oil. The trehalase activity during 0, 2, 4 and 6 days (before mating) and its activity on 7th day after mating are found to be respectively 1, 25, 45, 60 and 15 nmol/min. This significant reduction in the amount of trehalase activity in treated insects suggests that neem oil appears to interfere with the synthetic activity of this enzyme.

INTRODUCTION

In many invertebrates, sperm is passed from male to female within a spermatophore or sperm sac or seminal fluid (Mann, 1984). In the red cotton bug, *O. varicornis* seminal fluid contains the secretion of the male accessory reproductive glands. After adult ecdysis, there is a rapid accumulation of proteins in this gland involved in the formation of spermatophore (Happ et al., 1982; Shinbo et al., 1987), one major source of energy might be trehalase in the blood (Wyatt, 1967). MARGs might be expected to contain trehalase (α-glucoside-1-glucohydralase). In fact, high trehalase activity has been reported from the accessory glands of male American cockroach (Ogiso and Takahanshi, 1984; Ogiso et al., 1985). Further the neem oil has been shown to inhibit spermatozoal motility (Riar et al., 1990). The present investigation aims at finding out the trehalase activity of the normal and treated insects.

MATERIALS AND METHODS

The adults of *O. varicornis* were collected from their natural habitat. They were brought to the laboratory and maintained in insect cages. They were provided with *Sechium edule* (Chou-Chou) as their food. These adult insects during different mating periods such as 0, 2, 4 and 6 days, before mating and 7th day after mating were sacrificed and used for the estimation of trehalase activity in normal and experimental insects, each injected with 0. 01 ml of neem oil in the abdomen by a micro syringe fitted with 26 gauge needle. After 24 hours of treatment, the insects were sacrificed and used for enzyme assay. Distilled water injected insects served as controls.

Enzyme Preparation

MARGs were dissected from adults in ice-cold phosphate buffered saline (PBS 68.2 mM Na_2 HPo_4, 1.5 mM K_2HPo_4, 137 mM Nacl and 26 mM Kcl, pH 7.4) and blotted on filter paper (Shinbo et al., 1987). The glands were homogenized in 200 volumes of 25 mM NaH_2 Po_4-Na_2 HPo_4 buffer (pH 6.8 phosphate buffer) using a small glass homogenizer. The crude homogenate was used as an enzyme source.

Assay of Trehalase Activity

Conditions for routine enzyme assays were established from kinetic experiments as suggested by Sumida and Yamashita (1983). The standard reaction mixture contained: (1) 100 μl enzyme source (ca, 30–70 g protein and (2) 200 μl substrate solution (50 mM trehalase, 100 mM citrate Na_2 HPo_4. buffer, pH 5.6). The final reaction mixture (300μl) was at pH 5.75. The reaction was allowed to proceed at 30°C for 20 min and then stopped by boiling for 5 min. After cooling, 0.5 ml of distilled water was added and the solution was then spun at 1000 g for 15 minutes.

An aliquot of the supernatant was used for determination of glucose released from trehalase. Glucose content was determined by glucose oxidase (Hugget and Nixon, 1957) methods. Enzyme activity was expressed as μ mole or nmole of glucose formed per minute.

RESULTS

The kinetics of the trehalase enzyme in MARGs of normal and neem oil treated insects of 0 day, 2 days, 4 days, 6 days (before mating) and on 7 days of old adult insects are compared. The present observations indicate that the trehalase activity increases during the mating period of 0, 2, 4 and 6 days of adult male insects. The trehalase activity in MARGs homogenate is found to be 4, 100, 200 and 275 nmol/min. The activity of this enzyme in insects after mating has decreased significantly to about 100 nmol/min. (Table 1). The trehalase activity in MARGs homogenate of neem oil treated insect is found to be 1, 25, 45, 60 and 15 nmol/min in before and after mating. The activity of this enzyme in neem oil treated insect has decreased significantly (Fig. 1).

DISCUSSION

As a general rule, there are two types of trehalases in insects namely, 'soluble gut' type

Table 1. Trehalae activity in male accessory reproductive glands of normal and neem oil treated *O. varicornis*

Tissue	Age	Control (nmol/min)	Neem oil treated (nmol/min)
MARGs	0	4 ± 0.01	1± 0.03
Before mating	2	100 ± 0.31	25 ± 0.39
	4	200 ± 0.16	45 ± 0.53
	6	275 ± 0.31	60 ± 0.62
After mating	7	100 ± 0.01	15 ± 0.25

Mean of three individual observations ± S.D.

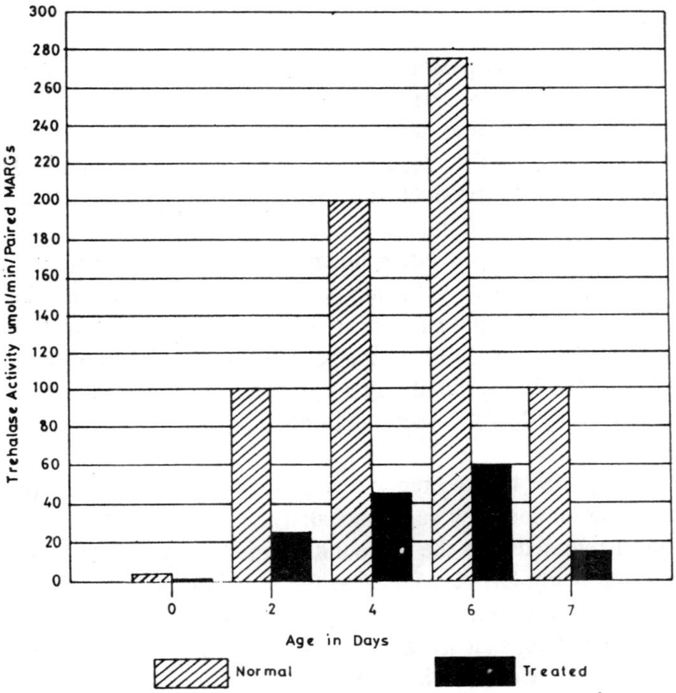

Fig. 1. Changes in Trehalase activity of MARGs per animal of normal and neem oil injected insects

and a 'peptide bound muscle' type which differ in their kinetic properties (Wyatt, 1967). Trehalase activity has also been found in muscles of *Melanoplus differentialis* (Derr and Randall, 1966) and high trehalase activity has been reported in the thoracic muscles of *Valanga nigricornis* and it was attributed to the function of flight (Teo and Heng, 1987). Other secretory trehalase activities have been reported from the salivary glands of *Chironomus thummi* (Laufer et al., 1963) and cocoon of *Bombyx mori* (Shimada et al., 1980).

In adult *O. varicornis*, histological observations on this gland show that it increases in its cell volume probably due to accumulation of proteins and other components (Selvi Sabhanayakam unpublished data). In control insects, the trehalase activity has also increased in relation to the development of MARGs during 0 to 6 days before mating period, followed by a reduction on 7th day after mating. In the neem oil treated insects, it has been reported that the trehalase activity has been reduced significantly from 0 to 6 days before mating period followed by a further reduction in trehalase activity during 7th day after mating.

It is evident from these observations that the enzyme trehalase appears to be transferred from the male to the female during the period of copulation. In several other instances, enzymes have been shown to have been transferred from male to female via a spermatophore or seminal fluids. The enzyme estrease is transferred to female *Drosophila melanogaster* in an ejaculate from the male species of this insect (Sheehan et al., 1979). Loher et al., (1981) have reported that prostaglandin synthetase is passed to the female in two crickets, *Acheta domesticus* and *Teleogryllus commodus*. In *T. commodus*, the prostaglandin synthetase originates from the testes (Tobe and Loher, 1983).

In *Bombyx mori*, arginase is passed from the male to the female as the spermatophore is formed within bursa copulatrix (Osanai et al., 1986). In this spermatophore, the arginase converts arginine into urea and ornithine. Ornithine then reacts with pyruvate derived from glycolysis, to produce alanine and 2-oxoglutarate. The latter is the preferred substrate for ATP production in the sperm (Osanai et al., 1987). Ogiso and Takahashi (1984) have noted that the male accessory glands of the American cockroach contains a trehalase of unknown functions. Happ and Yaginuma (1988) have reported that both trehalase and its substrate are found in MARGs and the spermatophore in *Tenebrio molitor*. Trehalase has been reported in the seminal fluids of other insects. Several small sugars, notably fructose have been reported to occur in mammals (Rodger, 1976). In insects, the seminal sugars are often mixtures (Tauber, 1977) and according to Happ et al. (1982) the seminal fluid of *T. molitor* contains glucose and trehalase but not fructose.

Gadzama and Happ (1974) have reported that trehalase plays some role in sperm nourishment and activation and they suggested that the trehalase of the spermatophore is soluble one and is released into the surrounding medium before the rupture of the sperm sac. The present study has revealed that the trehalase activity of control insects has increased gradully during premating period followed by a significant decrease in post mating period suggesting that trehalase is being utilized as an energy source for nourishment, motility and activation of sperms. On the other hand, insects treated with neem oil have shown the presence of a low level of trehalase activity in MARGs. It may be inferred from these observations that the neem oil seem to affect the synthetic activity of this enzyme as well as motility and activation of sperms. These findings are consistent with those of Riar et al. (1990) and Ashok Purohit and Dixit (1991) who have reported that the neem oil inhibits spermatozoal motility in rat and human.

ACKNOWLEDGEMENTS

The authors are grateful to Dr. N. Ramalingam, Professor and Head, Department of Zoology, Annamalai University for permission to publish this paper and for providing all facilities. Laboratory and Technical assistance for this work were provided by the National Institute of Nutrition, Government of India, Hyderabad, which is gratefully acknowledged.

REFERENCES

Ashok Purohit and V.P. Dixit. 1991. *Neem Newsletter* April–June, 1991. 8(2).

Derr R.F. and D.D. Randall. 1966. *J. Insect Physiol.* 12: 1105.

Gadzama, N.M. and G.M. Happ. 1974. The structure and evacuation of the spermatophore of *Tenebrio molitor* L. (Coleoptera: Tenebrionidae). *Tissue. Cell* 6: 95–108.

Happ G.M., C. Yuncker and P.J. Dailey. 1982. Cytodifferentiation in the accessory glands of *Tenebrio molitor.* VII Patterns of leucine incorporation by the bean shaped glands of males. *J. Expzool.* 220: 81–92.

Happ, G.M. and T.C. Yaginuma. 1988. Trehalase in accessory glands and spermatophores of mealworm beetle *Tenebrio molitor. J. Comp. Physiol.* 157: 765–770.

Hugget, A.S.G. and D.A. Nixon. 1957. Enzymatic determination of blood glucose. *Biochem J.* 66: 12.

Laufer, H., Y. Nagase and J. Vanberberg. 1963. Nucleo cytoplasmic interactions in the development of the salivary glands of *Chironomus thummi. Biol. Bull.* Mar Biol. Lab Woods Hole.

Loher, W., I.G. Ganyian, D. Kubol, Stanley-Samulson and S.S. Tobe. 1981. Prostaglandins: their role in egg-laying of the cricket, *Proc. Natl. Acad. Sci.* USA 78: 7835–7838.

Mann, T. 1984. Spermatophores. Development, structure, bio-chemical attributes and role in the transfer of spermatozoa, Springer, Berlin Heidelberg, New York, p. 217.

Osanai, M., T. Aigaki and H. Kasuga. 1987. Energy metabalism in the spermatophore of the silk moth, *Bombyx mori*, associated with accumulation of alanine derived from arginine. *Insect Biochem.* 17: 71–75.

Osanai, M., T. Aigaki, H. Kasuga and Yonezaway. 1986. Role of arginase transferred from the residue seminelis during mating and changes in amino acid pools of the spermatophore after ejaculation in silkworm, *Bombyx mori. Insect Biochem.* 16: 879–885.

Ogiso, M. and S.Y. Takahashi. 1984. Trehalases from the male accessory glands of the American cockroach, developmental changes and the humoral regulation of the enzymes. *Gen. Comp. Endocrinol.* 55: 387–392.

Ogiso M., Y. Shinohara, K. Hananoka, T. Kageyama and S.Y. Takahashi. 1985. Further purification and characterization of trehalaes from American cockroach, *Periplaneta americana. J. Comp. Physiol.* B155: 553–560.

Riar, S.S., C. Devakumar and G. Ilavazhagan. 1990. *Controception* 42(4): 479–487.

Rodger, J.C. 1976. Comparative aspects of the accessory sex glands and seminal biochemistry of mammals. *Comp. Biochem. Physiol.* 53-B: 1–8

Sheehan, K., R.C. Richmond and J.C. Cochrene. 1979. Studies of esterase 6 in *Drosophila melanogaster*. III. The developmental pattern and tissue distribution. *Insect Biochem.* 9: 443–450.

Shimada, S., A. Kamada, and S.C. Asano. 1980. The cocoon trehalase of the silkworm, *Bombyx mori. Insect Biochem.* 10: 49–52.

Shinbo, H., T. Yaginuma and G.M. Happ. 1987. Purification and characterization of a proline-rich secretory protein that is a precursor to a structural protein of an insect spermatophore. *J. Biol. Chem.* 262: 4794–4799.

Sumida, M. and O. Yamashita. 1983. Purification and some properties of soluble trehalase from midgut of pharate adult of the silkworm, *Bombyx mori. Insect Biochem.* 13: 257–265.

Tabuer, S.C. 1977. Semen of *Apis mellifera*. Fertility, chemical and physical characteristics. In: *Advances in Invertebrate Reproduction*, Eds. K.G. Adiyodi and R.G. Adiyodi, Vol I, John Wiley.

Tobe, S.S. and W.R. Loher. 1983. Properties of the prostaglandin synthetase complex in the cricket *Teleogryllus commodus. Insect Biochem.* B: 137–141.

Teo, L.H. and S.K. Heng. 1987. *Comp. Biochem. Physiol.* 87B: 373.

Wyatt, G.R. 1967. The Biochemistry of sugars and polysaccharides in insect. *Adv. Insect Physiol.* 4: 287–360.

34

Effect of Neem Kernel Extract and Neem Oil on Nutritive and Reproductive Physiology of *Heliothis armigera* Hub.

K. MURUGAN,* P. JAHANMOHINI AND R. BABU

Division of Entomology, Department of Zoology, Bharathiar University
Coimbatore 641 046, India

ABSTRACT

Neem kernel extract and neem oil are found to interfere with the food utilization and biochemical and enzyme profiles related to nutritive and reproductive physiology of *Heliothis armigera*. The amount of food consumed and the efficiency of ingestion and digestion of food were markedly inhibited by the neem products. Moths obtained from treated larvae failed to mature their oocytes, probably as a result of interference of azadirachtin with vitellogenin synthesis and/or its uptake by developing oocytes. Closer examination revealed decreased protein levels as well as fat body development in the larvae; enzymes involved in lipogenesis were suppressed by neem products in male and female reproductive tissues. Prospects for field control of this pest with neem are discussed in the light of these findings.

INTRODUCTION

Pesticide use has resulted in unprecedented chemical pollution (Matsumura, 1975) affecting non-target organisms (Pimental, 1971; Ware, 1980). The frequent and indiscriminate use of these chemicals may also have many undesirable consequences (Dunigan et al., 1972). Development of a safe insecticidal agent of plant origin has been the subject of intensive research efforts in recent years. The high biological activity of azadirachtin, a tetranortriterpenoid from seed kernels of the neem tree, *Azadirachta indica*, against insects is now well established. Though a strong antifeedant to the

* To whom all correspondence should be addressed.

locusts (Butterworth and Morgan, 1971), it also acts as a potent growth inhibitor when injected at microgram levels (Rembold and Sieber, 1981; Sieber and Rembold, 1983; Mordue et al., 1986). The most defined effects are (1) delay and/or inhibition of moult into the successive instar, (2) disturbance of the moulting process, and (3) delay, disturbance, or inhibition of ovarian development. Several studies have been made during the recent years to elucidate modifications in the endocrine mechanism induced by azadirachtin treatment that lead to the observed growth inhibiting effects.

Anti-reproductive effects of extracts from *A. indica* A. Juss. have been reported in a number of insects (see Schmutterer and Ascher, 1984). Oviposition deterrence was reported by Saxena and Rembold (1984) and Chiu et al. (1985), reduction in the fecundity by Ascher et al. (1984), and direct ovicidal action by Gaaboub and Hayes (1984). Azadirachtin also inhibits egg development of adult female *Locusta migratoria* (Rembold and Sieber, 1981). Chemical azadirachtin inhibits the synthesis (Sieber and Rembold, 1983) and/or release of ecdysone in some insects (Rembold, 1984). A direct effect of azadirachtin on ecdysone synthesis in *Bombyx mori* was demonstrated by Koul et al. (1987). The physiological events in the insect that lead to these observed effects have not, however, been fully examined and a number of them have been controversial.

Reproductive strategies of insects involve a wide range of complexities, various neural mechanisms, and accessory structures and secretions that are involved in the location of receptive partners and release of copulatory behaviour. The effect of neem and the role of hormones in the regulation of growth and development have been extensively studied, but many of the features of the 'fine tuning' of feeding strategies in relation to biochemical changes during the reproductive programming in pest species, particularly of lepidopteran pest species, have not been discussed.

The present study is an attempt to investigate the probable inhibitory role of neem seed kernel extract (NSKE) and neem oil on feeding and reproduction in relation to physiology of *Heliothis armigera* Hub. It is hoped that the study will provide background information for the development procedures which can disturb the biochemical machinery that operates during gonodotrophic period of male and female for the successful control of this pest at its reproductive stage.

MATERIALS AND METHODS

Mass Culture and Biological Studies

For mass culturing of insects in the laboratory, *H. armigera* was initially collected from the cotton field and cultured in cages measuring 25 × 25 × 30 cm. Mating pairs were isolated from the cages and provided a mud pot covered with muslin cloth for oviposition. Newly hatched, first-instar larvae were transferred into the artificial diet until fourth instar.

Preparation of Extracts

Fifty grams of neem seed kernel (NSK) powder was dissolved in one litre of water, left to stand for 24 hrs, and filtered through linen cloth to get neem seed kernel extract

(NSKE). The filtrate was used as a stock solution. The required concentration was prepared from the stock solution of five millilitres of neem oil mixed with 1000 ml water used as stock solution, from which the required concentration was prepared.

Treatment

The fifth-instar larvae were starved for 3 hrs and placed individually in plastic containers. Cotton leaves were dipped in 1 per cent NSKE and 0.5 per cent of neem oil, air dried and the larvae were allowed to feed until pupation. Pupae were collected, sexed and held in fresh pots to obtain viable adults which were then used for the fecundity studies and biochemical analysis.

Quantitative Food Utilization

Food utilization experiment was conducted only in the fifth instar. A gravimetric technique was used to determine weight gain, food consumption, and faeces produced; all weights were done using a monopan balance accurate to 0.1 mg. After the initial weight of the larvae was taken, they were introduced into separate containers and allowed to feed on weighed quantity of control and treated (1 per cent NSKE and 0.5 per cent neem oil coated) leaves of *G. hirsutum* for 24 hrs. Larvae were then weighed again and the difference in weight of the larvae gave the fresh weight gained during the period of study. The leaves remaining at the end of each day were oven dried and reweighed to establish a percentage dry weight conversion value to allow estimation of the dry weight of the diet given to the larvae. The quantity of food ingested was estimated by substracting the diet (dry weight) remaining at the end of each experiment from the total dry weight of diet provided. Faeces were collected daily and weighed, then oven dried and reweighed to estimate the dry weight of excreta. Consumption, growth rates, and post-ingestive food utilization efficiencies (all based on dry weight) were calculated in the traditional manner (Waldbauer, 1968; Slansky, 1985; Slansky and Scriber, 1985).

Efficiency of Digestion

To determine the efficiency of digestion on *H. armigera* larvae fed with cotton leaves coated with 1 per cent NSKE and 0.5 per cent neem oil (reared in separate containers), the time taken by the larvae to egest the first faecal pellet was noted in respect to the control and treated leaves.

Reproduction Studies

Freshly emerged adult male and female pairs which were formerly reared on the treated leaves were confined in mating cages and fed on sucrose solution. Mating pairs were isolated from the cage and provided separate mud pots covered with muslin cloth for oviposition. Fecundity was assessed on the basis of the pre-oviposition period and the number of eggs laid during the lifetime. The mean longevity of males and females which were reared on different treated leaves was also noted. Reproduction per day was calculated by dividing the fecundity by the reproductive period.

Biochemical Assays

Newly emerged adults of *H. armigera*, both control and experimental insects, were separated and allowed to mate. Reproductive organs, fat bodies, and gut were carefully removed and stored (deep frozen) for biochemical studies. Quantitative biochemical estimations were carried out for the total protein (Lowry et al., 1951), glycogen (Good et al., 1933), carbohydrates (Dubois et al., 1956), and phospholipid (Folch et al., 1957).

Enzyme Assays

The activities of protease, amylase and lipase were analysed by the method of Anson (1937), Bernfeld (1955), and Seligman and Kramer (1965) respectively.

The activity levels of malate dehydrogenase (MDH) (Ochoa, 1955), total lactate dehydrogenase (LDH) (King, 1965), acid phosphatase (ACP), and alkaline phosphatase (ALP) (Andersch and Szezypinski, 1947) were analysed in different reproductive tissues.

RESULTS

Effects on Reproductive Parameters

Comparative analysis of reproductive performances of *H. armigera* after the NSKE and neem oil treatment showed variations. The pre-reproductive and post-reproductive period and hatchability were lesser in treatment groups than in control; reproduction per day and average fecundity were also higher in control than in experimental groups (Table 1).

Table 1. Reproductive parameters of *H. armigera* treated with NSKE and neem oil

Biological parameters	Control	1% NSKE	0.5% Neem oil
Pre-reproductive period (days)	3.58 ± 0.066	2.78 ± 0.084***	2.12 ± 0.052***
Reproductive period (days)	6.06 ± 0.059	4.32 ± 0.038***	3.86 ± 0.026***
Post-reproductive period (days)	2.94 ± 0.017	2.13 ± 0.081***	1.89 ± 0.106***
Adult longevity (days)	13.94 ± 0.624	9.81 ± 0.285***	7.64 ± 0.447***
Reproduction per day	96.00 ± 5.69	84.7 ± 5.18	64.7 ± 3.17**
Average fecundity per generation (no. of eggs)	582.00 ±17.53	366.00 ± 9.70***	249.8 ± 10.01***
Hatchability (%)	87.8 ± 2.29	80.6 ± 1.97*	77.6 ± 2.22*

Each value is mean ± SEM of five replicates.
*P < 0.05, **P < 0.01, and ***P < 0.001

Effects on Quantitative Food Utilization

Data on the comparative food utilization of fifth instar of *H. armigera* are provided in Table 2. The consumption index, relative growth rate, and approximate digestibility were higher in the control than in NSKE and neem oil treatment. Larvae fed on leaves treated with NSKE and neem oil had higher efficiency of ingestion (ECI) and efficiency of digestion (ECD) values than control insects.

Effects on Efficiency of Digestion and Digestive Enzymes

Newly moulted final-instar larvae were segregated and were fed separately with control leaves and cotton leaves coated with NSKE and neem oil. Experimental larvae excreted fewer faecal pellets and had longer egestion period than control (Fig. 1). All the three digestive enzymes—protease, amylase, and lipase—were significantly inhibited in the midgut after the NSKE and neem oil treatment compared to control (Fig. 2).

Effects on Protein, Glycogen, Lipid and Enzyme Profiles in Subsequent Female Moths

Assays of protein, glycogen, and lipid profiles of fat body, ovary, oviduct, and spermatheca showed that at emergence, the levels were comparable in both treated and untreated insects. Activities of ACP, ALP, MDH, LDH and ATPase were adversely affected in both NSKE and neem oil treatments (Tables 3 and 4).

Effect on Protein, Glycogen, Lipid and Enzyme Profiles in Subsequent Male Moths

Tables 5 and 6 summarized the effects of NSKE and neem oil on protein, glycogen, and lipid of fat body, testis, seminal vesicle, and accessory reproductive glands of *H. armigera* respectively. The NSKE and neem oil treatment affected the protein, glycogen, and lipid profiles as well as caused considerable decrease in the profiles of ACP, ALP, MDH, and LDH, and in ATPase activities in the testis, seminal vesicle, and accessory glands of male insects with increase in the activity of glycogen level. The NSKE treatment caused marked reduction in spermatozoa and particularly affected the activities of glycolytic enzymes, LDH, and ATPase.

DISCUSSION

Food preference in lepidopteran larvae is known to be mostly influenced by physical, chemical, and biotic factors resulting in considerable variation. The choice of the food plant is again intricately associated with the diverse aspects of the feeding behaviour with considerable effect on the reproductive potential of the species (Slansky and Scriber, 1985; Slansky, 1992; Murugan and Ancy George, 1992). The utilization of a host plant by an insect is determined by its capacity to ingest, assimilate and convert the food into body tissues (Scriber and Slansky, 1981). In many insects, the development of the reproductive system is dependent on nutrients stored during larval feeding

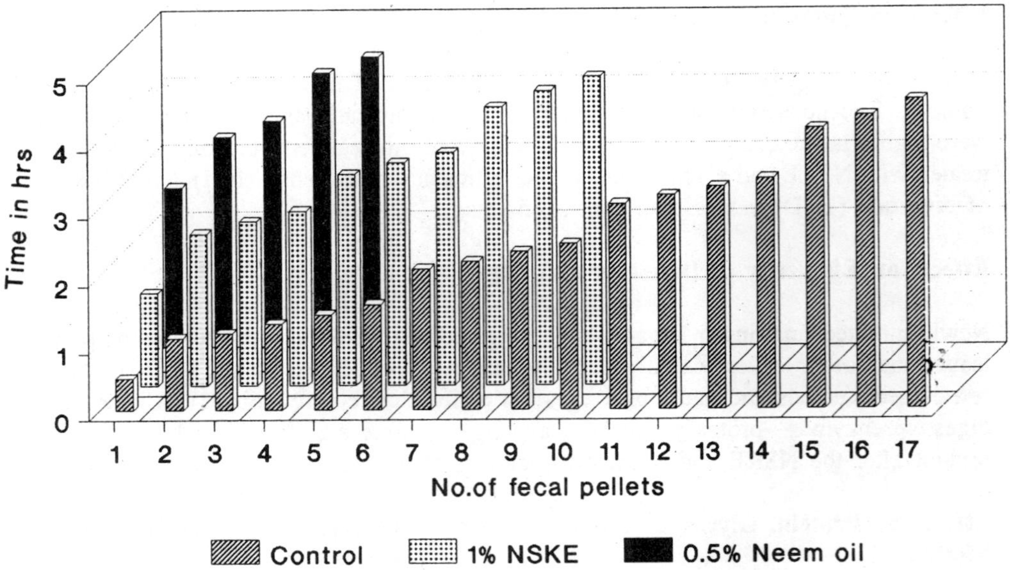

Fig. 1. Feeding efficiency of *H. armigera* treated with NSKE and neem oil

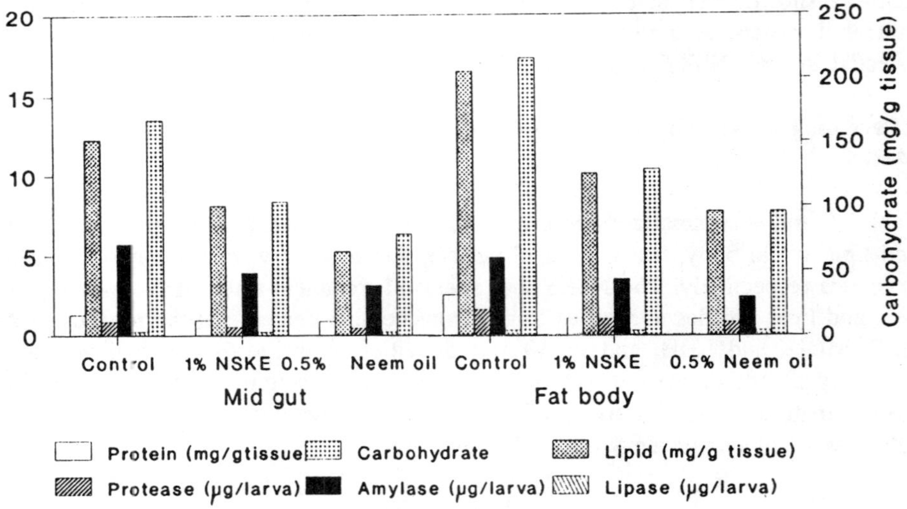

Fig. 2. Digestive enzyme activities of *H. armigera* treated with NSKE and neem oil

as well as those ingested by the adult (Johansson, 1964). The food utilization pattern of *H. armigera* showed considerable variation in NSKE and neem oil treatment compared with control insects.

In the present study, *H. armigera* treated with NSKE and neem oil were greatly affected in nutritional indices as well as digestive enzyme profiles. The neem treatment affected the rate of food utilization and also the efficiency process. The approximate digestibility estimates the proportion of ingested food that is digested and absorbed. So

Table 2. Quantitative food utilization of fifth instar of *H. armigera* treated with NSKE and neem oil

Biological parameters	Control	1% NSKE	0.5% Neem oil
Food intake (mg)	3849 ±226	2943 ±130.4**	1476 ± 50.99***
Faecal matter (mg)	2586 ±110.23	1965 ± 98.40**	898 ± 32.80***
Food digested (mg)	1263 ± 51.18	978 ± 34.56**	578 ± 35.09***
Body weight gain (mg)	478 ± 39.22	323 ± 29.08*	185 ± 34.27***
Consumption index (g)	1.842 ± 0.085	1.533 ± 0.086*	1.281 ± 0.047***
Growth rate (g)	0.207 ± 0.0017	0.176 ± 0.0022***	0.118 ± 0.0037***
Approximate digestability (%)	39.71 ± 1.059	33.60 ± 1.095**	28.612 ± 0.381***
Efficiency of Conversion of ingested food (%)	10.787 ± 0.070	12.25 ± 0.0273***	19.38 ± 1.375***
Efficiency of conversion of digest food	31.068 ± 1.286	36.281 ± 0.450**	47.97 ± 1.445***

Each value is mean ± SEM of five replicates.
*P < 0.05, **P < 0.01 and ***P < 0.001.

it was presumably reduced by allelochemical (NSKE and neem oil) inteferering with these processes (Reese, 1979).

In the present studies activities of digestive enzymes were markedly affected by NSKE and neem oil. This indicates that neem principle suppressed the nutritional value of the (protein) diet and in turn may have affected the digestive enzymes (Engelmann, 1969; Sibley, 1981).

Total egg output of *H. armigera* significantly affected the oocyte development. Decreased fecundity and oocyte development could thus be a consequence of impaired vitellogenin synthesis and its uptake by the developing oocytes (Ludlum and Sieber, 1988).

Neem treatment arrested the seminal secretion from the accessory reproductive gland of male and indirectly affected the neuroendocrine activity of the insect (Rembold, 1984). Decreased protein and lipid content as induced by NSKE and neem oil treatment may have affected the spermatophore moulding (Pickford et al., 1969; Friedel and Gilott, 1976).

NSKE and neem oil treatment caused a failure to transfer the spermatozoa to female and this may in turn result in the reduced fecundity. An increase of glycogen in the testis following NSKE treatment may indicate the arrest of spermatogenesis that leads to the accumulation of non-utilizable sugar in the testis, as this organ is the storehouse of glycogen from which spermatozoa get nourishment.

The decrease in the activities of ACP and ALP in the testis and seminal vesicle may be due to the inhibitory role of NSKE and neem oil on spermatogenesis and sperm survival (Schluter and Schulz, 1983). It is well known that spermatozoa derive their energy from ATPase and any interference in the production and the activity of ATPase would render the spermatozoa immobile, which occurred in this study after neem oil treatment.

In general LDH has been shown to play an important role in the mitochondrial lactate-pyruvate shuttle system for transferring the reducing equivalent from cytosol to mitochondria for the generation of energy necessary for motility and survival of germ

Table 3. Biochemical and enzyme profiles of female *H. armigera* treated with NSKE and neem oil

Biochemical parameters	Ovary			Oviduct		
	Control	1% NSKE	0.5% Neem oil	Control	1% NSKE	0.5% Neem oil
Protein (mg/g)	55.04 ± 1.34	40.6*** ± 0.98	29.44*** ± 1.02	28.1 ± 0.71	18.0*** ± 0.56	14.9*** ± 0.56
Glycogen (mg/g)	1.006 ± 0.406	1.439 ± 0.453	2.252 ± 0.573	1.335 ± 0.509	1.427 ± 0.564	1.295 ± 0.486
Lipid (mg/g)	1.081 ± 0.385	0.990 ± 0.384	0.864 ± 0.453	0.801 ± 0.312	0.721 ± 0.321	0.618 ± 0.309
Phospholipid (mg/g)	0.352 ± 0.0019	0.286*** ± 0.0015	0.261*** ± 0.0016	0.256 ± 0.0017	0.197*** ± 0.0023	0.171*** ± 0.0023
MDH (μg/mg protein/min)	1.644 ± 0.039	1.272*** ± 0.037	0.998*** ± 0.049	0.762 ± 0.029	0.686 ± 0.039	0.598** ± 0.032
LDH (μg moles/g/hr)	25.16 ± 3.159	20.94 ± 2.658	12.7** ± 1.936	23.06 ± 2.726	16.02 ± 2.245	10.56** ± 1.937
ACP (μg moles/g/hr)	22.15 ± 2.254	18.93 ± 2.191	12.61* ± 1.98	17.72 ± 2.574	12.87 ± 2.119	10.56* ± 1.623
ALP (μg/moles/g/hr)	26.94 ± 2.124	22.29 ± 1.635	16.42** ± 1.577	19.28 ± 1.948	15.87 ± 1.886	12.8* ± 1.855
ATPase (μg moles/g/hr)	115.91 ± 2.49	82.38*** ± 2.152	70.23*** ± 2.536	96.70 ± 2.158	71.06*** ± 2.098	52.91*** ± 2.346

Each value is mean ± SEM of five replicates.
*P < 0.05, **P < 0.01 and ***P < 0.001.

Table 4. Biochemical and enzyme profiles of female *H. armigera* treated with NSKE and neem oil

Biochemical parameters	Spermatheca			Fat body		
	Control	1% NSKE	0.5% Neem oil	Control	1% NSKE	0.5 % Neem oil
Protein (mg/g)	20.78 ± 0.78	14.58** ± 0.99	10.53*** ± 0.95	47.38 ± 1.059	38.25*** ± 1.86	31.97*** ± 1.445
Glycogen (mg/g)	1.286 ± 0.313	1.337 ± 0.481	1.471 ± 0.392	0.938 ± 0.073	1.095 ± 0.086	1.215* ± 0.070
Lipid (mg/g)	0.889 ± 0.327	0.747 ± 0.311	0.653 ± 0.325	1.497 ± 0.103	1.382 ± 0.065	1.295 ± 0.089
Phospholipid (mg/g)	0.281 ± 0.0016	0.217*** ± 0.0028	0.188*** ± 0.0026	0.496 ± 0.0026	0.410*** ± 0.0041	0.384*** ± 0.0025
MDH (ug/mg protein/min)	0.578 ± 0.022	0.528 ± 0.025	0.454** ± 0.026	1.31 ± 0.028	0.97** ± 0.095	0.732*** ± 0.071
LDH (μ moles/g/hr)	50.52 ± 3.478	42.58 ± 3.095	34.8** ± 2.669	23.27 ± 1.059	18.28* ± 1.288	15.60** ± 1.695
ACP (μ moles/g/hr)	19.65 ± 2.289	15.02 ± 1.819	11.03* ± 1.575	19.04 ± 1.006	16.6 ± 1.337	13.26** ± 1.273
ALP (μ moles/g/hr)	15.14 ± 1.803	12.1 ± 1.653	9.75 ± 1.526	23.38 ± 1.375	18.78* ± 1.070	15.28** ± 1.274
ATPase (μ moles/g/hr)	80.44 ± 2.239	67.62** ± 2.022	55.38*** ± 2.123	105.61 ± 2.381	89.97** ± 2.445	80.28*** ± 2.450

Each value is mean ± SEM of five replicates.
*$p < 0.05$, **$p < 0.01$ and ***$p < 0.001$.

Table 5. Biochemical and enzyme profiles of male *H. armigera* treated with NSKE and neem oil

Biochemical parameters	Testis			Seminal vesicle		
	Control	1% NSKE	0.5% Neem oil	Control	1% NSKE	0.5% Neem oil
Protein (mg/g)	100.48 ± 1.49	72.92*** ± 1.52	40.86*** ± 1.97	96.18 ± 1.37	77.86*** ± 1.35	52.16*** ± 1.99
Glycogen (mg/g)	0.96 ± 0.047	1.53*** ± 0.057	1.73*** ± 0.065	0.63 ± 0.078	0.698 ± 0.065	0.739 ± 0.082
Lipid (mg/g)	0.756 ± 0.045	0.648 ± 0.031	0.595 ± 0.068	0.718 ± 0.026	0.672 ± 0.059	0.591 ± 0.084
Phospholipid (mg/g)	0.243 ± 0.0074	0.191*** ± 0.0028	0.168*** ± 0.0021	0.238 ± 0.0022	0.183*** ± 0.0032	0.154*** ± 0.0018
MDH (μg/mg protein/min)	1.47 ± 0.266	1.09 ± 0.134	0.956 ± 0.187	0.507 ± 0.063	0.483 ± 0.087	0.442 ± 0.078
LDH (μ moles/g/hr)	29.26 ± 1.324	21.12** ± 1.303	12.84*** ± 1.093	49.32 ± 1.274	36.96*** ± 1.301	27.3*** ± 1.254
ACP (μ moles/g/hr)	20.04 ± 1.227	13.14** ± 1.036	7.88*** ± 0.99	9.092 ± 1.126	6.028 ± 1.119	6.2 ± 1.121
ALP (μ moles/g/hr)	27.84 ± 1.273	19.9** ± 1.476	16.0*** ± 1.197	8.02 ± 1.133	5.718 ± 1.173	5.32 ± 1.243
ATPase (μ moles/g/hr)	129.0 ± 5.797	103.0* ± 6.782	91.0** ± 6.711	102.46 ± 4.145	89.16* ± 3.90	72.36*** ± 2.60

Each value is mean ± SEM of five replicates.

*$p < 0.05$, **$p < 0.01$ and ***$p < 0.001$.

Table 6. Biochemical and enzyme profiles of male *H. armigera* treated with NSKE and neem oil

Biochemical parameters	Accessory gland			Fat body		
	Control	1% NSKE	0.5% Neem oil	Control	1% NSKE	0.5% Neem oil
Protein (mg/g)	83.23 ± 1.80	62.18*** ± 1.89	47.13*** ± 1.30	95.37 ± 2.346	80.15** ± 2.752	68.78*** ± 2.925
Glycogen (mg/g)	0.488 ± 0.091	0.545 ± 0.58	0.625 ± 0.079	0.856 ± 0.132	0.972 ± 0.186	1.127 ± 0.159
Lipid (mg/g)	0.818 ± 0.102	0.708 ± 0.098	0.658 ± 0.082	0.675 ± 0.098	0.562 ± 0.049	0.508 ± 0.051
Phospholipid (mg/g)	0.266 ± 0.0021	0.205*** ± 0.0021	0.184*** ± 0.0031	0.224 ± 0.0025	0.187*** ± 0.0022	0.167*** ± 0.0024
MDH (μg/mg protein/min)	0.963 ± 0.073	0.663* ± 0.097	0.608* ± 0.87	1.18 ± 0.096	1.05 ± 0.042	0.927 ± 0.065
LDH (μ moles/g/hr)	36.09 ± 1.341	29.11** ± 1.526	23.45*** ± 1.313	24.86 ± 1.492	20.15* ± 1.359	17.31** ± 1.456
ACP (μ moles/g/hr)	10.93 ± 1.042	7.95 ± 1.334	5.446** ± 0.977	16.54 ± 1.128	12.18* ± 1.235	9.75** ± 1.63
ALP (μ moles/g/hr)	13.12 ± 1.259	10.36 ± 1.221	7.84* ± 0.975	23.95 ± 1.370	19.71 ± 1.268	15.57** ± 1.324
ATPase (μ moles/g/hr)	91.31 ± 2.387	70.38*** ± 2.828	57.80*** ± 2.074	112.87 + 3.478	98.63** + 2.095	84.56*** ± 2.669

Each value is mean ± SEM of five replicates.
*P < 0.05, **P < 0.01 and ***P < 0.001.

cells. Besides, less lactate will be available for the protein synthesis and also for the oxidative phosphorylation. Malate dehydrogenase is an enzyme of the TCA cycle. It is present in both mitochondria and cytoplasm. In the present study the neem products alter the activity of the enzyme and lead to the accumulation of ammonia in various tissues leading to inhibition of MDH activity. The ATPase synthesis, being an essential requirement for protein synthesis, may if hampered, lead to an arrest of spermatogenesis. Further, depletion of protein, lipid, and phosphomonoesterase activities after the NSKE and neem oil treatment may suggest that the reproductive function is severely inhibited by the neem product.

CONCLUSION

This study illustrates that treatment with NSKE and neem oil results in certain metabolic changes in feeding and reproduction of *H. armigera* by interfering with the efficiency of feeding and causing a significant decline in the protein and lipid concentration together with enzyme systems in the testis and ovary as well as accessory reproductive glands. Thus, neem kernel extract may be considered as a chemosterilant agent for the insect, and as a highly promising candidate for an integrated pest management strategy. No ultrastructural studies especially for changes in the specific enzymes classes as well as metabolic pathwasy (protein, carbohydrate, and lipid) were carried out which could have thrown additional light on the probable nature and mechanism of antireproductive activity of *Azadirachta indica.*

ACKNOWLEDGEMENT

The authors wish to express their thanks to Dr. G. Vanithakumari, Bharathiar University (India) for critically reading the manuscript.

REFERENCES

Andersch, M.A. and A.J. Szezypinski. 1947. Use of P⁻. nitro phenyl phosphate as the substrate in determination of serum acid phosphatase. *Amer. J. Clin. Pathol.* 17: 571–573.

Anson, M.L. 1937. The estimation of pepsin, trypsin, papian and cathapsin with haemoglobin. *J. Gen. Physiol.* 22: 79–89.

Ascher, K.R.S., M. Eliyahu, N.E. Nemny and J. Meisner. 1984. Neem kernel extract as inhibitor of growth and fecundity in *Spodoptera littoralis. Proc. 2nd Int. Neem Conf.,* Eds. H. Schmutterer and K.R.S. Ascher. GTZ, Eschborn, pp. 331–334.

Baker, J.E. 1986. Amylase proteinase ratios in larval midguts of ten stored-products insects. *Entomol. Exp. Appl.* 40: 41–46.

Baumont, M.A. 1988. Stabilizing selection and metabolism. *Heredity* 61: 433–438.

Bernfeld, P. 1955. Amylases. In: *Methods in Enzymology.* Eds. S.P. Colowick and N.O. Kaplan. Academic Press, New York, pp. 149–158.

Butterworth, J.H. and E. Morgan. 1971. Investigation of the locust feeding inhibition of the seeds of the neem tree *Azadirachta indica. J. Insect Physiol.* 17: 974–977.

Chapman, R.F. 1985. Coordination of digestion. In: *Comprehensive Insect Physiology, Biochemistry and Pharmacology,* Vol. 4. Eds. G.A. Kerkut and L.I. Gilbert. Pergamon Press, Oxford, pp. 213–240.

Chen, P.S. 1984. The functional morphology and biochemistry of insect male accessory glands and their secretions. *Annu Rev. Entomol.* 29: 233–255.

Chiu, S.F., B.Q. Hyang and M.Y. Hu. 1985. Deterrent effects of seed oils and extracts of some meliaceaous plants on rice gall midge oviposition. *IRRI Newsletter* 10: 25.

Cripps, C., G.J. Blomquist and M. De Renobales. 1988. Changes in lipid biosynthesis during development of the house cricket, *Acheta domesticus* (Orthoptera: Gryllidae). *Bull. Entomol. Soc. Amer.* 127.

Dadd, R.H. 1970. Digestion in insects. In: *Chemical Zoology* Vol. 5. Eds. M. Florkin and B.T. Scheer. Arthropoda, Pare. A. Academic Press, New York, pp. 117–145.

Davey, K.G. 1985. The male reproductive tract. In: *Comprehensive Insect Physiology, Biochemistry and Pharmacology*, Vol. 1. Eds. G.A. Kerkut and L.I. Gilbert. Pergamon Press, Oxford, pp. 1–92.

Dubois M., K. A. Gilles, J.K. Hamilton, P.A. Robers and F. Smith. 1956. Colorimetric determination of sugars and related substances. *Analyt. Chem.* 28: 351–356.

Dunigan, E.P., F.J. Patric, L.D. Allen Jr. and Z. McMahon. 1972. Herbicidal effects on the nodulation of *Glycine max. Agron. J.*, 65: 806–808.

Engelmann, F. 1969. Food-stimulated synthesis of intestinal proteolytic enzymes in the cockroach *Leucophaea maderae. J. Insect Physiol.* 15: 217–235.

Engelmann, F. 1970. *The Physiology of Insect Reproduction*. Pergamon Press, London.

Folch, J., M. Less and G.H. Sloan-Stanley. 1957. A simple method for the isolation and purification of total lipids from animal tissue. *J. Biol. Chem.* 226: 497–506.

Friedel, T. and C. Gillott. 1976. Male accessory gland substance of *Melanoplus sanguinipes*: An oviposition stimulant under the control of the corpus allatum. *J. Insect Physiol.* 22: 489–495.

Gaaboub, A.I. and D.K. Hayes. 1984. Effect of larval treatment with azadirachtin, a moulting inhibitory component of the neem tree on reproductive capacity of the face fly, *Musca autumnalis* (Diptera: Muscidae). *Environ. Entomol.* 13: 1639–1643.

Good, C.A., H. Kramer and M. Somogyi. 1933. The determination of glycogen. *J. Biol. Chem.* 100: 488–491.

Hartl, D.L., D.E. Dykhuizen and A.M. Dean. 1985. Limits of adaptation: The evolution of selective neutrality. *Genetics* 111: 655–674.

Johansson, A.S. 1964. Feeding and nutrition in reproductive processes in insects. In: *Insect Reproduction*. Ed. K.C. Highnam. *Symp. Roy. Entomol. Soc. London*, 2: 43–52.

Kaeser, I. I. 1983. The control of enzyme systems *in vivo*: Elasticity analysis of the steady state. *Trans. 601st Meeting Biochemistry Society Aberdeen.*

King, J. 1965. *Practical Clinical Enzymology*. D. Van. Norstrand Co., London.

Koul, O., K. Amanai and Y. Ohtaki. 1987. Effect of azadiractin on the endocrine events of *Bombyx mori. J. Insect. Physiol.* 33: 103–108.

Lehane, M.J. 1976. Digestive enzyme secretion in *Stomoxys calcitrans* (Diptera: Muscidae). *Cell. Tiss. Res.* 170: 275–287.

Leopold, R.A., A.C. Terranova, B.J. Thorson and M.E. Degrugillier. 1971. The biosynthesis of the male housefly accessory secretion and its fate in mated female. *J. Insect Physiol.* 17: 987–1003.

Lowry, O.H., N.G. Rosebrough, A.L. Farr and R.G. Randall. 1951. Protein measurements with Folin phenol reagent. *J. Biol. Chem.* 193: 265–275.

Ludlum, C.T. and K.P. Seiber. 1988. Effect of azadirachtin on oogenesis in *Aedes aegypti. Physiol. entomol.* 13: 177–184.

Matsumura, F. 1975. *Toxicology of Insecticides*. Plenum Press, New York, 1975.

Mordue, A.J., K.A. Evans and M. Charlet. 1986. Azadirachtin, ecdysteroids and ecdysis in *Locusta migratoria. Comp. Biochem. Physiol.* 85C: 297–301.

Murugan, K. and Ancy George Sr. 1992. Feeding and nutritional influence on the growth and reproduction of *Daphnis nerii* Linn. (Lepidoptera: Sphingidae). *J. Insect Physiol.* 38: 961–968.

Ochoa, S. 1955. In: *Methods in Enzymology*, Vol. I. Eds. S.P. Colwick and N.O. Kaplan. Academic Press, New York, p. 739.

Pickford, R., A.B. Ewen and C. Gillott. 1969. Male accessory gland substance and egg-laying stimulant in *Melanoplus sanguinipes* (F.) (Orthoptera: Acrididae). *Can. J. Zool.* 47: 1199–1203.

Piemental, D. 1971. Ecological effects of pesticides on nontarget species. Office of Science Technology Documents, Washington, DC, 220 pp.

Rembold, H. 1984. Secondary plant products in insect control, with special reference to the azadirachtins. In: *Advances in Invertebrate Reproduction*, Vol. 3. Ed. W.E. Engels. Elsevier, Amsterdam pp. 481–491.

Rembold, H. and K.P. Sieber. 1981. Inhibition of oogenesis and ovarian ecdysteroid synthesis by azadirachtin in *Locusta migratoria migratorioides* (R + F). *Z. Naturforsch.* 36c: 466–469.

Reese, J.C. 1979. Interactions of allelochemicals with nutrients in herbivore food. In: *Herbivores: Their Interaction with Secondary Plant Metabolites*. Eds. G.A. Rosenthal and D.H. Janzen. Academic Press, New York, pp. 309–330.

Saxena, K.A. and H. Rembold. 1984. Orientation and oviposition responses of *Heliothis armigera* to certain neem constituents. *Proc. 2nd Int. Neem Conf.* Eds. H. Schmutterer and K.R.S. Ascher. GTZ, Eschborn, pp. 119–210.

Schluter, U. and W.D. Schulz. 1983. Structural damages caused by neem in *Epilachna varivestis*: A summary of histological and ultrastructural data. Tissues affected in larvae. *Proc. 2nd Int. Neem Conf.* Rauischholzhausen, 1983. GTZ, Eschborn, pp. 227–236.

Schmutterer, H. and K.R.S. Ascher, eds. 1984. *Proceedings of the 2nd International Neem Conference*, Rauischolzhausen, 1983. GTZ, Eschborn, pp. 587.

Scriber, J.M. and F. Slansky. 1981. The nutritional ecology of immature insects. *Annu. Rev. Entomol.* 26: 183–211.

Seigler, D.S. and P.W. Price. 1976. Secondary compounds in plants: Primary functions. *Amer. Nat.* 110: 101–105.

Seligman, A.M. and S.P. Kramer. 1965. *Methods in Enzymatic Analysis*. Academic Press, New York and London.

Sibley, R.M. 1981. Strategies of digestion and defecation. In: *Physiological Ecology: An Evolutionary Approach to Resource Use*. Eds. C.R. Townsend and P. Calow. Blackwell, Oxford, pp. 109–139.

Sieber, K.P. and H. Rembold. 1983. The effects of azadirachtin on the endocrine control of moulting in *Locusta migratoria*. *J. Insect Physiol.* 29: 523–527.

Slansky, F. 1985. Food utilization by insects: Interpretation of observed differences between dry weight and energy efficiencies. *Entomol. Exp. Appl.*. 39: 47–60.

Slansky, F. Jr. 1992. Allelochemical nutrient interactions in herbivore nutritional ecology. In: *Herbivores: Their Interactions with Secondary Plant Metabolites*, Vol. II, *Evolutionary and Ecological Process*. Ed. G.A. Rosenthal. Academic Press, San Diego, pp. 135–174.

Slansky, F., Jr. and J.M. Scriber. 1985. Food consumption and utilization. In: *Comprehensive Insect Physiology, Biochemistry and Pharmacology*. Eds. G.A. Kerkut and L.I. Gilbert. Pergamon Press, Oxford, pp. 87–164.

Terranova, A.C., R. A. Leopold, M.E. Degrugiller and J.R. Johnson. 1972. Electrophoresis of the male accessory secretion and its fate in the mated female. *J. Insect Physiol.* 18: 897–911.

Waldbauer, G.P. 1968. The consumption and utilization of food by insects. *Adv. Insect Physiol.* 5: 229–288.

Ware, G.W. 1980. Effects of pesticides on non-target organisms. *Residue Rev.* 173: 546.

35

Insecticide, Antifeeding and Juvenilizing Effects of Neem (*Azadirachta indica* A. Juss.) Oil against *Corcyra cephalonica* Staint. and *Epilachna vigintioctopunctata* F.

Y.K. MATHUR AND SUMAN NIGAM

Department of Entomology, C.S.A. University of Agriculture & Technology
Kanpur 208 002, India

ABSTRACT

The effects of neem oil as insecticidal, antifeedant and juvenomimetic agents were tested against a stored products pest, *Corcyra cephalonica* and phytophagous pest, *Epilachna vigintioctopunctata*. The results showed that the phytophagous insect, *E. vigintioctopunctata* was more susceptible than the stored product insect (*C. cephalonica*) in respect of the insecticidal value of neem oil in terms of Lc_{50} values. The data for the antifeeding properties of neem oil clearly indicate that higher concentrations of neem oil, i.e., 4.0 and 5.0 per cent offered the best results giving 100 per cent antifeeding effects against *C. cephalonica*. Higher concentration of 5 per cent neem oil when used against *E. vigintioctopunctata* gave maximum (89.16 per cent) leaf area protection over control. However, lower concentration (4.0 per cent) resulted in 16.24 per cent protection over control. Neem oil has also elicited different juvenomimetic effects against both the test insects (*C. cephalonica* and *E. vigintioctopunctata*), infesting wheat grain and brinjal, respectively. The most spectacular morphogenetic effects appeared were 1-p-intermediates, deformed pupae and adults of the treated insects. Percentage inhibition of adult emergence was used as the criterion for evaluation of juvenomimetic efficacy of neem oil against the test insects.

INTRODUCTION

The promiscuous use of synthetic insecticides has resulted in mammalian hazards and led to develop a high degree of resistance in insects. Besides, it has also posed the serious problems like elimination of naturally occurring biocontrol agents and other non-target organisms. Resurgence is also a common phenomenon following treatments of crops with synthetic insecticides. This dilemma has made it essential to identify such alternative methods that are comparatively safer and do not upset the agro-ecosystem. Among the alternative bio-insecticides, the future of botanicals in the modern concept of integrated pest management system appears to be very promising as they are bio-degradable in nature and much safer to higher animals including humans because of their low mammalian toxicity.

Keeping the aforesaid facts in view, the present authors have undertaken the study on insecticidal, anti-feedant and juvenomimetic properties of neem oil against a stored products pest, *Corcyra cephalonica* and a phytophagous pest, *Epilachna viginctioctopunctata*.

MATERIALS AND METHODS

The shade-dried seeds of neem (*A. indica*) were crushed and powdered in a common domestic grinder. The powdered material was extracted with petroleum ether in Soxhlet apparatus and this extract was transferred to different volumetric flasks to make up a known volume. From the above preparation a known quantity was taken in a weighed porcelain dish and the solvent was completely evaporated on a steam water bath. The dish was then kept in a dessicator for 15–20 minutes and its weight was recorded and this procedure was repeated till a constant weight was obtained. The extract (neem oil) thus obtained was used for preparing different concentrations. A stock solution of 10 per cent was prepared in acetone with 1 per cent Tween 60 (Sigma) as emulsifier. The solution was diluted with distilled water to get the desired concentration of neem oil.

The techniques employed to investigate the insecticidal, antifeedant and juvenomimetic effects of neem oil against the experimental insects are described here.

Insecticidal Test

The insecticidal effects of neem oil against *C. cephalonica* were studied by mixing 5 ml of each concentration in 20 gm of wheat grains. After evaporating the solvent grains were transferred into three glass tubes (4" × 1") covered with muslin cloth. Following this, ten newly emerged adults starved for 24 hours were released inside the tubes and observations on mortality counts were made 24 hrs after treatment. The grains treated with solvent only were taken as control. These tubes were kept at controlled room temperature (25 ± 1 °C). In order to study the insecticidal effects of neem oil against *E. vigintioctopunctata*, brinjal leaves were dipped in different concentrations of neem oil for two seconds and the solvent was evaporated under an electric fan for 30 minutes. One treated leaf bit was kept in each petri dish and ten newly emerged adults starved for 24 hours were released on treated leaf material. Each treatment was replicated three times. Observations on mortality counts were made after 48 hours of treatment. The observed mortality of insects in the treatment was adjusted with mortality occurring in control by Abbots' formula (1925). Data thus obtained were subjected to Probit analysis (Busvine, 1957). Solvent emulsified water served as control.

Antifeeding Test

The antifeeding effects of the extract (neem oil) against *C. cephalonica* were studied by mixing each concentration in 20 gm of wheat grains. There were three replications and ten larvae were released in each replicate including control (grains mixed with solvent only). After five days of experimentation, the grains left in the tubes were weighed and the amount of consumed food was estimated. Finally, antifeeding index (AFI) was calculated by the following formula as adopted by Abivardi and Benz (1984).

$$\text{Antifeeding index (AFI)} = \frac{C - T}{C} \times 100$$

Where C = consumed grains in control

T = consumed grains in treatment

The antifeeding effects of neem oil were studied in the case of *Epilachna vigintioctopunctata* by dipping the measured leaf bits of brinjal in different concentrations of the extract and the solvent was evaporated under an electric fan for 30 minutes. One treated leaf bit was kept in each petri dish and ten larvae starved for eight hours were released on the treated leaf bit in each petri dish. Observations were recorded after 48 hours of treatment and the area of leaf bit left over was measured. In control, leaves were treated only with solvent or emulsified water. The percentage feeding and leaf area protected over control was calculated by the following formula

$$\text{Per cent feeding} = \frac{\text{Leaf area given} - \text{corrected leaf area}}{\text{Leaf area given}} \times 100$$

Percentage reduction in feeding due to treatment was worked out by the formula as given by Abbot (1925).

$$\text{Leaf area protected over control} = \frac{\text{Percentage protection in treated} - \text{percentage protection in control}}{100 - \text{percentage protection in control}}$$

Juvenomimetic Test

Morphogenetic changes in larval-pupal intermediates, pupal adult intermediates, adult deformities, etc., due to treatment of food with different concentrations of the used plant extract in early last instar larvae of the experimental insects were noted at their different stages of growth. The feeding method was employed for both the test animals and the experiments were set up in the same manner as described in the antifeedant study.

RESULTS AND DISCUSSION

Insecticidal Effects of Neem Oil

The toxicity of neem oil against *C. cephalonica* and *E. vigintioctopunctata* was determined by bioassay method and the data thus obtained were subjected to probit analysis so as to work out the Lc_{50} value. The heterogenity, regression equation and fiducial limits of neem oil against both the test insects are presented in Table 1.

Table 1. Regression equation and Lc_{50} values of neem oil to the adults of *Corcyra cephalonica* and *Epilachna vigintioctopunctata*

Sl. No.	Name of the insect species	Heterogenity (x^2)	Regression equation	Lc_{50} value	Fiducial limit
1.	*C. cephalonica*	0.152	y = 0.807x +3.192	1.737	0.970 3.111
2.	*E. viginctioctopunctata*	0.238	y = 1.224x +2.748	0.692	0.389 0.367

It is amply documented from Table 1 that phytophagous insect (*Epilachna vigintioctopunctata*) was more susceptible than the stored product insects of *Corcyra cephalonica* as Lc_{50} observed was 0.692 and 1.737, respectively. No doubt, both the experimental insects could effectively be controlled by neem oil treatment. Jairaj (1976) while studying the effectiveness of neem oil, found it toxic against pulse beetle-infesting grains in storage. The toxic action of neem oil has also been reported by Savitri and Rao (1976) against *Sitophilus oryzae*, *Rhizopertha dominica* and *Sitotroga cerealella*. Later, Choudhary (1989) found the seed treatment of chickpea with neem oil as most effective treatment in reducing the number of eggs laid, adult emergence and seed damage due to *C. chinesis*. The toxicity of neem oil has also been reported by Mukherjee et al. (1990) against *S. oryzae*. Therefore, the toxic action of neem oil reported in the present study is in conformity with the results arrived at by earlier workers.

Saxena and Khan (1984) found neem oil highly effective in reducing the survival of *Nephotettix virescens* on rice seedlings. Rao and Srivastava (1984) also reported toxicity of neem oil against sorghum ear-head worm, *Heliothis armigera*. Logiswara et al. (1988) reported the toxic action of neem oil (2 per cent) against rice gall midge, rice leaf folder and found it to increase the yield of rice crop. The spraying of neem oil at 1.5 per cent concentration was found quite effective in killing the larvae of mustard web worm, *Crocidolomia binotalis* and nymphs and adults of *Lipaphis erysimi* (Mani et al., 1990). Recently, Chandramohan and Nanjan (1992) established the insecticidal property of neem oil against the diamond black moth, *Plutella xylostella* infesting cabbage in the field. In the present study too, neem oil has been found to be effective in killing *E. vigintioctopunctata* feeding on brinjal leaves.

Antifeedant Effects of Neem Oil

The data generated on neem oil in respect of its antifeeding properties clearly reveal that higher concentrations of neem oil i.e., 4 and 5 per cent offered the best results giving 100 per cent antifeeding effects against *C. cephalonica* on treated wheat grains, of course, the antifeedant index (AFI) at lowest concentration (0.5 per cent) was also as high as 87.52 (Table 2). While considering the effect of neem oil against *E. vigintioctopunctata*, it was found that the maximum leaf area protected over control (89.16 per cent) was obtained when brinjal leaves were treated with 5.0 per cent neem oil and it was closely followed by its lower concentration (4.0 per cent) which resulted in 86.24 per cent protection over control. The lowest concentration of 0.5 per cent was also rated quite effective as it registered 78.55 per cent leaf protection over control.

Table 2. Antifeeding index (AFI) of *Corcyra cephalonica* and leaf protection over control against *Epilachna vigintioctopunctata* by neem oil treatment

Sl. No.	Neem oil concentration (%)	Antifeeding index of *C. cephalonica*	Leaf area protected over control against *E. vigintioctopunctata*
1.	0.5	4.47 (87.52)	78.55
2.	1.0	4.46 (86.61)	81.94
3.	2.0	4.47 (89.42)	85.08
4.	4.0	4.61 (100.00)	86.24
5.	5.0	4.61 (100.00)	89.16
	CD at 5%	0.67	

Angularly transformed values (%) are given in parentheses

The oil has also been found to provide protection against *R. dominica* and *S. cerealella* in stored paddy at 1 per cent concentration. The effectiveness of neem against *Sitophilus oryzae* (Mukherjee et al., 1990) and against *Tribolium castaneum* (Kripa Shankar, 1992).

Neem oil has also been reported to deter feeding of several insects (Narayan et al., 1980; Kareem, 1984; Rao et al., 1986; Mala et al., 1987; Mani et al., 1990; and Kripashankar, 1992).

Juvenomimetic Effects of Neem Oil

The results of the juvenilizing effects of neem oil against a storage pest (*C. cephalonica*) and a phytophagous pest (*E. vigintioctopunctata*) are presented in Table 3 and 4, respectively. It is amply documented from the data that *A. indica* inducted deformities in the developing larvae, pupae and adults of treated last instar larvae of *C. cephalonica*. The most spectacular morphogenetic effects appeared were L-P intermediates, deformed pupae and the adults of the test insect. A critical look of the table reveals that L-P intermediates were recorded at higher concentration of 0.24, 0.12 and 0.06 per cent which resulted in 36.67, 23.33 and 3.33 per cent L-P intermediates, respectively. As regards deformed pupae, they were also recovered at higher concentration of the test chemical (Table 3). However, at lower concentrations (0.03 and 0.015 per cent) maximum percentage of normal pupation was noticed and their actual percentages were 96.67 and 100, respectively. The similar trend was also observed in the case of eclosion of normal adults when treated with different concentrations of neem oil.

While considering the formation of deformed adults, it was found that highest concentration (0.24 per cent) resulted in 30 per cent deformed adults, whereas the percentage of deformed adults decreased with lowering the concentration of neem oil. In control, no such deformities could be noticed. It may further be seen from Table 3 that maximum inhibition of adult emergence was recorded at the highest concentration

Table 3. Juvenilizing effects of neem oil on freshly moulted last instar larvae of *Corcyra cephalonica*

Sl. No.	Conc.	L-P[a] intermediate	Pupal deformity	Normal pupation	Deformed adults	Normal adults	Inhibition of adult emergence
1.	0.24	36.50 (36.67)	18.31 (20.0)	43.31 (43.33)	29.67 (30.00)	10.24 (13.33)	89.76 (86.67)
2.	0.12	23.18 (23.33)	22.45 (23.33)	53.35 (53.33)	23.18 (23.33)	30.00 (30.00)	70.00 (70.00)
3.	0.06	1.96 (3.33)	22.45 (23.33)	74.56 (73.33)	19.31 (20.00)	53.30 (53.33)	46.50 (46.67)
4.	0.03	0.25 (0.00)	1.96 (3.33)	98.04 (96.67)	16.36 (16.67)	80.68 (80.00)	19.31 (20.00)
5.	0.015	0.25 (0.00)	0.25 (0.00)	99.75 (100.00)	7.56 (10.00)	92.44 (90.00)	7.56 (10.00)
6.	Conrol	0.25 (0.00)	0.25 (0.00)	99.75 (100.00)	0.25 (0.00)	99.75 (100.00)	0.25 (0.00)
C.D. at 5%		0.101	0.140	0.144	0.196	0.212	0.212

Figures within parentheses represent the transformed back values in percentage.
Number of insects used per treatment were 30.

(0.24 per cent) of neem oil. The other concentrations of 0.12, 0.06, 0.03, 0.015 elicited 70.00, 46.67, 20.00 and 10.00 per cent inhibition of adult emergence, respectively. The statistical calculations showed significant difference at 5 per cent level of significance.

The data on juvenilizing effects of neem oil against *E. vigintioctopunctata* reveal three types of deformities, L-P intermediates were obtained at the higher concentrations ranging from 0.24 to 0.03 per cent (Table 4). Similarly, maximum pupal deformity (26.67 per cent) was recorded at 0.24 per cent concentration. However no such deformity could be seen at the lower most concentration of neem oil (0.015 per cent).

As regards the emergence of deformed adults, they were recorded at the concentrations of 0.12, 0.06 and 0.03 per cent only and maximum deformity (16.67 per cent) in adults was noticed when the grubs were treated with 0.12 per cent concentration of neem oil. The rest of the treatments gave intermediary result in respect of emergence of deformed adults. It would be pertinent to mention here that deformed adults were short-lived and non-viable and the adults with incomplete genitalia were unable to copulate and thus failed to produce their progeny. In the control set of experiment no deformities were seen in grubs, pupae and adults of the test insect.

Manifestations of morphogenetic effects of neem oil used in the present study may be due to the presence of azadirachtin as an active principle as also reported by Ibrahim et al. (1984) and Singh (1986). However, contrary results were reported by Redfern et al. (1984) who found that azadirachtin was least effective in inducing juvenomimetic effects against the nymphs of milkweed bug and the pupae of yellow meal worm. The juvenomimetic activity of neem based formulation has also been reported by Mehrotra and Gujar (1984) and Singh (1986) against the stored products and the phytophagous pests.

The validity of the juvenilizing effects of neem oil in the present set of experiment may be compared with the findings of Srivastava and Mathur (1982) and other workers

Table 4. Juvenilizing effect of neem oil on the freshly moulted last instar larvae of *Epilachna vigintioctopunctata*

Sl. No.	Conc.	L - P intermediate	Pupal deformity	Normal pupation	Deformed adult	Normal adult	Inhibition of adult emergence
1.	0.24	50.00 (50.00)	26.52 (26.67)	23.18 (23.33)	0.25 (0.00)	23.18 (23.33)	76.82 (76.67)
2.	0.12	16.36 (16.67)	13.01 (13.33)	70.00 (70.00)	15.72 (16.67)	53.35 (53.33)	46.65 (46.67)
3.	0.06	7.56 (10.00)	7.56 (10.00)	84.87 (80.00)	5.25 (6.67)	75.00 (73.33)	25.00 (26.67)
4.	0.03	0.25 (0.00)	1.96 (3.33)	98.04 (96.67)	10.00 (10.00)	86.89 (86.67)	13.01 (13.33)
5.	0.015	0.25 (0.00)	0.25 (0.00)	99.75 (100.00)	7.56 (10.00)	92.44 (90.00)	7.56 (10.00)
6.	Control	0.25 (0.00)	0.25 (0.00)	99.75 (100.00)	0.25 (0.00)	99.75 (100.00)	0.25 (0.00)
C.D. at 5%		0.129	0.161	0.175	0.185	0.208	0.208

Figures within parentheses represent the transformed back values per percentage.
Number of insects used per treatment were 30.

as quoted by Bambarkar (1990) who reported that azadirachtin acts as a metamorphosis and growth disrupter in *S. litura* and *Heliothis armigera*. It may, therefore, be inferred that presence of azadirachtin in neem oil is responsible for different types of interactions between juvenile hormone and moulting hormone because of that morphogenetic manifestations occur in the treated insects. A similar view is also held by Mathur et al. (1990) and Kripa Shankar (1992).

From the overall studies made, it is obvious that neem oil does possess insecticidal, antifeedant and juvenomimetic properties against both stored products pest, *Corcyra cephalonica* and the phytophagous pest, *Epilachna viginioctopunctata*. Therefore, it may suitably be well fitted in the integrated management of both the experimental insects, i.e., *C. cephalonica* and *E. vigintioctopunctata.*

ACKNOWLEDGEMENTS

The authors wish to record their sincere thanks to the Vice-Chancellor, C.S. Azad University of Agriculture and Technology, for providing all necessary facilities during the course of the study. Help rendered by Dr. S.K. Nigam, Utilisation Research Laboratory, National Botanical Research Institute, Lucknow in preparing the neem extract (oil) is also gratefully acknowledged.

REFERENCES

Abott, W.S. 1925. A method of computing the effectiveness of an insecticide. *J. Econ. Entomol.* 18: 265–267.

Abivardi, C. and G.C. Benz. 1984. Test with the extract of 21 medicinal plants for antifeedant activity against larvae of *Pieris brassicae. Mitt. Schweiz. Entomol. Besell* 57: 383–392.

Bambarkar, Sunil. 1990. Neem—A vast potential for agro-chemicals. *Pesticides* 24(1): 36–38.

Busvine, J.R. 1957. A critical review of techniques for testing insecticides. Commonwealth Instt. Entomol., London 1–5.

Chandramohan, N. and K. Nanjan. 1992. Effect of plant product spray on the ovipositional behaviour of diamond black moth, *Plutella xylostella*. *Neem Newsletter* 9(1): 8–9.

Chaudhary, B.S. 1989. Residual effect of eight vegetable oils in chickpea against pulse beetle. *Indian J. Plant. Prot.* 18: 89–91.

Devi, D.A. and N. Mohan Das. 1982. Relative efficacy of some antifeedants and deterrents against insect pests of paddy. *Entomon*. 7(3): 261–264.

Ibrahim, A., Baaboub and K.H. Dora. 1984. *Environ. Entomol*. 13: 803–812.

Jairaj. 1976. Neem products—*Entomological Trials*. Utilisation report. Department of Entomology, T.N.A.U., Coimbatore.

Kareem, Abdul A. 1984. Neem products: their uses in insect pest control. *Neem Newsletter* 1(1): 6

Kripa Shankar. 1992. Evaluation of plant products for their insecticidal, antifeedant and juvenomimetic properties against *Antigastra catalaunalis* Dup. and *Tribolium castaneum* Herbst. Ph.D. thesis, C.S.A. Univ. of Agri. and Tech. Kanpur, India.

Logiswaran, G., V K.R. Sathiyanandan and P.C. Sundarababu. 1988. Effect of neem oil on rice gall midge and leaf folder. *Neem Newsletter* 5: 18–20.

Mala, S., N.S. Nalini Sundri and S. Theinozhi. 1987. Antifeeding activity of neem oil extractive on a lepidopteran pest *Pericallia ricinie* (Meeting Abstract) I.A.R.I. *Neem Newsletter* 4(3): 36.

Mani, A., Kumudnathan and C.A. Jagdish. 1990. Antifeeding property of certain new products against red hairy Caterpillar, *Amsacta albistriga* Wlk. *Neem Newsletter* 7(2): 16.

Mathur, Y.K., J.P. Srivastava, S.K. Nigam and R. Banerji. 1990. Juvenomimetic effect of karanjin on the larval development of fleshfly, *Sarcaphaga ruficornis* Fabr. (Cyclorrhapha:Diptera). *J. Ent. Res.* 14(1): 44–51.

Mehrotra, K.N. and G.T. Gujar. 1984. Neem as insect growth regulator *Neem Newsletter* 1(1):6.

Mukherjee, U., A.K. Singh and S.F. Hameed. 1990. Evaluation of neem (*Azadirachta indica* A. Juss) products as wheat grain protectants against *Sitophilus oryzae* Linn. *Neem Newsletter* 7(1): 2–4.

Narayanan, C.R., R.P. Singh and D.O. Sawalkar. 1980. Phagodeterrency of various fractions of neem oil against *Schistocerca gregaria* Forsk. *Indian J. Entomol.* 42: 469–472.

Rao, D.R., S.S. Thakur, B.H.K. Rao and G. Srimannarayana. 1986. Neem seed oil residue and its essential oil as feeding inhibitors against fourth instar larvae of *Spodoptera litura*. *Natl. Symp. on Insect Physiology, Ecology and Behaviour*. Deptt. of Zoology, Univ. of Kerala, Trivandrum.

Rao, Satyanarayan, R.V. and K.P. Srivastava. 1984. Evaluation of neem formulations against sorghum earhead worm. *Neem Newsletter* 1(4): 37–38.

Redfern, R.E., D.K. Hayes, J.D. Wasdhen, Jr. A.B. Demio and T.P. McGovern. 1984. *Ann. Rev. Chronopharmcol.* 1: 239–242.

Savitri, P. and S.C. Rao. 1976. Studies on the admixture of neem seed kernel powder with paddy on the control of important storage pests of paddy. *Andhra Agric. J.* 23: 137–143.

Saxena, R.C. and Z.R. Khan. 1984. Neem oil disturbs *Nephotettix virescens* feeding. *Neem Newsletter* 1(3): 28.

Singh, R.P. 1986. Neem (*Azadirachta indica* A. Juss) an effective substitute to synthetic insecticide. *Proc. Natn. Conf. key Pests Agric. Crops*. pp. 322–333.

Srivastava, J.P. and Mathur, Y.K. 1982. Effect of juvenile hormone administration with larval flesh food on the metamorphosis of *Sarcophaga ruficornis* Fabr. (Sarcophagidae:Diptera). *Z. Angew. Entomol.* 69: 487–494.

36

Olfactory, Oviposition and Developmental Response of Four Species of Pulse Beetles to Four Neem Oil Treated Legumes

T.D. YADAV

Division of Entomology, Indian Agricultural Research Institute
New Delhi 110 012, India

ABSTRACT

The efficacy of neem oil against three species of bruchids infesting legumes has been reported yet information on the precise concentration needed for the treatment of legumes for safe storage is not available. Neem oil was extracted from powder of freshly harvested, dried kernel using petroleum ether. One per cent solution in carbon tetrachloride was prepared and further diluted to treat 100 g seeds of chickpea, cowpea, soybean and French bean. Seven doses varying between 20 and 500 ppm were tested. Seeds were stored in polyethylene bag (15 × 10 cm). The observations were made with a 24 chamber host-choice apparatus on per cent olfactory attraction of adult, oviposition, hatch and emergence of adult.

Tests were conducted on treated seeds after 24 hours. The results indicated that 200–500 ppm concentrations were ovicidal and *Zabrotes subfasciatus* (Boheman) was the most susceptible bruchid on French bean. *Callosobruchus maculatus* Fab. and *C. analis* Fab. were least susceptible. Larval development was also affected at 200–500 ppm concentrations and adult emergence was arrested. Adults with deformed wing and prepupal abnormal behaviour were also observed. The results have shown absence of olfactory response, dose dependent ovicidal and larval toxicity and juvenile hormone activity. The results have value in planning large-scale tests.

INTRODUCTION

The efficacy of neem (*Azadirachta indica* A. Juss.) oil and neem kernel powder (NKP), has been bioassayed by several workers. Jotwani and Sircar (1967) reported the efficacy of 1 and 2 per cent (w/w) NKP as seed treatment on four legumes against *Callosobruchus maculatus* Fab. during the 8 to 11 months of storage. Yadav (1984) also evaluated neem kernel 1, 2 and 3 per cent as seed treatment of seven legumes against *C. maculatus* and *C. chinensis* (Linn.) during 12 months of storage. Yadav (1985) evaluated the response of varying concentration from 200–5000 ppm of neem oil on mungbean against *C. maculatus*, *C. chinensis* and *C. analis* Fab. aiming at assaying relative efficacy among the species of bruchids. *Callosobruchus chinensis* was found most susceptible and *C. maculatus* least susceptible in reduction in oviposition and emergence of adult at a concentration level of 400 ppm. The level of efficacy was also influenced by the type of seed. Lentil displayed best efficacy at 1 per cent concentration of NKP among the seven legumes and mungbean was the least (Yadav, 1984). In the present study the response of neem oil at concentration of 20–500 was evaluated for olfactory attraction, oviposition, hatchability and emergence using a host-choice chamber involving five pulse beetles. Bioefficacy of oil treatment was assayed at the initial stage of treatment.

MATERIALS AND METHODS

To obtain neem oil, fresh neem kernels were obtained, decorticated and dried at 40°C for 24 hours, powdered in an electric mixer, and passed through a 40 mesh sieve. To 100 g powder held in stopper conical flask, 50 ml of petroleum ether (40–60°C) was added and stirred. It was kept for 24 hours and solvent was decanted through Whatman filter paper. The process was repeated once more with 20 ml of solvent. The solvent was allowed to evaporate and the oil left after evaporation of solvent was transferred to a volumetric flask and kept in a refrigerator. Subsequently 1 per cent stock solution of oil was prepared in carbon tetrachloride (CT).

Callosobruchus maculatus, C. chinensis and *C. analis* were reared on mungbean and *Zabrotes subfasciatus* on French bean at 27 ± 1.5°C and 60–65 per cent relative humidity. Clean 100 g seeds of chickpea, cowpea, soybean, and French bean were treated with 500, 200, 150, 100, 50, 40 and 20 ppm of oil using 5 ml of carbon tetrachloride. To treat seeds, 250 ml conical flask was used and shaking was done manually for two minutes. Only solvent treated seeds were used as control. Seeds were dried and transferred to polyethylene bags (15 × 10 cm). Randomly selected six seeds were removed from each concentration of treated seeds of chickpea, cowpea and soybean and 1 seed was placed in each chamber of 24 food chamber host-choice apparatus (15 × 15 × 5 cm) (Yadav and Tanwar, 1985). In case of French bean only four seeds were removed and individually placed in the chamber.

Three replications were maintained. In host-choice tests, five pairs of one-to-two days old adults were released for 24 hours. Olfactory attraction was based on number of adults found settled in host chamber. Insects remaining in central chamber or moving out side were counted as wanderer. Per cent attraction was calculated on the basis of total insects. Total number of oviposited eggs was basis of per cent oviposition in each concentration. Similarly hatchability and emergence of F1 adults were also calculated. Presence/absence of pin hole on the seeds was adjudged as criterion for larvicidal action of treatment.

RESULTS

As seen from Table 1, in chickpea choice test against *C. chinensis* at initial stage, 500 ppm was found to check hatching and F1 emergence. Results were similar in *C. analis* also. Ovicidal and larvicidal activity at 500 ppm dose was exhibited against both bruchids.

In cowpea, choice test at initial stage, *C. maculatus, C. chinensis* and *C. analis* displayed oviposition deterrency at 500 ppm dose followed by total check of hatching and emergence. *C. chinensis* was found most susceptible and at 50 ppm F1 emergence was 5.9 per cent.

In soybean, *C. analis* exhibited ovicidal activity and F1 emergence was checked at 500 ppm in the beginning.

In French bean, choice test against *Z. subfasciatus* at initial stage of treatment, indicated effectiveness as larvicidal and F1 was susceptible at 100 and 200 ppm doses. Thus *Z. subfasciatus* was susceptible at 100 ppm dose which could be due to the thick seed coat of French bean and round shape of egg.

The eggs of *Z. subfasciatus* deposited on hard seed coat of French bean were found more susceptible to neem oil treatment. The eggs laid in cluster are also associated in this beetle. The first stage larva was observed to develop in normal period in all the four species at effective concentration but entry in the cotyledon was delayed followed by mortality.

Olfactory attraction in the case of these beetles was dependent on the choice of female to select oviposition site. However, seeds of legumes treated with 20–500 ppm of neem oil did not display repellent response on oviposition of four species of pulse beetles involving four legume seeds specially during 24 hours of test except in cowpea treated at 500 ppm dose. The response to higher doses might have been camouflaged due to ovipositional preference to a lower dose in the first few hours. The ovicidal, larvicidal and juvenile hormone imbalance leading to deformity in adults were observed. The deformity was associated at 200–500 ppm and adult emergence was affected. Delayed emergence of adult was found in cowpea seeds treated at 500 ppm. The last instar larva failed to prepare emergence hole, a concurrent activity of abnormal development.

In another test, French bean treated with 50, 100, 200, 300 and 400 ppm doses of neem oil exhibited oviposition by *Z. subfasciatus* after four-and-a-half months but hatching was arrested at 400 ppm and F1 emergence at 100 ppm. However, after 7 months 300 ppm was needed to arrest F1 emergence.

Coefficient of regression values were worked out between oviposition and hatch, hatch and emergence of *Z. subfasciatus* and the values were 0.9149–0.9179. Thus a high degree of correlation was observed in these three parameters. Coefficient of regression were also worked for olfactory attraction and oviposition and hatch and emergence in case of *C. chinensis* initially. The value were between 0.897 and 0.936.

DISCUSSION

Jotwani and Sircar (1967) found 2 per cent neem kernel powder effective against *C. maculatus* developing in pea, chickpea, cowpea and mungbean up to 8 to 11 months

Table 1. Per cent olfactory attraction, oviposition, hatchability and F1 emergence of four pulse beetles on neem oil-treated seeds

Insect/ Period	Dose (ppm)	Choice	Attraction	Oviposition	Hatchability	Emergence (F1)
Chickpea						
C. chinensis						
Initial	500	Choice	0.0	17.5	2.8	0.0
(24 hrs)	50		17.4	29.9	31.8	23.5
	20		30.4	19.8	24.3	30.9
	0		21.8	32.8	41.1	45.6
	Wanderer		30.4	—	—	—
	Total number		23	137	107	68
	Mean		17.4	25.0	25.0	25.0
	±		12.8	7.5	16.3	19.0
C. analis						
Initial	500	Choice	12.1	25.0	5.1	6.4
	50		12.1	20.5	27.8	35.5
	20		30.3	25.0	26.6	32.2
	0		30.3	29.5	40.5	25.8
	Wanderer		15.2	—	—	—
	Total number		33	112	79	31
	Mean		21.2	26.5	25.0	25.0
	±		10.5	2.6	14.7	13.0
Cowpea						
C. maculatus						
Initial	500	Choice	8.3	12.8	0.0	0.0
	50		8.3	12.8	15.7	30.0
	20		16.7	23.1	25.0	40.0
	0		8.3	51.3	59.3	30.0
	Wanderer		58.4	—	—	—
	Total number		12	39	32	10
	Mean		10.4	25.0	25.0	25.0
	±		4.2	18.2	25.1	17.3
C. chinensis						
Initial	500	Choice	9.5	12.1	0.0	0.0
	50		19.0	39.7	24.4	5.9
	20		32.1	16.3	24.4	23.5
	0		19.1	31.1	51.2	70.6
	Wanderer		14.3	—	—	—
	Total number		21	141	82	34
	Mean		21.4	25.0	25.0	25.0
	±		12.0	12.9	20.9	32.0
C. analis						
Initial	500	Choice	5.3	17.3	0.0	0.0
	50		23.7	27.2	35.7	20.0
	20		36.8	30.2	32.1	50.0
	0		7.8	25.3	32.2	30.0
	Wanderer		26.3	—	—	—
	Total number		38	202	140	20
	Mean		18.4	25.0	25.0	25.0
	±		14.7	5.5	16.7	20.8

Contd.

Table 1. contd.

Insect/ Period	Dose (ppm)	Choice	Attraction	Oviposition	Hatchability	Emergence (F1)
Soybean						
C. analis						
Initial	500	Choice	10.3	29.4	0.0	0.0
	50		10.4	28.3	31.0	36.4
	20		13.8	27.6	46.0	45.4
	0		10.3	14.7	23.0	18.2
	Wanderer		54.2	—	—	—
	Total number		29	170	87	11
	Mean		11.2	25.0	25.0	25.0
	±		1.7	6.9	19.2	20.1
French bean						
Z. subfasciatus						
Initial	200	Choice	4.8	0.0	0.0	0.0
	150		0.0	10.5	7.0	12.0
	100		14.2	0.0	0.0	0.0
	40		0.0	12.4	9.3	16.0
	20		4.8	17.5	20.9	16.0
	0		4.8	59.6	62.8	56.0
	Wanderer		71.4	—	—	—
	Total number		21	57	43	25
	Mean		4.8	16.7	16.7	16.7
	±		5.2	22.2	23.9	20.6

of storage. Treated with NKP cowpea was protected up to 8 months against *C. maculatus* (Sowunmi and Arkinnusi, 1983). Yadav (1984) also found this dose effective on seven legumes against releases of *C. maculatus* and *C. chinensis* up to six months. Das (1986) reported 6 ml of neem oil/kg of chickpea antiovipositant against *C. chinensis* but 4 ml/kg (4000 ppm) dose was found ineffective. Neem kernel powder at the rate of 2 g/100 g of seed is expected to contain 0.6 g oil and equivalent dose is 6000 ppm. In the present tests dose of 500 ppm is 12 fold low.

Doses of 200, 400, 1000, 2000, 3000, 4000 and 5000 ppm of neem oil were applied on mungbean (*Vigna radiata* Wilcz.) seeds and tested against adult of *C. maculatus* Fab., *C. chinensis* and *C. analis*. Oviposition was arrested at 4000 ppm dose in *C. chinensis* and *C. analis* and at 5000 ppm dose in *C. maculatus*. The emergence of adults was checked at 1000 ppm in *C. chinensis*, 2000 ppm in *C. analis* and 400 ppm in *C. maculatus* (Yadav, 1985). After three months 5000 ppm remained effective against *C. maculatus*.

However, the traditionally used vegetable oils are cheap and acceptable due to long years of usage. Recently concerted efforts to develop suitable formulation of neem at the farm level to provide safe storage of legumes has been initiated. The immediate usage in cooking for human consumption is also an impediment because of the unacceptability of neem oil flavour and taste. Oil can be used as seed treatment for storage of the legumes provided germination is unaffected.

In the present study, very low concentrations were found effective against four species of pulse beetles infesting four legumes. The comprehensive information on parameters of olfactory attraction and associated oviposition, hatching and F1 emergence have brought out the explicit advantages of low dose application.

CONCLUSIONS

1) Neem oil possessed ovicidal, larvicidal and juvenile hormonal imbalancing properties at concentration of 200–500 ppm.
2) *Z. subfasciatus* was most susceptible infesting French bean.
3) *C. chinensis* was most susceptible than *C. maculatus* and *C. analis*.
4) Suitable formulation of neem oil is suggested to protect edible legume at farm level.

REFERENCES

Das, G.P. 1986. Effect of different concentrations of neem oil on the adult mortality and oviposition of *Callosobruchus chinensis* (Linn.) (Bruchidae:Coleoptera). *Indian Journal of Agricultural Science* 56: 793–794.

Jotwani, M.G. and P. Sircar. 1967. Neem seed as protectants against bruchid, *Callosobruchus maculatus* (Fabricius) infesting some leguminous seeds. *Indian Journal of Entomology* 29: 21–24.

Sowunmi, O.E. and O.A. Arkinnusi. 1983. Studies on the use of the neem kernel in the control of stored cowpea beetle (*Callosobruchus maculatus* F.). *Tropical Grain Legume Bulletin* 27: 28–31.

Yadav, T.D. 1984. Efficacy of neem (*Azadirachta indica* A. Juss) kernel powder as seed treatment against pulse beetles. *Neem Newsletter* 1: 5–6.

Yadav, T.D. 1985. Antiovipositional and ovicidal toxicity of neem (*Azadirachta indica* A. Juss) oil against three species of *Callosobruchus*. *Neem Newsletter* 2: 5–6.

Yadav, T.D. and R.K. Tanwar. 1985. Olfactory response of *Lasioderma serricorne* (Fab.) to different spices and dry fruit. *Annals of Agricultural Research* 6: 80–86.

37

Effectivity of Certain Plant Extracts against Larvae of *Mythimna separata* Wlk. and Adults of *Callosobruchus chinensis* Linn.

T.G.M. SINGH AND T.B. GOUR

Department of Entomology, College of Agriculture, Rajendranagar
Hyderabad 500 030, India

ABSTRACT

Seeds of neem (*Azadirachta indica* A. Juss.), custard apple (*Annona squamosa* Linn.), and *Cleome viscosa* (Linn.), pods of *Tephrosia villosa* Pers., debarked stem of *Milletia racemosa* and rhizomes of turmeric (*Curcuma longa* Linn.) were extracted in pertoleum ether and ethanol, and the extracts were tested against larvae of oriental armyworm, *Mythimna separata* Wlk. and adults of pulse beetle, *Callosobruchus chinensis* Linn.

Neem petroleum ether extract 0.1per cent was found to be repellent to the larvae of *M. separata,* while at lower concentration (0.06per cent) it showed antifeedant activity. Similar results were obtained with ethanol extract. When 0.1per cent petroleum ether extract was applied to sixth instar larvae, larval period was slightly extended but adults failed to emerge from pupae. Pupal mortality was observed at as low as 0.02per cent. Other plant extracts were less effective.

Potter's tower studies showed petroleum ether extract of neem to be more toxic (LC_{50} : 0.0165) than the ethanol extract (LC_{50} : 0.1421). LC_{50} value of DDVP was 0.0155. Custard apple seed petroleum extract (LC_{50} : 0.8288) was close to neem seed petroleum extract.

Redgram grains mixed with petroleum ether extract of neem at 0.06 % and 0.4% ethanol extract kept the adult pulse beetle, *C. chinensis* away. Growth index on treated grain was observed to be low. Toxicity to adult pulse beetle with these extracts was very low. These preliminary studies may open up avenues for neem utilization in the pest management of *M. separata* and *C. chinensis.*

INTRODUCTION

Synthetic organic insecticides have proven to be the most effective tools in pest control, however, the evils attendant on their use have led the biologists to seek replacement with safer and biodegradable ones. In this direction a great hope is being placed on plant products, because these possess a wide spectrum of biological properties which can hopefully be exploited in pest control programmes. These were established to be antifeedants (Pandey et al., 1977), repellents (Meisner et al., 1981), growth inhibitors (Steffens and Schumutterer, 1982), or as insecticides (Ali et al., 1983) against several crops as well as stored product pests. Further, naturally occurring phytochemicals are less harmful to the environment and hence, offer a great potential as safe, effective and economic pesticides.

Hence it was contemplated to evaluate the neem and other plant extracts to establish their repellent, antifeedant and insecticidal properties against oriental armyworm, *Mythimna separata* Wlk. (Noctuidae: Lepidoptera) a most destructive and formidable pest with a wide host range (rice, maize, barley, millets etc.) and pulse beetle *Callosobruchus chinensis* Linn. (Bruchidae: Coleoptera) a field as well as pest of stored pulses.

MATERIALS AND METHODS

Seeds of neem (*Azadirachta indica* A. Juss.), custard apple (*Annona squamosa* Linn.) and *Cleome viscosa* Linn., pods of *Tephrosia villosa* Pers., debarked stem of *Milletia racemosa* and rhizomes of turmeric (*Curcuma longa* Linn.) were shade dried, pulverised and extracted separately using a Soxhlet apparatus with solvents like petroleum ether, ethanol or methanol. Nine such extracts obtained were preserved at 5°C.

The cultures of the test insects, viz., *M. separata* and *C. chinensis* were maintained in the laboratory on sorghum leaves (var. CSH 5), and pigeonpea grains (var. LRG 30), respectively.

Repellent and Antifeedant Studies against *M. separata*

Leaf discs of sorghum leaves (4.5 cm dia.) were washed with distilled water, and after drying were dipped in 0.1 per cent concentration of various extracts for five minutes. Various dilutions of plant extracts were prepared using the respective solvents, used initially for extraction. Triton X – 100 at 0.04 per cent was used as an emulsifier. After air-dying, the treated leaf discs were placed randomly over a glass sheeth in a circular manner, along with an untreated leaf disc. Forty third instar larvae of *M. separata* starved for 24 hours were released in the centre. Leaf discs were then covered with a glass bell jar. The experiment was replicated thrice. After two hours, the number of leaves on each disc were counted, and the unconsumed leaf area was determined using a leaf-area meter.

Insecticidal Activity against *M. separata* Larvae

One ml of each of the diluted extracts was sprayed on petridishes (10 cm dia.) using a Potter's tower (0.28 kg per cm^2). Treated petridishes were dried and 10 two-day old third instar larvae of *M. separata* starved for 4 hours, were released. Three replications were maintained. The larvae after an exposure of 30 minutes were transferred into clean glass jar and were provided with sorghum leaves as food, and maintained at 25 ± 2°C and 70 ± 5% R.H. Mortality was recorded 24 hours later, and LC_{50} values calculated. DDVP was chosen as a standard insecticide.

Repellent Studies against *C. chinensis*

Twenty grams of pigeonpea, treated with six various concentrations of neem seed extracts of petroleum ether and its residues extracted with ethanol, were introduced into six arms of the olfactometer, while in one arm untreated grains were kept. Outer openings of the olfactometer were closed with rubber corks. Forty adults (2 days old) of *C. chinensis* were introduced in the olfactometer through the open mouth. The number of beetles reaching the pigeonpea grains in each of the seven arms of olfactometer were recorded after 30 minutes and 12 hours. The data were analysed statistically with three replications, adopting randomised block design.

Insecticidal Activity against *C. chinensis* Adults

The insecticidal activity of the nine plant extracts and malathion (standard) was evaluated against one-day adults of *C. chinensis*. LC_{50} values were determined following the same procedure as for larvae of *M. separata* described earlier.

Protection of Pigeonpea Grains from *C. chinensis* Damage

Samples of 100 gm grains of pigeonpea (var. LRG 30) were mixed with each of the plant extracts at various concentrations viz., 0.03, 0.06, 0.12, 0.2, 0.4 and 0.8 per cent (w/w). Ten pairs of one-day old beetles were introduced in each treatment. The number of undamaged grains by the grubs of subsequent progeny were recorded after 30 days of first release. In treatments showing no damage, a fresh batch of beetles were introduced on the following day and observations were recorded 30 days later. The process was continued in uninfested treatments and a maximum of seven such releases were made. Release of beetles was discontinued, if grain damage was noticed in any of the treatments. In treatments where the grains remained uninfested, final releases were made 180 days after treatments, and hence the last observations were made, in such situations, 210 days after the first release.

RESULTS AND DISCUSSION

Repellent and Antifeedant Properties against *M. separata* Larvae

Sorghum leaf discs treated with neem seed petroleum ether extract did not show presence of any larvae, while the untreated leaves had as many as nine larvae feeding, out of 40 larvae released (Table 1). Other extracts were intermediary with one to eight larvae on a single disc. Apart from mere repelling, the extracts also seem to have a pronounced influence, on larval feeding. Leaf disc treated with neem seed petroleum ether/ethanol extract showed one larvae, but it failed to feed on this disc. Other extracts showed considerably less feeding in comparison to untreated disc. Neem seed extracted

Table 1. Repellent and antifeedant effects of certain plant extracts against *M. separata* larvae

Plant extract (0.1% conc.)	Larvae found on leaf disc	Leaf area consumed (cm^2) per larva
Neem seed (Petroleum ether)	0	0.000
Neem seed (Petroleum ether/ethanol)	1	0.000
Custard apple seed (Petroleum ether)	2	0.082
Custard apple seed (Petroleum ether/methanol)	3	0.082
Tephrosia villosa pod (Petroleum ether)	4	0.145
T. villosa pod (Petroleum ether/methanol)	4	0.147
Milletia racemosa stem (Methanol)	5	0.180
Cleome viscosa seed (Petroleum ether)	4	0.166
Curcuma longa rhizome (Methanol)	8	0.182
Untreated control	9	0.943

with petroleum ether had a strong repellent effect on larvae of *M. separata*, and the residue thereof extracted with ethanol had the feeding prevention effects as well. Peirce (1981) reported that treated musk melon seedlings with azadirachtin successfully kept striped cucumber beetle, *Acalymna vitallum* Fab., away for 3 to 10 days. Shankar (1984) observed no feeding by larvae of *Achoea janata* L. when castor leaves treated with neem oil were offered.

Insecticidal Activity against *M. separata* Larvae

Among the nine plant extracts tested, neem seed petroleum ether extract appeared to be most effective (LC_{50}: 0.0165), against larvae of *M. separata* (Table 2). The value was close to the synthetic insecticide, DDVP, (LC_{50}: 0.0155). *Tephrosia* pod petroleum ether/methanol extract was least toxic (LC_{50}: 1.5699). Shankar (1984) has reported the superiority of neem seed oil over custard apple extract against *Spodoptera litura* F. larvae. Babu and Beri (1969) have extracted the neem seeds with alcohol, water, chloroform and acetone and reported alcohol as the best solvent in protecting castor leaves from damage by *Euproctis lunata* Wlk. Earlier, Pajni (1964), had indicated the effectivity of neem seed petroleum ether extract against red pumpkin beetles. It is not uncommon to observe a variation in toxicity of a chemical with different pest species. However, in a number of instances (Teotia and Pandey, 1979; Chauhan and Qadri, 1989), petroleum ether has been reported to be a better solvent for a number of plants.

Repellent Effects on Adults of *C. chinensis*

Pigeonpea grains treated with neem seed extracts were tested in an olfactometer at various concentrations. During the first half hour, there appeared to be some inconsistency with regard to the presence of population and increasing concentration of the extracts, especially at lower concentration levels (Table 3). Probably, during the period, the insects were still in the process of selecting their choice. Twelve hours after the release, the insect population decreased with increase in concentration. In petroleum ether extract at and beyond 0.04 per cent concentration, and in petroleum ether/ethanol

Table 2. Insecticidal activity of certain plant extracts against larvae of *M. separata*

Plant extract	Heterogenity		Regression equation	LC_{50}	Fiducial limits	Relative toxicities
	df	x^2				
Neem seed (Pet. ether)	4	4.5690	y = 4.7433x + 13.4490	0.0165	0.0147 0.0180	0.9393
Neem seed (Pet. ether/ethanol)	9	7.8538	y = 1.1754x + 5.9961	0.1421	0.1026 0.1967	0.1090
Custard apple seed (Pet. ether)	3	1.9960	y = 1.9731x + 8.4089	0.0187	0.0144 0.0243	0.8288
Custard apple seed (Pet. ether/ methanol)	6	1.8347	y = 1.3546x + 6.1345	0.1454	0.1075 0.1965	0.1066
T. villosa pod (Pet. ether)	6	2.0020	y = 1.2784x + 6.1622	0.1233	0.0891 0.1706	0.1257
T. villosa pod (Pet. ether/methanol)	5	4.1298	y = 1.3411x + 4.7373	1.5699	1.1380 2.1658	0.0098
C. viscosa seed (Pet. ether)	5	5.3084	y = 1.2745x + 4.8540	1.3019	0.9144 0.1856	0.0119
M. racemosa stem (Methanol)	5	6.9758	y = 1.2825x + 4.9672	1.0607	0.7424 1.5154	0.0146
C. longa rhizome (Methanol)	5	6.2718	y = 1.2493x + 4.8599	1.2946	0.9034 1.8552	0.0120
DDVP	5	3.7723	y = 1.7108x + 8.0962	0.0155	0.0117 0.0205	1.0000

Table 3. Population of *C. chinensis* adults present on pigeonpea grains treated with extract of neem seed (Mean of 3 replications)

Petroleum ether extract			Petroleum ether/ethanol		
% conc. w/w	Hours after release		% conc. w/w	Hours after release	
	0.5	12		0.5	12
0	58.65 (50.04)	70.00 (56.80)	0	61.88 (51.88)	63.12 (52.63)
0.01	11.25 (19.57)	15.00 (22.75)	0.1	13.12 (21.18)	27.50 (31.62)
0.02	7.50 (15.79)	10.00 (18.37)	0.2	10.62 (19.01)	9.38 (17.73)
0.04	13.75 (21.75)	5.00 (12.71)	0.4	14.38 (22.23)	0.00 (0.00)
0.06	8.75 (17.17)	0.00 (0.00)	0.6	0.00 (0.00)	0.00 (0.00)
0.08	0.00 (0.00)	0.00 (0.00)	0.8	0.00 (0.00)	0.00 (0.00)
0.10	0.00 (0.00)	0.00 (0.00)	1.0	0.00 (0.00)	0.00 (0.00)
CD at 5%	(1.85)	(2.22)	CD at 5%	(2.57)	(2.04)

Figures in parentheses are arcs in values.

extract of neem seed at 0.4 per cent and beyond there were no adults found after 12 hours of release. This clearly showed that neem seed petroleum ether extract has a strong repellent effect of *C. chinensis* adults.

In the other experiment pigeonpea grains treated with various extracts were tested for their period of prevention of infestation of *C. chinensis* (Table 4). Petroleum ether extracts of neem seeds and *T. villosa* pods (with seeds) at concentrations as low as 0.06 per cent effectively protected pigeonpea grains from pulse beetle infestation upto three months. At higher concentrations extracts from seeds of custard apple, *C. viscosa* and stem of *M. racemosa* offered protection upto 210 days. *C. longa* rhizome extract failed to offer any protection. Petroleum ether extract of neem seed at very low concentrations appears not only to have a strong repellence to *C. chinensis* but it also prevents the development of subsequent population by preventing the egg laying by the beetles. Protection to grains treated with neem products or extracts through antifeedant activity or preventing oviposition has previously been reported by Ivbijaro (1983), Pereria (1983), Sowunmi and Akinnusi (1983), Das (1986), Tanzubil (1986), Makanjuola (1989).

Insecticidal Activity against *C. chinensis* Adults

Petroleum ether extract of neem seed showed some toxicity against adults of *C. chinensis* (LC_{50} : 1.8974), though its relative toxicity (Table 5), in comparison to the standard insecticide, malathion (LC_{50} : 0.0447) was only 0.0236. Shankar (1984) against *Poecilocerus pictus* Fab. Pandey et al., (1981) against *Bagrada cruciferarum* Krik. and Chauhan and Qadri (1989) have reported similar observations. Neem extract has been reported to be toxic to *C. chinensis* (Naqvi et al., 1990), and chickpea seeds treated with neem oil were toxic to the beetle (Das, 1987). It would however be interesting to study the susceptibility of the malathion resistant strains of insects, to extracts of neem seeds which may provide useful information when control of stored grain insects is aimed at.

Table 4. Protection of pigeonpea grains against pulse beetle infestation, treated with various plant extracts

Plant extract	Minimum conc. % (w/w)	Period of protection (in days)
Neem seed (Petroleum ether)	0.06	90
Neem seed (Petroleum ether/ethanol)	0.20	90
Custard apple seed (Petroleum ether)	0.20	210
Custard apple seed (Petroleum ether/methanol)	0.20	210
T. villosa pod with seed (Petroleum ether)	0.06	90
T. villosa pod with seed (Petroleum ether/methanol)	—	—
M. racemosa stem (Methanol)	0.80	210
Cleome viscosa seed (Petroleum ether)	0.80	210
Curcuma longa rhizome (Methanol)	—	0
Untreated control	—	0

Table 5. Insecticidal activity of certain plant extracts against adults of *C. chinensis*

Plant extract	Heterogenity		Regression equation	LC_{50}	Fiducial limits	Relative toxicities
	df	x^2				
Neem seed (Pet. ether)	6	1.7152	y = 1.7903x + 4.5020	1.8974	1.5392 2.3389	0.0236
Neem seed (Pet. ether/ ethanol)	4	3.9250	y = 2.6680x + 4.0487	2.2728	1.9225 2.6968	0.0197
Custard apple seed (Pet. ether)	8	8.4976	y = 3.5681x + 2.6398	4.5865	4.1728 5.0412	0.0098
Custard apple seed (Pet. ether/ methanol)	8	7.9270	y = 3.1693x + 2.9699	4.3708	3.9370 4.8525	0.0102
T. villosa pod (Pet. ether)	3	6.2653	y = 3.0339x + 3.9323	2.2486	1.9274 2.6235	0.0199
T. villosa pod (Pet. ether/ methanol)	8	9.2512	y = 3.9394x +1.6409	7.1232	6.5461 7.7512	0.0063
Milletia racemosa stem (Methanol)	9	7.3287	y = 3.3361x + 2.5191	5.5417	5.0448 6.0874	0.0081
Cleome viscosa seed (Pet. ether)	8	6.4923	y = 4.2929x + 1.5250	6.4488	5.9689 6.9672	0.0069
Curcuma longa rhizome (Methanol)	8	5.3126	y = 3.8008x + 1.7196	7.2958	6.6827 7.9651	0.0061
Malathion	3	0.0466	y = 4.0736x + 10.4974	0.0447	0.0394 0.0507	1.0000

REFERENCES

Ali, S.I., O.P. Singh and U.S. Misra. 1983. Effectiveness of plant oils against pulse beetle, *Callasobruchus chinensis* Linn. *Indian Journal of Entomology* 45: 6–9.

Babu, T.H. and Y.P. Beri. 1969. Efficacy of neem (*Azadirachta indica* A. Juss.) seed extracts in different solvents as a deterrent to the larvae of *Euproctis lunata* Wlk. *The Andhra Agricultural Journal* 16(4): 107–111.

Chauhan, I.S. and S.M.M. Qadri. 1989. Studies on antifeedant and insecticidal activity of indigenous plant extacts against fruit borer in okra. *Pesticides* 23(1): 40–42.

Das, G.P. 1986. Pesticidal efficacy of some indigenous plant oils against the pulse beetle, *Callosobruchus chinensis* Linn. (Coleoptera: Bruchidae). *Bangladesh J. Zool.* 14(1): 15–18.

Das, G.P. 1987. Efficacy of neem oil on the egg and grub mortality of *Callosobruchus chinensis* Linn. (Coleoptera: Bruchidae). *Tropical Grain Legume Bull.* 34: 14–15.

Ivbijaro, M.F. 1983. Preservation of cowpea, *Vigna unguiculata* (L.) Walp., with the neem seed, *Azadirachta indica*. A. Juss. *Protection Ecology* 5(2): 177–182.

Makanjuola, W.A. 1989. Evaluation of extracts of neem (*Azadirachta indica* A. Juss) for control of some stored product pests. *J. Stored Products Res.* 25(4): 231–237.

Meisner, J., K.R.S. Ascher, R. Aly and J.D. Warthen. Jr. 1981. Response of *Spodoptera littoralis* Boisd and *Earias insulana* Boisd to Azadirachtin and Salanin. *Phytoparasitica* 9(1): 27–32.

Naqvi, S.N.H., R. Tabassum, R. Zia and S.M. Nurulain. 1990. Toxicity and residual effect of neem extract (factor C) against stored grain pest, *Callosobruchus analis*. *Pakistan J. Zool.* 22(3): 271–277.

Pajni, H.R. 1964. Some observations on the insecticidal activity of the fruits of Dharek. Research Bulletin in Punjab University 15th Part (3–4): 345–364.

Pandey N.D., Mahendra Singh and G.C. Tewari. 1977. Antifeeding repellent and insecticidal properties of some indigenous plant material against mustard saw fly. *Indian Journal of Entomology* 39(1): 60–63.

Pandey, U.K. and Mamta, Pandy and S.P.S. Chauhan. 1981. Insecticidal properties of some plant products against *Bagrada cruciferarum. Indian Journal of Entomology* 43(4): 404–407.

Pereria, J. 1983. The effectiveness of six vegetable oils as protectants of cowpeas, banana and groundnut against infestation by *Callosobruchus maculatus* F. *J. Stored Products Res.* 19: 57.

Pierce, R. 1981. Natural products repel cucumber beetle. *Journal of Agricultural Research* 30(2): 12–14.

Shankar, M. 1984. Efficacy of certain plant extracts in the control of certain leaf feeding insects. M.Sc. (Ag.) thesis submitted to Andhra Pradesh Agricultural University, Rajendranagar, Hyderabad.

Steffens, R.J. and M. Schmutterer. 1982. The effect of crude methanolic neem (*Azadirachta indica*) seed kernel extract on metamorphosis and quality of adults of the Mediterranean fruit fly, *Ceratetis capitata* (Diptera: Tephritidae). *Zeitschrift fur Angewandte Entomologie* 94(1): 98–103.

Sowunmi, O.E. and O.A. Akinnusi. 1983. Studies on the use of neem kernel in the control of stored cowpea beetle (*Callosobruchus maculatus* F.) *Tropical Grain Legume Bull.* 27: 28–31.

Tanzubil, P.B. 1986. The use of neem products in controlling the cowpea weevil, *Callosobruchus maculatus Proc. 3rd Int. Neem Conf.* Nairobi, Kenya pp. 517–523.

Teotia T.P.S. and G.P. Pandey. 1979. Insecticidal properties of rhizome of sweet flag (*Acorus calamus*) against rice weevil, *Sitophilus oryzae. Indian Journal of Entomology* 41: 91–94.

38

Neem Effects on *Spodoptera litura* (Fabr.): A Holistic Study

P.J. RAO, SHASHI GUPTA, D. MOHAN RAJ AND K.R. KRANTHI

Division of Entomology, Indian Agricultural Research Institute, New Delhi 110 012, India

ABSTRACT

A detailed study on the effects on neem (*Azadirachta indica* A. Juss.) on *Spodoptera litura* was carried out. Hexane and methanol extracts of neem seed kernels proved to be larval repellents and oviposition deterrents. Injection of a single dose of azadirachtin (1 µg/g body weight) into sixth-instar larvae affected food consumption and utilization, midgut enzymes, haemolymph constituents (haemocytes, proteins, trehalose, uric acid, nin-hydrin positive substances and cations), secretions of the median neurosecretory cells of the brain, corpora allata size, and ecdysteroid titres. It reduced nucleic acid and protein concentrations of the larval fat body, reduced pupal weight, and disrupted ovarian development. One-time exogenous application of methoprene (20 µg/larva) and injection of ecdysterone (4 µg/g) 24 hrs after azadirachtin injection into sixth-instar larvae were not able to overcome azadirachtin effects on ecdysteroid titres and pupation. Adults emerged from azadirachtin-treated larvae showed poor ovarian weight and reduced ovarian proteins and vitellogenin leading to reduced fecundity, hatchability, pupation, and adult emergence.

INTRODUCTION

The neem tree (*Azadirachta indica* A. Juss.) has been known for its plant protection properties from time immemorial. The introduction of synthetic insecticides in the late 1930s eliminated the use of botanicals in plant protection. With the realization of inherent drawbacks associated with synthetic insecticides such as development of insect resistance, environmental pollution, and ecological disturbance, the scientific community connected with plant protection research was forced to look for alternatives. Pradhan et al. (1962) demonstrated the antifeedant properties of neem seed kernel extracts against locusts. Since then renewed interest in neem started. The active ingredient from neem, azadirachtin, was isolated and characterized (Butterworth and Morgan, 1968; Zanno et al., 1975; Broughton et al., 1986).

Azadirachtin, a tetranortriterpenoid, was tested against a number of insects and was shown to be a strong feeding inhibitor and growth regulator (Butterworth and Morgan, 1968; Warthen, 1989; Schluter et al., 1985; Schmutterer, 1990). Though crude extracts of neem seed kernels are in use in Asia and Africa, azadirachtin use is being considered in several countries (Rembold, 1984). A neem-based product, Margosan-O, has been approved by the US Environmental Protection Agency for selective use (Larson, 1989).

Information on varied effects of neem extracts and azadirachtin is widespread over a number of insect species. Detailed studies on neem effects confined to a single insect species are relatively lacking. The paper reports the effects of certain neem extracts on larval and adult behaviour, and the effects of a single dose of injected azadirachtin (1 μg/g body weight) on the last-instar larvae of *Spodoptera litura*, and its carryover effects on the succeeding generation. Such detailed studies will facilitate understanding of not only the fuller potential of neem as a plant protection agent but also the various events leading to growth disruption in insects.

MATERIALS AND METHODS

The rearing of the test insect, *S. litura*, in the laboratory, preparation of neem seed kernel extracts and their testing, injection of azadirachtin (1 μg/g body weight) into 0–6 hr sixth-instar larvae, determination of food utilization indices, midgut enzymes, sampling of haemolymph, and the estimation of its constituents have been given elsewhere (Ayyangar and Rao, 1989a, b, 1990). The *in situ* staining of neurosecretory cells and estimation of corpora allata (CA) volume have been described (Shashi Gupta and Rao, 1990).

The chitin percentage of the cuticle was determined by the method of Hackman and Goldberg (1971) and the cuticle proteins were extracted according to Anderson and Hojrup (1987).

Ecdysteroid titres of haemolymph were determined by radioimmunnoassay according to Borst and Connor (1972) as described by Warren and Gilbert (1988).

Nucleic acid estimations were according to modified method of Munro and Fleck (1966).

The ovary samples were prepared for histological studies following the method of Feber and O'Brien (1968) and stained with 1 per cent toludine blue according to Schulz (1981).

Antiserum to larval haemolymph storage proteins and sex-specific proteins were prepared by immunizing healthy albino rabbits using affinity column purified haemolymph and ovary extract, respectively. Ouchterlony agar double diffusion (Ouchterlony, 1948) and ELISA (Clark and Adams, 1977) methods were used for detection and estimation of storage proteins and vitellogenin.

The protein profiles of haemolymph were carried out by SDS-PAGE gradient gel electrophoresis according to the method of Laemmli (1970) on a LKB-vertical electrophoretic system.

The data were subjected to statistical analyses using split-plot design according to Panse and Sukhatme (1957).

RESULTS AND DISCUSSION

Neem as Larval Repellent

Olfactometric tests of neem seed kernel extracts proved to be larval repellents. Methanolic extract at 0.02 per cent concentration repelled 63.9 per cent third instars larvae. It was superior to 0.046 per cent hexane extract. The effect of methanolic extract increased to 70.6, 81.9 and 92.1 per cent in fourth, fifth and sixth instar larvae respectively (Table 1). The increase in repellence with each succeeding instar could be due to fuller development of sensory system leading to greater discrimination.

Neem as Oviposition Deterrent

Demarcated areas of filter papers treated with methanol and hexane extracts of neem seed kernels along with control were folded and offered for egg laying in jars containing adults. The deterrence of the hexane extract (0.02 per cent) lasted for four days, whereas that of methanol extract (0.01 per cent) continued for five days as indicated by the absence of egg laying in the treated areas. During the five-day period, the total number of eggs laid in control was 1828. Only on the fifth day was there egg laying in the area treated with hexane extract (Table 2).

Table 1. Effect of methanol extract of neem seed kernels (0.02 per cent) on larval behaviour of *S. litura*

Larval stage	% larvae repelled
III	63.9
IV	70.6
V	81.9
VI	92.1

Source: Ayyangar and Rao (1989a).

Table 2. Effect of neem seed kernel extracts on oviposition behaviour of *S. litura*

Treatment	Daily egg laying				
	I	II	III	IV	V
Control	351.8 ± 10.4	346.8 ± 8.1	327.5 ± 25.1	344.5 ± 9.4	457.5 ± 21.9
Hexane extract (0.02%)					433.50 ± 22.14
Methanol extract (0.01%)					

Source: Ayyangar and Rao (1989a).

Azadirachtin Effects on 6th Larval Stage

Consumption and utilization of food and midgut enzymes

Azadirachtin reduced food intake and approximate digestibility (AD). The dry matter, carbohydrate, and nitrogen intake in treated larvae were lower and the values were 160, 15.09 and 8.83 mg/larva as against 190, 18.43 and 10.42 mg/larva in control. Similarly, AD values of dry matter, carbohydrates and nitrogen in treated larvae were 64.5, 90.8 and 87.9 per cent as compared to 67.7, 94.1 and 91.9 per cent in control, respectively (Ayyangar and Rao, 1989b). These lower AD values suggested that azadirachtin affected digestibility. This proved to be correct as there was reduction in the midgut enzyme activity (Table 3).

Table 3. Effect of azadirachtin (1 μg/g) on midgut enzyme activity of sixth-instar larvae of *S. litura*

Enzyme	Control	Azadirachtin	% reduction
Alkaline phosphatase	4.427	2.675	36.63
Alkakline protease[2]	2.652	1.806	31.69
amylase[3]	17.383	13.498	22.32
Invertase[4]	14.953	12.195	18.39

[1] Activity based on casein hydrolysed/hr/mg protein.
[2] Activity based on μmol p-nitrophenol released/hr/mg protein.
[3] Activity based on μmol maltose released/hr/mg protein.
[4] Activity based on μmol glucose released/hr/mg protein.

Source: Ayyangar and Rao (1989b).

Haemolymph constituents

Total haemocyte count provides an indication of normal functioning of an insect as haemocytes perform transport, synthesis, storage, detoxication, and other functions. Azadirachtin significantly reduced the total haemocyte count with the maximum reduction at 48 hrs after the treatement (Table 4). Since total haemocyte count is dependent on ecdysone titres (Crossley, 1975; Prasada Rao et al., 1984), the drastic reduction may be attributed to the interference of azadirachtin with hormone system.

The trend in protein concentration in both control and treated larvae remained the same with lower values in the latter. The highest concentration in both was at 72 hrs (control: 6.523 g/100 ml; treated: 6.260 g/100 ml). However, in the case of storage proteins (Fig. 1) the maximum concentration in control was at 48 hrs (4.191 g/100 ml), while in treated larvae was at 72 hrs (1.22 g/100 ml). Since storage proteins constitute the bulk of haemolymph protein during the last instar in lepidopteran insects, lack of concentration buildup in the treated larvae will affect food reserves in pupal stage with consequent effects on development and reproduction.

In contrast to protein concentration, nin-hydrin positive substance concentration was higher in the treated larvae, suggesting greater breakdown of protein. The trend in uric acid concentration in both control and treated larvae was the same with highest concentration at 72 hrs, which coincided with gut purging (Fig. 2). Uric acid accumulation was observed in *Manduca sexta* at prepupal and pupal stages, leading to higher values (Buckner and Caldwell, 1980).

Table 4. Effect of azadirachtin (1 μg/g) on total haemocyte count of sixth-instar larvae of *S. litura*

Hours after treatment	Number of haemocytes/mm³	
	Control	Treatment
24	20,400 ± 674	12,775 ± 396
48	40,725 ± 847	6625 ± 593
72	17,775 ± 1344	6300 ± 1042
Mean	26,300	8567

CD between main treatments: 2883
CD between two subtreatments in the same main treatments: 5538
CD between two main treatments in the same subtreatments: 6208

Source: Ayyangar and Rao (1990).

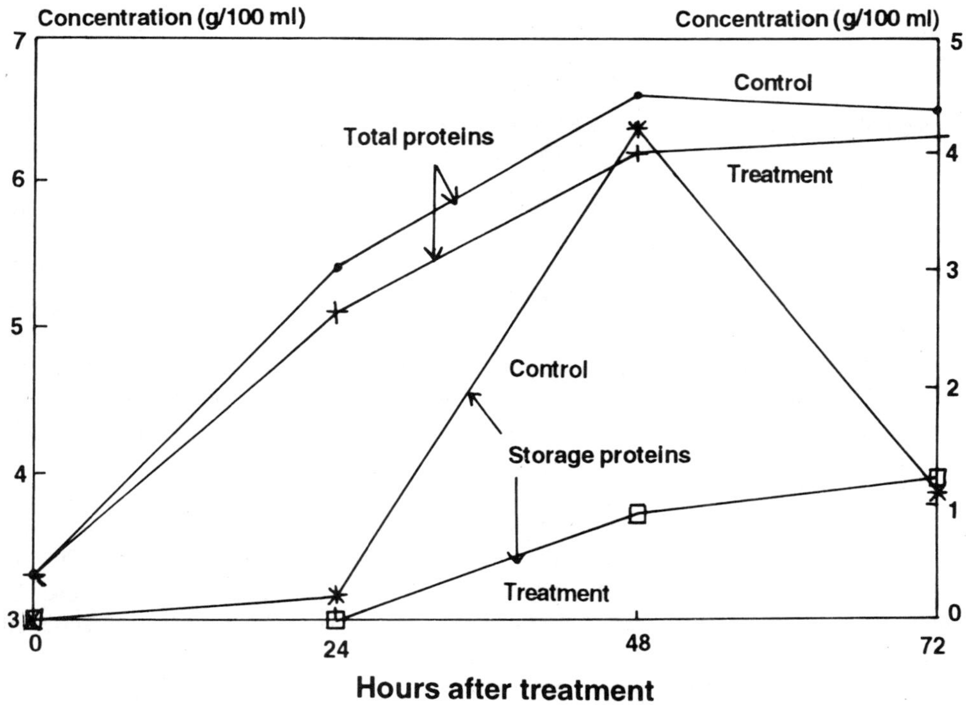

Fig. 1. Effect of Azadiractin (1 μg/g) on haemolymph proteins of sixth-instar larvae of *S. litura*

The electrophoretic pattern of larval haemolymph proteins showed the presence of three storage proteins with molecular weight 144 kDa (SP1), 68 kDa (SP2), and 36 kDa (SP3) in high concentration on day 3 of sixth-instar larvae, besides many other proteins

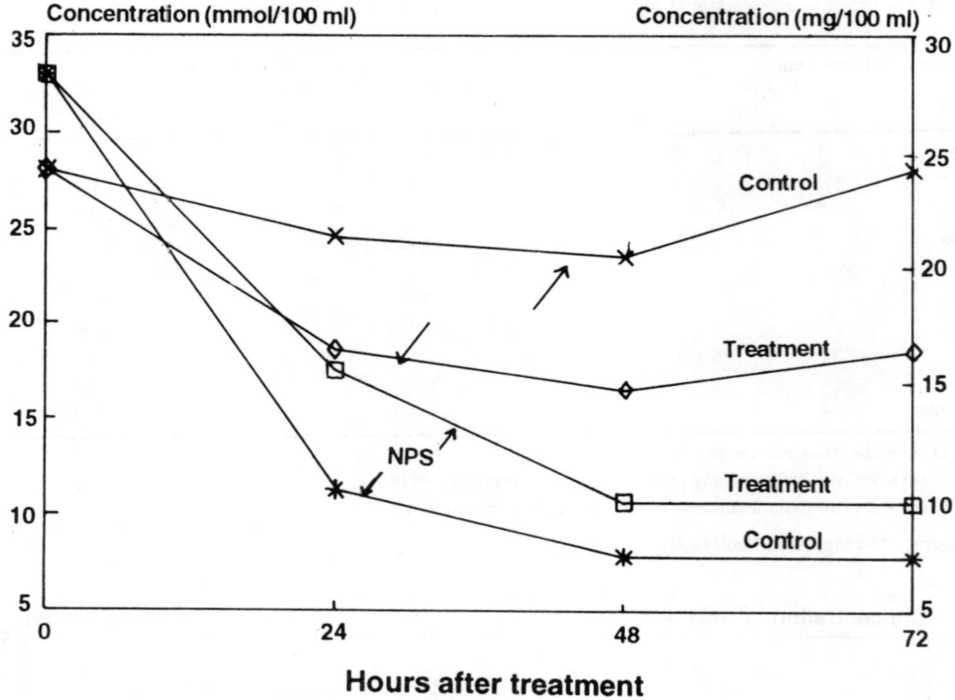

Fig. 2. Effect of azadirachtin (1μg/g) on haemolymph NPS (mmol/100 ml) and uric acid (mg/100 ml) of sixth-instar *S. litura* (Ayyangar and Rao, 1990)

(Fig. 3). It is interesting to note that SP3 band was missing in treated larvae at 72 hrs (Mohan Raj, 1990).

The accumulation pattern of the storage proteins was similar to the findings of Tojo et al. (1985), who reported three storage proteins on the third day of the last-instar larvae of this insect. Electrophoretic pattern when considered with storage proteins concentration showed that the total storage protein concentration in treated larvae was that of SP1 and SP2 only. The molecular weight of storage proteins of *S. litura* are in good agreement with those of some other insects (Tojo et al., 1980; Miller and Silhacek, 1982).

Although trehalose increased in both control and treated larvae, the rate of increase in treated larvae was lower (Fig. 4). As compared to an increase of 977.33 mg/100 ml in control at 72 hrs after moulting, the increase in treated larvae was only 615.00 mg/ 100 ml. The reduction may be due to reduced synthesis or utilization of trehalose for energy under stress. Reduced concentrations of trehalose were observed after insecticidal treatment in *Achaea janata* (Ramdev and Rao, 1989).

Azadirachtin increased sodium concentration throughout and potassium only up to 48 hrs and decreased calcium and magnesium concentrations (Fig. 5). Cations play an important role in nerve activity and other metabolic functions. Azadirachtin effect, on cations, thus may have an important bearing on physiology.

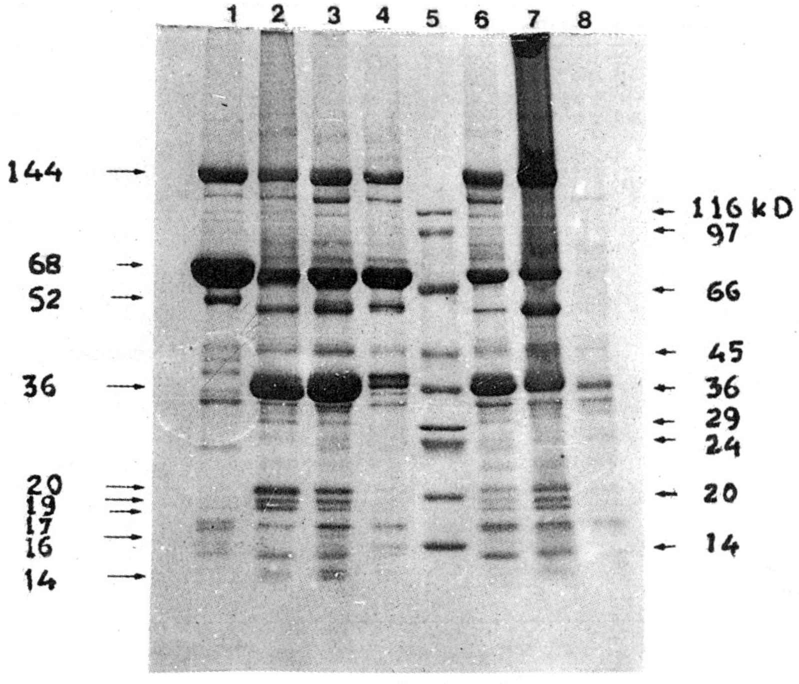

Fig. 3. Effect of azadirachtin on larval haemolymph proteins:
Lanes 8, 7, 4, and 2: sixth-instar larvae 0, 24, 48 and 72 hrs (Control); Lanes 6, 3, and 1: Sixth-instar larvae 24, 48 and 72 hrs after treatment; Lane 5: Standard protein molecular weight markers.

Cuticle

There was no significant difference in the overall chitin and protein percentage both in the treatment and in control (Table 5). Chitin percentage increased in control between 48 and 72 hrs in contrast to a decrease in treated larvae and the reverse was true with protein. Since the larvae enter prepupal condition after 72 hrs, the significance of these changes in prepupal condition is difficult to suggest (Shashi Gupta, 1988).

Neuroendocrine system

In situ staining of median neurosecretory cells revealed the presence of four A1 and five to six A2 cells in sixth-instar larvae. The staining intensity of A1 cells increased up to 48 hrs and thereafter it decreased. The decreased intensity in control in contrast to increased intensity in treated larvae at 72 hrs suggests the discharge of neurosecretory material in control against the accumulation in treated larvae (Fig. 6). Greater accumulation of neurosecretory material in adults at 72 hrs after azadirachtin treatment was reported in *Epilachna varivestis* (Schulz and Schluter, 1984) and *Schistocerca gregaria* (Subrahmanyam and Rao, 1986).

The mean corpora allata (CA) volume was also lower in treated larvae than in control (Table 6), suggesting reduced CA activity. Azadirachtin was reported to reduce CA size (Dorn et al., 1986) and may lead to its disintegration (Schulz and Schluter, 1984), leading to reduced juvenile hormone (JH) titres (Rembold et al., 1987).

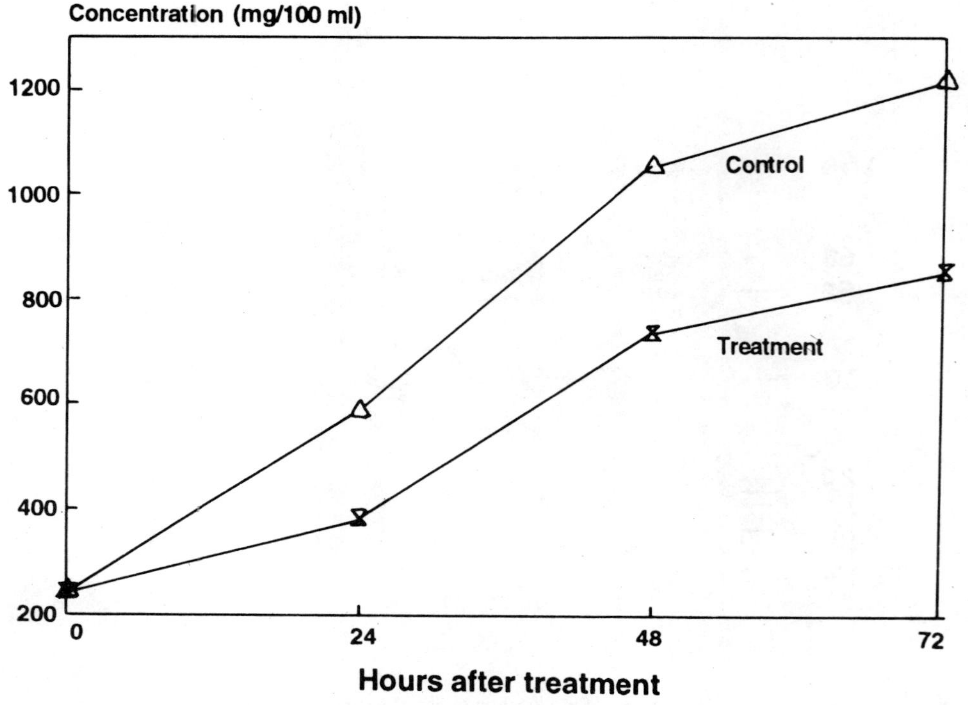

Fig. 4. Effect of azadirachtin on haemolymph trehalose of sixth-instar larvae of *S. litura* (Ayyangar and Rao, 1990)

Table 5. Effect of azadirachtin (1 μg/g) on cuticular chitin and protein content of sixth-instar larvae of *S. litura*.

Hours after treatment	Chitin (%)		Protein (%)	
	Control	Treatment	Control	Treatment
0	37.36 ± 0.80	37.36 ± 0.80	62.64 ± 0.78	62.64 ± 0.78
24	27.56 ± 0.60	29.50 ± 0.44	72.44 ± 0.60	70.50 ± 0.44
48	27.13 ± 0.49	30.96 ± 0.21	72.87 ± 0.49	69.04 ± 0.21
72	30.20 ± 0.74	27.13 ± 0.87	69.80 ± 0.74	72.87 ± 0.87
Mean	30.56	31.24	69.44	68.76
CD between main treatments		1.44		1.44
CD between two subtreatments in the same main treatments		2.09		2.06
CD between two main treatments in the same subtreatments		2.11		2.09

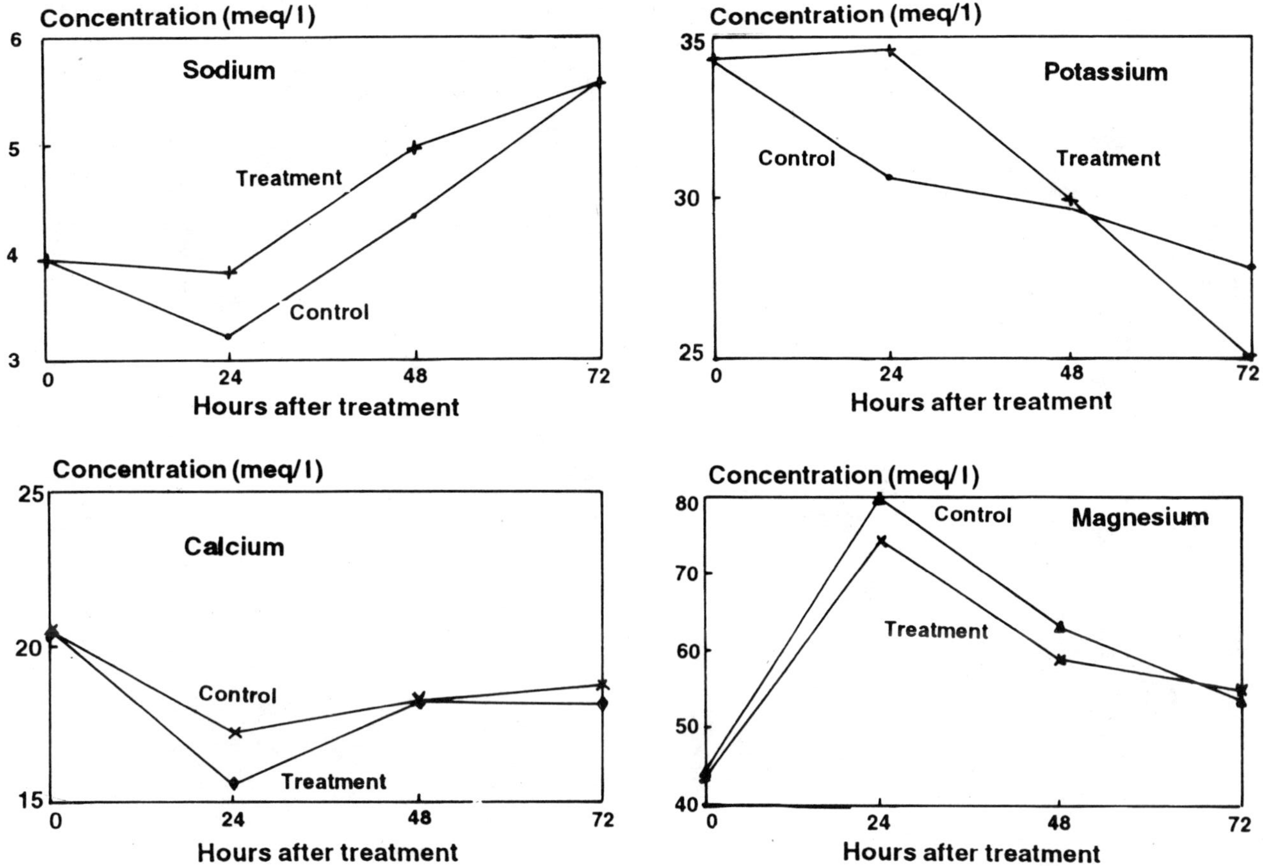

Fig. 5. Effect of azadirachtin on haemolymph cations of sixth-instar larvae of *S. litura* (Ayyangar and Rao, 1990)

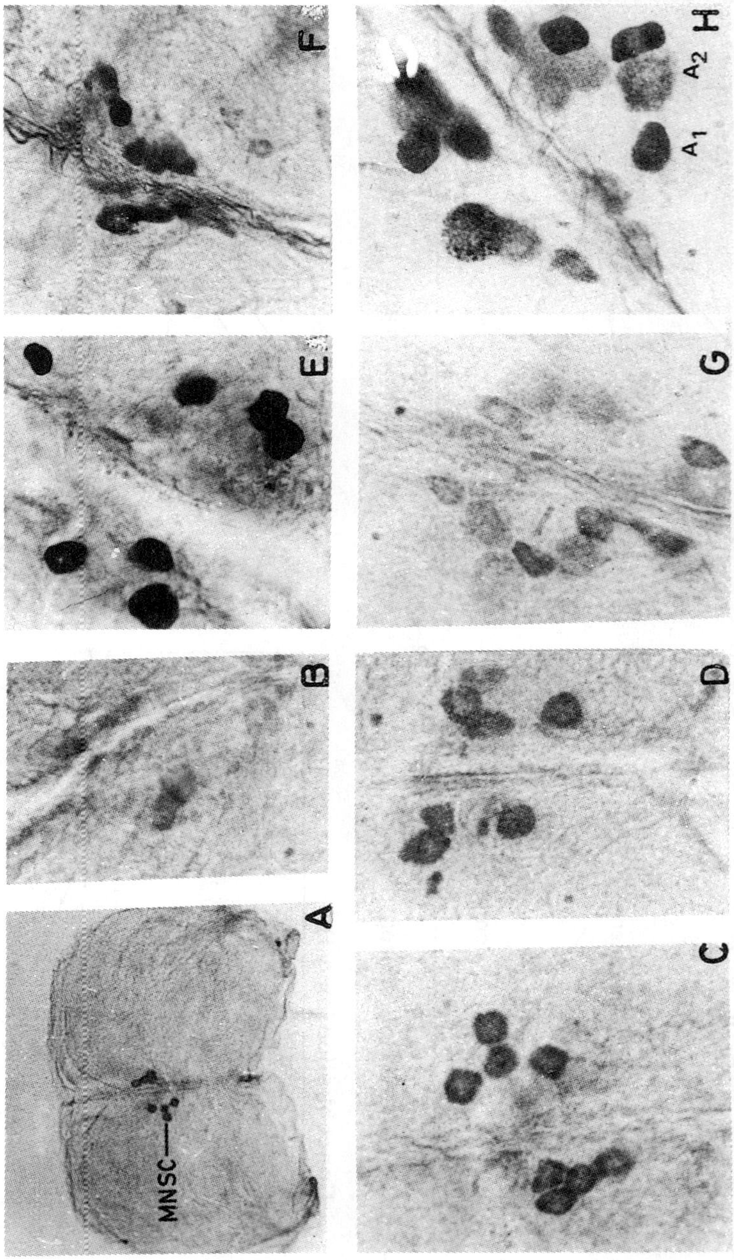

Fig. 6. Effect of azadirachtin on median neurosecretory cells (MNSC) of sixth-instar *S. litura*: A: Whole mount of brain showing MNSC. B: MNSC in 0–6 hr brain. C: MNSC in 24 hr control. D: MNSC in 24 hr treated. E: MNSC in 48 hr control. F: MNSC in 48 hr treated. G: MNSC in 72 hr control. H: MNSC in 72 hr treated. Magnification × 1120 (Shashi Gupta and Rao, 1990).

Table 6. Effect of azadirachtin of CA volume ($mm^3 \times 10^{-4}$) of sixth-instar of *S. litura*

Hours after treatment	Number of haemocytes/mm^3	
	Control	Treatment
0	1.057 ± 0.005	1.057 ± 0.005
24	1.852 ± 0.005	1.164 ± 0.082
48	1.757 ± 0.003	1.400 ± 1.133
72	2.769 ± 0.109	1.992 ± 0.059
Mean	1.859	1.403

CD between main treatments: 0.128
CD between two subtreatments in the same main treatment: 0.711
CD between two main treatments in the same subtretment: 0.626

Source: Shashi Gupta and Rao (1990).

Radioimmunoassay of haemolymph of sixth-instar larvae revealed the presence of two ecdysteroid peaks. The first peak appeared at 24 hrs after the moult with a titre of 242 ± 31 ng/ml and the second one at 84 hrs had a titre of 2465 ± 280 ng/ml in control. Larvae treated 24 hrs after moult showed two peaks, one at 84 hrs (1768 ± 315 ng/ml) and the other at 120 hrs (2177 ± 473 ng/ml), with pupation following thereafter. The sharp build up in the titre between 48 and 72 hrs as in control was lacking in the treated larvae. Thus, azadirachtin affected the attainment of the titre required for moult induction, thereby delaying the pupation. Therefore, studies were taken up to see whether exogenous application of methoprene (topical, 20 µg/larva) and 20-OH ecdysone (injection, 4 µg/g), 24 hrs after azadirachtin treatment (i.e., 48 hrs after the moult), would overcome the delay in the titre. Contrary to our expectations, both the treatments further delayed pupation (Fig. 7) as the titre required for moult induction was delayed (Figs. 8, 9) Interestingly, in all the hormonal treatments, the ecdysteroid titre required for moult induction was delayed (Kranthi, 1991).

Azadirachtin is known to affect juvenile hormone (JH) and ecdysteroid titres (Rembold et al., 1987; Min-Li and Shin-Foon, 1987). Administration of JHA in *Rhodnius* was shown to overcome moult inhibitory effect of azadirachtin (Garcia and Rembold, 1984) but failed to eliminate azadirachtin effect in *Galleria mellonella* (Malczewska et al., 1988). Similarly, Schluter et al. (1985) reported that application of hormone to larvae of *Manduce sexta* failed to counteract the azadirachtin effects. Recently, it was suggested that the mechanism of controlling JH titre by azadirachtin in lepidopterans may be different from others (Koul and Isman, 1991).

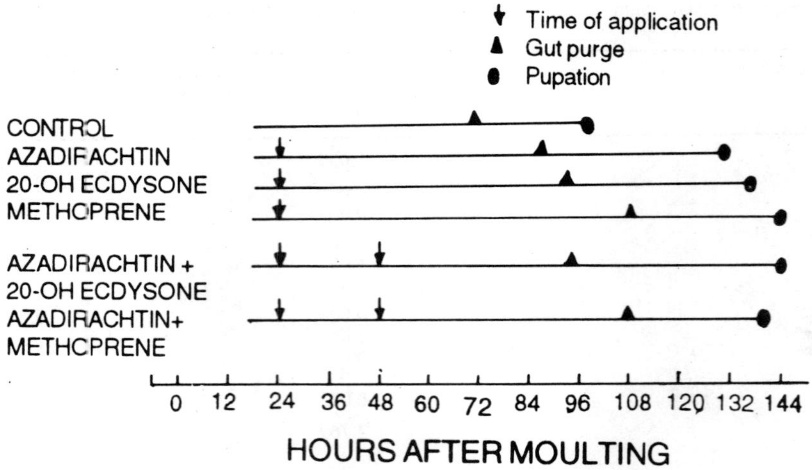

Fig. 7. Effect of treatments on metamorphosis of sixth-instar larvae of *S. litura*

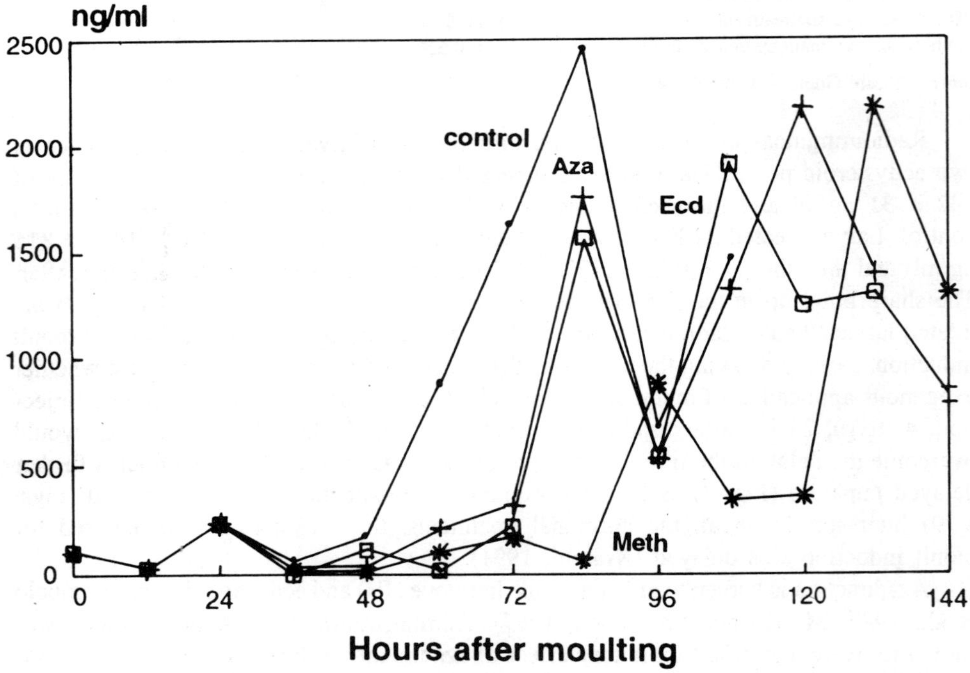

Fig. 8. Effect of treatments on haemolymph 20-OH ecdysone titres (ng/ml) of sixth-instar *S. litura*

Larval fat body

Azadirachtin significantly reduced DNA, RNA, and protein concentrations (Fig. 10) and these concentrations were the highest at 24, 48, and 72 hrs respectively in control. The reduced protein concentration was due to the effect of azadirachtin on nucleic acid concentrations (Shashi Gupta, 1988).

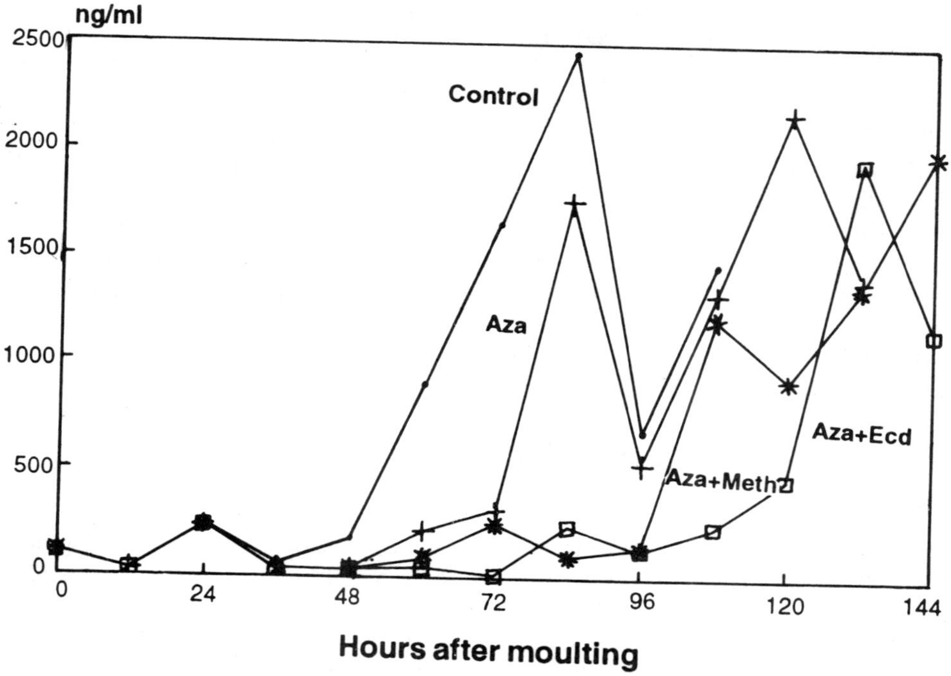

Fig. 9. Effect of treatments on haemolymph 20-OH ecdysone titres (ng/ml) of sixth-instar *S. litura*

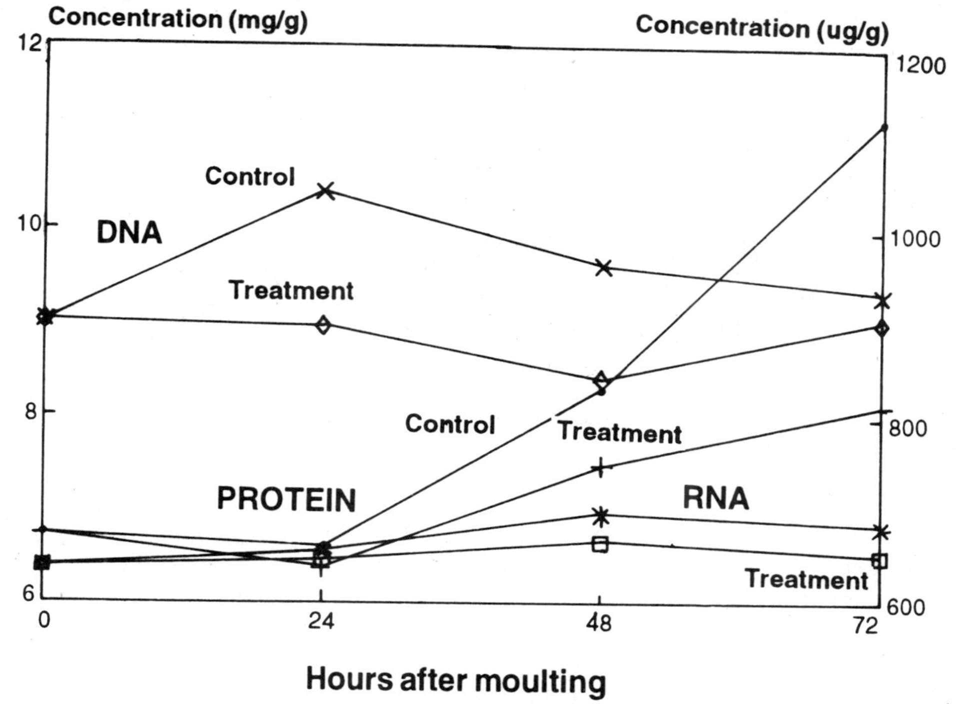

Fig. 10. Effect of azadirachtin (1 μg/g) on larval fat body RNA, DNA and proteins of sixth-instar *S. litura*

Effect of Azadirachtin on Pupal Stage

The pupae obtained from 0–6 hr treated sixth-instar larvae showed decreased pupal weight. The control pupae weighed 337.20 ± 19.10 mg/pupa as compared to 264.10 ± 10.26 mg/pupa in treated insects on the day of pupation and both of them lost weight during the pupal period (Shashi Gupta, 1988). The decrease in pupal weight was due to decrease in food reserves as azadirachtin affected food consumption and utilization.

The five-day-old treated female pupae showed lower fat body storage protein concentration (0.04 mg/g) than control (4.36 mg/g), indicating lack of sequestration of it into fat body from the larval stage (Mohan Raj, 1990). As sequestration into fat body is dependent on ecdysteroid titre, lack of it may be responsible for lower storage protein in fat body of the treated (Tojo et al., 1985).

Effect of Azadirachtin on Adult Stage

The adults obtained from 0–6 hr treated sixth-instar larvae revealed a number of carryover effects of azadirachtin. Azadirachtin reduced the ovary weights, ovary protein, and vitellogenin concentration (Fig. 11). The initial weight of the ovary in treated adults was 35.0 ± 7.70 mg just after adult emergence as compared to 66.7 ± 2.0 mg in control. There was continuous increase in the ovary weight in control up to 72 hrs (103.7 ± 3.3 mg), whereas in treated insects the increase was only up to 48 hrs (63.7 ± 2.8 mg). Subsequently, the weight decreased conspicuously to 47.3 ± 6.60 mg at 72 hrs in treated insects. The increase in ovary weight was due to greater yolk deposition in oocytes and

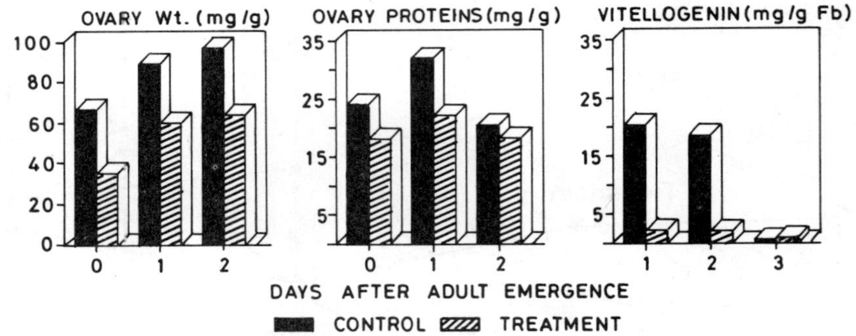

Fig. 11. Carryover effects of azadirachtin on *S. litura*

the reduction in weight in treated insects could be due to resorption of oocytes or inhibition in the uptake and deposition of yolk. The high vitellogenin values of the fat bodies in control suggest greater synthesis of sex-specific proteins which was lacking in the treated insects (Shashi Gupta, 1988; Mohan Raj, 1990).

The synthesis of yolk was shown to be dependent on JH and 20-OH ecdysone (Handler and Postlethwait, 1978; Jowett and Postlethwait, 1980) and their uptake on JH (Rembold, 1989). Azadirachtin appears to have affected both these processes. Indeed, inhibition of oogenesis and ovarian ecdysteroid synthesis were observed in *L. migratoria* after a single dose of injected azadirachtin (Sieber and Rembold, 1983).

Histological studies on ovary development revealed that azadirachtin drastically affected the follicular epithelium, yolk protein concentration, and vitellin membrane formation (Fig. 12). The damage to the follicular epithelium increased with the age of the ovary and the interfollicular tissue was disrupted. Greater vacuolization and sparse yolk granules were seen in the ooplasm of the treated insects (Shashi Gupta, 1988). Schulz (1981) and Schluter (1984) working with *Epilachna varivestis* reported histological disturbances in developing ovary in treated adults. It was established that growth and maturation of oocytes are largely dependent on reserve material available in the pupae of lepidopterans (Englemann, 1971) which was lacking in the treated insects.

The carryover effects of azadirachtin in the succeeding generation were reduced fecundity (39.51 per cent), hatchability (53.77 per cent), pupation (42.62 per cent) and adult emergence (41.92 per cent) over control (Ayyanar and Rao, 1991).

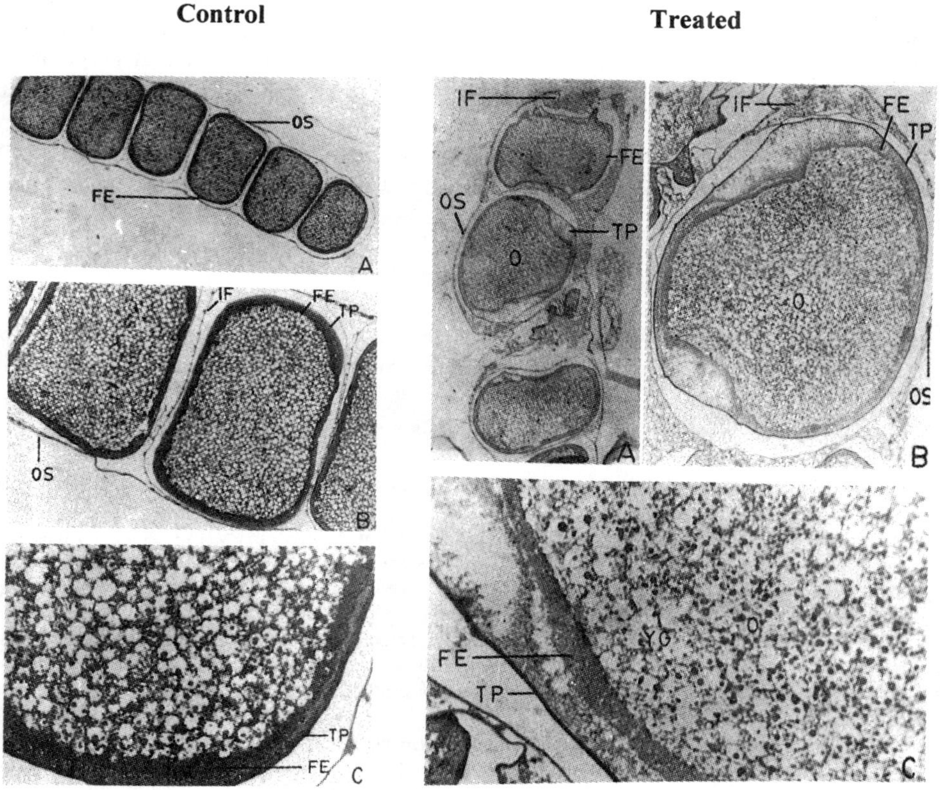

Fig. 12. Effect of azadirachtin on (24-hr-old) ovary. A: Portion of longitudinal section ovariole × 280. B: A single oocyte × 700. C: Part of oocyte magnified from B × 1120. OS: Ovariole sheath. IT: Interfollicular tissue. TP: *Tunica propria*. FE: Follicular epithelium. O: Ooplasm. YG: Yolk granules.

CONCLUSIONS

Neem affected *S. litura* at olfactory, gustatory, and physiological levels. Both methanol and hexane extracts of neem seed kernels proved to be larval repellents and oviposition deterrents.

Injection of single dose of azadirachtin into the sixth-instar larvae caused a number of immediate and delayed effects, some of which were hormone mediated. Histological studies in the larval stage revealed that the neuroendocrine system was affected by azadirachtin as reflected by the delayed buildup of neurosecretion, reduced size of CA, and shifting of ecdysteroid peak. The effect of azadirachtin on secretions of median neurosecretory cells led to hormonal disturbance in the larval stage resulting in delayed pupation and this could not be overcome with the exogenous application of hormones like methoprene and 20-OH ecdysone.

Reduced food intake and protein levels led to reduced food reserves at pupal stage. The adults emerged from such pupae showed lesser ovary weight, protein, and vitellogenin concentrations. Histological studies of ovary revealed azadirachtin effects on oogenesis and yolk deposition. Such adults were found to be biologically weak in perpetuating their progeny as their fecundity, hatchability, pupation, and adult emergence were reduced in the succeeding generation.

If the olfactory, gustatory, and growth disruptive properties of neem observed on *S. litura* are exploited, neem could be a good candidate for integrated pest management programmes.

ACKNOWLEDGEMENTS

We are grateful to Prof. H. Rembold of Max-Planck Institute for Biochemistry, Munich, Germany, for a free sample of azadirachtin and to Prof. Tojo, Saga University, Japan, for providing storage protein antiserum.

REFERENCES

Anderson, S.O. and P. Hojrup. 1987. Extractable protein from abdominal cuticle of sexually mature locusts *Locusta migratoria*. *Insect Biochem*. 17: 45–51.

Ayyangar, G.S.G. and P.J. Rao. 1989a. Neem (*Azadirachta indica* A. Juss.) extracts as larval repellents and ovipositional deterrents to *Spodoptera litura* (Fabr.). *Indian J. Entomol*. 51: 121–124.

Ayyangar, G.S.G. and P.J. Rao. 1989b. Azadirachtin effects on consumption and utilization of food and midgut enzymes of *Spodoptera litura* (Fabr.). *Indian J. Entomol* 51: 373–376.

Ayyangar, G.S.G. and P.J. Rao. 1990. Changes in haemolymph constituents of *Spodoptera litura* (Fabr.) under the influence of azadirachtin. *Indian J. Entomol*. 52: 69–83.

Ayyangar, G.S.G. and P.J. Rao. 1991. Carry over effects of azadirachtin on development and reproduction of *Spodoptera litura* (Fabr.) *Ann. Entomol.* 9: 55–57.

Borst, D.W. and J.D.O. Connor. 1972. Arthropod moulting hormone: Radioimmune assay. *Science*, 178: 418–419.

Broughton, H.B., S.V. Ley, M.Z. Slawin, D. Williams and E.D. Morgan. 1986. X-ray crystallographic structure determination of detigloyldihydroazadirachtin and reassignment of the structure of the limonoid, insect antifeedant azadirachtin. *J. Chem. Soc. Chem. Comm*. 46–47.

Buckner, J.S. and J.M. Caldwell. 1980. Uric acid levels during the last larval instar of *Manduca sexta* an abrupt transition from excretion to storage in fat body. *J. Insect Physiol*. 26: 27–32.

Butterworth, J.H. and E.D. Morgan. 1968. Isolation of a substance that suppresses feeding in locusts. *J. Chem. Soc. Chem. Comm*. 23–24.

Clark, M.F. and A.N. Adams. 1977. Characteristics of the microplate method ELISA for the detection of plant viruses. *J. Gen. Virol*. 34: 475–483.

Crossley, A.C.S. 1975. The cytophysiology of insect blood. *Adv. Insect Physiol*. 11: 117–221.

Dorn, A., J.M. Rademacher and E. Sehn. 1986. Effect of azadirachtin on the moulting cycle, endocrine system and ovaries in the last instar larvae of the milkweed bug, *Oncopeltus fasciatus*. *J. Insect Physiol.*, 32: 231–238.

Engelmann, F. 1971. Juvenile hormone controlled synthesis of female specific protein in the cockroach, *Leucophaea maderae*. *Arch. Biochem. Biophys.* 145: 439–447.

Feder, N. and T.P. O'Brien. 1968. Plant microtechnique: Some principles and new methods. *Amer. J. Bot.* 55: 123–142.

Garcia, E.S. and H. Rembold. 1984. Effect of azadirachtin on ecdysis of *Rhodnius prolixus*. *J. Insect Physiol.* 12: 939–941.

Hackman, R.H. and M. Goldberg. 1971. Proteins of the larval cuticle of *Agrianome spinicollis*. *J. Insect Physiol.* 17: 335–341.

Handler, A. and J. Postlethwait. 1978. Regulation of vitellogenin synthesis by ecdysone and juvenile hormone. *J. Exp. Zool.* 247–254.

Jowett, T. and T. Postlethwait. 1980. Regulation of vitellogenin synthesis in *Drosophila* ovaries and fat body by 20-OH ecdysone and a juvenile hormone analogue. *Dev. Biol.* 80: 225–234.

Koul, O. and M.B. Isman. 1991. Effect of azadirachtin on the dietary utilization and development of the variegated cutworm, *Peridroma saucia*. *J. Insect Physiol.* f37: 591–598.

Kranthi, K.R. 1991. Effect of hormones on the azadirachtin induced changes in *Spodoptera litura* (Fabr.). Ph.D. thesis, IARI, New Delhi.

Laemmli, U.K. 1970. Cleavage of structural proteins during the assembly of the head of bacteriophage T4. *Nature* (Lond.) 227: 680–685.

Larson, R.O. 1989. The commercialization of neem. In: *Focus on Phytochemical Pesticides*, Vol. 1. *The Neem Tree*. M. Jacobson. (ed.), CRC Press, Boca Raton, pp. 155–168.

Malczewska, M., D.B. Gelman and B. Cymbrowski (1988). Effect of azadirachtin on development, juvenile and ecdysteroid titres in chilled *Galleria mellonella* larvae. *J. Insect Physiol.* 34: 725–732.

Miller, S.G., and D.L. Silhacek. 1982. The synthesis and uptake of haemolymph storage proteins by the fatbody of the greater wax moth, *Galleria mellonella* (L.). *Insect Biochem.* 12: 293–300.

Min-Li, Z. and C. Shin-Foon. 1987. The effects of azadirachtin on the ecdysteroid titre in the larvae of *Ostrinia furnicalis* Guenee. *J. Appl. Entomol.* 103: 355–359.

Mohan Raj, D. 1990. Effect of azadirachtin on major proteins associated with ovarian development in *Spodoptera litura* (Fabr.), Ph.D. thesis, IARI, New Delhi.

Munro, H.N. and A. Fleck. 1966. Recent developments in the measurements of nucleic acids in biological materials. *Analyst* 91: 78–88.

Ouchterlony, O. 1948. *In vitro* method for testing the toxin producing capacity of diptheria bacteria. *Acta Path. Microbiol. Scand.* 25: 186.

Panse, V.G. and P.V. Sukhatme. 1957. *Statistical Methods for Agricultural Workers*. ICAR, New Delhi.

Pradhan, S., M.G. Jotwani and B.K. Rai. 1962. The neem seed deterrent to locusts. *Indian Fmg.* 12: 7–11.

Prasada Rao, C.G., Aparna Ray and P.S. Ramamurthy. 1984. Effect of ligation and ecdysone on total haemocyte count of the tobacco caterpiller, *Spodoptera litura*. *Can. J. Zool.* 62: 1461–1463.

Ramdev, Y.P. and P.J. Rao. 1989. Effect of insecticides on *Achaea janata* Linn. III. Haemolymph proteins, ninhydrin positive substances, trehalose and uric acid. *Proc. Indian Nat. Sci. Acad.* B55: 35–40.

Rembold, H. 1984. Secondary plant products in insect control with special reference to the azadirachtins. In *Advances in Invertebrate Reproduction* Vol. 3. W. Engles. (ed.), Elsevier, Amsterdam, 481 pp.

Rembold, H. 1989. Isomeric azadirachtins and their mode of action. In: *Focus on Phytochemical Pesticides*, Vol. 1, *The Neem Tree*. M. Jacobson. (ed.), CRC Press, Boca Raton, pp. 48–67.

Rembold, H., M. Uhl and T. Muller. 1987. Effect of azadirachtin-A on hormone titres during the gonadotrophic cycle of *Locusta migratoria*. In: *Natural Pesticides from Neem Tree and other Tropical Plants*. H. Schmutterer and K.R.S. Ascher (eds.), GTZ, Eschborn, pp. 289–298.

Schluter, U. 1984. Post-embryonal entwicklung des fettkorpis von *Epilachna varivestis* (Coleoptera, Coccinellidae). *Zool. Jb. Anat.* III: 343–355.

Schluter, U., H.J. Bidman and S. Grewe. 1985. Azadirachtin affects growth and endocrine events in larvae of the tobacco hornworm, *Manduca sexta*. *J. Insect Physiol.* 31: 773–777.

Schmutterer, H. 1990. Properties and potential of natural pesticides from the neem tree. *Azadirachtra indica*. *Ann. Rev. Entmol.* 35: 271–279.

Schulz, W.D. 1981. Pathological alterations in the ovaries of *Epilachna varivestis* induced by an extract from neem kernels. *Proc. 1st Inter. Neem Conf.* H. Schmutterer and K.R.S. Ascher (eds.), GTZ, Eschlorn, pp. 81–97.

Schulz W.D. and U. Schluter. 1984. Structural damages caused by neem in *Epilachna varivestis*. A summary of histogical and ultrastructural data. II. Tissues affected in adults. *Proc. 2nd int. Neem Conf.* H. Schmutterer and K.R.S. Ascher (eds.), GTZ, Eschborn, pp. 237–252.

Shashi Gupta. 1988. Effect of azadirachtin on proteins and nucleic acids of *Spodoptera litura* (Fabr.) Ph.D. thesis, IARI, New Delhi.

Shashi Gupta and P.J. Rao. 1990. Effect of azadirachtin on neuroendocrine system of *Spodoptera litura* (Fabr.). *Indian J. Entomol.* 52: 589–594.

Sieber, K.P., and H. Rembold. 1983. The effects of azadirachtin on the endocrine control of moulting in *Locusta migratoria. J. Insect Physiol.* 29: 523–527.

Subrahmanyam, B. and P.J. Rao. 1986. Azadirachtin effects on *Schistocerca gregaria* Forskal during ovarian development. *Curr. Sci.* 55: 534–538.

Tojo, S., M. Morita, N. Agui and K. Hirumai. 1985. Hormonal regulation of phase polymorphism and storage protein fluctuation in the common cutworm, *Spodoptera litura. J. Insect Physiol.* 31: 283–292.

Tojo, S., M. Nagata and M. Kobayashi. 1980. Storage proteins in the silkworm, *Bombyx mori. Insect Biochem.* 10: 289–303.

Warren, J.T. and L.I. Gilbert. 1988. Ecdysone metabolism and distribution during pupal-adult development of *Manduca sexta. Insect Biochem.* 17: 65–82.

Warthen, J.D., Jr. 1989. Neem (*Azadirachta indica* A. Juss.): Organisms affected and reference list update. *Proc. Entomol. Soc. Wash.* 91: 367–388.

Zanno, P.R., I. Miura, K. Nakanishi and D.L. Elder. 1975. Structure of insect phagorepellent azadirachtin. Applications of PRFT CWD Carbon-13 nuclear magnetic resonance. *J. Amer. Chem. Soc.* 97: 1975–1977.

39

Effect of Neem Extracts on Brown Planthopper, *Nilaparvata lugens* and Its Symbionts

S. RAGURAMAN, S. JAYARAJ AND R.C. SAXENA[1]

Assistant Professor (Entomology), Agricultural Research Station
Paramakudi 623 707, Ramnad District, Tamil Nadu, India
[1]International Centre of Insect Physiology and Ecology
P.O. Box 30, Mbita, Kenya

ABSTRACT

In two separate experiments, the neem seed derivatives — neem oil, aqueous seed kernel extract, cake extract, neem coated urea and neem seed bitters — were tested against brown planthopper (BPH) nymphs by exposing them on treated plants to know whether the different extracts also affect the yeast-like symbionts (YLS). The YLS population was significantly lower in BPH that fed on neem-treated rice plants than on untreated ones. The decrease may be due either to the direct effect of the compounds present in the extracts on the YLS or through BPH fat-bodies which store YLS. Nymphs that developed on treated plants weighed much less than those on control plants. Significantly higher mortality occurred of neem-treated plants than on controls, mainly because of the growth-disrupting effect of neem extracts.

INTRODUCTION

Intracellular symbionts of Homoptera, including bacteria and yeasts have been studied for their structural characters, host control on development, transovarial transmission and role of nutrition (Houk and Griffiths, 1980; Chen et al., 1981a; Koyama, 1986). The rice brown planthopper (BPH), *Nilaparvata lugens* (stal) harbour yeast-like symbionts (YLS) in the abdominal fat bodies and transmit them to the next generation through ovary (Nasu, 1963; Noda, 1977). Antibiotics such as polyoxin S and chloramphenicol, are reported to reduce the population of YLS of BPH (Chen et al., 1981 b). However, no earlier report on the effect of plant products on endosymbionts of insects is available, hence this study.

MATERIALS AND METHODS

Neem Oil (NO) and Custard Apple Oil (CAO) were obtained from the local market. For neem seed kernel extract (NSKE), the required quantity of NSK powder was loosely bundled in a muslin cloth and kept soaked overnight. The kernel powder inside the cloth was squeezed on the next day and harder particles left in the cloth were rejected. For neem cake extract (NCE), the required quantity of NC was taken and soaked in water overnight. The solution was filtered through a muslin cloth. Neem seed bitters (NSB), a lyophilized extract of neem seed kernel by water was also used, Experiments were conducted in glasshouse. The plants were sprayed with hand-operated pneumatic sprayer after adding Teepol at 0.1 per cent concentration. Neem cake coated urea NCU (1:1 W/W) was added at the time of planting the seedlings in pot.

Experiment I

Twenty-day-old IR 20 rice plants were sprayed with No, No + CAO, NSKE and NCE. NCU was applied at the time of planting the seedlings in pots. Twenty-five first instar nymphs were allowed per replication to feed on each of the treated and untreated control plants and the survival was observed once in four days. Each treatment was replicated four times. The design of the experiment was completely randomized block design.

The surviving fifth instar nymphs were collected for counting the intracellular YLS. All the nymphs from individual treatments were weighed separately. The insects were then homogenized using 0.8 per cent saline solution and made up to known volume. The homogenate was taken using Thoma pipette. An improved Neubauer haemocytometer (Weber, England) was used under a phase contrast microscope for counting symbionts. The symbionts were counted from the five one-millimetre squares of the homocytometer. The total number of symbionts per insect was calculated as per the formula given (Noda, 1977).

Total Number of symbionts = $a\ (x + m)\ /\ n.v.$ where,
a = number of symbionts in v,
m = total weight of insects homogenized
n = number of insects homogenized,
v = volume of the insect homogenate sampled in the haemocytometer, and
x = volume of the saline solution (0.8%) used to homogenize the insects.

Experiment II

In addition to the observations in the first experiment, this experiment was conducted to know the effect of neem extracts on honeydew excretion and its free aminoacids. Two sets of 30-day-old IR 20 plants having eight tillers/pot and one tiller/pot in each treatment were sprayed with neem products in four replications. The first set of plants (eight tillers/pot) were infested with 30 of four-day-old nymphs per replication, whereas, the second set of plants were protected by polyester cages. The number of survivors were counted on the fourth and eight day after release. The survivors (12-day-old insects) were collected and transferred to second set of treated plants containing honeydew chamber. Here four insects per replication were maintained with four such replication in each treatment. The insects were collected in vials after 24 hours feeding on plants. The honeydew spots were developed as described by Sogawa and Pathak (1970) and Bharathi (1982), and the spots were eluted in 0.1 per cent copper sulphate in 70 per cent ethanol. The total free amino acids content was estimated by the method of block et al. (1955) and expressed as glutamic acid standard in μg in honeydew excreted by four BPH. Soon after collection, the insect's size (length and breadth) and weight were assessed. The symbiont count was carried out as explained in Experiment I. In addition, symbiont's size (length and breadth) was assessed.

RESULTS

Effects of Neem Products on Mortality of BPH

The data have clearly brought out the mortality of BPH nymphs with different treatments (Table 1). In the first experiment, NSKE 5 per cent caused a high mortality of 73.47 per cent followed by NO + CAO 3 per cent (67.35) and NO 3 per cent (65.31), on the fourth day of observation. NCE 10 per cent and NCU basal application recorded 56.12 and 48.98 per cent mortality, respectively. Even after eight days of application a substantial increase in mortality of BPH nymphs was noticed in all the neem treatments.

In the second experiment, NSB 2500 ppm recorded a mortality of 48.47 per cent and 66.10 per cent, on the fourth and eighth day of treatment, respectively (Table 1).

Effect of Neem Products on Honeydew Excretion and Total Free Amino Acids

The results of the second experiment revealed that honeydew excretion and total free amino acids in honeydew were greatly reduced as compared to control (Table 2). The area of honeydew spots ranged from 114.00 to 208.00 mm^2 in neem treatments as against 663.25 mm^2 in control. Similarly, the total free amino acids content expressed as glutamic acid standard was very much reduced (110.03–116.39 μg in honeydew excreted by 4 BPH) in neem treatments compared to untreated check (208.22 μg in honeydew execreted by 4 BPH).

Table 1. Effect of neem products on mortality of BPH

Treatments	Experiment I % corrected mortality		Experiment II % corrected mortality	
	4 Dat*	8 Dat	4 Dat*	8 Dat
No 3%	65.31^b	74.49^b	70.44^a	86.09^a
No + CAO (2:1) 3%	67.35^b	77.55^{ab}	68.71^a	83.48^a
NSKE 5%	73.47^a	79.59^a	57.40^b	74.79^b
NCE 10%	56.12^c	63.27^c	68.70^a	84.35^a
NCU (Basal Application)	48.98^d	63.27^c	—	—
NSB 2500 ppm	—	—	48.71^c	66.10^c

* DAT—Days after treatment
In a column, means followed by same alphabet are not significantly different (p = 0.05) by DMRT.

Table 2. Effect of neem products on honeydew excretion and total free amino acids of honeydew of BPH

Treatments	Mean area of Honeydew spots (mm^2)	Mean of total free amino acid (μg in honeydew excreted by 4 BPH
NO 3%	185.00[a]	110.03[a]
NO + CAO (2:1) 3%	208.00[a]	129.87[a]
NCE 10%	173.00[a]	123.13[a]
NSKE 5%	114.00[a]	115.87[a]
NSB 2500 ppm	183.25[a]	116.39[a]
Control	663.25[b]	200.22[b]

* Total free amino acid is expressed in terms of glutamic acid.
In a column, means followed by same alphabet are not significant statistically (P = 0.05) by DMRT

Effect of Neem Products on Body Size and Weight of BPH

In the first experiment, weight of BPH recorded in different treatments showed no significant difference, they however differed significantly from control (Table 3).

In the second experiment, the weight of BPH was low on 5 per cent NSKE-treated plants (0.93 mg) while it was 1.88 mg in control. The length of BPH did not vary significantly among the neem products tested (2.095 to 2.416 mm) but was significantly shorter as compared to control (3.585 mm). And, BPH nymph measured a minimum breadth of 0.844 mm in NSKE 5 per cent and was followed by NO + CAO 3 per cent, NCE 10 per cent NSB 2500 ppm and NO 3 per cent, whereas in control it was 1.169 mm. All the insects in control reached adult stage while in neem treatments they were nymphs.

Table 3. Effect of neem products on body size and weight of BPH

Treatments	Experiment I	`Experiment II		
	Mean weight of BPH (mg)	Mean weight of BPH (mg)	Mean size of BPH (mm)	
			Length	Breadth
No 3%	1.05[a]	1.10[ab]	2.213[a]	1.053[cd]
No + CAO (2:1) 3%	1.12[a]	1.07[ab]	2.354[a]	0.904[ab]
NSKE 5%	1.02[a]	1.09[ab]	2.091[ab]	0.944[abc]
NCE 10%	1.12[a]	0.93[a]	2.075[a]	0.844[a]
NCU (Basal application)	1.16[a]	—	—	—
NSB 2500 ppm	—	1.38[b]	2.416[a]	0.988[abc]
Control	1.42[b]	1.88[c]	3.585[b]	1.169[d]

In a column, means followed by same alphabet are not significantly different (P = 0.05) by DMRT.

Effect of Neem Products on Endosymbionts of BPH

In the first experiment, significantly decreased number of YLS was recorded in all the neem treatments (24.72 to 31.08 × 10^4/BPH) when compared to untreated control (40.34 × 10^4/BPH (Table 4). And this was confirmed by the results of the second experiment

Table 4. Effect of neem products on endosymbionts of BPH

Treatments	Experiment I	Experiment II			
	Mean number of symbionts (× 10^4/BPH)	Mean number of (× 10^4/BPH)	Mean % of Budding symbionts	Size of symbionts	
				Length (μ)	Breadth (μ)
No 3%	24.72[a]	20.95[ab]	44.31[a]	13.02[b]	3.26[ab]
No + CAO (2:1) 3%	27.45[a]	19.63[ab]	35.08[a]	19.63[ab]	3.29[ab]
NSKE 5%	24.91[a]	21.04[ab]	32.59[a]	13.03[b]	3.39[ab]
NCE 10%	27.42[a]	17.21[a]	36.05[a]	13.33[b]	3.45[b]
NCU (Basal application)	31.08[a]	—	—	—	—
NSB 2500 ppm	—	26.80[a]	33.72[a]	12.65[b]	3.30[ab]
Control	40.35[b]	42.32[c]	31.56[a]	11.92[a]	3.14[a]

In a column, means followed by same alphabet are not significantly different (p = 0.05) by DMRT.

where, NSKE 5 per cent recorded the least YLS population of 17.21 × 10^4/BPH, and was followed by other treatments, whereas in control it was 42.32 × 10^4/BPH. Interestingly, the mean percentage of budding YLS did not vary significantly. Again, there was no appreciable reduction in the size of YLS in insects fed on neem-treated plants and, in contrast, the size was found to be reduced in insects fed on untreated plants.

DISCUSSION

The mortality of early instars was more than 70 per cent in NSKE 5 per cent, NO + CAO 3 per cent and NO 3 per cent treatments. The probable reasons are of poor growth and development of BPH in addition to various other physiological and behavioural effects of neem. The dramatic decrease may probably due to the direct effect of the principles of neem products on YLS or on the fat bodies or imbalanced hormonal metabolism thus leading to indirect effect as reduced size and weight of BPH. It has been shown by Bidmon et al. (1986) that several stages tissues are sensitive including brain and the fat body to azadirachtin in *C. vicina*. Since the store house of YLS is the fat bodies more detailed studies have to be made on the exact mode of action of azadirachtin against the YLS in BPH.

It was observed that the size (particularly length) of YLS decreased in the control insects. This might probably be due to more number of young cells and due to competition the size was diminished. Nasu et al. (1981) reported that even in the normal population of BPH two morphologically different YLS were isolated namely, oval-shaped cells (8–10 μ) and sheath-like cells (15–20 μ) in both egg and fat body. Hence, further investigations are necessary on the different morphs of YLS both in egg and fat bodies.

It is evident that the nutritional requirement of hemipterous insects in less than others because the symbionts supply many essential nutrients which inturn control the rate of reproduction, adult longevity, size and weight etc. Thus, the nutritional aspects of the insect host and its dependence on endosymbionts could critically be exploited with the use of neem products. It might then be possible to use specific antibiotics or metabolic inhibitors or growth disruptors (specific plant products) in the control of symbiont-dependent insects.

REFERENCES

Bharathi, M. 1982. Moderate resistance in rice to the brown planthopper *Nilaparvata lugens* (Stal) (Homoptera: Delphacidae). M.Sc., (Ag.) thesis (unpub.), Tamil Nadu Agric. Univ., Coimbatore, 95 p.

Bidmon, H.J.G. Kauser and J. Koolman. 1986. Action of azadirachtin on blowfly larvae and pupae. Third International Neem Conference, July 10–15, 1986. Abst. p. 32, ICIPE, Nairobi, Kenya.

Block, R.J., E.L. Durrum and G. Zweig. 1955. *A Manual of Paper Chromatography and Paper Electrophoresis*. Academic Press Inc., New York, 70 p.

Chen, C.C., L.L. Cheng and R.F. Hou. 1981 a. Studies on the intracellular yeast-like symbiote in the brown planthopper, *Nilaparyata lugens* (Stal) II. Effect of antibiotics and elevated temperature on the symbiotes and their host. *Z. Angew, Entomol.* 92: 440–449.

Chen, C.C., L.L. Cheng, C.C. Kuan and R.F. Hous. 1981 b. Studies on the intracellular yeast like symbiots in the brown planthopper *Nilaparvata lugens* (Stal) I. Histological observations and population changes of the symbiote *Z. Angew. Entomol.* 91: 321–327.

Houk, E.J. and C.W. Griffiths. 1980. Intraceallular symbiotes of the homoptera. *Ann. Rev. Entomol.* 25: 161–187.

Koyama, K. 1986. Nutritional physiology of the brown planthopper, Nilaparvata lugens (Stal) (Homoptera: Belphacidae) III. Essential vitamins for nymphal development. *Appl. Ent. Zool.* 21: 252–257.

Nasu, S. 1963. Studies on some leathoppers and planthoppers which transmit virus disease of plant in Japan. *Bull. Kvushu Agr. Expt. Sta.* 8: 153–349.

Nasu, S., T. Kusumi, Y. Suwa and H. Kita. 1981. Symbiotes of planthopper II. Isolation of Intracellular symbiotic microorganism from thr brown planthopper, *Nilaparvata lugens* (Stal) and immunological comparison of symbiotes associated with rice planthoppers (Homoptera: Delphacidae). *Appl. Ent. Zool.* 16: 88–93.

Noda, H. 1977. Histological and histochemical observation of intracellular yeast-like symbiotes in the fat body or the smaller brown planthopper, *Laodelphax striatellus* (Homoptera: Dephacidae). *Appl. Ent. Zool.* 12: 134–141.

Sogawa, K. and M.D. Pathak. 1970. Mechanisms of brown planthopper resistance in Mudgo variety of rice (Hemiptera: Delphacidae). *Appl. Entomol. Zool.* 5: 145–158.

40

Identification of Effective and Inexpensive Neem (*Azadirachta indica* A. Juss.) Seed Kernel Extract for Management of *Helicoverpa armigera* (Hubner)

B.R. JHANSI* AND R.P. SINGH

Division of Entomology, Indian Agricultural Research Institute
New Delhi 110 012, India

ABSTRACT

Sensitivity of insects to neem's different biological effects varies greatly. With a view to know the sensitivity of *H. armigera* to different biological effects of neem seed kernel extracts as also to find effective and inexpensive extract for management of this pest, six neem seed kernel extracts viz., aqueous (AE), ethanolic (EtoHE), hexane (HE), ethanol soluble part of hexane extract (EtoHSHE), ethanol insoluble part of hexane extract (EtoHIHE), chloroform extract of deoiled kernel (CHEDK) and combinations of CHEDK+HE (1:1) and CHEDK + EtoHSHE (1:1) were screened for their effects on egg hatching, oviposition, feeding and growth and development.

All extracts at 2 per cent concentration under choice condition significantly reduced egg laying by the moths on treated chickpea. The maximum deterrence was caused by CHEDK (90.4 per cent). The order of effectiveness was: CHEDK+EtoHSHE > CHEDK > EtoHE > EtoHSHE > CHEDK+HE > HE > EtoHIHE > AE.

Except EtoHIHE all the other extracts significantly reduced egg hatching. The percentage of egg hatched in different treatment ranged from 27.0 to 55.0 as against 61.0 to 66.0 in controls. EtoHSHE and CHEDK caused maximum inhibition in egg hatching. At 4 per cent egg hatchability was found to range from 24.0 to 39.0 per cent whereas it was 60.0 to 64.0 per cent in controls. The order of activity was: EtoHSHE > CHEDK > CHEDK + EtoHE = EtoHE > AE > HE > HE + CHEDK > EtoHIHE.

The percentage of feeding by 5, 10 and 15-day-old larvae on chickpea treated with 4 per cent concentration of different extracts ranged from 6.5 to 28.0 (controls 29.0 to 33.5), 5.0 to 30.0 (controls 38.0 to 44.0) and 10.0 to 34.0 (controls 44.0 to 51.0) respectively. None of the

* Present Address: Scientist, Division of Entomology, I.I.H.R., Hessarghatta Lake, Bangalore 560 089.

extract could completely inhibit feeding by larvae. The efficacy of the extracts was as follows: CHEDK > EtoHSHE+CHEDK > EtoHE > AE > EtoHSHE > CHEDK+HE > HE > EtoHIHE.

Five-day-old larvae when sprayed with 4.0, 2.0, 1.0 and 0.5 per cent concentration of different extract produced varied effects on their development. Significantly higher larval mortality (36.6 to 100 per cent) was recorded in different treatments of the extracts in comparison with control mortality (6.6 to 13.3 per cent). CHEDK, CHEDK+EtoHSHE and CHEDK+HE caused 100 per cent mortality at 4 per cent concentration. Extended larval period (20 to 45 days) were observed in these treatments and EtoHE.

The sizes and weights recorded on 5th, 10th and 15th days after treatment showed marked differences. The sizes and weights recorded on 15th day ranged from 5.0 to 22.0 mm and 5.6 to 176.4 mg as against 30.0 to 31.6 mm and 244.5 to 251.0 mg in controls respectively. The maximum reduction in size and weight was affected by CHEDK+EtoHSHE, CHEDK+HE and CHEDK.

The percentage of pupal mortality in different treatment ranged from 6.1 to 33.4 as against 7.4 to 8.1 in controls. The highest pupal mortality was found in aqueous extract.

EtoHIHE recorded highest percentage of adult emergence (30.3 to 56.7) while CHEDK the least (0.0 to 5.9) in different treatments. The adult emergence in control ranged from 79.3 to 85.4 per cent. The overall effectiveness of different extract was as follows: CHEDK > CHEDK+EtoHSHE > CHEDK+HE > EtoHE > EtoHSHE > AE > HE > EtoHIHE.

INTRODUCTION

The current plant protection methods followed for managing insect pests depend heavily on synthetic insecticides. Their uninterrupted and massive use have created several serious problems, making it essential to search for new methods or to revive old practices which are cheap, effective and environmentally safe. One such rationale which has gained world-wide attention is the use of plant products.

Numerous plant species are reported to possess pest control chemicals, but only a handful of them seem to be ideally suited for practical utilisation. Neem (*Azadirachta indica* A. Juss.) is one such perennial tree which grows widely in India and has come up as single most important source of pesticide. Ever since the discovery of extra ordinary antifeedant property of neem seed kernel against desert locust (Pradhan et al., 1962) the tree has been subjected to numerous investigations with a view to utilising the tree for management of insect pests of crops. Evaluations of seed kernel extracts and pure compounds against numerous species of insect pests have demonstrated neem's diverse biological effects: repellence (Jacobson et al., 1978; Quadri, 1973), feeding deterrence (Pradhan et al., 1962; Singh, 1987) reduced growth and abnormal development (Koul, 1985; Schmutterer and Rembold, 1980), oviposition deterrence (Saxena et al., 1981; Singh and Srivastava, 1983) reduced egg laying due to sterilizing effect (Rembold and Sieber, 1981; Gujar and Mehrotra, 1984) and also direct toxicity (Goyal, 1971; Singh et al., 1988).

Grampod borer, *Helicoverpa armigera* (Hubner) is a serious pest of several cultivated crops and has attained global importance as alarming pest. The insect is recorded from 96 crops and 61 weeds and wild plant species in India. It caused extensive damage in cotton in 1987-88 crop season in Guntur district of Andhra Pradesh. The loss was estimated to be to the tune of 416 crores in cotton alone in one season. In Tamil Nadu this pest was found very wide spread in summer cotton and *rabi* crop during the same year. The estimated loss to the crop by *H. armigera* was 92.8 crores (Jayaswal, 1988).

Various groups of insecticides including the potent synthetic pyrethroids have failed to contain the damage caused by the pest.

Despite well established potential of neem derivatives against insects only a few studies have been conducted to elucidate the efficacy of different neem seed kernel extracts against *H. armigera*. The present study aims at screening of different solvent extracts of neem seed kernel with a view to identify the most economical and biologically active extract(s) as also to know the most vulnerable stage(s) of the pest to hit upon.

MATERIALS AND METHODS

Seed Collection and Processing

Ripened fruits which had dropped to the ground were collected at New Delhi in the month of July. Fruits were depulped and the seeds washed with water and sun dried for 2–3 days. Kernels obtained after decortication were ground to powder in an electric blender. Neem seed kernel (NSK) powder thus obtained were used for extraction with solvents of different polarity as described below.

Extraction

Aqueous Extract (AE)

To obtain aqueous neem seed kernel extract 100 g of powdered NSK was stirred with 100 ml of distilled water with a magnetic stirrer for 3h and filtered through Whatman No. I filter paper using vacuum pump. The marc was stirred again for 1h followed by filtration. The combined extracts were freed of water in evaporating dishes at 60°C using a hot air blower.

Ethanolic Extract (EtoHE)

A similar procedure was adopted to obtain an ethanolic extract of NSK except that the combined extracts were freed of ethanol under reduced pressure at 50°C in a rotary vacuum evaporator.

Hexane Extract (HE)

To get hexane extract 500 g of NSK powder was stirred with 1 litre hexane in a blender for 10 minutes and filtered. Marc was restirred with 500 ml portion of hexane. The process was repeated until complete deoilation. Pooled extracts were freed of hexane at 50°C as in case of ethanolic extract. The residue left after extraction was dried under fan and collected as deoiled neem seed kernel powder (DNKP).

Chloroform Extract of DNKP (CHEDK)

One hundred gram of deoiled neem seed kernel powder, left after the extraction of NSK powder with hexane, was stirred 3 times with 400 ml portion of chloroform in a magnetic stirrer and filtered. Each extraction was done for 1h. Chloroform was evaporated as in case of ethanol to get the extract.

Ethanol Soluble (EtoHSHE) and Insoluble (EtoHIHE) Part of Hexane Extract

Hexane extract (100 ml) was partitioned with 100 ml of ethanol to get soluble and insoluble parts. The ethanol soluble part was freed of ethanol as described earlier.

Extract Yield

The yield of each extract was recorded and stored at 4°C in glass phials until used. The percentage of yield of each extract is given in Fig. 1.

Formulation of the Extracts

The above six extracts and also two combinations of extracts i.e. CHEDK + HE (1:1) and CHEDK + EtoHSHE (1:1) were formulated on a w/v basis to obtain desired concentrations as follows: the aqueous extract was used as solution in water while other extracts and combinations of extracts were formulated as emulsion using Triton X-100 as emulsifier. The extracts were first dissolved in solvents and then diluted with water containing emulsifier. The level of solvents and emulsifier are given in Table 1.

Rearing Culture of *H. armigera*

The culture of *H. armigera* was reared at 25°C ± 1°C and 65 ± 5 per cent relative humidity under sterilised conditions. The laboratory culture was initiated from field collected larvae infesting chickpea (*Cicer arietinum*). The larvae were fed on chickpea till pupation. Stock culture was later reared on artificial diet as given by Nagarkatti and Satyaprakash (1974).

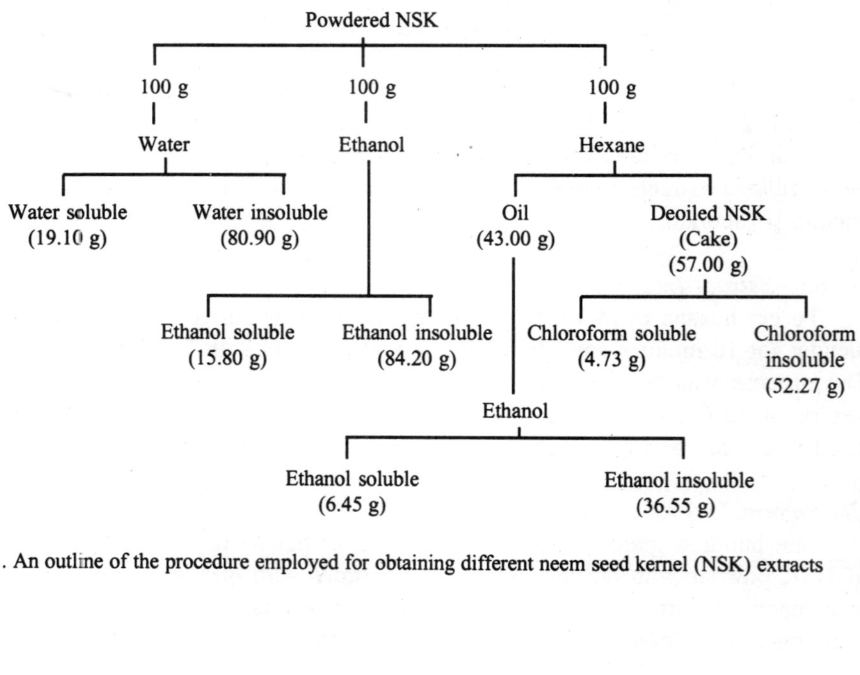

Fig. 1. An outline of the procedure employed for obtaining different neem seed kernel (NSK) extracts

Table 1. Various solvent and emulsifier levels used for formulating different NSK extracts

Extract	Solvent	Solvent Level (%)	Emulsifier
Aqueous extract	water	—	0.5
Ethanolic extract	ethanol	10	0.5
Hexane extract (oil)	benzene	5	1.0
Chloroform extract	acetone	5	0.5
Ethanolic extract of Hexane extract	ethanol	10	0.5
Insoluble part of hexane extract	benzene	5	1.0
hexane extract + chloroform extract (1:1)	acetone	5	1.0
Ethanolic extract of hexane extract + chloroform extract (1:1)	acetone	5	1.0

Biological Testing

All bioassays were conducted in the laboratory at the above temperature and relative humidity. Each extract was tested for oviposition deterrent, ovicidal, feeding deterrent and growth disrupting effect. Concentrations used and method employed for testing the effects of different extracts are detailed below.

Oviposition

The ovipositional response of mated female moth to different extracts was studied under choice conditions in a cage almost similar to the one designed by Saxena and Rembold (1984). The extracts were tested at 2 per cent concentration. Chickpea twigs of similar size were obtained from insecticide free crops and sprayed by hand automiser to run-off stage. Twigs were dried under fan and fixed in plastic boxes and transferred in cage along with control as shown in Fig. 7. Four mated females were released in the cage. The experiment was conducted in the evening. Cage was kept in uniform darkness to avoid differential effects of light. Adults were removed after 16 h and the number of eggs laid on treated and control twigs were counted. The experiments were repeated 3 times under identical conditions. The percentage of eggs deposited on the treated twigs was worked out from the total number of eggs laid on both treatment and control. The percentage was transformed into angular values.

Ovicidal

Eggs laid during the previous night by the mated females were collected in the morning and used for assessing ovicidal effect of the different extracts. Twenty five eggs were pasted on egg card with 5 per cent gum arabica. The cards after drying were placed individually in Petri dishes and sprayed with 1 ml of 2.0 and 4.0 per cent of different extracts in Potter's tower at 24 cm Hg pressure. Treated cards and controls were dried under fan and transferred into plastic boxes (8.5 × 5 cm). Boxes were covered with black cloth. Each treatment had four replications. The ovicidal effects of NSK extracts were determined by the percentage of eggs hatched.

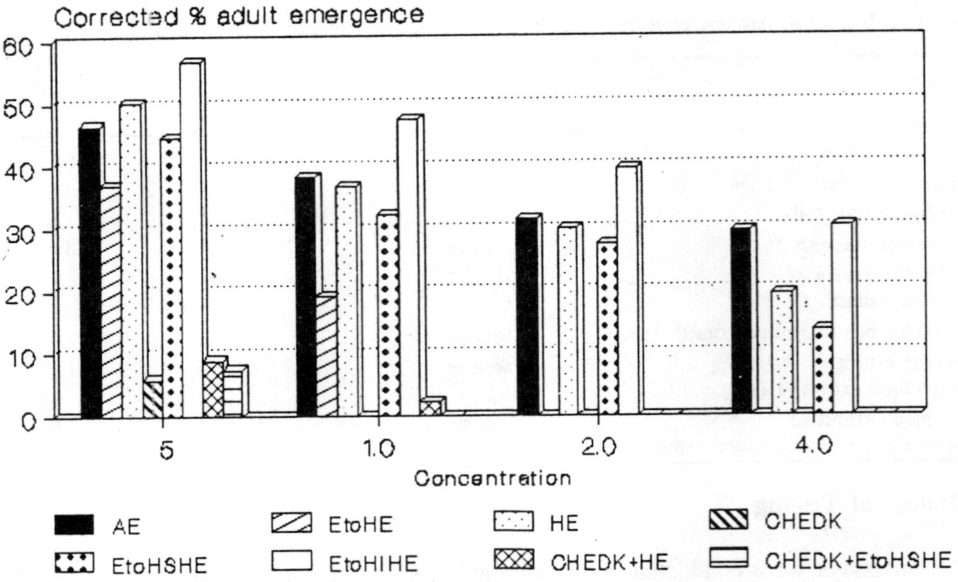

Fig. 2. Effect of neem seed kernel extracts on adult emergence of *H. amigera* (calculated from initial number of larvae used)

Feeding deterrent

Extract for feeding deterrent effects were evaluated at 4 per cent concentration against 5, 10 and 15-day-old larvae and at 2 per cent against only 10-day-old larvae. Chickpea leaves having ten leaflets of almost equal sizes were selected for treatment. Two leaves (20 leaflets) were dipped at a time in the formulated extracts for 5 seconds. Treated leaves were then dried under the fan and transferred (two leaves per box) into plastic boxes (8.5 × 5 cm). One larva earlier starved for 12 hours was introduced in each box. Each treatment and control were replicated ten times.

Observation on the area eaten by each larva was recorded after 24 h. The extent of feeding per leaflet in treatments and controls was rated on a 1–4 scale (1=76–100 per cent; 0.75=51–75 per cent; 0.5=26–50 per cent; 0.25=0–25 per cent). The per cent feeding recorded in treatment and control was transformed into angular values and subjected to statistical analysis.

Growth and Development

The growth-disrupting effect of NSK extracts was assayed at 4.0, 2.0, 1.0 and 0.5 per cent concentration on 5-day-old larvae in the following way.

Batches of ten larvae were taken in Petri dishes and sprayed with one ml formulation of each extract in Potter's tower as described earlier. The sprayed larvae were dried under fan and transferred singly into plastic tubes (7.5 x 2.5 cm) containing approximately 10 g diet. Tubes were plugged with non absorbent cotton. Ten treated larvae were considered as one replication. Three replications were made for each treatment. Observations were taken on larval and pupal durations, mortality, size and weight of larvae, percentage adult emergence, production of larval-pupal intermediates, pupal-adult intermediate and malformed adults.

RESULTS

Effect on Oviposition

The ovipositional response of mated female moths to 2.0 per cent NSK extracts and solvent treated twigs is given in Table 2. It is observed that all extract deterred oviposition of moth as the total number of eggs laid on treated twigs were much less than their respective controls. Of all the extracts tested chloroform extract (CHEDK) alone and in combination with ethanolic extract of hexane extract (EtoHSHE) caused maximum reduction in oviposition followed by ethanolic extract of NSK. The percentage of eggs laid on the twigs treated with these extracts were 9.54, 9.45 and 14.51 respectively. No significant difference was found in these treatments. EtoHSHE, hexane extract (HE) and hexane extract + chloroform also caused substantial reduction (22.93–28.73 per cent) in oviposition. Aqueous extract (AE) and ethanol insoluble part of hexane extract (EtoHIHE) exhibited low activity.

Table 2. Effect of 2.0 per cent concentration of different NSK extracts on oviposiotn of *H. armigera**

Extract	Total number of eggs laid[a]		Per cent eggs[b] laid on treatment
	Treatment	Control	
Aqueous extract	442	641	41.18 (39.88)**
Ethanolic extract	104	612	14.51 (22.23)
Hexane extract (oil)	342	842	28.73 (32.32)
Chloroform extract	97	911	9.54 (17.09)
Ethanolic extract of hexane extract	194	651	22.93 (28.56)
Insoluble part of hexane extract	277	528	40.83 (39.45)
Hexane extract + chloroform extract	292	726	27.95 (31.84)
Ethanolic extract of oil + chloroform extract	94	787	9.45 (16.87)
SEm ±			3.66
CD at 5%			7.85
CD at 1%			10.89

a = Total of 3 replications, 4 females per replicate
b = Average of 3 replications
* Cage kept under uniform darkness of controlled temperature and humidity
** Figures in parentheses are angular transformed

Effect on Egg Hatching

Table 3 shows that all the extracts except EtoHIHE at 2 per cent concentration significantly reduced egg hatching. The percentage of eggs hatched in treatments ranged from 27.0 to 55.0 as against 61.00 to 66.0 in different controls. The maximum inhibition in hatching (27.0 per cent) was observed in ethanol soluble part of oil (EtoHSHE) followed by chloroform extract (28.00 per cent) of DNKP (CHEDK). Both treatments differed significantly from rest of the treatments. At 4.0 per cent concentration the mean percentage of eggs hatched ranged from 24.00 to 39.00. All the treatments differed significantly from control. CHEDK and EtoHSHE were the most effective in inhibiting egg hatching. The lowest activity was shown by EtoHIHE (46.00 per cent).

Effect on Feeding

The percentage of feeding by 5, 10 and 15-day-old larvae at 4 per cent concentration on feeding rating scale is given in Table 4. It is seen that all extracts except EtoHIHE significantly deterred feeding by larvae but none could prevent feeding. The percentage of feeding by 5, 10, and 15-day-old larvae ranged from 4.0 to 28.0, 5.0 to 30.0 and 10.0 to 34.0 as against 29.0 to 33.5, 38.0 to 44.0 and 44.0 to 51.0 in controls respectively. It is evident from the table that feeding suspectibly decreased with increasing age of larvae. The order of activity was as follows: CHEDK > EtoHSHE + CHEDK > EtoHE > AE > EtoHSHE > HE + CHEDK > HE > EtoHIHE.

Table 3. Effect of different NSK extracts on hatchability of eggs of *H. armigera*

Treatment	% eggs hatched	
	Concentration	
	2.0	4.0
Aqueous extract	42 (42.37)*	36 (36.85)
Ethanolic extract	38 (38.02)	38 (37.94)
Hexane extract (oil)	43 (40.95)	39 (38.60)
Chloroform extract	28 (31.29)	24 (29.29)
Ethanolic extract of hexane extract	27 (31.29)	26 (30.48)
Insoluble part of hexane extract	55 (47.88)	46 (42.68)
Hexane extract + chloroform extract	44 (41.52)	38 (38.03)
Ethanolic extract of oil + chloroform extract	38 (38.02)	37 (36.79)
Acetone control	62 (51.97)	60 (50.79)
Benzene control	63 (52.61)	60 (50.85)
Ethanol control	61 (51.43)	62 (52.08)
Emulsified control	65 (53.77)	62 (52.02)
Water control	66 (54.37)	64 (53.40)
SEm ±	3.104	2.388
CD at 5%	8.60	6.62
CD at 1%	11.94	9.18

Average of 4 replications, 25 eggs per replicate
* Figures in parentheses are angular transformed values

Table 4. Effect of different NSK extracts on feeding of various age group larvae of *H. armigera*

Treatment	Feeding rating[a]			
	Concentration (%)			
	2.0	4.0		
	10 day	5 day	10 day	15 day
Aqueous extract	1.25 (2.06)*	0.65 (1.47)	1.00 (1.84)	1.80 (2.50)
Ethanolic extract	2.05 (2.69)	0.50 (1.36)	0.95 (1.79)	1.50 (2.27)
Hexane extract (oil)	3.00 (3.31)	1.60 (2.35)	2.60 (3.12)	3.00 (3.32)
Chloroform extract	1.05 (1.88)	0.40 (1.14)	0.50 (1.28)	1.25 (2.07)
Ethanolic extract of hexane extract	1.75 (2.46)	1.20 (2.20)	0.90 (1.74)	2.20 (2.79)
Insoluble part of hexane extract	4.55 (4.24)	2.80 (3.19)	3.00 (3.32)	3.40 (3.56)
Hexane extract + chloroform extract	1.55 (2.31)	1.60 (2.35)	2.00 (2.65)	2.40 (2.93)
Ethanolic extract of oil + chloroform extract	1.15 (1.97)	0.60 (1.41)	0.75 (1.58)	1.00 (1.83)
Acetone control	5.00 (4.50)	2.90 (3.25)	4.00 (3.92)	4.90 (4.44)
Benzene control	5.90 (5.02)	3.00 (3.32)	3.80 (3.80)	4.40 (4.15)
Ethanol control	5.65 (4.87)	3.20 (3.44)	4.25 (4.06)	4.60 (4.27)
Emulsified control	5.85 (4.99)	3.35 (3.53)	4.40 (4.09)	5.10 (4.55)
SEm ±	0.56	0.45	0.42	0.49
CD at 5%	1.58	1.26	1.19	1.38
CD at 1%	2.09	1.65	1.56	1.82

Average of 10 replications, 10 larvae per replicate
a = Based on the scale of 0.25 – 1.0, where, 0.25= 1–25%, 0.5= 26–50%, 0.75=51–75%, 1.0= 76–100%
* Figures in parentheses are angular transformed.

Effect on Growth and Development

Extracts when sprayed at 4.0, 2.0, 1.0 and 0.5 per cent concentration on 5-day-old larvae produced varied effects on different stages of development. Significantly higher larval mortality (36.6 to 100 per cent) in comparison with solvent controls (6.6 to 13.3 per cent) was recorded in different treatments.

Larvae succumbing to extracts at 0.5 and 1.0 per cent concentrations varied between 36.6 to 73.3 (Fig. 3) and 48.8 to 86.6 respectively (Fig. 4). At 2.0 per cent concentration CHEDK caused 100 per cent larval mortality while the mortality in other treatments ranged from 53.3 to 96.6 per cent (Fig. 5). At 4.0 per cent concentration all extract caused significantly high larval mortality which ranged from a minimum of 60.0 in AE to 100.0 per cent in CHEDK, CHEDK + EtoHSHE and CHEDK + HE (Fig. 6).

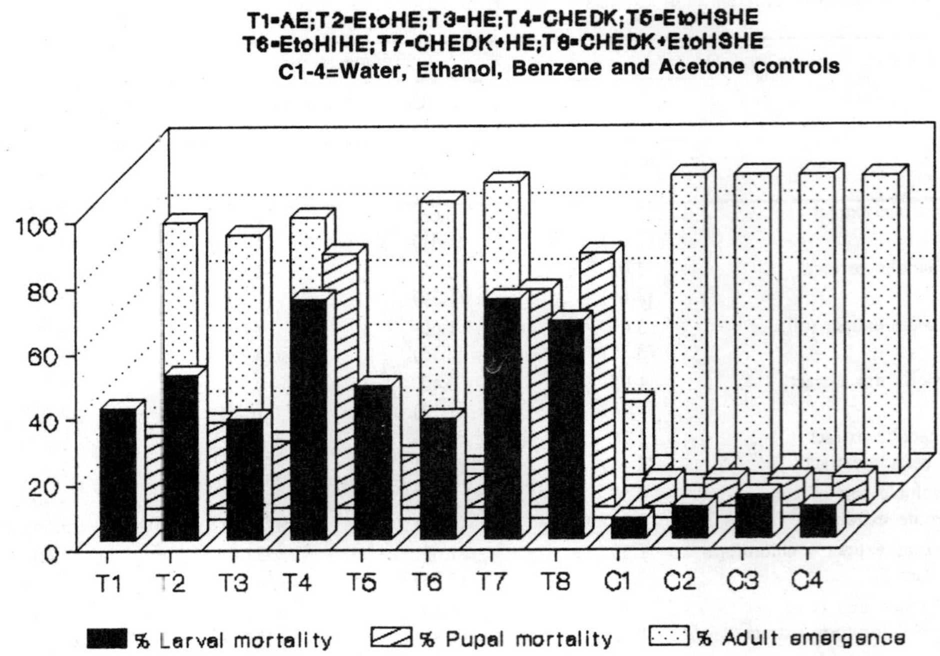

Fig. 3. Effect of seed kernel extracts at 0.5% on larval and pupal mortality and adult emergence of *H. armigera*

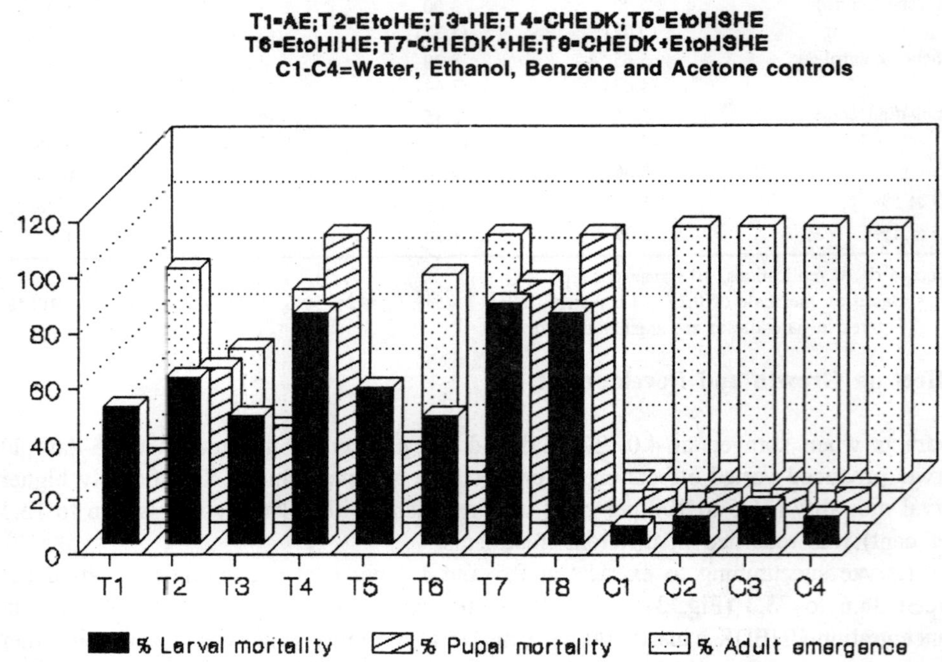

Fig. 4. Effect of seed kernel extracts at 1.0% on larval and pupal mortality and adult emergence of *H. armigera*

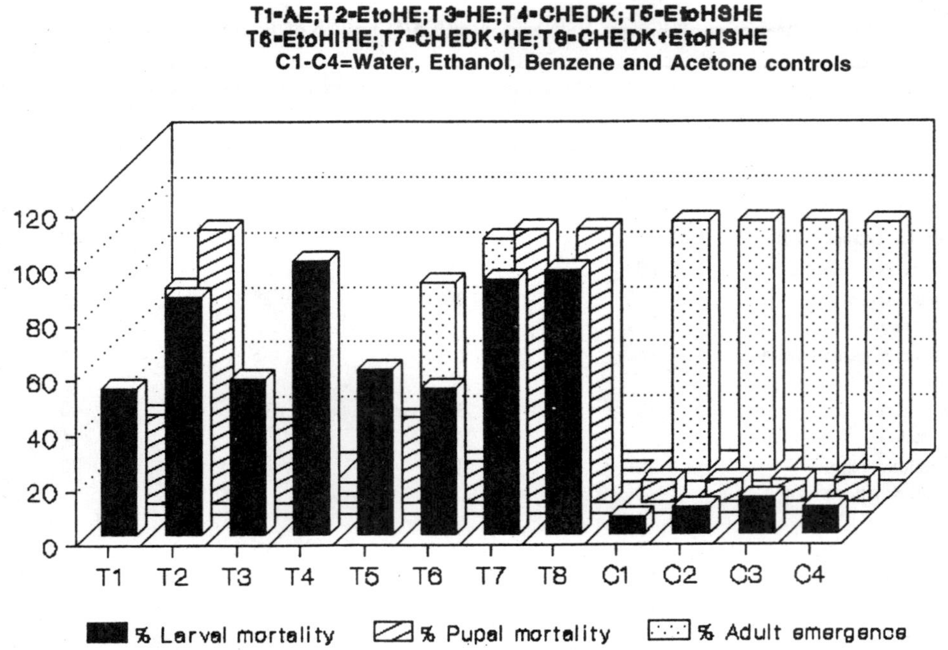

Fig. 5. Effect of seed kernel extracts at 2.0 % on larval and pupal mortality and adult emergence of *H. armigera*

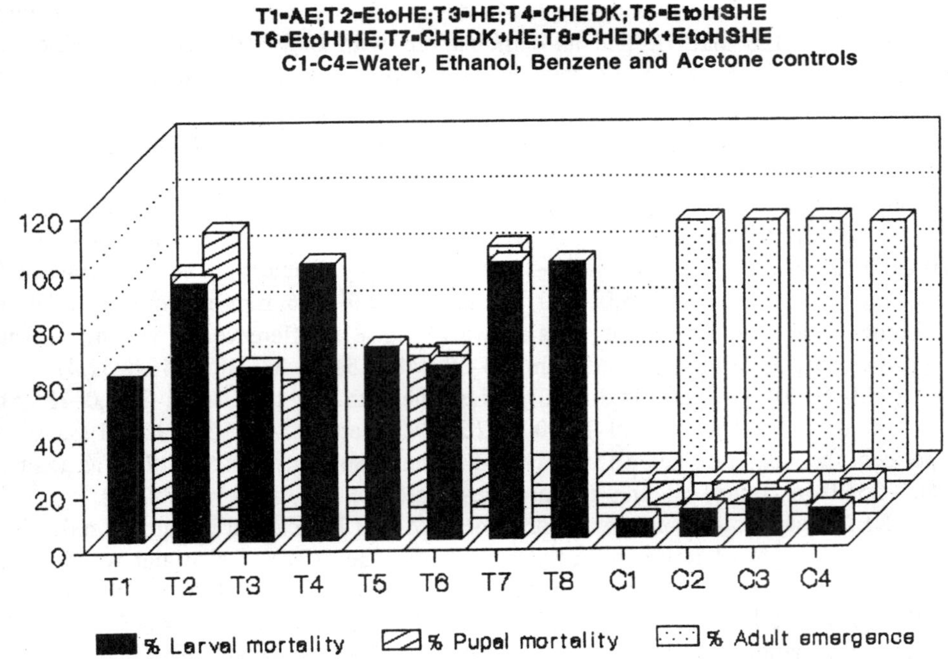

Fig. 6. Effect of seed kernel extracts at 4.0 % on larval and pupal mortality and adult emergence of *H. armigera*

Table 5. Effect of treatment of different NSK extracts on larval duration of *H. armigera*

Treatment	Larval duration (days) at different concentrations*			
	0.5	1.0	2.0	4.0
Aqueous extract	15	16	18	24
Ethanolic extract	20	30	36	30
Hexane extract (Oil)	16	18	20	20
Chloroform extract	28	40	36	45
Ethanolic extract of hexane extract	24	36	28	24
Insoluble part of hexane extract	16	17	16	15
Hexane extract + chloroform extract	30	38	36	40
Ethanolic extract of oil + chloroform extract	28	40	36	28
Acetone control	17	16	15	17
Benzene control	16	16	17	16
Ethanol control	16	18	15	16
Water control	15	16	17	16

* period required by all larvae in each treatment either to die or pupate

Prolongation of larval periods (20–45 days) was observed in larvae treated with EtoHE, EtoHSHE, CHEDK and its combinations with EtoHSHE and HE (Table 5). Normal larval durations (15–18 days) were found in the rest of the treatments and controls.

Marked differences were found in size and weight of larvae among different treatments. The size of the larvae on 5th day in 0.5, 1.0, 2.0 and 4 per cent concentrations ranged from 3.6–9.8 (10.2–10.8), 3.4–10.0 (10.0–11.0), 5.3–9.5 (9.8–10.4), 3.0–8.0 (8.5–10.0) and on the 10th day 6.2–21.4 (24.6–25.6), 6.2–20.4 (24.4–24.8), 6.0–20.0 (25.0–25.6), 5.1–20.0 (27.2–30.4) respectively (Table 6). The size range at 4 per cent on 15th day was 5.0–22.0 as against 30.0–31.6 mm in control (relative difference in average size of larvae is shown in Fig. 8). Differences in weight (mg) in the same order on 5th day ranged from 4.0–39, (48.0–51.2), 3.0–38.0 (46.8–51.4), 2.7–34.4 (46.0–50.0), 1.8–32.0 (44.6–50.0) and on the 10th day 17.4–152.6 (182.0–185.5), 12.4–147.0 (182.0–186.0), 10.0–136.0 (192.2–204.0) and 4.5–125.5 (192.0–195.0). At 4 per cent the weight range on 15th day in different treatments was 5.6–176.4 as against 244.5–251.0 in control (figures in parentheses are control values) (Table 7).

CHEDK and its combinations with EtoHSHE and HE caused maximum reduction in size and weight of larvae followed by EtoHE. The effect of size and weight was concentration dependent. It was also more pronounced on the 10th and 15th day than on the 5th day following treatment of larvae. Most of the larvae in EtoHE, CHEDK, and its combinations and few cases of EtoHSHE and HE remained small and failed to enter next instar. Many larvae died just before pupation while other attempting to pupate entered into larval-pupal intermediates in the following manner. After transformation of larval abdomen into pupal abdomen, a puff appeared just below thorax. After

Table 6. Effect of spraying of different NSK extracts on size of larvae of *H. armigera*

Treatment	Size (mm)* days after treatment								
	5th day concentration (%)				10th day				15th day conc. (%)
	concentration (%)				concentration (%)				concentration (%)
	0.5	1.0	2.0	4.0	0.5	1.0	2.0	4.0	4.0
Aqueous extract	9.20	9.20	8.30	6.00	12.60	13.00	12.60	12.00	21.60
Ethanolic extract	5.80	5.40	4.20	4.50	6.60	6.80	6.30	6.30	11.80
Hexane extract (oil)	8.30	8.00	7.00	6.50	12.00	12.00	10.90	10.80	20.40
Chloroform extract	4.10	4.00	3.60	3.00	6.20	6.00	6.00	5.10	6.00
Ethanolic extract of hexane extract	8.60	8.00	7.40	6.00	10.80	11.00	11.00	10.30	14.80
Insoluble part of hexane extract	9.80	10.00	9.50	8.00	21.40	20.40	20.00	20.00	22.00
Hexane extract + chloroform	4.00	4.00	3.80	3.50	7.00	6.80	6.50	6.00	5.00
Ethanolic extract of oil + chloroform	3.60	3.40	3.30	3.00	6.40	6.20	6.00	5.30	5.00
Acetone control	10.20	10.00	9.80	8.50	24.60	25.20	25.00	27.20	30.40
Benzene control	10.40	10.20	10.10	10.00	24.80	24.40	25.00	27.40	30.00
Ethanol control	10.40	10.60	9.90	9.00	24.80	25.20	24.40	29.80	30.40
Water control	10.80	11.00	10.40	10.00	25.60	25.80	25.60	30.40	31.60
SEm ±	0.33	0.46	0.23	0.26	0.42	0.42	0.43	0.43	0.54
CD at 5%	0.79	1.09	0.55	0.62	0.99	1.01	1.02	1.04	1.28
CD at 1%	0.94	1.31	0.66	0.74	1.19	1.21	1.23	1.25	1.54

* Average of 5 replications, one larva per replicate

Fig. 7. Cage for testing ovipositional response of *H. armigera* female, moths in the act of oviposition on treated and untreated twigs.

sometime the puff bursted and a viscous fluid oozed resulting in formations of cavity at the junction of thorax and abdomen. The anterior part remained as larva (Fig. 9).

Pupal Mortality and Adult Emergence

Larvae which survived and entered into pupal stages also suffered significantly higher mortality in different treatments except EtoHIHE and the lowest concentration in EtoHSHE (Fig. 3–6). Pupae in different treatments were much smaller than control and EtoHIHE (Fig. 10). The ranges of mortality in different concentrations in AE, EtoHE, HE, CHEDK, EtoHSHE, EtoHIHE, CHEDK + EtoHSHE and CHEDK + HE and controls were 22.5–32.8, 26.4–100, 20.6–47.2, 77.81–100.0, 16.2–55.1, 10.3–17.3, 66.7–100.0 and 77.78–100.0 and 8.3–8.9 per cent respectively. The percentage of adult emergence in the same order ranged from 67.3–77.6, 0.0–73.9, 52.8–79.4, 0.0–22.2, 44.9–83.8, 82.7–89.7, 0.0–33.3, 0.0–22.2, as against 91.1–91.7 in controls. Several pupae failed to emerge while others ended up as pupal-adult intermediates, deformed (Fig. 11) and miniature adults (Fig. 12). CHEDK and its combinations offered highest pupal mortality and had lowest percentage of adult emergence. This was followed by EtoHSHE, AE and HE were almost at par in activity.

When the percentage of emergence was calculated from initial number of larvae it ranged from 29.6–46.5, 0.0–37.0, 19.4–50.3, 0.0–5.9, 13.5–44.7, 30.0–56.8, 0.0–8.9, 0.0–7.4, in AE, EtoHE, HE, CHEDK EtoHIHE, CHEDK + HE, and CHEDK + EtoHSHE

Table 7. Effect of spraying of different NSK extracts on weight of larvae of *H. armigera*

Treatment	Weight (mg)* days after treatment								
	5th day				10th day				15th day
	concentration (%)				concentration (%)				concentration (%)
	0.5	1.0	2.0	4.0	0.5	1.0	2.0	4.0	4.0
Aqueous extract	12.40	10.80	9.20	9.00	72.60	64.00	66.50	44.00	175.00
Ethanolic extract	5.80	5.60	5.20	5.00	20.00	14.50	11.00	9.60	40.50
Hexane extract (oil)	11.40	10.00	9.70	8.50	61.40	56.00	58.50	34.00	80.00
Chloroform extract	4.80	3.20	3.00	2.50	17.00	12.50	10.00	5.20	6.50
Ethanolic extract of hexane extract	10.00	9.00	7.80	7.50	56.00	49.00	50.00	25.80	68.00
Insoluble part of hexane extract	39.00	38.00	34.40	32.00	152.60	147.00	136.60	125.50	176.40
Hexane extract + chloroform	4.80	4.60	2.80	2.00	21.00	14.60	15.00	5.00	6.00
Ethanolic extract of oil + chloroform	4.00	3.00	2.70	1.80	17.40	12.40	10.40	4.50	5.60
Acetone control	48.00	46.80	46.00	44.60	184.00	182.00	196.80	192.00	244.50
Benzene control	49.80	49.20	47.60	46.40	183.50	184.00	197.00	189.50	249.50
Ethanol control	50.20	49.60	48.00	45.00	182.00	182.00	196.20	192.50	248.00
Water control	51.20	51.40	50.00	50.00	185.50	186.00	204.00	195.00	251.00
SEm ±	0.69	0.66	0.45	1.27	2.96	2.40	4.09	2.72	2.85
CD at 5%	1.59	1.57	1.09	3.03	7.05	5.71	9.74	6.47	6.78
CD at 1%	1.91	1.88	1.30	3.64	8.45	6.85	11.68	7.76	8.14

* Average of 5 replications, one larva per replicate

Fig. 8. Effect of different extracts (at 4%) on size of larvae (15th day after treatment). From left to right: Control, EtoHIHE, AE, EtoHSHE, EtoHE, HE + CHEDK, EtoHSHE + CHEDK

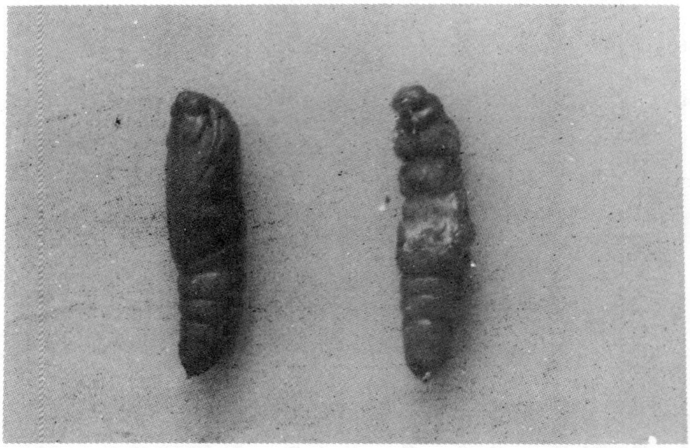

Fig. 9. Right larval-pupal intermediate showing wound following treatment with NSK extracts (except EtoHIHE). Left: A control pupa

respectively as against 79.5–85.4 in controls. All the treatments including EtoHIHE differed significantly (Fig. 2).

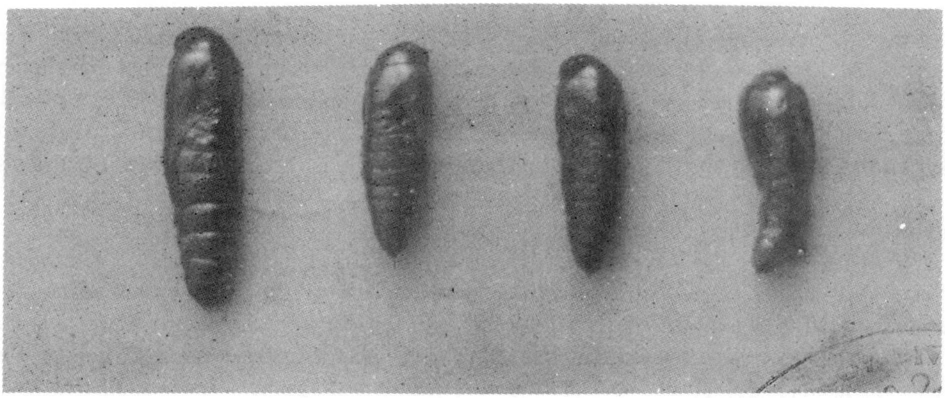

Fig. 10. Three malformed pupae which failed to emerge; extreme left a control pupa.

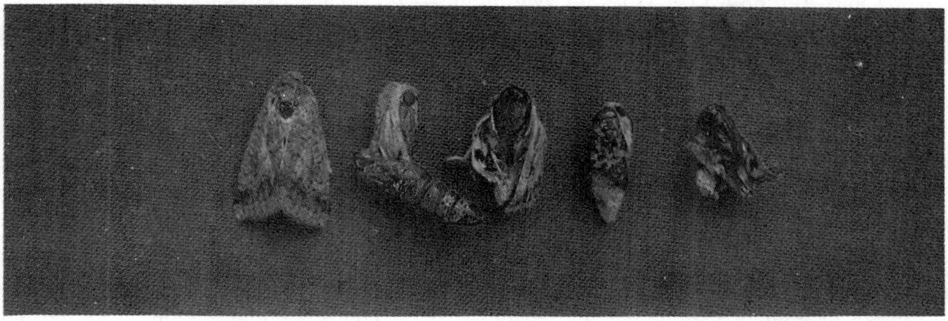

Fig. 11. Effect of NSK extracts on adult emergence: Control adult (a), eclosion problem (b), half covered abdomen (c), pupal-adult intermediate (d), malformed adult with twisted wings (e)

Fig. 12. Two miniature adults. Extreme left a control adult

DISCUSSION

All parts of neem are biologically active. The maximum bioactivity against insect pest is in seed kernel. Extract and pure compounds derived from seed kernel have been found to affect the insects on different steps of host selection. About 40 compounds have been isolated and identified from different parts of the tree. Most of these are tetranortriterpenoids and show varying degree of biological activity against insect pests.

Effect of oviposition

Neem seed kernel extract are reported to deter oviposition of several insects. Ethanolic extract of NSK when sprayed on cabbage affected the oviposition of *Crocidolomia binotalis* (Fagoonee, 1981). Ethanolic extract of hexane extract of seed kernel was reported to deter egg laying by *Dacus cucurbitae* on bittergourd (Singh and Srivastava, 1983). Aqueous extract of neem seed kernel at 2 per cent was found oviposition deterrent against *Spodoptera litura* infesting tobacco crop (Joshi and Sitaramaiah, 1979). Saxena and Rembold (1984) tested azadirachtin and oil for oviposition deterrent effect against *H. armigera.* Azadirachtin failed to cause any effect on oviposition while oil deterred the oviposition of the moths.

Azadirachtin and salannin are the major compounds present in the seed kernel. Azadirachtin is a highly polar tetranortoriterpenoid present in the deoiled part of the kernel and consequently unextractable with non-polar solvent. Salannin on the other hand is a less polar and hence extractable both by less polar solvent like chloroform as well as polar solvents.

CHEDK and CHEDK + EtoHSHE have exhibited highest oviposition deterrent activity (Table 2). Chloroform extract of deoiled neem kernel (CHEDK) is rich in azadirachtin content (Schneider and Ermel, 1987). The other major compound extractable with chloroform in DNKP is salannin. Since azadirachtin is ineffective as oviposition deterrent against *H. armigera* (Saxena and Rembold, 1984), salannin may be the possible compound responsible for deterring oviposition. The combination of CHEDK + EtoHSHE has also offered almost similar activity. This may be due to the fact that EtoHSHE too possess salannin and is extractable with ethanol. Ethanolic extract (EtoHE) has ranked third in order of activity. EtoHE being polar can extract salannin from both oil and deoiled part alongwith other inactive materials present in the seed kernel. Inactive materials must have diluted the extract and hence lower activity than CHEDK and CHEDK+ EtoHSHE. Water being most polar solvent can extract salannin only from deoiled part of the kernel and a lot of inactive polar materials and thereby diluting the extract. Except EtoHIHE all the other remaining extract also contain salannin. Their oviposition deterrent efficacy can also be correlated with qualitative difference in the salannin content. Thus, it appears that there exist some correlation between salannin and oviposition deterrent effect. We could not confirm it due to non-availability of salannin.

Effect on Egg Hatching

Very few references are available on the effects of extracts and pure compounds on eggs of insects. Saxena et al. (1981b) reported that hatchability of eggs of *Cnaphalocrocis medianalis* was strongly inhibited when dipped in 12, 25 and 50 per cent oil. Neem oil as spray however failed to cause any effect on eggs of *Nilaparvata lugens* (Saxena et al., 1981a). Neem oil is also reported to reduce egg hatching of *Callosobruchus* spp. (Luca, 1982; Yadav, 1985; Ali et al., 1983).

Of the different extracts tested only CHEDK and EtoHSHE caused substantial mortality of eggs while others had little effect (Table 3). Azadirachtin has no toxicity (Schluter et al., 1985) but has been reported to effect egg hatching (Gill and Lewis, 1971) presumably by arresting the development of embryo. Salannin and two other compounds present in the ethanolic extract of oil have appreciable toxicity (Singh et al., 1988). Oil contain many biologically active compounds (Lavie et al., 1971) extractable with ethanol. Of these salannin is a major compound. The high activity of EtoHSHE therefore may be due to this compound and of CHEDK due to both salannin and azadirachtin as both are extracted with chloroform. Non-significant differences among other extract is largely due to presence of inactive materials in HE, AE and EtoHE. EtoHIHE contains mainly glycerides and lack biologically active compounds and hence has shown lowest activity. Susceptibility of eggs is influenced by several factors such as permeability of chorion (Smith and Wagenknecht, 1959). The role of solvent thus cannot be eliminated.

Effect on Feeding

It is evident from Table 4 that none of the extract could completely inhibit feeding of *H. armigera* even at 4 per cent concentration. It is also noted that effect on feeding was age dependent. Of the different extract tested CHEDK, EtoHSHE + CHEDK, EtoHE and AE consistently proved better antifeedant against all the three age group larvae. Several compounds have been isolated from seed kernel which show antifeedant activity against insects (Kraus et al., 1987; Siddiqui et al., 1988). Azadirachtin and related compounds present in defatted part of seed kernel are extractable with chloroform and are considered more potent than those present in the seed oil (HE). The higher activity of CHEDK is due to presence of azadirachtin and also some other polar tetranortriterpenoids. Its higher activity than ethanolic (EtoHE) and aqueous extracts (AE) is due to concentration of active material especially azadirachtin per unit weight of the extract. Schneider and Ermel (1987) have reported that quantity of azadirachtin in chloroform extract of defatted seed kernel has highest concentration followed by ethanolic and aqueous extract. A similar observation was made by Singh (1987).

Pure compounds such as Salannin, 3-deacetyl salannin, salannol, salannol acetate present in the seed oil has low antifeedant activity than azadirachtin (Kraus et al., 1987). Oil (HE) is almost devoid of azadirachtin. The above less polar compounds which are extractable with ethanol when combined with CHEDK proved as effective as CHEDK. The lower activity of other extract is ascribed to the presence or absence of active compounds and dilution factor due to presence of inactive materials as explained earlier.

Effect on Growth and Development of Larvae

There are numerous reports that azadirachtin when applied topically prolongs larval period, arrests growth and development and cause mortality (Ruscoe, 1972; Rembold et al., 1981; Ladd et al., 1984). These effects are attributed mainly to azadirachtin's interference with endocrine system. Azadirachtin has no direct toxicity (Schluter et al., 1985).

Figures 3–6 show that CHEDK, CHEDK + EtoHSHE and CHEDK + HE were at par in causing larval mortality. These extracts were also significantly better than the rest of the extracts including ethanolic extract (EtoHE) of neem seed kernel. HE, AE, EtoHIHE were at par in activity. CHEDK, CHEDK + EtoHSHE and CHEDK + HE also caused maximum reduction in size and weight of 5, 10 and 15-day-old larvae. The larvae which pupated in these treatments were much smaller and abnormal (Fig. 10). Several larvae attempted to pupate produced larval-pupal intermediate. In these larvae cuticle was formed only on the abdomen. A wound appeared at the junction of abdomen and thorax. The anterior part retained larval character (Fig. 9).

The percentage of adult emergence in different treatments when compared showed no significant difference among CHEDK, CHEDK + HE and CHEDK + EtoHSHE at 4, 2, and 1.0 per cent concentration. However, at the lowest concentration of 0.5 per cent CHEDK produced more adult suggesting thereby lower activity than combination of the two extract.

The activity of CHEDK is mainly due to the presence of higher quantity of azadirachtin (see effect on feeding). The higher activity of CHEDK + EtoHSHE than CHEDK inspite of having only half of azadirachtin content is suggestive of joint action of azadirachtin with the compounds present in the HE extract. It may also be due to the presence of oil (HE) which might have helped in penetration of compound or a combination of the two.

The low activity of aqueous extract (AE) was due to lack of peneterability because when it was fed to larvae (not included in this paper) its IGR activity was almost at par with ethanolic extract of neem seed kernel. Influence of other extract on growth and development of larvae was based on quantitative and qualitative variation of active compounds as already explained under different biological effects.

CONCLUSION

H. armigera a polyphagus pest has developed resistance to practically all synthetic insecticide. The present investigations revealed that it is highly sensitive to IGR activity and oviposition deterrent effect of neem extracts viz. chloroform extract of deoiled neem seed kernel (CHEDK) and ethanol soluble part of hexane extract (EtoHSHE + chloroform extract (CHEDK) (50:50) and chloroform extract (CHEDK) + hexane extract (HE) (50:50) of NSK.

Aqueous, ethanolic and hexane extracts commonly used for management are thus not very promising for use against this pest. Though CHEDK has consistently given best results, yet its use is not feasible because of low yield. If however, it is combined with hexane extract (Fig. 1), it would not only be very economical but also give almost the same degree of effectiveness as did CHEDK and at the same time be much more economical for the management of this pest.

ACKNOWLEDGEMENT

The senior author is grateful to the Post-Graduate School, Indian Agricultural Research Institute, New Delhi for providing Senior Fellowship for conducting research work leading to Ph.D. Degree.

REFERENCES

Ali, S.I., O.P. Singh and V.S. Mishra. 1983. Effectiveness of plant oils against pulse beetle, *Callosobruchus chinensis* Linn. *Indian J. Ent.* 45: 6–9.

Fagoonee, I. 1981. Behavioural response of *Crocidolomia binotalis* to neem. In: *Natural Pesticides from Neem Tree* (*Azadirachta indica* A. Juss), H. Schmutterer, K.R.S. Ascher and H. Rembold, (eds.), G.T.Z., Eschborn, FRG, pp. 109–120.

Gill, J.S. and C.T. Lewis. 1971. Systemic action of an insect feeding deterrent. *Nature* 232: 402–403.

Goyal, R.S. 1971. *Studies on better utilization of neem* (*Azadirachta indica*) *seed cake—Biological activity of alcoholic extract and its isolates*. M.Sc. Thesis, Post Graduate School, Indian Agricultural Research Institute, New Delhi-110012, India.

Gujar, G.T. and K.N. Mehrotra. 1984. Inhibition of growth and development of the tobacco caterpillar, *Spodoptera litura* Fab. due to azadirachtin and other neem product. *Indian J. Ent.* 45(4): 431–435.

Jacobson, M., D.K. Reed, M.M. Crystal, D.S. Moreno and E.L. Soderstorm. 1978. Chemistry and Biological activity of insect feeding deterrents. *Entomologia Exp. Appl.* 24: 248–257.

Jayaswal, A.P. 1988. Changing pest complex in cotton in relation to cropping systems. National Seminar on Changing Pest Situation in Current Agriculture Scenario of India, IARI, New Delhi, 14–16 June, Seminar Abstract, 23 pp.

Joshi, B.G. and S. Sitaramaiah. 1979. Neem kernel as an ovipositional repellent for *Spodoptera litura* (Fab.) moths. *Phytoparasitica* 7: 199–202.

Koul, O. 1985. Azadirachtin interaction with development of *Spodoptera litura* (Fab.). *Indian J. Exp. Biol.* 23: 160–163.

Kraus, W., S. Baumann, M. Bokel, U. Killer, A. Klenk, M. Klinglee, M. Pohne and M. Schwinger. 1987. Control of insect feeding and development by constituents of *Melia azedarach* and *Azadirachta indica*. In: *Natural Pesticides from the Neem Tree and Other Tropical Plants*, H. Schmutterer and K.R.S. Ascher (eds.), G.T.Z., Eschborn, FRG, pp. 111–125.

Ladd, T.L., J.R. Warthen and M.G. Jr. Klein. 1984. Japanese beetle (Coleoptera: Scarabidae): The effect of azadirachtin on the growth and development of the immature forms. *J. Econ. Ent.* 77: 903-905.

Lavie, D., E.C. Levy and M.K. Jain. 1971. Limonoids of biogenetic interest from *Melia azedarach* L. *Tetrahedron* 27: 3927–3939.

Luca, Y. De. 1982. Products of vegetable origin that can be used against bruchids (Coleoptera) (Attractant, antifeedant, deterrent, repellent and lethal). *Frustula Entomologica* 2: 19–29.

Nagarkatti, S. and Satyaprakash. 1974. Rearing of *Heliothis armigera* (Hub.) on artificial diet. *Tech. Bull. Commonw. Inst. Biol. Control* 17: 169–173.

Pradhan, S., M.G. Jotwani and B.K. Rai. 1962. The neem seed deterrent to locusts. *Indian Farming* 12: 7–11.

Quadri, S.S.H. 1973. Some new indigenous plants repellent for storage pests. *Pesticides* 7: 18–19.

Rembold, H., G.K. Sharma and C.H. Czopett. 1981. Growth regulating activity of azadirachtin in two hemimetabolous insects. In: *Natural Pesticides from the Neem Tree* (*Azadirachta indica* A. Juss.), H. Schmutterer, K.R.S. Ascher and H. Rembold (eds.), GTZ Eschborn, FRG, pp. 121–126.

Rembold, H. and K.P. Sieber. 1981. Inhibition of oogensis and ovarian ecdysteroid synthesis by azadirachtin in *Locusta migratoria* (R and F). *Z. Naturforschc.* 36: 466-469.

Ruscoe, C.N.E. 1972. Growth-disrupting effects of an insect antifeedant. *Nature* 236: 159–160.

Saxena, K.N. and H. Rembold. 1984. Orientation and ovipositional response of *Heliothis armigera* to certain neem constituents. In: *Natural pesticides from Neem Tree and Other Tropical Plants*, H. Schmutterer and K.R.S. Ascher, (eds.), GTZ Eschborn, FRG, pp. 199–210.

Saxena, R.C., N.J. Liguido and H.D. Justo. 1981a. Neem oil a potential antifeedant for the control of rice brown planthooper, *Nilaparvata lugens*. In: *Natural Pesticides from the Neem Tree* (*Azadirachta indica* A. Juss), H. Schmutterer, K.R.S. Ascher and H. Rembold (eds.), GTZ Eschborn, FRG, pp. 171–188.

Saxena, R.C., G.P. Waldbauer, N.J. Liguido and B.C. Puma. 1981b. Effect of neem seed oil on the rice leaffolder, *Cnaphalocrocis medianalis*. In: *Natural Pesticides from Neem Tree* (*Azadirachta indica* A. Juss), H. Schmutterer, K.R.S. Ascher and H. Rembold, (eds.), GTZ, Eschborn, FRG, pp. 189–204.

Schluter, V., H.J. Bidmon and S. Grewe. 1985. Azadirachtin effects growth and endocrime events in larvae of the tobacco horn worm, *Manduca sexta*. *J. Insect Physiol.* 31: 773–779.

Schmutterer, H. and H. Rembold. 1980. Zur Wirkungeininger Reinfraklionen aus saman von *Azadirachta indica* and Fraßaktivitat und metamorphose von *Epilachna varivestis* (Col.: Coccinellidae). *Z. angew Ent.* 89: 179–188.

Schneider, B.H. and K. Ermel. 1987. Quantitative determination of azadirachtin from neem seeds using high performance liquid chromatography. In: *Natural Pesticides from the Neem Tree and Other Tropical Plants*, H. Schmutterer and K.R.S. Ascher (eds.), GTZ, Eschborn, FRG, pp. 161–170.

Siddiqui, S., B.S. Siddiqui, S. Faizi and T. Mahmood. 1988. Tetracyclic triterpenoids and their derivatives from *Azadirachta indica*. *J. Nat. Prod.* 51: 30–43.

Singh, R.P. 1987. Comparison of antifeedant efficacy and extract yields from different parts and ecotypes of neem (*Azadirachta indica* A. Juss) trees. In: *Natural Pesticides from the Neem Tree and Other Tropical Plants*, H. Schmutterer and K.R.S. Ascher (eds.), GTZ, Eschborn, FRG, pp. 185–194.

Singh, R.P., C. Devakumar and S. Dhingra. 1988. Activity of neem (*Azadirachta indica* A. Juss.) seed kernel extracts against the mustard aphid, *Lipaphis erysimi*. *Phytoparasitica* 16: 225–229.

Singh, R.P. and B.G. Srivastava. 1983. Alcohol extract of neem (*Azadirachta indica* A. Juss.) seed oil as oviposition deterrent for *Dacus cucurbitae* (Coq.). *Indian J. Ent.* 45: 497–498.

Smith, E.H. and A.C. Wagenknecht. 1950. The ovicidal action of organophosphate insecticides. *Can. J. Biochem. Physiol.* 37: 1135–1144.

Yadav, T.D. 1985. Antiovipositional and ovicidal toxicity of neem (*Azadirachta indica* A. Juss.) oil against three species of *Callosobruchus*. *Neem Newsl.* 2: 5–6.

5

FIELD EVALUATION

41

Effects of Neem Oil on the Grasshopper *Zonocerus variegatus* L. in the Field

MARTIN BAUMGART

Institut für Phytopathologie und Angewandte Zoologie, Justus-Liebig-Universität,
6300 Giessen, Germany

ABSTRACT

The variegated grasshopper *Zonocerus variegatus* (L.) is known as an important pest in the Republic of Benin since 1985. The dry season generation attacks various crops especially cassava (*Manioc esculenta*, L.), but also corn (*Zea mais*, L.) and teak (*Tectona grandis*, L.) are regularly endangered. The negative side effects for the environment to control this pest with chemical insecticides has prompted the studies of a new biocontrol system of *Zonocerus variegatus* with oil from seeds of the neem tree (*Azadirachta indica* A. Juss). The effect of neem oil on migration, feeding activity, mortality and morphogenetic defects on *Zonocerus variegatus* is described. The results proved to be very effective to reduce grasshopper populations on reasonable levels.

INTRODUCTION

The secondary ingredients of the neem tree e.g., azadirachtin (AZA) are proved to be very effective to increase the mortality, to disturb the metamorphosis or to reduce the general activity of locusts and grasshoppers (Schmutterer and Freres, 1990; Nicol and Schmutterer, 1991; Wilps et al., 1992) Azadirachtin, the mean active principle, was first identified by Butterworth and Morgan (1968), using *Schistocerca gregaria* Forsk., the desert locust, as test insect. Azadirachtin is also responsible for many biological effects on *Locusta migratoria migratorioides* R. & F., the African migratory locust (Rembold and Sieber, 1980; Mordue, 1985) This present study was carried out, to study the effects of pure neem oil, which contain low levels of azadirachtin and other biological active ingredients, on the first instar nymphs of the variegated grasshopper *Zonocerus variegatus*

(L.) in a field trial. The aggregation behaviour of the first and second instar grasshoppers were of most interest as target organisms for neem applications. After the synchronic hatching from egg-sites the young larvae are feeding in groups on different wild plants such as *Chromolaena odorata* L. and *Ageratum conyzoides* L. These groups are easily identified and treated with neem products.

Zonocerus variegatus is known as an important pest in the Republic Benin since 1985. Due to its polyphagus status, the dry season generation attacks various crops especially cassava (*Manioc esculenta*, L.), corn (*Zea mais*, L.) and teak (*Tectona grandis* L.). In 1986, an infested area of about 150,000 ha covered the four southern provinces of the country. Between 1986 and 1988 several control actions took place with the assistance of international donation groups. About 20,000 ha of heavily infested area were sprayed during three control campaigns with the insecticides propoxur and fenitrothion. Many problems associated to this campaigns have been observed including health hazards during the pesticide application process, contamination of drinking water, and elimination of non-target organisms from the agro-ecosystem. The side effects of this plant protection campaigns and the dependency of non-industrialized countries like Benin on financial and technical aids to control large damaging grasshopper populations has prompted the studies of potential ways for biological systems of *Z. variegatus*. One alternative could be the treatment of larvae aggregates with oil from seeds of the neem tree (*Azadirachta indica* A. Juss). Neem oil can be applied by ultra low volume (ULV) technique and has nearby technical advantages as farmers in West Africa are familiar with this technology. The neem tree is a locally available in Benin. The content of the azadirachtin ranged between 2.1 and 6.1 mg/g fm (n = 50), (Baumgart, 1991).

MATERIALS AND METHODS

From October to December 1990 a field trial was carried out in a six-month-old teak plantation of the Office National du Bois (ONAB) in the Lama forest, District Zou. The area is known for a high population density of *Zonocerus variegatus* during the dry season (Nov. to May). The following objectives were pursued in this trial:

(a) To study the migration behaviour of neem treated hoppers of *Zonocerus variegatus*. Background of this objective was to disturb the aggregation behaviour of young hoppers and thereby to reduce the level of damage due to the group feeding behaviour.

(b) To verify, if pure neem oil has an effect on mortality and morphogenetic defects on larvae.

(c) To verify, if topical neem oil application has an influence on the feeding activity.

NEEM PRODUCTS

Pure neem oil was produced by a hot-press-procedure at 80°C with an expeller which had 0.013 per cent azadirachtin.

EXPERIMENT

To study the long-term efficiency of the treatment, defined grasshopper populations with a known number of first instar were used. A 200 ha zone of a six-month old teak population was chosen after proving the absence of any hoppers. In this zone, 9 × 100 m^2 plots (10 m × 10 m) were laid. The distance between single plots was kept at app. 200 m. Each plot contained 15 teak plants of 1 m height. The test insects were captured with butterfly nets at the edge of gravel roads and in older teak plantations from young shoots of *Chromolaena odorata*. 18,000 hoppers were identified for their first instar status. 2000 first instar hoppers were then transferred in a net cage and released on the central teak plant in each of the experimental plots. The neem oil was applied in two application rates with 5 and 10 l/ha ULV. The check plot was untreated. Each treatment was replicated three times.

EVALUATION

The plant sociology in every plot was recorded and the migration, the attraction to different plants and the feeding behaviour of hoppers observed. Numbers and sizes of groups were counted. Counts were made in the morning when the insects were roosting on tall vegetation and so were readily visible.

The feeding activity was measured in a laboratory trial. Ten days after treatment, samples of app. 50 larvae were captured in every plot and transported to the laboratories of the Plant Protection Service in Porto Novo. Twenty larvae were transmitted in jars (Ø = 10 cm, 20 cm height) and covered with screen (Ø = 2 mm). To reduce mortality due to uniform diet, hoppers were offered leaves of *Chromolaena adorata* and *Tectona grandis* alternatively. The leaf area consumption was visual evaluated daily. Dead larvae were removed at 24-h intervals and replaced with new larvae from the sample stock to keep the pressure of feeding equal. Mortality due to the treatment was determined and was corrected using Schneider-Orelli's (1947) formula.

Morphogenetic defects on antennae and legs were recorded on samples of hoppers in every plot 4 and 8 weeks after treatment.

RESULTS

A behaviour study was conducted after neem oil application. During the following eight days after treatment it was observed that the insect groups with neem applications subsequently emigrated out of the plots leaving behind small number in the plot. The larvae groups in the check plots remained inside the designed area (Figure 1). The effort to leave the treated area was partly enormous. First instar larvae were found up to 20 m from the edge of the test plots. Ten days after treatment an immigration into the plots for the treated larvae was observed. Treated nymphs had a general south orientation, that means against the wind direction. Nymphs in the control plots moved with the wind direction.

In the treated plots no feeding occurred at the upper part of the vegetation. Some larvae fed only on herbs at the ground, where less numbers of drops and a smaller drop density of neem oil were counted on oil sensitive spray cards (Ciba-Geigy). The reduced feeding is probably a result of the strong perceptibility of *Z. variegatus* to neem treated plants (Olaifa and Akingbohungbe, 1987; Dreyer, 1986). The population movement out of the plots could be an answer due to an unavailable food source. Larvae in the check plots fed normal. All leaves of the central teak tree were damaged after two days. Then, the larvae moved to other food sources inside the plot and fed most on flowers of *A. conyzoides* and on leaves of *C. odorata*.

The originally defined population of 2000 hoppers per plot was splitted in smaller population in all cases. However, in the check plots significant more groups with a higher number of individuals were counted. The distribution maps (Fig. 1) demonstrates that the refounded larvae in the neem treated plots were splitted in groups with a lower number of individuals. These groups were spreaded over a larger area and increased therefore the infestation zone.

The results of the conducted feeding test are given in Figure 2. It is seen, that the neem oil treatments in both application rates caused an inhibition of feeding between 10 and 40 per cent compared to the check. The reduced feeding activity was accompanied with an increasing mortality for the neem oil applications during the trial period. Only one dead hopper was observed in the control. This result could also explain the lower presence of larvae in the neem treated plots.

Both application rates of neem oil in the field affected the antennae and legs of larvae effectively (Fig. 3). Four weeks after the start of the trial a significant higher rate of larvae with defected antennae was obtained. The application rate of 10 l neem oil/ha affected the loss and deformations of antennae stronger than the 5 l neem oil application/ha. The defects increased about 20 per cent at the second evaluation date, 8 weeks after the experiment started. Other morphogenetic defects beside on antennae were not significant.

CONCLUSION

The experimental evidence allows the conclusion that neem oil tested on *Zonocerus variegatus* is a suitable material of an effective control to this pest under field conditions in Benin. It was demonstrated that neem oil application disturbs the aggregation behaviour of young instar nymphs effectively. The so-called roosting larvae were splitted in more smaller groups and they have left the neem applied area, so that the intensity of infestation decreased and thereby the damage to the young teak trees. This result is considerable. Only the group feeding behaviour of roosting larvae is responsible for damages and/or loss of single teak trees. This change of behaviour will be additional reinforced if the mortality, the feeding activity, malformations on legs, wings and probably also the chemosensory activity due to missing antennae, were also affected by the treatment. So far all reported effects of neem oil on different species of locusts and grasshoppers e.g., phagorepellent effects (Bernays and Chapman, 1978), morphogenetic defects (Schmutterer and Freres, 1990), effects on the phase status (Langewald and Schmutterer, 1992) and mortality and fitness reduction (Wilps et al., 1992) were also observed on *Zonocerus variegatus* in this study, more or less well marked. From a practical point of view the gained results with pure neem oil in the field are encouraging and should be considered as an alternative control measure to this grasshopper pest. It is advisable to reduce the application rate because of the neem oil

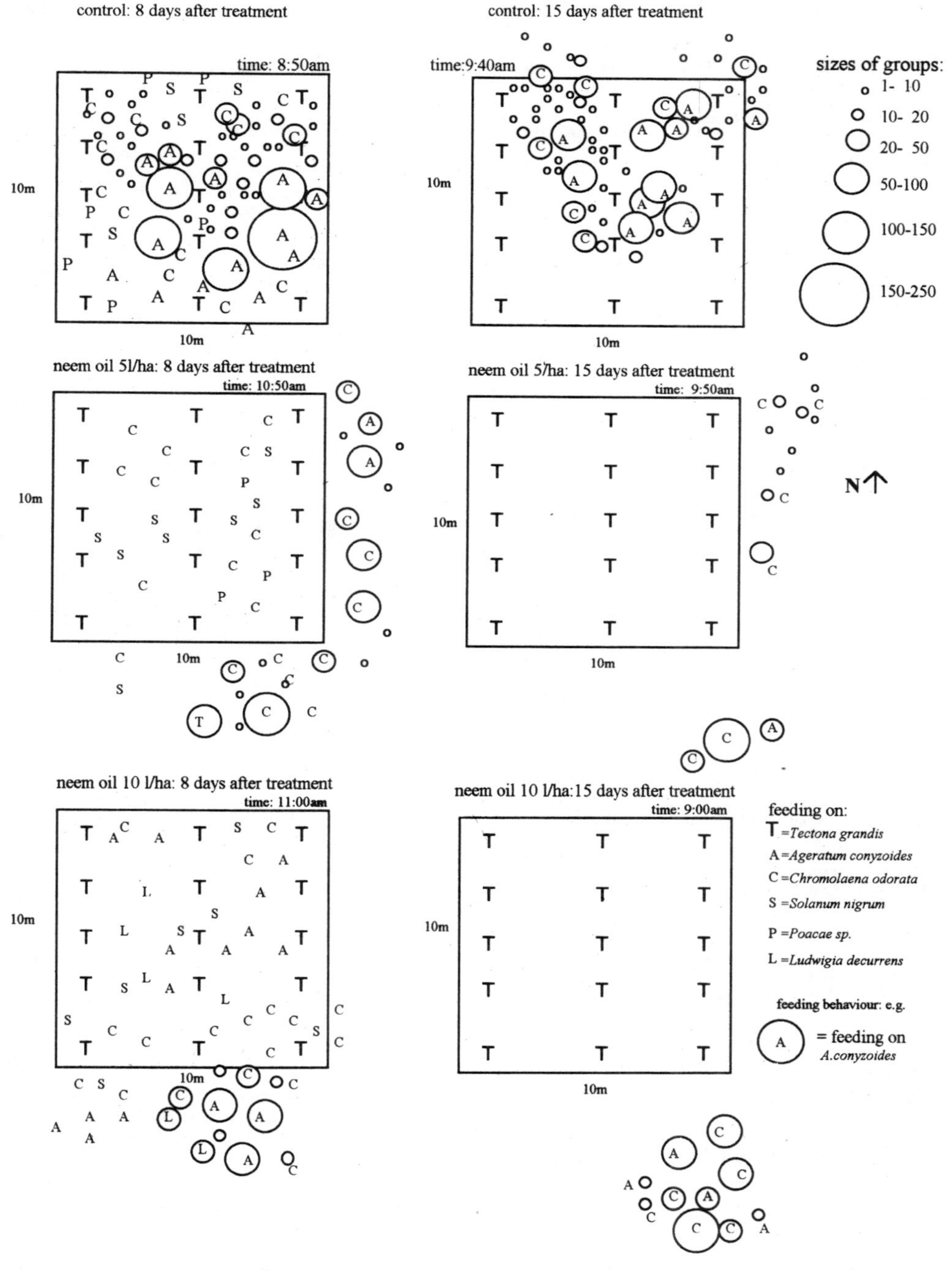

Fig. 1. Migration, feeding behaviour, and number and sizes of groups of *Zonocerus variegatus* larvae following neem oil application (5l/ha and 10l/ha) of a defined population released on the central teak plant of each plot (n = 6000 L1/treatment, 3 replicates/treatment).

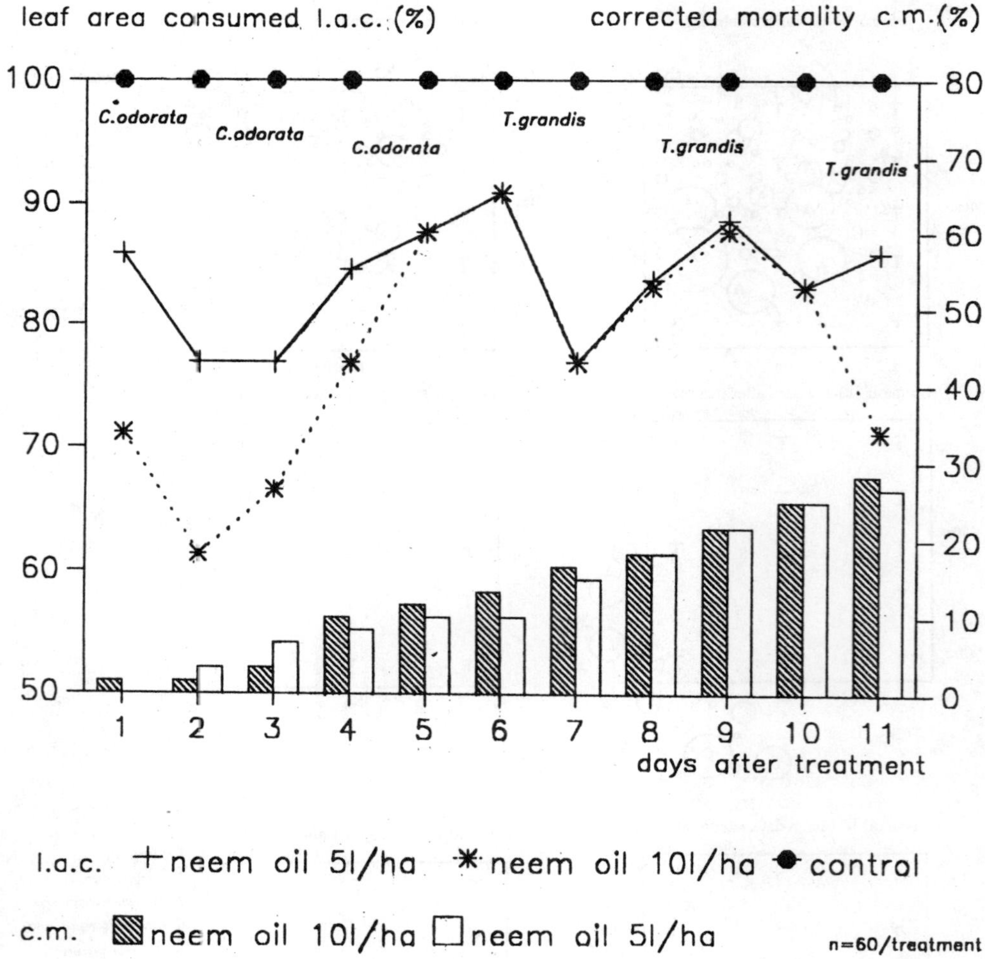

Fig. 2. Leaf area consumed by and corrected mortality of *Zonocerus variegatus* larvae on un-treated leaves of *Chromolaena odcrata* and *Tectona grandis* following application with neem oil (5l/ha and 10l/ha in the field.

production costs (Ostermann, 1992) and the probable phytotoxical damages on leaves on certain crops. There was no burning observed on teak leaves in this trials, but cassava leaves are more seriously affected. If applied in lower rates, additives to the oil such as azadirachtin may be considered as ingredient for the desired effect.

ACKNOWLEDGEMENTS

These studies were conducted at the National Plant Protection Service of Benin in Porto Novo and the Office National du Bois in Kota, Benin, in collaboration with the University of Giessen, Germany. Financial support of my studies and logistical assistance by the Deutsche Gesellschaft für Technische Zusammenarbeit (GTZ) in Eschborn and the Europäische Freiwilligenprogramm (EFP) in Berlin, both Germany, is acknowledged.

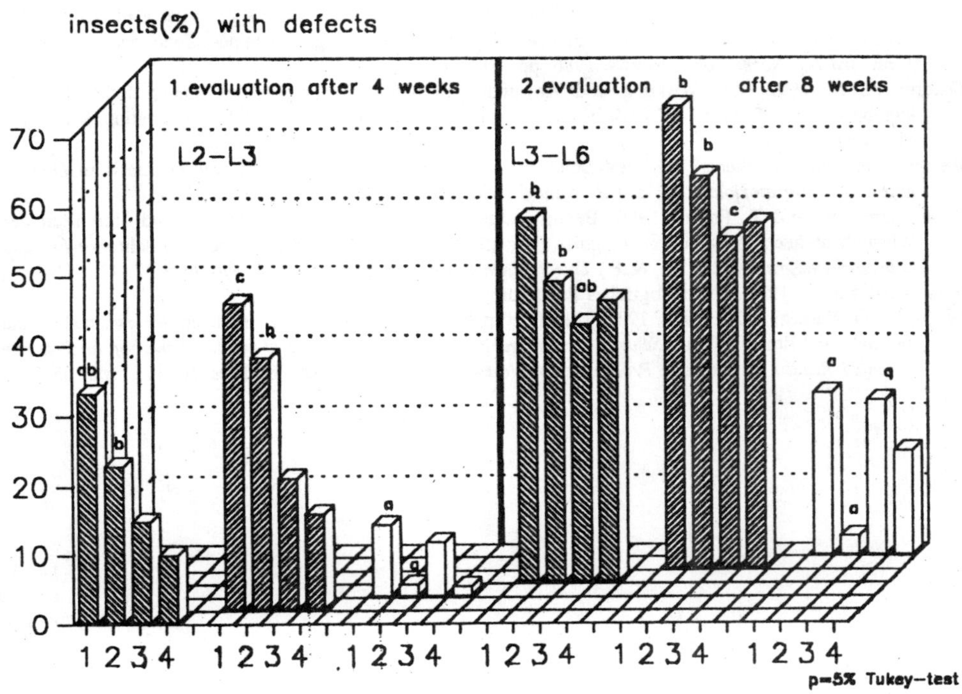

1. total loss of antennae
2. deformed antennae
3. missing segments in antennae
4. deformed legs

Treatments
neem oil 5l/ha   neem oil 10l/ha   check

Fig. 3. Morphogenetic defects of *Zonocerus variegatus* larvae following application of defined populations (n = 2000 L1 per treatment) with neem oil (5l/ha and 10l/ha) in a field experiment, 3 replications, per treatment.

REFERENCES

Baumgart, M. 1991. Das Niemvorkommen in der Republik Benin, Deutsche Gesellschaft für Technische Zusammenarbeit, 1991, Eschborn, 61 pp.

Bernays, E.A. and R.F. Chapman. 1978. Plant chemistry and acridoid feeding behaviour. In: *Biochemical aspects of plant and animal co-evolution*. Harborue, J.B. (ed.),—Phytochem. Soc. Europe Symp. 15: 99–141.

Butterworth, J.H. and E.D. Morgan. 1968. Isolation of a substance that suppresses feeding in Locusts. *Chem. Comm.* 23–24.

Dreyer, M. 1986. Untersuchungen zur Wirksamkeitvon Wasserextrakten und anderen Produkten aus Niemsamen gegen Schädlinge an Gemüse- und Feldkulturen in Togo. Ph.D. Dissertation, University of Giessen, Germany.

Langewald, J. and H. Schmutterer. 1992. Effects of neem oil treatment on the phase status of desert locust, *Schistocerca gregaria*. Proceedings of the Workshop : *Biological Control of Locusts and Grasshopper*, Cotonou, Republic of Benin. C.J. Lomer and C. Prior. (eds.)., pp. 142–154.

Mordue, A.J., 1985. Azadirachtin: its effect on gut motility, growth and moulting in *Locusta*. *Physiol. Ent.* 10: 431–437.

Nicol, C.M.Y. and H. Schmutterer, 1991. Kontakwirkungen des Samenöls des Niembaumes *Azadirachta indica* (A. Juss.) bei gregären Larvender Wüstenheu-schrecke *Schistocerca gregaria* (Forsk.). *J Appl. Ent.* 111: 197–205.

Olaifa, J.I. and A.E. Akingbohungbe. 1987. Antifeedant and insecticidal effects of extracts of *Azadirachta indica*, *Petiveria alliacea* and *Piper guineense* on the variegated grasshoppper, *Zonocerus variegatus*. *Proc. 3rd Int. Neem Conf.*, Nairobi 1986, pp. 405–418.

Ostermann, H. 1992. Zur Wirkung und Anwendung einfacher Niemprodukte gegen Schädlinge in kleinbäuerlichen Tomaten-, Vignabohnen- und Amaranthkulturen in Niger. Ph.D. Dissertation, University of Giessen, Germany.

Rembold, H. and K.P. Sieber. 1981. Inhibition of oogenesis and ovarian ecdystereoid synthesis in *Locusta migratoria migratorioides*. R. & F.Z. *Naturforsch.* 43c: 903-907.

Schmutterer, H. and T. Freres. 1990. Beeinflussung von Metarmorphose, Färbung und Verhalten der Wüstenheuschecke *Schistocerca gregaria* (Forsk.) und der Afrika-nischen Wander-heuschrecke *Locusta migratoria migratorioides* (R. & F.) *Durch Niemsamenöl.-Z. f.Pflkrankh. u. Pflschutz* 97: 431–438.

Schneider-Orlli, O. 1974. Entomologisches Praktikum. Aarau: Verlag Sauerländer. 237 pp.

Wilps, H., O. Nasseh, and S. Krall. 1992. The effect of Neem oil, Neem Azal-F and neem oil enriched on the mortality and fitness of adult *Schistocerca gregaria* (Forskal) *Proceeding of the 1st Workshop on Practice Oriented Results on Use and Production of Neem-Ingredients*, Wetzlar, H. Kleeberg (ed.), pp. 65–78.

42

Environmental Impact and Bioefficacy of Neem Products against Insect Pests in Rice

N.V. KRISHNAIAH AND M.B. KALODE

Directorate of Rice Research, Rajendranagar, Hyderabad 500 030
Andhra Pradesh, India

ABSTRACT

Neem oil 50 per cent EC and crude neem oil were 221 and 102 times respectively safer than malathion to black mirid bug *Tytthus parviceps* (Reuter) as compared to green leafhooper (GLH) *Nephotettix virescens* (Dist.) by LC_{50} method. Resistance development studies revealed that after eight generations, the LC_{50} of neem oil to GLH was 2.929 per cent as compared to 1.308 per cent in the beginning, while the corresponding values for the brown planthopper (BPH) *Nilaparvata lugens* (Stal.) were 1.54 per cent and 2.242 per cent respectively. A field study on BPH resurgence for three seasons showed that BPH population in plots treated with neem oil (1per cent and 4 per cent) was similar to untreated control while plots sprayed with deltamethrin (0.005 per cent) recorded higher population and caused complete hopper burn. Among 12 botanical formulations, Repelin RD 9, Neemark, and Margoside CK at 0.5 and 2.0 per cent exhibited significantly better growth inhibition against nymphs of BPH and GLH. Repelin at 0.5 per cent significantly enhanced the toxicity of synthetic insecticides like ethofenprox and monocrotophos against BPH nymphs.

INTRODUCTION

In pest management programmes, use of conventional insecticides has caused problems in some rice-growing countries. Hence, there is growing interest among entomologists the world over to search for alternatives which would minimize adverse effects on the ecosystem. In this context the use of insecticides of plant origin is a novel approach. The bioefficacy of extracts and compounds from the neem tree, *Azadirachta indica* A. Juss, against various insect pests has been well documented (Schmutterer, 1990). However, the environmental impact of these products has not been well investigated, with is most essential before they are put to large-scale use. The present investigations were carried out to study the adverse effect of neem products on natural enemies, development of target pest resistance, and pest resurgence. In addition, some studies on efficacy of ready-to-use botanical formulations based on neem were carried out.

MATERIALS AND METHODS

All glasshouse experiments were conducted in the DRR glasshouse at Hyderabad at a temperature of 30° ± 5°C and relative humidity 60 ± 10 per cent. All field experiments were carried out in the DRR research farm at Hyderabad.

Toxicity of Neem Oil and Malathion against Black Mirid Bug, *Tytthus parviceps* (Reuter)

Black mirid bug (BMB) *Tytthus parviceps* (Reuter) was reared on plants oviposited by brown planthopper (BPH) *Nilaparvata lugens* (Stal) from egg to adult stage. Three-day-old adults of BMB were used in toxicity studies. Potted 40-day-old TN_1 plants which were pre-oviposited by BPH were used in the tests. Neem oil emulsified with 1 per cent Teepol, neem oil 50 per cent EC (supplied by Tamil Nadu Agricultural University, Coimbatore) and malathion 50 per cent EC (Cynamid India Ltd.) were sprayed at suitable concentrations up to runoff stage. One day later, BMB adults were confined to treated plants at 10 per replication with the help of mylar cages and mortality was assessed after 24 hrs exposure. There were four replications. For comparison, neem oil emulsified with 1 per cent Teepol, neem oil 50 per cent EC, and malathion 50 per cent EC were tested against green leafhopper (GLH) *Nephotettix virescens* (Dist.) These formulations were sprayed on potted 40-day-old TN_1 plants at suitable concentrations up to runoff stage. One day later, seven to eight day old GLH nymphs were caged on treated plants with the help of suitable mylar cages and observations on mortality were recorded after 72 hrs exposure. There were four replications. Percentage mortalities were calculated and subjected to probit analysis (Finney, 1971).

Effect of Neem Oil Spray on Development of Resistance in BPH and GLH

Populations of BPH and GLH maintained in the Directorate of Rice Research glasshouse on TN_1 plants for 20 control without any exposure to insecticides or neem products were used as untreated control.

Baseline data on the susceptibility of BPH and GLH to neem oil were determined. For this purpose 40-day-old TN_1 plants were sprayed with suitable concentrations for neem oil emulsified with 1 per cent Teepol up to runoff stage. One day after spraying, seven- to eight-day old nymphs of each species were confined separately on treated plants with the help of suitable mylar cages. Observations on mortality were recorded after 96 hrs of exposure. Twenty BPH nymphs per replication and 10 GLH nymphs per replication were confined. There were four to six replications. The regression lines between log concentrations and probits were fitted and LC_{50} valves were calculated.

After the baseline data were established, 2000 individuals of each species were confined to separate cages. The caged GLH and BPH were sprayed with neem oil (emulsified with 1 per cent Teepol) at LC_{50} concentrations and the surviving insects were transferred to untreated plants. These were reared up to the next generation, which was again subjected to neem oil treatment at the same concentrations. The process was repeated for eight generations. The LC_{50} values for treated populations were determined after fourth and eighth treatment by the procedure described earlier.

The possibility of development of resistance in BPH against neem oil was also studied. Separate plots were treated with 4 per cent and 1 per cent neem oil at 10-day intervals. An untreated control was also maintained. The BPH nymphs from these plots were collected and reared for three generations under glasshouse conditions without exposure to neem oil. The LC_{50} values were determined by the procedure described earlier.

Effect of Intensive Spraying of Neem Oil on Resurgency of BPH

To study the effect of intensive spraying of neem oil on resurgence of BPH, field experiments were carried out for three seasons—wet seasons of 1991 and 1992 and dry season 1992—in Randomized Complete Block Design with six replications. The variety was RP 4–14 with a plot size of 12 m^2 and spacing of 10 × 10 cm. The following were the treatments:

T_1: Neem oil 4 per cent + Teepol 1 per cent to provide high level of BPH control.
T_2: Neem oil 1 per cent + Teepol 1 per cent to provide low level of BPH control.
T_3: Deltamethrin 50 active ingredient/ha (a known resurgence-causing insecticide).
T_4: Monocrotophos 500 g a.i./ha (a recommended insecticide for BPH control).
T_5: Untreated control.

From one week after transplanting, BPH gravid females were released thrice at weekly intervals in batches of 200 each at the centre of the plot to serve as initial inoculum. The treatments were applied six times at 10-day intervals starting from 30 days after treatment (DAT). Observations on BPH, mirid bugs *Cyrtorhinus lividipennis* (Reuter), and spiders were recorded by visual counting on 10 random hills one day before and two days after each treatment from the third application onwards. The damage due to stem borer and gall midge was also assessed by enumerating total tillers along with dead hearts and silver shoots on 10 random hills. The data were statistically analysed after suitable transformations and the means were separated by least significant difference (LSD) method (Cochran and Cox, 1957).

Influence of Neem Formulations on Survival of BPH and GLH

Botanical formulations received from various organizations—Wellgro, Godrej neem antifeedant, Neemark, Margoside CK, Margoside OK, Nimbosol, Ind NE-20, Repelin, Neemta 2100, and Neemgaurd (each at 0.5 and 2.0 per cent conc.), NOCIL formulation (2 and 4 per cent conc.), and Nimda (IARI) (0.1 and 0.5 per cent conc.) were sprayed on potted 40-day-old TN_1 plants up to runoff stage. The next day after spraying, eight-day-old nymphs of BPH and GLH were caged separately on treated plants and survival was observed seven days later.

In another experiment the combined effect of Repelin (0.5 per cent) and insecticides ethofenprox (5 and 20 ppm) and monocrotophos (20 and 100 ppm) on survival of BPH and GLH was studied by spraying, releasing the insects, and observing the survival after seven days' exposure as described in the earlier experiment. In these experiments percentage survival was worked out and the data were statistically analysed after suitable transformations (Cochran and Cox, 1957).

RESULTS AND DISCUSSION

Toxicity of Neem Oil and Malathion against *T. parviceps* (Reuter)

The LC_{50} values for crude neem oil emulsified with Teepol, neem oil 50 per cent EC, and malathion against BMB were 2.559 per cent, 2.876 per cent, and 8.407 ppm, respectively. The corresponding LC_{50} values for GLH were 1.308 per cent, 0.680 per cent and 438.915 ppm, respectively. It could be worked out that neem oil emulsified with 1 per cent Teepol and neem oil 50 per cent EC were 102 and 221 times safer than malathion to BMB as compared to GLH (Table 1). Abdul Kareem et al. (1989) Table 1 observed in field trials that neem seed kernel extract as seedling root dip and spray was safer to predatory mirid bug *C. lividipennis* and wolf spider *Lycosa pseudoannulata* than monocrotophos. These two predators were not affected by topically applied neem oil at doses up to 20 mg/adult and 50 mg/adult respectively (Saxena et al., 1993). According to Jayaraj (1992), 5.3 and 15.8 per cent mortality of *C. lividipennis* was observed at 0.3 and 0.6 per cent concentrations of neem oil at 48 hrs exposure. These results corroborate the present findings.

Effect of Neem Oil Spray on Development of Resistance in BPH and GLH

When GLH was treated with neem oil each generation, the LC_{50} value was 2.929 per cent after eight generations as compared to 1.308 per cent in the beginning of the experiment. However, this tendency needs to be confirmed in further generations. When BPH was exposed to neem oil in a similar manner, the LC_{50} value did not increase even after eight generations (Table 2). The BPH collected from resurgence experiment where six sprays of neem oil were given at 10-day interval showed no appreciable change in susceptibility to neem oil (Table 3). Vollinger (1986) reported that the larvae of diamondback moth *Plutella xylsotella* (L) showed no sign of resistance to neem seed kernel extract up to 42 generations when the larvae were treated during third and fourth instar with LC_{50} dosage of fourth instar. However, deltamethrin-treated insects developed 20- to 35-fold resistance.

Table 1. Relative toxicity of neem oil and malathion against BMB and GLH

Material	LC_{50} for BMB	LC_{50} for GLH	Safety index
Neem oil 50% EC	2.876%	0.680%	4.2294 (221)
Neem oil + 1% Teepol	2.559%	1.308%	1.9564 (102)
Malathion	8.407 ppm	438.915 ppm	0.0191 (1)

Table 2. Toxicity of neem oil against glasshouse strains of GLH and BPH under treatment at every generation

Generation	GLH		BPH	
	LC_{50} (%)	'b' slope	LC_{50} (%)	'b' slope
1	1.308	2.376 ± 0.120	2.242	1.838 ± 0.100
4	1.167	2.241 ± 0.096	0.769	2.023 ± 0.087
8	2.929	3.208 ± 0.295	1.540	2.984 ± 0.125

Table 3. Toxicity of neem oil to field-collected populations of BPH in resurgence experiment

Population from	LC_{50} (%)	'b' slope
4% neem oil treated	1.212	1.210 ± 0.082
1% neem oil treated	0.967	1.353 ± 0.082
Control (untreated)	1.161	1.156 ± 0.081

Effect of Intensive Spraying of Neem Oil on Resurgence of BPH

Wet Season 1991

The data on BPH and mirid bugs are furnished in Tables 4 and 5, respectively. Table 6 gives the ratios in treatments. After application of treatments two times (53 DAT), deltamethrin-treated plots recorded a significantly highest population of BPH (1277 insects/10 hills), while monocrotophhos-treated plots harboured a population of 419 insects/10 hills, which was significantly higher than the remaining three treatments. The BPH population in neem oil treatments remained low (152 to 161 insects/10 hills) and was on par with untreated control (151 insects/10 hills). At 74 DAT hopper burn occurred in the entire plot of deltamethrin, while in monocrotophos-treated plots, the population of BPH was low (1052 insects/10 hills), followed by untreated control (531/10 hills) and neem oil treatments (265 to 305/10 hills).

At 53 DAT, *C. lividipennis* population was relatively higher in deltamethrin followed by monocrotophos, neem oil treatments, and untreated control. However, there were no significant differences later on. Ratios of BPH/to mirid bugs were highest in

Table 4. Occurrence of BPH, stem borer and gall midge in BPH resurgence trial, wet season 1991

Treatment	BPH/10 hills at				Stem borer (% dead hearts) 46 DAT	Gall midge (% silver shoot) 46 DAT
	53 DAT ↓	54 DAT	74 DAT ↓	75 DAT		
Neem oil 4%	152.3^{c}	169.3^{c}	305.8^{b}	272.5^{bc}	8.2^{a}	21.4^{a}
Neem oil 1%	161.7^{c}	297.8^{c}	265.5^{b}	169.3^{c}	7.1^{a}	21.5^{a}
Deltamethrin 50 g a.i./ha	1277.5^{a}	1970.8^{a}	Hopper burn	Hopper burn	7.5^{a}	22.9^{a}
Monocrotophos 500 g a.i./ha	419.2^{b}	475.0^{b}	1052.1^{a}	812.5^{a}	8.0^{a}	18.9^{a}
Untreated control	151.2^{c}	220.7^{c}	531.0^{ab}	435.2^{ab}	9.7^{a}	21.4^{a}

↓ indicates application of treatments

Table 5. Mirid bug population in different treatments in BPH resurgence trial, wet season 1991

Treatment	Mirid bugs/10 hills at			
	53 DAT ↓	54 DAT	74 DAT ↓	75 DAT
Neem oil 4%	25.0[c]	23.5[a]	132.5[a]	78.5[a]
Neem oil 1%	25.8[bc]	41.2[a]	109.5[a]	80.3[a]
Deltamethrin 50 g a.i./ha	70.0[a]	51.8[a]	Hopper burn	Hopper burn
Monocrotophos 500 g a.i./ha	43.0[ab]	42.2[a]	88.3[a]	91.3[a]
Untreated control	27.0[bc]	30.5[a]	123.3[a]	115.3[a]

↓ Indicates application of treatments.

Table 6. Brown planthopper/mirid bug ratios in treatments in BPH resurgence trial, wet season 1991

Treatment	BPH/MB AT			
	53 DAT ↓	54 DAT	74 DAT ↓	75 DAT
Neem oil 4%	6.9[bc]	7.4[b]	2.9[b]	3.6[b]
Neem oil 1%	6.3[bc]	7.1[b]	1.9[b]	2.0[b]
Deltamethrin 50 g.a.i./ha	21.2[a]	48.8[a]	H.B.	H.B.
Monocrotophos 500 g.a.i./ha	9.3[b]	14.3[b]	13.2[a]	9.3[a]
Untreated Control	5.9[c]	7.4[b]	4.6[b]	4.2[b]

↓ Indicates time of application of treatments.
DAT = Days after transplanting.

Table 6a. Brown planthopper/mirid bug ratios in different treatments in brown planthopper resurgence trial, Dry Season, 1992

Treatment	BPH/MB AT DAT			
	58 DAT ↓	61 DAT	83 DAT ↓	87 DAT
Neem oil 4%	1.7[b]	2.0[a]	1.6[b]	2.8[b]
Neem oil 1%	1.7[b]	1.3[a]	1.8[b]	2.9[b]
Deltamethrin 50 g.a.i./ha	2.8[a]	2.8[a]	9.6[a]	16.5[a]
Monocrotophos 500 g.a.i./ha	1.6[b]	1.5[a]	1.5[b]	3.2
Untreated Control	2.4[a]	1.3[a]	2.0[b]	3.9[b]

↓ Indicates time of application of treatments.
DAT = Days after transplanting.

deltamethrin (21.2 to 48.8) followed by monocrotophos (9.3 to 14.3), as compared to 2.9 to 7.4 in plots treated with neem oil and 4.2 to 7.4 in untreated control, indicating greater safety of neem oil to mirid bug. The spider populations were, however, low in the experiment (maximum of 17/10 hills).

Deltamethrin recorded complete hopper burn with no grain yield, while hopper burn in neem oil treatments, monocrotophos, and untreated control was 15.5 to 16.1, 43.5, and 24.3 per cent without significant differences in grain yield.

Table 6b. Ratio of brown planthopper and mirid bug in different treatments in BPH resurgence trial, wet season, 1992

Treatments	50 DAT ↓	53 DAT	73 DAT ↓	76 DAT	84 DAT ↓	87 DAT
Neem oil 4%	2.93[b]	2.33[bc]	1.07[b]	1.27[b]	1.88[ab]	1.08[a]
Neem oil 1%	2.18[b]	1.85[c]	0.92[b]	1.55[b]	0.68[bc]	1.00[a]
Deltamethrin 50 g.a.i./ha	22.37[a]	12.12[a]	21.67[a]	22.92[a]	H.B.	H.B.
Monocrotophos 500 g.a.i./ha	5.35[b]	3.30[b]	2.30[b]	3.12[b]	2.38[a]	1.23[a]
Untreated control	2.80[b]	2.07[c]	0.58[b]	0.87[b]	0.62[c]	0.85[a]

↓ Indicates time of application of treatments.
DAT = Days after transplanting.

Table 6c. Influence of brown planthopper resurgence on grain yield in rice

Treatment	Wet Season 1991	Dry Season 1992	Wet Season 1992
Neem oil 4%	2553[a]	5959[c]	2460[b]
Neem oil 1%	3002[a]	7482[b]	2525[b]
Deltamethrin 50 g.a.i./ha	0[b]	6765[bc]	0[c]
Monocrotophos 500 g.a.i./ha	3181[a]	10170[a]	3172[a]
Untreated Control	3225[a]	9409[a]	2849[ab]

Dry Season 1992

The data on BPH and mirid bugs are presented in Tables 7 and 8 respectively. The BPH population at 61 DAT was highest in deltamethrin (223 insects/10 hills) as compared to 73/10 hills in monocrotophos, 93 to 120/10 hills in neem oil, and 116/10 hills in untreated control. Highest population of BPH was observed in deltamethrin (2135 insects/10 hills) up to 87 DAT, followed by gradual reduction.

The mirid bug population (62 to 106/10 hills) was highest in deltamethrin as compared to 34 to 68/10 hills in other treatments. However, ratios of BPH to mirid bugs were highest (2.8 to 16.5) in deltamethrin as compared to 9.0 to 3.9 in other treatments. Spider population was generally low, the maximum being 18/10 hills. Deltamethrin registered 30.5 per cent hopper burn, while it was negligible in all other treatments. Monocrotophos recorded a grain yield of 10.1 kg/ha followed by untreated control (9.4 kg/ha, while other treatments registered 6.0 to 7.5 kg/ha.

Wet Season 1992

The data on BPH and mirid bugs are presented in Tables 9 and 10 respectively. As in the wet season 1991, maximum population of BPH was observed in deltamethrin (1473/10 hills) followed by monocrotophos (542 insects/10 hills), neem oil (322 to 422 insects/10 hills), and untreated control (410 insects/10 hills) by 50 DAT (before third application of treatments). By 76 DAT, the population in deltamethrin reached a peak of 4166 BPH/10 hills followed by monocrotophos (729 insects/10 hills). Neem oil treatments registered 401 to 494 insects/10 hills as compared to 215 insects/10 hills in

Table 7. Occurrence of BPH in resurgence trial, dry season 1992

Treatment	BPH/10 hills at			
	58 DAT ↓	61 DAT	83 DAT ↓	87 DAT
Neem oil 4%	92.3[bc]	93.6[bc]	93.3[b]	188.0[bc]
Neem oil 1%	90.3[bc]	120.3[b]	118.0[b]	166.0[c]
Deltamethrin 50 g.a.i./ha	176.7[a]	223.2[a]	955.3[a]	2134.8[a]
Monocrotophos 500 g.a.i./ha	66.8[c]	72.8[c]	102.7[b]	211.3[bc]
Untreated Control	130.3[ab]	116.0[b]	128.5[b]	284.0[b]

↓ Indicates application of treatments.

Table 8. Occurrence of mirid bug population in BPH resurgence trial, dry season 1992

Treatment	Mirid bugs/10 hills at			
	58 DAT ↓	61 DAT	83 DAT ↓	87 DAT
Neem oil 4%	55.0[a]	48.5[bc]	60.2[b]	68.2[b]
Neem oil 1%	51.7[a]	60.5[ab]	67.2[b]	59.7[b]
Deltamethrin 50 g a.i./ha	62.2[a]	81.2[a]	96.5[a]	106.3[a]
Monocrotophos 500 g a.i./ha	41.2[a]	35.7[c]	65.5[b]	66.2[b]
Untreated Control	53.8[a]	62.3[ab]	64.0[b]	71.5[b]

↓ Indicates application of treatments.

Table 9. Occurrence of BPH in resurgence trial, wet season 1992

Treatment	BPH/10 hills at						Yield
	50 DAT ↓	53 DAT	73 DAT ↓	76 DAT	84 DAT ↓	87 DAT	(kg/ha)
Neem oil 4%	321.8[b]	266.5[b]	377.2[bc]	400.5[bc]	4378[a]	188.7[a]	2460[b]
Neem oil 1%	421.7[b]	231.5[b]	234.7[c]	493.5[bc]	204.8[a]	189.5[a]	2525[b]
Deltamethrin 50 g a.i./ha	1473.0[a]	1077.5[a]	3095.0[a]	4165.8[a]	Hopper burn	Hopper burn	0[c]
Monocrotophos 500 g a.i./ha	541.5[b]	349.0[b]	525.3[b]	728.5[b]	600.8[a]	189.7[a]	3172[a]
Untreated control	410.0[b]	298.5[b]	145.2[c]	214.7Pc	194.7[a]	182.5[a]	2849[ab]

↓ Indicates time of application of treatments.

untreated control. Complete hopper burn resulted in deltamethrin by 84 DAT while other treatments recorded negligible hopper burn.

Unlike in the wet season 1991, mirid bug population was lowest (67 MB/10 hills) in deltamethrin followed by monocrotophos (109/10 hills), neem oil treatments (127 to 190 MB/10 hills), and untreated control (163 MB/10 hills) by 50 DAT. A similar trend continued up to 73 DAT. As expected, ratios of BPH to mirid bugs were highest in deltamethrin (12.13 to 45.52) followed by monocrotophos (2.30 to 5.35), neem oil

Table 10. Mirid bug population in different treatments in BPH resurgence trial, wet season 1992

Treatment	MB/10 hills at					
	50 DAT ↓	53 DAT	73 DAT ↓	76 DAT	84 DAT ↓	87 DAT
Neem oil 4%	127.3[ab]	118.3[a]	297.7[a]	338.5[a]	248.7[a]	186.2[a]
Neem oil 1%	190.3[a]	123.3[a]	245.3[a]	290.3[a]	280.7[a]	190.8[a]
Deltamethrin 500 g a.i./ha	67.2[c]	91.8[a]	149.5[b]	195.5[a]	Hopper burn	Hopper burn
Monocrotophos	108.5[bc]	104.5[a]	255.0[a]	260.3[a]	229.0[a]	158.7[a]
Untreated control	162.5[ab]	141.5[a]	253.7[a]	271.7[a]	313.8[a]	232.7[a]

↓ Indicates time of application of treatments.

treatments (0.92 to 2.80), and untreated control (0.58 to 2.80). Spider population in general was low with a maximum of 5.3 to 10.7/10 hills in deltamethrin, while it ranged from 4.7 to 6.8/10 hills in other treatments. Higher grain yield was obtained in monocrotophos treatment (3.2 t/ha) followed by untreated control (2.8 t/ha), which was on par with neem oil treatments (2.5 t/ha). No grain yield was recorded in deltamethrin treatment due to complete hopper burn (Table 9).

The above results clearly showed that neem oil treatments at 1 and 4 per cent did not cause BPH resurgence when deltamethrin resulted in complete hopper burn. Neem oil was safer to mirid bug predator than monocrotophos. Further, it can be said that mirid bugs acted as passive dependants on BPH rather than active predators in checking the pest population. Jayarai (1992) also reported that neem oil did not cause resurgence of white-backed planthopper even when marked resurgence in insects treated with methyl parathion, deltamethrin, quinalphos, and fenvalerate was noticed. Krishnaiah and Kalode (1992) observed no resurgence of BPH in neem oil treatments even when deltamethrin caused complete hopper burn.

Influence of Neem Formulations on Survival of BPH and GLH

Among the ready-to-use botanical formulations tested against BPH and GLH, Repelin registered 7 and 18 per cent survival of BPH and 8 and 37 per cent survival of GLH at 2.0 and 0.5 per cent concentrations respectively, while 93 and 88 per cent insects survived in untreated control (Table 11). Among the other formulations, Neemark at 2.0 and 0.5 per cent concentrations exhibited good efficacy against GLH (6.7 and 21.7 per cent survival respectively) and BPH (7.5 and 24.2 per cent survival respectively) as compared to more than 90 per cent survival in untreated control. Margoside CK evaluated only against GLH resulted in lowest survival of nymphs (1.7 and 5.0 per cent at 2.0 and 0.5 per cent concentrations respectively), while untreated control recorded 92 per cent survival (Tables 13 and 14). The adverse effect of neem formulations on growth and survival of insect pests is well documented (Saxena, 1988; Schmutterer, 1990). However, the better efficacy of some of the formulations of neem mentioned above could be due to higher concentration of active principles (Schmutterer, 1990).

Table 11. Effect of formulations on survival of BPH and GLH

Formulation	Conc. (%)	Per cent survival*	
		BPH	GLH
Repelin	0.5	18[ef]	37[c]
Repelin	2.0	7[f]	8[e]
Neemta 2100	0.5	33[de]	42[c]
Neemta 2100	2.0	18[ef]	5[f]
Neemguard	0.5	66[bc]	73[b]
Neemguard	2.0	18[ef]	23[d]
Control		93[a]	88[a]

*Mean of six replicates.

Table 12. Effect of formulations on the survival of GLH and BPH

Formulation	Conc. (%)	Per cent survival*	
		BPH	GLH
Wellgro	0.5	98.3[a]	91.7[a]
Wellgro	2.0	85.0[b]	73.3[bc]
Godrej neem antifeedant	0.5	90.0[b]	84.2[ab]
Godrej neem antifeedant	2.0	83.3[b]	88.3[ab]
Nimba (IARI)	0.1	40.0[c]	63.0[c]
Nimba (IARI)	0.5	23.3[c]	21.7[d]
Untreated control		98.3[a]	91.7[a]

*Mean of six replicates.

Table 13. Effect of formulations on survival of GLH

Formulation	Conc. (%)	Per cent survival*
Neemark	2.0	6.7[a]
Neemark	0.5	21.7[b]
NOCIL formulation	4.0	58.3[d]
NOCIL formulation	2.0	50.0[cd]
Margoside CK	2.0	1.7[a]
Margoside CK	0.5	5.0[a]
Margoside OK	2.0	10.0[a]
Margoside OK	0.5	33.3[bc]
Untreated control		

*Mean of six replicates.

Effect of Repelin and Synthetic Insecticides on Survival of BPH and GLH

The efficacy of ethofenprox (a non-ester pyrethroid insecticide) at 5 ppm and monocrotophos (an organophosphate) at 20 ppm was significantly enhanced in combination with Repelin at 0.5 per cent as compared to insecticide alone against BPH nymphs (Table 15). However, such an effect could not be observed with GLH adults. Sitaramaiah et al. (1986) observed better efficacy of diflubenzuron (0.0035 per cent) in combination with 1 per cent neem seed kernel suspension in tobacco against *Spodoptera litura*. Higher mortality and lower oviposition were observed in desert locust *Schistocerca gregaria* by topical treatment of 1 per cent neem seed kernel extract in combination with the insecticides fenitrothion, dichlorvos, or chlorpyriphos (Brajendra Singh, 1987). In greenhouse experiments at IRRI, TN_1 rice plants treated with neem cake and carbofuran mixture had lowest rate of rice tungro virus infection as compared to either neem cake or carbofuran alone, probably because of better control of GLH (Abdul Kareem *et al.*, 1989). These observations support the present findings.

Table 14. Effect of formulations on survival of BPH

Formulation	Conc. (%)	Per cent survival*
Nimbosol	2.0	7.5[a]
Nimbosol	0.5	49.2[cd]
Ind NE-20	2.0	10.0[a]
Ind NE-20	0.5	30.8[bc]
Neemark	2.0	7.5[a]
Neemark	0.5	24.2[ab]
NOCIL formulation 2%	2.0	65.8[d]
NOCIL formulation 2%	0.5	62.5[d]
Untreated control		94.2[e]

*Mean of six replicates.

Table 15. Effect of Repelin and synthetic insecticides on survival of BPH and GLH

Treatment	Per cent survival*	
	BPH	GLH
Control	95[a]	97[a]
Repelin 0.5%	75[bc]	53[b]
Ethofenprox 5 ppm	83[ab]	50[b]
Ethofenprox 5 ppm + Repelin 0.5%	63[cd]	48[b]
Ethofenprox 20 ppm	45[def]	5[c]
Ethofenprox + Repelin 0.5%	33[fg]	7[c]
Monocrotophos 20 ppm	81[b]	28[b]
Monocrotophos 20 ppm + Repelin 0.5%	54[de]	12[c]
Monocrotophos 100 ppm	27[g]	5[c]
Monocrotophos 100 ppm + Repelin 0.5%	39[efg]	0[c]

*Mean of six replicates.

ACKNOWLEDGEMENT

The authors are highly grateful to Dr. E.A. Siddiq, Project Director, Directorate of Rice Research, for constant encouragement and for providing facilities.

REFERENCES

Abdul Kareem, A., R.C. Saxena and M.E.M. Bonocodin. 1989. Neem carbofuran mixtures against *Nephotettix virescens* (Distant) (Homoptera; Cicadellidae) and its transmission of tungro associated viruses. *J. Appl. Entomol.* 108(1): 68–71.

Brajendra Singh. 1987. Effect of neem kernel extract mixed with insecticides on the oviposition and mortality of the desert locust *Schistocerca gregaria* (Forsk). *Indian J. Plant Prot.* 15(2): 184–187.

Cochran, W.G. and G.M. Cox. 1957. *Experimental Designs,* 2nd ed., John Wiley & Sons Inc., New York, 611 pp.

Finney, D.J. 1971. *Probit Analysis*, 3rd ed. Cambridge Univ. Press., Cambridge, pp. 57–59.

Jayaraj, S. 1992. Studies on IPM in rice based cropping systems with emphasis on use of botanicals, their safety and socioeconomic considerations. In: *Botanical Pest Control Project Phase II. Proc. IRRI-ADB final Workshop on Botanical Pest Control*, 28-31 July, IRRI, Los Banos, Laguna, Phillippines, pp. 63–81.

Krishnaiah, N.V. and M.B. Kalode. 1992. Environmental impact studies on neem. In: *Botanical Pest Control Project Phase II. Proc. IRRI-ADB final Workshop on Botanical Pest Control*, 28-31 July, IRRI, Los Banos, Laguna, Philippines, pp. 50–53.

Saxena, R.C., P.B. Epino, Chen-Wen Tu, and B.C. Puma. 1983. Neem, chinaberry and custard apple: Antifeedant and insecticidal effects of seed oils on leafhopper and planthopper pests of rice. Proc. 2nd Int. Neem Conf., Rauischholzhausen, pp. 403–412.

Saxena, R.C. 1988. Neem—A source of natural insecticide. Final workshop of IRRI-ADB-EWC project on Botanical Pest Control in Rice-based Cropping Systems. 12-16 Dec., IRRI, Los Banos, Laguna, Philippines.

Schmutterer, H. 1990. Properties and potential of natural pesticides from the neem tree, *Azadirachta indica. Ann. Rev. Entomol.* 35: 271–297.

Sitaramaiah, S., G. Ramaprasad and B.G. Joshi (1986). Efficacy of diflubenzuron and neem seed kernel suspension against the tobacco caterpillar, *Spodoptera litura* in tobacco nurseries. *Phytoparasitica* 14(4): 265–271.

Vollinger, M. 1986. The Possible development of resistance against neem seed kernel extract and deltamethrin in *Plutella xylostella. Proc. 3rd Int. Neem Conf.*, Nairobi, pp. 543–554.

43

Potential of Neem in Insect Pest Management in Rice

G.S. DHALIWAL, JASWANT SINGH AND V.K. DILAWARI

Punjab Agricultural University, Ludhiana 141 004, India

ABSTRACT

In view of the environmental hazards of synthetic organic insecticides, the potential of neem was exploited for management of major insect pests of rice during 1987-1988 and 1990–1992. Neem oil (4 per cent) proved effective against rice stem borer, *Scirpophaga incertulas* (Walker) resulting in 5.99 per cent whiteheads as against 3.58 per cent in monocrotophos (0.05 per cent) and 13.86 per cent in untreated control. Neem oil (1, 2, 3, 4, per cent) was also effective against rice leaf folder, *Cnaphalocrocis medinalis* (Guenee) (20.56–26.39 per cent folded leaves) as compared to monocrotophos (20.71 per cent) and untreated control (33.32 per cent). Among neem-based insecticides, RD-9 Repelin (4 kg/ha) was most effective against stem borer with 10.60 per cent whiteheads, followed by Neemrich II and Neemrich I (both 2 kg/ha) with 11.21 and 12.79 per cent whiteheads as against 12.20 per cent in phosphamidon (0.527 kg active ingredient/ha) and 18.06 per cent in untreated control. All the treatments were at par with respect to yield, but proved significantly better than neem oil and control. RD-9 Repelin also reduced growth index of white-backed planthopper, *Sogatella furcifera* (Horvath), to 2.29 followed by Neemrich II (2.35), Neemrich I (3.08) Neemark (3.49), and untreated control (8.6).

Neem-based insecticides in combination with synthetic organic insecticides provided excellent control of rice leaf folder. The percentage of folded leaves in phosphamidon-Repelin-monocrotophos-Repelin, phosphamidon-Neemrich I-monocrotophos-Neemrich I, phosphamidon-Neemrich II-monocrotophos-Neemrich II, and phosphamidon-Neemark-monocrotophos-Neemark was 7.01, 6.18, 5.33, and 4.33 per cent respectively as against 32.15 per cent in untreated control. RD-9 Repelin, Neemrich II, Neemark, and Neemrich I resulted in 18.91, 20.64, 21.08, and 28.17 per cent folded leaves as against 3.13 per cent in synthetic organic insecticides phosphamidon-monocrotophos-chlorypriphos-phosphamidon sprayed at 10-day intervals alternatively at 0.6 kg a.i./ha. Similarly, schedule including Neemrich II controlled stem borer as effectively (10.51 per cent whiteheads, 5490 kg/ha yield) as synthetic insecticides (5.91 per cent, 5745 kg/ha).

INTRODUCTION

Most of the synthetic organic insecticides currently in use disrupt the agroecosystem. For this reason, interest in botanical insecticides has grown in the last few years. Among the numerous plant products, the ones from neem, *Azadirachta indica* A. Juss, have attracted the attention of scientists the most because of its availability and insecticidal properties (Saxena, 1989). It is in this context that neem products are being evaluated against insect pests of various crops. The rice agroecosystem receives, next to cotton, the highest amount of insecticides in India. Rice is severely attacked by over 20 species of insect pests. Among these, rice stem borer *Scirpophaga incertulas* (Walker), leaf folder, *Cnaphalocrocis medinalis* (Guenee), and white-backed planthopper, *Sogatella furcifera* (Horvath), are quite serious problems in Punjab. It was therefore decided to explore the possibility of use of neem products for the management of major insect pests in the rice agroecosystem.

MATERIALS AND METHODS

Field and laboratory studies were conducted using neem-based products on the insect pests of rice from 1987 and 1988 and 1990 to 1992. During the *kharif* seasons of 1987 and 1988, field trials were conducted at Punjab Agricultural University (PAU), at Ludhiana, on rice varieties PR 106 (1987) and Basmati 370 (1988) transplanted in a randomized block design. Neem oil at 1, 2, 3 and 4 per cent with 0.4 per cent Teepol as emulsifier, Teepol 0.4 per cent, monocrotophos 0.05 per cent, formed treatments with an untreated control applied in three replications. The treatments applied only once, i.e., at 80 days after transplanting (DAT) during 1987, and repeated three times at 20-day intervals starting from 35 DAT during 1988. Observations were recorded on per cent leaves folded by rice leaf folder in 1987 and on per cent white earheads due to stem borer in 1988, and yield data were recorded in both years. Based on these observations on neem, the neem-based insecticide Neemark was tested along with the other recommended insecticides for the control of rice stem borer at farmers' fields during 1990.

During 1991, a field trial was laid out in a randomized block design using variety PR 110 in farmers' fields. RD-9 Repelin at 4.0 kg/ha, Neemrich I, Neemrich II, and Neemark at 2.0 kg/ha each, neem oil at 6.0 kg/ha, phosphamidon at 0.527 kg/ha, and unsprayed control were applied randomly in three replications. Three applications were made at 12-day intervals and observations were recorded on per cent whiteheads due to rice stem borer and yield data were also recorded. In a greenhouse study during 1991, first-instar nymphs of white-backed planthopper were caged on 30-day-old seedlings of rice variety TN-1 treated with different concentrations of neem-based products at the rate of 10 individuals per cage with one untreated check. The observations on the number of nymphs becoming adults and time taken to become adults were recorded, and growth index was worked out (Saxena et al., 1981).

A field trial in a randomized block design was laid out at the PAU Research Farm, Ludhiana, during 1992 on rice variety PR 106. Ten treatments–four neem-based products individually and in combination with recommended insecticides, a combination of synthetic insecticides and neem oil were replicated three times. Each treatment was applied five times at 10-day intervals starting 37 DAT. The observations were recorded on the incidence of leaf folder after three and four replications, and on stem borer incidence after five replications, and yield was recorded from all the treatments.

RESULTS AND DISCUSSION

Effect of Neem Oil on Rice Leaf Folder and Stem Borer

Neem oil at 1, 2, 3 and 4 per cent was equally effective in controlling the leaf folder (Table 1). The percentage of folded leaves was on par among plants treated with neem oil at all the concentrations and the one treated with monocrotophos. The percentage of folded leaves was the highest in untreated control (33.32 per cent) and was significantly higher than in all other treatments (20.56 to 26.39 per cent). There was, however, no significant difference in yield among all the treatments.

In case of stem borer, performance of neem oil at 3 and 4 per cent was on par with monocrotophos 0.05 per cent, since percentage of dead hearts did not differ significantly among these three treatments; however, their differences over control were significantly superior. Similarly, whitehead emergence in case of monocrotophos (0.05 per cent) was on par with neem oil (4 per cent) and significantly less than that in untreated control. Performance of neem oil at all the concentrations was significantly superior over control. The highest yield of paddy was recorded from monocrotophos-treated plots (2027 kg/ha) followed by neem oil in order of decreasing concentrations, while untreated control recorded the lowest (1523 kg), which was significantly different from that in case of neem oil 4.0 per cent (1792 kg) and 3.0 per cent (1679 kg/ha).

Similar results have been obtained in coordinated trials conducted at other locations of India. Neem oil at higher concentration (4 per cent) was found to be moderately effective against stem borer and leaf folder (DRR, 1987, 1988). The

Table 1. Evaluation of neem oil against rice leaf folder and yellow stem borer on PR 106 (*kharif* 1987) and Basmati 370 (*kharif* 1988)

Treatment	Leaf folder (1987)		Stem borer (1988)		
	Folded leaves (%)	Yield (kg/ha)	Dead hearts (%)	Whiteheads (%)	Yield (kg/ha)
Neem oil 1%	23.72[a]	4951	6.76[bc]	10.97[cd]	1630
Neem oil 2%	20.56[a]	4827	6.27[bc]	9.78[c]	1622
Neem oil 3%	26.39[a]	5048	5.50[ab]	8.99[bc]	1679
Neem oil 4%	24.98[a]	5039	5.26[ab]	5.99[ab]	1792
Teepol 0.4%	26.91[ab]	4863	6.67[bc]	11.32[cd]	1637
Monocrotophos (Nuvacron 36 WSC at 0.05 %)	20.71[a]	4678	4.15[a]	3.58[a]	2027
Untreated control	33.32[b]	4360	7.18[c]	13.86[d]	1523

In a column, means followed by the same letter do not differ significantly at 5 per cent level by DMRT.

yield increase due to neem oil was also lower than that due to monocrotophos (DRR, 1987, 1988).

Efficacy of Neem-based Insecticides Against Stem Borer

Minimum number of whiteheads (10.60 per cent) were observed in case of RD-9 Repelin applied at 4 kg/ha, which was on par with that with Neemrich I (12.79 per cent) and Neemrich II (11.21 per cent) at 2 kg/ha each and phosphamidon at 0.527 kg/ha (12.20 per cent) (Table 2). Percentage of whitehead emergence in case of Neemark at 2 kg/ha (14.10 per cent) and neem oil at 6 kg/ha (14.43 per cent) was on par with Neemrich I, Neemrich II and phosphamidon but significantly higher than with RD-9 Repelin and lower than with untreated control (18.06 per cent). The yield of paddy in all the treatments (6000 to 6500 kg/ha) was on par except with neem oil (5150 kg) and control (4500 kg), where it was on par but significantly lower than all other treatments. Efficacy of Neemark in reducing yellow stem borer incidence was also observed in a trial conducted by Dilawari et al., (1991). Such effects have also been observed from the use of other neem based preparations which have been reviewed by Saxena (1989). The commercial products may contain several bioactive compounds, out of which azadirachtin exhibits a wide variety of biological effects on insects including growth disruption, male sterility, ovipositional inhibition, feeding deterrence and mortality (Schmutterer et al., 1981). Most commercial preparations have been standardized on the basis of amount of azadirachtin because it is the principal compound. The reduction in damage due to leaf folder and yellow stem borer among the treated plants over control can, therefore, be attributed to the effects of this principal compound.

Effect of Neem-based Insecticides on the Growth and Development of White-backed Planthopper

In the screenhouse, when nymphs of white-backed planthopper were reared on TN-1 plants treated with neem-based insecticides, the growth index of white-backed planthopper was the highest on untreated plants (8.6), which was significantly higher than the growth index on all other treatments (Table 3). The growth index was the lowest on plants treated with RD-9 Repelin 1 per cent (2.29) followed by Neemrich II 0.1 per cent (2.35), neem oil 1.0 per cent (2.59), Neemrich I 0.4 per cent (3.08), Neemrich II 0.05

Table 2. Efficacy of neem-based products in controlling yellow stem borer on PR 110 (*kharif* 1991)

Treatment	Dose (kg/ha)	Whiteheads (%)	Yield(kg/ha)
RD-9 Repelin	4.0	10.60[a]	6500[a]
Neemrich I	2.0	12.79[ab]	6100[a]
Neemrich II	2.0	11.21[ab]	6050[a]
Neemark	2.0	14.10[b]	6150[a]
Neem oil	6.0	14.43[b]	5150[b]
Phosphamidon	0.527	12.20[ab]	6000[a]
Control	—	18.06[c]	4500[b]

In a column, means followed by the same letter do not differ significantly at 5 per cent level by DMRT.

Table 3. Growth index of white-backed planthopper as affected by neem-based insecticides

Insecticide	Concentration (%)	Growth index
Neemark	0.05	3.49[abc]
	0.25	4.22[bcdef]
	0.125	5.11[f]
Neemrich I	0.4	3.08[ab]
	0.2	3.41[abcd]
	0.1	4.41[cde]
Neemrich II	0.1	2.35[a]
	0.05	3.20[ab]
	0.025	4.71[cdef]
RD-9 Repelin	1.0	2.29[a]
	0.5	3.66[bcde]
	0.25	4.69[def]
Neem oil	1.0	2.59[a]
	0.5	4.75[def]
	0.25	5.33[f]
Control	—	8.60[g]

In a column, means followed by the same letter do not differ significantly at 5 per cent level by DMRT

per cent (3.2), Neemrich I 0.2 per cent (3.41), and Neemark 0.5 per cent (3.49), which were on par with each other but superior to Neemrich I 0.10 per cent (4.41), Neemrich II 0.025 per cent (4.71), RD-9 Repelin 0.25 per cent (4.69), Neemark 0.125 per cent (5.11), and neem oil 0.25 per cent (5.33). Thus, the growth index was adversely affected by these preparations. The mortality of early instars and prolongation of nymphal period of survivors reduced the growth index among the individuals reared on treated plants. The effect of neem products on reduction of white-backed planthopper population has also been observed by Ramaraju and Sundra Babu (1989).

Combination of Neem-based Products with Synthetic Insecticides

The results on incidence of leaf folder after three applications of neem-based insecticides and synthetic insecticides individually and in combination indicated that Neemrich I and Neemark (14.82 and 15.49 per cent) were on par with but lower than control (21.04 per cent). RD-9 Repelin showed 10.08 per cent damaged leaves and Neemrich II showed 13.98 per cent. In combination with synthetic insecticides, RD-9 Repelin was the best (2.68 per cent damaged leaves). It was on par with synthetic insecticides alone (3.76 per cent) and all other combined treatments (4.05–5.37 per cent) but resulted in better control of leaf folder than other combined treatments. A similar trend was observed after four applications, where the incidence ranged from 3.13 to 32.15 per cent damaged leaves.

Results on incidence of stem borer (Table 4) indicated that a schedule of Neemrich II and synthetic insecticides (10.51 per cent whiteheads) was as good as a schedule of synthetic insecticides alone (5.91 per cent). All other schedules including neem-based insecticides were significantly superior (15.73 to 19.97 per cent) over unsprayed control (30.63 per cent). The yield in the Neemrich II schedule (54.90) was as high as that in the synthetic insecticide schedule (57.43), the best treatment. The schedule containing

Table 4. Evaluation of synthetic insecticides with neem-based insecticides for the control of yellow stem borer and leaf folder on rice variety PR 106 (*kharif* 1992).

Treatment	Leaf folder damage		Whiteheads (%)	Yield (kg/ha)
	Before 4th spray	Before 5th spray		
R-R-R-R-R	10.08[bc]	18.91[bc]	23.47[cd]	3750[c]
NI-NI-NI-NI-NI	14.82[cd]	28.17[cd]	24.55[cd]	3785[cd]
NII-NII-NII-NII-NII	13.98[c]	20.64[bc]	22.83[cd]	3636[c]
NA-NA-NA-NA-NA	15.49[cd]	21.08[bcd]	21.48[cd]	3330[c]
D-R-Nu-R-C	2.68[a]	7.01[a]	15.73[bc]	4245[bc]
D-NI-Nu-NI-C	4.52[ab]	6.18[a]	18.23[c]	4245[bc]
D-NII-Nu-NII-C	5.37[ab]	5.33[a]	10.51[ab]	5490[ab]
D-NA-Nu-NA-C	4.05[ab]	4.32[a]	19.97[c]	4245[bc]
D-Nu-C-D-Nu	3.76[ab]	3.13[a]	5.91[a]	5745[a]
Untreated control	21.04[d]	32.15[d]	30.63[d]	3540[c]

R = RD-9 Repelin, NI-Neemrich I, NII = Neemrich II, NA= Neemark, Nu = Nuvacron(monocrotophos), D= Dimecron (phosphamidon), C = Coroban (chlorpyriphos)).

In a column, means followed by the same letter do not differ significantly at 5 per cent level by DMRT.

Neemrich II was effective against both pests and yield from this treatment was on par with that from synthetic insecticide treatment. All other schedules having neem-based insecticides alone or in combination with insecticides were significantly inferior and were on par with each other and with control.

The studies on combination of neem-based products and synthetic insecticides against pests of rice indicated that such combinations have a great potential. The efficacy of combined treatments was found to be comparable with that of synthetic insecticides against leaf folder and stem borer. The neem products, thus, appear to reduce reliance on synthetic insecticides as the applications of these insecticides could be cut down to the extent of 40 per cent in case of stem borer and 50 per cent in case of leaf folder. This is particularly applicable in Punjab, where a schedule of five insecticidal sprays has recently been recommended for the control of stem boer on Basmati rice (PAU, 1992). Insecticides of botanical origin are environmentally safe and their effectiveness against a wide variety of insects makes them suitable in integrated pest management programmes.

REFERENCES

Dilawari, V.K., Jaswant Singh and G.S. Dhaliwal. 1991. Efficacy of Neemark against rice stem borer, *Scirpophaga incertulas*. *Neem Newsl.* 8(2): 18–19.

DRR (Directorate of Rice Research). 1987. *Progress Report*. All India Co-ordinated Rice Improvement Programme, Vol. 2, Entomology and Pathology, Kharif, 1987, Hyderabad.

DRR (Directorate of Rice Research). 1988. *Progress Report*. All India Co-ordinated Rice Improvement Programme, Vol. 2. Entomology and Pathology, Hyderabad.

PAU (Punjab Agricultural University). 1992. *Package of Practices for Crops of Punjab—Kharif 1992*, Ludhiana.

Ramaraju, K. and P.C. Sundra Babu. 1989. Effect of plant derivatives on brown planthopper (BPH) and whitebacked planthopper (WBPH) nymph emergence on rice. *Int. Rice Res. Newsl.* 14(5): 30.

Saxena, R.C. 1989. Insecticides from neem. In: *Insecticides from Plants*. J.T. Arnason, B.J.R. Philogene and P. Morand (eds.), ACS Symp. Ser., 387, Washington, 110–135.

Saxena, R.C., N.J. Liquido and H.D. Justo. 1981. Neem oil, a potential antifeedant for the control of the rice brown planthopper, *Nilaparvata lugens*. In: *Natural Pesticides from the Neem Tree Azadirachta indica* A. Juss). *Proc. 1st Int. Neem Conf.* Rottachegern, Germany, June 1980, pp. 171–188.
Schmutterer, H., K.R.S. Ascher and H. Rembold, Eds. 1981. *Natural Pesticides from Neem Tree* (*Azadirachta indica* A. Juss.) *Proc. 1st Int. Neem Conf.*, Rottachegern. GTZ, Eschborn.

44

Evaluation of Neem Products and Standard Chemicals for the Management of Bean Fly *Ophiomyia phaseoli* (Tryon) in French Beans

P.N. KRISHNA MOORTHY AND K. SRINIVASAN

Indian Institute of Horticultural Research
Hessaraghatta Lake P.O., Bangalore, India

ABSTRACT

Bean fly, *Ophiomyia phaseoli* (Tryon) is the most destructive pest of French beans (*Phaseolus vulgaris* L.) causing heavy mortality of plants in many tropical countries. During 1990 and 1991, two field trials were conducted to evaluate 2, 4 and 5 per cent concentrations of neem seed kernel extract (NSKE) in water, 1per cent neem oil and 1per cent Repelin (commercial botanical containing neem as one of the constituents) along with 0.07 per cent endosulfan and untreated control. Application of 5 per cent NSKE was found to give the best control of bean fly in both the trials. In 1992, an additional experiment was conducted to standardize the time of application of 5per cent NSKE. It was found that foliar application of NSKE at the time of appearance of puncture marks in the unifoliate leaf stage and again at petiole mining stages was the most appropriate.

Bean fly, *Ophiomyia phaseoli* (Tryon) is a very serious pest of several legumes like French beans (*Phaseolus vulgaris* L.), cowpea (*Vigna unguiculata* (L.) Walp.), mung bean (*Vigna radiata* (L.)), soy bean (*Glycine max* Merr.) etc., in many countries (Talekar, 1990). Scheduled application of insecticides either to soil or to foliage are very effective in controlling this pest (Spencer, 1973; Krishna Moorthy and Tewari, 1987) Spraying of endosulfan based on diagnostic characters of bean fly at unifoliate stage has also been found effective against this pest (Krishna Moorthy and Srinivasan, 1989 & 1991). This method requires two sprays of endosulfan timed at unifoliate leaf stage within a gap of one week. This works well even at times of heavy insect pressure. Considering the fact that sprays are required at very short interval, it was felt that botanical insecticides with a short residual toxicity could be substituted in place of endosulfan. To this end, therefore, experiments were conducted to evaluate foliar sprays of promising botanicals like neem seed kernel extract, neem oil and Repelin (a commercial botanical containing neem as one of the constituents) against bean fly and the results are communicated in this paper.

MATERIALS AND METHODS

Field experiments were conducted in randomised block design with three replications during 1990, 1991 and 1992 at the experimental farm of Indian Institute of Horticultural Research, Bangalore. Each plot measured 15 m^2. French beans variety 'Arka komal' was sown on June 2, 1990, June 17, 1991 and June 28, 1992.

During 1990 and 1991, the following treatments, namely, NSKE at 2, 4, 5 per cent; neem oil and Repelin both at 1 per cent were compared with 0.07 per cent endosulfan and untreated control.

Bean fly lays eggs in the unifoliate leaf stage of the plant. Puncture marks appear on the leaf as a result of oviposition. Thereafter the larvae mine into petioles and enter the stem. Treatments were directed at puncture mark stage (P stage) and again when petiole mining symptoms (PM) stage appeared on the unifoliate leaf stage.

During 1992, only 5 per cent NSKE was chosen from among the botanicals tried during 1990 and 1991. It was compared with 0.07 per cent endosulfan sprays. Different stages of bean fly incidence was taken as criteria for application of sprays. The following treatments were evaluated.

(i) NSKE at 7, 15, and 20 days after sowing.
(ii) NSKE at P and PM stage
(iii) NSKE at P and 5 PMT stage (when on an average 5/10 unifoliate leaves had petiole mining symptoms).
(iv) Endosulfan at P stage.
(v) Endosulfan at P 5 PMT stage.
(vi) Endosulfan at P and NSKE at 5 PMT stage.
(vii) Endosulfan at 7, 14 and 21 DAS and
(viii) Untreated control.

Plants were monitored every day, soon after germination to diagnose P, PM, and 5 PMT stages of bean fly incidence. Ten plants were observed at random in untreated plots to begin with. As soon as puncture marks were observed in any leaf it was assumed that P stage had reached. On the same day NSKE was prepared and kept overnight. Treatments were applied the following day. Thereafter, ten random plants were monitored in rest of the plots for the appearance of petiole mining symptoms. As and when, atleast one leaf had petiole mining symptoms, it was taken as PM stage. NSKE was also prepared on the same day, kept overnight and applied the following day. For monitoring 5 PMT stage the method given by Krishna Moorthy and Srinivasan (1989) was followed. In each plot, ten plants were chosen at random. When on an average, 5 unifoliate leaves had petiole mining symptoms/10 leaves examined, it was assumed that 5 PMT had reached and treatments were imposed on the following day.

Preparation of NSKE

Fifty grammes of neem seed kernel was ground with 200 ml water for 2 minutes. The slurry was kept overnight and strained on the following day by pouring over a coarse cloth. Then it was made upto 1 litre to get 5 per cent concentration. The 5 per cent stock solution was serially diluted to get 2 and 4 per cent. Quantity of kernels for preparation of extract was determined based on the area to be covered for spraying. All treatments were applied using a high volume foot sprayer delivering 700 1 of spray fluid/ha.

Counts of larvae and pupae and the number of plants with stem mines in 10 plants selected at random were taken at 25 DAS by carefully uprooting the plants. In 1992 experiment, observations were taken at 30 DAS. By that time, adults had already emerged from pupae. Some pupae were also dislodged from the plant while uprooting. Hence, data on the pest incidence could not be recorded. Marketable yield was recorded only for 1992 experiment. Heavy rains received during 1990 and 1991 caused extensive mortality of plants. Hence yield could not be recorded during these days.

RESULTS AND DISCUSSION

During 1990, the treatments were applied at 13 and 19 DAS for P and PM stages, respectively. Spray intervention for P and PM stages were made at 14 and 16 DAS, respectively, during 1991. The variation in the spray dates during both the years was basically due to differential insect pressure.

In 1990, all the treatments were effective in reducing bean fly incidence and foliar application of endosulfan reduced the pest to a minimum. However, application of 4 and 5 per cent NSKE, neem oil and Repelin were at par with endosulfan treatment in reducing bean fly incidence. During 1991, all the concentrations of NSKE, Repelin and endosulfan were significantly superior to rest of the treatments in reducing bean fly incidence. Minimum bean fly incidence (in terms of larval and pupal counts) was recorded only in 5 per cent NSKE sprayed plots (Table 1).

Table 1. Bean fly incidence and number of plants with stem mines in different management programmes (1990 and 1991)

Treatment	Conc.	larva + pupa/10 plants		No. of plants with stem mines	
	(%)	1990	1991	1990	1991
Neem seed kernel (NSKE)	2	14.00b	18.33a	8.33abc	10
NSKE	4	8.33ab	3.33a	8.67abc	10
NSKE	5	8.00ab	1.67a	6.67a	10
Neem oil	1	9.33ab	53.00bc	7.33ab	10
Repelin	1	10.33ab	29.00ab	9.67bc	10
Endosulfan	0.07	0.67a	3.00a	9.00abc	10
Control	—	40.00c	83.33c	10.00c	10
C.D. (P = 0. 05)	—	12.88	33.27	2.56	N.S.

Means in the same column with letters in common are not significantly different.

NS : Not significant

Larvae of bean fly usually mine into the petiole and enter the stem. During this process, the plant vigour is reduced. During 1990, when the pest incidence was relatively low, treatment with 5 per cent NSKE recorded the least number of plants with stem mines. However, none of the treatments were effective in reducing stem mines in 1991, when insect pressure was quite high. This is because, even a single larva reaching the stem is capable of causing mining symptom. This is also the reason, why, 5 per cent NSKE treatment was not consistently effective in reducing stem mining at times of heavy insect pressure.

NSKE 5 per cent was chosen for further evaluation during 1992, based on the general superiority of its performance in reducing pest incidence and stem mines during trials in 1990 and 1991 (Table 1).

During 1992, P and PM stages occurred at 11 and 15 DAS, respectively. There was a gap of 5 days in between appearance of these stages. There was only 2 days gap in 1991, which further supported our conclusion of heavy pest incidence during that year.

Data on stem mining recorded at 30 DAS in 1992 was not significantly different. However, marketable yield recorded in this experiment showed that neem application at 7, 15, 20 DAS; P and PM stages and endosulfan at four different stages as described in Table 2 were not significantly different but superior to control. The data also shows that 5 per cent NSKE sprays were equally effective as compared to endosulfan sprays for the management of bean fly.

Earlier, Mansuetus and Karel (1984) tried foliar applications of alcoholic and aqueous extracts of neem leaves and NSKE at 600 g a.i./ha in Tanzania. They had applied the treatments after 20 days after planting. Talekar (1990) opined that these applications of neem were not very encouraging due to the hidden nature of the immature stages of bean fly (in leaf lamina and stem). In the present study, we had applied NSKE at the critical stage of bean fly incidence i.e. unifoliate leaf stage

Table 2. Number of plants with stem-mines and yield of beans in different management programmes (1992)

Treatment	No. of plants with stem mining/10 plants	Yield (t/ha)
*NSKE at 7, 15, 20 DAS	7.7	6.6ab
NSKE at P & PM	5.0	7.8a
NSKE at P & 5 PMT	9.0	5.9bc
**Endosulfan at P	9.7	7.9a
Endosulfan at P & 5 PMT	4.0	8.2a
Endosulfan at P & NSKE at 5 PMT	7.7	8.1a
Endosulfan at 7, 14, 21 DAS	5.3	7.7ab
Control	9.2	4.5c
C.D. (P = 0.05)	N.S.	1.8

Means in the same column with letters in common are not significantly different.

NSKE : 5% Neem seed extract
DAS : Days after sowing
P : Puncture mark stage
PM : Petiole mining stage
PMT : Petiole mining threshold

(Krishna Moorthy and Srinivasan 1989) based on the symptoms of bean fly incidence and the treatments were effective in reducing bean fly incidence. Our empirical observations, however, indicate that NSKE deposit could be washed away at times of heavy rains. Hence, we recommend the first spray with endosulfan at P stage and followed by NSKE spray at 5 PMT. During dry months, NSKE alone could be used for bean fly management.

ACKNOWLEDGEMENTS

The technical assistance given by Sri. M.B. Munirathnaiah, Sri. S. Venkateshaiah and Sri. D. Ramaiah is gratefully acknowledged. This research was financed in part by a grant made by the United States Department of Agriculture under U.S. India Fund, FG-IN-760.

REFERENCES

Krishna Moorthy, P.N. and K. Srinivasan. 1989. Petiole mining thresholds and potassium chloride sprays for management of bean fly (Diptera : Agromyzidae) in snap beans and cowpea. *J. Econ. Entomol.* 82 (1): 246–250.

Krishna Moorthy, P.N. and K. Srinivasan. 1991. Diagnosis and management of bean files in beans. In: *Horticulture—New Technologies and Applications.* J. Prakash and R.L. M. Pierik (eds.), Kluwer Academic Publishers. The Netherlands. pp. 363–366

Krishna Moorthy, P.N. and G.C. Tewari. 1987. Management of stemfly *Ophiomya phaseoli*(Tryon) in beans with reduced insecticidal doses. *Entomon* 12: 363–366.

Mansuetus, S.B.A. and A.K. Karel. 1984. Effect of various neem (*Azadirachta indica* A. Juss.) extracts on insect pests of common beans. In: *Proceedings of the third workshop on bean research in Tanzania.* A.N. Minjas and M.P. Salema (eds.) CRSP, Washington State University, Pullman. pp. 52-56.

Spencer, K.A. 1973. *Agromyzidae (Dipteria) of Economic Importance*. Junk, The Hague, 418 pp.

Talekar, N.S. 1990. *Agromyzid Flies of Food Legumes in the Tropics.* Wiley Eastern Limited, New Delhi. 287 pp.

45

Neem in Integrated Management of *Helicoverpa armigera* Hubn. Infesting Chickpea

S.N. SINHA

Indian Agricultural Research Institute, Regional Station
Karnal 132001, India

ABSTRACT

Monitoring of the male *Helicoverpa* (*Heliothis*) *armigera* Hubn., was carried out from January 1986 to December 1988 at the IARI, Regional Station, Karnal (Haryana), using synthetic sex pheromone traps. Maximum activity was recorded during the spring and summer (7th to 28th standard week) while a minor peak activity was observed in autumn (42nd to 50th standard week). Moth catches in pheromone traps were found positively correlated with the larval count of *H. armigera* on chickpea made after one, two and three weeks of trap catches.

In a four year field efficacy trial, neem oil, neem seed kernel extract (NSKE), neem emulsion (NE) and neem wettable powder (NWP) along with insect growth regulators (IGR) and synthetic insecticides were evaluated against *H. armigera* infesting chickpea seed crop. At a level of 20.5 per cent (1986–87) and 12.5 per cent (1988–1989) overall infestation, diflubenzuron (0.05 per cent), neem oil and NSKE each at 5.0 per cent sprayed twice at an interval of 10 days, gave two-fold reduction in the pod borer infestation over untreated control and were found comparable to endosulfan (0.07 per cent) an insecticide. All these treatments gave significantly higher yield of seed in both years. A 39 to 47 per cent increase in seed yield was registered with diflubenzuron 0.05 per cent and endosulfan 0.07 per cent over untreated control.

During 1989–90 and 1990–91, some new neem-products and chitin biosynthetic inhibitors were evaluated along with endosulfan. A high level of infestation, 35.6 per cent (1989–90) and 37 per cent (1990–91) of the pest was recorded. Flufenoxuron irrespective of its dosages (0.0125 and 0.025 per cent) proved to be the most effective treatment and reduced pod borer infestation by 45–50 per cent in 1989–90 and 77–78 per cent in 1990–91, over control, respectively. Neem emulsion and neem wettable powder afforded 40 to 60 per cent whereas fenvalerate and endosulfan gave 72 to 73 per cent reduction in the infestation over untreated control (1990–91). The NSKE at 5.0 per cent gave 40 per cent reduction in the infestation and was comparable to endosulfan at 0.07 per cent (1989–90). Neemguard was found ineffective. There was no significant difference in the seed yield obtained from NE (0.125 per cent) NSKE (5.0 per cent) flufenoxuron and endosulfan (0.07 per cent) treatments.

INTRODUCTION

Among the botanical insecticides neem (*Azadirachta indica* A. Juss) occupies an important place. It contains many biologically active substances, but to get the desired results it is essential to apply it when the insect is in its early stage of development. In recent years *Helicoverpa armigera* Hubn. has posed a serious threat to chickpea and other field crops because it develops resistance against insecticides very fast. To overcome this problem, studies were made to determine early infestation of *H. armigera* on chickpea crop using synthetic sex pheromone traps. Field assay of neem products against the pest was also carried out.

MATERIALS AND METHODS

Monitoring of *H. armigera* and Its Infestation

The activity of *H. armigera* was monitored by moth catches in pheromone traps and by larval count in chickpea crop at the IARI, Regional Station, Karnal. Sleeve traps (Pawar et al., 1984) baited with *H. armigera* lure, a synthetic sex pheromone comprising Z-11-hexadecenal and Z-9-hexadecenal in 97 : 3 ratio impregnated into rubber septa at a rate of 5 mg/septa (supplied by Pheromone India, Tenali, Andhra Pradesh), were used in monitoring *H. armigera* moth population. The septa was changed every three weeks to maintain the catch efficiency of the trap. The traps were set 1.5 m above ground in and around chickpea fields and 10 m away from one another to avoid intra-trap interferences. The traps were emptied and moths caught were recorded daily. Larvae of all sizes were counted in all chickpea fields at weekly intervals starting from the first standard week of January till the maturity of the entire crop (15th standard week). An area of 1 sqm was randomly selected at 10 different spots in a field of around 1 ha to count the larvae.

Field Assay of Neem Products

The experiment was conducted from 1986–87 to 1990–91. Seed crop of chickpea cv Pusa 256 (bold seeded) was sown in November using tractor-drawn seeddrill with a flow rate of 60 kg/ha. After complete germination net plots measuring 3.3 × 3.7 m (1986), 3.0 × 2.6 m (1988), and 5.0 × 2.0 m (1989 and 1990) were demarcated leaving sufficient space between the plots to avoid cross-contamination by sprays. All the experiments were laid out in the randomized block design with four replications.

In 1986–87 and 1988–89, neem oil and neem seed kernel aqueous extract (NSKE) each at 2.5 per cent and 5.0 per cent were evaluated along with endosulfan (Thiodan 35EC) at 0.07 per cent, an insecticide check. To prepare spray solution of neem oil and NSKE, Sandovit (an emulsifier) was added at 2 ml/l. The NSKE was obtained by soaking measured quantity of freshly crushed neem seed kernel in water and then the decoction was taken out thoroughly. The desired concentration was obtained by dilution.

In 1989–90, neem emulsion (NE) 25EC (supplied by Head, Division of Agricultural Chemicals, IARI, New Delhi) at 0.03, 0.063, and 0.125 per cent Neemguard

(Gharda Chemicals, Bombay) at 2.5 and 5.0 per cent, NSKE at 2.5 and 5.0 per cent, diflubenzuron at 0.025 per cent, flufenoxuron (Cascade 10DC) at 0.0125 and 0.025 per cent, and endosulfan at 0.07 per cent were evaluated along with untreated control.

In the 1990–91 experiment, lower doses of NE (0.03 and 0.06 per cent) and Neemguard (2.5 per cent) were dropped. The NSKE treatment was replaced with neem wettable powder (NWP) 25 per cent (supplied by the Head, Division of Agricultural Chemicals, IARI, New Delhi). In addition to diflubenzuron, flufenoxuron, and endosulfan, the synthetic pyrethroid insecticide fenvalerate was also included.

The crop was sprayed twice at 750 l/ha at an interval of 10 days. The first application of spray was given at the time of appearance of *Heliothis* on the crop as determined by the pheromone trap catches.

Observation on pod borer infestation was recorded at maturity of the crop. Five plants were randomly taken from the plot and number of pods damaged out of total pods on them was recorded for working out the percentage. Seed yield was also recorded. Statistical analysis was done according to Panse and Sukhatme (1967) and the F-test was carried out after Duncan (1955).

RESULTS AND DISCUSSION

Monitoring of *H. armigera*

The pattern of pheromone trap catches was similar during 1986, 1987, and 1988 with two distinct periods of activity (Fig. 1). Major activity was recorded in the spring and summer (7th to 28th standard weeks), while a minor activity period was recorded in the autumn (42nd to 50th standard weeks). The maximum number of moths (52 moths/week) were captured between the 12th and 15th standard weeks (end of March to mid-April) in all the three years. During the entire spring and summer seasons, three to four peaks of catches indicated three to four generations in the season. This pattern of moth activity is in close conformity with the finding of Singh and Singh (1975) from Punjab and Lal et al. (1986) from Kanpur (Uttar Pradesh).

Moth catch was low during the winter (48th to 6th standard week) and increased after the 7th standard week. Larval population, on the other hand, was high during the early stage of chickpea growth presumably due to egg laying from the autumn population. It declined gradually, reaching its lowest level in January-February. Larvae were found to infest leaves and the apical portion of the plant without causing any economic damage. Further increase in larval population was observed after the 10th standard week or two to three weeks after the increase in the population of moths. Larval population at this stage coincided with the pod formation in chickpea causing severe damage. Moth catches in pheromone traps were found to be positively correlated with the larval counts of *H. armigera* on chickpea made after one, two, and three weeks of trap catches (Table 1). Tingle and Mitchell (1981) reported significant correlation between pheromone trap catch of *H. virescens* and larval infestation and damage level in tobacco when trap catch was compared with larval and plant damage count made one and two weeks later.

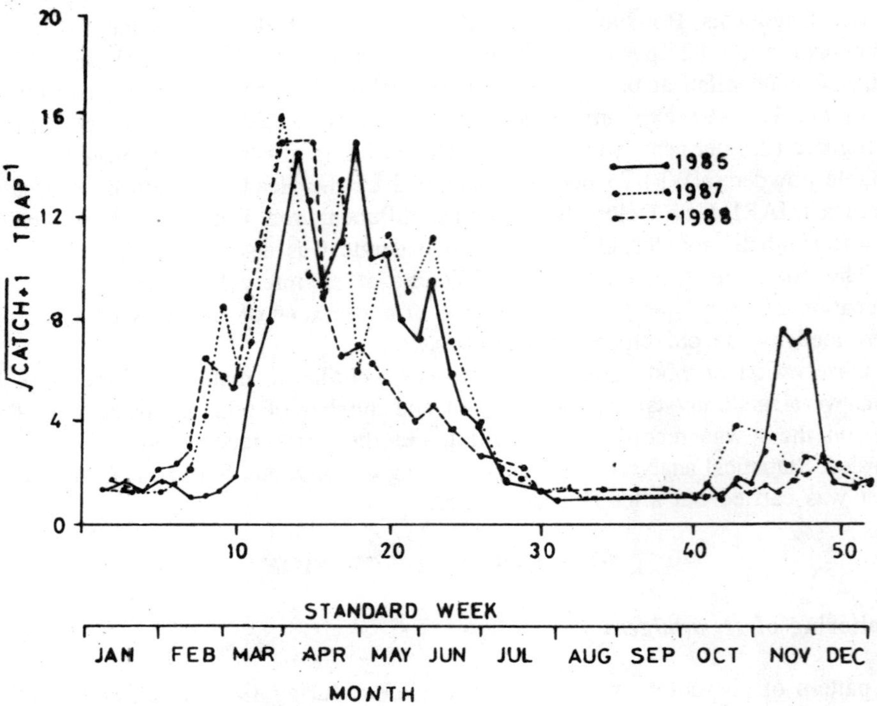

Fig. 1. Pheromone trap catches of *H. armigera* male moth at IARI Regional Station, Karnal

Table 1. Correlation coefficient between current √(moth catch + 1)/trap/week and number of *H. armigera* larvae/m² on chickpea crop during different time intervals

Year	Date of sowing	*r* values between moth catch (current) and larval count at the following time intervals (weeks after)			
		same	1	2	3
1986	20.10.85	0.757*	0.902**	0.989**	0.994**
1987	25.10.86	–0.505*	–0.491*	–0.359	–0.245
	3.11.86	0.330	0.233	0.161	0.438
	12.11.86	0.769**	0.695**	0.532	0.670
1988	16.11.87	0.782**	0.887**	0.920**	0.869**
1989	30.10.88	0.458	0.645**	0.834**	0.840**
	3.11.88	0.691*	0.851**	0.990**	0.869**
	12.11.88	0.550	0.687*	0.851**	0.972**

* Significant ($P < 0.05$).
** Significant ($P < 0.01$).

Pod Borer Infestation

The overall incidence of the pod borer in untreated plots was 20.5 per cent in 1986-87 and 12.5 per cent in 1988–89 (Table 2). In 1986–87, except diflubenzuron at the lowest dose (0.0125 per cent), all treatments were at par with each other and significantly superior to untreated control. The NSKE and neem oil, both at 5 per cent concentration, and diflubenzuron at 0.05 per cent brought down the borer infestation

Table 2. Effect of diflubenzuron and neem products on the incidence of *H. armigera* infesting chickpea

Treatment	Conc. (%)	Pod borer infestation (%)		Reduction over control (%)	
		1986–87	1988–89	1986–87	1988–89
Diflubenzuron	0.0125	15.3 (22.96)[bc]	8.3 (16.73)[a]	25.4	33.6
Diflubenzuron	0.025	12.7 (20.93)[abc]	7.5 (15.85)[a]	38.0	40.0
Diflubenzuron	0.05	11.6 (19.90)[ab]	5.6 (13.60)[a]	43.4	55.2
Neem oil*	2.5	13.7 (21.75)[abc]	8.1 (16.43)[a]	33.2	35.2
Neem oil	5.0	11.9 (20.05)[ab]	6.2 (14.37)[a]	42.0	50.4
NSKE	2.5	12.8 (20.85)[abc]	7.4 (15.69)[a]	37.6	40.8
NSKE	5.0	09.7 (17.89)[a]	6.0 (14.09)[a]	52.7	52.0
Endosulfan	0.07	08.7 (17.12)[a]	5.9 (14.05)[a]	57.6	52.8
Untreated control		20.5 (25.32)[c]	12.5 (20.69)[b]		

* Spreader at 2 ml/l.
Figures with same letter(s) are non-significant at $P < 0.05$ (Duncan's multiple range test).

from 20.5 per cent (untreated control) to 9–12 per cent and were comparable to endosulfan (8.7 per cent), an insecticidal check.

In 1988–89, all the treatments were at par with each other and significantly superior to untreated control. However, the lowest incidence was recorded with endosulfan (5.9 per cent) followed by diflubenzuron (5.6 per cent) at 0.05 per cent dose and NSKE (6 per cent) and neem oil (6.2 per cent), both at 5.0 per cent dose level.

In 1989–90, there was a high incidence (36.5 per cent). All the treatments except Neemguard at low concentration (2.5 per cent) reduced pod borer damage significantly when compared with untreated control (Table 3). Flufenoxuron at both levels (0.0125 and 0.025 per cent) was found to be the most effective against *H. armigera*. Neem emulsion (0.125 per cent), NSKE (5.0 per cent), and endosulfan (0.07 per cent) were at par with each other and gave 40–50 per cent reduction in borer incidence over untreated control. Also, they were found comparable to flufenoxuron in their field efficacy against the pest. Other treatments were relatively less effective.

In 1990–91, there was again a high pest incidence (37.0 per cent). Flufenoxuron once again proved most effective against the pest at both doses (Table 4). It gave 76–78 per cent reduction in pod borer infestation over untreated control (37.0 per cent). Flufenoxuron, an acylurea compound, is reported to be highly toxic towards larval stages of important Lepidopterous pests, including *Heliothis zea* (Anderson et al., 1986). At both the doses it was found superior to endosulfan (0.07 per cent) and fenvalerate (0.01 per cent), the insecticidal checks. Poor performance of fenvalerate is

Table 3. Effect of neem products and IGRs on *H. armigera* incidence on chickpea during 1989–90

Treatment	Conc. (%)	Pod borer infestation (%)	Reduction over control (%)	Seed yield (q/ha)
Neem emulsion 25EC	0.031	25.9 (30.56)[cd]	27.2	12.5[cde]
Neem emulsion 25EC	0.063	24.0 (29.28)[bc]	32.6	13.3[cde]
Neem emulsion 25EC	0.125	20.4 (26.83)[ab]	42.7	17.2[abc]
Neemguard	2.5	29.2 (32.68)[de]	18.0	10.4[de]
Neemguard	5.0	26.8 (31.30)[cd]	24.7	12.3[cde]
NSKE	2.5	27.2 (31.38)[cd]	23.6	10.4[de]
NSKE	5.0	20.5 (26.82)[ab]	42.4	15.2[abcd]
Diflubenzuron	0.025	23.4 (28.88)[bc]	34.3	14.2[bcde]
Flufenoxuron	0.0125	19.7 (26.35)[ab]	44.7	16.0[abc]
Flufenoxuron	0.025	17.9 (25.01)[a]	49.7	19.0[ab]
Endosulfan	0.07	20.6 (27.00)[ab]	39.3	19.2[a]
Untreated control		35.6 (36.63)[e]		00.97[e]

Figures with same letter(s) are non-significant at $P < 0.05$ (Duncan's multiple range test)

Table 4. Effect of neem products and IGRs on the incidence of *H. armigera* infesting chickpea crop in 1990–91

Treatment	Dose (%)	Pod borer infestation (%)	Reduction over control (%)
Neem emulsion 25EC	0.125	19.1 (25.88)[bc]	48.4
Neem emulsion 25EC	0.250	17.1 (24.40)[bc]	53.8
Neem wettable Powder 25 WP	0.125	24.1 (28.95)[cd]	34.9
Neem wettable powder	0.250	13.6 (21.57)[ab]	63.2
Neemguard	5.0	28.0 (31.77)[d]	24.3
Diflubenzuron	0.025	12.9 (20.98)[ab]	65.1
Flufenoxuron	0.0125	08.7 (17.01)[a]	76.5
Flufenoxuron	0.025	08.0 (16.39)[a]	78.4
Fenvalerate	0.01	10.4 (18.70)[a]	71.9
Endosulfan	0.07	09.9 (18.21)[a]	73.2
Untreated control		37.0 (38.50)[e]	

Figures with same letter(s) are non-significant at $P < 0.05$ (Duncan's multiple range test)

attributed to the growing ineffectiveness of insecticides in controlling *H. armigera* larvae. Neem wettable powder (0.25 per cent) and diflubenzuron (0.025 per cent) were found at par with endosulfan and flufenoxuron. Neem emulsion (0.25 per cent) was found relatively less effective than NWP. However, NE (0.25 per cent) and NWP (0.25 per cent) caused 54–63 per cent reduction in pod borer infestation over control whereas flufenoxuron, diflubenzuron, fenvalerate, and endosulfan reduced incidence by 65–78 per cent. These results are in close conformity with the earlier observations on field efficacy of diflubenzuron, neem oil, and NSKE against *H. armigera* (Sinha and Mehtrotra, 1988; Sachan and Lal, 1990).

Seed Yield

In 1986–87, diflubenzuron (0.05 per cent), neem oil (5.0 per cent) and endosulfan (0.07 per cent) were at par with each other and gave significantly higher yield of seed when compared with the seed yield (12.6 q/ha) of untreated control (Table 5). The seed yield increase over control was recorded as 40 and 47 per cent, respectively, during 1986–87. Other treatments were at par with the control.

All treatments were found significantly superior to the untreated control (19.9 q/ha) in 1988–89 except diflubenzuron at 0.0125 per cent, NSKE at 2.5 per cênt (Table 5). Neem oil at (2.5 and 5.0 per cent) and NSKE at 5.0 per cent gave as much seed yield as endosulfan, an insecticide check.

Results obtained in 1989–90 (Table 3) revealed that endosulfan gave the highest yield (19.2 q/ha) of seed followed by flufenoxuron at 0.025 per cent (19.0 q/ha), NE at 0.125 per cent (17.2 q/ha), flufenoxuron at 0.0125 per cent (16.0 q/ha), and NSKE at 5.0 per cent (15.2 q/ha). They were all at par and were significantly superior to untreated control (0.97 q/ha).

These results suggested that at a low level (< 20 per cent) of infestation, neem oil and NSKE were as effective as endosulfan in the control of *H. armigera*, while at a high level (> 30 per cent) of infestation neem products afforded only 50 to 60 per cent reduction in the infestation, necessitating use of some potent insecticides such as flufenoxuron. Flufenoxuron consistently gave good results in the present trials. It is reported to be very effective against resistant strains of diamond back moth, *Plutella xylostella* (Peter and Sundarajan, 1991). It is also evident that neem can replace prophylactic treatments which many a time trigger the insecticide-resistance process in resistance prone strains.

Table 5. Effect of diflubenzuron and neem products on the seed yield of chickpea cv. BG 256

Treatments	Dose (%)	Seed yield (q/ha)		Increase over control (%)	
		1986–87	1988–89	1986–87	1988–89
Diflubenzuron	0.0125	15.0^{bc}	23.1^{bc}	19.1	16.1
Diflubenzuron	0.025	16.0^{abc}	26.1^{ab}	27.0	31.1
Diflubenzuron	0.05	18.0^{ab}	27.8^{a}	42.9	39.7
Neem oil*	2.5	15.3^{abc}	23.9^{abc}	22.2	20.0
Neem oil*	5.0	17.6^{ab}	24.8^{ab}	39.7	24.6
NSKE*	2:5	15.2^{abc}	22.3^{bc}	20.6	12.1
NSKE*	5.0	16.1^{abc}	25.3^{ab}	27.8	27.1
Endosulfan	0.07	18.5^{a}	27.7^{a}	46.8	39.2
Untreated control		12.6^{c}	19.9^{c}		

*Spreader used at 2 ml/l.
Figures with same letter(s) are non-significant at $P < 0.05$ (Duncan's multiple range test).

ACKNOWLEDGEMENT

Thanks are due to the Head, IARI Regional Station, Karnal, for providing necessary facilities.

REFERENCES

Anderson, M., J.P. Fisher and J. Robinson. 1986. Flufenoxuron an acaricide/insecticide with novel properties *Brit. Crop Prot. Conf.—Pests and Diseases* 2B-1, pp. 89–96.

Duncan, D.B. 1955. Multiple range test and multiple F-test. *Biometrics* 11: 1–42.

Lal, S.S., C.P. Yadav and J.N. Sachan. 1986. Strategies for the development of an integrated approach to control gram pod borer, *Heliothis armigera* (Hubner) infesting chickpea. *Pesticides* 20 (5): 39–51.

Panse, V.G. and P.V. Sukhatme. 1967. *Statistical Methods for Agricultural Workers*, ICAR, New Delhi, 381 pp.

Pawar, C.S., S. Sithanatham, H.C. Sharma, S.L. Taneja, P.W. Amin, K. Leuschner and W. Reed. 1984. Use and development of insect traps at ICRISAT. In: *Proc. Natn. for Pest/Vector Research and Control*, 10–11 March, BCKVV, Kalyani, West Bengal, pp. 133–155.

Peter, C. and R. Sundarajan. 1991. Field tests with two new insect growth regulators for the control of diamond back moth, (*Plutella xylostella* L.). *Indian J. Plant Prot.* 19: 161–166.

Sachan, J.N. and S.S. Lal 1990. Role of botanical insecticides in *Helicoverpa armigera* management in pulses. *Natn. Symp. Problem and Prospects of Botanical Pesticides in Integrated Pest Management*, 21–22 Jan., CPPTI, Rajamundry (A.P.).

Sinha, S.N. and K.N. Mehrotra. 1988. Diflubenzuron and neem (*Azadirachta indica*) oil in control of *Heliothis armigera* infesting chickpea (*Cicer arietinum*). *Indian J. Agric Sci.* 58: 238–239.

Singh, H. and G. Singh. 1975. Biological studies on *Heliothis armigera* at ICRISAT, India, *Trop. Pest Mgmt.* 34(1): 39–43.

Tingle, F.C. and E.R. Mitchell. 1981. Relationship between pheromone trap catches of male tobacco budworm, larval infestation and damage level in tobacco. *J. Econ. Entomol.* 74(4): 437–440.

46

Evaluation of Neem Products and Other Standard Chemicals for the Management of Major Pest Complex on Cabbage: Comparison between Standard Spray Regime and I.P.M. Involving Mustard as a Trap Crop

K. SRINIVASAN AND P.N. KRISHNA MOORTHY

Indian Institute of Horticultural Research
Hessaraghatta Lake Post, Bangalore 560089, India

ABSTRACT

Two field experiments were conducted to evaluate 2, 3, 4 and 5 per cent neem seed kernel in water, 0.1 per cent cartap hydrochloride, 0.07 per cent endosulfan and 1 per cent endosulfan and 1 per cent of Neemark and Repelin against major pest complex of cabbage. These were evaluated both under standard spray regime at 10 days internal and in the integrated pest management (I.P.M.) involving Indian mustard as a trap crop. Five sprays were required as against three in the I.P.M. plot until the cabbage growing period. In both these trials, neem seed kernel extract (NSKE) at 4 and 5 per cent consistently recorded significant reduction of diamondback moth (DBM) *Plutella xylostella* larvae with á consequent increase in per cent marketable heads as compared to rest of the treatments. Sprays with NSKE did not affect the dry matter and size of cabbage heads. In addition, DBM larvae sampled from these plots, had higher levels of parasitism by *Cotesia plutellae*, a dominant parasitoid. Long term use of neem seed kernel extract for the management of insecticide resistant strain of *P. xylostella* under the I.P.M. programme is indicated in this paper.

Insect pest complex affecting cabbage (*Brassica olèracea* var. *capitata*) are diamondback moth (DBM), *Plutella xylostella*, (Linn.); leafwebber, *Corcidolomia pavonana* (*Zeller*); webworm, *Hellula undalis* Zeller and aphids, *Brevicoryne brassicae* L. (Nagarkatti and Jayanth, 1982; Chelliah and Srinivasan, 1986; Srinivasan and Veeresh,

1986). Of these pests, DBM is the most important one since it has developed resistance to several classes of insecticides including synthetic pyrethroids (Srinivasan, 1988; Srinivasan and Krishna Moorthy, 1992). An integrated pest management (IPM) which utilises Indian mustard as a trap crop has been successfully developed for the management of pest complex on cabbage (IIHR, 1991; Srinivasan and Krishna Moorthy, 1992). The technology consists of growing paired mustard rows at the beginning and after every 25 cabbage rows. Among the paired rows, the first is sown 15 days prior to cabbage planting while the second is sown 25 days after cabbage planting. This method of sowing ensures adequate maintenance of mustard foliage throughout the cabbage growing period. More than 80 per cent of DBM and almost entire population of leafwebber and webworm are attracted to mustard since it is a preferred host. Aphids are also attracted with the onset of flowering. Adoption of IPM technology ensures a saving of 6–9 sprays as compared to farmer's practice of spraying any commercially available insecticide or their combinations at weekly intervals to control pest complex on cabbage. This technology has presently spread to major cabbage areas in Karnataka and Tamil Nadu states ever since it was demonstrated through onfarm and adaptive trials around Bangalore during 1989 and 1990 (Srinivasan and Krishna Moorthy, 1992). During these onfarm and adaptive trials, two or three sprays with 0.1per cent cartap hydrochloride was recommended to control limited DBM infestation on the intercropped cabbages. However, beginning from 1991, our empirical observations in the grower's field with pure stand of cabbage (Non IPM plots) indicated reduced effectiveness of cartap hydrochloride in controlling DBM. DBM has an uncanny ability to develop resistance to any newly introduced insecticide with two or three years (Leibee and Savage, 1992; Shelton and Wyman, 1992; Sun, 1992; Syed, 1992). This shows that prolonged use of cartap hydrochloride even under the IPM may not give adequate control of DBM over extended period of time. DBM is also known to migrate long distances (Lokki et al., 1978; Bretherton, 1982; Chu, 1986). This characteristic will also enable infiltration of resistant population of DBM from pure cabbage into IPM plots making their control more difficult.

To overcome insecticide resistance problem, it is imperative that never chemicals should be constantly evaluated even under the IPM programme. Use of neem seed/seed kernel extracts for the management of insecticide resistant DBM population is indicated (Schmutterer, 1992). Schmutterer (1992) also opined that water extracts of neem could remove the waxy coating on cabbage leaves which results in increased transpiration and consequent reduction in size of head. Keeping these reports in view, this study was taken up to:

i) Evaluate different concentrations of need seed kernel extract (NSKE) in water along with commercially available neem formulations and standard chemicals for the control of pest complex both under the standard spray regime and IPM involving mustard as a trap crop.

ii) Study the extent of parasitisation by *Cotesia* (*Apanteles*) *plutellae* Kurdj., an important larval parasitoid on DBM sampled from different treatments in standard spray and IPM.

iii) Determine the dry matter and size of cabbage heads in plots treated with NSKE and standard chemical.

MATERIALS AND METHODS

Two field experiments in randomised block designs with three replications were conducted with F_1 cabbage hybrid Kuber during the rainy season (June–August) and with Maharani in late rainy season (August–October) 1991 to evaluate the following treatments with concentration mentioned against each in percentage.

i) NSKE in water 2
ii) NSKE in water 3
iii) NSKE in water 4
iv) NSKE in water 5
v) Cartap hydrochloride (Padan 50 SP) 0.1
vi) Endosulfan (Thiodan 35 EC) 0.07
vii) Repelin 1
viii) Neemark 1
ix) Control (unsprayed check)

Each plot measured 15 m^2 and 45 plants were maintained per plot.

Preparation of NSKE: Seed kernels were obtained by pounding good quality neem seeds. Quantity of kernels for preparation of extract was determined based on the concentration and area to be covered for spraying. The entire quantity of kernel required for a particular concentration was then ground in a mixie with a little water and the emulsion kept overnight. On the following day, it was filtered by pouring over a coarse cloth and made upto required volume.

FIELD EXPERIMENTS

Evaluation in Pure Cabbage Stand: This experiment was done during the rainy season (June–August). Five sprays with respective chemicals were applied at 10 days interval, the first spray beginning from 15 days after planting. Pre-treatment and post-treatment counts of insects at 5 and 9 days after each spray were recorded.

Evaluation in IPM: This was done in the late rainy season (August–October). Mustard was raised in paired rows at 13 m distance on either side of the cabbage experimental plot. Among the paired rows, the first sowing was done at 15 days prior to cabbage planting while in the second row, it was done at 25 days after planting (DAP).

Intercropped cabbages received three sprays with respective chemicals at 26, 36 and 46 DAP as per IPM recommendations. Pre-treatment and post-treatment counts of DBM larvae were taken at 5 and 9 days after each spray. Mustard was sprayed five times with 0.1 per cent DDichlorvos at 10 days interval beginning from 21 days after first sowing. Population counts of insects colonising mustard was recorded at 20, 30, 40, 50, 60 and 70 days after first mustard sowing.

Effect on Natural Enemies: Five DBM larvae were sampled at random from each plot at 50 and 60 DAP. The larvae were confined in petridishes and observed for emergence of *C. plutellae*.

Dry Matter and Size of Cabbage Heads: In both the experiments, five plants at random were sampled from each plot treated with different concentrations of NSKE and cartap hydrochloride. It was done at 23, 33, 43 DAP and 43, 53, 63 DAP during the rainy season and late rainy season, 91 trials, respectively. These samples were oven dried and weight recorded. Besides, length and width of five cut cabbage heads were measured at harvest.

Marketable Yield: When plants were ready to harvest, per cent marketable heads were determined from each plot. Shallow damage to head (< 2 cm deep damage) was also considered marketable since these are invariably trimmed to suit Indian marketing conditions.

RESULTS

Evaluation in Pure Cabbage Stand

Efficacy of NSKE, commercial neem formulations and standard chemicals for the control of DBM in pure cabbage stand is given (Table 1). It was found that application of 4 and 5 per cent NSKE consistently provided the best control of DBM. Both these treatments recorded significantly higher percentage of marketable heads as compared to rest of the treatments. Counts of DBM larvae in the plots treated with standard chemicals were significantly higher than NSKE treated plots. Commercial neem formulations gave effective control of DBM larvae only upto 9 days after third spray. With general increase in the population level at later stages, their efficacy declined. Webworm was observed upto preheading stages and plots treated with NSKE and standard chemicals significantly reduced its population at 5 and 9 days after first spray. Significantly higher percentage of marketable heads was recorded in treatments with 4 and 5 per cent NSKE, following overall decrease in the DBM and webworm population (Table 2). The cabbage heads harvested from the NSKE sprayed plots were free from smell at the time of harvest. Further, organoleptic tests conducted on fresh leaf of cabbage head indicated that there was no bitterness. In addition to DBM and webworm, the control plots supported larvae of leafwebber, especially during post heading stages. On an average 1.0, 1.2, 9.2, 8.1, 10.4, 21.8, 6.1 and 12.1 larvae/plant was recorded at 30, 39, 45, 49, 55, 59, 65 and 69 DAP, respectively, which together complemented to drastic yield reduction in the control plot.

Significant difference did not exist for the dry weight of cabbages recorded in plots treated with NSKE and cartap hydrochloride at 43, 53 and 63 DAP. The size of cut cabbages recorded in these treatments at harvest also did not differ (Table 3).

Evaluation in IPM: There was negligible incidence of insect pests on the intercropped cabbage until 24 DAP and after 55 DAP. Three sprays with respective chemicals were applied at 10 days interval from 26 DAP on the intercropped cabbages. Consistent control of DBM was obtained only in 4 and 5 per cent NSKE treated plots. Commercial neem formulations and standard chemicals were only better than check in reducing DBM population. Highest percentage of marketable heads was obtained in 5 per cent NSKE and it was followed by 4 per cent NSKE treated plot. Marketable heads recorded

Table 1. Larvae of diamondback moth in different treatments (Rainy season 91)

Treatment	Conc. (%)	Pre treatment count	Mean larvae/plant on different days after: I spray		II spray		III spray		IV spray		V spray	
			5	9	5	9	5	9	5	9	5	9
Neem seed kernel extract (NSKE)	2	0.20	0.00^{a}	0.10^{ab}	0.50^{ab}	0.60^{bc}	0.60^{bc}	0.47^{bc}	0.37^{a}	0.30^{b}	0.33^{c}	0.40^{b}
NSKE	3	0.17	0.00^{a}	0.03^{b}	0.83^{ab}	0.57^{ab}	0.50^{b}	0.37^{ab}	0.53^{ab}	0.37^{b}	0.27^{bc}	0.33^{b}
NSKE	4	0.23	0.00^{a}	0.03^{a}	0.53^{ab}	0.33^{a}	0.23^{a}	0.30^{a}	0.33^{a}	0.27^{ab}	0.13^{ab}	1.17^{a}
NSKE	5	0.17	0.00^{a}	0.03^{a}	0.20^{a}	0.37^{a}	0.17^{a}	0.30^{a}	0.33^{a}	0.12^{a}	0.03^{a}	0.10^{a}
Cartap hydro-chloride	0.1	0.17	0.13^{b}	0.33^{b}	0.70^{ab}	0.87^{bcd}	0.57^{b}	0.53^{c}	0.83^{cd}	0.90^{d}	0.27^{bc}	0.63^{c}
Endosulfan	0.7	0.17	0.16^{b}	0.37^{b}	0.77^{ab}	1.17^{de}	1.37^{e}	0.37^{ab}	1.03^{d}	0.53^{c}	0.50^{d}	0.73^{c}
Neemark	1	0.17	0.13^{b}	0.20^{ab}	1.03^{b}	0.90^{cd}	1.13^{de}	0.37^{ab}	0.70^{bc}	0.60^{c}	0.67^{e}	0.87^{d}
Repelin	1	0.27	0.13^{b}	0.27^{ab}	1.00^{b}	1.37^{e}	0.80^{c}	1.17^{d}	0.80^{c}	0.53^{c}	0.50^{d}	1.03^{e}
Control	—	0.27	0.50^{c}	0.77^{c}	2.63^{c}	4.00^{f}	5.53^{f}	3.80^{e}	2.53^{e}	2.57^{e}	4.13^{f}	3.10^{f}
CD 5%		NS	0.10	0.28	0.65	0.30	0.22	0.10	0.20	0.15	0.14	0.10

N.S. : Non Significant
Means in the same column with letters in common are not significantly different

Table 2. Larvae of webworm and percentage of marketable heads recorded in different treatments (Rainy season, 91)

Treatment	Conc (%)	Pre-treatment count	Mean larvae/plant on different days after I spray		% marketable heads
			5	9	
Neem seed kernel extract (NSKE)	2	0.00	0.00[a]	0.03[a]	83.6 (66.3)[bc]
NSKE	3	0.07	0.10[a]	0.03[a]	84.2 (66.8)[b]
NSKE	4	0.03	0.03[a]	0.03[a]	92.2 (74.2)[a]
NSKE	5	0.07	0.03[a]	0.03[a]	92.6 (73.4)[a]
Cartap hydrochloride	0.1	0.03	0.10[a]	0.10[ab]	85.9 (68.2)[b]
Endosulfan	0.07	0.03	0.00[a]	0.03[a]	82.4 (65.5)[bc]
Neemark	1	0.03	0.23[b]	0.23[c]	82.4 (65.5)[bc]
Repelin	1	0.03	0.20[b]	0.17[bc]	79.9 (63.5)[c]
Control	—	0.03	0.27[b]	0.47[d]	0.0 (4.0)[d]
CD 5%		NS	0.08	0.12	3.9 (3.1)

NS : Non significant
Means in the same column with letters in common are not significantly different
Figures in parentheses are angular transformed values
Negligible population of webworm was recorded in all the treatments after II spray. Hence, the data was not analysed

Table 3. Effect of neem seed kernel extract (NSKE) on dry matter and size of cabbage at different crop stages (Rainy season, 91)

Treatment	Conc. (%)	Mean dry matter/plant (g)			Mean diameter of head (cm) at harvest	
		23 DAT	33 DAT	43 DAT	Length	Width
NSKE	2	6.3	55.8	151.3	18.3	18.9
NSKE	3	6.1	51.9	152.0	18.7	19.4
NSKE	4	6.2	52.5	149.0	18.9	18.9
NSKE	5	5.4	52.1	149.3	19.7	20.0
Cartap hydrochloride	0.1	6.0	52.3	148.3	19.6	19.4
CD 5%		NS	NS	NS	NS	NS

DAT : Days after transplanting
NS : Non-significant

Table 4. Larvae of diamondback moth recorded in different treatments under the Integrated Pest Management (IPM) involving mustard as a trap crop (Late rainy season, 1991)

Treatment	Conc. (%)	Pre-treat-ment count	Mean larvae/plant on different days after						% heads
			I spray		II spray		III spray		
			5	9	5	9	5	9	
Neem Seed Kernel Extract (NSKE)	2	0.03	0.03^{ab}	0.33^{ab}	0.23^{b}	0.23^{abc}	0.60^{ab}	0.77^{bc}	86.6 $(68.5)^{def}$
NSKE	3	0.10	0.07^{bc}	0.03^{ab}	0.20^{b}	0.27^{b}	0.70^{bc}	0.87^{c}	91.1 $(72.6)^{c}$
NSKE	4	0.10	0.00^{a}	0.23^{ab}	0.07^{a}	0.17^{ab}	0.47^{ac}	0.63^{ab}	94.0 $(75.8)^{b}$
NSKE	5	0.07	0.00^{a}	0.20^{a}	0.00^{a}	0.13^{a}	0.43^{a}	0.53^{a}	96.6 $(79.3)^{a}$
Cartap hydrochloride	0.1	0.13	0.07^{bc}	0.43^{bc}	0.43^{c}	0.53^{d}	0.77^{bc}	1.43^{d}	85.7 $(67.8)^{ef}$
Endosulfan	0.07	0.13	0.13^{cd}	0.60^{cd}	0.60^{d}	0.50^{d}	0.83^{c}	1.43^{d}	86.6 $(67.3)^{f}$
Neemark	1	0.13	0.03^{ab}	0.30^{ab}	0.27^{b}	0.33^{e}	0.63^{b}	0.90^{c}	88.9 $(70.5)^{cde}$
Repelin	1	0.17	0.17^{d}	0.80^{d}	0.47^{c}	0.60^{d}	1.53^{d}	2.13^{e}	89.0 $(70.7)^{cd}$
Control	—	0.10	0.30^{e}	1.50^{e}	2.37^{e}	4.27^{e}	5.87^{e}	5.47^{f}	44.9 $(42.0)^{g}$
CD 5%		NS	0.07	0.20	0.07	0.10	0.17	0.16	4.1 (2.7)

NS: Mustard sown 15 days prior to cabbage planting helped to attract early arrival of pests resulting in continuously low incidence on intercropped cabbage upto 24 days after planting

*Pre-treatment count and first spray commenced at 25 and 26 days after cabbage planting respectively

in plots with 2 and 3 per cent NSKE and commercial neem formulations were not significantly different but were superior to standard chemicals and untreated check (Table 4). In addition to DBM, control plots supported larvae of leafwebber and gram caterpillar, *Helicoverpa armigera* (Hubner). On an average, 4.0, 3.0 and 0.6 leafwebber and 3.3, 1.3 and 6.0 gram caterpillar larvae/plant, were recorded at 45, 51 and 55 DAP, respectively.

Mustard grown as a trap crop on either side of the experimental plot supported DBM, leafwebber and webworm beginning from 25 days after I mustard sowing (DAIMS) (10 days after cabbage planting (DACP)) until 71 DAIMS (56 DACP), aphids, *B. brassicae* appeared on mustard at 41 DAIMS, when it started flowering (Table 5). The

Table 5. Population count of insect pests on mustard grown on either side (13 m) of experimental plot with cabbage. (Late rainy season, 1991)

DAIMS/DACP	Row	Insect pest/plant				Stem borer Larvae	Aphids
		Diamondback moth		Leafwebber			
		Egg	Larvae	Egg mass	Larvae		
25/10	I	0.3	0.1	0.1	0.0	0.1	0.0
	II	0.3	0.0	0.1	0.0	0.1	0.0
	I	0.4	0.2	0.0	3.0	0.2	0.0
31/16	II	0.5	0.2	0.0	3.0	0.2	0.0
	I	0.9	0.4	0.0	0.8	0.0	0.0
41/26	II	0.6	0.4	0.2	0.5	0.0	0.0
	I	2.3	0.2	0.0	2.0	0.1	26.0
51/36	II	2.5	0.4	0.0	1.6	0.1	25.0
	I	1.0	0.9	0.0	2.5	0.0	406.0
61/46	II	2.2	1.0	0.0	1.6	0.1	125.0
	I	2.3	1.6	0.0	1.3	0.0	370.0
71/56	II	2.6	0.7	0.2	1.9	0.1	330.0

DAIMS : Days after I Mustard Sowing
DACP : Days after Cabbage Planting

Table 6. Effect of Neem Seed Kernel Extract (NSKE) on dry matter and size of cabbage at different crop stages (Late Rainy season, 1991)

Treatment	Conc. (%)	Mean dry matter/plant (g) DAT			Mean diameter of head (cm) at harvest	
		43	53	63	Length	Width
NSKE	2	19.8	73.4	115.2	12.6	18.4
NSKE	3	22.7	77.4	123.7	12.7	19.5
NSKE	4	22.9	74.0	128.3	12.2	18.8
NSKE	5	22.9	67.1	130.6	12.5	19.0
Cartap hydro-chloride	0.1	22.7	71.5	116.4	12.8	18.3
CD 5%		NS	NS	NS	NS	NS

NS = Non Significant
DAT = Days After Transplanting

dry weight and size of cut cabbages did not differ for samples taken at 43, 53, 63 DAP and at harvest, respectively (Table 6).

Effect on natural enemies: There was no significant difference in the parasitism levels for the DBM larvae sampled at 50 DAP, both in the standard spray and IPM plots. However, samples taken at 60 DAP showed significant variation in the parasitism levels between treatments in the standard spray and IPM plots. Parasitism in NSKE and control plots was significantly high (26.7 to 50.0 per cent) at 60 DAP as compared to rest of the treatments in both the experiments. Commercial neem formulations and standard chemicals treated plots registered 16.7 to 26.7 and 3.3 to 16.6 per cent parasitism at 60 DAP, respectively (Table 7).

DISCUSSION

The water extracts of neem seed kernel at 4 and 5 per cent were uniformly effective in checking DBM population in both the experiments. Commercial formulations of neem, though effective in early stages, did not show the same consistency with build up of population at later crop growth stages. Efficacy of water extracts from seed kernels (25 and 50 g seed kernel/1 of water) for the control of DBM and pyralid moth, *Hellula philidealis* (Walker) in the Dominican Republic of South America is on record

Table 7. Parasitisation by *Cotesia plutellae* on DBM under the standard spray regime and IPM.

Treatment	Conc (%)	Per cent parasitism for larvae sampled on difft. DAP			
		Standard Spray		IPM	
		50	60	50	60
NSKE	2	26.7 (30.7)	30.0 (32.9)[bc]	33.3 (34.9)	33.3 (34.8)[b]
NSKE	3	40.0 (39.1)	40.0 (39.0)[ab]	26.7 (30.7)	26.7 (30.9)[b]
NSKE	4	40.0 (38.8)	40.0 (39.0)[ab]	33.3 (34.9)	33.3 (34.9)[b]
NSKE	5	33.3 (34.5)	46.7 (42.9)[a]	26.7 (30.7)	33.3 (35.1)[b]
Cartap hydrochloride	0.1	13.3 (17.7)	10.0 (15.0)[d]	20.0 (26.5)	13.3 (21.1)[c]
Endosulfan	0.07	6.7 (8.8)	3.3 (6.1)[e]	13.3 (17.7)	16.6 (23.8)[c]
Neemark	1	33.3 (29.9)	26.7 (30.7)[bc]	26.7 (30.7)	16.7 (23.8)[c]
Repelin	1	13.3 (17.7)	20.0 (26.5)[c]	20.0 (26.5)	16.7 (23.8)[c]
Control	—	46.7 (42.9)	50.0 (44.9)[a]	46.7 (42.9)	50.0 (44.9)[a]
CD 5%		NS	9.0	NS	7.0

DAP : Days after planting; NS : Non significant
Figures in parentheses are angular transformed values
Means in the same column with letters in common are not significantly different

(Dreyer, 1986). Schmutterer (1992) compiled information on the use of neem for the control of DBM in Africa, Asia and Latin America. He reported that in all the three continents, water extracts from seed kernels or seeds gave the best results and were superior to methanolic formulated seed kernel extract. Neem has deterrent, antifeedant and growth disrupting properties. Seed kernels contain azadirachtin isomers AZA to AZG as the main component which has all these properties (Schmutterer, 1990) and numerous other active principles contributing to antifeedant and insect growth regulatory properties (Jones et al., 1989; Kraus et al., 1987). These principles easily come out in water within 4-6 hours (Schmutterer, 1992). However, commercial formulations are made from solvent extracts. It is likely that few or many of the active principles may be rendered inactive during the fractional distillation. This could be one of the reason, why, the commercial formulations tried were not as effective as water extract.

Schmutterer (1992) also opined that water extracts of seed kernels might remove the waxy layer on the leaves, leading to increased transpiration and consequent reduction in size of head. Our results did not indicate any such effects for all the concentrations tried in both the experiments.

It may be due to the fact that both the experiments were conducted in rainy season and adequate water was available to the plants during the entire period of crop growth. This might have compensated for increased transpiration, if any, following neem applications.

First spray commenced at 15 DAP in the standard spray regime experiment. However, in the IPM experiment, mustard attracted incoming pests (early arrivals) until 24 DAP. This is the reason, why, preheading spray was not required on the intercropped cabbage. Spray after 55 DAP was also not required because, immigrant and resident insects readily colonised on the green foliage of second mustard sown at 25 days after cabbage planting resulting in significant reduction of overall pest population on the intercropped cabbage. In addition, its reduction was also helped by parasitisation and treatments with respective chemicals.

Relative safety of NSKE to important natural enemies belonging to hymenoptera parasitising eggs, larvae and pupae of pests have been documented (Joshi et al., 1982; Schauer, 1985). Our results also show significantly high parasitism by *C. plutellae* for DBM larvae sampled from NSKE plots.

A versatile insect like DBM can even adapt to natural product like neem if continuous and indiscriminate application is resorted to on cabbage (Schmutterer, 1990). It is in this context, limited sprays of neem as required under the IPM involving mustard as a trap crop assumes importance.

REFERENCES

Bretherton, R.F. 1982. Lepidoptera immigration to the British Isles. 1969 to 1977. *Proc. Trans. British Entomol. Nat. Hist. Soc.* 15: 98–110.

Chelliah, S. and K. Srinivasan. 1986. Bio-ecology and management of diamondback moth in India. In: *Proc. I. Int. Workshop on Diamondback Moth Management*, Tainan, Taiwan, 11-15 Mar. 1985, N.S. Talekar (ed.), AVRDC Publication No. 86–248, pp. 63–76.

Chu, Y.I. 1986. The migration of diamondback moth. In: *Proc. I. Int. Workshop on Diamondback Moth Management,* Tainan, Taiwan, 11–15 Mar. 1985, N.S. Talekar, (ed.), AVRDC, Publication No. 86–248, pp. 77–82.

Dreyer, M. 1986. Untersuchungen zur Wirksamkeit von Wassereextrakten und anderen Produkten aus Niemsamen gegen Schadlinge an Gemuse-und Feldkulturen in Togo. Ph.D. Thesis, University of Giessen, 138 pp.

I.I.H.R. 1991. Integrated Pest Management on Cabbage Using Indian Mustard as a Trap corp. Extension folder 33. Indian Institute of Horticultural Research, Bangalore, 6 pp.

Jones, P.S., S.V. Ley, E.D. Morgan and D. Santafianos. 1989. The Chemistry of the Neem Tree. *Focus on Phytochemical Pesticides*, Vol. I, *The Neem Tree*, Boca Raton, CRC Press, pp. 19–45.

Joshi, B.G., G. Ramaprasad and S. Sitaramaiah. 1982.. Effect of neem seed kernel suspension on *Telenomus remus* an egg parasite of *Spodoptera litura. Phytoparasitica* 10 : 61–63.

Kraus, W., S. Baumann, M. Bokel, U. Keller, A. Klenk, M. Klingele, H. Pohnl and M. Schwinger. 1987. Control of insect feeding and development by constituents of *Melia azedarach* and *Azadirachta indica. Proc. III. Int. Neem Conf.*, Nairobi, 1986, Eschborn, GTZ, pp. 111–125.

Leibee, G.L. and K.E. Savage. 1992. Insecticide resistance in diamondback moth in Florida. In: *Proc. II. Int. Workshop on Diamondback Moth and other Crucifer Pests*, op. cit. pp. 427–436.

Lokki, J., K.K. Malmstrom and E. Sumalainen. 1978. Migration of *Vanessa cardui* and *Plutella xylostella* (Lepidoptera) to Spitsbergen in the summer 1978. *Nat. Entomol.* 58 : 121–123.

Nagarkatti, S. and K.P. Jayanth. 1982. Population dynamics of major insect pests of cabbage and of their natural enemies in Bangalore district (India): In: *Proc. Int. Conf. Plant Prot. Tropics*, Malaysian Plant Protection Society, Kuala Lumpur, Malaysia, pp. 325–347.

Schauer, M. 1985. Die Wirkung von Nieminhaltsstoffen auf Blaltlause und die Rubenblattwanze. Ph.D. Thesis. Univ. Giessen, FRG.

Schmutterer, H. 1990. Properties and potential of natural pesticides from the neem tree *Azadirachta indica. Ann. Rev. Entomol.* 35: 271–297.

Schmutterer, H. 1992. Control of diamondback moth by application of neem extracts. In: *Proc. II. Int. Workshop on Diamondback moth and other Crucifer Pests*, op. cit., pp. 325–332.

Shelton, A.M. and J.A. Wyman. 1992. Insecticide resistance in diamondback moth in North America. In: *Proc. II. Int. Workshop on Diamondback Moth and other Crucifier Pests*, pp. 447–454.

Srinivasan, K. 1988. Changing pest complex in relation to cropping systems—vegetables. National Seminar on "Changing Pest Situation in the Current Agriculture Scenario of India", Indian Agricultural Research Institute, New Delhi, India, Abstract Volume, 27 pp.

Srinivasan, K. and P.N. Krishna Moorthy, 1992. Development and adoption of integrated pest management for major pests of cabbage using Indian mustard as a trap crop. In: *Proc. II. Int. Workshop on Diamondback moth and other Crucifer Pests*, op. cit., pp. 511–522.

Srinivasan, K. and G.K. Veeresh. 1986. The development and comparison of visual damage thresholds for the chemical control of *Plutella xylostella* and *Crocidolomia binotalis* on cabbage in India. *Insect Sci. Applic.* 7: 559–563.

Sun, C.N. 1992. Insecticide resistance in diamondback moth. In: *Proc. II. Int. Workshop on Diamondback moth and other Crucifer Pests*, op. cit., pp. 419–426.

Syed, A.R. 1992. Insecticide resistance in diamondback moth in Malaysia. In: *Proc. II. Int. Workshop on Diamondback moth and other Crucifer Pests,* op. cit., pp. 437–442.

47

Neem in *Helicoverpa* [*Heliothis*] *armigera* Management: Pulses Ecosystem

J.N. SACHAN AND GURURAJ KATTI

Indian Institute of Pulses Research [ICAR]
Kanpur, India

ABSTRACT

Helicoverpa armigera Hubner is a key pest of chickpea and pigeonpea in India. Till now, use of insecticides has been the only tool for its effective management. The need has been felt to minimize the reliance on use of insecticides in the pulses ecosystem. Strategies have been formulated under the All India Co-ordinated Pulses Improvement Project (AICPIP] to evolve suitable integrated pest management (IPM) systems incorporating the use of safe and effective botanical insecticides including neem products along with other components of IPM such as host plant resistance, cultural practices and biological control.

Large-scale multilocational field trials have revealed that neem seed kernel extract (NSKE) and neem leaf extract (NLE) at 5 per cent each provided significant and cost-effective control of *H. armigera* compared to conventional insecticides in chickpea. Laboratory studies on commercial formulations of neem showed that Repelin, Neemguard, neem 25 EC, and neem 25 WDP provided leaf area protection up to 75 to 90 per cent from *Helicoverpa* damage in chickpea. Another commercial formulation, Neemark, proved quite effective against *H. armigera* in chickpea.

In pigeonpea, two sprays of NSKE 5 per cent and a single spray of cypermethrin 0.01 per cent gave significantly better cost benefit ratio of (1:3:81) and endosulfan 0.07 per cent (1:3:39) applied alone, against *H. armigera*. Soil treatment with phorate at 20 kg/ha at preflowering stage followed by nuclear polyhedrosis virus application and NSKE 5 per cent spray was effective in reducing *Helicoverpa* larval population to the extent of 4.16 per cent and was statistically on par with monocrotophos 0.04 per cent followed by endosulfan 0.07 per cent and quinalphos 0.05 per cent treatments.

Further efforts are required to develop efficient neem formulations and for their thorough testing with a view to integrate them along with other components of IPM against *H. armigera*.

INTRODUCTION

Helicoverpa armigera Hubner, commonly known as gram pod borer, is a highly polyphagous and widely distributed pest in India. It is an important pest of several cultivated crops like cotton, pigeonpea, tomato and tobacco. Among pulses, *H. armigera* is a key pest of chickpea and pigeonpea. Based on pest damage maps drawn in case of chickpea and pigeonpea in India, Virmani et al. (1991) have concluded that *H. armigera* is one of the most serious biotic stresses affecting both these crops. Losses to the extent of $ 300 million (Reed, 1983) or Rs. 2486 million (Lal et al., 1985; Lal and Yadava, 1987) have been reported in India alone.

Varied numbers of generations of this pest have been reported from different agroclimatic regions of the country. The pest passes through four generations in Punjab (Singh and Singh, 1975), while seven to eight have been reported in Andhra Pradesh (Bhatnagar, 1980). In Uttar Pradesh, five generations have been recorded during December to March on chickpea and pigeonpea, one during April to May on tomato, one on okra during June to July, and one on groundnut, sorghum, millets, and other crops during August to September (Tripathi, 1985). The pest remains active throughout the year in peninsular India due to uninterrupted supply of host plants. It thus causes damage to chickpea at both vegetative and reproductive stages and to pigeonpea of different maturity groups. In North India, it inflicts serious crop losses only at reproductive phase of chickpea and pigeonpea. However, there are several gaps in our knowledge of the seasonal incidence and number of generations. This is primarily because rapid changes in the cropping patterns during the last decade have facilitated the survival of *H. armigera* on different host plants throughout the year (Sachan and Yadava, 1991).

Intensive studies have been carried out under the All India Co-ordinated Pulses Improvement Project (AICPIP) on various components of *Helicoverpa* management in both chickpea and pigeonpea (Sachan, 1983–1992 a, b). Initially, the emphasis was on selection of suitable insecticides for its effective control. However, in the past decade an integrated approach has been adopted to tackle this pest. Systematic research work on host plant resistance has led to the identification of promising resistant genotypes such as PDE 2 and PDE 5 in chickpea and ICPL 6, PDE 54-2, ICP 1903, and MA 1 in pigeonpea (Sachan, 1990). However, only limited success has been achieved in controlling *H. armigera* through host plant resistance, though pseudoresistance (host avoidance phenomenon) has shown good promise particularly in northern India. The peak activity period of *H. armigera* in this region is usually in March and April (Fig. 1). Through selection of plant types of both chickpea and pigeonpea which mature earlier, it is possible to escape *Helicoverpa* damage (Sachan, 1992). Among the biocontrol agents of *H. armigera*, an ichneumonid parasite, *Campoletis chlorideae* Uchida, has been found very effective in regulating *Helicoverpa* population in chickpea. However, it has been observed that there is limited scope for large-scale field releases, as it is uneconomical (Nagarkatti, 1981). Nuclear polyhedrosis virus (NPV), a microbial pathogen has been found effective alone or in combination with insecticides in the management of *H. armigera* in both chickpea and pigeonpea. However, no commercial NPV formulation is yet available in India.

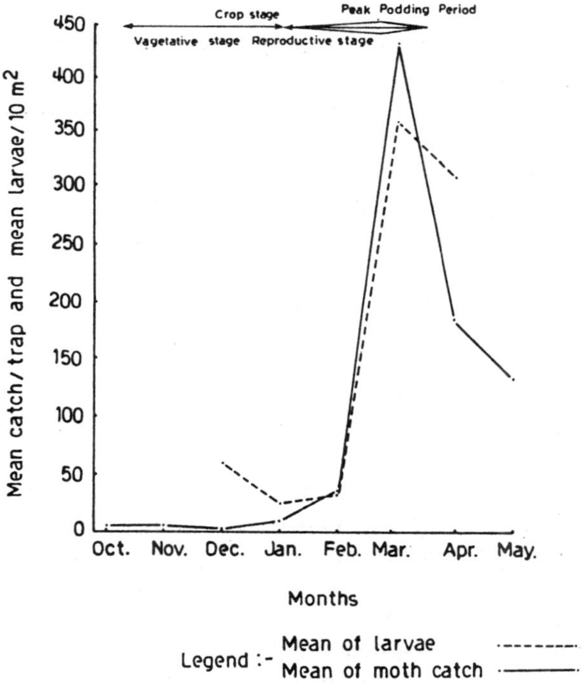

Fig. 1. The mean monthly catch of males per trap in sex pheromone traps and mean monthly number of larvae/ $10m^2$ on chickpea (Mean of three years i.e. 1982–83 to 1984–85)

Chemical control has been one of the most practical methods available to the farmer for the effective control of *H. armigera*. Insecticides like endosulfan, quinalphos, monocrotophos, dimethoate, fenvalerate, and cypermethrin have been found to be cost effective and are commonly recommended against *H. armigera* in both chickpea and pigeonpea. However, in the past few years, there have been increasing reports of development of insecticide resistance in *H. armigera*, particularly to synthetic pyrethriods. Resistance to conventionally used insecticides in *H. armigera* has been extensively reported in cotton ecosystem in Andhra Pradesh (Dhingra et al., 1988), Punjab, Haryana, and Rajasthan (Sachan et al., 1991). High levels of resistance to fenvalerate in pigeonpea have also been recorded (King and Sawicki, 1990). In addition, most of the insecticides are not compatible with the biocontrol agents because of their toxic nature. Thus, the indiscriminate use of insecticides has led to problems of environmental pollution and upsurge of secondary pest outbreaks resulting in ecological imbalance.

Although such a situation is yet to arise in pulse ecosystem, safer, more effective, and cheaper methods of managing *H. armigera* in chickpea and pigeonpea are needed. Keeping this in view, safe and effective botanical insecticides such as neem and its products have been intensively tested under the AICPIP.

Studies on Use of Neem Products under AICPIP

Studies have been carried out at various locations throughout the country to evaluate the efficacy of neem for possible inclusion in the management of *H. armigera* in chickpea and pigeonpea. The emphasis has been on testing of crude products of neem such as neem seed kernel extract (NSKE), neem leaf extract (NLE) and neem oil. In the last few years commercial formulations such as Neemark, Neemguard and RD 9 Repelin have been evaluated for their effectiveness against *H. armigera*. The salient findings obtained so far are discussed below:

Efficacy and Economics of Crude Products of Neem

Chickpea

Field trials were carried out at several locations to find out relative performance of neem products alone as well as in combination with the recommended insecticides against *H. armigera*.

Use of neem products provided significantly good control of *H. armigera* in comparison to untreated control (Fig. 2). Among these NSKE 5 per cent and NLE 5 per cent performed on par with quinalphos (0.05 per cent spray) but were slightly inferior to fenvalerate (0.2 per cent), cypermethrin (0.006 per cent) and endosulfan (0.07 per cent spray and 4 per cent dust).

The cost benefit ratios (CBRs) obtained in case of both NSKE and NLE were quite encouraging and comparable to that of quinalphos 1.5 per cent dust, endosulfan 4 per cent dust and fenvalerate 0.02 per cent spray (Fig. 3). Thakur et al., (1988) reported similar results on the basis of yield and gain profitability in case of NSKE. Trials conducted with neem leaf and NSKE at Sehore under AICPIP also reported increase in yields up to 24 per cent with the use of neem products compared to that of control (Table 1).

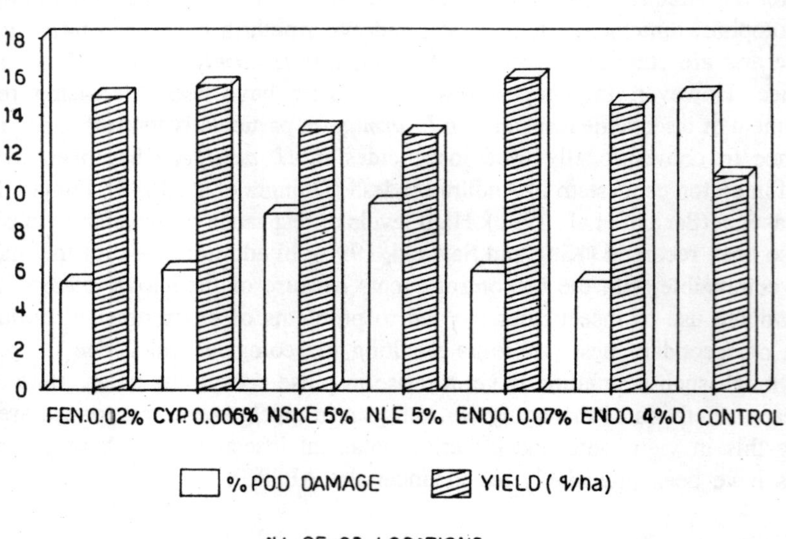

Fig. 2. Efficacy of neem products vs insecticides against *H. armigera* in chickpea (1982–86)

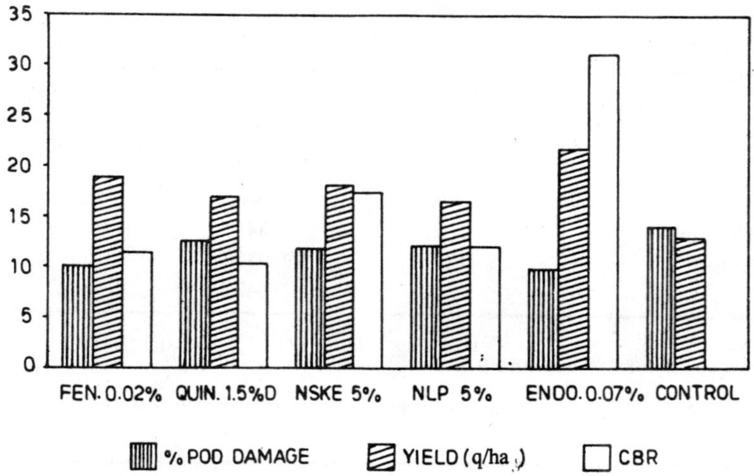

Fig. 3. Economics of neem products vs insecticides against *H. armigera* in chickpeas (S.K. Nagar, 1982–1985)

Pigeonpea

Field trials were carried out over several years to evaluate the efficacy of neem products against pod borers of pigeonpea, mainly *H. armigera*. The results revealed that NSKE, NLE and neem oil could not provide effective control of *H. armigera* as in chickpea (Fig. 4). The prolonged reproductive phase of pigeonpea may have diluted the effect of neem products, which are known to have short residual toxicity.

The CBRs obtained in case of neem products were less favourable than those of the synthetic insecticides, endosulfan and cypermethrin. However, encouraging results were obtained through combination of neem product sprays and cypermethrin and fenvalerate against *H. armigera* (Table 2). Degaonkar et al. (1988) also reported that use of NLE alone is less economical than BHC and endosulfan.

Compatibility of Neem Products with Other Components of Management

Efforts are on under AICPIP to integrate neem products in the IPM trials on pigeonpea.

Studies conducted at Coimbatore revealed that NSKE 25 per cent proved very effective in combination with NPV and endosulfan and significantly reduced pod damage to 10 to 11 per cent compared to 39 per cent in control (Table 3).

At Gulbarga (1991 to 1992), application of NSKE 5 per cent alone as well as in combination with NPV and phorate reduced *Helicoverpa* larval populations up to 35 to 40 per cent compared to that of control (15 to 61 per cent larval reduction).

Testing of Commercial Formulations of Neem

Chickpea

Laboratory studies carried out at Kanpur revealed that RD 9 Repelin (1 per cent), Neemguard (0.4 per cent), Neem 25 EC (0.15 per cent) and Neem 25 WDP (0.15 per cent) provided 75 to 90 per cent leaf area protection from *Helicoverpa* damage (Table 4).

Table 1. Effect of neem products on pod damage due to *H. armigera* and yield in chickpea*

Treatment	Pod damage (%)	Yield (kg/ha) (kg/ha)	Increase in yield over control (%)
NSKE 5%	8.54	757.5	15.5
NLE 5%	13.76	696.5	8.1
NSKE 5% + 1% soap	6.51	841.5	23.9
NLE 5% + 1% soap	11.04	722.5	11.4
Control	16.75	640.0	

Location : Sehore.
*Average of two years: 1988-89 and 1990-91.

Table 2. Efficacy of NSKE alone and in combination with synthetic insecticides against *H. armigera* in pigeonpea (1989–90)

Treatment *	Pod damage (%)	Yield (Kg/ha)	Cost benefit ratio (CBR)
NSKE 5% + 1% soap	40.36 (42.03)	1481	1:2:38
Endosulfan 0.07%	40.16 (45.15)	1512	1:3:39
Cypermethrin 0.01%	43.80 (48.01)	1467	1:3:81
Fenvalerate 0.04% dust at 25 kg/ha	43.58 (47.53)	1495	1:1:45
Endosulfan 0.07% + NSKE 5% + NSKE 5%	40.30 (43.73)	1353	1:2:70
Endosulfan 0.07% + Endosulfan 0.07% + NSKE 5%	44.02 (48.30)	1353	1:1:27
NSKE 5% + Endosulfan 0.07%+ NSKE 5%	39.09 (39.77)	1552	1:3:30
Cypemethrin 0.01% – NSKE 5% + NSKE 5%	37.82 (37.88)	1695	1:6:18
NSKE 5% + Cypermethrin 0.01% + NSKE 5%	37.20 (46.44)	1753	1:6:91
Fenvalerate 0.4% dust + NSKE 5% + Fenvalerate 0.4% dust	38.13 (38.30)	1695	1:4:62
Fenvalerate 0.4% dust + 0.4% dust + NSKE 5%	32.93 (29.73)	1923	1:6:98
NSKE 5% + Fenvalerate 0.4% dust + NSKE 5%	36.47 (35.57)	1713	1:4:17
Control	45.50 (50.89)	1153	
CD at 1%	9.53	269	
CV %	10.50	7.60	

Location : Gulbarga.
* Three sprays/dust were given.

Field study conducted at Rahuri in 1991 to 1992 revealed that Neemark 5 per cent proved quite effective against *H. armigera* and significantly reduced pod damage to 6.10 per cent, compared to 13.35 per cent in control and its performance was on par with that of monocrotophos 0.04 per cent (4.24 per cent damage). However, its perfor-

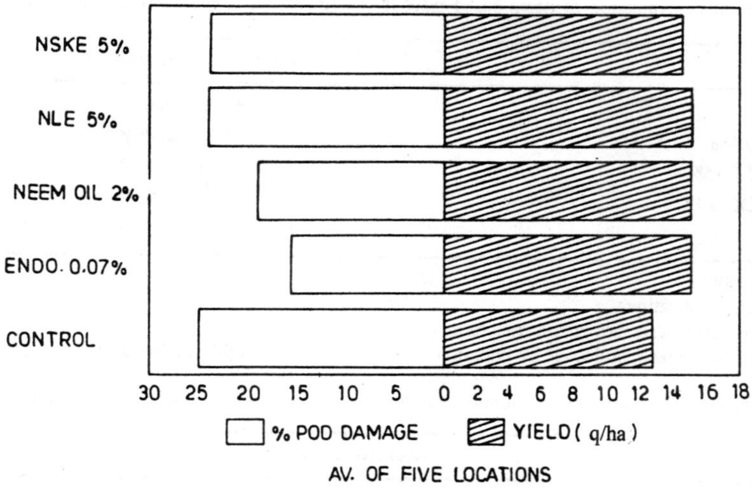

Fig. 4. Efficacy of neem products against *H. armigera* in pigeonpea (1983–84)

Table 3. Efficacy of NSKE with nuclear polyhedrosis virus (NPV) and Endosulfan on *H. armigera* in pigeonpea (1988–89).

Treatment *	Pod bore damage (%)	Yield (kg/ha)
NPV 250 LE/ha	13.36 (21.44)[ab]	480[ab]
NPV 250 LE + NSKE 24%	11.23 (19.56)[ab]	502[ab]
NPV 250 LE + NSKE 25% + Endosulfan 219 g a.i./ha	9.87 (18.18)[a]	521[a]
NPV 250 LE + NSKE 25% + Crude sugar 20% + Endosulfan 219 g a.i./ha	14.49 (22.37)	491[b]
NSKE 25%	21.57 (27.60)[c]	433[b]
Endosulfan 437.5 g a.i./ha	11.41 (19.76)[ab]	497[a]
Control	38.96 (38.62)[d]	338

Location: Coimbatore.
* Applied as ULV spray using helisprayer.
Figures followed by common letters are not significantly different at $p = 0.05$.

mance was inferior to that of synthetic pyrethroids (1.95 to 2.51 per cent) and *Bacillus thuringiensis* (2.86 per cent).

Simwat and Dhawan (1992) carried out field and laboratory tests and reported that Neemark (0.05 and 0.75 per cent) and RD 9 Repelin (1 and 2 per cent) proved less effective than quinalphos against third-instar larvae of *H. armigera*.

Pigeonpea

Under AICPIP, Neemark 5 per cent was evaluated for its effectiveness against *H. armigera* at Rahuri, during 1990 to 1992. The results revealed that its performance was on par with that of monocrotophos 0.04 per cent (Table 5).

Table 4. Laboratory evaluation of neem-based insecticides against *H. armigera* in chickpea (1991–92).

Treatment	Protection of leaf area* (%)
Repelin 1%	86.73 (68.68)[ab]
Neemguard 0.4%	75.69 (60.49)[c]
Neem 25 EC 0.15%	89.54 (71.18)[b]
Neem 25 WDP 0.15%	83.96 (66.46)[c]
Control	44.59 (41.88)[d]

Location : Kanpur.
* Mean of four replications.
Figures in the parentheses are arcsin transformed values.
Figures in the parentheses followed by common letter are not significantly different at $p = 0.05$.

Table 5. Efficacy of Neemark (R) vis-a-vis synthetic insecticides against *H. armigera* in pigeonpea*

Treatment	Pod damage (%)	Yield (kg/ha)
Endosulfan 0.07%	31.74	910
Monocrotophos 0.04%	39.98	866
Decamethrin 0.003%	30.67	806
Neemark 5%	38.28	726
Fenpropathrin 0.002%	37.42	764
Acephate 0.07%	34.14	733
Control	46.75	525

Location: Rahuri.
* Average of two years: 1990–91 and 1991–92.

CONCLUSION

Trials conducted at several locations throughout the country in the past ten years have revealed that use of neem products has good scope in *Helicoverpa* management in chickpea and pigeonpea. In case of pigeonpea, their effect is diluted due to the prolonged reproductive stage of the crop; however, they can be very effectively used alternatively or in combination with other components of IPM.

ACKNOWLEDGEMENTS

The authors are thankful to AICPIP entomologists for their cooperation in generating the above information through the conduct of coordinated trials.

REFERENCES

Armes, N.J, G.S. Bond and A.B.S. King. 1989. Report on insecticide resistance in *Helicoverpa armigera* (Hubner) (Lepidoptera: Noctuidae) in Southern India. Annual crops programme management of flower seed and fruit pests (unpublished).

Bhatnagar, V.S. 1980. A report on research on *Heliothis* complex at ICRISAT (India) 1974–79. *Proc. All India Workshop on Consolidation of Pest Management. Recommendations and Guidelines of Research*, April 24–26, 1980, Udaipur, India.

Degaonkar, A.M., J.S. Kurtadikar and P.V. Rekawar. 1988. Economic control measures against arhar pod borer complex in Marathwada region. *Pesticides* 22(8) : 15–16, 27.

Dhingra, S., A. Phokela and K.N. Mehrotra. 1988. Cypermethrin resistance in population of *Heliothis armigera* (Hubner). *Nat. Acad. Sci. Letters* 11: 123–125.

King, A.B.S. and R.M. Sawicki. 1990. Insecticides resistance of *Helicoverpa* and its management. In: *Chickpea in the Nineties : Proc. 2nd Int. Workshop on Chickpea Improvement*, 4–8 Dec., 1989, ICRISAT Centre, Patancheru, Andhra Pradesh, pp. 195–201.

Lal. S.S. and C.P. Yadava. 1987. Estimate of crop losses in pigeonpea caused by the pod borer complex. *FAO Plant Prot. Bull.* 35(3): 93–98.

Lal, S.S., C.P. Yadava and C.A.R. Dias. 1985. Assessment of crop losses in chickpea caused by *Heliothis armigera*. *FAO Plant Prot. Bull.* 33(1): 27–35.

Nagarkatti, 1981. The utilization of biological control in *Heliothis* management in India. *Proc. Int. Workshop on Heliothis Management*, 15–20 Nov. 1981, ICRISAT Centre, Patancheru.

Reed, W. 1983. Crop losses caused by insect pests in the developing world. *Proc 10th Int. Cong of Plant Protection* pp. 74–84.

Sachan, J.N. 1983–1992a. All India Co-ordinated Pulses Improvement Project, Kharif Workshop, Consolidated Reports on Kharif Pulses (Entomology), 1983–92.

Sachan, J.N. 1983–1992b. All India Co-ordinated Pulses Improvement Project, Rabi Workshop, Consolidated Reports on Rabi Pulses (Entomology), 1983–92.

Sachan, J.N. 1990. Progress in host plant resistance work in chickpea and pigeonpea against *Helicoverpa armigera* (Hubner) in India. *Summary Proc. of 1st Consultative Group Meeting on Host Selection Behaviour of H. Armigera.* 5–7 Mar., 1990, ICRISAT Centre, Patancheru, pp. 19–22.

Sachan, J.N. 1992. Present status of *Helicoverpa armigera* on pulses and strategies for its management. In: *Helicoverpa Management: Current Status and Future Strategies. Proc. 1st Nat. Workshop*, J.N. Sachan (ed.) Directorate of Pulses Research, Kanpur, 30–31, Aug., 1990, pp. 7–23 (in press).

Sachan, J.N., S.S. Lal and O.P. Dubey. 1991. Outbreak of *Heliothis armigera* in northern states of India during 1990–91. *Pesticides Information* 17(3): 15.

Sachan, J.N. and C.P. Yadava. 1991. *Heliothis* ecology in relation to different agro-ecosystems. Paper presented at ICAR/IOPRM/USDA/Joint Development Group Meeting on Managing Resistance with Focus on *Heliothis* Resistance Management in India, 16–17 Oct. 1991, Directorate of Rice Research, Hyderabad.

Simwat. G.S. and A.K. Dhawan. 1992. Use of plant insecticides against *Heliothis armigera* (Hubner) *Proc. 1st Nat. Symp.: Allelopathy in Agroecosystems (Agriculture and Forestry)*, Feb. 12–14. 1992, CCS, Haryana Agricultural University, Hisar. P. Tauro and S.S. Narwal. (eds.), Indian Society of Allelopathy, CCS Haryana Agricultural University, pp 158–159.

Singh, H. and G. Singh. 1975. Biological studies on *Heliothis armigera* (Hubner) in Punjab. *Indian J. Entomol.*, 37(2): 154–164.

Thakur, R.C., K.K. Nema and K.N. Kango. 1988. Comparative efficacy of neem seed kernel and some insecticidal formulations against the gram pod borer, *Heliothis armigera* (Hubner). *Legume Res.* 11(3): 114–116.

Tripathi, S.R. 1985. Final Report of ICAR ad-hoc scheme, biology, food preference and extent of damage by *Heliothis armigera* on different varieties of gram in trial belt of eastern U.P. Department of Zoology, University of Gorakhpur.

Virmani, S.M., D.G. Faris and G. Johansen, eds. 1991. Agroclimatology of Asian grain legumes (chickpea, pigeonpea and groundnut). *Res. Bull. No. 14.* International Crops Research Institute for the Semi-Arid Tropics, Patancheru.

48

Use of Neem in Management of Pest Complex in Cotton

G.P. GUPTA

Division of Entomology, Indian Agricultural Research Institute
New Delhi 110 012, India

ABSTRACT

Field trials were conducted with upland cotton, *Gossypium hirsutum* var. Pusa 31 during 1990, 1991 and 1992 crop seasons at the Indian Agricultural Research Institute, New Delhi, to evaluate the bioefficacy of neem products and their possible combinations with synthetic insecticides in a spray schedule for the management of pest complex in cotton with ultimate aim to reduce the load of synthetic insecticides and producing cost effective quality cotton. Spraying was initiated at 50 per cent flowering and applied 5 round with high volume knapsack sprayer at an interval of 15 days. Observations were recorded on jassid (*Amrasca devastans* Dist.) and whitefly (*Bemisia tabaci* Genn.) population on 30 leaves per plot basis, incidence of bollworms viz. spotted (*Earias vittella* Fab., *E. insulana* Boisd.), pink (*Pectinophora gossypiella* Saunders), American bollworm (*Heliothis armigera* Hubn.) in shed reproductives, green bolls and open bolls on loculi basis, and total yield of seed cotton. Result shows that application of neem products alone or in combination with biopesticides (Bt.) or conventional synthetic insecticides failed to suppress pest complex in cotton under field conditions. However, neem products in combination with synthetic pyrethroids proved effective in managing pest complex in cotton and increasing quality cotton produce. The seed cotton yield in such a schedule was more (1905 kg/ha and 2476 kg/ha) as compared to recommended spray schedule (1861 kg/ha and 2405 kg/ha) during 1991 and 1992 respectively, in which conventional insecticides alternated with synthetic pyrethroids, and is in vogue in northern India. It shows that neem can replace conventional insecticides in a spray schedule and can effectively manage the pest complex in cotton, and will also minimise the exposure of environment to toxic pesticides.

INTRODUCTION

Cotton is assailed by a multitude of pests as it evolves through various production levels. Of these, the sucking pests jassid (*Amrasca devastans* Dist.) and white-fly

(*Bemisia tabaci* Genn.) and the bollworm complex viz., spotted (*Earias vittella* Fab. *Earias insulana* Boisd.), pink (*Pectinophora gossypiella* Saund.) and American bollworm (*Heliothis armigera* Hubn.) are the key pests and pose a serious threat in limiting the cotton production in India. The main thrust of cotton growers has so far been towards the application of synthetic insecticides for combating these pests because of effectiveness and quick knockdown action. This crop alone consumes more than 50 per cent of the total insecticides produced or imported in the country. The rising consumption of synthetic insecticides in developing countries like India has led to a number of problems, such as insect resistance, high residue in market produce, ecological imbalances, and health hazards. Thus development of alternate plant protection technologies has become imperative, and botanical pest control materials have been under investigation.

Intensive research on the products of neem trees (*Azadirachta indica* A. Juss.) has indicated that it is a nearly ideal model for the use of plant products in agriculture and other areas. Results of field trials, still somewhat scarce, have demonstrated the great potential in neem for pest control. These results give rise to the hope that in developing countries, where sufficient neem trees are available, the use of synthetic pesticides could be replaced partly by the use of locally, produced neem (Schmutterer and Ascher, 1987). It has now been proved worldwide that neem affects insects in many ways and lepidopterous larvae are very sensitive to neem derivatives under field conditions (Meisner et al., 1981; Schmutterer, 1990). Since lepidopterous larvae of the bollworm complex are very serious pests of cotton and have defied all practical control measures in India, field trials were conducted to evaluate the bioefficacy of neem products and their possible combinations with synthetic insecticides in a spray schedule for the management of pest complex in cotton with the ultimate aim of minimizing the use of synthetic insecticides and producing cost-effective quality cotton.

MATERIALS AND METHODS

Synthetic insecticide alone and in combination with neem-based-insecticide is used. Field trials conducted during 1990 to evaluate the different spray schedules based on different economic threshold levels against pest complex in cotton showed promising results of those schedules in which two neem products Neemrich I (Margocide OK and Neemrich II (Margocide CK), were included. This gave rise to the hope that the use of synthetic insecticides could be replaced partly by the use of neem products. Exhaustive field trials were conducted at the Indian Agricultural Research Institute, New Delhi, during the 1991 and 1992 crop seasons with upland cotton variety Pusa 31 (*Gossypium hirsutum* L.) and different neem products. These products were supplied by National Chemical Laboratory, Pune (Neemrich I, Neemrich II), National Organic Chemical Industries Ltd. (Neem 2 per cent, Neem A_1, A_2), Gharda Chemicals Ltd. (Neemguard) and T. Stanes and Co. Ltd. (Nimbecidine) and were used alone as well as in different combinations with synthetic insecticides against cotton pest complex. The details of the treatments and dosages are given in Table 1. The experiments were laid out in 42–45 m^2 in randomized block design with three replicates. All the agronomical practices recommended for growing irrigated cotton in northern India were followed. Spray materials were initiated at 50 per cent flowering and applied five times with high-

Table 1. Details of spray schedule

1990			1991			1992		
Spray schedule	Name of insecticide	Dose (g a.i./ha)	Treatment	Name of insecticide	Dose (g a.i./ha)	Treatment	Name of insecticide	Dose (g a.i./ha)
S_1	Monocrotophos 36% WSC	500	T_1	Neemrich II 20% EC	5 ml/l water	T_1	Neem A_1	5 l/ha
	Cypermethrin 25% EC	60		(Margocide CK)				
	Phosalone 35% EC	750						
	Deltamethrin 2.8% EC	12.5	T_2	Neem 2%	10 l/ha	T_2	Neem A_2	2.5 l/ha
	Qunialphos 25% EC	750						
S_2	Endosulfan 35% EC	750	T_3	Monocrotophos 36% WSC	500	T_3	Neemrich II 20% EC	5 ml/l water
	Deltaphos 36% EC	250		Endosulfan 35% EC	750		Dipel (BT) 8L	750
	Phosalone 35% EC	750		Quinalphos 25% EC	750		Neemrich II 20% EC	5 ml/l water
	Fenvalerate 20% EC	75		Triazophos 40% EC	600		Dipel (BT) 8L	750
	Quinalphos 25% EC	750		Endosulfan 35% EC	750		Neemrich I 80% EC	10 ml/l water
S_3	Deltaphos 36% EC	250	T_4	Monocrotophos 36% WSC	500	T_4	Monocrotophos 36% WSC	500
	Triazophos 40% EC	600		Deltamethrin 2.8% EC	12.5		Deltamethrin 2.8% EC	12.5
	Cyhalothrin 5% EC	50		Quinalphos 25% EC	750		Quinalphos 25% EC	750
	Quinalphos 25% EC	750		Cypermethrin 25% EC	60		Cypermethrin 25% EC	60
	Phosalone 35% EC	750		Triazophos 40% EC	600		Triazophos 40% EC	600
S_4	Monocrotophos 36% WSC	500	T_5	Deltaphos 36% EC	250	T_5	Deltaphos 36% EC	250
	Quinalphos 25% EC	750		Neem 2%	10 l/ha		Neemrich II 20% EC	5 ml/l water
	Alphamethrin 10% EC	25		Alphamethrin 25% EC	25		Neemrich I 80% EC	10 ml/l water
	Endosulfan 35% EC	750		Neemrich II 20% EC	5 ml/l water		Cypermethrin 25% EC	60
	Phosalone 35% EC	750		(Margocide CK)	5 l/ha		Triazophos 40% EC	600
				Neem 2% +	+ 30 g			
				Cypermethrin 25% EC				
S_5	Margocide OK 80% EC	10 ml/l water	T_6	Pyraclophos 50% EC	500	T_6	Deltaphos 36% EC	250
	Endosulfan 35% EC	750		Neemrich I 80% EC	10 ml/l water		Water	Tap water
	Margocide CK 20% EC	5 ml/l water		(Margocide OK)	100		Cypermethrin 25% EC	60
	Phosalone 35% EC	750		Trebon 10% EC				

				Neem 2% +	5 l/ha		Water	Tap water
	Quinalphos 25% EC	750		Cypermethrin 25% EC	+ 30 g		Neemguard	5 ml/l water
				Trebon 10% EC	100			
S_6	Margocide OK 80% EC	10 ml/l water	T_7	Neem 2% +	5 l/ha	T_7	Water	Tap water
	Cypermethrin 25% EC	60		Monocrotophos 36% WSC	+ 250 g		Deltamethrin 2.8% EC	12.5
	Margocide CK 20% EC	5 ml/l water		Endosulfan 35% EC	750		Water	Tap water
	Deltamethrin 2.8% EC	12.5		Neemrich II 20% EC	5 ml/l water		Cypermethrin 25% EC	60
	Quinalphos 25% EC	750		(Margocide CK)	600		Nimbecidine	5 ml/l water
				Triazophos 40% EC				
				Neemrich I 80% EC	10 ml/l water			
				(Margocide OK)				
S_7	Half control		T_8	Full control	No protection	T_8	Full control	No protection
	Protection against jassid							
	Dimethoate	0.03%						
S_8	Full control							
	No protection							
Dates of spraying: Jul. 31, Aug. 17, Sep. 1, Sep. 18, Oct. 4			Aug. 8, Aug. 22, Sep. 10, Sep. 27, Oct. 10			Aug. 13, Aug. 28, Sep. 10, Sep. 25, Oct. 9		

a.i. = active ingredients

volume knapsack sprayer at an interval of 15 days as suggested by Gupta and Agarwal (1985) against the bollworm complex. The quantity of spray fluid was calculated at each stage of the crop growth by applying a known quantity of water on control plot. As insects are able to differentiate between treated and untreated parts of their host plants, spraying was done with utmost care to cover the entire foliage. Care was also taken to avoid drift of insecticides. The last spray was stopped two to three weeks before first picking. During early stage of crop growth, dimethoate 0.03 per cent was applied as blank coverage to check the population of jassid below prescribed economic threshold level (one nymph per leaf).

Observations were recorded on the intermittent population of jassid and white-fly on 10 randomly selected plants in each plot at regular intervals. The nymphal population of jassid and adult population of white-fly were recorded by visual observation on the lower side of the leaves; three leaves from each plant leaving crown from the top were screened. These observations were recorded between 7 a.m. and 9 a.m.

Observations were also recorded on the incidence of bollworm complex in reproductives, shed parts (squares, flowers, both) and on locule basis in green bolls and open bolls on randomly selected plants in each plot. Twenty randomly selected green bolls in each plot were dissected and observed visually and per cent damage was worked out on the basis of incidence of bollworms in each locule. The incidence in open bolls was recorded on the basis of loculi damaged in all the open bolls on randomly selected plants in each plot. Data were also recorded on the incidence of bollworms in leftover green bolls after harvest. After last picking, all the green bolls left on five randomly selected plants in each plot were collected and dissected to ascertain the incidence of bollworms and their larval population. Data were also recorded on quality and quantity of seed cotton based on two pickings. The percentage incidence of bollworms was subjected to angular transformation for statistical analysis.

RESULTS AND DISCUSSION

Evaluation of Insecticides against Pest Complex of Cotton (1990)

Field trial conducted during 1990 showed 160 kg/ha yield losses due to sucking pests but the avoidable losses due to bollworm complex varied from 800 to 1470 kg/ha. The maximum avoidable losses of seed cotton were recorded in spray schedule S_3 (deltaphos, triazophos, cyhalothrin, quinalphos, phosalone) minimum in S_5 (Margocide OK, endosulfan, Margocide CK, phosalone, quinalphos). However, the spray schedule S_6, in which Margocide OK and Margocide CK were applied in combination with synthetic pyrethroids, proved effective in reducing avoidable losses to 1350 kg/ha. The mean population of overwintering pink bollworm larvae in open and leftover green bolls was also minimum in this schedule as compared to half control and full control (Table 2). This result gave rise to the hope that neem products can be used in spray schedules for the management of pest complex in cotton. Exhaustive field trials were conducted during 1991 and 1992 with neem products alone as well as in different combinations with synthetic insecticides for the management of pest complex in cotton to minimize the use of synthetic insecticides.

Table 2. Impact of different spray schedules on the avoidable losses caused by sucking pests, bollworm complex, and total pest complex and population of diapausing pink bollworm larvae during 1990

Spray schedule	Seed cotton yield (kg/ha)	Net gain in yield over half control (kg/ha)	Net gain in yield over full control (kg/ha)	Av. no. of hibernating pink bollworm larvae per boll	
				Open boll	Left over green boll
S_1	2560	1380	1540	0.34 (0.83)	1.20 (1.30)
S_2	2450	1270	1430	0.39 (0.89)	1.04 (1.26)
S_3	2650	1470	1630	0.30 (0.89)	1.16 (1.29)
S_4	2410	1230	1390	0.44 (0.97)	1.19 (1.30)
S_5	1980	800	960	0.77 (1.13)	1.05 (1.24)
S_6	2530	1350	1510	0.35 (0.92)	1.00 (1.23)
Half control	1180		160	1.08 (1.26)	4.32 (2.16)
Full control	1020			0.83 (1.15)	3.98 (2.12)
SEM ±				0.12	0.34
CD (p = 0.05)				0.28	0.78
CD (p = 0.01)				0.42	1.09

S_1–S_6Crop was protected against sucking pests and bollworm complex.
Half control: Crop was protected against sucking pests only.
Full control: Crop was fully unprotected.
Figures in parenthesis are angular transformed values.

Evaluation of Different Insecticidal Schedules against Bollworms

Incidence of bollworms complex (1991–1992)

The incidence of bollworms in shed reproductives and on locule basis in green bolls, open bolls, and leftover green bolls during 1991 and 1992 are presented in Tables 3 and 4 respectively. Table 3 shows that, during 1991 there was no significant difference among the treatments in the incidence values of bollworms in shed reproductives; however, significant differences were noticed in green bolls and open bolls after completion of four rounds of insecticidal sprays. Observations recorded on October 4 and 11 showed that spray schedule T_5, in which neem was used in combination with synthetic pyrethroid, was significantly effective in reducing the incidence of bollworms in green bolls (13.0 per cent, 8.1 per cent) and was statistically at par with T_4, in which conventional insecticides were used in combination with synthetic pyrethroides (19.4 per cent, 12.7 per cent) on October 4 and 11 respectively. Repeated application of neem alone (T_1, T_2) or its use in a spray schedule in combination with conventional insecticides (T_3) did not prove effective in reducing the incidence of bollworms in green bolls. The same trend in the incidence values of bollworms in open bolls on November 13

Table 3. Effect of spray schedules of different combination of insecticides on the incidence of bollworms and yield of seed cotton during 1991

Treat-ment	% incidence of bollworm complex in reproductives							% stained seed cotton		Total yield of seed cotton (kg/ha)
	Shed parts		Green bolls (loculi basis)		Open bolls (loculi basis)		Leftover green bolls (loculi basis)	1st picking	2nd picking	
	Sept. 9	Sept. 19	Oct. 4	Oct. 11	Nov. 13	Dec. 18	Dec. 31			
T_1	87.8 (67.7)	94.4 (78.8)	50.6 (45.3)	30.6 (33.4)	67.6 (55.6)	78.4 (67.2)	32.6 (34.7)	28.7 (32.4)	38.8 (38.2)	515
T_2	85.6 (68.4)	86.5 (68.6)	35.8 (36.6)	27.8 (31.8)	42.9 (40.9)	65.6 (54.5)	29.4 (32.7)	25.1 (30.4)	24.2 (35.8)	1213
T_3	88.5 (70.5)	94.0 (76.1)	36.5 (36.9)	25.0 (29.7)	38.3 (38.2)	50.7 (45.4)	25.5 (30.3)	24.3 (29.4)	25.4 (30.2)	1539
T_4	91.0 (72.9)	89.5 (74.4)	19.4 (25.8)	12.7 (20.2)	23.6 (28.9)	26.1 (30.6)	11.6 (17.2)	18.4 (22.5)	15.4 (22.7)	1861
T_5	89.5 (71.3)	93.6 (78.0)	13.0 (20.7)	8.1 (16.4)	17.0 (24.2)	22.3 (28.0)	16.1 (22.9)	12.6 (20.7)	15.2 (22.6)	1905
T_6	84.8 (67.7)	88.5 (71.1)	33.8 (35.4)	20.6 (26.5)	41.1 (41.8)	32.7 (34.7)	24.7 (29.7)	22.8 (28.4)	24.4 (29.5)	1033
T_7	88.1 (70.5)	83.5 (72.1)	50.3 (45.2)	25.0 (29.7)	42.7 (40.7)	63.9 (52.9)	28.9 (32.5)	28.9 (32.5)	31.5 (34.0)	763
T_8 (control)	92.0 (74.2)	94.2 (79.2)	70.9 (58.1)	43.2 (49.4)	67.6 (55.4)	86.7 (70.1)	35.6 (35.7)	37.3 (37.6)	40.8 (39.7)	424
SEM ±	NS	NS	4.6	4.5	2.5	6.6	6.1	2.0	2.6	165
CD ($p = 0.05$)			14.0	13.8	7.7	19.8	9.4	6.2	7.9	502
CD ($p = 0.01$)					10.7			8.6	10.9	696

Figures in parentheses are angular transformed values.

before first picking and December 18 before second picking was noticed (Table 3.) Data recorded on bollworms incidence in leftover green bolls showed little difference between schedules T_4 and T_5, but both the schedules were significantly effective in reducing the incidence as compared to other treatments.

Table 4 shows that during 1992, there was significant variation in the incidence levels of bollworms in shed reproductives, green bolls and open bolls. The per cent incidence values in shed reproductives (76.4, 54.4), green bolls (8.4, 9.2) open bolls (15.5, 19.7), and leftover green bolls (12.0) were significantly less in T_5, in which neem was used in combination with synthetic pyrethroids, closely followed by T_4, in which conventional insecticides were used in combination with synthetic pyrethroids. During 1992 also, application of neem alone (T_1, T_2) or in combination with Dipel (T_3) did not prove effective in reducing the incidence of bollworms. This shows that a spray schedule in which all the conventional insecticides were used or one of them replaced with neem products did not prove effective in reducing the incidence of bollworms in cotton under field conditions. These treatments were at par with untreated control and have lost importance in the present-day management strategy in cotton pest complex. It was also noted by Thangavel et al. (1975) that application of neem oil alone was least effective in controlling bollworm complex in cotton. However, a neem product registered in the United States (Margosan-O) gave good result against larvae of *Earias insulana* (Meisner et al., 1990).

Quality of Cotton (Per cent Stained Seed Cotton)

There were significant differences among treatments in the quality of seed cotton at the time of first and second picking during 1991 and 1992. During 1991, significantly lower percentages of stained seed cotton (12.6, 15.2) were recorded at first and second picking respectively in spray schedule T_5, which was closely followed by T_4 (18.4, 15.4) than in untreated control (37.3, 40.8) (Table 3).

During 1992, significantly less stained seed cotton was recorded at first and second picking in spray schedule T_4 (7.5 per cent, 8.7 per cent) closely followed by T_5 (7.7 per cent, 9.3 per cent) (Table 4). It is clearly evident that neem or conventional insecticides alternated in combination with synthetic pyrethroids would be effective in reducing the stained seed cotton. Maximum bad quality cotton was recorded from the plots which were protected with repeated application of neem alone or in combination with conventional insecticides or Dipel (BT).

Yield of Seed Cotton

In general, seed cotton yield was greater in 1992 than in 1991. During 1991, except T_1 and T_7 all the treatments were significantly superior to untreated control. Neem 2 per cent was also found more effective than Neemrich II in increasing the Yield of seed cotton. During 1992, except T_1 (Neem A_1) all the treatments were significantly superior to the untreated control, but neem A_2 was more effective than A_1 in increasing the cotton yield. It is evident from Tables 3 and 4 that seed cotton yield was significantly higher in the plot treated with spray schedule T_5, in which neem products were applied alternately in combination with synthetic pyrethroids (1905, 2476 kg/ha), closely followed by T_4, in which conventional insecticides were used alternately with synthetic

Table 4. Effect of spray schedules of different combination of insecticides on the incidence of bollworm complex and yield of seed cotton during 1992

Treatment	% incidence of bollworm complex in reproductives							% stained seed cotton		Total yield of seed cotton (kg/ha)
	Shed parts		Green bolls (loculi basis)		Open bolls loculi basis		Leftover green bolls (loculi basis)	1st picking	2nd picking	
	Aug. 31	Sept. 10	Sept. 28	Oct. 19	Oct. 31	Dec. 8	Dec. 31			
T_1	74.7 (60.6)	74.5 (60.1)	26.6 (30.6)	31.0 (33.7)	37.1 (37.3)	41.7 (40.1)	32.0 (34.4)	20.6 (26.7)	26.0 (30.4)	1492
T_2	70.4 (57.2)	61.5 (51.7)	23.8 (28.9)	33.6 (35.4)	34.0 (35.6)	37.5 (37.5)	34.2 (35.7)	18.8 (25.6)	27.6 (31.6)	1723
T_3	82.8 (65.5)	61.6 (51.9)	32.0 (34.3)	29.7 (30.0)	34.3 (35.7)	42.7 (40.7)	37.7 (37.9)	18.4 (25.3)	25.7 (30.4)	1516
T_4	53.1 (46.8)	47.9 (43.9)	8.1 (15.8)	8.2 (16.4)	16.9 (24.3)	18.3 (24.9)	12.0 (20.2)	7.5 (15.8)	8.7 (17.1)	2405
T_5	76.4 (61.2)	54.4 (57.5)	8.4 (15.9)	9.2 (16.6)	15.5 (23.0)	19.7 (26.3)	12.0 (20.2)	7.7 (16.0)	9.3 (17.6)	2476
T_6	68.8 (56.2)	51.5 (45.8)	7.4 (15.2)	16.1 (23.3)	21.7 (27.6)	23.7 (29.1)	14.1 (21.2)	8.9 (17.1)	12.3 (20.5)	2202
T_7	69.0 (56.2)	47.4 (43.4)	12.6 (20.4)	11.8 (19.9)	19.3 (25.9)	20.0 (26.5)	16.8 (23.8)	8.8 (17.0)	10.2 (18.6)	2335
T_8 (control)	85.9 (68.3)	81.3 (64.5)	36.9 (37.4)	40.4 (39.5)	44.4 (41.9)	51.9 (46.1)	36.4 (30.1)	26.9 (31.2)	35.9 (36.8)	896
SEM ±	3.0	3.7	3.4	2.5	3.3	3.8	2.7	1.9	2.1	202
CD (p = 0.05)	9.1	11.4	10.4	7.7	10.1	11.7	8.3	5.7	6.3	613
CD (p = 0.01)				10.7			11.5	7.9	8.8	851

Figures in parentheses are angular transformed values.

pyrethroids (1861, 2405 kg/ha) during 1991 and 1992 respectively. The yield of seed cotton in untreated control was only 424 kg/ha in 1991 and 896 kg/ha in 1992.

During 1992, in spray schedules T_6 and T_7, instead of neem tap water was sprayed to ascertain the potentiality of neem against bollworms and to find out whether, without use of neem in these schedules, only two rounds of synthetic pyrethroids would be effective in realizing good yield of seed cotton. It was observed that these schedules were also effective in reducing the incidence of bollworm and increasing the good quality yield of seed cotton. In both schedules, two sprays of synthetic pyrethroids and a final spray of neem product were given, which were also significantly superior to T_1, T_3, and untreated control, but these schedules were inferior to T_5 and T_4. The result shows that use of neem alone is not effective in increasing the yield of seed cotton, but its alternate application in combination with synthetic pyrethroids is effective in increasing good quality seed cotton. Rao et al. (1990) found that neem oil alone was less effective against *Helicoverpa armigera*, and consequently yield of seed cotton was also reduced in Andhra Pradesh. Samuthiruvelu and David (1990) found effective control of bollworms with neem in combination with deltamethrin.

Effect on Jassid

Though spray schedules were targeted to control bollworm complex their impacts were also noted on the intermittent population of jassid which is also a major pest of cotton and is responsible for limiting the yield. Jassid population was recorded regularly after 24 hrs, one week, and two weeks of spraying of different insecticides, and the results are presented in Tables 5 and 6. Data recorded after 24 hrs during 1991 showed significant variation in population of jassid among treatments (Table 5). It is clearly evident from the table that jassid nymphal population per 30 leaves after 24 hrs of first round of spray, i.e., on August, 9 was significantly low in plots treated with monocrotophos (1.5, 2.1) and deltaphos (1.5), whereas significant variation in neem-treated plots (29.2, 39.8) and untreated control (48.8) was not observed. After a second round of spraying, jassid population was significantly less in plots treated with deltamethrin (1.2) and endosulfan (1.5, 2.2). At this occasion also there was no significant variation in jassid population between neem-treated plots (26.2, 39.8, 21.2, 38.2) and untreated control plot (36.5). The same trend was observed in jassid population after 24 hrs of subsequent rounds of spraying. Table 5 shows that spray schedules in which neem was used as tank-mix with monocrotophos and cypermethrin (T_5, T_6, T_7) were effective in suppressing jassid population below prescribed economic threshold level. These schedules were also more potent against jassid than those schedules in which neem was used alone.

During 1992, jassid population was recorded after 24 hrs, one week, and two weeks of spraying to observe the long-term effect of different insecticides used at different stages of crop growth (Table 6). Data recorded on August 14, 24 hrs after application of first round of spraying, showed significant variation in jassid population among treatments. Significantly lower population was recorded in plots treated with monocrotophos (1.0) and deltaphos (1.7. 6.0), whereas neem failed to check jassid population even after 24 hrs of its application. The population in neem-treated plots was 40.7 and 36.7, which was at par with untreated control (42.3). Data recorded after 24 hrs of subsequent round of spraying revealed the same trend in variation in jassid

Table 5. Effect of spray schedules of different combination of insecticides on intermittent population of jassid after 24 hrs of spraying during 1991

Treatment		Jassid nymphal population/30 leaves				
		Aug. 9	Aug. 23	Sept. 11	Sept. 28	Oct. 11
T_1	Neemrich II	29.2	26.2	51.8	44.0	41.6
T_2	Neem 2%	39.8	39.8	28.2	49.3	34.0
T_3	Monocrotophos	1.5				
	Endosulfan		1.5			
	Quinalphos			4.8		
	Triazophos				15.0	
	Endosulfan					16.3
T_4	Monocrotophos	2.1				
	Deltamethrin		1.2			
	Quinalphos			2.8		
	Cypermethrin				13.0	
	Triazophos					9.0
T_5	Deltaphos	1.5				
	Neem 2%		21.2			
	Alphamethrin			2.5		
	Neemrich II				13.3	
	Neem 2% + cypermethrin					15.7
T_6	Pyraclofos	18.8				
	Neemrich I		38.2			
	Trebon			0.5		
	Neem 2% + cypermethrin				6.3	
	Trebon					7.7
T_7	Neem 2% + Monocrotophos	22.2				
	Endosulfan		2.2			
	Neemrich II			21.5		
	Triazophos				19.0	
	Neemrich I					39.3
T_8	Control	48.8	36.5	41.8	36.3	42.0
SEM ±		6.7	5.8	3.5	6.9	7.8
CD (p = 0.05)		20.3	17.7	10.6	21.0	23.6
CD (p = 0.01)			24.7	14.7	29.3	32.8

population. On all occasions, neem products failed to suppress jassid population below economic threshold level and were at par with untreated control (Table 6).

Observations recorded after one week of spraying showed the same trend in the effectiveness of insecticides against jassid. The effect of all the insecticides was also noted after two weeks of each round of spraying. Table 6 shows significant variation in jassid population among the treatments on two occasions, i.e. August 26 and September 9, and thereafter there was no significant variation. The minimum population of jassid was recorded from plots treated with monocrotophos (9.7), deltamethrin (14.3), and deltaphos (28.3) and these insecticides were only effective up to two weeks in reducing the population of jassid. Neem products were not effective in suppressing jassid population either after 24 hrs or after one week or two weeks of their application

Table 6. Effect of spray schedules of different combination of insecticides on intermittent population of jassid after spraying during 1992

Treatment		Jassid nymphal population/30 leaves														
		After 24 hrs of spraying					After 1 week of spraying					After 2 weeks of spraying				
		Aug. 14	Aug. 29	Sep. 11	Sept. 26	Oct. 10	Aug. 20	Sept. 4	Sept. 10	Oct. 1	Oct. 15	Aug. 26	Sept. 9	Sept. 24	Oct. 8	Oct. 22
T_1	Neem A_1	40.7	69.3	66.7	114.7	26.7	58.7	95.3	97.3	82.7	22.7	64.3	82.3	85.0	54.0	16.3
T_2	Neem A_2	36.7	72.0	73.7	131.3	34.3	53.7	107.3	78.0	89.7	32.7	78.0	77.3	101.7	58.0	37.7
T_3	Neemrich II	34.3					69.0					58.0				
	Dipel (BT)		75.0					89.0					68.7			
	Neemrich II			49.7					89.0					104.3		
	Dipel (BT)				127.0					75.7					51.3	
	Neemrich I					50.0					37.3					13.7
T_4	Monocrotophos	1.0					8.0					9.7				
	Deltamethrin		6.3					33.3					14.3			
	Quinalphos			10.0					68.3					80.7		
	Cypermethrin				44.3					55.3					44.7	
	Triazophos					10.7					19.7					17.0
T_5	Deltaphos	1.7					30.0					28.7				
	Neemrich II		39.7					66.0					60.0			
	Cypermethrin			17.0					76.3					50.7		
	Neemrich I				84.0					73.3					36.3	
	Triazophos					23.0					18.7					18.7
T_6	Deltaphos	6.0					39.0					61.7				
	Water		79.0					102.0					81.7			
	Cypermethrin			36.7					92.0					84.3		
	Water				129.7					87.0					65.3	
	Neemguard					48.3					48.7					25.7
T_7	Water	46.0					64.0					93.0				
	Deltamethrin		11.7					79.0					34.0			
	Water			40.3					63.3					118.3		
	Cypermethrin				43.0					60.7					54.3	
	Nimbecidine					37.7					25.3					27.7
T_8	Control	42.3	84.3	108.0	123.0	48.7	67.0	165.7	130.3	74.7	31.3	85.3	68.3	121.3	58.0	29.7
SEM ±		7.4	10.0	11.6	14.3	NS	8.2	13.7	10.1	NS	NS	8.9	10.1	NS	NS	NS
CD (P = 0.05)		22.4	30.3	32.4	43.5		24.8	41.5	30.6			27.1	30.6			
CD (P = 0.01)			42.1	44.9				57.6				37.7	42.5			

and were significantly at par with untreated control. Dhawan et al. (1992) observed that neemark 95 EC at 750 ml/ha was effective in reducing jassid population during vegetative phase under Punjab conditions, which differs from the finding of the author.

Effect on White-fly

The impact of insecticides of different spray schedules on the population buildup of white-fly was recorded after termination of all the five rounds of spraying targeted towards bollworm complex during 1991 and 1992. The spraying was terminated on October 10 and October 9 respectively during both years. White-fly populations recorded after 24 hrs of last round of spraying and thereafter at regular intervals of about a week are presented in Tables 7 and 8. During 1991, significant variation in white-

Table 7. Effect of spray schedules of different combination of insecticides on population buildup of white-fly after termination of last spray during 1991

Treatment	White-fly adult population/30 leaves				
	Oct. 11	Oct. 21	Oct. 28	Nov. 4	Nov. 11
T_1	60.3	105.0	112.3	102.0	60.0
T_2	48.7	84.0	95.7	58.0	37.6
T_3	30.7	89.0	157.0	86.3	59.3
T_4	54.0	98.3	159.3	92.7	44.0
T_5	88.0	150.6	222.7	111.3	58.7
T_6	29.3	55.3	122.7	62.7	49.7
T_7	55.0	76.3	169.3	78.0	46.7
T_8 (Control)	44.3	51.0	42.7	33.0	24.7
SEM ±	7.4	NS	26.4	NS	NS
CD ($p = 0.05$)	22.5		79.9		
CD ($p = 0.01$)					

Table 8. Effect of spray schedules of different combination of insecticides on population buildup of white-fly after termination of last spray during 1992

Treatment	White-fly adult population/30 leaves				
	Oct. 10	Oct. 15	Oct. 22	Oct. 28	Nov. 5
T_1	20.3	22.0	25.7	10.3	4.7
T_2	11.0	15.3	18.3	8.7	1.7
T_3	8.7	14.7	19.3	11.7	8.7
T_4	26.3	44.7	44.0	37.3	12.0
T_5	16.7	27.3	20.0	20.7	2.0
T_6	15.7	15.0	14.0	14.3	3.0
T_7	25.0	27.3	19.0	10.7	2.0
T_8 (control)	11.0	12.0	10.3	9.0	2.0
SEM ±	NS	4.6	5.8	2.9	2.0
CD ($p = 0.05$)		13.9	17.6	8.9	6.1
CD ($p = 0.01$)				12.4	

fly population was recorded on two occasions, October 11 and 28. Significantly higher population of white-fly was noted on October 11 in T_5 (88.0) as compared to untreated control (44.3). The population buildup was more in plots treated with neem 2 per cent and cypermethrin tank-mix as last spray; all other treatments were at par with untreated control. The white-fly population buildup was minimum in T_6 (29.3) and T_3 (30.7), in which a final spray of trebon and endosulfan respectively was given. There was heavy buildup in white-fly population on October 28, after 18 days of last round of spray. Significantly higher population of white-fly was recorded in all the plots treated with different insecticidal schedules except T_1 and T_2, in which repeated application of neem was given. The maximum population of white-fly was recorded in T_5 (222.7), in which neem 2 per cent was used as tank-mix with cypermethrin. Though white-fly population buildup in T_1 and T_2 was not significantly higher than in control, population was greater in those plots (112.3 and 95.7 respectively) than in untreated control (42.7).

During 1992, there was less buildup in white-fly population than in control mainly because synthetic pyrethroids were not used either alone or in tank-mix as last spray (Table 8). There was no significant variation in white-fly population after 24 hrs of termination of last spraying, but significant differences were noticed after a week. Table 8 shows lower population of white-fly in untreated control plots and higher population in plots treated with different insecticides. At all times maximum population was recorded in T_4 (44.7, 44.0, 37.3), in which two sprays of synthetic pyrethroids were applied alternately with three sprays of conventional insecticides, followed by T_5 (27.3, 20.0, 20.7), in which two sprays of synthetic pyrethroids were used in combination with two sprays of neem and one spray of conventional insecticide. White-fly population was also greater in plots treated with neem alone than in untreated control but the tendency for population to build up was less than in other treatments. Neem products thus were also not effective in reducing the population of white-fly under field conditions. Phadke et al. (1988) also observed less buildup in white-fly population in neem-treated plot than in fenvalerate-treated plot. They have observed that neemark 0.4 per cent was as effective as endosulfan against white-fly and bollworms, which differs with the present findings. The finding of Prabhakar et al. (1989) is in conformity with the present work that neem seed extract was least toxic to *Bemisia tabaci*. Deminik and Mohansundaram (1992) found neem seed kernel extract 5 per cent moderately effective against white-fly.

CONCLUSION

On the basis of three years of field trials, it is concluded that application of neem alone or in combination with conventional insecticides failed to check pest complex in cotton under field conditions. However, its combination with synthetic pyrethroids proved effective in managing pest complex in cotton and increasing good quality yield. Such a spray schedule was even more effective than recommended spray schedule, in which conventional insecticides are used in combination with synthetic pyrethroids and which is in vogue for the control of pest complex in cotton under north Indian conditions. The findings show that neem can replace conventional insecticides in a spray schedule and can effectively control the pest complex in cotton. It will also minimize the exposure of the environment to toxic pesticides and will postpone the development of resistance.

Neem products can also be mixed with synthetic insecticides or other by-products based on *Bacillus thuringiensis* to increase efficacy, if necessary. Fagoonee (1987) achieved improved performance by using a synthetic pyrethroid and neem alternately for the protection of vegetable crop from the attack of *Heliothis* in Mauritius. Schmutterer (1990) also recommended use of neem product alternately or mixed with other products in order to increase efficacy and to prevent or at least postpone the development of resistance.

ACKNOWLEDGEMENT

The author is thankful to Dr. Kirti Sharma for her help in compilation and calculation of data for preparation of the manuscript.

REFERENCES

Deminik, S.J. and M. Mohansundaram. 1992. Susceptibility of *Bemisia tabaci* Genn. (Hom: Aleyrodidae) to insecticides. *Pestology* 16(2): 5–7.

Dhawan, A.K., G.S. Srimannarayana, G.S. Simwat and K. Nagaiah. 1992. Management of cotton pests on upland cotton, *Gossypium hirsutum* with Navneem 95 per cent EC. *Proc. 1st Nat. Symp. Allelopathy in Agro-ecosystems (Agriculture and Forestry)*, HAU Hissar, 12–14 Feb. Seminar Paper, pp. 156–157.

Gupta, G.P. and R.A. Agarwal. 1985. Effect of insecticides for the control of bollworms in relation to phenological characters in cotton. *Pesticides* 19(10): 58-60.

Fagoonee, I. 1987. Use of neem in vegetable crop protection in Mauritius. *Natural Pesticides from the neem tree (Azadirachta indica* A. Juss) *and other tropical plants.* H. Schmutterer and K.R.S. Ascher eds., Eschborn, pp. 419–429.

Meisner, J., K.R.S. Ascher and R. Aly. 1981. The residual effect of some products of neem seeds on larvae of *Spodoptera littoralis* in laboratory and field trials. *Proc. 1st Int. Neem Conf.*, Rottach-Egern, pp. 157–170.

Meisner, J., M. Klein and S. Koren. 1990. Effect of Margosan-O on the development of *Earias insulana. Phytoparasitica* 18(4): 287–297.

Phadke, A.D., V.S. Khandal and S.R. Rahlkar. 1988. Use of neem product in insect resistance management (IRM) in cotton. *Pesticides* 22(4): 36–37.

Prabhakar, N., N.C. Toscano and D.L. Coudriet. 1989. Susceptibility of immature and adult stages of the sweet potato white-fly to selected insecticides. *J. Econ. Entomol.* 82(4): 983–988.

Rao, N.V., A.S. Reddy and P.S. Reddy. 1990. Relative efficacy of some new insecticides on insect pests of cotton. *Indian J. Plant Prot.* 18(1): 53–58.

Samuthiravelu and B.V. David. 1990. Bioefficacy of neem oil and deltamethrin against spotted bollworm *Earias vittella* (Fab.); Noctuidae, Lepidoptera on cotton (MCU-5). *Madras Agric. J.* 77(7–8): 294–298.

Schmutterer, H. 1990. Properties and potential of natural pesticides from the neem tree, *Azadirachta indica. Annu. Rev. Entomol.*, 35: 271–297.

Schmutterer, H. and K.R.S. Ascher (eds.). 1987. *Natural pesticides from the neem tree (Azadirachta indica* A. Juss.*) and other tropical plants*. GTZ, Eschborn, 703 pp.

Thangavel, P., T.R. Subramaniam and S. Parameshwaran. 1975. Efficacy of certain insecticides against the incidence of cotton bollworms. *Pesticides* 9(10): 37–38.

49

Field Evaluation of Some Botanical Insecticides Alone and in Combination with Other Insecticides for Management of Bollworm Complex on Cotton

ASHOK K. DHAWAN AND G.S. SIMWAT

Department of Entomology, Punjab Agricultural University
Ludhiana 141 004, India

ABSTRACT

Cotton crop in Punjab is attacked mainly by pink bollworm and spotted bollworms. Three botanical insecticides viz. RD-9 Repelin, Neemark, and Ind-Ne 20 EC were tested alone and in combination with conventional insecticide (quinalphos) and synthetic pyrethroid (cypermethrin) for the control of bollworm complex on cotton in three different experiments. Botanical insecticides when used alone, significantly reduced the bollworm damage and increased the seed cotton yield than control. All except RD-9 Repelin were significantly inferior to synthetic insecticides. RD-9 Repelin was as effective as quinalphos for reducing bollworm damage and increasing seed cotton yield. Botanical insecticides gave better bollworm control and higher yield when used in combination than quinalphos and cypermethrin alone. Mixing of these botanicals with conventional insecticides make them as effective as synthetic pyrethroids. These combinations can be substituted for synthetic pyrethroids for bollworm management.

INTRODUCTION

Cotton crop during the flowering phase in northern *hirsutum-arboreum* zone is attacked mainly by bollworm complex (*Pectinophora gossypiella* Saund.) and spotted bollworms (*Earias vittella* F. and *E. insulana* Boisd.) Sometimes American bollworm (*Heliothis armigera* Hubner) also does considerable damage. Dhawan et al. (1988) estimated a loss of 44.5 per cent due to bollworms in upland cotton. In the absence of resistant varieties and effective natural enemies, the main reliance is on the use of insecticides to reduce the losses due to bollworms. The excessive use of these insecticides has resulted in serious outbreak of white-fly and American bollworm especially in West and South India in the recent past (Jayaraj and Rangarajan, 1987). Many workers have observed neem-based formulations to be effective against cotton pests (Dhawan and Simwat, 1992; Lal et al., 1986; Natarajan and Sundramurthy, 1990; Subramanian, 1992; Uthamasamy and Gajendran, 1992). In the present studied some neem-based formulations were evaluated alone and in combination with other recommended insecticides for the control of bollworm complex of cotton during 1991.

MATERIALS AND METHODS

Three botanical insecticides—Neemark (West-Coast Herbochem, Bombay), RD-9 Repelin (ITC and ILTD, Guntur), and Ind-Ne 20EC (Indofill Chemicals, Bombay)—were tested alone and in combination with recommended organophosphatic insecticide (quinalphos 20EC at 2 l/ha) and synthetic pyrethroid (cypermethrin 25EC at 0.2 l/ha) in three different experiments on upland cotton *Gossypium hirsuthum* var. LH-900. Neemark and Ind-Ne are neem-based formulations while RD-9 Repelin contains products of neem (*Azadirachta indica*), karanjia (*Pongamia glabra*), and mahua (*Madhuca indica*). The crop was raised as per recommended agronomic practices. The three experiments were in randomized block design with three replications. In these experiments six sprays were done at 10-day intervals starting from the first week of August with manually operated knapsack sprayer using 250 l of water/ha.

In the first experiment, the treatments were: RD-9 Repelin at 2.50 and 3.75 l/ha; quinalphos alone at 2 l/ha and in combination with RD-9 Repelin at 2.50 l/ha; and cypermethrin alone at 0.2 l/ha and in combination with RD-9 Repelin at 2.5 l/ha. In the second experiment, Neemark was tested alone at 0.625 and 1.25 l/ha; quinalphos alone at 2 l/ha and in combination with Neemark at 0.625 and 1.25 l/ha; and cypermethrin alone at 0.2 l/ha and in combination with Neemark at 0.625 and 1.25 l/ha. In the third experiment, Ind-Ne 20EC was tested alone at 0.625 and 1.25 l/ha; quinalphos alone at 2 l/ha and in combination with Ind-Ne at 0.625 l/ha; and cypermethrin alone at 0.2 l/ha and in combination with Ind-Ne at 0.625 l/ha. The plot size in these three experiments was 150, 200, and 100 sq m, respectively. Incidence of pink and spotted and bollworms was recorded among the pickable bolls on boll and loculi basis. For this purpose 10 plants were tagged in the middle two rows of each plot after the last spray. All the opened bolls on these plants were collected and dissected to determine the bollworm damage. The shedding of floral forms due to bollworms was determined by marking a quadrat of 1.2 × 1.8 m. All the shed fruiting bodies from this quadrat were collected and dissected to determine the shedding due to bollworms. The yield of seed cotton was recorded on whole-plot basis.

The effect of different treatments was also observed on the non-target sucking pests, jassids (*Amrasca biguttula* Ishida) and white-fly (*Bemisia tabaci* Genn.); on the leaf eating caterpillars cotton leaf rollers (*Sylepta derogata* F.) green semi-looper (*Anomis flava* F.), and cotton grey weevil (*Myllocerous maculosis* Desbr); and on predators such as adults and larvae of Coccinellids (*Coccinella septumpunctata* Linn. and *Menochilus sexmaculatus* Fab.), predatory bugs (*Geocoris* and *Zelus* spp.), *Chrysopa*, and spiders.

The population counts of jassid nymphs and white-fly adults were taken from three fully opened leaves in the upper half of the 10 plants selected at random in the middle two rows, on the eighth day after each spray. The population of leaf-eating caterpillars and predators was also recorded from these 10 plants.

RESULTS AND DISCUSSION

RD-9 Repelin

The per cent shedding of floral forms due to bollworms and boll and loculi damage among treated plots varied from 1.4 to 8.6, 4.1 to 18.4, and 2.0 to 7.2 against 11.1, 28.4, and 16.8 in control. The seed-cotton yield in treated plots varied from 12.86 to 18.43 q/ha against 11.23 q/ha in control. The bollworm damage was significantly low and seed-cotton yield was high in all treated plots. The addition of RD-9 Repelin to quinalphos significantly lowered damage and raised yield as compared to quinalphos alone. The combination of RD-9 Repelin and cypermethrin did not significantly reduce the bollworm damage but seed cotton yield was increased significantly over cypermethrin alone. The RD-9 Repelin at 3.75 l/ha was as effective as quinalphos (Table 1). The population of jassid and white-fly was significantly low in all the treated plots. However, the population of leaf-eating caterpillar did not differ significantly in different plots. The predator population was significantly high in control. Among treated plots predator population was greater with RD-9 Repelin (Table 2).

Table 1. Field evaluation of RD-9 Repelin alone and in combination with other insecticides for bollworm control (1991)

Insecticide	Dose (l/ha)	Shedding of floral forms due to bollworms (%)	Boll damage (%)	Loculi damage (%)	Seed-cotton yield	
					q/ha	% increase over control
RD-9 Repelin + cypermethrin 25 EC	2.500 + 0.200	1.4 (6.87)	4.1 (11.74)	2.2 (8.45)	18.43	64.11
RD-9 Repelin + quinalphos 25 EC	2.500 + 2.000	3.2 (10.32)	7.2 (15.60)	2.5 (9.15)	15.96	42.12
Cypermethrin 25 EC	0.200	3.4 (10.62)	9.6 (18.08)	2.0 (8.06)	17.09	52.18
Quinalphos 25 EC	2.000	5.3 (13.26)	13.7 (21.73)	5.6 (13.64)	14.03	24.93
RD-9 Repelin	2.500	8.6 (17.07)	18.4 (25.40)	7.2 (15.52)	12.86	14.51
RD-9 Repelin	3.750	4.2 (11.83)	14.4 (22.27)	5.5 (13.52)	13.80	22.89
Control		11.1 (19.40)	28.4 (32.19)	16.8 (24.17)	11.23	
CD (p = 0.05)		(0.60)	(0.61)	(0.45)	0.51	

Figures in parentheses are angular transformations.

Table 2. Effect of RD-9 Repelin alone and in combination with other insecticides on population of non-target insect pests and predators (1991)

Insecticide	Dose (l/ha)	Jassid nymphs/ 30 leaves*	Whitefly adults/ 30 leaves*	Leaf-eating caterpillar/ 10 plants	Predators/ 10 plants**
RD-9 Repelin + cypermethrin 25 EC	2.500 + 0.200	15.68 (3.96)	12.96 (3.60)	1.34	3.53 (2.13)
RD-9 Repelin + Quinalphos 25 EC	2.500 + 2.000	11.15 (3.34)	16.00 (4.00)	1.66	1.99 (1.73)
Eypermethrin 25 EC	0.200	10.75 (3.28)	20.43 (4.52)	2.50	2.39 (1.84)
Quinalphos 25 EC	2.000	13.61 (3.69)	20.34 (4.51)	2.84	3.20 (2.05)
RD-9 Repelin	2.500	16.81 (4.10)	18.74 (4.33)	1.34	5.35 (2.52)
RD-9 Repelin	3.750	27.87 (5.28)	17.55 (4.19)	1.66	4.11 (2.26)
Control		49.28 (7.02)	76.73 (8.76)	6.67	10.02 (3.32)
CD (p = 0.05)		(1.28)	(1.57)	NS	(0.58)

* Figures in parentheses are $\sqrt{n}$ transformation

** Figures in parentheses are $\sqrt{n+1}$ transformation

Mean of four observations

NS : Not significant

Neemark

The shedding of floral forms and boll and loculi damage varied from 3.3 to 12.7, 3.6 to 17.1, and 1.6 to 9.2 per cent in treated plots against 12.8, 22.3, and 11.3 per cent in control, being significantly low in all the treated plots. The seed-cotton yield was also significantly high in treated plots. Among the latter, Neemark alone was significantly inferior to all other treatments. The addition of Neemark at 1.25 l/ha to quinalphos or cypermethrin was more effective than these insecticides alone. The quinalphos when used in combination with Neemark was as effective as cypermethrin alone (Table 3.) So the combination can be substituted for synthetic pyrethroid (cypermethrin), which are being used in excess and have resulted in resurgence of many secondary pests and also insecticide resistance especially of American bollworm (Jayaraj and Rangarajan, 1987). The population of jassid and white-fly was significantly lower in treated plots than in control. Among the treated plots the population was comparatively high in Neemark-treated plots. The predator population was significantly high in control and Neemark at 0.625 l/ha. The population of leaf-eating caterpillar was comparatively low in all the treated plots but the difference among various treatments was not significant (Table 4).

Table 3. Field evaluation of Neemark alone and in combination with other insecticides for bollworm control (1991)

Insecticides	Dose (l/ha)	Shedding of floral forms due to bollworms (%)	Boll damage (%)	Loculi damage (%)	Seed-cotton yield	
					q/ha	% increase over control
Neemark + quinalphos 25 EC	0.625 + 2.000	5.8 (13.95)	12.0 (20.26)	4.9 (12.84)	15.47	42.84
Neemark + quinalophos 25 EC	1.250 + 2.000	4.1 (11.68)	9.8 (18.24)	4.6 (12.44)	15.91	46.90
Neemark + cypermethrin 25 EC	0.625 + 0.200	4.9 (12.78)	6.7 (14.96)	3.1 (10.18)	16.98	56.79
Neemark + cypermethrin 25 EC	1.250 + 0.200	3.3 (10.49)	3.6 (10.88)	1.6 (7.39)	17.37	60.38
Quinalphos 25 EC	2.000	7.0 (15.32)	11.8 (20.08)	5.2 (13.23)	14.91	37.58
Cypermethrin 25 EC	0.200	5.2 (13.12)	8.5 (16.95)	4.0 (11.54)	15.90	46.81
Neemark	0.625	12.7 (20.87)	17.1 (24.46)	9.2 (17.65)	11.29	4.24
Neemark	1.250	9.9 (18.37)	14.4 (22.25)	7.6 (16.01)	11.75	8.49
Control		12.8 (20.95)	22.3 (28.20)	11.3 (19.60)	10.83	
CD ($p = 0.05$)		(0.65)	(3.28)	(1.84)	0.67	

Figures in parentheses are angular transformations

Ind-Ne

The boll and loculi damage among treated plots varied from 6.6 to 18.9 and 3.1 to 8.1 per cent against 28.3 and 13.4 per cent in control, being significantly less in treated plots. The shedding of floral forms also showed the same trend. The seed-cotton yield was significantly high in treated plots (12.37 to 16.88 q/ha.) Among treated plots Ind-Ne alone at 1.25 l/ha was as effective as quinalphos in reducing bollworm damage but yield was significantly low. The mixing of Ind-Ne with cypermethrin had no significant effect on bollworm damage but mixing with quinalphos significantly reduced the loculi damage. The addition of Ind-Ne to different synthetic insecticides significantly increased the seed-cotton yield (Table 5). The bollworm damage and seed-cotton yield did not differ significantly between quinalphos + Ind-Ne and cypermethrin alone. This indicates that this combination can be used as an alternative to synthetic pyrethroids. The population of different sucking pests, leaf-eating caterpillar, and predators was low even in untreated control.

Table 4. Effect of Neemark alone and in combination with other insecticides on population of non-target insect pests and predators (1991)

Insecticide	Dose (l/ha)	Jassid nymphs/ 30 leaves*	Whitefly adults/ 30 leaves*	Leaf-eating caterpillar/ 10 plants	Predators/ 10 plants**
Neemark + quinalphos 25 EC	0.625 + 2.000	7.37 (2.71)	21.62 (4.65)	1.66	1.34 (1.53)
Neemark + quinalphos 25 EC	1.250 + 2.000	5.47 (2.34)	14.44 (3.80)	1.33	3.00 (2.00)
Neemark + cypermethrin 25 EC	0.625 + 0.200	12.11 (3.48)	19.36 (4.40)	1.66	0.99 (1.41)
Neemark + cypermethrin 25 EC	1.250 + 0.200	5.90 (2.43)	18.66 (4.32)	1.00	1.99 (1.73)
Quinalphos 25 EC	2.000	10.56 (3.25)	20.34 (4.51)	2.02	6.34 (2.71)
Cypermethrin 25 EC	0.200	5.90 (2.43)	17.55 (4.19)	2.69	7.64 (2.94)
Neemark	0.625	29.59 (5.44)	42.25 (6.50)	4.60	12.03 (3.61)
Neemark	1.250	18.49 (4.30)	30.14 (5.49)	2.67	9.30 (3.21)
Control		55.20 (7.43)	74.13 (8.61)	6.67	13.67
CD (p = 0.05)		(1.15)	(1.15)	NS	(0.36)

* Figures in the parentheses are $\sqrt{n}$ transformation

** Figures in the parentheses are $\sqrt{n+1}$ transformation

Mean of four observations

NS : Not significant

CONCLUSIONS

The present studies indicate that botanical insecticides like RD-9 Repelin can be used for the management of bollworm complex on cotton. The use of insecticides in combination with botanicals will give better bollworm control and seed-cotton yield. The mixing of botanicals like Neemark and Ind-Ne with conventional insecticides like quinalphos make them as effective as synthetic pyrethroids. So these combinations can be used to decrease the use of synthetic pyrethroids.

Table 5. Field evaluation of Ind-Ne alone and in combination with other insecticides for bollworm control (1991)

Insecticide	Dose (l/ha)	Shedding of floral forms due to bollworms (%)	Boll damage (%)	Loculi damage (%)	Seed-cotton yield	
					q/ha	% increase over control
Ind-Ne 20 EC + cypermethrin 25 EC	0.625 + 0.200	1.4 (6.87)	6.6 (14.91)	3.1 (10.10)	16.88	48.85
Ind-Ne 20 EC + quinalphos 25 EC	0.625 + 2.000	2.0 (8.19)	11.2 (19.55)	3.8 (11.27)	15.98	43.32
Cypermethrin 25 EC	0.200	2.3 (8.65)	7.7 (16.13)	3.1 (10.12)	15.31	37.31
Quinalphos 25 EC	2.000	4.7 (12.43)	12.2 (20.46)	5.7 (13.87)	14.28	28.07
Ind-Ne 20 EC	0.625	7.5 (15.85)	18.9 (25.72)	8.1 (16.57)	12.37	10.94
Ind-Ne 20 EC	1.250	6.9 (15.28)	12.8 (20.92)	6.3 (14.55)	12.87	15.43
Control		8.8 (17.26)	28.3 (32.25)	13.4 (21.48)	11.15	
CD ($p = 0.05$)		(0.61)	(3.20)	(2.41)	0.52	

Figures in parentheses are angular transformations

REFERENCES

Dhawan, A.K. and G.S. Simwat. 1992. Field evaluation Neemrich 20EC for management of insect pest on cotton during reproductive phase. *Proc. 1st Nat. Symp. Allelopathy in Agroecosystems*. Eds. P. Tauro and S.S. Narwal. Indian Society of Allelopathy, Hissar, pp. 152–153.

Dhawan, A.K., G.S. Simwat and A.S. Sidhu. 1988. Assessment of avoidable loss in cotton due to sucking pests and bollworms. *Indian J. Agric. Sci.* 58: 290–292.

Jayaraj, S. and A.V. Rangarajan. 1987. Trends in insect control pyrethroids. *Pesticides* 21(8): 1–16.

Lal, S.S., S.P. Yadav and J.P. Sachan. 1986. Strategies for development of an integrated approach to control gram pod borer infesting okra. *Pesticides* 20(5): 39–51.

Mehrotra, K.N. 1991. Current status of pesticides resistance in insect pest in India. *J. Insect Science* 4(1): 1–14.

Natarajan, K. and V.T. Sundramurthy. 1990. Effect of neem oil on cotton white-fly. *Indian J. Agric. Sci.* 60(4) 290–291.

Subramanian, T.S. 1992. Neem and RD-9 Repelin in pest control. *Proc. Nat. Sem. on Changing Scenario in Pests and Pest Management in India.* CPPTI, Hyderabad, Jan. 31–Feb. 1. 15 pp.

Uthamasamy, S. and G. Gajendran. 1992. Efficacy of neem products in management of cotton pest. *Proc. Nat. Sem. on Changing Scenario in Pests and Pest Management in India*, CPPTI, Hyderabad, Jan. 31–Feb. 1. 45 pp.

50

Evaluation of Botanical Insecticides Alone and in Combination with Carbaryl against Insect Pest Complex of Okra

S. VENKATESWARA RAO AND B. ROSAIAH

Andhra Pradesh Agricultural University, Regional Agricultural Research Station
Lam, Guntur 522 034, India

ABSTRACT

Botanical insecticides such as Neemguard 0.5 per cent, Repelin 0.5 per cent, neem seed kernel extract (NSKE) 5 per cent, and nicotine sulphate 0.5 per cent were evaluated in the field individually and in combination with carbaryl 0.15 per cent against the major pest complex of okra. Nicotine sulphate alone or in combination with carbaryl 0.15 per cent was effective in suppressing jassids. Similarly, Repelin followed by nicotine sulphate with and without carbaryl was equally effective in reducing aphid populations and fruit borer incidence due to *Heliothis*. Significant seed yields were realized with the use of nicotine sulphate and Repelin both with and without carbaryl, carbaryl alone, Neemguard + carbaryl, and NSKE + carbaryl. The botanical insecticides did not exert any adverse effect on the natural beneficial predatory fauna such as coccinellids, syrphids, the spiders and hence can be safely used for the management of the key pest problems on okra.

INTRODUCTION

Okra (*Abelmoschus esculentus*), the most popular vegetable crop in India is subject to the ravages of the sucking pest complex of jassids, aphids, mites and fruit borers, resulting in economic yield loss to the growers.

Overuse and misuse of highly toxic, persistent, and broad spectrum insecticides at short intervals is a common practice among vegetables growers. The fruits harvested either every day or on alternate days contain insecticide residues over and above the tolerance limits (Narasimha Rao, 1990) posing unwarranted health hazards to the consumer. The use of safer insecticides of botanical origin to contain the pest problems of okra will help to overcome the problems related to harmful residues, environmental pollution, the conservation of beneficial insects.

MATERIALS AND METHODS

The botanicals Neemguard, neem seed kernel extract (NSKE), Repelin, and nicotine sulphate were evaluated alone and in combination with carbaryl against the pest complex of okra at the Regional Agricultural Research Station, Lam, during *kharif* season 1991–92. The field testing was done in a randomized block design with 10 treatments replicated three times using Parbhani Kranthi as the test variety. A plot size of 20 m^2 per treatment and the normal agrotechniques were adopted. Five foliar spray applications of the test treatments were given at 10-day intervals. The incidence of jassids, aphids, and *Helicoverpa armigera* fruit borer and the seed yield were recorded. The data analysed statistically are presented in Table 1.

RESULTS AND DISCUSSION

The results of the investigation revealed significant variations among the test treatments in respect of the pest, natural enemy complex, and seed yield parameters.

Effect of Treatments on Pest Complex and Yield

Foliar application of nicotine sulphate either in combination with carbaryl (2.56/5 plants) or alone (3.03/5 plants) was either at par or superior to untreated check (4.0), Neemguard (3.34), Repelin (4.84), and NSKE (3.94) with and without carbaryl in keeping the jassid populations to a minimum (Table 1).

Nicotine sulphate followed by Repelin with and without carbaryl (4.5 to 6.0/5 plants) and Neemguard with carbaryl (6.6) were superior to the remaining treatments (7.96 to 10.1/5 plants) and check (10.0) in reducing aphid populations to a minimum (Table 1).

Helicoverpa armigera that occurred as fruit borer was effectively checked with the use of Repelin + carbaryl (13.48 per cent), Repelin alone (16.97 per cent), nicotine sulphate (18.32 per cent), and nicotine sulphate + carbaryl 19.39 per cent. These treatments were at par and superior to untreated control (27.88 per cent), Neemguard (24.05 per cent), and other treatments (21.60 to 22.74 per cent) (Table 1).

A maximum seed yield of 1510 kg/ha was realized with nicotine sulphate + carbaryl and it was at par with Repelin + Carbaryl (1300 kg/ha), nicotine sulphate (1245 kg/ha), carbaryl (1230 kg/ha), Repelin (1145 kg/ha), Neemguard + carbaryl (1125 kg/ha), and NSKE + carbaryl (1105 kg/ha) but superior to Neemguard alone (925 kg/ha), NSKE alone (940 kg/ha), and control (525 kg/ha) (Table 1).

Effect of Botanicals on Predators

Foliar spray applications with botanical protectants did not exert any adverse effect on natural occurrence of predatory populations and contributed to the biological control of the pest.

Table 1. Effect of botanical insecticides and carbaryl on the pest complex of okra and associated natural enemies

Treatment	Conc. dose (%)	Population/5 plants*		*Helicoverpa* damage**(%)	Seed yield (kg/ha)	Natural enemy population/5 plants*		
		Jassids	Aphids			Spiders	Syrphids	Coccinellids
Neemguard	0.5	11.2 (3.34)	103.3 (10.12)	16.6 (24.05)	925	0.7 (0.83)	2.0 (1.50)	12.3 (3.33)
Repelin	0.5	22.8 (4.84)	36.1 (6.02)	8.7 (16.97)	1145	0.3 (0.73)	1.0 (1.20)	12.0 (3.43)
NSKE	5.0	15.3 (3.94)	65.8 (8.33)	15.0 (22.74)	940	2.0 (1.50)	1.0 (1.20)	11.7 (3.20)
Nicotine sulphate	0.5	8.7 (3.03)	25.0 (4.91)	10.5 (18.32)	1245	0.7 (0.83)	1.0 (1.20)	9.0 (2.93)
Carbaryl	0.15	13.3 (3.67)	98.1 (9.88)	15.0 (22.68)	1230	2.0 (1.50)	5.3 (2.63)	0.6 (1.40)
Neemguard + Carbaryl	0.5 0.15	12.6 (3.58)	58.1 (7.46)	14.7 (22.05)	1125	1.0 (1.20)	3.0 (1.77)	13.7 (3.70)
Repelin + Carbaryl	0.5 0.15	12.9 (3.63)	33.3 (5.33)	5.6 (13.48)	1300	0.3 (0.73)	1.0 (1:20)	3.7 (1.90)
NSKE + Carbaryl	5.0 0.15	11.9 (3.52)	68.1 (7.96)	13.7 (21.60)	1105	0.7 (0.83)	4.7 (2.13)	1.3 (1.33)
Nicotine sulphate + Carbaryl	0.5 0.15	6.3 (2.56)	20.4 (4.50)	11.1 (19.39)	1510	1.7 (1.48)	1.3 (1.34)	11.3 (3.10)
Control		14.2 (4.08)	99.9 (10.02)	21.9 (27.85)	525	1.3 (1.17)	1.7 (1.44)	19.7 (4.20)
CD (p = 0.05)		(0.64)	(3.12)	(6.37)	(445)	(NS)	(0.35)	(1.95)

*Figures in parentheses are means of $\sqrt{x + 0.5}$ values
**Figures in parentheses are arcsin values

Untreated check (4.2°/5 plants) followed by Repelin (3.43), Neemguard (3.33), NSKE (3.20), and nicotine sulphate (2.93) alone and with carbaryl (3.10) conserved the coccinellid predators significantly in maximum numbers (Table 1).

The activity of the syrphid predators was significantly high in carbaryl treatment (2.63/5 plants) followed by NSKE + carbaryl (2.13) and Neemguard + carbaryl (1.77) compared to the remaining treatments (1.20 to 1.44/5 plants) (Table 1).

In predatory spiders significant variations between the treatments could not be observed. However, the spider population was high with NSKE (2.0/5 plants) and carbaryl (2.0), followed by nicotine sulphate + carbaryl (1.7) and control (1.3) as against 0.3 to 1.05/5 plants in the remaining treatments (Table 1).

Evidently the predatory populations of coccinellids, syrphids, and spiders were conserved and their populations prevailed at significant levels with the use of botanical protectants and played an active role in the natural suppression of the injurious insect pests on okra crop.

These results are in contradiction to the observations made by Singh et al. (1985) and Bhamburkar (1989), who reported beneficial effects with the use of NSKE in reducing the incidence of *Heliothis* on cotton and red gram crops respectively. However, the use of NSKE alone exerted beneficial effect in conserving the spider populations, while NSKE with carbaryl preserved syrphids and reduced pest incidence of jassid, aphid, and *Heliothis* pests to a moderate level, besides improving seed yields.

The beneficial effects achieved with the use of Repelin and nicotine sulphate alone and in combination with carbaryl in reducing the *Heliothis* incidence are in consonance with the observations made earlier at the Regional Agricultural Research Station, Lam (Anon., 1991–92).

Similarly, results reported by Raman et al. (1992) on antifeedant behavioural aspects of *Earias vitella* and *H. armigera* larvae on okra fruits with the use of neem insecticides and by Rajasekharan and Jayaraj (1990) on the inhibition of larval feeding in *H. armigera* with sprays of neem bitters support the present findings in respect of *Heliothis* on okra.

These results suggest that the botanical protectants, nicotine sulphate and Repelin either alone or in combination with carbaryl could best be used for the management of the insect pest problems on okra as they are safe, effective and promising in conserving coccinellids, syrphids, and spiders which suppress the pests.

REFERENCES

Anonymous. 1992. Annual Report. *Heliothis* Management Scheme. Regional Agricultural Research Station. Lam.

Bhamburkar, M.W. 1989. Studies on the antifeedant and insecticidal properties of various plant extracts for the control of bollworms in cotton. *Proc. Nat. Sem. on Futuristic Approaches in Cotton Improvement.* 42 pp.

Rajasekharan, B. and S. Jayaraj. 1990. Use of plant products in the management of *Heliothis armigera. Proc. Nat. Workshop on Heliothis Management*, Coimbatore, pp. 230–236.

Raman, K., B.N. Vyas and S. Ganesan, 1992. Pest control using neem antifeedent. *Proc. Nat. Sem. on Changing Scenario in Pest and Pest Management in India,* 62 pp.

Narasimha Rao, B. 1990. Pesticide contamination. Bulletin by A.P. Agricultural University, Hyderabad, pp. 1–18.

Singh, R.P., Yasbhir Singh and S.P. Singh. 1985. Field evaluation of neem (*Azadirachta indica*) seed kernel extracts against the pod borers of pigeon pea *Cajanus cajan. Indian J. Entomol.* 47(1): 111–112.

51

Evaluation of Neem Products against Insect Pests of Rice and Pulse Crops

B.C. SHUKLA, U.K. KAUSHIK AND R. GUPTA

I.G.A.U., Raipur 492 012 (M.P.), India

ABSTRACT

Rice, the major food crop of eastern Madhya Pradesh, is subjected to low productivity due to many constraints and insect pests are one of them. Gall midge and leaf and plant hoppers are major insects pests found reducing farmers harvests of the region. Among pulse crops pigeonpea and chickpea suffer damage mainly by pod borers up to 40–50 per cent on pod and grain basis. Farmers depend on conventional synthetic insecticides to protect their crops from ravages of insect pests without knowing their ill-effects on ecosystems. With the increasing hazards of these chemicals, the need to search and use materials environmentally safe, socially acceptable and economically viable has now been increased. Neem based insecticides were tested against rice, pigeonpea and chickpea insect pests. The field trials were conducted with neem oil (NO 3 per cent), neem seed kernel extract (NSKE 5 per cent) along with conventional insecticides moncrotophos, endosulfan and cypermethrin. The population/damage was assessed after need based HV applications. The population of natural enemies (NE) like spiders was also sampled to assess effects of these chemicals. The neem reduced population/damage in rice as well as in pulse crops but lesser than the conventional insecticides which were deleterious to spiders. Grain yield in rice was 13.71 and 13.43 per cent more in NSKE and NO, while in pigeonpea it was 26.29 and 4.29 per cent and chickpea 12.55 and 6.61 per cent more, respectively over control.

INTRODUCTION

Rice is the major food crop of eastern Madhya Pradesh (Area 3.92 mha; Production 3.5 mt). It is subjected to low productivity (1.16 t/ha) due to many bio-physical and socio-economical constraints (Shukla, 1988). Among bio-physicals, insects pests are important one. Gall midge and leaf and plant hoppers are major insect pests found reducing farmer's harvests of the area. Under rice based cropping systems, pulse crops pigeonpea and chickpea are occupying about 50 and 100 thousand hectares in this region. Both the crops suffer damage mainly by pod borers up to 40–50 per cent on pod and grain basis (Shukla 1990). Farmers depend on conventional synthetic insecticides to protect their crops from insect pest without knowing their ill-effects due to illiteracy and unawareness. In India, continuous attempts are being made to generate information on different aspects of plant protection technology leading to integrated pest management in different states. In view of several drawbacks of synthetic chemicals, there is a growing interest among the entomologists world over to search for alternatives which would minimise the adverse effects on the ecosystems (Kalode and Krishnaiah, 1987), including "safe guard" role of natural enemies against major pests (Lamb and Saxena 1988; Shukla et al., 1987; Shukla et al., 1991).

The use of plant derivatives especially of neem *Azadirachta indica* A. Juss., is gaining importance as an innovative method of pest management in recent years. The neem products used for insect control include dried leaves, neem oil, cake, aqueous and organic solvent extracts from neem kernels etc. (Jayraj, 1991). Effects of neem against the rice and pulse pests has successfully been demonstrated by several workers (Saxena et al., 1981; Samalo et al., 1987; Gupta et al., 1988) and they reported that the insects were found to feed far less, grow poorly, lay fewer eggs and to have lower longevity. Out of the seeds of 14 million neem trees in India only 25 per cent are collected for industrial use (Ketkar, 1976) and negligible quantity is used as pesticides. In country like India where neem is available in plenty, its products like neem oil and neem seed kernel extracts can easily and successfully be utilized for pest control as a low cost technology for the poor farmers. Keeping these in views field studies were conducted on the use of neem products against insect pests of rice, pigeonpea and chickpea at Indira Gandhi Agricultural University, Raipur (M.P.) India.

MATERIALS AND METHODS

Three separate field trials were conducted on rice and pigeonpea during wet season and on chickpea during dry season 1991–92.

Rice: The trial on Kranti rice transplanted on Aug. 1991 with 30 days old seedlings in 8 × 5 m plots, was carried out in randomised block design (RBD) with following treatments replicated 5 times.

1) Neem Oil (NO) 3 per cent + 0.1 per cent Teepol
2) Neem Seed Kernel Extract (NSKE) 5 per cent + 0.1 per cent Teepol
3) Monocrotophos 0.5 kg a.i./ha
4) Untreated Control

The treatments were applied 3 times, first at 10 days after transplanting (DT) for gall midge (GM) and the other two on need basis. Insect population of green leafhopper (GLH), white backed planthopper (WBPH) and spiders on 10 random hills/plot and GM damage in terms of per cent silver shoots (per cent ss) out of total tillers in 20 random hills/plot at different stages along with grain yield/plot at harvest were recorded (Table 1).

Pigeonpea: This trial was conducted under RBD with cultivar NO. 148 sown on July 2, 1991 in 5 × 4 m plots replicated four times with following treatments:

1. Cypermethrin 0.02%
2. Endosulfan 0.07%
3. Monocrotophos 0.05%
4. Neem seed kernel extract 5%
5. Neemark 2%
6. Neem oil 3%
7. Untreated control —

The treatments were applied thrice at 20 days interval starting from 15 days after flowering initiation. Observations on pod and grain damage were recorded along with grain yield of five random plants/plot (Table 2).

Chickpea: This trial on chickpea (Variety JG 74) was sown on December 17, 1991 after rice in 5 × 4 m plots in RBD. The following treatments replicated four times were applied twice at 15 days interval starting from 15 days after flowering initiation :

Table 1. Field efficacy of neem seed kernel extract (NSKE), neem oil (NO) and monocrotophos on insect pests, predators and yield of rice (Kranti), IGAU, Raipur (India) WS 1991

Treatment	WBPH	GLH	Spider	LF	GM		Yield kg/ha	Increase over control
	(No./10 hills)			(%)	(%)			
	46 DT	45 DT	45 DT	65 DT	35 DT	65 DT		
NSKE 5%	15.4 b	10.8 b	4.8 b	9.5 a	12.6 a	14.7 a	3549 a	13.71
NO 3%	13.2 b	8.4 ab	4.4 ab	9.5 a	11.3 a	13.9 a	3540 a	13.43
Monocrotophos 0.5 kg a.i./ha	5.0 a	4.0 a	3.4 a	7.4 a	12.8 a	14.4 a	3592 a	15.09
Untreated control	28.4 a	22.6 c	7.8 c	12.4 b	18.0 b	17.2 b	3121 b	—

1. Mean of 5 replications
2. In a column means followed by same letters are not significantly different

Table 2. Efficacy of NSKE, NO and synthetic insecticides against borer complex and yield of pigeonpea (No. 148), IGAU, Raipur (India) WS 1991–92

Treatment	Damage		Yield (kg/ha)	Increase over control (%)
	Pod	Grain		
Cypermethrin	22.02 a	9.97 a	4626 a	47.04
Endosulfan	27.12 b	11.85 ab	4420 ab	40.50
Monocrotophos	28.77 b	13.30 b	4203 ab	33.60
NSKE	40.97 c	17.15 c	3973 bc	26.29
Neemark	41.30 c	17.02 c	3563 cd	13.25
Neem oil	41.80 c	17.47 c	3300 d	4.90
Control (Untreated)	57.05 d	25.50 d	3146 d	—

Mean of 4 replications.
Means followed by same letters are not significantly different

1.	NSKE	5%
2.	NO	3%
3.	Neemark	3%
4.	Endosulfan	0.07%
5.	Cypermethrin	0.002%
6.	NPV (Raipur-R)	250 LE (Larval equalent) (Prepared at Raipur)
7.	NPV (Hyderabad-H)	250 LE (Larval equalent) (Prepared at Hyderabad)
8.	Untreated Control	—

Observations on larval population and pod damage on 10 random plants/plot were recorded. The grain yield at harvest was also recorded (Table 3).

In all the trials the data were analysed following suitable standard statistical method suggested for RBD (Steel and Torrie, 1960).

RESULTS AND DISCUSSION

Rice

The results (Table 1) reveal that the population of GLH and WBPH was significantly lower in treated plots, however, it was lowest (5.0 and 4.0 per 10 hills) in monocrotophos. The population in NSKE and NO plots was on par. Kalode and Krishnaiah (1987) also observed reduced survival of BPH nymphs in NO treated plots. Similarly, Saxena and Barrion (1987) found disrupted growth of plant and leafhoppers on NS bitters solution treated plants. Baqui et al. (1988) from Bangladesh reported lower infestation of GLH in NC + NO treated plots at BRRI Farm and in NO treated plots at farmers field. The spiders population varied from 3.4 to 7.8 in different treatments and was least (3.4/10 hills) in monocrotophos plots. In neem treated plots, the high population on par with that of untreated plots prove them to be safer to this predator. Chiu Shin Foon (1984) from China and Jayaraj (1991) from India reported that there was no difference in population of spiders in rice fields between neem or other botanicals and the control

Table 3. Effect of neem and other non conventional insecticides on pod borer and grain yield of chickpea (JG 74), IGAU, Raipur (India) DS 1991–92

Treatment	Larvae/ 10 plants	Damage Pod (%)		Yield (Kg/ha)	Increase over control (%)
		I	II		
NSKE 5%	2.45 bc	1.13 b	1.06 ab	1874 a	12.55
NO 3%	2.77 bc	1.53 ab	1.11 ab	1775 c	6.61
Neemark 3%	2.50 bc	1.03 a	1.43 bc	1808 c	8.59
Endosulfan 0.07%	1.09 a	1.14 ab	1.06 ab	1775 c	6.61
Cypermethrin 0.002%	1.22 a	1.25 ab	1.01 ab	1940 a	16.52
NPV (R) 250 LE	2.41 b	1.33 ab	0.78 a	1777 c	6.73
NPV (H) 250 LE	2.77 bc	1.32 ab	0.96 a	1852 b	11.23
Untreated control	3.00 c	2.20 c	1.76 c	1665 d	—

Source (R) - Raipur; (H) - Hyderabad

Mean of 4 replications. In a column means followed by same letters are not significantly different

and that these plant products were safer to these potential predators of rice insect pests. At the moderate level of infestation (11.3 to 18.0 per cent SS) the GM damage as compared to untreated control was less in treated plots. In NO 31.7 per cent and in NSKE 30 per cent less as were recorded (Table 1). Samalo and Dash (1988) also recorded 35 per cent less gall midge in NO combined with chlorpyriphos or monocrotophos. Similarly leaf-folder damage was also lowest (7.47 per cent) in monocrotophos followed by NSKE and NO but were on par.

The grain yield (3540 to 3592 kg/ha) was significantly higher in treated plots (Table 1) and there was 13.71, 13.43 and 15.09 per cent increase over control in NSKE, NO and monocrotophos, respectively. Thus, neem available at door-steps of poor farmers can be profitable low cost technique in IPM in rice sparing natural enemies of insect pests.

Pigeonpea

The results presented in Table 2 indicated that there was significant variation in pod and grain damage by the pod borer complex (including *Heliothis, Exelastis* and *Melanagromyza*) of pigeonpea with the treatments. The lowest damage was in cypermethrin (22.02 per cent pod and 9.97 per cent grain) followed by endosulfan and monocrotophos. Among neem products, plots treated with NSKE had the lowest (40.97 per cent pod) damage but it was on par with NO and Neemark. Grain damage was also on par in all the three treatments. Similarly, Abdul Kareem (1978a) found NSKE to be effective against pod borers of pigeonpea. Shrivastava et al. (1984) also reported that NSKE 8 per cent significantly reduced the damage by *Melangromyza obtusa* and *Heliothis armigera* on pigeonpea. The grain yield was highest in cypermethrin (4626 kg/ha) followed by endosulfan. Among neem products the yield was on par in NSKE 5 per cent, Neemark 3 per cent and NO 3 per cent, however, it was 26.29, 13.25 and 4.90 per cent higher over control in NSKE, Neemark and NO respectively. Thus, neem products can effectively be used for the control of pod borers of pigeonpea, also and higher yields can be obtained.

Chickpea

The results on pod borer *Heliothis armigera* population/damage and grain yield presented in Table 3 reveal that the population was very moderate but the conventional synthetic insecticides controlled it at early stage of crop growth. Later on, the damage was on par with other neem products and NPV. The grain yield was the highest (1940 kg/ha) in cypermethrin which was 16.52 per cent higher over control. Among neem products NSKE 5 per cent recorded highest (1874 kg/ha) grain yield followed by Neemark (1808 kg/ha) and NO (1775 kg/ha) and these were 12.55, 8.59 and 6.61 per cent higher over that in untreated control. Abdul Kareem (1978 b) recorded NSKE effective against gram pod borer in chickpea while efficacy of NSKE and NO against *Heliothis armigera* was also reported through laboratory tests by Rao and Shrivastava (1985) and Gupta et al. (1988). Sundra Babu and Rajsekaran (1984) found spraying of neem oil at 3 and 5 per cent effective for reduction of pod borer (*Heliothis armigera*) damage in chickpea.

SUMMARY

The field trial studies reveal the effectiveness of neem products like NSKE 5 per cent and NO 3 per cent against insect pests of rice, pigeonpea and chickpea sparing the potent predator-spiders in rice fields and giving higher yields, although, these products were not as effective as synthetic insecticide. These studies suggested neem products to be pest effective, safer to predators, economically viable and socially acceptable, especially, by the poor farmers of the region.

ACKNOWLEDGEMENT

The authors are grateful to the authorities of Indira Gandhi Agricultural University, Raipur for providing facilities for these studies.

REFERENCES

Abdul Kareem, A. 1978a. Control of pod borers in red gram, cowpea and lablab. Kharif Report, 1978. All India Coordinated Project for Intensification of Research on the Improvement of Pulses. TNAU, Coimbatore, India.

Abdul Kareem, A. 1978b. Efficacy of antifeedants in the control of pod borers. Rabi Report 1978, Ibid.

Baqui, M.A., N. Quraishi and R. Karim. 1988. Efficacy of neem derivatives against some insect pests of rice in Bangladesh. Paper presented in the *Final Workshop of IRRI-ADB-EWC Project on Botanical Pest Control in Rice Based Cropping Systems*. IRRI, Dec. 12–16, 1988, Los Banos, Laguna, Philippines.

Chiu Shin Foon. 1984. Recent research findings in meliaceae and other promising botanical insecticides in China. In: *Research Planning Workshop on Botanical Pest Control Project*, IRRI, Los Banos, Laguna, Philippines. 6–10

Gupta, R., B.C. Shukla, S.K. Srivastava and A.D. Gupta. 1988. Efficacy of neem oil against *Heliothis armigera* Hubner. Paper presented at the *Final Workshop of IRRI-ADB-EWC Project*, op. cit.

Heyde, J.V.D., R.C. Saxena and H. Schmutterer. 1984. Neem oil and neem extracts as potential insecticides for control of hemipterous rice pests. *Proc. 2nd Int. Neem Conf.,* Rauischholzhausen 1983, pp. 377–390.

Jayraj, S. 1991. An overview of botanical pest control research at TNGDNAU in Tamil Nadu, India. *Proc. of the Midterm Review Meeting of Botanical Pest Control Project*, Phase II held at Dhaka, Bangladesh, IRRI, Philippines, 1991, pp. 36–38.

Kalode, M.B. and N. Krishnaiah. 1987. Evaluation of botanicals against rice pests. *Proc. of the Midterm Review Workshop on Botanical Pest Control in Rice Based Cropping Systems,* TNAU, Coimbatore, India. IRRI, Philippines. 48 pp.

Ketkar, C.M. 1976. *Utilization of neem (Azadirachta indica* A. Juss) *and its by products*. Directorate of Non-edible oils and soap industry, Khadi and Village Industries Commission, Bombay, India.

Lamb, R. and R.C. Saxena. 1988. Effects of neem seed derivatives on rice leaf folders (Lepidoptera: Pyralidae) and their natural enemies. *Final Workshop of IRRI-ADB-EWC Project*, op. cit.

Rao, S.R.V. and K.P. Shrivastava. 1985. Relative efficacy of neem formulations against gram pod borer. *Neem Newsletter* 2: 28–29.

Samalo, A.P., N. Panda and B.K. Sontakke. 1987. Effect of neem cake based nitrogen application on the incidence of rice gall midge and yield. *Neem Newsletter* 4(4): 41–43.

Samalo, A.P. and A.C. Dash. 1988. Phytoproducts on pest control in Rice Based Cropping Systems. *Final Workshop of IRRI-ADB-EWC Project*, op. cit.

Saxena, R.C. and A.A. Barrion. 1987. Cytogenetic effects of neem seed bitters (NSB) on green leafhoppers (GLH) males. *Int. Rice Res. Newsl.* 12(5): 25–6.

Saxena, R.C., N.J. Liquido and H.D. Justo Jr. 1981. Neem seed oil, a potential antifeedant for the control of the rice brown planthopper *Nilaparvata lugens*. Proc. *1st Int. Neem Conf.* Rottach-Egern. 1980, pp. 171–188.

Saxena, R.C., P.B. Epino, Tu Cheng-Wen and B.C. Puma. 1984. Neem, chinaberry and custard apple; antifeedant and insecticidal effects of seed oils on leafhopper and planthopper pests of rice. *Proc. 2nd Int. Neem Conf.*, op. cit, pp. 403–412.

Schmutterer, H. 1988. Potential of azadirachtin containing pesticides for integrated pest control in developing and industrialized countries. *J. Insect Physiol.* 34: 713–719.

Shrivastava, K.P., N.P. Agnihotri, V.T. Gajbhiye and H.K. Jain. 1984. Relative efficacy of fenvalerate, quinalphos and neem kernel extracts for control of pod fly, *Melanagromyza obtusa* and pod borer *Heliothis armigera* infesting red gram *Cajanus cajan* together with their residues. *J. Ent. Res.* 8: 1–4.

Shukla, B.C. 1990. Insect pests problem in the rice based cropping systems in Chhatisgarh (M.P.) India. *IRRI-ADB Res. Planning Workshop on Botanical Pest Control*, TNAU, Coimbatore March, 26–29, 1990 (Mimeo). 13 pp.

Shukla, B.C., S.K. Shrivastava and R. Gupta. 1987. Evaluation of NO against WBPH and gall midge pests of rice. *Proc. Midterm Review Workshop on Botanical Pest Control in Rice Based Cropping Systems* held at TNAU, Coimbatore, India, June 1–4, 1987, IRRI Philippines. 48 pp.

Shukla, B.C., U.K. Kaushik and R. Gupta. 1991. Field evaluation of botanicals for management of rice insect pests. *Proc. Midterm Review Workshop on Botanical Pest Control Phase II*, Bangladesh, 28–31 July, 1991, IRRI, Philippines, 110 pp.

Shukla, V.P. 1988. Perspectives for Chhattisgarh Agriculture Development. *Proc. Symposium on Chhattisgarh Agril. Systems Accomplishments, Present Status and Future Strategies*, IGAU, Raipur Jan. 20–23, 1988. I.G. Agril. Uni. Raipur (M.P.) India. 207 pp.

Steel, R.G.D. and J.H. Torrie. 1960. *Principles and Procedures of Statistics.* Mc-Graw Hill Book Co, New York, 481 pp.

Sundara Babu, P.C. and B. Rajsekaran. 1984. Evaluation of certain synthetic pyrethroids and vegetable products for the control of Bengal gram pod borer *Heliothis armigera*. *Pesticides* 18: 58–59.

52

Efficacy of Neem and *Melia* in Overcoming Tolerance of *Plutella xylostella* to Some Commonly Used Synthetic Insecticides

S. FACKNATH

Faculty of Agriculture, University of Mauritius, Reduit, Mauritius

ABSTRACT

Extracts of neem (*Azadirachta indica* A. Juss.) were found to be successful in breaking down the resistance developed by *Plutella xylostella* to some commonly used pyrethroids, namely deltamethrin and cypermethrin, and to have a synergistic effect with the organophosphate prothiofos, both in laboratory bioassays as well as in field trials. Comparable studies on (*Melia azederach* L.) extracts did not give significant results.

INTRODUCTION

Awareness of the harmful side effects and hazards of an overuse and/or misuse of synthetic pesticides from the 1960s onwards has revived interest in the potential of several plants as sources of pest control chemicals. Of the various plant species studied so far, *Azadirachta indica* (neem) is among those showing the highest bioactivity against a range of invertebrate pests of cultivated crops and animals. While records of the pesticidal and medicinal properties of parts of the neem tree and crude extracts of neem date back to ancient times, the accelerated pace of scientific investigation during the past decade or so has resulted in a flood of scientific and popular literature on the chemistry, ecology, mode of action, toxicity and application of neem and neem extracts for a variety of purposes. A large number of studies have been carried out in many parts of world on the biological activity of neem against a range of crop and animal pests (Warthen, 1979; Gujar and Mehrotra, 1983; Meisner et al., 1981; Kubo and Klocke, 1982; Schmutterer, 1986). While not as exhaustively studied as neem, the related *Melia azederach* is also known to possess strong pesticidal properties against many economically important pests (Chiu Shin-Foon, 1986; Ascher, 1986; Kraus et al., 1986).

This paper studies the biological activity of Neem and *Melia* in connection with one of the most important pests of cruciferous crops in Mauritius, the diamondback moth, *Plutella xylostella*. This notorious insect pest causes serious damage to cabbage, cauliflower and Chinese cabbage, necessitating intensive application of insecticides. The situation is further aggravated by the fact that in Mauritius, as in many other parts of the world, *P. xylostella* has developed tolerance to the majority of insecticides used in the recent past. Farmers therefore are constantly looking to the agrochemical firms and Extension Services of the Ministry of Agriculture for newer, more effective chemicals for the control of this pest. However, the increasingly stringent legal and safety requirements necessary for the granting of an import permit prior to the introduction of a new pesticide into the country has drastically reduced the number of new products entering the local market. While waiting for a better alternative, farmers continue to use the available insecticides at higher and more frequent doses, which not only compounds the problem of resistance, but also poses serious environmental and health hazards.

The Entomology Division of the Ministry of Agriculture is currently working on developing an integrated pest management programme for the control of *P. xylostella*. Its main thrust is towards the establishment of biological control using parasites and microbial agents.

A decisive factor in the eventual success of this programme is a considerable reduction in the total amount of pesticides being used presently in vegetable cultivation, with preferably a shift towards the use of less persistent and less toxic insecticides. The present study addresses both pesticide tolerance in the pest and pesticide pollution.

Three of the insecticides used extensively in the recent past and no longer found to be effective in the control of *P. xylostella* are the pyrethroids Decis (deltamethrin) and Cymbush (cypermethrin) and the organophosphate Tamaron (methamidophos). The efficacy of neem and *melia* in overcoming the resistance developed by *P. xylostella* to these three insecticides was studied. Admixtures of neem and *melia* extracts with the synthetic insecticides were tested in the lab. for antifeedance and mortality in fourth-instar larvae of *P. xylostella*. Trials were also conducted in order to study the effectiveness of the mixtures in protecting cabbage plants under field conditions.

MATERIALS AND METHODS

Preparation of Extracts

Simple, easy to prepare mixtures were made by crushing leaves of the appropriate plant species in an electric blender in the desired quantity of water. The mixture was left overnight on a shaker for maximum extraction, followed by filtration and dilution to a concentration of 10 per cent. This was further diluted with water to obtain a 4 per cent aqueous leaf extract. Commercially available deltamethrin (Decis), cypermethrin (Cymbush) and methamidophos (Tamaron) were diluted in water to the recommended dose of 1 ml/l in each case, and then mixed with the 4 per cent leaf extract in a 1:5 ratio. The mixture was stirred with a magnetic stirrer for five minutes in order to ensure thorough mixing, and then used for the laboratory bioassays as well as for the field sprays. One per cent Teepol was added as a wetting agent to all the mixtures prior to use.

Culturing of Test Insects

The culture of *Plutella xylostella* was bred and reared on cabbage leaves in wooden cages (0.75 × 0.60 × 0.75 m) under normal laboratory conditions of 20–25°C and 65–75 per cent relative humidity. Adults were fed 10 per cent honey solution.

Laboratory Bioassays

Leaf discs of 2.5 cm diameter were cut out from tender, untreated cabbage leaves, dipped in the treatment solutions for 5 seconds air-dried and weighed. The discs were then placed in a 12 cm diameter glass petri dish lined with filter paper. There were eight discs per dish. Fifteen fourth-instar larvae of *P. xylostella* from the lab culture were starved for two hours in order to equilibratre their level of hunger, introduced into each petri dish, and allowed to feed on the discs for 24 hours. The discs were reweighed at the end of 24 hours (D1) and then again at the end of 48 hours (D2). The amount eaten in mg/larva was calculated on a fresh weight basis. Each treatment as well as the control (water + teepol only) was replicated five times. Another petridish was set up under similar conditions but without larvae for the estimation of the desiccation factor. After 48 hours the larvae were transferred to fresh untreated cabbage leaves and maintained until they moulted or died. Total number of normal adults that survived were noted in each case.

Field Experimentation

Field trials were conducted on cabbages grown in a randomized block design with five replicates. Each plot measuring 3.2 × 4.5 m was sprayed weekly with one of the treatments as from the third week of transplantation. One day prior to the spray treatment, the number of larvae found on 10 plants chosen at random in each plot was recorded. Furthermore, the total number of plants in each plot having leaves damaged by *Plutella* larvae, i.e., the total number of infested plants per plot, was noted. Manual weeding was done regularly and the cabbages harvested two-and-a-half months from the time of transplantation. The harvested cabbages were graded according to their quality into three categories, and the total yield in terms of number of heads per treatment and weight in kg/treatment was noted.

RESULTS AND DISCUSSION

Laboratory Bioassays

Figure 1 presents the amount of leaf tissue consumed in mg/larva during the first 24 hours (D1) and the following 24 hours (D2). Both the plant species tested in this study exhibited a strongly negative effect on the extent of feeding by *P. xylostella* on a preferred host plant like cabbage, with neem showing greater activity. In the first 24 hours one larva consumed 10.3 mg of leaf tissue that had been treated with a 4 per cent

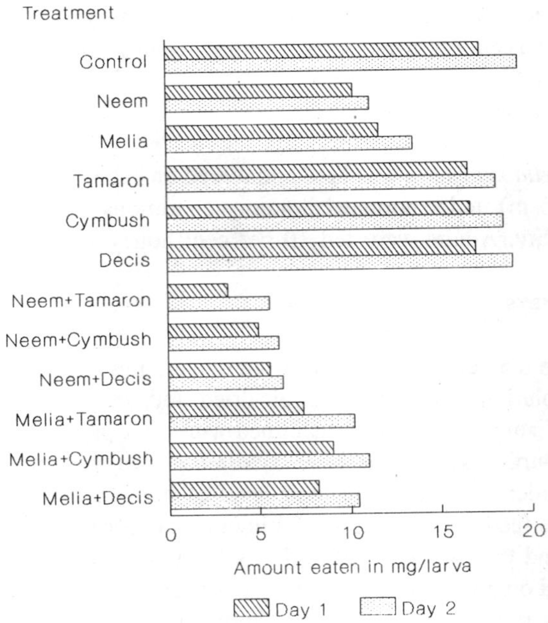

Fig. 1. Antifeedant effect of Neem and *Melia*, alone and with insecticides

neem leaf extract as compared to 17.3 mg untreated leaf discs, i.e., an antifeedant effect of 40.5 per cent (Table 1). Treatment with Tamaron, Cymbush and Decis at the recommended dose of 1 ml/l did not result in significant antifeedance in the larva, being 4.1 per cent, 3.5 per cent and 1.7 per cent respectively, as compared to control. The various admixtures of plant extracts and synthetic insecticides were found to be highly effective in inhibiting feeding. The mixture of neem + Tamaron produced a 80.9 per cent reduction in feeding, whereas Tamaron alone, at the recommended dose, caused a 4.1 per cent feeding inhibition. The antifeedance produced by Cymbush increased from 3.5 per cent to 71.1 per cent when mixed with neem while Decis, which by itself did not exhibit significant antifeedant effect (1.7 per cent), caused 67.6 per cent feeding inhibition when combined with neem. Reduction in feeding due to the 4 per cent leaf extract of *Melia* (32.4 per cent) was almost as high as that due to neem (40.5 per cent). The combined treatments of *Melia* with Tamaron, Cymbush, or Decis also resulted in significant antifeedance when compared to control, however the feeding inhibition due to the combination of synthetic insecticide + *Melia* was not significantly different statistically from that caused by *Melia* alone (Table 1).

In the majority of cases, the antifeedant activity of the plant extracts, as well as that of the mixtures decreased after the first 24 hours whereas that of the synthetic insecticides increased (Table 1). This is not surprising in view of the fact that most natural extracts of plant origin are photo-unstable and lose activity rapidly on exposure to light. The commercially formulated synthetic insecticides, on the other hand, being relatively more stable demonstrate a short-term cumulative effect, particularly under lab. conditions.

Table 1. Efficacy of neem and *Melia* leaf extracts in overcoming tolerance to some synthetic insecticides in larvae of *P. xylostella*

Treatment	Mean antifeedant effect at D1 (%)	% larval mortality (D1)	Mean antifeedant effect at D2 (%)	% larval mortality (D2)
Control	0a	0a	0a	0a
Neem	40.5bc	16.0b	42.3bc	17.3a
Melia	32.4b	14.7a	29.9b	16.0a
Tamaron	4.1a	10.7a	6.7a	16.0a
Cymbush	3.5a	6.7a	4.6a	10.7a
Decis	1.7a	10.7a	2.1a	9.3a
Neem + Tamaron	80.9g	45.3c	71.1d	42.7b
Neem + Cymbush	71.1fg	37.3c	68.6d	40.0b
Neem + Decis	67.6efg	40.0c	67.5d	44.0b
Melia + Tamaron	57.2def	44.0c	47.4c	42.7b
Melia + Cymbush	48.0bcd	33.3c	43.3bc	40.0b
Melia + Decis	52.6cde	30.7c	46.4bc	36.0b

Means followed by the same letter within a column are not significantly different at P = 0.05 with Duncan's Multiple Range Test.

Table 1 also lists larval mortality at D1 and D2 resulting from the different treatments. Comparison of larval mortalities at D1 and D2 resulting from treatment with synthetic insecticides alone and in combination with neem or *Melia* showed significant differences. Larval mortality rates for the synthetic insecticides, which varied from 6.7 per cent for Cymbush (at D1) to a maximum of 16 per cent for Tamaron (at D2), increased by factors ranging from 3 to 10 when combined with either neem or *Melia* (Table 1).

Comparison of feeding reduction and larval mortality for the different treatments after 48 hours demonstrates maximum antifeedance and highest mortality resulting from the neem + Tamaron combination (Fig. 2). Total antifeedant effect of neem at the end of 48 hours was 41.4 per cent, feeding inhibition due to Tamaron alone was 5.4 per cent, whereas the combined antifeedant effect of Neem + Tamaron was 75.7 per cent. In combination with *Melia* (antifeedant effect = 31.1 per cent) Tamaron exerted a feeding inhibition of 52.0 per cent, Cymbush which inhibited larval feeding to the extent of 4.1 per cent, reduced the amount eaten by 69.8 per cent when mixed with neem by 45.5 per cent when mixed with *Melia*. Antifeedant effect of Decis increased from 1.9 per cent to 67.6 per cent when combined with neem and to 49.3 per cent in combination with *Melia*.

Tamaron which produced 29.3 per cent larval mortality, killed 88.0 per cent of the larvae when mixed with neem and 86.7 per cent when mixed with *Melia*. Larvicidal effect of Cymbush was enhanced from 17.3 per cent to 77.3 per cent with neem and 73.3 per cent with *Melia*, while the effectiveness of Decis was improved from 24 per cent to 84 per cent when combined with neem and to 66.7 per cent in combination with *Melia*.

In this study a 4 per cent aqueous neem leaf extract was found to cause 33.3 per cent larval mortality after 48 hours whereas Sombatsiri and Tigvattanot (1983) reported 78.6 per cent larval mortality in *P. xylostella* with a 5 per cent methanolic extract of neem leaves. The joint action of neem oil or neem extracts with synthetic insecticides has been tested against a number of insect pests, with varying results (Sontakke, 1989; Mani et al.,

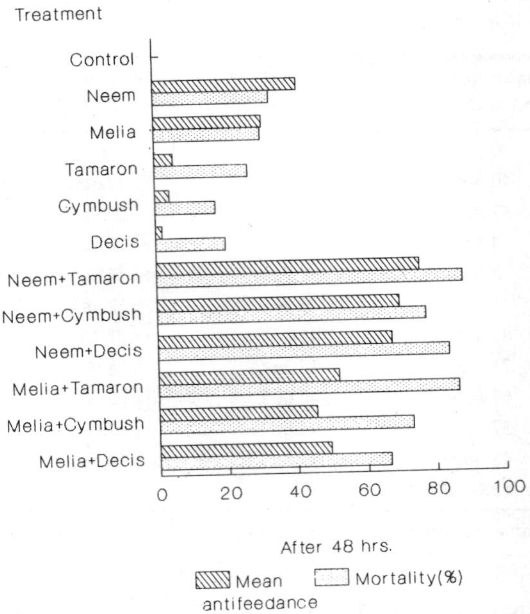

Fig 2. Effect of neem and *Melia* on DBM larvae, alone and with insecticides

1990; Abdul Kareem et al., 1988; Singh and Singh, 1987). The growth regulating properties of Neem (Warthen, 1979; Jacobson, 1989; Maurer, 1983) and *Melia* (Saxena et al., 1983) on several important pests have been documented. This was further confirmed in the present study, with only 9 and 22 pupae developing from the 75 larvae originally treated with neem and *Melia* extracts respectively, as compared to 72 in the untreated control (Table 2). Moreover treatment with neem resulted in only 12.0 per cent of the larvae developing into normal adults, while the *Melia* treatment led to complete and normal development of only 16.0 per cent larvae, as compared to 93.3 per cent in the control set. While treatment with Tamaron, Cymbush and Decis resulted in 53.3, 57.3 and 66.7 per cent of the insects developing normally to the adult stage, the combined treatments of a plant extract and a synthetic insecticide allowed normal development in a considerably lower percentage of insects (Table 2). Furthermore it was observed that larvae treated with the neem or *Melia* extracts or with the combined treatments showed signs of abnormal development, particularly in the wings. The wings were frizzled and/or curled. Moreover several of the adults were observed to be smaller than the ones in the control branch.

Field Trials

As was expected, *P. xylostella* was the most damaging pest encountered in the cabbage field. The small number of *Crocidolomia binotalis* larvae found in the plots were hand-picked at an early stage in the crop cycle and did not cause significant damage.

Table 2. Effect of plant extracts and/or synthetic insecticides on development of *P. xylostella*

Treatment	% larval survival after 48 hours	Total no. of pupae	Total no. of normal adults (%)
Control	100.0	72	93.3
Neem	66.7	20	12.0
Melia	69.3	22	16.0
Tamaron	70.7	47	53.3
Cymbush	82.7	50	57.3
Decis	76.0	55	66.7
Neem + Tamaron	12.0	5	4.0
Neem + Cymbush	22.7	10	9.3
Neem + Decis	16.0	7	8.0
Melia + Tamaron	13.3	8	8.0
Melia + Cymbush	26.7	14	12.0
Melia + Decis	33.3	15	13.3

Progress of infestation by *P. xylostella* was rapid in the control plot with 100 per cent of the cabbages showing some degree of damage by the 10th week (Fig. 3). While the plots treated with *Melia* and Tamaron treated plots had comparable levels of infestation in the sixth week, development of infestation was more rapid in the Tamaron treated plot, with 86 per cent of the plants being attacked by the 10th week. Neem extract slowed down the rate of infestation somewhat, with 62 per cent of the plants showing some degree of damage by the 10th week, as compared to 92 per cent damaged plants in the Cymbush-treated plots and 93 per cent in the Decis-treated plots. The combined treatments were relatively more effective with damage incidences of 44 per cent in the neem + Tamaron plots, 59 per cent in the neem + Cymbush plots and 62 per cent in the neem + Decis plots at the end of week 10. The recommended organophosphate, Tamaron at 1 ml/l gave the highest degree of protection from pest attack. Figure 2 charts the pest incidence in the different plots for a three-week period. The mean larval count was highest in the control plot at any given time. Development of pest incidence was more rapid in cabbage plants treated with Tamaron, Cymbush, and Decis, while neem and *Melia* were effective in slowing the progress of pest incidence to some extent. Of the various combined treatments neem + Tamaron, followed by neem + Cymbush and neem + Decis, was observed to be fairly efficacious in lowering the rate of development of pest incidence. The Tokuthion treated plot had the lowest larval density at any time—a result which is reflected in the figures for the post harvest yield and quality of cabbages (Table 3). Of a total of 104 cabbage heads harvested from the Tokuthion-treated plot all were of marketable quality (Grade A) while only 3.8 per cent were of Grade B, whereas in the untreated control plot more than half of all harvested cabbages could not be marketed. Treatment with Tamaron, Cymbush and Decis provided 65–69 per cent protection on the cabbage crop and resulted in 32–35 per cent loss while treatment with the combination of neem + a synthetic insecticide gave significantly better results, with more than 50 per cent Grade A cabbages and less than 10 per cent loss. Tamaron, Cymbush and Decis when combined with *Melia* offered 82–91 per cent protection from *Plutella* damage.

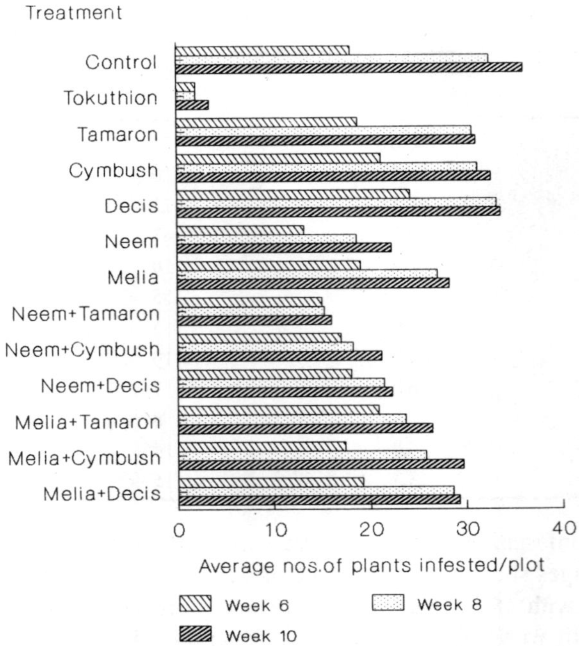

Fig. 3. Effect of Neem and *Melia*, alone and with insecticides, on damage incidence

Table 3. Yield and quality of cabbages treated with synthetic insecticides and plant extracts, alone and in combination

Treatment	Marketable*		Unmarketable	Total yield of cabbages	
	Grade A (%)	Grade B (%)	Grade C (%)	No of heads harvested	Wt of heads harvested (kg/plot)
Untreated control	0	45.6	54.4	103	116.39
Tokuthion control	96.2	3.8	0	104	212.16
Neem	42.4	40.4	17.2	99	179.19
Melia	22.8	46.5	30.7	101	164.63
Tamaron	14.3	54.3	31.4	105	130.20
Cymbush	10.7	58.2	31.1	103	123.60
Decis	5.7	59.4	34.9	106	126.14
Neem + Tamaron	54.5	39.4	6.1	99	195.03
Neem + Cymbush	63.7	34.3	2.0	102	192.78
Neem + Decis	58.3	33.3	8.3	108	189.00
Melia + Tamaron	47.6	43.7	8.7	103	182.31
Melia + Cymbush	50.5	32.0	17.5	103	177.16
Melia + Decis	49.0	37.7	13.3	98	166.60

* Grade A: Superior quality.
Grade B: Marketable after removal of infested outer leaves.

CONCLUSION

The laboratory and field trials in this study demonstrate that the synthetic insecticides Tamaron, Cymbush and Decis, which by themselves gave poor control of *Plutella xylostella*, performed significantly better when mixed with a 4 per cent aqueous extract of neem leaves. The non-significance of the antifeedance and mortality test values confirmed the tolerance developed by *P. xylostella* to the recommended dosage of Tamaron, Cymbush and Decis. Addition of neem leaf extract to these synthetic insecticides improved the efficacy of these products in *Plutella* control. Neem kernel suspension has been reported to have a potentiating effect on cypermethrin as well as on carbaryl and on several organophosphates (Singh and Singh, 1987), whereas Mariappan and Saxena (1984) demonstrated that neem oil enhances the pesticidal effect of custard-apple oil in the control of the leaf-hopper transmission of the rice tungro virus. Conversely, a number of substances have been reported as potentiating the antifeedant, insecticidal and growth-regulating effects of neem extracts, for example oil of *Petiveria alliacea* (Olaifa and Akingbohungbe, 1986), sesame oil and plant lecithins (Schauer, 1983).

Tolerance and eventually resistance in an insect to an insecticidal compound is due to an increase in enzymatic degradation of the toxicant (Sun et al., 1978) and according to Vollinger (1986), the mixed function oxidases are responsible for the development of resistance in *Plutella xylostella* to neem products. Compounds that inhibit mixed function oxidises or other enzymes involved in insecticide breakdown therefore exhibit a synergistic effect when combined with neem preparations, for example piperonyl butoxide (Lange, 1983; Sombatsiri and Temboonkeat, 1986).

The present study demonstrating the synergistic and/or resistance breakdown effect of a neem leaf extract on an insect species known to be resistant or tolerant to several organophosphate and pyrethroid insecticides, is in keeping with the results obtained by Saxena et al., (1989) on the synergistic and DDT-resistance knockdown effect of neem oil on *Tribolium casteneum*. The synergistic effect of neem oil on several organophosphates such as quinalphos and monocrotophos has also been reported (Parmar and Dutta, 1986). In the present study, the neem leaf extract used was observed to breakdown tolerance in *P. xylostella* to methamidophos, cypermethrin and deltamethrin with a greater effect on the organophosphate than on the two pyrethroids. The effect of *Melia*, on the other hand, was not found to be statistically significant.

At the grassroot level, the results obtained here are important economically as well as from the environmental and health point of view. Weekly application of deltamethrin for the protection of vegetable crops is a routine practice widespread among Mauritian farmers, and one that is not easy to discourage. In spite of the easy availability of Tokuthion which is recommended specifically for the control of leaf-eating caterpillars such as *Plutella*, planters often continue to use Decis at much higher doses in the hope of controlling this particularly damaging pest, in addition to the other insect pests occurring in cultivations of cruciferous crops. Admittedly, Tokuthion is a more expensive product, 100 ml costing 78 Mauritian rupees, whereas the same quantity of Decis cost Rs. 58.40. Cymbush and Tamaron are even cheaper, costing Rs. 53.30/100 ml and Rs. 30/100 ml respectively.

Farmers who grow various types of vegetables are usually unwilling to buy different insecticidal products for different crops, and perhaps therein lies the popularity of a broad-spectrum formulation such as Decis. Hitherto the effectiveness of deltamethrin justified such a course of action, but the development of tolerance in *Plutella* necessitates an alternative approach. Results of this study provide an interesting option, whereby farmers may be able to continue using Decis at the recommended dose even for the control of insect pests of cruciferous crops. Further studies are indicated in order to investigate the effect of such combined treatments on other pests, as well as the exact proportion of the plant extract in relation to the synthetic insecticide for optimum efficacy and minimum side effects.

REFERENCES

Abdul Kareem, A., R.C. Saxena and M.E.M. Boncodin. 1988. Mixtures of neem seed kernel or neem cake powder and carbofuran granules for controlling green leaf-hopper (GLH) and rice tungro virus (RTV). *Neem Newsletter* 5(2): 15–17.

Ascher, K.R.S. 1986. Notes on Indian and Persian lilac pesticide research in Israel. *Proc. 3rd Int. Neem Conf., Nairobi,* Eschborn 1987, pp. 45–53.

Chiu Shin-Foon. 1986. Experiments on the practical application of china berry, *Melia azederach* and other naturally occurring insecticides in China. Ibid. pp. 661–668.

Gujar, G.T. and K.N. Mehrotra. 1983. Inhibition of growth and development of the tobacco caterpillar *Spodoptera litura* Fabr. due to azadirachtin and other neem products. *Ind. J. Ent.* 45(4): 431–435.

Jacobson, M. 1989. Botanical Pesticides—Past, Present and Future. In : *Insecticides of Plant Origin.* Arnason, Philogene, Morand (eds.), ACS Symposium Series 387: 1–10.

Kraus, W., S. Baumann, M. Bokel, U. Keller, A. Klenk, A.M. Klingele, H. Pohnl and M. Schwinger. 1986. Control of insect feeding and development by constituents of *Melia azederach* and *Azadirachta indica. Proc. 3rd Int. Neem Conf., Nairobi,* op. cit. pp. 111–125.

Kubo, T. and J.A. Klocke. 1982. Azadirachtin, insect ecdysis inhibitor. *Agric. Biol. Chem.* 46(7): 1951–1953.

Lange, W. 1983. P B.O. : Synergistic effects on different neem seed extracts and influence on degradation of an enriched extracts and influence on degradation of an enriched extract by U.V. light. *Proc. 2nd Int. Neem Conf.* (Rauischholzhausen, 1983, pp. 129–140.

Mani, A., K. Kumudanathan and C.A. Jagadish. 1990. Relative efficacy of neem oil and endosulfan against insect pests of mustard. *Neem Newsletter* 7(2): 19–20.

Mariappan, V. and R.C. Saxena. 1984. Custard apple oil, neem oil and their mixtures: effect on survival of *Nephotettix virescens* and rice tungro virus transmission. *Proc. 2nd Int. Neem Conf.* op. cit., pp. 413–424.

Maurer, G. 1983. Effect of a methanolic extract of neem and seed kernels on the metamorphosis of *Ephestia kuehniella.* Ibid., pp. 365–376.

Meisner, J., K.R.S. Ascher, R. Aly and J.D. Warthen. 1981 Response of *Spodoptera littoralis* (Boisd.) and *Earias insulana* (Boisd.) larvae to azadirachtin and salannin. *Phytoparasitica* 9(1): 27–32.

Olaifa, J.I. and A.E. Akingbohungbe. 1986. Antifeedant and insecticidal effects of extracts of *Azadirachta indica, Periveria alliacea* and *Piper guineense* on the variagated grasshopper, *Zonocerus variegatus. Proc. 3rd Int. Neem Conf.,* op. cit., pp. 405–418.

Parmar, B.S. and S. Dutta. 1986. Neem oil as a synergist for insecticides. *Neem Newsletter* 3: 3–5.

Saxena, R.C., P.B. Epino, Tu Cheng-wen and B.C. Puma. 1983. Neem, chinaberry and custard-apple : antifeedant and insecticidal effects of seed oils on leafhopper and plant hopper pests of rice. *Proc. 2nd Int. Neem Conf.,* op. cit., pp. 403–412.

Saxena, V.S., C. Devakumar, D.B. Saxena and R.P. Singh. 1989. DDT resistance breakdown by neem oil. *Neem Newsletter* 6(1). pp. 1–3.

Schauer, M. 1983. Effects of variously formulated neem seed extracts on *Acyrthosiphon pisum* and *Aphis fabae. Proc. 2nd. Int. Neem Conf.,* op. cit., pp. 141–150.

Schmutterer, H. 1986. Fecundity reducing and sterilizing effects of neem seed kernel extracts in the Colorado potato beetle, *Leptinotarsa decemlineata. Proc. 3rd Int. Neem Conf.,* op. cit., pp. 351–360.

Singh, B. and A.P. Singh. 1987. Joint action of neem kernel suspension with some insecticides against *Schistocerca gregaria* Forsk. *Neem Newsletter* 6(4): 38–40.

Sombatsiri, K. and S. Tigvattanot. 1983. Effects of neem extracts on some insect pests of economic importance in Thailand. *Proc. 2nd Int. Neem Conf.,* op. cit., pp. 95–100.

Sombatsiri, K. and K. Temboonkeat. 1986. Efficacy of an improved neem kernel extract in the control of *Spodoptera littoralis* and *Plutella xylostella* under laboratory conditions and in field trials. *Proc. 3rd Int. Neem Conf.*, op. cit. pp. 195–203.

Sontakke, B.K. 1989. Comparative efficacy of insecticides alone and in combination with neem oil against rice plant hopper. *Neem Newsletter* 6(4) : 38–40.

Sun, C.N., H. Chi and H.T. Feng. 1978. Diamond back moth resistance to diazinon and methomyl in Taiwan. *J. Econ. Ent.* 71 pp. 551–554.

Vollinger, M. 1986. The possible development of resistance against neem seed kernel extracts and deltamethrin in *Plutella xylostella. Proc. 3rd Int. Neem Conf.*, op. cit. pp. 543–554.

Warthen, J.D. 1979. *Azadirachta indica* : A source of insect feeding inhibitors and growth regulators. *U.S. Dept. Agric. ARM-NE-4*. 21 pp.

53

Bioefficacy of Neem Derivatives against the Insect Pest Complex of Rice

U.K. NANDA, B. PARIJA, N.C. PRADHAN, B. NANDA AND D.D. DASH

Department of Entomology, Orissa University of Agriculture and Technology
Bhubaneswar 751 003, India

ABSTRACT

Field experiments were conducted at the Central Research Station, OUAT., Bhubaneswar over five seasons to evaluate the bioefficacy of the neem derivatives against the rice insect-pest complex, because to explore the feasibility of inclusion of neem derivatives in rice pest management schedule as an approach to low cost technology, to restore the environment quality and to study their effects on biocontrol agents. A susceptible rice variety, Jaya was included in all the field tests. Locally available neem derivatives like neem oil, neem seed kernel extract and neem cake extract alongwith commercial formulations such as Repelin, Wellgrow at 2 per cent; and Biosol and Neemguard at 0.5 per cent concentration; and mixture of karanja and neem oils (1 per cent + 1 per cent) utilised as foliar sprays at 20, 40 and 70 DAT were tested. All these products were supplemented with 0.1 per cent detergent. From the field investigations over five seasons, it was indicated that the natural pest pressure of gall midge *Oresolia oryzae* (Wood-Mason) and stem borers [*Scirpophaga incertulas* (Walk) and *Chil suppressalis* (Walk)] ranged from 9.4 per cemt to 15.3 per cent silver shoot; 11.3 per cent to 11.9 per cent deadheart and 7.5 per cent to 12.7 per cent whitehead respectively whereas the damage level in various treatments of neem pesticides exhibited 5.1 per cent to 8.8 per cent silvershoot; 5.82 per cent to 9.2 per cent deadheart and 4.8 per cent to 8.0 per cent whitehead and thus proved moderately effective against them. These botanicals showed moderate level of effectiveness against rice green leafhoppers, [*Nephotettix virescens* (Distant) and *Nephotettix nigropictus* (Stal)] and leaffolders (*Cnaphalocrosis medinalis* Guenee). Among the neem derivatives neem seed kernel extract, neem oil and Repelin (Commercial product) at 2 per cent concentration were found more promising. Neem kernel extract and neem oil were found less effective against the pests of summer rice which may be possibly for photodegradation of pesticidal agents due to relatively high temperature. Neem kernel extract and oil were found relatively safer than the insecticides like monocrotophos and ethofenprox to predators like spiders (*Lycosa pseudoannulata* and *Phidippus* sp) and mirid bug (*Cyrtorhinus lividipennis*) in field condition.

INTRODUCTION

Presently there is a global call for rice technologists to explore the feasibility of inclusion of neem derivatives in integrated pest management of rice. The aims are to find a low-cost technology for crop protection, to restore the environmental quality, and to use the benefit of biocontrol agents. Jacobson (1986) reported that as many as 120 insect pest species, three species of mites, and five species of nematodes are adversely affected by neem derivatives. Saxena et al. (1981), Saxena and Khan (1984), Saxena (1986), Mariappan et al. (1982), Saxena and Khan (1985a, b), Saxena and Barrion (1987a, b) Nanda and Pradhan (1990), and Nanda and Nanda (1991) noted the adverse effects of neem derivatives on various insect pests of rice crop.

MATERIALS AND METHODS

Field experiments were conducted at the Central Research Station, Orissa University of Agriculture and Technology, Bhubaneswar, Orissa, over five seasons from *kharif* 1987 to *kharif* 1991 in randomized block design with three replications to find out the bioefficacy of neem derivatives against the rice pest complex. Attempts have been made during the field investigations of 1990–91 to observe the effects of such phytoproducts on the biocontrol agents particularly on the predators such as spiders and mirid bug. Neem oil, neem seed kernel extract, and neem cake extract have been tested as foliar sprays at concentration 1 per cent during the first phase of the field experiment, and subsequently at 2 per cent concentration along with 0.1 per cent detergent at 10 or 20 days after transplanting (DAT), 40 DAT, and 70 DAT. Commercial products from neem such as Repelin, Wellgrow, Biosol, and Neemguard developed by various agencies in India have also been tested for their effectiveness against rice insect pests over seasons. A prescheduled management, i.e., maximum protection, need-based application of the conventional insecticides monocrotophos and ethofenprox and untreated check, have been included as treatments for the comparison of results. A susceptible rice variety, Jaya was included in all the field tests under report.

RESULTS AND DISCUSSION

The results obtained over five seasons commencing from *kharif* 1987 to *kharif* 1991 are presented in Tables 1 to 12. The natural pest load as revealed from the untreated check plots over five seasons was found to be moderate with respect to rice gall midge, stem borers, leaf folders, green leafhoppers, and whorl maggot (Tables 1 to 10). The level of injury in the untreated check plots (control) over five seasons ranged from 9.4 to 15.3 per cent silver shoot due to the attack by gall midge; 10.73 to 17. per cent dead heart and 7.5 to 12.7 per cent whitehead due to the incidence of stem borers; 10 to 30.7 per cent leaf damage by leaf folder; and 3.9 to 12.2 per cent leaf damage by whorl maggot. The green leafhopper population as estimated by sweeping methodology was found over two seasons as 285.6/sweep and 280.3/sweep. In the other two seasons, the population estimation was 22 and 24.4/hill. The grasshopper damage was quite negligible. The maximum protection treatment which provided full protection umbrella to the crop comprising the seedling root dip treatment with chlorpyriphos solution fol-

Table 1. Incidence of dead heart and white earhead percentage in different treatments (*kharif*, 1987)

Treatment	Dead heart percentage		White earhead percentage at 130 DAT	Average grain yield (q/ha)
	At 30 DAT	At 50 DAT		
Maximum protection	2.1	3.2	2.2	28.0
Monocrotophos at 0.5 kg a.i./ha	8.2	10.5	3.1	26.0
Neem oil 2%	8.7	10.0	4.8	26.5
Repelin 1%	7.4	8.2	4.9	27.0
Neem kernel extract 1%	5.7	7.0	2.9	26.0
Neem cake extract 1%	10.1	11.4	7.1	24.6
Wellgrow 1%	7.8	8.5	4.8	25.5
Control	15.1	17.9	12.7	22.7
'F' test	S	S	S	S
SEM ±	0.917	0.438	0.379	0.355
CD (0.01)	3.858	1.843	1.595	1.497
CD (0.05)	2.779	1.328	1.149	1.078

Table 2. Incidence of silver shoot percentage and green leafhopper population per sweep in different treatments (*kharif* 1987)

Treatment	Silver shoot percentage		Average green leafhopper population/sweep at 90 DAT
	At 30 DAT	AT 50 DAT	
Maximum protection	0.8	7.0	70.3
Monocrotophos at 0.5 kg a.i./ha	1.7	10.1	173.3
Neem oil 2 %	4.1	11.3	115.3
Repelin 1 %	5.1	7.6	183.0
Neem kernel extract 1%	2.2	8.2	105.3
Neem cake extract 1%	3.8	13.0	145.6
Wellgrow 1%	3.1	10.0	111.0
Control	9.4	8.8	285.6
'F' test	S	S	S
SEM ± ·	1.503	1.109	9.19
CD (0.01)	6.329	4.667	38.69
CD (0.05)	4.56	3.363	27.87

lowed by three round applications of granular insecticides proved to be the best of all the treatments. Neem oil and neem kernel extract tested over five seasons afforded a moderate level of control against gall midge, recording 4.03 to 11.3 per cent silver shoot and 3.4 to 8.2 per cent silver shoot respectively as against 9.4 to 15.3 per cent silver shoot in control plots. Similarly, the dead heart percentage in those two neem products was restricted within 3.9 to 10.0 per cent compared to 10.73 to 17.9 per cent dead heart in control plots. A similar trend was obtained with respect to stem borer damage at heading phase of the crop. The neem oil and neem seed kernel extract were

Table 3. Silver shoot percentage and green leafhopper population per sweep in different treatments (*kharif* 1988)

Treatment	Silver shoot percentage		Average number of green leafhopper sweep at 70 DAT
	At 30 DAT	AT 50 DAT	
Maximum protection	2.3	2.6	69.3
Biosol 0.5% as foliar spray	5.0	7.0	123.0
Neem oil 2% as foliar spray	6.4	8.1	114.6
Repelin 2% as foliar spray	5.8	7.1	143.3
Neem seed kernel extract 2% as foliar spray	5.1	6.8	105.0
Need-based application of monocrotophos	4.7	3.5	108.3
Wellgrow 2% as foliar spray	5.5	6.6	121.3
Untreated control	13.3	15.3	280.3
'F' test	S	S	S
SEM ±	0.77	0.382	8.19
CD (0.01)	3.278	1.612	34.502
CD (0.05)	2.362	1.161	24.859

Table 4. Percentage of dead heart and whitehead and the grain yield in different treatments (*kharif* 1988)

Treatment	Dead heart percentage		Whitehead percentage at 90 DAT	Average grain yield (q/ha)
	AT 30 DAT	AT 50 DAT		
Maximum protection	2.9	2.8	2.9	33.54
Biosol 0.5% as foliar spray	4.6	4.9	5.2	28.28
Neem oil 2% as foliar spray	4.4	5.9	6.0	31.27
Repelin 2% as foliar spray	5.9	6.2	6.6	29.28
Neem seed kernel extract 2% as foliar spray	3.8	5.2	4.0	31.84
Need-based application of monocrotophos	3.5	4.1	3.1	29.76
Wellgrow 2% as foliar spray	4.5	5.8	5.2	30.25
Untreated control	12.0	14.6	11.6	27.44
'F' test	S	S	S	
SEM ±	0.927	0.387	0.52	0.743
CD (0.01)	3.903	1.630	2.200	3.131
CD (0.05)	2.812	1.174	1.585	2.256

found less effective against the insect pests of summer rice, which may be due to photodegradation of pesticidal agents at relatively high temperature.

Chin et al. (1987), Saroja (1986), Panda (1987), Nanda and Parija (1988), Nanda and Pradhan (1990), Balasubramanian et al. (1987), and Gunathilagaraj and Sunderbabu (1987) have found similar results on the bioefficacy of neem derivatives against the rice stem borers and gall midge.

Table 5. Percentage of dead heart, whitehead, leaf damage by leaf folder and whorl maggot in different treatments (*rabi* 1988–89)

Treatment	Dead heart percentage		Leaf folder percentage		Whorl maggot damage at 30 DAT	Whitehead percentage at 75 DAT
	AT 30 DAT	At 50 DAT	At 30 DAT	At 50 DAT		
Maximum protection	3.6	2.9	11.1	9.9	1.0	1.6
Biosol 0.5% as foliar spray	6.9	7.4	18.8	20.3	2.5	3.3
Neem oil 2% as foliar spray	7.7	9.2	16.5	21.4	2.8	2.7
Repelin 2% as foliar spray	7.1	8.5	19.8	19.2	3.0	3.2
Neem seed kernel extract 2% as foliar spray	6.4	6.7	15.3	17.5	2.1	2.2
Need-based application of monocrotophos	5.5	5.3	13.2	10.7	1.2	1.9
Wellgrow 2% as foliar spray	6.2	7.1	19.8	18.6	3.3	3.6
Untreated control	9.8	14.3	23.3	30.7	3.9	7.5
'F' test	S	S	S	S	S	S
SEM ±	0.428	0.525	0.864	0.635	0.294	0.580
CD (0.01)	1.802	2.214	3.637	2.673	1.239	2.442
CD (0.05)	1.298	1.595	2.620	1.926	0.892	1.759

Table 6. Percentage of silver shoot, dead heart, and whitehead in different treatments (*kharif* 1990)

Treatment	Silver shoot percentage		Stem borer damage in percentage		
	30 DAT	50 DAT	Dead heart at 30 DAT	Dead heart at 50 DAT	Whitehead at preharvest
Maximum protection	3.46	1.26	0.25	2.0	0.50
Neem oil 2% as foliar spray with 0.1% detergent	4.03	2.0	0.86	5.23	3.33
Neemguard 0.5% as foliar spray	3.22	2.20	0.27	4.70	2.40
Neem oil 1% + karanja oil 1% with 0.1 detergent	2.46	5.50	0.90	3.80	1.50
Neem seed kernel extract 2% with 0.1% detergent	3.40	2.86	0.46	3.90	2.0
Need-based application of monocrotophos as foliar spray	4.03	3.10	0.80	4.0	0.96
Ethofenprox at 0.075 kg a.i./ha	5.53	3.03	0.56	5.50	2.53
Untreated control	12.93	13.96	3.00	10.73	7.56
	S	S	S	S	S
CD (0.01)	8.50	4.69	1.42	0.41	3.68
CD (0.05)	6.12	3.38	1.02	0.30	2.66

Table 7. Percentage of leaf damage of leaf folders, whorl maggot, and grasshopper in different treatments (*kharif* 1990)

Treatment	Leaf folder damage percentage		Whorl maggot damage percentage at 50 DAT	Grasshopper damage percentage at 30 DAT
	30 DAT	50 DAT		
Maximum protection	0.63	4.23	5.93	1.63
Neem oil 2% as foliar spray with 0.1% detergent	1.33	5.96	4.20	1.93
Neemguard 0.5% as foliar spray	1.06	6.36	5.06	1.76
Neem oil 1% + karanja oil 1% with 0.1% detergent	1.53	4.10	5.43	1.96
Neem seed kernel extract 2% with 0.1% detergent	0.90	5.44	7.70	1.63
Need-based application of monocrotophos as foliar spray	0.63	3.20	5.53	1.86
Ethofenprox at the rate of 0.075 kg a.i./ha	0.83	4.45	7.03	1.13
Untreated control	3.03	10.0	12.20	2.30
	NS	S	S	NS
CD (0.01)		0.46	5.65	
CD (0.05)		0.33	4.08	

Table 8. Green leaf hopper population per hill at 70 DAT in different treatments (*kharif* 1990)

Treatment	Green leaf hopper population/hill at 70 DAT
Maximum protection	5.8
Neem oil 2% as foliar spray with 0.1% detergent	8.0
Neemguard 0.5% as foliar spray	7.1
Neem oil 1% + karanja oil 1% with 0.1% detergent	9.7
Neem seed kernel extract 2% with 0.1% detergent	6.6
Need-based application of monocrotophos as foliar spray	7.0
Application of Ethofenprox at the rate of 0.075 kg a.i./ha	7.8
Untreated control	22.0
	S
CD (0.01)	0.58
CD (0.05)	0.42

Table 9. Percentage of silver shoot, deadheart and whitehead in different treatments (*kharif* 1991)

Treatment	Gall midge damage (silver shoot percentage)		Stem borer damage in percentage		
	30 DAT	50 DAT	Dead heart at 30 DAT	Dead heart at 50 DAT	Whitehead at preharvest
Maximum protection	2.65	3.75	0.60	2.50	1.60
Neem oil 2% as foliar spray with 0.1% detergent Teepol	5.27	5.39	0.80	5.82	5.20
Karanja oil 2% as foliar spray with 0.1% Teepol	7.74	9.75	1.20	7.34	6.78
Neem oil at 1% + karnaja oil 1% with 0.1% Teepol	8.29	8.88	0.90	6.57	8.00
Neem seed kernel extract 2% with 0.1% Teepol	6.39	7.26	0.72	4.95	4.90
Need-based application of monocrotophos as foliar spray at 0.5 kg a.i./ha	3.96	4.74	0.85	3.90	2.30
Ethofenprox at 0.075 kg a.i./ha	5.59	5.54	0.79	5.00	5.50
Untreated check	10.29	14.27	5.20	11.37	9.70
	S	S	S	S	S
CD (0.01)	3.360	4.509	3.735	3.319	2.598
CD (0.05)	2.255	3.026	2.505	2.228	1.743

Table 10. Percentage of leaf damage by leaf folders, whorl maggot and grasshoppers in different treatments (*kharif* .991)

Treatment	Leaf folder damage percentage		Whorl maggot damage percentage at 50 DAT	Grasshopper damage percentage at 30 DAT	Green leaf hopper population/hill at 70 DAT
	30 DAT	50 DAT			
Maximum protection	0.82	4.20	4.80	1.05	6.00
Neem oil 2% as foliar spray with 0.1% Teepol	1.59	6.98	5.24	1.60	7.20
Karanja oil 2% as foliar spray with 0.1% Teepol	3.00	9.97	6.63	1.77	12.30
Neem oil 1% + karanja oil 1% with 0.1% Teepol	2.80	10.00	6.82	1.80	10.70
Neem seed kernel extract 2% with 0.1% Teepol	2.10	7.00	5.00	1.37	7.10
Need-based application of monocrotophos as foliar spray at 0.5 kg a.i./ha	2.00	4.90	4.90	1.54	5.30
Enthofenprox at 0.075 kg a.i./ha	1.94	5.91	7.29	1.40	8.70
Untreated check	4.47	11.98	10.29	2.47	24.40
	S	S	S	S	S
CD (0.01)	2.571	2.185	2.693	1.231	4.048
CD (0.05)	1.725	1.466	1.807	0.826	2.717

Table 11. Population of predators/hill at pre and post treatment periods in various treatments (39 and 45 DAT), *kharif*, 1990.

Treat ment No.	Details of the treatment	Treatment period	No. of Wolf spider	No. of Lynx spider	No. of Jumping spider	No. of Dwarf spider	No. of Orb spider	Total No. of spider	No. of Damselflies	No. of mirid bug	Total no. of predators
T_1	Maximum Protection	Pre post	—	—	—	—	—	—	2	—	2
		post	—	—	—	—	—	—	—	—	0
T_2	Neem oil at 2% concentration as foliar spray with 0.1% detergent	Pre	—	2	—	3	—	5		2	7
		post	—	—	1	2	—	3	—	2	5
T_3	Neemguard at 0.5% concentration as foliar spray	Pre	—	—	—	2	—	2	—	3	5
		post	—	—	—	—	—	—	—	—	0
T_4	Neem oil 1% + karanja oil 1% with 0.1% detergent	Pre	4	—	—	—	—	4	—	—	4
		post	—	—	—	—	—	—	—	—	0
T_5	Neem seed kernel extract at 2% concentration with 0.1% detergent	Pre	2	—	—	—	—	2	—	2	4
		post	1	—	1	—	—	2	—	—	2
T_6	Need based application of monocrotophos as foliar spray	Pre	—	—	—	2	—	2	—	1	3
		post	—	—	—	—	—	—	—	—	0
T_7	Application of ethofenprox at the rate of 0.075 kg a.i./ha	Pre	—	—	—	—	—	—	—	2	2
		post	—	—	—	—	—	—	—	—	0
T_8	Untreated control	Pre	3	—	—	—	2	5	—	4	9
		post	4	—	—	—	—	4	—	3	7

Wolf Spider - *Lycosa pseudoannulata*
Lynx Spider - *Oxyopes* sp.
Jumping Spider - *Phidippus* sp.
Dwarf Spider - *Callitrichia formosana*
Orb Spider - *Argiope* sp.
Damselflies - *Agriocnemis* sp.
Mirid bug - *Cyrtorhinus lividipennis*

Table 12. Population of predators/hill at pre and post treatment periods in various treatments (at 39 and 45 DAT) *kharif*, 1991

Treat ment No.	Details of the treatment	Treat- ment period	No. of Wolf spiders	No. of Dwarf spiders	No. of Orb spiders	Total No. of spiders	No. of Damselflies	No. of mirid bugs	Total no. of predators
T_1	Maximum protection	pre	—	—	—	—	2	—	2
		post	—	—	—	—	—	—	—
T_2	Neem oil at 2% concentration as foliar spray with 0.1% detergent (Teepol)	pre	—	3	2	5	—	1	6
		post	—	2	1	3	—	—	3
T_3	Karanja oil at 2% concentration with 0.1% detergent (Teepol)	pre	—	—	—	—	1	2	3
		post	—	—	—	—	—	—	—
T_4	Neem oil at 1% + karanja oil at 1% concentration with 0.1% detergent (Teepol)	pre	1	—	—	1	—	—	1
		post	—	-	-	—	—	—	—
T_5	Neem seed kernel extract at 2% concentration with 0.1% detergent (Teepol)	pre	2	—	—	2	—	2	4
		post	1	—	—	1	—	2	3
T_6	Need based application of monocrotophos as foliar spray	pre	—	2	—	2	—	—	2
		post	—	—	—	—	—	—	—
T_7	Application of Ethofenprox at the rate of 0.075 kg a.i./ha	pre	2	—	2	4	—	—	4
		post	1	—	1	2	—	—	2
T_8	Untreated check	pre	2	2	2	6	—	—	6
		post	4	—	2	6	—	2	8

Wolf Spider - *Lycosa pseudoannulata*
Dwarf Spider - *Callitrichia formosana*
Orb Spider - *Argiope* sp.
Damselflies - *Agriocnemis* sp.
Mirid bug - *Cyrtorhinus lividipennis*

Neem oil and neem seed kernel extract were also found moderately effective against the rice leaf folders in field conditions registering 5.44 to 21.4 per cent leaf damage as against 10.0 to 30.7 per cent leaf damage recorded in control plots over three seasons. Saxena (1986) and Schmutterer et al. (1983) have given similar views. Green leafhopper was also suppressed to a fairly high level by neem oil and neem seed kernel extract.

Repelin 2 per cent, Wellgrow 2 per cent, Biosol 0.5 per cent, and Neemguard at 0.5 per cent concentration applied to the rice crop as foliar sprays at 20, 40, and 70 DAT provided a moderate level of control against the gall midge, stem borers, leaf folders, and green leafhoppers (Table 1 to 10).

Neem kernel extract and oil were found relatively safer than monocrotophos and ethofenprox to spiders and mirid bug in field conditions (Tables 11 and 12).

REFERENCES

Balasubramanian, G.S., Jayaraj and M. Balasubramanian. 1987. Effect of slow release fertilizer nitrogen and neem products in the control of stem borer of rice. *Indian J. Plant Prot.* 15(2): 132–135.

Chin, S.F., B.Q. Huang, Z.X. Huang and M.Y. Hu. 1987. Evaluation and utilization of neem cake, custard apple oil, neem oil and china berry oil against BPH, WBPH, thrips and leaffolder pests of rice in China. *Research Planning Workshop on Botanical Pest Control in Rice-Based Cropping System,* Tamil Nadu Agriculture University, Coimbatore, June 1–6, 6 pp.

Gunathilagaraj, K. and P.C. Sunderbabu. 1987. Role of plant products in pest management in rice. *Neem Newsletter* 4(3): 35.

Jacobson, M. 1986. The neem tree : Natural resistance par excellence. *ACS Symp. on Natural Resistance of Plants to Pests, Roles of Allelochemicals*, 189th meeting of ACS, Florida, 1985. pp. 221–232.

Mariappan, V., R.C. Saxena and K.C. Long. 1982. Effect of custard apple and neem oil on the life span of rice tungro virus transmission by *Nephotettix virescens*. *IRRN* 7(3): 13–14.

Nanda, U.K. and B. Parija. 1988. Bioefficacy of plant products against the rice pests complex. M.Sc. (Ag) thesis, Orissa University of Agriculture and Technology, Bhubaneswar, Orissa, pp. 38–41.

Nanda, U.K. and N. Pradhan. 1990. Bioefficacy of phytoproducts against the rice pest complex. M.Sc. (Ag) thesis, Orissa University of Agriculture and Technology, Bhubaneswar, Orissa, pp. 44–46.

Nanda, U.K. and B. Nanda. 1991. Effects of botanical pesticides on the insect pest complex of kharif rice and their natural enemies. M.Sc. (Ag) thesis, Orissa University of Agriculture and Technology, Bhubaneswar, Orissa, pp. 53–54.

Panda, N. 1987. Use of botanicals against pests in rice-based cropping system. *Research Planning Workshop on Botanical Pest Control in Rice-Based Cropping System,*Tamil Nadu Agriculture University, Coimbatore, June 1–4.

Saroja, R. 1986. Effect of neem product foliar sprays on rice plants. *IRRN* 11(4): 34.

Saxena, R.C. 1986. Effect of custard apple oil, neem oil and neem cake on green leafhopper population and on tungro infection. *IRRN*, 11(2): 25.

Saxena, R.C. and A.A. Barrion. 1987a. Cytogenetic effects of neem seed 'bitters' (NSB) on green leafhopper (GLH) males. *IRRN*, 12(5): 24.

Saxena, R.C. and A.A. Barrion. 1987b. Cytogenetic effects of neem seed kernel extract (NSKE) on brown planthopper (BPH) *Nilaparvata lugens* (Stal.) spermatocytes. *IRRN*, 12(5): 25.

Saxena, R.C. and Z.R Khan. 1984. Neem oil disturbs *Nephotettix virescens* (Distant.) feeding. *Neem Newsletter* 1(3): 28–29.

Saxena, R.C. and Z.R. Khan. 1985a. Electronically recorded disturbances in feeding behaviour of *Nephotettix virescens* (Distant.) (Homoptera: Cicadellidae) on neem oil treated plants. *J. Econ. Entomol.* 78(1): 222–226.

Saxena, R.C. and Z.R. Khan. 1985b. Effect of neem oil on survival of *Nilaparvata lugens* (Stal.) (Homoptera: Delphacidae) and on grassy stunt and ragged stunt virus transmission. *J. Econ. Entomol.* 78(3): 647–651.

Saxena, R.C., G.P. Waldbauer, N.J. Liquido and B.C. Puma. 1981. Effects of neem seed oil on the rice leaf folder, *Cnaphalocrosis medinalis* (Guenee). In: *Natural Pesticides from Neem tree*. H. Schmutterer, KRS Ascher and H. Rembold eds.,GTZ: Eschborn FRG, pp. 189–204.

Schmutterer, H., R.C. Saxena and J.V. Heyde. 1983. Morphogenetic effects of some partially purified fractions and methanolic extracts of neem seeds on *Mythimna separata* (Walker.) and *Cnaphalocrosis medinalis* (Guenee.) *Z. Angew. Entomol*. 95(3): 230–237.

54

Evaluation of Neem Products for the Control of Vectors of Japanese Encephalitis Virus in Rice Fields—A Five-year Study (1987–1992)

D. RAGHUNATHA RAO[1], R. REUBEN[1] AND C. HELLPAP[2]

[1]Centre for Research in Medical Entomology (ICMR), Madurai 625 002, India
[2]Institut für Phytopathologie und Angewandte Zoologie, Germany

ABSTRACT

Application of neem products to rice fields was evaluated with the dual objective of controlling the mosquito vectors of Japanese encephalitis virus and enhancing grain yield. A bioassay was developed and used for screening neem cake powder (NCP) against mosquito to larvae of *Culex (Culex) quinquefasciatus* Say. Only samples giving over 90 per cent bioassay mortality were used in field trials. When NCP was applied at the rate of 500 kg/ha, either alone or coated over urea, to rice fields immediately after transplantation, there was a significant reduction in the breeding of culicines. Two relatively stable formulations, Neemrich-I (lipid rich) and Neemrich-II (azadirachtin rich), also significantly reduced breeding. Neemrich-I was compared with a commercial product Nimin alone or coated over urea. Both gave significant reduction in the abundance of culicine immatures. Nimin costs Rs. 75 (US$ 2.7) for 1.5 l/ha. Neemrich-I costs Rs. 250 (US$ 9) for 2.5 l/ha. Neemrich-II costs Rs. 300 (US$ 11) for 1.5 l/ha. Treatments with urea coated with Nimin and Neemrich-I gave significantly higher grain yield than the formulations used alone.

INTRODUCTION

Standing water in rice fields supports the breeding of vectors of Japanese encephalitis virus (JEV) *Culex (Culex) tritaeniorhynchus* Giles, C. (C.) *pseudovishnui* Colless, and C. (C.) *vishnui* Theobald. Use of synthetic insecticides is neither cost-effective nor desirable because of the enormous area involved in rice-growing rural tracts (Reuben et al., 1992). In addition, indiscriminate use of insecticides in rice agro-ecosystem has led to the development of resistance to pesticides and resurgence in the population of rice pests (Saxena et al., 1984) and also in mosquito populations (Kamimura and Maruyuma, 1983). Therefore, there is a revival of interest in integrating traditionally used natural products in pest control programme.

Studies on the effect of neem products on Diptera, and in particular on mosquitoes, are very limited (Steffens and Schmutterer, 1982). Some laboratory studies are available on the larvicidal and growth regulatory activity of crude neem extracts and semi-purified azadirachtin-rich fractions against *Culex* sp., *Anopheles* sp., and *Aedes* sp., (Attri and Prasad, 1980; Singh, 1984; Zebitz, 1984; Rao et al., 1988, 1989; Mwangi and Rembold, 1988; and others summarized by Ascher and Meisner, 1988). A small-scale field trial was carried out by Kalyanasundaram and Babu (1982) with petroleum ether extracts of neem leaves for the control of mosquitoes, and the practical applicability of neem products was demonstrated by Reuben et al. (1992), who found that application of neem cake powder (NCP) in rice fields at the rate of 500 kg/ha reduced the abundance of late instars by 81 per cent and pupal abundance by 40 per cent.

In the studies described in this paper applications of neem products directly to the field were compared with applications of urea coated with crude neem products and neem-based formulations. Coating urea with NCP inhibits nitrifying soil bacteria and thus prolongs the useful life of urea, leading to higher grain yield (Chhonkar, 1984; Grant et al., 1984).

MATERIALS AND METHODS

Laboratory Bioassay

Eight NCP samples from different places were screened for their quality. The samples from Virudhunagar, Karnataka, Tanjore, and Mixed were obtained from oil mills near Madurai, while four other NCP samples were made from the seeds collected from four different localities in Madurai, South India during the main fruiting season (April to July). The fruits were peeled, washed, and shade-dried. Kernels were separated by removal of the hard outer shell. Four kilograms of neem kernels from each locality were crushed by rotary method in Virudhunagar, in Tamil Nadu, South India, and NCP and oil were obtained. Azadirachtin content of eight NCP samples was analysed at the Institut für Phytopathologie and Angewandte Zoologie, Giessen, Germany. All eight samples were soaked in distilled water to obtain dilutions of 2, 1, and 0.5 per cent. Three replicates each of 25 early fourth-instar larvae of laboratory-reared *C. quinquefasciatus* Say were exposed to each concentration and parallel controls were maintained. Observations on larval mortality were made at 24 and 48 hrs.

Field Trial 1

An area of 4500 m^2 (0.45 ha) was divided with elevated dykes into nine plots, 500 m^2 each, with separate inlets and outlets. Neem cake-coated urea (NCU) was prepared by mixing 62.5 kg urea with 500 kg of NCP/ha using vegetable gum as a binding agent. Immediately after transplantation of rice seedlings, NCP and NCU were broadcast at the rates of 500 kg and 562.5 kg/ha respectively. Both treatments were triplicated.

Parallel untreated control plots were maintained. Ponni, a traditional long-term variety (130–140 days including nursery) was transplanted and the usual farming practices were followed in all the treatment plots. Observations on immature population were made as described under Sampling.

Field Trial 2

Two relatively stable neem-based formulations were tested: Neemrich-I, a lipid-rich formulation, and Neemrich-II, an azadirachtin-rich formulation, obtained from the National Chemical Laboratory (courtesy Dr. B.A. Nagasampagi), Pune, India. An area of 4800 m^2 (0.48 ha) was divided as before into 12 equalized plots, each 400 m^2. A new medium-term variety ADT 39 (125–130 days) was planted and routine agronomical practices were followed. Neemrich-I 80 EC at 8000 ppm, Neemrich-II 20 EC at 1000 ppm, and at 500 ppm were appropriately diluted in water and sprayed immediately after transplantation, each formulation in triplicate with a manually operated knapsack sprayer.

Field Trial 3

In the third field trial Neemrich-I and Nimin were tested alone and coated over urea.

An area of 8400 m^2 (0.84 ha) was divided into 21 plots, each 400 m^2. All treatments were laid out in triplicate in a randomized block design. Seedlings of a short-term variety ADT 36 (110–115 days) were transplanted. Neemrich-I and Nimin were applied as in the second field trial. Samples of urea coated with Neemrich-I at the rates of 6000 and 8000 ppm were obtained from Dr. B.A. Nagasampagi, National Chemical Laboratory. For Nimin-coated urea, 1.25 l of Nimin was mixed with 62.5 kg/ha of urea. Neemrich-I coated urea and Nimin-coated were broadcast immediately after the transplantation. The treatments were as follows:

T1) Control
T2) Nimin (1.25 l/ha)
T3) Nimin-coated urea
T4) Neemrich-I (6000 ppm)
T5) Neemrich-I-coated urea (6000 ppm)
T6) Neemrich-I (8000 ppm)
T7) Neemrich-I-coated urea (8000 ppm)

Sampling

Sampling of the immature mosquitoes was carried out thrice weekly from the day of transplantation up to harvest in the first two field trials, but only for six weeks in the third field trial. In the first field trial 10 dips were taken in each plot, while in the second and third field trials 12 and 20 dips per plot respectively were taken, using standard 350 ml dippers. Late-instar larvae (third/fourth) were collected, classified by genus, counted, and returned to the same plots. Pupae were brought to the laboratory and identified to species eclosion. Counts on non-target organisms were also made during the observation period in all the field trials.

Assessment of Grain Yield Parameters

In all the field trials grain yield was assessed in each plot by pulling out five plant bunches from five hills at random from four corners of the plot (20 samples/replicate). In each bunch productive tillers and grains from a tiller selected at random from the bunch were counted, and weight was recorded.

Data Analysis

Data collected in all the three field trials on late instars and pupae in different weeks were transformed to log (n + 1) and then subjected to factorial randomized block design analysis. Complete randomized block design was used for analysing the yield parameters. Data on non-target organisms were analysed by Student's *t* test.

RESULTS

Laboratory Bioassay

Full expression of mortality in the test larvae was observed only after 48 hrs after exposure to the NCP samples. Of the three doses tested, 2 per cent gave a reasonable degree of correspondence between larval mortality and the azadirachtin content of the samples tested. Therefore, 2 per cent was used as the discriminating dose for routine use. The azadirachtin content of the eight samples varied from 0.4 to 3 mg/g and the biological activity ranged from 32 to 95 per cent.

The samples (1–4) collected from the oil expeller showed lower azadirachtin content (0.4–1.2 mg/g) with lower biological activity (32–95 per cent) than that of the samples (5–8) of different localities from Madurai (azadirachtin content, 2.7–3.0 and biological activity 76–95 per cent). However, poor correspondence between azadirachtin content and the biological activity was observed (Table 1). Therefore, for field appli-

Table 1. Azadirachtin content of neem cake samples and their biological activity against *C. quinquefasciatus* larvae

Sample no.	Locality	Azadirachtin content (mg/g)	% Bioassay mortality at 2%
1	Virudhunagar	1.3	87
2	Karnataka	0.4	83
3	Tanjore	0.5	32
4	Mixed	1.1	95
5	TVS Nagar Madurai	2.8	95
6	Satya Sai Nagar Madurai	2.7	91
7	Alagappa Nagar Madurai	3.0	91
8	Muthupatty Madurai	2.8	76

cation, only NCP samples which showed at least 90 per cent bioassay mortality were selected.

Field Trial 1

The NCP used in the first field trial gave 100 per cent bioassay mortality in the laboratory. The predominant culicine species during this field trial were *C. tritaeniorhynchus* (68 per cent), *C. pseudovishnui* and *C. vishnui* (31 per cent), and *C.* (*C.*) *bitaeniorhynchus* Giles (1 per cent).

In the control plots culicine breeding was high during the first five weeks immediately after transplantation. The abundance of late-instar larvae was high in the second, and again during the fourth and fifth weeks. Similarly, the abundance of pupae was high during the fourth week. From the sixth week onwards there was a natural decline and the immature population during the rest of the crop period (7–13) weeks) was negligible (Fig. 1). The breeding of culicines was significantly reduced in NCP and NCU-treated plots in comparison with control plots as shown in Figs. 1 and 2 and ANOVA table (Table 2). Further analysis showed significant reduction in the breeding of culicine immatures in NCP- and NCU-treated plots, which were on par with each other.

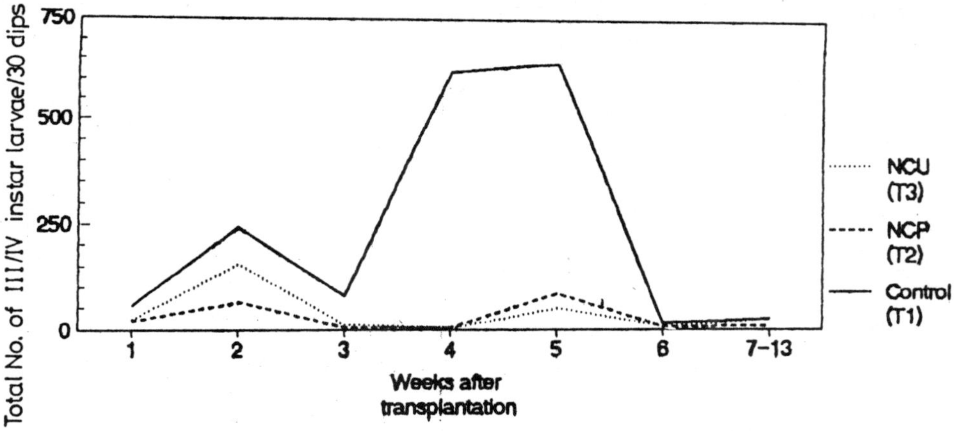

Fig. 1. Neem cake powder (NCP) and neem cake coated urea (NCU) against culicines in rice fields

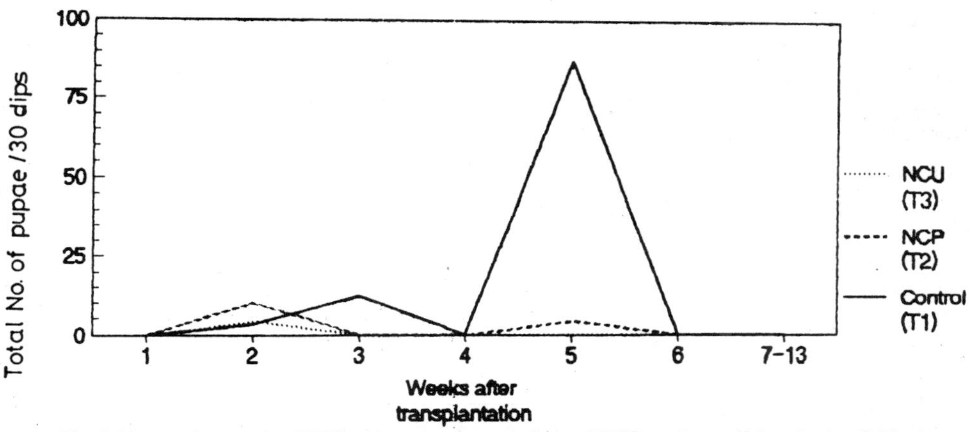

Fig. 2. Neem cake powder (NCP) and neem cake coated urea (NCU) against culicines in rice fields

There was no significant increase in grain yield components in the neem-treated plots over control. Significantly higher numbers of productive tillers were recorded in the NCP-treated plots than in control and NCU-treated plots, which were on par with each other. As regards the average number of grains per panicle, significantly lower number of grains per panicle was observed in NCP- and NCU-treated plots than in control plots (Table 3). This was because of severe damage due to leaf foler *Cnaphalocrosis medinalis* (*Guenee*), which seriously affected both treatments.

Field Trial 2

The second field trial compared the efficacy of two relatively stable neem-based formulations against culicines.

The culicine species identified during the crop period were *C. tritaeniorhynchus* (71 per cent), *C. vishnui* (19 per cent), *C. pseudovishnui* (1 per cent), and *C. (C. fuscocephala* Theobald (2 per cent).

Late-instar larvae in control plots were more abundant during the second, fourth, and fifth weeks. Pupal abundance showed a small peak in the second week and a large peak during the fourth week (Figs. 3 and 4). All the three treatments with neem-based formulations—Neemrich-I at 8000 ppm (T2), Neemrich-II at 1000 ppm, and Neemrich-II at 500 ppm—produced striking reduction in the abundance of late-instar larvae and

Table 2. Effect of NCP and NCU on breeding of culicines in rice fields (ANOVA Table)

Source	D.F.	M.S		S.S		F	
		III/IV	Pupae	III/IV	Pupae	III/IV	Pupae
Replicates	2	1.70	0.08	0.85	0.04	14.55*	1.70
Treatments	2	5.50	0.47	2.75	0.23	47.18*	10.52*
Periods	12	32.02	3.28	2.66	0.27	45.71*	12.33*
Interaction	24	8.20	3.62	0.34	0.15	5.05*	6.80*
Error	76	4.44	1.69	0.06	0.02	1.00	1.00

*$p < 0.05$.

Factorial randomized block design analysis

	Control	T2	T3	CD value
X log (n + 1) III/IV larvae	0.87^a	0.38^b	0.45^b	0.11
X log (n + 1) pupae	0.17^a	0.06^b	0.02^b	0.07

Values followed by the same letter do not differ significantly from each other.

Table 3. Effect of NCP and NCU on rice yield components

Treatments	Yield components (av. + SE)		
	No. productive tillers	Weight of plant	No. grains/ panicles
T1 Control	11.78^a + 0.03	52.32^a + 0.07	202.6^a + 0.12
T2 NCP	13.56^b + 0.03	77.34^b + 0.09	171.45^b + 0.09
T3 NCU	11.03^a + 0.03	51.04^a + 0.07	188.91^b + 0.12
CD value	1.25	8.04	17.67

Values followed by the same letter do not differ significantly from each other

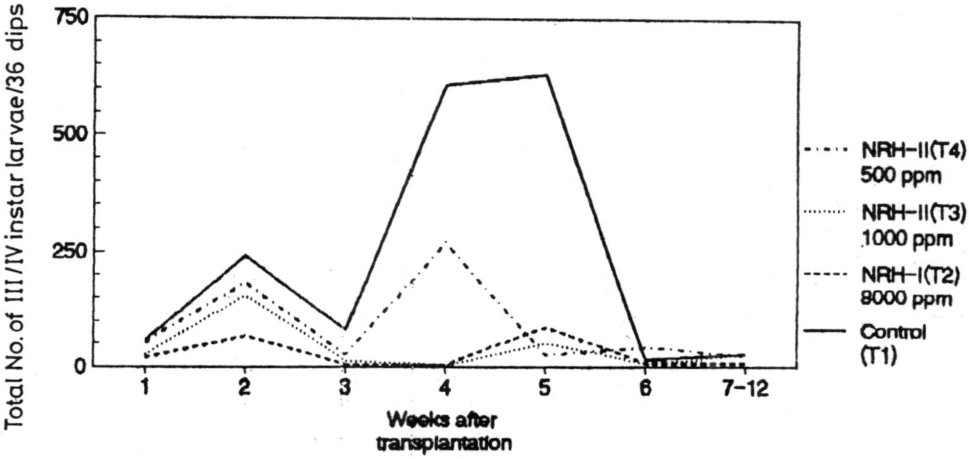

Fig. 3. Neemrich-I (NRH-I) and II formulations against culicines in rice fields

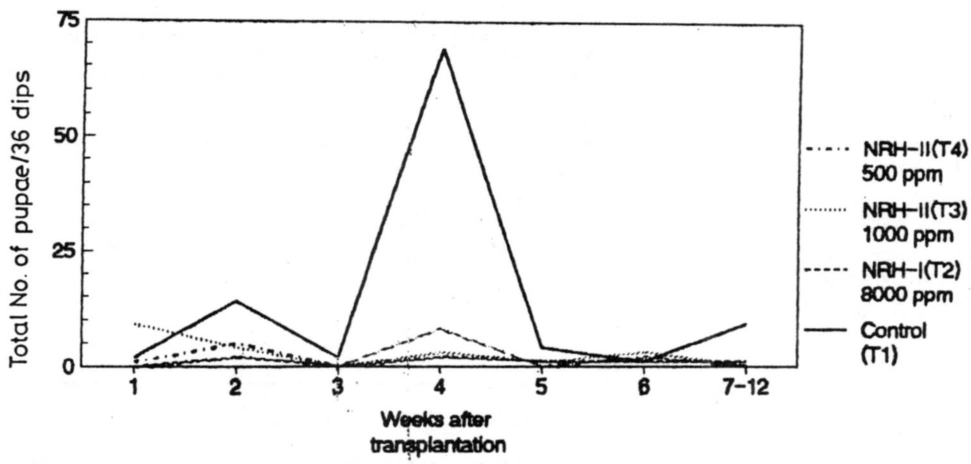

Fig. 4. Neemrich-I (NRH-I) and II formulations against culicines in rice fields

pupae as compared with the control. Factorial randomized block design analysis showed that the reduction in the abundance of late-instar larvae and pupae of culicines in the treated plots (T2–T4) was significantly different from control (Table 4). Further analysis showed that Neemrich-I and Neemrich-II plots were on par with each other.

Significantly higher grain yield was observed in Neemrich-II at 1000 ppm (T3) and 500 ppm (T4) and the treatments were at par with each other. There was no significant increase in the grain yield observed in the plots treated with Neemrich-I (T2), which was at par with the control (Table 5).

Field Trial 3

In the third field trial neem-based formulations were applied either directly or coated over urea. The culicine species identified during this trial were *C. tritaeniorhynchus* (77

Table 4. Effect of Neemrich-I and Neemrich-II on the breeding of culicines in rice fields (ANOVA Table)

Source	D.F.	S.S		M.S		F	
		III/IV	Pupae	III/IV	Pupae	III/IV	Pupae
Replicates	2	2.06	0.26	1.03	0.13	8.09	2.97
Treatments	3	6.02	1.63	2.03	0.54	15.89*	12.32*
Periods	5	10.69	2.32	2.14	0.46	16.73*	10.54*
Interaction	15	1.87	2.36	0.12	0.16	0.97*	3.57*
Error	46	5.88	2.02	0.13	0.04	1.00	1.00

*p < 0.05

Factorial randomized block design analysis

	Control	T2	T3	T4	CD value
X log (n + 1) III/IV larvae	1.77[a]	1.03[b]	1.10[b]	1.23[b]	0.24
X log (n + 1) pupae	0.48[a]	0.12[b]	0.13[b]	0.14[b]	0.14

Values followed by the same letter do not differ significantly from each other

Table 5. Effect of Neemrich-I and Neemrich-II on rice yield components

Treatments	Yield components (av. + SE)		
	No. productive tillers	Weight of plant	No. grains/ panicles
T1 Control	10.00[a] + 0.53	36.77[a] + 3.72	152.13[a] + 9.25
T2 Neemrich-I at 8000 ppm	12.0[a] + 0.95	41.23[a] + 7.23	166.8[b] + 16.42
T3 Neemrich-II at 1000 ppm	11.93[a] + 1.01	63.97[a] + 11.65	188.97[c] + 11.06
T4 Neemrich-II at 500 ppm	12.13[a] + 0.74	45.17[a] + 4.14	180.6[c] + 14.0
CD value	2.6	24.3	9.2

Values followed by the same letter do not differ significantly from each other.

per cent), *C. vishnui* (14 per cent), *C. pseudovishnui* (8 per cent), and *C. bitaeniorhynchus* (2 per cent).

As in the previous field trials maximum numbers of late-instar larvae and puape of culicines were recorded during the second fourth weeks in control plots (Figs. 5 and 6). The lipid-rich formulations Neemrich-I and Nimin, used alone or coated over urea, significantly reduced the abundance of late-instar larvae and pupae as compared with the control. The reduction in the abundance of late-instar larvae of culicines was at par in all the treatments (T2–T7) (Table 6). Lowest pupal production was observed in Neemrich-I coated urea at 8000 ppm (T7) and the rest of the treatments (T2–T6) were on par.

All the treatments (T2–T7) were significantly better than the control as regards the average number of productive tillers, weight of the plant, and number of grains per panicle. Grain yield obtained with Neemrich-I coated urea at 8000 ppm was significantly higher than with the rest of treatments. Nimin-coated and NRH-I-coated urea gave significantly higher grain yield than the same formulations used alone (Table 7).

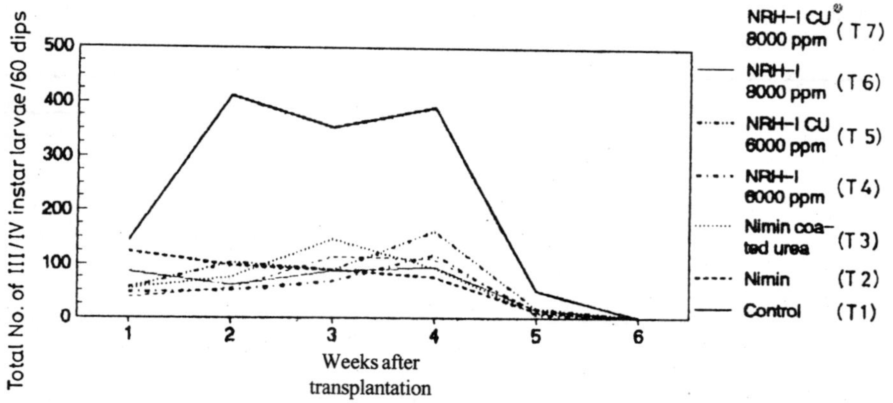

Fig. 5. Neemrich-I (NRH-I) and Nimin either alone or coated over urea against culicines in rice fields

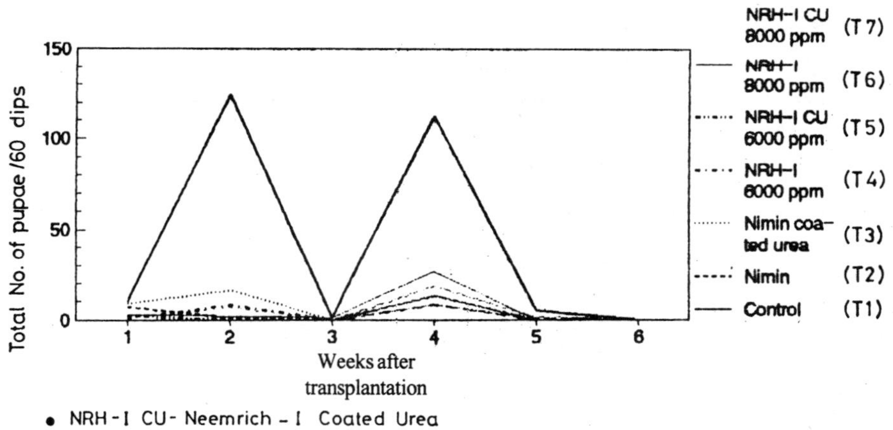

Fig. 6. Neemrich-I (NRH-I) and Nimin either alone or coated over urea against culicines in rice fields

Effect of Neem Products on Non-target Biota

In all the on-farm trials various non-target organisms were recorded: anisopteran and zygopteran naiads, *Corixa* sp., *Chironomus* larvae, *Dytiscus* sp., ephimeropteran naiads, frog tadpoles, *Laccobius* sp., notonectids, *Ranatra* sp., plant spiders, and various other water beetles. There were no significant differences for most of the non-target biota, except that there were less zygopteran naiads in NCU-treated plots than in controls ($t = 4.1$, $p < 0.05$) in the first field trial. In the second field trial more *Dytiscus* sp. were recorded in untreated control plots than with Neemrich-I, 1000 ppm ($t = 3.313$, $p \leq 0.05$) and 500 ppm ($t = 2.89$, $p < 0.05$). The tadpole population was significantly higher in Neemrich-II (1000 ppm) than in control. In the third field trial there was no significant difference in the population of non-target organisms between the control and treated plots.

Table 6. Effect of neem-based formulations and urea coated with them on culicine breeding in rice fields (ANOVA table)

Source	D.F.	M.S		S.S		F	
		III/IV	Pupae	III/IV	Pupae	III/IV	Pupae
Replicates	2	0.41	0.09	0.21	0.21	1.98	0.16
Treatments	6	3.62	3.82	0.60	0.64	5.84*	17.13*
Periods	5	42.02	11.12	8.40	2.22	81.26*	59.82*
Interaction	30	1.80	5.21	0.06	0.17	0.58	4.67*
Error	82	8.48	3.05	0.10	0.04	1.00	1.00

*$p < 0.05$

Factorial randomized block design analysis

	Control	T2	T3	T4	T5	T6	T7	CD value
X log (n + 1) III/ IV larvae	1.47[a]	1.05[b]	1.0[b]	1.0[b]	0.97[b]	0.93[b]	1.10[b]	0.21
X log (n + 1) pupae	0.70[a]	0.21[bc]	0.32[b]	0.24[bc]	0.21[bc]	0.24[bc]	0.11[c]	0.13

Values followed by the same letter do not differ significantly from each other

Table 7. Effect of Nimin, Neemrich-I and urea coated with them on rice yield components

Treatments	Yield components (av. + SE)		
	No. productive tillers	Weight of plant	No. grains/ panicles
T1 Control	8.9[a] + 1.3	31.1[a] + 4.0	104.5[a] + 9.0
T2 Nimin	10.3[b] + 0.9	37.9[bc] + 0.41	113.1[a] + 6.2
T3 Nimin-coated urea	12.7[d] + 1.1	53.2[d] + 5.6	127.5[c] + 7.3
T4 Neemrich-I 6000 ppm	10.5[c] + 0.8	36.2[b] + 4.3	119.1[b] + 9.9
T5 Neemrich-I-coated urea 6000 ppm	10.7[c] + 0.8	40.0[c] + 3.0	116.2[b] + 11.3
T6 Neemrich-I 8000 ppm	9.7[c] + 0.8	39.5[bc] + 3.6	107.7[ab] + 7.1
T7 Neemrich-I-coated urea 8000 ppm	12.3[d] + 1.3	40.5[c] + 3.9	141.6[d] + 15.7
CD value	0.7	2.8	6.9

Values followed by the same letter do not differ significantly from each other

DISCUSSION

These studies have demonstrated that application of good quality NCP and NCU to rice fields can significantly reduce the breeding of culicines while simultaneously increasing grain yield by inhibiting nitrification of urea by soil bacteria. Farmers in India are already using NCU in this way. However, the amount of crude neem required to control culicine breeding was at least twice the normal rate recommended by the agricultural department (Rao et al., 1992). Crude neem products vary in their biological activity and quality control is difficult in large-scale use. There are also problems of stability, bulk, and transportation to the fields. We have shown that relatively stable neem-based formulations Neemrich-I, Neemrich-II and Nimin are as effective as crude products in reducing culicine breeding at the rates recommended for agricultural use. These formulations enhanced grain yield when they were coated over urea. The neem-based formulations have all the advantages of crude NCP and avoid the problems associated with its use.

According to Dr. B.A. Nagasampagi, production costs for Neemrich-I, Rs 100/l (US$ 3.5), are lower than for Neemrich-II, Rs 200 (US$ 7). For 1 ha the cost of Neemrich-I is Rs 250 (US$ 9) and the cost of Neemrich-II is Rs 300 (US$ 11). Meanwhile, Nimin, a lipid rich formulation, has been commercialized and put on the market in 0.5 l sachets at Rs 30 (US$ 1.1) to be used at 0.5 l coated over urea for 1 acre (0.4 ha), i.e., Rs 75 (US$ 2.7) per hectare. According to Vyas et al., (1991), Nimin has been tested at various research stations of the Indian Council of Agricultural Research and agricultural universities. The cost of the Neemrich fractions may come down when produced on an industrial scale.

The farmer collaborator showed keen interest in our studies, and also demonstrated the method for making Nimin-coated urea to fellow farmers. The utility of neem-based formulations in rice fields is well recognized by the farming community and their availability commercially at an affordable price will promote large-scale use. The use of neem-based formulations in rice aquatic system is safe as in our studies no adverse effects on the non-target population were recorded. Neem products are also safe to fish (Ragunatha Rao, 1992) and can be used in rice-fish culture. Therefore, we feel neem-based formulations could be used as a component of a community-based integrated vector control programme.

ACKNOWLEDGEMENTS

This study was supported in part by WHO\FAO\UNEP Panel of Experts on Environmental Management for Vector control grant through contract no. HQ/89/484162. The authors thank Drs. H. Schmutterer and K. Ermel, Institut für Phytopathologie und Angewandte Zoologie, Justus Liebig-Universitat, Giessen, Germany, for analysing azadirachtin content of neem samples. We thank Dr. B.A. Nagasampagi, National Chemical Laboratory (CSIR), Pune, India, for providing Neemrich-I and II samples. Thanks are due to the technical staff of Control Section, CRME, for their excellent assistance in the field.

REFERENCES

Ascher, K.R.S. and J. Meisner. 1988. The effects of neem on insects affecting man and animal. In: *Focus on Phytochemical Pesticides*, Vol. 1, *The Neem Tree* M. Jacobson (ed.), CRC Press, Boca Raton, pp. 113–131.

Attri, B.S. and G. Ravi Prasad. 1980. Neem oil extractive—An effective mosquito larvicide. *Indian. J. Entomol.* 42: 371–374.

Chhonkar, P.K. 1984. Control of nitrification of urea in an alluvial soil by neem *Azadirachta indica* cake. *Neem Newsletter* 1(2): 18–19.

Grant, I.F., R. Seegers and I. Watanabe. 1984. Increasing biological nitrogen fixation in flooded rice using neem. *Proc. 2nd Int. Neem Conf.*, Rauischholzhausen, 1983, GTZ, Eschborn, pp. 493–505.

Kalyanasundaram, M. and C.J. Babu. 1982. Biologically active plant extracts as mosquito larvicides. *Ind. J. Med. Res.* 76: 102–106.

Kamimura, K. and Y. Maruyuma. 1983. Appearance of highly resistant strain of *Culex tritaeniorhynchus* to organophosphorus insecticides. *Jpn. J. Sanit. Zool.* 34: 33–37.

Mwangi, R.W. and H. Rembold. 1988. Growth inhibiting and larvicidal effects of *Melia volkensii* extracts on *Aedes aegypti* larvae. *Entomol. Exp. Appl.* 46: 103–108.

Raghunatha Rao, D., R. Reuben, Y. Gitanjali and G. Srimannarayana. 1988. Evaluation of four azadirachtin rich fractions from neem, *Azadirachta indica* A. Juss. (Fam. Meliaceae), as mosquito larvicides. *Indian. J. Malariol.* 25: 67–72.

Raghunatha Rao, D., R. Reuben and R.C. Saxena. 1989. Larvicidal activity of neem seed bitters (NSB) against *Culex quinquefasciatus* Say in flooded rice fields. *Int. Rice Res. Newsletter* 14: 28.

Raghunatha Rao D. 1992. Effect of neem, *Azadirachta indica* A. Juss. on the mosquito fish, *Gambusia affinis*. *Neem Newsletter* 9(2): 28–29.

Raghunatha Rao, D., R. Reuben, M.S. Venugopal, B.A. Nagasampagi and H. Schmutterer. 1992. Evaluation of neem, *Azadirachta indica*, with and without water management, for the control of culicine mosquito larvae in rice fields. *Med. Vet. Entomol.* 6: 318–324.

Reuben, R., D. Raghunatha Rao, S. Anthony, Philip S. Corbet, Wu Neng and Liao Guhou. 1970. Biological control methods suitable for community use. In: *Appropriate Technology in Vector Control,* C.F. Curtis (ed.), CRC Press, Boca Raton, pp. 139–158.

Reuben, R., R. Rajendran and D. Raghunatha Rao. 1992. Appropriate technology for the control of rice field breeding vectors of Japanese encephalitis. 22: 87–91.

Steffens, R.J. and H. Schmutterer. 1982. The effect of a crude methanolic neem (*Azadirachta indica*) seed kernel extract on metamorphosis and quality of adults of the mediterranean fruit fly, *Ceratitis capitata* Wied. (Diptera, Tephritidae). *Z. angew. Ent.* 94: 98–103.

Saxena, R.C., P.B. Epino, Tu Cheng-Wen and B.C. Puma. 1984. Neem, chinaberry, and custard-apple: Antifeedant and insecticidal effects of seed oils on leafhopper and planthopper pests of rice and their natural enemies. *Proc. 2nd Int. Neem Conf.*, Rauischolzhausen, 1983, GTZ, Eschborn, 403–412.

Singh, R.P. 1984. Effect of water extract of deoiled neem kernel on second instar larvae *Culex fatigans* Weidemann. *Neem Newsletter* 1: 16–17.

Vyas, B.N., N.B. Godrej and K.B. Mistry. 1991. Development and evaluation of neem extract as a coating for urea fertiliser. *Fertil. News* 19–25.

Zebitz. 1984. Effect of some crude and some azadirachtin enriched neem (*Azadirachta indica*) seed kernel extract on larvae of *Aedes aegypti*. *Entomol. Exp. Appl.* 35: 11–16.

55

The Performance of an IPM Package Including Neem for the Control of Potato Pests in Farmers Fields

S.A. SIDDIQ AND A. KHALAFALLA

Shambat Research Station, PO Box 30, Khartoum, North Sudan

ABSTRACT

Following the termination of various back-up research and researcher-managed trials (Siddiq 1983–1988), an integrated pest management (IPM) package including neem and wider spacing of potato plant was tested in farmer fields near Khartoum for the control of potato pests under light soil conditions.

The first season of the trial (1988–89) witnessed an outbreak of cutworms (*Agrotis ipsilon* and *A. segetum*) and negligible infestations by other pests. The IPM package at an on-island site reduced cutworm damage and increased yield from 16.2 to 25.6 tons/feddan (4200 m^2) compared to a package housing farmer practices (no spraying and closer spacing). However, at an out-of-island site, the IPM package significantly reduced damage and increased the yield from 12.7 to 20.5 tons/feddan.

Similar results were also obtained in the 1989–90 season for the control of white-fly (*Bemisia tabaci*) and jassid (*Empoasea lybica*) and in 1990–91 for the control of white-fly.

INTRODUCTION

In the Sudan, potato crop harbours, under cracking clay soil conditions, the potato tuber worm *Phthorimaea operculella* as major soil pest and leafhoppers (*Bemisia tabaci*, *Jacobiasea lybica*, and *Aphis gossypii*) as major foliage pests. Under light soil conditions, cutworms (*Agrotis ipsilon*, *A. spenifera*, and *A. segetum*) and leafhoppers prevail as soil and foliage pests respectively.

Based on the results of intensive back-up research trials (Siddiq, 1983–1986) and researcher-managed trials (Siddiq, 1987–1989), an integrated pest management (IPM) package for the control of potato pest complex under light soil conditions and a second one for heavy soil conditions were identified.

The IPM for light soil conditions consisted of:

1) Three sprays at 10-day intervals against early larval stages and leafhoppers with neem seed water extract at 1 kg seed powder/40 l water.
2) Potato seed cut into two longitudinal halves, each sown at a spacing of 20 m, as a control measure against older larval stages of cutworms.

The IPM for cracking heavy soil conditions consisted of:

1) Spraying leafhoppers with neem seed water extract.
2) Control of tuber worms through early sowing of tubers in November, deep sowing at 3 inch depth, and early harvesting of the crop.

The present paper reports on the results of the first package in farmers fields at Shehenab, a major potato production area 20 km north of Khartoum.

MATERIALS AND METHODS

A farmer-managed, on-farm trial to combat cutworms and leafhoppers on potato crop by the suggested IPM package for light soils was executed at on-island and out-of-island sites in 1988–89 and 1990–91. The trial was repeated in 1989–90 at an out-of-island site. Five plots measuring 3 × 3 m were selected at random in a farmer field for implementation of the IPM package. Another five plots in the rest of the farmer field were considered for comparison as package of farmer methods, whereby no spraying was done and tubers were cut into more than 10 eye pieces spaced at less than 20 cm spacing.

The first spray was applied when insect population suddenly built up. Counts of white-fly scales and adults, all stages of jassid, and aphids per 100 leaves of potato were made one day before the subsequent spray. Counts of cutworm damage on tubers (in percentage transformed to degrees) were made at harvesting. Furthermore, yield of marketable tubers in tons/feddan (4200 m^2) was determined for evaluation of the IPM package.

RESULTS AND DISCUSSION

All major pests of potato (except cutworms) appeared in the 1988–89 season at such low infestation levels that no counts were done for them.

At the on-island site mean per cent cutworm damage transformed to degrees (Table 1) was found to be 4.8 degrees in the IPM package compared to 7.3 degrees in the package of farmer methods. Accordingly, the yield of marketable tubers was significantly increased from 16.2 tons/feddan in the package of farmer methods to 25.6 tons/feddan in the IPM package.

At the cut-of-island site the IPM package showed significantly lighter damage by cutworms than the package of farmer methods. As a result, yield of marketable tubers was significantly increased in the IPM package to 20.5 tons/feddan compared to 1.27 tons/feddan in the package of farmer methods.

In the 1989–90 season the incidence of white-flies and jassid was alarming but that of cutworms and aphids was negligible.

At an out-of-island site the number of white-fly scales, white-fly adults, and all stages of jassid per 100 leaves of potato was significantly lower in the IPM package than in the package of farmer methods (Table 2). As a result, yield of marketable tubers per feddan was significantly increased in the IPM package to 7.1 tons compared to 3.8 tons in the package of farmer methods.

In the 1990-91 season only white-flies infested potato crop as major pest. At an on-island site the first count of white-fly scales per 100 leaves showed significantly lighter infestation in the IPM package than in farmer methods (Table 3). Similarly, in the subsequent two counts infestation was light in the IPM package but differences were not significant. However, all three counts of adult white-flies showed lighter infestation in the IPM package but the difference was also not significant.

Marketable tuber yield in tons/feddan was significantly increased in the IPM package to 7.87 tons compared to 1.64 tons in the package of farmer methods.

At the out-of-island site, results (Table 4) showed that the IPM package received significantly lighter infestation with white-fly scales and adults in all counts made compared to the package of farmer methods. Consequently, yield was significantly increased from 1.51 tons in the package of farmer methods to 4.4 tons in the IPM package.

Table 1. Testing of IPM package at Shehenab area for protection against potato pests (1988-89)

Type of package	On-island site		Out-of-island site	
	Mean cut-worm damage in degrees	Yield of marketable tuber in tons/feddan	Mean cut-worm damage in degrees	Yield of marketable tuber in tons/feddan
	SE ± 3.0	SE ± 1.6	SE ± 2.5	SE ± 1.5
IPM package	4.8	25.6	4.6	20.5
Package of farmer methods	7.3	16.2	27.7	1.27

Table 2. Testing of an IPM package at Shehenab area for protection against potato pests (1989–90 out of island)

Type of package	Mean number of insects per 100 leaves			Marketable in tons/feddan
	White-fly scales	White-fly adults	Jassid (all stages)	
	SE ± 36.7	SE ± 20.6	SE ± 14.0	SE ± 0.5
IPM package	105	75	17	7.1
Package of farmer methods	258	247	67	3.8

Table 3. Testing of an IPM package at Shehenab area for protection against potato pests (1990-91, on island)

Type of package	White-fly adults per 100 leaves			White-fly scales per 100 leaves			Yield tons/feddan
	1st count	2nd count	3rd count	1st count	2nd count	3rd count	
	SE ± 120.4	SE ± 119.9	SE ± 561.9	SE ± 38.6	SE ± 102.5	SE ± 751.4	SE ± 0.38
Package of recommended treatments	684	716	1772	184	495	2896	7.87
Package of farmers methods	1040	856	2532	352	492	3700	1.64

Table 4. Testing of an IPM package at Shehenab area for protection against potato pests (1990-91, out of island)

Type of package	White-fly adults per 100 leaves		White-fly scales per 100 leaves		Yield tons/feddan
	1st count	2nd count	1st count	2nd count	
	SE ± 88.0	SE ± 17.1	SE ± 38.6	SE ± 27.6	SE ± 0.54
Package of recommended treatments	160	80	184	188	4.48
Package of farmer methods	528	192	352	368	1.51

REFERENCES

Siddiq, S.A. (1983–1986). Evaluation of neem and cultural practices for the control of potato pest. *Ann. Rep.* Shambat Res. Sta., Khartoum North, Sudan.

Siddiq, S.A. (1987). Integrated pest management programme including neem treatment for combating potato pests in Sudan. In: *Natural Pesticides from Neem Tree and other Tropical Plants, Proc. 3rd Int. Neem Conf.*, Nairobi. H. Schmutterer and K.R.S. Ascher (eds.), GTZ, Eschborn, FRG, pp. 449-549.

Siddiq, S.A. 1989. Integrated pest management for potato in Sudan. *Proc. DLG Int. Conf.*, Bad Durkheim, FRG Vol. 1. pp. 281-285.

56

Potential of Neem for Controlling Leaf-eating Caterpillar Complex on Cabbage

K.A.N.P. BANDARA[1] AND C. KUDAGAMAGE[2]

[1]Research Officer,
[2]Head and Entomologist, Division of Entomology, Central Agricultural Research Institute, Peradeniya, Sri Lanka

ABSTRACT

Methanol, petroleum ether, and water extracts of neem seed kernel, neem oil, and several commercial insecticides with different modes of action were evaluated for their efficacy and economy against leaf-eating caterpillar complex on cabbage. The effect of neem seed kernel water extract (NSKWE), chlorfluazuron (Atabron^R 5 EC) and methamidophos (Tamaron^R LC 60) on the quality of treated cabbage and on the larval parasitoid *Apanteles plutellae* Kurdjumov (Hymenoptera: Braconidae) was also investigated. Further, two action threshold levels were compared using NSKWE and chlorfluazuron.

Methanol, petroleum ether, and water extracts of neem seed kernel, and neem oil were found to be as effective as the commercial insecticides, chlorfluazuron, ethofenprox, and triflumuron and more effective than methamidophos, cartap, *Bacillus thuringiensis*, and diflumuron against leaf-eating caterpillar complex on cabbage. Neem seed extract did not cause any harmful effect on *A. plutellae* and the quality of cooked cabbage. Use of NSKWE and neem oil was found to be more cheaper than methamidaphos and chlorfluazuron; they should, however, be applied at a lower action threshold to avoid significant loss.

Neem products such as water, methanol, and petroleum ether extracts of neem seed kernels can be used effectively and economically for controlling the leaf-eating caterpillar complex on cabbage.

INTRODUCTION

The steadily increasing cost of synthetic pesticides make them unaffordable for farmers in developing countries. However, due to heavy insect pest outbreaks, especially on vegetable crops, their control is needed to obtain reasonable yields. Thus, investigations are urgently needed on alternative control strategies from locally available renewable resources with simple and inexpensive processing.

Numerous studies in many countries have demonstrated the effectiveness of neem (*Azadirachta indica* A. Juss.) extracts against 198 species of insects, mites, and nematodes (Saxena, 1988). Furthermore, unlike ordinary insecticides based on a single active ingredient, the bioactive principles of neem are a complex array of compounds with diverse behavioural and physiological effects on insects.

Although neem is not commercially grown, it is abundant in most of the developing countries in the tropical region, including Sri Lanka. The introduction of a simple and cheap technology to use this resource as a pesticide will benefit the agricultural sector, providing low cost insecticides with less environment impact in terms of insecticidal hazards.

In the present studies, the efficacy of different neem extracts and commercial insecticides on the cabbage caterpillar complex *Plutella xylostella, Plusia eriosoma, Crocidolomia binotalis*, and *Spodoptera litura* and their effects on larval parasitoid *Apanteles plutellae* were evaluated. In addition, the effects of neem formulations on the quality of cooked cabbage were studied.

MATERIALS AND METHODS

Laboratory and field experiments were conducted at the Central Agricultural Research Institute (CARI), Gannoruwa, and Open Prison Camp (OPC), Palekelle, from 1989 to 1991.

Laboratory Experiments

Effect of Neem Oil, Root Extract of H. augustifolia, B. thuringiensis and Conventional Insecticides on Egg Hatchability and Larval Mortality of Diamond Back Moth (DBM)

The adult female DBM, *Plutella xylostella* (L.), were obtained from a culture maintained on radish plants in the laboratory. They were allowed to oviposit on one-month-old potted cabbage plant (of 15 cm diam. and 11 cm height). When these eggs were one day old, they were counted and sprayed with different insecticides and neem oil emulsion (Table 1) up to the point of runoff with an atomizer (Quickfit, England) and a rotating table. The neem oil sample was purchased from Ayurvedic Pharmacy Ltd., Kandy. A conventional wetting agent, Teepol®, was added at a rate of 0.5 ml/l for better adherence. The egg hatchability was recorded. In another experiment, cabbage leaf disks were dipped in different insecticides, air dried, and placed in petri dishes (9 cm diam. & 1.5 cm height) lined with moistened filter papers. The second-instar larvae of DBM obtained from the laboratory culture were allowed to feed on treated leaf disks. In order to obtain second-instar larvae of DBM, female DBM moths were allowed to oviposit on potted radish plant for 6 hrs and the oviposited radish plant were transferred into another cage. When the eggs hatched and first-instar larvae shed their exuviae (one to two days), they were considered second-instar larvae of DBM. These cabbage leaf disks were replaced with fresh untreated cabbage leaf disks 48 hrs after introduction. Ten larvae were used for each treatment and experiment was rep-

Table 1. Effect of different pest control agents on egg hatchability and larval/pupal mortality of diamond back moth in the laboratory

Treatment	Dosage (ppm)	Egg hatchability (%)	Pupation (%)	Adult emergence (%)
Diflubenzuron 25 EC	200 ai*	53 b	70 a	40 b
Chlorfluazuron 5 EC	100 ai	19 c	40 b	00 d
Triflumuron 480 SC	200 ai	70 ab	60 ab	10 cd
Methamidophos 60 LC	672 ai	00 d	00 c	00 d
Neem oil emulsion	20000	24 c	40 b	10 cd
Hortonia angustifolia root methanol extract	4000		10 c	00 d
Bacillus thuringiensis	2000		00 c	00 d
Untreated control		96 a	70 a	60 a
CV (%)		11.5	14.0	8.4

*ai = active ingredient
In each column, the values followed by the same letter do not differ significantly according to DMRT at p = 0.05.

licated three times. The larval and pupal mortalities and adult emergence were recorded.

Toxicity of Neem Seed Kernel Water Extract (NSKWE) and Insecticides on A. plutellae

To prepare the NSKWE, 25 g of decorticated neem kernels were ground to a fine powder using an iron mortar and pestle. One litter of water was added to this powder and the mixture was agitated for 10 min. and allowed to stand for 24 hrs. The mixture was filtered through a fine gauze (American jeorjet cloth[R]) and 0.5 ml of Teepol® was added and stirred well with a glass rod to make the emulsion. One-day-old adults of *A. plutelle* obtained from the laboratory culture were exposed to the recommended concentrations of methamidophos, chlorfluazuron, and NSKWE using the method described by Mani and Krishnamoorthy (1984). Ten adults were used for each treatment and they were fed on bee's honey. The treatments were replicated five times. The mortality counts were recorded 6, 24, and 48 hrs after the introduction of the adults.

To test the toxicity on pupal stage, one-day-old cocoons were placed on a petri dish lined with a filter paper and sprayed with 5 ml of the same insecticides and NSKWE using the Potters Spray Tower (Burkard Manufacturing Co. Ltd., England). These treated cocoons were dried in the shade and kept in 15 × 2.5 cm glass valves. The mouths of the valves were plugged with cotton wool. Water was used on the untreated control. Each treatment was replicated at 20° ± 2°C and 70 ± 10 per cent RH.

Field Experiments

Four field experiments were carried out (three at CARI, Gannoruwa, and one at OPC, Pallekelle) to study the effectiveness of neem formulations on the cabbage caterpillar control. In all these experiments, cabbage was planted in 4 × 3 m plots at a spacing of 40 × 50 cm. The plots were separated from one another by a 1 m wide strip. The experiments were laid out in a randomized complete block design.

Evaluation of Neem Oil, H. augustifolia, B. thuringiensis and Conventional Insecticides

To study the comparative efficacy of neem oil and other insecticides, two field evaluations were carried out during *yala* 1989 and *maha* 1989–90 at the CARI field using the cabbage cv SD Cross. Six commercial insecticides with various modes of action and neem oil emulsion were sprayed on individual cabbage plots at 10-day intervals commencing two weeks after transplanting (Table 2). The treatments were replicated three times. Insecticides were applied by using high volume knapsack sprayer (Cooper Pegler and Co. Ltd., England). The number of caterpillars in 10 randomly selected and labelled plants in the net plot (6.0 m^2) immediately before and after each insecticide application was recorded. Before the harvest, percentage leaf damage was assessed using 1–4 scale. After harvesting, the weights of whole heads and marketable heads after removing the damaged leaves were recorded. The yield loss was determined by subtracting the weight of the cabbage head after removing the damaged leaves, from the weight of the whole head.

Evaluation of Different Neem Seed Kernel Extracts

To determine the most effective extract, four different neem seed kernel extractions were field tested against this caterpillar complex at CARI. They were neem seed kernel oil, NSKWE, NSK methanol extract, and NSK petroleum ether extract. Neem seeds collected from *maha* Illuppallam area were used for the preparation of different neem extracts while neem oil was purchased from the Ayurvedic Pharmacy Ltd., Kandy. Neem seed kernel extracts were obtained by extracting finely ground decorticated neem seed kernel using a Soxhlet apparatus. Chlorfluazuron was used as check for comparison. The different leaf-eating caterpillar species and the number of *A. plutellae* cocoons were separately recorded. Each treatment was replicated three times.

Comparison of Two Action Thresholds with NSKWE and Chlorfluazuron

Two action thresholds, 2 and 5 cabbage looper equivalents (CLE) per 10 plants, were compared using NSKWE and chlorfluazuron (Shelton et al., 1983). The other treatments in this trial consisted of an application of chlorfluazuron and NSKWE at 10-day intervals and an untreated control. This experiment was conducted at the OPC, Pallekelle, during *maha* 1990–91. The cabbage cultivar Exotic F1 was used. The treatments were replicated four times. The number of leaf-eating caterpillar larvae was recorded once in every 10 days from transplanting to harvest in 10 randomly labelled plants per plot. Leaf damage incidence was recorded using a 1–4 scale. Percentage yield loss due to caterpillar damage was also obtained at harvesting. Insecticides were sprayed using a high volume knapsack sprayer.

At harvest the cabbage samples obtained from different treatments were tested at the Food Technology Division, CARI, Gannoruwa, for their taste, smell, flavour, fibre content, and boiling time.

Table 2. The efficacy of different pest control agents against leaf-eating caterpillars of cabbage (*yala* 1989)

Treatment	Rate/ha	No. of larvae	Leaf damage (1–4 scale)	Yield loss (%)*
Triflumuron (Alsystine 480 SC)	200–400 g ai	4.22 a	1.33 bc	25.5 ab
Bacillus thuringiensis (Thuricide HP)	500–1000 g	5.44 ab	2.00 d	24.6 ab
Diflubenzuron (Dimilin WP 25)	75–100 g ai	6.22 ab	1.66 cd	28.3 ab
Chlorfluazuron (Atabron 5 EC)	50–100 g ai	2.89 a	1.00 ab	27.3 ab
Neem oil	11–16 l	12.11 a	0.66 a	18.4 a
Hortonia angustifolia root methanol extract	2200–4000 g	4.78 a	1.66 cd	28.3 ab
Methamidophos (Tamaron LC 60)	420–500 g ai	5.55 ab	2.66 de	28.1 ab
Untreated control		8.89 b	3.00 e	28.7 ab
CV (%)		26.0	19.2	13.6

In each column, the values followed by the same letter do not differ significantly according to DMRT at p = 0.05.

$$\text{*Yield loss (\%)} = \frac{\text{Weight of whole head} - \text{Weight of head after removing damaged leave}}{\text{Weight of whole head}}$$

RESULTS AND DISCUSSION

Laboratory Evaluation

Effect on Egg Hatchability and Larval Mortality of DBM

Larvae of DBM did not emerge from eggs treated with methamidophos (Table 1). Neem oil and chlorfluazuron also reduced egg hatchability. Inhibition of egg hatchability of *Cnaphalocrosis medinalis* by neem oil reported by Saxena et al., (1980). The present results confirm the potential ovicidal activity of neem on DBM.

Larvae fed on leaves treated with methamidophos and *B. thuringiensis* could not pupate. Very few larvae fed on leaves treated with root methanol extract of *Hortonia angustifolia* (Monomiaceae) pupated and no adults emerged. Though 40 per cent of larvae fed on leaves treated with neem oil and chlorfluazuron could pupate, very little or no adult emergence was observed. Triflumuron also reduced the adult emergence considerably but diflumuron showed comparatively poor activity. Inhibition of DBM larval development by neem oil and extracts has been reported by several authors (Kareem, 1980; Sombatsiri and Tigvattanont, 1984). In the present study the survival rate of second-instar larvae of DBM fed on leaves treated with 2 per cent neem oil emulsion was observed to be 10 per cent.

Toxicity of NSKWE and Insecticides on *A. plutellae*

There was no adverse effect of NSKWE (25 g/l) on the adult stage of *A. plutellae* (Table 3). Methamidophos and chlorfluazuron drastically reduced their longevity and induced rapid mortality within 24 hrs after introduction. But neither of these two insecticides nor NSKWE caused any significant reduction in adult emergency, when the cocoons were sprayed with them. These results indicate that NSKWE (25 g/l) does not have any adverse effect on adult and cocoon stages of *A. plutellae*, a natural enemy of DBM. Mansour et al., (1986) reported that neem extracts were considerably much more toxic to the pests than to the predacious mite *Phytoseiulus persimilis* and the predatory spider *Chiracanthium mildei*. Saxena et al., (1980) reported that parasitization of *Crocidolomia binotalis* larvae increased in fields sprayed with 50 per cent neem oil.

Field Evaluation of Neem Oil, *H. angustifolia*, *B. thuringiensis* and Conventional Insecticides

In *yala* 1989, plots treated with neem oil, chlorfluazuron, triflumuron, and root methanol extract of *H. angustifolia* gave significantly lower number of leaf-eating caterpillar larvae compared to the untreated control and other treatments (Table 2). Significantly low leaf damage was observed in plots treated with neem oil and chlorfluazuron. However, no significant difference in yield reduction was observed among different treatments.

From the results of the *maha* 1989–90 evaluation, where each leaf-eating caterpillar species was recorded separately, all treated plots except methamidophos reported significantly low number of DBM larvae than the untreated control (Table 4). But there was no significant difference among different insecticide treatments. In the case of *Plusia eriosoma*, a significantly low number of larvae was reported only in plots treated with neem oil. The larval counts of the *Crocidolomia binotalis* and *S. litura* did not show any significant difference among treatments. Significantly low leaf damage incident and yield reduction were observed in plots treated with neem oil, chlorfluazuron, ethofenprox, and triflumuron than other treatments.

Adhikary (1980) reported that methanol extract of NSK powder at 2 per cent and 4 per cent on cabbage reduced the DBM damage and led to higher yield with better

Table 3. Effect of insecticides and neem seed kernel water extract on *A. plutellae*

Insecticide	Dosage	Adult mortality (%)	Adults emerged from treated cocoons (%)
Methamidophos (Tamaron LC 60)	672 ppm ai	100 a	72 a
Chlorfluazuron (Atabron 5EC)	100 ppm ai	98 a	72 a
NSKWE	25 g/l	38 b	75 a
Untreated control		34 b	80 a
CV (%)		7.4	10.5

In each column, the values followed by the same letter do not differ significantly according to DMRT at p = 0.05.

Table 4. The efficacy of different pest control agents against leaf-eating caterpillars of cabbage (*maha* 1989/90)

Insecticide	Rate/ha	No. of larvae				Leaf damage	% Yield loss
		P. xylostella	*P. eriosoma*	*C. binotalis*	*S. litura*		
Triflumuron (Alystine 480 SC)	200–400 g ai	0.22 a	2.22 a	0.00	13.3	1.00 a	0.25 cd
B. thuringiensis (Thuricide HP)	500–1000 g	0.00 a	1.66 a	0.11	3.33	1.66 ab	4.48 bcd
Chlorfluazuron (Atabron 5 EC)	50–100 g ai	0.11 a	1.77 a	0.00	7.22	1.00 a	0.09 d
Neem oil 2% emulsion	11.0–16.0 l	0.00 a	0.55 b	0.00	0.22	1.00 a	0.00 d
Ethofenprox (Tribon 10 EC)	75–150 g ai	0.11 a	2.25 a	0.00	6.11	1.00 a	0.65 cd
Cartap (Padan 50 SP)	500–750 g ai	0.44 a	2.86 a	0.11	50.55	2.00 b	5.29 abc
Methamidophos (Tamaron 60 LC)	420–500 g ai	1.33 b	4.11 a	0.00	8.77	2.00 b	6.33 ab
Untreated control		4.88 c	4.00 a	0.00	3.44	1.66 ab	14.25 a
CV %		25.1	29.8		111.0	15.2	35.5

In each column, the values followed by the same letter do not differ significantly according to DMRT at $p = 0.05$.

quality than mevinphos at 0.05 per cent. Ramprasad et al., (1983) reported NSK suspension to be as effective as some synthetic pyrethroids and some other insecticides against *S. litura*. Sombatsiri and Tembookeat (1986) reported that improved aqueous NSK extract is as effective as cypermethrin against the same pest. Fagoonee (1986) reported that acetone extract of NSK was as effective as deltametrin (Decis) against DBM and *C. binotalis*.

The present results indicate that 2 per cent emulsion of NSK oil is as effective as the recommended concentrations of chlorfluazuron, ethofenprox, and triflumuron and more effective than diflumuron, *B. thuringiensis*, cartap, and methamidophos against leaf-eating caterpillar complex on cabbage where DBM and *Plusia eriosoma* are predominant.

Neem Seed Kernel Extracts: Different Neem Formulations

Chlorifluazuron and all four NSK formulations gave significantly low number of DBM, *C. binotalis*, and total caterpillar larvae than the untreated control (Table 5). But no significant difference was observed among these different treatments. The plots treated with NSKWE, chlorfluazuron, and NSK methanol extracts showed lowest leaf damage incidence. However, all treatments showed significantly lower yield reduction due to caterpillar damage when compared with the untreated control. Ascher et al., (1983) reported that extracts prepared with polar solvents such as methanol, ethanol, and water were the most effective against *S. litura*. Feuerhake (1983) reported that water, methanol, ethyl-ethylketone, and azerotropic mixture of methyl tertiary butyl ether and methanol were the most efficient in the extraction of active compounds of neem seeds.

Present results showed that all neem formulations used were as effective as chlorfluazuron for controlling leaf-eating caterpillar complex and obtaining quality cabbage. But NSKWE could be selected as the most suitable formulation for use by small-scale farmers because they need very simple processing with low cost.

Significantly low number of *A. plutellae* cocoons were observed in plots treated with all neem formulations and chlorfluazuron when compared with the untreated control (Table 5). However, the results presented in Table 3 indicated that NSKWE had no detrimental effect on either the adult of the cocoon stage of *A. plutellae* while chlorfluazuron and methamidophos had an adverse effect on the adult stage but not on the cocoon stage. Thus, the significant reduction of *A. plutellae* population on cabbage plants treated with neem formulations should be further investigated.

Comparison of Two Action Threshold with NSKWE and Chlorfluazuron

All treatments reported significantly low number of CLE, low leaf damage, low percentage yield loss, and low number of heads destroyed by the larvae when compared with the untreated control (Table 6).

The lowest yield reduction was observed in all chlorfluazuron-treated plots and plots treated with NSKWE at > 2 CLE. However, spraying of NSKWE at > 5 CLE showed a higher yield loss. All chlorfluazuron treatments and treatment of NSKWE at > 2 CLE level gave the lowest number of heads destroyed by the larvae. Plots treated with NSKWE at 10-day intervals and NSKWE > 5 CLE level showed more heads

Table 5. The effect of different neem seed kernel extracts and chlorfluazuron against leaf-eating caterpillars and *A. plutellae* on cabbage

Treatment	Rate/ha conc.	No. of caterpillars/10 plants/count				Leaf damage[a]	No. of *A. pleutellae* cocoons	% Yield loss
		P. xyl tella	*C. binotalis*	*P. eriosoma*	*S. litura*			
NSK[b] methanol extract	1.0%	3.40 bc	0.40 b	0.00	0.20	1.23 cd	0.00 b	1.80 b
NSK petroleum extract	1.0%	4.4 bc	0.20 bc	0.00	0.66	1.45 bc	0.20 b	7.10 b
NSK water extract	25 g/l	2.7 bc	0.20 bc	0.06	1.50	1.20 d	0.00 b	6.80 b
NSK oil	11.0–16.0 l	5.0 b	0.00 c	0.00	0.30	1.46 b	0.0 b	16.50 ab
Chlorfluazuron (Atabron 5 EC)	50–100 g ai	1.90 c	0.00 c	0.06	1.40	1.26 bcd	0.00 b	0.00 b
Untreated control		14.4 a	1.66 a	0.06	4.70	2.86 a	1.73 a	64.30 a
CV %		13.49	43.5	ns	142.0	4.8	82.6	40.0

[a]1–4 scale; 1 = 0–5 per cent leaf damage, 2 = 6–20 per cent leaf damage, 3 = 21–40 per cent leaf damage, 4 > 40 per cent leaf damage
[b]NSK-Neem seed kernel

Table 6. Effect of different action thresholds with neem seed kernel extract and chlorfluazuron on leaf-eating caterpillar population, their leaf damage, and yield loss of cabbage

Treatment	Average no. of of larvae CLE*	Leaf damage (1–4 scale)	Yield loss (%)	Heads destroyed (%)
Chlorfluazuron at > 2 CLE/10 plants	10.30 b	1.17 b	0.69 c	1.25 d
Chlorfluazuron at > 5 CLE/10 plants	9.52 b	1.25 b	1.14 c	0.83 d
Chlorfluazuron at 10 day intervals	3.70 b	1.10 b	0.63 c	0.00 d
NSKWE at > 2 CLE/10 plants	5.82 b	1.25 b	0.78 c	2.08 cd
NSKWE at > 5 CLE/10 plants	9.47 b	1.37 b	5.45 b	9.58 b
NSKWE at 10 day intervals	4.80 b	1.20 b	5.75 bc	6.66 bc
Untreated control	52.05 a	2.82 a	51.32 a	61.25 a
CV (%)	30.2	15.1	22.5	35.1

*CLE: Cabbage looper equivalents.
In each column, the values followed by the same letter do not differ significantly according to DMRT at $p = 0.05$.

destroyed by the larvae. These results seem to indicate that it is advisable to use a CLE > 5 threshold for chlorfluazuron and a little lower threshold for NSKE.

Effect of NSKWE on Quality of Cabbage

There was no difference in cooking time, fibre content, texture, moisture content, taste, and smell between NSKWE-treated and untreated cabbage (Table 7). Adhikary (1980) also reported no adverse effect of neem on the taste or smell of the cooked cabbage.

CONCLUSIONS

Neem products such as NSKWE, NSK oil, NSK methanol, and NSK petroleum ether extracts are effective as recommended rates of the commercial insecticides chlorfluazuron, ethofenprox, and triflumuron and more effective than methamidophos for controlling the leaf-eating caterpillar complex of cabbage. Further, neem products such as NSKWE and NSK oil are very cheap. Their use in cabbage caterpillar control gives some additional benefits to the farmer by not harming natural enemies of cabbage caterpillars. Furthermore, neem application does not bring any undesirable taste, flavour or texture to the treated boiled cabbage.

ACKNOWLEDGEMENTS

The authors wish to thank Mr. R.P. Wimalasena and Mr. R.G. Wimalasena for their technical assistance and Mr. S. Ekanayake and his staff for their help in quality testing work.

Table 7. Effect of neem seed kernel water extract on the quality of treated, cabbage

Treatment	Moisture (%)	Cooking time (min.)	Fibre (%)	Texture	Remarks
NSKWE 25 g/l	91.9	9.50	0.712	Soft	No undesirable taste of smell
Untreated control	92.1	11.25	0.627	Soft	
CV %	15.1	18.4	12.0		

REFERENCES

Adhikary, S. 1980. The Togo experience in moving from neem research to its practical application for plant protection. In: *Natural Pesticides from the Neem Tree, Proc., 1st Int. Neem Conf.,* H. Schmutterer, K.R.S. Ascher and H. Rembold (eds.), GTZ. Eschborn, West Germany, pp. 215–222.

Ascher, K.R.S., M. Eliahu, N.E. Nemny and J. Meisner. 1983. Neem seed kernel extract as an inhibitor of growth and fecundity in *Spodoptera litura*. In: *Natural Pesticides from the Neem Tree and Other Tropical Plants, Proc., 2nd Int. Neem Conf.,* H. Schmutterer and K.R.S. Ascher (eds.), GTZ, Eschborn, West Germany, pp. 223–250.

Fagoonee, I. 1987. Use of neem in vegetable crop protection in Mauritius. In: *Natural Pesticides from the Neem Tree and Other Tropical Plants, Proc. 3rd Int. Neem Conf.,* H. Schmutterer and K.R.S. Ascher (eds.), GTZ, Eschborn, West Germany, pp. 95–100.

Feuerhake, K.J. 1983. Effectiveness and selectivity of technical solvents for the extraction of neem seed compounds with insecticidal activity. In: *Natural Pesticides from the Neem Tree and Other Tropical Plants, Proc. 2nd Int. Neem Conf.,* H. Schmutterer and K.R.S. Ascher (eds.), GTZ, Eschborn, West Germany, pp. 103–114.

Kareem, A. 1980. Neem as an antifeedant for certain phytophagous insects and bruchid on pulses. In *Natural Pesticides from the Neem Tree, Proc. 1st Neem Conf.,* H. Schmutterer, K.R.S. Ascher and H. Rembold (eds.), GTZ, Eschborn, West Germany, pp. 223–250.

Mani, M. and A. Krishnamoorthy. 1984. Toxicity of some insecticides to *Apanteles plutellae*, a parasite of the diamond back moth. *Trop. Pest Mgmt.* 30(2): 130–132.

Mansour, F., K.R.S. Ascher and N. Omari. 1986. Effect of neem seed kernel extracts from different solvents on the predacious mite *Phytoseiulus persimilis* and the phytophagous mite *Tetranychus cinnabarinus* as well as on the predatory spider *Chiracanthium mildei.* In: *Natural Pesticides from the Neem Tree and Other Tropical Plants, Proc. 3rd Neem Conf.,* H. Schmutterer and K.R.S. Ascher (eds.), GTZ, Eschborn, West Germany, pp. 577–587.

Ramprasad, G., S. Sitaramaiah, B.G. Joshi and R.S.N. Rao. 1984. Relative efficacy of neem seed kernel suspension, some synthetic pyrethroids and insecticides as plant protection agents against *Spodoptera litura* in tobacco nurseries. *Neem Newsletter* 1: 6.

Saxena, R.C. 1988. Insecticides from neem. In: *Insecticides of Plant Origin*. J.T. Arnason, B.J.R. Philogene and P. Moranc (eds.), ACS Symposium Series No 387. American Chemical Society, Washington, D.C., pp. 220–232.

Saxena, R.C., G.P. Waldbauer, N.J. Liquido and B.C. Puma. 1980. Effects of neem seed oil on the rice leaf folder, *Cnaphalocrocis medinalis*. In: *Natural Pesticides from Neem Tree. Proc. 1st Int. Neem Conf.,* H. Schmutterer, K.R.S. Ascher and H. Rembold (eds.), GTZ, Eschborn, West Germany, pp. 189–204.

Shelton, A.M., M.K. Sears, J.A. Wyman and T.C. Quick. 1983. Comparison of action thresholds for lepidopterous larvae on fresh marker cabbage. *J. Econ. Entomol* 76(1): 196–199.

Sombatsiri, K. and K. Tembookeat. 1986. Efficacy of an improved neem kernel extract in the control of *Spodoptera litura* and of *Plutella xylostella* under laboratory conditions and in field trials. In: *Natural Pesticides from the Neem Tree and Other Tropical Plants, Proc. 3rd Int. Neem Conf.,* H. Schmutterer and K.R.S. Ascher (eds.), GTZ, Eschborn, West Germany, pp. 195–203.

Sombatsiri, K. and Tigvattanont. 1984. Effects of neem extracts on some insect pests of economic importance in Thailand. In: *Natural Pesticides from the Neem Tree and Other Tropical Plants, Proc. 2nd Int. Neem Conf.,* H. Schmutterer and K.R.S. Ascher (eds.), GTZ, Eschborn, West Germany, pp. 95–100.

57

Neem Oil as Possible Biorational Insecticide for the Management of Spotted Pod Borer (*Maruca testulalis* Geyer) in Early Pigeon Pea

R.K. AKHAURI, M.M. SINHA AND R.P. YADAV

Department of Entomology, Rajendra Agricultural University, Pusa, Samastipur 848 125, Bihar, India

ABSTRACT

Since the cultivation of early pigeon pea in Bihar is almost impossible without protection against the spotted pod borer (*Maruca testulalis* G.) and because its control with chemical insecticides is ecologically undesirable and unaffordable, field trials in randomized block design with the early pigeon-pea cv UPAS-120 as test crop were conducted during 1990–91 and 1991–92 to compare the efficacy to two indigenous natural plant products. Neem oil (2.0 per cent) proved better than mahua oil (2.0 per cent) spray in significantly reducing the mean pod damage (27.2 to 18.9 per cent) while increasing the mean number of matured pods (67.4 to 85.8/plant) and grain yield (13.5 to 17.4 g/plant) in comparison to untreated check. Neem oil, therefore, might be exploited more efficiently as a biorational insecticide in pod borer management in early pigeon-pea cultivation, hitherto uneconomical, in Bihar.

INTRODUCTION

The spotted pod borer (*Maruca testulalis* G.) is one of the most serious insect pests of early pigeon pea, both in geographical distribution as in damage potential. It causes direct losses by injuring both flowers and pods. According to an estimate, the extent of damage to pods alone varies widely from 5.7 to 89.4 per cent under Bihar conditions (Sinha et al., 1979). It is almost impossible to grow the early pigeon pea crop in this state without protecting against the ravages of pod borer. Although a good number of chemical insecticides are known to be effective (Saxena, 1978, 1981; Samalo and Patnaik, 1986; Bhalani and Parsana, 1987), very little work seems to have been done on management of the pod borer by botanical products (Singh et al., 1985). In view of the ecological implications as well as costs of insecticides, plant products need to be exploited. The present study, was therefore, undertaken to evaluate the field efficacy of two indigenously available non-edible plant products—neem, (*Azadirachta indica* A. Juss.) and *mahua* (*Bassia latifolia* L.) oils—as biorational insecticides and possible alternatives or substitutes for chemical insecticides against *M. testulalis* in early pigeon-pea.

MATERIALS AND METHODS

Field trials in randomized block design were conducted for evaluating the relative efficacy of neem and mahua oils against spotted pod borer (*M. testulalis*) in early pigeon pea cv UPAS-120 during the 1990–91 and 1991–92 crop seasons at the research farm of Tirhut College of Agriculture, Dholi (Muzaffarpur), Bihar. Neem and *mahua* oils each at 2.0 per cent concentration and untreated control were included in the trial. Each treatment was replicated six times. The crop was sown on 21 July in both years and the experimental plots received uniform agronomic practices. Plot size was 3.0×1.8 m^2. Row-to-row and plant-to-plant spacings were 0.3 m and 0.5 m, respectively. The phytoproducts under study were applied twice, once at 50.0 per cent flowering and again at 50.0 per cent podding.

Observation on number of larvae of *M. testulalis* on 10 randomly selected plants were recorded on each plot, at one day before and seven days after the first as well as second round of treatments by exposing and closely examining webbed flowers (racemes) and damaged pods. The data on the average number of larvae/10 plants for each plot and year as well as the pooled data were subjected to statistical analysis and the treatment effect was compared at 5.0 per cent probability level. The reduction in natural population of the pest after each round of treatment application was worked out by using the formulae of Handerson and Tilton (1955). The data thus obtained were further subjected to statistical analysis and the differential efficacy of the treatments was compared at 5.0 per cent probability level.

Overall performance of the non-edible oils under study was judged by average number of pods per plant, mean per cent damaged pods, mean per cent reduction in pod damage over control (i.e., preventable damage), and mean grain yield (g) per plant, recorded on five randomly selected plants in each plot at harvest. The data pertaining to each of these parameters were subjected to statistical analysis for both the seasons separately and also on pooled basis.

RESULTS AND DISCUSSION

From the data summarized in Table 1, it became quite obvious that both neem oil and *mahua* oil each at 2.0 per cent concentration as spray were able to reduce the larval population of *M. testulalis* on early pigeon pea cv UPAS-120. Although the two oils under study consistently proved superior to control in suppressing the pest population in either crop season, neem oil gave the best results when compared at 5 per cent level of significance. During 1990–91, at seven days after the first and second sprayings of neem oil, the mean number of larvae/10 plants was 30.7 to 21.0, respectively, as against

Table 1. Relative efficacy of *mahua* and neem oils in controlling the field population of *Ṁaruca testulalis* G. in early pigeon pea cv UPAS-120

Crop season	Treatment		Mean no. of pest larvae/10 plants 7 days after			Mean % reduction in pest population 7 days after		
			1st spray	2nd spray	Mean	1st spray	2nd spray	Mean
1990-91	Neem oil	(2.0%)	30.7 (5.54)*	21.0 (4.57)*	25.8	48.6 (44.18)**	72.2 (58.20)**	60.4
	Mahua oil	(2.0%)	36.7 (6.03)	33.0 (5.78)	34.8	44.7 (41.93)	64.2 (53.31)	54.4
	Control		72.7 (8.53)	130.7 (11.45)	ſ01.7			
	SE (Tr.)	Mean	(0.51)	(0.24)		(1.54)	(1.76)	
	CD at 5%		(0.77)	(0.37)		(2.34)	(2.68)	
1991–92	Neem oil	(2.0%)	41.0 (6.44)	22.3 (4.78)	31.6	56.1 (48.50)	64.2 (53.28)	60.1
	Mahua oil	(2.0%)	55.0 (7.45)	37.7 (6.15)	46.3	42.0 (40.33)	56.7 (48.84)	49.3
	Control		91.0 (9.56)	163.7 (12.81)	127.3			
	SE (Tr.)	Mean	(0.15)	(0.19)		(1.56)	(1.53)	
	CD at 5%		(0.23)	(0.29)		(2.36)	(2.33)	
Pooled	Neem oil	(2.0%)	35.8 (6.02)	21.7 (4.70)	28.7	52.4 (46.36)	68.2 (55.71)	60.3
(1990–91 and	*Mahua* oil	(2.0%)	45.8 (6.80)	35.3 (5.98)	40.5	43.4 (41.15)	60.5 (51.04)	51.9
1991–92)	Control		81.8 (9.06)	147.2 (12.15)	114.50			
	SE (Tr.)	Mean	(0.23)	(0.14)		(1.24)	(1.27)	
	CD at 5%		(0.34)	(0.21)		(1.89)	(1.93)	

* Figures in parentheses are the values of $\sqrt{x}$ transformation.
** Figures in parentheses are the values of angular transformation.

Table 2. Effect of *mahua* and neem oils on pod damage due to *Maruca testulalis* G. and grain yield in early pigeon pea cv UPAS-120

Crop season	Treatments		Mean % of damaged pods at maturity	Mean % reduction in pod damage	Mean no. of pods/ plant at harvest	Mean grain yield (g)/plant	Mean % increase in grain yield
1990–91	Neem oil	(2.0%)	19.1 (25.93)*	33.2	98.4 (9.94)**	19.9	19.2
	Mahua oil	(2.0%)	25.2 (30.14)	11.9	96.7 (9.85)	18.6	11.4
	Control		28.6 (32.28)		77.4 (8.82)	16.7	
	SE (Tr.)	Mean	(0.82)		(0.27)	1.24	
	CD at 5%		(1.24)		(0.41)	1.86	
1991–92	Neem oil	(2.0%)	18.7 (25.57)	27.5	73.1 (8.54)	14.8	43.7
	Mahua oil	(2.0%)	24.3 (29.53)	5.8	64.3 (8.02)	11.5	11.7
	Control		25.8 (30.54)		57.3 (7.60)	10.3	
	SE (Tr.)	Mean	(0.84)		(0.44)	1.32	
	CD at 5%		(1.26)		(0.66)	1.98	
Pooled (1990–91 and 1991–92)	Neem oil	(2.0%)	18.9 (25.77)	30.4	85.8 (9.27)	17.4	31.5
	Mahua oil	(2.0%)	24.8 (29.85)	8.9	80.5 (8.99)	15.0	11.6
	Control		27.2 (31.42)		67.0 (8.24)	13.5	
	SE (Tr.) Mean		(0.57)		(0.26)	0.95	
	CD at 5%		(0.86)		(0.39)	1.43	

* Figures in parentheses are the values of angular transformation.

** Figures in parentheses are the values of $\sqrt{x}$ transformation.

72.7 and 130.7 larvae/10 plants in untreated control. A similar trend was observed during the next crop season. When the results of both seasons were pooled together, the neem oil, closely followed by *mahua* oil, registered the best performance in suppressing the field population of *M. testulalis*. Neem oil gave better performance than *mahua* oil consistently in both years. The former reduced the pest population by more than 60 per cent, whereas the latter reduced it by 49.3 to 54.4 per cent only over the control. A critical examination of the data would also reveal that the pest population in the untreated control continued to increase between the two sprayings, whereas both neem and *mahua* oils continued to suppress the pest in the corresponding periods, the former being more persistent and effective than the latter.

In protection against pod damage from *M. testulalis* neem oil established its superiority over *mahua* oil in both years (Table 2). The performance of *mahua* oil was statistically at par with the untreated control. The spraying of neem oil (2.0 per cent) resulted 27.5 to 33.2 per cent reduction in pest-caused pod damage over control, whereas the *mahua* oil (2.0 per cent) spray prevented only 5.8–11.9 per cent damage. The crop treated with neem oil recorded a significantly greater number of pods per plant at harvest than the untreated one. Similarly, the mean grain yield per plant was found to increase by 19.2–43.7 per cent with neem oil, as against 11.5 per cent with *mahua* oil.

The present findings, thus, gave an indication that neem oil had a better potential than *mahua* oil as a botanical insecticide in controlling the population of *M. testulalis*, reducing pod damage, and increasing the number of pods as well as the grain yield in early pigeon-pea. More studies are, however, required for determining the frequency of application and dosage of this botanical insecticide for its effective exploitation in the management of the pod borer, in order to ensure the profitable cultivation of early pigeon-pea in Bihar. Since no attempt seems to have been made to ascertain the efficacy of this oil against *M. testulalis*, the present findings assume greater importance. According to an earlier report, ethanolic extract of neem seed kernel extract (2.0 per cent) is known to be effective against pod borer (Singh et al., 1985).

ACKNOWLEDGEMENT

The authors are thankful to the Principal, Tirhut College of Agriculture, Dholi (Muzaffarpur), Bihar, for providing the facilities to carry out the present investigation.

REFERENCES

Bhalani, P.A. and G.J. Parsana. 1987. Relative toxicity of certain newer insecticides to *Maruca testulalis* (Geyer) attacking pigeonpea. *Pesticides* 21(4): 23–25.

Handerson, C.F. and E.W. Tilton. 1955. Tests with insecticides against brown wheat mite. *J. Econ. Ent.* 48: 158–161.

Samalo, A.P. and H.P. Patnaik. 1986. Efficacy of different insecticides and their frequency of application on the pod borers infesting pigeonpea. *Madras Agric. J.* 73(6): 352–354.

Saxena, H.P. 1978. Pests of grain legumes and their control in India. In: *Pests of Grain Legumes, Ecology and Control*. S.R. Singh, H.F. Van Emden and T.A. Taylor (eds.), Academic Press, London pp. 15–23.

Saxena, H.P. 1981. Insect pests of arhar. *Indian Fmg.* 31(12): 16–17.

Singh, R.P., Yashbir Singh and S.P. Singh. 1985. Field evaluation of neem (*Azadirachta indica* A. Juss.) seed kernel extracts against the pod borers of pigeonpea, *Cajanus cajan* (L.) Millsp. *Indian J. Ent.* 47(1): 111–112.

Sinha, M.M., R.P. Yadav and Ashok Kumar. 1979. Multi-directional approach for pest management in arhar (*Cajanus cajan*) in Bihar. *Pesticides* 13(11): 14–16.

58

Effects of the Extracted Substances from Thai Neem Seed, *Azadirachta indica* var. *siamensis*, on the Insect Pests of Chinese Kale

KWANCHAI SOMBATSIRI

Deptt. of Entomology, Kasetsart University, Bangkok 109 000, Thailand

ABSTRACT

Thai neem seed extracts were tested for the control of insect pests on Chinese kale, namely, *Plutella xylostella*, *Spodoptera litura*, and *Hellula undalis*. The experiment was carried out on Chinese kale field in randomized complete block design with five treatments and three replications. The small plot is 3 × 3.5 m^2. Various neem formulations were sprayed on the plots right on the 10th day after sowing and continued once every five days until seven days before harvest. All plants treated with neem extracts appeared to have withstood *P. xylostella* and *S. litura* and even showed better resistance to *H. undalis*. Methanolic neem extract at 5 per cent concentration probably was the best among the neem formulations in controlling insects and resulted in greater marketable yield.

INTRODUCTION

Diamond back moth (DBM), *Plutella xylostella*, is one of the serious pests of cruciferous plants in Thailand. This insect has developed rapid resistance to many conventional insecticides. In areas with heavy infestations, total yield loss may be expected. *Spodoptera litura* and *Hellula undalis* are also found to cause a great deal of damage to crucifers all year round.

New control methods and synthetic insecticides have been used to solve the problem of DBM outbreak, but DBM is still a serious pest of crucifer. Neem extract is one of the promising agents to control DBM and other insect pests of crucifer. Adhikary (1981) applied 4 per cent neem seed methanolic extract to control DBM and the result was better than with the synthetic insecticide mevinhos (0.05 per cent). Sombatsiri and Temboonkeat (1987) showed a similar effect in controlling DBM and tobacco cutworm (*S. litura*) in both laboratory and field trial. Moreover, no sign of resistance development to neem extract after treatment up to 42 generations was encountered, whereas larvae treated with deltamethrin developed 20 and 30 times resistance in feeding and fecundity tests respectively (Vollinger, 1987).

A study of neem extracts sprayed on Chinese kale field, where DBM is a serious problem, will provide information on its potential as an insecticide in controlling insect pests of Chinese kale.

MATERIALS AND METHODS

Preparation of the Extracts

The first treatment is untreated control. The second treatment is methanolic neem seed extract 5 per cent (MNSE) prepared by blending 500 g powder of Thai neem depulped seed in methanol for 1 min. The mixture is left to stand for one night, then filtered and condensed by evaporation to a volume of approximately 100 ml. When used, the extract is diluted with 6 l of water to obtain 5 per cent concentration. The third treatment consisted methanolic neem seed extract (p) 0.25 per cent (MNPE) prepared by the same procedure as in the second treatment (the extract is evaporated to dryness as paste) plus neem oil. Fifty grams of paste is dissolved in methanol and then mixed with 20 ml of neem oil and 10 ml emulsifier to make 100 ml solution. The desired concentration is 0.25 per cent prepared by diluting 30 ml of the solution mentioned in 6 l of water. The fourth treatment is methanolic neem seed extract (p) 0.5 per cent (MNPE) prepared as in the third treatment, but it is not mixed with neem oil. Thirty grams of neem paste is dissolved in methanol and diluted in 6 l of water to obtain 0.5 per cent concentration. The fifth treatment is the conventional insecticide Argona (*Bacillus thuringiensis*) mixed with carbosulfan.

Field Layout and Data Collection

The experiment is designed in randomized completed block with plots of 3 × 3.5 m^2. Each treatment is replicated three times. The first spray begins on the 10th day after sowing and is continued once every five days until seven days before harvest. The control plot is left unsprayed. The number of *P. xylostella*, *S. litura*, and *H. undalis* larvae are visually counted at random of 15 plants per replication on the day proceeding spray. The total yield and yield quality (marketable yield) are also recorded.

Results and Discussion

The larval populations of *P. xylostella* and *S. litura* were relative high, but that of *H. undalis* was low. All neem extracts (MNSE 5 per cent, MNPE 0.25 per cent and 0.50 per cent) could control *P. xylostella*, *S. litura*, and *H. undalis* larvae in low level in all seven counts (Figs. 1, 2, and 3), whereas the larvae in untreated control gradually increased and reached the peak at the fifth count. The larval population among all

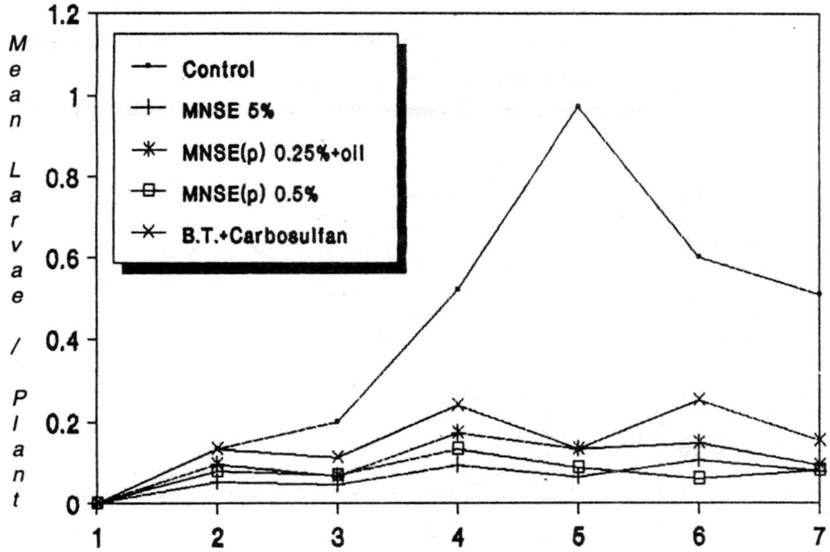

MNSE=Methanol neem seed extract
MNSE(p)=Methanol seed extract as paste
B.T=Bacillus thuringiensis

Fig. 1. The population of *P. xylostella* larvae in each count.

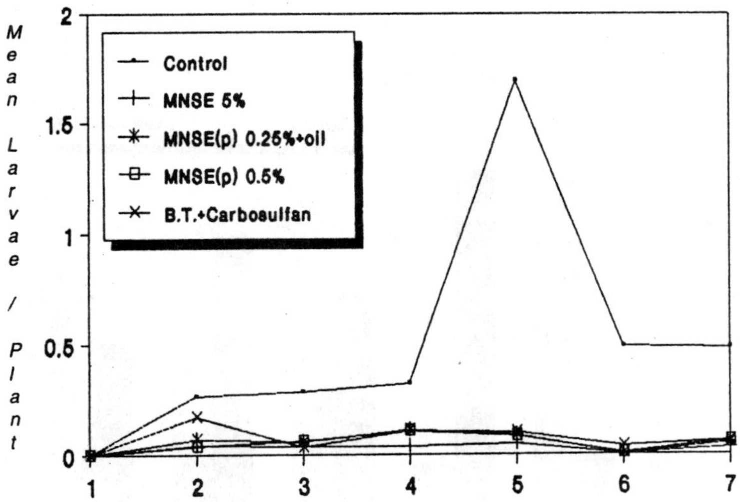

MNSE=Methanol neem seed extract
MNSE(p)=Methanol seed extract as paste
B.T=Bacillus thuringiensis

Fig. 2. The population of *S. litura* larvae in each count.

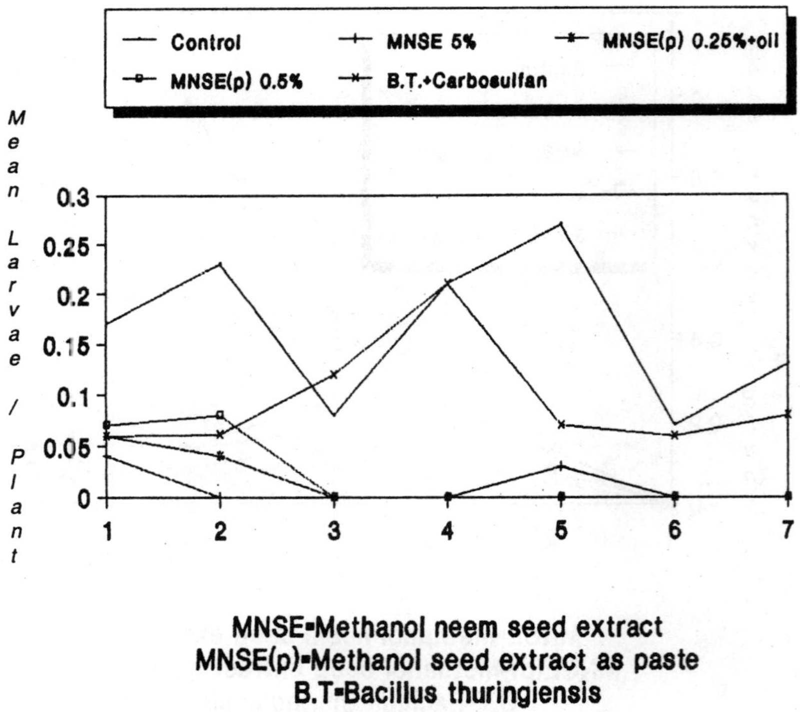

Fig. 3. The population of *H. undalis* larvae in each count

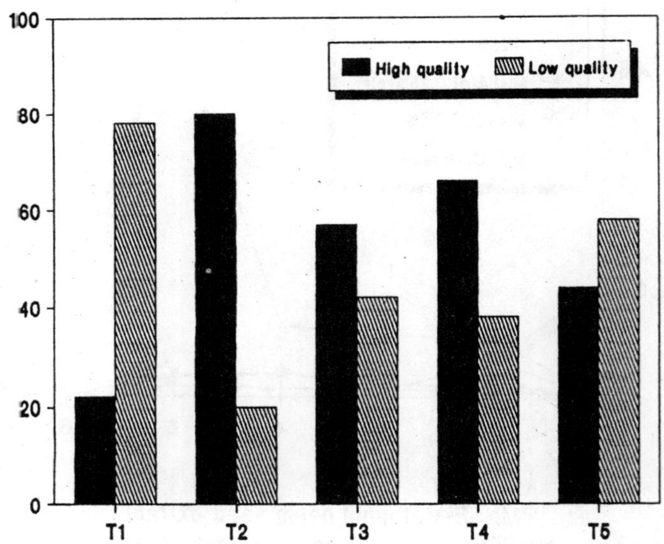

T_1 = Control T_2 = MNSE 5% T_3 = MNPE 0.25 % + oil
T_4 = MNPE 0.5% T_5 = B.T. + Carbosulfan

Fig. 4. Yield quality of Chinese kale.

treatments was not of statistical significance, but it was significantly different from the untreated control. Methanol neem extracts showed better control of *H. undalis* larvae than the B.T. and insecticide (Fig. 3). Although all neem extracts offered the same effectiveness in the control of these three insect species, plants treated with MNSE 5 per cent produced more marketable yield (high quality) with significant difference from those treated with MNPE 0.25 and 0.5 per cent (Fig. 4).

The same result was reported by Adhikary (1981) that application of alcoholic neem extract of Indian seed kernel at 2 and 4 per cent controlled *P. xylostella* larvae and resulted in high yield and better quality than spraying with mevinphos at 0.05 per cent. Fagoonee (1981) has also found that neem seed kernel extract from Indian variety was as effective as the commercial insecticide deltamethrin in the control of major insect pest of Chinese cabbage, cauliflower, and tomato.

REFERENCES

Adhikary, S. 1981. The Togo experience in moving from neem research to its practical application for plant protection. In: *Natural Pesticides from Neem Tree* (*Azadirachta indica* A. Juss.), H. Schmutterer, K.R.S. Ascher and H. Rembold (eds.) GTZ, Eschborn, FRG, Germany, pp. 215–222.

Fagoonee, I. 1981. Behavioral response of *Crocidolomia binotalis* to neem. In: *Natural Pesticides from Neem Tree* (*Azadirachta indica* A. Juss.). H. Schmutterer, KRS Ascher and H. Rembold (eds.), GTZ, Eschborn, Germany, pp. 543–554.

Sombatsiri, K. and K. Temboonkeat. 1987. Efficacy of an improved kernel extract in the control of *Spodoptera litura* and of *Plutella xylostalla* under laboratory conditions and in field trials. In: *Natural Pesticides from Neem Tree, and Other Tropical Plants,* H. Schmutterer and KRS Ascher (eds.) GTZ. Eschborn, FRG. pp. 195–203.

Vollinger, M. 1987. The possible development of resistance against neem seed kernel extract and deltamethrin in *Plutella xylostella*. In: *Natural Pesticides from Neem Tree* (*Azadirachta indica* A. Juss.) *and Other Tropical Plants*, H. Schmutterer and KRS Ascher (eds.), GTZ, Eschborn, FRG., pp. 543–554.

59

Exploratory Evaluation of Neem Extract Sprays on Maize for Control of Stalk Borers in Coastal Kenya

M. KIARIE[1] AND R.C. SAXENA[2]

[1]Regional Research Centre—Mtwapa, PO Box 16, Kikambala, Kenya
[2]ICIPE Field Station, Mbita, PO. Box 30, Mbita, Kenya

ABSTRACT

Neem extract at 3, 6, and 12 per cent was sprayed on maize (cv Coast Composite) for the control of maize stalk borers (*Chilo pertellus*) Swinh, at the Regional Research Centre—Mtwapa in Coastal Kenya during the 1992 long rain season. Dipterex (5 per cent G at 10 kg/ha) served as insecticide check and unprotected control plots were maintained. Extracts were sprayed thrice at three, four, and five weeks after plant emergence on plants which had been infested at five larvae (one to two days old) per plant one day before the first spray application. Neem reduced the leaf damage, tunnelling, and larval population. Control efficacy increased with the concentration. Neem extract 12 per cent was at par with dipterex. Yield increases with 3, 6, and 12 per cent neem extract were 17, 24, and 28 per cent respectively as compared with 30 per cent increase with insecticide.

INTRODUCTION

Maize stem borer *Chilo pertellus* Swinh is a serious constraint to maize production in Kenya (Ampofo et al., 1986). Studies carried out in the country indicate that 20 per cent of maize produced in the highlands is lost to stem borers. The losses increase to 40 per cent in the hot and humid coastal zone of Kenya and Kiarie. Presently, dipterex is found to be effective and is recommended for control of the pest (Warui and Kuria, 1983).

However, the use of insecticide is to be minimized and if possible avoided in view of its undesirable effect on the agricultural ecosystem (Sohati and Sithanantham, 1990).

In recent years there has been a rapid increase in interest in evaluating botanical insecticides in view of their relative safety to the environment (Schmutterer, 1990).

In the present study, the optimum dose of neem seed extract required in control of maize stem borers was evaluated. The potential of neem in replacing dipterex 5G was also assessed and the results are discussed herein.

MATERIALS AND METHODS

The field trials were conducted at Regional Research Centre—Mtwapa, at the Kenya coast, during the 1992 long rains. Coast Composite maize variety was used. Basal dosage of triple superphosphate fertilizer at 200 kg/ha was applied at the time of planting. Two seeds per hill were planted at a spacing of 75 cm between rows and 50 cm within the row. Five treatments consisting of neem extract 3, 6, and 12 per cent and control were tested while dipterex was used as a check for comparison. The trial was laid out in randomized block design with six replications in plots of six rows 4 m long.

Neem kernels were ground to fine powder and soaked overnight. To make a good emulsion 20 ml of Teepol liquid was added to the solution. The filtered extract was diluted to give the required concentrations for treatments at 3, 6, and 12 per cent. Extracts were sprayed thrice at three, four, and five weeks after emergence on plants which had been infested with five larvae (one to two days old) one day before the first spray application.

Infestation of maize by *Chilo* sp. was determined at five weeks by evaluating the leaf damage on a scale of 1–5 (where 1 represented no damage and 5 dead hearts). A sample of 20 random plants per plot was also taken and the plants dissected to count the number of larvae and pupae and to determine the tunnel length per stem. Larva and pupa count and tunnel length were also taken at the time of harvesting.

RESULTS

Table 1 shows that leaf damage was significantly less in neem extract 6 and 12 per cent. dipterex treatments were on par with 12 per cent neem extract. Unprotected plots showed the highest leaf stem borer damage. The number of larvae and pupae recovered from dissected stems was lowest in plots treated with neem extract 12 per cent and dipterex. The plots treated with neem extract 3 and 6 per cent and the unprotected plots showed no significant difference in the number of larvae recovered. Neem extract 12 per cent spray and dipterex 5G showed superior performance in reduction of tunnel length as compared to control plots. However, there was no significant difference in tunnel length between neem extract at 3, 6, and 12 per cent and dipterex 5G.

The highest grain yields were recorded in plots treated with neem extract 12 per cent and dipterex 5G. Yields in treatments with neem extract 3 and 6 per cent were still superior to those in unprotected plots.

Table 1. Assessment of the influence of neem extract sprays on plant damage, larval/pupal population and maize yield per hectare

Treatment	Leaf damage score	No. of larvae/ pupae	Tunnel length	Grain yield/ha
Neem extract 3%	2.53 b	0.55 a	4.19 ab	1664 b
Neem extract 6%	2.03 bc	0.5 a	3.82 ab	1725 b
Neem extract 12%	1.42 c	0.1 b	2.21 b	2934 a
Dipterex	1.38 c	0.28 ab	2.22 b	2799 a
Control	3.37 a	0.48 a	5.27 a	1617 b
Mean	2.146	0.382	3.542	2147.8
CV %	19.22	56.74	30.52	17.52
Probability	0.01	0.01	0.01	0.01
LSD	0.6562	0.3450	2.057	598.8

DISCUSSION

These preliminary results have shown that foliar spray of neem seed extract at 12 per cent can control maize stem borer *Chilo* sp. as well as the presently recommended insecticide dipterex 5G. Neem spray reduced leaf damage, larval and pupal population, and length of the stem tunnelled. Neem extract 12 per cent increased yields by 11 and 4 per cent over unprotected and dipterex 5G treatments respectively. However, the difference in yield between neem extract 12 per cent and dipterex 5G was not significantly different.

Neem spray, therefore, appears to be a potential substitute for chemical use in management of maize stem borer.

Neem is found in abundance on the Kenya coast and is widely spreading to the rest of the country. Hence, there is a great potential for future adoption of neem seed extract by farmers not only in the coastal zone but also in the rest of the country.

ACKNOWLEDGEMENTS

This paper was submitted to the World Neem Conference with permission of the Director, Kenya Agricultural Research Institute. A special tribute is paid to the Director. UNESCO (MAB) Dakar, for financing the authors, attendance at the conference, and for MAB's contribution in encouraging the use of natural resources. Thanks are also due to Dr. Sithanantham and Mr. Malinga, Director, RRC Mtwapa, for their invaluable consultation.

REFERENCES

Ampofo, J.K.O., K.N. Saxena, J.G. Kibua and E.O. Nyangiri. 1986. Evaluation of some maize cultivars for resistance to stem borer *Chilo partellus* (Swinhoe) in Western Kenya. *Maydica* XXXI(1986): 379–389.

Schmutterer, H. 1990. Properties and potential of natural pesticides from the neem tree, *Azadirachta indica.*, *Anr. Rev. Entomol.* 35: 271–297.

Sohati, P.H. and S. Sithanantham. 1990. Preliminary field evaluation of extracts of plants as sprays for controlling the bean maggot in Zambia. Food Legume Research Team, Masekera Regional Research Station. Special reports.

Warui, C.M. and J.N. Kuria. 1983. Population incidence and the control of maize stalkborers *Chilo partellus* (Swinh), *C. orichalcociliellus* strand and *Sesamia calamistis* Hmps, in Coast Province, Kenya. *Insect Sci. Appl.* 4(1/2): 11–18.

60

Repelin—An Effective Botanical Insecticide against Ber Fruitfly *Carpomyia vesuviana* Costa

V.V.S. REDDY, T. RAMESH BABU AND B. NARASIMHA RAO

Department of Entomology, College of Agriculture
Andhra Pradesh Agricultural University, Hyderabad 500 030, India

ABSTRACT

The incidence of fruitfly, *Carpomyia vesuviana* Costa, on ber was initiated from middle of October (13 per cent) and the peak activity (45 per cent) was observed during December, 1988 in APAU Orchard, Hyderabad. Studies were conducted to evaluate six insecticides—dichlorvos 0.05 per cent, malathion 0.05 per cent, carbaryl 0.15 per cent, phosalone 0.07 per cent, monocrotophos 0.05 per cent and Repelin 2 per cent—against this important pest. Repelin 2 per cent was the most effective with 6.2 per cent fruit damage followed by monocrotophos 0.05 per cent (8.1 per cent damage) and were on par as against 38.1 per cent fruit damage in control. The lesser incidence in Repelin treatment might be due to prevention of oviposition by the repellent activity. The tree that received monocrotophos realized highest fruit yield (42.3 kg) and was on par with the tree treated with Repelin (40.8 kg). Hence, Repelin 2 per cent can be recommended during fruiting season as it has the additional advantage of ecological safety.

INTRODUCTION

The ber fruitfly, *Carpomyia vesuviana* Costa, is an important pest of jujube (*Zizyphus* spp.) and is found to occur everywhere in India where ber is grown (Batra, 1953). In Haryana, it was reported to cause 16.8 per cent damage (Lakra et al., 1979). Basha (1952) reported 80–82 per cent combined infestation of fruitfly and fruit borer *Meridarchis scyrodes* Meyr. at Panyam, Andhra Pradesh. Jejurkar (1986) recorded 73 per cent damage in Maharastra. The influence of different insecticides against ber fruitfly and fruit borer was documented (Patel et al., 1989, Patel and Patel, 1979; Lakra, 1981;

Jejurkar, 1986). In view of the undesirable effects of unilateral dependence on synthetic chemicals, research for environmentally sound and eco-friendly methods to manage insect pests resulted in the encouragement of non-conventioanl methods of pest control. The adverse effects of neem products on pests and the successful use of neem from the control of crop pests were reported (Schmutterer, 1990). Repelin, a combination of neem and other plant products, was successfully used for the control of various crop pests (RARS, 1988). Hence, it was contemplated to establish the efficacy of Repelin against fruitfly and fruit borer of ber, since ber cultivation is gaining importance in dryland horticulture.

MATERIALS AND METHODS

The field survey to record the extent of damage of fruitfly and fruit borer on ber was conducted at three different locations around Hyderabad at fortnightly intervals. A hundred fruits were collected from each earmarked tree at the time of bearing at each observation to record the fruitfly and fruit borer infestation.

A field experiment was laid out at the Agricultural Research Institute, Rajendranagar, in a randomized block design with three replications for evaluating the insecticides. Each replication was represented by one tree. Six insecticides were included. The common, trade, and chemical names of the insecticides used and their concentrations and sources are listed in Table 1. A total of five sprayings were given at 15-day intervals. Five litres of spray fluid was used per tree. The ber fruitfly and fruit borer damage was recorded separately by collecting 100 fruits at random from harvested produce at each picking. The data were analysed statistically.

RESULTS AND DISCUSSION

The fruitfly damage was initiated (10–13 per cent) from the middle of October and a maximum infestation of 42–45 per cent was recorded during December, with a gradual decline from then onwards. Though fruit borer damage was noticed from October

Table 1. Particulars of the test chemicals

Common name (Trade name)	Formulation	Concentration (per cent)	Source
Dichlorvos (Nuvan)	100 EC	0.05	Ciba-Geigy India Ltd.
Malathion (Cythion)	50 EC	0.05	Cynamid India Ltd.
Carbaryl (Sevin)	50 WP	0.15	Union Carbide (I) Ltd.
Phosalone (Zolone)	35 EC	0.07	Voltas India Ltd.
Monocrotophos (Nuvacron)	36 EC	0.05	Ciba-Geigy India Ltd.
Repelin	100 EC	2.00	ITC

second fortnight, the peak infestation (40–45 per cent) was observed during second fortnight of January to first fortnight of February (Fig. 1).

Repelin 2 per cent and monocrotophos 0.05 per cent were the most effective treatments which recorded 6.2 and 8.1 per cent fruitfly infestation respectively followed by dichlorvos 0.05 per cent (13.2 per cent infestation), phosalone 0.07 per cent (16.5 per cent infestation), carbaryl 0.15 per cent (19.7 per cent infestation), and malathion 0.05 per cent (22.5 per cent infestation) as against 38.1 per cent in control (Table 2). Patel and Patel (1979) recorded 100 per cent control of ber fruitfly with monocrotophos. Similar results with moncrotophos were also reported by Lakra (1981) and Patel et al. (1989). The least per cent infestation with Repelin 2 per cent recorded in the present study may be attributed to the formation of a thin coating of the chemical over the ber fruit ultimately preventing the oviposition. However, no information is available to corroborate the present findings with Repelin.

Monocrotophos 0.05 per cent was found to be effective in checking the fruit borer infestation as it recorded only 6.9 per cent infestation as against 24.4 per cent in control. However, phosalone 0.07 per cent (8.7 per cent) and Repelin 2 per cent (9.9 per cent) were found to be statistically at par with monocrotophos 0.05 per cent. The tree that received monocrotophos 0.05 per cent realized highest fruit yield (42.3 kg) which recorded 112 per cent increase in yield over control followed by Repelin 2 per cent (40.8 kg) with 104.7 per cent increase over control and were found to be on par (Table 2). Patel et al. (1989) reported around 91 per cent increased yield over control with monocrotophos 0.04 per cent.

Monocrotophos 0.05 per cent was adjudged the most effective chemical, as it could control fruitfly and fruit borer besides contributing highest fruit yield. Repelin was

Table 2. Insecticidal efficacy against fruitfly and fruit borer of ber and yield

Treatment	Concentration %	Per cent infestation		Yield	
		Fruitfly	Fruit borer	Mean yield kg/tree	Per cent increase over control
Dichlorvos	0.05	13.2 (21.2)	12.5 (20.7)	36.4	82.6
Malathion	0.05	22.5 (28.3)	13.3 (21.3)	31.5	58.2
Carbaryl	0.15	19.7 (26.3)	14.3 (22.1)	34.0	70.6
Phosalone	0.07	16.5 (23.9)	8.7 (17.0)	37.8	89.7
Monocrtophos	0.05	8.1 (16.4)	6.9 (15.1)	42.3	112.0
Repelin	2.00	6.2 (14.2)	9.9 (18.3)	40.8	104.7
Control		38.1 (38.1)	24.4 (29.6)	19.9	0.0
CD (p = 0.05per cent)		4.96	5.16	4.28	

Figures in parentheses are angular transformed values.

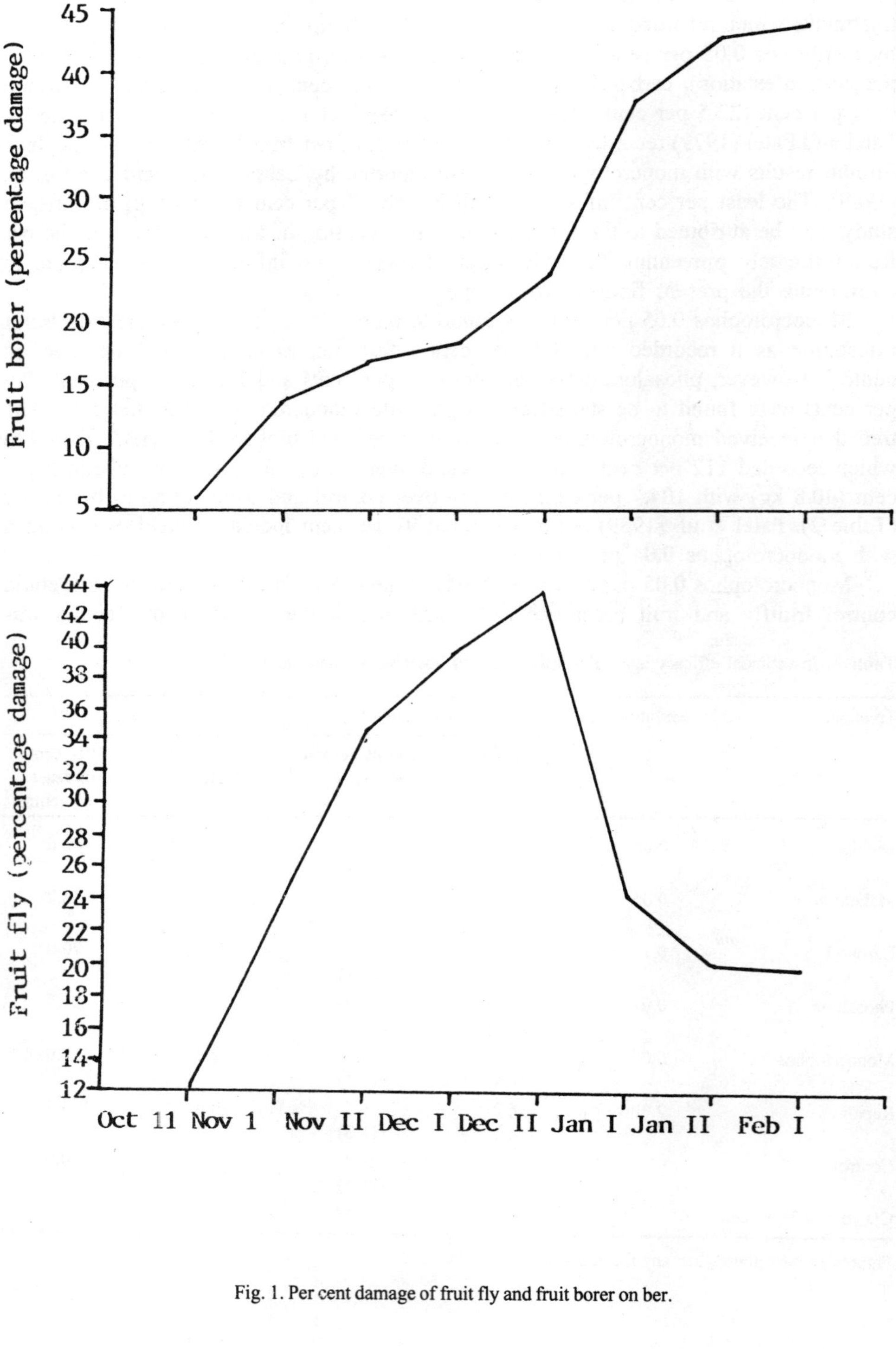

Fig. 1. Per cent damage of fruit fly and fruit borer on ber.

found to be effective against ber fruitfly and hence can be recommended during early fruiting season to avoid oviposition as it has the additional advantage of ecological safety over other chemicals tested. However, further multilocational investigations are necessary to confirm and to establish the mechanism action of this botanical insecticide.

REFERENCES

Basha, J.M.G. 1952. Experiments on the control of the fruit borer of *Jujube* (*Ziziphus* spp.) *Carpomyia vesuviana* Costa and *Meridarchis scyrodes* Meyr. in South India. *Indian J. Entomol.*, 14: 229–238.

Batra, H.N. 1953. Biology and control of *Dacus diversus* Coquillett and *Carpomyia vesuviana* Costa and important notes on other fruit flies in India. *Indian J. Agric. Sci.*, 23: 87–112.

Cherian, M.C. and C.V. Sundaram. 1941. The ber fruit fly *Carpomyia vesuviana* Costa and its control. *Proc. 28th Indian Science Congress III*, 191 p.

Jejurkar, N.K. 1986. Studies on ber fruit borer (*Meridarchis scyrodes* Meyr.) and ber fruitfly (*Carpomyia vesuviana* Costa) in respect of losses caused susceptibility to pesticides and harvest time residues of promising pesticides. M.Sc. (Ag) Thesis, Mahatma Phule Agricultural University, Maharashtra.

Lakra, R.K. 1981. Bionomics and control of ber fruit fly *Carpomyia vesuviana* Costa. Ph.D. Thesis, Haryana Agricultural University, Hissar.

Lakra, R.K., W.S. Kharub and Zile Singh. 1979. Relationship of shedding of ber fruits and the incidence of ber fruit fly and bird damage. *Indian J. Ecol.*, 6: 68–72.

Patel, B. and V.J. Patel. 1979. Impact of chemicals on ber (*Zizyphus mauritiana* Lamk.). *Pesticides*, 13: 28–30.

Patel, B.H., V.R. Upadhyay, C.M. Muralidharan and G.S. Judal. 1989. Comparative efficacy of different insecticides against ber fruit fly *Carpomyia vesuviana* Costa. *Indian J. Plant Prot.*, 17: 39–45.

Regional Agricultural Research Station, Lam. 1988. Annual Report. Effect of different Repelin treatments against cotton pests. pp. 43–48.

Schmutterer, H. 1990. Properties and potential of natural pesticides from the neem tree *Azadirachta indica*. *Annu. Rev. Entomol.*, 35: 271–294.

61

Evaluation of Enriched Neem (*Azadirachta indica* A. Juss.) Seed Extracts against Mango Hoppers, *Idioscopus nitidulus* Walker and *Amritodus atkinsoni* Leth.

R.P. SRIVASTAVA, R.P. SINGH*, M. HASEEB AND S.R. ABBAS

Central Institute of Horticulture for Northern Plains, B-217, Indira Nagar
Lucknow 226 016, India
*Division of Entomology, Indian Agricultural Research Institute
New Delhi 110 012, India

ABSTRACT

Two cheap neem seed enriched extracts, an oil-based and a cake-based were identified. Both the extracts proved very effective against mango-hoppers, *Amritodus atkinsoni* and *Idioscopus nitidulus* serious pests of mango.

In laboratory tests the oil-based concentrate at 1 per cent and the cake-based concentrate of 0.2 per cent concentration caused 100 per cent and 88.6 per cent mortality respectively of nymphs of *I. nitidulus*. The efficacy of the extract was also determined under field conditions against *I. nitidulus*. Oil-based concentrate offered the same percentage kill of nymphs as in the laboratory tests. The cake-based, however, did not show much promise.

Both the enriched extracts at the above concentrations, when tested for settling response of adults of *A. atkinsoni* under choice condition, only one fourth adults settled on plants treated with extracts in comparison to control.

INTRODUCTION

Growing public alarm over dangerous side effects of pesticides on public health and serious concern of natural imbalance have made it essential to find alternative methods for managing pests which are more economical, more lasting and less harmful to public health and the environment. Keeping in view of this fact different ways are being devised to be used in insect pest management. One such rationale is the use of plant extracts and other natural products which may be of immense help for combating insects.

Neem tree, *Azadirachta indica* is a unique source of insecticide as unlike synthetic insecticides, it has diverse biological effects on insects: repellent, feeding and oviposition deterrent, growth and development disruptor (IGR), sterilant etc. (Singh, 1990). Because of this unique property, not possessed by any plant, the tree has attracted attention of entomologist the world over. Neem extracts and pure compounds have been evaluated against more than 300 species of insect pests. There is, however, no report of any evaluation of neem against mango hoppers. We therefore envisaged the use of neem derivatives for the control of mango hoppers, *Idioscopus nitidulus* and *Amritodus atkinsoni*, the serious pests of mango, an important fruit crop of our country. Presently these pests are being controlled by spraying of different synthetic insecticides. Their use in 1987 caused resurgence of these hoppers. In the present study, two enriched extracts were evaluated for their effects on orientation, growth and development and survival against both the noxious pests.

MATERIALS AND METHODS

Two neem seed enriched extracts an oil-based and a cake-based were isolated. Both extracts were formulated as emulsion and evaluated against mango-hoppers. Oil-based formulation contained a non-polar solvent and polysorbate as emulsifier while cake-based formulation had a polar solvent and polysorbate as emulsifier. Formulation were diluted with water to get desired concentration. The concentrations were based on weight of the extracts.

Laboratory Trial

Test 1: This test was conducted for evaluation of effect of above neem extracts on nymphs of mango-hopper, *I. nitidulus*. Fresh mango panicles along with the nymphs of hopper were sprayed separately with 1 per cent oil-based and 0.2 per cent cake-based extracts. Panicles were kept in reagent bottles containing water to keep them fresh. Three replications were made for each formulation and each replicate had 30 nymphs. Similarly a check sprayed with emulsified water served as control. Reagent bottle were transferred in cages and kept at 25 ± 2°C. Observations were recorded on mortality, abnormal development and adult emergence, on 2nd, 5th and 9th day after treatment.
Test 2: : This test was conducted to see the effect of neem treatments on settling and ovipositional response of adults of *A. atkinsoni* under choice condition. For this test, fresh panicles of mango were treated with 1 per cent oil and 0.2 per cent cake-based extract and kept in a tube containing water to maintain its turgidity and freshness. Treated and control panicles were kept together at a distance of 30 cm in big glass cage. Two such sets were prepared for each treatment. Twenty five adults of *A. atkinsoni* were released in each cage. Control were same as in Test 1. Observations were recorded on their settling response at 10, 20 and 40 minutes and ovipositional response after 24 hours.

Field trials

Two introductory field trials were conducted to evaluate the efficacy of neem extracts for the control of mango hopper. In one trail 1 per cent oil-based extracts and in the other cake-based extract was used. In the first trial half of a mango tree (Var. Dashehari) with precounted population was treated with oil-based extract and the other half was kept as control which was treated with emulsified water. Panicles were bagged after spraying. Observations were recorded on population of hopper on 1st, 4th, 10th day after spraying. Similarly in the second trial 0.2 per cent cake-based extract was sprayed on a heavily infested mango tree while one control tree was sprayed with emulsified water. Observation were recorded on population of hoppers before spray and after ten days of spray. Ten panicles were selected randomly for this purpose. The mango hopper nymphs were counted by jerk method, panicle was given a gentle jerk thrice and the nymphs fallen on the paper were counted on a white sheet of paper. Average population of nymphs per panicle was calculated for interpretation of data.

On the basis of results of laboratory and preliminary field trials (1989), a large scale field trial (1990) was conducted in mango orchard at Budhadia, Lucknow to evaluate the effect of two neem extracts (2 per cent oil-based and 0.3 per cent of cake based) against mango hoppers, *I. nitidulus*. Heavily infested trees of even age and growth were selected for the experiment. Above two treatments and a control was taken for comparison. Each treatment and control was replicated thrice in RBD. One tree was treated as one replicate. From each tree 5 panicles were selected randomly for counting of population before and after treatment as described above. Plants were sprayed thoroughly with the respective chemicals. Control tree was not sprayed. Data were recorded on population of hoppers after 48 hours and at weekly intervals and subjected to statistical analysis.

RESULTS

Laboratory trials

Table 1 presents data on development and survival of mango hoper nymphs, *I. nitidulus*. From the data it is evident that treatment with oil-based concentrate resulted in 100 per cent mortality of nymphs within 48 hours. Cake-based concentrate caused 46.6 per cent mortality during the same period which increased to 66.6 per cent by 9th day where as only 17.7 per cent mortality was recorded in control. Among dead nymphs 2.2 per cent nymphs suffered from imperfect moulting. As observed at 5th day of treatment. Adult emergence in treatment was 22.2 where as 78.8 per cent adult emerged in control. Adult emergence was higher i.e. 55.5 per cent and 78.8 per cent on 5th and 9th day respectively in control while in treatment, adult emergence was 15.5 per cent and 22.2 per cent at 5th and 9th day respectively.

Table 2 represents data on settling and ovipositional response of adults of mango-hopper, *A. atkinsoni*. The data revealed that in both the treatments the percentage of hoppers settled on treated panicles was low as compared to control. The difference in settling response initially (10 minutes after treatment) was however more in case of cake-based extract as only 7.5 per cent hoppers settled on treated panicle as against 60

Table 1. Effect of neem extracts on the development of nymphs of mango hopper, *I. nitidulus* (1989)

Treatment	Post-treatment observation days after								
	% Mortality			% Malformed nymph			% Adult emergence		
	2	5	9	2	5	9	2	5	9
Oil-based extract (1%)	100								
Cake-based extract	46.6	60.0	66.6	0	2.2	2.2	6.6	15.5	22.2
(0.2%)	3.8	12.2	17.7	0	0.0	0.0	7.7	55.6	78.8
Control									

Table 2. Effect of neem extracts on the settling and ovipositional response of adults of mango hopper, *Amritodus atkinsoni* (1989)

Treatment	Settling pattern (%)			Ovipositional activity after 24 hours*
	10	20	40	
Oil-based extract (1%)	10.0	4.0	4.0	0.0
Control	22.0	16.0	16.0	0.0
Cake-based extract (0.2%)	7.5	12.5	13.0	0.0
Control	60.0	45.0	46.0	0.0

*All adults died by this time.

per cent on control. In oil-based extract the percentage of hoppers settled in treatment and control during the same period was 10 and 22 respectively. The percentage settlement ratio in treatment and control in both extracts decreased after, 20 minutes and remained almost same even after 40 minutes. The above observations show that both extracts have deterrent effect on the hopper, *A. atkinsoni*. No oviposition was noticed in both the treatments and controls as observed after 24 hours. The hoppers died by that time.

Field Trials

Evaluation of neem derivatives for control of mango hopper, I. nitidulus

Results of introductory field trial with oil extract revealed 100 per cent mortality of nymphs of mango hopper as found in laboratory tests. After four days of treatment it was found that the number of dead hoppers were more (18.13 per panicle) as compared to initial number (15.71 per panicle) at the time of spray. It is evident from these observations that the nymphs hatched later were also affected besides causing 100 per cent mortality of the initial population (Table 3).

Results of field trial (Table 4) with 0.2 per cent cake extract revealed that there was not much difference in population of hoppers on control and treated panicles. These figures are not in Table 3 (30.7 and 34 hoppers per panicle respectively) as observed after ten days of treatment. But when compared with the initial population it was found that number of hoppers increased by 2.4 fold in treatment and 3.57 fold in control which shows that the spraying could exert some check on population growth of hoppers. Hormonal action of the chemical was also observed as a few hoppers with imperfect moulting were also seen in bagged panicles.

Table 3. Field evaluation of oil-based extract for control of mango hopper, *I. nitidulus* (1989)

Treatment	Mean No. of hopper/panicle*							
	Pre-treatment		Post- treatment					
	Live	Dead	1 day		4 days		10 days	
			Live	Dead	Live	Dead	Live	Dead
Oil-based extract (1%)	15.71	0	5.42	12.71	13	18.43	11.29	20.93
Control	12	0	15	0	17.86	0.57	18.43	0

*Average of 10 panicles

Table 4. Field evaluation of cake-based neem extract for control of mango hopper, *I. nitidulus*

Treatment	Mean No. of hoppers per panicle*		% malformed nymphs
	Pre treatment	Post treatment	
Cake-based extract (0.2%)	14.2	34	2.35
Control	8.6	30.7	0

*Average of 10 panicles

Results of large scale field trial in which 2 per cent oil-based extract and 0.3 per cent cake-based extract were used, revealed that both these extracts on the above concentrations were highly significant in bringing down the population of mango hopper i.e., nil and 2.7 hoppers per panicle respectively in above treatment as compared to 16.50 hoppers per panicle in the control as observed at 14th day of treatment. The trend was nearly same even at 23rd day after treatment (Table 5).

DISCUSSION

Crude neem extracts are known to deter feeding and oviposition, disrupt growth and development and also cause mortality of insects. Neem oil has been found very effective against a number of hemipterous pests including plant hoppers. Neem oil at concentrations of 1, 2, 3 and 4 per cent was found highly significant in reducing the hopper, *Nephotettix virescens* population as compared to control (Nigam and Sen, 1989). Singh et al. (1988) reported that hexane extract of seed kernel and alcoholic extract of hexane extract has LC 50 of 0.674 per cent and 0.328 per cent respectively against mustard aphid, *Liaphis erysimi*, and exhibited toxic effect as seen in our experiments. Results of laboratory and field trials show that treatments of mango panicles with 1per cent alcohol extract resulted in 100 per cent mortality of nymphs and also caused knock down effect (Table 1 and 5). This shows that neem oil-based extract has insecticidal property. Santakke (1989), in his field experiments showed that 3 per cent neem oil treatment of rice seedlings was at part with B.P.M.C. and significantly superior over chloropyriphos after 7 days of treatment, in reducing mixed population of rice plant hopper. Saxena and Khan (1985) reported that crude oil expelled from decorticated neem seeds at 50 per cent concentration could cause significant reduction in survival

Table 5. Evaluation of neem extracts for control of mango hopper, *I. nitidulus* in the field (1990)

Treatments	Average pre count population	Averages	Post count hopper population at indicated dates								
		3	5	7	10	12	14	16	19	21	23
Oil-based extract (2%)	10.80 (3.427)	0.00 (1.000)	0.15 (0.070)	0.05 (1.024)	0.00 (1.000)	0.20 (1.086)	0.00 (1.000)	0.0 (1.000)	0.05 (1.024)	0.25 (1.104)	0.00 (1.00)
Cake-based (0.3%)	13.95 (3.823)	1.85 (1.603)	3.25 (2.031)	4.50 (2.316)	3.65 (2.081)	4.05 (2.181)	2.70 (1.851)	2.65 (1.864)	1.40 (1.527)	2.10 (1.704)	1.25 (1.46)
Control	12.90 (3.707)	19.80 (4.547)	27.80 (5.356)	25.10 (5.075)	24.40 (4.987)	22.90 (4.810)	16.50 (4.144)	16.15 (3.966)	10.05 (3.258)	12.95 (3.657)	9.20 (3.16)
CD at 5% =		0.438	0.510	0.884	1.288	1.019	0.949	0.841	0.932	0.966	0.067
SEM =		0.157	0.184	0.318	0.464	0.367	0.342	0.303	0.336	0.348	0.219

(Figures in parentheses are transformed mean values)

percentage of brown plant hopper, *Nilaparvata lugens* and ragged stunt virus transmission decreased on neem oil treated seedlings. Likewise, use of neem oil and its formulations have been found very effective against a number of hemipterous pests viz., safflower aphid, *Dactynotus carthami*; cotton white fly, *Bemisia tabaci*; aphid, *Aphis craccivora* and other such insects (Patel and Srivastava, 1989; Mani et al., 1990; Natrajan et al., 1990; Dhanarkar et al., 1990). There is however, no report of evaluation of any neem products against mango hopper. In our laboratory studies it was found that less number of hoppers, *A. atkinsoni* were oriented to the panicle of mango treated with 1 per cent oil based enriched extract. Similar observations were made by Saxena et al. (1981) in case of *N. lugens* where rice plants treated with 3 per cent neem oil attracted less number of hoppers i.e., about 36 per cent. As the concentration of oil increased, to 100 per cent the insect arrivals decreased to only 6 per cent.

Cake-based extracts are reported to affect feeding and growth and development of insects (Singh, 1990, Jhansi and Singh, 1993). In our laboratory studies also few nymphs with imperfect moulting were seen, also less number of nymphs developed to adult stage in comparison to control (Table 1). In field experiments also the rate of growth of hopper population on mango trees treated with 0.2 per cent cake-based extract was low as compared to control (Table 4). Similar observations were reported by Saxena et al. (1984) in case of *N. lugens* when they were released on rice plants treated with 3:10 neem cake urea mixture. But there was no significant difference in settling pattern of the *N. lugens* females when a free choice among the treated and control plants were given. However in our studies four times more hoppers were found settled on control panicles of mango than the panicles treated with 0.2 per cent cake extract. Like-wise foliar application of neem oil and neem extracts also disrupt growth and development of *N. lugens*, *N. virescens* and *Sagatella furcifera* the plant hoppers (Saxena et al., 1981; Saxena et al, 1984; Heyde et al., 1984). The antifeedant, deterrent and disruption of growth and development are such properties of neem extract which leads to self destruction of insect pests and hence their use will be more cheaper and safer for plant protection in horticulture.

ACKNOWLEDGEMENTS

Thanks are due to Dr. S.S. Negi, Director, Central Institute of Horticulture for Northern Plains, Lucknow and Director, I.A.R.I., New Delhi for their constant encouragement to carry out this work. Thanks are also due to Shri R.P. Shankhwar and Shri Ram Dayal Technicians for their help.

REFERENCES

Dhanarkar, B.K., S.N. Puri and B.S. Thakur. 1990. Efficacy of neemark against cotton white fly, *Bemisia tabaci* Gannadius. *Neem Newsl.* 7:39.

Jhansi, B.R. and R.P. Singh. 1993. Identification of effective and inexpensive neem (*Azadirachta indica* A. Juss.) seed kernel extract for management of *Helicoverpa armigera* Hubner. *World Neem Conference*, Bangalore, 24–28 February 1993. Abstract page 7.

Heyde, J.V.D., R.C. Saxena and H. Schmutterer. 1984. Neem oil and neem extracts as potential insecticides for control of hemipterous rice pests. In: *Natural Pesticides from the Neem Tree and other Tropical Plants*, H. Schmutterer and K.R.S. Ascher (eds.), GTZ, Eschborn. pp. 263–290.

Mani, A., K. Kumudanatham and C.A. Jagdish. 1990. Efficacy of neem oil emulsion for the control of safflower aphid, *Dactynotus carthami*. *Neem Newsletter* 7:25.

Natrajan, K. and V.T. Sundaramurthy. 1990. Effect of neem oil on cotton white fly, *Bemisia tabaci*. *Indian J. Agric. Sci.* 60: 290–291.

Nigam, P.M. and R. Sen. 1989. Efficacy of neem oil against *Nephotettix virescens* and *Mythimna separata* (Lepidoptera: Noctuidae) as pests of rice. *Neem Newsl.* 6: 23.

Patel, M.B. and K. Srivastava. 1989. Evaluation of a neem formulation against cowpea aphid (*Aphis craccivora*). *Neem Newsl.* 6: 41–44.

Santakke, B.K. 1989. Comparative efficacy of insecticides alone and in combination with neem oil against rice plant hopper. *Neem Newsl.* 6: 38–40.

Saxena, R.C., N.J. Liquido, H.D. Justo. 1981. Neem seed oil, a potential antifeedant for the control of the rice brown plant hopper, *Nilaparavata lugens*. In: *Natural Pesticides from the Neem Tree* (*Azadirachta indica* A. Juss.) H. Schmutterer, K.R.S. Ascher and H. Rembold (eds.), G.T.Z. Eschborn, West Germany. pp. 171–188.

Saxena, R.C., B.P. Epino, Tucheng-Wen and B.C. Puma. 1984. Neem, chinaberry and custard apple: antifeedant and insecticidal effects of seed oils on leaf hopper and planthopper pests of rice. In: *Natural pesticides from the neem tree and other Tropical Plants*. H. Schmutterer and K.R.S. Ascher (eds.) G.T.Z., Eschborn, West Germany, pp. 403–412.

Saxena, R.C. and Z.R. Khan. 1985. Effect of neem oil on survival of *Nilaparvata lugens* (Homoptera: Delphacidae) and on grassy stunt and ragged stunt virus transmission. *J. Econ. Entomol.* 78: 647–651.

Singh, R.P., C. Devkumar and S. Dhingra. 1988. Activity of neem (*Azadirachta indica* A. Juss) seed kernel extract against the mustard aphid, *Liaphis erysimi*. *Phytoparasitica* 16: 225–230.

Singh, R.P. 1990. Neem in agriculture—Indian Scenario. *Proc. Symp. Botanical Pesticides in IPM*, Rajahmundry, pp. 78–113.

62

Evaluation of Neem Derivatives and *Lantana camara* L. against Cardamom Pests

B. GOPAKUMAR, D. KUMARESEN, S.S. CHANDERSEKAR AND S. VARADARASAN

Indian Cardamom Research Institute, Myladumpara
Kailasanadu P.O., 685 553, Kerala, India

ABSTRACT

Application of neem derivatives in seven rounds per year showed no satisfactory control over cardamom thrips. Neem oil 1 per cent when sprayed with wetting agent has been found effective against cardamom whitefly nymphs. A 1 per cent aqueous suspension of the extract of *Lantana camara* produced larval and pupal mortality in the cardamom hairy caterpillar *Eupterote undata* Blanchard. This extract showed insecticidal property against larvae, pupae and adults of cardamom shoot borer *Conogethes punctiferalis* (Guen.).

INTRODUCTION

Pest control using synthetic chemicals generally entails unforeseen hazards. This urged research on the lookout of pesticides of plant origin with the presumption that bioproducts are less hazardous. With the growing interest on the use of plant products for pest control, a large number of plant species have been screened for their pest controlling properties and nearly 2500 plant species have been reported to possess the property (Grainge and Ahmad, 1988). Among these plants the neem *Azadirachta indicia* A. Juss. has been widely recognized. Azadirachtins in the neem products have been found to act as repellents, antifeedants, affect food consumption and utilisation and interfere with growth regulation and ovarian development in insects (Subrahmanyam, 1990). The insecticidal property of a neem seed derivative (Margoside C K) has also been shown on cardamom root grubs, *Basilepta fulvicorns* by Varadarasan et al., (1992). *Lantana camara* L, a straggling weed plant has captured the attention of many plant protection-ists due to its insecticidal property (Feuntebella et al., 1980; Pandey et al., 1982, 1983a, b and 1987), repellent property (Islam et al., 1990), juvenile hormone like property (Jaipal et al., 1983) and antifeedant property (Singh, 1983). This paper briefs our preliminary attempts on the use of neem derivatives and *Lantana camara* for the management of a few pests of cardamom.

MATERIALS AND METHODS

Field Evaluation of Neem Derivatives against the Cardamom Thrips, *Sciothrips cardamomi* (Ramk.)

Three neem-based insecticides viz., Neemine (0.025 per cent ai, 0.038 per cent ai and 0.05 per cent ai) (these are fractions of neem kernel extract supplied by Prof. G. Srimannarayana, Osmania University, Hyderabad), Neem SSF F.21 (0.004 per cent, 0.006 per cent and 0.008 per cent) and Margoside C K (0.04 per cent and 0.06 per cent ai) and quinalphos 0.025 per cent as standard chemical and an absolute control were imposed in seven rounds per year in two different blocks of the ICRI farm. There were 12 plants per plot and three replications were maintained. The treatments were laidout in a RBD. Thrips damage was assessed from harvested capsules. The data obtained from five harvests were pooled, and pooled data was analysed for variance.

Field Evaluation of Neem Oil against Cardamom Whitefly, *Dialeurodes cardamomi* David & Subr.

A field trial with the following treatments: Ethion 0.1 per cent, phosalone 0.07 per cent, Acephate 0.07 per cent, methylaprathion 0.05 per cent, pyridaphenthion 0.04 per cent, 0.06 per cent, phosphamidon 0.05 per cent, neem oil 0.5 per cent, methyldemeton 0.05 per cent, organo 80 0.05 per cent and control, replicated thrice was laidout in a RBD in the infested area at Vandiperiyar. Wetting agent @ 50 ml/10 l. was mixed with spray fluid. The plants were of Cv. Vahukka at a spacing of 10' × 10'. The plot size was 12 plants. Treatments were imposed four times at an interval of 10 days during January to February. After standardisation of the population of nymphs on different leaves of a tiller, fourth leaf from the top was selected for assessment of nymphal population. Five clumps were selected at random in each treatment and the fourth leaf from 10 per cent of the total number of tillers were chosen for assessment of population of live nymphs before and 3 days after each round of spray.

Laboratory Evaluation of the Crude Extract of Stem of *Lantana camara* against the Cardamom Hairy Caterpillar, *Eupterots undata*

Stem of *Lantana camara* red variety was chopped into fine pieces and shade dried. Acetone extract of the dry material was prepared using a soxhlet apparatus and the acetone was evaporated off using a rotary vacuum evaporator. A one per cent aqueous suspension of the extract was prepared using Tween 80 as emulsifier. The suspension was sprayed on 15 fully grown cardamom leaves and allowed to dry. When dry, the leaf bases were inserted into a conical flask containing water to prevent wilting and ten eighth instar larvae of *E. undata* were released over it. The larvae with the leave were covered with a polythene cover having numerous small holes. Mouth of the polythene cover was tied with the mouth of the conical flask to prevent escape of the larvae. When the leaves offered as food were fed by the larvae, fresh unsprayed leaves were supplemented as food. The larvae were observed every day for mortality, moulting and when pupated the pupae were transferred to polythene jars for adult emergence. In the control units, leaves sprayed with water and dried off were offered as food for the larvae. The experiment was replicated six times. Number of larvae which pupated and number of pupae which became adults were noted.

Laboratory Evaluation of the Crude Extract of the Stem of *Lantana camara* L. against Cardamom Shoot Borer, *Conogethes punctiferalis (Guen.)*

The extract was prepared as mentioned above and 1 per cent suspension of the extract was applied topically to second instar larvae of *C. punctiferalis* using an atomiser. Each larva was allowed to enter into a stem-cutting through a previously made hole. On the larvae of the control group, water, instead of the extract was sprayed. Fifteen larvae were maintained like this in both groups and the experiment was replicated 6 times. The stem-cuttings were opened every day to note larval mortality and the live larvae were allowed to re-enter a fresh stem-cutting. The data was statistically analysed. Similarly the extract was applied on fourth instar larvae, mortality was recorded and the data statistically analysed.

RESULTS AND DISCUSSION

Effect of Neem Derivatives on Cardamom Thrips

Data on impact of seven rounds of application of neem derivatives on thrips damage is appended in Table 1. Treatments of quinalphos recorded the lowest thrips damage. With all the other treatments more than 50 per cent of the capsules received thrips damage.

Effect of Neem Oil on Cardamom Whitefly

Data on impact of the different treatments on mortality of whitefly nymphs is presented in Table 2. It indicates that population of nymphs which was uniform before spraying reduced significantly in all treatments except control. However, with neem oil treatment the nymphal population was drastically reduced to 12.2, which is highly significant over that of control and some other insecticides. It therefore, becomes obvious that neem oil 0.5 per cent is effective against whitefly nymphs.

Effect of 1 per cent Suspension of the Extract of *Lantana camara* on *E. undata*

Released larvae did not feed the treated leaves on the 1st day, however, by the second day all the treated leaves were fed on. Results on larval mortality (Table 3) show that 40 per cent of the treated larvae died, while in the control only 20 per cent of the larvae

Table 1. Effectiveness of some neem derivatives against cardamom thrips

S.No.	Treatments	Percent thrips damage on harvested capsules (Mean of 5 harvests)
1.	Neemine 0.025%	73.03 (60.50)
2.	Neemine 0.038%	62.92 (52.20)
3.	Neemine 0.05%	54.91 (47.90)
4.	Neem SSF F. 21 0.004%	59.74 (50.64)
5.	Neem SSF F. 21 0.006%	48.44 (44.11)
6.	Neem SSF F. 21 0.008%	50.41 (45.22)
7.	Margoside C K 0.04%	61.30 (51.54)
8.	Margoside C K 0.06%	62.21 (52.88)
9.	Quinalphos 0.025%	12.34 (20.39)
10.	Control	79.93 (60.45)
CD (p = 0.05)		14.395

Values in the parentheses are angular transformed values

Table 2. Effectiveness of certain insecticides against cardamom whitefly

S. No.	Insecticide	Dosage (%)	No. of live nymphs/leaf	
			Precount	Postcount
1.	Ethion	0.10	90.67	8.00
2.	Phosalone	0.07	90.00	59.41
3.	Acephate	0.075	91.00	3.13
4.	Methylparathion	0.05	87.33	34.67
5.	Pyridaphenthion	0.04	91.00	23.00
6.	Pyridaphenthion	0.06	89.33	33.73
7.	Dimethoate	0.05	91.00	50.80
8.	Phosphamidon	0.05	90.67	69.73
9.	Neem oil	0.50	87.33	12.20
10.	Methyldemeton	0.05	92.00	41.67
11.	Organo 80	0.50	87.67	20.13
12.	Control		90.00	85.13
	C.D. (p = 0.05)		NS	23.11

died before pupation. Among the pupae which received the treatment at the larval stage, 56.9 per cent died before reaching adult stage, whereas in the control insects pupal mortality was 35.08 per cent, accounting a high mortality difference with the treatment of the extract of *Lantana camara*. Also all the adults from the treated larvae were deformed whereas in the control all the adults were normal. These clearly show that 1 per cent suspension of the extract of *Lantana camara* produces mortality in the larvae and pupae of *E. undata* and deformity in the adults.

Reluctance of the larvae of *E. undata* to feed on the treated leaves on the day of release ascribes to the extract some repellent/antifeedant property which did not persist long. Singh (1983) has found out antifeedant property in the extract of *L. camara*

Table 3. Effect of extract of *Lantana camara* L. on *Conogethes punctiferalis* and *Eupterote undata*

Pest species	Stage	% Mortality	
		treated	control
C. punctiferalis	Early stage larvae	80.00	30.00
	Pupa	7.5.00	0.00
	Adult	48.33	0.00
	Late stage larvae	40.00	20.00
	Pupa	9.12	3.95
E. undata	Larvae	40.00	20.00
	Pupa	56.91	35.08

against the desert locust *Schistocerca gregaria* Forskal. Repellent property of the plant has been exhibited towards *Phthorimaea operculella* (Islam et al., 1990). The crude extract of stem of *L. camara* produced deformity in the adults of *E. undata*, a phenomenon usually associated with hormonal imbalance and physiological disorder. Juvenilizing effect on *Dysdercus koenigii* due to hormonal imbalance has been observed in response to the application of extract of *L. camara* on it (Jaipal et al., 1983). It is also observed that 1 per cent suspension of the stem extract of *L. camara* produced different levels of mortality in the larvae and pupae of *E. undata* and *C. punctiferalis*. Similar insecticidal property of *L. camara* has also been reported on *Bagrada cruciferarum* (Pandey et al., 1982), *Heliothis armigera* (Pandey et al., 1983a) *Aphis gossypi* (Pandey et al. 1983b) and *Lipaphis erysimi* (Pandey et al., 1987).

Effect of 1 per cent Suspension of the Extract of *L. camara* on *C. punctiferalis*

Second instar larvae of *C. punctiferalis* which received the treatment suffered heavy mortality (Table 3). Nearly 80 per cent of the larvae which received the treatment and nearly 30 per cent of the control larvae died showing a higher mortality due to the treatment. When the treatment was imposed on fourth instar larvae of *C. punctiferalis* mortality of the treated larvae was less (40 per cent) than that observed with second instar larvae indicating higher tolerance towards the extract with advanced stage of the larva. However, mortality observed in fourth instar larvae was higher than that of the control. It is noted that pupae from the recipients of the treatment at second instar stage had higher mortality than control (Table 3), whereas treatment that fourth instar stage did not alter the mortality of its pupae to a great extent. It is also seen that adults from the larvae that received the treatment at second instar stage were seriously affected (Table 3) while the treatment that fourth instar stage had no impact over it. These clearly indicate a tolerance towards the plant extract with advancing age of the insect.

ACKNOWLEDGEMENTS

The authors are grateful to Dr. R. Naidu, Director (Research), ICRI, for encouragement and facilities afforded for conducting the work, and to Kum. S. Sreelatha for neatly typing the manuscript.

REFERENCES

Fuentebella, F. and B. Morallo-Rejesus. 1980. The insecticidal activity of fancy buttons (*Lantana camara* L.) flower extracts to several insect species. Youth Res. Apprenticeship Action Rep. Society for the Advancement of Research, May 1980 UPLB College Laguna 16 pp.

Grainge, M. and S. Ahmad. 1988. *Hand book of Plants with Pest Control Properties*. Wiley, New York.

Islam, M.N., M.A. Karim and Z. Nessa. 1990. Control of the potato tuber moth *Phthorimae operculella* (Zeller) (Lepidoptera; Gelechidae) in the store houses for seed and warehouse potatoes in Bangladesh. *Bangladesh J. Zool.* 18 (1): 41–52.

Jaipal, S., Z. Singh and R. Chauhan. 1983. Juvenile hormone like activity of some common Indian plants. *Indian J. Agric. Sci.* 53(8): 30–733.

Pandey, U.K., G.S. Verma, C. Lekha and A.K. Singh. 1982. Note on the use of some insecticides against *Bagrada cruciferarum* Kirk. *Indian J. Agric Sci.* 52(3): 205–206.

Pandey, U.K., A.K. Srivastava, A.K. Singh and M. Pandey. 1983a. Evaluation of some plant origin insecticides against gram caterpillar, *Heliothis armigera* Hubn. *Indian J. Ent.* 45(2): 211–212.

Pandey, U.K., A.K. Srivastava, C. Lekha and A.K. Singh. 1983b. Efficacy of certain plant extracts against brinjal aphid, *Aphis gossypi* Glover. *Indian J. Ent.* 45(3) 313–314.

Pandey, N.D., L. Singh, Y.P. Singh and Tripathi. 1987. Effect of certain plant extracts against *Lipaphis erysimi* Kalt under laboratory conditions. *Indian J. Ent.* 49(2): 238–242.

Singh, R.P. 1983. Search for antifeedants in some botanicals for desert locust, *Schistocerca gregaria* Forskal. *Zeitschrift fur Angewandte. Entomologie* 96(3): 316–319.

Subrahmanyam, B. 1990. Azadirachtin—A naturally occurring insect growth regulator. *Proc. Indian Acad. Sci.* (Anim. Sci.) 99(3): 277–288.

Varadasrasan, S., T. Sivasubramonian and R. Manimegalai. 1992. Evaluation of some insecticides on cardamom root grub, *Basilepta fullvicorne* Jacoby in field condition. *Pestology* 16(8): 5–7.

63

Bioefficacy of Some Neem Formulations against *Spodoptera litura* F.

M.S. CHARI, G. RAMAPRASAD, S. SITARAMAIAH AND P.S.N. MURTHY

Central Tobacco Research Institute, Rajahmundry 533 105, India

ABSTRACT

The bio-efficacy of six new neem formulations viz. Repelin A (RD-9 Repelin) at 1 per cent, a product of Neem (*Azadirachta indica*), Karanjia (*Pongamia pinnata*) and Mahua (*Madhuca indica*) containing 300 ppm of azadirachtin in its natural form along with 500 ppm of pongamin and pongamol, Neemgold (0.003 per cent and 0.0015 per cent Azt.), Vepacide-I and II (0.25 per cent), Dichloroethane (DCE) extract of deoiled neem cake (DOC) and Navnem (0.1 per cent) was studied against *Spodoptera litura* F. on tobacco in comparison with aqueous kernel suspension of neem (NSKS) (2 per cent), DCE extract of neem seed kernel (NSK) 1 per cent and Neemrich II (0.1 per cent) during 1990 and 1991. The results indicated that Repelin A (1 per cent), NSKS, Neemgold (0.003 per cent), DCE of NSK (1 per cent) and DCE of deoiled neem cake (DOC) (1 per cent) effectively controlled *S. litura* as compared with Neemgold (0.0015 per cent), Vepacide-I and II (0.25 per cent) and Neemrich-II (0.1 per cent). Cent per cent mortality was obtained in the laboratory with DCE of NSK within 5 days, followed by Neemgold (0.003 per cent) in 6 days, NSKS and Neemrich-II in 7 days, Vepacide-II in 8 days, DCE of DOC, Repelin A, Neemgold (0.0015 per cent) and Vepacide-I in 9 days, Repelin A (1 per cent) when compared with Repelin B (RD-9 Repelin) containing 3000 ppm azadirachtin in 2 per cent, 1.25 per cent and 0.66 per cent concentrations and NSKS (2 per cent) for its bio-efficacy against 8-day old *S. lilura* larvae on castor in laboratory during 1992, it was found that Repelin B 2 per cent and 1.25 per cent concentrations were equally effective as NSKS and they were superior to Repelin A (1 per cent) and Repelin B (0.66 per cent) which had shown similar antifeedant effect. Subsequent laboratory tests during 1992 revealed that Repelin B (1 per cent, 2 per cent and 3 per cent) alone and in combination with half the recommended dose of endosulfan (0.025 per cent) were most effective in preventing the damage on tobacco as shown by almost nil consumption of leaf and it remained on par with NSKS (2 per cent), whereas the recommended dose of endosulfan (0.05 per cent) showed slight consumption of leaf by 8-day old *S. litura* larvae after 48 hours from treatment. Cent per cent mortality of larvae was observed under no-choice situation in petri dishes in 2 per cent and 3 per cent concentrations of Repelin B, endosulfan and all combination treatments after 48 hours from the treatment. When Repelin B and other neem formulations viz., Nimbecidine, Godrej Achook and Neemox were tested for their bio-efficacy against *S. litura* on tobacco in comparison with NSKS and other neem formulations found promising earlier viz., DCE of NSK and DOC both at 1 per cent, Repelin A (1 per cent), Neemgold (0.003 per cent Azt.) and Navnem (0.1 per cent), the results indicated that tobacco seedlings can be effectively protected from the damage caused by larvae of *S. litura* by spraying Repelin B (RD-9 Repelin) at 1 per cent (0.003 per cent Azt.) alone or in combination with half the recommended dose of endosulfan (0.025 per cent) or 2 per cent aqueous NSKS or Neemgold (0.003 per cent Azt.) or Repelin A (1 per cent) or DCE of NSK or DOC at 1 per cent. Further it can be inferred from the above that neem formulations containing azadirachtin were highly effective at 0.003 per cent and higher concentrations in protecting the tobacco foliage from the *Spodoptera* damage.

INTRODUCTION

Spodoptera litura F., a polyphagous pest popularly known as tobacco caterpillar or tobacco cutworm, is a major pest of tobacco in Andhra Pradesh, India, causing serious damage by defoliation in tobacco nurseries and also in planted crop. It causes damage to an extent of 80 to 100 per cent in tobacco nurseries (Chari and Ramaprasad, 1987). This Institute has recommended several insecticides viz., endosulfan, chlorpyrifos, qunialphos, acephate, monocrotophos and carbaryl (Anon., 1989) to combat this pest. However, because of possible dangers to the environment through use of insecticides and in view of restrictions imposed for use of pesticides on tobacco (Wittekindt, 1978 and 1986) and development of insecticide resistance in the pest population (Srivastava and Joshi, 1965; Krishnamurthi, 1982 and Ramakrishnan et al., 1984) search has begun in early seventies for an easily available and safe substitute for recommended insecticides.

Neem (*Azadirchta indica* A. Juss) seed kernel suspension (NSKS) in water was found highly effective as an antifeedant and ovipositional repellent against *S. litura* in tobacco nurseries (Joshi and Ramaprasad, 1975; Joshi et al., 1978b; Joshi and Sitaramaiah, 1979). Based on the results of phytotoxicity studies (Ramaprasad et al., 1979), a schedule of NSKS sprays viz., 1 per cent NSKS on 3-week-old seedlings and subsequently three sprays of 2 per cent NSKS at 10-day intervals was evolved and recommended for protecting tobacco seedlings from the *Spodoptera* damage (Joshi et al., 1978 a and Anon., 1979). Neem products were also reported to be effective as antifeedants against *S. litura* on various crops like sweet potato (Abdul Kareem, 1981); groundnut (Koshiya and Ghelani, 1990); and on cotton (Ramachandrarao et al., 1990). Opender Koul (1987) reported feeding deterrency and growth inhibition with neem oil in early 3rd instar larvae on castor in laboratory. In order to recommend a ready to use neem formulation for easy adoption by the farmers investigations were carried out earlier in Central Tobacco Research Institute, Rajahmundry which revealed that apart from 2 per cent aqueous NSKS, Repelin 1 per cent and dichloroethane extract (DCE) 1 per cent of neem seed kernel (NSK) were most effective against *S. litura* in tobacco nurseries followed by margosan-O (0.0018 per cent) and neem-rich II (0.1 per cent) (Chari et al., 1990). In the present investigations the bio-efficacy of repelin A and B and some new neem formulations was studied against *S. litura* in comparison with the neem formulations found promising earlier viz., 2 per cent NSKS and DCE extract of NSK 1 per cent and the results are reported in this paper.

MATERIALS AND METHODS

The bio-efficacy of Repelin A (RD-9 Repelin) a product of neem, *A. indica,* karanjia, *Pongamia pinnata* and mahua (*Madhuca indica*) containing 300 ppm of azadirachtin in its natural form along with 5000 ppm of pongamin and pongamol, supplied by M/s ITC Limited and some new neem formulations viz., Neemgold, a natural and biological antifeedant developed by M/s SPIC Scientific Foundation, Madras containing 3000 ppm of azadirachtin at 0.003 per cent and 0.0015 per cent, Navnem 50 per cent E.C. a product developed at Osmania University at 0.1 per cent, Vepacide I and II of Indian Institute of Chemical Technology both at 0.25 per cent concentration and DCE extract of deoiled neem cake (DOC 1 per cent formulated at the Central Tobacco Research Institute was studied against *S. litura* in comparison with 2 per cent NSKS, DCE of NSK 1 per cent and Neemrich II 0.1 per cent both in the laboratory and also in the nursery during 1990 and 1991 nursery seasons.

Laboratory Tests

The bio-efficacy of neem formulations was studied both under choice and no-choice situations by leaf dip method. One litre solution of each formulation was prepared in beakers. Aqueous NSKS 2 per cent was prepared according to Pradhan and Jotwani (1971). Leaves from 6-week old seedlings of Jayasri 'MR' (*Nicotiana tabacum* L.) variety of flue-cured Virginia tobacco were dipped in respective treatment solutions and allowed them to dry under fan.

In choice situation the leaves were kept in plastic tray of 40 cm × 28 cm × 12 cm size containing wet cotton wool covered by filter paper. The treated leaves were kept in 2 rows @ 1 leaf for each treatment. In choice situation the leaves were kept in 2 rows @ one leaf per each treatment on the filter paper at the bottom of the tray and the respective treatments labelled. Laboratory bred 3rd instar larvae of 8-day old starved for 2 hours were released @ 1 larva per treatment in the centre of each tray (replication) and given the choice for feeding on leaves treated with different treatments. The trays were covered with black cloth pieces and tightened by an elastic rope.

In no-choice situation one tobacco leaf from 6-week old seedlings was kept for each treatment and replication in a petri dish (15 cm diameter) containing wet cotton wool covered by filter paper disc. In each petri dish one 3rd instar larva of 8-day old starved for two hours was released for feeding on the leaf. The treatments were replicated 3 times.

Observations on leaf area consumed by *S. litura* larvae were recorded both in choice and no-choice situations by measuring the area on the outlines of leaves drawn on graph sheets prior to their dipping in the treatment solutions. In no-choice situation weights of the larvae were recorded initially and at 48 hours after introduction of the larvae into petri dishes. Data on leaf area consumed in both the situations and decreased/increase in weight of larva at 48 hours after treatment under no-choice situation were statistically analysed using analysis of variance technique.

Studies on the effect of neem formulations on larval mortality and pupation were also carried out in the laboratory in glass troughs of 20 cm diameter and 10 cm ht. in size. For each treatment and replication 20 laboratory bred 3rd instar larvae of eight-day-old were released into a glass trough and fed with treated leaves daily. The glass troughs were covered with cloth piece and tightened by an elastic rope. Observations were recorded on mortality of larvae, malformation of larvae and number of larvae pupated.

Nursery Experiment

Six nursery beds of 6.0 m × 1.0 m size were sown with seed of Jayasri 'MR' variety of fcv tobacco @ 3.0 kg/ha. The seedlings were grown upto 4 weeks without applying any insecticides, however fungicides were used as and when necessary to control diseases like damping off, leaf blight etc. Four weeks old healthy seedlings were transplanted on 1 m^2 beds @ 400 on each bed and the experiment was laid out with the above treatments which were replicated three times. After establishment of seedlings and when they were six weeks old, sprayings were carried out thoroughly with the above neem formulations using 0.7 l spray fluid per each bed to ensure full coverage. Just before spraying 30 laboratory bred third instar larvae were released in each bed. Observations were recorded on the number of seedlings damaged at 2 and 9 days after spraying separately and the data were statistically analysed using analysis of variance technique.

Studies on Effect of Repelin B versus Repelin A and NSKS against *S. litura* on Castor in Laboratory

Repelin B (RD-9 Repelin) the new neem formulation developed very recently by M/s ITC Limited, containing 3000 ppm of azadirachtin, was tested at 2 per cent, 1.25 per cent and 0.66 per cent concentrations for establishing its antifeedant effect against *S. litura* larvae on castor in comparison with Repelin A 1 per cent and aqueous 2 per cent NSKS and control (water) treatment. Two laboratory tests were conducted during April–May, 1992 under choice and no-choice situations by following the same method as adopted above for testing the neem formulations excepting that instead of tobacco leaves, castor leaf circles of 5.5 cm diameter were cut from fresh castor leaves and they were dipped in the treatment solutions and given for feeding by eight day-old third-instar larvae. The treatments were replicated four times in both the tests. In first test, six of the third-instar larvae, starved for two hours i.e. @ one larva per treatment were released in the centre of each tray (replication) while in no-choice situation one larva per each treatment and replication was released in a petri dish. Observations on leaf area consumed on graph sheets and mortality of larvae were recorded at 48 hours after introducing of the larvae. In second test 12 third-instar larvae i.e. @ two larvae per treatment were released in the centre of each tray (replication) under choice situation while in no-choice situation two larvae were released into a petri dish for each treatment and replication. Observations on leaf area consumed by the larvae were recorded 24 hours after introduction of larvae both under choice and no-choice situations. Data on leaf area consumed were statistically analysed.

Studies on Effect of Repelin B Alone and in Combination with Endosulfan against *S. litura (1992)*

Repelin B at 1, 2 and 3 per cent concentration alone and in combination with half the recommended dose of endosulfan (0.025 per cent) were tested in comparison with recommended dose of endosulfan (0.05 per cent) and aqueous 2 per cent NSKS and control (water) treatment against *S. litura* on tobacco both under choice and no-choice situations during 1992 by following exactly the same method as was adopted for testing the neem formulations during 1990 and 1991. Observations on leaf area consumed at 48 hours after feeding were recorded on graph sheet in the choice and no-choice situations and the data were subjected to statistical analysis. The effect of above treatments on larval mortality was also studied in glass troughs in laboratory. For each treatment and replication 10 laboratory bred 3rd instar larvae of 9-day-old were released into a glass trough and the larvae were fed with treated leaves daily. Observations were recorded on larval mortality daily.

Bio-efficacy of Repelin B and Other New Neem Formulations against *S. litura* on Tobacco (1992)

The bio-efficacy of Repelin B at 1 per cent (0.003 per cent azadirachtin) and some other new neem formulations viz., Godrej Achook, a neem oil based water soluble powder containing 0.03 per cent azadirachtin of Godrej Agrovet Limited, Bombay at 0.5 per cent (0.00015 per cent azadirachtin), Nimbecidine, a neem oil based emulsion containing 0.03 per cent azadirachtin of M/s. T. Stanes and Company Limited, Coimbatore at 0.00006 per cent (azadirachtin) and Neemox an emulsion concentrate containing 80 per cent neem biomass of M/s. Dinkal Agro Chemicals Private Limited, Bombay at 0.4 per cent was studied against *S. litura* on tobacco in comparison with aqueous 2 per cent NSKS, DCE of NSK and DOC at 1 per cent concentration, Repelin A at 1 per cent and Neemgold at 0.003 per cent (azadirachtin) and Navnem at 0.1 per cent both under choice and no-choice situations in the laboratory and also in tobacco nursery using exactly the same methodology as was adopted in 1990 season. Observations of leaf area consumed under choice and no-choice situations and mean increase/ decrease in weight of larvae at 48 hours after treatment under no-choice situation in the laboratory and number of seedlings damaged at 2 and 9 days after spraying were recorded. The data were subjected to statistical analysis using analysis of variance technique. Studies on the effect of neem formulations on larval mortality were also carried out in glass troughs as in 1990 by releasing 20 larvae in each glass trough (replication) per treatment and observations were recorded on larval mortality and pupation.

RESULTS AND DISCUSSION

All the neem formulations tested during 1990 and 1991 nursery seasons significantly protected tobacco from the *Spodoptera* damage both in laboratory and nursery tests (Tables 1 and 2).

Laboratory Tests

Under choice situation, DCE extract of NSK, Neemgold and NSKS were highly effective and showed significantly less area of leaf consumed by the *Spodoptera* larvae as compared to Neemrich II and Navnem in 1990 and Neemrich II, Vepacide I and II in 1991 season (Table 1). However, these formulations did not differ from DCE of DOC, Repelin A, Vepacide I and II and Neemgold (0.0015 per cent) during 1990 season whereas in 1991 season they remained on par with DCE of DOC, Repelin A, Navnem and Neemgold (0.0015 per cent). During 1990 Neemrich II was equally effective as Vepacide II but it was inferior to all other formulations except Navnem which was least effective whereas in 1991 Neemrich II was on par with DCE of DOC, Repelin A, Neemgold (0.0015 per cent) and also Vepacide I and II. The latter two treatments were least effective in 1991 (Table 1).

Under no-choice situation all the neem formulations excepting Navnem were equally effective and showed significantly less area of leaf consumed during 1990. Among these formulations, Repelin A, DCE extracts of NSK and DOC, Neemgold (0.003 per cent) and NSKS were highly effective and showed less area of leaf consumed though the differences were non-significant. During 1991, DCE extract of NSK, NSKS, Repelin A and Neemgold (0.003 per cent and 0.0015 per cent) were highly effective and superior to Neemrich II and Vepacide II. Navnem was equally effective as DCE of NSK and DOC, Neemgold (0.003 per cent and 0.0015 per cent) and Vepacide I and II, and superior to Neemrich II which was least effective in 1991 (Table 1).

Three neem formulations viz., DCE of NSK, Repelin A and NSKS showed maximum antifeeding effect against *S. litura* during both the years as shown by decrease in weight in 1990 and significantly less increase in weight in 1991. Among other neem formulations Navnem though significantly showed less increase in weight than in controls but it had shown significantly more increase in weight than the rest of the neem formulations in 1990. However, in 1991 Navnem was on par with all other neem formulations. The better antifeedant effect exhibited by Navnem in 1991 might be due to extrinsic and intrinsic factors including physiological condition of the larvae which needs to be further probed. Among others, NSKS, DCE of NSK and Vepacide I were superior to Neemrich II as shown by decrease or significantly less increase in weight of larvae.

Laboratory tests on larval mortality revealed that cent per cent mortality of larvae was obtained in DCE extract of NSK within 5 days followed by Neemgold 0.003 per cent in 6 days, NSKS and Neemrich II in 7 days, Vepacide II in 8 days, DCE of DOC, Repelin A, Neemgold (0.0015 per cent) and Vepacide I in 9 days. Navnem gave 90 per cent mortality of larvae within 9 days and rest of the larvae pupated. Out of 60 larvae released in each treatment maximum number of malformed larvae were obtained in Repelin A treatment i.e., 11 followed by Vepacide I and II (7), Neemgold 0.003 per cent (6), Neemgold 0.0015 per cent (5), Neemrich II (4), DCE extract of DOC (3), DCE extract of NSK and Navnem (2 each) (Table 3). Malformed larvae did not pupate and they are considered as dead. Forty two larvae in control (water spray) and forty nine larvae in control (no spray) were pupated.

Table 1. Effect of neem formulations on leaf area consumed and larval weight in 3rd instar larvae of *S. litura* 48 hours after treatment.

S. No.	Treatment	Leaf area consumed in sq cm				Mean increase/decrease in larval weight (mg)	
		Choice		No-choice			
		1990	1991	1990	1991	1990	1991
1.	NSKS 2%	0.37	1.00	1.55	1.00	−1.67	−7.33
2.	DCE extract 1% of NSK	0.28	1.00	1.48	1.38	−4.67	0.50
3.	DCE extract 1% of deoiled neem cake	0.70	1.11	1.33	1.85	11.00	16.50
4.	Repelin A 1%	1.12	1.15	1.25	1.13	−4.33	Dead
5.	Neemrich II 0.1%	5.75	2.18	4.17	2.09	28.67	18.33
6.	Vepacide I 0.25%	1.03	2.26	3.00	1.67	14.33	0.00
7.	Vepacide II 0.1%	1.30	2.74	3.27	1.88	20.67	9.00
8.	Navnem 0.1%	7.50	1.76	11.08	1.50	79.33	6.50
9.	Neemgold 0.003% (Azadirachtin)	0.38	1.05	1.52	1.23	13.33	4.00
10.	Neemgold 0.0015% (Azadirachtin)	1.23	1.11	3.25	1.21	17.33	15.33
11.	Control (Water Spray)	15.60	5.45	23.00	5.26	157.00	90.67
12.	Control (No Spray)	18.33	6.92	18.58	5.19	137.67	88.00
	C.V.%	59.10	28.10	37.90	13.79	35.05	31.74
	C.D. at 5%	4.47	1.12	3.93	0.49	22.69	16.26

Table 2. Effect of neem formulations on number of seedlings damaged by larvae of *S. litura*

Treatment	Number of seedlings damaged			
	2 days after spraying		9 days after spraying	
	1990	1991	1990	1991
NSKS 2%	2.63 (6.93)	2.87 (8.24)	3.18 (10.13)	2.76 (7.61)
DCE extract 1% of NSK	2.69 (7.23)	2.93 (8.59)	3.39 (11.50)	2.70 (7.28)
DCE extract 1% of deoiled neem cake	3.86 (14.90)	3.73 (13.89)	3.58 (12.84)	3.81 (14.54)
Repelin A 1%	3.35 (11.24)	3.04 (9.22)	3.61 (13.03)	3.45 (11.90)
Neemrich II 0.1%	4.65 (21.63)	4.54 (20.61)	4.96 (24.56)	5.62 (31.56)
Vepacide I 0.25%	4.28 (18.29)	3.91 (15.31)	4.64 (21.26)	4.45 (19.83)
Vepacide II 0.25%	3.83 (14.64)	4.03 (16.29)	4.43 (19.59)	4.19 (17.54)
Navnem 0.1%	5.77 (33.25)	3.10 (9.59)	5.50 (30.24)	3.65 (13.30)
Neemgold 0.003%(Azadirachtin)	3.36 (11.30)	3.25 (10.56)	3.34 (11.15)	3.40 (11.57)
Neemgold 0.0015% (Azadirachtin)	4.12 (16.96)	3.73 (13.95)	4.86 (23.62)	4.23 (17.88)
Control (Water Spray)	9.19 (84.48)	9.52 (90.60)	10.24(104.91)	9.41 (88.56)
Control (No Spray)	9,34 (87.16)	9.70 (94.18)	9.84 (96.85)	9.62 (92.65)
C.V.%	5.94	7.66	8.74	7.68
C.D. at 5%	0.48	0.59	0.76	0.62

Figures in parentheses are retransformed values.

Nursery Experiments

Two days after spraying, NSKS and DCE extract of NSK were highly effective and showed significantly less number of seedlings damaged by *S. litura* as compared to all other neem formulations in 1990 whereas they did not differ from Repelin A, Navnem and Neemgold (0.003 per cent) in 1991 (Table 2). Among others, Repelin A and Neemgold while not differing among themselves were superior to Neemrich-II, Vepacide-I, Neemgold 0.0015 per cent and Navnem in 1990 season. Vepacide II and DCE of DOC did not differ from Neemgold (0.003 per cent and 0.0015 per cent) and they were superior to Neemrich II. Vepacide I and II were on par with each other among which Vepacide I was superior to Neemrich II. Nine days after spraying, NSKS and DCE of NSK were highly effective during both the seasons and showed significantly less number of seedlings damaged by *S. litura* larvae than all other neem formulations except Repelin A, DCE of DOC and Neemgold (0.003 per cent), in 1990 season which remained on par with each other. Among other Vepacide I and II while not differing from Neemrich II and Neemgold 0.0015 per cent, they were superior to Navnem in 1990 season where as in 1991 season Neemgold 0.003 per cent, Navnem and Repelin A were equally effective and superior to Neemrich II and Vepacide I but they did not also differ from DCE of DOC, Vepacide-II and Neemgold (0.0015 per cent). Vepacide I and II were equally effective and superior to Neemrich II which was least effective and showed maximum number of seedlings damaged.

Effect of Repelin B versus Repelin A and NSKS against *S. litura* on Castor in Laboratory

All the treatments of Repelin and 2 per cent NSKS were found significantly more effective than control (water) treatment in protecting the castor leaf from damage caused by 3rd instar larvae of *S. litura* in both the tests under choice and no-choice situations in laboratory (Table 4).

Choice Situations

In both the tests Repelin B at 2 per cent and 1.25 per cent concentrations and 2 per cent NSKS in water had shown significantly better antifeedant and repellent activity and showed significantly less area of leaf consumed by larvae of *S. litura* as compared to Repelin A 1 per cent and Repelin B 0.66 per cent at 48 hours after treatment. Among the latter two treatments Repelin B 0.66 per cent was superior to Repelin A 1 per cent.

No-choice Situation

As in choice situation Repelin B at 2 per cent and 1.25 per cent concentrations and 2 per cent NSKS in water had shown significantly better antifeedant activity than Repelin A 1 per cent and Repelin B 0.66 per cent. But unlike choice situation the latter two treatments did not differ among themselves. Cent per cent mortality of larvae was obtained in Repelin B 2 per cent and 1.25 per cent concentrations while 50, 25 and 25 per cent mortality was observed in Repelin B 0.66 per cent, Repelin A 1 per cent and 2 per cent NSKS in water, respectively at 48 hours after treatment in test I.

Table 3. Percentage morality and number of malformed *S. litura* larvae in the laboratory (1990)

S. No.	Treatment	Per cent mortality of larvae days after treatment									No. of malformed larvae	Remarks
1.	NSKS 2% in water	6.67	20.00	76.67	93.33	96.67	98.33	100.00			0	
2.	DCE extract of NSK 1%	15.00	38.33	73.33	88.33	100.00	—	—			2	
3.	DCE extract (DOC) 1%	16.67	36.67	70.00	81.67	86.67	88.33	91.67	95.00	100.00	3	
4.	Repelin A 1%	20.00	35.00	55.00	73.33	85.00	88.33	95.00	96.67	100.00	11	
5.	Neemrich II 0.1%	33.33	48.33	71.67	78.33	98.33	98.33	100.00			4	
6.	Vepacide I 0.25%	15.00	36.67	75.00	78.33	88.33	95.00	95.00	98.33	100.00	7	
7.	Vepacide II 0.25%	31.67	55.00	68.33	76.67	90.00	95.00	98.33	100.00		7	
8.	Navnem 0.1%	25.00	33.33	53.33	63.66	80.00	85.00	85.00	90.00	90.00	2	6 larvae pupated
9.	Neemgold (0.003%) (Azadirachtin)	83.33	55.00	73.33	91.67	93.33	100.00				6	
10.	Neemgold (0.0015%) (Azadirachtin)	45.00	58.33	71.67	81.67	88.33	91.67	95.00	98.33	100.00	5	
11.	Control-water spray	0.00	0.00	0.00	15.00	30.00	30.00	30.00	30.00	30.00	0	42 larvae pupated
12.	Control-No spray	0.00	0.00	0.00	8.33	18.33	18.33	18.33	18.33	18.33	0	49 larvae pupated

Table 4. Effect of Repelin A and B and NSKS on leaf area consumed in castor by 3rd instar larvae of *S. litura* F. (1992).

Sl. No.	Treatment	Leaf area consumed in sq cm 48 hrs after treatment			
		Choice situation		No choice situation	
		Test I	Test II*	Test I	Test II*
1.	Repelin A 1%	2.98 (7.86)	2.73 (6.45)	1.60 (1.57)	1.78 (3.17)
2.	Repelin B 2%	1.02 (0.05)	1.04 (0.07)	1.00 (0.00)	1.08 (1.16)
3.	Repelin B 1.25%	1.07 (0.15)	1.11 (0.22)	1.00 (0.00)	1.71 (2.92)
4.	Repelin B 0.66%	1.79 (2.19)	1.41 (0.99)	1.80 (2.24)	2.05 (4.18)
5.	NSKS 2% in water	1.00 (0.00)	1.06 (0.11)	1.12 (0.26)	0.90 (0.81)
6.	Control (Water)	4.35 (17.88)	4.80 (22.00)	3.01 (8.09)	4.23 (17.89)
	C.V. %	19.56	7.63	14.67	13.90
	CD at 5%	0.60	0.23	0.35	0.41
	CD at 1%	0.83	0.32	0.49	0.57

*Leaf area consumed in sq cm 24 hours after treatment

Effect of Repelin B alone and in Combination with Endosulfan against Tobacco Caterpillar, *S. litura* on Tobacco

Laboratory tests revealed that Repelin 'B' alone at 1 per cent, 2 per cent and 3 per cent concentrations and in combination with endosulfan at half the recommended dose (0.025 per cent) gave significantly superior protection over endosulfan (0.05 per cent) against *S. litura* as shown by leaf area consumed at 48 hours after treatment both under choice and no-choice situations in the laboratory (Table 5). However, they remained on par with 2 per cent aqueous NSKS, the present recommended practice. Cent per cent mortality of larvae was observed under no-choice situation in petri dishes in 2 per cent and 3 per cent concentrations of Repelin 'B', endosulfan (0.05 per cent) and in all combination treatments 48 hours after introduction.

Bio-efficacy of Repelin B and Other New Neem Formualtions against *S. litura* (1992)

Repelin B and all other new neem formulations tested were significantly more effective than two controls in protecting tobacco from the *Spodoptera* damage both under choice and no-choice situations in the laboratory and also in the nursery (Table 6).

Laboratory tests

Choice Situation

Repelin B, Neemgold, DCE extracts of both NSK and DOC, and NSKS prevented the feeding completely by 3rd instar larvae of *S. litura* as shown by leaf area consumed. These formulations and Repelin A gave significantly better protection to tobacco from *S. litura* as compared to others. Among other neem formulations, Neemox was superior to Navnem which in turn was superior to Godrej Achook and Nimbecidine which did not differ among themselves and showed maximum leaf area consumed (Table 6).

Table 5. Effect of Repelin B alone and in combination with endosulfan on leaf area consumed by larvae of *S. litura* (1992)

		Leaf area consumed in sq cm	
S. No.	Treatment	Choice situation	No choice situation
1.	Repelin B 1%	1.00 (0.0)	1.04 (0.08)
2.	Repelin B 2%	1.00 (0.0)	1.00 (0.0)
3.	Repelin B 3%	1.00 (0.0)	1.00 (0.0)
4.	Repelin B 1% + Endosulfan 0.025%	1.00 (0.0)	1.00 (0.0)
5.	Repelin B 2% + Endosulfan 0.025%	1.00 (0.0)	1.00 (0.0)
6.	Repelin B 3% + Endosulfan 0.025%	1.00 (0.0)	1.00 (0.0)
7.	Endosulfan 0.05%	1.88 (2.54)	1.42 (1.02)
8.	NSKS 2% in water	1.00 (0.0)	1.14 (0.29)
9.	Control (Water spray)	4.93(23.27)	3.99(14.92)
	C.V. %	17.04	7.66
	C.D. at 5%	0.45	0.19

Figures in parentheses are retransformed means

Table 6. Effect of neem formulations on leaf area consumed and number of seedlings damaged in sq cm by larvae of *S. litura* (1992)

Treatment	Leaf area consumed in sq cm		Number of seedlings damaged (Days after treatment)	
	Choice	No choice	2	9
DCE extract of NSK 1%	1.00 (0.0)	1.00 (0.0)	3.15 (9.95)	3.00 (8.98)
NSKS 2%	1.00 (0.0)	1.01 (0.01)	2.82 (7.98)	2.57 (6.61)
DCE extract of DOC 1%	1.00 (0.0)	1.05 (0.10)	3.05 (9.33)	2.82 (7.98)
Repelin A 1%	1.25 (0.57)	1.19 (0.43)	3.69 (13.62)	3.65 (13.30)
Repelin B 1% (0.003% Azadirachtin)	1.00 (0.0)	1.00 (0.0)	3.59 (12.91)	3.16 (9.98)
Navnem 0.1%	3.15 (8.93)	2.95 (7.71)	4.83 (23.32)	4.90 (23.99)
Godrej Achook 0.5%	4.10 (15.77)	2.22 (3.94)	5.13 (26.32)	5.29 (27.96)
Nembicidine 0.00006% (Azadirachtin)	4.04 (15.29)	2.81 (6.87)	5.83 (33.98)	5.88 (34.57)
Neemox 0.4%	2.19 (3.81)	1.52 (1.31)	4.12 (16.93)	4.39 (19.28)
Neemgold 0.003 (Azadirachtin)	1.00 (0.0)	1.27 (0.62)	3.26 (10.63)	3.31 (10.98)
Control (Water spray)	5.28 (26.83)	5.59 (30.29)	10.17(103.52)	10.91(119.00)
Control (No spray)	5.83 (32.97)	5.67 (31.15)	9.77 (95.54)	10.11(102.20)
C.V. %	8.20	10.43	5.88	4.98
C.D. at 5%	0.36	0.40	0.49	0.42

Figures in parentheses are retransformed values.

No-choice Situation

Repelin B and DCE extract of NSK showed maximum antifeeding effect which has prevented the larvae completely from feeding on treated leaves as shown by area of leaf consumed. However, these formulations did not differ from NSKS, DCE of DOC, Repelin A and Neemgold which were superior to all other neem formulations excepting Neemox. Neemox remained on par with Repelin A and gave better protection than Godrej Achook, Nimbecidine and Navnem. Godrej Achook was superior to both Nimbecidine and Navnem which were on par with each other (Table 6).

Three neem formulations viz., DCE extracts of NSK and DOC and NSKS showed maximum antifeeding effect as shown by decrease in larval weight at 48 hours after treatment. Even though all other neem formulations showed increase in larval weight but the mean increase in larval weight is significantly less than in control treatments exhibiting their antifeeding effect against *S. litura*. Repelin A and B did not differ either from NSKS or DCE extracts of NSK and DOC, and they were superior to all other neem formulations. Among others, Neemgold showed better antifeeding effect as shown by less increase in weight of larvae, but it was superior to only Navnem (Table 7).

Studies on the effect of neem formulations on the larval mortality and pupation revealed that none of the larvae pupated in treatments of NSKS, DCE extracts of both NSK and DOC, Repelin A and B and Neemgold whereas few larvae pupated in treatments of Navnem, Godrej Achook, Nimbecidine and Neemox. Cent per cent mortality was observed after 6 days in Neemgold, 7 days in NSKS, DCE extracts of NSK and DOC and after 8 days in Repelin A treatment. The mortality in other treatments was ranging from 85 to 91.67 per cent. All the larvae were pupated in control treatments after 9 days from treatment (Table 7).

Nursery Tests

All neem formulations gave superior protection to tobacco seedlings from the *Spodoptera* damage as shown by number of seedlings damaged (Table 6).

Two days after spraying DCE extracts of NSK and DOC and NSKS were highly effective and showed significantly less number of seedlings damaged by *S. litura* as compared to Repelin A, Neemox, Nimbecidine, Godrej Achook and Navnem; however they did not differ from Neemgold. Among others Repelin B remained on par with DCE extract of NSK, Repelin A and Neemgold. Neemox was superior to Navnem and Godrej Achook which were on par with each other and superior to Nimbecidine which had shown maximum number of seedlings damaged (Table 6).

Nine days after spraying NSKS and DCE extract of DOC were most effective and showed minimum number of seedlings damaged by *S. litura* which were superior to all other neem formulations excepting that DCE extract of DOC did not differ from DCE extract of NSK and Repelin B. Next best treatments were DCE extract of NSK, Repelin B, Neemgold and Repelin A. Among others Neemox was superior to Navnem, Godrej Achook and Nimbecidine. In the latter group, Navnem and Godrej Achook were equally effective and superior to Nimbecidine which showed maximum number of seedlings damaged by *S. litura* larvae (Table 6).

The performance of DCE extracts of NSK and DOC against *S. litura* on tobacco in the present studies is in agreement with the previous findings (Murthy et al., 1989

Table 7. Effect of neem formulations on larval weight of 3rd instar larvae of *S. litura* 48 hours after treatment and mean percentage of mortality of larvae (1992)

S. No.	Treatment	Mean Increase /decrease in larval weight (mg.)	Mean percentage mortality									Remarks
			Days after treatment									
			1	2	3	4	5	6	7	8	9	
1.	DCE extract of NSK 1%	–9.00	18.33	25.0	55.0	75.0	80.0	88.33	100.00			
2.	NSKS 2%	–5.00	13.33	40.0	60.0	85.0	86.67	96.67	100.00			
3.	DCE extract of DOC 1%	–1.50	3.33	31.67	60.0	78.33	88.33	98.33	100.00			
4.	Repelin A 1%	4.33	20.00	28.33	45.0	63.33	71.67	85.00	96.67	100.00		
5.	Repelin B 1%	0.50	5.00	26.67	50.0	68.33	75.00	88.33	100.00			
6.	Navnem 0.1%	45.00	8.33	20.00	35.0	60.00	66.67	88.33	91.67	91.67	91.67	Rest pupated
7.	Godrej Achook 0.5%	39.00	8.33	18.33	30.0	60.00	68.33	78.33	85.0	85.0	85.0	Rest pupated
8.	Nembicidine 0.00006%	27.67	11.67	26.67	48.33	58.33	70.00	76.67	85.0	85.0	85.0	Rest pupated
9.	Neemox 0.4%	42.00	10.00	28.33	61.67	73.33	78.33	86.67	88.33	88.33	88.33	Rest pupated
10.	Neemgold 0.003%	21.33	15.00	16.67	50.0	88.33	93.33	100.00				
11.	Control (Water spray)	146.67	0.0	0.0	0.0	0.0	0.0	0.0	0.0	0.0	0.0	20 larvae pupated
12.	Control (No spray)	188.67	0.0	0.0	0.0	0.0	0.0	0.0	0.0	0.0	0.0	20 larvae pupated
	C.V. %	25.13										
	C.D. at 5%	21.25										

and 1992). According to HPLC analysis carried out by Murthy (1992) using external standard azadirachtin, the neem formulations of 2 per cent NSKS, DCE extract of NSK 1 per cent and DCE extract of DOC 1 per cent contains 3700, 1270 and 260 ppm azadirachtin respectively, whereas the above results indicate that tobacco seedlings can be effectively protected from the *Spodoptera* damage by spraying Repelin B at 1 per cent (0.003 per cent azadirachtin) alone or in combination with half the recommended dose of endosulfan (0.025 per cent) or Neemgold (0.003 per cent azadirachtin) or Repelin A at 1 per cent besides 2 per cent aqueous NSKS or DCE extracts of NSK and DOC. From this it can be inferred that neem formulations containing azadirachtin were effective at 0.003 per cent and higher concentrations in protecting the foliage from the *Spodoptera* damage.

REFERENCES

Abdul Kareem, A. 1981. Neem as an antifeedant for certain phytophagous insects and a bruchid on pulses. *Proc. 1st Int. Neem Conf.* Rottach-Egern. GTZ, Escborn. 1980. pp. 223–50.

Anonymous, 1979. Lab to Land. Good news to the tobacco growers. Neem seed kernel suspension gives good protection to your tobacco nurseries from tobacco caterpillar damage. Pamphlet issued by the Central Tobacco Research Institute, Rajahmundry, India.

Anonymous, 1989. Tobacco Production Technology. Central Tobacco Research Institute, Rajahmundry, India.

Chari, M.S. and G. Ramaprasad. 1987. Integrated Control of Pests on tobacco. *Proc. National Symp.* on *Integrated Pest Control—Progress* and *Perspectives*, November, 1988. Association for Advancement of Entomology, Trivandrum, India pp. 42–52.

Chari, M.S., G. Ramaprasad, S. Sitaramaiah and P.S.N. Murthy. 1990. Bio-efficacy of neem formulations against *Spodoptera litura* F. in tobacco nurseries. In: *Botanical Pesticides in Integrated Pest Management,* M.S. Chari and G. Ramaprasad (eds.), *Proc. Symp. Botanical Pesticides in IPM*, January, 21–22, 1990, ISTS, Rajahmundry, India, pp. 145–153.

Joshi, B.G. and G. Ramaprasad. 1975. Neem kernel as an antifeedant against the tobacco caterpillar (*Spodoptera litura* F.) *Phytoparasitica* 3(1): 59–61.

Joshi, B.G., G. Ramaprasad and S. Sitaramaiah. 1978a. Neem kernel suspension protects tobacco nurseries. *Indian Fmg.* 28(9): 17–18.

Joshi, B.G. and G. Ramaprasad and S.V.V. Satyanarayana. 1978b. Relative efficacy of neem kernel, fentin acetate and fentin hydroxide as antifeedants against tobacco caterpillar, *Spodoptera litura* Fabricius in the nursery. *Indian J. Agric. Sci.* 48(1): 19–22.

Joshi, B.G. and S. Sitaramaiah. 1979. Neem kernel as an ovipositional repellent for *Spodoptera litura* (F). *Phytoparasitica* 7(3): 199–202.

Koshiya. D.J. and A.B. Ghelani. 1990. Antifeedant activity of different plant derivatives against *Spodoptera litura* (Fabricius) on goundnut. In: *Botanical Pesticides in Integrated Pest Management,* M.S. Chari and G. Ramaprasad (eds.), *Proc. Symp. Botanical Pesticides in IPM*, January 21–22, 1990, ISTS, Rajamundry, India, pp. 270–75.

Koul, O. 1987. Antifeedant and growth inhibitory effects of Calamus oil and neem oil on *Spodoptera litura* under laboratory conditions. *Phytoparasitica* 15(3): 169–80.

Krishnamurti, C.R. 1982. Integrated Pest Management. Dr. T.V. Ramakrishna Ayyar Centenary Memorial Lecture-II. Entomology Research Institute, Loyola College, Madras. pp. 13–27.

Murthy, P.S.N. 1992. HPLC analysis of neem formulations using external standard Azadirachtin (Personal Communication).

Murthy, P.S.N., G. Ramaprasad, S. Sitaramaiah, B.V. Kameswararao, S.R. Prabhu and S. Nageswara Rao. 1989. Studies on the enrichment of neem fractions and their relative efficacy against tobacco caterpillar (*Spodoptera litura* F.) on lanka and f.c.v. tobaccos. *Tob. Res.* 15(1): 65–70.

Murthy, P.S.N., G. Ramaprasad and S. Sitaramaiah. 1992. Efficacy of deoiled neem cake extracts against *Spodoptera litura* on tobacco. Souvenir 1st. National Group Discussion "Quest for Quality and Productivity of Tobacco", Rajahmundry, 21–22 January, 1992, 20 p.

Pradhan, S. and M.G. Jotwani. 1971. Neem kernel as an antifeedant for locust. *Sneha Sandesh* (India) 13:1–5.

Ramachandra Rao, G., G. Raghavaiah and B. Nagalingam. 1990. Effect of Botanicals on certain behavioural responses and on the growth inhibition of tobacco caterpillar, *Spodoptera litura* Fabricius. In: *Botanical Pesticides in Integrated Pest Management*. M.S. Chari and G. Rama Prasad (eds.), *Proc. Symp. Botanical Pesticides in IPM*, January 21–22, 1990, ISTS, Rajahmundry, India, pp. 175–82.

Ramakrishnan, N., V.S. Saxena and S. Dhingra. 1984. Insecticide resistance in population of *Spodoptera litura* F. on tobacco. *Pesticides,* 8: 23–27.

Ramaprasad, G., S.V.V. Satyanarayana and S. Sitaramaiah. 1979. Studies on phytotoxic effect of neem kernel suspension on tobacco seedlings. *Indian J. Plant Prot.* 7:100–101.

Srivastava, B.K. and H.C. Joshi. 1965. Occurrence of resistance to BHC in *Prodenia litura* F. *Indian J. Ent.* 27: 102–104.

Wittekindt, W. 1978. Current West German regulations on maximum pesticide residues in tobacco products. *Tob. J. International* No. 4: 222–29.

Wittekindt, W. 1986. Recommended Plant Protecting Agents—Current Strategies in 22 important tobacco growing countries *Beit. Zur. Tabak. International* 13(5): 271–80.

64

Bioefficacy of Neem Extracts against *S. litura* on Tobacco

P.S.N. MURTHY[1], G. RAMAPRASAD[2], S. SITARAMAIAH[3], C.V. NARASIMHA RAO[4], B.V. KAMESWARA RAO[5], S.R. PRABHU[6] AND S. NAGESWARA RAO[7]

[1, 5]Principal Scientists
[4, 6]Sr. Scientists, Division of Crop Chemistry and Soil Science
[2]Principal Scientist,
[3]Sr. Scientist
[7]Technical Officer (T-5), Division of Crop Protection (Entomology)
Central Tobacco Research Institute Rajahmundry 533 105, India

ABSTRACT

Investigations were carried out to study the bioefficacy of various extracts of neem kernel and deoiled neem cake (DOC) against *S. litura* on tobacco in terms of putative ingredient Azadirachtin (AZT). In the laboratory tests under choice situation, dichloroethane (DCE) extract of NSK (1 per cent), NSKS (Neem Seed Kernel Suspension) (2 per cent) and DCE extract of DOC (1 per cent) were highly effective and showed significantly less leaf area consumed as compared to DCE extract at 0.5 per cent and methanolic extract of DOC (1 per cent). Same trend was observed under no-choice situation, however the difference among the treatments were not significant.

In the nursery trials, DCE extract of NSK (1 per cent) and NSKS (2 per cent) were equally effective both at 2 and 9 days after spraying and superior to alcoholic extract of NSK (1 per cent) and DCE of NSK (0.5 per cent). Next best were DCE extract of DOC and methanolic extract of NSK (1 per cent). The results indicated that NSKS (2 per cent), DCE of NSK (1 per cent) and DCE of DOC were most effective in protecting the nursery from *S. litura* damage, followed by methanolic extract of NSK, methanolic extract of DOC, DCE extract of NSK (0.5 per cent) and alcoholic extract of NSK (1 per cent).

INTRODUCTION

The bioefficacy of neem extracts against various insect pests has been documented in three international neem conferences. However, the trials conducted against *S. litura* are sparse. Much interest has been focussed by CTRI, on the activity of 2 per cent NSKS (Neem Seed Kernel Suspension) against *S. litura* on tobacco (Joshi et al., 1984). It was recommended that 25 kg of neem seed kernel was sufficient for complete protection of nursery in 400 sq m area. In order to facilitate easy application, attempts were made to enrich the active component azadirachtin from neem seed kernel which will obviate the problems of procurement on large scale. The present paper deals with the study of various neem extracts including enriched product and the bioactivity has been correlated in terms of putative reagent Azadirachtin (AZT).

MATERIALS AND METHODS

Neem seed kernel powder (1 kg) was extracted with alcohol, methanol, dichloroethane (DCE) six times at room temperature. The combined filtered extracts were concentrated in vacuo to give crude extracts. Fractionation scheme of crude extracts and the bioefficacy tests were carried out by adopting the method described earlier (Murthy et al., 1989). The neem seed kernel suspension (NSKS) was prepared according to Pradhan and Jotwani (1971). The extract of deoiled cake obtained from expellers were also tested.

Analysis of azadirachtin was determined by reverse phase HPLC, consisted of RP-18 column, using a linear isocratic system with 60 per cent aqueous methanol at flow rate of 1 ml/min. Absorbance was measured at 217 nm. Using this system, azadirachtin has retention time of 3.9 min.

RESULTS AND DISCUSSION

All the neem extracts viz., deoiled cake extracts, DCE extract (enriched) were significantly more effective than controls in protecting tobacco from *S. litura* damage, both in the laboratory and nursery.

Choice Situation

It is evident from Table 1, that enriched DCE extract, DCE extract of DOC and 2 per cent NSKS were superior to other fractions, which were on par. Among the extracts tested under no-choice situation, the significant superiority and enriched DCE extract was reflected. However the same trend was observed among other fractions.

Number of seedlings damaged 2 days after spraying in nursery, (Table 1) indicated that 2 per cent NSKS, DCE extract of NSK (1 per cent) and methanolic extract (1 per cent) of fresh seed kernel exhibited marked effect on the bio-efficacy of *S. litura* followed by DCE of DOC and DCE extract of NSK (0.5 per cent). Alcoholic extract of NSK showed least effectiveness compared to all other extracts. The order of bioefficacy of extracts remained same even after 9 days treatment.

Table 1. Effect of neem extracts against tobacco caterpillar, *Spodoptera litura* F.

Sl. No.	Treatment	Leaf area consumed 48 hours after treatment (Sq cm)		Number of seedlings damaged per bed, days after treatment in the nursery*	
		Choice situation	No choice situation	2 Days	9 Days
1.	Alcoholic *ext.* of Neem seed Kernel (1%)	4.13	3.10	20.33	19.33
2.	Methanolic *ext.* of Neem seed Kernel (1%)	3.18	1.57	12.33	16.00
3.	DCE *ext.* of NSK (1%)	0.73	0.00	8.67	9.33
4.	DCE *ext.* of NSK (0.5%)	5.85	2.53	18.33	18.67
5.	DCE *ext.* of DOC (1%)	0.75	1.17	12.67	15.00
6.	Methanolic *ext.* of DOC (1%)	6.73	3.15	16.33	17.33
7.	Neem seed kernel suspension (2%)	0.98	0.83	7.00	10.67
8.	Control (Solvent)	36.00	25.72	82.33	85.67
9.	Control (Water)	31.82	21.40	93.33	96.00
	CV%	21.61	40.96	15.63	10.16
	CD at 5%	3.75	4.68	8.16	5.63
	CD at 1%	5.17	6.45	11.24	7.75

*In the nursery, plot (bed) size was kept as 1 sq m. In each plot 400 seedlings were examined.

DISCUSSION

Relationship between Bioefficacy and Content of AZT against *S. litura*

The bioefficacy of NSKS (2 per cent) has been established over the years. NSKS (2 per cent) contained 10–15 per cent oil besides sediment, water soluble emulgators, and azt. It was segregated by partition into different fractions. Water solubles had azadirachtin (azt.) content of 0.37 per cent (Table 2), which was extracted into hexane and dichloromethane (CH_2Cl_2). Hexane contained only (0.055 per cent of AZT) while CH_2Cl_2 fraction had 0.262 per cent azt. Most of the azt. (85 per cent) was present in CH_2Cl_2. It is revealed from the analysis that NSKS (2 per cent) is a mixture of oil, sediment and water solubles (azt. and natural emulgators, sulphur compounds, sugars and protein) which exhibited better activity and persisted upto 9 days. It was found earlier that oil and sediment showed less bioefficacy than water solubles of NSKS (2 per cent) (Anon., 1987). However, when water solubles of NSKS (2 per cent) alone was tested for its bioefficacy against *S. litura* it was inferior to NSKS (2 per cent) (Anon., 1988) and the activity did not persist upto 9 days due to the possible degradation of azadirachtin. The soil and sediment present in NSKS (2 per cent) may help in spreading water solubles thereby enhancing the bioefficacy of 2 per cent NSKS. Because of the above reasons the activity of NSKS 2 per cent was on par with enriched DCE extract (1 per cent). It may be deduced that oil and sediment will synergise the activity of water solubles in 2 per cent NSKS.

The extraction of AZT. with different solvents had been envisaged to find out the suitable solvent, with maximum recovery in order to exploit on commercial scale.

Methanol extraction could provide an efficient method for extraction of AZT (1.75 to 3.39 per cent). It has been observed that the bioefficacy of methanolic extract at 1 per cent concentration exhibited activity on par with enriched DCE extract but there was no consistent activity over the years.

Reverse phase HPLC was used for the quantitative determination of azadirachtin. Crude and enriched neem seed kernel extracts were made with different solvents and their AZT was investigated. The solvent methanol had highest extraction capacity. The crude extracts contained 22–35 per cent oil, besides tetra nor-triterpenoids like salannin, nimbin and azadirachtin. They were tested against *S. litura* larvae both in the laboratory (under choice and no-choice) as well as in nursery, along with available formulations AISKE and margosan-O. Differences in bioactivity were not observed in the treatments except 2 per cent NSKS (Murthy et al., 1989).

The percentage of azadirachtin in various extracts and their enriched products is given in Table 2. With two further purification steps, the concentration increases to a higher level in any of the crude extracts. Extracts made by stirring the ground neem seed kernels in an appropriate solvent without heating is the practical method. Purification steps to enrich AZT were carried out by partitioning with dichloroethane, which is considered to be efficient, as it extracted most of AZT. The yield of enriched extracts are ranging from 3.35 to 6.5 per cent, depending on the quality of seeds, while AZT contents in the enriched extracts ranged from 8.3 to 12.72 per cent. The bioactivity of the enriched extract at 1 per cent concentration exhibited superior activity over others. The purity could not be further improved by merely partitioning in different solvents. The methods adopted by Schroeder and Nakanishi (1987) involving number of crystallisations from CCl_4 could afford AZT of 70 per cent purity. In order to fix the optimum levels of enriched extracts, different concentrations were tried. It was observed that less than 1 per cent concentration did not exhibit good efficacy against *S. litura*, when tested in both laboratory and nursery. The efficacy of enriched extract (AZT 8.8 to 12.72 per cent) was also on par with 2 per cent NSKS. A higher content of AZT could be achieved by liquid/liquid extraction of methanol/water extract with hexane and dichloroethane (DCE).

Table 2. Percentage of Azadirachtin (AZT) in various extracts

1. Alcoholic extracts of fresh neem cake	0.70
2. Methanolic extracts of fresh neem cake	1.75
3. Me OH/Eto AC extract	0.795
4. Acetonitrile extract	1.48
5. Enriched DCE extract from Alcoholic extract	12.72
6. Enriched DCE extract from Methanolic extract	8.8
7. Methanolic extract of Deoiled neem cake (DOC)	2.4
8. DCE extract of DOC	2.6
Fractions from 2% NSKS	
1. Hexane solubles of 2% NSKS	0.055
2. CH_2Cl_2 solubles of 2% NSKS	0.262
3. Neem seed kernel extract (2% NSKS)	0.37

It is evident from the azadirachtin content, the bioefficacy of enriched product has been increased, which affords complete protection of nursery beds upto 9 days after spraying. This was also compared with known formulations available in the market and exhibited superiority over others. Inspite of high bioeffiacy, the yields of enriched product are low and involved number of partitions which affects exploitation on commercial scale.

To circumvent the operations with concurrent increase in the bioefficacy, fresh deoiled neem cakes obtained from the expellers have been chosen to evaluate against *S. litura*. The extracts of DOC were compared with the promising fractions of fresh neem seed kernel as well as with 2 per cent NSKS. The results indicated that the efficacy of DCE of DOC and methanolic extract of DOC were on par with enriched extracts. These extracts served better proposition as they contained only 10 per cent neem oil besides active ingredients, sufficient for bioactivity. The deoiled cake extracts eliminated the partition steps, thus helping commercialisation of the process, and diversifying oil for suckericides and insecticides. The superior action of methanolic and DCE extracts of DOC is reflected with increasing amounts of azadirachtin with the concomitant decrease in leaf area consumed. The crude DOC extracts were analysed for AZT by HPLC following external standard method and they were found to range from 1.65 to 3.4 per cent.

From the above results, it can be deduced that bioactivity persists for a longer time, if the amount of azadirachtin present in the extracts is high. But it may be difficult to process the enriched extracts on a large scale as number of operations are involved. To make the process feasible, one can resort to easily available deoiled neem cakes in order to have significant bioactivity. The compounds other than AZT like salannin or nimbin may show synergistic activity, which may not deter the antifeedant activity when they are present in combination.

REFERENCES

Anonymous. 1987. *Annual Report.* Central Tobacco Research Institute, Rajahmundry pp. 36 & 37.

Anonymous. 1988. *Annual Report.* Central Tobacco Research Institute, Rajahmundry. pp. 41 and 42.

Joshi, B.G., G. Ramaprasad and S. Nageswara Rao. 1984. Neem seed kernel suspension as an antifeedant for *Spodoptera litura* in plated flue-cured tobacco crop. *Phytoparasitica.* 12(1): 3–12.

Murthy, P.S.N., G. Ramaprasad, S. Sitaramaiah, B.V. Kameswara Rao, S.R. Prabhu and S. Nageswara Rao. 1989. Studies on the enrichment of neem fractions and their relative bioefficacy against tobacco caterpillar (*S. litura* F.) on lanka and FCV tobacco. *Tob. Res.* 15(1): 65–70.

Murthy, P.S.N., S. Sitaramaiah, G. Ramaprasad, S. Nageswara Rao, B.V. Kameswara Rao and S.R. Prabhu. 1990. In: *Botanical Pesticides in Integrated Pest Management*, M.S. Chari and G. Ramaprasad (eds.), *Proc. Symp. Botanical Pesticides in IPM*, January, 21–22, 1990, ISTS, Rajahmundry, India, pp. 306–317.

Pradhan S. and M.G. Jotwani. 1971. Neem kernel as antifeedant for locust. *Sneha Sandesh* (India) 13:1–3.

Daniel Schroeder and Koji Nakanishi. 1987. A simplified isolation procedure for azadirachtin. *J. Natl. Products* 50(2): 241–244.

Neem and Environment

Volume 2

Editors

R.P. Singh • M.S. Chari
A.K. Raheja • W. Kraus

SCIENCE PUBLISHERS, INC.

SCIENCE PUBLISHERS, INC.
10 Water Street, # 310
Lebanon, New Hampshire 03766
United States of America

ISBN Set 1-886106-32-0
ISBN Vol. 1 1-886106-33-9
ISBN Vol. 2 1-886106-34-7

1-L6-SP.2

Foreword

For centuries, neem has been held in high esteem in India because of its medicinal and insecticidal properties. Neem tree, also known as margosa or the Indian lilac, grows extensively in many Asian and African countries. This multipurpose tree lives for 200 years and can grow well even under semi-arid and subhumid conditions up to 700 m above sea-level. It is tolerant to soil salinity, alkalinity and acidity and thrives well on all types of soils, besides establishing well even in the drought-prone areas. It is known to arrest desertification and soil erosion. It is used as a pesticide, nitrification inhibitor and is an effective component in the pharmaceuticals, and cosmetics. Research on neem is an excellent example of blending traditional wisdom with scientific knowledge.

The World Neem Conference was held for the first time in India at Bangalore from 24 to 28 February, 1993. I understand that this was the largest gathering on this subject, which comprised over 400 delegates from 30 countries who deliberated on the various aspects of neem. Research work done around the world in the last two decades has confirmed the immense value of neem. This tree will not only usher in a new era in pest management and medicine, but is also likely to play a significant role in preventing environmental degradation by reducing soil erosion and deforestation. Thus, the theme of the conference, 'Neem and Environment' had been in keeping with the spirit of the Earth Summit held at Rio de Janeiro in 1992.

It has been a matter of great privilege and pleasure for me to be associated with the hosting of the World Neem Conference in India. I congratulate the editors for their painstaking efforts in reviewing the papers presented at the Conference and bringing out the Proceedings. I am sure, the information contained in these Proceedings will be of immense value to all those concerned about the conservation and improvement of our environment. Indeed, the information is a valuable compilation of the available literature and knowledge on neem — the wonder tree of the World.

R.S. Paroda
Director General, ICAR and
Secretary, Department of Agricultural Research
and Education, Government of India

Preface

All efforts to increase the productivity of crops go waste if plants are not protected. Pests and diseases are important components of the adverse environmental factors. Putting of tonnes of toxic material in the environment to manage them is certainly not a right approach of solving pest problems. Cost development of resistance etc. are the other drawbacks associated with the use of synthetic pesticides. Neem (*Azadirachta indica* A. Juss.) has emerged as a single most important safe and cheap alternative to synthetic pesticides. We sincerely extend our compliments to the Indian Council of Agricultural Research and Indian Tobacco Company for sponsoring, and Indian Society of Tobacco Science for organising the World Neem Conference in the homeland of the tree, in 1993 at Bangalore.

The Conference received overwhelming response from all those who are directly or indirectly associated with neem. In the 5-day-deliberation, about 400 delegates from 30 countries exchanged their views on different aspects of neem. It was highly satisfying that almost all eminent researchers such as Prof. H. Schmutterer, Dr. R.C. Saxena, Prof. H. Rembold and Dr. Martin Rice participated in the conference and contributed papers.

Because of large number of papers presented in the Conference, it was decided to take out proceedings in two volumes. In preparing the articles, the editors have fully exercised their privilege of altering and reducing some of the manuscripts. It is likely that in doing so certain parts in the manuscript may have been altered unintentionally. For that editors tender their sincere apologies.

We express our deep sense of gratitude to Dr. Balram Jakhar, then Hon'ble Minister of Agriculture, Govt. of India, New Delhi for his wholehearted support and blessing for the Conference.

We express our sincere thanks to Mr. Verrapa Moily then Chief Minister of Karnataka for kindly inaugurating the Conference and Dr. V.L. Chopra, then Director-General, Indian Council of Agricultural Research, New Delhi and Secretary, DARE, Govt. of India, New Delhi for presiding over the Conference. We are also equally grateful to Shri G.K.P. Reddi, Director, ITL Ltd. for his keen interest and encouragement.

We sincerely thank Dr. R.S. Paroda, Chairman of the Organizing Committee for taking initiative in organizing the Conference and providing constant supervision and guidance.

The editors are grateful to Dr. T.S. Subramaniam, Senior Scientist, ITC Ltd. who played a very important role in conducting the conference seminar and providing all the necessary facilities to delegates.

The editor (Dr. R.P. Singh) is grateful to Prof. R.B. Singh, Director, Indian Agricultural Research Institute for providing facilities and permission for editing the proceedings.

We express thanks to the eminent persons of the Steering Committee and Members of the Executive Committee for providing guidance and help in organizing the Conference.

Finally we would like to express our great appreciation to all those who were directly or indirectly involved in making the Conference a success and delegates who responded to our invitation and participated in the Conference.

The Editors

WORLD NEEM CONFERENCE, FEB. 24–28, 1993, BANGALORE

Recommendations

I. *Sustainable Agriculture*

- While we need to produce more the environment is often harmed for immediate economic gains, sustainability of agro-ecosystem should, therefore, be the guiding policy for future research activity.
- Increased awareness of the potential of neem in agricultural sustainability — mass media, farming community and group meetings.
- Create favourable policy environment — at national, regional and international levels for careful and effective utilization of neem as sustainable substitute to synthetic agro-chemicals.

II. *Management of Pests and Diseases of Crops*

- For wider adoptibility and effective management, the pest control efficiency of neem-based pesticides should be evaluated for diverse biological effects using innovative bioassays and application methods against major insect pests and diseases of crops through on-farm trials.
- Relative safety to pests' natural enemies and other non-target organisms especially invertebrates needs to be determined.
- Evaluate the compatibility with synthetic pesticides especially plant derived pesticides and biocontrol agents.
- Encourage the traditional use of neem at farm level. Educate farmers about the advantages of using neem.
- Data on effect on seed germination, vegetative vigour phytotoxicity data etc. should be generated.
- Regulatory agencies should re-examine the existing pesticide registration requirements to encourage the use of neem formulations.
- Investigate systemic activity of bioactive principles and its effect on plant.
- Establish standard method for residue analysis of end product.
- Evaluate suppressive role of neem constituents on aflatoxins.

III. *Botany, Silviculture and Production*

- Genetic variability needs to be assessed keeping in view different utilization objectives especially bioactivity against pests.
- Understand neem biology and tree requirements to determine the adaptability of neem ecotypes/genotypes under diverse ecosystems, soil nutrients, etc.
- Develop techniques for seed collection preservation to increase the viability and deterioration of the active components.
- Investigate neem's nursery management including protection against pest and diseases.
- Investigate the value of monoculture versus mixed plantation with other economic trees including its role as wind breaker and in soil erosion.
- Promote neem cultivation especially in degraded land and investigate the impact of large-scale plantation on environment.

IV. *Neem Chemistry*

- Identify bioactive principles including non-terpenoids content from leaves, seed coat kernel, bark and flowers.
- Evaluate the biological effects of neem volatiles especially against stored grain pests.
- Investigate effect of temperature and photo-degradation on active compounds especially azadirachtin.
- Standardization of bio-assays to determine pest control effectiveness of neem-based products.
- To match synthetic pesticides develop as many neem formulations.
- Increase the shelf-life of neem formulations.
- Study structure-activity relationships of neem compounds.

* Identify metabolic pathways to production of bioactive principles.

V. *Neem Industry; Commercialization*

- Standardization of neem products.
- Improved shelf-life and stability.
- Cost effectiveness.
- Manufacturers to submit pertinent data as per requirement of the government. Compatibility with other pesticides should be declared.
- Aflatoxin content to be established.
- Azadirachtin should be declared.

VI. *Neem in Human and Veterinary Medicines and Public Health*

- Evaluate antibacterial, antiviral and fungicidal activities of neem.
- Evaluate neem products against human, poultry livestock diseases.
- Evaluate neem products against vectors and parasites affecting human, livestock and poultry.
- Evaluate immunostimulating properties of neem.
- Evaluate neem for its possible application in population control.
- Test neem for its efficacy against human disease (malaria, tuberculosis, leprosy, etc.).

Contents

Foreword iii
Preface v

VOLUME 1

1. INVITED LECTURES AND GENERAL PAPERS

1. Neem and Its General Development for Pest Control 3
Heinz Rembold

2. Interspecific Differences in the Susceptibility of Noctuid Larvae to Behavioural and Physiological Effects of Azadirachtin 11
Murray B. Isman

3. Repellent Action of Neem Oil on Insects of Public Health Importance 19
V.P. Sharma

4. Neem and Transfer of Technology 27
M.S. Chari

5. The Use of Neem by Farmers in the Dominican Republic 39
Carsten Hellpap

6. Neem in Holland 47
Arnold A. Ruiter

7. Neem as a Safe Trend for Pest Control in the Future 51
Nadia Z. Dimetry

8. Neem Pest Control Research in West Africa 61
Joanny O. Zongo

9. Improving the Productivity of Neem Trees 69
Narayan G. Hegde

10. Neem for Vegetable IPM in Nicaragua: Experiments and Small Farmers' Perspective — 81
D. Gómez, F. Miranda, B. Guerrero, I. Rivas, J. Mercado and F. Guharay

11. Potential for Neem (*Azadirachta indica* A. Juss.) in Armyworm Control in Africa — 91
Paul B. Tanzubil

12. Strategies in Management of Insect Pests with Neem (*Azadirachta indica* A. Juss.) — 103
R.P. Singh and A.K. Raheja

2. BOTANY

13. An Assay of Genetic Variability through Phenological Studies on Neem (*Azadirachta indica* A. Juss.) Plantation — 123
M.G. Gogate and D.R. Gujar

14. Engineering Properties of Neem Nut — 129
P.T. Palanisamy, R. Vishwanathan, S.S. Sivakumar, L. Gothandapani and K.R. Swaminathan

15. Location of Neem Triterpenoids and Other Secretory Structures — 141
P. Dayanandan, A. Stephen and B. Muruganandam

16. Studies on the Resin Canals and Ducts of *Azadirachta indica* A. Juss. (Meliaceae) — 151
K. Arumugasamy, K. Udaiyan, S. Manian, C. Balasubramanian and P.S. Mohan

17. Neem Seedlings — Growth Potentials of Different Organs — 159
A. Stephen, B. Muruganandam, P. Ravichandran and P. Dayanandan

18. Variation Studies and Heritable Components of Seed Parameters in Neem (*Azadirachta indica* A. Juss.) — 167
K. Kumaran, C. Surendran and R.S.V. Rai

19. *De novo* Synthesis of Secondary Metabolites in Cultured Cells of Neem — 173
K. Ramesh and M.A. Padhya

3. NEEM CHEMISTRY

20. Chemistry of Neem (*Azadirachta indica*), a Sustainable Source of Natural Pesticides — 187
Bina Shaheen Siddiqui, Shaheen Fazi, Ghiasuddin and Salimuzzaman Siddiqui

21. Photo-oxidation of Teranortriterpenoids — 199
S.R. Rojatkar, S.P. Joshi and B.A. Nagasampagi

22. New Tetra and Pentanortriterpenoids from *Azadirachta indica* A. Juss. — 203
S.R. Rojatkar, D.D. Sawaikar and B.A. Nagasampagi

23. Azadirachtin Content and Bioactivity of Some Neem Ecotypes of India 207
Seeni Rengasamy, Nutan Kaushik, Jitendra Kumar, Opender Koul and Balraj Singh Parmar

4. PHYSIOLOGICAL AND BEHAVIOURAL EFFECTS

24. The Effects of Various Neem Formulations on Mortality Rate and Morphogenetic Defects Upon *Schistocerca gregaria* (Forskal) Larvae 221
H. Wilps, O. Nasseh and S. Krall

25. Antifeedant and Insecticidal Activity of Some Neem Fractions 237
M. Narayana Rao, K. Nagaiah and G. Srimannarayana

26. Antifeedant and Growth-inhibiting Effects of Azadirachtin-rich Neem Fractions on Rice White-backed Planthopper *Sogatella furcifera* Horvath, *Spodoptera litura* Fb. and *Helicoverpa armigera* Hb. 243
S.J. Nelson, P.C. Sundarababu, D.S. Rajvel, G. Sreemannarayana and Y. Geethanjali

27. Effects of Extracts from Neem on Aphids (Homoptera: Aphididae) and Their Natural Enemies 253
D. Thomas Lowery and Murray B. Isman

28. Differential Sensitivity of *Spodoptera litura* (Fabricius) (Lepidoptera: Noctuidae) Larvae to Multifarious Activity of Azadirachtin and Possible Implications in Integrated Pest Management 265
S.N. Mukherjee and R.N. Sharma

29. Effect of Neem-based Insecticides on Growth and Development of White-backed Planthopper, *Sogatella furcifera* (Horvath) 281
S.S. Sandhu, G.S. Dhaliwal and V.K. Dilawari

30. Toxic, Developmental and Reproductive Effects of AZT-VR-K and Some Neem Products against Mustard Aphid, *Lipaphis erysimi* (Kaltenbach) 287
Surinder S. Bhathal and Darshan Singh

31. Effects of Azadirachtin on Neural Regulation of Midgut Peristalsis in *Locusta migratoria* 297
A. Dorn and P. Trumm

32. Fate of Azadirachtin in the Variegated Cutworm, *Peridroma saucia* 307
Opender Koul

33. Effect of Neem Oil on Trehalase Activity in Accessory Reproductive Glands of the Male *Odontopus varicornis* (Dist.) Hemiptera: Pyrrhocoridae 315
N. Selvi Sabhanayakam, K. Ramalingam, Senthamil Selvi and R. Krishnakumari

34. Effect of Neem Kernel Extract and Neem Oil on Nutritive and Reproductive Physiology of *Heliothis armigera* Hub. 321
K. Murugan, P. Jahanmohini and R. Babu

35. Insecticide, Antfieeding and Juvenilizing Effects of Neem (*Azadirachta indica* A. Juss.) Oil against *Corcyra cephalonica* Staint. and *Epilachna vigintioctopunctata* F. 335
Y.K. Mathur and Suman Nigam

36. Olfactory, Oviposition and Developmental Response of Four Species of Pulse Beetles to Four Neem Oil Treated Legumes 343
T.D. Yadav

37. Effectivity of Certain Plant Extracts against Larvae of *Mythimna separata* Wlk. and Adults of *Callosobruchus chinensis* Linn. 349
T.G.M. Singh and T.B. Gour

38. Neem Effects on *Spodoptera litura* (Fabr.): A Holistic Study 357
P.J. Rao, Shashi Gupta, D. Mohan Raj and K.R. Kranthi

39. Effect of Neem Extracts on Brown Planthopper, *Nilaparvata lugens* and Its Symbionts 375
S. Raguraman, S. Jayaraj and R.C. Saxena

40. Identification of Effective and Inexpensive Neem (*Azadirachta indica* A. Juss.) Seed Kernel Extract for Management of *Helicoverpa armigera* (Hubner) 381
B.R. Jhansi and R.P. Singh

5. FIELD EVALUATION

41. Effects of Neem Oil on the Grasshopper *Zonocerus variegatus* L. in the Field 405
Martin Baumgart

42. Environmental Impact and Bioefficacy of Neem Products against Insect Pests in Rice 413
N.V. Krishnaiah and M.B. Kalode

43. Potential of Neem in Insect Pest Management in Rice 425
G.S. Dhaliwal, Jaswant Singh and V.K. Dilawari

44. Evaluation of Neem Products and Standard Chemicals for the Management of Bean Fly *Ophiomyia phaseoli* (Tryon) in French Beans 433
P.N. Krishna Moorthy and K. Srinivasan

45. Neem in Integrated Management of *Helicoverpa armigera* Hubn. Infesting Chickpea 439
S.N. Sinha

46. Evaluation of Neem Products and Other Standard Chemicals for the Management of Major Pest Complex on Cabbage: Comparison between Standard Spray Regime and I.P.M. Involving Mustard as a Trap Crop 447
K. Srinivasan and P.N. Krishna Moorthy

47. Neem in *Helicoverpa* [*Heliothis*] *armigera* Management: Pulses Ecosystem 459
J.N. Sachan and Gururaj Katti

48. Use of Neem in Management of Pest Complex in Cotton 469
G.P. Gupta

49. Field Evaluation of Some Botanical Insecticides Alone and in Combination with Other Insecticides for Management of Bollworm Complex on Cotton 485
Ashok K. Dhawan and G.S. Simwat

50. Evaluation of Botanical Insecticides Alone and in Combination with Carbaryl against Insect Pest Complex of Okra 493
S. Venkateswara Rao and B. Rosaiah

51. Evaluation of Neem Products against Insect Pests of Rice and Pulse Crops 497
B.C. Shukla, U.K. Kaushik and R. Gupta

52. Efficacy of Neem and *Melia* in Overcoming Tolerance of *Plutella xylostella* to Some Commonly Used Synthetic Insecticides 505
S. Facknath

53. Bioefficacy of Neem Derivatives against the Insect Pest Complex of Rice 517
U.K. Nanda, B. Parija, N.C. Pradhan, B. Nanda and D.D. Dash

54. Evaluation of Neem Products for the Control of Vectors of Japanese Encephalitis Virus in Rice Fields — A Five-year Study (1987–1992) 529
D. Raghunatha Rao, R. Reuben and C. Hellpap

55. The Performance of an IPM Package Including Neem for the Control of Potato Pests in Farmers Fields 541
S.A. Siddiq and A. Khalafalla

56. Potential of Neem for Controlling Leaf-eating Caterpillar Complex on Cabbage 547
K.A.N.P. Bandara and C. Kudagamage

57. Neem Oil as Possible Biorational Insecticide for the Management of Spotted Pod Borer (*Maruca testulalis* Geyer) in Early Pigeon Pea 559
R.K. Akhauri, M.M. Sinha and R.P. Yadav

58. Effects of the Extracted Substances from Thai Neem Seed, *Azadirachta indica* var. *siamensis*, on the Insect Pests of Chinese Kale 565
Kwanchai Sombatsiri

59. Exploratory Evaluation of Neem Extract Sprays on Maize for Control of Stalk Borers in Coastal Kenya 571
M. Kiarie and R.C. Saxena

60. Repelin — An Effective Botanical Insecticide against Ber Fruitfly *Carpomyia vesuviana* Costa 575
V.V.S. Reddy, T. Ramesh Babu and B. Narasimha Rao

61. Evaluation of Enriched Neem (*Azadirachta indica* A. Juss.) Seed Extracts against Mango Hoppers, *Idioscopus nitidulus* Walker and *Amritodus atkinsoni* Leth. 581
R.P. Srivastava, R.P. Singh, M. Haseeb and S.R. Abbas

62. Evaluation of Neem Derivatives and *Lantana camara* L. against Cardamom Pests 589
B. Gopakumar, D. Kumaresen, S.S. Chandrasekar and S. Varadarasan

63. Bioefficacy of Some Neem Formulations against *Spodoptera litura* F. 595
M.S. Chari, G. Ramaprasad, S. Sitaramaiah and P.S.N. Murthy

64. Bioefficacy of Neem Extracts against *S. litura* on Tobacco 613
P.S.N. Murthy, G. Ramaprasad, S. Sitaramaiah, C.V. Narasimha Rao, B.V. Kameswara Rao, S.R. Prabhu and S. Nageswara Rao

VOLUME 2

6. EVALUATIONS AGAINST PHYTONEMATODES

65. Use of Neem Cake in Control of Lesion Nematode, Cyst Nematode and Root Knot Nematode 621
P. Gupta, S. Sebastian, L.R. Sobita and S.M. Mishra

66. Efficacy of Neem Cake against Soil Nematodes and Fungi on Chickpea 627
Sartaj A. Tiyagi and M. Mashkoor Alam

67. Efficacy of Cakes in the Management of Root Knot Nematode (*M. arenaria*) in Groundnut 643
M.U. Vaishnav, I.U. Dhruj and R.P. Lakhataria

68. Integrated Management of the Citrus Nematode, *Tylenchulus semipenetrans*, Using Neem Cake and the Parasitic Fungus *Paecilomyces lilacinus* 647
P. Parvatha Reddy, R.M. Khan and M.S. Rao

69. Effective Use of Wellgro and RD-9 Repelin for the Management of Root Knot Nematode on Tomato 651
M.S. Rao and P. Parvatha Reddy

70. Nematicidal Efficacy of Decomposed Neem Products on *Meloidogyne incognita* Infecting Mungbean and Their Residual Effect in Chickpea 657
S.D. Mishra and Vijayalakshmi Mojumder

71. Effect of Neem Cake, Neem Oil, Repelin and Carbofuran on Control of Soybean Nematodes 663
S.C. Vyas

72. Evaluation of Nematicidal Potential in Neem Allelochemicals 669
Mansoor A. Siddiqui and M. Mashkoor Alam

73. Management of Root Knot Nematode in Betelvine by Neem Cake 679
K. Sitaramaiah, A. Acharya, E.I. Jonathan, A.R. Wasnikar, R.V. Nakat, D. Subba Rao and Satyabrata Maiti

74. Nematode-controlling Efficiency of *Paecilomyces lilacinus* in Relation to Neem Products 685
Anjum Ahmad and M. Mashkoor Alam

75. Management of *Meloidogyne incognita* Infesting Pulse Crops through Seed Treatments with Neem Products 699
Vijayalakshmi Mojumder and S.D. Mishra

76. Neem Cake for Management of Root Knot Disease in Green Gram and Tomato Nursery 705
D.J. Patel, S.K. Patel and H.V. Patel

7. PLANT DISEASE MANAGEMENT

77. Neem and Control of Aflatoxin Contamination 713
Hampden J. Zeringue, Jr. and Deepak Bhatnagar

78. Management of Pest and Diseases of Cotton and Soybean Using Neem Products 729
F. Vidya, K. Sethuraman, V. Mariappan, K.R. Karunakaran and P. Vidyasekaran

79. Efficacy of Plant Derivatives on Rice Blast Pathogen *Pyricularia oryzae* cav., Blast Infection and Enzymes of Rice Plants 735
E. Rajeswari and V. Mariappan

80. Evaluation of *Azadirachta indica* and *Lawsonia inermis* for Antifungal Activity and Studies on Neem Products for Control of Damping-off in Tobacco Nurseries 745
M.M. Shenoi, S.M. Abdul Wajid and S.S. Sreenivas

81. Management of Pests and Diseases of Rice Crop with Neem Products 755
V. Mariappan, K. Sethuraman, F. Vidya, K.R. Karunakaran and P. Vidhyasekaran

82. Neem for the Control of Aphid and Whitefly Vectors and Virus Diseases of Plants 763
R. Roychoudhury and R.K. Jain

83. Effect of Neem and Other Plant Extracts on *Xanthomonas capestris citri*, the Incitant of Citrus Canker 771
B. Chendramohan and G.J. Moses

84. Effect of Plant Derivatives on the Transmission of Potato Virus Y Infecting Chilli and Its Aphid Vectors 777
L. Duleep Kumar Samuel and V. Mariappan

85. Efficacy of Neem Products in the Control of Rice Bacterial Blight 783
S. Eswaramurthy, V. Mariappan, M. Muthusamy, M.N. Alagianagalingam and K.S. Subramaniam

86. Effect of Non-edible Seed Oils in Aphid Transmitted Chilli Mosaic Virus in Tamil Nadu 787
V. Mariappan and L. Duleep Kumar Samuel

87. Use of Neem in Integrated Management of Rot Diseases of Small Cardamom (*Elettaria cardamomum* Maton.) 793
K. Dhanapal, Thomas Joseph and R. Naidu

88. Antimycotic Activity of a Methanolic Extract from *Azadirachta indica* A. Juss. 799
B. Steinhauer

89. Management of Chilli (*Capsicum annuum* L.) Diseases by Neem-based Preparations 809
R.P. Maharishi

8. ROLE IN INCREASING FERTILIZER NITROGEN EFFICIENCY

90. Potential Use of Urea Coated with Neem Cake to Improve Nutrient Uptake and N Recoveries and to Reduce N Losses in a Perennial Aromatic Grass, Java citronella (*Cymbopogon winterianus* Jowitt.) 815
E.V.S. Prakasa Rao

91. Relative Agronomic Efficiency of Neem Extract-coated Urea versus Prilled Urea for Rice 821
A.L. Bhandari, K.N. Sharma, R.S. Brar and D.S. Rana

92. Development of Neem-coated Urea for Increasing Nitrogen Use Efficiency and Crop Yields 827
B.N. Vyas, N.B. Godrej and K.B. Mistry

93. Use of Different Oil Cakes as Fertilizers: Comparison of Processes of Immobilization, Mineralization and Nitrification, Inhibition in Soil and Nitrogen Uptake 835
B.F. Hulagur

94. Comparison of Nimin-coated Urea with Other Sources at Graded Level of Nitrogen in Lowland Rice 843
M.K. Porwal, G.S. Bhatnagar and P.C. Chaplot

95. Use of Neem as Nimin to Increase N-use Efficiency in Transplanted Rice 847
S.S. Tomar and P.K. Gupta

96. Influence of Neem Oil Coating on Prilled Urea on Volatilization Loss from Swell-shrink Soil Amended with Organic Matter under Different Moisture Regimes 851
Muneshwar Singh and P.N. Takkar

9. COMMERCIAL NEEM INSECTICIDES

97. RD-9 Repelin: An Effective Botanical for Integrated Pest Management 857
T.S. Subramaniam, B. Sivaprasad and V.V.L.N. Prasad

98. Development of a Neem Formulation and Its Evaluation for Control of Crop Pests 865
K. Raman, S. Ganesan, B.N. Vyas, N.B. Godrej and K.B. Mistry

99. The NeemAzal Conception: Future Possibilities of the Use of Neem in Biological and Integrated Pest Management 875
Hubertus Kleeberg

100. The Use of Margosan-O® in the Greenhouse Environment 883
J.F. Walter and J.F. Knauss

101. Efficacy Profile of a Commercial Neem Insecticide 897
John Immaraju, Timothy Wood, Reese Nelson, William Ruggero and Samuel Wells

10. TECHNOLOGICAL DEVELOPMENT

102. Machineries for Neem Processing 909
S.S. Sivakumar, P.T. Palanisamy, N. Varadharaju, L. Gothandapani and K.R. Swaminathan

103. Technology for Production of Insecticides of Plant Origin at Rural Level 921
A. Dakshinamurthy

104. Cold Processing of Neem Seed 931
G. Ramakrishna, N.B.L. Prasad and G. Azeemoddin

11. PROPAGATION

105. Site Suitability for Neem in Queensland 941
Axel Bosselmann

106. Somatic Embryogenesis and Organogenesis in Neem (*Azadirachta indica* A. Juss.) 955
R. Islam, N. Joarder, O.I. Joarder, M. Hossain, S.M. Rahman and A.T.M. Naderuzzaman

107. Micropropagation of Neem through Axillary Bud Culture 961
O.I. Joarder, A.T.M. Naderuzzaman, R. Islam, M. Hossain, N. Joarder and B.K. Biswas

108. Potential Application of *in vitro* Methods for Propagation of Neem (*Azadirachta indica* A. Juss.) 967
M.S. Joshi and S.R. Thengane

109. High Frequency Regeneration of Neem *in vitro* (*Azadirachta indica* A. Juss.) from Immature Embryo Explants 975
M. Thiagarajan and P.M. Murali

110. Tissue Culture Studies in Neem (*Azadirachta indica* A Juss.) 981
A. Nirmalakumari, N.M. Ramaswamy and S.R. Sree Rangaswamy

111. Investigations on Neem and *Melia* for Sustainable Production 993
M.V.R. Prasad, A.R.G. Ranganatha, M. Sujatha, Mukta Dhingra M. Ramachandram and G. Nagaraj

112. Clonal Propagation of Neem by Tissue Culture 999
R.A. Drew

113. Twig Blight — A New Disease of Neem Seedlings 1007
J.C. Kaushik, S.S. Bisla and D.P. Singh

114. Vegetative Propagation of *Azadirachta indica* A. Juss. 1013
K.P. Srivasuki, N.S. Philomina and R. Vijaya Kumar

12. OTHER INSECTICIDAL PLANTS AND MISCELLANEOUS

115. Juvenomimetic Activity of Benzene Extracts of Twenty Plants from Kerala on the Red Cotton Bug *Dysdercus cingulatus* (Fb.) 1019
K. Saradamma, D. Dale and N.M. Das

116. Biological Effects of Certain Oils and Terpene Fractions of Plant Origin of Insects 1029
B. Subrahmanyam and P.J. Rao

117. Antifeedant Effects of *Jatropha curcas* L. Seed Oil and Extracts on the Variegated Grasshopper *Zonocerus variegatus* and the Eggplant Defoliator, *Selepa docilis* 1037
J.R. Cobbinah

118. Phytophagous Insects Living on *Ajuga* Species *A. bracteosa, A. chamaepitys, A. genevensis, A. pyramidalis. A. reptans* var. *reptans* and *A. reptans* var. *atropupurea* 1043
Béla Darvas, László Alfonz Polgár, Abdel-Monem Mokthar, Piroska Szabo, Mária Torma-Gazdag, Zoltán Ilovai, Ede Petro, Chih-Hua Tsou, Yaw-Huei Lin and Arild Anderson

119. Biological Activity of the Constituents of *Dalbergia sissoo* and Their Derivatives 1057
P. Venkat Kumar, D.B. Saxena and G.T. Gujar

120. Effects of *Acorus calamus* Oil and Its Main Compound β-Asarone on the Larger Corn Borer, *Prostephanus truncatus* (Horn) the Main Pest of Corn in Central Africa 1069
Gerhard H. Schmidt and Martin Streloke

121. Efficacy of Botanical Pesticides in the Control of Citrus Butterfly *Papilio demoleus* Linn. (Papilionidae: Lepidoptera) 1081
P. Kameshwara Rao and Ch. Balagangadhar Tilak

122. Efficacy of *Ajuga* (*A. chamaepitys, A. reptans*, var. *reptans*, and var. *atropurpurea*) Extracts on a Wide Variety of Non-adapted Insect Species 1085
Béla Darvas, László Alfonz Polgár, Ahmad Saber Bream, Imre Csatlos, Ali Ibrahim Farag, Mária Torma-Gazdag, Zoltán Ilovai, Maria Pia Calcagno and Josep Coll Toledano

123. Utility of Botanicals and Biocides as Additives to Insecticides in *Helicoverpa armigera* Hübner Management on Cotton 1101
B. Rosaiah and A.S. Reddy

124. Performance of Some Plant Products as Additives to Insecticides against *Helicoverpa armigera* Hübner (Noctuidae: Lepidoptera) 1109
N. Venugopal Rao, P. Raja Sekhar, M. Venkataiah and A.S. Reddy

125. Effect of Botanicals on Certain Parasitoids 1117
K. Srinivasa Babu, M.S.N. Murthy and T. Ramesh Babu

126. Effectiveness of Neem-based Products (Azadirachtin) against *Tetranychus macfarlanei* (Acarina: Tetranychidae) on Okra and Their Safety to Predatory Mites 1127
A.B. Rai, G.R. Bhanderi and C.B. Patel

127. Use of Botanical Formulations as Potential Repellents in Bird Pest Management 1135
P. Syamsunder Rao, K. Seshachalam Babu, M. Anand Rao and B.P. Vardhani

13. THERAPEUTIC POTENTIAL AND OTHER MEDICAL APPLICATION

128. *Azadirachta indica* A. Juss. Stem Bark as an Anti-Leprosy Source 1143
K.S. Subramanian and K.K. Lakshmanan

129. Identification and Characterization of the Immunomodulatory Fraction from Neem Seed Extract Responsible for Long-term Antifertility Activity 1151
Sanjay Garg, G.P. Talwar and S.N. Upadhyay

130. Clinical Studies with Praneem Polyherbal Cream in Chlamydial Cervicitis 1155
Sanjay Garg, A. Mittal, S.K. Kapur, S. Gupta, S.N. Upadhyay and G.P. Talwar

131. Rat Toxicity Studies with Neem Oil 1159
P. Badri Sriman Narayanan, R. Sampathraj and G. Vanitha Kumari

132. Alteration in Testicular Histoarchitecture Following Neem Oil Administration in Albino Rats 1165
M.P. Manoranjitham, A. Prem Anandhi, R. Sampathraj and C. Vanithakumari

133. Effect of Neem Oil: Structural and Functional Changes in the Epidymis of Rats — 1173
R. Sampathraj, P. Badri Sriman Narayanan and G. Vanithakumari

134. An Experimental Evaluation of *Azadirachta indica* (A. Juss.) for Adaptogenic Potential — 1183
P. Sen, P. Mediratta and A. Ray

135. Thyroid Hormones and Intermediary Metabolism in Fish: Influence of Neem Kernel Extract — 1189
M.C. Subash Peter

Appendix: List of Contributors and Keynote Speakers — 1199

6

EVALUATIONS AGAINST PHYTONEMATODES

65

Use of Neem Cake in Control of Lesion Nematode, Cyst Nematode and Root Knot Nematode

P. GUPTA, S. SEBASTIAN, L.R. SOBITA AND S.M. MISHRA

Allahabad Agricultural Institute, Allahabad 211 007, India

ABSTRACT

With a view to reduce dependence on use of chemical nematicides for minimizing crop losses due to phytonematodes, the efficacy of neem cake against *Pratylenchus thornei* on chickpea; *Heterodera cajani* on pigeonpea, black gram, green gram, and cowpea; and *Meloidogyne incognita* on soybean was tested. Cake was applied in soil/or as seed dressing, at recommended dosages, in fields infested with nematode species. The effect of these cake applications was evaluated on nematode population in roots and soil, crop growth and yield, and rhizobium nodulation on roots. Efficacy of neem cake extracts (water, ether, ethanol, acetone) was also evaluated on *P. thornei* and *H. cajani*. In all experiments, the population of nematode species in host roots and soil was significantly reduced in neem cake application with a corresponding increase in crop growth and yield. Number of rhizobium nodules were significantly greater with neem cake application. All neem cake extracts significantly reduced larval emergence from egg sacs of *H. cajani* and increased nematode mortality of *P. thornei*. It is concluded, that neem cake can successfully be used in control of these nematode species.

INTRODUCTION

Phytonematodes are responsible for huge losses annually to agricultural produce all over the world. Various chemical formulations (pesticides) used to prevent these losses are not only expensive and cumbersome to apply but also hazardous to humans and their environment. With a view to reduce dependence on pesticides, alternatives to chemical pest control are being sought.

Neem plant (*Azadirachta indica*), is known for its value in human health and hygiene, medicine, and protection of stored grains. Lately this plant has emerged as a source of several potent botanical insecticides. Some neem parts and products are reported to be antagonistic to plant parasitic nematodes.

Lesion nematode (*Pratylenchus* sp.), cyst nematode (*Heterodera* sp.), and root knot nematode (*Meloidogyne* sp.) are important phytonematodes causing significant losses to their host crops. In the present study the nematicidal efficacy of neem cake was tested against *P. thornei* on chickpea, *H. cajani* on pigeonpea cowpea, green gram, and black gram, and *M. incognita* on soybean in glasshouse, field, and also in neem extract tests. Its effect on plant growth and root nodulation was also assessed.

MATERIALS AND METHODS

In glasshouse tests finely powdered neemcake was mixed with autoclaved soil at the rate of 5 per cent w/w and after 15 days distributed in 15 cm earthen pots (500 g soil per pot) and each pot planted with two seeds of the host species. The test nematodes of known density (2 larvae/g soil) were inoculated into the rhizosphere of seven-day-old seedlings. Nematode inocula were obtained by blending and extraction of infected roots in case of *P. thornei* and *M. incognita* and by crushing freshly collected cysts in water for *H. cajani*. Observations were recorded on final nematode population in soil and roots for *P. thornei* and *H. cajani* and root gall index and root larval population for *M. incognita*.

Field tests were conducted at sites naturally infested with test nematode species. Finely ground neem cake was either incorporated into the upper 10 cm layer of soil in field 15 days prior to seed sowing at the rate of 400 kg/ha or applied as seed dressing along with seed sowing at the rate of 2g/10g seed. Population of nematodes per 500 g soil in the respective field experiments was recorded before cake application and 15 days after cake application in soil application, at sowing time in seed application, and at the end of the experiments.

In addition to the above, root and shoot weight, number of root nodules were recorded.

In the neem extract tests, 30 g of finely powdered neem cake was soaked in water, ether, acetone and alcohol solvents for 24 hours. The filtrate was evaporated to dryness (at room temperature) and then reconstituted in 50 ml water (N) or mixed with water to get required dilution of extracts. The test nematode species were exposed to these various extracts for 24 and 48 hours in small cavity blocks and per cent motality of *P. thornei* and larval emergence from egg sacs of *H. cajani* recorded.

The methods followed for nematode study were as per Southey (1970).

RESULTS AND DISCUSSION

The population of *P. thornei* per gram of root in chickpea was significantly reduced with soil application of neem cake in both glasshouse and field tests with a corresponding increase in plant growth parmeters (Table 1a).

Water, alcohol and acetone extracts of neem cake when evaluated for their efficacy against *P. thornei* population showed 82~100 per cent mortality (Control 0 per cent) in various extracts (Table 1b).

Table 1a. Evaluation of neem cake against *Pratylenchus thornei* on chickpea

Type of treatment	Glass house study Field study	Characters studied	Treatments		Statistical significance
			Control	Neem cake	
Soil application of cake	Glass house study	Nematode population/g of roots	305	56	S
		Shoot weight (g)	33	44	S
		Root weight (g) @ 90 DAS	5.9	13.6	S
-do-	Field study	Initial nematode population in soil/ 500 g soil	320		
		Final nematode population in soil/ 500 g soil @ 120 DAS	44776	13034	S
		Nematode population /g root @ 120 DAS	405	153	S

Table 1b. Evaluation of neem extracts against *Pratylenchus thornei*

Characters studied	Type of extracts			Control
	Water extract	Alcohol extract	Acetone extract	
Nematode mortality	100.0	86.6	82.2	0.0

In tests on *H. cajani*, both seed and soil application of cake significantly reduced the cyst population in all crops besides increasing the plant growth and root nodulation in both glass house and field trials (Table 2a).

Neem cake extracts in various solvents (water, alcohol, and ether) completely inhibited the larval emergence from egg sacs of *H. cajani* at both dilutions (N and N/2) till 48 hours exposure (Table 2b).

Neem cake application in *M. incognita* infested soil gave significantly reduced level of root knot index and root larval population with a corresponding increase in plant growth and root nodulation in soybean (Table 3).

CONCLUSIONS

From the above results it can be concluded that neem cake can successfully be used in control of lesion nematode (*P. thornei*), cyst nematode (*H. cajani*), and root knot nematode (*M. incognita*). This can be applied in field soils infested with the above nematode species either 15 days before seed sowing, or at the time of seed sowing as seed dressing. Both methods of cake application were effective in reducing nematode population and increasing plant growth. Besides, in reference to its use in pulse crop cultivation our results on rhizobium nodulation have added significance. While several reports exist in literature on the use of neem cake against *Meloidogyne* sp. (Parmar,

Table 2a. Evaluation of neem cake against *Heterodera cajani* on black gram, green gram, pigeonpea and cowpea

Type of treatment	Glass house study/Field study	Crop	Characters studied	Treatments		Satistical significance S/NS
				Control	Neem cake	
Soil application of cake	Glass house study	Black gram	Cyst population	64	32	S
			Shoot weight (g)	4.5	5.5	NS
			Root weight (g)	1.2	2.6	S
			@ 60 DAS			
-do-	-do-	Green gram	Cyst population	493	61	S
			Nodule number	13	12	NS
			Shoot weight (g)	0.5	0.8	S
			Root weight (g)	0.4	1.0	S
			@ 55 DAS			
-do-	-do-	Cowpea	Cyst population	1450	282	S
			Nodule number	43	60	S
			Shoot weight (g)	5.1	18.4	S
			Root weight (g)	2.8	6.8	S
			@ 45 DAS			
-do-	Field study	Black gram	Cyst population	171	33	S
			Shoot weight (g)	5.8	20.5	S
			Root weight (g)	1.9	3.8	S
			yield (g)	70	284	S
			@ 60 DAS			
-do-	-do-	Pigeonpea	Cyst population	152	34	S
			Shoot weight (g)	14.3	39.1	S
			Root weight (g)	0.9	1.9	S
			@ 90 DAS			
-do-	Glass house study	Black gram	Cyst population	506	35	S
			Shoot weight (g)	21.2	23.7	NS
			Root weight (g)	1.2	5.1	S
			@ 60 DAS			
-do-	-do-	Pigeonpea	Cyst population	465	132	S
			Shoot weight (g)	21.2	109.2	S
			Root weight (g)	5.2	11.8	S
			@ 90 DAS			
-do-	Field study	Black gram	Cyst population	167	52	S
			Shoot weight (g)	5.5	14.7	S
			Root weight (g)	0.7	2.8	S
			@ 60 DAS			
-do-	-do-	Pigeonpea	Cyst population	276	37	S
			Shoot weight (g)	14.2	92.9	S
			Root weight (g)	2.8	20.7	S
			@ 90 DAS			

1987) and a couple of them against some *Pratylenchus* sp. (Rössner and Zebitz, 1987), our results with *H. cajani* and *P. thornei* are new reports.

Table 2b. Evaluation of neem extracts against *Heterodera cajani.*

Characters studied	Concentration	Exposure period	Type of extracts			Control
			Water extract	Ether extract	Alcohol extract	
Larval emergence from 4 eggs sacs	N	24 hours	0	0	0	90~109
	N/2	24 hours	0	0	0	90~109
	N	48 hours	0	0	0	149~162
	N/2	48 hours	0	0	0	149~162

Table 3. Evaluation of neem cake against *Meloidogyne incognita* on Soybean.

Type of treatment	Glass house study Field study	Characters studied	Treatments		Statistical significance S/NS
			Control	Neem cake	
Soil application of cake	Glass house study	Gall index	3.8	1.2	S
		Larval population in roots	5792	550	S
		Nodule number	7	11	NS
		Shoot weight (g)	0.5	11.5	S
		Root weight (g) @ 90 DAS	1.8	4.0	S

S = Significant
NS = Non-significant
DAS = Days after sowing

REFERENCES

Parmar, B.S. 1987. An overview of neem research and use in India during the year 1983–86. In: *Proc. 3rd Int. Neem Conf.*, Nairobi, Kenya, 10–15 July, 1986. H. Schumutterer and K.R.S. Ascher (eds.), Eschborn, 1987, pp. 55~80.

Rossner, J. and C.P.W. Zebitz. 1987. Effect of neem products on nematodes and growth of tomato (*Lycopersicum esculentum*) plants. In: *Ibid.*, pp. 611–621.

Southey, J.F. 1970. *Laboratory Methods for Work with Plant and Soil Nematodes.* Min., Agr. Fish., Food, HMSO, London, 148 pp.

66

Efficacy of Neem Cake against Soil Nematodes and Fungi on Chickpea

SARTAJ A. TIYAGI AND M. MASHKOOR ALAM

Department of Botany, Aligarh Muslim University, Aligarh 202 002, India

ABSTRACT

Efficacy of oilseed cakes of neem (*Azadirachta indica*), castor (*Ricinus communis*), mustard (*Brassica campestris*), and duan (*Eruca sativa*) was evaluated against plant-parasitic nematodes and fungi infesting chickpea and also the subsequent crop, mung bean. The parameters recorded were plant weight, pollen fertility, pod numbers, chlorophyll content and nitrate reductase activity in leaves, and root nodulation. Significant reduction was observed in the population of plant-parasitic nematodes such as *Meloidogyne incognita*, *Rotylenchulus reniformis*, and *Tylenchorhynchus brassicae*, and in the frequency of pathogenic fungi such as *Fusarium oxysporum* f. *ciceri*, *Macrophomina phaseolina*, and *Rhizoctonia solani*. Frequency of saprophytic fungi was, however, increased. Manifold improvement in growth parameters was observed. The beneficial effects of different cakes persisted in the subsequent crop (mung bean), i.e., after a lapse of six months.

INTRODUCTION

Pulse crops contribute 10–13 million tonnes grain annually from 22–23 million ha of cultivable land in India. Pests and pathogens are major factors responsible for low production and uncertainty in yields of pulses (Grewal, 1988; Tiyagi, 1991). In chickpea and mung bean, plant-parasitic nematodes cause severe damage to plant growth (Tiyagi and Alam, 1986, 1987, 1988, 1989). Sasser and Freckman (1987) have estimated 13.7 and 10.9 per cent yield losses in these crops respectively. Parasitic fungi also cause considerable damage to these crops (Vidhyasekharan et al., 1977; Nene et al., 1984). About 10 per cent loss in the yield of chickpea was caused by the wilt fungus, *Fusarium oxysporum* f. *ciceri* (Singh and Dahiya, 1973). It is therefore imperative that these pathogens be managed in order to recover more economic mass and better quality of grains and to increase soil fertility. Pesticides play a vital role in this respect but they entail setbacks such as pesticide residues, environmental hazards and disruption of natural ecosystem.

The use of plant products has shown promise in solving this problem to some extent. The aim of the present investigation is to evaluate the efficacy of oilseed cakes in combination with mode of ploughing against soil-inhabiting plant-parasitic nematodes and fungi infecting chickpea. The residual effect of cake treatments on the subsquent crop mung bean has also been studied.

MATERIALS AND METHODS

The experimental field situated at Aligarh Muslim University Agricultural Farm was ploughed and small beds measuring 6 m^2 prepared, with a 0.5 m wide buffer zone between them. The beds were treated with oilseed cakes of neem (*Azadirachta indica* A. Juss.), castor (*Ricinus communis* L.), mustard (*Brassica campestris* L.), and duan (*Eruca sativa* Mill.) at 110 kg N/ha. The inorganic fertilizers—urea at 110 kg N/ha, superphosphate at 55 kg P/ha, and murate of potash at 55 kg K/ha—were applied before sowing. Untreated beds and those receiving inorganic fertilizers served as control. The treatments were randomized with five replications. Immediately after the soil was treated with oilseed cakes, the beds watered and after one week, bacterized seeds of chickpea cv K-850 were sown. This experiment was conducted under normal ploughing (20 cm deep) and deep ploughing (40 cm deep) conditions. Weeding and watering were done whenever required during the four month period. Plant growth parameters, nematode populations and frequency of soil fungi were determined.

Plant Growth and Other Parameters

At chickpea harvest, shoot weight, root weight, pods number per plant, root-nodule index (on 0–5 scale) were determined.

Pollen fertility (in percentage) at flowering stage was estimated by the method of Brown (1949) using stainability of pollen grains in 1 per cent acetocarmine solution. The pollen grains which took up stain and had irregular shape were considered sterile. Nitrate reductase activity in the leaves was estimated by the process of Jaworski (1971). The leaves were collected in the early morning, and cut into small pieces. Freshly harvested leaf tissue (200 mg) suspended in 5 ml mixture of 0.1 M phosphate buffer (pH 7.5), 0.02 M KNO_3, 5 per cent propanol, and two drops of chloram phenicol (0.5 mg/ml) in sealed tube was incubated at 25°C in dark for 60 min. Nitrate reductase activity was measured by NO_2^- production, which was detected by treating 0.4 ml of the incubation mixture with 0.3 ml each of 1 per cent sulfanilamide in 3 M HCl and 0.02 per cent N-1-nephthyl-ethylenediamine hydrochloride for 20 min. After appropriate dilution, the absorption peak at 540 nm was measured with the help of Bausch & Lomb spectronic-20 spectrophotometer.

Chlorophyll content of leaf was estimated by the method of Hiscox and Israelstam (1979). One hundered milligram of leaf pieces were placed in vial containing 7 ml dimethyl sulphoxide (DMSO) and the chlorophyll was extracted into the fluids at 65°C by incubating it for 60 min. The extract was transferred to a graduated tube and made up to 10 ml with DMSO and assayed immediately. A sample of 3.0 ml chlorophyll extract was transferred in covette and the Optical Density (O.D.) values at 645 and 663 nm were read in Spectronic-1001 spectrophotometer against DMSO blank.

Extraction of Nematodes

Nematode populations in each bed were determined before treating the soil as well as after terminating the experiment by processing soil samples with Cobb's sieving and decanting and Bearmann's funnel techniques (Southey, 1986), and examining the nematode aliquot under sterioscopic microscope. Numbers of root galls per plant were also counted.

Frequency of Soil Fungi

The frequency of soil fungi from the rhizosphere soil (on dry weight basis) was also determined, before the soil was treated as well as after the experiment was terminated by the dilution plate method of Dickinson and Pugh (1965), using peptone dextrose agar medium (Martin, 1950).

For each treatment 20 petriplates were used. Petri plated were incubated at 28°C and the fungi which developed after one week were examined and identified. The frequency of fungi was calculated by the formula of Mclean and Cook (1957).

$$\frac{\text{No. of plates containing a particular fungus}}{\text{Total plates poured}} \times 100$$

The residual effect of different treatments was also investigated on the succeeding mung been crop. The field was given normal ploughing treatment for all the beds, mung bean seeds were bacterized and sown. Weeding and watering were done whenever required during the 100 days crop period. Plant growth parameters were measured as with chickpea. Nematode populations and frequency of soil fungi were also determined as described above.

RESULTS

Effect of Oilseed Cakes on Chickpea

Plant Growth Parameters

Plant growth parameters, per cent pollen ferility, root nodulation, nitrate reductase activity, and chlorophyll content of chickpea greatly improved in all the treatments compared with control. The increase was more pronounced in deep ploughed beds than in normal-ploughed ones. In normal-ploughed beds, highest improvement was observed in plant weight (63.84 g), pollen fertility (89.50 per cent), number of pods (47.92), root nodulation (4.9), nitrate reductase activity (0.711) and chlorophyll (2.889 mg/g in the beds treated with neem cake compared with control and other oilseed cakes. In control it was observed that plant weight (25.38 g), pollen fertility (40.17 per cent), number of pods (19.34), root nodulation (2.2), nitrate reductase activity (0.396) and chlorophyll

(1.676 mg/g) were greatly reduced by the presence of pathogens. Significant improvement was noted in castor cake, followed by duan and mustard cakes. Deep ploughing enhanced the efficacy of various treatments (Table 1).

Root Galling

There was great reduction in the number of root galls caused by *M. incognita* with oilseed cakes, especially with neem cake (Table 1). As compared to 152.07 galls per plant in untreated bed, there were only 56.77 galls in the bed treated with neem cake, and 70.00 in duan cake. In case of inorganic fertilizers, though the reduction in galling was statistically significant, the reduction was not as much as in other treatments. Deep ploughing reduced root galls in all the treatments including control.

Population of Nematodes

Population of all the stylet bearing nematodes increased in untreated beds and beds treated with inorganic fetilizers. *Meloidogyne incognita*, *Rotylenchulus reniformis*, and *Tylenchorhynchus brassicae* were dominant species (Table 2). In normal-ploughed beds, the population of nematodes increased from the initial level of 1327/200 g soil to 3098 in untreated beds and 2216 in beds treated with inorganic fertilizers, whereas treatments with cakes brought reduction in nematode population, highest reduction being in neem cake (482), followed by castor cake (822), mustard cake (858) and duan cake (968). Deep ploughing further reduced the population of these nematodes in beds treated with oilseed cakes (Table 2).

Frequency of Rhizosphere Fungi

The frequency (percentage) of fungi (parasites as well as saprophytes) was greatly influenced by application of oilseed cakes (Table 3). In inorganic fetilizers the frequency increased in all types of fungi. The frequency of most of the pathogenic fungi, namely, *Macrophomina phaseolina*, *Botrytis cinerea*, *Curvularia tunata*, *Alternaria altrnata*, *Fusarium solani*, *Fusarium oxysporum*, *F. oxysporum* f. *ciceri*, *Rhizoctonis solani*, and *Sclerotium rolfsii*, greatly reduced in various treatments in normal as well as deep-ploughed beds. Neem cake proved most efficacious, followed by castor, mustard, and duan cakes. The reduction in the frequency of pathogenic fungi was greater in deep-ploughed beds. On the other hand, the frequency of saprophytic fungi—*Aspergillus niger*, *A. flavus*, *A. fumigatus*, *A. terreus*, *Rhizopus oryzae* and *Mucor* spp.—and of antagonistic fungi—*Trichoderma viridae* and *Penicillium degetatum*—increased in all the treated beds with oilseed cakes, the highest being in neem cake followed by castor, duan and mustard cakes.

Residual Effect on the Succeeding Mung Bean Crop

Plant Growth Parameters

Plant growth parmeters of mung bean significantly improved in all the treatments compared with control. However, it was more pronounced in deep-ploughed beds than

Table 1. Effect of oilseed cakes in combination with tillage on chickpea in field

Treatment	Plant weight (g)			Pollen fertility (%)	No. pods/ plant	Root nodule index (0–5)	No. root galls/ plant	Nitrate reductase activity*	Chlorophyll content (m/g)		
	Shoot	Root	Total						Chl.a	Chl.b	Total
Normal ploughing											
Untreated	17.88	7.50	25.38	40.17	19.34	2.2	152.07	0.396	1.298	0.378	1.676
Inorg. fert.	24.56	9.23	33.79	50.67	26.64	3.3	132.40	0.404	1.504	0.423	1.927
Neem cake	50.34	13.50	63.84	89.50	47.92	4.9	56.77	0.711	2.344	0.543	2.889
Castor cake	43.39	12.27	55.66	86.76	40.58	4.6	56.80	0.668	2.249	0.530	2.780
Mustard cake	32.42	10.65	43.07	59.67	38.75	3.8	76.34	0.414	1.681	0.428	2.109
Duan cake	35.08	11.00	46.08	68.72	39.62	4.0	70.00	0.437	1.891	0.485	2.379
CD ($P = 0.05$)			5.89	6.79	3.72	0.42	8.43	0.059	0.136	0.042	1.183
CD ($P = 0.01$)			8.03	9.26	5.07	0.57	11.49	0.080	0.185	0.057	0.250
Deep Ploughing											
Untreated	33.53	9.67	43.20	62.52	36.42	3.7	118.27	0.442	1.808	0.489	2.231
Inorg. fert.	39.96	10.53	50.49	73.44	44.31	4.0	67.90	0.491	2.033	0.471	2.504
Neem cake	62.84	14.80	77.64	93.60	55.14	4.9	37.47	0.742	2.583	0.625	3.213
Castor cake	47.69	14.78	62.47	91.37	48.94	4.6	45.20	0.728	2.578	0.647	3.229
Mustard cake	42.71	12.00	54.71	76.42	48.72	4.5	47.56	0.514	2.084	0.512	2.598
Duan cake	46.84	14.74	58.58	88.26	50.64	4.7	47.00	0.516	2.178	0.504	2.586
CD ($P = 0.05$)			5.39	5.81	4.07	0.38	6.24	0.058	0.083	0.036	0.131
CD ($P = 0.01$)			7.35	7.92	5.55	0.52	8.51	0.079	0.113	0.049	0.178

* Unit of nitrate reductase activity (NRA) = μ mole NO_2/h/g/fresh wt.

Table 2. Effect of oilseed cakes in combination with tillage on the population of nematodes on chickpea in field

Treatment	Population of plant-parasitic nematodes/200 g soil										
	Hop	Hel	Rot	Trh	Tyl	Mel	Dor	Hem	Pra	Others	Total parasitic
Normal ploughing											
Untreated	96	115	650	223	61	1728	114	50	37	24	3098
Inorg. fert.	60	67	275	233	59	1339	91	47	30	15	2216
Neem cake	25	34	104	57	18	170	38	20	10	09	482
Castor cake	35	52	212	113	32	269	56	23	17	13	822
Mustard cake	30	51	206	129	35	288	61	27	20	11	858
Duan cake	37	68	179	124	66	362	71	33	18	10	968
Initial Population	50	72	302	146	41	513	101	45	39	18	1327
Deep ploughing											
Untreated	81	93	548	197	88	1585	111	45	52	13	2813
Inorg. fert.	60	66	310	174	48	1154	70	69	22	10	1983
Neem cake	15	20	112	100	24	125	41	08	06	05	456
Castor cake	22	40	192	150	20	230	60	18	13	10	755
Mustard cake	25	41	127	71	40	205	45	18	04	05	581
Duan cake	26	50	133	82	35	245	58	11	06	02	648
Initial populations	50	72	302	146	41	513	101	45	39	18	1327

Hop = *Hoplolaimus indicus*, Hel = *Helicotylenchus indicus*, Rot = *Rotylenchulus reniformis*, Trh = *Tylenchorhynchus brassicae*, Tyl = *Tylenchus filiformis*, Mel = *Meloidogyne incognita*, Hem = *Hemicriconemoides mangiferae*, Pra = *Pratylenchus coffeae*, Others = Other Tylenchids, Dor = Dorylaims *longidorus elongatus, Xiphinema basiri* and *Trichodorus mirzai.*

Table 3. Effect of oilseed cakes in combination with tillage on the frequency of fungi in the rhizosphere of chickpea in field

Treatment	1	2	3	4	5	6	7	8	9	10	11	12	13	14	15	16	17
Normal ploughing																	
Untreated	50	45	35	45	45	35	30	35	30	25	20	20	25	35	50	30	10
Inorg. fert.	75	75	55	55	70	60	50	50	25	35	40	30	30	45	60	30	20
Neem cake	65	60	45	60	60	55	55	60	20	—	10	—	15	20	30	5	—
Castor cake	60	45	40	55	60	50	50	50	10	15	15	—	15	25	35	5	—
Mustard cake	50	60	40	45	50	55	45	45	15	10	20	20	15	25	40	10	5
Duan cake	55	55	50	40	50	55	35	50	15	-	25	15	15	20	40	15	5
Initial population	20	20	25	15	25	30	15	20	15	15	10	15	15	15	25	10	5
Deep ploughing																	
Untreated	55	30	30	35	35	45	40	30	25	15	10	20	25	15	35	15	5
Inorg. fert.	75	60	50	55	55	50	60	50	20	20	25	25	30	35	50	25	15
Neem cake	65	45	55	45	60	55	65	55	5	—	—	—	10	5	15	—	—
Castor cake	45	35	45	45	60	50	55	45	10	—	5	5	10	5	20	—	—
Mustard cake	60	35	40	45	50	40	50	45	10	5	5	10	15	10	25	10	—
Duan cake	55	40	40	35	45	40	50	45	5	5	—	10	15	10	20	15	—
Initial population	20	20	25	15	25	30	15	20	15	15	10	15	15	15	25	10	5

1 = *Aspergillus niger*; 2 = *A. flavus*; 3 = *A. fumigatus*; 4 = *A. terreus*; 5 = *Rhizopus oryzae*; 6 = *Mucor* spp.; 7 = *Trichoderma viridae*; 8 = *Penicillium degetatum*; 9 = *Macrophomina phaseolina*; 10 = *Botrytis cinerea*; 11 = *Curvularia tunata*; 12 = *Alternaria alternata*; 13 = *Fusarium solani*; 14 = *F. oxysporum*; 15 = *F. oxysporum* f. *ciceri*; 16 = *Rhizoctonia solani*; 17 = *Sclerotium rolfsii*.

normal-ploughed ones. In normal-ploughed bed, the highest increase in plant weight (71.18 g), pollen fertility (97.50 per cent), number of pods (56.62), root nodulation (5.0), nitrate reductase activity (0.624), and chlorophyll content (3.110 mg/g) of mung bean was observed with neem cake, as compared to control, where plant weight 20.97 g, pollen fertility 38.64 per cent, number of pods 16.84, root nodulation 2.2, nitrate reductase activity 0.296 and chlorophyll content 1.538 mg/g were recorded (Table 4).

Root Galling

Root galling caused by *M. incognita* greatly reduced due to the application of oilseed cakes in the preceding crop. As compared to 156.50 root galls per plant in untreated plants, there were only 50.37 galls with neem cake followed by 52.64 with castor cake, 68.65 with mustard cake, and 73.30 with duan cake. In case of inorganic fertilizers the reduction in galling was statistically significant but not as great as with other treatments. Deep ploughing brought further stress on root knot development (Table 4).

Population of Nematodes

With normal ploughing, the population of *M. incognita* slightly increased in beds previously treated with oilseed cakes (neem, castor, mustard, duan). The highest reduction in the population of nematodes was observed with neem cake, followed by castor cake, duan cake, and mustard cake. A similar trend was noted in *R. reniformis* and *T. brassicae*. Deep ploughing of the preceding crop had adverse effect on nematode populations, which was less than in normal ploughing. It is evident that oilseed cakes applied to the preceding crop can successfully contain the population of nematodes in the succeeding crop (Table 5).

Frequency of Rhizosphere Fungi

In the second crop (mung bean) the frequency of most of the saprophytic fungi further increased over the frequency in the first crop (chickpea) in untreated, normal-ploughed beds, the highest being with *A. flavus*, *T. viridae*, and *P. degetatum*. Soil application of oilseed cakes to the first crop also had a favourable effect on the frequency of saprophytic fungi in the present crop. The residual effect of neem cake was found highest in increasing the frequency of some saprophytic fungi, e.g. *A. flavus*, *A. fumigatus*, *T. viridae*, and *P. degetatum*, whereas the residual effect of mustard cake was found least effective in promoting frequency in saprophytic forms. However, these oilseed cakes showed some differential behaviour towards saprophytic fungi (Table 6). The residual effect of inorganic fertilizers was also seen on some fungi.

Parasitic fungi were also influenced by the residual effect of oilseed cakes applied to the previous crop. But their efficacies differed (Table 6). Neem cake and castor cake residues were found effective in continuously suppressing parasitic fungi in most cases. Mustard and duan cakes did not influence them appreciably. Some parasitic fungi were totally eliminated by the application of neem and castor cakes.

Table 4. Residual effect of oilseed cakes on the succeeding mung bean in field

Treatments	Plant weight (g)			Pollen fertility (%)	No. pods/ Plant	Root-nodule index (0–5)	No. root galls/ plant	Nitrate reductase activity	Chlorophyll content (mg/g)		
	Shoot	Root	Total						Chl.a	Chl.b	Total
Normal ploughing											
Untreated	15.57	5.40	20.97	38.64	16.84	2.2	156.50	0.296	0.909	0.628	1.538
Inorg. fert.	26.20	8.50	34.70	65.89	25.27	3.2	100.00	0.392	1.301	0.714	2.015
Neem cake	55.40	15.78	71.18	97.50	56.62	5.0	50.37	0.624	1.751	1.351	3.110
Castor cake	50.52	14.62	65.14	96.64	49.50	4.9	52.64	0.593	1.654	1.304	2.961
Mustard cake	44.72	13.36	58.08	92.50	48.39	4.7	68.65	0.532	1.592	1.004	2.598
Duan cake	47.57	14.27	61.84	88.21	44.20	4.7	73.30	0.500	1.423	1.085	2.509
CD (P = 0.05)			5.96	6.17	4.75	0.27	5.09	0.058	0.098	0.081	0.182
CD (P = 0.01)			8.13	8.41	6.48	0.37	6.94	0.079	0.133	0.110	0.248
Deep ploughing											
Untreated	26.64	7.84	34.48	59.37	24.74	3.2	100.43	0.334	1.352	0.824	2.177
Inorg. fert.	34.42	9.87	43.29	77.57	33.49	3.9	63.17	0.407	1.567	0.933	2.500
Neem cake	69.44	17.00	86.44	97.68	61.52	5.0	38.50	0.705	2.110	1.376	3.267
Castor cake	57.72	15.64	73.36	96.50	56.40	4.9	40.75	0.674	1.922	1.342	3.267
Mustard cake	48.36	14.14	62.50	63.66	44.83	4.5	39.62	0.604	1.587	1.239	2.828
Duan cake	49.36	15.62	64.98	92.40	50.24	4.5	48.77	0.583	1.471	1.304	2.777
CD (P = 0.05)			4.87	5.77	4.65	0.24	6.29	0.061	0.082	0.070	0.155
CD (P = 0.01)			6.64	7.87	6.34	0.33	8.58	0.083	0.112	0.095	0.211

Table 5. Residual effect of oilseed cakes on the population of nematodes on the succeeding mung bean in field

Treatment	Population of plant parasitic nematodes/200 g soil										
	Hop	Hel	Rot	Trh	Tyl	Mel	Dor	Hem	Pra	Others	Total parasitic
Normal ploughing											
Untreated	52	127	413	175	77	1816	184	65	22	37	2968
Inorg. fert.	51	77	312	200	60	1407	127	42	35	10	2321
Neem cake	30	40	124	42	22	200	20	15	10	05	508
Castor cake	37	45	237	75	35	255	44	34	20	15	797
Mustard cake	25	65	220	118	30	302	47	30	23	06	866
Duan cake	40	57	196	109	68	462	65	37	20	10	1064
Deep ploughing											
Untreated	60	100	372	150	66	1550	160	50	24	33	2565
Inorg. fert.	44	59	348	180	52	1149	100	32	26	10	2000
Neem cake	27	44	110	34	27	185	17	16	10	06	476
Castor cake	26	43	214	85	32	240	40	27	16	20	743
Mustard cake	20	52	196	97	27	256	37	26	17	05	733
Duan cake	30	48	200	85	38	334	44	29	16	08	832

Table 6. Residual effect of oilseed cakes on the frequency of fungi in rhizosphere of succeeding mung bean in field

Treatments	1	2	3	4	5	6	7	8	9	10	11	12	13	14	15	16	17	18
Normal ploughing																		
Untreated	50	60	40	25	45	20	40	40	35	35	10	40	20	15	20	10	20	35
Inorg. fert.	75	70	45	50	50	35	65	35	40	40	15	50	35	25	30	15	25	30
Neem cake	65	70	65	45	45	20	65	50	15	15	—	20	—	—	10	—	—	10
Castor cake	65	65	60	45	50	15	65	45	15	25	—	25	5	—	15	—	5	15
Mustard cake	55	65	55	40	50	20	55	45	25	20	—	25	10	—	15	—	—	20
Duan cake	55	70	55	45	45	20	50	50	30	20	5	30	5	10	10	5	10	20
Deep ploughing																		
Untreated	45	35	45	35	30	30	35	30	20	20	10	30	10	10	10	10	10	20
Inorg. fert.	65	60	65	45	45	45	60	40	35	30	10	35	25	15	20	10	15	30
Neem cake	60	65	75	45	40	15	60	35	15	5	—	10	—	—	—	—	—	5
Castor cake	50	55	45	35	40	15	55	35	10	10	—	15	—	—	5	—	5	5
Mustard cake	55	50	50	35	35	25	55	30	20	10	—	15	—	—	—	—	—	15
Duan cake	50	55	55	30	35	20	45	30	20	15	—	20	10	10	5	—	5	10

1 = *Aspergillus niger*; 2 = *A. flavus*; 3 = *A. fumigatus*; 4 = *Rhizopus oryzae*; 5 = *R. nigricans*; 6 = *Mucor* spp.; 7 = *Trichoderma viridae*; 8 = *Penicillium degetatum*; 9 = *Macrophomina phaseolina*; 10 = *Rhizoctonis solani*; 11 = *Alternaria alternata*; 12 = *Phyllosticta phaseolina*; 13 = *Sclerotium rolfsii*; 14 = *Botrytis cinerea*; 15 = *Septoria leguminum*; 16 = *Fusarium solani*; 17 = *F. oxysporum* f. *ciceri*; 18 = *F. oxysporum* f. *lycopersici*.

In deep-ploughed beds, irrespective of the treatments of the first crop, there was a reduction in the frequency of some of the saprophytic as well as parasitic fungi; the reduction was greater in parasitic fungi (Table 6).

DISCUSSION

In the present study, amending the soil with oilseed cakes (neem, castor, mustard, duan) has caused significant reduction in the population of plant-parasitic nematodes, including *M. incognita*, *R. reniformis*, *T. brassicae*, *Hoplolaimus indicus*, *Tylenchus filiformis*, *Helicotylendus indicus*, *Hemicriconemoides mangiferae*, and *Pratylenchus coffeae*, on chickpea and mung bean (Tables 2, 5). Root galling due to *M. incognita* was also reduced by these treatments. These results are, in a way, in agreement with those of Singh and Sitaramaiah (1970), Alam and Khan (1974), and Siddiqui (1986) with other crops.

The frequency of parasitic fungi was also found reduced in the beds amended with oilseed cakes. However, the amendments supported the growth of saprophytic fungi. Oilseed cake amendments have been found effective in suppressing the soil population of pathogenic fungi like *R. solani*, and *Colletotrichum* sp. and *Fusarium* sp. in the rhizosphere of eggplant, okra, and tomato (Khan et al., 1974b), *F. udum* on pigeonpea (Singh and Singh, 1982), and *F. oxysporum* f, *ciceri* and *Sclerotium rolfsii* on gram (Sinha and Prasad, 1986).

There may be sevral reasons for the suppression of parasitic fungi with oilseed cake amendments. Khan et al. (1974a) have found water soluble fractions of oilseed cakes and bitter principles of neem, e.g., nimbidin and thionimone, inhibitory to the growth of pathogenic fungi. Neem extracts were also inhibitory to *F. oxysporum f. ciceri*, *R. solani*, *S. rolfsii* and *Sclerotinia sclerotiorum* infecting gram crop (Singh et al., 1980).

Several fatty acids (Toussoun et al., 1968), aldehyde, ketones, amino acids, carbohydrates, and free sulphur (Ahmad et al., 1972) are released during decomposition of organic amendments. Some of these have been reported to be toxic to pathogenic fungi (Singh and Pandey, 1965; Khan et al., 1974 a,b). The volatile degradation products of oilseed cakes and meals also inhibit the population of *Fusarium* spp. (Zakaria et al., 1980).

The suppressant effects of oilseed cakes on parasitic fungi may also be due to stimulation of microbial activity and biocontrol agents (Lloyed and Lockwood, 1966; Singh and Singh, 1981). In the present study frequency of both the saprophytes and antagonistic fungi such as *A. niger*, *A. flavus*, and *T. viridae*, and *P. degetatum* increased due to the application of oilseed cakes in both the crops. Evans (1955), Saxena (1960) and Palakshappa et al. (1986) have also reported that high population of *T. viridae* and other antagonistic fungi may suppress the population of parasitic fungi. In the present study this might be one of the reasons for reduction in the frequency of parasitic fungi.

Thus, it appears that improvement in plant growth of chickpea and mung bean in amended soil was due to reduction in the populations of plant-parasitic nematodes and frequency of pathogenic fungi and also due to their manurial effect. Increased microbial activity in amended soil is known to bring about increased conversion of N to nitrate form (Schmidt, 1954; Gunner, 1963), which in turn appears to be responsible for stimulation of nitrate reductase activity, as has been evident in the present study. Plants utilize the nitrate form of nitrogen, which, however, must be reduced to ammonia before incorporation into the nitrogenous compounds of plants. This reaction is mediated by enzymes such as nitrate reductase. The application of oilseed cakes provides more and more inducing substrate (nitrate) for the enzyme (nitrate reductase) to accelerate its activity, which results ultimately in increased motabolic activity of plants and then plant growth. The chlorophyll content was also increased by amendments with oilseed cakes. A similar study was reported by Ahmad et al. (1990) with the application of oilseed cakes on black gram.

Deep ploughing resulted in more reduction in nematode population in the present study. Similar effect was also reported by Thomas (1978), Khan et al. (1990), Alam (1991) and Siddiqui and Alam (1991). Most probably the ecological set-up of nematodes is distrubed by deep ploughing. Frequency of fungi was also found to be aversely affected by deep ploughing. As a consequence of these adverse effects on pathogens, plant growth of chickpea and mung bean improved manifold in deep-ploughed beds treated with oilseed cakes.

The residual effects of oilseed cakes and ploughing on the population of plant-parasitic nematodes and frequency of fungi remained below the initial levels. This clearly shows that the suppressant effects of cake treatments to first crop persisted even in the subsequent crop. These findings on plant-parasitic nematodes are in conformity with those of Singh and Sitaramaiah (1966) and Alam et al. (1977).

ACKNOWLEDGEMENT

The senior author is thankful to the Council of Scientific and Industrial Research (C.S.I.R), New Delhi for financial assistance.

REFERENCES

Ahmad, A., S.A. Tiyagi and M. Mashkoor Alam. 1990. Efficacy of oilseed cakes and nematicides agianst root-knot nematode attacking black-gram. *Science Khyber* 3: 165–170.

Ahmad, R., A.M. Khan and S.K. Saxena. 1972. Changes resulting from amending the soil with oil-cakes and analysis of oil-cakes. Proc. 59th *Sess. of Indian Science Cong.*, Calcutta Part III, pp. 164.

Alam, M.M. 1991. Influence of ploughing on the efficacy of aldicarb and carbofuran against ecto-parasitic nematodes. *Int. Pest Control* 33: 39–41.

Alam, M.M. and A.M. Khan. 1974. Control of phytonematodes with oil-cake amendments in spinach field. *Indian J. Nematol.* 4: 239–240.

Alam, M.M., A.M. Khan and S.K. Saxena. 1977. Presistent action of oil-cakes and nematicides on the population of nematodes in field. *Botyu Kagaku* 40: 159–161.

Brown, G.T. 1949. *Pollen Slide Studies*. Charles C. Thomas Publ., Springfield, Illinois.

Dickinson, C.H. and G.J.F. Pugh. 1965. Use of a selective cellulose agar for isolation of soil fungi. *Nature* (London) 207: 440–441.

Evans, E. 1955. Survival and recolonization by fungi in soil treated with formalin or carbon disulphide. *Trans. Brit. Mycol. Soc.* 38: 335–346.

Grewal, J.S. 1988. Diseases of pulse crops—An overview. *Indian Phytopathol.* 41: 1–14.

Gunner, H.B. 1963. Nitrification by *Arthrobacter globiformis*. *Nature* (London) 197: 1127–1128.

Hiscox, J.D. and G.F. Israelstam. 1979. A method for the extraction of chlorophyll from leaf tissue without maceration. *Canadian J. Bot.* 57: 1332–1334.

Jaworski, E.G. 1971. Nitrate reductase assay in intact plant tissues. *Biochem. Biophy. Res. Comm.* 43: 1274–1279.

Khan, M.A.A., M.A. Siddiqui and M.M. Alam. 1990. Influence of yellow vein mosaic virus of okra on the incindence of rhizosphere fungi and nematodes under two tillage regimes. *Indian Bot. Soc.* 69: 457–461.

Khan, M.W., M.M. Alam, A.M. Khan and S.K. Saxena. 1974a. Effect of water soluble fractions of oil-cakes and bitter principles of neem on some fungi and nematodes. *Acta Bot. Indica* 2: 120–128.

Khan, M.W., A.M. Khan and S.K. Saxena. 1974b. Rhizosphere fungi and nematodes of eggplant as influenced by oil-cake. *Indian Phytopathol.* 27: 480–484.

Lloyed, A.B. and J.L. Lockwood. 1966. Lysis of fungal hyphae in soil and its possible relation to autolysis. *Phytopathology* 56: 595–602.

Martin, J.P. 1950. Use of acid, rose bengal and streptomycin in the plate method for estimating soil fungi. *Social Sci.* 69: 215–233.

Mclean, R.C. and W.R.I. Cook. 1957. *Practical Field Ecology*. George Allen and Unwin Ltd., London.

Nene, Y.L., V.K. Sheilla and S.B. Sharma. 1984. A world list of chickpea (*Cicer ariethinum* L.) and pigeonpea (*Cajanus cajan* L.) Millsp.) pathogens. ICRISAT Pulse Pathology Progress Report, 32: 19 pp.

Palakshappa, C.M., S. Kulkarni and R.K. Hegde. 1986. Studies on footrot of betelvine caused by *Sclerotium rolfsii* Sacc. in Karnataka. *Plant Pathol. Newsletter* 4: 17.

Sasser, J.N. and D.W. Freckman. 1987. World Perspective on Nematology: the role of the society. In: *Vistas on Nematology*, a commeration of the Twenty-fifth Anniversary of the Society of Nemalogists. J.A. Veech and D.W. Dickson. Society of Nematologists. Inc. Hyattsville, Maryland, pp. 7–14.

Saxena, S.B. 1960. Effect of carbon disulphide fumigation on *Trichoderma viride* and other soil fungi. *Trans. Brit. Mycol. Soc.* 43: 111–116.

Schmidt, E.L. 1954. Nitrification by *Aspergillus niger*. *Science* 119: 187.

Siddiqui, M.A. 1986. Studies on the effect of organic soil amendments on plant-parasitic nematodes. Ph.D. thesis, Aligarh Muslim University, Aligarh, 402 pp.

Siddiqui, M.A. and M.M. Alam. 1991. Efficacy of oil-seed cakes and nematicides against plant-parasitic nematodes attacking tomato in relation to ploughing. *J. Bangladesh Acad. Sci.* 15: 23–29.

Singh, K.B. and B.S. Dahiya. 1973. Breeding for wilt-resistance in chickpea. *Sym. on Wilt Problems and Breding for Wild Resistance in Gram.*, Sept. 23, IARI, New Delhi, pp.13–14.

Singh, N. and R.S. Singh. 1981. Lysis of mycelium of *Fusarium oxysporum* f. *udum* in soil amended with organic matters. *Plant and Soil* 59: 9–15.

Singh, N. and R.S. Singh. 1982. Effect of oil-cake amended soil atmosphere on pigeonpea wilt pathogen. *Indian Phytopathol.* 35: 300–305.

Singh, R.S. and K.R. Pandey. 1965. Inhibitory and stimulatory effects of certain oil-cakes on *Pytnium aphanidermatum* in soil. *Sci. Cult.* 31: 534–535.

Singh, R.S. and K. Sitaramaiah. 1966. Incidence of root-knot of okra and tomatoes in oil-cake amended soil. *Plant Dis. Reporter* 50: 668–672.

Singh, R.S. and K. Sitaramaiah. 1970. Control of plant-parsitic nematodes with organic soil amendments. *Pest Artic. News Summary* 16: 287–297.

Singh, U.P., H.B. Singh and R.B. Singh. 1980. The fungicidal effect of neem (*Azadirachta indica*) extracts on some soil-borne pathogens of gram (*Cicer arietinum*). *Mycologia* 72: 1077–1092.

Sinha, D.C. and R.K. Prasad. 1986. Effect of the organic amendments in soil on foot and root-rot of gram caused by *Fusarium oxysporum* f. sp. *ciceri* and *Sclerotium rolfsii*. *Indian Phytopathol.* 39: 145.

Southey, J.F. 1986. *Laboratory Methods for Work with Plant and Soil Nematodes*. Ministry of Agriculture, Fish & Food, HMSO, London, 202 pp.

Thomas, S.H. 1978. Population densities of nematodes under seven tillage regimes. *J. Nematol.* 10: 24–27.

Tiyagi, S.A. 1991. Pathogenicity and control of *Meloidogyne incognita* and *Rotylenchulus reniformis* in relation to *Macrophomina phaseolina* on greengram and *Fusarium oxysporum* f. sp. *ciceri* on chickpea. Ph.D. thesis, Aligarh Muslim University, Aligarh, 478 pp.

Tiyagi, S.A. and M.M. Alam. 1986. Root-knot development and plant growth of chickpea in relation to inoculum density of *Meloidogyne incognita*. *Int. Chickpea Newsletter* 14: 30–32.

Tiyagi, S.A. and M.M. Alam. 1987. Pathogenicity of *Rotylenchulus reniformis* to two cultivars of chickpea. *Int. Nematol. Network Newsletter* 4: 19–21.

Tiyagi, S.A. and M.M. Alam. 1988. Pathogenicity of root-knot nematode on mungbean. *Int. Nematol. Network Newsletter* 5: 22–24.

Tiyagi, S.A. and M.M. Alam. 1989. Effect of different inoculum densities of stunt nematode, *Tylenchorhynchus brassicae* on growth of chickpea. *Int. Chickpea Newsletter* 20: 14–15.

Toussoun, T.A., A.R. Weinhold, R.G. Linderman and Z.A. Patrick. 1968. Nature of phytotoxic substances produced during plant residues decomposition in soil. *Phytopathology* 58: 41–45.

Vidhyasekharan, P., G. Arjunan, V. Mariappan, K. Ranganath and S. Kolandisamy. 1977. Field tolerance of some bengal gram types to wilt and root-rot. *Indian Phytopathology* 30: 537–538.

Zakaria, M.A., J.L. Lockwood and A.B. Filonow. 1980. Reduction in *Fusarium* population density in soil by volatile degradation products of oil-seed meal amendments. *Phytopathology* 70: 495–499.

67

Efficacy of Cakes in the Management of Root Knot Nematode (*M. arenaria*) in Groundnut

M.U. VAISHNAV, I.U. DHRUJ AND R.P. LAKHATARIA

Department of Plant Pathology, College of Agriculture, Junagadh 362 001, India

ABSTRACT

The root knot nematode (*Meloidogyne arenaria*) causes considerable damage to the groundnut crop in some parts of western Gujarat. A field experiment was conducted to test the efficacy of different oil cakes in controlling nematodes for three years in nematode-prone area. It comprised nine treatments: neem, castor, mustard, and groundnut oil cake applied at 1 and 2 t/ha and check in quadruplicate. The three-year pooled analysis of data of final nematode population and root knot index showed significant reduction in all treatments as compared to check. Similarly, groundnut pod yield also significantly increased, except with groundnut cake treatment at 1 t/ha. The minimum final population of nematode and root knot index, and maximum pod yield were observed under neem cake treatment applied at 2 t/ha followed by 1 t/ha. However, the net incremental cost-benefit ratio was maximum when neem cake was applied at 1 t/ha. Hence, neem cake is recommended for the control of root knot nematode in groundnut.

INTRODUCTION

Root knot nematode (*Meloidogyne arenaria*) causes considerable damage to groundnut crop in some parts of western Gujarat (Saurashtra region). It was first observed in Sanala village near Dhoraji of Rajkot district (Vaishnav and Dhruj, 1981). Groundnut is a main oilseed crop in Saurashtra. Organic soil amendments including oilseed cakes have been used for the control of plant parasitic nematodes (Muller and Gooch, 1982). The effectiveness of castor, mustard, neem, and groundnut cakes for the control of root knot nematode was also reported (Alam, 1989). But no systematic work has been done on management of root knot nematode of groundnut with cakes. Therefore, the present investigation was carried out to compare the efficacy of various cakes in controlling root knot nematode.

MATERIALS AND METHODS

To study the efficacy of oil cakes in the control of root knot nematode, a field experiment was conducted from 1986 to 1989 in a highly infested field. There was a sevre drought during 1987–88; hence, the experiment was not conducted for that period. The treatments consisted of two doss of castor (*Ricinus communis* L.), mustard (*Brassica campestris* L.), neem (*Azadirachta indica* A. Juss.), and groundnut (*Arachis hypogaea* L.) cakes applied at 1 and 2 t/ha and check. The powdered cakes were applied in furrow before sowing seeds. The experiment was conducted in quadruplicate by randomized block design.

Observations were recorded on final population of nematode, root knot index, and pod yield from each treatment. Soil samples of 250 ml were collected for estimating nematode population from each treatment at the time of harvesting and processed according to Cobb's sifting and gravity method (Cobb, 1918). Root knot index on 0–5 scale for nematode infection was taken at the time of harvest on five random palnts per treatment. Pod yield was also recorded. The data were analysed statistically and the net incremental cost-benefit ratio (ICBR) was calculated for castor and neem cake treatments.

RESULTS AND DISCUSSION

It is clear from the data that all the cake applications were effective in controlling root knot nematode on groundnut (Table 1 and 2). The three-year pooled root knot index as presented in Table 1 was minimum (1.85) in neem cake applied at 2 t/ha, followed by neem cake at 1 t/ha and castor cake applied 1 and 2 t/ha. But neem cake is significantly more effective than the other cakes. Which in turn also significantly reduced root knot index as compared to check. A similar trend was noticed in the individual years. The final second-stage larval population of root knot nematode was reduced significantly in all treatments as compared to check, the minimum being in

Table 1. Effect of different cakes on root knot nematode index on groundnut

Treatment	Root knot index*			Mean of pooled root knot index
	1986–87	1988–89	1989–90	
Castor 1 t/ha	2.45	2.26	1.80	2.17
Castor 2 t/ha	2.38	2.10	1.55	2.01
Groundnut 1 t/ha	2.72	2.55	2.70	2.66
Groundnut 2 t/ha	2.56	2.41	2.62	2.53
Mustard 1 t/ha	2.58	2.36	2.58	2.51
Mustard 2 t/ha	2.39	2.33	2.41	2.38
Neem 1 t/ha	2.28	2.00	1.50	1.93
Neem 2 t/ha	2.23	1.98	1.35	1.85
Check	4.60	4.48	4.50	4.53
SEM ±	0.14	0.18	0.13	0.13
CD at 5 per cent	0.40	0.52	0.39	0.39
CV per cent	10.10	14.13	11.36	11.94

* 0 = Free, 5 = maximum disease intensity.

neem cake applied at 2 t/ha (Table 2). However, it was statistically at par with the rest of the cake applications. A similar trend was more or less observe for all the treatments in individual years in nematode populatiton.

Three-year pooled analysis of the data on pod yield as presented in Table 3 showed that the maximum pod yield was obtained with neem cake applied at 2 t/ha, closely followed by neem cake at 1 t/ha and castor cake at 2 t/ha. The remaining cake applications also increased pod yield significantly as compared to check, except groundnut cake (1 t/ha). The maximum net ICBR 1:1.89 was obtained with neem cake applied at

Table 2. Effect of different cakes on final root knot nematode population

Treatment	Nematodes population/250g soil			Mean of pooled data
	1986–87	1988–89	1989–90	
Castor 1 t/ha	14,345	10,800	11,000	12,048
Castor 2 t/ha	13,470	9,000	10,230	11,200
Groundnut 1 t/ha	18,790	13,202	11,715	14,569
Groundnut 2 t/ha	17,845	11,898	12,215	13,986
Mustard 1 t/ha	15,575	8,165	11,560	11,767
Mustard 2 t/ha	14,335	9,090	10,994	11,473
Neem 1 t/ha	12,895	9,050	10,005	10,650
Neem 2 t/ha	12,100	8,071	10,560	10,244
Check	22,395	14,301	20,145	18,947
SEM +	293	513	278	758
CD at 5 per cent	854	1,500	811	2,275
CV per cent	3.7	10	4.61	6.26

Table 3. Effect of different cakes on groundnut pod yield (kg/ha)

Treatment	Mean yield kg/ha			Yield, mean of pooled data kg/ha	Net ICBR, Rs.
	1986–87	1988–89	1899–90		
Castor 1 t/ha	1,193	1,060	900	1,051	1:1.06
Castor 2 t/ha	1,294	1,103	912	1,099	1:0.22
Groundnut 1 t/ha	1,035	877	740	884	
Groundnut 2 t/ha	1,111	912	766	929	
Mustard 1 t/ha	1,134	1,115	790	1,013	
Mustard 2 t/ha	1,174	1,191	800	1,055	
Neem 1 t/ha	1,455	1,082	930	1,155	1:1.89
Neem 2 t/ha	1,478	1,175	937	1,197	1:0.61
Check	1,010	815	557	794	
SEM +	37	31	8	41	
CD at 5 %	112	90	25	125	
CV %	8.4	16	2.11	5.68	

1 t/ha. The second best net ICBR (1:1.06) was obtained with castor cake at 1 t/ha. Hence, neem or castor at 1 t/ha is recommended for control of root knot nematode of groundnut.

The present investigation did not pin point the exact mechanism of nematode control by addition of oilseed cakes, but some theories put forward for other cases could be relevant to the present case. The control of nematodes may be due to the toxicity of the decomposition products (Khan et al., 1974; Alam et al., 1978, 1979), increase in the predacious and parasitic activity of soil biota (Linford, 1937), or change in physical and chemical properties of the soil inimical to the nematodes (Ahmad et al., 1972), or the cakes themselves may have some nematotoxic principles.

REFERENCES

Ahmad, R., A.M. Khan and S.K. Saxena. 1972. Changes resulting from amending the soil with oilcakes and analysis of oilcakes (Abstr.) *Proc. 59th Sess. Indian Sci. Cong.*, Calcutta, Part III, 164 pp.

Alam, M.M. 1989. control of root knot and stunt nematodes with horn meal, bone meal and oilseed cakes. *Indian J. Nematol.* 19: 166–170.

Alam, M.M., A.M. Khan and S.K. Saxena. 1978. Mechanisms of control of plant parasitic nematodes as a result of the application of organic amendments to the soil. IV. Role of formaldehyde and acetone. *Indian J. Nematol.* 8: 172–174.

Alam, M.M., A.M. Khan and S.K. Saxena. 1979. Mechanisms of control of plant parasitic nematodes as a result of the application of organic amendments to the soil. V. Role of phenolic compounds. *Indian J. Nematol.* 9: 136–142.

Cobb, N.A. 1918. Estimating the nematode population of the soil. *Agric. Tech. Circ. I. Bur. Plant Indust.* USDA, 48 pp.

Knan, A.M., M.M. Alam and S.K. Saxsena. 1974. Mechanism of the control of plant parasitic nematodes as a result of the application of oilcakes to the soil. *Indian J. Nematol.* 4: 93–96.

Linford, M.B. 1937. Stimulated activity of natural enemies of nematodes. *Science* 85: 123–124.

Muller, R.R. and P.S. Gooch. 1982. Organic amendments in nematode control. An examination of the literature. *Nematropica* 12: 319–326.

Vaishnav, M.U. and I.U. Dhruj. 1981. Occurrence of root knot nematode in groundnut in Gujarat. Paper presented at the *Nematology Symposium* 13–15 Feb. 1981. Tamilnadu Agricultural University, Coimbatore, India.

68

Integrated Management of the Citrus Nematode, *Tylenchulus semipenetrans*, Using Neem Cake and the Parasitic Fungus *Paecilomyces lilacinus**

P. PARVATHA REDDY, R.M. KHAN AND M.S. RAO

Division of Entomology and Nematology, Indian Institute of Horticultural Research
Hessaraghatta, Bangalore 560 089, India

ABSTRACT

Three oil cakes—castor (*Ricinus communis*), karanj (*Pongamia glabra*), and neem (*Azadirachta indica*)—at 20 g per plant and the parasitic fungus *Paecilomyces lilacinus* at 8 g per plant alone and in combination at half the dosages were evaluated for the management of the citrus nematode *Tylenchulus semipenetrans* infecting acid lime under glasshouse conditions. The results revealed that the application of neem cake in combination with *P. lilacinus* gave maximum shoot length and weight and root weight and was most effective in reducing the citrus nematode population both in soil and in roots.

INTRODUCTION

The citrus nematode, *Tylenchulus semipenetrans* Cobb, 1913, is one of the important limiting factors in citrus production and is widely distributed throughout the citrus-growing regions of the world. The nematode causes a disease called "slow decline" and is also considered to be one of the factors responsible for dieback disease of citrus trees (Chona et al., 1965). This nematode infects most of the commercial citrus rootstocks (Parvatha Reddy and Singh, 1979) and is responsible for 19 to 69 per cent loss in citrus fruit yield (Mukhopadhyaya and Suryanarayana, 1969; Mukhopadhyaya and Dalal,

* Contribution No. 122/92.

1971; Baghel and Bhatti, 1983). The present investigations were carried out for integrated management of the citrus nematode infecting acid lime (*Citrus aurantifolia* L.) by rational combination of oil cakes and a biocontrol agent, *Paecilomyces lilacinus* (Thom.) Samson.

MATERIALS AND METHODS

Three oil cakes—castor (*Ricinus communis* L.), karanj (*Pongamia glabra* Vent.), and neem (*Azadirachta indica* Juss.)—at 20 g per plant and the parasitic fungus, *Paecilomyces lilacinus* grown on paddy seeds at 8 g per plant alone and in combination at half the dosages were evaluated for the management of the citrus nematode infecting acid lime under glasshouse conditions.

Three-month-old acid lime plants were planted singly into pots of 15 cm diameter containing 2 kg sterilized soil-compost mixture (3:1). One month after transplanting about 5,000 larvae of *T. semipenetrans* were inoculated per plant. All the treatments were replicated five times and randomized on a glasshouse bench. Ten months after nematode inoculation, observations were recorded on plant growth characters, nematode population both in soil and in roots, percentage egg masses parasitized, percentage eggs infected, and density of spores in soil. Ten egg masses collected at random from infected roots were treated with 4 per cent sodium hypochlorite solution for 2–3 min. and transferred on a glass slide and crushed with cover slip for observing the presence of fungal mycelium as an indicator of parasitization. Presence of fungal hyphae and spores inside the eggs was taken as indication of infection. The fungus was reisolated from egg masses and eggs.

RESULTS AND DISCUSSION

The results presented in Table 1 revealed that integration of neem cake with *P. lilacinus* was most effective in increasing plant growth characters such as shoot length, shoot weight, and root weight, while karanj cake with *P. lilacinus* gave maximum root length of citrus.

The maximum reduction in the citrus nematode population both in soil and in roots was obtained with neem cake and *P. lilacinus*, which was on par with castor cake and *P. lilacinus* (Table 2). Combination of neem cake with *P. lilacinus* gave highest parasitization of egg masses and eggs of *T. semipenetrans* and increased spore density of *P. lilacinus* in soil, which differed significantly over other treatments.

Neem cake in combination with *P. lilacinus* proved significantly effective over all other treatments in increasing shoot length, shoot weight, and root weight and in reducing the citrus nematode population both in soil and in roots. This may be due to increased percentage of infected egg masses and parasitized eggs of the citrus nematode by *P. lilacinus* in soil amended with neem cake. There were significantly higher densities of spores of *P. lilacinus* in the soil amended with neem cake when compared to castor and karanj cakes, suggesting that neem cake encourages the growth and multiplication of *P. lilacinus*. Similar results have been reported by Rao and Parvatha Reddy (1992) working with *Meloidogyne incognita* infecting brinjal. Agar media mixed with neem cake extract were reported to encourage the growth of *P. lilacinus* (Mani and Anandam, 1989).

Table 1. Effect of oil cakes and *P. lilacinus* on plant growth of acid lime infected with *T. semipenetrans* (mean of five replicates)

Treatment (g/plant)	Dose	Shoot		Root	
		Length (cm)	Weight (g)	Length (cm)	Weight (g)
1. Castor cake	20	38.4	17.6	24.6	10.0
2. Karanj cake	20	31.4	15.5	25.5	8.5
3. Neem cake	20	33.2	15.8	20.5	7.7
4. *P. lilacinus*	8	35.2	17.0	24.7	8.8
5. Castor cake + *P. lilacinus*	10 + 4	40.5	18.1	24.9	12.9
6. Karanj cake + *P. lilacinus*	10 + 4	36.0	18.0	26.1	11.4
7. Neem cake + *P. lilacinus*	10 + 4	43.8	25.6	22.3	15.0
8. Control	28.1	12.0	15.2	7.6	
C.D. at 5%		5.34	4.34	4.02	3.06

Table 2. Effect of oil cakes and *P. lilacinus* on *T. semipenetrans* infecting acid lime (means of five replicates)

Treatment	Dose (g/plant)	Nematode population/plant		% egg masses parasitized	% eggs infected	Spore density (cfu/g soil)
		Soil	Root			
1. Castor cake	20	3980	2342	—	—	—
2. Karanj cake	20	5040	2932	—	—	—
3. Neem cake	20	3840	2320	—	—	—
4. *P. lilacinus*	8	4680	2074	52.8	56.3	5556
5. Castor cake + *P. lilacinus*	20 + 4	2500	1606	58.0	65.0	6471
6. Karanj cake + *P. lilacinus*	20 + 4	4240	1650	57.0	59.0	5997
7. Neem cake + *P. lilacinus*	20 + 4	2120	1330	67.5	72.4	7123
8. Control		16080	4498	—	—	—
C.D. at 5%		995.3	449.5	5.58	5.5	576

The volatile fatty acids (formic, acetic, propionic, and butyric acids), ammonia, formaldehyde, hydrogen sulphide, phenols, and amino acids that are released during the decomposition of oil cakes have been reported to be toxic to the nematodes (Khan et al., 1974; Alam et al., 1978, 1979; Parvatha Reddy et al., 1975). These factors seem to be responsible for the effectiveness of oil cakes and P. lilacinus in reducing the citrus nematode population.

ACKNOWLEDGEMENT

The authors are grateful to Dr. R.M. Pandey, Director, Indian Institute of Horticultural Research, Bangalore, for providing necessary facilities and encouragement.

REFERENCES

Alam, M.M., A.M. Khan and S.K. Saxena. 1978. Mechanism of control of plant parasitic nematodes as a result of the application of organic amendments to the soil. IV—Role of formaldehyde and acetone. *Indian J. Nematol.* 8: 172–174.

Alam, M.M., A.M. Khan and S.K. Saxena. 1979. Mechanism of control of plant parasitic nematodes as a result of the application of organic amendments to the soil. V—Role of phenolic compounds. *Indian J. Nematol.* 9: 136–142.

Baghel, P.P.S. and D.S. Bhatti. 1983. Evaluation of pesticides for the controls of phytonematodes on citrus. *Third Nematol. Symp.*, H.P. Agri. Univ., Solan, pp. 38–39.

Chona, B.L., C.L. Sethi, G. Swarup and J.S. Gill. 1965. Citrus nematodes (*Tylenchulus semipenetrans* Cobb, 1913) associated with die-back disease of citrus. *Indian J. Hort.* 22: 370–373.

Khan, A.M., M.M. Alam and R. Ahmad. 1974. Mechanism of the control of plant-parasitic nematodes as a result of the application of oil-cakes to the soil. *Indian J. Nematol.* 4: 93–96.

Mani, A., and R.J. Anandam. 1989. Evaluation of plant leaves, oil cakes and agro-industrial wastes as substrates for mass multiplication of the nematophagous fungus *Paecilomyces lilacinus*. *J. Biol. Control* 3: 56–58.

Mukhopadhyaya, M.C., and M.R. Dalal. 1971. Effect of two nematicides on *Tylenchulus semipenetrans* and on sweet lime yield. *Indian J. Nematol.* 1: 95–97.

Mukhopadhyaya, M.C., and D. Suryanarayana. 1969. Citrus decline in Haryana. Role of *Tylenchulus semipenetrans* and its control. *Indian Phytopathol.* 22: 495–497.

Parvatha Reddy, P., H.C. Govindu and K.G.H. Setty. 1975. Further studies on the action of DL-methionine on Meloidogyne incognita. *Indian J. Nematol.* 5: 200–206.

Parvatha Reddy, P., and D.B. Singh. 1979. Nematode problems of citrus in India. *PANS* 25: 409-419.

Rao, M.S., and P. Parvatha Reddy. 1992. A method for delivery of a bio-agent *Paecilomyces lilacinus* to soil for the management of root-knot nematode on egg plant. *Nematol. Medit.* (in press).

69

Effective Use of Wellgro and RD-9 Repelin for the Management of Root Knot Nematode on Tomato

M.S. RAO AND P. PARVATHA REDDY

Division of Entomology and Nematology, Indian Institute of Horticultural Research
Hessaraghatta Lake P.O., Bangalore 560 089, India

ABSTRACT

Studies were conducted to standardize the methods of efficient use of commercially available botanicals Wellgro and RD-9 Repelin for integrated management of root knot nematode *Meloidogyne incognita* on tomato. Wellgro suspension and Repelin at 2.5 and 5 per cent were used for seed treatment, nursery bed drenching, and seedling dip. These botanicals were also evaluated for their ability to support the growth of promising bio-agent *Paecilomyces lilacinus*. Nursery bed drenching with Wellgro suspension and Repelin both at 2.5 and 5 per cent concentrations significantly reduced the population of *M. incognita* in the nursery beds. Seed treatment and seedling dip in these botanicals at both concentrations significantly reduced the root knot nematode infestation on tomato. Wellgro suspension at both concentrations was found to support the growth of *P. lilacinus*. Further use of Wellgro suspension mixed with spores of *P. lilacinus* for seed treatment, seedling root dip, and nursery bed drenching was found highly effective in the management of *M. incognita* on tomato due to their combined and complementary interactive effect. These treatments have also increased plant growth parameters of tomato, revealing the great scope for use of these botanicals.

INTRODUCTION

Keeping in view the hazards of excessive chemicalization of agro-ecosystems researchers (Muller and Gooch, 1982; Nandal and Bhatti, 1986; Parvatha Reddy, 1985; Rao and Parvatha Reddy, 1992; Singh et al., 1980; Sitaramaiah and Singh, 1978) have attempted the management of various nematodes by means of certain indigenous plant products. Ready availability of the material is an important consideration. This paper presents the efficient use of the botanicals Wellgro and RD-9 Repelin (which are readily available in the market) for the integrated management of root knot nematode, *Meloidogyne incognita* on tomato.

MATERIALS AND METHODS

Wellgro and Repelin suspensions (2.5 and 5 per cent concentrations) were evaluated for their efficacy by drenching the 3 kg sick soil (1 larva of *M. incognita*/g of soil) in the seed pans. Two hundred ml of each botanical of each concentration was used for drenching each seed pan. Tomato (Cv Pusa Ruby) seeds were sown in the seed pans. In control, soil was drenched with ordinary water. Each treatment was replicated five times. Simultaneously, for another experiment, the seeds of tomato were treated with these botanicals (2.5 and 5 per cent concentration) for 20 minutes, air dried, and sown the next day in seed pans containing 3 kg sick soil. Five replicates were maintained for each treatment.

Thirty-five days later, 10 seedlings from each seed pan were pulled at random and observations on plant growth parameters and number of galls on the roots were recorded.

In another experiment, soil drenching and seed treatment with these botanicals were done as mentioned above. But here the seed pans were filled with sterilized soil. Thirty-day-old seedlings of these experiments were transplanted in pots filled each with 2 kg sterilized soil. Each treatment was replicated five times with appropriate controls. One week after transplanting the seedlings were inoculated each with 2,000 larvae of *M. incognita*. Observations on plant growth parameters, root knot index (on 1–5 scale), and soil and root (final) population were recorded.

The botanicals (at 2.5 and 5 per cent conc.) were also used for root dip treatment of nursery seedlings (raised in sterilized soil) for 20 minutes. The seedlings were transplanted and inoculated and observations were taken as mentioned above.

Another experiment was conducted to evaluate the feasibility of integration of these botanicals along with promising biological control fungus, *Paecilomyces lilacinus*. Wellgro and Repelin (5 and 10 per cent concentration) were evaluated for their ability to support the growth of *P. lilacinus*. Wellgro at both concentrations was found to support the growth of *P. lilacinus*, whereas Repelin did not support the growth of this useful nematophagous fungus. Hence, Wellgro suspensions (2.5 and 5 per cent concentration) mixed with the spores (2×10^4 spores/ml of suspension) of *P. lilacinus* were used along with *P. lilacinus* spore suspension (2×10^4 spores/ml of sterilized water) for soil drenching, seed treatment in sick soil and sterilized soil, and root dip treatment as mentioned above. Observations were made on plant growth parameters, nematode infestation, spore density in soil, and extent of colonization of *P. lilacinus* fungus on tomato.

RESULTS AND DISCUSSION

It is evident from the data that soil drenching with Wellgro suspension (2.5 and 5 per cent concentration) has proved significantly effective in reducing the number of root galls on seedlings and favouring the growth of seedlings (Table 1). In case of seed treatment, Repelin at 2.5 and 5 per cent has proved significantly more effective than Wellgro in reducing the number of galls on the roots (Table 1) and also in reducing the gall index (Table 2) but Repelin at 5 per cent conc. has not affected the plant growth parameters favourably (Tables 1, 2).

When the seedlings grown in sterilized soil drenched with either Wellgro suspension or Repelin were transplanted and inoculated with *M. incognita*, those raised in sterilized soil drenched with Wellgro (2.5 and 5 per cent concentration) suspensions were least affected by *M. incognita* (Table 2).

It is evident that drenching of soil with Wellgro suspension and seed treatment with Repelin are effective in the management of root knot nematode on tomato.

Reasons for the differential efficacies of these botanicals were not clearly understood. It is possible that Wellgro suspension encourages more of useful fungus and due to fungal antibiosis it is effective for drenching of soil. It is also possible that absorption of this botanical by seeds protects against nematodes. Increased efficacy of Repelin as seed treatment raises a doubt whether it triggers resistance in tomato to *M. incognita*. Higher contents of phenols were reported in tomato grown in soils amended with oil cake (Alam and Khan, 1980).

Wellgro suspensions (2.5 and 5 per cent concentration) were found to support the growth of *P. lilacinus*. The spore densities in Wellgro suspensions of 2.5 and 5 per cent concentrations were 4×10^8/ml and 4×10^{10}/ml respectively 25 days after inoculation.

Table 1. Effect of Wellgro, Repelin and Wellgro + *P. lilacinus* on tomato seedling growth and management of *M. incognita* in nursery stage

Botanical and conc. (+ bio-agent)	Mean seedling length (cm)	Mean seedling wt (g)	No. of root galls in 10 seedlings
Soil Drenching			
Wellgro 2.5%	14.8	2.1	36
Wellgro 5.0%	15.2	2.0	30
Repelin 2.5%	12.0	1.8	48
Repelin 5.0%	12.3	2.0	45
Wellgro 2.5% + *P. lilacinus*	18.5	2.5	22
Wellgro 5.0% + *P. lilacinus*	20.3	2.8	17
P. lilacinus	14.2	1.9	31
Control	9.2	1.6	67
Seed Treatment			
Wellgro 2.5%	13.0	1.9	46
Wellgro 5.0%	14.6	2.0	42
Repelin 2.5%	15.0	2.2	34
Repelin 5.0%	11.3	2.4	27
Wellgro 2.5% + *P. lilacinus*	16.2	2.3	25
Wellgro 5.0% + *P. lilacinus*	17.8	2.5	22
P. lilacinus	13.5	1.8	30
Control	10.5	1.7	63
CD 5%	2.64	0.53	6.57

Table 2. Effect of Wellgro, Repelin and Wellgro + *P. lilacinus* on tomato growth and management of *M. incognita*

Botanical and conc. (+bio-agent)	Plant height (cm)	Plant wt (g)	Root knot index	Final population of nematodes	% colonization on the roots	% parasitization of egg masses
Soil Drenching						
Wellgro 2.5%	60	21	2.9	2410		
Wellgro 5%	65	23	2.7	2122		
Repelin 2.5%	53	20	3.5	2954		
Repelin 5.0%	57	19	3.6	2727		
Wellgro 2.5% + *P. lilacinus*	71	28	2.4	1856	50	56
Wellgro 5.0% + *P. lilacinus*	77	32	2.2	1530	53	58
P. lilacinus	57	22	2.8	2047	46	51
Control	46	15	4.3	5572		
Seed Treatment						
Wellgro 2.5%	56	20	3.3	3241		
Wellgro 5.0%	55	18	3.4	3086		
Repelin 2.5%	62	23	2.8	2437		
Repelin 5.0%	69	20	2.8	2143		
Wellgro 2.5% + *P. lilacinus*	64	26	2.6	2140	49	50
Wellgro 5.0% + *P. lilacinus*	62	28	2.5	2078	48	52
P. lilacinus	53	23	2.7	2349	45	48
Control	48	14	4.4	5848		
CD 5%	7.92	3.32	0.32	364.72	5.2	6.1

Table 3. Effect of root dip treatment with Wellgro, Repelin, and Wellgro + *P. lilacinus* on tomato growth and management of *M. incognita*

Botanical and conc. (+bio-agent)	Plant height (cm)	Plant wt (g)	Root knot index	Final population of nematodes	% colonization on the roots	% parasitization of egg masses
Wellgro 2.5%	62	24	3.2	2847		
Wellgro 5.0%	66	27	3.1	2826		
Repelin 2.5%	67	29	3.0	2594		
Repelin 5.0%	65	30	2.7	2640		
Wellgro 2.5% + *P. lilacinus*	79	29	2.4	2143	54	58
Wellgro 5.0% + *P. lilacinus*	84	34	2.1	1738	57	60
P. lilacinus	65	22	2.8	2349	46	50
Control	50	18	4.3	6349		
CD 5%	7.53	3.16	0.35	412.14	5.1	6.0

Repelin did not support the growth of this biocontrol fungus. Hence, Wellgro suspension integration with *P. lilacinus* proved significantly more effective than Wellgro suspension (2.5 and 5 per cent concentration) treatments in reducing the gall index and root knot nematode final population and increasing plant growth parameters, per cent parasitization of egg masses, and colonization of *P. lilacinus* on the roots (Table 2). Though root-dip of the seedlings in these botanicals increased the growth of the plants, it did not reduce the nematode infestation to the desirable extent, whereas root-dip treatment in Wellgro suspension mixed with the spores of *P. lilacinus* proved significantly effective in reducing the nematode infestation and increased plant growth parameters (Table 3).

REFERENCES

Alam, M.M. and A.M. Khan. 1980. Effect of organic amendment on the growth and chemical composition of tomato, egg plant and chilli and their susceptibility to attack by *Meloidogyne incognita*. *Plant and Soil* 57: 231–236.

Muller, R. and P.S. Gooch. 1982. Organic amendments in nematode control: An examination of the literature. *Nematropica* 12: 319–326.

Nandal, B. and D.S. Bhatti. 1986. Influence of four plant extracts on the hatching of *Meloidogyne javanica* and invasion of host root. *Nematol. Medit.* 14: 291–294.

Parvatha Reddy, P. 1985. Integrated management of nematode pests in vegetable crops. In: *Integrated Pest and Disease Management.* S. Jayaraj (ed.), Tamil Nadu Agricultural University, Coimbatore, pp. 202–207.

Rao, M.S. and P. Parvatha Reddy. 1992. Studies on the comparative efficacy of certain plant leaves and carbofuran in the management of *Meloidogyne incognita* on tomato. *Curr. Nematol.* 3(1): 5–6.

Singh, S.P., M. Ahmed, A.M. Khan and S.K. Saxena. 1980. Effect of seed treatment with certain oil cakes of nematicides on the growth of tomato and rhizosphere population of nematodes and fungi. *Nematol. Medit.* 4: 222–224.

Sitaramaiah, K. and R.S. Singh. 1978. Effect of organic amendments on phenolic content of soil and plant response of *Meloidogyne javanica* and its host to related compounds. *Plant and Soil* 50: 671–679.

70

Nematicidal Efficacy of Decomposed Neem Products on *Meloidogyne incognita* Infecting Mungbean and Their Residual Effect in Chickpea

S.D. MISHRA AND VIJAYALAKSHMI MOJUMDER

Division of Nematology, Indian Agricultural Research Institute
New Delhi 110 012, India

ABSTRACT

Neem seed, seed kernel, and seed coat at 1, 5, and 10 per cent (w/w) after decomposing for two months were effective in reducing *Meloidogyne incognita* in mungbean when directly used or added to soil in 1:3 ratio. At 10 per cent some phytotoxicity was noticed but at 5 per cent dose and addition to soil in 1:3 ratio were found to be most effective in reducing root knot galls and improving plant growth. In the second crop, chickpea, which was raised after mungbean in the next season in the same pots, there was reduction in root knot incidence and improvement of plant growth indicating a residual effect of the decomposed neem materials.

INTRODUCTION

Neem cake, leaf, and other neem products as soil amendments are known to reduce plant parasitic nematodes and are often used in nematode management. These amendments were found to be more toxic when applied after decomposition than when applied directly. Not much work has been done, however, on the period of decomposition. The present investigations were designed to observe the effect of properly decomposed neem seed, neem kernel, and seed coat individually in soil against *Meloidogyne incognita*, which causes a severe root knot nematode disease in most agricultural crops, including pulse crops.

MATERIALS AND METHODS

Two experiments were conducted in 20 cm earthen pots with mungbean (*Vigna radiata* cv Pusa Baisakhi) in kharif season and with chickpea (*Cicer arietinum* cv Pusa 240) in rabi season. Before the start of the first experiment, neem seed, seed kernel, and seed coat were added in soil at three doses—1, 5, and 10 per cent (w/w) and left to decompose. The pots were watered frequently. After two months, the first experiment was laid out with two sets of treatments: (1) amending the decomposed material directly, and (2) amending it with soil in the ratio of 1:3. In either case the materials were thoroughly mixed and filled in earthen pots. All the treatments and nematode-infested and uninfested checks were replicated thrice.

In the first experiment, five mungbean seeds were sown in each pot and their germination percentage was recorded up to seven days. The addition of neem seed kernel at 10 per cent and its addition in soil at 1:3 ratio provided no germination of seed at the initial stage, so these treatments were discontinued. After observing the effect of treatments on germination, in other treatments and checks only one plant was allowed to grow in each pot. These plants were inoculated with 1000 second-stage juveniles of *M. incognita* per pot in respect of all the treatments excepting those maintained as uninfested control. Observations were recorded 60 days after sowing.

The soil of experimental pots maintained for mungbean experiment had the same treatments without watering so that it may become free of nematodes in dry conditions. At the time of chickpea sowing, the soil of each pot was thoroughly mixed and refilled for the experiment planned to observe the residual effect of the treatments.

Seeds of chickpea were sown in each pot and their germination was noted up to 10 days. After that only one plant per pot was allowed to grow and 1000 freshly hatched juveniles of *M. incognita* per pot were inoculated. Observations on plant growth characters and development of nematodes were recorded after 90 days of sowing.

RESULTS

The data presented in Table 1 indicated that in the mungbean experiment, the seed germination was delayed and the percentage germination was reduced in treatments of neem seed and seed coat at 10 per cent, while their application in 1:3 ratio and all other treatments showed 100 per cent germination. A significant reduction in number of root knot galls in all treatments was observed at 1, 5, and 10 per cent doses and in their 1:3 ratios (Table 2). Maximum reduction was recorded in neem seed at 10 per cent followed by neem seed coat at 10 per cent. At 5 per cent doses and at their 1:3 ratios, the reduction was 50 per cent or more; at 1 per cent doses and at their 1:3 ratios also, the reduction was significant. The data on plant growth (Table 3) revealed that the improvement in plant growth characters was best at 5 per cent doses and their 1:3 ratios in all three treatments. At 10 per cent doses, though there was less nematode attack, the plant growth was not that good, indicating a slight phytotoxicity. However, at their 1:3 ratios plant growth was significantly improved. Even at lowest dose of 1 per cent and their 1:3 ratios, there was improvement in plant growth.

In the chickpea experiment also, there was significant reduction in root knot galls indicating the residual effect of the treatments in the next season.

Table 1. Effect of decomposed neem seed, neem seed kernel and neem seed coat on the germination of mungbean and chickpea subsequently

Treatments	Dose % (w/w)	Percentage germination	
		Mungbean	Chickpea
Neem seed	1	100	100
	5	100	100
	10	20	100
	1 (1:3)	100	100
	5 (1:3)	100	100
	10 (1:3)	100	100
Neem seed kernel	1	100	100
	5	100	100
	10	0	—
	1 (1:3)	100	100
	5 (1:3)	100	100
	10 (1:3)	0	—
Neem seed coat	1	100	100
	5	100	100
	10	60	100
	1 (1:3)	100	100
	5 (1:3)	100	100
	10 (1:3)	100	100
Check (infested)		100	100
Check (uninfested)		100	100

Table 2. Effect of decomposed neem seed, neem seed kernel and neem seed coat on *M. incognita* in mung bean

Treatments	Doses % (w/w)	No. of galls/plant	% reduction in galls over check
Neem seed	1	32.0	40.7
	5	25.0	53.7
	10	6.0	88.9
	1 (1:3)	37.0	31.5
	5 (1:3)	27.0	50.0
	10 (1:3)	23.3	56.9
Neem seed kernel	1	31.3	42.0
	5	24.0	55.6
	1 (1:3)	35.3	35.2
	5 (1:3)	26.0	51.9
Neem seed coat	1	33.0	38.9
	5	26.0	51.9
	10 18.3	66.1	
	1 (1:3)	40.0	25.9
	5 (1:3)	32.0	40.7
	10 (1:3)	30.6	43.5
Control		54.0	
CD 5%		5.39	

Table 3. Effect of decomposed neem seed, neem cake kernel and neem seed coat on growth of mungbean infested with *M. incognita*

Treatments	Doses %(w/w)	Root length	Shoot length	Root weight	Shoot weight	Pod number
Neem seed	1	14.0	31.0	3.0	5.2	3.0
	5	17.0	35.6	4.0	7.3	4.0
	10	10.7	27.5	2.5	3.8	3.3
	1 (1:3)	12.0	31.6	2.4	4.2	2.5
	5 (1:3)	16.0	34.0	3.8	6.1	4.1
	10 (1:3)	17.3	36.3	4.5	7.3	5.0
Neem seed kernel	1	15.0	33.0	3.0	5.8	3.8
	5	17.0	36.0	4.5	8.1	4.8
	1 (1:3)	12.7	31.0	2.5	5.0	2.5
	5 (1:3)	16.5	34.7	3.8	6.6	4.0
Neem seed coat	1	14.3	31.0	2.5	4.8	3.0
	5	16.5	34.6	3.5	6.6	4.0
	10	12.3	30.7	3.0	6.0	4.2
	1 (1:3)	11.7	25.6	2.1	3.8	2.0
	5 (1:3)	15.0	33.0	3.2	5.6	3.5
	10 (1.3)	15.3	33.7	3.0	6.0	3.8
Control						
Infested		10.7	22.7	1.65	2.5	1.0
Uninfested		14.3	34.0	3.50	6.6	3.3
CD 5%		1.84	2.75	0.43	0.38	NS

Table 4. Residual effect of decomposed neem seed, kernel and seed coat on *M. incognita* in chickpea

Treatments	Doses % (w/w)	No. of galls/plant	% reduction in galls over check
Neem seed	1	57.3	35.9
	5	46.0	47.7
	10	36.7	58.7
	1 (1:3)	64.3	26.9
	5 (1:3)	53.3	39.4
	10 (1:3)	50.3	42.8
Neem seed kernel	1	55.7	36.8
	5	45.0	48.9
	1 (1:3)	61.3	31.3
	5 (1:3)	46.7	47.0
Neem seed coat	1	59.3	32.6
	5	49.0	44.3
	10	42.3	51.9
	1 (1:3)	70.3	20.1
	5 (1:3)	56.7	35.6
	10 (1:3)	54.0	38.6
Control		88.0	
CD 5%		6.67	

Table 5. Residual effect of decomposed neem seed, kernel and seed coat on growth of chickpea infested with *M. incognita*

Treatments	Doses %(w/w)	Root length	Shoot length	Root weight	Shoot weight	Pod number
Neem seed	1	20.5	37.0	2.8	4.5	2.5
	5	24.5	41.3	3.3	5.7	3.3
	10	31.5	43.3	3.7	7.4	3.5
	1 (1:3)	19.0	37.0	2.4	4.1	2.0
	5 (1:3)	23.5	40.6	2.9	5.2	2.8
	10 (1:3)	28.5	42.6	2.9	5.3	2.8
Neem seed kernal	1	21.3	39.6	2.7	5.2	2.5
	5	25.3	42.0	3.6	6.0	3.5
	1 (1:3)	20.8	37.0	2.8	4.1	2.0
	5 (1:3)	23.5	40.0	3.2	5.7	3.0
Neem seed coat	1	20.5	37.5	2.7	4.5	2.0
	5	23.0	40.0	3.2	5.5	3.0
	10	28.6	43.3	3.6	6.3	3.5
	1 (1:3)	18.0	34.3	2.3	3.6	1.5
	5 (1:3)	21.8	38.6	2.7	4.9	2.5
	10 (1:3)	22.5	40.0	2.8	5.2	2.7
Control						
Infested		17.6	29.8	1.9	2.7	1.0
Uninfested		38.6	42.5	3.0	4.2	3.0
CD 5%		2.33	1.82	0.43	NS	NS

The data on the root knot galling (Table 4) showed that the trend of reduction due to different treatments was similar to that of mungbean. Maximum reduction was obtained with neem seed at 10 per cent followed by seed coat at 10 per cent. There was significant reduction when 10 per cent decomposed material was mixed with soil in the ratio of 1:3 or at 5 and 1 per cent and their 1:3 ratios. There was improvement in plant growth characters in all the treatments (Table 5). In case of chickpea, even at 10 per cent dose, plants were healthy, indicating that at this stage there was no phytotoxic effect.

DISCUSSION

Neem products, particularly neem cake, had been reported to manage plant parasitic nematodes as soil amendments. Mojumder and Mishra (1992) used neem seed coat as soil amendment against *M. incognita*. Neem products were found to be more toxic to nematodes after some period of decomposition. Prasad et al. (1972) found maximum reduction in plant parasitic nematode population at 75 days of application of neem cake though reduction was noticed after 45 days. Goswami and Vijayalakshmi (1987) found maximum nematicidal effect of neem cake against *M. incognita* juveniles three weeks after adding the cake to the soil. Use of pre-decomposed materials of neem seed, seed kernel, and seed coat has been tried successfully against root knot nematode for the first time in the present studies.

REFERENCES

Goswami, B.K. and K. Vijayalakshmi. 1987. Effect of period of decomposition of oil seed cakes in soil on *Meloidogyne incognita* juveniles. *Indian J. Nematol.* 17: 84–86.

Mojumder, V. and S.D. Mishra. 1992. Management of root knot nematode, *Meloidogyne incognita* infesting chickpea with neem seed coat. *Ann. agric. Res.* 13.

Prasad, S.K., S.D. Mishra and A.C. Gour. 1972. Effect of soil amendments on nematodes associated with wheat followed by mung and maize. *Indian J. Ent.* 34: 307–311.

71

Effect of Neem Cake, Neem Oil, Repelin and Carbofuran on Control of Soybean Nematodes

S.C. VYAS

Department of Plant Pathology, J.N. Agricultural University
Indore, Madhya Pradesh 452 001, India

ABSTRACT

Studies were conducted to evaluate the effect of neem cake, neem oil, Repelin and carbofuran applied 2g/kg of soil infested with *Helicotylenchus* sp., *Pratylenchus* sp., *Xiphenema* sp., and *Aphelenchus* sp. at the number which produces disease in an epidemic form. Significant reduction in the nematode population was recorded. Repelin application was comparable to carbofuran. The application significantly increased saprophytic soil biota, especially soil fungi which parasitized the pathogenic nematodes. The application significantly controlled soil-borne plant fungal pathogens such as *Sclerotium rolfsii* and *Rhizoctonis bataticola*, which cause collar and root rots respectively. The infectivity of fungal pathogens is associated with nematode infection. The treatment increased yield significantly (22 q/ha) over control (12 q/ha). These results indicate the role of Repelin in the integrated control of nematodes and fungal plant pathogens in soybean.

INTRODUCTION

Soybean (*Glycine max*) is one of the major oil-yielding crops and is grown in 29 million ha in kharif season (July through September) in India. Its area is further increasing in the states of Tamil Nadu, Karnataka and Andhra Pradesh. The crop has been cultivated for the last 30 years as a sole crop continuously in the same fields. This has resulted in the occurrence of several fungal nematode populations which constrain efforts to increase the yield, which is very low (6–7 q/ha). Systematic surveys conducted to ascertain the yield losses due to nematodes indicated that the crop suffers in some places loss of 10 to 50 per cent yield and a major share of that loss is attributed to nematodes. Attempts to control nematodes with chemicals and resistant cultivars have not been fruitful; hence, the present studies were conducted to ascertain the effects of neem cake, neem oil, Repelin and carbofuran on the nematode population and on yield of soybean.

MATERIALS AND METHODS

Composite samples of soil (250 cc) along with root pieces (5 g) were taken from 10 soybean fields. Soil was washed by gravity and Baermann funnel method (Christie and Perry, 1952) and roots were washed by blending and sieving method (Ayub, 1977). Estimation of nematode population and quantifying and qualifying of the various genera associated with soybean were in accordance with the keys developed by Mail and Lyon (1975) and Taylor (1967).

Pathogenicity tests were carried out by comparing the growth of inoculated and uninoculated plants of 20-day-old soybean seedling raised in earthen pots of 15 cm diameter containing 1 kg steam-sterilized soil, and an equal number of uninoculated plants were kept as check.

Earthen pots were also used for experiments on the control of nematode by soil application of various treatments. Sterilized soil was filled in pots. Composite samples of nematodes were collected by the gravity and Baermann funnel method and the soil was inoculated with 5000 nematode propagules per kg. Soybean seeds of CV-JS-72-44, which has been grown for the last 20 years and shows high susceptibility to a range of nematodes were used. When the nematodes were developed in the soil in sufficient number, the soil was amended with neem cake, neem oil, Repelin, and carbofuran at the recommended rates. The treatments were replicated four times. The effect of soil amendments was recorded in terms of shoot and root length and dry weight. The nematode population reduction percentage over control was also recorded. A similar type of treatment was made in the 3 × 2 m plots to simulate the pot experiment for recording growth and yield. The data were analysed.

RESULTS AND DISCUSSION

Surveys conducted in soybean-growing fields at J.N. Agricultural University Experimental Farm, Indore and its vicinity indicate that several plant parasitic nematodes are associated with soybean and are a potential threat to the crop. The data (Table 1) indicate that in soil samples *Helicotylenchus multicinctus* and *Aphelenchus* sp. occur in maximum number as compared to *Xiphenema index*, *Pretylenchus coffee* and *Hetenodera glycine*. A considerable number of unidentified nematodes were also encountered. Pathogenicity test was carried out by mixed inoculum and it was observed that almost all the prevailing varieties such as JS-72-44, PK-472, Punjab-1 and JS-74-05 were found susceptible to highly susceptible. Soybean variety JS-72-44 was found highly susceptible. Nematode damage has a variety of symptoms, although it is difficult to single out symptoms produced by the single nematode species associated with soybean because of overlapping and border line differences (Table 2). By and large, the aerial symptoms comprise stunting, dwarfing, chlorosis, yellowing necrosis, and leaf spotting and mottling associated with ground symptoms such as root lesions and root knots. Nematode damage will likely show up as circular areas in the infested fields with stunted yellow plants at the centre grading out to green normal plants at the perimeter.

Table 1. Average density of different plant parasitic nemotode species

Soil	Density of nematodes in 250 cc soil/5 g root						
	1	2	3	4	5	6	Total
Sample 1	700	100	175	300	200	500	1975
Sample 2	500	150	150	300	150	700	1950
Sample 3	300	200	200	300	150	600	1750

1 = *H. multicinctus* 2 = *X. index*
3 = *P. coffee* 4 = *Aphelenchus* sp.
5 = *H. glycine* 6 = *others*

Table 2. Effect of several species of nematodes on the occurrence of plant symptoms in cultivar JS-72-44

Aphelenchus sp.	*H. multicinctus*	*P. penetran*	*X. index*	*H. glycine*
Chlorosis, dwarfing, stunting, sickness	Nematode causes symptoms in cortex of roots and is easily introduced in fresh area with infested soil and plant tissues	Nematode is associated with soil sickness and root lesion nematode; they are parasites of cortex but may damage vascular tissues in soybean	Migratory root ectoparasite; they also transmit viruses in grapes, and cause necrosis, terminal swelling, cellular hypertrophy, and multinucleate condition of cortical cells near the feeding site	Diseases plants with nematode start yellowing as early as one month after planting; careful removal of soil from roots reveals cream coloured cysts

The above ground symtoms of plants infected by nematodes are usually conspicuous because of the reduced unthrifty growth and a tendency to wilt on warm days. Heavy infection results in a severe loss of yield, reduced quality of produce, and deficiency symptoms. It was observed that nematode infection was associated with increased susceptibility to foliage diseases and soil-borne diseases such as collar rot and dry root rot.

The data presented in Table 3 indicate that Repelin at 2g/kg soil was significantly superior to other treatments in reducing nematode populations. The data indicate that carbofuran was less effective than Repelin but more effective than neem cake and neem oil. However, neem cake and oil application was significantly superior to control in reducing nematode population. Repelin gave significantly higher yield than other treatments. This indicates that besides nematode control, Repelin has growth promoting effects. Highest yield was recorded in the Repelin application, that is 22 q/ha as compared to 12 q/ha in control (Table 4).

It is suggested that neem cake and oil and Repelin control nematodes by encouraging many saprophytic soil biota which parasitize the nematodes and decomposition of organic matter *in situ* which releases toxic metabolites which directly kill nematodes (Ayoub, 1977). The control of soil-borne fungal pathogens was also observed, which may be due to competitive saprophytic ability of the saprophytes, production of anti-

Table 3. Effect of soil application of neem cake, neem oil, Repelin, and Carbofuran on nematode population in soybean pots

Treatment	Nematode			
	Conc. (%)	Population count	Population reduction percentage over control	Disease Index (%)
Neem cake	1.00	2610a	65.25	15.75b
	2.00	1900d	47.50	13.75b
Neem oil	1.00	2510e	62.75	14.25b
	2.00	1350c	33.75	10.00b
Repelin	1.00	0900b	22.50	08.00a
	2.00	0300a	07.50	05.00a
Carbofuran	1.00	1600c	40.00	37.50c
	2.00	1300c	32.00	35.00c
Check		4000f		40.00c

Column means followed by the same letter are not significantly different according to Duncan's multiple range test (p = 0.05).

Table 4. Effect of soil application of neem oil, neem cake Repelin, and Carbofuran on the growth parameters and yield of soybean

Treatment	Conc. (%)	Shoot		Root		Yield (q/ha)
		Length (cm)	Dry wt. (g)	Length (cm)	Dry wt. (g)	
Neem cake	1.00	36.91	10.91	13.00	2.64	15.00c
	2.00	38.91	11.92	14.00	3.62	16.00c
Neem oil	1.00	47.43	12.91	14.07	2.86	15.50c
	2.00	49.23	12.00	14.50	3.50	15.50c
Repelin	1.00	51.99	13.58	17.81	5.63	22.00a
	2.00	57.80	17.81	19.82	5.83	23.00a
Carbofuran	1.00	49.00	11.68	14.07	4.64	18.00b
	2.00	50.00	12.00	15.00	4.90	18.00b
Check		29.00	08.80	10.75	2.64	12.00d

Column means followed by the same letter are not significantly different according to Duncan's multiple range test (p = 0.05).

biotics, and lastly direct penetration by the saprophytic fungi in the pathogenic fungi (Vyas, 1982).

The present investigation indicates that application of Repelin reduced nematode population and pathogens and thus increased the yield significantly. The information generated can be used for effective management of the parasitic nematodes of soybean.

REFERENCES

Ayoub, S.M. 1977. Plant nematology—An Agricultural Training Aid. Department of Food and Agriculture, Division of Plant Industry, Laboratory Services—Nematology, California, 157 pp.

Christie, J.R. and V.G. Perry. 1952. Removing Nematodes from Soil. *Proc. Helminthol. Soc. Wash.* 18: 9–13.

Mail, W.F. and H.H. Lyon. 1975. *Pictorial Key to Geneva of Plant Parasitic Nematodes*. Comstock Publ. Assoc., Ithaca, New York, 219 pp.

Taylor, A.L. 1967. *Introduction to Reserch on Plant Nematology, An FAO Guide to the Study and Control of Plant Parasitic Nematodes*. FAO, Rome, 133 pp.

Vyas, S.C. 1992. Studies on biological control of soil-borne pathogens of soybean and wheat. Scheme financed by M.P. Council of Science and Technology, Bhopal. Department of Plant Pathology, J.N. Agric. University, Indore, 134 pp.

72

Evaluation of Nematicidal Potential in Neem Allelochemicals*

MANSOOR A. SIDDIQUI AND M. MASHKOOR ALAM

Department of Botany, Aligarh Muslim University, Aligarh 202 002, India

ABSTRACT

Considerable work has been carried out on nematode control with neem (*Azadirachta indica*) products such as oil cakes and leaf. However, such information is meagre with respect to the nematotoxicity of neem allelochemicals. Hence, the nematicidal potential of azadirachtin, nimbidic acid, nimbin, kaempferol, and quercetin has been evaluated against *Meloidogyne incognita*, *Hoplolaimus indicus*, *Helicotylenchus indicus*, *Rotylenchulus reniformis*, *Tylenchorhynchus brassicae*, and *Tylenchus filiformis*. All the allelochemicals were found to be highly toxic to the test nematodes, however, to varying degree. The mortality of the test nematodes increased with an increase in the concentration of the allelochemicals and the exposure period. These allelochemicals also caused a significant inhibition in the hatching of *M. incognita*. This effect increased with an increase in the chemical concentration. The present findings may help understand the mechanism of action of neem oil cake and leaf when used as soil amendment to control plant parasitic nematodes.

INTRODUCTION

Margosa or neem (*Azadirachta indica* A. Juss.) is a tropical tree with several uses in crop protection. In India, it has been used since time immemorial as medicine for a large number of ailments. Much work has been carried out with respect to its insecticidal and insect repellent properties. Although neem derivatives have shown great nematode control potential, there is a need for more research. Water-soluble fractions of different plant parts and products and root exudates of neem have been found to be toxic to plant-parasitic nematodes (Khan et al., 1974; Alam et al., 1975, 1982; Siddiqui and Alam, 1985 a, b, 1989; Devakumar et al., 1985). The nematicidal property of neem

* This paper formed part of the Ph.D. thesis of the senior author.

may be due to toxic chemicals in it (Khan et al., 1974). The present study was undertaken to investigate the nematotoxicity of some triterpenoids and flavenoids occurring in neem.

MATERIALS AND METHODS

Some commonly occurring triterpenoids (azadirachtin, nimbidic acid, and nimbin) and flavenoids (kaempferol and quercetin) from neem were selected for the present study. One hundred milligrams of each chemical was dissolved in 1 ml of 90 per cent alcohol. The solution was further diluted to 1.00, 0.01, and 0.01 per cent concentrations with distilled water. In control sets, the same concentrations of the solvent were used in addition to the main control of distilled water alone.

For the mortality experiment, separate suspensions of freshly hatched second stage juveniles of the root knot nematode, *Meloidogyne incognita* (Kofoid & White, 1919) Chitwood 1949, and freshly isolated specimens of the lance nematode *Hoplolaimus indicus* Sher 1963, the spiral nematode *Helicotylenchus indicus* Siddiqi 1963, the stunt nematode *Tylenchorhynchus brassicae* Siddiqi 1961, and the filiform nematode *Tylenchus filiformis* Butschli 1873 were transferred separately to petri dishes (40 mm diameter) containing 10 ml of the various dilutions of the neem chemicals following the procedure described by Alam (1985). There were three replicates of each treatment.

For the hatching experiment, five average-sized and fresh egg masses of *M. incognita* were transferred to 40 mm diameter petri dishes containing 10 ml neem chemicals of different concentrations. Each treatment was replicated three times. Total hatched juveniles were counted after five days. Per cent inhibition in larval hatching over control was calculated.

RESULTS

Different chemicals of neem were highly toxic to the test nematodes. The mortality of the test nematodes increased with increase in concentration and exposure period. In the solvent, 0.01 per cent concentration was least effective in all cases, as no nematode was killed, even after 48 hrs exposure, and in 1 per cent concentration mortality of the test nematodes ranged from 26 to 20 per cent after 48 hrs of exposure.

All juveniles of *M. incognita* were killed in 0.10 per cent azadirachtin and nimbidic acid and 1 per cent nimbin after 48 hrs exposure. The same level of mortality (100 per cent) was achieved in 1 per cent nimbidic acid and quercetin after 24 hrs of exposure. The lowest concentration (0.01 per cent) of all the chemicals except nimbidic acid caused 50 per cent or more nematode kill after 24 hrs. exposure. The same concentration of nimbidic acid caused a similar mortality after 48 hrs (Table 1).

All *R. reniformis* were killed in 0.10 per centt azadirachtin and nimbidic acid after 48 hrs in 1 per cent azadirachtin, nimbidic acid, and nimbin after 24 hrs, and kaempferol and quercetin after 48 hrs. Fifty per cent mortality was observed in the lowest concentration (0.01 per cent) of azadirachtin, kaempferol, and quercetin after 12 hrs and nimbidic acid and nimbin after 24 hrs exposure (Table 2).

All *T. brassicae* were killed in 0.01 per cent azadirachtin, nimbidic acid, and nimbin and 1 per cent concentration of kaempferol and quercetin after 48 hrs. The

Table 1. Effect of different chemicals of margosa on the mortality of *M. incognita* juveniles *in vitro*

Chemicals	Exposure period (hr)	Per cent mortality in different concentrations				Regression equations
		1.00%	0.10%	0.01%	Distilled water (control)	
Azadirachtin	12	87.20 (87.98)	58.80 (59.36)	34.20 (30.74)	0.00 (02.12)	Y = 45.05 + 28.62 (X–1.50)
	24	100.00 (108.57)	83.10 (75.59)	53.30 (42.61)	0.00 (09.63)	Y = 49.10 + 32.98 (X–1.50)
	48	100.00 (116.21)	100.00 (82.43)	62.20 (48.66)	0.00 (14.88)	Y = 65.55 + 33.77 (X–1.50)
Nimbidic acid	12	80.30 (82.53)	60.50 (54.54)	21.40 (26.56)	0.00 (–01.43)	Y = 40.55 + 27.99 (X–1.50)
	24	100.00 (104.78)	79.50 (69.84)	30.00 (34.90)	0.00 (–00.40)	Y = 50.37 + 34.94 (X–1.50)
	48	100.00 (109.80)	100.00 (77.09)	51.20 (43.65)	0.00 (–08.20)	Y = 61.37 + 35.44 (X–1.50)
Nimbin	12	70.00 (73.69)	55.50 (49.64)	25.00 (25.60)	0.00 (01.54)	Y = 37.62 + 24.05 (X–1.50)
	24	90.00 (95.71)	67.70 (67.57)	56.30 (39.43)	0.00 (11.29)	Y = 53.50 + 28.14 (X–1.50)
	48	100.00 (109.80)	83.40 (77.91)	64.50 (46.03)	0.00 (14.13)	Y = 61.97 + 31.88 (X–1.50)
Kaempferol	12	66.00 (68.20)	50.00 (45.40)	20.00 (22.60)	0.00 (–00.20)	Y = 34.00 + 22.80 (X–1.50)
	24	78.30 (84.28)	61.20 (59.67)	50.00 (35.07)	0.00 (–10.46)	Y = 47.37 + 24.60 (X–1.50)
	48	83.50 (92.67)	70.00 (66.85)	62.30 (41.02)	0.00 (15.22)	Y = 53.95 + 25.81 (X–1.50)
Quercetin	12	80.00 (87.55)	70.00 (59.78)	33.33 (31.88)	0.00 (4.10)	Y = 45.83 + 27.81 (X–1.50)
	24	100.00 (108.09)	81.20 (75.61)	56.30 (43.13)	0.00 (10.65)	Y = 59.37 + 32.48 (X–1.50)
	48	100.00 (113.48)	90.00 (81.96)	74.80 (50.44)	0.00 (18.92)	Y = 66.20 + 31.52 (X–1.50)
Solvent	12	10.00 (9.15)	5.40 (5.61)	0.00 (2.09)	0.00 (–01.45)	Y = 3.85 + 3.53 (X–1.50)
	24	13.50 (13.43)	10.00 (8.39)	0.00 (3.35)	0.00 (–01.69)	Y = 5.87 + 5.04 (X–1.50)
	48	23.40 (22.57)	15.50 (14.00)	0.00 (5.44)	0.00 (–03.13)	Y = 9.72 + 8.57 (X–1.50)

Values calculated from regression equations are given in parentheses.

Table 2. Effect of different chemicals of margosa on the mortality of *R. reniformis in vitro*

Chemicals	Exposure period (hr)	Per cent mortality in different concentrations				Regression equations
		1.00%	0.10%	0.01%	Distilled water (control)	
Azadiractin	12	91.30 (102.78)	83.30 (72.61)	55.55 (42.45)	0.00 (12.29)	Y = 57.53 + 30.16 (X–1.50)
	24	100.00 (116.21)	92.10 (81.55)	64.80 (46.89)	0.00 (12.23)	Y = 64.22 + 34.66 (X–1.50)
	48	100.00 (117.55)	100.00 (85.10)	75.50 (52.65)	0.00 (20.20)	Y = 68.87 + 32.45 (X–1.50)
Nimbidic acid	12	85.50 (94.49)	75.30 (65.84)	45.30 (37.20)	0.00 (08.54)	Y = 51.52 + 28.65 (X–1.50)
	24	100.00 (110.59)	88.88 (76.75)	50.50 (42.93)	0.00 (09.08)	Y = 59.84 + 33.83 (X–1.50)
	48	100.00 (116.00)	100.00 (82.00)	60.00 (48.00)	0.00 (14.00)	Y = 65.00 + 34.00 (X–1.50)
Nimbin	12	83.50 (88.76)	65.80 (61.13)	40.00 (33.51)	0.00 (05.87)	Y = 47.32 + 27.63 (X–1.50)
	24	100.00 (107.61)	80.30 (75.07)	54.90 (42.53)	0.00 (09.99)	Y = 58.80 + 32.54 (X–1.50)
	48	100.00 (111.29)	87.10 (79.03)	64.50 (46.77)	0.00 (14.51)	Y = 62.90 + 32.26 (X–1.50)
Kaempferol	12	71.50 (81.56)	65.00 (59.12)	55.55 (36.68)	0.00 (14.24)	Y = 47.90 + 22.44 (X–1.50)
	24	79.20 (90.08)	70.80 (65.58)	63.33 (41.08)	0.00 (16.57)	Y = 53.33 + 24.50 (X–1.50)
	48	100.00 (114.15)	90.10 (83.26)	81.20 (52.38)	0.00 (21.48)	Y = 67.82 + 30.89 (X–1.50)
Quercetin	12	73.33 (82.32)	64.50 (59.07)	52.00 (35.83)	0.00 (12.57)	Y = 47.45 + 23.24 (X–1.50)
	24	85.50 (95.09)	73.33 (68.03)	59.20 (40.97)	0.00 (13.90)	Y = 54.50 + 27.06 (X–1.50)
	48	100.00 (114.48)	91.20 (83.36)	80.00 (52.24)	0.00 (21.12)	Y = 67.80 + 31.12 (X–1.50)
Solvent	12	11.10 (7.73)	0.00 (4.43)	0.00 (1.11)	0.00 (–02.21)	Y = 2.77 + 3.22 (X–1.50)
	24	14.44 (12.14)	5.10 (7.30)	0.00 (2.46)	0.00 (–02.38)	Y = 4.88 + 4.84 (X–1.50)
	48	22.30 (19.65)	10.10 (11.95)	0.00 (4.25)	0.00 (–03.45)	Y = 8.10 + 7.70 (X–1.50)

Values calculated from regression equations are given in parentheses.

lowest concentration (0.01 per cent) of kaempferol and quercetin caused 50 per cent mortality after 24 hrs, and of the remaining chemicals after 48 hrs exposure (Table 3).

All *Hoplolaimus indicus* were killed in 0.01 per cent concentration of azadirachtin, nimbidic acid, and nimbin and 1 per cent kaempferol and quercetin after 48 hrs exposure. In nimbidic acid and nimbin this level of mortality was also achieved with 1 per cent concentration in 50 per cent nematode killing after 48 hrs of exposure in nimbidic acid and quercetin, and after 24 hrs in the remaining chemicals (Table 4).

All *Helicotylenchus indicus* were killed in 0.01 per cent azadirachtin and nimbidic acid and 1 per cent kaempferol and quercetin after 48 hrs exposure. This mortality level was achieved in 1 per cent azadirachtin, nimbidic acid, and nimbin after only 24 hrs exposure (Table 5).

All *T. filiformis* were killed in 1 per cent concentration of all the test chemicals after 48 hrs exposure. After the same period in 0.10 per cent concentration the nematode kill was greatest in nimbidic acid where 96 per cent nematodes were killed and minimum (85 per cent) in quercetin. The lowest concentration (0.01 per cent) caused 50 per cent mortality in kaempferol and quercetin ater 24 hrs and in the remaining chemicals after 48 hrs of exposure (Table 6).

The neem chemicals also caused a significant inhibition in the hatching of eggs of *M. incognita*. There was decrease in hatching with an increase in the concentration of the chemicals. The average number of juveniles hatched in 1, 0.10, and 0.01 per cent concentrations of azadirachtin was 12, 30 and 36 respectively. The corresponding figures for different dilutions of nimbidic acid were 18, 45, and 100; for nimbin 19, 35, and 90; for kaempferol 14, 40, and 70; for quercetin 15, 30 and 88; and for the solvent 260, 300, and 350 respectively. In distilled water control, 380 juveniles were hatched.

DISCUSSION

The results clearly show that all the neem chemicals were toxic to all the test nematodes. The toxicity of the extracts of different plant parts of margosa to plant-parasitic nematodes has already been reported (Siddiqui and Alam, 1985 a, b). The root exudates of margosa seedlings also brought about considerable mortality to the nematodes (Alam et al., 1975; Siddiqui and Alam, 1989). Many active principles like nimbidine and thionimone from margosa have been reported to be highly nematicidal and nematostatic (Khan et al., 1974). Limonoids belonging to the β furanotriterpenoids have also been found to be nematotoxic (Devakumar et al., 1985). The results of the present study are in general agreement with those of others as discussed above. Neem is known to contain over 34 bitter principles (Rao and Parmar, 1984). The nematicidal nature of any of them is, however, still to be investigated. The present finding may help us to understand the mechanism of action of neem oil cake and leaf when used as soil amendment to control plant-parasitic nematodes.

Table 3. Effect of different chemicals of margosa on the mortality of *Tylenchorhynchus brassicae in vitro*

Chemicals	Exposure period (hr)	Per cent mortality in different concentrations				Regression equations
		1.00%	0.10%	0.01%	Distilled water (control)	
Azadirachtin	12	56.30 (59.40)	41.20 (41.60)	33.33 (23.80)	0.00 (06.00)	Y = 32.70 + 17.80 (X–1.500)
	24	87.50 (90.88)	62.40 (63.07)	46.80 (35.27)	0.00 (07.46)	Y = 49.17 + 27.80 (X–1.50)
	48	100.00 (116.93)	100.00 (83.56)	67.50 (50.18)	0.00 (16.81)	Y = 66.87 + 33.37 (X–1.50)
Nimbidic acid	12	63.33 (70.19)	55.55 (49.29)	36.50 (28.39)	0.00 (07.48)	Y = 38.84 + 20.90 (X–1.50)
	24	91.00 (94.77)	66.66 (65.13)	43.50 (35.49)	0.00 (05.84)	Y = 50.31 + 29.64 (X–1.50)
	48	100.00 (115.67)	100.00 (81.57)	58.10 (47.47)	0.00 (13.37)	Y = 64.52 + 34.10 (X–1.50)
Nimbin	12	55.60 (63.60)	51.60 (45.81)	40.50 (28.03)	0.00 (10.23)	Y = 36.92 + 17.79 (X–1.50)
	24	93.40 (99.61)	74.50 (68.59)	44.44 (37.57)	0.00 (06.45)	Y = 53.08 + 31.02 (X–1.50)
	48	100.00 (115.12)	100.00 (80.25)	51.30 (45.39)	0.00 (10.51)	Y = 62.82 + 34.87 (X–1.50)
Kaempferol	12	48.50 (54.01)	41.30 (38.88)	35.50 (23.76)	0.00 (08.68)	Y = 31.32 + 15.13 (X–1.50)
	24	79.20 (91.34)	73.40 (66.79)	65.50 (42.25)	0.00 (17.75)	Y = 54.52 + 24.55 (X –1.50)
	48	100.00 (109.80)	81.20 (79.03)	73.40 (48.27)	0.00 (17.50)	Y = 63.65 + 30.77 (X–1.50)
Quercetin	12	56.10 (61.70)	45.80 (44.61)	41.20 (27.13)	0.00 (09.83)	Y = 35.77 + 17.28 (X–1.50)
	24	78.50 (88.64)	69.60 (63.99)	58.60 (39.35)	0.00 (14.70)	Y = 51.67 + 24.64 (X–1.50)
	48	100.00 (112.57)	89.30 (80.53)	68.80 (48.51)	0.00 (16.47)	Y = 64.52 + 32.03 (X–1.50)
Solvent	12	5.10 (4.80)	3.10 (2.96)	0.00 (1.14)	0.00 (–00.70)	Y = 2.05 + 1.83 (X–1.50)
	24	15.00 (12.69)	5.50 (7.64)	0.00 (2.60)	0.00 (–02.45)	Y = 5.12 + 5.04 (X–1.50)
	48	20.00 (18.99)	12.50 (11.74)	0.00 (4.50)	0.00 (–02.75)	Y = 8.12 + 7.24 (X–1.50)

Values calculated from regression equations are given in parentheses.

Table 4. Effect of different chemicals of margosa on the mortality of *Hoplolaimus indicus in vitro*

Chemicals	Exposure period (hr)	Per cent mortality in different concentrations				Regression equations
		1.00%	0.10%	0.01%	Distilled water (control)	
Azadirachtin	12	63.33 (57.75)	54.44 (46.82)	25.80 (24.96)	0.00 (03.09)	Y = 35.89 + 21.86 (X–1.50)
	24	76.10 (73.28)	67.80 (61.04)	51.30 (36.56)	0.00 (12.08)	Y = 48.80 + 24.48 (X–1.50)
	48	100.00 (115.18)	100.00 (82.07)	62.10 (48.97)	0.00 (15.86)	Y = 65.52 + 33.10 (X–1.50)
Nimbidic acid	12	71.20 (77.85)	63.50 (52.75)	26.10 (27.65)	0.00 (02.55)	Y = 40.20 + 25.10 (X–1.50)
	24	100.00 (105.07)	76.50 (71.93)	45.00 (38.81)	0.00 (05.66)	Y = 55.37 + 33.13 (X–1.50)
	48	100.00 (115.12)	100.00 (80.24)	51.20 (45.36)	0.00 (10.48)	Y = 62.80 + 34.88 (X–1.50)
Nimbin	12	61.30 (67.90)	55.00 (47.03)	30.10 (26.17)	0.00 (05.29)	Y = 36.60 + 20.87 (X–1.50)
	24	100.00 (106.33)	78.00 (73.63)	51.40 (41.03)	0.00 (08.36)	Y = 57.35 + 32.65 (X–1.50)
	48	100.00 (116.22)	100.00 (82.45)	62.30 (48.91)	0.00 (14.91)	Y = 65.57 + 33.76 (X–1.50)
Kaempferol	12	55.55 (59.91)	44.44 (42.06)	32.60 (24.22)	0.00 (06.36)	Y = 33.14 + 17.84 (X–1.50)
	24	80.00 (92.53)	74.30 (67.91)	68.10 (43.29)	0.00 (18.67)	Y = 55.60 + 24.62 (X–1.50)
	48	100.00 (115.31)	92.50 (84.38)	83.20 (53.46)	0.00 (22.52)	Y = 68.92 + 30.93 (X–1.50)
Quercetin	12	47.50 (48.43)	35.50 (31.64)	10.00 (14.86)	0.00 (–01.93)	Y = 23.25 + 16.79 (X–1.50)
	24	73.50 (82.22)	71.60 (55.14)	21.30 (28.06)	0.00 (00.98)	Y = 41.60 + 27.08 (X–1.50)
	48	100.00 (98.88)	90.00 (72.82)	50.40 (46.84)	0.00 (–20.82)	Y = 59.85 + 26.02 (X–1.50)
Solvent	12	8.50 (5.94)	0.00 (3.39)	0.00 (0.85)	0.00 (–01.70)	Y = 2.12 + 2.55 (X–1.50)
	24	15.00 (14.61)	10.30 (9.08)	0.00 (3.56)	0.00 (–01.97)	Y = 6.32 + 5.52 (X–1.50)
	48	23.80 (22.69)	15.10 (14.04)	0.00 (5.40)	0.00 (–03.25)	Y = 9.72 + 8.64 (X–1.50)

Values calculated from regression equations are given in parentheses.

Table 5. Effect of different chemicals of margosa on the mortality of *Helicotylenchus indicus in vitro*

Chemicals	Exposure period (hr)	Per cent mortality in different concentrations				Regression equations
		1.00%	0.10%	0.01%	Distilled water (control)	
Azadirachtin	12	71.30 (81.83)	58.20 (57.26)	44.44 (32.70)	0.00 (08.13)	Y = 44.98 + 24.56 (X–1.50)
	24	100.00 (89.76)	75.20 (73.81)	56.30 (41.93)	0.00 (10.04)	Y = 57.87 + 31.88 (X–1.50)
	48	100.00 (99.67)	100.00 (82.89)	64.50 (49.35)	0.00 (15.80)	Y = 66.12 + 33.54 (X–1.50)
Nimbidic acid	12	81.30 (92.16)	76.30 (64.88)	47.40 (37.61)	0.00 (10.33)	Y= 51.25 + 27.27 (X–1.50)
	24	100.00 (108.88)	83.60 (75.97)	54.50 (43.07)	0.00 (43.07)	Y = 59.52 + 32.91 (X–1.50)
	48	100.00 (114.52)	100.00 (75.97)	65.33 (43.59)	0.00 (11.11)	Y = 61.32 + 33.47 (X–1.50)
Nimbin	12	80.90 (83.62)	56.40 (58.14)	44.30 (32.66)	0.00 (07.18)	Y = 45.40 + 25.48 (X–1.50)
	24	100.00 (110.29)	81.30 (79.93)	77.77 (49.59)	0.00 (19.23)	Y = 64.76 + 30.35 (X–1.50)
	48	100.00 (114.65)	91.60 (83.53)	80.30 (52.41)	0.00 (21.29)	Y = 67.97 + 31.12 (X–1.50)
Kaempferol	12	60.30 (69.59)	53.33 (50.23)	48.60 (30.87)	0.00 (11.50)	Y = 40.55 + 19.36 (X–1.50)
	24	75.50 (85.74)	68.50 (61.74)	55.00 (37.76)	0.00 (13.75)	Y = 49.75 + 23.99 (X–1.50)
	48	100.00 (113.02)	90.00 (81.05)	70.30 (49.09)	0.00 (17.11)	Y= 65.07 + 31.96 (X–1.50)
Quercetin	12	63.50 (72.03)	59.00 (51.09)	40.00 (30.15)	0.00 (09.20)	Y = 40.62 + 20.94 (X–1.50)
	24	81.20 (92.48)	72.00 (67.76)	68.40 (43.04)	0.00 (18.32)	Y = 55.40 + 24.72 (X–1.50)
	48	100.00 (114.46)	93.33 (82.27)	71.40 (50.08)	0.00 (17.89)	Y = 66.18 + 32.19 (X–150)
Solvent	12	16.00 (15.30)	10.30 (9.48)	0.00 (3.66)	0.00 (–02.16)	Y = 6.57 + 5.82 (X–1.50)
	24	21.30 (21.14)	15.60 (13.19)	0.00 (5.25)	0.00 (–02.70)	Y = 9.22 + 7.94 (X–1.50)
	48	25.50 (27.72)	23.40 (18.24)	5.10 (8.76)	0.00 (–00.72)	Y = 13.50 + 9.48 (X–1.50)

Values calculated from regression equations are given in parentheses.

Table 6. Effect of different chemicals of margosa on the mortality of *Tylenchus filiformis in vitro*

Chemicals	Exposure period (hr)	Per cent mortality in different concentrations				Regression equations
		1.00%	0.10%	0.01%	Distilled water (control)	
Azadirachtin	12	66.50 (64.23)	37.70 (43.11)	26.00 (21.99)	0.00 (00.87)	Y = 32.55 + 21.12 (X–1.50)
	24	83.33 (84.47)	55.55 (57.83)	39.20 (31.30)	0.00 (04.56)	Y = 44.52 + 26.63 (X–1.50)
	48	100.00 (112.55)	89.30 (80.46)	68.40 (48.38)	0.00 (16.28)	Y = 64.42 + 32.08 (X–1.50)
Nimbidic acid	12	70.00 (71.62)	60.00 (50.74)	31.20 (29.86)	0.00 (08.98)	Y = 40.30 + 20.88 (X–1.50)
	24	94.50 (99.38)	72.00 (68.28)	44.44 (37.18)	0.00 (06.07)	Y = 52.73 + 31.10 (X–1.50)
	48	100.00 (113.73)	95.50 (79.71)	55.30 (45.69)	0.00 (11.67)	Y = 62.70 + 34.02 (X–1.50)
Nimbin	12	59.20 (61.41)	41.20 (43.05)	35.10 (24.69)	0.00 (06.33)	Y = 33.87 + 18.36 (X–1.50)
	24	83.60 (91.04)	70.10 (63.46)	45.00 (35.88)	0.00 (08.29)	Y = 49.67 + 27.58 (X–1.50)
	48	100.00 (112.70)	93.50 (78.68)	53.20 (44.66)	0.00 (10.64)	Y = 61.67 + 34.02 (X–1.50)
Kaempferol	12	50.00 (54.82)	43.33 (37.99)	25.00 (21.17)	0.00 (04.33)	Y = 29.58 + 16.83 (X–1.50)
	24	75.80 (85.58)	65.70 (60.76)	51.90 (35.94)	0.00 (11.11)	Y = 48.35 + 24.82 (X–1.50)
	48	100.00 (111.93)	85.30 (81.22)	78.20 (30.52)	0.00 (19.80)	Y = 65.87 + 30.70 (X–1.50)
Quercetin	12	41.80 (46.84)	37.60 (33.08)	25.40 (19.32)	0.00 (05.56)	Y = 26.20 + 13.76 (X–1.50)
	24	73.50 (83.82)	64.50 (59.90)	55.80 (37.00)	0.00 (14.07)	Y = 48.45 + 22.91 (X–1.50)
	48	100.00 (111.53)	84.70 (80.72)	76.60 (49.92)	0.00 (19.10)	Y = 65.32 + 30.81 (X–1.50)
Solvent	12	5.00 (3.49)	0.00 (1.99)	0.00 (0.51)	0.00 (–00.99)	Y = 1.25 + 1.49 (X–1.50)
	24	11.00 (9.90)	5.55 (6.05)	0.00 (2.21)	0.00 (–01.64)	Y = 4.13 + 3.85 (X–1.50)
	48	20.30 (18.19)	100.00 (11.11)	0.00 (4.03)	0.00 (–03.05)	Y = 7.57 + 7.08 (X–1.50)

Values calculated from regression equaiions are given in parentheses.

REFERENCES

Alam, M.M. 1985. A simple method for *in vitro* screening of chemicals for neem-toxicity. *Nematol. Network Newsletter* 2(1): 6.

Alam, M.M., A.M. Khan and S.K. Saxena. 1982. Relative toxicity of decomposed and undecomposed oil-cakes to plant parasitic nematodes. *Acta Bot. Indica* 10: 124 –127.

Alam, M.M., A. Masood and S.I. Husain. 1975. Effect of margosa and marigold root-exudates on mortality and larval hatch of certain nematodes. *Indian Biol.* 13: 412–414.

Devakumar, C., B.K. Goswami and S.K. Mukherjee. 1985. Nematicidal properties from neem (*Azadirachta indica* A. Juss.) Part I. Screening of neem kernel fractions against *Meloidogyne incognita*. *Indian J. Nematol.* 15: 121–124.

Khan M.W., M.M. Alam, A.M. Khan and S.K. Saxena. 1974. Effect of water soluble fractions of oil-cakes and bitter principles of neem on some fungi and nematodes. *Acta. Bot. Indica* 2: 120–128.

Rao, K.N. and B.S. Parmar. 1984. A compendium of chemical constituents of neem. *Neem Newletter* 1: 39–46.

Siddiqui, M.A. and M.M. Alam. 1985a. Evaluation of nematicidal properties of different parts of margosa and Persian lilac. *Neem Newsletter* 2: 1–4.

Siddiqui, M.A. and M.M. Alam. 1985b. Further studies on the nematode toxicity of margosa and Persian lilac. *Neem Newsletter* 2: 43–47.

Siddiqui, M.A. and M.M. Alam. 1989. Effect of root-exudates of neem and Persian lilac on plant parasitic nematodes. *Anzeiger fur Schalingskunde Pflanzenschutz Umweltschutz* 62: 33–35.

73

Management of Root Knot Nematode in Betelvine by Neem Cake

K. SITARAMAIAH[1], A. ACHARYA[2], E.I. JONATHAN[3], A.R. WASNIKAR[4], R.V. NAKAT[5], D. SUBBA RAO[6] AND SATYABRATA MAITI[7]

[1]Associate Professor, APAU
[2]Assistant Professor, OUAT
[3]Assistant Professor, TNAU
[4]Assistant Professor, JNKVV
[5]Assistant Professor, MPAU
[6]Assistant Professor, APAU
[7]Project Co-ordinator

All India Coordinated Research Project on Betelvine, Indian Institute of Horticultural Research, Hessaraghatta Lake, Bangalore 560 089, India

ABSTRACT

The root knot nematode (*Meloidogyne incognita*) is one of the major constraints for successful betelvine cultivation throughout India. Therefore, field experiments were conducted across the locations under AICRP on Betelvine in newly established gardens.

Results revealed that neem cake at 0.5 t/ha + carbofuran at 0.75 kg active ingredient per ha + NPK (150 : 100 : 50) treatment was most effective in increasing the leaf yield and reducing the root galls and soil nematode population in most of the locations.

INTRODUCTION

Betelvine (*Piper betle* L.) is a perennial creeper cultivated in India for its leaves, which are chewed unprocessed along with betelnut. Betelvine cultivation is highly input-intensive and grown on about 40,000 ha area with an annual turnover of about Rs 700 crores.

The crop is attacked by a number of fungal, bacterial, and nematode diseases. Among the 12 genera of nematodes reported to be associated with this crop, the root knot nematode (*Meloidogyne incognita*) is considered to be a major parasite prevalent in all betelvine-growing areas of the country (AICRPB Annual Report, 1989–90). Since betel leaves are consumed fresh, it is not possible to use any chemical pesticides which may leave a residue. Neem cake is known to reduce nematode population in soil (Singh and Sitaramaiah, 1973) and it is extensively used as a source of organic manure in this crop. Therefore, several field experiments were conducted across the locations under the All India Coordinated Research Project on Betelvine for management of root knot by application of neem cake in soil.

MATERIALS AND METHODS

Experiments were conducted in the experimental fields of state agricultural universities at different centres during 1989–90 using randomized block design with seven treatments replicated thrice. Treatments were (1) sawdust at 2 t/ha + NPK; (2) neem cake at 1 t/ha + NPK; (3) neem cake at 0.5 t/ha + carbofuran 3 per cent G at o.75 kg/ha active ingredient (ai) + NPK; (4) sawdust at 1 t/ha + carbofuran 3 per cent G at 0.75 kg/ha ai + NPK; (5) carbofuran 3 per cent G at 1.5 kg/ha + NPK; (6) NPK alone; and (7) control. NPK was applied in 150:100:50 ratio. N was applied as four split doses where as P and K were applied at the time of land preparation. Observations on gall index and nematode population in soil were recorded up to 180 days. Vegetative growth parameters, such as weight of 100 leaves, and number of leaves per vine were also recorded and analysed statistically.

RESULTS AND DISCUSSION

Experiments conducted at Chinthalapudi revealed that highest reduction in nematode population was obtained in carbofuran + NPK treatment followed by neem cake at 1 t/ha + NPK and neem cake at 0.5 t/ha + carbofuran 0.75 kg ai + NPK. Similar results were recorded for root gall infection. Yield data showed that maximum leaves per vine (115) were harvested from neem cake at 0.5 t/ha + carbofuran at 0.75 ha ai/ha + NPK treatment. However, highest 100 leaf weight was recorded with sawdust at 2 t/ha treatment (Table 1).

Similar results were also obtained from Sangli where the lowest nematode population was noted from carbofuran 1.5 kg/ha + NPK treatment followed by neem cake at 0.5 t/ha + carbofuran 0.75 kg ai/ha + NPK treatment. However, highest leaf yield per vine was recorded on neem cake at 0.5 t/ha + carbofuran 0.75 ai/ha + NPK treatment (Table 2).

In Bhubaneshwar nematode population was at par in neem cake at 1 t/ha + NPK, carbofuran at 1.5 kg ai/ha + NPK, sawdust at 1t/ha NPK, carbofuran at 1.5 Kg ai/ha + NPK, and neem cake 0.5 t/ha + carbofuran at 0.75 kg ai/ha + NPK treatments. Similar results were also observed for root gall indices. However, highest leaf yield was recorded in neem cake at 0.5 t/ha + carbofuran 0.75 kg ai/ha + NPK treatment. Weight of 100 leaves was also highest in the same treatment (Table 3).

Table 1. Effect of neem and sawdust as soil supplement for root knot management at Chinthalapudi

Treatments	Nematode population (180 days)		No. of leaves/ vine	Wt of 100 leaves (g)
	Soil (250 g)	Root (10 g)		
Sawdust at 2 t/ha + NPK (150:100:50)	88	75	95	188
Neem cake at 1 t/ha + NPK (150:100:50)	84	41	76	157
Carbofuran 3% G at 1.5 kg ai/ha + NPK (150:100:50)	75	37	79	167
Sawdust at 1 t/ha + carbofuran 3% G at 0.75 kg ai/ha + NPK (150:100:50)	133	46	90	177
Neem cake at 0.5 t/ha + Carbofuran 3% G at 0.75 kg ai/ha + NPK (150:100:50)	89	48	115	158
NPK (150:100:50) alone	162	159	92	158
Control	244	152	68	154
CC (P = 0.05)	64	55	19	23

ai : active ingredient.

Table 2. Effect of neem and sawdust as soil supplement for root knot management at Sangli

Treatments	Nematode population (180 days)		No. of leaves/ vine	Wt of 100 leaves (g)
	Soil (250 g)	Root (10 g)		
Sawdust at 2 t/ha + NPK (150:100:50)	480	—	145	—
Neem cake at 1 t/ha + NPK (150:100:50)	355	—	181	—
Carbofuran 3% G at 1.5 kg ai/ha + NPK (150:100:50)	305	—	190	—
Sawdust at 1 t/ha + carbofuran 3 % G a 0.75 kg ai/ha + NPK (150:100:50)	340	—	151	—
Neem cake at 0.5 t/ha + Carbofuran 3% G at 0.75 kg ai/ha + NPK (150:100:50)	307	—	205	—
NPK (150:100:50) alone	415	—	148	—
Control	985	—	137	—
CD (P = 0.05)	76	—	17.4	—

Table 3. Effect of neem and sawdust as soil supplement for root knot management at Bhubaneshwar

Treatments	Nematode population (180 days)		No. of leaves/ vine	Wt of 100 leaves (g)
	Soil (250 g)	Root (10 g)		
Sawdust at 2 t/ha + NPK (150:100:50)	410	64	33	208
Neem cake at 1 t/ha + NPK (150:100:50)	385	58	40	225
Carbofuran 3% G at 1.5 kg ai/ha + NPK (150:100:50)	395	60	38	218
Sawdust at 1 t/ha + carbofuran 3% G at 0.75 kg ai/ha + NPK (150:100:50)	365	55	45	235
Neem cake at 0.5 t/ha + carbofuran 3% G at 0.75 kg ai/ha + NPK (150:100:50)	348	50	52	250
NPK (150:100:50) alone	438	68	30	202
Control	1145	138	22	185
CD (0.05)	102	5.05	7.3	3.56

Table 4. Effect of neem and sawdust as soil supplement for root knot management at Sirugamani

Treatments	Nematode population (180 days)		No. of leaves/ vine	Wt of 100 leaves (g)
	Soil (250 g)	Root (10 g)		
Sawdust at 2 t/ha + NPK (150:100:50)	383.33	19.58	95	185.00
Neem cake at 1 t/ha + NPK (150:100:50)	388.33	19.68	96	190.00
Carbofuran 3% G at 1.5 kg ai/ha + NPK (150:100:50)	300.00	16.31	110	205.00
Sawdust at 1 t/ha + carbofuran 3% G at 0.75 kg ai/ha + NPK (150:100:50)	418.00	20.44	90	180.00
Neem cake at 0.5 t/ha + Carbofuran 3% G at 0.75 kg ai/ha + NPK (150:100:50)	262.33	16.19	107	200.00
NPK (150:100:50) alone	531.00	23.04	75	170.00
Control	645.67	25.39	68	153.33
CD (P = 0.05)	60.18	NS	10	40.36

Table 5. Effect of neem and sawdust as soil supplement for root knot management at Jabalpur

Treatments	Nematode population (180 days)		No. of leaves/ vine	Wt of 100 leaves (g)
	Soil (250 g)	Root (10 g)		
Sawdust at 2 t/ha + NPK (150:100:50)	133.3	68.3	18.33	242.64
Neem cake at 1 t/ha + NPK (150:100:50)	188.6	67.9	21.33	289.44
Carbofuran 3% G at 1.5 kg ai/ha + NPK (150:100:50)	150.6	132.0	25.00	339.77
Sawdust at 1 t/ha + NPK carbofuran 3% G at 0.75 kg ai/ha + NPK (150:100:50)	125.3	91.0	20.66	283.98
Neem cake at 0.5 t/ha + carbofuran 3% G at 0.75 kg ai/ha + NPK (150:100:50)	160.6	115.0	21.75	297.40
NPK (150:100:50) alone	136.0	129.3	23.33	317.1
Control	229.3	158.6	16.00	210.50
CD (P = 0.05)	60.08	50.59	NS	43.79

In Sirugamani the lowest nematode population was recorded in soil at 180 days in carbofuran at 1.5 kg ai/ha + NPK treatment which was statistically at par with neem cake at 0.5 t/h + carbofuran at 0.75 kg ai/ha + NPK treatment. A similar trend was obtained in case of leaf yield and 100 leaf weight (Table 4).

In Jabalpur root gall index was observed to be minimum in neem cake at 1 t/ha + NPK treatment followed by sawdust at 1 t/ha + carbofuran at 0.75 kg ai/ha + NPK treatment. Leaf yield per vine was found to be non-significant (Table 5).

The role of toxic decomposition products in the suppression of plant parasitic nematodes in soil has been emphasized by several workers (Sitaramaiah, 1990). Singh and Sitaramaiah 1973 in a more detailed study extracted different fractions from soil amended with neem cake and found that the extracts obtained at different stages of decomposition up to three weeks contained nematotoxic substances. This phenomenon has been observed in this case also. Nematode population in soil was, therefore, reduced in neem-amended soil.

Increase in yield in neem cake plus fertilizer treatments may be due not only to reduction of root knot but also to amount and speed of availability of plant nutrients as observed by Singh and Sitaramaiah (1973) in tomato and okra. However, the actual mechanism involved in this case needs further investigation.

AKNOWLEDGEMENTS

The authors are thankful to the Indian Council of Agricultural Research, New Delhi, for financial assistance to the project. The technical assistance of Mr. B.E. Uma Shankar is also gratefully acknowledged.

REFERENCES

Annual Report. 1989–90. All India Coordinated Research Project on Betelvine. IIHR, Bangalore.

Singh, R.S. and K. Sitaramaiah. 1973. Nematode control. Res. Bull. 6. Expt. Stn., G.B. Pant Univ. Agric. Technology Pantnagar, pp. 147–169.

Sitaramaiah, K. 1990. Mechanism of reduction of plant parasitic nematodes in soils amended with organic materials. In: *Progress in Plant Nematology*. M.W. Khan, A. Rashid and S.K. Saxena (eds.), CBS Publication, New Delhi, pp. 263–295.

74

Nematode-controlling Efficiency of *Paecilomyces lilacinus* in Relation to Neem Products

ANJUM AHMAD AND M. MASHKOOR ALAM

Department of Botany, Aligarh Muslim University
Aligarh 202 002, India

ABSTRACT

The efficiency of nematode biocontrol fungus *Paecilomyces lilacinus* was evaluated in the presence of neem (*Azadirachta indica*) oilseed cake and leaf. The combined effect of the fungus and amendments in suppressing root knot nematode (*Meloidogyne incognita*) and thus improving different growth parameters of tomato, basil, and castor was more than that of either of the treatmetns alone. The study thus reveals the compatibility of these treatments in efficient control of the test nematode. Neem products besides being directly effective against the nematode have also served as substrate for enhancing the inoculum load of the fungus.

INTRODUCTION

Neem (*Azadirachta indica*) is supposed to be an inexpensive naturally insecticidal plant that causes no environmental damage. A neem tree becomes fully productive in about 10 years and produces 30–50 kg seed every year (Ketkar, 1976). Thirty kilograms of seeds yield 6 kg of neem oil and 24 kg of neem cake. Tonnes of non-edible neem cake are produced annually and huge quantities of it are not disposed in any organized way. Recently scientists have started using neem cake as soil amendment for nematode control.

Paecilomyces lilacinus is a nematode-parasitic fungus. Its efficiency as a biocontrol agent of nematodes has been tested in many countries with fruitful results (Jatala, 1986). Since the suppression of nematode population with biotic organisms is manifested slowly over a period of time, it was thought worthwhile to use *P. lilacinus* as a component of integrated nematode control strategy with neem cake and other organic amendments.

MATERIALS AND METHODS

Autoclaved soil, amended with oilseed cakes and chopped leaves of margosa and castor (*Ricinus communis*) and rice polish at the rate of 1 g/kg soil, was filled in 15 cm clay pots. The soil was watered for 10 days to ensure proper decomposition of the additives. Thereafter, three-week-old seedlings of tomato, *Lycopersicon esculentum*, and basil, *Ocimum basilicum* (raised in autoclaved pots), were transplanted into the treated pots singly. In the case of castor, four to five seeds were sown directly in the pots. After germination seedlings were thinned to keep one plant per pot. At three-week stage the plants were inoculated with 5000 freshly hatched second-stage juveniles of *Meloidogyne incognita*. The uninoculated plants served as control. In a subsequent experiment 8 g *P. lilacinus* inoculum on rice seeds was added simultaneously with *M. incognita*. Each treatment was replicated six times.

At the termination of the experiment (in tomato and basil 70 days and castor 100 days after inoculation) plants were carefully uprooted and gently washed free of soil, and the water absorption efficiency of the roots was determined by the method of Alam et al. (1974). Lengths and weights of shoots and roots were determined, and number of fruits or inflorescence per plant was counted. For obtaining the leaf surface area of castor the method of Alam (1989) was adopted. Root galling caused by *M. incognita* was assessed on the 0–5 scale of Taylor and Sesser (Sasser et al., 1984). Data were statistically analysed for critical difference (CD) at 0.05, 0.01 (Pansey and Sukhatme, 1978).

RESULTS

The root knot nematode *M. incognita* caused significant reduction in the growth parameters of tomato, basil, and castor in inoculated and unamended soil (Tables 1, 2, 3). Water absorption capacity of roots was also decreased because of the infection (Tables 1, 2, 3). In the uninoculated set plants grew better in amended soil than in unamended soil.

In the nematode-inoculated set root knot index was highest in unamended soil, being 4.2, 4.0, and 2.5 in tomato, basil, and castor respectively. As a result of the application of oilseed cakes, chopped leaves, and rice polish, root galling was significantly inhibited. The highest inhibition was in neem cake followed by castor cake, neem leaf, rice polish, and castor leaf in order of efficiency. In different treatments root knot index ranged from 2.2 to 3.3, 2.0 to 3.0, and 1.3 to 2.2 in tomato, basil, and castor respectively (Tables 1, 2, 3). Growth retardation caused by the nematode gradually decreased in correspondence with decreased root galling and water absorption capacity of roots improved (Tables 1, 2, 3).

Paecilomyces lilacinus effectively reduced the damage caused by the nematode on all the three test crops (Tables 4, 5, 6). The combined effect of the fungus and the

Table 1. Effect of organic amendments on root knot development, plant growth and water absorption of tomato cv Pusa Ruby

Treatment		Length (cm)				Fresh weight (g)				Dry weight (g)				No. of fruits/ plants	Root knot index (0–5 scale)	Water absorbed/ plant (g/day)
Organic amendment	Un/In	Shoot	Root	Total	% reduction	Shoot	Root	Total	% reduction	Shoot	Root	Total	% reduction			
Untreated	Un	30.0	15.0	45.0		16.1	9.2	25.3		4.0	1.1	5.1		5		17.0
	In	18.0	8.0	26.0	42.22	9.2	5.4	14.6	42.29	2.2	0.5	2.7	47.05	2	4.2	12.5
Neem cake	Un	45.0	22.0	67.0		25.7	14.3	40.0		7.1	2.2	9.3		9		19.5
	In	33.2	17.0	50.2	25.07	21.8	9.4	31.2	22.00	5.4	1.6	7.0	24.73	7	2.2	17.2
Castor cake	Un	40.0	20.0	60.0		23.0	12.0	35.0		6.5	1.6	8.1		8		18.7
	In	29.0	12.5	41.5	30.83	15.1	10.0	25.1	28.28	4.6	1.1	5.7	29.62	5	2.7	15.9
Neem leaf	Un	35.0	18.2	53.2		21.0	10.0	31.0		5.8	1.2	7.0		7		18.0
	In	20.0	15.4	35.6	33.08	13.4	8.1	21.5	30.64	4.0	0.7	4.7	32.85	4	3.0	15.0
Castor leaf	Un	33.0	16.1	49.1		18.0	8.5	26.5		4.9	0.8	5.7		6		17.1
	In	20.0	10.6	30.6	37.67	12.0	5.0	17.0	35.84	3.0	0.6	3.6	36.84	3	3.3	13.4
Rice polish	Un	37.0	19.2	56.2		20.0	9.0	29.0		5.7	1.0	6.7		7		17.9
	In	23.0	13.8	36.8	34.51	14.0	6.0	20.0	31.03	3.6	0.8	4.4	34.32	4	3.0	14.4
CD (p = 0.05)				4.18				1.40				0.62		1.15	0.89	0.91
(p = 0.01)				5.42				1.81				0.80		1.50	1.16	1.18

Un = Uninoculated; In = Inoculated.

Table 2. Effect of organic amendments on the root knot development, growth and water absorption of basil

Treatment		Length (cm)				Fresh weight (g)				Dry weight (g)				No. of inflore-scence/ plants	Root knot index (0-5 scale)	Water absorbed plant (g/day)
Organic amendment	Un/In	Shoot	Root	Total	% reduction	Shoot	Root	Total	% reduction	Shoot	Root	Total	% reduction			
Untreated	Un	45.5	24.0	69.5		19.8	4.7	24.5		10.0	1.9	11.9		12		40.0
	In	27.8	12.0	39.8	42.73	11.4	3.0	14.4	41.22	5.8	0.8	6.6	44.53	7	4.0	27.5
Neem cake	Un	80.0	38.0	118.0		34.0	8.0	42.0		14.0	4.0	18.0		20		48.1
	In	70.0	22.1	92.1	21.94	29.5	5.7	35.2	16.19	11.3	3.1	14.4	20.00	16	2.0	42.0
Castor cake	Un	60.2	28.0	88.2		30.0	7.0	37.0		12.0	3.0	15.0		18		44.1
	In	46.3	18.7	65.0	26.30	23.4	5.4	28.8	22.16	9.5	1.9	11.4	24.00	13	2.3	36.0
Neem leaf	Un	58.7	24.0	82.7		30.0	5.0	35.0		12.0	2.3	14.3		16		42.7
	In	42.1	18.1	60.2	27.20	21.7	3.8	25.5	27.14	8.2	1.9	10.1	29.37	11	2.7	31.5
Castor leaf	Un	61.3	30.5	91.8		29.6	4.4	34.0		11.6	1.3	12.9		14		41.2
	In	42.2	19.4	61.6	32.89	20.3	3.0	23.3	31.47	7.9	1.0	8.9	31.00	9	3.0	29.0
Rice Polish	Un	61.0	38.0	99.0		28.8	6.1	34.9		12.2	2.5	14.7		16		45.3
	In	48.5	21.6	70.1	29.19	21.0	4.4	25.4	27.22	9.3	1.2	10.5	28.57	11	2.7	34.1
CD (p = 0.05)				6.95				3.51				1.00		1.99	0.83	1.20
(p = 0.01)				9.01				4.55				1.30		2.59	1.07	1.55

Table 3. Effect of organic amendments on the root knot development caused by *M. incognita* plant growth and water absorption of castor *R. communis* cv. Local

Treatment		Length (cm)				Fresh weight (g)				Dry weight (g)				Leaf surface area (sq. cm/leaf)	% reduction	Root knot index (0–5 scale)	Water absorbed/plant (g/day)
Organic amendment	Un/In	Shoot	Root	Total	% reduction	Shoot	Root	Total	% reduction	Shoot	Root	Total	% reduction				
Untreated	Un	54.0	14.5	78.5		28.4	21.7	50.1		4.9	1.9	6.8		168			39.1
	In	40.0	16.0	56.0	28.66	21.0	15.5	36.5	27.14	3.8	1.0	4.8	29.41	130	22.61	2.5	25.4
Neem cake	Un	69.0	27.9	96.9		32.1	28.8	60.9		9.2	3.5	12.7		217			44.2
	In	58.0	23.5	81.5	15.89	29.2	23.8	53.0	12.97	8.1	2.6	10.7	15.74	194	10.59	1.3	36.7
Castor cake	Un	57.0	24.1	81.1		30.1	27.0	57.1		7.2	2.2	9.4		190			41.0
	In	46.0	20.2	66.2	18.37	26.4	21.5	47.9	16.11	5.9	1.8	7.7	18.08	163	14.21	1.7	31.9
Neem Leaf	Un	76.5	35.5	112.0		43.2	28.5	71.7		12.9	4.1	17.0		260			48.9
	In	63.2	25.2	88.4	21.07	38.0	19.3	57.3	20.08	9.7	3.5	13.2	22.35	215	17.30	2.0	36.0
Castor leaf	Un	66.7	31.2	97.9		44.0	32.1	76.1		13.6	4.8	18.4		300			52.3
	In	55.0	19.4	74.4	24.00	35.4	23.8	59.2	22.20	10.4	3.7	14.1	23.36	241	19.66	2.2	37.0
Rice polish	Un	69.4	30.7	100.1		34.5	30.4	64.9		12.1	4.0	16.1		248			46.1
	In	58.0	22.0	80.0	20.07	29.2	23.3	52.5	19.10	9.7	3.2	12.9	19.87	210	15.32	1.8	34.5
CD (p = 0.05)				3.37				2.64				1.92		8.76		0.30	1.90
(p = 0.01)				4.37				3.43				2.49		11.36		0.39	2.47

Table 4. Combined effect of organic amendments and the fungus *P. lilacinus* on the root knot development plant growth and water absorption of tomato

Treatment		Length (cm)				Fresh weight (g)				Dry weight (g)				No. of fruits /plants	Root knot index (0-5 scale)	Water absorbed /plant (g/day)
Organic amendment	Un/In	Shoot	Root	Total	% reduction	Shoot	Root	Total	% reduction	Shoot	Root	Total	% reduction			
P. lilacinus																
Untreated	Un	24.6	1.0	41.6		14.2	7.1	21.3		3.1	1.3	4.4		8		16.5
	In	15.0	9.5	24.5	41.10	9.1	2.2	11.3	46.94	1.7	0.5	2.2	50.00	3	4.2	11.7
P. lilacinus	Un	25.1	17.3	42.4		14.8	7.9	22.7		3.2	1.5	4.7		8		16.3
	In	22.5	10.3	32.8	22.64	11.4	4.4	15.8	30.39	2.2	1.0	3.2	31.91	5	2.8	13.0
Neem cake	Un	38.7	22.0	60.7		20.4	12.1	32.5		5.5	2.1	7.6		12		19.1
+ *P. lilacinus*	In	35.5	19.1	54.6	10.04	19.2	9.6	28.8	11.38	5.0	1.6	6.6	13.15	9	1.5	17.5
Castor cake	Un	37.1	17.0	54.1		18.5	10.5	29.0		5.1	1.4	6.5		11		18.0
+ *P. lilacinus*	In	28.8	14.4	43.2	20.14	15.2	7.6	22.8	21.37	3.8	1.1	4.9	24.61	8	2.2	15.4
Neem leaf	Un	31.4	16.8	48.2		16.4	9.3	25.7		4.5	1.2	5.7		10		17.1
+ *P. lilacinus*	In	25.8	14.1	39.9	17.21	13.9	6.8	20.7	19.45	3.6	0.9	4.5	21.05	7	1.8	14.8
Castor leaf	Un	28.1	16.0	44.1		14.4	7.1	21.5		3.7	1.3	5.0		9		16.5
+ *P. lilacinus*	In	21.2	11.8	33.0	25.17	10.2	5.0	15.2	29.30	2.9	0.6	3.5	30.00	6	2.7	13.2
Rice polish	Un	33.2	17.6	50.8		15.2	8.5	23.7		4.0	1.5	5.5		10		17.0
+ *P. lilacinus*	In	26.4	14.1	40.5	20.27	12.0	6.2	18.2	23.20	3.0	1.1	4.1	25.45	7	2.2	14.4
CD (p = 0.05)				6.28				1.77				0.81		1.48	1.05	1.11
(p = 0.01)				8.49				2.39				1.09		2.00	1.43	1.50

Table 5. Combined effect of organic amendments and the fungus *P. lilacinus* on the root knot development by plant growth and water absorption of basil

Treatment		Length (cm)				Fresh weight (g)				Dry weight (g)				No. of fruits /plants	Root knot index (0–5 scale)	Water absorbed /plant (g/day)
Organic amendment *P. lilacinus*	Un/In	Shoot	Root	Total	% reduction	Shoot	Root	Total	% reduction	Shoot	Root	Total	% reduction			
Untreated	Un	38.5	20.7	59.2		15.2	3.9	19.1		8.0	1.1	9.1		15		37.5
	In	27.1	9.0	36.1	39.02	7.9	2.9	10.8	43.45	4.5	0.6	4.9	46.15	7	4.0	25.1
P. lilacinus	Un	39.0	21.1	60.1		15.0	4.2	9.2		8.2	1.1	9.3		16		37.9
	In	34.0	10.4	44.4	26.12	10.4	3.8	14.2	26.04	5.5	1.2	6.7	27.95	10	2.5	29.5
Neem cake	Un	66.0	39.0	105.0		30.6	6.9	37.5		12.6	4.1	16.7		24		44.8
+*P. lilacinus*	In	60.1	34.2	94.3	10.19	28.4	4.9	33.3	11.20	11.8	2.9	14.7		20	1.2	40.3
Castor cake +	Un	57.7	27.3	85.0		27.6	4.2	31.8		10.0	2.9	12.9		22		40.9
P. lilacinus	In	48.0	18.5	66.5	21.76	22.7	2.7	25.4	20.12	8.5	1.6	10.1	21.70	16	2.0	33.9
Neem leaf +	Un	49.3	20.0	69.3		24.2	3.4	27.8		9.8	2.1	11.9		21		39.0
P. lilacinus	In	40.1	18.0	58.1	16.16	19.9	2.6	22.5	19.06	7.9	1.7	9.6	19.32	17	1.8	32.7
Castor leaf +	Un	51.3	24.0	75.3		22.9	3.0	25.9		7.8	1.6	9.4		19		38.2
P. lilacinus	In	39.2	16.3	55.5	26.29	16.8	2.4	19.2	25.86	5.6	1.2	6.8	27.65	12	2.5	30.0
Rice pollish +	Un	60.2	30.2	90.4		23.6	3.4	27.0		10.1	2.4	12.5		21		41.5
P. lilacinus	In	48.4	20.2	68.6	24.11	17.9	2.8	20.7	23.33	8.0	1.5	9.5	24.00	15	2.2	33.6
CD (p = 0.05)				6.13				2.28				0.30		4.01	1.11	0.70
(p = 0.01)				8.29				3.08				0.40		5.43	1.50	0.95

Table 6. Combined effect of organic amendments and the fungus *P. lilacinus* on the root knot development by plant growth and water absorption of castor

Treatment		Length (cm)				Fresh weight (g)				Dry weight (g)				Leaf surface area (sq. cm/ leaf)	% reduc- tion	Root knot index (0–5 scale)	Water absorbed/ plant (g/day)
Organic amendment/ *P. lilacinus*	Un/In	Shoot	Root	Total	% reduc- tion	Shoot	Root	Total	% reduc- tion	Shoot	Root	Total	% reduc- tion				
Untreated	Un	46.2	19.0	65.2		25.1	16.5	41.6		4.0	2.0	6.0		162			35.8
	In	32.0	14.2	46.2	29.14	19.5	10.8	30.3	27.16	2.8	1.5	4.3	28.33	123	24.07	2.5	22.5
P. lilacinus	Un	45.3	19.4	64.7		24.0	17.2	41.2		3.9	2.3	6.2		164			36.0
	In	37.9	15.7	53.6	17.15	20.7	14.3	35.0	15.04	3.2	1.8	5.0	19.35	135	17.68	2.0	26.2
Neem cake	Un	55.3	22.3	77.6		32.6	18.4	51.0		8.3	2.7	11.0		210			40.2
P. lilacinus	In	50.1	21.2	71.3	8.11	30.3	16.0	46.3	9.21	7.7	2.2	9.9	10.00	192	8.57	0.8	35.3
Castor cake +	Un	50.0	20.0	70.0		30.9	16.1	47.0		7.9	2.1	10.0		183			37.1
P. lilacinus	In	41.8	17.5	59.3	15.28	25.7	14.6	40.3	14.25	6.9	1.4	8.3	17.00	160	12.56	1.3	29.6
Neem leaf +	Un	60.5	33.1	93.6		32.0	21.0	53.0		9.2	3.1	12.3		247			44.2
P. lilacinus	In	52.3	29.0	81.3	13.14	28.2	18.3	46.5	12.26	7.8	2.6	10.4	15.44	221	10.52	1.0	36.6
Castor leaf +	Un	50.5	28.5	79.0		32.3	26.7	59.0		10.6	3.4	14.0		285			47.0
P. lilacinus	In	43.5	22.3	65.8	16.70	27.6	22.3	49.9	15.42	8.5	2.9	11.4	18.57	238	16.59	1.7	35.2
Rice polish +	Un	61.0	20.0	81.0		30.0	24.0	54.0		9.2	3.7	12.9		233			41.5
P. lilacinus	In	52.0	17.1	69.1	14.69	26.8	19.3	46.1	14.62	8.2	2.6	10.8	16.27	207	11.15	1.3	32.3
CD (p = 0.05)				4.79				2.65				0.52		8.88		0.49	1.30
(p = 0.01)				6.48				3.58				0.70		12.00		0.67	1.76

organic amendments was better for nematode control and improvement in plant growth than either of the treatments singly (Table 4, 5, 6).

The root knot index was 4.2 (on 0–5 scale) in tomato and was reduced to 2.8 by *P. lilacinus*. Further reduction was noted in the presence of various organic additives along with *P. lilacinus* (Table 4). Minimum root knot index was noted in plants treated with neem cake + *P. lilacinus* (1.5) followed by combinations of *P. lilacinus* with neem leaf (1.8), castor cake, rice polish (2.2), and castor leaf (2.7).

In basil and castor root knot index was 4.0 and 2.5 respectively in untreated-inoculated plants (Tables 5, 6). It was minimum (1.2) in basil plants treated with neem cake + *P. lilacinus* and maximum (2.5) in castor leaf + *P. lilacinus* and *P. lilacinus* alone (Table 5) while it ranged from 0.8 to 2.0 in castor; the sequence of efficiency was similar to that given above (Table 6).

DISCUSSION

It is clear from the results that neem cake proved to be most beneficial for improvement in plant growth and reduction in the disease incidence on all the three test crops (Figs. 1, 2, 3). It was followed in order of efficiency by castor cake, neem leaf, rice polish, and castor leaf. When *P. lilacinus* was added in amended soil, damage to plant growth was further reduced and root galling was also less severe than in the plants treated with organic amendments alone. The sequence of efficiency of different treatments was the combination of *P. lilacinus* with neem cake, followed by neem leaf, castor cake, rice polish, and castor leaf (Figs. 1, 2, 3). Organic matter when incorporated in the soil changed the physical and chemical properties of the soil, which may be inimical to the nematodes. Organic matter itself may be toxic to the nematodes.

The high efficiency of neem products is understandable because they comprise a pool of biologically active components. The neem tree is known to contain a large number of chemicals. These include proto-meliacins, meliacins (limonoids or tetranortriterpenoids, tetranortriterpenoid-γ-hydroxybutenolides, ring C Seco-tetranortriterpenoids, and ring C Seco-tetranortriterpenoid-γ-hydroxybutenolides), pentanortriterpenoids, a hexanortriterpenoid, and non-triterpenoidal constituents (Siddiqui et al., 1988). During the decomposition of oilseeds cakes many compounds and chemicals are released, such as phenols, aldehydes, and different gases including ammonia (Khan et al., 1974; Alam et al., 1978, 1979, 1982). These chemicals have been found to be highly deleterious to many plant-parasitic nematodes (Eno et al., 1955; Walker et al., 1967; Alam et al., 1979). If absorbed by the roots, phenols increase the resistance of plants to the parasites. Predacious and parasitic activity of the soil microorganisms is increased in the presence of organic amendments. Metabolites of these microbes are also found to be toxic to plant-parasitic nematodes. Plant growth promoting substances are also released into soil which may play some role in helping plants indirectly to escape nematode attack.

The treatment with organic amendments improved water absorption capacity of roots as compared to untreated control. This appears to be due to better root and shoot development. Better root development provides more surface area for water absorption and better shoot development increases transpirational pull which in turn improves the

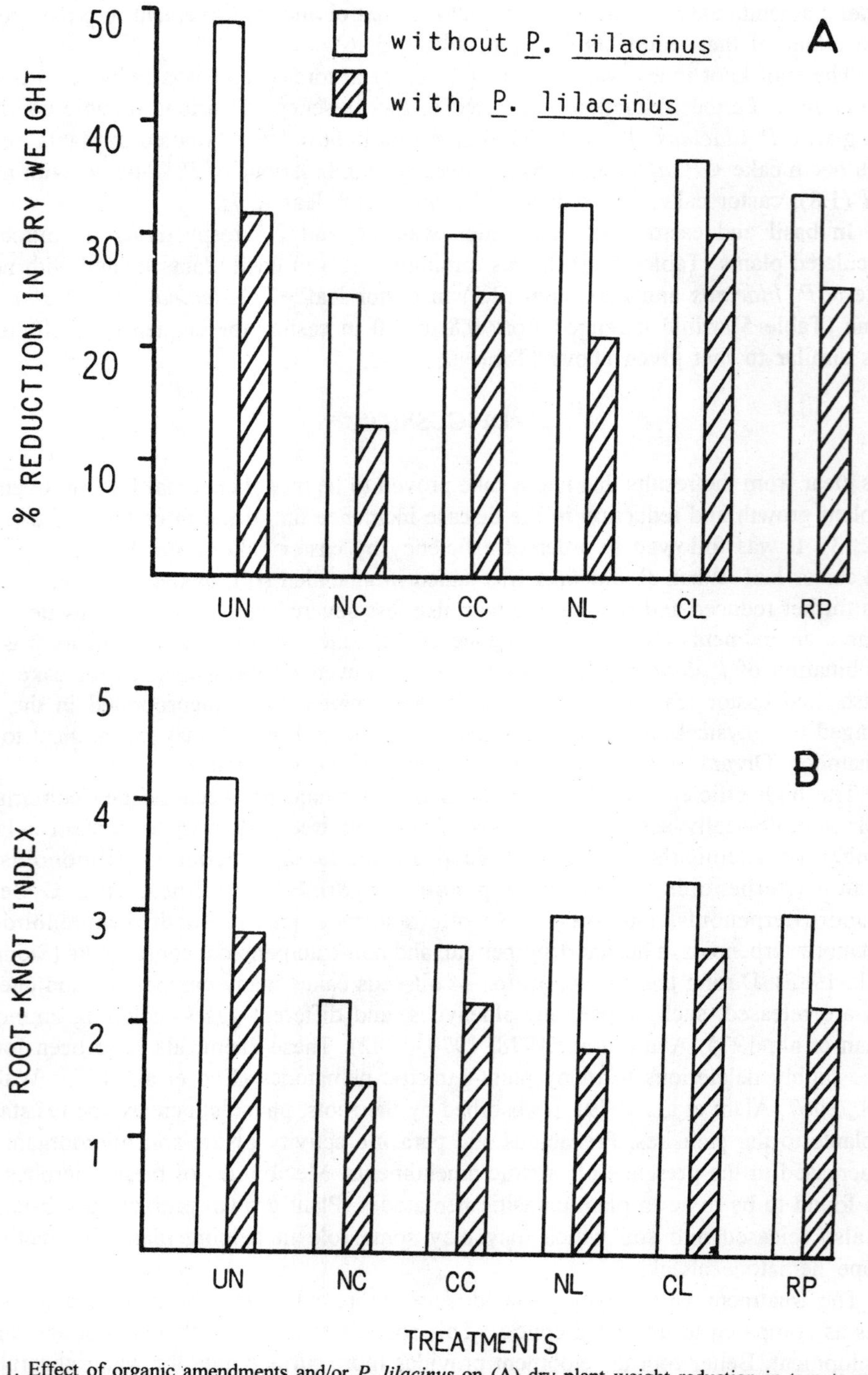

Fig. 1. Effect of organic amendments and/or *P. lilacinus* on (A) dry plant weight reduction in tomato and (B) root knot development. UN = untreated; NC = neem cake; CC = castor cake; NL = neem leaf; CL = castor leaf; RP = rice polish.

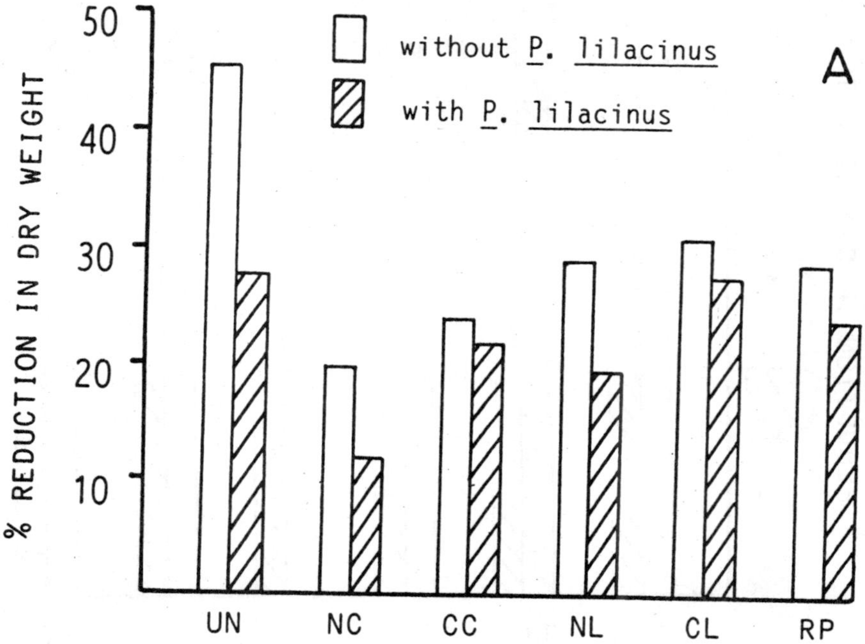

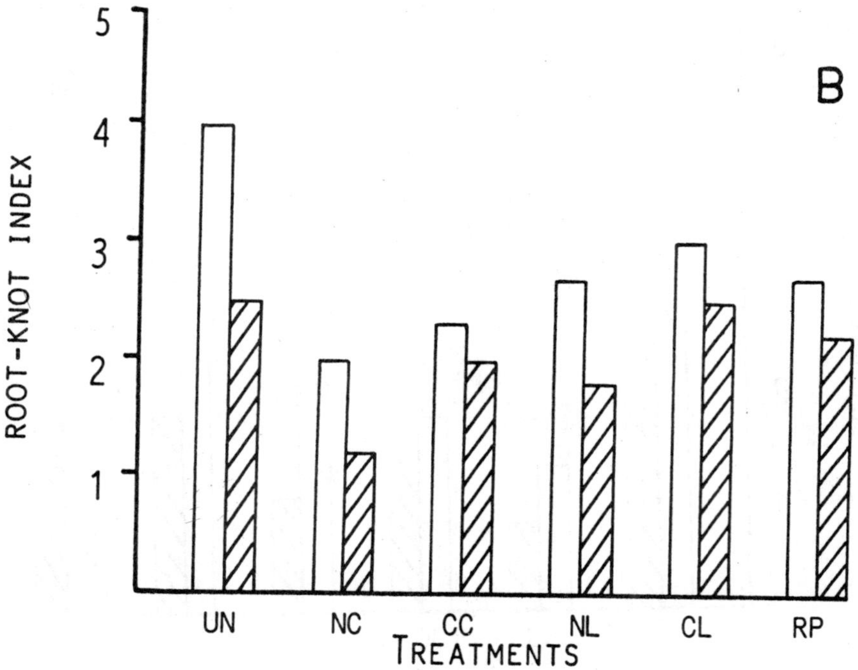

Fig. 2. Effect of organic amendments and/or *P. lilacinus* on (A) dry plant weight reduction in basil and (B) root knot development. UN = untreated; NC = neem cake; CC = castor cake; NL = neem leaf; CL = castor leaf; RP = rice polish.

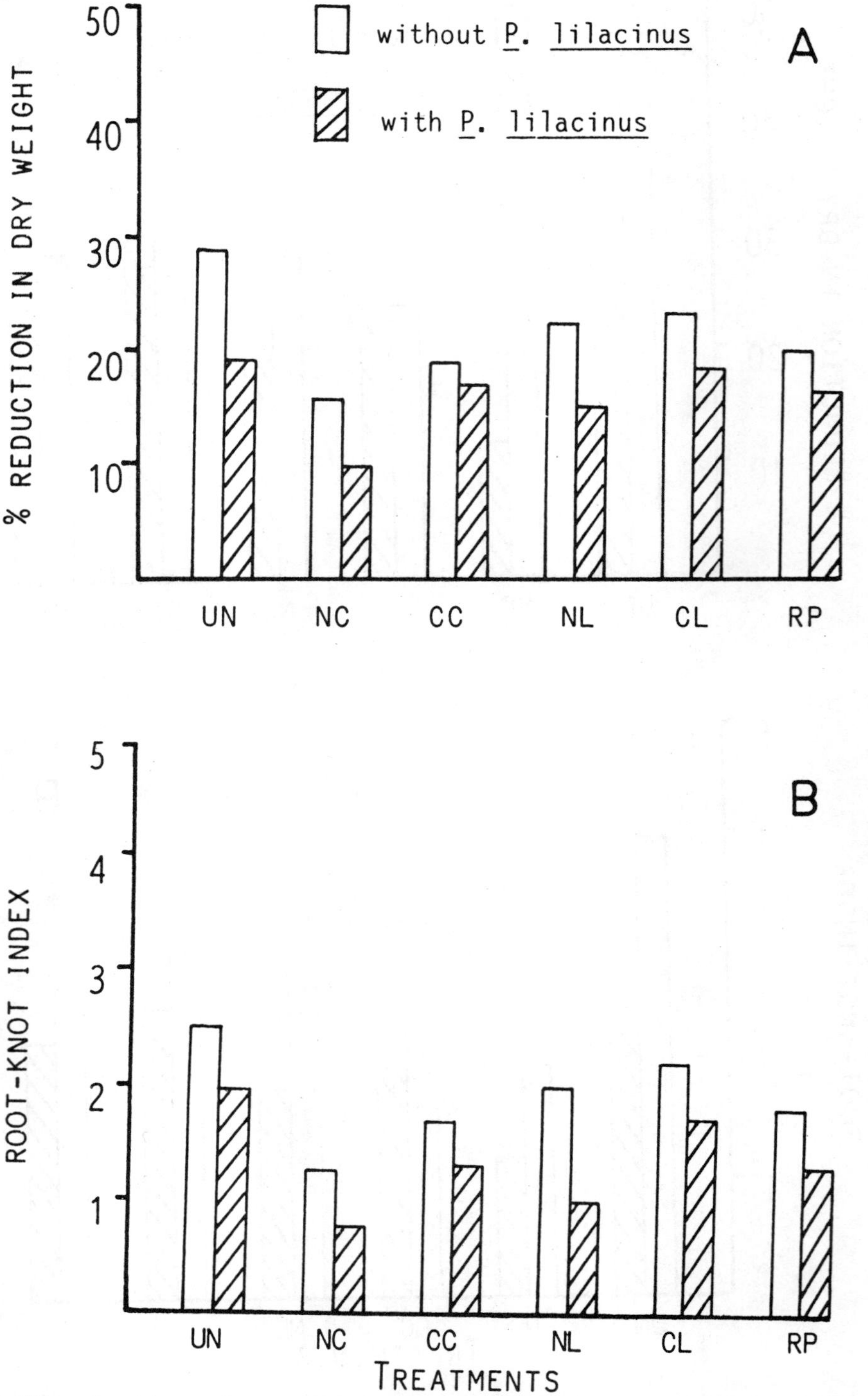

Fig. 3. Effect of organic amendments and/or *P. lilacinus* on (A) dry plant weight reduction in castor and (B) root knot developement. UN = untreated; NC = neem cake; CC-castor cake; NL = neem leaf; CL = castor leaf; RP = rice polish.

absorption capacity of roots. Moreover, inhibited multiplication of nematodes reduces damage to the conducting tissue, thereby improving the water absorption capacity.

The results of the present experiment clearly indicate that *P. lilacinus* was not pathogenic to the test plants. It was found to be effective in reducing plant damage caused by *M. incognita*. Jatala et al. (1979) and Morgan-Jones and Rodriguez-Kabana (1984) have reported that reduction in population of root knot results by way of fungal attack on larvae or death of females before egg laying, thus reducing their fecundity, and if eggs are produced these are colonized by the fungus and destroyed. When *P. lilacinus* and organic amendments were used together the level of nematode control increased. Being an opportunistic fungus, *P. lilacinus* is capable of colonizing and destroying females and eggs of nematodes and is also a good competitor of soil microbes, growing easily on organic substrates in the absence of its natural host (e.g., nematodes). Thus, it may well have first multiplied on organic substrates before attacking the nematodes. *Paecilomyces lilacinus* showed compatibility with organic matters used for soil amendments, favouring integrated management approach of plant parasitic nematodes.

REFERENCES

Alam, M.M. 1989. A simple method for estimating leaf area. *J. Indian Bot. Soc.*, 68: 427–428.

Alam, M.M., N. Hasan and S.K. Saxena. 1974. Effect of root-knot nematode, *Meloidogyne incognita* on water absorption capability of tomato roots. *Indian J. Nematol.*, 4: 244–246.

Alam, M.M., A.M. Khan and S.K. Saxena. 1978. Mechanism of control of plant-parasitic nematodes as a result of the application of organic amendments to the soil. IV. Role of formaldehyde and acetone. *Indian J. Nematol.*, 8: 172–174.

Alam, M.M., A.M. Khan and S.K. Saxena. 1979. Mehanism of control of plant-parasitic nematodes as a result of the application of organic amendments to the soil. V. Role of phenolic compounds. *Indian J. Nematol.*, 9: 136–142.

Alam, M.M., A.M. Khan and S.K. Saxena. 1982. Relative toxicity of decomposed and undecomposed oil-cakes to plant-parasitic nematodes. *Acta Bot. Indica*, 10: 124–127.

Eno, C.F., W.G. Blue and J.M. Good. 1955. The effect of anhydrous ammonia on nematodes, fungi, bacteria, and nitrification in some Florida soils. *Proc. Soil. Sci. Amer.*, 19: 55–58.

Jatala, P. 1986. Biological control of plant-parasitic nematodes. *Ann. Rev. Phytopathol.*, 24: 453–489.

Jatala, P., R. Kaltenbach and M. Bocangel. 1979. Biological control of *Meloidogyne incognita acrita* and *Globodera pallida* on potatoes. *J. Nematol.*, 11: 303.

Ketkar, C.M. 1976. *Utilization of Neem* (Azadirachta indica) and *Its By-products*. Directorate of Non-edible Oils and Soap Industry, Khadi and Village Industries Commission, Pune.

Khan, M.W., M.M. Alam, A.M. Khan and S.K. Saxena. 1974. Effect of water soluble fractions of oil-cakes and bitter principles of neem on some fungi and nematodes. *Acta Bot. Indica*, 2: 120–128.

Morgan-Jones, G. and R. Rodriguez-Kabana. 1974. Species of *Verticillium* and *Paecilomyces* as parasites of cyst and root-knot nematodes. *Phytopathology*, 74: 831.

Pansey, V.G. and P.V. Sukhatme. 1978. *Statistical Methods for Agricultural Workers*. Rev. P.V. Sukhatme and V.V. Amble. Indian Council of Agricultural Research, New Delhi, 347 pp.

Sasser, J.N., C.C. Carter and K.M. Hartman. 1984. *Standardization of Host Suitability Studies and Reporting of Resistance to Root-knot Nematodes*. Coop. Pub., Dept. Plant Pathol., N.C. State Univ. and USAID, Raleigh, North Carolina, 7 pp.

Siddiqui, S., B.S. Siddiqui, S. Faizi and T. Mahmood. 1988. Tetracyclic triterpenoids and their derivatives from *Azadirachta indica*. *J. Natural Products*, 51: 30–43.

Walker, J.T., C.H. Specht and A.U. Maurodine. 1967. Reduction of lesion nematodes in soybean meal and oil amended soil. *Plant Dis. Reptr.*, 51: 1021–1024.

75

Management of *Meloidogyne incognita* Infesting Pulse Crops through Seed Treatments with Neem Products

VIJAYALAKSHMI MOJUMDER AND S.D. MISHRA

Division of Nematology, Indian Agricultural Research Institute
New Delhi 110 012, India

ABSTRACT

Seed coating of mung bean seed with neem seed kernel and neem seed coat at 1, 5, 10, and 20 per cent (w/w) reduced penetration of second-stage juveniles of *Meloidogyne incognita* after seven days and root knot incidence at 45 days. In a microplot trial, 20 per cent dose of neem seed kernel almost completely controlled the nematode and seed coat was also effective.

Seed soaking of chickpea seed in aqueous extracts of neem cake, neem seed kernel, and seed coat also reduced penetration and development of the nematode. At 90 days, there was more than 50 per cent reduction in number of galls in almost all treatments at both S and S/2 doses.

INTRODUCTION

The nematicidal efficacy of neem and management of nematode crop pests through neem products are well established (Mojumder and Mishra, 1993). The use of neem products as seed treatments may prove more economical than any other method of nematode management. Seed coating of chickpea with neem seed kernel and seed coat was effective against root knot nematode, *Meloidogyne incognita* (Mojumder and Mishra, 1991a), while seed soaking in aqueous extracts of neem products was effective against penetration of *M. incognita* in chickpea (Mojumder and Mishra, 1991b) and mungbean (Mojumder and Mishra, 1992, 1993).

Root knot nematode is one of the major problems of pulse crops in India and causes a general yield loss of 10–12 per cent. The present investigations were designed to manage the infestation of *M. incognita* by using neem products as seed treatments through seed coating and seed soaking.

MATERIALS AND METHODS

(A) Seed Coating of Mung bean

Seed coating trials were conducted with neem seed kernel and neem seed coat in mung bean (*Vigna radiata* cv PS 16). Seeds of mung bean were coated with dried, powdered neem seed kernel and seed coat at four doses—1, 5, 10, and 20 per cent (w/w) using gum as adhesive and chalk powder for drying. With these treated seeds, three experiments—preliminary trial, greenhouse trial, and microplot trial were conducted.

In a preliminary trial, four treated seeds were sown in respect of each treatment in icecream cups having soil infested with root knot nematode (four second-stage juveniles/g soil). For check, seeds without any treatment were sown. All the treatments and check were replicated thrice. Germination was noted up to seven days, after which only one plant was allowed to grow. Observations on penetration of second-stage juveniles of *M. incognita* were taken after seven days.

The greenhouse trial was conducted in 20 cm diameter earthen pots containing infested soil (four second-stage juveniles/g soil). Four treated seeds per pot were sown and untreated seeds were sown in check. All the treatments were replicated thrice. After the germination only one plant was allowed to grow. Observations on nematode galling were recorded after 45 days of sowing.

The third experiment was conducted in microplots of 0.1 sq m size, infested with root knot nematode at 1 second-stage juvenile/g soil. Seed coating of mung bean was done with only the highest dose of neem seed kernel and seed coat, i.e., 20 per cent (w/w). As in the previous experiment, untreated seed was used for check. All the treatments were replicated six times. Fifteen seeds were sown in two rows in each plot but after germination only 10 plants per row were kept. Final observations on the development of nematode were taken after 90 days.

(B) Seed Soaking of Chickpea

Seed soaking trials were conducted by soaking chickpea (*Cicer arietinum* cv Pusa 240) seed for 6 hours in aqueous extracts in two concentrations of neem cake, seed kernel, and seed coat. The standard aqueous extracts (S) of these materials were prepared by soaking 25 g in 100 ml water for 24 hours and then filtering through muslin cloth. The other concentration of S/2 for all three extracts were prepared by adding water to standard extracts.

RESULTS

(A) Seed Coating of Mungbean

A perusal of the data obtained from three trials shows that seed coating was effective against penetration and development of *M. incognita* without affecting seed germina-

Table 1. Effect of seed coating of mungbean seed with neem seed kernel and seed coat against *M. incognita*

Treatments	Doses (% w/w)	Penetration after 7 days		Root knot galling after 45 days	
		No. of juveniles/ plant	% reduction over check	No. of galls/ plants	% reduction over check
Neem seed kernel	1	13.3	36.9	22.3	31.2
	5	9.3	52.7	21.0	34.5
	10	5.7	68.5	18.3	43.8
	20	2.0	89.5	13.3	59.5
Neem seed coat	1	15.0	21.1	25.3	21.9
	5	13.0	36.9	23.0	28.2
	10	9.7	47.4	20.0	37.5
	20	6.0	68.5	15.6	50.0
Check		19.0		32.3	
CD at 5%		2.59		2.92	

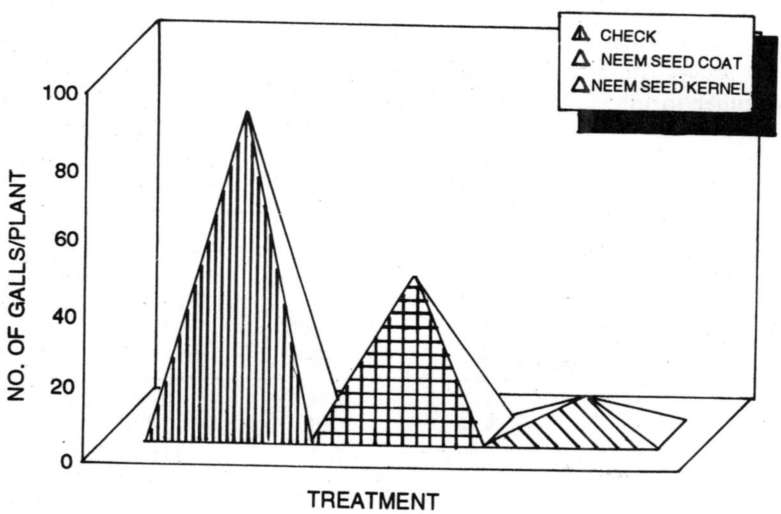

Fig. 1. Effect of neem seed kernel and neem seed coat as seed coating of mungbean on the number of root knot galls.

tion (Table 1, Fig. 1). There was significant reduction in penetration in the preliminary trial. In the greenhouse trial, there was reduction in number of root knot galls in both the treatments. The percentage reduction increased with increasing dosage. In the microplot trial, neem seed kernel showed very small and few galls and the root system was healthy, while seed coat also resulted in smaller and fewer galls than the check, which had large and multiple root knot galling. The reduction in number of root knot galls was 88.6 and 51.0 per cent respectively in neem seed kernel and seed coat at 20 per cent (w/w).

(B) Seed Soaking of Chickpea

In the chickpea experiments, seed soaking in aqueous extracts of neem cake, seed kernel, and seed coat at S and S/2 concentrations for six hours reduced penetration of second-stage juveniles and development of *M. incognita* in roots, without affecting seed germination (Table 2). Maximum reduction in number of juveniles penetrated was observed in neem cake at higher dose. In the greenhouse trial, maximum reduction in number of galls was obtained in higher dose of neem cake while other treatments also significantly reduced the number of galls.

DISCUSSION

Seed treatments with neem cake, seed kernel, and seed coat as seed coating or soaking are reported to be effective against *M. incognita* in both mungbean and chickpea (Mojumder and Mishra, 1991a, b, 1992). The present studies also confirm the earlier studies that seed coating or seed soaking of mungbean and chickpea reduce root knot incidence either in infested soil or in later inoculation (Fig. 2).

Table 2. Effect of seed soaking of chickpea in aqueous extracts of neem cake, seed kernel and seed coat against *M. incognita*

Treatments	Doses	Penetration after 7 days		Root knot galling after 45 days	
		No. of juveniles/ plant	% Reduction over check	No. of galls/ plant	% Reduction over check
Neem cake	S	5.00	82.6	10.0	71.0
	S/2	9.70	66.3	13.8	60.0
Neem seed kernel	S	6.70	76.8	12.0	65.2
	S/2	10.00	65.1	16.0	53.6
Neem seed coat	S	12.70	55.8	13.8	60.0
	S/2	14.00	51.2	17.5	48.6
Check		28.70		34.5	
CD at 5%		2.43		2.23	

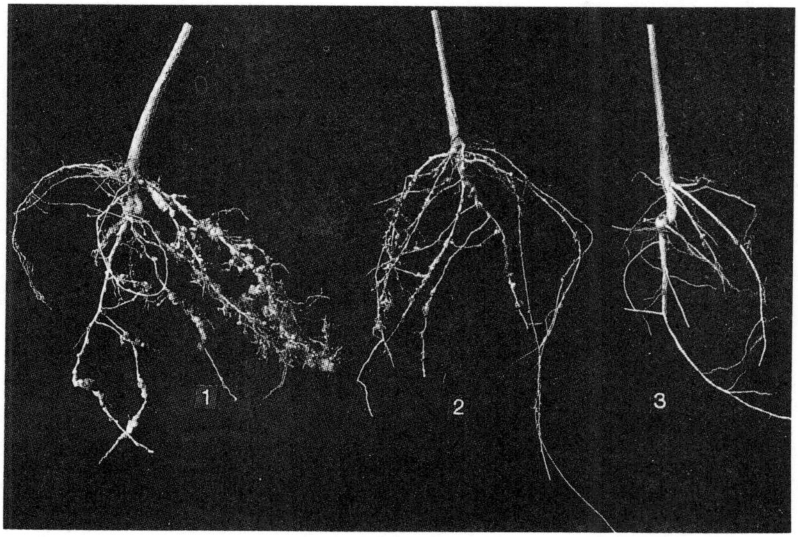

Fig. 2. Effect of neem seed kernel and neem seed coat as seed treatments on the development of root knot galls on mungbean 1. Check, 2. Neem seed coat, 3. Neem seed kernel.

REFERENCES

Mojumder, V. and S.D. Mishra. 1991. Seed treatment of chickpea with powdered neem seed kernel and cover for the management of root knot nematode, *Meloidogyne incognita*. Abstract. Souvenir and Abstracts, *Nat. Sem. on Agriculture in Nineties: Challenges and Research Needs*, IARI, New Delhi, Oct. 22–24, pp. 49–50.

Mojumder, V. and S.D. Mishra. 1991b. Nematicidal efficacy of some plant products and management of *Meloidogyne incognita* in pulse crops by soaking seeds in their aqueous extracts. *Curr. Nematol.* 2: 27–32.

Mojumder, V. and S.D. Mishra. 1992. Effect of aqueous extracts of neem seed as seed soaking on germination of mung bean and penetration of second-stage juveniles of *Meloidogyne incognita. Ann. Agric. Res.* 13: 297–299.

Mojumder, V. and S.D. Mishra. 1993. Management of nematode pests. In: I.A.R.I. Research Bulletin No. 40 "*Neem in Agriculture*", B.S. Parmar and R.P. Singh (eds.), New Delhi, pp. 40–48.

76

Neem Cake for Management of Root Knot Disease in Green Gram and Tomato Nursery

D.J. PATEL, S.K. PATEL AND H.V. PATEL

Department of Nematology, Gujarat Agricultural University, Anand Campus
Anand 388 110, Gujarat, India

ABSTRACT

Experiments were conducted to study bioefficacy of the amendments with castor, piludi and neem cakes and poultry and farm yard manures each at 0.5 and 1.0 t/ha along with controls with normal fertilization (20:40:0 NPK kg/ha), and no fertilization, in green gram and bioefficacy of azolla (*Azolla pinnata* L.), neem, mustard, karanj and piludi cakes each at 1 and 2 t/ha along with controls with normal fertilization (180:0:0 NPK kg/ha) and no fertilization in tomato nursery in root knot nematodes infected soils.

Pooled observations recorded on plant height, grain yield and root knot disease on green gram indicated neem cake applied at 1.0 t/ha caused significant increase in grain yield by 30.3 per cent with 22.3 per cent reduction in root knot disease over control with normal fertilization, giving net realization of Rs. 1806/ha. Next best was castor cake at 1.0 t/ha which gave 24.7 per cent higher grain yield with 21.8 per cent reduction in root knot disease, giving net realization of Rs. 1266/ha.

Pooled observations on number of transplantable seedlings and root knot index on tomato indicated that piludi, neem and mustard cakes applied a week prior to tomato seeding increased number of transplants by 37.7, 44.9 and 22.5 per cent and decreased root knot disease by 27.3, 31.7 and 30.8 per cent at 1.0 t/ha, whereas application of these cakes at 2.0 t/ha increased number of transplants by 62.3, 12.3 and 76.8 per cent and decreased root knot disease by 33.7, 25.0 and 23.4 per cent over control respectively with normal fertilization.

INTRODUCTION

Plant parasitic nematodes cause 12.3 per cent loss in crop yields amounting to US $ 77.7 billion world wide. The tomato crop accounts for 20.6 per cent loss (Sasser and Freckman, 1987). Bhatti and Jain (1977) estimated 46 per cent yield loss due to *M.*

incognita in tomato crop in Hissar. Green gram yield due to infection of *Rotylenchulus reniformis* in combination with *M. incognita acrita* was reduced by 24 per cent (Castillo et al., 1977). Loss as high as 60 per cent in green gram yields was also recorded (Anon., 1985). At Anand, pooled data of five years on yield loss study in green gram due to root knot nematodes indicated 12.7 per cent loss (Anon., 1989–91). Several systemic and non-systemic chemicals tried for management of root knot disease in crops (Prasad, 1978; Patel et al; 1981; Anon., 1987) were effective and increased crop yields, but they are either unavailable or hazardous to health and cumbersome in application, leaving residue problem too. Therefore, organic amendments are now used for management of plant parasitic nematodes. Besides their nematicidal properties, they enrich soil fertility. Linford and Oliveira (1938) demonstrated for the first time the reduction of damage in plants caused by plant-parasitic nematodes through decomposition of organic matter. Singh and Sitaramaiah (1989); Bora and Phukan (1983) and Jagdale et al. (1985) reported that decomposition of organic matter brings about physical, chemical and microbial changes in soil which are generally beneficial for managing plant-parasitic nematodes. The present study was conducted to locate an effective and cheap organic amendment for control of root knot nematodes in green gram and tomato nursery.

MATERIALS AND METHODS

Green Gram

Castor (*Ricinus communis* L.), Piludi (*Salvadora persica* L.), and neem (*Azadirachta indica* A. Juss.) cakes and poultry and farmyard manures each at 0.5 and 1.0 t/ha were tested in soils (course sand 1.3 per cent, fine sand 63 per cent, silt 15 per cent and clay 20 per cent) infested with root knot nematodes at 228 second-stage juveniles per 200 g, soil along with controls with normal fertilization (20:40:0 NPK kg/ha) and no fertilization. Amendments were applied under the crop row 10 days prior to seeding. Green gram cv Gujarat 2, susceptible to root knot, was sown at 20 kg/ha 45 cm apart in a 3.00 × 2.25 m plot. Randomized block design with three replications was used. Observations on plant height, gain yield and root knot index (1–5 scale) were recorded and data were statistically analysed.

Tomato Nursery

Five organic amendments azolla (*Azolla pinnata* L.), neem, mustard (*Brassica juncea* Coss), karanj (*Pongamia glabra* vent.), and piludi cakes—each at 1 and 2 t/ha along with controls with normal fertilization (180:0:0 NPK kg/ha) and no fertilization, were tried for control of root knot disease and production of tomato transplants. Amendments were incorporated in soil infested with root knot nematodes (100 second-stage juveniles per 200 g soil) prior to tomato seeding in 1.2 × 1.2 m bed. Treatments were replicated three times in randomized block design. Root knot susceptible tomato cv Pusa Ruby was sown at 20 kg/ha. Recommended practices were adopted for raising tomato nursery. Observations were recorded on plant height, number of transplantable seedlings and root knot index (1–5 scale) and data were statistically analysed using appropriate procedures.

RESULTS AND DISCUSSION

Green Gram

Data pooled for three years, 1988 through 1990 kharif season, revealed that all the organic amendments tried improved plant height, increased crop yield and reduced root knot disease over controls with normal fertilization and no fertilization (Table 1). Castor, piludi and neem cakes as well as poultry manure significantly increased plant height and thereby the grain yield of green gram over other treatments including controls. For increased yield and root knot control, neem cake applied at 1.0 and 0.5 t/ha gave maximum of 30.3 and 20.0 per cent higher grain yield with reduction of root

Table 1. Effect of organic amendments for management of root knot nematodes in green gram field (pooled for 1988 to 1990)

Organic amendment (Dose, t/ha)	Plant height (cm)	Yield (kg/ha)	Root knot index (1–5) *
Castor cake 0.5	69.9	972 (+ 21.3)	1.62 (– 13.8)
Castor cake 1.0	73.5	999 (+ 24.7)	1.47 (– 21.8)
Piludi cake 0.5	70.3	961 (+ 20.0)	1.73 (– 8.0)
Piludi cake 1.0	70.9	913 (+ 14.0)	1.62 (–13.8)
Neem cake 0.5	68.8	961 (+ 20.0)	1.45 (–22.9)
Neem cake 1.0	67.9	1044 (+ 30.3)	1.46 (–22.3)
Poultry manure 0.5	70.4	954 (+ 19.1)	1.50 (–20.2)
Poultry manure 1.0	71.8	921 (+ 15.0)	1.59 (–15.4)
Farmyard manure 0.5	66.8	903 (+ 12.7)	1.65 (–12.2)
Farmyard manure 1.0	66.3	873 (+ 9.0)	1.58 (–16.0)
Normal fertilization (20:40:0 NPK kg/ha)	65.3	801	1.88
Control (no cake/no fertilization)	65.4	735	1.91
S Em	1.6	54	0.07
C D 0.05	4.4	149	0.20
Year × treatment	NS	NS	NS
Year effect	s	s	s
CV %	7.3	17.2	13.2

* 1 = free 5 = maximum disease intensity.

Figures in parentheses indicate per cent increase (+) or decrease (–) over normal fertilization.

Table 2. Effect of organic amendments on production of transplantable seedlings and control of root knot disease in tomato nursery (pooled for 1987–1989)

Organic amendment (dose, t/ha)	Seedling height (cm)	No. of trans-plantable seedlings/ m^2 Log x trans.	% increase over normal fertilization	Root knot index (1–5), * $\sqrt{x}$ trans.	% decrease over normal fertilization
Dry Azolla 1 (*Azolla pinnata* L.)	19.0	2.2510 (178)	29.0	1.61 (2.60)	16.6
Dry Azolla 2	19.5	2.1850 (153)	10.9	1.45 (2.09)	33.0
Neem cake 1	17.9	2.3020 (200)	44.9	1.46 (2.13)	31.7
Neem cake 2	18.0	2.1910 (155)	12.3	1.53 (2.34)	25.0
Mustard cake 1	20.2	2.2280 (169)	22.5	1.47 (2.16)	30.8
Mustard cake 2	21.6	2.3870 (244)	76.8	1.55 (2.39)	23.4
Karanj cake 1	19.7	2.2810 (191)	38.4	1.55 (2.39)	22.7
Karanj cake 2	20.8	2.2010 (191)	38.4	1.64 (2.70)	13.5
Piludi cake 1	23.6	2.2770 (190)	37.7	1.51 (2.27)	27.3
Piludi cake 2	19.6	2.3510 (224)	62.3	1.44 (2.07)	33.7
Normal ferti-lization (180:0:0 NPK kg/ha)	18.0	2.1390 (138)		1.77 (3.12)	
Control (no cake/no fertilizer)	14.8	2.1060 (128)		1.84 (3.37)	
S Em	1.3	0.05		0.07	
C D 0.05	3.6	0.13		0.19	
Year x treat-ment	ns	ns		ns	
Year effect	s	s		s	
C V %	20.2	6.1		13.1	

* 1 = free; 5 = maximum disease intensity.
Figures in parentheses are retransformed values.

knot disease by 22.3 and 22.9 per cent respectively over normal fertilization. Next best in efficacy was castor cake at 0.5 and 1.0 t/ha, which gave 21.3 and 24.7 per cent increased yield with 13.8 and 21.8 per cent reduction in root knot disease. Other organic amendments, though better than control, had mediocre effect on yield and root knot control. Controls had minimum plant height with least grain yield and highest root knot disease.

Tomato Nursery

All organic amendments showed significant influence on plant height, production of transplantable seedlings and reduction of root knot disease over controls with normal fertilization and no fertilization (Table 2). Mustard, piludi, karanj and neem cakes each at 1 and 2 t/ha were more effective than dry azolla. Mustard cake at 2 t/ha produced 76.8 per cent more transplantable seedlings followed by piludi cake at 2 t/ha (62.3 per cent), neem cake at 1.0 t/ha (44.9 per cent), karanj cake at 1 and 2 t/ha (38.4 per cent each) and piludi cake at 1.0 t/ha (37.7 per cent) over normal fertilization. Azolla and neem cake at 2 t/ha, and mustard cake at 1.0 t/ha, though better than controls, were mediocre in effect on production of transplantable seedlings. For root knot control, piludi cake at 2 t/ha gave maximum reduction of 33.7 per cent over control with normal fertilization followed by dry azolla at 2 t/ha (33.0 per cent), neem cake at 1 t/ha (31.7 per cent), mustard cake at 1 t/ha (30.8 per cent) and piludi cake at 1.0 t/ha (27.3 per cent). Least production of transplantable seedlings with highest root knot disease was recorded in control treatments. Thus, neem cake was observed to be more effective for control of root knot disease in green gram than in tomato nursery. An increase in shoot and root weights by amendment of the soil with neem cake was recorded in jute by Jagdale et al. (1985), in french bean by Padhi and Mishra (1987), and in okra and tomato by Singh and Sitaramaiah (1989). Similarly poultry manure at 1 or 2 t/ha was also found effective for control of root knot nematodes on green gram by Bora and Phukan (1992) and on tomato by Babatola (1989).

REFERENCES

Anonymous. 1985–86. Consolidated Biennial Report, AICRP on Plant Parasitic Nematodes with Integrated Approach for Their Control, ICAR, New Delhi, pp. 28–29.

Anonymous. 1987. Biennial Report. AICRP on nematodes, Gujarat Agricultural University, Anand Centre, pp. 36–55.

Anonymous. 1989–91. Consolidated Biennial Report, AICRP on Plant Parasitic Nematodes with Integrated Approach for Their Control, ICAR, New Delhi, pp. 22–24.

Babatola. J.O. 1989. Effect of some organic manures on nematodes in tomato cultivation. *Pak. J. Nematol.* 7: 39–46.

Bhatti D.S. and R.K. Jain. 1977. Estimation of loss in okra, tomato and brinjal yields due to *Meloidogyne incognita*. *Indian J. Nematol.* 7: 37–41.

Bora B.C. and P.N. Phukan. 1983. Organic amendment for the control of root knot nematode *M. incognita* on jute. *J. Res. Assam Agric. Univ.* 4: 50-54.

Bora, A. and P.N. Phukan. 1992. Effect of different organic amendments for the management of *Meloidogyne incognita* on green gram. *Currt. Nematol.* 3: 49–50.

Castillo, M.B., M.S. Alejar and J.A. Litsinger. 1977. Pathological reactions and yield loss of mungbean to known population of *Rotylenchulus reniformis* and *Meloidogyne incognita* acrita. *Philippines Agriculturist* 61: 12–24.

Jagdale, G.B., A.B. Pawar and K.S. Darekar. 1985. Effect of organic amendments on root knot nematodes infecting betelvine. *Int. Nematol. Network Newsletter* 2: 7–10.

Linford, N.B.F. and J.M. Oliveira. 1938. Reduction of soil populations of the root knot nematodes during decomposition of organic matter. *Soil Sci.* 45: 127–140.

Padhi, N.N. and R.P. Mishra. 1987. Control of *Rotylenchulus reniformis* on french bean (*Phaseolus vulgaris*) *Indian J. Nematol.* 17: 130–131.

Patel, G.J., H.M. Shah, D.J. Patel and G.B. Valand. 1981. Effective duration of some proven nematicides in control of root knot disease in bidi tobacco nursery. *Indian J. Nematol.* 11: 74–77.

Prasad D. 1978. Studies on DBCP in relation to diffusion, persistence in soil and uptake by plants. Ph.D. Thesis, IARI, New Delhi.

Sassar, J.N. and D.W. Freckman. 1987. A world perspective on nematology: The role of the society. In: *Vistas on nematology*. J.A. Veech and D.W. Dickson. (eds.), Society of Nematologists, Inc. Hyattsville, Maryland, pp. 7–14.

Singh R.S. and K. Sitaramaiah. 1989. Control of root knot nematodes through organic and inorganic amendments of soil. 1, Effect of oil cakes and saw dust. All India Nematology Symp. IART, New Delhi, 64 pp.

7

PLANT DISEASE MANAGEMENT

77

Neem and Control of Aflatoxin Contamination

HAMPDEN J. ZERINGUE, JR. AND DEEPAK BHATNAGAR

USDA/Agricultural Research Service
Southern Regional Research Center
New Orleans, Louisiana 70124, USA

ABSTRACT

Aflatoxins are carcinogenic secondary metabolites of the fungi *Aspergillus flavus* and *A. parasiticus*. The effect of neem (*Azadirachta indica*) extracts on *Aspergillus* growth and aflatoxin biosynthesis was investigated. The leaf extracts were prepared by blending 100 g (wet weight) of fresh leaves in one liter of water. The formulation (1 to 40 per cent, v/v), when added to fungal growth medium prior to inoculation, did not affect fungal growth but essentially blocked aflatoxin biosynthesis in both *A. flavus* (100 per cent) and *A. parasiticus* (> 95 per cent) at concentrations greater than 10 per cent (v/v). Neem seed aqueous extracts were not as effective as the leaf extracts, whereas neem oil had no effect. Injection of the neem leaf extract followed by an *A. flavus* spore suspension (48 hrs later) onto the surface of locks of developing cotton bolls (30-day post-anthesis) did not affect fungal growth in the bolls, but the seeds from the locules exhibited > 98 per cent inhibition in aflatoxin production. *In vitro* studies with the fungi suggest that non-volatile neem leaf constituents inhibit aflatoxin biosynthesis in the early stages of the biosynthetic pathway. The neem-mediated inhibition appears to involve regulation of secondary metabolism, because once secondary biosynthesis was initiated the inhibitory effect of the neem leaf constituents was lost. Practical applications of this observation are being developed to eliminate the preharvest contamination of crops with aflatoxin.

INTRODUCTION

Aflatoxins are extremely potent, naturally occurring carcinogens. They occur in feed for livestock and in food for human consumption. The two aflatoxigenic fungi, *Aspergillus flavus* and *A. parasiticus*, can grow and produce aflatoxins on a number of substrates, but aflatoxin contamination is a serious concern on diverse substrates such as corn, peanuts, cottonseed, and tree nuts. *Aspergillus flavus* appears to be the primary aflatoxin-producing fungus on these commodities; *A. parasiticus* also occurs frequently on peanuts. Both fungi produce a family of related aflatoxins; those most commonly produced by *A. flavus* are B1 and B2 (Fig. 1) and *A. parasiticus* produces two additional aflatoxins, G1 and G2. Aflatoxin B1 is the most carcinogenic of the aflatoxins and is thus receiving the most attention in mammalian toxicology. The aflatoxin family of compounds can form 'adducts' with animal and human DNA (Hsu et al., 1991; Bressac et al., 1991) and can cause primary liver cancer in certain animal systems (Groopman and Sabbioni, 1991; Wogan, 1991). The carcinogenicity of aflatoxins in human systems is less clear (Campbell et al., 1990), but certain associations between aflatoxin intake by human populations and primary liver cancer have been reported (Groopman and Sabbioni, 1991; Yeh et al., 1989). Although aflatoxin contamination occurs worldwide (Jelinek et al., 1989), the extent of contamination in many countries is unknown because of reluctance to report its occurrence. With the implications to human and animal health worldwide, intense efforts are underway to remove these compounds from animal and human food supplies. The US Food and Drug Administration (FDA) prohibits interstate commerce of dairy feed grain containing more than 20 ppb aflatoxin and the sale of milk containing more than 0.5 ppb aflatoxin.

Several conventional agronomic practices reduce preharvest aflatoxin contamination of crops; these include use of pesticides, altered cultural practices (such as irrigation), and use of resistant crop varieties (Widstrom, 1987; Scott and Zummo, 1988; Darrah and Barry, 1991; Lillehoj, 1991). However, such procedures have exhibited only a limited potential for reducing aflatoxin levels in the field. Detoxification (Park et al., 1988) and absorptive removal (Phillips et al., 1991) of aflatoxins from already contaminated foods and feeds are two promising methods currently under extensive investigation; these should prove to be important control measures for immediate application, since less progress has been made in the development of practical methods to prevent

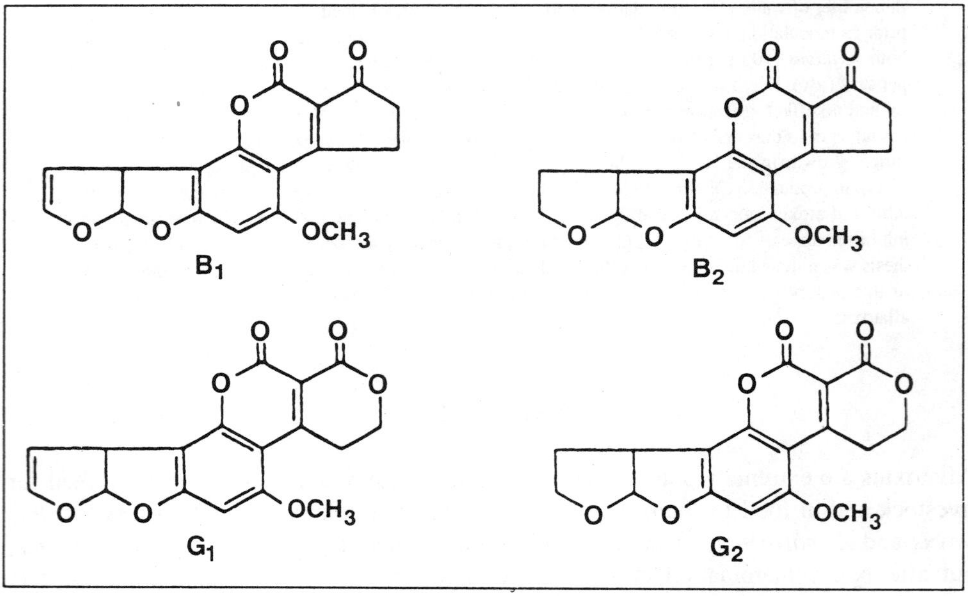

Fig. 1. Chemical structures of the major aflatoxins B1, B2, G1 and G2.

preharvest aflatoxin contamination of foods and feeds. However, prevention is probably the best long-term approach; this strategy would obviate the need to detoxify large quantities of aflatoxin-contaminated seed material and avoid the uncertainties of gaining approval from regulatory agencies for the use of detoxified seed for animal or human food.

Growers and food processors worldwide are being placed under increased pressure from consumer groups, merchants, and regulatory agencies to eliminate mycotoxins from food and feed. Therefore, there is an increasing need to develop new technology to reduce and eventually eliminate preharvest aflatoxin contamination. Elimination of preharvest contamination might be achieved with the development of novel biotechnological approaches (Cleveland and Bhatnagar, 1992; Payne, 1992). To develop such approaches, additional knowledge is required in two broad areas: (1) fundamental molecular and biological mechanisms that regulate the biosynthesis of aflatoxin by the fungus and the ecological and biological factors that influence toxin production in the field and (2) biochemistry of host-plant resistance to aflatoxin and/or aflatoxigenic moulds. Knowledge in these areas will aid in development of novel methods to manipulate the chain of events in aflatoxin contamination; extensive research in these areas is being conducted in this lab and elsewhere (for reviews see Bhatnagar et al., 1991; Keller et al., 1991; Cleveland et al., 1990).

There is also a need to identify short-term strategies for managing aflatoxin contamination at a nominal cost. Plant products, like neem extracts, could be used in design of ecologically safe 'soft' pesticides to achieve this goal. Neem, *Azadirachta indica* A. Juss., is a subtropical tree native to the drier areas of Pakistan, India, Indonesia, Thailand, Burma, Sri Lanka, Malaysia, and Africa. Several tetranortriterpenoids, mainly found in neem seeds, have been described to be quite active as feeding deterrents, toxicants, and/or disruptants of growth and development against a large variety of insect species and nematodes (Koul et al., 1990). Neem components are also reputed for their medicinal, spermicidal, antiviral, antibacterial, antiprotozoal, and antifungal properties (Singh et al., 1980; Jacobson, 1989; National Research Council, 1992). Several active principles from different parts of the neem tree have been reported. The neem leaves are known to contain desactylimbin, quercetin, and sitosterol. Neem oil yields various acids, sulphur, and other products (for review see National Research Council, 1992; Koul et al., 1990).

The current study was conducted to assess the inhibitory effect of neem components on *A. flavus* and *A. parasiticus* growth and aflatoxin synthesis by these fungi both in submerged culture and in preharvest cottonseed.

MATERIALS AND METHODS

Fungal Cultures

Wild-type aflatoxigenic isolates of *A. flavus* (SRRC 1100) and *A. parasiticus* (SRRC 143) were cultured on potato dextrose agar (PDA) petri plates. The spores were harvested after seven days' incubation at 28°C, and spore suspensions of these cultures (10^6 spores/ml in sterile, distilled water with 1 per cent Triton X-100) were used in the submerged culture and cotton boll inoculation studies.

Neem Extract Preparations

Fresh neem leaves were obtained from Dr. Robert Knight, ARS/USDA Subtropical Horticulture Research Station, Miami, Florida. After the leaves were washed thoroughly with sterile, distilled water, extracts were prepared by either blending or boiling 100 g (wet weight) of fresh leaves in one litre sterile, distilled water. Extracts were filtered through several layers of cheesecloth, and the filtrate was centrifuged for 15 min. at 7000 × g. The pellet was discarded, the volume adjusted to one litre, and the supernatant extract stored at 4°C. Soluble extracts were sterilized by passing through a millipore filter (0.22 μm pore size). Also, one-half of both leaf formulations (blended and heat-extracted) was autoclaved to achieve sterilization.

Neem seed extracts were prepared similarly to leaf extract except that 10 g of powdered seed was used per 100 ml of water. Neem oil and seeds were generous gifts of J.C. Locke, USDA/ARS/Florist & Nursery Crops Lab, Beltsville, Maryland, and J.F. Walter, W.R. Grace & Co.-Conn, Columbia, Maryland, respectively.

Fungal Treatments in Submerged Cultures and Metabolite Extractions

The fungus was grown on growth medium (Adye and Mateles, 1964) by inoculation with 0.1 ml of the spore suspension (10^6 spores/ml) to 100 ml of the medium containing 0 to 40 per cent v/v neem leaf formulations in 250 ml flasks. The flasks were incubated for two to four days on a shaker incubator (Lab-Line Instruments, Inc.) at 150 rpm and 28°C. Growth of the fungus was recorded after four days by harvesting the mycelial pellets through filtration on oven-dried Whatman No. 42 filter papers. The filter papers containing the fungal growth mass were oven-dried at 70°C for 24 hrs and the dry weight of the fungus was determined as an index of fungal growth. For the study of the conversion of secondary metabolites to aflatoxins, mycelia were washed thoroughly and transferred to low sugar replacement media (LSRM) as described earlier (Bhatnagar et al., 1987). After the desired incubation, mycelial pellets and media were extracted with aqueous acetone, followed by methylene chloride (Bhatnagar et al., 1987). Aflatoxins were separated on silica gel thin-layer chromatographic (TLC) plates in ether:methanol:water (96:3:1, v/v/v). The toxins were quantitated by fluorometric scans (360 nm) of TLC plates containing the extracted samples and comparison with aflatoxin standards run on the same plate (Bhatnagar et al., 1987). Aflatoxin precursors from the mycelia were identified by the HPLC procedure described by McCormick et al. (1988).

Determination of Volatile Components of Neem Leaves

Volatiles from neem extracts were collected onto Tenax GC (60–80 mesh) tubes as described earlier (Bhatnagar and McCormick, 1988). The volatiles absorbed on the Tenax tubes were heat-desorbed onto a capillary chromatographic column and were analysed with a Finnigan MAT GC/MS 4000 system interfaced with an external closed inlet device (Bhatnagar and McCormick, 1988). Data acquisition and analysis were accomplished with a Finnigan-Incos data system and the compounds were identified on the basis of computer-assisted library searches.

Effect of Volatile Components of Neem Leaves

Activity of volatiles from the blended neem leaf extracts on growth and aflatoxin production in *A. flavus* was determined on PDA plates (Fig. 2) by the procedure described earlier (Bhatnagar and McCormick, 1988). The radial growth was recorded each day over a four-day test period. At the end of the four days, petri plate contents were extracted for aflatoxin analysis.

Fungal Treatments in Cotton Bolls and Aflatoxin Extractions

Developing cotton bolls (30 days post-anthesis) obtained from approximately four-month-old Acala SJ-2 cotton plants growing under greenhouse conditions were used throughout the experiment. For inoculation of all fungal cultures or neem extract, two 3 mm holes were produced on the surface of all treated bolls, simulating the exit holes of the pink bollworm, *Pectinophora gossypiella*, one of the main sources of *A. flavus* cotton boll infection (Lee et al., 1987). Ten microlitres of *A. flavus* spore suspension and 50 μl neem leaf extract, each inoculated in separate bored holes, were used in the treatments. The neem extract was inoculated either simultaneously with fungal spore suspension or 48 hrs prior to the spore suspension. The controls included inoculations of 10 μl *A. flavus* spore suspension and 50 μl sterile, distilled water. The developing bolls were harvested 13 days after spore inoculation. The seeds in the treated locks from bolls harvested 13 days after treatment were delinted and extracted for aflatoxin B1 according to established procedures (AOAC, 1984).

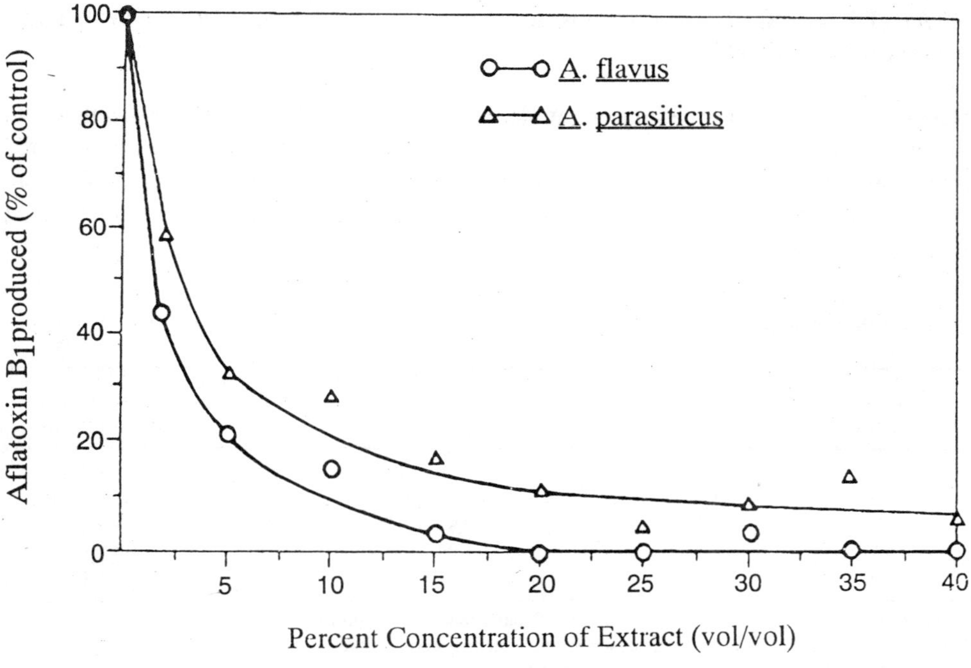

Fig. 2. Effect of various concentrations of neem leaf extracts on aflatoxin synthesis in submerged cultures of *A. flavus* and *A. parasiticus*. 100% refers to 38.4 μg aflatoxin B1 per g mycelial dry weight in *A. flavus* and 28.6 μg aflatoxin B1 in *A. parasiticus*.

Cottonseeds were also separated from either locks treated with neem leaf extract and *A. flavus* or locks treated with *A. flavus* only to assay the fungal growth within the seeds. These seeds were acid-delinted with concentrated sulphuric acid (1 min.), rinsed in distilled H_2O, treated in 10 per cent NaOCl (1 min.), and rinsed several times with sterile, distilled H_2O. The surfaced-sterilized seeds were positioned on PDA petri plates and after incubation for four days at 28°C, the radial growth of emerging developing cultures was compared to radial growth of control seeds (Zeringue and McCormick, 1990).

RESULTS

Effect of Neem Leaf Formulations on *Aspergillus* Growth and Aflatoxin Production in Submerged Cultures

Neem leaf extracts (0 to 40 per cent, v/v) did not affect fungal growth (i.e. mycelial dry weight), but essentially blocked (> 95 per cent) aflatoxin synthesis at concentrations greater than 10 per cent (v/v) (Table 1, Fig. 2). The effect of the neem leaf extracts was more pronounced in *A. flavus* than in *A. parasiticus*; total inhibition of aflatoxin synthesis was not achieved in *A. parasiticus*.

The inhibitory property of neem extracts was considerably reduced after the leaf extracts were heated by either boiling the leaves or autoclaving the leaf extracts (Bhatnagar and McCormick, 1988); a 10-fold higher amount of heated extract is required to obtain the inhibitory effect of the unheated extract (Fig. 3).

Effect of Neem Seed Formulations and Neem Oil on *A. parasiticus* Growth and Aflatoxin Production

Neem seed extracts, at the concentrations tested, did not inhibit aflatoxin synthesis to the extent observed with neem leaf extracts (Table 2). Extracts of neem seeds prepared by heating were unaffected by the process in the ability to inhibit aflatoxin production. Neem oil, on the other hand, did not exhibit any inhibition, but may have even enhanced fungal growth and consequently aflatoxin production (Table 3).

Table 1. Effect of neem leaf extracts on *A. flavus* and *A. parasiticus* growth in submerged cultures[a]

Concentration of extract (v/v)[b]	Mycelial dry weight (g)[c]	
	A. flavus	*A. parasiticus*
0	0.96 ± 0.15	1.11 ± 0.16
5	0.94 ± 0.11	1.04 ± 0.21
10	0.98 ± 0.09	1.19 ± 0.12
25	1.09 ± 0.18	1.08 ± 0.10
40	1.01 ± 0.11	1.13 ± 0.22

[a] The results are the means and standard errors of four experiments with two replicates each.
[b] The neem extracts were prepared as described in Materials and Methods.
[c] The mycelial dry weights were determined after four days of fungal growth.

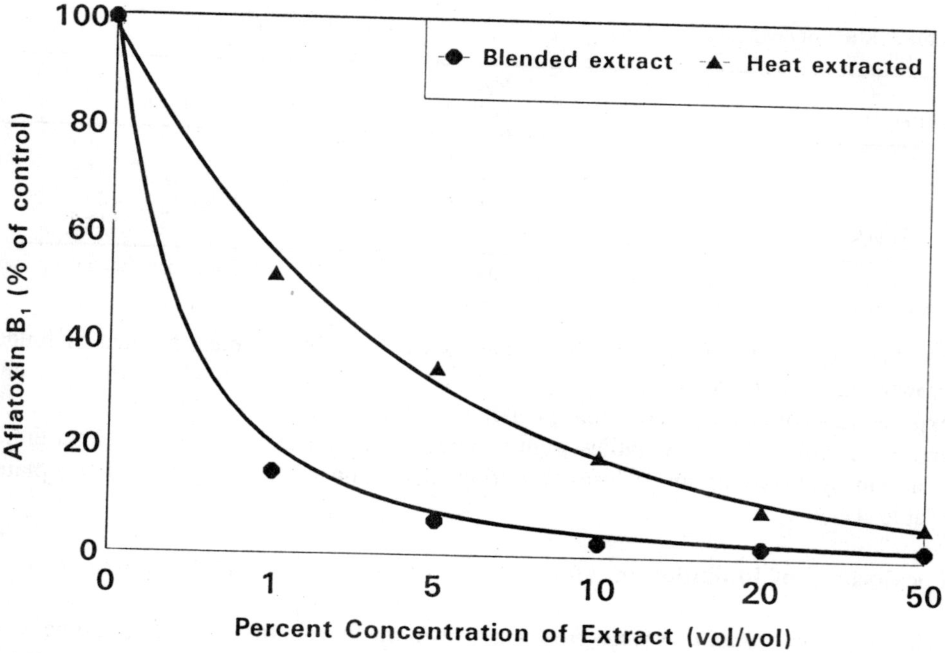

Fig. 3. Effect of various concentrations of unheated and heated neem leaf extracts on aflatoxin synthesis in submerged cultures of *A. flavus.* 100% refers to 20.6 μg aflatoxin produced per g mycelial dry weight in *A. flavus.* The results are the pooled means of three experiments, the mean standard error was ± 12.4%.

Table 2. Effect of neem seed extract on *A. parasiticus* growth and aflatoxin production[a]

Extract[b] (ml/100 ml culture media)	Mycelial dry weight (g)	Aflatoxin B1 (μg/g mycelial dry wt)
Non-boiled		
0	0.33	59.8
1	0.38	59.6
2	0.44	54.7
4	0.35	43.3
5	0.41	30.2
Boiled		
0	0.38	52.4
1	0.41	57.1
2	0.35	43.1
4	0.46	40.2
5	0.41	38.6

[a] The results are mean of three experiments, the standard deviation was less than 15 per cent.
[b] The extracts were prepared as described in Materials and Methods.

Effect of Neem Leaf Volatiles on Fungal Growth and Aflatoxin Production on Agar

The content of 10 of 13 major volatile components was significantly higher in the blended neem leaf extract than in the autoclaved, heated extract (Bhatnagar and McCormick, 1988), a major volatile component being 3-methyl-2-buten-1-ol (Fig. 4).

Table 3. Effect of neem oil on *A. parasiticus* growth and aflatoxin production

Oil[a] (μl/100 ml culture media)	Mycelial dry weight (g)	Aflatoxin B1 (μg/g mycelial dry wt)
0	0.38	10.2
250	0.73	13.8
500	0.61	26.0
1000	0.50	13.7

[a] Neem oil was added to culture flasks prior to fungal inoculation and Triton X-100 was included in the growth medium at a final concentration of 0.03 per cent to disperse the oil.

The bioactivity of the neem leaf volatiles was assessed by measurement of fungal growth and aflatoxin production by *A. flavus* grown on agar medium (Fig. 5) and exposed to an atmosphere containing volatiles from blended neem leaf extract (Bhatnagar and McCormick, 1988). Volatiles from blended neem leaf extracts did not affect either aflatoxin synthesis or fungal growth during incubation of the fungus on PDA plates (Table 4).

Localization of Inhibition of Aflatoxin Biosynthesis by Neem Leaf Extracts

When mycelia of *A. flavus* were grown in growth medium in the presence of 20 per cent neem extract and transferred to a resting medium (without neem) in the presence of several aflatoxin precursors, the results (Table 5) indicated that although aflatoxin synthesis was inhibited by the neem leaf extract, the affected fungal mycelia were able to convert aflatoxin precursors to aflatoxin B1. When the extracts of neem-treated mycelia from these experiments (with no aflatoxin precursor added) were analysed for the presence of aflatoxin precursors by the HPLC procedure described earlier (McCormick et al., 1988), none of the metabolites detected in extracts of control mycelia were found in the treated mycelial extracts (data not shown). However, when the mycelia grown in the absence of neem extracts for 72 hrs (duration required to initiate aflatoxin synthesis) were transferred to a resting medium containing no neem extract, there was no inhibition of aflatoxin synthesis (Table 6).

Effect of Neem Leaf Formulations on *A. flavus* Growth and Aflatoxin Production in Cotton Bolls

Treating developing cotton bolls (30 day post-anthesis) with the neem leaf extract 48 hrs prior to infecting the bolls with a spore suspension of *A. flavus* produced almost total (> 98 per cent) inhibition of aflatoxin synthesis measured in cottonseed from treated cotton boll locks 13 days after spore inoculation (Table 7). Similar cotton bolls which received a simultaneous application of both *A. flavus* spore suspension and neem leaf extract treatment exhibited only a 16 per cent inhibition of aflatoxin production in the cottonseeds after the 13-day incubation period.

When delinted, surface-sterilized seeds from control or treated locks were placed on PDA plates and the fungus was allowed to grow for four days, the radial fungal growth in the case of seeds from treated locks was 93–96 per cent of that of controls (Table 7).

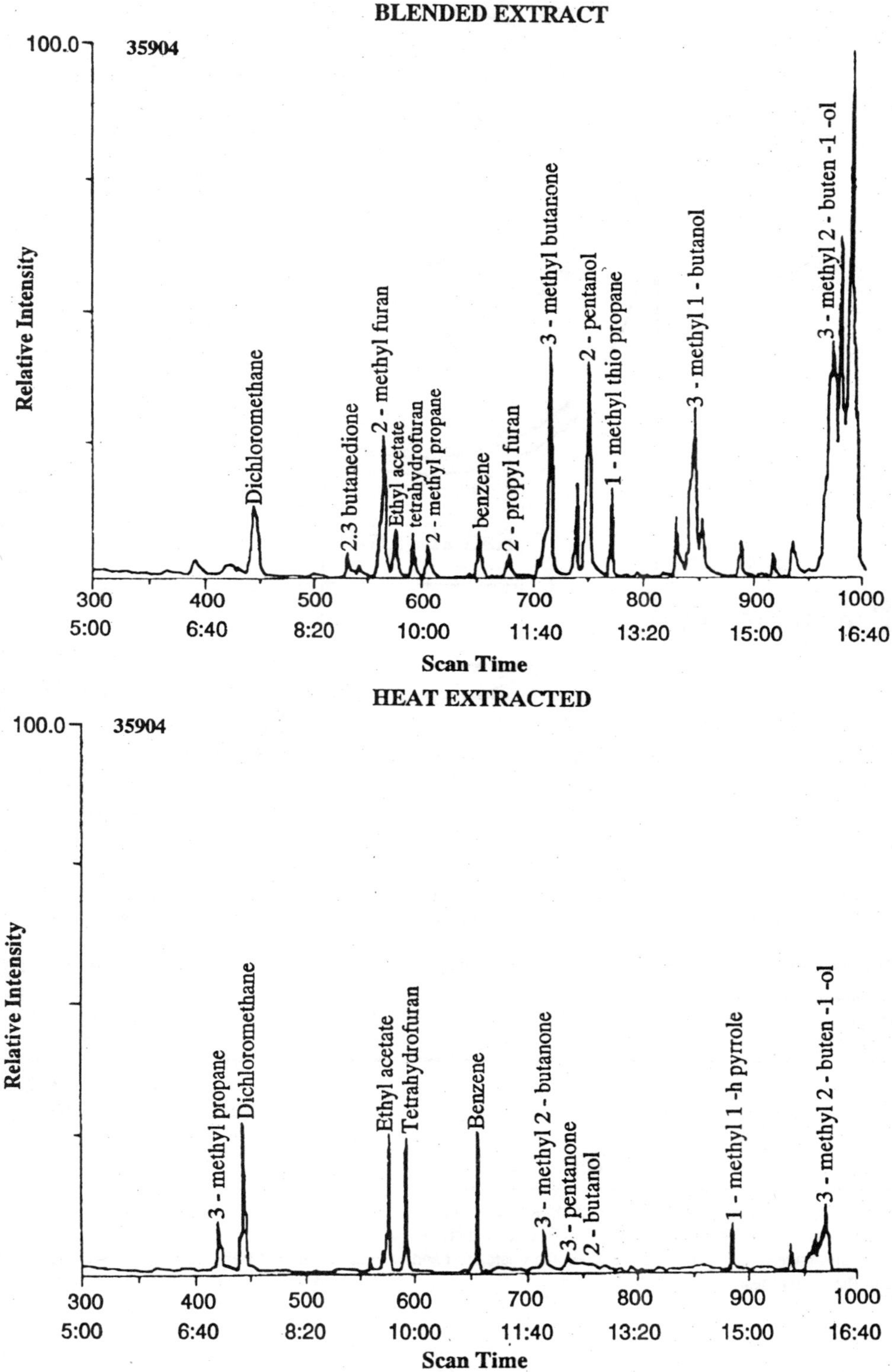

Fig. 4. Profile of the major volatile components in blended neem leaf extract (top panel) and the autoclaved, heated extract (bottom panel). The extract preparation and the determination of profile of volatiles was carried out as described in Material and Methods.

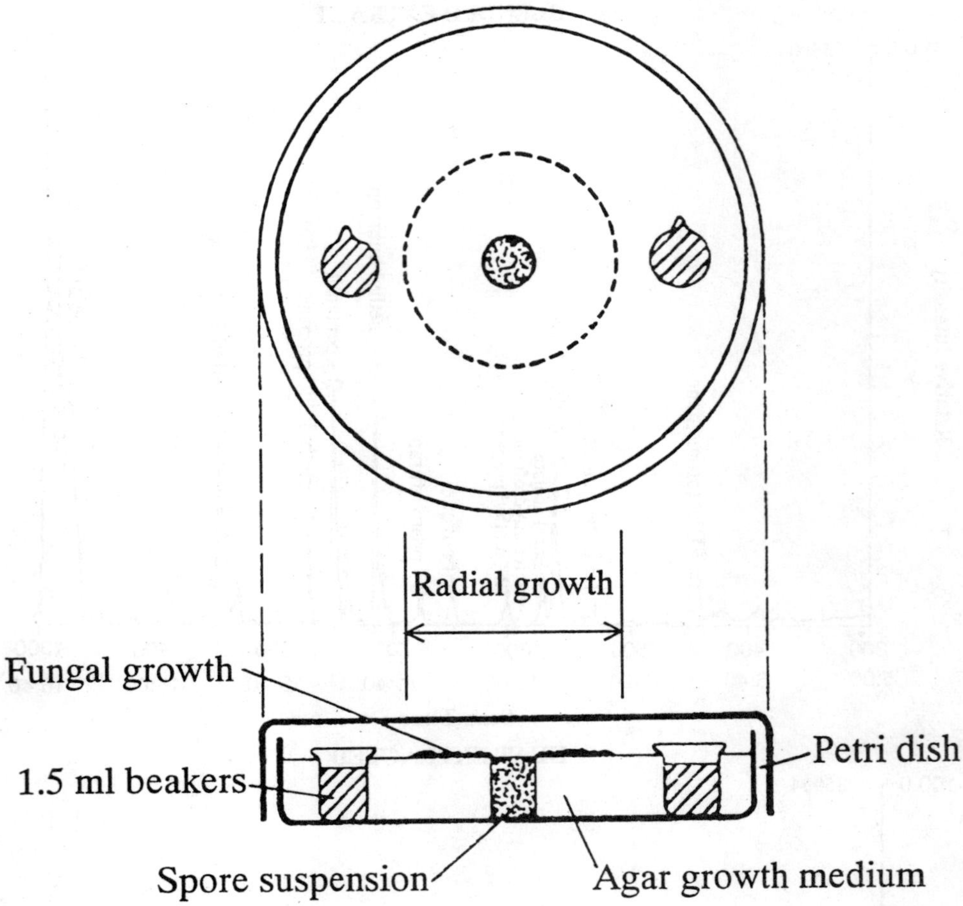

Fig 5. A diagrammatic representation of the experimental set-up for the determination of the effect of volatile component of neem leaves on fungal growth and aflatoxin production on a PDA plate.

Table 4. Effect of volatiles in neem leaf extracts on fungal growth and aflatoxin production by *A. flavus*[a]

Duration of incubation	Treatment[b]	Aflatoxin B1 produced (μg)	Fungal growth (cm)
2 days	Control	4.9 ± 1.5	2.8 ± 0.3
	Neem extract	3.8 ± 1.0	2.9 ± 0.4
4 days	Control	16.3 ± 2.0	4.8 ± 1.1
	Neem extract	17.1 ± 1.6	5.2 ± 1.3

[a] Fungal growth and aflatoxin production were determined as described in Material and Methods. The results represent the means and standard errors of three experiments.

[b] 1.5 ml of neem-leaf extract was added in the treated test and an equal volume of sterile, distilled water was added in the control test.

Table 5. Ability of *A. flavus* mycelia grown in the presence of neem leaf extract to utilize aflatoxin precursors

Precursor added	Amount added (μg)	Aflatoxin B1 produced in LSRM (μg)[a]	
		Control mycelia	Neem-treated mycelia
None		8.3 ± 1.8	ND
Norsolorinic acid	40	11.1 ± 2.6	3.6 ± 1.1
Averantin	30	13.3 ± 1.9	5.9 ± 1.2
Versicolorin A	20	17.4 ± 1.1	8.8 ± 2.2
Sterigmatocystin	10	15.2 ± 2.1	6.4 ± 2.0
O-methylsterigmatocystin	10	17.6 ± 2.0	8.1 ± 1.1

[a] One-gram fractions of wet 48-hr-old mycelia were obtained from fermentation carried out either in the absence (control mycelia) or presence (neem-treated mycelia) of 20 per cent (v/v) of neem leaf extract in growth media. The ability of these mycelia to utilize aflatoxin precursors in resting media was determined as described in Material and Methods. The values represent the means and standard errors of two experiments with two replicates each.

Table 6. Direct effect of neem leaf extract on aflatoxin B1 synthesis by resting *A. parasiticus* mycelia

Incubation (hr)	Aflatoxin B1 produced (μg)[a]	
	Control mycelia	Neem-treated mycelia
2	2.7 ± 0.4	2.7 ± 0.4
24	5.8 ± 0.7	5.5 ± 0.6
48	7.1 ± 1.1	6.8 ± 1.0

[a] One-gram fractions of wet 72-hr-old mycelia were obtained from fermentations in GM and transferred to 10 ml LSRM containing either 0 (control mycelia) or 20 per cent (neem-treated mycelia) v/v of neem-blended extract. Aflatoxin content was determined after 2, 24, and 48 hrs of incubation.

Table 7. Effect of neem leaf extracts on *A. flavus* growth and aflatoxin B1 production in developing cottonseed[a]

Treatment	Aflatoxin B1/g seed (ng)	*A. flavus* growth (cm)[b]
A. flavus inoculated	38,692 ± 3,864	25.1 ± 0.53
A. flavus and neem extract inoculated at the same time	30,799 ± 13,397	24.0 ± 0.48
A. flavus inoculated, 48 hrs after neem extract injection	8,435 ± 367	23.5 ± 0.71

[a] Assayed 13 days after fungal spore inoculation. The results represent the means and standard deviations of at least three experiments.
[b] Four-day radial growth of emerging *A. flavus* from seeds of treated locks.

DISCUSSION

In an effort to obtain a botanically derived inhibitor of *A. flavus* or *A. parasiticus* growth and aflatoxin production by these fungi, the efficacy of extracts of neem leaves was investigated in our laboratory (Bhatnagar and McCormick, 1988; Zeringue and Bhatnagar, 1989, 1990). Aqueous extracts of neem leaves, when added to fungal growth media prior to fungal spore inoculation, did not affect fungal growth (i.e. mycelial dry weight) in submerged cultures, but essentially blocked aflatoxin biosynthesis in both *A. flavus* and

A. parasiticus. This effect was not a pH effect because the neem extract did not alter the pH of the growth medium (data not shown) and neem extracts prepared in 50 mM potassium phosphate buffer (pH 7.0) also inhibited aflatoxin synthesis (Bhatnagar and McCormick, 1988).

The inhibitory effect of these extracts was also investigated under *in vivo* conditions, i.e. in developing cotton bolls. Fungal growth in cottonseed was unaffected by the treatments, but the seeds from locules receiving both neem leaf extracts and *A. flavus* spores simultaneously exhibited one-sixth the inhibition of aflatoxin production as compared to the treatment where the locules received the *A. flavus* spores 48 hrs after the neem extract was injected into the locules. It appears, therefore, that neem leaf extracts contain an aflatoxin inhibiting factor. However, this factor may need to translocate from the fibrous locule surface of the cotton boll to the seed, prior to fungal mycelia growth in the seed, for maximal effect (Zeringue and Bhatnagar, 1989, 1990).

Neem seed aqueous extracts were not inhibitory to *A. flavus* growth, but inhibited aflatoxin synthesis by nearly 50 per cent. However, this inhibition was not as pronounced as that observed with neem leaves. Neem oil, on the other hand, enhanced fungal growth and consequently aflatoxin production.

It was also observed (Bhatnagar and McCormick, 1988) that the inhibitory activity of the neem leaf extracts on aflatoxin production by the fungi was considerably reduced after the extracts were heated by either boiling the leaves in the buffer or autoclaving the leaf extracts (blended and heat-extracted neem leaf preparations). It was, therefore, postulated that the inhibitory component may be a volatile compound in leaf extracts. The differences in the volatile components of the blended and the autoclaved, heated neem leaf extracts were examined. The major volatile component present, 3-methyl-2-buten-1-ol, was nearly 400-fold greater in the blended extract than in the heated extract. It has been demonstrated (Zeringue and McCormick, 1990) that a similar compound, 3-methyl-1-butanol, one of the main characteristic odours of living colonies of *A. flavus* and also a volatile component of cotton leaves, inhibited *A. flavus* growth by 20 per cent but increased production of aflatoxin B1 two-fold. Volatiles from blended neem leaf extracts, however, did not inhibit or stimulate either fungal growth or aflatoxin production in *A. flavus* (Table 4) or *A. parasiticus* (Bhatnagar and McCormick, 1988), even though the amount of inoculum and neem extract added in the bioassay was in the same ratio as that added in submerged cultures.

The mode of action of neem leaf extracts in the inhibition of aflatoxin synthesis was investigated by determining the effect of the aqueous leaf extracts on the process of secondary metabolism (i.e., on aflatoxin biosynthetic pathway). The ability of resting mycelia to carry out secondary metabolism was used as a test system to monitor the effects of leaf extracts on the process in *A. flavus* (Table 5) and *A. parasiticus* (Bhatnagar and McCormick, 1988; Cleveland et al., 1990). The results demonstrated that aflatoxin biosynthesis was irreversibly inhibited in the mycelia of both fungi by neem leaf constituents; removal of mycelia from exposure to leaf extracts did not restore aflatoxin synthesis. Enzymes required for aflatoxin biosynthesis were, however, apparently intact in the treated mycelia because all the aflatoxin precursors (norsolorinic acid, averantin, versicolorin A, sterigmatocystin, and O-methysterigmatocystin) when fed to the treated mycelia were converted to aflatoxin B1 with nearly the same efficiency as control mycelia not exposed to leaf extracts. Metabolite precursors of aflatoxin biosynthesis were also not detected in mycelial extracts of treated mycelia, but these compounds were present in extracts of the control mycelia. Therefore, neem leaf extracts inhibit aflatoxin biosynthesis by preventing the production of secondary metabolites rather than the regulation of the synthesis or inactivation of enzymes specific to secondary metabolism. This inhibition of aflatoxin biosynthesis in fungal cells apparently occurs at a very early stage of the biosynthetic pathway because (1) norsolorinic acid, a very early aflatoxin precursor (Bhatnagar et al., 1991), was not detected in the treated mycelia and (2) once the secondary metabolism was initiated, the inhibitory effect of the neem leaf constituents was lost (Bhatnagar and McCormick, 1988).

CONCLUSION

Non-volatile, somewhat heat-labile neem leaf constituents irreversibly and almost totally inhibit aflatoxin biosynthesis in *A. flavus* and *A. parasiticus*, without affecting fungal growth, in both *in vitro* and *in vivo* conditions. Inhibition of aflatoxin biosynthesis appears to occur in the early stages of the biosynthetic pathway. Experiments are underway for analysing separated components from neem leaf extract in order to determine the components responsible for the bio-activity described. If the inhibitory factor in neem leaf extracts could be effective in field studies, these natural products could be

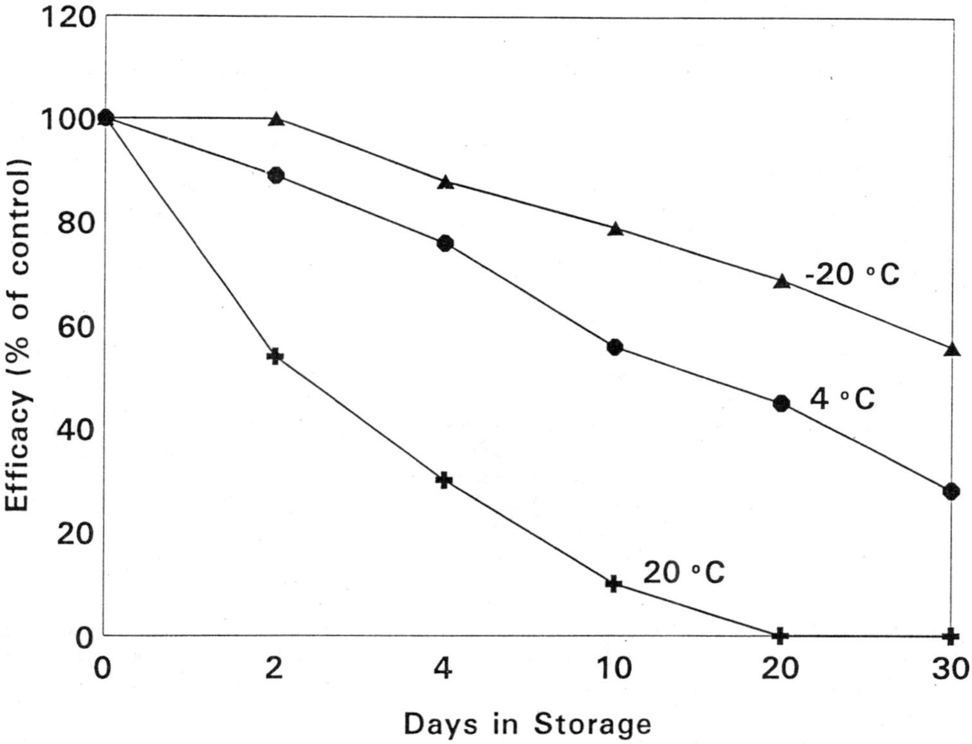

Fig. 6. Loss of efficiency of neem leaf formulation during storage at −20°C, 4°C, and 20°C. 20 ml fractions of leaf extracts were added to 100 ml of growth medium prior to inoculation with *A. flavus* spore suspension. Analysis of inhibition of aflatoxin synthesis was carried out after three days of fungal growth. A control experiment was carried out with no neem extract at day 0 and represented the ability of neem extract (20% v/v) to completely inhibit (100%) aflatoxin synthesis in *A. flavus*.

used in controlling the preharvest aflatoxin contamination of food and feed commodities. In the meantime, however, neem-growing countries could easily prepare fresh neem leaf extracts and use them to manage aflatoxin contamination of crops, basically at zero cost. The delivery system for these extracts has yet to be developed and careful attention needs to be paid to storage conditions because the extracts are highly unstable (Fig. 6) under normal storage conditions.

ACKNOWLEDGEMENTS

We are grateful to Dr. S.P. McCormick for neem volatile bioassays, and Troy S. Lewis and Brian Monette for excellent technical assistance.

REFERENCES

Adye, J. and R.I. Mateles. 1964. Incorporation of labelled compounds into aflatoxins. *Biochim. Biophys. Acta* 86: 418–420.

Association of Official Analytical Chemists. 1984. *Official Methods of Analysis*, 14th ed. Arlington, Virginia. pp. 26.052–26.060.

Bhatnagar, D., K. C. Ehrlich and T.E. Cleveland. 1991. Oxidation-reduction reactions in biosynthesis of secondary metabolites. In: *Mycotoxins in Ecological Systems*. D. Bhatnagar, E.B. Lillehoj and D. K. Arora (eds.), Marcel Dekker, Inc., New York, pp. 255–286.

Bhatnagar, D. and S.P. McCormick. 1988. The inhibitory effect of neem (*Azadirachta indica*) leaf extracts on aflatoxin synthesis in *Aspergillus parasiticus*. *J. Am. Oil Chem. Soc.* 65: 1166–1168.

Bhatnagar, D., S. P. McCormick, L.S. Lee and R.A. Hill. 1987. Indentification of O-methylsterigmatocystin as an aflatoxin B1 and G1 precursor in *Aspergillus parasiticus*. *Appl. Environ. Microbiol.* 53: 1028–1033.

Bressac, B., M. Kew, J. Wands and M. Ozturk. 1991. Selective G to T mutations of p53 gene in hepatocellular carcinoma from southern Africa. *Nature* 350: 429–431.

Campbell, T. C., J. Chen, C. Liu, J. Li and B. Parpia. 1990. Nonassociation of aflatoxin with primary liver cancer in a cross-sectional ecological survey in the People's Republic of China. *Cancer Res.* 50: 6882–6893.

Cleveland, T.E., D. Bhatnagar, P.J. Cotty and H.J. Zeringue. 1990. Aflatoxin reduction—a molecular strategy. In: *Biotechnology and Food Safety*. D.D. Bills and S.D. Kung (eds.), Butterworth-Heinemann, Boston, pp. 117–138.

Cleveland, T.E. and D. Bhatnagar. 1992. Molecular strategies for reducing aflatoxin levels in crops before harvest. In: *Molecular Approaches to Improving Food Quality and Safety*. D. Bhatnagar and T.E. Cleveland (eds.), Van Nostrand Reinhold, New York, pp. 205–228.

Darrah, L.L. and B. D. Barry. 1991. Reduction of preharvest aflatoxin in corn. In: *Pennington Center Nutrition Series, Mycotoxins, Cancer and Health*, Vol. I. G.A. Bray and D.H. Ryan (eds.), Louisiana State University Press, Baton Rouge, pp. 283–310.

Groopman, J.D. and G. Sabbioni. 1991. Detection of aflatoxin and its metabolites in human biological fluids. In: *Mycotoxins, Cancer and Health*, Vol. I. G.A. Bray and D.H. Ryan (eds.), Pennington Center Nutrition Series, Louisiana State University Press, Baton Rouge, pp. 18–31.

Hsu, I.C., R.A. Metcalf, T. Sun, J.A. Welsh, N.J. Wang and C.C. Hams. 1991. Mutational hotspot in the p53 gene in human hepatocellular carcinomas. *Nature* 350: 427–428.

Jelinek, C.F., A.E. Pohland and G.E. Wood. 1989. Review of mycotoxin contamination. Worldwide occurrence of mycotoxins in food and feeds—An update. *J. Assoc. Of. Anal. Chem.* 72: 223–230.

Jacobson, M. (ed.). 1989. *Phytochemical Pesticides*, Vol. I. *The Neem Tree*. CRC Press, Boca Raton, 178 pp.

Keller, N.P., T.E. Cleveland and D. Bhatnagar. 1991. A molecular approach towards understanding aflatoxin production. In: *Mycotoxins in Ecological Systems*. D. Bhatnagar, E.B. Lillehoj and D.K. Arora (eds.), Marcel Dekker, Inc., New York, pp. 287–310.

Koul, O., M.B. Isman and C.M. Ketkar. 1990. Properties and uses of neem, *Azadirachta indica*. *Can. J. Bot.* 68: 1–11.

Lee, L.S., P.E. Lacey and W.R. Goynes. 1987. Aflatoxin in Arizona cottonseed: A model study of insect vectored entry of cotton bolls by *Aspergillus flavus*. *Plant Dis.* 71: 997–1000.

Lillehoj, E.B. 1991. Aflatoxin: Genetic mobilization agents. In: *Mycotoxins in Ecological Systems*. D. Bhatnagar, E.B. Lillehoj and D.K. Arora (eds.), Marcel Dekker, Inc., New York, pp. 1–22.

McCormick, S.P., E. Bowers and D. Bhatnagar. 1988. An improved HPLC procedure for determining the profiles of aflatoxin precursors in wildtype and mutant strains of *Aspergillus parasiticus*. *J. Chromatog.* 441: 400–405.

National Research Council. 1992. *Neem: A Tree for Solving Global Problems*. National Academy Press, Washington, D.C.

Park, D.L., L.S. Lee, R.L. Price and A.E. Pohland. 1988. Review of the decontamination of aflatoxins by ammoniation: Current status and regulation. *J. Assoc. Of. Anal. Chem.* 71: 685–703.

Payne, G.A. 1992. Aflatoxin in maize. *Crit. Rev. Plant Sci.* 10: 423–440.

Phillips, T.D., B.A. Sarr, B.A. Clement, L.F. Kubena and R.B. Harvey. 1991. Prevention of aflatoxicosis in farm animals via selective chemisorption of aflatoxin. In: *Pennington Center Nutrition Series, Mycotoxins, Cancer and Health*, Vol. I. G.A. Bray and D.H. Ryan (eds.), Louisiana State University Press, Baton Rouge, pp. 223–237.

Scott, G.E. and N. Zummo. 1988. Sources of resistance in maize to kernel infection by *Aspergillus flavus* in the field. *Crop Sci.* 28: 504–507.

Singh, U.P., H.B. Singh and R.B. Singh. 1980. The fungicidal effect of neem (*Azadirachta indica*) extracts on some soil-borne pathogens of gram (*Cicer arietinum*). *Mycologia* 72: 1077–1093.

Widstrom, N.W. 1987. Breeding strategies to control aflatoxin contamination of maize through host plant resistance. In: *Aflatoxin in Maize: A Proceedings of the Workshop*. M.S. Zuber et al. (eds.), CIMMYT, Mexico City, Mexico, pp. 212–220.

Wogan, G.N. 1991. Aflatoxins as risk factors for primary hepatocellular carcinoma in humans. In: *Mycotoxins, Cancer and Health*, Vol. 1. G.A. Bray and D.H. Ryan (eds.), Pennington Center Nutrition Series, Louisiana State University Press, Baton Rouge, pp. 3–17.

Yeh, F.-S., M.C. Ju, C.-C. Mo, S. Lou, M.J. Tong and B.E. Henderson. 1989. Hepatitis B virus, aflatoxins, hepatocellular carcinoma in Southern Guangxi, China. *Cancer Res.* 49: 2506–2709.

Zeringue, H.J., Jr. and D. Bhatnagar. 1989. Inhibition of aflatoxin production in *Aspergillus flavus* infected cotton bolls after treatment with neem (*Azadirachta indica*) leaf extracts. *Plant Physiol.* (Supp.) 89: 200.

Zeringue, H.J., Jr. and D. Bhatnagar. 1990. Inhibition of aflatoxin production in *Aspergillus flavus* infected cotton bolls after treatment with neem (*Azadirachta indica*) leaf extracts. *J. Am. Oil Chem. Soc.* 67: 215–216.

Zeringue, H.J., Jr. and S.P. McCormick. 1990. Aflatoxin production in cultures of *Aspergillus flavus* incubated in atmospheres containing selected cotton leaf derived volatiles. *Toxicon* 28: 445–448.

78

Management of Pest and Diseases of Cotton and Soybean Using Neem Products

F. VIDYA, K. SETHURAMAN, V. MARIAPPAN,
K.R. KARUNAKARAN, AND P. VIDHYASEKARAN

Tamil Nadu Agricultural University
Coimbatore 641 003, India

ABSTRACT

The neem seed kernel extract (NSKE), neem oil (NO) and endosulfan were effective in reducing the aphid (*Aphis gossypii*) while NSKE and NO were effective in reducing the whitefly (*Bemisia tabaci*) population. However there was no effect on leaf hopper (*Amrasca biguttula biguttula*). The yield recorded by NO was significantly higher (2036 kg/ha).

Soybean crop was affected by leaf miner (*Aproaerema modicelli*) and yellow mosaic virus transmitted by *B. tabaci*. Leaf miner incidence was reduced significantly by NO, NSKE and monocrotophos while yellow mosaic was reduced significantly by monocrotophos and NSKE.

INTRODUCTION

Neem (*Azadirachta indica* A. Juss.) products like seeds, seed oil and leaves are known to contain compounds with biological activity against insect pests and pathogens (Butterworth and Morgan, 1968; Zanno et al., 1975; Kraus et al., 1981). Neem oil and seed kernel derivatives have made their debut as soft pesticides and possible components in integrated management of plant protection (Mariappan et al., 1987; Saxena, 1987; Locke, 1990).

The whitefly (*Bemisia tabaci* Genn.) has assumed serious proportion on cotton causing 10 to 45 per cent losses in seed cotton (Reddy, 1986). Epidemic outbreaks of whitefly has been reported during 1984, 1985 and 1988 in Andhra Pradesh and Maharashtra (Nene, 1986) and farmers used all sorts of pesticides but failed to get the desired results. The insect whitefly (*B. tabaci*) being vector of yellow mosaic virus disease inflicted 61 to 86 per cent loss to soybean crop (Singh et al., 1971; Sutteri et al., 1974).

The sucking pests of cotton like aphid (*Aphis gossypii* Glov.) and leaf hopper (*Amrasca biguttula biguttula* Ishida.) are serious threats to cotton crops (Hussain and Lal, 1940; Painter, 1954). The inadequate control of insect pests and diseases by the chemical pesticides; the increasing cost of pesticides and awareness of the environment pollution due to the excess usage of chemical pesticides have prompted the authors to test the efficacy of neem products as the alternative for the chemical pesticides.

MATERIALS AND METHODS

Field trials were conducted in farmers field to demonstrate the efficacy of neem products like neem oil (NO) at 3 per cent, neem seed kernel extract (NSKE) at 5 per cent concentration in comparison with insecticide (Monocrotophos 500 ml/ha; Endosulfan 2 l/ha) for the control of insect pests and diseases of cotton and soybean crops.

Neem oil (3 per cent) was prepared by mixing 30 l neem oil with 1000 l of water containing 1000 ml of liquid detergent 'teepol'.

Neem seed kernel extract 5 per cent (NSKE) was prepared by powdering 50 kg of neem seed kernel and soaking in 200 l water for 12 hrs and filtered. The filtrate was made up to 1000 l and added with 1000 ml of liquid detergent 'teepol'.

The treatments on Cotton (*Gossypium hirsutum* L.) (Variety MCU 11) were

1. NO (3 per cent) @ 1000 l/ha
2. NSKE (5 per cent) @ 1000 l/ha
3. Endosulfan 2 l/ha @ 1000 l/ha

and 4. Control

The treatments on Soybean (*Glycine max* (L.) Merril.) (Variety CO l) were

1. NO (3 per cent) @ 500 l/ha
2. NSKE (5 per cent) @ 500 l/ha
3. Monocrotophos @ 500 l/ha

and 4. Control

The experiment was laid out in plots of 40 m^2 each with six replications in randomized block design. Spraying with NO, NSKE and insecticide was given using high volume sprayer on crossing of Economic Threshold Level (ETL) by any one insect pest or disease.

In cotton crop observations on aphid whitefly and jassid population were taken on the leaves of 10 plants in a plot and averages per leaf were worked out.

In soybean crop observations were taken on leaf miner and yellow mosaic on the number of plants affected in each plot and their per cent worked out.

RESULTS

Cotton

The crop was affected by aphid (*A. gossypii*), whitefly (*B. tabaci*) and leaf hopper (A. *biguttula biguttula*). There were no significant differences between treatments. However there was reduction in aphid population in plots treated with NSKE, NO and endosulfan. Whitefly population was also reduced in NSKE and NO treatments when compared to control. There was no effect on leaf hopper. The yields recorded from NO and insecticides were significantly higher (2236 and 2226 kg/ha respectively) than NSKE and control treatments. The yield recorded by NSKE (1939 kg/ha) was also higher than control (1607 kg/ha). Among the neem products, NO application has been found to be significantly superior than NSKE (Table 1).

Soybean

The crop was affected by leaf miner (*Aproaerema modicelli.* Deventer) and yellow mosaic virus disease transmitted by whitefly (*B. tabaci*). The incidence of leaf miner was reduced significantly in plots treated with NO, NSKE and monocrotophos (7.30, 7.32 and 6.55 per cent respectively) when compared to 15.63 per cent incidence in control treatment. Neem products (NO, NSKE) were as effective as insecticides.

The yellow mosaic disease was reduced significantly by monocrotophos and NSKE (11.01 and 12.59 per cent infection respectively) when compared to 13.09 per cent incidence in NO treatment and 20.39 per cent in control. Among the neem products, NSKE was more effective than neem oil.

The yield recorded by NO, NSKE and insecticide treatments were higher than control (Table 2).

DISCUSSION

Neem products like neem oil and neem seed kernel extract were found to reduce the aphid and whitefly population on cotton and leaf miner damage and yellow mosaic virus disease in soybean. Jayashree (1984) reported that a mixture of neem oil + Karanji oil

Table 1. Field efficacy of neem products in management of cotton pests

Treatment	Aphid No./leaf	Whitefly No./leaf	Leafhopper No./leaf	Yield kg/ha	Yield value Rs.	Treatment cost Rs./ha
NSKE 5%	8	1.6	1.8	1939^{b}	34902	440
NO 3%	8	1.9	0.7	2236^{a}	40248	365
Endosulfan	8	2.0	0.7	2226^{a}	40068	210
Control	12	2.0	0.7	1607^{c}	28926	—

Season: August to December 1991

Table 2. Field efficacy of neem products in management of soybean pests and diseases

Treatment	Leaf miner %	Yellow mosaic %	Yield kg/ha	Yield value Rs.	Treatment cost
NSKE	7.32^{a}	12.59^{b}	1038.75^{a}	9342	440
NO	7.30^{a}	13.09^{b}	978.75^{a}	8811	365
Monocrotophos	6.55^{a}	11.01^{a}	993.75^{a}	8946	235
Control	15.63^{b}	20.39^{c}	656.25^{b}	5904	—
CD	3.43	1.78	—	—	—

Season: March to June 1992

(at 5:1 and 10:1 proportions) at 3 per cent or neem oil 3 per cent were comparable to insecticide endosulfan (700 g a.i/ha in reducing the aphid (*A. gossypii*) and jassid (*A. biguttula biguttula*) population in cotton. Uthamasamy and Parameswaran (1989) reported that neem oil 0.5 per cent or neem seed kernel extract 5 per cent were effective in controlling the cotton whitefly (*B. tabaci*). The neem seed kernel extract 5 per cent controlled *B. tabaci* population by 80 per cent in bhendi followed by neem oil 3 per cent spray with 78 per cent reduction (Kathirvel, 1988).

Spraying of NSKE 10 per cent or neem cake extract 10 per cent or neem leaf extract 10 per cent or neem oil 2 per cent on 35th and 45th day after sowing has been reported to check the leaf miner, *A. modicelli* (Chandramohan and Sivasubramanian, 1987). Mariappan et al. (1987), observed that neem cake extract, Iluppai (mahua) seed kernel extract, neem seed kernel extract, pinnai oil and Iluppai oil spray reduced the yellow mosaic virus disease of blackgram (*Vigna mungo*) transmitted by whitefly (*B. tabaci*). Ragupathi (1989) reported a significant reduction in the yellow mosaic virus infected plants in soybean crop sprayed with neem oil and also recorded higher crop yield.

The present observations are in confirmity with the results obtained earlier and confirm the usefulness of neem products like neem oil (3 per cent) and neem seed kernel extract (5 per cent) in containing the pests and diseases of cotton and soybean. In both crops the yields recorded from neem oil (3 per cent) and insecticide (endosulfan/monocrotophos) treatments were on par and comparable with each other. Neem seed kernel extract treatment also has been comparable to neem oil in enhancing the yield of soybean.

REFERENCES

Butterworth, J.H. and E. Morgan. 1968. Isolation of a substance that suppresses feeding insects. *J. Chem. Society Commun.*: 23–24.

Chandramohan, N. and P. Sivasubramanian. 1987. Evaluation of neem products against leaf miner, *Aproaerema modicelli* Deventer. *Neem Newsl.* 4(4): 44.

Hussain, M.A. and K.B. Lal. 1940. The bionomics of *Empoasca devastans* distant on some varieties of cotton in Punjab. *Indian J. Ent.* 2(2): 123–136.

Jayashree, S. 1984. *Studies on the efficacy of neem products against pests of cotton* M.Sc. (Ag.) Thesis, Tamil Nadu Agricultural University, Coimbatore, 90 pp.

Kathirvel, M. 1988. *Studies on the management of pests and nematodes of bhendi with botanicals and insecticides*. M.Sc. (Ag.) Thesis, Tamil Nadu Agricultural University, Coimbatore, 135 pp.

Kraus, W., R. Cramer, M. Bokel and G. Sawitzti. 1981. New insect antifeedant from *Azadirachta indica* and *Melia azederach.* In: *Proc. of the 1st Int. Neem Conference.* H. Schumutterer, K.R.S. Asher and H. Rembold (eds.), Eschborn, FRG, GTZ, pp. 507–518.

Locke, J.C. 1990. Activity of extracted neem seed oil against fungal plant pathogens. In: *Proc. of USDA Neem Workshop*, J.C. Locke and R.H. Lawson (eds.), Beltsville, Maryland, USA, April 16–17, 1990, pp. 132–136.

Mariappan, V., M. Gopalan, V. Narasimhan and S. Suresh. 1987. Effect of neem and other plant products on yellow mosaic virus (YMV) disease in blackgram (*Vigna mungo* L.). *Neem Newsl.* 4: 9–10.

Nene, S.R. 1986. An appraisal of cotton whitefly problem for further research, pp. 90–94. In: *Resurgence of sucking pests.* S. Jayaraj (ed.), Tamil Nadu Agricultural University, Coimbatore, India, 230 pp.

Painter, R.H. 1954. *Insect Resistance in Crop Plants*. Macmillan, New York.

Ragupathi, N. 1989. *Studies on yellow mosaic virus disease of soybean* (*Glycine max* (L.) Merrill) M.Sc.(Ag.) Thesis, Tamil Nadu Agricultural University, Coimbatore, 91 pp.

Reddy, A.S. 1986. Whitefly, the emerging pest. *Seminar on strategies for raising production and productivity in southern cotton zone*, Central Institute of Cotton Research, Coimbatore, India, Feb. 20–21, 1986.

Saxena, R.C. 1987. Neem seed derivatives for management of rice insect pests, a review of recent studies. In: *Natural pesticides from the neem tree, Azadirachta indica and other tropical plants. Proc. 3rd Int. Neem Conference*, Nairobi, 1986. H. Schmutterer and K.R.S. Asher (eds.), Deutsche Gesellschaft fur Technische Zusummenarbeit, Eschborn, pp. 81–83.

Singh, H., G.S. Sandhu and G.S. Mavi. 1971. Control of yellow mosaic virus in soybean, *Glycine max* (L.). *Indian J. Entomology* 33: 272–278.

Sutteri, B.D. 1974. Occurrence of soybean yellow mosaic virus in Uttar Pradesh. *Curr. Sci.* 43: 689–690.

Uthamasamy, S. and S. Parameswaran. 1989. *Cotton insect pest management* (in Tamil Nadu), Centre for Plant Protection Studies, Tamil Nadu Agricultural University, Coimbatore, 61 pp.

Zanno, P.R., I. Miura, K. Nakanishi and D.E. Elder. 1975. Structure of the phago repellent, azadirachtin. *J. American Chem. Society* 97: 1957–1977.

79

Efficacy of Plant Derivatives on Rice Blast Pathogen *Pyricularia oryzae* cav., Blast Infection and Enzymes of Rice Plants

E. RAJESWARI AND V. MARIAPPAN

Centre for Plant Protection Studies
Tamil Nadu Agricultural University
Coimbatore 641 003, India

ABSTRACT

Leaf extracts (in cold water, hot water and ethanol) of *Prosophis juliflora* (sw.) DC. adversely affected the spore germination, mycelial growth and biomass production of rice blast pathogen *Pyricularia oryzae* cav and reduced blast disease of rice. *Adhatoda vasica* Nees. (Acanthaceae), *Vitex negundo* Siv. and Mold. (Verbanaceae), *Azadirachta indica* A. Juss. (Meliaceae), and *Polyalthea longifolia* Sonner (Annonaceae) exhibited similar efficacy. The leaf extracts of these plant species contained protein, glycoprotein, total phenol, free amino acids, and tannin at higher proportions than the least effective *Caesalpinia pulcherrima* (L.) Sw. (Caesalpiniaceae) leaf extract. The enzymes peroxidase, polyphenol oxidase, and phenyl alanine ammonia lyase worked faster in *P. oryzae* inoculated rice plants treated with *A. vasica* and *P. juliflora* than in those treated with *C. pulcherrima*. Leaf extracts were more effective than seed extracts or seed soil emulsions.

INTRODUCTION

Plants are an unlimited source of phytochemicals. The last two decades have witnessed an increase in investigation on plants as a source of pest control materials and as a possible alternative to synthetic pesticides. The literature available today is mostly on the usefulness of neem tree derivatives for insect pest control (Schmutterer, 1990; Warthen, 1989) and there are relatively few references to activity against fungus pathogens (Locke, 1990). Screening programmes have been initiated on other plant materials occasionally, in spite of centuries of tradition in most of the world's cultures of controlling microbial infection by plant materials.

Rice blast disease caused by *Pyricularia oryzae* Cav. is a serious disease in India and many other rice-growing countries.

Lapis and Dumancas (1979) screened 93 plant species for their fungicidal property against rice blast pathogen *P. oryzae* and found 19 of them inhibited the growth of the fungus. Tewari and Premalatha Dath (1984) observed that media containing leaf extracts of *Ocimum sanctum*, *Lawsonia inermis*, and *Nyctanthes arbortristis* suppressed the growth of *P. oryzae*. Muthusamy et al. (1988) found that plant extracts from *Pepper cubeba* and *Andrographis paniculatus* inhibited the biomass production of *P. oryzae* for more than 60 per cent.

The objective of this research was to test the efficacy of the leaf extracts, seed extracts, and seed oils on the rice blast pathogen *P. oryzae*, their effect on disease development, and their influence on the host.

MATERIALS AND METHODS

Pathogen

The test fungus *P. oryzae* was isolated from leaves of IR50 rice variety affected by blast disease. Purified culture from single spore was maintained on apple agar medium (125 g potato/125 g apple/20 g dextrose/15 g agar/1000 ml water, pH 6.0 to 6.5).

Plant Extracts/Products

Extracts of leaves of 15 plant species belonging to 14 families, two non-edible seeds of plants belonging to two families, and three on edible seed oils of plants from three families were screened.

Leaf Extracts

Fresh, mature leaves were washed, ground with cold sterile water at 1:1 ratio (w/v) in a sterile pestle and mortar, and filtered, and the filtrate was taken as 100 per cent cold water extract. Similarly, leaves immersed in hot water (80°C) for 10 min. were ground and filtered and the filtrate was taken as hot water extract.

Fresh leaves (100 g) were ground in 100 ml of ethanol (95 per cent) and filtered, and filtrate evaporated to dryness at laboratory temperature, 32°C. The residue was dissolved in 100 ml of sterile water. These three extracts were heated up to 45° to 50°C for 10 min. to avoid contamination (Jaganathan and Narasimhan, 1988). Leaf extract in cold water autoclaved at a pressure of 1.2 kg/sq cm was kept separately.

Seed Extracts

Dried seeds (neem, *Azadirachta indica* and pinnai, *Calophyllum inophyllum* L) were washed, air dried, ground in sterile water at 1:2 (w/v) and filtered and the filtrate was taken as 50 per cent solution of seed extract.

Seed Oil Emulsion

The non edible seed oils (*A. indica, C. inophyllum,* and mahua, *Madhuca longifolia* Koen. macbr. var. *latifolia* Roxb. Cheval) were mixed with water (containing liquid detergent Teepol 0.1 per cent) at 3:100 (v/v).

Spore Germination

Spore suspension of *P. oryzae* (at a concentration of 106 cells/ml) was mixed in appropriate proportions of cold water, hot water, and ethanol extracts and with autoclaved cold water extracts of leaves to get 10 per cent concentration; with seed extracts to get 5 per cent concentration; and with seed oil emulsions to get 3 per cent concentration, in cavity slides. Spore germination per cent was recorded after 8 hrs. Spores mixed with sterilized water were kept as control.

Mycelial Growth

The above plant preparations were mixed with apple agar medium in appropriate proportions (to maintain the concentrations mentioned above) in petri plates and inoculated with the mycelial discs of *P. oryzae* of 8 mm size in the centre. Plain apple agar served as control. The radial growth measurement was taken on the 12th day when the growth in control reached the periphery.

Biomass Production

The plant preparations were mixed with 'Richard's broth' (Fahmy, 1923) to maintain the concentration mentioned above and inoculated with mycelial discs of *P. oryzae*. Richard's broth inoculated with mycelial disc served as control. After 15 days the content was filtered and residue dried in a hot-air oven at 45°C till two consequent weighings remained constant.

Blast Disease Development

Leaf extracts of *Adhatoda vasica, Azadirachta indica, Prosophis juliflora, Polyalthea longifolia* and *Vitex negundo* (10 per cent), seed extracts of *A. indica* and *C. inophyllum* (5 per cent) and seed oil emulsions of *A. indica, C. inophyllum* and *M. latifolia* (3 per cent) were tested for their efficacy on blast disease development. Ten-day-old IR50 rice seedlings were sprayed with these preparations of plant derivatives 3 hrs before inoculation with *P. oryzae* spore suspension (10^6 spores/ml) using hand atomizer. Another set of rice plants were inoculated with *P. oryzae* spores and treated with the preparations of plant derivatives after 3 hrs. The above operation was done in the evening. The treated plants were kept in cages covered with alkathene sheets and maintained in glasshouse at 19° and 21°C. For each treatment 10 pots containing 100 seedlings were maintained and each treatment was replicated thrice. Disease rating was scored on seventh day and 15th day following Standard Evaluation System of 0–9 scale and per cent disease index was worked out using the formula (McKinney, 1923):

$$\text{PDI} = \frac{\text{Sum of all ratings}}{\text{Total no. of leaves observed}} \times \frac{100}{\text{Maximum disease rating}}$$

Biochemical Constituents of Plants

Leaf extracts of *A. vasica, C. pulcherrima* and *P. juliflora* were analysed for the estimation of total sugars (Somoyogi, 1952), reducing sugars (Somoyogi, 1952), non-reducing sugars, protein (Lowry, et al., 1951), glycoprotein (Keller and Block, 1960), total phenols (Spies, 1955), free amino acids (Spies, 1955), and tannin (AOAC, 1965).

Enzymes in Rice Plants

The peroxidase (Srivastava, 1987), polyphenol oxidase (Srivastava, 1987) and phenyl alanine ammonia lyase (Zucker, 1965) activities in rice plants (inoculated with *P. oryzae* and treated with plant derivatives) were estimated.

RESULTS

Effect of Plant Derivatives on *P. oryzae*

The plant derivatives adversely affected the spore germination, mycelial growth, and biomass production of *P. oryzae*.

Leaf extracts (cold water, hot water, ethanol and autoclaved) of *P. julifora* (Mimosaceae) were found to be highly inhibitory to spore germination, mycelial growth, and biomass production of *P. oryzae* followed by *A. vasica* (Acanthaceae), *V. negundo* (Verbenaceae), *A. indica* (Meliaceae), and *P. longifolia* (Annonaceae) among the leaf extracts of 15 species tested. The efficacy of the different leaf extracts was found to vary. The autoclaved extracts were least effective (Table 1).

The seed extracts and seed oil emulsions were relatively less effective than the leaf extracts (Table 2).

Effect of Plant Derivatives on Blast Disease Development

Plant derivatives were effective in reducing the blast disease development and also the lesion size to a considerable extent.

Pre-inoculation application (prophylactic) has been more effective than post-inoculation (curative) application. The reduction in the per cent disease index recorded by plant derivatives has been considerably significant. The reduction in size of the blast lesion due to application of leaf extract has been impressive. *Adhatoda vasica* and *P. juliflora* were able to contain the disease and reduce the lesion size effectively, followed by *A. indica, V. negundo,* and *P. longifolia*. The seed extracts and seed oil emulsions were relatively less effective than the leaf extracts (Table 3).

Table 1. Effect of plant extracts on germination, mycelial growth, and biomass production of *P. oryzae*

Extract	Inhibition % over control							
	Spore germination			Mycelial growth				Biomass
	C	H	E	C	H	E	C A	
Acanthaceae								
Adhatoda vasica (Nees)	76.7b	78.7a	82.5b	87.4b	80.3a	89.1b	65.1a	89.7b
Rutaceae								
Aegle marmelos (L) (Corr)	34.2g	35.6b	35.5i	35.0h	28.4f	34.5h	20.4e	42.5h
Polygonaceae								
Antigonan leptopus (Hook + Arn)	3.5i	4.7i	8.9l	4.8i	3.7h	3.6k	2.4g	4.1i
Meliaceae								
Azadirachta indica (A. Juss.)	55.6d	58.2c	61.9d	66.5d	61.6b	71.5c	60.2ab	60.8d
Caesalpiniaceae								
Caesalpinia pulcherrima L. SN.	7.7h	5.9i	9.8kl	6.1i	6.2g	7.3ij	4.8g	5.7i
Apocynaceae								
Catharanthus roseus (L.) G. Don	46.5ef	41.5e	43.8g	46.1f	43.6de	47.3fg	22.8de	45.9g
Solanaceae								
Datura metel L.	6.3h	8.3h	12.3g	5.5i	4.9gh	9.6i	5.3f	6.0i
Convolvulaceae								
Ipomoea carnea Jack	35.3g	37.3f	39.2h	46.8f	40.4e	43.6g	27.2d	41.6h
Lythraceae								
Lawsonia inermis L.	45.2f	45.0d	51.4e	48.1f	49.3c	52.1e	47.1e	47.3g
Labiaceae								
Ocimum sanctum	45.0g	42.8de	45.5g	40.3g	46.1cd	49.6ef	42.8c	51.9b
Euphorbiaceae								
Phyllanthus fraternus, Web.	47.3e	41.7e	47.9b	53.3e	43.5de	47.9fg	48.4c	46.6g
Annonaceae								
Polyalthea longifolia Sonnen	54.6d	56.9c	61.4d	64.5d	44.1de	68.4d	55.8b	55.2e
Mimosaceae								
Prosophis juliflora (SWDC.)	80.5a	80.8a	84.5a	93.3a	78.5a	92.1a	62.7a	100.0a
Solanaceae								
Solanum nigrum L.	6.4h	10.6g	10.8h	4.2i	3.0h	6.7j	4.2g	5.4i
Verbanaceae								
Vitex negundo Siv and Mold	59.4c	64.3b	65.0c	72.4c	64.5b	73.3c	55.2b	71.4c

C = cold water extract, H = hot water extract, E = ethanol extract, CA = Cold water extract autoclaved.

Table 2. Effect of seed extracts and seed oils on germination and mycelial growth of *P. oryzae*

Plant products	Inhibition % over control	
	Spore germination	Mycelial growth
Seed Extracts (5%)		
Meliaceae		
Azadirachta indica	18.9b	27.9b
Gutiferae		
Calophyllum inophyllum	9.2c	9.2d
Seed Oils (3%)		
Meliaceae		
A. indica	19.7b	27.7b
Gutiferae		
C. inophyllum	9.0c	17.5c
Sapotaceae		
Madhuca latifolia	61.20a	51.8a

Table 3. Effect of plant products on blast disease*

Plant species	Per cent reduction over control in blast infection (PDI values)				Per cent reduction in lesion size	
	7* DAI		15* DAI			
	a	b	a	b	a	b
Leaf Extract (10%)						
Adhatoda vasica	57.0a	47.3b	50.2a	43.9b	78.8a	62.8b
Azadirachta indica	43.3c	40.5c	31.7c	40.3c	43.4bc	37.7de
Polyalthea longifolia	30.8e	34.3d	28.6d	30.3d	35.5cd	28.9e
Prosophis juliflora	54.6b	51.4a	46.4b	45.1a	78.5a	78.4a
Vitex negundo	38.9d	35.6d	3.5cd	33.3d	49.1b	55.8bc
Seed Extract (5%)						
A. indica	21.8f	20.9ef	18.8e	20.8e	40.1c	53.0e
Calophyllum inophyllum	10.3h	15.0g	10.4g	17.4f	10.7e	10.2f
Seed Oil Emulsion (3%)						
A. indica	20.4f	19.5f	15.3f	15.5g	31.2d	35.9de
C. inophyllum	12.6g	13.8e	10.9g	10.5h	11.1e	10.5f
Madhuca latifolia	21.9f	21.2e	20.2e	18.7ef	39.8c	43.7d

* Mean of three replications
DAI = days after inoculation
a = pre-inoculation spray, b = post-inoculation spray.

Biochemical Constituents of the Plant Extracts

The leaf extracts of *A. vasica, A. indica, P. longifolia, P. juliflora* and *V. negundo* that were effective in reducing the mycelial growth, spore germination and biomass production of *P. oryzae* and blast disease development contained more total sugar, reducing sugar, non-reducing sugar protein, glycoprotein, free amino acids, phenols, and tannin contents than the least effective *C. pulcherrima* (Table 4).

Table 4. Biochemical contituents of leaf extracts**

Leaf extracts	Total sugar* dry wt	Reducing sugar*	Non reducing sugar	Protein*	Glyco protein	Total phenol*	Free amino acids*	Tannin
Adhatoda vasica	10.2c	6.9b	3.3c	24.4c	1.5b	1.2b	2.7b	28.3ab
Azadirachta indica	11.4b	5.5c	5.9b	25.4b	1.1d	0.8c	5.1a	26.1c
Polyalthea longifolia	14.5a	8.3a	6.3a	15.4d	1.1d	3.1a	2.3d	27.1b
Prosophis juliflora	9.3d	6.1b	3.3c	29.4a	1.9a	2.9a	1.9e	29.8a
Vitex negundo	11.5b	8.5a	3.1c	29.5a	1.3c	3.5a	2.4cd	22.3d
Caesalpinia pulcherrima	6.1e	4.2d	1.9d	12.9e	0.8e	1.4b	1.2f	17.6e

* mg/g of leaf tissue fresh wieght

** Mean of six replications

Enzyme Activities in Rice Plants Treated with Plant Derivatives

The enzyme (peroxidase, polyphenol oxidase, phenyl alanine ammonia lyase) activities accelerated in *P. oryzae* inoculated rice plants when treated with leaf extracts of *A. vasica* and *P. juliflora* rather than the leaf extracts of *C. pulcherrima* (Table 5).

DISCUSSION

Plants synthesize several self-defense chemicals in addition to their food materials. The pesticidal properties of these chemicals have been demonstrated by many in recent years. Extracts of *Allium cepa, A. indica* (Krishnamohan et al., 1988), *Impatiens balsamina, Pseudocalymma alliaceum, Allium sativum* and *Tagetes erecta* (Lapis and Dumancas, 1979) were reported to inhibit the growth of *P. oryzae*. Similarly, many plant species were reported to exert adverse influence on many plant pathogens. The leaf extracts of *A. indica* reported to inhibit the growth of *Curvularia lunata* (Bhowmick and Vardhan, 1981) and spore germination of *Pestalotia psidii* (Pandey et al., 1983). Leaf extracts of *A. indica* and *V. negundo* reduced the spore germination of *Alternaria alternata* (Bhowmick and Choudhary, 1982). Leaf extracts of *V. negundo* inhibited the growth of

Table 5. Effect of plant derivatives on rice plants**

Leaf extracts	Changes in enzymes in rice plants							
	PO		PPO		PAL		Total phenol*	
	H	I	H	I	H	I	H	I
Adhatoda vasica	90.0	110.0	6.0	8.0	9.0	13.3	2.7	3.0
Caesalpinia pulcherrima	78.0	80.0	5.6	6.3	8.7	9.3	2.7	2.8
Prosophis juliflora	80.0	123.0	7.0	8.0	8.0	15.33	2.7	3.1
Control (untreated rice plants)	75.0	69.0	5.0	3.0	8.7	2.0	2.6	2.0

PO = peroxidase activity, PPO = polyphenol oxidase activity, PAL = phenyl alanine ammonia lyase activity, 0.01 OD = 1 unit for PO, PPO and PAL activities.

H = healthy, I = inoculated

* mg/g of leaf tissue (fresh weight)

** Mean of six replications

Drechslera turcica (Bhowmick and Vardhan, 1982). Kishore et al. (1982) observed the leaf extracts of *P. longifolia* were toxic to *Colletotrichum falcatum* and contained several antimicrobial compounds (Annapurna et al., 1983). Charya et al. (1979) reported the root extracts of *P. juliflora* completely inhibited the spore germination of *D. rostrata* and *C. lunata*. The extracts of *A. vasica* inhibited the spore germination of *Fusarium equiseti*, *F. semitectum* and *C. lunata* (Ojha and Prasad, 1986).

Several plant species have been reported to contain substances which may either act on the pathogen or induce resistance in host plants resulting in reduction in disease development (Datar, 1988; Bhowmick and Vardhan, 1981; Devi et al., 1982; Chaturvedi et al., 1987; Doubrava et al., 1988). The biochemical constituents of *P. juliflora, A. vasica, V. negundo, A. indica* and *P. longifolia* were more than those of *C. pulcherrima*. Plant extracts and chemicals have been reported to act as elicitors or induce defense mechanism (Langcake and Wickins, 1975; Arimoto et al., 1976; Ohashi, 1980). In the present study application of the leaf extracts of *P. juliflora* and *A. vasica* to *P. oryzae* inoculated rice plants resulted in an acceleration of the enzymes peroxidase, polyphenol oxidase, and phenyl alanine ammonia lyase, and phenols which are the precursors of resistance. Josefa et al. (1984) reported that crude extracts of cassava, poinsettia and ipil-ipil sprayed either as a protectant or as an eradicant controlled the *Cercospora* leaf spot of Mungbean (*Vigna radiata*). Natarajan and Lalithakumari (1987) observed the *Drechslera oryzae* infection in rice seedlings reduced by application of the leaf extracts of *L. inermis*. Doubrava et al. (1988) reported that spray of spinach and rhubarb leaf extracts induced systemic resistance in cucumber to anthracnose caused by *Colletotrichum lagenarium*.

Among the preparations tested, *P. juliflora* and *A. vasica* significantly reduced the rice blast disease. The results demonstrate that the leaf extracts of these two plant species adversely affect the pathogen and also induce host-mediated resistance.

In many rice-growing countries of South and East Asia, marginal and submarginal farmers with small land holdings can easily use these leaf extracts as an inexpensive method to save the rice crop from blast disease as the species are commonly available. Identification of the templates of the phytochemicals of these plant species will be helpful in synthesizing novel pesticides.

REFERENCES

Annapurna, Y., S. Mitra, D.S. Iyengar, S.N. Rao and U.T. Bhalerao. 1983. Antimicrobial activity of leaf extracts of *Polyalthea longifolia*. *Phytopathol.* 106(2): 183–185.

Association of Agricultural Chemists. 1965. *Official methods of analysis.* 10th ed. AOAC, Washington, DC, 501 pp.

Arimoto, V.N., T. Homna, T. Chtsu and T. Misato. 1976. Studies on chemically induced resistance of plants to diseases. 1. The effect of soaking of rice seeds in dodecyl DL alanine hydrochloride on seedling infection by *Pyricularia oryzae*. *Ann. Phytopath. Soc.* 42: 392–400.

Bhowmick, B.N. and S.K. Choudhary. 1982. Antifungal activity of leaf extracts of medicinal plants of *Alternaria alternata* (Fr.) Keissler. *Indian Bot. Rep.* 1(2): 164–165.

Bhowmick, B.N. and N. Vardhan. 1981. Antifungal activity of some leaf extracts of medicinal plants on *Curvularia lunata*. *Indian Phytopathol.* 34: 385–386.

Bhowmick, B.N. and N. Vardhan. 1982. Antimycotic activity of leaf extracts of some medicinal plants on *Drechslera turcicum* (Pats) Subram and Jain. *Biol. Bull. India* 4(1): 58–60.

Charya, M.A.S., S.M. Reddy, N.P. Kumar and S.R. Reddy. 1979. Laboratory evaluation of some medicinal plant extracts against two pathogenic fungi. *New Botanist* 6: 171–174.

Chaturvedi, R., D. Anupam, C. Dikshit and S.N. Dixit. 1987. *Adenocalyma allicea* a new source of natural fungitoxicant. *Trop. Agric.* 64(4): 318–382.

Datar, V.V. 1988. Antifungal activity of plant extracts against *Fusarium* sp. causing rhizome rot of ginger. *Nat. Sem. on Management of Crop Diseases with Plant Products/Biological Agents*. Agricultural College and Res. Inst., Madurai, 88 pp. (Abst.)

Devi, S.B., A. Naseema and M.C. Nair. 1982. *In vitro* effect of lemon grass oil on the mycelial growth of *Rhizoctonia solani*. *Indian Phytopathol*. 35: 714–716.

Doubrava, N.S., A. Ralph and Deanh Kuc. 1988. Induction of resistance to anthracnose caused by *Colletotrichum lagenarium* in cucumber by oxalate and extracts from spinach and rhubarb leaves. *Physiol. Mol. Plant Pathol*. 33(1): 69–79.

Fahmy, T. 1923. The production by *Fusarium solani* of a toxic excretary substance capable of causing wilting in plants. *Phytopathology* 13: 543–550.

Jaganathan, R. and V. Narasimhan. 1988. Effect of plant extracts/products on two fungal pathogens of finger millet. *Indian J. Mycol. Plant Pathol*. 18: 250–254.

Josefa, C., M. Maroon, O.S. Opina and A.B. Moling. 1984. Evaluation of crude plant extracts as fungicides for the control of *Cercospora* leaf spot of mungbean (*Vigna radiata*, L. Wilzek.). *Philippines Phytopathol*. 20: 27–38.

Keller, S. and R.J. Block. 1960. The separation and isolation of proteins. *A Laboratory Manual of Analytical Methods of Protein Chemistry*, Vol. I, Pergamon Press, London, 250 pp.

Kishore, N., N.K. Dubey, R.D. Tripathi and S.K. Singh. 1982. Fungitoxicity of leaf extracts of some higher plants. *Natl. Acad. Sci. Letters* 5(1): 9–10.

Krishnamohan, G., P. Narayanasamy, K. Pillayarsamy and R. Jeyarajan. 1988. Efficacy of certain products on the rice blast pathogen *Pyricularia oryzae*. *Nat. Sem. on Management of Crop Diseases with Plant Products/ Biological Agents*, Agricultural College and Res. Inst. Madurai, 88 pp. (Abst.).

Langcake, P. and S.G.A. Wickins. 1975. Studies on the action of dichlorocyclo propanes on the host-parasite relationship in rich blast disease. *Physiol. Plant Pathol*. 7: 113–126.

Lapis, D.B. and E.E. Dumancas. 1979. Fungicidal activity of crude plant extracts against *Helminthosporium oryzae*. *Philippines Phytopathol*. 14: 23–37.

Locke, J.C. 1990. Activity of extracted neem seed oil against fungal pathogens. *Neem's Potential in Pest Management Programs. Proc. USDA Neem Workshop. Natl. Tech. Inf. Service 5285 Port Royal Road, Spring field VA 22161, U.S.A.* Beltsville, Maryland, pp. 132–136.

Lowry, O.H., N.J. Roserbrough, A.L. Farr and R.J. Randall. 1951. Protein measurement with Folin-Phenol reagent. *J. Biol. Chem*. 193: 268–275.

McKinney, H.H. 1923. A new system of grading plant diseases. *J. Agric. Res*. 26: 195–218.

Muthusamy, M., K. Pillayarsamy and V. Mariappan. 1988. Efficacy of plant products against blast and brown spot diseases of rice. *Nat. Sem. on Management of Crop Diseases with Plant Products/Biological Agents*, Agricultural College and Res. Inst., Madurai, 88 pp. (Abst.).

Natarajan, M.R. and D. Lalithakumari. 1987. Antifungal activity of leaf extract of *Lawsonia inermis on Drechslera oryzae*. *Indian Phytopathol*. 40: 390–395.

Ohashi, T. 1980. Oryzemate; Novel systemic antibacterial and antifungal agrochemical. Jap. Pesticide Infor., 37: 37–39.

Ojha, N.L. and A.K. Prasad. 1986. Antifungal evaluation of leaf extracts for the control of some cucurbitaceous fruit rot diseases. *Indian Phytopathol*. 39: 153.

Pandey, R.S., S.N. Bhargava, D.N. Shukla and D.K. Dwivedi. 1983. Control of *Pestalotia* fruit rot of guava by leaf extracts of two medicinal plants. *Revista Mexicana de Fitopathologia* 2(1): 15–16.

Schmutterer, H. 1990. Properties and potential of natural pesticides from neem tree. *Azadirachta indica*. *Annu. Rev. Entomol*. 35: 271–297.

Somoyogi, M. 1952. Notes on sugar determination. *J. Biol. Chem*. 200: 145–154.

Spies, J.R. 1955. In: *Methods in Enzymology*, Vol. III. S.P. Clockwick and W.O. Kapalan (eds.), Academic Press, New York, U.S.A. pp. 467–468.

Srivastava, S.K. 1987. Peroxidase and polyphenol oxidase in *Brassica juncea* plants infected with *Macrophomina phaseolina* (Tassi) Goid. and their implication in disease resistance. *J. Phytopathol*. 120: 244–254.

Tewari, S.N. and A. Premalatha Dath. 1984. Effect of leaf extract media of some plants on the growth of fungal pathogen of rice. *Indian Phytopathol*. 37(3): 458–461.

Warthen, J.D., Jr. 1989. Neem (*Azadirachta indica* A. Juss.): Organisms affected and reference list up to date. *Proc. Entomol. Soc. Wash*. 91:367–388.

Zucker, M. 1965. Induction of PAL by light and its relation to chlorogenic acid synthesis in potato tuber tissue. *Plant Physiol*. 40: 779–781.

80

Evaluation of *Azadirachta indica* and *Lawsonia inermis* for Antifungal Activity and Studies on Neem Products for Control of Damping-off in Tobacco Nurseries

M.M. SHENOI, S.M. ABDUL WAJID AND S.S. SREENIVAS

CTRI Research Station, Hunsur 571 105, Karnataka, India

ABSTRACT

Ten botanicals were evaluated *in vitro* by paper disc and aqueous leaf extract methods against *Pythium aphanidermatum*, *Phytophthora parasitica* var. *nicotianae*, *Colletotrichum tabacum*, and *Alternaria alternata*, which are fungal pathogens of tobacco. *Azadirachta indica* and *Lawsonia inermis* effectively inhibited the radial growth of mycelia, and sporulation and germination of spores, indicating broad spectrum fungicidal potentiality.

In sick microplots, all the five neem products tested controlled damping-off caused by *P. aphanidermatum* up to 25–35 days after sowing (DAS). Of the five, Neemguard was most promising. A replicated nursery trial also indicated efficacy of Neemguard, Ashirwad, and neem cake alone against damping-off for 25 DAS and subsequently in combination with Ridomil MZ 72 WP. A preliminary trial in sick nursery on neem leaf and neem cake amendments before soil solarization with and without subsequent sprays of Ridomil MZ gave promising indication of the feasibility of evolving an integrated schedule for overall management of tobacco nurseries.

INTRODUCTION

The integrated approach to the control of pests and diseases involving an optimal combination of physical, biological, and chemical means is gaining importance in view of the overwhelming concern for indiscriminate use of pesticides which result in residual toxicity, development of resistance in pathogen, imbalance of biological equilibrium, and atmospheric pollution (Elad et al., 1980). Economically and environmentally safe methods such as botanical pesticides (Chari et al., 1990), organic amendments

(Singh, 1983), and soil solarization (Katan, 1980, 1981; Katan et al., 1983) are used either alone or in combination with chemical control for management of pests and diseases in field crops. In the present study, an attempt has been made initially to evaluate certain botanicals against four major fungal pathogens of tobacco. Neem and its products are preliminarily evaluated as organic amendments or straight products in combination with soil solarization and also with and without chemical control to explore the feasibility of evolving an integrated schedule for control of two major diseases in tobacco nursery, damping-off caused by *Pythium aphanidermatum* and black shank caused by *Phytophthora parasitica* var. *nicotianae*. Overall management of Flue cured Virginia tobacco nurseries using these alternate methods of disease control is also discussed.

MATERIALS AND METHODS

In vitro Evaluation of Botanicals

Crude leaf extracts of 10 botanicals were evaluated by paper disc method, poisoned food technique, and aqueous extract method (Tripathi et al., 1983) for testing their antifungal properties such as inhibiting mycelial growth and sporulation of four major fungal pathogens of tobacco, *P. aphanidermatum*, *P. parasitica* var. *nicotianae*, *Colletotrichum tabacum*, and *Alternaria alternata* which cause damping-off, black shank, and anthracnose in nursery, and brown spot in field crop respectively. The effect of extracts on spore germination was studied as described by Natarajan and Lalithakumari (1987).

Evaluation of Neem Products in Sick Microplots and Nursery

Five neem products—Neemguard, Ashirwad, Neemark, Neemox, and Neemorate—along with neem cake amendment and untreated check were evaluated under glasshouse conditions in microplots of 0.5 sq m. Soil was made sick by artificially inoculating *P. aphanidermatum* grown on sand oat media. A higher seed rate of 0.3 g/sq m was used to create disease pressure. Neem products were applied as pre-sowing soil drenching at selected doses (Table 1). Damping-off disease initiation and subsequent incidence was recorded periodically and per cent disease control was calculated. In a sick nursery, Neem cake, Neemguard, Neemark, and Ashirwad were evaluated singly as well as in combination with Ridomil MZ. A total of 10 treatments including recommended effective check and untreated check were tested (Table 2). Randomized block design was adopted with four replications. Incidence of damping-off, blight, and black shank diseases was recorded at regular intervals and per cent disease control was estimated by recording number of transplants pulled in three instalments.

Integrated Management of Nursery Diseases

A total of 10 treatment schedules involving soil solarization, soil amendment with neem cake or shredded fresh neem leaf, and foliar spray of Ridomil MZ in different combinations were evaluated preliminarily for the purpose of assessing the integrated schedules. Recommended Ridomil MZ schedule and untreated check were also included. Soil

Table 1. Preliminary evaluation of neem products in sick microplots against damping-off disease

Neem products	Av. germination count/100 sq cm at 15 DAS	% disease incidence at regular intervals			
		20 DAS	25 DAS	30 DAS	35 DAS
Neem cake 400 g/sq m	66.4	0.0 (100)	0.0 (100)	0.0 (100)	25.0 (75)
Neemguard 0.5%	62.2	0.0 (100)	0.0 (100)	0.0 (100)	0.0 (100)
Ashirwad 0.5%	67.0	0.0 (100)	0.0 (100)	12.0 (87)	60.0 (40)
Neemark 5 g/sq m	54.0	0.0 (100)	0.0 (100)	2.0 (98)	25.0 (75)
Neemox 0.5%	69.2	0.0 (100)	5.0 (89)	30.0 (67)	60.0 (40)
Neemorate	60.0	0.0 (100)	0.0 (100)	5.0 (94)	25.0 (75)
Check (untreated)	60.8	5.0	45.0	90.0	100.0

Figures in parentheses indicate per cent disease control over check

Table 2. Evaluation of neem products against damping-off caused by *P. aphanidermatum* in tobacco seed beds*

Pre-sowing treatment	Av. germination count/100 sq cm at 15 DAS	% damping-off incidence at regular intervals		
		At 30 DAS without foliar sprays of Ridomil MZ	After 30 DAS without foliar sprays of Ridomil MZ	After 30 DAS with foliar sprays of Ridomil MZ at 30 and 50 DAS
Neem cake 400 g/sq m	26.5	5.0 (46)	99.3 (0)	15.3 (85)
Neemguard 0.5% at 2 l solution/sq m	25.6	0.8 (91)	83.3 (16)	0.5 (99)
Neemark 5 g/sq m	20.1	1.1 (88)	98.3 (1)	2.5 (98)
Ashirwad 0.5% at 2 l solution/sq m	25.5	1.0 (89)	95.0 (4)	2.6 (97)
Ridomil MZ 72 WP 0.1% at 500 ml solution/sq m	23.1	0.0(100)	0.3 (99)	0.3 (99)
Untreated check	21.3	9.3	99.0	99.0

*Average of four replications
Figures in parentheses indicate per cent disease control over check

solarization was done as described by Katan (1980) for six weeks using clear, transparent polyethylene sheets of 50 gauge density. Neem leaf and neem cake was incorporated into soil at 3 kg and 400 g per sq m respectively, prior to soil solarization. Ridomil MZ schedules were given as per the recommendations. Observations on germination were recorded by counting number of sprouts in random sample area of 100 sq cm at 15 DAS. Disease incidence and production of transplants were determined as described earlier.

RESULTS AND DISCUSSION

In vitro Evaluation of Botanicals

The effect of aqueous extracts of botanicals on growth and sporulation of fungal pathogens *in vitro* and on spore germination is presented in Table 3. The 10 leaf extracts varied in antifungal activity. *Lawsonia inermis* completely inhibited mycelial growth and greatly affected sporulation and germination of spores of all the four pathogens. The strong antifungal activity of this plant is reported to be due to lawsone, a high phenol content in its leaf extract (Natarajan and Lalithakumari, 1987). *Azadirachta indica* was the only other plant extract which exhibited similar inhibitory effect prominently on all the four pathogens. Active principles such as meliantriol and azadirachtin isolated and identified as reported by Jotwani and Srivastava (1981b) could be responsible for this fungitoxicity. Ketkar and Ketkar (1990) have reported that neem and its products are effective against six species of fungi. Antifungal property of neem on *Aspergillus niger* (Sinha and Saxena, 1987), *Puccinia arachidis* (Muthusamy, 1988), *Macrophomina* (Dwivedi and Dubey, 1986), *Alternaria alternata* (Bhowmick and Chowdary, 1982), *Solerotium rolfsii* (Singh and Dwivedi, 1987), and seed microflora (Khan and Shah, 1992) has been reported in recent years. The strong broad spectrum fungicidal potentiality registered *in vitro* by *L. inermis* and *A. indica* opens up a new chapter on the research related to their use in control of fungal diseases in field situations, particularly in tobacco nurseries. The crude leaf extracts of *Datura fastuosa*, *Ipomea hispida*, *Ocimum canum*, and *Polyalthia longifolia* were also effective against these pathogens, except against *P. aphanidermatum*, while the leaf extract of *Bougainvillaea spectabilis* was inhibitory only against *P. aphanidermatum*.

Evaluation of Neem Products in Sick Microplots and Nursery

The results of a preliminary evaluation of neem products in sick microplots and nursery are presented in Tables 1 and 2. No adverse effect on germination was noticed except in Neemark treatment, wherein the germination was less and toxicity on young seedlings in the form of stunted growth was also observed. Up to 25 DAS, incidence of damping-off in the treatments was nil to 5 per cent as against 45 per cent in untreated check (Table 1). The efficacy of neem products gradually decreased subsequently to 40–75 per cent disease control by 35 DAS, except Neemguard, which exhibited 100 per cent disease control. The disease buildup after 25 DAS indicated that effect of treatments was mitigated with passage of time. When these neem products were further tested in sick nurseries either alone or in combination with Ridomil MZ (Table 2), their limited efficacy up to 25 to 30 DAS was confirmed. Combining these treatments with supportive foliar sprays of Ridomil MZ at 30 and 50 DAS increased efficacy to the level obtained in recommended schedule of Ridomil MZ presowing drench to be followed by foliar sprays. Thus, the two initial trials have given positive results on the effectiveness of neem products for control of damping-off caused by *P. aphanidermatum* in tobacco nurseries at initial stages, and incorporation of chemical fungicide to the soil can be avoided. The trend of results suits well with the present-day stress on the significance of using pesticides from naturally occurring plant products as part of developing envi-

Table 3. *In vitro* evaluation of botanicals against four fungal pathogens of FCV tobacco

Botanicals	Family	Percentage inhibition of mycelial growth				Percentage inhibition of sporangia/conidia germination			
		Pa	Pp	Ct	Aa	Pa	Pp	Ct	Aa
* *Azadirachta indica*	Meliaceae	65.0	59.0	37.0	52.0	100.0	100.0	66.1	100.0
Bougainvillea spectabilis	Nyctagenaceae	100.0	0.0	0.0	0.0	0.0	0.0	0.0	0.0
Carica papaya	Caricaceae	0.0	0.0	20.0	10.0	0.0	0.0	16.7	94.0
Datura fastuosa	Solanaceae	0.0	31.0	36.0	28.0	0.0	100.0	31.4	52.3
Ipomea hispida	Solanaceae	0.0	18.0	7.5	6.6	0.0	31.1	28.2	76.2
Lantana camera	Verbenaceae	0.0	0.0	0.0	0.0	0.0	0.0	0.0	0.0
* *Lawsonia inermis*	Lythraceae	100.0	100.0	100.0	100.0	31.7	86.0	100.0	100.0
Ocimum canum	Labiatae	0.0	33.0	10.5	40.0	0.0	100.0	10.8	17.6
* *Polyalthia longifolia*	Annonaceae	0.0	34.0	34.0	38.0	10.3	44.2	100.0	79.3
Tridax procumbens	Compositae	0.0	0.0	0.0	0.0	0.0	0.0	0.0	0.0

* Extracts of these three botanicals have also inhibited prominently the production of sporangia of *P. aphanidermatum* and *P. parasitica* var. *nicotianae*.
Pa = *P. aphanidermatum*, Pp = *P. parasitica* var. *nicotianae*, Ct = *C. tabacum*, Aa = *A. alternata*

ronmentally sound methods of pest and disease control (Jotwani and Srivastava, 1981 a). Products from *A. indica* for control of pests have attained worldwide prominence in recent years (Govindachari, 1992). According to Bindra (1992), more than 200 insect species as well as mites, nematodes, fungi, and bacteria are reported to be affected by neem. Nearly 30 pests on agricultural crops are reported to be controlled by neem products (Govindachari, 1992). However, existing reports on fungicidal activity of neem and neem-based products are very few. The present study attempts to highlight the wider scope for use of neem-based products.

Integrated Management of Nursery Diseases

Data presented in Table 4 indicate that soil amendments by application of neem cake or chopped fresh neem leaves alone or in combination with solarization for six weeks control initial damping-off very effectively in highly sick soils. Combination treatments like solarization + neem leaf and solarization + neem cake without Ridomil MZ were effective up to 30 DAS (0.5 to 2.0 per cent disease as against 20 per cent incidence in untreated check). This further supports the finding that application of Ridomil MZ could be delayed up to 25 to 30 DAS if neem products are incorporated to the soil alone or in combination with soil solarization. However, sudden buildup in disease pressure after 30 DAS indicates their limited efficacy and the necessity of using chemical fungicides like Ridomil MZ subsequently for management of damping-off and other diseases. Nevertheless, the trend of results points to the feasibility of evolving an integrated schedule consisting of soil solarization, soil amendments, and fungicide sprays for overall management of tobacco nurseries. In the present trial, the best treatments which produced maximum healthy transplants include the combination schedules such as soil solarization + neem leaf + Ridomil MZ spray (748/sq m) and soil solarization + neem cake + Ridomil MZ (716/sq m) as compared to presently recommended schedule of Ridomil MZ presowing drench + sprays which yields 540 transplants per sq m.

Soil solarization per se has been reported to be successful in controlling several diseases (Pullman et al., 1981; Katan, 1980; and Katan et al., 1976). Soil amendment with products like neem cake are also reported to result in ultimate reduction of inoculum density and disease suppression (Singh, 1983). Most organic amendments are reported to work excellently at higher concentration and higher temperature causing maximum suppression of fungi (Singh, 1983; Narendra Singh and Singh, 1982). Results of the present study on soil amendments with neem leaf and neem cake before soil solarization support this view.

The various studies reported in this paper highlight the importance of using neem products such as Neemguard and also soil amendment with neem cake before soil solarization for initial control of diseases and for overall management of tobacco nurseries in combination with delayed application of chemical fungicides. The higher yield of healthy transplants obtained in combination schedules is indicative of other influences of the components of the schedule such as (1) neem cake amendment promoting the rapid, luxuriant growth of seedlings to overcome disease attack, regulating nitrogen nutrition (Attri and Raviprasad, 1981), and improving physical characteristics of soil, and (2) soil solarization inhibiting weed growth in general and suppressing to a certain extent the perennial weed nutgrass, which is common in tobacco nurseries. These two

Table 4. Integrated approach for the management of damping-off, blight, and black shank diseases in tobacco nurseries *

Schedule of treatments			Av. germination count at 15 DAS/ 100 sq cm	% damping-off incidence			% Blight and black shank incidence at end	Healthy transplants/ sq m from 3 pullings	Total healthy seedlings/ sq m at end
Organic soil amendments	Solarization + for 6 weeks	Ridomil Mz + foliar sprays 0.2% at 25 and 40 DAS		25 DAS	30 DAS	40 DAS			
Chopped neem leaf 3 kg/sq m			23.7	2.6 (71)	10.0 (50)	100.0 (0)		32	32
Chopped neem leaf 3 kg/sq m	+		23.0	1.0 (89)	2.0 (90)	100.0 (0)		35	35
Chopped neem leaf 3 kg/sq m		+	23.7	2.5 (72)	3.0 (85)	10.0 (90)	4.6	388	528
Chopped neem leaf 3 kg/sq m	+	+	23.0	1.0 (89)	2.0 (90)	5.0 (95)	0.7	748	1088
Neem cake 400 g/sq m			21.6	1.3 (86)	6.0 (70)	100.0 (0)		30	30
Neem cake 400 g/sq m	+		21.6	0.0 (100)	0.5 (98)	100.0 (0)		20	20
Neem cake 400 g/sq m		+	22.6	0.0 (100)	0.5 (98)	10.0 (90)	4.7	196	246
Neem cake 400 g/sq m	+	+	23.0	0.0 (100)	0.0 (100)	0.5 (99)	0.8	716	1326
		Ridomil MZ Pre-sowing drench + two foliar sprays (recommended check)	23.0	0.0 (100)	0.0 (100)	1.0 (99)	0.7	540	880
Untreated check			23.3	9.0	20.0	100.0		0	0

*Mean of three replications

Figures in parentheses indicate per cent disease control over untreated check

components of integrated schedule, thus might even help in avoiding application of top dressing with nitrogen.

ACKNOWLEDGEMENTS

The authors are grateful to Dr. M.S. Chari, Director, Tobacco Research, Central Tobacco Research Institute, Rajahmundry, India, for encouraging us to take up these studies and for providing the facilities. Our sincere thanks to Sri J.S.L. Moses, Officer-in-Charge, CTRI Research Station, Hunsur, for his cooperation and for critical comments on the paper. The supply of neem product samples by various firms is gratefully acknowledged.

REFERENCES

Attri, B.S. and Raviprasad. 1981. Preliminary studies on the inhibition of carbofuran degradation in soil by neem cake and neem oil extractives. *Pesticides* 15: 6–7.

Bhowmick, B.N. and B.K. Chowdary. 1982. Antifungal activity of leaf extracts of medicinal plants on *Alternaria alternata* (Fr.) Keissler. *Indian Bot. Reptr.* 1: 164–165.

Bindra, S. 1992. Neem, back to the roots—The West discover's the tree's time-tested virtues. *India Today* 31 July: 146–151.

Chari, M.S., S.R. Prabhu and G. Ramaprasad. 1990. Role of botanicals in pest management of tobacco in India. Souvenir, *Nat. Symp. on Problems and Prospects of Botanical Pesticides in Integrated Pest Management*. CTRI, Rajahmundry, 73 pp.

Dwivedi, R.S. and R.C. Dubey. 1986. Effects of volatile and non-volatile fractions of two medicinal plants on germination of *Macrophomina phaseolinasclerotia. Trans. Brit. Mycol. Soc.* 87: 326–328.

Elad, Y., J. Katan and I. Chet. 1980. Physical, biological and chemical control integrated for soil borne diseases in potatoes. *Phytopathology* 70: 418–422.

Govindachari, T.R. 1992. Chemical and biological investigations in *Azadirachta indica* (the neem tree). *Curr. Sci.* 63: 117–122.

Jotwani, M.G. and K.P. Srivastava. 1981a. Neem Insecticide of future—I. As protectant against stored grain pests. *Pesticides* 15: 19–23.

Jotwani, M.G. and K.P. Srivatsava. 1981b. Neem: Insecticide of the future—II. Protection against field pests. *Pesticides* 15: 40–47.

Katan, J. 1980. Solar pasteurization of soils for disease control: Status and prospects. *Plant Dis.* 64: 450–454.

Katan, J. 1981. Solar heating (solarization) of soil for control of soil-borne pests. *Annu. Rev. Phytopathol.* 19: 211–236.

Katan, J., A. Greenberger, H. Alan and A. Grinstein. 1976. Solar heating by polyethylene mulching for control of diseases caused by soil-borne pathogens. *Phytopathology* 66: 683–688.

Katan, J., G. Fishler and A. Grinstein. 1983. Short and long term effects of soil solarization and crop sequence on *Fusarium* wilt and yield of cotton in Israel. *Phytopathology* 73: 1215–1219.

Ketkar, C.M. and Mohan S. Ketkar. 1990. Versatile neem. A source for plant protection. Souvenir, *Nat. Symp. on Problems and Prospects of Botanical Pesticides in Integrated Pest Management* CTRI, Rajahmundry, 73 pp.

Khan, M.I. and Nazeer H. Shah. 1992. Antifungal activity of leaf extract of neem on seed mycoflora of wheat. *Bioved.* 2: 209–210.

Muthusamy, M. 1988. Evaluation of neem products against rust disease of groundnut. *Neem Newsletter* 5: 48.

Narendra Singh and R.S. Singh. 1982. Effect of oil cake amended soil atmosphere on pigeon pea wilt pathogen. *Indian Phytopathol.* 35: 300–305.

Natarajan, M.R. and D. Lalithakumari. 1987. Leaf extracts of *Lawsonia inermis* as antifungal agent. *Curr. Sci.* 56: 1021–1022.

Pullman, G.S., J.E. Devay and R.H. Garber. 1981. Soil solarization and thermal death: A logarithmic relationship between time and temperature for four soil-borne plant pathogens. *Phytopathology* 71: 959–964.

Singh, R.S. 1983. Organic amendments for root disease control through management of soil microbiota and the host. *Indian J. Mycol. Plant Pathol.* 13: 1.

Singh, R.K. and R.S. Dwivedi. 1987. Fungitoxicity of different plant spp. against *Sclerotium rolfsii* Sacc. a root-rot pathogen of barley. *Nat. Acad. Sci. Letters* 10: 89–91.

Sinha, Purnima and S.K. Saxena. 1987. Effect of neem powder and extract on development of fruit rot caused by *Aspergillus niger. Neem Newsletter* 4: 45–47.

Tripathi, N.N., N.K. Dubey, R.D. Tripathi and S.N. Dixit. 1983. Growth stages vis-a-vis fungitoxicity in *Iberis amara*. *Indian Phytopathol*. 36: 284–287.

81

Management of Pests and Diseases of Rice Crop with Neem Products

V. MARIAPPAN, K. SETHURAMAN, F. VIDYA, K.R. KARUNAKARAN AND P. VIDHYASEKARAN

Tamil Nadu Agricultural University
Coimbatore 641 003, India

ABSTRACT

The efficacy of neem (*Azadirachta indica* A. Juss.) oil at 3 per cent and neem seed kernel extract (NSKE) at 5 per cent was compared with that of the insecticide phosphamidon against pests and diseases of rice crop in farmer's fields. Rice seedlings raised in nursery, treated with neem cake (at 150 kg/ha as basal application), and seedlings raised with carbofuran 3G (at 3.5 g/m^2 at 10 days after sowing), were planted in main field in separate plots, with suitable controls. Neem oil (NO) and NSKE were sprayed at 500 l/ha on plants that had received neem cake in nursery separately. Phosphamidon (500 ml/ha) was sprayed on plants that had received carbofuran in nursery. Spraying was started immediately after the crossing of Economic Threshold Level (ETL) by any insect or disease. The damage caused by the insect pests leaf hopper (*Cnaphalocrocis medinalis* Guenee) and brown plant hopper (*Nilaparvata lugens* Stal.) and sheath blight disease (*Rhizoctonia solani* Kuhn.) to rice crop were significantly reduced by NO and NSKE and the reduction was comparable to that of phosphamidon. Neem oil and NSKE were not harmful to the beneficial predators lycosa (*Lycosa pseudoannulata* Boes et Str.) and mirid bug (*Cyrtorhinus lividipennis* Reuter.) as was the insecticide phosphamidon. Stem borer (*Scirphophaga incertulas* Walk.), earhead bug (*Leptocorisa acuta* Thumb.), and brown spot (*Helminthosporium oryzae* Dreshler) were below ETL in all treatments and there were no differences in the level of incidence between treatments. The yields obtained from neem oil and NSKE treatments were significantly higher than from the control treatment and either comparable or higher than the yield of phosphamidon-treated plants. Similarly, the net gain obtained from neem oil and NSKE treatments was higher than or similar to that obtained from phosphamidon treatment.

INTRODUCTION

The bioefficacy of neem (*Azadirachta indica* A. Juss.) products against several crop pests has been reported in recent years (Schmutterer, 1990; Warthen, 1989). The neem products have been demonstrated as selective to targetted insect pests and pathogens. Neem formulations also were reported to be least harmful to beneficial organisms that survive exclusively on crop pests. Neem oil, neem fractions or extracts caused no obvious harm to natural enemies of hoppers and the endoparasitic hymenopterous parasites of *Cnaphalocrocis medinalis* larvae (Saxena et al., 1981a; 1981b; Schmutterer et al., 1983). Neem products have been reported to encourage the growth of antagonistic organisms that adversely affected the pathogen. Wajidkhan et al. (1974) reported that the application of neem cake and castor cake increased the population of fungi in the rhizosphere that adversely affected the pathogenic *Rhizoctonia solani*. Krishnamoorthy (1987) found that in soil amended with neem cake the population of beneficial antagonistic fungus *Trichoderma viride* increased by 300 per cent over control. These results have been encouraging the use of neem products instead of chemical pesticides for management of pests and diseases of crop plants in farmer's field. Rice crop is affected by many insect pests and diseases around Coimbatore. Farmers in this region frequently used to apply synthetic pesticides which were relatively unselective, affecting not only the insect pests and pathogens, but also the beneficial organisms. Hence field trials were conducted 1) to verify the efficacy of neem cake, neem oil and need seed kernel extract (NSKE) for the control of insects pests and diseases of rice in farmer's field, 2) to compare the performance of neem cake, neem oil and NSKE with that of recommended pesticide, and 3) to determine whether neem products can improve on the existing farmer practice of pest control and whether it is cost effective.

MATERIALS AND METHODS

Three field trials were conducted at Coimbatore (July–October 1991), Pasur (July–October 1991) and Mathampatti (February–May 1992) using IR 50, ADT 36 and J 13 rice varieties respectively.

Treatments

In the nursery neem cake was applied at the rate of 150 kg/ha as basal application during the last ploughing (in four plots of 40 m^2 each). Carbofuran 3G (Furadon) at 3.5 g/m^2 was applied 10 days after sowing (in two plots of 40 m^2 each). Two plots of 40 m^2 each made up the untreated control.

The experiment in the main field was laid out with the following treatments in a randomised block design with six replications and with the plot size of 40 m^2 each. The treatments were as follows:

1. NSKE (5 per cent) at 500 l/ha
2. Neem oil (3 per cent) at 500 l/ha
3. Phosphamidon at 500 l/ha
4. Untreated control

The seedlings from neem cake-treated plots in nursery were planted in neem oil and in NSKE plots in the main field. The seedlings from carbofuran-treated plots were planted in phosphamidon treatment and the seedlings from untreated control plots in nursery were planted in control plots in main field.

Preparation of Neem Oil (3 per cent)

Neem oil 15 l were mixed with 500 l of water containing 500 ml of liquid detergent Teepol and stirred well. The emulsion was immediately used for spraying.

Preparation of NSKE (5 per cent)

Neem seed kernal 25 kg was powdered and soaked in 100 l water for 12 hrs. The content was filtered. The filtrate was made up to 500 l and added to 500 ml of Teepol for spraying.

Spraying with neem oil, NSKE, and phosphamidon was started immediately after the crossing of ETL by any one insect pest or disease was noticed.

Sampling of Pests and Diseases

The sampling of pests and diseases was as follows:

1) Stem borer : Per cent of dead hearts out of 20 plants in five microplots.
2) Leaf folder : Per cent damaged leaves in 20 hills each in five micro plots.
3) Brown plant hopper : Number per 10 hills by counting nymphs and adults present on two hills each from five microplots.
4) Earhead bug : Number on two hills in each of five microplots.
5) Lycosa : Number per hill from 20 hills each in five microplots.
6) Mirid bug : Number per 10 hills by counting from two hills each in five microplots.
7) Brown spot : Maximum grade value for each plant from 20 plants each from five microplots.
8) Sheath blight : Maximum grade value of 0–9 Scale (IRRI, 1988) for each hill and 10 hills each from five microplots were scored and per cent disease index was worked out using the formula (Mckinny, 1923):

$$\frac{\text{Sum of all grades multiplied by 100}}{\text{Number of hills observed multiplied by 9}}$$

RESULTS

Season and Incidence of Pests and Diseases

Incidence of stem borer was noticed at Coimbatore and Pasur during samba season (July–October), while leaf folder was observed at Coimbatore during samba and at

Mathampatti during summer (February–May). Brown plant hopper was observed at Pasur during samba season, while earhead bug was prevalent during summer at Mathampatti. Among the predators, mirid bug was prevalent during summer at Pasur, while lycosa was seen at both seasons at all three places.

Among the diseases, incidence of brown spot (Coimbatore) and sheath blight (Pasur) was noticed during samba season only.

Control of Insect Pests and Diseases and Cost-effective Considerations

At Coimbatore incidence of stem borer was first noticed crossing ETL (10 per cent dead hearts) from 45 days after sowing. Neem oil, NSKE and phosphamidon were immediately sprayed and spraying was repeated after 15 days. On the subsequent days incidence of leaf folder, lycosa, and brown spot was observed.

Stem borer damage was recorded up to 27.0 per cent in control treatment, while the damage was slightly reduced (23.1 to 25.5 per cent) in treated plots. However there was no significant difference between treatments.

The damage due to leaf folder was significantly reduced in neem oil and NSKE plots (4.3 and 3.6 per cent respectively) as compared to 5.2 and 5.5 per cent respectively in phosphamidon and control plots. The lycosa population was reduced in the plots treated with neem oil, NSKE and phosphamidon (3.5, 4.8 and 2.3 per hill respectively) compared with control (5.3/hill), but the reduction with neem oil and NSKE was not drastic as with phosphamidon. Brown spot incidence was less in all treated plots (grade 1) when compared to control (grade 3).

All the treated plots recorded a higher yield than control. The NSKE treatment recorded the maximum yield of 3,347 kg/ha, which was significantly higher than the 3,210 kg/ha obtained with phosphamidon and 2,974 kg/ha with neem oil. The increased net gain obtained from NSKE, neem oil and phosphamidon treatments were in the order of Rs 1,805, 667 and 1,465 over control respectively. The NSKE treatment yields a net gain of Rs 340/ha over phosphamidon treatment.

Table 1. Efficacy of neem products in management of rice pests and diseases at Coimbatore (India)

(Mean of six replications)

Treatment	Stem borer (% dead heart)	Leaf folder (% damaged leaves)	Lycosa (No./hill)	Brown spot (Grade value)	Yield (kg/ha)	Yield value (Rs)	Treatment cost (Rs)	Net gain due to treatment (Rs)
NC + NSKE	23.7	3.6a	4.8	1	3,347a	10,870	464	10,414
NC + NO	23.1	4.3ab	3.5	1	2,974c	9,665	389	9,276
C + phospha midon	25.5	5.2bc	2.3	1	3,210b	10,434	360	10,074
Control	27.0	5.5c	5.3	3	2,649d	8,609	—	8,609
C.D.	NS	S	NS	NS	S	—	—	—

Variety IR 20
Season July–October 1991

At Pasur brown plant hopper was first noticed crossing ETL (1/tiller) at 42 days after planting. Neem oil, NSKE and phosphamidon were sprayed immediately and spraying was repeated after 15 days. Stem borer, mirid bug, and lycosa were observed in the subsequent periods. Sheath blight appeared late at grain setting stage and hence no fungicide treatment was given.

The treatments with neem oil, NSKE and phosphamidon reduced the brown plant hopper population significantly (58.8, 59.5 and 56.0 per cent respectively) when compared to control (100.0 per cent) (Table 2). Incidence of stem borer was much less than ETL (10 per cent dead hearts) in all the treatments including control and there were no significant differences between the treatments. The predators mirid bug and lycosa were relatively larger in number in all the treated plots than in control. Sheath blight was reduced significantly in all the treatments (25.18 to 30.68) when compared to control (33.16). The neem oil and NSKE were on par and were more effective than phosphamidon.

The phosphamidon treatment gave a maximum yield (7,667 kg/ha) while neem oil and NSKE treatments yielded 5,476 and 5,684 kg/ha respectively. The net gain recovered by phosphamidon, NSKE, and neem oil was Rs 10,277, 3,403 and 2,802 per ha over control.

At Mathampatti incidence of insect pests was below ETL and there was no disease incidence. However, one spraying was given 45 days after planting. Leaf folder and earhead bug incidence were uniform in all the treatments (Table 3). Lycosa was observed in all the treatments.

The yields obtained from plots treated with NSKE, neem oil and phosphamidon (4,672, 4,236 and 4,725 kg/ha respectively) were significantly more than the yield obtained from the control plot (3,750 kg/ha). The net gains obtained from phosphamidon, NSKE and neem oil were in the order of Rs 3,008, 2,764 and 1,434 respectively over control.

Table 2. Efficacy of neem products in management of rice pests and diseases at Pasur, Coimbatore District

(Mean of six replications)

Treatment	BPH (no./10 hills)	Stem borer (%)	Mirid bug (no./10) hills)	Lycosa population (no./10 hills)	Sheath blight (PDI)	Yield (kg/ha)	Yield value (Rs)	Treatment cost (Rs)	Net Gain (Rs)
NC+NSKE	59.5a	5.5	8.0	5.5	25.18a	5,684b	18,473	464	18,009
NC+NO	58.0a	5.4	8.2	6.0	26.18a	5,476b	17,797	389	17,408
C+phosphamidon	56.0a	5.9	8.2	5.0	30.68b	7,667a	24,918	360	24,883
Control	100.7b	4.7	2.5	3.0	33.16b	4,494c	14,606	—	14,606
C.D.	S	NS	NS	NS	S	S	—	—	—

Variety ADT 36

PDI: Per cent disease index; Season July–October 1991.

Table 3. Efficacy of neem products on management of rice pests at Mathampatti, Coimbatore District

(Mean of six replications)

Treatment	Leaf folder (%)	Earhead bug (no.)	Lycosa (no.)	Yield (kg/ha)	Yield value (Rs)	Treatment cost (Rs)	Net gain (Rs)
NC + NSKE	2.6	0.1	1.5	4,672a	15,184	232	14,952
NC + NO	2.4	0.2	1.0	4,236a	13,767	145	13,622
C + phosphamidon	2.1	0.1	0.8	4,725a	15,356	180	15,176
Control	2.5	0.2	1.0	3,750b	12,188	—	12,188
C.D.	NS	NS	NS	NS	—	—	—

Variety J 13
Summer Season, February–May 1992

DISCUSSION

Neem oil and NSKE were significantly effective in reducing the damages caused by leaf folder and brown plant hopper to rice crop and were comparable to insecticide phosphamidon. Saxena et al. (1981b) reported that neem seed oil was highly effective against rice leaf folder from egg to adult stages. Brown plant hopper nymphal population were reported to have been reduced more in plots treated with neem urea mixture than in urea-treated plots by Saxena et al. (1983) at IRRI, Philippines. Prasad et al. (1984) reported that the application of neem cake at 125 kg/ha in an operational research trial to rain fed low land rice crop at the panicle initiation stage reduced the susceptibility to brown plant hopper.

Sheath blight disease intensity was significantly reduced by NSKE and neem oil treatments. Incidence and intensity of sheath blight disease of rice were reduced in rice plants applied with oil cakes of neem and Punnai (*Calophyllum inophyllum*: Rajan and Alexander, 1988). Rajan and Alexander also reported that the population of the saprophytes (fungi and bacteria) which are antagonistic to the pathogen increased in plots treated with oil cake and there was a negative correlation between saprophytes and pathogen population. In the present study the brown spot disease incidence also was reduced in plots treated with NSKE and neem oil. Thangamani and Narayanasamy (1988) reported that the rice brown spot incidence was reduced to 12.9 and 9.9 per cent respectively by neem oil (1 per cent) foliar spray, as against 18.5 per cent incidence in untreated control.

Among the two neem products tested, NSKE has been more effective than neem oil in reducing the leaf folder, brown plant hopper, and sheath blight.

The populations of the beneficial predators mirid bug and the predatory spider lycosa were not affected by NSKE and neem oil when compared to phosphamidon. Topical application of neem oil was not reported to affect lycosa with 50 μg neem oil/individual (Saxena et al., 1984a). Saxena et al. (1984b) reported that the natural enemies of rice pests, whorl maggot, stem borer and rice bugs were not affected by neem cake-urea mixture (2:10) in field.

The yields obtained from the plots treated with NSKE and neem oil were significantly higher than the control at all the three places. Saxena et al. (1983) observed from field trials conducted at IRRI during 1980 that plots treated with neem cake-urea mixture (2:10) gave significantly higher yield than the control. Saxena and Justo (1986) reported that the application of custard apple oil and neem oil mixture alone or in combination with neem cake-urea mixture has resulted in lower incidence of tungro (RTV) and higher yields in three consecutive croppings of rice from 1984-85 at IRRI, Philippines. Saxena (1986) reported that the yields from rice field sprayed with neem oil + carbofuran, neem oil + latex, and neem oil were increased (3.85, 3.65 and 3.70 t/ha respectively from unsprayed control, which yielded only 3.45 t/ha. From the present study it is evident that the yield from NSKE-treated plots was significantly greater than (Coimbatore) or on par with yield from plots treated with phosphamidon spray (Mathampatti).

The application of NSKE on rice crop has been found to be as effective as phosphamidon in reducing the loss caused by insect pests and diseases of rice crop and also in conserving the beneficial predators. The use of neem oil and NSKE which are commonly available to the native rice farmers, is ideally suited for crop protection in developing countries.

REFERENCES

International Rice Research Institute. 1988. *Standard Evaluation System for Rice*, 3rd ed. The International Rice Testing Programme, Los Banos, Philippines, 54 pp.

Krishnamoorthy, A.S. 1987. Biological control of damping off disease of tomato caused by *Pythium indicum*. M.Sc. (Ag.) Thesis, Tamil Nadu Agricultural University, Coimbatore, India, 135 pp.

McKinny, H.H. 1923. A new system of grading plant diseases. *J. Agric. Res.* 26: 195–218.

Prasad, C., J.P. Kulshresta and V.R.P. Sinha. 1984. Operational research projects on rice in India. *Proc. Int. Conf. on Rainfed Lowland Rice*. Bhubaneswar, India (mimeo).

Rajan, K.M. and S. Alexander. 1988. Oilcakes as tools of biological control of sheath blight of rice. *Proc. Nat. Sem. on Management of Crop Diseases with Plant Products/Biological Agents*. Agricultural College and Research Institute, Madurai, India, pp. 11–12 (Abst.).

Saxena, R.C. 1986. Neem seed derivatives for management of rice insect pests a review of recent studies. *Proc. 3rd Int. Neem Conf.*, Nairobi, pp. 81–93.

Saxena, R.C., P.B. Epino, Tu Cheng Wen and B.C. Puma. 1984a. Neem, chinaberry and custard-apple: Antifeedant and insecticidal effects of seed oils on leaf folder and planthopper pests of rice and their natural enemies. *Proc. 2nd Neem Conf.*, Rauischholzhausen, pp. 403–412.

Saxena, R.C. and H.D. Justo, Jr. 1986. Effect of custard-apple oil, neem oil and neem cake on green leaf hopper (GLH) population and on tungro (RTV) infection. *Int. Rice Res. Newsl.* 11(2): 25.

Saxena, R.C., H.D. Justo, Jr. and P.B. Epino. 1983. Evaluation and utilization of neem cake against the brown planthopper *Nilaparvata lugens*. *Proc. 2nd Int. Neem Conf.*, Rauischholzhausen, pp. 391–402.

Saxena, R.C., N.J. Liquido and H.D. Justo, Jr. 1981a. Neem seed oil, a potential antifeedant for the control of rice brown planthopper (*Nilaparvata lugens*). *Proc. 1st Int. Neem Conf.*, Rottach-Egern, pp. 171–188.

Saxena, R.C., G.P. Waldbauer, N.J. Liquido and B.C. Puma. 1981b. Effects of neem seed oil on the rice leaf folder (*Cnaphalocrocis medinalis*). *Proc. 1st Int. Neem Conf.*, Rottach-Egern, 189–204.

Schmutterer, H. 1990. Properties and potential of natural pesticides from the neem tree, *Azadirachta indica*. *Annu. Rev. Entomol.* 35: 271–297.

Schmutterer, H., R.C. Saxena and J.V.D. Heyde. 1983. Morphogenetic effects of some partially purified fractions and methanolic extracts of neem seeds on *Mythimna separata* Walker and *Cnaphalocrocis medinalis* Guenee. *Z. Angew. Ent.* 95: 230–237.

Thangamani Narayanasamy and P. Narayanasamy. 1988. Role of neem products in the control of fungal diseases of rice. *Proc. Nat. Sem. on Management of Crop Diseases with Plant Products/Biological Agents*, Agricultural College and Research Institute, Madurai, India, pp. 6–7 (Abst.).

Wajidkhan, M., M. Abrar, M. Khan and S.K. Saxena. 1974. Rhizosphere fungi and nematodes of egg plants as influenced by oil cake amendments. *Indian Phytopath.* 27: 480–484.

Warthen, J.D., Jr. 1989. Neem (*Azadirachta indica* A. Juss.): Organisms affected and references list upto date. *Proc. Entomol. Soc.* Washington 91: 367–388.

82

Neem for the Control of Aphid and Whitefly Vectors and Virus Diseases of Plants

R. ROYCHOUDHURY AND R.K. JAIN

Department of Botany
University of Lucknow, Lucknow 226 007, India

ABSTRACT

A major strategy for the control of plant virus diseases is to control the insect vectors by synthetic insecticides, but virus transmission by aphid and whitefly vectors cannot be prevented by these insecticides. Neem, which possesses insecticidal, antifeedant, and antiviral properties, offers immense potential as a virus disease control agent. Direct sprays with neem oil (2 per cent) were toxic to the apterous adult and nymphal stage of aphids *Aphis gossypii* and *Myzus persicae* and to whitefly *Bemisia tabaci* adults and pupae. Imago emergence from mature pupae of whiteflies was also inhibited. On leaves treated with neem oil the mortality of the aphids varied with the species and the development stage. The oviposition by whiteflies was also affected. Neem oil sprays on plants also affected soybean mosaic and bhindi yellow vein mosaic virus transmission by its vectors and transmission of tobacco mosaic virus by mechanical inoculation. Sprays on *Abelmoschus esculentus* (bhindi) in experimental plots reduced bhindi yellow vein mosaic virus disease incidence to 21 per cent as compared to 44 per cent in control and 30 per cent in oxydemeton-treated plants.

INTRODUCTION

Aphids and whiteflies (Order: Hemiptera) are important pests and plant virus vectors. The pest infestation is largely controlled by synthetic insecticides but this measure does not prevent virus transmission as the insect can transmit the virus before being killed (Robertson and Klostermeyer, 1961). Neem (*Azadirachta indica* A. Juss.) has shown immense potential as a pest and vector control agent because of its insecticidal, antifeedant, antiecdysteroidal, and antiviral properties (Ketkar, 1976). However, studies on the virus transmission inhibiting activity of neem are limited. Neem seed oil has potential for being developed as a pest control agent due to its effectiveness against pests, safety to the environment, low cost, and easy availability (Saxena et al., 1980). Neem oil has been reported to reduce rice tungro virus transmission by leaf hopper vector *Nephotettix virescens* (Mariappan et al., 1982; Aiyanathan and Narayanswamy, 1987; Srinivasulu and Jeyarajan, 1988), grassy stunt virus by *Nilaparvata lugens* (Saxena and Khan, 1985), cucumber mosaic virus by aphid *Aphis gossypii* (Srivastava et al., 1986), and bhindi yellow vein mosaic virus by whitefly *Bemisia tabaci* (Roychoudhury and Jain, 1991). Neem oil and neem leaf extract have also been reported to inhibit lesion production by mechanically transmitted viruses when mixed with virus inoculum or when applied to test plants (Verma, 1974; Chowdhuri and Saha, 1985; Zaidi et al., 1988). This work was undertaken to explore the potential of neem oil for control of virus transmission by vector and of virus diseases.

MATERIALS AND METHODS

Crude unrefined oil obtained by pressing from neem seed kernel was mixed with 1 per cent soap solution, which was used as emulgator. The mixture was shaken vigorously to obtain an emulsion. Preliminary trials had indicated 2 per cent neem oil emulsion to be most effective on insects and least phytotoxic (Roychoudhury and Jain, 1991). This concentration was used in all experiments.

Oxydemeton methyl (Metasystox c-2, Bayer India Ltd.) and Dimecron (Phosphamidon) were used at 0.02 per cent concentration in water for sprays.

The vectors were raised and maintained on their natural host plants: *A. gossypii* (Koch) on *Cucumis sativus* and *Lagenaria siceraria*; *M. persicae* (Sulzer) on *Nicotiana tabacum* var. N.P. 31; and *B. tabaci* (Genn.) on *Crotalaria juncea* and *Lycopersicon esculentum*.

Progeny obtained from single adult aphids were used for the studies. Nymphs which had undergone final ecdysis were used as apterous adults. Whiteflies were caged on host plants for three to four days and then removed. The eggs were allowed to develop to obtain the preimaginal stages.

The leaves were sprayed on both surfaces and excess liquid was allowed to drip off. Ten nymphs or adults per replicate were placed on treated leaves and percentage mortality was observed after 24 hrs. Fecundity of aphids was calculated from the number of progeny produced by a single adult aphid during its reproductive period.

In the direct treatment, 100–150 aphids present on host plant were treated with sprays and the mortality of adults and nymphs was determined 24 hrs after treatment.

Virus transmission by whiteflies was tested on *Abelmoschus esculentus* with bhindi yellow vein mosaic virus, which is transmitted semipersistently by the vector. The acquisition and inoculation feeding time was 24 hrs each. *Myzus persicae* was tested for transmission of soybean mosaic virus, which is transmitted non-persistently. The aphids were starved for 1 hr before feeding. Acquisition and inoculation feeding time was 10 min. each. Test sprays were done on plants before acquisition or inoculation feeding by the vectors. Tobacco mosaic virus was tested for lesion production on local lesion hosts *Nicotiana glutinosa*, *N. tabacum* var. samsun NN, *Chenopodium amaranticolor*, and *Datura stramonium*. The test plants were sprayed soon before virus inoculation. Virus inoculum infective sap was freshly prepared from diseased tobacco plants. The treated leaves were rinsed with water before virus inoculation.

RESULTS

Effect of Pretreatment of Test Plants on Aphids and Whitefly

Pretreatment of leaves with neem oil (2 per cent) emulsion showed varied effect on aphids. The mortality of adult *A. gossypii* was greater (73.33 per cent) than that of nymphs (52.38 per cent) after 24 hrs, but after 48 hrs both stages were equally affected (93.33 and 95.20 per cent respectively). The nymphs of *M. persicae* showed 100 per cent mortality after 24 hrs while the adult aphids were not much affected even after 48 hrs (16.67 and 30 per cent mortality). The fecundity of the adult was, however, suppressed and 23 progeny per adult were produced on treated leaves as compared to 160 on the untreated leaves. The progeny on the treated leaves did not survive and died after one day (Table 1).

Pretreatment of leaves with neem oil or with Dimecron caused 100 per cent mortality of whiteflies and also suppressed oviposition, as indicated by nil imago emergence on treated leaves. Oxydemeton methyl caused 85.71 per cent mortality, and also inhibited oviposition (Table 2).

Effect of Direct Treatment on Aphids and on Whitefly Pupae

Neem oil caused 100 per cent mortality of the nymphs of *A. gossypii* and *M. persicae* but was less effective on the apterous adult stages causing 60 and 98 per cent mortality respectively. The effect of neem on nymphs was comparable to that by Dimecron (Table 3). Mature pupae of whiteflies were killed by neem oil treatment with no imago emergence, while oxydemeton methyl treatment caused 22.82 per cent imago emergence (Table 4).

Table 1. Mortality of *Aphis gossypii* and *Myzus persicae* nymphs and apterous adults placed on plants treated with neem oil

Treatment	Hrs after treatment	Percentage mortality*				Avg. no. of progeny per adult
		A. gossypii		*M. persicae*		
		Nymph	Adult	Nymph	Adult	
Control (water)	24	5.5 (13.56)	20 (26.56)	6.6 (14.89)	0	160
	48	5.5 (13.56)	20 (26.56)	10 (18.44)	0	200
Neem oil (2%)	24	52.38 (48.38)[a]	73.33 (58.89)[a]	100 (90)[a]	16.67 (24.12)[a]	
	48	95.20 (77.34)[a]	93.33 (75)[a]		30 (33.21)[a]	23

Figures in parentheses are transformed values.
* From 30 nymphs/apterous adults.
[a] Values significantly different from control at 5 per cent level by Dunnett's test for comparison of control mean to each group mean.

Table 2. Mortality and oviposition of *B. tabaci* on *Crotalaria juncea* plants treated with neem oil and with synthetic insecticides

Treatment	Percentage mortality*	Per cent imago emergence after 20 days
Control (water)	28.57 (21.78)[a]	63.8
Neem oil (2%)	100 (90)[b]	0
Dimecron (0.02%)	100 (90)[b]	0
Oxydemeton methyl (0.02%)	85.71 (67.78)	0

Figures in parentheses are transformed values.
For each treatment the values followed by the same letter are not significantly different at 5 per cent level by SNK's multiple range test.
* From 40 whiteflies after three days on treated plants.

Table 3. Mortality of *Aphis gossypii* and *Myzus persicae* nymphs and apterous adults treated directly with neem oil and Dimecron

Treatment	Percentage mortality after 24 hrs			
	A. gossypii		*M. persicae*	
	Nymph	Adult	Nymph	Adult
Control (water)	5.2 (15.23)	8.5 (12.92)	8.9 (15.23)	3.35 (10.63)
Neem oil (2%)	100 (90)[a]	60 (55.77)[a]	100 (90)[a]	98 (81.87)[a]
Dimecron (0.02%)	100 (90)[a]	100 (90)[a]	100 (90)[a]	100 (90)[a]

Figures in parentheses are transformed values.
[a] Values significantly different from control at 5 per cent level by Dunnett's test for comparison of control means to each group mean.

Table 4. Percentage imago emergence from mature pupae of *B. tabaci* on *Crotalaria juncea* treated with neem oil and oxydemeton methyl

Treatment	Total no. of pupae	Percentage of imago emergence*
Control (water)	130	96
Neem oil (2%)	56	0
Oxydemeton methyl (0.02%)	92	22.82

* Ten days after treatment.

Effect of Treatment on Virus Transmission by Vectors

Neem oil sprayed on healthy soybean plants before inoculation feeding with viruliferous (soybean mosaic virus) *M. persicae* caused 50 per cent disease incidence in treated plants as compared to 100 per cent in control plants. Neem oil did not completely suppress acquisition or inoculation feeding by whiteflies and the test plants showed 66.6 per cent disease incidence in both cases. By comparison, oxydemeton methyl completely suppressed acquisition feeding and on plants treated before inoculation feeding the disease incidence was 33.3 per cent (Table 5).

In tests with mechanically transmitted tobacco mosaic virus, spray with neem oil emulsion on the hypersensitive hosts 24 hrs prior to virus inoculation inhibited lesion production on *D. stramonium* only (Table 6), indicating a host-specific action of neem oil on virus infection.

On bhindi plants grown in experimental plots and sprayed with neem oil and oxydemeton methyl at six-day interval, the disease incidence was 21.05 and 29.72 per cent respectively compared to 44.11 per cent in control plants (Table 7).

Table 5. Effect of treatments on virus transmission by *M. persicae* and by *B. tabaci.*

Virus/host/vector	Treatment	Percentage disease incidence*	
		Before acquisition feeding	Before inoculation feeding
Soybean mosaic/ soybean/*M. persicae*	Control (water)	ND	100
	Neem oil (2%)	ND	50
Bhindi yellow vein mosaic/bhindi/*B. tacaci*	Control (water)	75	50
	Neem oil (2%)	66.60	66.60
	Oxydemeton methyl (0.02%)	0	33,30

* From 10 plants
ND: Not done.

Table 6. Lesion production by tobacco mosaic virus on plants treated with neem oil

Virus hosts	Per cent decrease(–)/increase in lesion number*
Nicotiana glutinosa	27
N. tabacum Samsun NN	–34.2
Chenopodium amaranticolor	11
Datura stramonium	84.34**

* Lesions from 30 leaves. Control treated with water.
** P > 0.05.

Table 7. BYMV disease incidence in *Abelmoschus esculentus* (bhindi) plants sprayed at six-day intervals with neem oil and oxydemeton methyl

Treatment	Per cent disease incidence*, Days after first spray			
	6	12	34	47
Control (water)	14.70	23.52	44.11	44
Neem oil (2%)	10.52	13.15	21.05	21
Oxydemeton methyl (0.02%)	8.10	13.51	29.72	30

* From 30 plants in each experimental plot.

DISCUSSION

Neem has been reported to possess insecticidal, antifeedant, and antiviral properties (Ketkar, 1976). It therefore has potential for use in control of plant virus diseases as it may affect plant viruses and virus vectors and prevent virus transmission by vectors. These possibilities have been tested using neem oil against aphid vectors *A. gossypii* and *M. persicae* and whitefly *B. tabaci* and on their virus transmission.

The insecticidal properties of neem oil tested by direct sprays on the aphids showed it to be more toxic to the nymphs, causing 100 per cent mortality, than to the adult stages, in which it caused 60–98 per cent mortality. The toxicity to nymphs was at par with Dimecron, a systemic organophosphorus insecticide, although Dimecron was equally toxic to the nymphs and adult stages. Similar results have been reported in tests with neem oil formulation against nymphs and apterous adults of *A. craccivora* (Patel and Srivastava, 1989), the formulation being more toxic to the nymphs than to the adults and the effect being at par with phosphamidon. On plants treated with neem oil the toxicity varied with the aphid species, and also between nymphs and adults. The mortality of *A. gossypii* nymphs was 52.38 per cent after 24 hrs and 95.20 per cent after 48 hrs. But *M. persicae* nymphs showed 100 per cent mortality after 24 hrs. Similarly, adult *A. gossypii* showed 73.3 per cent mortality after 24 hrs and 93.33 per cent after 48 hrs, while *M. persicae* adults showed 16.67 and 30 per cent mortality on the first and second day respectively. The fecundity of adult *M. persicae* was also inhibited. Schauer (1983) has reported differences in the LC_{50} value of *Acyrthosiphum pisum* and *Aphis fabae* nymphs on plants treated with neem seed kernel extract. The fecundity of *A. fabae* was also reduced. The effect of neem treated plants on insects could be due to its antifeedant effect, and the variation in the effect on different species could be due to the differences in the feeding behaviour of the aphids. The degree of penetration by neem oil to the leaf tissue could also influence its uptake and its subsequent moult-inhibiting effect on the insects. In the experimental results reported by Schauer (1983), *A. fabae*, which showed low mortality on treated plants, did not show morphological abnormalities, while *A. pisum* nymphs were most affected. These results would imply that apart from being antifeedant or phagodeterrent, the toxic principle from neem may also be ingested and cause abnormalities and mortality of the aphids. Thus, the acquisition and inoculation feeding by aphids and the virus transmission may also vary with the nymph and adult, with aphid species, and with the host plant being tested. The differences in the degree of penetration of neem oil to various plants could also be responsible for the variation in the effect of oil on lesion production by tobacco mosaic virus on various host plants.

Neem oil has been reported to suppress the growth and development of nymphs and adult whitefly population on cotton (Dreyer, 1983; Natarajan and Sundaramurthy, 1990). A similar effect has been observed with neem oil on sunnhemp (*Crotalaria juncea*) (Roychoudhury and Jain, 1991). Neem oil was better than oxydemeton methyl in reducing disease incidence on bhindi. Neem oil can be used as substitute for synthetic insecticide and can provide a base for formulation of an effective plant-based insecticide for the control of vector and virus transmission, provided its ability to penetrate the leaf tissue can be improved.

REFERENCES

Aiyanathan, K.E.A. and P. Narayanswamy. 1987. Effect of neem oil on rice tungro virus (RTV) infection in susceptible and resistant varieties. *Int. Rice Res. Newsletter* 12: 23–24.

Chowdhury, A.K. and N.K. Saha. 1985. Inhibition of urd bean leaf crinkle virus by different plant extracts. *Indian Phytopath.* 38: 566–568.

Dreyer, M. 1983. Effect of aqueous neem extracts and neem oil on the main pests of *Cucurbita pepo* in togo. *Proc. 2nd Int. Neem Conf.*, Rauischholzhausen. GTZ, Eschborn, pp. 435–444.

Ketkar, C.M. 1976. *Utilization of Neem (Azadirachta indica* A. Juss.) *and Its By-products*. Directorate of Non-edible oils and Soap Industries Commission, Bombay, 234 pp.

Mariappan, V., R.C. Saxena and K.L. Ling. 1982. Effect of custard apple oil and neem oil on the life span of and rice tungro virus transmission by *Nephotettix virescens*. *Int. Rice Res. Newsletter* 7: 13–24.

Natarajan, K. and V.T. Sundaramurthy. 1990. Effect of neem oil on cotton fly (*Bemisia tabaci*). *Indian J. Agric. Sci.* 60: 290–291.

Patel, M.B. and K.P. Srivastava. 1989. Evaluation of a neem oil formulation against cowpea aphid (*Aphis craccivora*, Koch). *Neem Newsletter* 6: 41–44.

Robertson, R.S. and E.C. Klostermeyer. 1961. Aphid transmission of bean viruses in field beans in central Washington. *J. Econ. Ent.* 54: 414–416.

Roychoudhury, R. and R.K. Jain. 1991. Effect of neem oil and citronella oil on the development of whitefly (*Bemisia tabaci*) and virus transmission. *Recent Advances in Medicinal, Aromatic and Spice Crops*, Vol. I, Today and Tomorrow's Printers and Publishers, New Delhi, pp. 159–161.

Saxena, R.C., N.J. Liquido and H.D. Justo. 1980. Neem seed oil, a potential antifeedant for the control of the rice brown plant hopper, *Nilaparvata lugens*. *Proc. 1st Int. Neem Conf.* Rottach-Egern, GTZ, Eschborn, pp. 171–181.

Saxena, R.C. and Z.R. Khan. 1985. Effect of neem oil on survival of *Nilaparvata lugens* (Homoptera: Delphacidae) and on grassy stunt and ragged stunt virus transmission. *J. Econ. Entomol.* 78: 647–651.

Schauer, M. 1983. Effects of variously formulated neem seed extracts on *Acyrthosiphon pisum* and *Aphis fabae*. *Proc. 2nd Int. Neem Conf.*, Rauischholzhausen. GTZ, Eschborn, pp. 141–150.

Srinivasulu, B. and R. Jeyarajan. 1988. Effect of seed oil on tungro transmission and survival of green leaf hopper of rice crop. *Indian J. Plant Prot.* 16: 171–175.

Srivastava, K.M., N.S. Rana, V.C. Dwadash Shreni and B.P. Singh. 1986. Mechanism of inhibition of CMV by crude oil from margosa (*Azadirachta indica*) during inoculation with single apterous *Aphis gossypii*. *Indian Phytopath.* 39: 20–25.

Verma, V.S. 1974. Chemical compounds from *Azadirachta* as inhibitors of potato virus X. *Acta Microbiol. Polonica* 6: 9–13.

Zaidi, Z.B., V.P. Gupta, A. Samad and Q.A. Naqvi. 1988. Inhibition of spinach mosaic virus by extracts of some medicinal plants. *Curr. Sci.* 57: 151–153.

83

Effect of Neem and Other Plant Extracts on *Xanthomonas campestris citri*, the Incitant of Citrus Canker*

B. CHENDRAMOHAN AND G.J. MOSES**

Department of Plant Pathology, S.V. Agricultural College
Andhra Pradesh Agricultural University, Tirupati 517 502,. India

ABSTRACT

Extracts from 15 plant species and neem cake were tested for their inhibitory effect on *Xanthomonas campestris citri in vitro* as a prelude to using them under field conditions for citrus canker management.

There was no difference in the efficacy of crude and centrifuged extracts.

The water extracts of neem cake, *Allium sativum*, *Polyalthia longifolia*, *Tephrosia purpurea* and *Lantana camara* were found significantly highly superior to others with a mean inhibition area from 62 to 137 mm^2. Those of *Croton sparsiflorus*, *Bougainvillea spectabilis*, *Ficus bengalensis*, *Datura stramonium*, *Parthenium hysterophorus*, *Azadirachta indica*, *Cassia auriculata*, *Vinca rosea* and *Eucalyptus* sp. had moderate effect with a mean inhibition area from 25 to 56 mm^2, while those of *Calotropis gigantia* and *Tridax procumbens* had the least inhibitory effect.

The active principle was extracted in ether, which was found more inhibitory than water extract.

The effect of the plant extracts was lost or reduced when heated to 50°C for 10 minutes.

INTRODUCTION

Citrus canker, incited by *Xanthomonas campestris citri*, is a major disease on acid lime and sweet orange in Andhra Pradesh. Though effective chemical control measures are available, they are not widely used by the citrus growers for obvious reasons. Neem cake

* Part of M.Sc. (Ag.) thesis submitted by the senior author to A.P. Agricultural University.

** Professor and Head

extract recommended by Dakshinamurthy and Koteswara Rao (1959) widely used by the nursery growers. Many workers have shown that natural products are useful phytochemotherapeutants in plant disease management (Skinner and Seghal, 1961; Thapliyal and Nene, 1967; Nariani, 1971; Ward et al., 1975). To find out the *in vitro* efficacy of plant extracts of the locally available flora on the causal bacterium investigations were carried out with selected plant species and the results are published in this paper.

MATERIALS AND METHODS

Fifteen plant species were selected to get leaf extracts besides neem cake.

Xanthomonas campestris citri was isolated from acid lime on nutrient agar medium and pathogenicity proved.

For crude extracts, standard leaf extracts were prepared by grinding 1g leaf material in 1 ml distilled water and filtering through Whatman No. 1 filter paper.

To make neem cake extract, 10 g of neem cake was soaked in 200 ml water for seven days. The material was thoroughly agitated and the supernatant decanted, centrifuged (5000 rpm for 10 min.) and filtered throughly Buchner funnel. The filtrate was used for the study.

For centrifuged extracts, standard leaf extracts were centrifuged at 30,000 rpm for 30 min. in a REMI R-24 centrifuge. The supernatant was filtered through a double layer of Whatman No. 1 filter paper.

For ether extracts, the leaf sample was shade dried, ground into a fine powder, and subjected to extraction in petroleum ether (BP 40°–60°C) in a Soxhlet apparatus.

To check thermostability, the extracts were heated to 50°C for 10 min. in a thermostatically controlled water bath.

In vitro inhibition was tested by filter paper disc impregnated with the test extract and placed on seeded nutrient agar plates and incubated. Streptocycline (100 ppm) and water served as controls. The area of inhibition was calculated by the formula

$$\frac{\Pi}{4}(D^2-d^2)$$

where D = diameter of inhibition zone and
d = diameter of filter paper disc.

RESULTS AND DISCUSSION

The results of *in vitro* effect of crude and centrifuged extracts of 15 selected plant species are presented in Table 1.

The data indicate that there was no difference in the efficacy of crude and centrifuged extracts. The following plant extracts were significantly more effective *Allium sativum* (134 and 135 mm²), *Polyalthia longifolia* (131 and 131 mm²), *Tephrosia purpurea* (89 and 89 mm²), and *Lantana camara* (62 and 60 mm²); moderate effect was found with *Croton sparsiflorus* (56 and 58 mm²), *Bougainvillea spectabilis* (54 and 56 mm²), *Ficus bengalensis* (46 and 42 mm²), *Datura stramonium* (31 and 28 mm²), *Parthenium hysterophorum* (31 and 31 mm²), *Azadirachta indica* (28 and 35 mm²)

Table 1. *In vitro* effect of crude and centrifuged plant extracts of *Xanthomonas campestris citri*

Plant species	Mean area of inhibition (mm^2)	
	Crude extract	Centrifuged extract
Cassia auriculata	25	31
Ficus bengalensis	46	42
Azadirachta indica	28	35
Bougainvillea spectabilis	54	56
Calotropis gigantia	13	16
Parthenium hysterophorum	31	31
Tridax procumbens	10	12
Tephrosia purpurea	89	89
Croton sparsiflorus	56	58
Lantana camara	62	60
Allium sativum	134	135
Vinca rosea	25	28
Datura stramonium	31	28
Eucalyptus sp.	25	20
Neem cake	137	137
Polyalthia longifolia	131	131
Streptocycline (100 ppm)	378	369
Distilled water	0	0

Cassia auriculata (25 and 31 mm^2), *Vinca rosea* (25 and 28 mm^2) and *Eucalyptus* sp. (25 and 20 mm^2). Least inhibitory effect was observed with *Calotropis gigantia* (13 and 16 mm^2) and *Tridax procumbens* (10 and 12 mm^2). Neem cake extract was the most effective with 137 mm^2 in both crude and centrifuged conditions.

Seven treatments found effective in crude and centrifuged conditions were tested with their ether extracts and the results are presented in Table 2.

All the ether extracts tested were more inhibitory than either crude or centrifuged extracts. The ether extracts of *Polyalthia longifolia, Croton sparsiflorus* and *Azadirachta indica* were far more effective (305, 134 and 131 mm^2) than their crude extracts (131, 56 and 28 mm^2).

The thermostability of the crude extracts was tested and the results are presented in Table 3.

All the crude extracts lost their inhibitory effect on heating at 50°C for 10 min.

In the present study, neem cake extract was found to be significantly far superior to other plant extracts in inhibiting the growth of the test bacterium. Similar effect of its antimicrobial activity was reported by Singh and Pande (1965) against *Pythium aphanidermatum* and Damodar Naidu and John (1984) against the rice pathogens, *Rhizoctonia solani* and *Sclerotium oryzae*.

In the present study, the plant extracts from *Allium sativum, Polyalthia longifolia, Azadirachta indica, Lantana camera* and others were found effective against the test bacterium. The extract of *Allium sativum* was found effective against X.c. oryzae (Grainge et al., 1985), *Pseudomonas solanacearum* (Hatagalung, 1988), and the fungi *Cephalosporium sacchari* and *Curvularia lunata* (Misra and Dixit, 1977a, b). Extract

Table 2. *In vitro* effect of ether extracts on *X. compestris citri*

Plant species	Mean area of inhibition (mm^2)
Cassia auriculata	58
Ficus bengalensis	104
Azadirachta indica	131
Bougainvillea spectabilis	80
Parthenium hysterophorum	84
Croton sparsiflorus	134
Polyalthia longifolia	305
Streptocycline (100 ppm)	360
Distilled water	0

Table 3. Thermostability of crude plant extracts and their *in vitro* effect on *X. campestris citri*

Plant species	Mean area of inhibition (mm^2)
Lantana camara	13
Tephrosia purpurea	0
Croton sparsiflorus	25
Bougainvillea spectabilis	13
Cassia auriculata	0
Allium sativum	38
Neem cake extract	46
Streptocycline	378
Distilled water	0

of *Polyalthia longifolia* was effective against *Alternaria brassicicola* (Mukadam et al., 1976) and *Glomerella tucumanensis* (Kishore et al., 1982). Extract of *Azadirachta indica* was effective against *Botryodiplodia theobromae* (Manoharachary and Reddy, 1978), and *Alternaria alternata* (Bhowmick and Chaudhury, 1982), and that of *Lantana camara* against *Leviellula taurica* (Srivastava and Rawat, 1982) and *Curvularia lunata* (Bhowmick and Vardhan, 1981).

Ether extracts of *Polyalthia longifolia, Croton sparsiflorus* and *Azadirachta indica* were found more effective than their water extracts in inhibiting the test bacterium. Mahadevan and Mohan (1978) showed that alcoholic extract of *Catharanthus roseus* was found inhibitory to *Helminthosporium oryzae*, *Cladosporium herbarium* and *Curvularia* sp. Janardhan and Narsimhan (1988), observed that acetone extracts of *Azadirachta indica* and *Parthenium hysterophorum* were inhibitory to *Helminthosporium nodulosum*. Thus, the active principle may be extracted with any of the organic solvents from different plant species.

The plant extracts lost their efficacy on heating at 50°C for 10 min. However, Misra and Dixit (1978), Interjit Singh and Verma (1981), and Verma and Dwivedi (1984) found that the active principle involved in their studies was thermostable. The loss of inhibitory effect on heating in the present study shows that the active principle may be proteinaceous in nature. Misra et al. (1978) and Narayanaswamy and Ramaiah (1983) reported that the active principle in their studies was proteinaceous in nature.

ACKNOWLEDGEMENTS

The authors are grateful to the Principal, S.V. Agricultural College, Tirupati and Andhra Pradesh Agricultural University for providing necessary facilities and a Research Fellowship to the senior author.

REFERENCES

Bhowmick, B.N. and B.K. Choudhury. 1982. Antifungal activity of leaf extracts of medicinal plants on *Alternaria alternata* (Fr) Keissler, *Indian Bot. Reptr.* 1: 164–165.

Bhowmick, B.N. and V. Vardhan. 1981. Antifungal activity of some leaf extracts of medicinal plants on *Curvularia lunata*. *Indian Phytopathol.* 34: 385–386.

Dakshinamurthy, V. and D. Koteswara Rao. 1959. Preliminary studies on the control of citrus canker on acid lime. *Andhra Agric. J.* 6: 145–148.

Damodar Naidu, V. and V.T. John. 1984. Antifungal activity of degraded oil cake extracts. *J. Res. Assam Agric. Univ.* 2: 209–210.

Grainge, M., L. Berger and S. Ahmed. 1985. Effect of extracts of *Artabotrys uncinatus* and *Allium sativum* on *Xanthomonas campestris* pv *oryzae*. *Curr. Sci.* 54: 90.

Hantagalung, L. 1988. Garlic bulb as a material for suppressing the incidence of bacterial wilt (*Pseudomonas solanacearum* E.F. Smith) on tomatoes. *Bull. Penelition Hortic.* 16: 84–93.

Interjit Singh and J.P. Verma. 1981. Virus inhibitor from *Datura metel*. *Indian Phytopathol.* 34: 452–457.

Janardhan, R. and V. Narsimhan. 1988. Effect of plant extracts/products on two fungal pathogens of finger millet. *Indian J. Mycol. & Plant Pathol.* 18: 250–254.

Kishore, N., N.K. Dubey, R.D. Tripathi and S.K. Singh. 1982. Fungitoxic activity of leaves of some higher plants. *Nat. Acad. Sci. Letters* 5: 9–10.

Mahadevan, A. and N. Mohan. 1978. Prohibitions in *Catheranthus roseus*. *Curr. Sci.* 47: 57–58.

Manoharachary, C. and M.M. Reddy. 1978. Efficacy of some leaf extracts on *Botryodiplodia theobromae*. *Geobios* 5: 190–191.

Misra, S.B. and S.N. Dixit. 1977a. Fungicidal spectrum of the leaf extract of *Allium sativum*. *Indian Phytopathol.* 29: 448–449.

Misra, S.B. and S.N. Dixit. 1977b. Antifungal spectrum of some plant extracts. *Geobios* 4: 29–30.

Misra, S.B. and S.N. Dixit. 1978. Fungicidal properties of *Clemetis gouriana*. *Indian Phytopathol.* 30: 577–579.

Misra, A., R. Sinha, P. Rani and M. Sinha. 1978. Antifungal principle in *Cassia*. *Med. Fac. Landhouww. Rijka Univ. Gent.* 43: 1043–1049.

Mukadam, D.S., L.D. Balakhande and G.V. Umalkar. 1976. Antifungal activities in deproteinised leaf extracts of weeds and non-weeds. *Indian J. Microbiol.* 16: 78–79.

Narayanaswamy, P. and M. Ramaiah. 1983. Effect of leaf extracts of millet crops on lesion production of groundnut ring mosaic strain of tomato spotted wilt virus. *Proc. Nat. Sci. Acad.* 378: 332–340.

Nariani, A. 1971. Control of some fungal diseases by antibiotics and crude plant extracts. *Proc. Indian Nat. Sci. Acad.* 378: 332–340.

Singh, R.S. and K.R. Pandey. 1965. Inhibitory and stimulatory effect of certain oil cakes on *Pythium aphanidermatum* in soil. *Sci. & Cult.* 31: 534–535.

Skinner, F.A. and J.M. Seghal. 1961. Antimicrobial substances from flowering plants 1. Antifungal substances. *Hindusthan Antibiotics* 4: 3–29.

Srivastava, S.L. and R. Rawat. 1982. Efficacy of some pesticides and leaf diffusates of *Lantana camara* on the control of powdery mildew of chilli. *Indian J. Mycol. & Plant Pathol.* 12: 127–128.

Thapliyal, P.N. and Y.L. Nene. 1967. Inhibition of plant pathogens by higher plant substances. *J. Sci. Indust. Res.* 26: 289–290.

Verma, H.N. and S.D. Dwivedi. 1984. Properties of a virus-inhibiting agent isolated from plants which have been treated with leaf extracts from *Bougainvillea spectabilis*. *Physiol. Plant Pathol.* 25: 93–101.

Ward, E.W.B., C.H. Unwin and A. Stoessi. 1975. Experimental control of late blight of tomatoes with capsidol and phytoalexin from pepper. *Phytopathol.* 65: 168–169.

84

Effect of Plant Derivatives on the Transmission of Potato Virus Y Infecting Chilli and Its Aphid Vectors

L. DULEEP KUMAR SAMUEL AND V. MARIAPPAN

Centre for Plant Protection Studies
Tamil Nadu Agricultural University, Coimbatore, 641 003, India

ABSTRACT

A mosaic virus was isolated from naturally infected chilli (*Capsicum annuum* L.) plants collected from severly infected field at Madurai, Tamil Nadu. The virus isolate was readily transmitted by mechanical inoculation and by aphids (*Aphis craccivora* Koch and *Myzus persicae* Sulz), and reacted positively with potato virus Y antiserum. The seed oil from neem (*Azadirachta indica* A. Juss), its derivatives, and derivatives from other plant species were aphicidal and effective in reducing aphid transmission of the virus.

INTRODUCTION

Chillies (*Capsicum annuum* L.), commonly known as bell pepper are grown all over India. The crop is affected by 51 pathogens and 14 of them are viruses causing mosaic (Saha and Singh, 1988). Some of them are reported to be transmitted by aphids and a few by mechanical inoculation. The mosaic disease noticed around Madurai is spreading to all areas in Tamil Nadu and control of the disease by application of insecticides has not yielded desirable results. The seed oil from neem (*Azadirachta indica* A. Juss.) and derivatives from other plant species were tested for their effect on the transmission by aphids (*Aphis craccivora* Koch and *Myzus persicae* Sulz) as they were known for their antifeedant and insecticidal properties (Schmutterer et al., 1981; Schmutterer and Ascher, 1984, 1986). The observations are described herein.

MATERIALS AND METHODS

Virus Source and Maintenance

The virus culture was isolated from a naturally infected chilli plant and transmitted to K 1 chilli variety (10 days old) by mechanical inoculation. All plants were kept in insect-proof glass house.

Transmission by Insects

Aphids (*A. craccivora* and *M. persicae*) collected from virus-free colonies maintained in the glass house were allowed different acquisition and inoculation access feeding periods ranging from sec. to min. Infected chilli plants served as virus source plants while healthy plants of K 1 chilli variety (10 days old) were used as test plants.

Serology

The reaction of the virus isolate to the antiserum of potato virus Y (PVY) received from Central Potato Research Institute, Simla was tested by agar gel double diffusion tests in 0.85 per cent Difco Agar gels in 0.85 per cent Sodium chloride/0.25 per cent sodium azide/0.01 M. Tris HCl pH 8.0. Reactant wells were 4 mm in diameter and were spaced 2 mm apart.

Preparation of Plant Derivatives

Freshly extracted neem oil was obtained from local oil expeller mill. The seed oil derivatives were provided by Dr. Sremannarayana, Osmania University, Hyderabad, Azadirachtin rich fractions by Dr. B.A. Nagasampagi, National Chemical Laboratory, Poona, India. The seed oil formulation RD 9 repelin was provided by Dr. T.S. Subramaniam, India Tobacco Company, Hyderabad, India. Emulsifiable Neem concentrates were obtained from the local market. The oil derivatives were emulsified in water (containing 1 per cent liquid detergent, Teepol) to obtain preparations of 5 per cent concentration. Mosaic infected chilli plants were sprayed with these preparations. Three hours later non viruliferous starved adults of both *A. craccivora* and *M. persicae* were allowed to feed on these diseased but sprayed plants. On completion of the acquisition feeding the efficacy of the plant products, in reducing the acquisition of the virus was tested by allowing these aphids to inoculate healthy seedlings. To study the efficacy of the plant products in reducing the transmission of the virus, healthy chilli seedlings were sprayed with the preparations and viruliferous adults of both *A. craccivora* and *M. persicae* were allowed to feed on these healthy but sprayed seedlings. The Aphid mortality was observed 24 hrs after feeding on the treated plants (Table 1).

RESULTS

Sap Transmission

The virus was transmitted readily by mechanical inoculation. The inoculated plants developed symptoms in eight days.

Table 1. Plant derivatives and emulsifiable concentrates.

Plant species and family	Common name	Oil Derivative	Code Name
PLANT DERIVATIVES			
Ailanthus excelsa Roxb. (Simaroubaceae)	Tree of heaven	Fruit oil	Ailanth
Butea monosperma (Lamk) Taubert (Papilionaceae)	Flame of the forest	Seed oil	Bufro
Cassia fistula L. (Cesalpiniaceae)	Indian laburnam	Seed oil	Casili
Cleome liscosa L. (Cleomaceae)	Sticky cleome	Seed oil	Clsvisc
Azadirachta indica A. Juss (Meliaceae)	Neem	Hexane fraction of neem oil	Comnhex
A. indica A. Juss	Neem	Purified neem oil	Comnol
A. indica A. Juss	Neem	Neem seed powder	Nekhex
Parkia biglandulosa Wight and Arn. (Mimosaceae)	Parkia tree	Seed oil	Parkbi
EMULSIFIABLE CONCENTRATES			
Azabin-NSB			
Crop Guard			
Neemark			
Nemex			
Nemocide			
SEED OIL PREPARATION			
RD9 Repelin			
AZADIRACHTIN-RICH FRACTION			
Margocide CK			
Margocide OK			

Insect Transmission

Aphids transmitted the disease successfully. The maximum number of seedlings were infected when the aphids were given a preacquisition starvation period for 180 min. an acquisition period of 180 sec. on virus source plants and an inoculation feeding period of 24 hrs on healthy chilli CV. K. 1 plants.

Serology

The virus reacted positively with the PVY antiserum in agar gel double diffusion test and formed a distinct precipitation line.

Transmission to Plants Treated with Plant Derivatives

Preparations of the Plant derivatives were less effective in reducing the virus transmission in mechanical inoculation, when compared to aphid transmission (Table 2). The acquisition of the virus by *A. craccivora* was reduced to 54.7, 43.1 and 43.1 per cent respectively by Neem oil, Ailanth and Margocide Ok treatments. Similarly an inhibition of 46.9, 43.1, 41.2 and 41.2 per cent respectively was observed in respect of *M. persicae* for Neem oil, Ailanth, Bufro, and Margocide Ok. The transmission was reduced by Neem oil up to 58.8 and 46.9 per cent respectively in respect of *A. craccivora* and *M. persicae*. This was followed by Ailanth, Margocide Ok and Margocide Ck. The plant derivatives were effective in reducing the survival of aphids. Neem oil spray recorded 69.5 per cent mortality of *A. craccivora* and 91.6 per cent mortality of *M. persicae* followed by Margocide Ok (68.6 and 68.6 per cent) and Margocide Ck (67.7 and 67.7 per cent). Among the remaining treatments neem derivatives Comnhex, Comnol, and Nekhex were relatively more effective.

DISCUSSION

Control of aphid-transmitted non-persistent or semi persistent viruses by insecticides does not work, as the insects acquire and transmit the virus in a short period before they are killed. Also, none of the methods including insecticides prevent or reduce the feeding activities of the insect vector and even a short period of feeding has been sufficient to transmit the disease. Therefore, methods that would reduce the feeding activity are considered to be more useful in the control of viruses that are spread by aphid vectors. Paraffin oil was reported to impede the transmission of PVY (Bradley, 1963; Vanderveken, 1977). Earlier reports also emphasize the use of mineral oil with

Table 2. Effect of plant derivatives on mechanical and aphid transmission of chilli mosaic virus and on the aphid survival*

Plant	Inhibition					Insect Mortaility %	
	Sap transmission	Acquisition		Transmission		Ac	Mp
		Ac	Mp	Ac	Mp		
Neem oil	27.7a	54.7a	46.9a	58.8a	46.9a	69.5a	91.6a
Ailanth	27.7a	43.1bcde	43.1abc	46.9bcd	45.8ab	57.5c	57.5ef
Bufro	19.9c	39.2def	41.2abcd	41.2defg	43.1abc	56.8e	57.5ef
Casili	19.9c	35.3g	33.2de	39.2efgh	37.3bcde	56.8e	58.9def
Clsvisc	21.4bc	23.3efg	39.2bcde	39.2efgh	39.2abcde	56.1e	56.1f
Comnhex	24.1ab	33.2fg	35.3cde	41.2defg	39.2abcde	62.7d	63.4c
Comnol	21.4bc	39.2def	33.2de	37.3fghi	41.2abcd	64.2bcd	61.1cd
Nedhex	19.9c	33.2fg	21.4g	37.3fghi	37.3bcde	63.4bcd	60.4cde
Parkbi	24.1ab	41.2cde	39.2bcde	33.2hi	35.3cde	62.7d	62.7c
Azabin	18.4c	33.2fg	33.2de	33.2hi	31.1e	41.2gh	43.7h
Crop guard	18.4c	31.1g	31.1ef	35.3ghi	33.2de	43.1fg	45.0h
Neemark	24.1ab	33.2fg	35.3cde	41.2defg	35.3cde	45.0f	45.6gh
Nemex	21.4bc	24.1h	24.1fg	35.3hi	31.1e	37.9h	44.4gh
Nemocide	19.9c	24.1h	24.1fg	31.1i	31.1e	42.5fg	45.6gh
RD9 Repelin	25.4a	41.2cde	39.2bcde	43.1cdefg	33.2de	44.4fg	48.2g
Margocide CK	24.1ab	39.2def	39.2bcde	45.0bcdef	41.2abcd	67.7ab	67.7b
Margocide OK	26.6a	43.1bcde	41.2abcd	45.0bcdef	43.1abc	68.6a	68.6ab

* Mean of three replications
Ac : *Aphis craccivora*
Mp: *Myzus persicae*

naturally occurring antifeedants derived from plants for prevention of vector-transmitted plant viruses (Simons, 1981). The potential of seed oils for use against rice leafhopper and planthoper has been demonstrated in laboratory and field trails (Heyde et al., 1984; Saxena et al., 1984; Saxena and Justo, 1986; Saxena and Khan, 1986). Neem oil and custard apple oil or their mixtures and other non-edible oils reduced feeding and survival and leafhopper (*Nephotettix virescens* Distant) and its transmission of rice tungro virus (Mariappan and Saxena, 1983a, b; Mariappan et al., 1988). From the results it may be seen that neem oil and its derivatives were more effective than the derivatives from other plant species in reducing the survival of aphid vectors and their transmission of chilli mosaic virus.

REFERENCES

Bradley, R.H.E. 1963. Some ways in which a paraffin oil impeded aphid transmission of PVY, *Canadian J. Microbiol.* 9: 369–380.

Heyde J.V.D., R.C. Saxena and H. Schumutterer. 1984. Neem oil and neem extracts as potential insecticides for control of hemipterous rice pests. In: *Natural Pesticides from the Neem Tree (Azadirachta indica* A. Juss) *and other Tropical Plants*. Proc. 2nd Int. Neem Conf., Rauischholzhausen, 1983. H. Schmutterer and K.R.S. Ascher (eds.), GTZ, Eschborn pp. 377–390.

Mariappan, V. and R.C. Saxena. 1983a. Effect of custard apple oil and neem oil on survival of *Nephotettix virescens* (Homoptera: Cicadellidae) and on rice tungro virus transmission. *J. Econ. Entomol.* 76(3): 573–576.

Mariappan., V. and R.C. Saxena. 1983b. Effect of mixtures of custard apple oil and neem oil on survival of *Nephotettix virescens* (Homoptera: Cicadellidae) and on rice tungro virus transmission. *J. Econ Entomol*. 77: 513–521.

Mariappan, V., S. Jayaraj and R.C. Saxena. 1988. Effect of non-edible seed oils on survival of *Nephotettix virescens* (Homoptera: Cicadellidae) and transmission of rice tungro virus. *J. Econ. Entomol.* 81(5): 1369–1372.

Saha, L.R. and H.B. Singh. 1988. Diseases of Chilli and their management. *Int. J. Trop. Plant Diseases.*, 6: 135–143.

Saxena, R.C., P.B. Epino, W. Tucheng and B.C. Puma. 1984. Neem Chinaberry and Custardapple: Antifeedant and insecticidal effect of seed oils in leafhopper and planthopper pests of rice. In: *Natural pesticides from the neem tree* (*Azadirachta indica* A. Juss) *and other tropical plants. Proc. 2nd Int. Neem Conf.*, Rauischholzhausen, 1983. H. Schmutterer and K.R.S. Ascher (eds.), GTZ, Eschborn pp. 403–412.

Saxena, R.C. and H.D. Justo Jr. 1986. Effect of Custard apple oil, Neem oil and Neem cake on green leaf hopper (GLH) population and on tungro (RTV) infection. *Int. Rice Res. Newsl.* 11(2): 25.

Saxena, R.C. and Z.R. Khan. 1986. Aberrations caused by neem oil odour in green leaf hopper feeding on rice plants. *Entomol. Exp. Appl.* 42: 279–284.

Schumutterer, H., K.R.S. Ascher and H. Rembold. 1981. *Natural Pesticides from the Neem tree* (*Azadirachta indica* A. Juss. *Proc. 1st Int. Neem Conf.*, 1983. GTZ, Eschborn.

Schumutterer, H. and K.R.S. Ascher. 1984. *Natural pesticides from the neem tree (Azadirachta indica* A. Juss). *Proc. 2nd Int. Neem Conf.*, Rauischholzhausen, 1983. GTZ, Eschborn.

Schumutterer, H. and K.R.S. Ascher. 1986. *Natural pesticides from the neem tree and other tropical plants*. Proc. 3rd Neem Conf., Nairobi, 1986. GTZ, Eschborn.

Simons, J.N. 1981. Innovative methods for control of insect transmitted plant viral diseases. In: *Vector of Disease Agents, Interaction with Plants, Animals and Man.* K. McKelvy Jr., B.F. Elridge. and K. Maramorosch (eds.), Praeger, New York, pp. 169–178.

Vanderveken, J.J. 1977. Oils and other inhibitors of non-persistent virus transmission. In: *Aphids as Virus Vectors.* K.F. Harris and K. Maramorosch (eds.), Academic Press, New York, pp. 435–454.

85

Efficacy of Neem Products in the Control of Rice Bacterial Blight

S. ESWARAMURTHY, V. MARIAPPAN, M. MUTHUSAMY, M.N. ALAGIANAGALINGAM AND K.S. SUBRAMANIAM

Agricultural College and Research Institute
Tamil Nadu Agricultural University, Killikulam 627 252, India

ABSTRACT

In the field trial conducted during rabi 1990 using eight treatments, neem cake extract 5 per cent, neem seed kernel extract 2 per cent, *Prosopis* leaf extract 10 per cent, *Acacia* leaf extract 10 per cent, plantomycin 500 ppm, plantomycin 1000 ppm, and *Bacillus subtilis* were sprayed twice at weekly intervals. *Xanthomonas campestris* pv. *oryzae* was inoculated 48 hrs prior to spraying. Neem cake extract 5 per cent, neem seed kernel extract 2 per cent and plantomycin 1000 ppm were very effective, recording bacterial blight intensity grade of 1, 1 and 1 respectively as against 3.5 in control and yield of 2650, 2950 and 3100 kg/ha respectively as against 2300 kg/ha in control. In the field trial conducted during rabi 1991 with the same treatments and procedure, neem cake extract 5 per cent and neem seed kernel extract 2 per cent controlled bacterial blight pathogen, exhibiting bacterial blight intensity grade of 5 and 5 respectively as against 8 in control and yield of 3207 and 3120/kg/ha respectively as against 1213 kg/ha in control.

INTRODUCTION

Rice bacterial blight caused by *Xanthomonas campestris* pv. *oryzae* is one of the most serious diseases of rice and causes heavy loss in yield. In addition to the chemical control of the disease, control with use of plant products and antagonists is of paramount importance. Experiments were conducted to assess the efficacy of neem products and other plant products and antagonists in the control of rice bacterial blight.

MATERIALS AND METHODS

Field trials were conducted during rabi 1990 and rabi 1991 using ADT 36 variety adopting the randomized block design with eight treatments and four replications. The plot size was 5m × 4m. Neem cake extract 5 per cent, neem seed kernel extract 2 per cent, *Prosopis juliflora* leaf extract 10 per cent, *Acacia* leaf extract 10 per cent, plantomycin 500 ppm, plantomycin 1000 ppm and *Bacillus subtilis* were sprayed twice at weekly interval. *Xanthomonas campestris* pv. *oryzae* was inoculated 48 hrs prior to spraying. Bacterial blight intensity was assessed using 1980 Standard Evaluation System for rice and the data are presented in Table 1.

RESULTS AND DISCUSSION

From the data in Table 1 it is evident that neem cake extract 5 per cent, neem seed kernel extract 2 per cent and plantomycin 1000 ppm were very effective, recording bacterial blight intensity grade of 1, 1 and 1 respectively as against 3.5 in control and registering yield of 2650, 2950 and 3100 kg/ha respectively as against 2300 kg/ha in control during rabi 1990 season.

During *rabi* 1991 neem cake extract 5 per cent and neem seed kernel extract 2 per cent controlled bacterial blight, recording intensity grade of 5 and 5 respectively as against 8 in control and yield of 3207 and 3120 kg/ha respectively as against 1213 kg/ha in control. Biological control of Rice bacterial blight with Azospirillum lipoferum application had been reported by Eswaramurthy et al. (1988). Sakthivel and Mew (1991) reported the control of *X. campestris* pv. *oryzae* with bacteriocinogenic strain while Huang and Chang (1975) controlled the pathogen with *B. subtilis*. The significance of biological control is well known (Baker and Cook, 1974). Swinburne et al. (1975, 1977) found that *B. subtilis* produced at least eight antibacterial antibiotics and two antifungal antibiotics. Since neem products are able to control the disease, neem products also can be extensively used to tackle the bacterial blight problem.

Table 1. Efficacy of neem products in the control of rice bacterial blight

Treatment	Mean bacterial blight intensity grade		Mean yield in (kg/ha)	
	Rabi 1990	Rabi 1991	Rabi 1990	Rabi 1991
Neem cake extract 5 per cent	1	5	2650	3207
Neem seed kernel extract 2 per cent	1	5	2950	3120
Prosopis leaf extract 10 per cent	1.75	4	2650	3315
Acacia leaf extract 10 per cent	1.75	4	2850	2395
Plantomycin 500 ppm	1	4.5	2850	3055
Plantomycin 1000 ppm	1	3.7	3100	2960
Bacillus subtilis	1	3	3000	2835
Control	3.5	8	2300	1213
C.D. (P = 0.05)	0.87	1.68		865

REFERENCES

Baker, K.F. and R.J. Cook. 1974. *Biological Control of Plant Pathogens*. W.H. Freeman, San Francisco, 433 pp.

Eswaramurthy, S., S. Muthusamy, M. Muthuswamy, V. Mariappan, M. Jeyasekar and S. Natarajan. 1988. Management of rice brown spot, sheath rot and bacterial blight with *Azospirillum lipoferum* application. *Proc. Nat. Sem. on Management of Crop Diseases with Plant Products/Biological Agents*, AC & RI Madurai, Jan. 11–12, 1988. 8 pp.

Huang and Chang. 1975. Rice Pathology. S. Kannaiyan, Tamil Nadu Agricultural University, Coimbatore.

Sakthivel, N., T.W. Mew. 1991. Efficacy of Bacteriocinogenic strains of *Xanthomonas oryzae* pv. *oryzae* on the incidence of bacterial blight disease of rice (*Oryza sativa*). *Can. J. Microbiol.* 377: 764–768.

Swinburne, T.R., J. Barr and A.E. Brown. 1975. *Trans. Brit. Mycol. Suc.* 65: 211–217.

Swinburne, T.R., J. Cartwright, N.J. Flace and A.E. Brown. 1977. *Ann. Appl. Biol.* 81: 61–73.

86

Effect of Non-edible Seed Oils in Aphid Transmitted Chilli Mosaic Virus in Tamil Nadu (India)

V. MARIAPPAN AND L. DULEEP KUMAR SAMUEL

Centre for Plant Protection Studies
Tamil Nadu Agricultural University
Coimbatore 641 003, India

ABSTRACT

A viral strain causing severe mosaic on Chilli (*Capsicum annuum* L.) and transmissable by sap and aphids (*Aphis craccivora* Koch and *Myzus persicae* Sulz.) was isolated. It was proved to be serologicaly related to potato virus Y by agar gel double diffusion tests. Non-edible seed oils of Neem (*Azadirachta indica* A. Juss) and Mahua (*Madhuca longifolia* Koen-Macbr. var *latifolia* Roxb. Cheval) both individually and as mixtures reduced the acquisition and transmission of the virus by the aphid vectors. These oils were also highly aphicidal.

INTRODUCTION

Chillies (*Capsicum annuum* L.) also known as bell peppers, are grown all over India in 826,000 ha with an annual production of 5,11,000 tonnes of dry chillies. The crop is affected by 14 viral mosaics (Saha and Singh, 1988). These mosaic diseases have been reported to be caused by Chilli mosaic virus (Kamara and Dubey, 1975), Potato virus Y (PVY) (Gowda and Reddy, 1985), Potato virus X (PVX) (Rishi and Dhawan, 1989) and a strain of tobacco mosaic virus (Singh, 1988). Multiple viral infections causing mosaic have been reported like that of cucumber mosaic virus and PVY (Conti and Masenga, 1977; Rishi and Dhawan, 1989), PVX and tobacco mosaic virus (Conti and Masenga, 1977) and PVY with cucumber mosaic virus or PVX (Dossa and Mungur, 1988).

During 1988 to 1990 a mosaic disease on chilli caused by a virus assumed serious proportions in southern districts of Tamil Nadu, India, and became a limiting factor for profitable chilli production. Control of this sap and aphid transmitted, non persistent virus by insecticides or by cultural practices was not successful. We present the results of a green house study in identifying the virus and comparing the effectiveness of non-edible seed oils against the survival of aphid vectors and their acquisition and transmission of this virus.

MATERIALS AND METHODS

The virus was isolated from naturally infected chilli plants and was maintained in K 1 Chilli variety by mechanical transmission. To test the host range 15 species were sap inoculated. Seed oils of Neem (*Azadirachta indica* A. Juss) and Mahua (*Madhuca longifolia* Koen. Macbr. Var *latifolia* Roxb. Cheval) were taken and mixtures in the ratios of 1:0, 1:1, 1:2, 1:4, 0:1, 2:1 and 4:1 were prepared. These mixtures were emulsified to 3 and 5 per cent concentrations in water containing 1 per cent Teepol. These emulsions were taken @ 0.1 ml and to them 1 ml of infective sap was added, and the volume was made upto 10 ml with water. This inoculum was mechanically transmitted to 10-day old chilli seedlings variety K 1 after dusting the leaves with 600 mesh carborandum powder. Already mosaic infected seedlings were sprayed with these oil mixtures and 5 nonviruliferous adults of *A. craccivora* and *M. persicae* which had been given a preacquisition fasting period of 2 h were allowed to have an acquisition feeding period of 120 sec. on the sprayed diseased seedling and on completion of it, were transferred to healthy chilli seedlings cv K1. Similarly 5 viruliferous adults of *A. craccivora* and *M. persicae* which had been given a preacquisition fasting period of 2 h, and an acquisition feeding period of 120 sec. were allowed to inoculate 10-day old healthy chilli seedlings sprayed with the oil mixtures at 3 and 5 per cent concentrations. The per cent mortality of *Aphis craccivora* and *Myzus persicae* was noted after 24 h. The antiserum to the virus was prepared as described by Mali et al. (1989). Agar gel diffusion tests were done in 0.85 per cent Difco Agar in 0.85 per cent sodium chloride, 0.25 per cent sodium azide 0.01M Tris HCL pH 8.0. Reactant wells were 4 mm in diameter and spaced 2 mm apart.

RESULTS

Symptoms in Infected Plants

Infected seedlings showed a severe mosaic on apical leaves, vein clearing, vein banding, chlorosis, mosaic mottling, and filiformity on older leaves and they often had a bushy appearance. Fruits were small and pale green with reduced pungency. All varieties K 1, MDU 1 and local samba were severely affected.

Host Range, Stability in Sap and Transmissibility

The virus was readily transmitted by sap inoculation but infected solanaceous hosts only (Table 1). The virus had a thermal inactivation point between 55°C and 60°C, a dilution

Table 1. Host range and symptomatology of the chilli mosaic virus

Family, genus, species	Symptoms
Amaranthaceae	
Gomphrena globsa L.	No Symptoms
Chenopodiaceae	
Chenopodium amaranticolor Coste & Reyn.	-do-
Cucurbitaceae	
Cucurbita moschata Duch.	-do-
Compositae	
Lagasca mollis Cav.	-do-
Labiatae	
Ocimum sanctum Linn.	-do-
Leguminosae	
Cyamopsis soraloides DC.	-do-
Dolichos lablab Linn.	-do-
Phaseolus aurous Roxb.	-do-
Solanaceae	
Capsicum annum L. K1	VC, VB, MM, C, F, RL, S
Nicotiana glutinosa L.	MM, F, RL
N. tabacum var. white Burley	MM
N. tabacum L. var, Xanthi.	MM, VC, CS
Nicandra physaloides (L) pers.	MM, C, S
Petunia hybrida Vilm.	MM, RL, F
Physalis floridana Rydberg	MM, RL, S

Abbreviations for symptoms: C = chlorosis; F = filiformity; MM = mosaic mottling; RL = reduction in leaf size; S = stunting; VB = vein banding; VC = vein clearing

point of 10^{-2} and infectivity was retained for five but not six days at room temperature (30°C $\pm$ 2°C). The aphids *A. craccivora* and *M. persicae* transmitted the virus (61.7 and 71.6 per cent) with a preacquisition fasting period of 120 min., an acquisition feeding period of 120 sec. and an inoculation feeding period of 24 hrs.

Impeded Transmission Due to Seed Oils

At 3 and 5 per cent concentration of Neem oil and Mahua oil the level of inhibition of sap transmission was low and ranged from 21 to 28 per cent (Table 2). Neem oil had the maximum inhibition of acquisition of the virus by 39, and 37 per cent and 55 and 45 per cent at 3 and 5 per cent concentration for *A. craccivora* and *M. persicae*. The effect of all oil mixtures in impeding the acquisition was on par at 3 per cent concentration. While at 5 per cent the effects of Mahua oil and Neem oil were on par for *A. craccivora* and for *M. persicae* the effects of NO + MO (1:1) were also on par.

The efficacy in reducing transmission of the virus ranged from 39 to 43 per cent and 37–41 per cent at 3 per cent concentration and between 41–47 per cent and 47 to 51 per cent at 5 per cent concentration for *A. craccivora* and *M. persicae* respectively. At 3 per cent concentration all oils and their mixtures were on par for their effect on both aphid species. At 5 per cent concentration Neem oil and NO + MO (2:1) were on par, for *A. craccivora*. The antitransmission effect of all mixtures at 5 per cent was on par for *M. persicae*.

Table 2. Effect of seed oils on survival of aphids *A. craccivora* (Ac) and *M. persicae* (Mp) and their transmission of chilli mosaic virus

Treatment	Inhibition per cent													
	Sap transmission		Acquisition				Transmission				Insect mortality			
			Ac		Mp		Ac		Mp		Ac		Mp	
	3	5	3	5	3	5	3	5	3	5	3	5	3	5
Mahua oil MO	21ab	28a	35a	55a	35ab	47a	43ab	51b	39ab	47a	52ab	67ab	53ab	62b
Neem oil NO	24a	28a	39a	55a	37a	45ab	45a	59a	41a	49a	54a	70a	54a	92a
NO + MO (1:1)	21ab	24ab	33abc	47b	33abc	45ab	39abc	49bc	39ab	45ab	49cd	68ab	50cd	62b
NO + MO (1:2)	21ab	21b	33abc	39c	33abc	39d	41abc	49bc	39ab	45ab	49bcd	68ab	50cd	69b
NO + MO (1:4)	18bc	21b	31abc	39c	33abc	41c	39abc	47bc	37abc	45ab	44c	67ab	49d	67b
NO + MO (2:1)	21ab	27a	33abc	47b	35ab	47a	41abc	53ab	39ab	47a	47d	66ab	51bcd	67b
NO + MO (4:1)	24a	28a	33abc	45b	35ab	43b	41abc	51c	41a	47a	49bcd	67ab	51bcd	69b

3 = Concentration of oil preparation 3 per cent
5 = Concentration of oil preparation 5 per cent

Maximum aphicidal activity was noted for Neem oil at 5 per cent concentration (70 and 92 per cent). The aphicidal activity was on par at 3 per cent concentration for Neem and Mahua oil while at 5 per cent concentration Neem oil was more aphicidal (92 per cent). At 5 per cent concentration the effect of all oil mixtures were generally on par.

Serology

The virus reacted positively with the antiserum produced and with the reference antiserum received from Central Potato Research Institute, Simla, India in gel diffusion tests and formed a distinct precipitation line and identified as a strain of PVY.

DISCUSSION

The virus isolate is determined to be a strain of PVY reported by Thakur et al. (1988) due to its similarity in physical properties, host range, transmission and also by serological tests. Vanderveken (1977) discussed how oils inhibited aphid transmission of viruses, the efficiency of oil treatments, and the possible field application of oils for the control of stylet-borne viruses, and Simons (1981) emphasized the use of oil formulations with naturally occurring antifeedants derived from plants. Bradley (1963) reported the control of PVY by use of paraffin oils. Peters (1977) suggested that oil droplets which collect at the tip of the labium while the aphids feed on the oil sprayed leaves were transferred to the infection site where they interfere with the development of infection. Seed oils from plants like neem (*A. indica*) and custard apple (*Annona squamosa* L.) either individually or in mixtures have been reported to possess insecticidal properties against the insect vector, rice green leaf hopper (*Nephotettix virescens* Distant) and its transmission of rice tungro virus (Mariappan and Saxena, 1983, 1984).

Qadri et al. (1977) reported that the mixtures of seed extracts increased the insecticidal potential. Mariappan et al. (1988) reported that non-edible seed oils of Karanj (*Pongamia pinnata* Pierre), mahua, and pinnai (*Callophylum inophyllum* L.) were more effective than the oil of neem in reducing the survival of *N. virescens* and its transmission of rice tungro virus. Extracts of leaves and flowers of *Peltophorum ferrugeneum Crassula indica*, *Achras sapota*, *Bougainvillaea spectabilis*, *Nerium odorum* and Neem possess antiviral properties against Chilli mosaic virus sap transmission (Sabitha and Ramakrishnan, 1988; Samuel, 1990). Results comparable to the above were obtained in the present study. However Neem oil and Mahua oil individually had the maximum effect in reducing the acquisition and transmission of PVY. Reduction in transmission was more than that of acquisition, this may be due to the longer contact with the oil. The aphicidal effect was very high upto 92 per cent. These triple factors of reducing the acquisition of the virus, reducing the transmission, and the high aphicidal nature of the oil mixtures open a new way to control the aphid stylet borne non-persistent viruses by non edible seed oils and their mixtures.

REFERENCES

Bradley, R.H.E. 1963. Some ways in which a paraffin oil impeded aphid transmission of PVY. *Canadian J. Microbiol.* 9: 369–380.

Conti, M. and V. Masenga. 1977. Identification and prevalence of pepper viruses in North-West Italy, *Phytopath Z.* 90: 212–222.

Dossa, M.I. and C. Mungur. 1988. The status of virus diseases of *Capsicum annuum* L. in Mauritius. *FAO Plant Prot. Bull.* 30: 151–156.

Gowda, K.T.P. and H.R. Reddy. 1985. Distribution of chilli mosaic virus in some parts of Kolar district. *Curr. Sci.* 16: 17.

Kamara, B.S. and G.S. Dubey. 1975. Studies on mosaic disease of chilli. *Indian Pytopath.* 28: 212–217.

Mali, V.R., G.E. Mundhe and W.R. Shain. 1989. Serodiagnosis of six cowpea seed borne viruses in India. *Indian J. Virol.* 5: 45–55.

Mariappan, V. and R.C. Saxena. 1983. Effect of custard apple oil and neem oil on survival of *Nephotettix virescens* (Homoptera: cicadellidae) and on rice tungro virus transmission. *J. Econ. Entomol.* 76: 573–576.

Mariappan, V. and R.C. Saxena. 1984. Effect of mixtures of custard apple oil and neem oil on survival of *Nephotettix virescens* (Homoptera: Cicadellidae) and on rice tungro virus transmission. *J. Econ. Entomol.* 77: 519–521.

Mariappan, V., S. Jayaraj and R.C. Saxena. 1988. Effect of non-edible seed oils on survival of *Nephotettix virescens* (Homoptera: Cicadellidae) and on rice tungro transmission *J. Econ. Entomol.* 81: 1369–1372.

Peters, D. 1977. The use of oil to prevent the spread of virus diseases and its mode of action. In: *Proc. British Crop Prot. Conf.—Pests and Diseases*, British Crop Prot Council, Brighton, England, Vol. III: 823–834.

Qadri, S.S., H. Rao and B. Brahmmanand. 1977. Effect of combining some indigenous plant seed extracts against house hold insects. *Pesticides* 11: 21–23.

Rishi, N. and P. Dhawan. 1989. Characterisation and identification of viruses causing mosaic disease in *Capsicum* in nature. *Indian Phytopath.* 42: 221–226.

Sabitha, D. and K. Ramakrishnan. 1988. Inhibitors of tobacco mosaic virus. In: *Proc. Natl. Sem. on Management of Crop Diseases with Plant Products/Biological Agents.* (Abstr). V. Mariappan. (ed.), Dept. Pl. Path, AC & RI, Madurai, India, 130 pp.

Saha, L.R. and H.B. Singh. 1988. Diseases of chilli and their management. *Int. J. Trop. Plant Diseases* 6: 135–143.

Samuel, D.K. 1990. Studies on chilli mosaic virus, M.Sc.(Ag) Dissertation, Tamil Nadu Agricultural University, Coimbatore, pp. 147.

Simons, J.N. 1981. Innovative methods of control for insect transmitted plant viral diseases. In: *Vectors of Disease Agents, Interaction with Plants, Animals and Man.* J.J. McKelly Jr., B.F. Elridge and K. Maramorosch (eds.), Praeger, New York, pp. 169–178.

Singh, M.N. 1988. Relationship of *Aphis gossypii* with Potato virus Y. *Indian J. Virol.* 41: 108–120.

Thakur, P.D., S.C. Chowfla and S.M.P. Khurana. 1988. Natural occurrence of an atypical strain of potato virus Y on bell pepper in Himachal Pradesh. *Indian J. Virol.* 1: 91–96.

Vanderveken, J.J. 1977. Oils and other inhibitors of non-persistent virus transmission. In: *Aphids as Virus Vectors*, K.F. Harris and K. Maramorosch (eds.), Academic Press, New York, pp. 435–454.

87

Use of Neem in Integrated Management of Rot Diseases of Small Cardamom (*Elettaria cardamomum* Maton.)

K. DHANAPAL,* THOMAS JOSEPH AND R. NAIDU

Indian Cardamom Research Institute
Myladumpara, Kailasanadu P.O. Pin 685 553
Idukki District, Kerala, India
* ICRI Regional Research Station
Thadiyankudisai 624 212, India

ABSTRACT

Extracts of neem and neem products were tested against the rot pathogens, *Rhizoctonia solani* and *Phytophthora meadii* under *in vitro* and field conditions. Aqueous extracts prepared from neem leaf, unripened neem fruits, neem seeds and also commercially available neem based formulations such as neen oil, kemisal, Azabin-NSB and Neem Gold were used for inhibiting the pathogen. *In vitro* studies showed 80 to 100 per cent growth inhibition on *R. solani* and *P. meadii* with all the products tested. The minimum inhibiting concentration causing above 80 per cent inhibition was applied in seedlings in pot culture and on mature plants in the field. The potentiality of neem products in controlling rot-diseases of cardamom is discussed in this paper.

INTRODUCTION

Rot diseases such as clump rot and capsule rot (Azhukal) of small cardamom (*Elettaria cardamomum* Maton.) are the most serious diseases causing considerable crop loss. These diseases are prevalent both in nurseries as well as in plantations. Rhizome rot and capsule rot (Azhukal) are caused by *Rhizactonia solani* Kohn. and *Phytophthora meadii* Mc Rac respectively. So far the diseases have been managed following plant sanitation coupled with fungicidal applications. However, fungicidal control has got its own limitations such as involvement of high costs, residual problems, environmental hazards etc.

In recent years scientists are on the look out of non-chemical means of plant disease control using biocontrol systems or antagonistic plant extracts, which are comparatively less expensive and non-hazardous. Some of the higher plants have recently been shown to be potential sources of fungitoxic substances. Avdhesh Narain and Satapathy (1977) have reported that extracts from *Vinca rosea* inhibited mycelial growth, sporulation and spore germination of the wilt pathogen, *Fusarium oxysporum*. Crude extracts from garlic (*Allium sativum*) has been shown to inhibit the mycelial growth of a number of phytopathogenic fungi (Shekhawat and Prasada, 1971, Misra and Dixit, 1976). So far no work has been done on the use of antagonistic plant extracts to control cardamom diseases. The present investigation was carried out to evaluate the antifungal properties of different parts of neem, garlic and commercially available neem based formulations against rot pathogens of cardamom under *in vitro* and field conditions.

MATERIALS AND METHODS

In vitro Studies

Aqueous extracts prepared from leaves, unripened fruits and seeds of neem (*Azadirachta indica*), garlic bulbs (*Allium sativum*) and commercially available neem based formulations such as neem oil, kemisal, Azabin-NSB and Neem Gold were used in the studies. Fresh samples were collected, washed in tap water and finally in three changes of sterile distilled water. Extracts were prepared in sterile distilled water by grinding the samples in a sterilized mortar and pestle at the rate of 0.5 g/ml on fresh weight basis. The extracts were strained through Whatman No.1 filter paper. Effect of the plant extracts on the radial growth of *R. solani* and *P. meadii* was studied following poisoned food techniques described by Grover and Moore (1962). These extracts were thoroughly mixed with sterilized molten potato dextrose agar to make different concentrations. Fifteen ml of the medium mixed with the extracts were poured in sterilized petriplates. Petriplates poured with pure PDA alone served as the control. Plates containing the amended and control media were inoculated at the centre with 5 mm discs of seven day old *R. solani* and *P. meadii* cultures. Inoculated plates were maintained at room temperature (28 ± 3°C). Each treatment was replicated four times. The colony diameter of *R. solani* and *P. meadii* in test and control plates were measured after 120 hours intervals and percentage of growth inhibition was calculated.

Pot Culture and Field Studies

The *in vivo* effects of plant extracts on *R. solani* infection on six months old seedlings was studied in pot culture under green house conditions. Cardamom seedlings were raised in the nursery and transplanted to earthernware pots (8" diameter), each pot containing five seedlings. Inoculum of *R. solani* was prepared by grinding 5 days old PDA cultures in sterile water (one plate of culture/100 ml water). In the case of *P meadii* infection the spray inoculation method (Peethambaran et al., 1988) was followed in six months old seedlings using 7 day old cultures on carrot agar. Zoospores were prepared by the method of Tsao and Garber (1970). The final spray fluid was prepared by filtering the suspension and adjusting the concentration to get 10^5 zoospores/ml. Seedlings were inoculated by spraying of zoospore (25 ml/seedling) suspension with a glass automiser. Five days later, the test seedlings of *R. solani* infection were drenched and sprayed for *P. meadii* infection with the minimum inhibiting concentrations of respective extracts at the rate of 100 ml/seedlings. The control plants were applied with equal quantity of water. Observations were recorded 15 days after the application of extracts and per cent disease incidence on leaves and seedling mortality were calculated.

In Field Conditions

Field evaluation of the antagonistic extracts and neem formulations was carried out in a plot severely affected with clump rot caused by *R. solani*. The trial was carried out using selected concentration (Minimum inhibiting concentration) of respective plant extracts with five replications. Plants were sanitered by removing the disease affected parts before application. The respective plant extracts were drenched at the plant base @ 2 to 3 litres/clump. Observations on disease incidence were taken by counting healthy and infected tillers. The data were analysed statistically.

RESULTS AND DISCUSSIONS

Results on *in vitro* inhibition of *R. solani* and *P. meadii* with plant extracts are presented in Table 1. All the concentrations of the extracts tested inhibited the pathogen in varying degrees. Maximum inhibition of mycelial growth was recorded at higher concentrations of the extracts. Growth inhibition started 48 hrs after inoculation of the pathogen and reached maximum after 120 hrs. No sclerotial fromation was observed in any of the treated cultures except in control plates. The concentrations causing 80 to 100 per cent inhibition were selected as the minimum inhibiting concentrations (MIC) and these were used for testing in the field. Complete inhibition of growth of *R. solani* was observed with extracts from neem seed (30 per cent), neem oil (4 per cent) and garlic bulb (4 per cent). Similar inhibitory properties of plant extracts from neem, garlic and *Bougainvillea spectabiles* on the mycelial growth and sclerotial germination of *Thanatophorus cucumeris,* the perfect state of *R. solani* has earlier been reported (Lakshmanan and Mohan, 1989). Dubey and Dwivedi (1991) also showed that extract of *A. sativum* was inhibitory to mycelial growth and sclerotial germination of *Macrophomina phaseolina*. In the case of *P. meadii* the mycelial growth was completely inhibited in different concentration of all extracts tested under *in vitro* condition. Root extract of garlic inhibited the germination of chlamidospores and encysted zoospores of *Phytophthora palmivora* in soil and reduce the damping-off of the pappaya seedlings (Paul and Ko, 1986). The inhibition of *R. solani* and *P. meadii* observed in the present study using extracts from garlic bulb may be attributed to the presence of inhibitory compounds such as Allicin as reported by Skinner (1955). The active principle i.e., Azadirachtin present in neem products may be responsible for the inhibition of the pathogen.

R. solani inoculated seedlings in the pots showed initiation of disease symptoms 10–15 days after the inoculation of the pathogen. Symptoms were in the form of rotting of leaf sheaths and leaves. Such leaves dried-off at a later stage. In severe cases, the collar portions showed decay symptoms and the entire seedling collapsed. In the case of *P.*

Table 1. *In vitro* effect of neem, neem formulations and garlic on mycelial growth of *R. solani* and *P. meadii*

Plant extracts	Conc. (%)	% Growth inhibition after 120 hours	
		R. solani	*P. meadii*
Neem leaf	5	50.34	45.00
	10	81.37	60.60
	20	86.01	94.23
	30	96.58	100.00
Neem seed	5	50.15	38.42
	10	76.21	100.00
	20	84.99	100.00
	30	100.00	100.00
Neem Unripend Fruit	5	49.22	25.25
	10	66.00	100.00
	20	71.01	100.00
	30	81.09	100.00
Kemisal	0.1	40.61	65.63
	0.2	82.59	100.00
	0.3	92.15	100.00
Azabin-NSB	0.5	57.81	45.61
	1.0	61.15	87.40
	2.0	96.82	100.00
Neem oil	1.0	50.10	34.60
	2.0	66.53	69.37
	3.0	76.13	100.00
	4.0	100.00	100.00
Garlic bulb	1.0	38.60	44.11
	2.0	51.56	57.45
	3.0	76.46	75.20
	4.0	100.00	90.19

Mean values of four replications

Table 2. Effect of neem, neem formulations and garlic on *R. solani* and *P. meadii* infection in cardamom seedlings

Sl. No.	Treatment	Conc. (%)	*R. solani* infection (%)		Conc. (%)	*P. meadii* infection (%)
			Infection on leaves	Mortality		
1.	Neem Leaf	20	12.37 (20.67)	13.33 (21.09)	20	31.50 (27.33)
2.	Neem seed	20	15.54 (23.00)	4.00 (9.32)	10	29.94 (24.92)
3.	Neem unripened fruit	20	12.53 (20.70)	8.00 (16.08)	10	32.43 (28.76)
4.	Kemisal	0.3	11.32 (19.65)	16.67 (23.16)	0.3	33.71 (30.82)
5.	Azabin - NSB	2.0	17.16 (24.06)	16.00 (23.19)	2	29.45 (24.24)
6.	Neem oil	3.0	8.13 (15.92)	9.33 (14.33)	3	26.92 (20.52)
7.	Garlic bulb	4.0	27.33 (31.47)	8.67 (16.75)	4	29.03 (23.56)
8.	Control	—	28.67 (32.15)	41.67 (40.00)	—	50.37 (59.20)
			S.E. = 2.17	S.E. = 4.38		S.E. = 0.97
			CD 5% = 6.59	CD 5% = 13.29		CD 5% = 2.95

Figures in parentheses are transformed values

Table 3. Evaluation of neem and garlic extracts against clump rot caused by *R. solani* in the field

Sl. No.	Treatment	Conc. (%)	Disease incidence (%)
1.	Neem Leaf	20	10.57 (18.62)
2.	Neem Seed	20	11.52 (16.80)
3.	Neem unripened fruit	20	14.05 (21.76)
4.	Kemisal	0.3	18.89 (25.60)
5.	Azabin - NSB	2.0	10.74 (18.26)
6.	Neem Gold	1.0	9.01 (17.13)
7.	Garlic bulb	4.0	12.33 (20.09)
8.	Control	—	26.46 (30.83)
			L.S.D. (5%) = 7.49 L.S.D. (1%) = 10.11

Figures in parentheses are transformed values

meadii sprayed seedlings, water soaked lesions were formed in the leaf and finally rotting occured by the association of the lesions. Results on the extent of foliar infection (Table 2) showed difference between control and treatments. However, significant difference has been observed in mortality rates of seedlings of treated and control pots. Among the neem products tested, it was found that neem seed extract (20 per cent) was most effective in reducing the seedling mortality rates followed by neem oil (3 per cent) and neem unripened fruit (20 per cent). Neem leaf extract, Azabin-NSB and kemisal also reduced the mortality rates as compared to untreated control. Garlic extract tested (4 per cent) also minimized the disease incidence to a great extent. Leaf infection caused by *P. meadii* was significantly reduced in neem oil 3 per cent followed by all the extracts tested in this study.

The effect of plant extracts on clump rot in the field showed encouraging results. Disease incidence in treated plots were lesser as compared to control plots. Among the extracts tested, Neem Gold (1 per cent) was found to be the best in reducing disease severity. Other extracts were also found to be on par with each other (Table 3). Garlic extract was known to inhibit a number of soil borne fungi and control damping-off disease of several crop plants (Powell and Ko, 1986). Application of neem cake and neem leaves in sick soil has been reported to reduce leaf infection of *Collectotrichum capsici* in betelvine (Marimuthu and Rabindran, 1991). Similar inhibitory effects of neem extracts on *R. solani* and their efficacy in controlling damping-off was reported in *Phaseolus aureus* (Lakshmana and Mohan, 1989). Neem extracts and other neem formulations have been largely used in controlling pest problem of several crop plants (Ahmed and Grainge, 1986, Subrahmanyam, 1990) but are being seldom used in controlling plant diseases. Our study shows encouraging results on the use of neem products in the control of soil borne pathogens. The potentiality of neem and neem based formulations cannot be ruled out in modern disease management strategies though not complete suppression of pathogens in possible.

ACKNOWLEDGEMENTS

Our sincere thanks are due to Miss. S. Sreelatha and Mr. N. Thulasidas for their help in statistical analysis and typing works. We also gratefully acknowledge M/s. SPIC LTD., Madras for kindly supplying Neem Gold for conducting the trials.

REFERENCES

Ahmed, S. and M. Grainge. 1986. Potential of the neem tree (*A. indica*) for pest control and rural developments. *Econ. Bot.* 40: 201–201.

Avdhesh Narain and J.N. Satapathy. 1977. Antifungal characteristic of *Vinca rosea* extracts. *Indian Phytopath.* 30: 36–40.

Carrie R. Powell and Ko. Win-hsium. 1986. Screening for Antagonistic Plants for Control of *Phytophthora palmivora* in soil. *Ann. Phytopath. Soc. Japan* 52: 817–824.

Dubey, R.C. and R.S. Dwivedi. 1991. Fungitoxic properties of some plant extracts against vegetative growth and sclerotial viability of *Macrophomina phaseolina. Indian Phytopath.* 44: 411–413.

Grover, R.K. and C.D. Moore. 1962. Toximetric studies of fungicides against brown rot organisms *Sclerotia fructicola* and *S. laxa. Phytopathology* 52: 876–880.

Lakshmanan, P. and S. Mohan. 1989. Antifungal properties of some plant extracts against collar rot of *Phaseolus aureus. Madras Agric. J.* 76(5): 266–270.

Marimuthu, T. and R. Rabindran. 1991. Non-chemical method of management of Betelvine anthracnose. *Madras Agric. J.* 78: 55–56.

Misra, S.B. and S.N. Dixit. 1976. Comparison of activity of active principle of *Clematis qourina* with commercial fungicides. *Indian Phytopath.* 29: 448.

Peethambaran, C.K., R. Suseela Bhai, J. Thomas and R. Naidu. 1988. Standardisation of screening techniques against 'Azhukal' diseaase of small cardamom (*Electtaria cadamomum* Maton). *J. Plantn. Crops.* 18(Suppl) 252–256.

Shekhawat, P.S. and R. Prasada. 1971. Antifungal properties of some plant extracts, 11 Growth inhibition studies. *Sci. and Cult.* 37: 40–41.

Skinner, F.A. 1955. Antibiotics, In: *Modern Method of Plant Analysis.* K. Peach and M.V. Tracey (eds.), Vol. III. Springer-Verlag, Vierter Band, pp. 626–725.

Subrahmanyam, B. 1990. Azadirachtin—A naturally occurring insect growth regulator. *Proc. Indian Acad. Sci.* (Ann. Sci.). 99: 277–288.

Tsao, P.H. and M.J. Garbar. 1970. Methods of soil infestation watering and assessing degree of root infection *in situ* ecological studies with citrus *Phytophthora Pi. Dis Reptr.* 44: 710–715.

88

Antimycotic Activity of a Methanolic Extract from *Azadirachta indica* A. Juss.

B. STEINHAUER

Institut für Phytopathologie und Angewandte Zoologie, Justus-Liebig-Universität
Bismarckstr. 16, D-6300 Giessen, Germany

ABSTRACT

Extracts of ground neem kernels were prepared by Soxhlet extraction with different organic solvents and with increasing polarity. After evaporation of the solvents, the residues were reconstituted in water and tested for their activity against phytopathogenic fungi. Tests were carried out by mixing pasteurized extracts into an agar medium, followed by incubation with fungal mycelium. The strongest growth inhibition was obtained with a methanolic extract. Further purification of the crude methanolic extract was carried out by liquid/liquid extraction and low-pressure column chromatography. The purity of the fractions was controlled by HPLC analysis and their activity subsequently tested. By using semipreparative HPLC the fungicidal activity of the extract was attributed to a single component. The structure of this component has to be elucidated.

INTRODUCTION

The ingredients of *Azadirachta indica* A. Juss. are well known for their broad spectrum of biological activities. Among these, antimycotic activities have been reported by Lal et al. (1980), Singh et al. (1980), Naidu and John (1982), Achimu and Schlosser (1992), Häberle (1992), Lehmann (1992), and Rovesti et al. (1992). Antimycotic effects of neem extracts can be due to direct effects of fungitoxic or fungistatic compounds on pathogens or to indirect effects like activation of antagonists on the phylloplane or in the rhizosphere. The active compounds responsible for the fungicidal activity are, however, still unknown and it is for that reason impossible to determine whether neem extracts had direct or indirect effects.

The aim of this investigation was to identify and purify substances from *A. indica* with direct fungitoxic or fungistatic effects. The procedures of extraction and cleanup will be described.

MATERIALS AND METHODS

Fungi for testing the activity of crude extracts and fractions were obtained from the culture collection of the Institute of Phytopathology and Applied Zoology, Justus-Liebig-University, Giessen. They were cultivated on glucose-yeast extract agar containing 17 g agar, 10 g glucose, and 2 g yeast extract in 1000 ml deionized water. Neem kernels for extraction originated from Niger, harvested during 1991. Solvents for extraction were of analytical grade. Solvents for chromatographic separations were of the quality gradient grade, and double-distilled water was produced with a Büchi Fontavapor apparatus. Low-pressure column chromatography (LPCC) was carried out with a Merck Lobar column (310 × 25 mm) filled with Lichroprep RP18 40–63 μm and connected to the HPLC system. Fractions were collected with a Frac100 fraction collector (Pharmacia Inc., Sweden). The HPLC equipment consisted of a constant flow pump 300 CS, a gradient former 250 B, and a photometer SP4 from Gynkotek. Chromatograms were recorded on a Merck D 2000 chromatointegrator. Samples were injected by hand or by a Marathon autosampler (Spark Inc., Germany). The columns were kept at 40°C in the oven of the autosampler. The detection wavelength was 215 nm. Semipreparative HPLC was performed on Lichrosorb RP18 5 μm with a column of 250 × 8 mm and a flow rate of 2.5 ml/min. Samples of 200 μl were injected and fractions at 1.5 min. intervals were collected. On the base of the chromatograms, single peaks of several injections were pooled. The eluent programme P1 was used for semipreparative separation of substances of a fraction from the LPCC and P2 was used for an overview of fractions from the semipreparative HPLC. Analytical HPLC was carried out on Lichrosorb RP18 5 μm with a column of 250 × 4 mm, at flow rates of 1.2 ml/min. and 1 ml/min. respectively.

The eluent programme A1 with 1.2 ml/min. was used for an overview of crude extracts and fractions from the LPCC, A2 with 1 ml/min. was used to control the purity of the fractions from semipreparative HPLC.

A1

Time (min.)	% water	% acetonitrile
5	100	0
in 40	0	100
for 15 until end	0	100

A2

Time (min.)	% water	% methanol
35	38	62

P1

Time (min.)	% water	% methanol
10	40	60
in 5 to	35	65
for 10	35	65
in 5 to	30	70
for 10	30	70
in 10	0	100
until the end	0	100

P2

Time (min.)	% water	% methanol
for 10	60	40
in 5	50	50
for 10	50	50
in 5	40	60
for 10	40	60
in 5	30	70
for 10	30	70
in 5	20	80
for 10	20	80
in 5	10	90
for 10	10	90
in 5	0	100
for 10	0	100

For extraction, neem kernels were ground in an electric mill and 30 g of the powder was filled in a Soxhlet extraction tube. The material was defatted with 300 ml petroleum benzine (boiling point 40°–60°C) for 8 hrs in a Soxhlet extractor. After defatting, the residue was extracted either (1) directly with 300 ml methanol ($MeOH^1$) for 8 hrs or (2) subsequently with 300 ml methyl tertiary butyl ether (MTB), tetrahydrofurane (THF), ethylacetate (EtAc), and methanol ($MeOH^2$) for 8 hrs respectively. The extracts were evaporated to dryness under vacuum at 40°C. For testing of crude extracts and fractions, the residues after evaporation of the solvents were resuspended in deionized water and reconstituted to the original volume.

RESULTS

Selection of Fungi

To select fungi for the testing of neem extracts and fractions, 43 different fungal species or strains were incubated on a medium containing 10 per cent of a crude methanolic extract from neem kernels. The reaction of the isolates differed widely. Some fungi were sensitive and their mycelial growth was inhibited, some were not affected, and a few species showed an increased mycelial growth on the agar medium amended with neem extract. Out of these, eight isolates were chosen for further investigations (Table 1).

SOLVENTS FOR EXTRACTION

After defatting of the ground neem kernels, the residues were extracted with different solvents and the resuspended residues were tested for their antimycotic activity (Table 2).

The highest fungicidal activity was found in $MeOH^1$, followed by $MeOH^2$. The extraction with THF also showed considerable activity. In the treatment $MeOH^{1*}$, even though the extract had been autoclaved (129°C, 25 min.) with the agar medium, the fungicidal activity was not reduced.

Table 1. Mycelial growth (mm) of fungi on a glucose-yeast extract agar amended with 10 per cent crude methanolic neem kernel extract

Fungus	2 days after inoculation		4 days after inoculation	
	Control	Treatment	Control	Treatment
Botrytis fabae 13	26.5	27.0 (101.9)	58.0	58.0 (100.0)
Drechslera teres 149	15.5	9.5 (61.3)	30.5	22.5 (73.8)
Fusarium graminearum 151	24.5	17.0 (69.4)	45.5	32.5 (71.4)
Alternaria porri 163	11.0	8.0 (72.7)	20.0	10.0 (50.0)
Ascochyta tritici 166	17.0	12.0 (70.6)	34.5	23.0 (66.7)
Phoma lingam 171	17.0	13.5 (79.4)	32.5	25.0 (76.9)
Rhizoctonia cereale 269	18.5	16.0 (86.5)	37.0	32.5 (87.8)
Gerlachia nivalis 283	23.5	17.0 (72.3)	47.0	39.0 (83.0)

In parentheses: per cent growth relative to control.

Table 2. Effect of different neem extracts on mycelial growth of eight fungal species four days after inoculation

Fungus (mm)	Treatment	Growth (mm)	Fungus	Treatment	Growth
G. nivale (100.0)	Control	82.0 (100.0)	*F. graminearum*	Control	45.0
	MeOH[1]	58.0 (70.7)		MeOH[1]	41.3 (91.9)
	MeOH[1]*	61.3 (74.8)		MeOH[1]*	40.3 (89.6)
	MeOH[2]	58.0 (70.7)		MeOH[2]	36.0 (80.0)
	MTB	80.3 (98.0)		MTB	50.7(112.6)
	THF	61.3 (74.8)		THF	39.0 (86.7)
	EtAc	74.7 (91.1)		EtAc	38.0 (84.4)
R. cereale	Control	37.7 (100.0)	*B. fabae*	Control	80.7(100.0)
	MeOH[1]	35.7 (94.7)		MeOH[1]	66.3 (82.2)
	MeOH[1]*	41.0 (108.8)		MeOH[1]*	73.0 (90.5)
	MeOH[2]	38.3 (101.8)		MeOH[2]	54.7 (67.8)
	MTB	44.3 (117.7)		MTB	81.3(100.8)
	THF	49.7 (131.9)		THF	68.0 (84.3)
	EtAc	47.0 (124.8)		EtAc	80.0 (99.2)
A. porri	Control	32.3 (100.0)	*A. tritici*	Control	26.3(100.0)
	MeOH[1]	22.0 (68.0)		MeOH[1]	24.0 (91.1)
	MeOH[1]*	19.3 (59.8)		MeOH[2]	23.7 (89.9)
	MeOH[2]	27.7 (85.6)		MTB	34.7(131.6)
	MTB	30.7 (94.8)		THF	27.7(105.1)
	THF	24.7 (76.3)		EtAc	28.7(108.9)
	EtAc	25.7 (79.4)	*P. lingam*	Control	47.3(100.0)
D. teres	Control	49.7 (100.0)		MeOH[1]	33.3 (70.4)
	MeOH[1]	15.7 (31.5)		MeOH[2]	36.3 (76.8)
	MeOH[1]	17.0 (34.2)		MTB	48.7(102.8)
	MeOH[2]	16.7 (33.6)		THF	39.3 (83.1)
	MTB	46.3 (93.3)		EtAc	42.7 (90.1)
	THF	17.0 (34.2)			
	EtAc	22.0 (44.3)			

In parentheses: per cent growth relative to control; MeOH[1]*=MeOH[1] autoclaved.

Liquid/Liquid Extraction

The most active extract MeOH[1] was chosen for further purification. As a first step, the reconstituted turbid extract which contained a lot of insoluble material (non-polar components suspended in the water) was shaken with dichloromethan (DCM) in a separation funnel. After separation of the phases in the funnel and by centrifugation at about 3000 × g, two clear and light brown phases were obtained, with only a very small white interphase which was discarded.

Analytical HPLC, under conditions A1, showed a sharp separation into a more polar and a more non-polar fraction. After evaporation of the solvents the two phases were reconstituted in water and the activity was tested against three fungal species (Table 3).

After liquid/liquid extraction, the activity of the crude extract MeOH[1] was mainly found in the DCM phase. This phase was used for further purification.

Low-pressure Column Chromatography

The DCM phase was evaporated and dissolved in 50 per cent aqueous methanol. Three ml were separated on the Lobar column at a flow rate of 2.5 ml/min. with the following gradient:

Time (min.)	% water	% methanol
for 20	50	50
in 10 for 20	40	60
in 10 for 20	30	70
in 10 for 20	20	80
in 10 for 20	10	90
in 10 for 20	0	100

When the four large fractions were pooled and tested, the most active one was found to be fraction 4 (Table 4).

Table 3. Antimycotic activity of polar and nonpolar fractions after liquid/liquid extraction with water and (DCM) against three fungal species (growth in mm)

Fungus	Treatment	Days after inoculation	
		3	5
D. teres	Control	18.0 (100.0)	33.3 (100.0)
	MeOH[1]	9.0 (50.0)	14.7 (44.4)
	Water phase	15.3 (85.2)	28.0 (84.0)
	DCM phase	11.7 (64.8)	20.0 (60.0)
A. porri	Control	19.0 (100.0)	26.7 (100.0)
	MeOH[1]	8.0 (42.1)	10.0 (37.5)
	Water phase	14.7 (77.2)	18.7 (70.0)
	DCM phase	9.7 (50.9)	13.0 (48.8)
A. tritici	Control	22.3 (100.0)	32.7 (100.0)
	MeOH[1]	16.3 (73.1)	24.7 (75.5)
	Water phase	21.3 (95.5)	30.0 (91.8)
	DCM phase	16.3 (73.1)	25.3 (77.6)

In parentheses: per cent growth relative to control.

Table 4. Antimycotic activity of fractions from the LPCC of the DCM phase against three fungal species (measured as growth in mm)

Fungus	Treatment	Days after inoculation	
		3	4
D. teres	Control	19.5 (100.0)	30.7 (100.0)
	F 1	16.0 (82.1)	25.3 (82.5)
	F 2	15.3 (64.5)	24.0 (64.5)
	F 3	13.0 (43.0)	18.7 (39.2)
	F 4	11.8 (26.1)	17.0 (21.7)
A. porri	Control	14.7 (100.0)	20.3 (100.0)
	F 1	14.3 (97.5)	18.8 (92.8)
	F 2	12.7 (84.0)	16.7 (76.2)
	F 3	9.3 (53.3)	12.0 (45.0)
	F 4	10.0 (36.3)	13.0 (28.8)
A tritici	Control	21.0 (100.0)	30.0 (100.0)
	F 1	20.7 (98.4)	29.7 (98.9)
	F 2	19.7 (92.2)	29.7 (97.8)
	F 3	15.3 (67.3)	24.3 (79.3)
	F 4	14.3 (45.9)	22.7 (59.9)

In parentheses: per cent growth relative to control.

Semipreparative HPLC

The most active fraction 4 of the LPCC contained only four main peaks (Fig. 1). This fraction was further purified via semipreparative HPLC under conditions P1. The fractions were collected and single peaks were pooled for testing (Fig. 2). The purity of these fractions was tested by analytical HPLC under conditions A2 (Fig. 3). Fraction 3 completely inhibited the growth of *D. teres* and *A. porri*. The growth of *A. tritici* was also strongly inhibited (Table 5).

DISCUSSION

The reaction of isolates of different fungi to a crude methanolic extract of neem kernels varied widely from sensitive over insensitive to a growth promoting effect. Besides the known varying sensitivity of fungi to a fungicidal compound, the qualitative and quantitative differences in the compounds extracted largely depend on the solvent used for extraction (Feuerhake, 1985). Among these there could be certain compounds that are nutritive to certain fungi and so promote the growth. This may lead to a nullifying of the fungicidal effects of the active compounds. Purification of crude extracts towards an isolation of active compounds should step by step lead to a stronger growth inhibition. This effect was clearly demonstrated in the present investigation. *Ascochyta tritici*, which was less sensitive to the crude neem extract MeOH[1], reacted with more than 50 per cent growth inhibition to the purified compound. Almost complete growth inhibition was found with *D. teres* and *A. porri*.

The use of different solvents for the Soxhlet extraction of the neem material showed that the highest fungicidal activity could be obtained by a direct extraction with methanol after defatting of the material. Extraction with other solvents yielded only com-

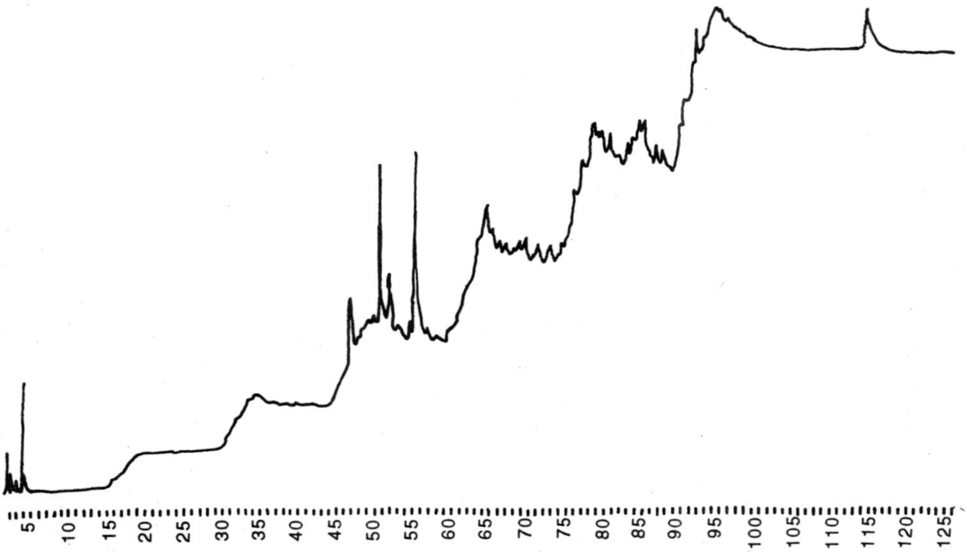

Fig. 1. Chromatogram of fraction 4 of the separation by LPCC under analytical conditions A1.

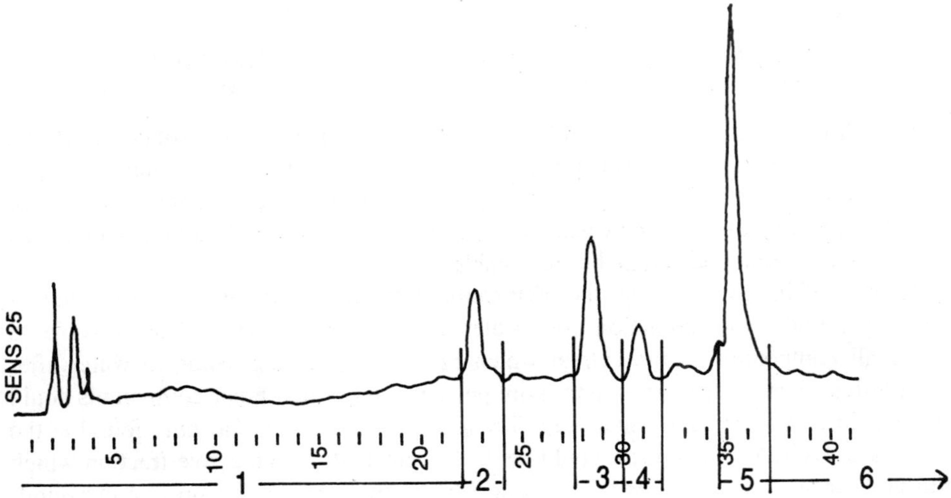

Fig. 2. Chromatogram of semipreparative separation and pooled fractions under conditions P1.

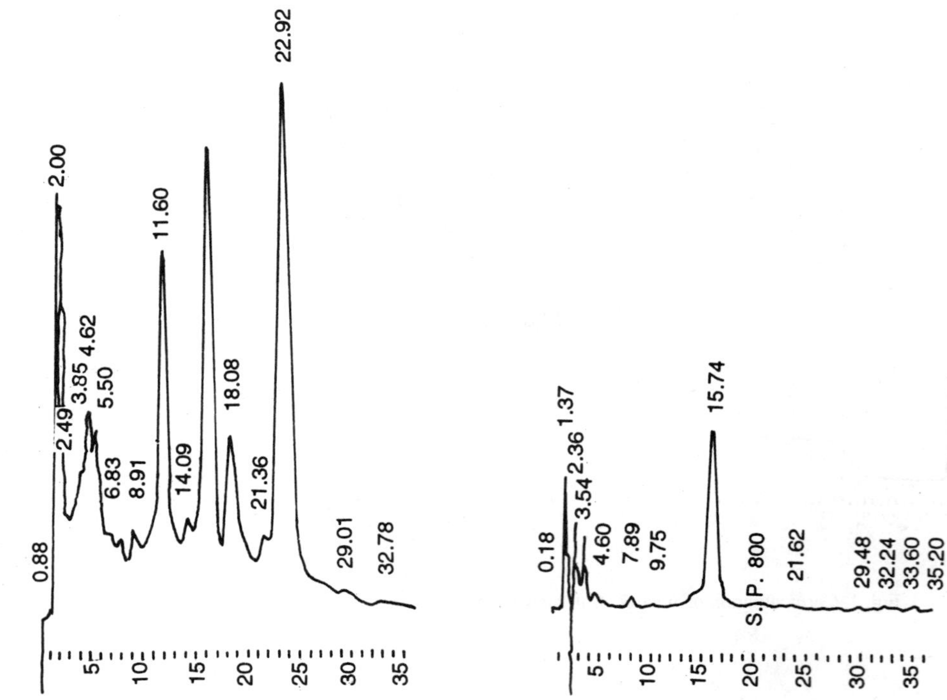

Fig. 3. Purity obtained by semipreparative HPLC, chromatograms under analytical conditions. A2: (a) fraction 4 of LPCC, (b) isolated active compound (corresponds to fraction 3 in Table 5)

pounds with lower activity. Other fungicidal compounds than the one isolated could be present in these extracts, and their purification will be the subject of further investigations. In the treatment MeOH[1*] the crude extract had been autoclaved for 25 min. The fungicidal activity was not diminished by this heat treatment, which indirectly indicated that the active compound might be heat-stable.

Because of its high activity, the methanolic extract was chosen for further purification. By liquid/liquid extraction with water and dichloromethan, it was possible to dissolve all compounds of the extract which had been in a suspension in water after reconstitution. A strict separation into more polar and more non-polar compounds could be obtained by this simple cleaning step. The antimycotic activity was only found in the DCM phase. By LPCC this phase could be further purified into an active fraction which contained only four main peaks. For a final isolation of the pure compound, semipreparative HPLC was used. It was possible to isolate three compounds with high purity while a fourth one was contaminated with one of the others. Testing of these isolated substances clearly showed that only one compound contained almost all of the fungicidal activity of the methanolic extract.

Table 5. Activity of fractions from the semipreparative HPLC against three fungal species (measured as growth in mm)

Fungus	Treatment	Days after inoculation		
		2	3	4
D. teres	Control	12.7 (100.0)	19.7 (100.0)	28.3 (100.0)
	F1	8.3 (65.8)	13.7 (69.5)	20.3 (71.8)
	F2	8.0 (63.2)	12.7 (64.4)	18.3 (64.7)
	F3	0.0 (0.0)	0.0 (0.0)	0.0 (0.0)
	F4	8.3 (65.8)	14.0 (71.2)	21.3 (75.3)
	F5	10.3 (81.6)	14.8 (75.4)	21.0 (74.1)
	F6	10.3 (81.6)	14.0 (71.2)	19.7 (69.4)
A. porri	Control	11.3 (100.0)	15.0 (100.0)	20.7 (100.0)
	F1	9.3 (82.4)	12.0 (80.0)	16.0 (77.4)
	F2	9.0 (79.4)	12.2 (81.1)	16.0 (77.4)
	F3	0.0 (0.0)	0.0 (0.0)	0.0 (0.0)
	F4	9.0 (79.4)	12.5 (83.3)	16.2 (78.2)
	F5	9.0 (79.4)	11.8 (78.9)	14.7 (71.0)
	F6	10.3 (91.2)	13.2 (87.8)	17.0 (82.3)
A. tritici	Control	17.0 (100.0)	25.0 (100.0)	32.3 (100.0)
	F1	15.7 (92.2)	23.7 (94.7)	30.7 (94.8)
	F2	14.3 (84.3)	24.3 (97.3)	30.0 (92.8)
	F3	8.8 (52.0)	10.8 (43.3)	14.3 (44.3)
	F4	14.0 (82.4)	22.8 (91.3)	30.0 (92.8)
	F5	15.3 (90.2)	23.3 (93.3)	30.0 (92.8)
	F6	15.0 (88.2)	22.2 (88.7)	29.7 (91.8)

In parentheses: per cent growth relative to control.

CONCLUSION

By the use of chromatographic methods it was possible to isolate a compound with high fungicidal activity from a crude methanolic extract of neem kernels. It is likely that other fungicidal compounds could be isolated from neem in a similar way which will be subject to further investigations. The activity of the isolated compound against a wider range of pathogens has to be tested.

After structure elucidation of the isolated compound and others that may be isolated, it will be possible to quantify their presence in different raw neem materials. This will be very useful to assess the quality of a neem material for extraction and use as a fungicide.

It will be possible to develop simple extraction procedures with a high yield of active ingredients for practical use.

REFERENCES

Achimu, P. and E. Schlösser. 1992. Effect of neem seed extracts (*Azadirachta indica* A. Juss.) against downy mildew (*Plasmopara viticola*) of grapevine. *Med. Fac. Landbouww. Univ. Gent* 57: 423–431.

Feuerhake, K.J. 1985. Untersuchungen zur Gewinnung and Formulierung von Sameninhaltsstoffen des Niembaumes (*Azadirachta indica* A. Juss.) in Hinblick auf ihre Verwendung als Schädlingsbekämpfungsmittel in den Entwicklungsländern. Dissertation, GieBen.

Häberle, R. 1992. *Untersuchungen zur Wirkung wäBriger Neemextrakte auf Gurkenmehltau.* Institut für Phytopathologie and Angewandte Zoologie, GieBen, Unveröffentlicht.

Lal, S., K. Nath and S.C. Saxena. 1980. Use of pesticides and natural products in control *Sclerospora sacchari* in maize. *Trop. Pest Mt.* 26: 286–292.

Lehmann, W. 1992. Fungizide Inhaltsstoffee aus Meliaceen-Neembaum: *Azadirachta indica A. Juss.*, Persischer Flieder: *Melia azedarach* L. - Ihre Gewinnung und Wirkung auf Wurzelbranderreger der Zuckerrübe, Dissertation, Gottingen.

Naidu, V.D. and V.T. John. 1982. *In vitro* toxicity of some fermented oilcake extract to *Rhizoctonia solani* and *Sclerotium oryzae*. *Int. Rice Res. Newsletter* 7: 15–16.

Rovesti, L., S. Di Marco and D. Pancaldi. 1992. Effect of neem kernel extract on some phytopathogenic fungi under greenhouse conditions. *Z. Pflkrankh. Pflschutz.* 99: 293–296.

Singh, U.P., H.B. Singh and R.B. Singh. 1980. The fungicidal effect of neem (*Azadirachta indica*) extracts on some soil-borne pathogens of gram (*Cicer arietinum*). *Mycologia* 72: 1077–1093.

89

Management of Chilli (*Capsicum annuum* L.) Diseases by Neem-based Preparations

R.P. MAHARISHI

Associate Professor (Plant Pathology)
Agricultural Research Station
(Rajasthan Agricultural University)
Mandor-Jodhpur (Rajasthan), India

ABSTRACT

Chilli (*Capsicum annuum* L.) is an important commercial crop and suffers considerable damage due to dieback, fruit rot, bacterial leaf spot, powdery mildew, and viral infections. The efficacy of neem (*Azadirachta indica* A. Juss.) preparations such as fresh neem leaf aqueous extract, Margocide OK 80 EC, and Margocide CK 20 EC *vis-a-vis* pesticides already in use against these diseases is presented.

INTRODUCTION

Chilli (*Capsicum annuum* L.) is a major commercial crop of Rajasthan (Anon., 1988). It is widely cultivated for green and red fruits in Mathania-Jodhpur area of the arid western plain agroclimatic zone of Rajasthan and suffers considerable damage due to dieback and fruit rot (*Colletotrichum capsici* (Syd.) Butler & Bisby), bacterial leaf spot (*Xanthomonas campestris* pv. *vesicatoria* (Doidge) Dye), powdery mildew (*Leveillula taurica* (Lev.) Arn.), and viral infections causing mosaic and leaf curl symptoms. Although chemical pesticides are available to manage these diseases, the possibility of using natural plant-based preparations, namely those of neem (*Azadirachta indica* A. Juss.), has been explored to reduce the chemical pesticide load in the environment. Chemical pesticides are now used indiscriminately in this area.

MATERIALS AND METHODS

The field evaluation studies in a replicated and randomized block design using recommended doses of fertilizers were conducted during 1990–91 and repeated in 1991–92 at Agriculture Research Station, Mandor-Jodhpur. Three neem-based preparations—fresh neem leaf aqueous extract 2.0 per cent, Margocide OK 80 EC 0.05 per cent, and Margocide CK 20 EC 0.1 per cent—were compared with pesticides carbendazim 0.1 per cent + streptocycline 100 ppm + dimethoate 0.1 per cent as spray applied, with a knapsack sprayer using 500 l water/ha. In all, three sprays were applied. The incidence of disease and per cent disease index were recorded and analysed statistically.

RESULTS AND DISCUSSION

The data depicted in Table 1 reveal that although the spray of carbendazim + streptocycline + dimethoate proved most effective, all the neem-based preparations also showed efficacy against various diseases as compared to control. The order of efficacy of preparations was Margocide OK 80 EC 0.05 per cent, Margocide CK 20 EC 0.1 per cent, and fresh neem leaf aqueous extract 2.0 per cent. The efficacy of carbendazim, streptocycline, and dimethoate against fungi, bacteria, and insect pests is well documented in literature. Carbendazim, besides being an effective fungicide, has been reported to slightly increase the chlorophylls in chilli plants infected with potato virus

Table 1. Evaluation of neem-based preparation against diseases of chilli at Agricultural Research Station, Mandor-Jodhpur

Treatment	Per cent disease Index						Per cent incidence	
	Dieback		Bacterial leaf spot		Powdery mildew		Mosaic/leaf curl	
	1990–91	1991–92	1990–91	1991–92	1990–91	1991–92	1990–91	1991–92
Spray of aqueous	5.83	13.75	2.50	12.37	10.83	35.62	18.83	25.00
Neem leaf extract	(13.63)	(21.7)	(7.33)	(20.6)	(19.03)	(36.6)	(19.03)	(30.00)
2.0%	9.79**		7.43		23.22		21.92	
Margocide OK 80EC	3.00	10.00	0.83	6.87	6.66	15.62	5.83	15.00
0.05%	(12.90)	(18.4)	(3.03)	(15.2)	(14.73)	(23.3)	(13.90)	(22.8)
	6.05		3.85		11.14		10.41	
Margocide CK 20EC	3.00	10.62	0.83	7.50	10.00	13.75	8.33	17.50
0.1%	(12.90)	(19.00)	(3.03)	(15.9)	(18.33)	(21.7)	(16.73)	(24.7)
	6.81		4.16		11.87		12.91	
Carbendazim 0.1%	2.50	8.75	0.83	5.62	7.50	10.00	3.33	15.00
+ Streptocycline	(7.33)	(17.03)	(3.03)	(13.7)	(15.73)	(18.4)	(10.36)	(22.8)
100 ppm +	5.62		3.22		8.75		9.16	
Dimethoate 0.1%								
Control	16.66	20.62	6.66	18.12	40.00	47.50	23.30	40.00
	(24.06)	(27.00)	(14.90)	(25.1)	(39.20)	(43.6)	(28.80)	(39.2)
	18.64		12.39		43.75		31.65	
SEM ±	2.36	1.17	1.61	1.04	1.52	2.18	1.66	3.17
CD 5%	7.73	3.62	5.28	3.23	4.98	6.72	5.43	9.79

* Angular value in parentheses.
** Average of 1990–91 and 1991–92.

(Suriachandraselvan and Narayanasamy, 1991), delaying the senescence of chloroplasts also. Derivatives of neem, even simple leaf aqueous extract, were shown to provide suitable reduction of diseases without causing environmental hazards.

ACKNOWLEDGEMENTS

Thanks are due to Dr. B.A. Nagasampagi, Scientist, Organic Chemistry, National Chemical Laboratory, Pune, Maharashtra, for making available the samples of Margocide OK 80 EC and Margocide CK 20 EC for evaluation, and to university authorities for facilitating the work.

REFERENCES

Anonymous. 1988. Status Report, NARP, Durgapura, Jaipur, 76 pp.

Suriachandraselvan, M. and P. Narayanasamy. 1991. Effect of systemic fungicides on the physiology of potato virus infected chilli plants. *Indian Phytopath* 44(1): 1–4.

8

ROLE IN INCREASING FERTILIZER NITROGEN EFFICIENCY

90

Potential Use of Urea Coated with Neem Cake to Improve Nutrient Uptake and N Recoveries and to Reduce N Losses in a Perennial Aromatic Grass, Java citronella (*Cymbopogon winterianus* Jowitt.)

E.V.S. PRAKASA RAO

Central Institute of Medicinal and Aromatic Plants
Regional Centre, GKVK Post Office, Bangalore 560 065, India

ABSTRACT

Large quantities of N fertilizers are required to sustain good yields of oil from Java citronella (*Cymbopogon winterianus* Jowitt.). However, N-use efficiency is very poor. A two-year study conducted to assess the nutrient uptake, N recoveries, and NH_3 volatilization losses showed that neem cake-coated urea (NCU) significantly increased the uptake of N, P, and K by 17, 15, and 25 per cent respectively over ordinary urea. It increased apparent recoveries by 90 per cent (from 8.4 to 16 per cent) and by 45 per cent (from 11.4 to 16.5 per cent) over ordinary urea at 300 and 400 kg N/ha/year respectively, by a two-year crop. It reduced NH_3 volatilization losses by 31 per cent (from 38 to 29 per cent) over ordinary urea, during a 25-day period, indicating that it is economical.

INTRODUCTION

Java citronella (*Cymbopogon winterianus*) is an important aromatic plant cultivated in many parts of India. The essential oil obtained by steam distillation of citronella grass is widely used in perfumes and mosquito repellent creams, and is a source of some aroma chemicals such as citronellal and hydroxy citronellal. In the past decade, substantial data were generated to show that the oil yields of Java citronella could be improved tremendously by applying large quantities of N fertilizers (Prakasa Rao et al., 1983,

1984, 1985, 1988a, Prakasa Rao and Singh, 1991); however, N recoveries by this crop remained quite poor (Prakasa Rao et al., 1983, 1988a). In an effort to improve the N-use efficiency, (NCU) tested for its effect on nutrient uptake, N recoveries, and NH_3 volatilization losses in a citronella field. Neem cake was reported to possess nitrification inhibitory properties (Reddy and Prasad, 1975; Devakumar, 1986).

MATERIALS AND METHODS

A field experiment was conducted for two years at the experimental farm of the Central Institute of Medicinal and Aromatic Plants, Regional Centre, Bangalore. The soil of the experimental field was a sandy loam with 8.4 pH (soil to solution ratio, 1:2.5), 0.4 per cent organic carbon, 0.05 per cent total N, 14 kg/ha Olsen's available P, and 170 kg/ha exchangeable K. The treatments consisted of all combinations of four levels of N (100, 200, 300, and 400 kg/ha/year) and two N sources (urea and NCU) and a control and were arranged in a randomized block design with three replications. Citronella slips were planted at a spacing of 60 × 45 cm. P_2O_5 and K_2O at 40 kg each per ha were applied basally and N was applied in four splits in each year. Five harvests were taken during the two-year period. At each harvest, plant samples were collected; Nitrogen, phosphorus, and potassium in these samples were analysed, and apparent N recoveries were calculated as described earlier (Prakasa Rao et al., 1983). NH_3 volatilization studies were made in the field as described earlier (Prakasa Rao and Bhat, 1984).

RESULTS AND DISCUSSION

Nitrogen Recoveries

At low levels of N (100 and 200 kg N/ha/year), urea and NCU were on par. However, when large quantities of N were applied (at 300 and 400 kg N/ha/year), NCU increased apparent N recoveries substantially by 45 to 90 per cent over ordinary urea (Fig. 1). When large quantities of N are applied, the potential for N losses through leaching and NH_3 volatilization are expected to be greater; thus, neem cake helped to reduce losses in such situations. Neem cake-coated urea was shown to reduce leaching losses of N (Prakasa Rao and Prasad, 1980).

NH_3 Volatilization Losses

Ammonia volatilization losses were measured over 25 days (Fig. 2). Most of the losses occurred within seven days after urea application and this confirms our earlier findings (Prakasa Rao and Bhat, 1984; Prakasa Rao and Puttanna, 1987). Neem cake-coated urea reduced the losses substantially, by 31 per cent (from 38 to 29 per cent) over ordinary urea. Soil pH at the end of 25 days was 7.8 and 7.7 (initial soil pH, 8.4) with urea and NCU respectively. Perhaps, due to lowering of pH, NCU reduced NH_3 volatilization losses. Neem cake-coated urea was shown to reduce NH_3 volatilization losses in some other studies also (Reddy and Mishra, 1983; Sharma and Gupta, 1989).

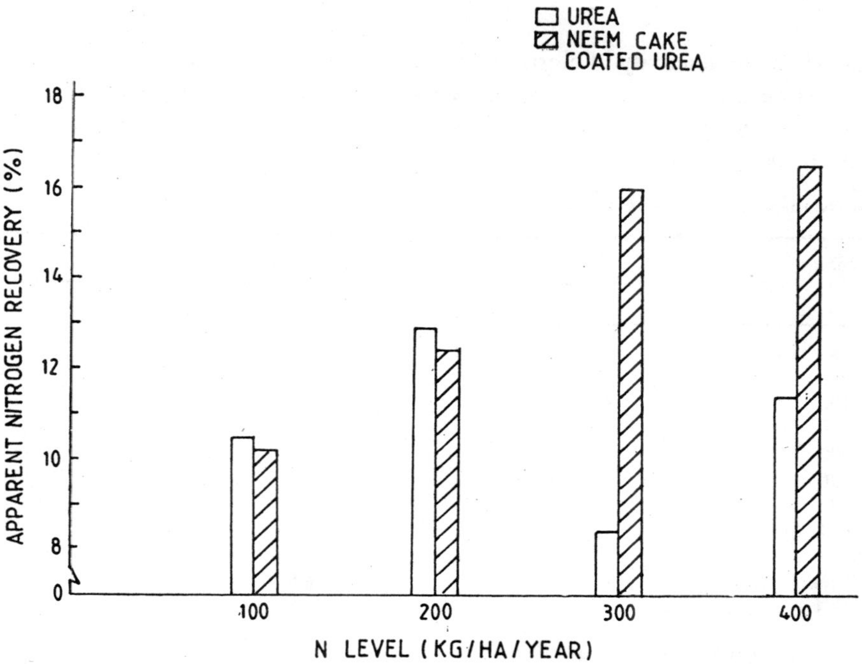

Fig. 1. N recoveries by Java citronella with urea and NCU at different N levels

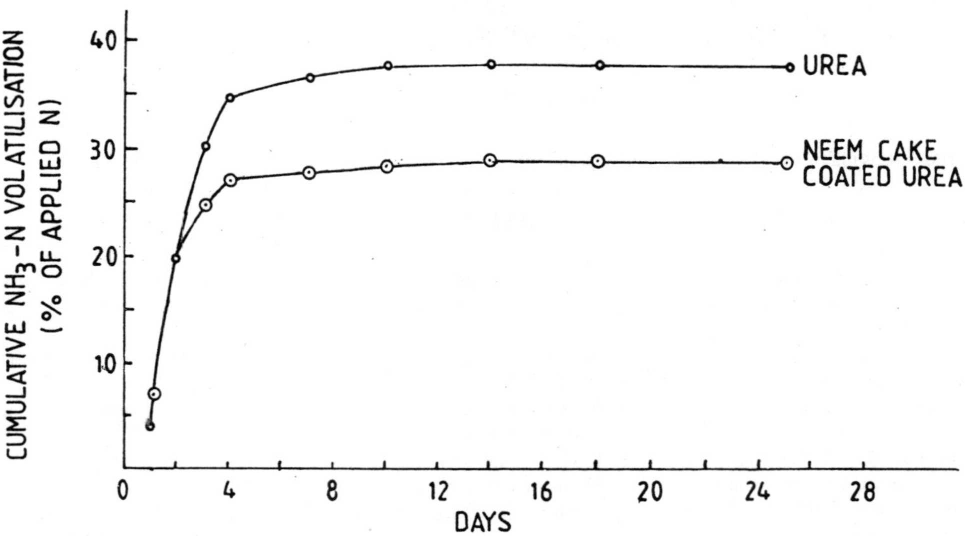

Fig. 2. Ammonia volatilization losses from urea and NCU applied to soil

Nutrient Uptake

Increase in the productivity of citronella was associated with increase in uptake of major nutrients (Prakasa Rao et al., 1988a,b). Neem cake-coated urea significantly increased the uptake of nitrogen, phosphorus, and potassium over ordinary urea by 17, 15 and 25 per cent respectively (Table 1).

Table 1. Nutrient uptake by Java citronella with urea and NCU

Treatment	Nutrient uptake (kg/ha/2 years)		
	Nitrogen	Phosphorus	Potassium
N level (kg/ha/year)			
0	64.2	15.1	82.5
100	84.9	16.1	110.8
200	115.5	19.9	143.3
300	137.4	20.6	169.4
400	175.6	24.2	206.6
S.Ed	9.65	1.48	16.49
LSD (p = 0.05)	20.46	2.84	34.97
N source			
Urea	118.2	18.8	136.1
NCU	138.6	21.6	170.1
S.Ed	6.82	0.95	11.66
LSD (p = 0.05)	14.46	2.01	24.72

CONCLUSION

The present study has shown that NCU has a good potential in the production of citronella oil by reducing N losses and improving nutrient utilization by the crop.

ACKNOWLEDGEMENT

I am thankful to the Director, Central Institute of Medicinal and Aromatic Plants, for facilities.

REFERENCES

Devakumar, C. 1986. Identification of nitrification retarding principles in neem (*Azadirachta indica* A. Juss.) seeds. Ph.D. thesis. Division of Soil Science and Agricultural Chemistry, Post Graduate School, IARI, New Delhi.

Prakasa Rao, E.V.S. and P.R. Bhat. 1984. Ammonia volatilization losses from prilled urea, urea-supergranules and sulphur coated urea when surface applied and deep placed. *Fert. News* 29: 31–32, 47.

Prakasa Rao, E.V.S. and R. Prasad. 1980. Nitrogen leaching losses from conventional and new nitrogenous fertilisers in lowland rice culture. *Plant and Soil* 57: 383–392.

Prakasa Rao, E.V.S. and K. Puttanna. 1987. Nitrification and ammonia volatilization losses from urea and dicyandiamide treated urea in a sandy loam soil. *Plant and Soil* 97: 201–206.

Prakasa Rao, E.V.S., M.R. Narayana, M. Singh and K. Puttanna. 1983. Effect of N, P and K fertilizers on growth, yield and nutrient uptake in Java citronella (*Cymbopogon winterianus* Jowitt.). *Z. Acker und Pflanz.* 152: 279–283.

Prakasa Rao, E.V.S., M. Singh, M.R. Narayana and G. Chandrasekhara. 1984. Relative efficiency of prilled urea and urea supergranules in Java citronella (*Cymbopogon winterianus* Jowitt.) *Fert. Res.* 5: 435–437.

Prakasa Rao, E.V.S., M. Singh, R.S. Ganesha Rao and S. Ramesh. 1985. Effect of urea and neem cake coated urea on yield and concentration and quality of essential oil in Java citronella (*Cymbopogon winterianus* Jowitt.). *J. Agric. Sci. Camb.* 104: 477–479.

Prakasa Rao, E.V.S., M. Singh and G. Chandrasekhara. 1988a. Effect of nitrogen application on herb yield, nitrogen uptake and nitrogen recovery in Java citronella (*Cymbopogon winterianus* Jowitt.). *Indian J. Agron.* 33(4): 412–415.

Prakasa Rao, E.V.S., M. Singh and R.S. Ganesha Rao. 1988b. Intercropping studies in Java citronella (*Cymbopogon winterianus*). *Field Crops Res.* 18: 279–286.

Prakasa Rao, E.V.S. and M. Singh. 1991. Long term studies on yield and quality of Java citronella (*Cymbopogon winterianus* Jowitt.) in relation to nitrogen application. *J. Essential Oil Res.* 3: 419–424.

Reddy, V.R.M. and B. Mishra. 1983. Effect of altered soil urease activity and temperature on ammonia volatilization from surface applied urea. *J. Indian Soc. Soil Sci.* 31: 143–145.

Reddy, R.N.S. and R. Prasad. 1975. Studies on mineralisation of urea and nitrification inhibitor treated urea in soil. *J. Soil Sci.* 26: 304–312.

Sharma, B.D. and I.C. Gupta. 1989. Effect of rate and source of nitrogen and moisture content of soil on ammonia volatilization for sandy soils. *J. Indian Soc. Soil Sci.* 37: 665–669.

91

Relative Agronomic Efficiency of Neem Extract-coated Urea versus Prilled Urea for Rice

A.L. BHANDARI, K.N. SHARMA, R.S. BRAR AND D.S. RANA

Department of Agronomy
Punjab Agricultural University, Ludhiana 141 004, India

ABSTRACT

To evaluate the agronomic effectiveness of neem extract-coated urea (NECU) vs prilled urea (PU) for rice, a field experiment was conducted for two consecutive years at Punjab Agricultural University Farm, Ludhiana, on an alkaline loamy sand (Typic Ustochrepts) soil inherited with heavy water percolation rate of 100 mm/day. In this study three levels of NECU (80, 100, 120 kg N/ha) with four methods of application—M_1 (single split of NECU), M_2 (two splits, 1/3 + 2/3), M_3 (two splits, 2/3 + 1/3), M_4 (three splits, 1/3 + 1/3 + 1/3)—were compared with PU application in three splits (1/3 + 1/3 +1/3) including N control in a randomized block design. The results revealed that rice yield increased significantly with N application from both NECU and PU in both years. The NECU in three splits as per recommended practice was significantly superior to the rest of the methods. There existed a highly significant positive relationship between different yield attributes and rice yield. At the same rate of N application NECU showed an average yield increase of 330 kg/ha over the PU. In fertilizer-use efficiency, NECU gave the maximum values of 41.2 at 80 kg N/ha followed by 39.2 and 34.7 at 100 and 120 kg N/ha in comparison to 31.9 kg grains per kg N by PU application at 120 kg N/ha. The data suggested a saving of nearly 20 kg fertilizer N/ha by the use of NECU as nitrogen source for rice. The application of NECU to rice also showed its residual effect on the grain yield of succeeding wheat crop.

INTRODUCTION

A recovery of applied N by rice crop under wetland conditions seldom exceeds 30–40 per cent even with the best management practices (Reddy and Prasad, 1977; DeDatta, 1987). Any method which would increase the N-use efficiency in rice even by 5–10 per cent could result in considerable saving of the fertilizer nitrogen. Prilled urea (PU) is the main source of N applied to soil; it hydrolyses to ammonium carbonate subsequently converted to nitrite, then to nitrate, by the action of nitrifying bacteria. The nitrate nitrogen in flooded rice is subject to losses either as elemental N or as nitrous oxide by the action of denitrifying bacteria or its loss in mass through percolation of irrigation water (Katyal et al., 1984; Mikkelsen et al., 1978). Therefore, to ensure continuous and optimal supply of N to rice crop, there is a need to regulate its supply by reducing the rate of either urea hydrolysis or nitrification or both.

Neem (*Azadirachta indica* A. Juss.) is a potential indigenous material which can effectively be used as N-inhibitor. In spite of encouraging results obtained with its by-product neem cake used in coating the urea (Reddy and Prasad, 1977; Sinha et al., 1980), this practice has not attracted farmers' attention mainly because of the crude and cumbersome method of preparing the coated material. Since neem bitters (tetranortriterpenoids) in the neem are reported to be responsible for nitrification inhibition (Hulagur and Shinde, 1984; Bringi, 1987), neem extract coating for PU in concentrated form performs better than cake coating. With these considerations a concentrated neem extract obtained from Godrej Soaps Ltd., Bombay, under the trade name Nimin (containing about 5 per cent neem bitter w/w) was used for preparing the neem extract-coated urea (NECU), the agronomic effectiveness of which in relation to PU was evaluated for rice grown in a sequence with wheat.

MATERIALS AND METHODS

A field experiment was conducted at Punjab Agricultural University Farm, Ludhiana, during the rainy season of 1988 and 1989 to evaluate the performance of (NECU) in rice under different methods and levels of N application in relation to standard application. The soil of the experimental field was loamy sand (Typic Ustochrepts) and had pH 8.2, low organic carbon, low potassium permanganate oxidizable nitrogen, and medium phosphorus (0.5 M $NaHCO_3$ extractable) and potassium (1 N neutral NH_4 OAC extractable) with average percolation rate of 100 mm/day. The experiment was laid out in a randomized block design with three replications. The treatments included three levels of NECU application (80, 100, 120 kg N/ha) through four methods—M_1 = single application at planting, M_2 = two splits (1/3 at planting + 2/3 after three weeks of planting), M_3 = two splits (2/3 at planting + 1/3 after six weeks of planting), and M_4 = three splits (1/3 at planting and remaining 2/3 in equal splits, three and six weeks after planting)—besides a N control and standard PU application (120 kg N/ha in three equal splits as in M_4). A basal dose of 30 kg each of P_2O_5 and K_2O along with 25 kg zinc sulphate/ha was also applied at the time of planting. Rice seedlings (40–45 days) of cultivar PR 106 were transplanted during the third week of July in both years. In order to prepare NECU, the required quantity of Nimin (1 per cent w/w) was thoroughly mixed with PU by hand for 10–15 min. to obtain a uniform coating and allowed to dry in air for about 3 hrs prior to its application.

Grain (unhusked rice) and straw yield was recorded at harvest and computed for 15 per cent moisture and oven-dry weight, respectively. Ancillary information on productive tillers, panicle length, grain per panicle, panicle weight, and 1000 grain weight was also recorded to work out simple correlation coefficients (r) between the yield attributes and grain and straw yield of rice. After harvest of rice crop the residual effect of N fertilizers was studied on the succeeding wheat crop (cv HD 2329) receiving uniform dressing of 60 kg N, 60 kg P_2O_5, and 30 kg K_2O/ha every year.

RESULTS AND DISCUSSION

Source and Methods of N Application

Application of N through either NECU or PU increased the rice yield significantly over the control yield of 2.66 t/ha (Table 1). The mean yield response to successive N addition through NECU was 3.10, 3.60, and 3.93 t/ha at 80, 100, and 120 kg N/ha respectively. The yield increase due to PU application at 120 kg N/ha was 3.87 t/ha. Among methods of NECU application the split application (M_4) as per recommended practice (three splits) was significantly superior to the other methods. Furthermore, NECU application as a single dose at all levels of N gave lower yield than two or three split application though the significant differences between M_1 and M_2/M_3 at all N levels were not observed.

The yield of rice with 100 kg N/ha application through NECU in three splits was nearly identical with that obtained with split PU application at 120 kg N/ha, whereas at the same rate of N the NECU gave a yield increase of 0.33 t/ha over PU. A similar trend was found in straw yield in both years (data not presented).

Yield Attributes vs Yield Relationship

The yield attributes of rice such as productive tillers, panicle length, grains per panicle, panicle weight, and 1000 grain weight also indicated a significant increase with successive N application by NECU and 120 kg N/ha with the PU over N control. However, these attributes were significantly higher with three N splits (M_4), and the other three methods of NECU application were at par with each other. For the sake of brevity the data of different yield attributes are not presented. Only coefficients of correlation (*r*) between the yield attributes and grain and straw yield of rice are given in Table 2.

There exists a highly significant positive relationship between yield parameters and grain and straw yield. Among different parameters *r* values were higher with number of product tillers, panicle length, and 1000 grain weight in that in order.

Table 1. Effect of application method and N levels of NECU on grain yield of rice (mean two years)

Method of N application	Yield (t/ha) at N levels (kg/ha)		
	80	100	120
M_1	5.55	6.05	6.37
M_2	5.72	6.19	6.64
M_3	5.82	6.21	6.49
M_4	5.95	6.58	6.86
PU in splits			6.53
N control			2.66
CD at 5%		0.31	

Fertilizer-use Efficiency

The fertilizer-use efficiency on computation as kg grains per kg of applied N indicated was undoubtedly higher with NECU application in three splits at all levels of N addition (Table 3). The values obtained with 80, 100, 120 kg N/ha were 41.2, 39.2, 34.7 kg grains per kg N in comparison to 31.9 kg grains with PU application at 120 kg N/ha. The data clearly suggest higher agronomic efficiency of the applied N with NECU over the PU and more so at 80 kg and 100 kg N/ha.

Residual Effect of N Fertilizers

The residual effect of N fertilizers applied to rice was also observed on the succeeding wheat crop. The average increase in grain yield of wheat was 0.12, 0.19, and 0.30 t/ha at 80, 100, 120 kg N/ha over the control yield of 4.07 t/ha. The yield differences were significant only at 120 kg N/ha through NECU over the N control (data not presented) during both years; however, wheat yields were comparable with PU application.

The large yield increase with nitrogen application was apparently due to low N fertility of the native soil (Prasad and Thomas, 1982; Narang and Bhandari, 1990). The significant increase in rice yield in three-split (M_4) application of NECU is ascribed to continuous supply of nitrogen at different growth stages of the crop and less susceptibility to losses through different pathways especially in light-textured soils with intermittent flooding. The better performance of NECU over PU application is obviously due to its inhibitory characteristics, which alleviate the rate of urea hydrolysis or nitrification or both (Prasad et al., 1971; Prasad and Thomas, 1982; Hulagur and Shinde, 1984). The production of more grains per kg of nitrogen with NECU in comparison to PU and

Table 2. Correlation coefficients (*r*) between different yield parameters vs grain and straw yield of rice

Parameter	Grain	Straw
Productive tillers/m^2	0.936	0.872
Panicle length (cm)	0.941	0.889
Grain/panicle	0.881	0.814
Panicle weight (g)	0.917	0.826
1000 grain weight (g)	0.928	0.884

Table 3. Effect of application methods and N levels of NECU on fertilizer-use efficiency (kg grains/kg N) by rice mean two years)

Method of N application	Fertilizer-use efficiency (kg grains/kg/N) at N levels (kg/ha)		
	80	100	120
M_1	36.16	34.13	31.93
M_2	38.22	35.27	33.14
M_3	39.53	35.58	31.96
M_4	41.16	39.23	34.73
PU in splits			31.92
CD at 5%		4.07	

highly significant positive *r* value between the yield parameters and grain and straw yield also confirmed the above observations.

CONCLUSIONS

The study concludes that NECU can be a better N source in producing more rice grains than PU application, besides improving fertilizer-use efficiency in sandy soils. Application of 100 kg N/ha to rice through NECU gave identical yields (6.58 t/ha) to 120 kg N/ha (6.53 t/ha) via PU application, resulting in net saving of 20 kg fertilizer N/ha.

REFERENCES

Bringi, N.V. 1987. *Nontraditional Oilseeds and Oils in India*. Oxford & IBH Publishing Co. Pvt. Ltd., New Delhi, 254 pp.

DeDatta, S.K. 1987. *Principles and Practices of Rice Production*. John Wiley and Sons, New York, 89–145 pp.

Hulagur, B.F. and J.E. Shinde. 1984. Effect of nonedible oil cakes and neem extract on urea hydrolysis and ammonia volatilization loss in flooded soils. *Bull. Indian Soc. Soil Sci.* 13: 138–144.

Katyal, J.C., Bijay Singh and E.T. Craswell. 1984. Fate and efficiency of urea supergranules applied to high percolating rice growing soil. *Bull. Indian Soc. Soil Sci.* 13: 229–237.

Mikkelsen, D.S., S.K. DeDatta and W.N. Oboemea. 1978. Ammonia volatilization losses from flooded rice soils. *Soil Sci. Soc. Am. J.* 42: 725–773.

Narang, R.S. and A.L. Bhandari. 1990. Integrated nitrogen management in rice-wheat system. *Rice-Wheat Workshop*, Directorate for Cropping System Research, Modipuram, Meerut, 15–16 Oct.

Prasad, R., G.B. Rajale and B.A. Lakhdive. 1971. Nitrification retarders and slow release nitrogen fertilizers. *Adv. Agron.* 23: 337–376.

Prasad, R. and J. Thomas. 1982. Nitrogen. In: *Review of Soil Research in India*, Part I, Proc. 12th Int. Cong. Soil Sci. Indian Soc. Soil Sci., New Delhi, 309–322 pp.

Reddy, R.N.S. and R. Prasad. 1977. Fertilizer management in rice-wheat rotation. Part 1. Nitrogen. *Fert. News* 22(7): 3–5.

Sinha, N.P., S.N. Singh and B.K. Jha. 1980. Effect of neem (*Azadirachta indica*) and its extractives on yield and nutrient uptake in rice. *J. Indian Soc. Soil Sci.* 28: 535–536.

92

Development of Neem-coated Urea for Increasing Nitrogen Use Efficiency and Crop Yields

B.N. VYAS, N.B. GODREJ AND K.B. MISTRY

Godrej Soaps Limited, Vikhroli, Bombay 400 079, India

ABSTRACT

A concentrated neem extract (Nimin) which contains 5–6 per cent neem bitters (tetranortiterpenoids) responsible for inhibition of nitrification in soils has been developed. Nimin is self-adhesive, and neem-coated urea (NCU) can be prepared by mixing with urea (Nimin/urea ratio being 1/100). For quantities of 50–100 kg, simple hand mixing would serve the purpose. Experiments on a laterite (oxisol) and a black soil (vertisol) indicated significantly lower levels of nitrification (nitrate + nitrate nitrogen) at least up to 30 days in NCU-treated soils in comparison with levels observed 10–15 days in soils treated with uncoated urea. Application of 75–80 kg N/ha as uncoated urea is equivalent in terms of nitrogen use efficiency and crop yields to the application of 100 kg N/ha as uncoated urea. Extensive field trials with NCU on paddy sugarcane, wheat, maize, and pepper have shown an increase in yield ranging from 4 to 280 per cent.

The cost-benefit ratio for the use of NCU ranged from 1:2.4 to 1:36.4 for the different crops studied.

INTRODUCTION

The primary objective of sustainable agriculture is to develop farming systems that meet the increasing primary demands of food, clothings, and shelter of the society by conserving the natural resource base and maintaining a safe environment in the long run. It is estimated that nearly 225 million of food grains, 30 million of sugar, 10 million of vegetable oil, 17 million bales of cotton, 64 million of milk, and 2 million of meat will be required to feed a projected Indian population of 1,000 million by the end of this century. With limited scope for horizontal geographical expansion in the net cultivable area, it is essential that highly efficient management practices and modern scientific techniques are adopted to meet the food and fodder demands of the society.

Efficient fertilizer management has played an extremely important role in increasing food grain production from 55 million t in 1951 to nearly 180 million t in 1992 and will continue to be the key factor in increasing production. Since the recovery of fertilizer nitrogen by kharif-grown rice seldom exceeds 35–40 per cent and that for the other crops is not more than 60–65 per cent under the best management, there is an urgent need to improve the efficiency of fertilizer nitrogen use by crops. This in turn relates to increasing the urea use efficiency, since urea accounts for more than 80 per cent of nitrogenous fertilizers used in India.

Scientists around the world are concentrating their efforts on regulating the nitrogen supply to crops by reducing the rate of either hydrolysis or nitrification or both, so as to ensure continuous and optimal supply of nitrogen to match the requirements of crops at different stages of growth. In this context, slow-release urea forms such as sulphur-coated urea, lac-coated urea, polymer-coated urea, and urea supergranules have been extensively investigated (Katyal et al., 1985; Prasad et al., 1971; Rao, 1987; Tandon, 1987; Thomas and Prasad, 1983; Vyas and Mistry, 1985; Youngdahl et al., 1986). However, for reasons of high cost of coating materials and other factors, none of these materials currently being used in India.

In order to improve nitrogen use efficiency of crops, synthetic chemicals such as phenylphosphorodiamidate (PPDA), thiourea (TU), N-serve, and dicyandiamide (DCD) have also been examined for inhibition of either urea hydrolysis or nitrification in soils. Again, owing to their cost, it is found difficult to use these synthetic chemicals in agricultural practice in India, and it is imperative to direct our efforts to the development of indigenous products.

A considerable volume of data is available on the potential of neem (*Azadirachta indica* A. Juss.) and karanj (*Pongamia glabra* Vent) constituents as effective nitrification inhibitors (Patil, 1972; Sahrawat, 1981; Sahrawat and Parmaar, 1975; Slanghen and Kerkhof, 1979). The Indian Council of Agricultural Research in its publication 'Technology for Rice Production' for different states has recommended the use of neem cake blended or coated urea. In spite of the encouraging results obtained with the use of urea coated with neem cake (Hulagur and Shinde, 1984; Panda et al., 1988; Thomas and Prasad, 1983), this practice has not attracted the attention of farmers on a large scale because the process of coating urea with neem cake is cumbersome and because neem cake is not readily available at the farm level.

DEVELOPMENT OF NEEM EXTRACT (NIMIN)

The neem bitters (tetranortriterpenoids) reported to be responsiblle for nitrification inhibition—nimbin, nimbinin nimbidin, nimbinimin, and others—are lipid associates and constitute about 2 per cent (w/w) of oil (Bringi, 1987; Mitra, 1963).

A concentrated neem extract, which contains as much as 5 per cent neem bitters responsible for nitrification inhibition in soils, has been prepared by processing industrial-grade neem oil. For this purpose, neem oil was successively treated in oder to remove the major part of triglycerides and free fatty acids. The resultant fraction was subsequently solvent-extracted to concentrate the desired neem alkaloids. This extract has been given the trade name 'Nimin'. Nimin is self-adhesive and no external adhesives have been added.

Nimin contains 6–7 per cent unsaponifiable matter, out of which nearly 80 per cent (about 5 per cent w/w, in the finished product) is composed of neem bitters responsible for nitrification inhibition in soils. It also contains 15–20 per cent of unrecovered free fatty acids. The elemental composition of Nimin is carbon 75–79 per cent, hydrogen 11–12 per cent, and nitrogen 0.45–0.55 per cent and the remainder consists of oxygen, sulphur, and traces of phosphorus and potassium.

The estimation of neem bitters responsible for nitrification inhibition was carried out by the scheme suggested by Mitra (1963). The constituents of Nimin were also analysed using advanced techniques such as infra-red spectrometry UV absorption spectrometry, high performance liquid chromatography, and carbon-hydrogen-nitrogen analyser.

In order to prepare neem-coated urea (NCU), Nimin (1 per cent w/w) needs to be mixed with urea. For small quantities of urea, say 50–100 kg, no mechnical mixer is required. A urea bag is emptied on a clean floor and Nimin is uniformly spread over it. The constituents are mixed thoroughly by hand to obtain a uniform coating. For larger quantities of urea, rotary mixers can be used. Urea is added to the mxing container and Nimin introduced from the top while mixing continues. Generally, coating requires 15–20 min., but mixing duration can vary depending on the type of mixer and the quantities of urea involved. Our extensive trials have shown that the uniform coating of Nimin over prilled urea is indicated by the golden yellow appearance of urea prills without the presence of dark brown spots due to unmixed Nimin in the product.

It is preferable to air dry NCU for 2–3 hrs before its application to soils. Dried NCU, if not required immediately, can be stored up to six to eight months without any deterioration in its appearance, physicochemical properties, or efficacy.

LABORATORY EXPERIMENTS

Nimin-coated urea was examined for hydrolytic degradation and nitrification in a laterite soil (Ultisol) from Maharashtra and a medium black soil (Vertisol) also from Maharashtra under submerged as well as upland conditions. Soils were incubated at field capacity moisture status in upland conditions, whereas 4 cm water head was maintained for submerged conditions. Soil samples were removed at different incubation periods of 1, 3, 7, 15, and 30 days and extracted with 2 M KCl. Ammoniacal and nitrite + nitrate nitrogen were determined separately by steam distillation using MgO and Devarda's alloy (Bremner, 1965).

GREENHOUSE EXPERIMENT

A greenhouse experiment at the Research Centre of Godrej Soaps was conducted on Ultisol from Raigad district of Maharashtra. Rice and maize were grown till the completion of vegetative phase as test crops. Nimin-coated urea as well as prilled urea at 100 kg N/ha were added as basal application as well as in three split doses. The data on dry matter yield, total nitrogen uptake, and apparent nitrogen recovery were collected and are presented in Table 1.

Table 1. Influence of nimim-coated urea on yield and N uptake by rice and maize plants under greenhouse conditions

Treatments	Dry matter yield (g)		Total N uptake (mg/pot)		% apparent N recovery	
	Rice	Maize	Rice	Maize	Rice	Maize
Control (O–N)	51.2	47.0	453.6	273.5		
Uncoated urea (100 kg N/ha) single basal application	57.5	56.7	541.1	362.3	34.2	35.5
Uncoated urea (100 kg N/ha) 3 split applications	59.4	55.0	596.3	384.4	57.1	44.4
Nimin-coated urea (100 kg N/ha) single basal application	66.9	63.8	669.0	414.5	86.1	56.4
Nimin-coated urea (60 kg N/ha) single basal application	57.4	56.9	512.6	341.9	39.1	45.6
Nimin-coated urea (70 kg N/ha) single basal application	60.3	59.4	544.5	375.6	51.9	58.2
Nimin-coated urea (80 kg N/ha) single basal application	65.5	58.0	581.6	398.3	64.0	58.4
LSD (P = 0.05)	4.3	3.8	38.8	30.5	5.4	3.0

$$\%\ \text{apparent N recovery} = \frac{\text{N uptake by plants from N-treated pots} - \text{N uptake by plants from control pots}}{\text{N applied/pot}} \times 100$$

FIELD EXPERIMENTS

Since NCU was found to perform well in respect of rate of nitrification, nitrogen recovery, and dry matter yield of rice and maize plants in laboratory and greenhouse experiments, a large number of well-replicated trials for rice, sugarcane, wheat, maize, and plantation crops were organized at agircultral universities and research stations employing standard protocols for evaluation of ferilizer materials.

RESULTS AND DISCUSSION

Laboratory Experiments

The data on the nitrite and nitrate nitrogen content in both the soils and water regimes indicated a gradual increase up to a period of one month when soils were treated with NCU. In contrast a sharp increase in NO_2 + NO_3-N levels within a period of 7 or 15 days was observed in the soils treated wth uncoated urea. The values on NO_2 + NO_3 nitrogen level were used to calculate nitrification inhibition in soils and are presented in Table 2. The data clearly indicate a significant inhibition of nitrification in NCU-treated soils up to a period of 30 days.

Table 2. Influence of nimin-coated urea on nitrification inhibition in two soils under two moisture regimes

	Percent nitrification inhibition* Days after incubation			
	3	7	15	30
Laterite soil (Ultisol)				
Aerobic conditions	74.4	63.5	53.3	46.2
Flooded conditions	22.1	53.7	25.2	10.5
Black soil (Vertisol)				
Aerobic conditions	86.9	78.9	57.6	39.7
Flood conditions	39.1	66.9	36.6	38.8

$$\text{*\% Nitrification inhibition} = \frac{(NO_2 + NO_3\text{– N in PU-treated soil}) - (NO_2 + NO_3\text{ – N in NCU-treated soil})}{NO_2 + NO_3\text{ – N in PU-treated soil}} \times 100$$

Table 3. Per cent hydrolysis of urea nitrogen in two soils

Days	1		3		7		15		30	
	A	F	A	F	A	F	A	F	A	F
Ultisol										
Prilled urea	65	60.6	100	100	81.7	92.3	69.3	72.3	46	64.6
Nimin-coated urea	27.7	82.3	85.7	97	94.7	98.5	79.3	73.2	73.4	61.8
Vertisol										
Prilled urea	94	60.4	94.3	94.3	73.1	64.7	72.2	77.8	62.0	72.7
Nimin-coated urea	58.0	49.8	93.3	69.3	100	66.8	99.4	82.9	93.3	77.7

A = Aerobic conditions (upland)
F = Flooded conditions (submerged)

Ammoniacal nitrogen content in soils was used to determine the per cent urea nitrogen hydrolysed and these data are presented in Table 3. The results did not reveal any notable difference between the rate of hydrolysis of the two urea forms, indicating that the Nimin coating had negligible influence on the activity of urease enzyme.

Greenhouse Experiments

The results of greenhouse experiments on rice and maize are presented in Table 1. Data in Table 1 clearly indicate that the maximum values of dry matter yield, nitrogen uptake, and apparent nitrogen recovery were obtained with NCU application, which was superior to three split applications of uncoated urea. In both cases there was apparent saving of nearly 30 per cent of applied N.

Field Experiments

The salient findings of the field trials on the evaluation of NCU for rice, wheat, maize, sugarcane, and plantation crops are summarized below.

Rice

Results obtained from 16 trials during 1987–89 at distant locations indicated that there was an increase in grain yield of rice ranging from 4.2 to 70.2 per cent in NCU-treated plots as compared to uncoated urea-treated plots. Nimin-coated urea was also included in the research trials of the All India Co-ordinated Rice Improvement Programme of the Indian Council of Agricultural Research in kharif 1990 and 1991. On the basis of those trials it was concluded that large granular urea and NCU applied in two split doses performed better in increasing grain yield and nitrogen use efficiency of rice (Anon., 1992).

In two field trials on rice conducted during the summer of 1989 and 1990 in a clay loam soil of West Bengal, three split doses of NCU gave significantly higher grain yield, the mean yield increase of 12.2 q/ha amounting to 50.6 per cent over control and 14.69 per cent over prilled urea treatment (De et al., 1992). Application of 70 kg N through NCU resulted in a statistically similar yield to that obtained with the application of 100 kg N through uncoated prilled urea.

Wheat and Maize

Two field experiments on wheat during rabi 1988 and an equal number of trials on maize in kharif 1989 demonstrate an increase in the grain yield of 16.7 and 36.9, respectively, in NCU-treated plots over plots treated with uncoated prilled urea.

Sugarcane

In a recently concluded study on sugarcane (Chitra, 1992), it was observed that there was a conspicuous increase in the cane yield with use of NCU at 100 per cent N (106.33 t cane/ha) and 80 per cent N (105.67 t cane/ha) to that over control (75.67 t/ha), prilled urea (98.3 t/ha), and urea coated with neem cake (97.3 t/ha). It was also observed that volatilization and leaching losses monitored for a period of 13 weeks were reduced to a greater extent on application of NCU at 45 and 90 days after planting. The reduction in nitrogen losses resulted in a saving of at least 25 per cent applied nitrogen. In this study as well as in earlier studies conducted at the farms of Nizam sugar and EID Parry, Tamil Nadu, it was found that values of Brix, Pol, and Commercial cane sugar were higher in NCU-treated plots than in plots treated with uncoated urea.

Plantation Crops

A field experiment was laid out in a laterite soil at the experimental farm of the National Research Centre for Spices, Calicut at Peruannamuzhi, Kerala, to evaluate the efficiency of slow-release nitrogenous fertilizers in black pepper (Sadanandan, 1991). The results indicated that application of NCU at 100 g N/vine increased the yield of Panniyur-1 black pepper by 280 per cent compared to uncoated urea. Laboratory incubation as well as field studies showed the superiority of NCU over the slow-release nitrogenous fertilizers, namely urea formaldehyde, cyclodi-urea, and urea pellets, as evidenced by the release patterns of ammoniacal, nitrite + nitrate N, and total N in the soil during different stages of crop growth.

DISCUSSION

It is well established that use efficiency of fertilizer urea nitrogen and in turn the crop productivity can be increased by delaying the rate of urea hydrolysis and/or the rate of nitrification. In this context, as indicated in the Introduction, neem cake blended or coated urea has been observed to be superior to prilled urea in terms of crop yields and apparent fertilizer nitrogen recovery in several research trials (Hulagur and Shinde, 1984; Katyal et al., 1985; Prasad et al., 1971; Tandon, 1987; Thomas and Prasad, 1983). The results of increased grain and cane yield ranging from 4.3 to 71.6 per cent, depending upon the crop, soil, and other environmental conditions, obtained with NCU generally conform with the results reported earlier on the use of urea coated with neem cake (Hulagur and Shinde, 1984; Panda et al., 1988; Prasad et al., 1971; Rao and Shinde, 1985; Thomas and Prasad, 1983).

COST : BENEFIT RATIO OF NCU APPLICATION

One kg of Nimin required to coat 100 kg urea is currently being marketed at Rs 55/kg. The cost:benefit ratios for the use of NCU were calculated considering the rate of fertilizer nitrogen application at which maximum produce was obtained in a given trial, and the current procurement price of the produce. The values for different crops were found to range from 1:2.4 to 1:36.3 in field trials conducted with NCU.

ACKNOWLEDGEMENTS

The authors are greatly thankful to all the scientists from universities and research stations who cooperated in the field trials for their unceasing interest. Help rendered by Mr. D. V. Shukla, Mr. S.J. Bhosale, and Ms. Savitha of GSL and Dr. C.D. Karve and Dr. Raman of GAVL is also appreciated.

REFERENCES

Anonymous. 1992. *Progress report of the All India Coordinated Rice Improvement Programme*, Vol. 3, Kharif 1991. Directorate of Rice Research, Hyderabad.

Bremner, J.M. 1965. Inorganic forms of nitrogen. In: *Methods of Soil Analysis*, Part 2, *Agronomy*, Vol. 9. American Society of Agronomy, Madison, Wisconsin, pp. 1179–1234.

Bringi, N.V. 1987. *Non-traditional Oil Seeds in India*. Oxford and IBH Pub. Co. Pvt. Ltd., New Delhi, 254 pp.

Chitra, S. 1992. Studies on the effect of sources, levels and time of nitrogen application in sugarcane. M.Sc. thesis, Tamil Nadu Agricultural University, Madurai, 191 pp.

De, G.C., S. Das and D. Pal. 1992. Efficiency of neem extract coated urea on transplanted summer rice. *Indian J. Agron*. 37(1): 163–165.

Hulagur, B.F. and J.E. Shinde. 1984. Effect of non-edible oilcakes and neem extract on urea hydrolysis and ammonia volatilisation loss in flooded soils. *Bull. Indian Soil Sci*. 13: 138–144.

Katyal, J.C., B. Singh, V.K. Sharma and E.T. Craswell. 1985. Efficiency of some modified urea fertilisers for wetland rice grown on permeable soil. *Fert. Res.* 8: 137–146.

Mitra, C.R. 1963. *Neem*. The Indian Central Oil Seeds Committee, Hyderabad, 190 pp.

Panda, D., R.N. Samantray and S. Patnaik. 1988. Management of fertiliser nitrogen for direct-sown, rainfed lowland rice. (*Oryza sativa* L.) *J. Agric. Sci.* (Camb.) 110: 475–479.

Patil, N.D. 1972. Neem (*Azadirachta indica* L.) oil as nitrification inhibitor. *Fert. News* 17(9): 37–38.

Prasad, R., G.B. Rajale and B.A. Lakhdive. 1971. Nitrification retarders and slow release nitrogen fertilisers. *Adv. Agron*. 23: 337–383.

Rao, D.L.N. 1987. Slow release urea fertilisers—Effect on flood water chemistry, ammonia volatilisation and rice grown in an alkaline soil. *Fert. Res*. 13: 209–221.

Rao, K.V. and J.E. Shinde. 1985. Transformation of ^{15}N labelled urea and its modified forms in tropical wetland rice culture. *Plant and Soil,* 85: 413–421.

Sadanandan, A.K. 1991. *Progress Report of the National Research Centre*, Caiicut, pp. 53–57.

Sahrawat, K.L. 1981. Karanja (*Pongamia glabra* Vent) as source of nitrification inhibitors. *Fert. News* 26: 29–33.

Sahrawat, K.L. and B.S. Parmar. 1975. Alcohol extract of 'neem' (*Azadirachta indica* L.) seed as nitrification inhibitor. *J. Indian Soc. Soil Sci.* 23: 131–134.

Slanghen, J.H.G. and P. Kerkhoff. 1979. Nitrification inhibitors in agriculture and horticulture: A literature review. *Fert. Res.* 5: 1–75.

Tandon, H.L.S. 1987. Status of fertiliser products under development in India. *Fert. Res.* 13: 181–189.

Thomas, J. and R. Prasad 1983. Mineralisation of urea, coated urea and nitrification inhibitor treated urea in different rice growing soils. *Zeitschrift fur pflanzenernahrung Dilingung und Bodenkunde*. 146: 341–347.

Vyas, B.N. and K.B. Mistry. 1985. Hydrolysis of prilled urea, sulphur-coated urea and urea supergranules in an Oxisol, Vertisol, and Inceptisol. *Indian J. Agric. Sci.* 55: 35–40.

Youngdahl, L.J., M.S. Lupin and E.T. Crasswell. 1986. New developments in nitrogen fertilisers for rice. *Fert. Res.* 9: 149–160.

93

Use of Different Oil Cakes as Fertilizers: Comparison of Processes of Immobilization, Mineralization and Nitrification, Inhibition in Soil and Nitrogen Uptake

B.F. HULAGUR*

Station Agronomique Laon (Inra), France

ABSTRACT

Immobilization, mineralization and nitrification inhibition processes as influenced by different cakes were investigated in incubation experiments. After incubation of neem, mahuva, karanja and castor cakes for 42 days at 35°C, the per cent mineralization of nitrogen was 23.2, –3.6, 42.1 and 78.5 respectively, by computing the NH_4 + NO_3 N values. Remineralization of immobilized nitrogen was clearly observed in case of neem and mahuva cake at the later period of incubation. None of the cakes or acetone extracts of them inhibited the process of nitrification even for the smallest period of incubation. Mahuva cake immobilized most and castor cake mineralized at a rapid rate. The same trend was exhibited by the yield and nitrogen uptake of rye grass grown in greenhouse.

INTRODUCTION

Ammoniacal nitrogen is subject to chemical and biochemical reactions in soils resulting in significant loss of nitrogen from the root zone. One approach to minimize these losses involves the formulation of ammoniacal fertilizers with certain chemicals to repress nitrification.

* Present address: Deputy Manager (Technical), National Bank for Agriculture and Rural Development (NABARD), RTC 'X' Roads, Musheerabad, Hyderabad-500 020.

Research findings form part of the post-doctoral research work. The views expressed in the paper are the author's personal views and do not reflect the policies of NABARD.

Acetone extract of neem seed crush could be used as a nitrification inhibitor (Bains et al., 1971). Two fractions of neem oil, total bitters and adourescent compounds, possessed nitrification inhibition property (Patil, 1972), conversion of ammonia to nitrite was checked presumably by selective effects of neem cake on the ammonium oxidizing bacteria (Mishra et al., 1975). It was as effective as sulphathiazole till one week of incubation (Reddy and Prasad, 1975). Lipid associates of neem check the conversion of ammonium to nitrite and the subsequent nitrate (Saharawat and Parmar, 1975). In all these studies the balance sheet of nitrogen was not worked out. Thus, it was difficult to make out whether the drop in nitrate production was due to nitrification inhibition or immobilization.

The low nitrifiability of mahuva cake was reported by Mishra et al. (1960). The saponin of mahuva cake was responsible for slow mineralization of nitrogen (Bains et al., 1971). A furanoflavonoid from karanja was responsible for nitrification inhibition (Saharawat and Mukerjee, 1977). Lipid associates at 20 ppm reduced the population of nitrifying bacteria (Ahmed et al. 1978). In this study, it was attempted to assess the immobilization, mineralization, nitrification inhibition capacities and nitrogen uptake of rye grass as affected by neem (*Azadirachta indica* Juss), mahuva (*Madhuca latifolia)* (syn, *Bassia latifolia*), karanja (*Pongamia pinnata*) and castor (*Ricinus communis*) cakes.

EXPERIMENTS

Experiment 1. Nitrification of Ammonium Sulphate in Presence of Cakes

A loamy soil containing 0.90 per cent carbon and 0.102 per cent nitrogen was used for the incubation tests. Non-edible oil cakes (neem, mahuva, karanja and castor having C/N ratios of 20.5, 19.3, 11.1 and 11.2 respectively) were applied at the rate of 1 per cent together with 3 mg ammoniacal nitrogen. The soil moisture content was maintained at 60 per cent of the maximum water holding capacity. The incubation was conducted for 21 days at 25°C. Ammoniacal, nitrate and organic nitrogen were estimated at intervals of 3, 7, 14 and 21 days. Ammonia nitrogen was trapped in 0.2 N, H_2SO_4 traps for 21 days only and excess acid was back-titrated using standard NaOH. Ammoniacal and NO_3-N were determined on the 0.5 M KCI extract by distillation in the presence of calcined MgO and further reduction of nitrates with Devarda's alloy. Organic nitrogen was determined after digestion and distillation.

Experiment 2. Immobilization and Mineralization of Nitrogen of Cakes in Soil

The procedure of Stanford and Smith (1972) was followed. Different cakes at the rate of 1 per cent were mixed thoroughly. The tubes were wrapped in parafilm and incubated at 35°C. At the end of 7, 14, 21, 28 and 42 days mineral N was recovered by leaching with 0.01 M $CaCI_2$ and nutrient solution devoid of nitrogen. Loss of water during incubation was made up by addition of water to make up to the initial weight. The determination of NH_4 N was done by the development of blue colour complex in the presence of phenol and hypochlorite. The colour was measured by molecular absorption spectrometry at 625 nμ. The nitrates reduced to nitrites by hydrazine sulphate to form a complex azoique rouge purple in the presence of sulphanilamide and N-napthyl-l-ethylene diamine dichloro hydrate. The intensity of the colour was measured by molecular absorption spectrometry at 520 nμ.

Experiment 3. Test for Nitrification Inhibition

A simple and rapid test for measuring the rate of nitrification in the laboratory was developed. One ml of nutrient solution without nitrogen, 1 ml of ammonium sulphate containing 50 mg N, 5 ml of freshly prepared soil suspension and 3 ml of nitrapyrin to supply 65×10^{-3}, 65×10^{-2}, 65×10^{-1}, 6.5 and 650 mg after dilution in distilled water were added to a 40 ml test tube and incubated at 25°C for 10 days and shaken twice daily. Ammoniacal and NO_3 N were estimated after extracting with 10 ml of molar KCI, so that the final concentration of KCI was 0.5 M. Acetone extracts of the cakes were employed in the same system. One ml of acetone extract of cake was diluted to 25 ml using 2 ml of Teepol as emulsifier. The dilutions equal to the application of 200, 20, 2 and 0.2 mg cakes were employed. Acetone with similar dilution was used as control.

Experiment 4. Effect of Different Levels of Cakes and Urea N on Dry Matter Yield and Nitrogen uptake of Rye Grass

A pot experiment to study the influence of different cakes on the utilization of urea nitrogen by rye grass was conducted. The four cakes at 0.1, 0.2 and 0.3 per cent and urea at 9, 18 and 36 mg nitrogen were added to each pot of 250 g soil. Each pot received 200 mg rye grass seeds, 71.5 mg P, and 90 mg K in the form of KH_2PO_4 followed by N. The distilled water was added to make up to 50 ml to obtain a uniform distribution. Pots were irrigated with distilled water when necessary. Plants were harvested after five weeks. Total nitrogen in the plant was determined after kjeldahl digestion and distillation. Uptake of nitrogen was computed after correcting the yield of dry matter for the moisture content.

RESULTS AND DISCUSSION

Effect of Oil Cakes on Immobilization and Mineralization of Nitrogen

The data presented in Table 1 indicated that the amount of mineralized nitrogen from neem cake increased up to seven days of incubation. Mahuva cake decreased the soil mineral nitrogen immediately. Karanja and castor cakes mineralized gradually. The mineral nitrogen level with mahuva cake did not improve appreciably even up to 42 days of incubation. Recovery of mineral nitrogen from neem cake diminished at 14th day of incubation and thereafter there was gradual increase. Karanja and castor cakes showed gradual increasing trend after exhibiting an initial decline. Percentage of nitrogen recovered from mahuva cake was negative even up to 42 days of incubation, indicating that immobilization was the major process. Neem cake had a declining effect on the 14th day, whereas others increased gradually. Mahuva cake immobilized fertilizer nitrogen to the extent of 60 per cent (19.17 ppm) at the end of 14 days of incubation. With neem, karanja and mahuva the recoveries on N^{15} N in organic fractions

Table 1. Production of mineral nitrogen as influenced by different oil cakes (means of duplicates in ppm-N)

Incubation (days)	Soil	Treatments			
		Neem 1%	Mahuva 1%	Karanja 1%	Castor 1%
0	33.01	34.33	28.86	31.15	35.92
7	33.82	78.83	2.59	126.90	249.47
14	20.87	4.88	14.10	59.22	93.80
21	8.95	16.51	18.76	27.94	25.01
28	6.72	12.41	14.00	24.07	18.44
42	12.20	22.10	27.30	33.86	29.83

were 11.69 (36 per cent), 10.76 (33 per cent) and 5.56 ppm (17 per cent) respectively. Slow nitrifiability of mahuva cake was observed by Mishra et al. (1960). They attributed this property to its wide C/N ratio apart from the presence of toxic materials in it. They further indicated that toxic materials present did not affect the general microbiological population of soil. The saponin in mahuva cake was responsible for slow mineralization of nitrogen (Bains et al., 1971).

Apart from C/N ratio, the type of carbonaceous material also played an important role in the decomposition of organic matter in the soil. In the present study, neem and mahuva cake had more or less identical C/N ratios. But their mineralization characteristics were different. Kanamari and Yasuda (1979) observed that the application of rice straw and peat moss with C/N ratio of 56 and 44 respectively to supply 2 per cent carbon retained 90–95 per cent of the applied nitrogen in organic form with peat moss, wheras rice straw carbon transformed almost completely into organic nitrogen within 15 days. This supported the hypothesis of differential mineralization of organic matter based on the nature of carbon apart from C/N ratio.

Effect of Nitrapyrin and Acetone Extracts of Cakes on Nitrification

Influence of acetone extracts of cakes and nitrapyrin on the nitrification inhibition of ammonium sulphate has been presented in Tables 2 and 3. As expected nitrapynin inhibited nitrification Acetone alone also inhibited nitrification, conserving NH_4+ ions. None of the cake extracts inhibited markedly the production of nitrates. In mahuva and karanja at 200 mg cake equivalents, almost the entire quantity of nitrogen was immobilized. With castor at 200 mg cake equivalents there was accumulation of ammonium from the cake in addition to its rapid mineralization as shown in the mineralization experiment.

As revealed from research reports from India, acetone extract of neem seed crush could be used as a nitrification inhibitor (Bains et al., 1971). Alcohol extract of neem effectively conserved the ammoniacal nitrogen. Lipid associates of neem check the conversion of ammonium to nitrite and the subsequent nitrate (Saharawat and Parmar, 1975). Misra and Chonkar (1978) observed the conversion of NH_4 by application of neem cake with urea.

The nitrification inhibition capacity of karanja was also investigated. The furan ring in karanja was essential for nitrification inhibition (Saharawat and Mukerjee,

Table 2. Distribution of ^{15}N in ammonium nitrogen, percentage nitrogen derived from fertilizer

Treatment	3 day	7 day	14 day	21 day	28 day
Neem Cake 0.1%	47.58 (38.48)	69.13 (13.70)	8.40 (0.21)	— (M)	— (M)
Neem Cake 1%	50.50 (57.41)	45.18 (30.50)	21.04 (0.95)	14.34 (0.46)	12.18 (0.48)
Castor Cake 0.1%	74.34 (62.84)	60.29 (12.32)	15.24 (0.56)	1.63 (0.03)	— (M)
Castor Cake 1%	52.72 (43.82)	27.23 (11.50)	12.72 (1.83)	5.64 (0.21)	5.10 (0.16)
Nitrapyrin (0.5% of N)	84.50 (72.30)	81.55 (30.41)	48.44 (6.16)	27.21 (12.21)	51.99 (28.82)

Figures in parentheses indicate percentage ^{15}N recovery values.

Table 3. Distribution of ^{15}N in nitrate nitrogen, percentage nitrogen derived from fertilizers

Treatment	3 day	7 day	14 day	21 day	28 day
Neem cake 0.1%	32.59 (2.90)	58.94 (18.97)	58.92 (78.20)	56.04 (74.56)	55.55 (75.86)
Neem cake 1%	12.51 (3.35)	32.46 (23.64)	36.78 (49.26)	34.88 (49.21)	34.00 (50.20)
Castor cake 0.1%	36.20 (16.22)	54.30 (23.85)	54.86 (75.20)	50.32 (66.74)	50.95 (78.48)
Castor cake 1%	30.35 (5.15)	26.32 (14.09)	24.95 (67.94)	23.39 (50.19)	23.94 (70.07)
Nitrapyrin	1.37 (25.00)	1.55 (0.32)	— (M)	1.39 (0.06)	— (M)

1977). Lipid associates from karanja applied at 20 ppm have reduced the number of Nitrosomonas and Nitrobacter four weeks after its application (Ahmad et al., 1978).

On the contrary, Dr. Ashworth (personal communication, 1979) reported that the neem oil cake and kernel materials were ineffective in inhibiting urease activity and nitrification. Only at 1 g material/10 g soil, nitrate concentration was small compared with control in any form of cakes. But much nitrogen was unrecovered. Denitrification or immobilization could have been responsible as easily as inhibition of nitrification. All the cake materials are reported to promote growth of fungus, which was observed in our studies also. Naphade and Wasnik (1975) attributed the reduction in the rate of nitrification at 100 kg N/ha as compared to 50 kg N/ha from applied urea to the higher concentration of ammonia which might have become toxic to the nitrifiers or to the possible loss of gaseous ammonia from the soil.

Results of our incubation studies have indicated that the major process responsible for disappearance of nitrates in soil was its incorporation into organic fraction.

Table 4. Effect of oil cakes and urea on yield, N content, and uptake by rye grass

Cakes		N_9 mg/pot			N_{18} mg/pot			N_{36} mg/pot		
		Dry matter g/pot	Content % N	Uptake mg/pot	Dry matter g/pot	Content % N	Uptake mg/pot	Dry matter g/pot	Content %N	Uptake mg/pot
Soil		1.27	1.13	14.33	1.52	1.51	23.48	1.64	1.87	30.61
Neem	0.1%	1.31	1.11	14.37	1.58	1.41	22.25	1.73	1.78	30.58
cake	0.2%	1.45	1.09	15.89	1.56	1.46	22.78	1.77	1.91	33.64
	0.3%	1.41	1.17	16.55	1.57	1.49	23.38	1.84	1.74	31.91
Mahuva	0.1%	1.07	1.16	12.43	1.27	1.29	16.43	1.42	2.04	29.16
cake	0.2%	0.90	1.14	10.31	1.31	1.16	15.25	1.40	1.87	26.16
	0.3%	0.80	1.24	9.93	1.27	1.23	15.59	1.49	1.58	23.76
Karanja	0.1%	1.42	1.26	17.93	1.56	1.69	26.32	1.49	2.24	33.36
cake	0.2%	1.44	1.36	19.32	1.52	1.99	30.49	1.61	2.23	37.32
	0.3%	1.38	1.50	20.71	1.41	2.30	32.33	1.58	2.79	44.05
Castor	0.1%	1.46	1.26	18.34	1.63	1.66	26.74	1.86	2.10	35.42
cake	0.2%	1.66	1.31	21.75	1.90	1.68	31.93	1.97	1.89	37.31
	0.3%	1.96	1.29	25.23	1.91	1.74	33.27	2.20	1.97	43.37

Effect of Different Levels of Oil Cakes and Urea Nitrogen on the Dry Matter Yield and Nitrogen Uptake by Rye Grass

The data on dry matter, nitrogen content and nitrogen uptake of rye grass grown in pots are presented in Table 4. The different levels of nitrogen and also the cakes influenced the yield significantly. Interaction of nitrogen levels and cakes was also significant. Yield, nitrogen content and nitrogen uptake were significantly increased with increase in nitrogen level. Mahuva cake depressed the yield and nitrogen uptake to a greater extent at 9 mg nitrogen per pot. Neem cake and karanja cake did not have adverse effect on the yield and improved the nitrogen uptake especially at higher level of application. Castor cake at all levels and with varying nitrogen levels increased the yield and uptake of nitrogen.

Apparent low nitrogen recoveries of cakes further confirmed the immobilizing characteristics of mahuva cake (Table 5). Figures for mahuva, neem, karanja and castor were minus 21 to minus 102, minus 6 to 26, 10 to 40 and 31 to 45 respectively.

Results of incubation test could offer adequate explanation for low yield and low nitrogen uptake in pots with treated mahuva cake. Hulagur and Shinde (1981) observed no panicle formation of rice grown in pots receiving mahuva cake. Ketkar (1974) reported increase in the yield of rice when neem cake blended with urea was applied. Saharawat (1988) reported the possibility of the effects of non-edible oil cakes like neem and karanja due to immobilization and remineralization of nitrogen particularly when the carbonaceous materials are added with fertilizers at high rates.

Table 5. Apparent coefficient of utilization of N from oil cakes by rye grass

Treatment	Per cent	N_9 mg/pot	N_{18} mg/pot	N_{36} mg/pot
Neem cake	0.1%	0.7	minus 21	minus 0.5
	0.2%	14	minus 6	26
	0.3%	13	minus 0.6	7
Mahuva cake	0.1%	minus 28	minus 102	minus 21
	0.2%	minus 29	minus 60	minus 32
	0.3%	minus 21	minus 38	minus 33
Karanja cake	0.1%	32	25	25
	0.2%	22	31	30
	0.3%	19	26	40
Castor cake	0.1%	38	31	45
	0.2%	35	39	31
	0.3%	34	30	40

$$\frac{\text{N uptake by treatment (minus) N uptake in control}}{\text{N added through (cakes) treatment}} = \text{Apparent coefficient of utilization of N}$$

CONCLUSION

From the data presented here, it can be concluded that neem, mahuva, karanja and castor oil cakes have great value as means of immobilizers, thus conserving the applied and soil nitrogen and mineralizing steadily over a longer period. They could aid in metered supply of nitrogen over a stipulated period of crop growth. Their advantage in conserving soil and fertilizer nitrogen could be attributed largely to immobilization. Castor cake could be recommended as the best organic manure as it contains a higher level of nitrogen. Mahuva cake should be avoided on N-deficient soil and should be accompanied by an adequate amount of inorganic nitrogen.

ACKNOWLEDGEMENTS

The author wishes to acknowledge the facilities provided by Institute National de la Recherche Agronomique LAON, France, expresses his gratitude to Prof. J. C. Remy for guidance, and thanks the Government of India for the award of French Government Fellowship.

REFERENCES

Ahmad, N., M. Rai, N.L. Yadav and B.P. Sahi. 1978. Studies on efficiency of karanja cake and its liquid associates on Nitrification Regulation under sub-merged rice cultivation in a calcareous soil of North Bihar. *J. Indian Soc. Soil Sci.* 26: 304.

Bains S.S., R. Prasad and P.C. Bhatia. 1971. Use of indigenous materials to enhance the efficiency of fertilizer nitrogen for rice. *Fertil. News* 16: 30.

Hulagur, B.F. and J.E. Shinde. 1981. Evaluation of non-edible oil cakes on the uptake of urea-n-by low land rice crop. *Madras Agric. J.* 68: 36.

Kanamari, T and T. Yasuda. 1979. Immobilisation, mineralisation and the availability of the fertiliser nitrogen during the decomposition of the organic matter applied to the soil. *Pl. Soil* 52: 219

Ketkar, C.M. 1974. Neem cake blended urea for nitrogen economy. *Fertil. News* 19: 25

Mishra M.M., S. Neelakantan, K.G. Narayan and S.R. Vyas. 1972. Studies on the de-composition of organic matter in soil and effects of sesbania and bajra stubble plus neem cake, J. Res. Haryana Agrl. University, 2: 150

Mishra M.M., S. Neelakantan, K.C. Khandelwal, S.K. Bharadwaj and S.K. Vyas. 1975. Margosa (neem) seed cake as an inhibitor of nitrogen. *Soil Biol. Biochem.* 7: 183.

Misra K.C., Chonkar, P.K. 1978. Possible utilisation of neem cake for inhibiting nitrification in soil. *J. Indian Soc. Soil Sci.* 26: 90.

Misra, V.K. A. Sen and N.D. Vyas. 1960. Utilisation of 'Mahua Cake' as manure. *J. Indian Soc. Soil Sci.* 8: 97.

Naphade, K.T. and H.M. Wasnik. 1975. Nagpur Univ. *Agric. Coll. Mag.* 47: 43.

Patil N.D. 1972. *Fertil. News* 17: 37.

Reddy, R.N.S. and Prasad, R. 1975. Studies on the mineralisation of urea, coated urea and nitrification inhibitor treated urea in soil. *J. Soil. Sci.* 26: 304.

Saharawat, K.L. 1988. *J. Nuclear Agric. Biol.* 17: 238.

Saharawat, K.L. and S.K. Mukherjee. 1977. Nitrification inhibitors 1. Studies with Karanjin—Furanal flavonoid from Karanja (*Pongamia glabra*) seeds. *Pl. Soil* 47: 27.

Saharawat, K.L. and B.S. Parmar. 1975. Alcohol extract of neem (*Azadirachta indica*). Seeds as nitrification inhibitor. *J. Indian Soc. Soil Sci.* 23: 131.

Stanford, G. and S.J. Smith. 1972. Nitrogen mineralisation potentials of soils. *Soil Sci. Soc. Amer. Proc.* 36: 465.

94

Comparison of Nimin-coated Urea with Other Sources at Graded Level of Nitrogen in Lowland Rice

M.K. PORWAL, G.S. BHATNAGAR AND P.C. CHAPLOT

Agricultural Research Station, Rajau, Banswara 327 001, India

ABSTRACT

A field experiment conducted during kharif 1989 and 1990 on transplanted rice revealed that increasing levels of nitrogen increased the grain yield but higher nitrogen use efficiency was observed with split application of low dose of nitrogen. Sources of urea coating did not influence paddy yields significantly at 100 kg N/ha. Further, 80 kg N/ha through nimin-coated urea was on par with 100 kg N/ha through prilled urea.

INTRODUCTION

Of the various governing factors of nitrogen use efficiency (NUE) in transplanted rice, the levels, source, and time of application of nitrogen play vital roles. Split application of nitrogen as prilled urea at specific growth stages, use of coating, and controlled release of N fertilizers are common features of techniques to increase NUE. Joseph et al. (1991) reported that split application of prilled urea was equally effective to basal application of neem cake-coated urea. Since availability of finally graded neem cake and coltar is not easy and also its coating process is somewhat difficult, an attempt was made to test the neem extract nimin as coating material for urea and the effect of different doses of nitrogen on the growth and yield of rice in the vertisols of southern Rajasthan.

MATERIALS AND METHODS

A field experiment was conducted in the kharif seasons of 1989 and 1990 at Agricultural Research Station, Banswara, Rajasthan. The soil was clay loam of moderate fertility. Twelve treatments (Table 1), comprising source of coating materials, doses, and time of application of nitrogen, were laid in randomized block design and replicated four times. The prilled urea was applied in three splits at specific stages, i.e., basal, tillering, and panicle initiation. Urea was coated with nimin at the rate of 1 per cent and used after it was air dried. A basal application of 40 kg P_2O_5 ha was made in all plots. Thirty-day-old seedlings of paddy cv BK 79 were transplanted. Continuous shallow submergence in about 5 cm water was maintained up to crop harvest. Economic analyses and NUE were worked out for treatment evaluation.

RESULTS AND DISCUSSION

Tillers/m row and panicle length did not differ significantly at 100 kg N/ha due to sources of urea coating and prilled urea but their doses and time of application did. Similar results of no variation of nitrogen sources were reported by Pandey and Singh (1987). In control plots (without Nitrogen), the number of tillers/m row and panicle length were significantly less than in N treated plots. Further, tillers/m row and panicle length at 80 and 120 kg N/ha through nimin-coated urea were significantly different but not at 100 kg N/ha. Similarly, basal application and three splits of nimin-coated urea influenced tillers/m row and panicle length significantly (Table 1). During two years,

Table 1. Effect of treatments on the grain yield, yield attributes of paddy, and nitrogen use efficiency.

Treatment			Tillers/m row		Height (cm)		Panicle length (cm)		Grain yield (q/ha)			NUE (kg grain/ kg N)
N level (kg/ha)	Source	Application	1989	1990	1989	1990	1989	1990	1989	1990	Pooled	
0	—	—	33.3	68.5	38.1	39.5	15.3	17.5	17.4	25.0	21.2	—
100	PU	3	39.3	80.3	43.5	45.2	16.5	19.7	27.8	32.5	30.2	30.2
100	NCCU	1	41.8	82.8	44.3	46.4	17.0	20.0	29.3	34.0	31.7	31.7
100	NCU	1	40.2	84.5	44.2	47.2	16.8	19.7	31.2	34.8	33.0	33.0
100	NCU	2	48.0	90.0	46.9	49.4	17.6	20.7	33.3	36.5	34.9	34.9
100	NCU	3	45.6	94.3	48.7	51.6	18.0	21.6	36.7	39.5	38.1	38.1
80	NCU	1	38.0	75.5	42.4	45.5	16.2	19.2	26.5	33.0	29.8	37.3
80	NCU	2	40.8	82.8	44.4	47.3	17.1	20.0	31.9	35.5	33.7	42.1
80	NCU	3	44.9	85.8	46.7	49.5	17.8	20.5	33.3	37.5	35.7	44.6
120	NCU	1	43.5	86.3	47.3	50.2	17.9	21.4	34.7	38.8	36.1	30.1
120	NCU	2	45.4	95.3	48.5	53.4	18.5	22.6	38.9	40.5	39.7	33.1
120	NCU	3	48.5	103.5	51.0	55.8	19.0	23.5	42.3	43.0	42.7	35.6
	C.D. at 5%		4.43	9.34	3.46	1.61	1.01	1.61	5.69	3.60	7.50	—

PU = Prilled urea
NCCU = Neem cake-coated urea
NCU = Nimin-coated urea

1 = Basal
2 = Basal + tillering
3 = Basal + tillering + panicle initiation

maximum tillers/m row (48.5 and 103.5) and panicle length (19.0 and 23.5 cm) were recorded with three splits of 120 kg N/ha through nimin-coated urea. Plant height of paddy followed a similar trend but, in 1990, nitrogen levels and time of application differed significantly.

The grain yield of paddy increased with increasing levels of nitrogen. The grain yield was maximum with three splits of 120 kg N/ha through nimin-coated urea (42.7 q/ha) and found statistically on par with 100 kg N/ha but superior to 80 kg N/ha (Table 1).

Basal application of 100 kg N/ha with two sources of coatings (31.7–33.0 q/ha) and three splits of prilled urea (30.7 q/ha) were found on par with respect to the grain yield of paddy. The results obtained are in close conformity to those reported by Pandey and Singh (1987) and Raju et al. (1987), but nimin-coated urea yielded 9.3 per cent higher over prilled urea. The better performance of nimin-coated urea might be due to slow rate of enzymatic hydrolysis (Raju et al., 1987) and slow and steady release of NH_4^+ in the soil and better utilization by the rice plants (Eswaramoorthy, 1983). Split application of nimin-coated urea did not influence the grain yield statistically, but three splits of 100 kg N/ha through prilled urea was on par with basal application of 80 kg N/ha through nimin-coated urea. Agarwal et al. (1990) and Joseph et al. (1991) reported similar effects with neem cake-coated urea.

Nitrogen use efficiency increased with split application of nitrogen. Maximum NUE (44.6) was observed at three splits of 80 kg N/ha. Though, grain yield increased with increase in nitrogen levels, NUE was decreased. Fertilized treatments were significantly superior over control with respect to the grain yield, tillers, and panicle length. A similar trend of results was reported by Ramasubha Reddy and Anand Reddy (1989).

Table 2. Economic analysis of different treatments

Treatments			Cost of treatment (Rs/ha)	Net return (Rs/ha)	Additional income over control (Rs/ha)	B:C over control
N Level (kg/ha)	Source	Application				
0	—	—	0	8480	—	—
100	PU	3	783	11297	+ 2817	+ 3.59
100	NCCU	1	978	11702	+ 3222	+ 3.29
100	NCU	1	847	12353	+ 3873	+ 4.57
100	NCU	2	869	13091	+ 4611	+ 5.31
100	NCU	3	891	14349	+ 5869	+ 6.59
80	NCU	1	678	11242	+ 2762	+ 4.07
80	NCU	2	700	12780	+ 4300	+ 6.14
80	NCU	3	722	13558	+ 5078	+ 7.03
120	NCU	1	1016	13424	+ 4944	+ 4.87
120	NCU	2	1038	14842	+ 6362	+ 6.13
120	NCU	3	1060	16020	+ 7540	+ 7.11

Input cost (Rs)
1 kg N through PU = 7.39
1 kg N through NCCU = 9.78
1 kg N through NCU = 8.47
Neem cake = 5.00/kg
Nimin = 50.00/kg
Application cost = 22.0/ha

Selling price
Paddy grain = Rs 400/q.

Three splits of 120 kg N/ha through nimin-coated urea yielded maximum (B:C) (+7.11) followed by three splits of 80 kg N/ha (+7.03). Economic analysis further revealed that 80 kg N/ha through nimin-coated urea yielded higher B:C than 100 kg N/ha through prilled urea (+3.59) (Table 2).

REFERENCES

Agarwal, S.K., H. Shankar and M.M. Agarwal. 1990. Effect of slow release nitrogen and nitrification inhibitor on rice wheat sequence. *Indian J. Agron.* 35: 337–340.

Eswaramoorthy, A. 1983. Efficacy of urea super granules over prilled urea in rice. M.Sc. (Ag.) thesis, TNAU, Coimbatore.

Joseph, K., P.K.G. Manon and K. Anilakumar. 1991. Studies on the comparative efficiency of nitrogen sources in low land rice. *Indian J. Agron.* 36: 122–124.

Pandey, J. and R.P. Singh. 1987. Effect of urea, slow release fertilizers and nitrification inhibiting on the yield of rainfed lowland rice. *Indian J. Agron.* 32: 303–305.

Raju, R.A., M.M. Hussain and M.N. Reddy. 1987. Relative efficiency of modified urea materials for lowland rice. *Indian J. Agron.* 32: 460–462.

Ramasubha Reddy, G. and K. Anand Reddy. 1989. Effect of levels and sources of nitrogen on rice. *Indian J. Agron.* 34: 364–366.

95

Use of Neem as Nimin to Increase N-use Efficiency in Transplanted Rice

S.S. TOMAR AND P.K. GUPTA

Rajasthan Agricultural University
Agricultural Research Station
Borkhera, Kota 324001, India

ABSTRACT

A field experiment was conducted at Rajasthan Agricultural University Research Centre, Kota, to evaluate the N-use efficiency of urea coated with nimin (neem extract) in comparison to prilled urea, Mussoorie rock phosphate-coated urea, and large granulated urea on transplanted rice in vertisols during kharif seasons of 1990 and 1991. Basal incorporation of all modified and coated urea materials gave significantly higher grain yield of rice than conventional prilled urea application either basal or in two splits. Yield components also showed a similar response. Nimin-coated urea produced highest grain yield as well as N-use efficiency, 16 and 50 per cent higher than split application of prilled urea, respectively. This contention is further supported by N-mineralization study. The concentration of NH_4-N was comparatively less under the nimin-coated urea, indicating that it releases NH_4-N slowly so that available N may be utilized by the rice plant more efficiently during the entire growth period.

INTRODUCTION

In intensive rice cultivation, high nitrogen losses by leaching and denitrification of ammonia occur mainly after waterlogging when rice seedlings are in the first stages of development. These losses can be reduced if the rapid nitrification of ammonia is prevented or if the fertilizer-N has a slow dissolution rate. Urea can be lost to the extent of 80 per cent through leaching and denitrification (Sharma and Ghosh, 1976). For increasing the fertilizer-N efficiency several agronomical practices, such as split application of urea, foliar spray, and use of coated and modified urea materials and nitrification inhibitors, have been tried.

In the present study, some indigenous materials such as nimin (neem extract) and Mussoorie rock phosphate-coated urea along with modified urea were used to increase the N-use efficiency in transplanted rice in vertisols of the Chambal Command Area of Rajasthan.

MATERIALS AND METHODS

A field experiment was carried out on rice (var. Jaya) at Rajasthan Agricultural University Research Farm, Kota (Rajasthan), during kharif 1990 and 1991 in randomized block design with three replications. The soil was clay loam having pH 8.0, organic carbon 0.62 per cent, available N 158 kg/ha, P_2O_5 24 kg/ha, and K_2O 280 kg/ha.

Nitrogen at the rate of 100 kg/ha was supplied to the rice crop in the form of prilled urea, Mussoorie rock phosphate-coated urea, large granulated urea, and nimin-coated (1 per cent w/w) urea as all basal and in two splits (two-thirds basal and one-third at panicle initiation). Phosphate and potash were applied at the rate of 60 and 40 kg/ha to all the plots.

Surface soil samples (0.15 cm) were collected from treatment plots under each replication after 2, 4, 6, 8, 10, and 14 weeks of transplanting. The wet soil samples were extracted with Morgan's reagent and shaken for 30 min. before extraction. In the extract, NH_4-N was estimated. Moisture in the soil samples was determined separately by gravimetric method.

RESULTS AND DISCUSSION

Yield Attributes

The data presented in Table 1 reveal that yield attributes such as number of panicles and panicle weight were influenced significantly by the use of different urea materials during both years. The maximum number of panicles/m^2 (432 and 401) and panicle weight (3.25 and 3.19 g) were observed with nimin-coated urea applied as basal during 1990 and 1991, which was higher by 19 and 15 per cent (average basis) respectively over split application of prilled urea. This is possibly due to higher availability of N under nimin-coated urea as a result of its slow dissolution and inhibitory effect on nitrification.

Significantly higher grain yield of rice was observed under all modified and coated urea materials as compared to prilled urea applied as basal. This is because of slow dissolution of modified and coated urea materials. This is corroborated by the results of Jena and Mishra (1987) and Panigrahi and Dixit (1991). The contention is also supported by yield attributes and NH_4-N studies carried out under the present study.

Nimin-coated urea applied all basal produced maximum grain yield (59.01 and 56.99 q/ha), which was also significantly higher than split application of urea during both years. The yield under nimin-coated urea was higher by 24.2 and 28.7 per cent over basal and 19.2 and 12.9 per cent over split application of urea during 1990 and 1991 respectively, probably because of the dual effect of nimin on N-mineralization. Being a coated material, it may slow down the dissolution of urea and at the same time the presence of nimbinin and nimbidin alkaloids may inhibit the activity of nitrifying

Table 1. Effect of modified and coated urea materials on grain yield, number of panicles and panicle weight of transplanted rice, kharif 1990 and 1991

Treatment	Grain yield (q/ha)		N-use efficiency (kg grain/kg N)	Panicles/m^2		Panicle wt (g)	
	1990	1991		1990	1991	1990	1991
Control (no nitrogen)	34.35	33.65	—	240	224	2.61	2.62
100 kg N/ha PU all basal	47.50	46.61	13.1	350	349	2.80	2.77
100 kg N/ha, MRPU all basal	52.03	51.56	17.8	360	351	2.95	2.90
100 kg N/ha, LGU all basal	56.21	54.89	21.6	393	386	3.00	3.01
100 kg N/ha, nimin-coated urea, all basal	59.01	56.99	24.0	432	401	3.25	3.19
100 kg N/ha, PU two splits 2/3 basal and 1/3 at panicle initiation	49.52	50.48	16.0	352	350	2.80	2.81
100 kg N/ha, MRPU two splits as above	53.00	50.74	17.9	365	350	2.97	2.89
100 kg N/ha, LGU two splits as above	56.42	55.83	22.1	394	390	3.11	3.10
100 kg N/ha, nimin-coated urea two splits as above	58.02	56.41	23.22	424	394	3.20	3.13
S. Em ±	1.62	1.91		6.11	7.1	0.09	0.11
CD at 5%	4.84	5.73		18.30	21.3	0.26	0.32

PU = Prilled urea, MRPU = Mussoorie rock phosphate-coated urea, LGU = Large granulated urea.

bacteria. These findings are in close conformity with results of Tomar et al. (1991) and Vyas et al. (1991).

NH_4-N Contents

From the data given in Table 2, it is seen that a higher amount of NH_4-N was released from control and basal application of urea in the second week of transplanting and then there was a gradual reduction in subsequent weeks. In case of modified and coated urea materials the production of NH_4-N was quite slow in the second week; thereafter, it maintained more or less the same level till eight weeks and then slowed down. This indicates that modified urea materials released nitrogen slowly so that it may be utilized by rice crop during the entire growth period (Surve and Daftardar, 1985). The contention is further supported by the grain yield obtained under the present study, which is higher under these treatments. The production of NH_4-N in split application of urea was decreased gradually from the second week but after six weeks it increased and attained

Table 2. Production of NH_4-N (ppm) in transplanted rice as influenced by modified and coated urea materials (average of two years)

Treatment	Weeks after transplanting					
	2	4	6	8	10	14
Control (no nitrogen)	52.4	48.2	39.0	34.3	32.6	33.0
100 kg N/ha, PU all basal	98.5	78.4	66.5	54.2	49.3	42.2
100 kg N/ha, MRPU all basal	78.6	84.7	76.0	71.0	53.2	50.1
100 kg N/ha, LGU all basal	76.2	83.5	79.6	73.1	56.0	52.0
100 kg N/ha, nimin-coated urea all basal	65.2	89.4	88.1	76.0	73.3	61.2
100 kg N/ha, PU two splits (2/3 basal and 1/3 at panicle initiation)	60.3	61.0	50.4	58.3	78.2	67.7
100 kg N/ha, MRPU two splits as above	67.5	70.0	65.2	58.5	68.4	64.5
100 kg N/ha, LGU two splits as above	69.0	70.3	66.6	62.0	70.4	63.7
100 kg N/ha, nimin-coated urea two splits as above	63.0	73.8	71.3	70.0	74.4	64.0

maximum value (78.2 ppm) in the 10th week probably due to application of second dose at the time of panicle initiation. However, modified urea materials when applied in split doses produced NH_4-N consistently during the entire growth period but the amount released was lower than with basal application at early stages of crop growth (up to eight weeks) and then it was higher at maturity, i.e., after eight weeks.

N-use Efficiency

Application of all modified and coated urea materials showed higher N-use efficiency than prilled urea application (Table 1). On an average it varied from 13 to 24 kg grain/ kg N with the maximum value under nimin-coated urea applied all basal. The per cent increase in N-use efficiency over split application of urea was 50 under this treatment.

REFERENCES

Jena, B.K. and C. Mishra. 1987. Effect of several urea based N sources. *Int. Rice Res. Neswletter* 12(3): 49–50.

Panigrahi, B.K. and L. Dixit. 1991. Relative efficiency of slow-release nitrogenous fertilizers in rainfed rice (*Oryza sativa*). *Indian J. Agron.* 36(3): 401–403.

Sharma, R.C. and A.B. Ghosh. 1976. A note on the leaching behaviour of urea and ammonium nitrate *J. Indian Soc. Soil Sci.* 24: 328.

Surve, S.P. and S.Y. Daftardar. 1985. Effect of some neem (*Azadirachta indica*) and karanj (*Pongamia glabra*) products on utilization of urea nitrogen by upland rice. *J. Indian Soc. Soil Sci.* 33(1): 182–186.

Tomar, S.S., S.C. Sharma and A. Sharma. 1991. Effect of neem extract coated urea on rice yield and nitrogen use efficiency of transplanted rice. *Int. Rice Res. Newsletter* 16(2): 20–21.

Vyas, B.N., N.B. Godrej and K.B. Mistry. 1991. Development and evaluation of neem extract as a coating for urea fertilizer. *Fert. News* 2: 19–25.

96

Influence of Neem Oil Coating on Prilled Urea on Volatilization Loss from Swell-shrink Soil Amended with Organic Matter under Different Moisture Regimes

MUNESHWAR SINGH AND P.N. TAKKAR

Indian Institute of Soil Science, Bhopal 462 011, India

ABSTRACT

The volatilization loss of nitrogen from calcareous soils is one of the major channels of nitrogen loss and is responsible for the low efficiency of urea. The coating of urea with neem oil, the production of which in India is around 30,000 t/year, as well as with lac holds promise to reduce the volatilization loss or increase the urea-N use efficiency, as these materials delay urea hydolysis. The ammonia loss from neem oil-coated urea (NOCU), lac-coated urea (LCU), prilled urea (PU), and urea super granule (USG) was studied under different moisture levels and added organic manure to a Vertic Ustochrept. Under submergence and field-capacity conditions, the ammonia loss from NOCU was 1.7 and 6.7 per cent, which was markedly less than that of 6.5 and 21.3 per cent from PU, respectively. However, it was at par with LCU. The addition of 2.0 per cent organic carbon at field capacity further reduced the ammonia loss from 6.7 to 4.0, 6.1 to 3.9, 15.3 to 9.5, and 21.3 to 11.6 per cent from NOCU, LCU, USG, and PU, respectively. Although the addition of organic manure decreased the ammonia loss, the effect was not as marked as that of the soil moisture regimes. The coating of neem oil or lac on PU proved most effective in reducing volatilization loss of N from calcareous soil. The loss of NH_4-N was the largest under field capacity and the least under submerged moisture conditions.

INTRODUCTION

Urea contributes nearly 80 per cent to the total nitrogenous fertilizers used in agriculture in India. But its use efficiency is quite low due to loss of N through ammonia volatilization, denitrification, and leaching depending upon soil and environmental condition. The loss of N through ammonia volatilization from urea has been reported to be fairly larger from the soils high in pH and $CaCo_3$, as the activity of urease or hydrolysis of applied urea increases with soil pH (McGarity and Meyers, 1967). There are several organic compounds and toxic heavy metals which reduce the ammonia volatilization loss by inhibiting urease activity or urea hydrolysis, but their high cost and pollution hazards prohibit their use. Another alternative is to modify the urea itself and identify the materials which reduce ammonia volatilization. In general Vertisols are high in pH and $CaCO_3$ and applied urea is susceptible to ammonia volatilization. The coating of urea with neem oil and lac and manipulation of soil moisture and organic matter content with manure may help in reducing volatilization loss from applied urea. In view of this, ammonia volatilization loss from neem oil-coated area (NOCU), lac-coated urea (LCU), prilled urea (PU), and urea super granule (USG) as influenced by soil moisture regime and build-up of organic matter levels in a Vertisol has been studied.

MATERIALS AND METHODS

Surface (0–15 cm) soil sample representing Islamnagar-3 and Vertic Ustochrept was collected from a cultivated field of Islamnagar village, Bhopal, India. The soil sample was air-dried, ground to pass through 2 mm sieve and analysed for physicochemical properties. It had clayey texture, pH 8.1 (1:2 soil:water, suspension), electrical conductivity 8.1 and 0.086 dS/m at 25°C (1:2 soil:water, suspension), organic carbon 0.48 per cent, $CaCo_3$ 2.12 per cent, and available N 192 kg/ha.

The processed soil (3.5 kg) was added to an 8 kg capacity plastic pot having two holes of 4 mm opposite to each other at 2 cm below the upper edge. Farmyard manure was added equivalent to 1.0 and 2.0 per cent organic carbon in the required number of pots. After the organic matter was mixed with the soil, the pots were subjected to two wet and dry cycles. Again the soil of the pot was processed and filled into the same pot for study. Calculated amounts of deionized water were added to bring the soil to field capacity, saturation, saturation with wetting and drying cycles, and submergence under 3 cm water and equilibrated for a period of 24 hrs. Then 131.1 mg N/kg was applied to soil surface through, PU, LCU, NOCU, and USG of 1 g size, which were placed at 5 and 10 cm below the surface. Thereafter, the lid was placed immediately on the pot and made airtight with a layer of putty in between the lid and the pot.

The inlet hole was connected to an oil-free air compressor and outlet hole through a PVC tube dipped in a 2 per cent solution of boric acid with mixed indicator. A constant flow volume (4.5–5.5 l/min.) of NH_3-free air was allowed to replace the air in the pot and, thus evolved, NH_3 was sorbed in boric acid and titrated against 0.02 NH_2SO_4. The study was conducted in batches. Two replications and the desired moisture levels were maintained by weighing the pots weekly upto four weeks.

RESULTS AND DISCUSSION

Effect of Neem Oil and Lac on NH_3 Volatilization

The influence of neem oil and lac coating on urea under different soil moisture regimes is presented in Table 1. At field capacity, the loss of N through ammonia volatilization was 21.3 per cent from PU, which reduced to 6.7 per cent on application of NOCU and

6.1 per cent on application of LCU under similar conditions. The coating of neem oil over prilled urea also performed better than the USG granule placed at 5 cm and 10 cm below the surface. A similar effect of neem oil coating was also noted under other soil moisture regimes. But the overall magnitude was less. The reduction in ammonia volatilization loss from NOCU and LCU resulted from a physical barrier between urea granule and soil solution provided by neem oil and lac. The neem oil and lac formed a thin layer around the urea granule and made it less accessible to urease for hydrolysis.

Effect of Soil Moisture Regimes

The data presented in Table 1 indicate significant influence of soil moisture regimes on ammonia volatilization. It is evident from the data that field-capacity soil moisture regime is most conducive to ammonia volatilization. For instance, from unamended soil the ammonia volatilization loss from applied PU was 21.3 per cent, which reduced to 15.4, 8.0, and 6.5 per cent respectively under alternate wet and dry, saturation, and submergence moisture regime. A similar behaviour of soil moisture was seen from the soil amended with organic matter (farmyard manure). However, the magnitude was less. The reduction in NH_3 under submergence and saturation soil moisture conditions seems to have resulted from the expected decrease in pH, which tends to approach neutrality (Ponnamperuma et al., 1986), as well as from reduced activity of urease under these conditions (Mikkelsen et al., 1978). The lesser loss of NH_3 under wet and dry cycle than in field-capacity moisture condition appears to be due to availability of larger surface area for sorption of ammonia ion due to churning of soil.

Effect of Organic Matter on Ammonia Volatilization

Application of farmyard manure significantly reduced ammonia volatilization loss from all the modified urea materials (Table 1). The effect of organic matter was more visible under wet and dry and field-capacity moisture conditions. Under wet and dry soil moisture condition, 15.4 per cent of applied nitrogen loss occurred from the soil having 0.48 per cent organic carbon (control), which reduced to 12.2 and 8.2 when the soil was amended with farmyard manure at 1 and 2 per cent organic carbon, respectively. The reduction in ammonia loss on addition of organic matter may be due to lowering in urease activity by binding of urease by functional group of humic substance, which resulted in reduction of urea hydrolysis and secondly the additional surfaces available for sorption of ammonium ions. Both the processes reduce the concentration of ammonium ions in the solution, which is susceptible to volatilization loss. Bramner and Douglas (1971) found that P Benzo quonons which exist in humic substances reduce the hydrolysis of urea, since there is an intimate relationship among hydrolysis rate of urea, concentration of ammonium carbonate, and ammonia volatilization.

CONCLUSION

From the study it may be concluded that ammonia volatilization loss from urea applied to a Vertic Ustochrept can be reduced by coating urea with neem oil. Ammonia vola-

Table 1. Per cent cumulative ammonia loss as influenced by neem oil coating on urea granule in a vertisol amended with organic matter under different moisture regimes

Urea material	Organic carbon (%)		
	0.48	1.0	2.0
Submergence			
PU	6.5	6.5	5.1
NOCU	1.7	1.8	1.3
LCU	2.2	2.3	1.7
USG 5 cm	3.3	2.9	2.3
USG 10 cm	3.2	2.9	2.3
CD 5% urea material x organic carbon, NS			
Saturation			
PU	8.0	4.4	2.8
NOCU	2.6	1.9	1.6
LCU	2.7	1.9	1.9
USG 5 cm	6.7	4.5	4.2
USG 5 cm	3.6	2.8	3.1
CD 5% urea material × organic carbon, 1.2			
Wet/Dry			
PU	15.4	12.2	8.2
NOCU	3.3	3.1	3.1
LCU	3.8	3.2	3.0
USG 5 cm	12.2	8.9	7.5
USG 10 cm	9.4	8.4 -	7.6
CD 5% urea material × organic carbon, 2.0			
Field capacity			
PU	21.3	15.7	11.6
NOCU	6.7	4.5	4.0
LCU	6.1	4.6	3.9
USG 5 cm	15.3	12.5	9.5
USG 10 cm	16.3	12.1	8.5
CD 5% urea material × organic carbon, 2.1			

tilization can be further reduced by building up organic matter content of soil with manure and by manipulating moisture content.

REFERENCES

Bramner, J.M. and L.A. Douglas. 1971. Decomposition of urea phosphate in soil. *Soil Sci. Soc. Am. Proc.* 35: 575–578.

McGarity, J.W. and M.G. Meyers. 1967. A survey of urease activity in soils of northern New South Wales. *Plant and Soil* 27: 217–238.

Mikkelsen, D.S., S.K. De Datta and W.N. Obcenia. 1978. Ammonia losses from flood rice soils. *Soil Sci. Soc. Am. J.* 42: 725–730.

Ponnamperuma, F.N., E. Martinez and T. Loy. 1966. Influence of redox potential and partial pressure of carbon dioxide in pH values and the suspension effect of flooded oil. *Soil Sci.* 101: 421–431.

9

COMMERCIAL NEEM INSECTICIDES

97

RD-9 Repelin: An Effective Botanical for Integrated Pest Management

T.S. SUBRAMANIAM, B. SIVAPRASAD AND V.V.L.N. PRASAD

Agricultural Research Centre, I.T.C. Ltd.—International Business Division
7th Floor, Amrutha Topaz, Punjagutta, Hyderabad 500 482, India

ABSTRACT

RD-9 Repelin, a neem-based EC formulation containing an azadirachtin concentration of 3000 ppm, was tested for its bioefficacy and was found effective in managing several key pests of major crops. The pests controlled included American bollworm and whitefly in cotton; stemborer and brown plant hopper in rice; pod borer, mites, and thrips in chillies; aphids and green leaf hopper in groundnut; and diamond-back moth in cabbage. RD-9 Repelin was found to be effective at a concentration of 1.0% either alone or in combination with recommended chemical pesticides. The dosage of the chemical pesticide could be reduced to half of the recommended dosage when used in combination with RD-9 Repelin. When RD-9 Repelin was used, the number of sprayings required during the crop duration was also reduced as it inhibited the build-up of pests. The LD_{50} values of RD-9 Repelin for acute and sub-acute dermal toxicity, oral toxicity and inhalation toxicity obtained from toxicological studies were very high revealing a high degree of its safety to mammals, fish and honeybees. RD-9 Repelin has been registered with Central Insecticides Board, Government of India.

INTRODUCTION

Pests management is an important aspect of crop production. Indiscriminate use of pesticides resulted in pest resurgence, pesticide resistance, residual toxicity, imbalances in ecological equilibrium and environmental pollution. An integrated approach to pest management is needed to reduce the crop losses due to pests and to make agriculture sustainable. Botanical pesticides, especially neem-based, offer immense scope in the integrated pest management.

ITC's research on neem dates back to 1969 when experiments were conducted to control the tobacco mosaic virus. Further research on neem resulted in the development of Wellgro, a product aimed at the protection of nurseries and production of healthy seedings. In 1985, ITC Ltd developed a potential neem-based formulation viz., RD-9 Repelin, enriched with the active ingredient, azadirachtin. It was extensively studied for its bioefficacy against a range of pests in several crops at different research centres of agricultural institutes, ICAR institutes and international institutes. This paper presents some of the results obtained from the collaborative trials on RD-9 Repelin.

TOXICOLOGICAL STUDIES

Toxicological studies of RD-9 Repelin were carried out by Indian Institute of Chemical Technology, Hyderabad. The data obtained are given below:

1. Acute dermal toxicity to rats:
   LD_{50} for single application by dermal route=>20000 mg/kg
2. Acute toxicity to honey bee by contact:
   LD_{50} in honey bee = 169.4 × 26.7 mg/kg
3. Acute oral toxicity (Route: P.O.):
   LD_{50} for rat = 18106.0 + 96.24 mg/kg
4. Acute toxicity to fish (Route: Exposure):
   LD_{50} for fish = 40.37 + 0.11 ppm
5. Acute inhalation toxicity (Route: Inhalation):
   LD_{50} in Rats = >750 ppm > (mist-aerosol of 1–5 micron particle size)
6. Sub-acute oral toxicity to rat:
   No significant damages were found (macroscopically and microscopically) in liver, spleen, lung, brain, kidney and heart tissues at different concentrations.
7. Sub-acute dermal toxicity to rat (skin application):
   No significant differences (changes) between treated and control tissues at different concentrations when observed macroscopically and microscopically.
8. Sub-acute inhalation toxicity to rat (inhalation for 14 days):
   Histopathological examination of various tissues treated with different concentrations failed to reveal any remarkable damage.
9. Acute oral toxicity to mice (Route: P.O.):
   LD_{50} for mice: 13545.36 + 456 mg/kg
10. Mucous membrane irritation in rabbits:
    There was no mortality in treated animals. They exhibited no adverse symptoms. Hence, the compound is practically non-toxic as mucous membrane irritant.
11. Primary skin irritation on rabbits (Draize Method):
    The compound did not cause any oedema or erythema in the treated animals. There was no mortality. Therefore, the compound does not appear to be a skin irritant.

BIOEFFICACY STUDIES

Trials on Rice

RD-9 Repelin was tested for its bioefficacy against different insect pests of rice viz., yellow stem borer (*Scirpophaga incertulus* Walker), brown plant hopper (*Nilaparvata lugens* Stal), gallmidge (*Orseolia oryzae* Wood Mason), leaf folder (*Cnaphalocrocis medinalis* Guinea), white backed plant hopper (*Sogatella furcifera* Horvath), green leaf hopper (*Nephotettix virescens*) and rice tungro virus (RTV). Studies conducted at Agricultural Research Station (APAU), Maruteru revealed that RD-9 Repelin (1.0 per cent) plus half the dose of monocrotophos (0.03 per cent) significantly reduced the population of yellow stem borer, brown plant hopper and leaf folder.

Krishnaiah et al. (1990) reported that RD-9 Repelin in combination with monocrotophos and ethofenprox at lower concentrations resulted in reduced survival of BPH and GLH nymphs than the application of insecticides alone. Studies on the bioefficacy of RD-9 Repelin against white backed plant hopper (WBPH) were conducted at Punjab Agricultural University, Ludhiana (Dhaliwal; 1993). The results obtained revealed that RD-9 Repelin repelled WBPH and also that it adversely affected its fecundity. It was also found that minimum number of nymphs attained the adult stage in case of RD-9 Repelin 1.0 per cent (28 per cent) while 98 per cent of nymphs became adults in untreated control. Gurubasawaraj et al. (1988) reported that neem seed bitter (NSB) and RD-9 Repelin significantly reduced the survival of green leaf hopper and its transmission of RTV (Table 1). On feeding for one day on plants sprayed with the preparations of NSB (1 per cent), NSB (2 per cent), RD-9 Repelin (1 per cent), RD-9 Repelin (2 per cent), and NSB (1 per cent) root dip, 80, 60 56, 73 and 50 per cent of the insects only survived as against 100 per cent survival in the untreated control. On the 5th day none survived in NSB 1 per cent root dip and RD-9 Repel in 2 per cent spray resulting in significantly lower infection levels of RTV.

Trials on Cotton

Field experiments were conducted at Regional Agricultural Research Station (APAU), Lam to evaluate RD-9 Repelin for controlling cotton insect pests viz., boll worm (*Helicoverpa armigera*), aphid (*Aphis gossypii* G.) and mite (*Tetranychus telarius*). Data were collected on the insect population and cotton yield. RD-9 Repelin @ 1.5 per cent sprayed 4–6 times and at 1.0 per cent sprayed 6–8 times was found to be signifcantly superior to application of insecticides comprising 8 sprayings of conventioanl insecticides and 2 sprayings of synthetic pyrethroids in controlling aphids, mites, and bollworm population in cotton (RARS, 1989).

Table 1. RD-9 Repelin for controlling GLH and RTV in Rice

Treatment	Days after treatment					
	1	2	3	4	5	6
GLH Survival (%)						
RD-9 Repelin 2%	50	40	26	13	0	0
Teepol 2%	100	100	100	90	80	80
Control	100	100	100	100	100	90
RTC Infected Seedlings (%)						
RD-9 Repelin 2%	36	23	20	10		
Teepol 2%	70	50	40	30		
Control	73	56	40	30		

Source: T.M. Gurubasawaraj et al. (1988).

Further studies at the same station during 1990–91 revealed that RD-9 Repelin (0.5 per cent) in combination with quinalphos (0.05 per cent) was significantly superior to lone use of quinalphos 0.05 per cent in controlling *Helicoverpa armigera* as evidenced by insect population in terms of egg and larval loads and damage to squares, flowers and bolls while RD-9 Repelin (0.5 per cent) application was on par with quinalphos (0.05 per cent) in terms of pest population, damage and cotton yield (Table 2). The yield obtained was highest when RD-9 Repelin (0.5 per cent) was combined with quinalphos (0.05 per cent) (RARS, 1991). In the same study, the combination treatment exhibited significantly superior control of jassids and aphids.

Large scale trials on cotton conducted in farmers fields in Guntur and Praskasam districts of Andhra Pradesh (AICCIP, 1988) also revealed that RD-9 Repelin sprayed plots showed significantly lower levels of Whitefly population, egg and larval loads of bollworm and per cent boll damage and gave higher yields than the plots sprayed with conventional insecticides (control) (Table 3).

Trials on Vegetable Crops

Studies carried out by Samalo et al. (1988) at Orissa University of Agriculture and Technology, Bhubaneshwar revealed that RD-9 Repelin (2.0 per cent) root dip treatment

Table 2. Effect of RD-9 Repelin on *Helicoverpa armigera* and cotton yield

Treatment	Number per five plants		% damage to bolls	Yield kg/ha
	Eggs	Larvae		
RD-9 Repelin (0.5%)	4.42 (2.10)	3.35 (1.94)	20.24 (26.40)	799
RD-9 (0.5%) + Quinal (0.05%)	4.53 (2.13)	1.39 (1.34)	10.60 (18.40)	1262
Quinalphos (0.05%)	6.34 (2.60)	3.29 (1.92)	22.11 (27.50)	819
Control	7.33 (2.76)	6.31 (2.88)	35.81 (38.06)	403
CD (0.05)	(0.21)	(0.19)	(3.39)	231

Figures in parentheses are transformed values
Source: Regional Agricultural Research Station, APAU, Lam.

Table 3. Incidence of whitefly and American bollworm in cotton as influenced by RD-9 Repelin

	Pedanandipadu		Parchuru	
	Sprayed	Control	Sprayed	Control
Whitefly				
No. of adults/leaf	11.4	34.1	16.9	23.2
Bollworm				
No. of eggs/plant	0.1	8.4	2.3	14.2
No. of larvae/plant	0.1	13.2	5.4	19.2
% Boll damage	1.1	27.1	9.1	37.7
Yield kg/ha	2625	820	1625	500

was on par with Carbofuran (2.0 per cent) in controlling root-knot nematode, *Meloidogyne incognita* on tomato (Table 4). There was a reduction of 50 per cent in the nematode population recorded in the RD-9 Repelin (2.0 per cent) root dip treatment over control. Field experiments conducted at Regional Agricultural Station, Lam (RARS, 1993) indicated that spray application of RD-9 Repelin (0.5 per cent) alone or in combination with carbaryl (0.15 per cent) reduced the aphid population and American boll worm damage on okra significantly as compared to Carbaryl (0.15 per cent) alone (Table 5).

The results of field trials conducted by Kareem et al. (1988) in Philippines revealed that the population of immature stages of *Plutella xylostella* Lemn on broccoli 10 days after 1st spraying was significantly lower in all the plots treated with botanicals including RD-9 Repelin than that of either untreated or of insecticides treated plots. The pest population remained at a low level. The data obtained on per cent leaf damage indicated that it was higher in monocrotophos treated plot than in RD-9 Repelin plot.

Table 4. RD-9 Repelin for controlling root-knot nematode of tomato

Treatment		Nematode population (thousand)	Reduction over control (%)
Neem Bitter	(0.5%)	2.3 b	35
	(1.0%)	1.9 a	47
RD-9 Repelin	(1.0%)	2.3 b	35
	(2.0%)	1.8 a	50
Wellgro	(1.0%)	2.5 b	31
	(2.0%)	1.7 a	53
Carbofuran	(1.0%)	2.8 b	24
	(2.0%)	2.0 a	46
Untreated		3.6 c	

Source: Samalo et al. (1988).

Table 5. RD-9 Repelin for controlling okra pests

Treatment	Number per five plants		Bollworm	Yield
	Jassid	Aphid	damage (%)	(kg/ha)
RD-9 Repelin (0.5%)	22.8 (4.84)	36.06 (6.03)	8.73 (16.97)	1145
Carbaryl (0.15%)	13.3 (3.67)	98.13 (9.88)	15.00 (22.68)	1230
RD-9 (0.5%) + Carbaryl (0.15%)	12.9 (3.63)	33.33 (5.33)	5.60 (13.48)	1300
Control	14.9 (4.08)	99.90 (10.02)	21.86 (27.85)	525
CD (0.05)	(0.64)	(3.12)	(6.37)	445

Source: Regional Agricultural Station, APAU, Lam.

Trials on Oildseed Crops

RD-9 Repelin was tested for controlling groundnut aphid (*Aphis craccivora* Koch) and hopper (*Empoasca kerri*) at Nimbar Agricultural Research Institute, Phaltan, (NARI, 1989a) (Tables 6 and 7). The best treatments for controlling these sucking pests were found to be spraying dimethoate or endosulfan as also diemthoate (0.25 per cent) plus RD-9 Repelin (1.0 per cent) or alternate treatment of dimethoate and RD-9 Repelin. The pod and green fodder yield obtained from RD-9 Repelin alone or in combination with other pesticides indicating its compatibility and safe use in groundnut.

At NARI experiments were conducted to evaluate RD-9 Repelin for controlling *Helicoverpa armigera* in sunflower crop (NARI, 1989b). Results revealed that RD-9 Repelin gave a good control of *H. armigera* when sprayed at the rate of 100 ml and 150 ml per 10 litres of water (Table 8). It appeared to be well compatible with fungicide, insecticide and fertilizer used. No phytotoxic symptoms were recorded from any of the plots sprayed with RD-9 Repelin, indicating its safety to sunflower crop. Data on seed yield showed that RD-9 Repelin at 1.0 per cent and 1.5 per cent alone and at 1.0 per cent in combination with endosulfan and dimethoate or spraying RD-9 Repelin alternated with them gave significantly higher yields than the untreated control.

Trials on stored product pests and household pests revealed that RD-9 Repelin was effective against flour beetle (*Tribolium confusum*) and cockroach (*Blatella germanica*). Trials conducted by All India Coordinated Research Project on Agricultural Ornithology revealed that RD-9 Repelin repelled bird pests on maize and sunflower.

Table 6. RD-9 Repelin for controlling groundnut aphid

Treatment	Aphid number per five plants		
	I spray	II spray	III spray
RD-9 Repelin 1.5%	19	14	14
Dimethoate 0.25%	17	3	3
RD-9 (1%) + Dimethoate	17	3	4
Control	56	71	89
CD (0.05)	5	6	8

Source: Nimbkar Agricultural Research Institute, Phaltan.

Table 7. RD-9 Repelin for controlling groundnut hopper

Treatment	Hopper number per five plants		
	I spray	II spray	III spray
RD-9 Repelin 1.5%	12	10	10
Dimethoate 0.25%	8	13	11
RD-9 (1%) + Dimethoate (0.25%)	15	7	7
Control	35	47	53
CD (0.05)	4	5	6

Source: Nimbkar Agricultural Research Institute, Phaltan.

Table 6. RD-9 Repelin for controlling *H. armigera* in sunflower

Treatment	Larval number per five plants		
	I spray	II spray	III spray
RD-9 Repelin 1.5%	10	8	8
Dimethoate 0.25%	14	12	11
RD-9 (1%) + Dimethoate (0.25%)	10	8	7
Control	39	51	56
CD (0.05)	4	5	6

Source: Nimbkar Agricultural Research Institute, Phaltan.

CONCLUSION

RD-9 Repelin was found to be effective against a wide range of pests on different crops. RD-9 Repelin is safe to mammals, fishes, and pollinators. It did not exhibit any phytotoxicity properties. It is compatible with other pesticides. The experimental results indicated that RD-9 Repelin formed an important component in the management of key pests of rice, cotton, vegetables and oil seed crops.

REFERENCES

AICCIP. 1988. Testing of new plant products and new pesticides. All India Coordinated Cotton Improvement Project in Andhra Pradesh, Lam. Annual Progress Report.

Dhaliwal, G.S. 1983. Effect of neem-based insecticides on growth and development of insect pests of rice (Personal communication).

Gurubasawaraj, T.M., V. Mariappan and S. Jairaj. 1988. Neem seed bitter and Repelin reduce GLH survival and RTV, Paper presented at the *Final Workshop of IRRI-ADB-EWC project on Botanical Pest Control in Rice-based Cropping Systems*, IRRI, Philippines.

Kareem, A.A., R.C. Saxena, E.L. Palaniginan, E.M. Boncodin and M.T. Malayba. 1988. Effects of neem seed derivatives and other botanical pesticide formulation against diamondback moth, *Plutela xylostella* Linn on Brocolli (Cruciferaceae), a vegetable crop. In: *Neem Derivatives: Effects of insect pests of rice and certain other crops in the Philippines* (Lab and Field Evaluations 1986–88), pp. 77–79.

Krishnaiah, N.V., M.B. Kalode and I.C. Fasalu. 1990. New approaches in utilization of botanicals in rice insect pest control. *Proc. Symp. Botanical Pesticides in IPM*, Rajahumundry, pp. 114–117.

NARI. 1989. Bioefficacy, compatibility and phytotoxicity of "RD-9 Repelin" of ITC Ltd., on groundnut cultivar "Phule Pragati". Nimbkar Agricultural Research Institute, Phaltan, Pune, Report.

NARI. 1989. Bioefficacy and phytotoxicity of "RD-9 Repelin" of ITC Ltd., on sunflower cultivar NSFH-9. Nimbkar Agricultural Research Institute Report.

RARS. 1989. Studies on the evaluation of Repelin on the control of cotton. A.P. Agricultural University, Regional Agricultural Research Station, Lam Report.

RARS. 1991. Evaluation of Repelin for controlling cotton pests. A.P. Agricultural University, Regional Agricultural Research Station, Lam Report.

RARS. 1993. Effect of RD-9 Repelin in the management of okra pests. Regional Agricultural Research Station, Lam Report.

Samalo, A.P.., A.N. Dash, B.K. Mishra and N.N. Padhi. 1988. Phytoproducts on pest control in rice-based cropping systems. Paper presented at the *Final Workshop of IRRI-ADB-EWC Project on Botanical Pest Control in Rice-based Cropping Systems*, IRRI, Los Banos, 12–16 December, 1988.

98

Development of a Neem Formulation and Its Evaluation for Control of Crop Pests

K. RAMAN[1], S. GANESAN[1], B.N. VYAS[2], N.B. GODREJ[2] AND K.B. MISTRY[2]

[1]Godrej Agrovet Ltd., Madras 600 098, India
[2]Godrej Soaps Ltd., Bombay 400 079, India

ABSTRACT

An enriched neem formulation (Godrej *Achook*), containing azadirachtin, azadiradione, nimbocinol, and epinimbocinol (totalling 2800 ppm), in the form of a water-soluble powder has been developed. The physiological effects of *Achook* were evaluated under laboratory conditions on two important lepidopterous pests: *Helicoverpa armigera* Hub. and *Earias vitella* Fab. Besides showing strong 'antifeedant effects', *Achook* also affected the larval and pupal survival and adult emergence in a dose-dependent manner in both the species tested. Further, *Achook* reduced the larval growth and total development growth, prolonged time for pupation, and lowered pupal weights, resulting in the formation of deformed individuals. Field evaluation of the said neem formulation against the pests of major crops such as cotton, okra, brinjal, cabbage, jute, and tea were undertaken at different locations in the country. Results of the studies indicated significant reduction in pest population and associated crop damage, leading to increase in crop yields.

INTRODUCTION

It is well known that synthetic chemical insecticides play an indomitable role in the field of crop protection due to their effective action against insect pests, long shelf-life, wider applicability, and ease of handling. However, the continuous use of synthetic chemicals has led to problems such as toxicity to non-target organisms, development of resistance and resurgence in several pests, environmental pollution, and health hazards to human beings. This has necessitated exploration of alternate pest management practices that are effective but relatively safe. The development and use of botanical insecticides for protecting crops from pest damage, as an alternative pest management strategy, has reached a stage of commercial exploitation. Among the most promising natural insecticides investigated, the extracts obtained from different parts of the neem tree, *Azadirachta indica* A. Juss, possessed diverse biological effects such as phagodeterrence and repellence (Pradhan et al., 1963; Joshi and Ramaprasad, 1979; Jacobson et al., 1978; Warthen et al., 1978), growth disruption and abnormal development (Leuschner, 1972; Meisner et al., 1978; Schmutterer and Rembold, 1980; Sharma et al. 1980), and oviposition suppression (Jacobson et al., 1978; Joshi and Sitaramaiah, 1979). The active principles present in neem seed kernels have been identified (Butterworth and Morgan, 1968; Nakanishi, 1975; Warthen et al., 1978) and the biological effects of a few individual components on insects have also been reported (Kraus et al., 1981; Schmutterer, 1981; Morgan, 1981; Schmutterer and Rembold, 1980; Schmutterer et al., 1983). However, it was only during the second International Neem Conference at Rauischholzhausen, Germany, that Sharma et al. (1983) first reported the concept of using 'enriched fractions (neemrich)' from neem for the behavioural and physiological changes which could be helpful for control of insects. While information is available on the effects of several neem components on insects, attempts to quantify the active ingredient(s) and commercially formulate a suitable 'neem formulation' for the use of Indian farmers have been wanting. A programme was initiated at Godrej Soaps Ltd., Bombay, to develop a suitable neem formulation for pest control. This paper reports the development of a neem formulation as well as its effects on feeding, growth, and control of certain major pests of crops under laboratory and field conditions.

MATERIALS AND METHODS

Product Development

The neem formulation was developed by successive treatment of industrial-grade neem oil to remove the major part of triglycerides and free fatty acids. The fraction remaining after separation of triglycerides was further processed to yield a water-soluble powder (fraction I). Further, neem seeds were separately solvent-extracted and neem triterpene constituents were concentrated to yield a product (fraction II). The final neem formulation was prepared by mixing fractions I and II in the ratio of 95:5. The individual neem constitutents were separated by solvent extraction and silica gel column chromatography as per the schemes described by Mitra (1963), Yamasaki et al. (1986), and Govindachari et al. (1990). The finished product was also analysed with the help of HPLC for its chemical constituents, namely azadirachtin (minimum 300 ppm), azadiradione (minimum 500 ppm), nimbocinol, and epinimbocinol (together totalling 2000 ppm). The product has been given the trade name 'Godrej *Achook*'.

Laboratory Studies

Bioassays under laboratory conditions were carried out on the larvae of two lepidopterous pests: *Helicoverpa armigera* Hub. and *Earias vitella* Fab. For antifeedant studies, three three-day-old larvae of the respective species were placed separately at the centre of a petri dish (10 cm diameter). Cut okra (bhendi) dipped in the test solution of *Achook* of desired concentrations (0.5, 1, 2, and 3 per cent) were air-dried and placed at opposite ends inside the petri dish. A control was also maintained in a similar manner with okra treated with distilled water. The entire set-up was left undisturbed for 24 hrs, after which the number of larvae feeding on control and *Achook*-treated okra were recorded separately for different concentrations and for each species. Each test was repeated 15 times using three three-day-old larvae each time. At the end of the experiment, the total number of larvae of *H. armigera* and *E. vitella* feeding on control as well as on different concentrations of *Achook* was determined and the 'antifeedant index' was computed using the method of Blaney et al. (1987). For growth and development studies, larvae of the two species were individually confined in plastic vials provided with cut okra fruits treated with different concentrations of the test solutions. A control was also maintained as described above. Observations were recorded regularly on the mortality of larvae, number of days to pupate, pupal duration, and adult emergence. In addition, the occurrence of any larval, pupal, and adult abnormalities due to *Achook* treatment were recorded. Each treatment was repeated 20 times using fresh larvae of the same age. The data were statistically analysed and means were compared using DMRT (Duncan's multiple range test).

Field Studies

The efficacy of *Achook* was also tested under field conditions against major pests of cotton, okra, brinjal, cabbage, paddy, jute, tea, and other crops. Trials were conducted at different locations in the country using the standard protocols employed for the evaluation of insecticidal formulations against crop pests. *Achook* was used at 0.5 per cent concentration (i.e., 2.5 kg in 500 l water per hectare). Insecticides were used as per the recommended package of practices. Control plots received water sprays. Each treatment was replicated three times. Observations were recorded on the incidence of sucking pests and borers. Results collected were statistically analysed using F-test and the data from different locations were pooled for interpretation.

RESULTS AND DISCUSSION

Laboratory Studies

Results from the study clearly indicate the antifeedant and growth-disruptive effects of *Achook* on larvae of *H. armigera* and *E. vitella*. Computation of the antifeedant indices for various concentrations revealed 50 per cent more larval feeding inhibition for treatments from 1 per cent and above for *E. vitella* and 0.5 per cent for *H. armigera*. Similarly, compared to control, the larval growth indices and total development growth indices showed a profound decrease in *Achook* treatments, indicating the effects of *Achook* on the growth and development of the two insect species tested. Also the antifeedant effects and growth inhibition effects varied according to the concentration of *Achook* tested, suggesting a dose-dependent response (Table 1). Similar dose-dependent effects leading to reduction in larval growth rates were reported with aqueous neem extracts and ethanolic neem extracts for *Spilosoma maculatus* Walk. (Umeh, 1988) and *Spodoptera frugiperda* (J.E. Smith) (Mikolajczak et al., 1989).

Table 1. Antifeedant and growth-inhibitory effects of *Achook* on larva of *H. armigera* and *E. vitellas*

Concentrations (%)	Antifeedant index		Larval growth index		Total development growth index		Pupal weight (g)	
	Ha	Ev	Ha	Ev	Ha	Ev	Ha	Ev
Control	—	—	8.47a	14.99a	6.08a	7.42a	0.355a	0.055a
0.5	33.33a	14.24a	6.12b	10.00b	2.08b	4.00b	NP	0.045b
1.0	53.85b	50.00b	NP	9.48b	NP	2.73c	NP	0.042b
2.0	73.33c	60.00c	NP	3.85c	NE	1.80c	NP	0.040b
3.0	86.67d	63.64c	*	1.25c	*	NE	*	NE

Three-day-old larvae of *H. armigera* (Ha) and *E. vitella* (Ev) were used for the study.

$$\text{Larval growth index} = \frac{\%\,\text{pupation}}{\text{larval period (days)}}$$

$$\text{Total development growth index} = \frac{\%\,\text{emergence}}{\text{Total develop. period (days)}}$$

NP: No pupation: Ne: No adult emergence; * Concentration not tested.
Means followed by a common letter are not different significantly at 5% level (DMRT).

Among the various active constituents present in neem seed, azadirachtin is considered to be the most potent compound in view of its showing very strong deterrent, antifeedant, and growth-disruptive properties on several insects (Schmutterer and Ascher, 1984; Schmutterer, 1990). Other active compounds such as salannin, salannol, azadiradione, gedunin, nimbin, nimbolide, nimbinic acid, and deacetylnimbinen are also reported to have diverse biological effects on insect pests (Jones et al., 1989; Warthen, 1989). Hence, the antifeedant and growth-disruptive effects of *Achook* observed in the present study could be attributed to the presence of such potent principles (in addition to azadirachtin) derived from neem oil and neem seed kernels.

Exposure of larvae of *H. armigera* and *E. vitella* to *Achook* treatment also reduced their survival and post-embryonic development significantly. Mortalities of larvae from 24 hrs to four days after treatment were observed. Younger larvae were more susceptible to *Achook* treatment than older larvae and their survival depended on the concentration tested. In both the species, prolongation of larval and pupal periods and reduction in pupal weights were evident due to *Achook* treatment (Tables 1, 2). Incidentally, in *H. armigera* there was 100 per cent larval mortalilty and hence no pupation was observed. In general, treated larvae of *H. armigera* and *E. vitella* either died within four days after treatment or transformed into abnormal larval-pupal intermediates. Likewise, affected pupae either remained unhatched or often produced adults with deformed wings, which failed to survive after emergence. We also observed certain characteristic symptoms, such as blackening of mouth parts and anal regions, development of black prolegs and spots on the body of dead larva and inability of larva to spin a cocoon, resulting in the formation of a fragile cocoon as observed in *E. vitella* (Fig. 1). The inability of larva to pupate or the formation of larval-pupal intermediate forms could be attributed to the interference of the neem constituents present in *Achook* with the ecdysteroid titres of treated larva (Seiber and Rembold, 1983; Barnby and Klocke, 1990). Similarly, the

Table 2. Effects of *Achook* on the larval and pupal developments of *H. armigera* and *E. vitella*

Particulars	Insect species	Age of larva (days)	Concentration of neem formulation (%)				
			Control	0.5	1.0	2.0	3.0
Mortality (%)	Ev	3*	—	41.18	59.41	64.71	88.24
		5*	—	33.33	40.00	46.67	93.30
	Ha	3*	—	100.00	100.00	100.00	**
		8*	—	25.00	100.00	100.00	**
Days for pupation	Ev	3*	5.67± 0.28	5.70+ 0.10	6.33± 0.10	6.50± 0.08	8.00± 0.01
		5*	5.00± 0.07	5.10± 0.22	5.30± 0.11	5.50± 0.12	7.00± 0.01
	Ha	3*	13.00±	NP	NP	NP	**
		8*	6.75± 0.17	7.00+ 0.01	NP	NP	**
Pupal duration (days)	Ev	3***	7.80± 0.21	10.00± 0.01	12.00± 0.19	12.00± 0.01	NE
		5***	8.00± 0.05	8.22± 0.05	8.50 0.01	9.00± 0.01	NE
	Ha	3*	8.00 0.01	NP	NP	NP	**
		8*	4.70± 0.01	9.00+ 0.01	NP	NP	**

Values indicate mean + SE. Each treatment is replicated 20 times.
NP = No pupation; NE = No adult emergence;
* Significant at 5% level; **concentration not tested; *** Significant at 5% level.

blackening of prolegs as well as appearance of certain black spots on the body of the larva are the effects of neem derivatives on the chitin synthesis in insects (Cassier et al., 1987). Such instances of developmental abnormalities leading to high mortalities have also been reported for several other lepidopterous pests (Saxena et al., 1981; Schmutterer et al., 1983; Mikolajczak et al., 1989; Tanzubil and McCaffery, 1990).

Field Studies

The results obtained from different field studies on cotton, okra, brinjal, paddy, and other crops are presented in Tables 3 and 4. Data compiled from field studies revealed lower incidence of major pests of cotton, vegetables, and paddy in *Achook*-treated plots than in untreated control. On cotton the mean number of sucking pests such as aphids, jassids, and whiteflies were significantly lower due to *Achook* spray compared to insecticide spray and untreated control (Table 3).

On okra and brinjal *Achook* spray controlled the incidence of *E. vitella* and *Epilachna* beetles, and fruit borer attack respectively, leading to higher yields. Similarly, on paddy the incidence of leaf roller, *Cnaphalocrocis medinalis* Guenee, was lower in *Achook*-treated plots than in control (Table 4). Spraying cabbage plots with *Achook* at 0.5 per cent concentration checked the multiplication of diamondback moth, *Plutella xylostella* L., and increased the marketable cabbage yield by 26.98 per cent compared to untreated control (Nimbkar Research Institute, 1990). In jute there was a considerable reduction of damage due to semi-looper (*Anomis sabulifera* Guen) and hairy caterpillar (*Spilosoma*

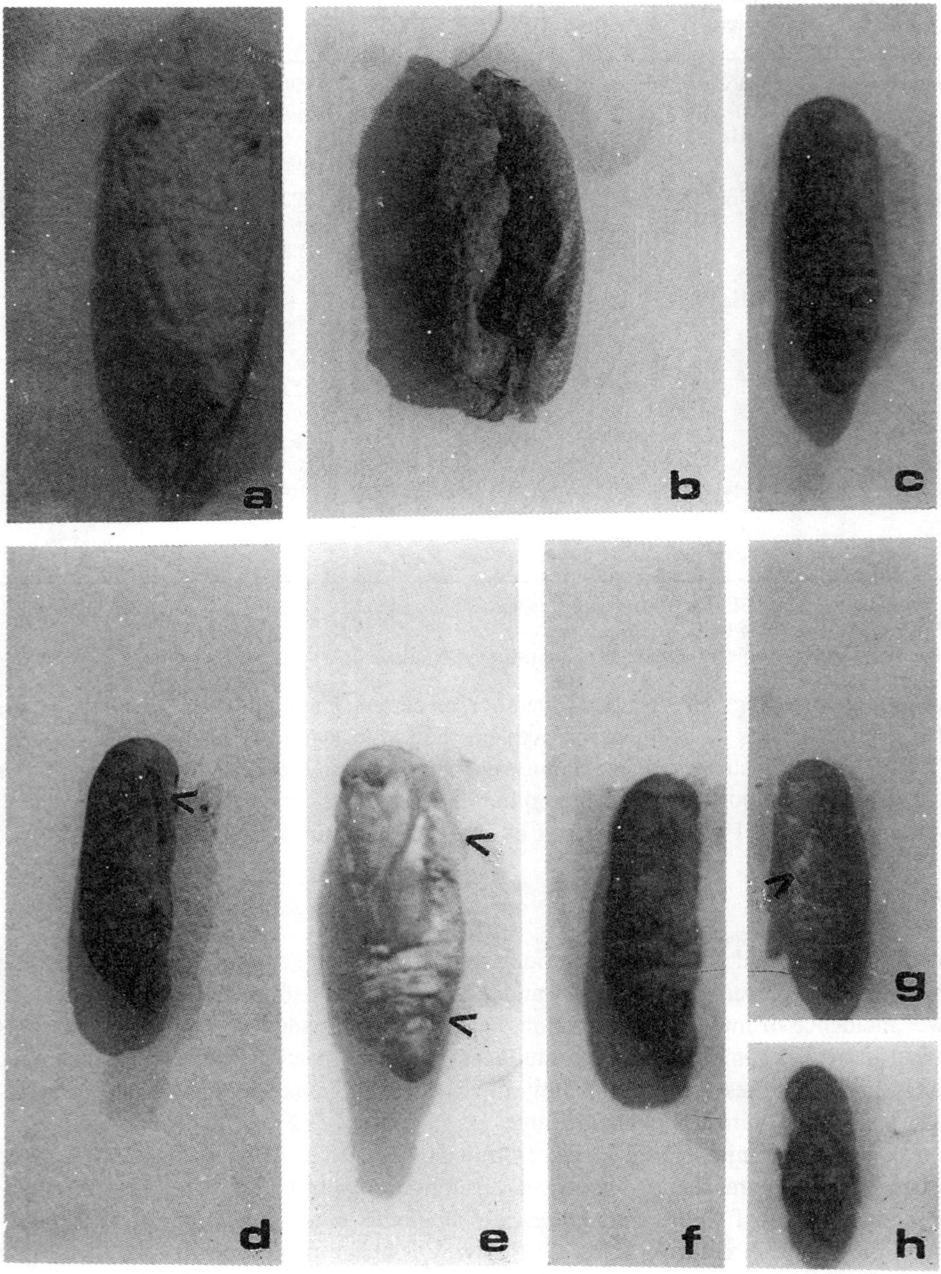

Fig. 1. Morphogenetic effects of *Achook* in *E. vitella* (a) Healthy pupal case. (b) Shrivelled pupal case. (c) Healthy pupa. (d–h) Pupal deformities of various forms. (Arrows indicate affected regions.)

Table 3. Effect of neem products and insecticide on cotton insects

Particulars (0.5%)	Neem oil (0.5%)	Godrej neem formula (0.07%)	Endosulfan	Control
Whiteflies	3.8 ab	3.8ab	4.0ab	4.5b
Aphids	6.1bc	3.6a	4.2ab	6.3c
Leafhoppers	4.1a	4.8a	4.5a	5.2a
Ash weevil	4.3b	3.9	3.9a	5.1b
Bollworm damage (%)	21.6a	22.7a	19.6a	27.6b
Bad kapas (%)	22.8ab	20.0a	23.7abc	27.3c
Kapas yield (Q/ha)	13.31bc	13.32bc	16.38a	12.13c

Source: Central Institute of Cotton Research, Coimbatore 1990a.
Data represent mean number of insects/leaf. Spraying was done four times at 15 days interval commencing from 45 days after germination. In a column means followed by a common letter are not significant at 5% level (DMRT).

Table 4. Efficacy of *Achook* and insecticide in controlling pests of crops

Treatment	Okra* % fruits damage by *E. Vitella*	Brinjal*		Rice** % reduction due to leaf roller	
		% reduction by *Epilachna* bettles	% fruits damaged by borer	2nd DAT	7th DAT
1. *Achook* @ 2.5 kg/ha	8.98 (45.63)	54.90	40.89 (22.29)	55.91	35.18 (55.29)
2. Recommended insecticide	7.27 (56.59	83.65	20.51 (30.91)	97.98	86.75 (60.86)
3. Recommended insecticide + *Achook* @ 2.5 kg/ha	1.92 (58.81)	89.70	18.87 (32.72)	99.91	90.89 (61.95)
4. *Achook* alternate with insecticide	2.13 (55.27)	82.67	27.62 (28.36)	88.94	78.77 (57.84)
5. Control (water spray)	(43.14)		(16.88)		(51.27)
C.D. (0.5%)	5.58	6.57	1.86	4.15	5.13

Sprays were given four times at 15 days' interval commencing from 20 days after sowing/transplanting.
Each treatment was replicated three times.
DAT: Days after sowing/transplanting.
Figures in bracket indicate yield (q/ha).
* BCKV (1991).
** Konkam Krishi Vidyapeeth (1991).

obliqua Walk.), when *Achook* was alternated with Endosulfan at recommended doses (BCKV, 1991). In tea the population of flushworms, bunch caterpillar (*Andraca bipunctata*), tea mosquito bug (*Helopeltis theivora* Fab.), and jassids (*Empoasca flavescens* Fabr.) showed significant reduction with *Achook* treatment (Tocklai Experimental Station, 1991).

Achook at 0.5 per cent concentration provided adequate control of major pests under field conditions. Satisfactory control of grain legume crops from attack of major lepidopterous pests was achieved by spraying wettable powder and emulsifiable formulations of neem (Parmar and Srivastava, 1986), spraying neem extracts alone or alternatively with chemical insecticide also provided adequate protection of tobacco (Kirsch, 1986; Joshi, 1986), cabbage (Kirsch, 1986), cauliflower, and tomato (Fagoonee, 1986) against attack of major caterpillar pests. Likewise, spraying neem oil or neem seed kernel extracts suppressed the growth and development of cotton whitefly (Natarajan and Sundaramurthy, 1990), rice brown planthopper and leaf folder (Saxena, 1987), and major sucking pests of potato (Siddig, 1986).

The present programme lead to successful development of an enriched neem formulation. Considering the effectiveness of *Achook* as an antifeedant and growth-disruptive agent in controlling insect pests of major crops, it could be successfully used as a substitute for chemical pesticide and the total number of pesticidal rounds could be effectively reduced (Kogan, 1988). It would thus be useful in an integrated pest management programme. Formulations based on neem are also reported to be environment-friendly due to their low toxicity to natural enemies of insects and humans (Saxena, 1987).

ACKNOWLEDGEMENT

Grateful acknowledgements are due to all concerned scientists from different agricultural universities and research stations in the country who have rendered help in organizing field experiments.

REFERENCES

Nimbkar Research Institute. 1990. Maharashtra Progress Report.

Tocklai Experimental Station. 1991. Tea Research Association communication.

Konkan Krishi Vidyapeeth. 1991. Project Report. Maharashtra.

Barnby, M.A. and J.A. Klocke. 1990. Effects of azadirachtin on levels of ecdysteroids and prothoracicotropic hormone-like activity in *Heliothis virescens* (Fab.) larvae. *J. Insect Physiol.* 36: 125–131.

Bidhan Chandra Krishi Viswavidyalaya. 1991. Project Report. Nadia, West Bengal.

Blaney, N.M., M.S. Simmonds, S.V. Ley and R.B. Katz. 1987. An electrophysiological and behavioural study of insect antifeedant properties of natural and synthetic drimane related compounds. *Physiol. Entomol.* 12: 281–291.

Butterworth, J.H. and E. Morgan. 1968. Isolation of a substance that suppresses feeding by locusts. *Chemical Comm.* 1: 23–24.

Cassier, P., D. Chalaye, N. De Basse, M. Papillon, L. Baldaia and A. Farcheron. 1987. Poster VIII (Abstract). In: *Ecdysone Workshop*, Marburgg, 39 pp.

Fagoonee, I. 1986. Use of neem in vegetable crop protection in Mauritius. *Proc. 3rd Int. Neem Conf.*, Nairobi, Dt. Ges. für Techn. Zusammenarbait (GTZ) GmbH, Eschborn, pp. 419–429.

Govindachari, T.R., G. Sandhya and S.P. Ganeshraj. 1990. Simple method for the isoloation of azadirachtin by preparative high performance liquid chromatography. *J. Chromatrog.* 513: 389–391.

Jacobson, M., D.K. Reed, M.M. Crystal, D.S. Morneo and E.L.C. Soderstrom. 1978. Chemistry and biological activity of insect feeding deterrents from certain weed and crop plants. *Entomol. Exp. Appl.* 24: 248–257.

Jones, P.S., S.V. Ley, E.D. Morgan and D. Santafiancos. 1989. The chemistry of the neem tree. In: *The Neem Tree: Natural Resistance par Excellence*. M. Jacobson (ed.), ACS Symp. Ser. 296, Washington, pp. 19–45.

Joshi, B.G. and G. Ramaprasad. 1979. Neem kernel as an antifeedant against the tobacco caterpillar (*Spodoptera litura* F.). *Phytoparasitica* 3: 59–61.

Joshi, B.G. and S. Sitaramaiah. 1979. Neem kernel as an ovipositional repellent for *Spodoptera litura* F. moth. *Phytoparasitica* 7: 199–202.

Joshi, B.G. 1986. Use of neem products in tobacco in India. *Proc. 3rd Int. Neem Conf.*, Nairobi Dt. Ges. für Techn. Zusammenarbeit (GTZ), GmbH, Eschborn, pp. 479–494.

Kirsch, K. 1986. Studies of the efficacy of neem extracts in controlling major insect pests in tobacco and cabbage. *Proc. 3rd Int. Neem Conf.*, Nairobi, Dt. Ges. für Techn. Zusammenarbeit (ZTZ) GmbH, GTZ, Eschborn, pp. 495–515.

Kogan, M. 1988. Integrated pest management theory and practice. *Entomol. Exp. Appl.* 49: 59–70.

Kraus, N., R. Cramer, M. Bokel and G. Sawitzki. 1981. New insect antifeedants from *Azadirachta indica* and *Melia azedarach. Proc. 1st Int. Neem Conf.*, Dt. Ges, für Techn. Zusammanarbeit (GTZ) GmbH. Rottach-Egern, pp. 53–62.

Leuschner, K. 1972. Effect of an unknown plant substance on a shield bug. *Naturwissenschaftern* 59: 217–218.

Meisner, J., M. Kehat, M.A. Zur and C. Elzick. 1978. Response of *Earias insulana* Bosid. larvae to neem (*Azadirachta indica* A. Juss.) kernel extract. *Phytoparasitica* 6: 85–88.

Mikolajczak, K.L., B.W. Zilkowski and R.J. Bartlet. 1989. Effect of Meliaceous seeds extracts on growth and survival of *Spodoptera frugiperda* (J.E. Smith). *J. Chem. Ecol.* 15: 121–126.

Mitra, C.R. 1963. *Neem.* The Indian Central Oilseeds Committee, Hyderabad, 190 pp.

Morgan, E.D. 1981. Strategy in the isolation of insect control substances from plants. *Proc. 1st Int. Neem Conf.*, Dt. Ges. für Tech. Zusammanar (GTZ) GmbH. Rottach-Egern, pp. 43–52.

Nakanishi, K. 1975. Structure of the insect antifeedant azadirachtin. In: *Recent Advances in Phytochemistry.* V.C. Runeckles (ed.), Plenum, New York, pp. 283–298.

Natarajan, K. and V.T. Sundaramurthy. 1990. Effect of neem oil on cotton whitefly (*Bemesia tabaci*). *Indian Agric. Sci.* 60: 290–291.

Parmar, B.S. and K.P. Srivastava. 1986. Development of some neem formulations and their evaluation for the control of *Spilosoma obliqua* in the laboratory and *Euchrysops cnejus* in the field. *Proc. 3rd Int. Neem Conf.*, Nairobi, Dt. Ges. für Tech. Zusammenarbeit (GTZ) GmbH, Eschborn, pp. 205–215.

Pradhan, S., M.G. Jotwani and B.K. Rai. 1963. The repellent properties of some neem products. *Reg. Res. Bull.* 11: 149–151.

Saxena, R.C., G.P. Waldbauer, N.J. Liquido and B.C. Puma. 1981. Effects of neem seed oil on the rice leaf folder, *Cnaphalocrocis medinalis. Proc. 1st Int. Neem Conf.*, Dt. Ges. für Tech. Zusammenarbeit (GTZ), GmbH, Eschborn. Rottach-Egern, pp. 189–204.

Saxena, R.C. 1987. Antifeedants in tropical pest management. *Insect Sci. Appl.* 8: 731–736.

Schmutterer, H. 1981. Some properties of components of the neem tree (*Azadirachta indica*) and their use in pest control in developing countries. *Meded. Fac. Landbouww. Rijksuniv. Gent.* 46: 39–47.

Schmutterer, H. and H. Rembold. 1980. Effects of some pure fractions of extracts from neem (*Azadirachta indica*) seeds on feeding activity and metamorphosis of *Epilachna varivestis* (Col. Coccinellidae). *Z. Ang. Ent.* 89: 179–188.

Schmutterer, H., R.C. Saxena and J.V.D. Heyde. 1983. Morphogenetic effects of some partially purified fractions and methanolic extracts of neem seeds of *Mythima separata* Walk. and *Cnaphalocrocis medinalis* Guenee. *Z. Ang. Ent.* 95: 230–237.

Schmutterer, H. and K.R.S. Ascher. 1984. Natural pesticides from the neem tree (*Azadirachta indica* A. Juss). *Proc. 1st Int. Neem Conf.*, Dt. Ges für Techn. Zusammenarbeit (GTZ) GmbH, Eschborn, 279 pp.

Schmutterer, H. 1990. Properties and potential of natural pesticides from the neem tree, *Azadirachta indica. Ann. Rev. Entomol.* 35: 271–297.

Seiber, K.P. and H. Rembold. 1983. The effects of azadirachtin on the endocrine control of moulting in *Locusta migratoria. J. Insect Physiol.* 29: 523–527.

Sharma, G.K., Ch. Czoppert and H. Rembold. 1980. Further evidence of growth-disruption by neem seed fractions. *Z. Angew. Ent.* 90: 439–444.

Sharma, R.N., B.A. Nagasampagi, A.S. Bhosale, M.M. Kulkarni and V.B. Tungikar. 1983. 'Neemrich': The concept of enriched fractions from neem for behavioural and physiological control of insects. *Proc. 2nd Int. Neem Conf.*, Rausichholzhausen, Dt. Ges. für Techn. Zusammenarbeit (GTZ) GmbH Eschborn. pp. 115–128.

Siddig, S.A. 1986. A proposed pest management programme including neem treatments for combating potato pests in the Sudan. *Proc. 3rd Int. Neem Conf.*, Nairobi, pp. 449–459.

Tanzubil, P.B. and A.R. McCaffery. 1990. Effects of azadirachtin and aqueous neem seed extracts on survival, growth, and development of the African armyworm, *Spodoptera exempta. Crop Protection* 9: 383–386.

Umeh, E.N. 1988. The effect of methanolic extracts of neem seed kernel on feeding, development, and survival of *Spilosoma maculatus* Walk. (Lep. Arctiidae). *J. Appl. Entomol.* 106: 494–499.

Warthen, J.D., Jr. 1989. Neem (*Azadirachta indica* A. Juss.): Organisms affected and reference list update. *Proc. Entomol. Soc. Wash.* 91: 367–388.

Warthen, J.D., Jr., R.E. Redfern, E.C. Uebel and G.D. Mills, Jr. 1978. An antifeedant for fall armyworm larvae from neem seeds. *U. S. Dept. Agric. Res. Results* ARR-NE-1.

Yamasaki, R.B., J.A. Klocke, S.M. Lee, G.A. Stone and M.V. Darlington. 1986. Isolation and purification of azadirachtin from neem (*Azadirachta indica* A. Juss.) seeds using flash chromatography and high performance liquid chromatography. *J. Chromatog.* 356: 220–226.

99

The NeemAzal Conception: Future Possibilities of the Use of Neem in Biological and Integrated Pest Management

HUBERTUS KLEEBERG

Trifolio-M-GmbH
D-35633 Lahnau, Sonnenstr. 22, Germany

ABSTRACT

The use of neem seed kernel preparations has large potential for biological or integrated pest management. Problems of standardization may be overcome by the use of proper standards. The shelf-life of neem extracts may be increased by a proper product design. The application of products based on neem oil does not look promising. Slow insecticidal effect is the most desirable property in neem preparations whereas repellent effects play a secondary role.

INTRODUCTION

The worldwide search for safer and more specific means of pest control has led to intensive investigations of the possibilities of the use of neem (*Azadirachta indica* A. Juss.) seed kernels (NSK). Since time immemorial the different parts of the neem tree have been efficiently used in India for many purposes. In more industrialized countries, the news about this 'wonder tree' is convincing increasingly more people, and has drawn the attention of pesticide and pharmaceutical companies.

The economic exploitation of the tremendous scope of Neem has already started. In this situation it may be useful to discuss critically the potential of neem products with respect to biologial and integrated pest management (IPM).

The 10 most important properties of future pesticides are:

1) Efficiency
2) Security in the success of application

3) Considerate influence on beneficial insects
4) Minimization of the appearance of resistancies
5) Toxicological safety for the user
6) Toxicological safety for the consumer
7) Environmental soundness (biodegradability, no tendencies to accumulate)
8) Toxicologically and environmentally safe production
9) Inexpensive production
10) Worldwide availability of standard quality

Points 1–7 are taken care of by the registration requirements in many countries. It has been demonstrated and discussed many times that neem fulfils these requirements most satisfactorily.

The last three points are usually not considered as requirements for pesticides. Toxicologically and environmentally safe production should be included in registration considerations, since at least in the event of disasters the society has to carry a large burden. In addition to a new, efficient, and ecologically as well as toxicologically sound insecticide with large market potential we should not neglect the benefits of afforestation with neem and of the creation of new socioeconomic centres in neem-growing areas. It would be too ambitious to include these aspects in this contribution, but the benefits of neem cannot be estimated highly enough and should not be considered only in terms of profitability.

As far as points 1–8 are concerned, present knowledge and technology clearly speak in favour of neem for pest control purposes in comparison to synthetic pesticides.

Presently, the largest problems of processing aspects of neem seem to arise from points 9 and 10. Very clearly, inexpensive production should not be achieved at the cost of the possible socioeconomic improvements. In other words, the people collecting neem seeds or fruit have to get adequate payment and for those working in larger plantations of neem advantageous social structures have to be organized.

PROBLEMS OF STANDARDIZING NEEM PRODUCTS

The possibilities of standardizing neem insecticides are discussed here with respect to their content of azadirachtin A, the most abundant and best studied single chemical compound of NSK with very high insecticidal activity. In Table 1 some analytical results of azadirachtin A standard substances are compared.

The chemical identity of our own internal standard (azadirachtin sup.) has been proved by ^{1}H and ^{13}C NMR. The analysis of the impurities of our azadirachtin A sup. amounts to about 2 per cent.

The error of the usual HPLC procedure amounts to about $\pm$ 5 per cent. Thus, deviations of at the most 10 per cent between analytical results of the same compound of different laboratories may be tolerated. Table 1 indicates that the azadirachtin A standards now available are lacking in quality. The last column indicates that the ratio between the analytically determined purity and the specified purity of the azadirachtin A contents varies between about 0.4 and 1.

Thus, it is not surprising that the azadirachtin A content of different neem products deviates considerably from the stated amount, as shown in Table 2.

Table 1. Azadirachtin A content of different commercial azadirachtin A standards

Source	Specified purity (%)	Determined purity (%)	Determined purity / Specified purity
Internal: supn.	≥ 98	Taken as 98	'1'
Internal: pur.	≥ 95	96 ± 3	1
A	≥ 90	72 ± 4	0.80
B	≥ 96	78 ± 3	0.81
C	≥ 95	39 ± 4	0.411

In addition to the uncertainty arising from several standards, degradation during storage is a problem of many products. Thus the exact determination of the initial azadirachtin A concentration of a product may be possible by the use of a proper standard. However shelf-life is an intrinsic property of each product and can only be increased by a better product design.

We have developed a reasonable compromise of the different aspects of neem products in "NeemAzal". We isolate a concentrate of the insecticidally active neem ingredients by a few simple extractions steps. The concentrate powder form is called 'NeemAzal' and usually contains 30 per cent azadirachtin A and about the same amount of other azadirachtins. The formulated product 'NeemAzal-F' or 'NeemAzal-T' contains 5 per cent azadirachtin A and has a shelf-life of about two years at about 20°C.

In order to quantify the amount of the different ingredients of an extract a simple semi-preparative chromatographic procedure may be applied. The total amount of extract has to be known (1.4 g NeemAzal in the example given in Fig. 1). Then different fractions are collected and evaporated to dryness and the weight of each fraction is determined. According to this procedure 33 per cent azadirachtin (Fraction No. 2; see Fig. 1) has been determined, in excellent agreement with the analytical value of 30 per cent.

COMPARATIVE DISCUSSION OF INSECTICIDAL NEEM PRODUCTS

For a general design of a neem insecticide there are various possibilities (see Fig. 2). The most simple and inexpensive possibility for insect control with neem, probably, is the use of NSK powder (NSKP) or neem seed cake (NSC). In this case rainfall or watering of the plants is a prerequisite in order to dissolve the active ingredients of

Table 2. Azadirachtin A content of some neem products

Source	Specified purity (%)	Determined purity (%)	Determined purity / Specified purity
D	11.36	2.4	0.21
E	0.3	0.23	0.77
F	10	8	0.8
G	3	2.5	0.83
H	1.2	0.3	0.25

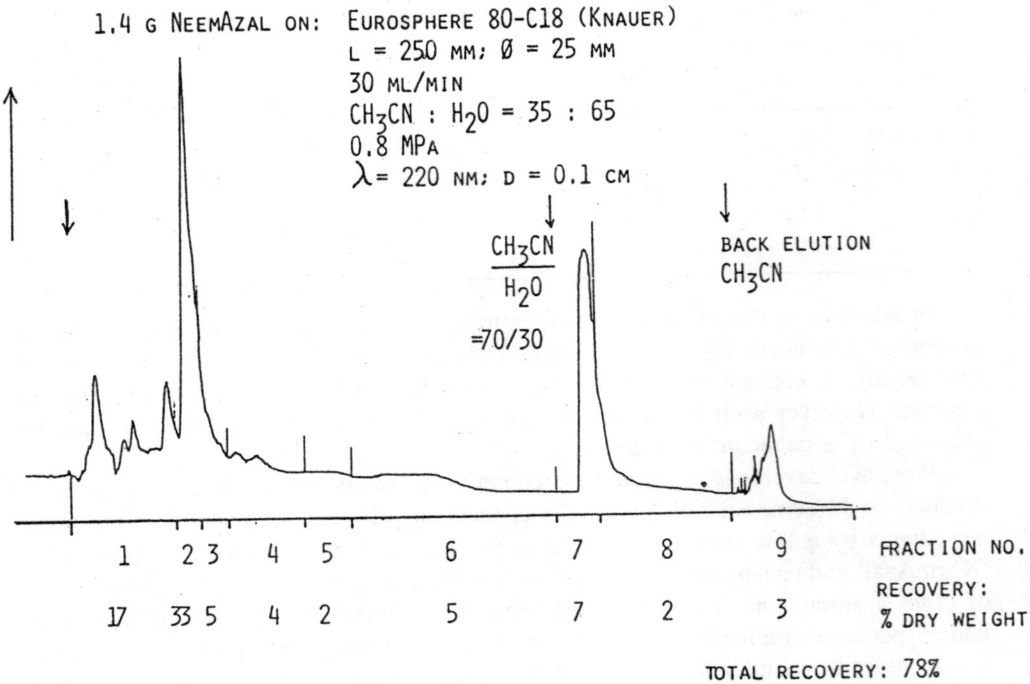

Fig. 1. Elution of neem ingredients from an RP-18 column. Injection is indicated by downward arrows; after fraction 6 the eluent was changed from $CH_3CN:H_2O$ = 35:65 to $CH_3CN:H_2O$ = 70:30. After fraction 8 the column was eluted in the opposite direction using CH_3CN only.

NSKP/NSC and make it available to the plant for systemic transportation inside the plant.

Under proper drying and storage conditions the azadirachtin content of NSK, NSKP, or NSC may decrease only by about 20 per cent per year, less than with many 'formulated' products. The preparation and application of aqueous NSKP/NSC extracts is simple, inexpensive and is successfully used in many countries. However aqueous extracts are not marketable because of the rapid degradation of the active ingredient and growth of microorganisms.

In comparison to commercial formulated products which currently cost US $ 0.5 to 30 per g azadirachtin A in bulk, the aqueous extracts are inexpensive even if the prices of raw material will increase.

For good quality (about 5 g azadirachtin A/kg) NSK about US $ 0.5/kg has to be paid in India at present. For a single treatment about 20 g azadirachtin A is necessary per hectare. Thus, the cost of labour for the preparation (grinding, stirring, filtration) of water extract (WEx) may be US $ 2/ha .

Thus, in a situation where the cost of labour is low, about US $ 5 will have to be paid for a single treatment of one hectare. This makes WEx a very competitive insecticide for all countries which have neem and where the cost of labour is low.

It may not be advisable for neem-growing countries to export the raw material. If they are looking for export, it should be in the form of highly efficient, good quality concentrates or their stable formulations. Processing neem in industrialized countries

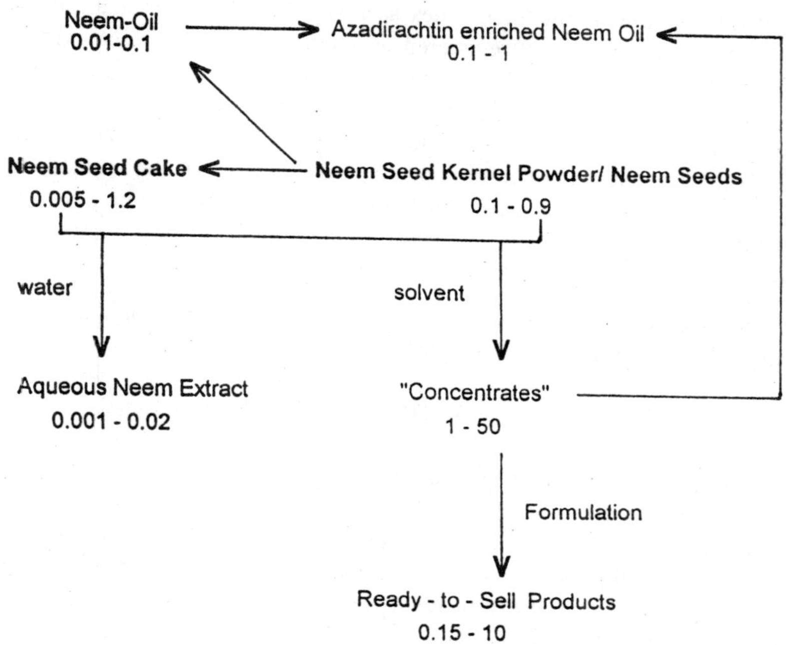

Fig. 2. Schematic representation of the production of different neem products. Ranges of the content of azadirachtin A determined in the various products is indicated in %.

should be politically prohibitive. Thus, not only the high cost of labour but also political reasons oppose the use of WEx in industrialized countries.

Technically, the most attractive possibility for a marketable azadirachtin-containing product is extraction with polar solvents. With few purification steps 'azadirachtin concentrates' containing about 50 per cent azadirachtin A are possible. In our opinion these extracts, if they contain clearly more azadirachtin than NSC or NSK (i.e., above 1 per cent azadirachtin A), are the only reasonable technical neem insecticide which will have a future on the world market. It will probably be better to sell or buy NSC or NSK rather than extracts which contain active ingredients in comparable amounts. The price of the natural product will be lower. It does not contain chemical additives and frequently its shelf-life is even longer.

Taking into account bulk prices for azadirachtin concentrates of about US $ 2/g azadirachtin content, the high expenditure for process development and registration, cost for formulation, a reasonable margin, and about 20 g azadirachtin to be applied for a single treatment of one hectare, the cost of the insecticide for the user will be of the order of US $ 60 to 100 per hectare. In cases this cost will be competitive with modern synthetic insecticides.

Finally, we may consider the market potential of neem oil or neem oil enriched in azadirachtin (see Fig. 2). There are serious limitations for marketing of neem insecticides based on neem oil:

1) It is difficult to enrich the content of active ingredient in neem oil above that in NSK.

2) Neem oil contains about three times as many compounds which have to be standardized, investigated, and registered as concentrates of solvent extracts.
3) Some of the neem oil ingredients possess unwanted properties for insecticidal use (e.g. fungicidal, contraceptive, or toxic properties).
4) Phytotoxic properties of neem oil limit its application to low concentration.
5) Due to the limited concentrations of neem oil which may be applied, the addition of synthetic insecticide is frequently proposed. This is contradictory to the concept of IPM and prevents the use of neem oil for biological pest control.
6) It is difficult to assure the absence of aflatoxins in neem oil preparations.

In view of these arguments against neem oil preparations, the application of the WEx or formulated concentrates of NSK/NSC seems to be the choice for the future.

THE MODE OF ACTION

In order to interpret the efficiency of neem ingredients it is important to know the time dependence of its mode of action.

The second instar larvae of *Croesus septentrionalis* were sprayed with an aqueous solution of a neem preparation containing 2 ppm azadirachtin A in the field. The lower part of Fig. 3 shows that the efficacy (per cent mortality) increased one day after treatment to 25 per cent and gradually reached 100 per cent after 10 days. This time dependence is very typical of the active ingredients on neem.

However, as shown in the upper part of Fig. 3, the larvae did not increase in weight after the second day after treatment. The larvae were obviously still on the leaves; they were still alive, since they reacted to various kinds of stress or irritation, but they did not feed any more, in opposition to the blank treated control group (see Fig. 3, upper part, indicated by x). In tests where the larvae were transferred one day after treatment from treated leaves to fresh leaves they did not take up leaves as the blank treated control did.

These results clearly indicate that neem ingredients act very slowly; for this reason the term “insectistat” (as mentioned by M. Rice) may be better than “insecticide”.

Especially at low concentrations, no ‘repellent effects’ are observed. At least as far as NeemAzal is concerned there is a slow insecticidal effect. The much faster effect of the reduction of food intake and fitness is very important since it may make the larvae an easy prey of beneficial animals.

Whatever the mode of action of neem ingredients may be, it seems advisable rather to record the reduction of the damage of the plants infested by a pest after neem application than to look for mortality, especially in comparative investigations with other insecticides.

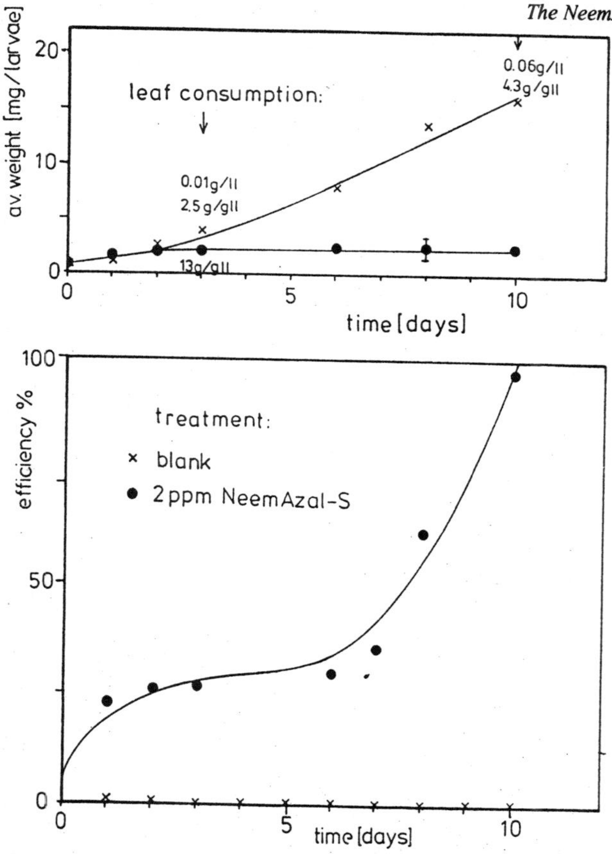

Fig. 3. Mortality (lower part) and change of the average body weight (upper part) of second-instar larvae after treatment with a neem preparation (corresponding to 2 ppm azadirachtin A). The larvae fed throughout the experiment on the treated leaves.

100

The Use of Margosan-O® in the Greenhouse Environment

J.F. WALTER[1] AND J.F. KNAUSS

[1]W.R. Grace & Co. -Conn.
Washington Research Center
7379 Route 32
Columbia, Maryland 21044, USA

ABSTRACT

The neem-based insecticide Margosan-O® was introduced into the horticultural market in 1990 by the Grace-Sierra Company. Because of the unique demands of this market, intensive insect pressure, high premium placed on plant quality, and a sensitivity to worker exposure, Margosan-O® has been readily accepted by this market and is used extensively for the control of whiteflies, aphids and thrips.

INTRODUCTION

Today, more than ever before, the floricultural and horticultural industries face a potential crisis in pest management. Two of the industry's most prominent pest management specialists, Lindquist (1992) and Parella (1990, 1992), indicate that the following primary factors influencing this potential crisis are:

1) The fact that current regulatory requirements, especially those for conventional plant protection products are becoming more stringent.
2) The fact that few new active ingredients are being registered.
3) The fact that greenhouse insects are rapidly developing resistance to conventional pest management products.
4) The fact that legislation governing worker safety, health regulations, and pollution affect this industry more strongly.

These factors highlight the uncertainty on how pest management will be handled in the future. Especially important is the phenomenon of insect resistance to pesticides. Although not always agreeing tactically, entomologists agree that alternation of chemical pesticide classes is a main step needed to manage this problem. Current pesticide classes are pyrethroids, organophosphates and carbamates, macrocyclic lactones, insect growth regulators, horticulture soaps, and horticultural oils.

Entomologists are vigorously stressing integrated pest management (IPM) principles, the need for new pesticide chemistry, and the movement towards people- and environment-friendly pesticide products.

The principles inherent in IPM should form the basis for all new pesticide development, whether the product be a novel conventional or a novel natural or biological one.

THE NEEM TREE

Background and Distribution

Currently, no other member of the green plant world holds greater promise as a natural source for pest management products than the neem or margosa tree, *Azadirachta indica* (Grainge and Ahmed, 1988). This fast-growing evergreen tree originated either in the Indo-Pakistan region (Schmutterer, 1990) or in the dry scrub forest in Burma and on the Siwalik hills (Jacobson, 1988). It grows only in frost-free areas, to heights of 30 m, and thrives in tropical regions having 400 to 800 mm of yearly rainfall in conjunction with an extended dry season. It is found in abundance in Africa, India and Pakistan. The tree is also found in greater or lesser numbers in Antigua, Argentina, Australia, Barbados, Belez, Bolivia, Brazil, Cuba, Dominican Republic, Ecuador, Fiji, Guana, Guatemala, Guyana, Haiti, Hawaii, Honduras, Indonesia, Jamaica, Kenya, Malaysia, Mexico, Nicaragua, Puerto Rico, Senegal, Sri Lanka, St. Lucia, Sudan, Suriname, The Philippines, Thailand, Tobago, Trinidad, The United States of America and Yemen (Jacobson, 1988).

Neem's Active Ingredient and Modes of Activity

The natural compounds responsible for the biological activities of neem were discovered in 1968 (Isman et al., 1990). Since then, azadirachtin has proven unquestionably to be neem's main active ingredient and is primarily responsible for its pest managemnet activities (Schmutterer, 1990; Vietmeyer, 1992). Azadirachtin is most concentrated in the neem fruit with normal yields in recently depulped and dried seed averaging 2 to 4 mg/g (Vietmeyer, 1992).

Azadirachtin, in crude neem seed extracts, has shown many modes of activity against pests. Prominent among these are insect growth regulation (IGR), repellence, antifeedant activity, anti-egg-laying, and antifecundity. It is the IGR activity of Margosan O* botanical insecticide that will be more thoroughly discussed in this paper.

Azadirachtin belongs to a general class of compounds termed "triterpenes", more specifically liminoids (Vietmeyer, 1992). Research, over the past 20 years, has indicated azadirachtin to be one of the most potent insect growth regulators and antifeedants ever assayed. It is very similar in chemical structure to ecdysone, the primary hormone that regulates insect moulting. Azadirachtin acts as an ecdysone blocker and inhibits ecdysone production and release, normal release being vital to insect moulting.

Azadirachtin's modes of action, both IGR and insect antifeedant activity, have been found to play a major role in the Margosan-O® botanical insecticide's activity. The IGR activity is prominent at the earliest larval instar through the developing pupal stages. This insecticide has no significant IGR effects on the insect eggs or adult stages. This fact is clearly demonstrated in data taken from the studies of Price and Schuster (1990) and illustrated in Table 1. These IGR activities, to be discussed further later, are particularly profound on insects belonging to the Lepidoptera, Coleoptera and some Homoptera. The relatively large spectrum of efficacy is a major asset of Margosan-O®.

Past research studies on neem can be found in over 1,300 research articles, the proceedings of three International Congress (Vietmeyer, 1992) and two reviews (Olkowski, 1987; Warthen, 1989).

MARGOSAN-O® BOTANICAL INSECTICIDE DEVELOPMENT

The early research culminating in the development of Margosan-O® is succinctly presented by Menn (1990). Active research, under the direction of Dr. M. Jacobson, was started at USDA-Beltsville's Biologically Active Natural Products Laboratory (now The Insect Chemical Ecology Laboratory) in 1975. This laboratory coordinated neem research for 15 other ARS coooperator laboratories. Research encompassed extract eficacy, insect host range, and the HPLC analysis, pioneering efforts so important to the developments to follow, particularly Margosan-O® registration and patenting in 1985 by R. Larson.

The early version of Margosan-O® had problems in shelf-life and product consistency that caused Larson to seek commercial help. This led to W.R. Grace & Co.'s purchase of the technology.

W.R. Grace Neem Development Objectives

Researchers at Grace's Washington Research Center in Columbia, Marlyland, undertook the task of developing the shelf-stable, reproducible product necessary for commercial-

Table 1. Sweet potato whitefly mortality on poinstettia leaves after one spray of Margosan-O® to the specific life stages. Data are averaged for four replications per treatment with 25 life forms for each replicate (J. Price, University of Florida)

Treatment	% mortality				
	Conc/100 gal	Eggs	1st instars	Pupae	Adults
Control		0	6	8	4
Talstar®	1 lb	0	98	12	95
Margosan-O®	5 pt[1]	0	96	74	8

[1] 5 pt/100 = 18.7 ppm azadirachtin.
Sprays were applied to poinsettia leaves and allowed to dry, then plants were introduced into cages containing whitefly.
Talstar® is the treatment of the FMC Agric. Chemicals Co. for bifenthirin.

ization. It was known from previous literature that azadirachtin was a relatively sensitive molecule that can easily degrade. Stokes and Redfern et al. (1982) showed its sensitivity to UV light. Stability was also affected by solution pH and the concentration of azadirachtin in solution (Isman et al., 1990). Furthermore, Schumutterer (1990), noted that azadirachtin showed varying degrees of sensitivity to temperature, UV light, pH, rainfall, and other environmental factors when applied to plant tissue. The Grace researchers recognized that besides product shelf life, reproductibility, and efficacy, the improved product must also be non-phytotoxic (non-injurious to plant tissue), especially for horticultural and floricultural applications.

A research team comprising an organic chemist, an analytical chemist, and a chemical engineer, formed in the subsequent nine months, identified and solved the critical parameters that needed resolution. In 1989, improved Margosan-O® insecticide was put into research and commercial cooperator hands for greenhouse and field evaluations.

1989 to 1990 Research and Grower Cooperator Trials

The initial research trials involved efficacy and phytotoxicity evaluations of four separate formulations to control the sweet potato whitefly, *Bemisia tabaci*, and/or the common greenhouse whitefly, *Trialeurodes vaporariorum*. The studies, perfomed mostly on poinsettias, resulted in selection of a formula for further evaluation on poinsettias by commercial growers at 16 different sites in Florida, Michigan, Missouri, and Ohio.

Based on the poinsettia results, a formulation was selected and subsequently evaluated under commercial conditions on a wide variety of bedding plants in the four states mentioned. These grower trials showed the selected formulation to be efficacious and non-phytotoxic, allowing the official launch of Margosan-O® by Grace-Sierra marketing in the fall of 1990.

Neem's Activity Spectrum

Crude neem extracts containing active azadirachtin have shown a medium to broad efficacy spectrum against insects in the order Orthoptera, Heteroptera, Homoptera, Coleoptera, Lepidoptera, and Diptera (Schmutterer, 1990). Adult insect stages within these orders are normally not killed by neem's active components. However, adults in Heteroptera, Homoptera, Orthoptera and Coleoptera may often be strongly influenced, showing significant reduction in fecundity or even sterility at higher doses. It was this known spectrum of activity that stimulated the commercial development of Margosan-O®.

Insect Management with Margosan-O®

Leafminers (Diptera)

The earliest studies on crude neem extracts and the early version of Margosan-O® were done on the leafminers *Liriomyza trifolii* and *L. sativae* (Larew et al., 1984, 1985; Lindquist et al., 1986; Webb et al., 1983). The former pest attacks numerous floricultural crops, particularly chrysanthemum, zinnia, and dahlia. In tests involving drench application (solution containing 9.9 ppm azadirachtin) to chrysanthemums, Margosan-O® proved non-phytotoxic and very effective in controlling emergence of the leafminer adult winged state (knodel et al., 1985, 1988). Although *L. trifolii* was controlled by drench applications to infested chrysanthemums, similar application to infested marigolds, zinnia, and snapdragon was ineffective. Since the spray application to marigolds and zinnia was effective, drench effectiveness appears tobe host dependent and possibly restricted to very few plant types.

Another leafminer affected by neem-based products was the birch leafminer, *Fenusa pusilla*. Both spray (Larew et al., 1987) and trunk microinjection (Marion et al., 1990) were effective on this pest.

More recent studies with the imroved Margosan-O® on *L. trifolii* showed that two foliar applications, 5 pt/100 gal, seven days apart, reduced the number of developing pupape by 35 per cent and emerging adults by 82 per cent (Lindquist and Casey, 1990b). Oetting, University of Georgia (unpublished), showed that this same formulation and concentration applied once to potted chrysanthemums infected with *L. trifolii* reduced recovered pupae by 59 per cent and completely eliminated emergence of adults from the recovered pupae.

In a recent study, the Grace formulation applied in a shipping carton to the growing media surrounding the roots of leafminer-infested chrysanthemum cuttings produced an 80.6 per cent reduction in emerging adults after subsequent unpacking and planting (Sanderson et al., 1989).

Lepidopterous Insects (Lepidoptera)

In general, lepidopterous insects have shown a remarkable sensitivity to neem-based products (Schmutterer, 1990; Vietmeyer, 1992) both by direct wetting of the larval form and by consumption of sprayed plant parts.

Work with the pre-Grace Margosan-O® showed no activity on the beet armyworm, *Spodoptera exigua*, where neem seed extract proved effective (Harris et al., 1986). Recent studies (unpublished) by Ali in Florida with the new Margosan-O® revealed significant mortality to instars of *S. exigua* (Table 2). This difference in activity is not surprising in light of the known instability of the earlier Margosan-O®.

Another lepidopterous pest controlled by the current Margosan-O® formulation is the mathogany webworm, *Macalla thyrsisalis*. Applications by Howard (1990) in Florida

Table 2. Mortality of beet armyworm first instars exposed to Margosan-O® sprayed and dried on chrysanthemum leaves (A.D. Ali Yoder Brothers, Florida)

Treatment	% mortality			
	Conc/100 gal	24 HAT	48 HAT	168 HAT
Control		0 b	2.5 b	20.0 b
Margosan-O®	1.5 pt	12.5 b	15.0 b	37.5 b
Margosan-O®	2.5 pt	12.5 b	15.0 b	45.0 b
Margosan-O®	4.0 pt	68.8 a	79.4 a	92.5 a
Margosan-O®	5.0 pt	52.5 a	77.5 a	90.0 a

HAT = Hours after application

to the foliage of the West Indies mahogany, *Swietenia thyrsisalis*, decreased young instar growth by 84.5 per cent and the mean number of live instars per tree by 94.2 per cent.

Whiteflies (Homoptera)

The insect group specifically targeted for initial evaluation with early Grace Margosan-O® formulations was the whitefly, specifically, the sweet potato whitefly and the common greenhouse whitefly. The key researchers cooperation on this project were J. Baker, North Carolina State University; R. Lindquist, Ohio State University; L. Osborne, University of Florida; M. Parella, University of California; J. Price, University of Florida; and J. Sanderson, Cornell University. Evaluations by these researchers (Table 3) led to the selection of the formulation to commercialize.

The effect of the current Margosan-O® on adult whitefly emergence has additionally been emphasized by Lindquist in studies where the labelled concentration applied to the greenhouse whitefly showed no emergence compared to 75 per cent to 85 per cent in the untreated controls (Lindquist and Casey, 1990b; Lindquist et al., 1990).

The effect upon the sweet potato whitefly's adult emergence showed a definite dose-related effect under Florida conditions in sprays to heavily infested salvia (Table 4).

Sanderson, in a trial on first and second instar and fourth instar stages of the sweet potato whitefly, demonstrated that Margosan-O®'s activity was comparable to most pest management products in current use for whitefly control on poinsettians (Table 5).

Table 3. Summary of results obtained with Margosan-O® selected for commercialization. Application concentration, 5 pt/100 gal

Researcher	No. spray	% instar mortality		% adult emerged	
	appl.[1]	Control	Margosan-O®	Control	Margosan-O®
J. Price	1	6.3	61.0	NE	NE
M. Parella	1	7.6	69.7	NE	NE
J. Sanderson	1	12.3	98.9	NE	NE
J. Baker	7	0.0	98.7	NE	NE
L. Osborne	5	0.0	77.5	100	6.4
R. Lindquist	3	0.0	66.5	100	0.7

[1]Repeated on seven-day intervals.
NE = Not evaluated.

Table 4. Effect of Margosan-O® sprays salvia upon sweet potato whitefly instars and subsequent adult emergence after one, two and three consecutive applications at seven-day intervals (L. Osborne, University of Florida)

Treatment	Conc/100 gal	1st Application		2nd Application		3rd Application	
		Live instars	Adults emerged	Live instars	Adults emerged	Live instars	Adults emerged
Control		173.4	45	120.4	45.0	45.0	124.6
Margosan-O®	2.5 pt	116.6	25	45.6	25.6	8.6	39.6
Margosan-O®	5.0 pt	180.6	24	11.6	11.6	9.2	14.2
Margosan-O®	10.0 pt	119.6	17	5.2	5.2	6.2	3.0

Data taken on average per plant at appropriate interval after spraying.

Thrips (Thysanoptera)

Activity against thrips is included by Lindquist and Casey (1990b) in their update on Margosan-O®. In their published studies on western flower thrips, *Frankleiniella occidentals* (Lindquist and Casey, 1990a), they showed that three sprays of Margosan-O®, at seven-day intervals, significantly reduced instar numbers compared to the controls. They also showed that a tank mix of Margosan-O® with Avid (abamectin) showed the greatest thrip reduction (83.2 per cent) among the tank mixes tried. In another unpublished trial, Lindquist found that three applications of Margosan-O® provided control of *F. occidentals* equal to Avid (Table 6).

Baker (unpublished), in North Carolina, showed that Margosan-O®, in applications to chrysanthemum flowers was equal in activity to Avid®, Carsol® (formetanate hydorchloride) and Dursban® (chlorpyrifos) when applied twice, at a seven-day interval (Table 7). Oetting, in Georgia (unpublished), recently showed that spray, and possibly drench application, significantly reduces the number of emerging western flower thrip adults from infested chrysanthemums. One foliar application to leaves infested with second-stage instars reduced emerging adults and live pupal stages by 66 per cent and 53 per cent, respectively. In another study, Oetting showed that placement of second-stage instars onto growing media drenched with Margosan-O® resulted in 89 per cent mortality.

Table 5. Comparison of activity against sweet potato of Margosan-O® to other pesticides. One spray application (J. Sanderson, Cornell University)

Treatment	(% mortality)		
	Conc	1st/2nd instar	4th instar
Control		12.7 e	2.3 e
Ultra-fine Spray Oil®	1: 50	100.0 a	98.0 a
Margosan-O®	1: 150	99.0 a	54.0 b
Avid®	1: 2000	98.0 a	51.0 b
Safer's Soap®	1: 100	78.0 bc	98.0 a
Tempo®	1: 8000	68.0 cd	3.0 e
Mavrik®	1: ·8000	83.0 bc	9.3 d

Ultra-fine Spray Oil®, Avid, Safer's Soap®, Tempo® and Mavrik® are the trademarks of the Sun Oil Company, Merck and Co., Ringer Corp., Sanex, Inc., and Sandoz Crop Protection Corp. for the common chemical compounds horticultural spray oil, abamectin, potassium salts of fatty acids, temephos, and fluvalinate, respectively.

Table 6. Western flower thrips management using repeated spray applications of Margosan-O® to chrysanthemum foliage (R. Lindquist, Ohio State University)

Treatment	Conc/gal	Average no. live thrips/leaf
Control		10.6 b
Margosan-O®	25.0 ml	0.4 a
Avid®	2.4 ml	0.0 a

Table 7. The effect of two Margosan-O® applications, seven days apart, to chrysanthemum flowers upon *F. occidentals* (J. Baker, North Caroline State University)

Treatment	Conc/100 gal	Average No. thrips/3 flowers	
		Pretreated	9 days after treatment
Control		27.1	173.4 a
Avid®	8 fl. oz.	9.0	15.4 b
Carzol	8 fl. oz.	19.4	41.7 b
Durshan®	64 fl. oz.	20.9	48.5 b
Margosan-O®	80 fl. oz.	11.1	73.5 b

Avid®, Carsol® and Dursban® are the trademark of Merck and Co., NOR-AM Chemical Co., and Dow Elanco for the common chemical compounds abamectic, formetanate hydrochloride, and chlorpyrifos, respectively.

Table 8. The influence of a single Margosan-O® spray on *E. americanus* infesting impatients (R. Oetting, University of Georgia)

Treatment	Conc/gal	Pretreatment number	3 DAT	7 DAT	21 DAT
Control		1.0	3.8	12.5	1.9
Margosan-O®	25 ml	2.2	0.5	1.7	0.3

DAT = Days after treatment.

In a trial on other thrips, Oetting (unpublished) showed that a single foliar appliaction of Margosan-O® to *Echinothrips americanus*-infested impatients significantly reduced the insect's population (Table 8).

MARGOSAN-O® APPLICATION

Application Technologies

Margosan-O® can be applied as a foliar spray in greenhouses and oher enclosed growing structures. It currently has limited use as a dip and a drench (Lindquist, 1988; Schmutterer, 1990). In application outdoors, it appears to have limited persistence and is affected adversely by high temperatures, UV light, pH of plant parts, and rainfall (Schmutterer, 1990). In outdoor applications, the product activity approximately has a maximum persistence of five to nine days.

The equipment used for foliar application within greenhouses has been actively studied by Lindquist (1988, 1991), and Lindquist and Powell (1990). Currently, this equipment consists of diverse variety of low concentration, high volume equipment that deposit approximately 2 to 6 per cent of the pesticide sprayed upon the target plant. More recently, new innovations in pesticide application equipment have swung towards the higher concentration, low and ultra low volume application devices. These are typified by air-assisted mist blowers, thermal pulse-jet foggers, cold foggers, rotary atomizers and air-assisted electrostatic sprayers (Lindquist, 1988, 1991; Lindquist and Powell, 1990).

Mathews (1971) has listed the following droplet ranges as optimum for effective application to the following targets: flying insects, 10 to 50 μ; insects on foliage, 30 to 50 μ; foliage coverage, 40 to 100 μ; and soil application, 250 to 500 μ.

Specific Application Studies

Lindquist (1991) evaluated Margosan-O®, at the same per acre, against the sweet potato whitefly on poinsettia using a high volume sprayer, an air-assisted electrostatic sprayer, an aerosol generator, and a thermal pulse-jet fogger. Two applications at a seven-day interval were made and initiated at the insect's first instar stage. Data were taken later on adult emergence. No significant difference from the water-sprayed control was noted for the thermal pulse-jet fogger and the aerosol generator. The air-assisted electrostatic spray treatment produced no adult emergence at all. These results clearly showed how important deposit of Margosan-O® is on the leaf undersurface, where the targeted insect instar resides.

Although this preceding trial indicated high volume application could be the best choice for whitefly control, the fact remains that in 1991 to 1992 commercial practice, any growers applying Margosan-O® using the other application devices mentioned have been able to achieve acceptable to excellent whitefly control. High greenhouse humidity and mechanically produced air movement within the greenhouse apparently move the spray droplets to leaf undersurfaces and maintain a film of moisture on the insects, thus providing greater activity through these low and ultra low-volume applications.

Drench applications of Margosan-O® have been proved active in leafminer control in chrysanthemums (Jacobson, 1988). Research indicated, however, that comparable spray applications produce a quicker instar kill and are as effective in adult emergence suppression as long as spray coverage is uniform over the infested foliage.

Price and Shuster (1990), of the University of Florida, demonstrated that drench applications of Margosan-O® at 20 ppm azadirachtin to poinsettia produced no significant control of sweet potato whitefly. It appears from the research to date that Margosan-O® has shown activity as a drench only against leafminers and then only on a few select plant types.

Dip application of Margosan-O® should help control whitefly-infested plant tissue. Lindquist (unpublished) showed that a 1:150 solution dip of salvia leaves infested with third-stage instars of the greenhouse whitefly produced 84.8 per cent mortality compared to 21.1 per cent for the water dip control. The results were statistically equal to Avid® and insecticidal soap dips and superior to results of leaves dipped in Talstar (bifenthrin), Morestan (exythioquinox), Danitol (fenproputhrin), Danitol + Orthene, and Orthene (acephate). This indicated that Margosan-O® dips hold promise, but past research on dips with conventional chemicals indicate this method can be more potentially phytotoxic than the comparable spray application.

MARGOSAN-O® SAFETY

Early in W.R. Grace's evaluation of the original Margosan-O® formula, it was found this formula produced phytotoxic reactions on some foliage, and especially on a variety of flowers grown in greenhouse. Grace's previously mentioned product development work solved this problem and resulted in the patented and EPA-registered product currently being sold for use on floricultural and ornamental plants in and around greenhouses, commercial nurseries, interiorscapes, and homes. The list of plants tried that showed non-phytotoxicity is too extensive to include in this article, but covers cut and potted flowering, tropical foliage, flowering bedding and nursery and vegetable transplants. Two flowers that show slight injury from Margosan-O® sprays are orchids and hibiscus. Some delicate ferns and succulents have also exhibited slight plant injury.

The few phytotoxicity complaints that did occur in 1991, when examined, invariably resulted from improper use in application or from the unlabelled mixing of Margosan-O® with other plant protection products or adjuvants. Growers can legally mix products together as long as it is at the labelled concentration for a labelled use, but it is not recommended and the grower assumes all risk for unlabelled tank mix applications.

MARGOSAN-O® INSECTICIDE IPM COMPATIBILITY

Neem IPM Compatibility

Neem extracts are stated by Schmutterer (1990) to have relatively weak insecticidal or acute effects on adult insects. In fact, due to azadirachtin's special modes of action, neem extracts have normally proven non-harmful, or only slightly harmful, to beneficial insects (predators and parasitoids). In addition, neem leaves or seed kernels incorporated in potting soil showed no ill effects on earthworms. In actuality, the number of young worms in the neem treatments increased over the control (Vietmeyer, 1992). Outdoor studies also showed no harmful effects on the worm populations.

Vietmeyer (1992) states neem to be remarkably benign to spiders, butterflies, bees, ladybugs and beneficial parasitic wasps. Mention is also made of an unpublished study in which collections of aphid enemies were made at seven different sites where neem was in trial. This study concluded there were no detrimental effects on either predators (coccinellids, chrysopids, and syrphids or paraditoids (ichenumonids, braconids). An interesting observation made in these studies was that more parasitoids and predators were found in the neem plots than in either the control or the pyrethrum-sprayed plots.

MARGOSAN-O® INSECTICIDE EFFECTS ON BENEFICIALS

It has been found that Margosan-O® shows no effects on worker honeybees through direct contact sprays up to 4.418 ppm azadirachtin (Schmutterer, 1990).

Sanderson (unpublished) at Cornell University found Margosan-O® to have no significant effects on the predatory mite, *Phytoseiulus persimilis*, when tested at 2.5 and 5.0 pt/100 gal concentrations. Neither mite survival or fecundity of female mites was affected.

The most extensive studies done on beneficials with Margosan-O® have been carried out in Florida and Hawaii. Specifically, trials were performed on predators and parasitoids of the cotton (melon) aphid and sweet whitefly in the greenhouse and laboratory by Hoelmer et al. (1990) and on parasitoids of fruit flies by Stark et al. (1990a, b).

The results in both studies showed no significant effects. Particularly significant in Hoelmer's work was the fact that parasitism, by *Eretmocerus californicus* and *Encarsia transvena*, in the greenhouse trials involving plants infested with sweet potato whitefly showed that even at up to four application at 20 ppm, azadirachtin had no effect on parasitism compared to the control. Also, parasitism in the Margosan-O® and control treatments was very prevalent while little was seen where conventional pesticides were used. Other beneficials evaluated by Hoelmer and found unharmed by Margosan-O® included the following: *Delphastus pusillus*, *Scymnus* sp., *Lysiphlebus testaceipes*, *Aphelinus asychis*, and *Encarsia formosa*. Furthermore, Stark and co-workers found azadirachtin-treated oriented fruit flies that are parasitized by *Psytallia incisi* and *Diachasmimorpha longicaudata* have little effect on the development of the parasitoids. Their subsequent unpublished studies showed exposure to elevated levels of azadirachtin contained in Margosan-O® applications did not affect reproduction rates or eclosion of the fruit fly parasitoids.

ENVIRONMENT FATE AND MECHANISM OF ACTION

One of the environmetal disadvantages and application problems associated with azadirachtin is its lack of persistence in the environment. Studies by Stokes and Redfern (1982) and Ley et al. (1989), show that azadirachtin is degraded by both UV light and acidic conditions. Furthermore, W.R. Grace associates (unpublished) have shown that azadirachtin is very sensitive to water hydrolysis. The instability of the molecule leads to short persistence in the environment. Studies show that azadirachtin readily degrades in water and has a half-life of less than one week under ambient conditions. Similar results are found in other moist environments such as in soils and on plant surfaces. Insect control studies confirm that the insecticidal activity of neem compounds rarely persists beyond two weeks.

This degradation has led to an understanding of the important structural functions of azadirachtin. Studies by Simmond et al. (1990) show that modifications to the azadirachtin structure, in general, reduce its IGR activity. Subsitution at Cl and C3 of the decalin ring significantly reduces IGR potency of the molecule, as do substitutes to C-22, 23 Ley et al. (1989) also showed that substitution at C-22, 23 caused an additional drop in the antifeedant effect of the molecule. However, modification at Cl and C3 had little effect on antifeedant effect.

Similarly, disruption of the epoxide growitn between C-13 and 14 results in a significant loss in both antifeedant and IGR activity.

It is apparent that the decalin ring structure is important to both IGR and antifeedant activities.

CONCLUSION

Neem-based insecticides, specifically those containing the compound azadirachtin, offer an ecologically minded approach to the control of insect pests. Over the last two decades, research and development of neem products have conclusively shown their value. The use of neem products to control major horticultural, agricultural and domestic pest is slowly becoming a reality.

REFERENCES

Grainge, M. and S. Ahmed. 1988. *Handbook of Plants With Pest Control Properties*. John Wiley and Sons, New York, 470 pp.

Harris, M., D. Warkentin and J. Begley. 1986. *Insecticide and Acaricide Tests* 42: 331.

Hoelmer, K.A., L.S. Osborne and R.K. Yokomi. 1990. Effects of neem extracts on beneficial insects in greenhouse culture. *Proc. USDA Neem Workshop*. ARS-86: 100–105.

Howard, F.W. 1990. Population suppression of mahogany webworm *Macalla thyrsisalis* (Lepidoptera: Pyralidae) with natural products. *Fla. Entomol*. 73(2): 225–229.

Isman, M.B., O. Kovl, D.T. Lowery, J.T. Arnason, D. Gagnor, J.G. Stewart and G.S. Salloom. 1990. Development of a neem-based insecticide in Canada. *Proc. USDA Neem Workshop* ARS-86: 32–39.

Jacobson, M. 1988. Phytochemical pesticides. *The Neem Tree*. CRC Press, Boca Raton, 178 pp.

Knodel, J.J., H.G. Larew and R.E Webb. 1988. Margosan-O, a commercial formulation of neem seed extract, control *Lirimyza trifolli* on chrysanthemums. *J. Agric. Entomol*. 3: 249–254.

Knodel, J.J. and H.G. Larew. 1985. Control of leafminer using neem (Azadirachtin induce A. Juss) products and other insecticides as foliar sprays. *Insecticides and Acaricide Tests*. 10: 282.

Larew, H.G., J.J. Knodel-Muntz, R.E. Webb and J.D. Warther. 1995. *Liriomyza trifolii* (Burgress) (Diptera: Agromyzidae) control on chrysanthemum by neem seed extract applied to soil. *J. Econ. Entamol*. 78: 80–84.

Larew, H.G., J.J. Knodel and D.F. Marion. 1987. Use of foliar-applied neem (*Azadirachta indica* A. Juss.) seed extract for the control of the birch leafminer, *Fenusa pusilla* (Lepeletier) (Hymenoptera: Teuthredinidae). *J. Environ. Hort*. 5(1): 17–19.

Larew. H.G., R.E. Webb and J.D. Warthen. 1984. Leafminer controlled on chrysanthemum by neem seed extract applied to potting soil. *Proc. 4th Conference on the Leafminer*, Society of American Florists, pp. 108–117.

Ley, S.V., D. Santafianos, W.M. Blarey and M.S.J. Simmonds. 1984. Synthesis of a hydroxydihydrofuranacetal related to azadirachtin; A potent insect antifeedant. *Tetrahedron* 45(16): 5175–5192.

Lindquist, R.K. 1985. Whitefly control and pesticide application methods. *Proc. 4th Conference on Insect and Disease Management on Ornamentals*. Society of American Florists, pp. 172–173.

Lindquist, R.K. 1991. Best pesticide application method depends on pests and chemicals. *Greenhouse Manager* 10(7): 89–90.

Lindquist, R.K. 1992. Ohio State Flower Growers Hotline, (1992) 7: 1–2.

Lindquist, R.K. and M.L. Casey. 1990a. *Ohio State Florist Assoc. Bull*. 727: 3–5.

Lindquist, R.K. and M.L. Casey. 1990b. Marogsan-O®—A new botanical insecticide. *Ohio State Florist Assoc. Bull*. 728: 1–3.

Lindquist, R.K. and C.C. Powell. 1990. *Proc. 6th Conference on Insect and Disease Management on Ornamentals*, Society of American Florists, pp. 98–117.

Lindquist, R.K., A.J. Adams, F.R. Hall and I.H.H. Adams. 1990. Laboratory and greenhouse evaluations of Margosan-O against bifeuthrium-resistant and susceptible greenhous whiteflies, *Trialeurodes vaporariorum* (Homoptera: Aleyrodiadae). *Proc. USDA Neem Workshop*. ARS-86: 91–99.

Lindquist, R,K., H.R. Krueger and M.L. Casey. 1986. Use of neem extract as systematic treatment for *Liriomyza trofolii* control on greenhouse chrysanthemum. *Brit. Crop Protection Conference. Pest and Diseases*. pp. 271–276.

Mathews, G.A. 1979. *Pesticide Application Methods*. Longman, London and New York, 336 pp.

Marion, D.F., H.G. Larew, J.J. Knodel and W. Natli. 1990. Systemic activity of neem extract against the birch learminer. *J. Arboric*. 16(1): 12–16.

Menn, J. 1990. Proc. USDA interest in neem research: Past and present. *USDA Neem Workshop*. ARS-86: 1–3.

Olkowski, W. 1987. *The IPM Practitioner* 9(10): 1–8.

Parella, M. 1990. *Proc. 6th Conference on Insect and Disease Management on Ornamentals*. Society of American Florists, pp. 150–155.

Parella, M. 1992. The future of pesticides in floriculture. *Proc. 8th Conference on Insect and Disease Management on Ornamentals*. Society of American Florists, pp. 1–11.

Price, J.F. and D.J. Schuster. 1990. Responses of sweet potato whitefly to azadirachtin extracted from neem tree seeds (*Azadirachta indica*). *Proc*. USDA *Neem Workshop*, ARS-86: 85–90.

Sanderson, K.C., R.D. Oetting and D.A. Smith. 1989. In-transit neem insecticide treatment of rooted chrysanthemum cuttings controls leafminers. *Hort. Sci*. 24(5): 856.

Schmutterer, H. 1990. Properties and potential of natural pesticides from the neem tree, *Azadirachta indica*. *Ann. Rev. Entomol*. 35: 271–297.

Simmond, M.S., W.M. Blaney, S.V. Ley, J.C. Anderson and P.L. Toogood. 1990. Azadirachtin: structural requirements for reducing growth and increasing mortality in lepidopterous larvae. *Entomol. Exp. Appl.* 55: 169–181.

Stark, J.D., R.I. Vargas and T.Y. Wang. 1990a. Effects of neem tree extracts on tephritid fruit flies (Diptera: Tephritidae) and their parasitoids in Hawaii. *Proc. USDA Neem Workshop* ARS-86: 106–111.

Stark, J.D., R.I. Vargas and R.K. Tyalman. 1990b. Azadirachtin effects on metamorphosis, longevity, and production of three tephritid fruit fly species (Diptera: Tephritidae) *J. Econ. Entomol.* 83(6): 2168–2173.

Stokes, J.B. and R.E. Redfern. 1982. Effect of sunlight on azadirachtin: Antifeeding potency. *J. Environ. Sci. and Health*, Health A 17(1): 57.

Vietmeyer, N.D. 1992. *A Tree for Solving Global Problems*. National Research Council—Neem Nat. Acad. Press, Washington D.C., 141 pp.

Warthen, J.D., Jr. 1989. Neem (*Azadirachta indica* A. Juss.): Organisms affected and references list update. *Proc. Entomol. Soc. Wash.* 91(3): 367–388.

Webb, R.E., M.A. Hinegaugh, R.K. Lindquist and M. Jacobson. 1983. Evaluation of aqueous solution of neem seed extract against *Liriomyza sativae* and *L. trifolii* (Diptera: Agromydidae) *J. Econ. Entomol.* 6(2): 357–362.

101

Efficacy Profile of a Commercial Neem Insecticide

JOHN IMMARAJU, TIMOTHY WOOD, REESE NELSON, WILLIAM RUGGERO AND SAMUEL WELLS

AgriDyne Technologies Inc., 2401 South Foothill Dr.
Salt Lake City, Utah 84109, USA

ABSTRACT

AgriDyne Tech ologies Inc. is introducing a family of neembased insecticides to hotricultural, agricultural, and turf care markets. One of the products, Azatin EC, is an emulsifiable concentrate containing 3 per cent azadirachtin. Over the past three years, this formulaiton has been tested against a wide range of insect pests. Results from these trials indicate that the product is broadly effective at rates of 50 g of active ingredients per hectare against most lepidopteran species including army-worms, cutworms, loopers, webworms, leaf skeletonizers, gypsy moth, and diamondback moth. Excellent efficacy has also been seen against Colorado potato beetle, leafhoppers, whiteflies, and some leafminers. Control of aphids, thrips and some cryptic pests has been inconsistent. To date, Azatin EC has been tested as a stand alone product. In coming years, AgriDyne's products will be tested in combinations (tank mixes and rotations) with a variety of other insecticide products to define the role of azadirachtin in comprehensive pest control programme.

INTRODUCTION

Neem has for centuries been used to control insect pests of agricultural and veterinary importance. Despite extensive research and documentation of the efficacy of neem extracts, only in the past decade or so has there been a significant interest in commercializing neem, particularly in the United States. The main deterrent has been the lack of patent protection of azadirachtin. AgriDyne Technologies Inc. (formerly Native Plants Inc.) has been active in neem research for over 10 years. The Environmental Protection Agency (EPA) has granted registration to Azatin EC for indoor and outdoor ornamental uses and to Turplex for turf pest control. Align, another azadirachtin formulation, is currently awaiting EPA approval for vegetable and fruit crop applications.*

All the formulations have 3 per cent (w/w) azadirachtin, contain no neem oil and emulsify well in a wide range of water types. The various formulations are with proprietary technology made from blended neem extract concentrates that contain 10 to 20 per cent azadirachtin. AgriDyne is the only company in the United States that holds registration for the neem extract concentrates as well.

Recently, the interest in neem has been fuelled further with the publication of *Neem—A Tree for Solving Global Problems* by the National Academy of Sciences panel on neem. Despite this tremendous tail wind, there are several aspects of commercialization that need to be looked into that will determine the magnitude of neem use for insect control in the near future. Since azadirachtin is the key compound that provides consistent insecticidal activity in neem, for the rest of the paper, reference will be made only to azadirachtin in the context of insect control.

FACTORS INFLUENCING THE COMMERCIAL SUCCESS OF AZADIRACHTIN

Consistency and Quality of Extracts

For any product to be commercially viable, it needs to have both consistency and quality. Neem extracts in general, depending on the extraction methodology, contain in excess of 25–30 compounds but only the azaadirachtin content of the extracts closely correlates to the insect growth regulator (IGR) activity. Neem extracts with higher azadirachtin content show higher potency. It is therefore understandable that the high degree of variation observed in the efficacy of various crude extracts from different laboratories around the world was subject to not only from where, when and how the seeds were procured, stored and extracted but to also how the extracts themselves were stored and tested. With tremendous advances in the development of analytical extraction, and formulation procedures of neem extracts, this hurdle has largely been overcome. It is now possible to measure azadirachtin content in crude extracts precisely, selectively remove aflatoxins to below the required 20 ppb range, and formulate to provide a shelf life of at least 18 months to EC formulations. Moreover, with tremendous advances in neem concentration formulation technology, and AgriDyne EC formulations have the consistency, texture, and compatibility profile of synthetic insecticides.

Product Positioning

Neem extracts are primarily recognized for their repellent and antifeedant properties but the IGR properties of azadirachtin are not universally recognized. However, it is only the IGR activity that provides a consistent and measurable yardstick for efficacy and the dosages required are typically much less than those required to obtain repellent, antifeedant or antiovipositional effects. In fact, much of the antifeedant activity is probably due to the IGR activity because of hormonal disruption. It is our opinion that

* On January 27 1993, EPA approved align for food crop use in the USA. Some state registrations are still pending.

commercialization of azadirachtin as an insecticide should focus primarily on the IGR properties because, when the azadirachtin content is accurately estimated and evaluated as an IGR, the efficacy is remarkably consistent against a given species of insect. However, this emphasizes the need for education of growers on the use of azadirachtin. Also, since azadirachtin is slow-acting, needing 3–14 days for mortality, the concept that there is no quick kill needs to be communicated effectively.

Supply and Availability

The challenge of supply is by far the most formidable. Adequate supplies of good quality (i.e, high azadirachtin content) neem kernels in a timely fashion is critical for the commercial success of azadirachtin. In countries such as India and Burma where oil is currently being extracted from neem seeds for soap manufacture, the opportunity for obtaining azadirachtin-rich extracts from preexisting facilities is available. However, unlike the oil, the azadirachtin content in kernels is subject to rapid degradation under harsh environmental conditions and therefore the procurement, storage, handling, and extraction procedures need to be followed closely to get the highest yield at the extraction step. Poor quality extracts generally yield poor quality formulations that perform inconsistently.

Increase in Environmental Awareness

Perhaps the main reason for the commercial opportunity for azadirachtin that exists today is the changing social and governmental attitude towards pest control. Until recently, insecticides were primarily synthetic, possessing quick knockdown, and the inherent toxicity of the molecules to mammals and non-target organisms was not of primary importance. With more and more stringent governmental regulations, worker safety issues, and increased data requirements for registration purposes, the availability of newer insecticides has been drastically reduced and the need for environmentally sound insecticides is greater than ever before. At the same time, older insecticides that have lost patent protection have not been supported by the parent companies. This has led to cancellation of numerous registrations and reduction in options available for insect control. Since Azadirachtin has been classified as a biological insecticide for registration purposes by the EPA, the date requirements have been less stringent based on the biological origin and safety profile of neem.

The Acceptance of Insect-Growth-Regulators

There are many synthetic IGRs such as kinoprene and hydroprene available in the market and it is only now that these chemicals have started to gain acceptance because of their unique modes of action. Azadirachtin is perhaps the only natural IGR that currently possesses commercial potential. By definition, IGRs act only on the immature stages of the insect-larvae or nymphs and pupae and do not kill adults. They have excellent potential in integrated pest management (IPM) programmes. Therefore, the marketing and positioning of azadirachtin becomes crucial for the success of the product in the market place, where quick knockdown and adulticidal activity are generally expected. The growers have to understand the unique mode of action and use it appropriately along with the other insecticides for maximum benefit.

AgriDyne has taken the approach of emphasizing the 'widening the window of opportunity' for insect control and positioning the product as an alternative insecticide with a new mode of action that is compatible in conventional or IPM programme. Azadirachtin controls insects not only at all stages of larvae/nymphs but also at the pupal stage, which is unaffected by traditional synthetic insecticides. Since beta-ecdysone, the key moulding hormone, is active in the pupal stage, the insecticidal activity of azadirachtin extends into this stage as well. Again, it is important that these essential differences be communicated to the growers for the commercial success of azadirachtin.

Cost of Raw Material

In the United States, from extensive efficacy trials conducted by both AgriDyne and key university researchers, it has been determined that depending on pest species, a use rate of 25 to 50 g (a.i.) active ingredients azadirachtin per hectare is necessary for consistent efficacy. Lepidopteran pests typically require 20 g a.i. (such as *Trichoplusia ni*) to 50 g a.i. (*Plutella xylostella*) per hectare. The spectrum of activity covers whiteflies, leafhoppers, and sawflies, loopers, gypsy moth, armyworms, cutworms, leafminers, fungus gnats, webworms, leafhoppers, Colorado potato beetle and the list continues to grow. Despite such a wide activity spectrum, the commercial feasibility and success of azadirachtin depends on (1) what the market is willing to pay for the product and (2) how much it will cost a company such as AgriDyne to produce and market it. Barring niche markets such as greenhouse ornamentals, home and garden, and perhaps turf, more than 90 per cent of the market opportunity today is in the high-value food or agriculture market where competitive products for caterpillar control (including the *Bacillus thuringiensis* products) sell in the range of \$25–50/ha. Given that for diamondback moth control 50 g a.i./ha of azadirachtin is required, it follows that in order to be fully competitive with other products, 1 g of formulated azadirachtin needs to cost around \$1 to the grower. This cost includes cost of procuring, processing, extracting, refining, formulating, shipping and handling, manufacturer's profit, and distributor margins. It is our estimate that in order to sell formulated azadirachtin at \$1/g a.i. to the grower, the cost of neem extract concentrate needs to be at a maximum of 15 to 20 cents per gram a.i. of azadirachtin.

IMPORTANT COMMERICAL CHARACTERISTICS OF AZADIRACHTIN

Many features of azadirachtin have relevant in today's pest control environment. Its advantages are:

1) Broad spectrum of activity
2) No known insecticide resistance mechanisms
3) Compatible with many commercial insecticides and fungicides
4) New mode of action with possible multiple sites of attack
5) Classified as a biological for registration purposes
6) Low use rates (< 50 g a.i./ha)
7) Not persistent in the environment

8) Compatible with biological control agents for IPM programmes
9) Minimal impact on non-target organisms
10) Fromulation flexibility
11) No re-entry restrictions
12) Supply available from preexisting infrastructure
13) Application flexibility—can be sprayed or drenched
14) Non-phytotoxic formulations available.

The disadvantages of azadirachtin are:

1) Slow action
2) No patent protection
3) Supply can be erratic
4) May be expensive for high-volume markets.

REPRESENTIVE EFFICACY DATA WTIH AZATIN/TURPLEX/ALIGN IN THE UNITED STATES

Figure 1 shows the dose response generated with Azatin EC for controlling sweet potato whitefly (*Bemisia tabaci*) at the AgriDyne Research Greenhouse in Salt Lake city. This was an eight-treatment, three replicate, single-factoral experiment set in a completely randomized block design. eight poinsettia plants (var. Annette Hegg) in each organdie cage were infested with 100 sweet potato whitefly adults after each treatment for up to three weeks. On the fourth week another 200 sweet potato whitefly adults were introduced and final evaluations were made after six weeks. Spraying was done with a

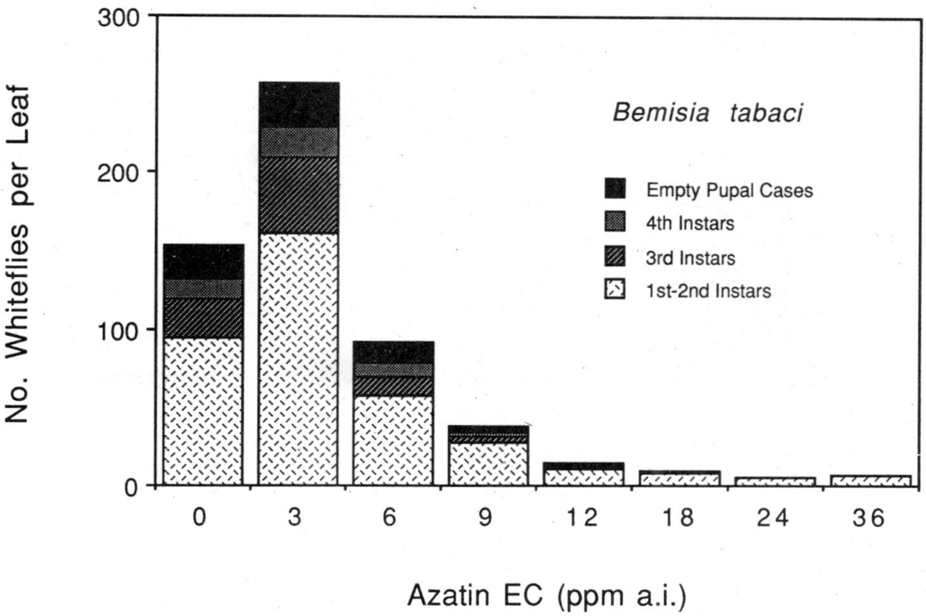

Fig. 1. Control of sweet potato whitefly on poinsettia, day 56.

portable electric sprayer and the plants were sprayed to runoff. Sampling was done by detaching 10 fully developed leaves from each cage (replicate) and counting the insects under a stereomicroscope. All surviving whiteflies were categorized as first and second instars, third instars, four instars, and empty pupal cases.

The results show a strong dose response that is very consistent when repeated. More than 15 ppm azadirachtin is required to control the late-instar and to disrupt the life cycle. Label recommendations for Azatin EC require 30–40 ppm a.i. for control of sweet potato whitefly.

This experiment depicted in Fig. 2 which was conducted at the University of Georgia, Griffin, demonstrates the competitiveness of azadirachtin against standard pest control measures against sweet potato whitefly. This trial was a three-treatment, three replicate, single-factorial experiment set in a completely randomized block design. A similar protocol was followed to that described for the first experiment. Azatin EC was applied at the 20 ppm and 30 ppm concentrations and the standard treatment consisted of bifenthrin, fenpropathrin plus acephate or endosulfan applied weekly up to six weeks. All applications were made to runoff with a portable hydraulic sprayer unit. Ten fully expanded leaves were sampled from each cage after six applications. Figure 2 summarizes week 6 data.

Results indicate that Azatin EC at both 20 and 30 ppm even when used alone, show excellent control of sweet potato whitefly which is on par with accepted grower pest control practices.

The experiment depicted in Fig. 3, which was conducted at the AgriDyne Research Greenhouse, Salt Lake City, shows the ability of azadirachtin (Azatin) to control ser-

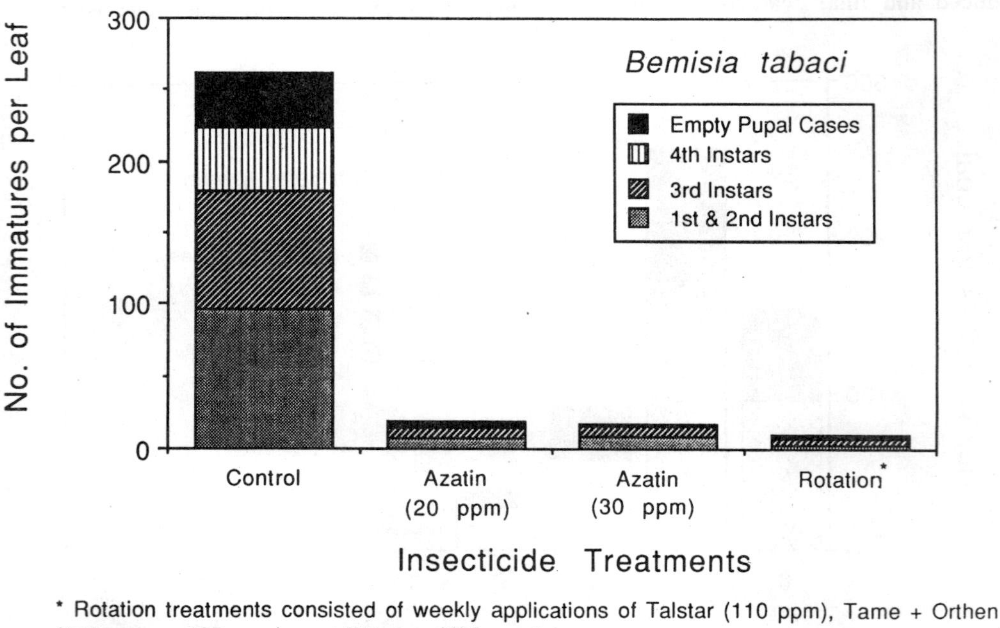

Fig. 2. Efficacy of Azatin and other insecticides for the control of sweet potato whitefly on poinsettia, day 42.

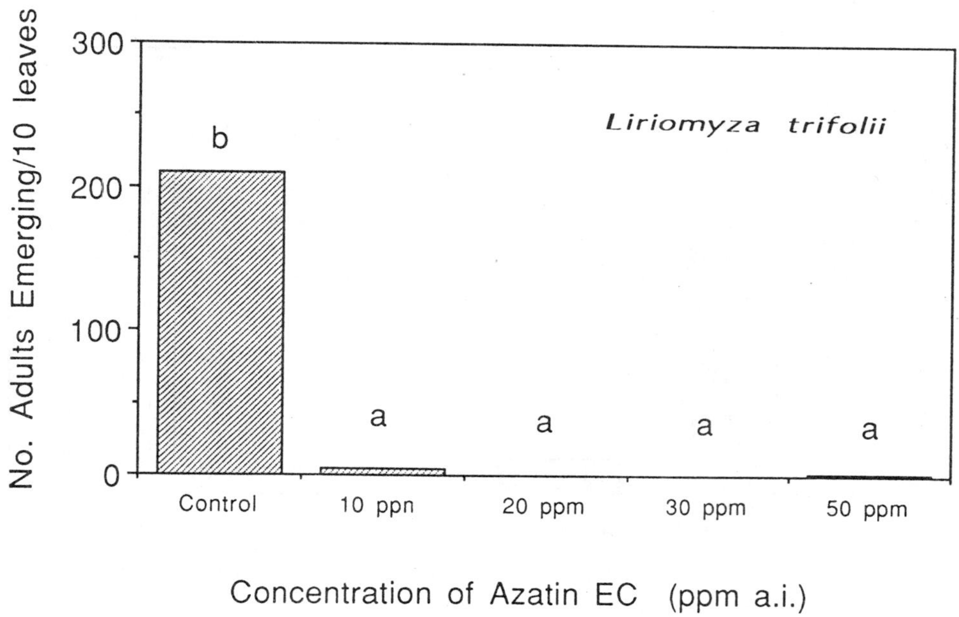

Fig. 3. Efficacy of Azatin EC as a drench for controlling serpentine leafminer on chrysanthemum, day 9.

pentine leafminer (*Liriomyza trifolii*) when applied as a drench. The trial was a five-treatment, three-replicate, single-factorial experiment laid out in a completely randomized block design. Eight chrysanthemum plants (var. Hurricane) each in a 3.8 l pot were arranged in a cage. Each pot was drenched with 250 ml of drench solution and, 48 hrs later, infested with 25 newly emerged leafminer adults. After nine days, 10 leaves per cage (replicate) were sampled and set over sand and the pupae were allowed to drop in a growth chamber. Later, pupae were collected, counted, and allowed to emerge in glass vials.

These results are very consistent and repeatable when sprayed or drenched. A significant and commercially acceptable level of control was achieved even at 10 ppm azadirachtin (2.5 mg azadirachtin per pot). Since azadirachtin is an IGR, Some mining damage will be present that may not be aesthetically acceptable in potted chrysanthemums but is fully acceptable for cut flower production. Nevertheless, pupal mortality is unique to azadirachtin and it widens the leafminer control window and should be used accordingly. The feature will be execellent for leafminer pest or resistance management programmes, particularly in conjunction with adulticides.

Thr trial (Fig. 4) was conducted on the Cocheco Country Club golf course in Dover, New Hamspshire, to test the efficacy to Turplex (3 per cent azadirachtin formulation) against native black cutworm (*Agrotis ipsilon*) larvae. This was a five-treatment, three-replicate, single-factorial experiment set in a completely randomized block design. Fifteen whole greens of bentgrass were selected and Turplex was applied at rates ranging from 25 to 125 g azadirachtin a.i./ha and compared with Dursban (chloropyrifos), an organophospate. All treatments were applied in 740 l/ha. Eight days after treatment, the greens were sampled by pouring weak soap solution to flush live larvae to the surface. Results of the day 8 sampling are summarized in Fig. 4.

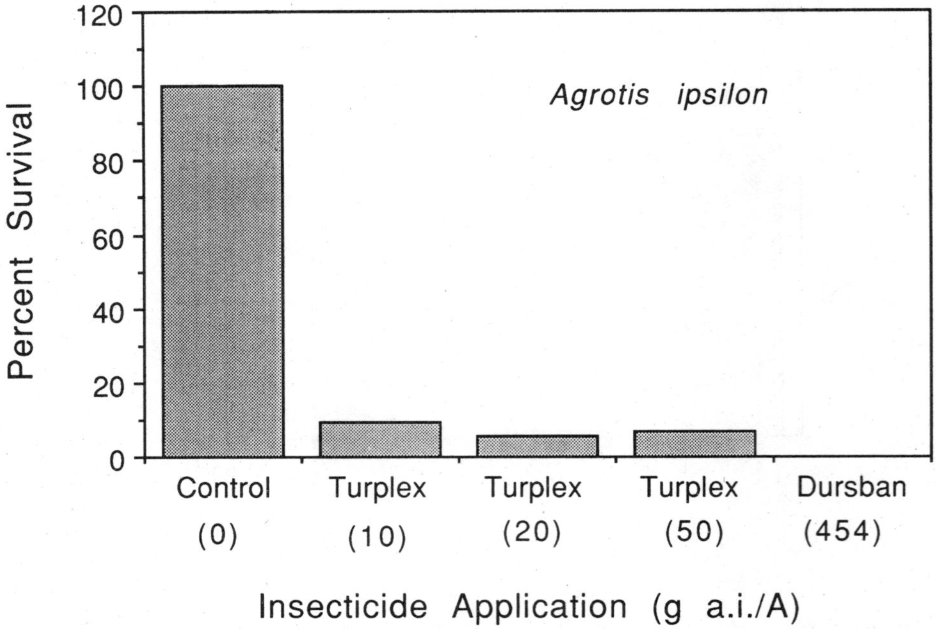

Fig. 4. Survival of black cutworms on golf course greens at 8 days post treatment: Dover, NH.

Results indicate execellent control of black cutworm larvae even at 25 g a.i./ha rate. Turplex has generally shown excellent efficacy against all the lepidopterous caterpillar complex on turf such as sod webworms and armyworms.

The trial (Fig. 5) was conducted at San Tormo Winery, Lodi, California, on Sauvignon Blanc variety of grapes to test the efficacy of Align (3 per cent azadirachtin) against grape leafhopper (*Erythroneura elegantula*). This was a four treatment, five-replicate, single-factorial experiment set in a completely randomized block design. Application of Align and the standard treatments were made weekly using a hand-gun sprayer at the rate of 950 l/ha. Sampling was conducted by harvesting 10 fully expanded leaves from the interior of the canopy of each replicate and the immature/adults counts were made. Data from the day 42 counts are summarized in Fig. 5.

There was significant control of both the adults and immature grape leafhoppers to levels below the economic threshold by day 42 even at the 20 g a.i./acre (50 g a.i./ha) populations and are therefore used sparingly on grapes. Align could be used to control the leafhopper populations without the risk of secondary pest resurgence.

The experiment (Fig. 6) was conducted in a commercial tomato field at Dondarro Farms in Stockton, California to test the efficacy of Align for the control of cabbage looper (*Trichoplusia ni*). The trial was four treatment, four replicate, single factorial experiment laid out in a completely randomized block design. Align at 20 and 40 g a.i./ acre (50 and 100 g a.i./ha, respectively) and a pyrethroid (Asana at 20 g a.i./acre or 50 g a.i./ha) were applied with a CO_2 sprayer calibrated to deliver 240 l/ha. Sampling was done weekly by examining all tomato plants in 5 ft linear row and counting all

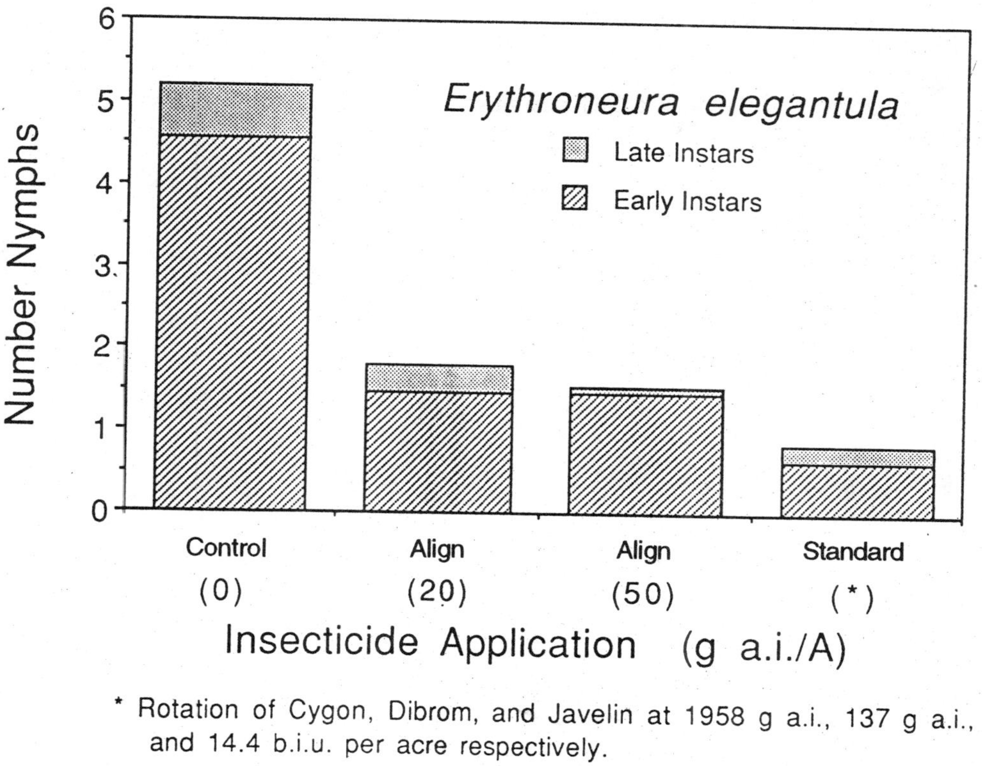

Fig. 5. Control of grape leafhoppers on wine grapes, San Tomo Winery, Lodi, CA: day 42.

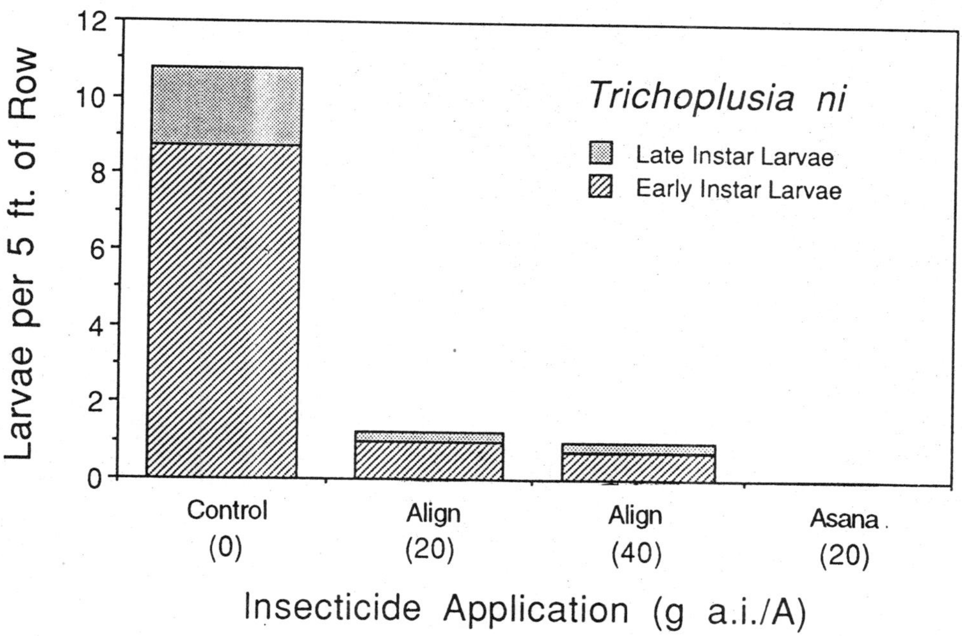

Fig. 6. Cabbage looper control on field tomatoes, Dondarro Farms, Stockton, CA: day 14.

live early- and late-instar *T. ni* larvae. Results of day 14 counts are summarized in Fig. 6.

Align even at the 50 g a.i./ha rate showed excellent efficacy against the cabbage looper in the field within 14 days (two applications).

REFERENCE

National Research Council. 1992. *Neem: A Tree for Solving Global Problems*. National Academy Press, Washington D.C., 139 pp.

10

TECHNOLOGICAL DEVELOPMENT

102

Machineries for Neem Processing

S.S. SIVAKUMAR, P.T. PALANISAMY, N. VARADHARAJU, L. GOTHANDAPANI AND K.R. SWAMINATHAN

Post Harvest Technology Scheme, College of Agricultural Engineering
Tamil Nadu Agricultural University, Coimbatore 641 003, India

ABSTRACT

In recent times neem products such as neem kernel suspension in water, neem oil, and neem kernel powder are being used for pest control. Neem kernel powder is used at the rate of 10–20 g per kg of seeds for storage pest control. The major active principle in neem, azadirachtin, is highly heat-sensitive. Hence, neem oil obtained will not contain much of the principle because the heat generated during extraction inactivates the principle. To get the maximum possible active principle, an attempt has been made to develop and test various neem processing equipments, such as neem decorticator, neem kernel crusher, and neem pulverizer in the Department of Agricultural Processing. College of Agricultural Engineering, Tamil Nadu Agricultural University. The cost analysis was also worked out for a village-level neem processing plant to prepare defatted neem kernel powder for pest control.

INTRODUCTION

It is an age-old practice in rural India to mix dried neem leaves with stored grains, and to place leaves in folds of cloths and woolen blankets to keep the insects away. Neem oil and neem powder are the major products now being considered as effective substitutes for chemical pesticides and research is going on in India and several other Asian countries on the use of neem products as plant protection agents.

The processing of neem involves shelling, crushing, defatting, oil processing, and other operations. These processes are available at large-scale units. At present, for pilot-scale application for the use of the farmer at his holdings, it is necessary to develop small-scale units for neem processing.

MATERIALS AND METHODS

Neem seeds are collected during April–July and are sun-dried and stored till October–December or even later. Deterioration of neem seed occurs when it is stocked without proper drying. The collected fruits are classified into four categories: dried fruit, semi-dried fruit, fresh fruit, and wet and/or damaged fruit.

Fresh fruit take more time to dry and it would be unwise to dry the fruit in the wet and/or damaged condition where the deterioration had already started and would proceed at an accelerated rate even during drying. In case of semi-dried and undamaged fruits, steps should be taken to dry them completely and as quickly as possible either in bright sunshine or in a suitable drier. The completely dried whole fruit can be decorticated later and kernels obtained for processing. The dried fruits can be dumped in big heaps or in tall bins temporarily made out of bamboo. The fresh and wet or damaged lots should never be stocked in big pyramidal heaps for more than one day.

Dried neem nut obtained from the local market at 11.6 per cent (wet basis) moisture content was used in the experiments. The experiments were conducted in the moisture range of 8.0 to 20.0 per cent (w.b.).

The samples of lower moisture content were obtained by drying in a mechanically ventilated oven at 45°C with 15 mm nut layer. The samples of higher moisture content were obtained by adding distilled water. The samples were stored in airtight containers at 5°C in the refrigerator. The moisture content of the samples was determined by drying the samples at 130°C $\pm$ 1°C for 12 hrs.

The decorticator (Fig. 1) consists of a feed hopper, rotor or shelling drum, sieve, outlet, and motor and pulley.

The feed hopper or inlet is made of mild steel and is conical for easy flow of the feed material into the threshing chamber.

The rotor or shelling drum is made of cast iron with a diameter of 20 cm and length of 15 cm. Six rubber pads of size 15 × 15 × 1 cm are fixed along the periphery of the rotor at 6 cm intervals. The motor is attached with a pulley by a central shaft.

The sieve is made of mild steel plate having oblong slots of 4.5 × 0.5 cm size with 0.5 cm slot-to-slot interval cut to the required size. The sheet is bent to semi-circular shape and fitted to the guide, which acts as a shelling platform. The clearance between the rubber pad and sheet is adjusted by raising and lowering of the rotor chamber. The rotor assembly is covered to avoid the scattering of the neem nut. An outlet is provided at the bottom of the sieve for collection of shelled neem kernel and husk.

The rotor is powered by a 1 hp motor through a V belt drive. The sheller was tested using a dynodrive unit by setting the speed of operation at 100, 150, 200, 300, and 350 rpm, with varying moisture contents of 8, 10, and 12 per cent (w.b.). The power required for shelling was noted using watt-hour meter or energy meter and the time of operation was noted for each test having the constant feed of 1 kg. The capacity and shelling efficiency were calculated for each case and are reported in Table 1.

$$\text{Capacity} = A/T \text{ kg/hr}$$

$$\text{Shelling efficiency} = (1 - B/A) \times 100 \text{ per cent}$$

where A = weight of the feed material, kg; B = wieght of the unshelled nut, kg; and T = time of operation, hr.

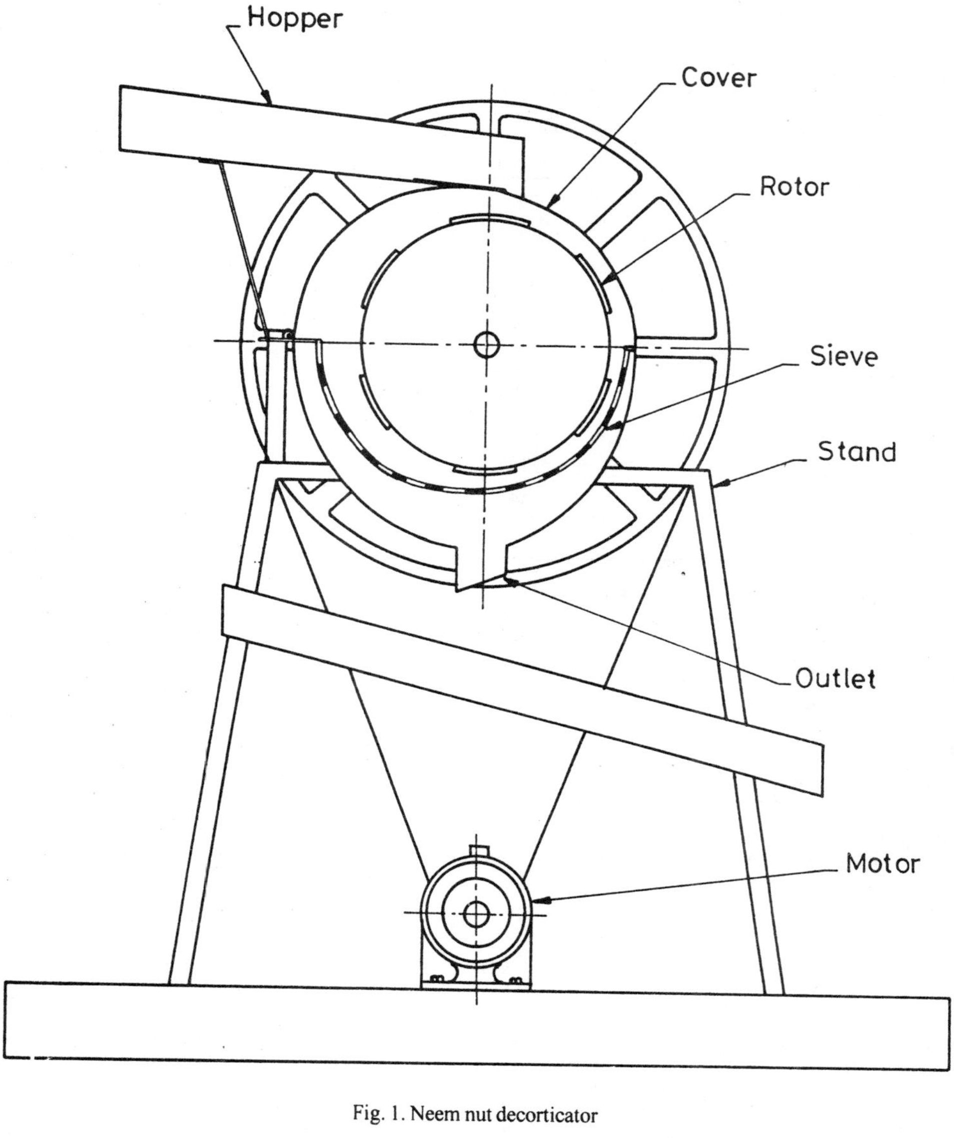

Fig. 1. Neem nut decorticator

The neem kernel crusher (Fig. 2) consists of a feed hopper, crushing rollers, outlet for kernel powder collection, outlet for oil collection, and motor and pulley.

The neem kernel crusher has two rollers of diameter 354 mm and length 97 mm, rotating in opposite directions, which yields the powder by squeezing out the oil. The rollers are powered by 2 hp electric motor. A known weight of sample was fed into the rollers through the hopper and the time taken for crushing, weight of powder recovered, and weight of uncrushed kernel were noted to calculate the capacity and crushing efficiency.

Commercially available pulverizers, ball mills mini flour mills, or mixing units are used for pulverizing the neem kernel. Then the powdered material is sieved and used as defatted neem kernel powder.

Table 1. Trials in Decorticator

$$\text{Capacity in kg/hr} = \frac{\text{Sample (kg x 60)}}{\text{Time (min.)}}$$

$$\text{Shelling efficiency} = \frac{\text{Wt. of unshelled nut}}{\text{Wt. of nut}}$$

I. Moisture content = 8 per cent (w.b.)

Sample (kg)	rpm	Kernel (g)	Husk (g)	Unshelled nut (g)	Time (min.)	Capacity (kg/hr)	Shelling efficiency (%)
1	100	350	620	30	4.0	15.0	97.0
1	150	323	605	72	2.5	24.0	92.8
1	200	310	578	112	1.5	40.0	88.8
1	300	288	559	153	1.0	60.0	84.7
1	350	276	542	182	0.7	85.7	81.8
II. Moisture content = 10 per cent (w.b.)							
1	100	340	580	80	5.0	12.0	92.0
1	150	320	570	110	2.0	30.0	89.0
1	200	300	550	150	1.5	40.0	85.0
1	300	280	520	200	1.0	60.0	80.0
1	350	270	500	230	0.7	86.0	77.0
III. Moisture content = 12 per cent (w.b.)							
1	100	345	560	95	4.5	13.3	90.5
1	150	320	570	110	1.8	33.3	89.0
1	200	305	545	150	1.5	40.0	85.0
1	300	285	515	200	0.9	66.6	80.0
1	350	275	490	235	0.7	86.0	76.5

RESULTS AND DISCUSSION

The decorticator was tested for its performance under different moisture levels (8, 10, and 12 per cent w.b.) and speeds of operation (100, 150, 200, 300, and 350 rpm). The capacity, shelling efficiency, and powder requirment are evaluated and given in the Table 1 and are represented in Figs. 3, 4, 5, and 6.

As shown in the table, for the constant feed of 1 kg, the maximum shelling efficiency is 97.0 per cent under 100 rpm. If the speed of operation increases, the shelling efficiency decreases. The same trend was noticed for all the moisture levels.

From Table 1, it is seen that the optimum moisture level to get maximum shelling efficiency is 8 per cent (w.b.) is under 100 rpm the efficiency is 97.0 per cent. If the moisture content increases, the shelling efficiency decreases.

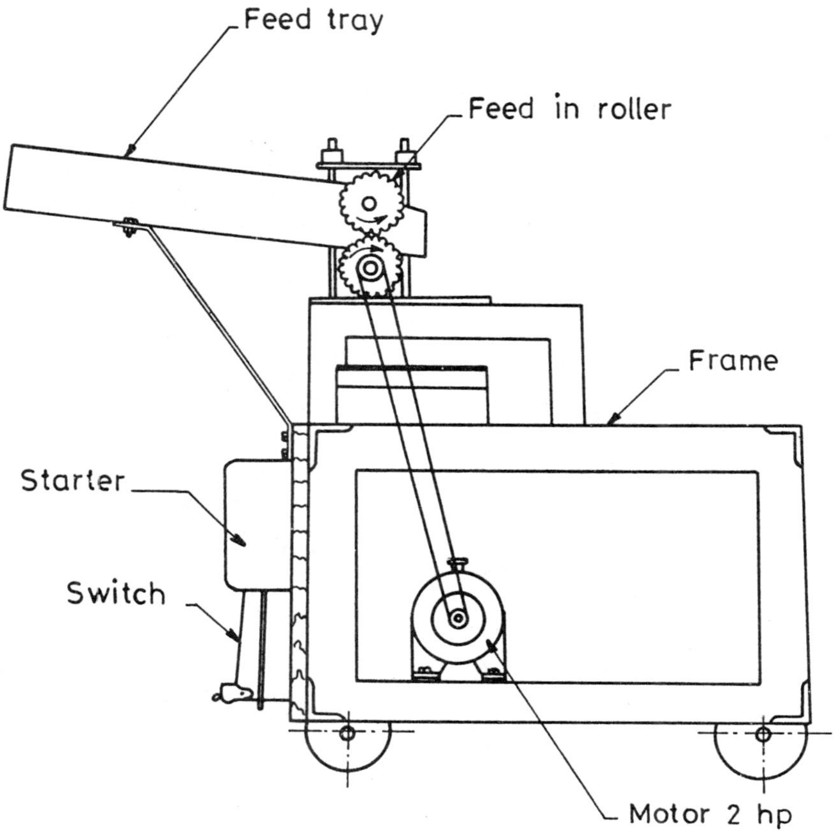

Fig. 2. Neem kernal crusher

The cost of shelling is Rs 40/quintal and the analysis given in Annexure 1.

Performance of the crusher is satisfactory when giving 95 per cent crushing at Rs 15/quintal. The calculation is given in Annexure 1. The roller length is 35.4 cm, diameter 9.7 cm. The top roller rotates at 95 rpm and the bottom roller at 100 rpm. The 2 hp motor runs at 1440 rpm.

The kernels measure 1.1 cm in length, 0.5 cm in diameter.

The performance is calculated as follows:

Weight of feed	= 1 kg
Time taken	= 104 sec
Capacity	= 34.0 kg/hr

Power Requirement:

Number of revolutions	= 2
Time taken	= 192 sec
60 revolutions	= 1 kwh

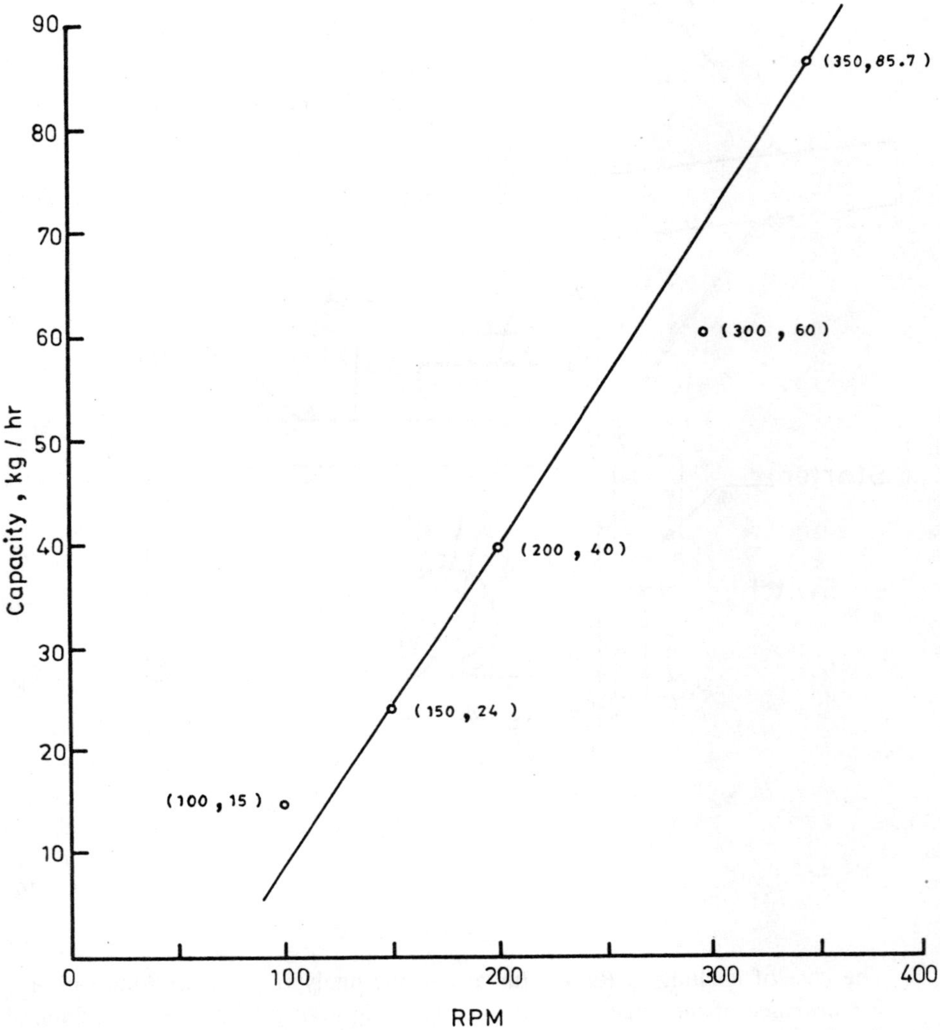

Fig. 3. Trials in decorticator, rpm vs capacity

Power required for 1 hr	= 2/192 × 3600 × 1/60
	= 0.625 kwh

Crushing efficiency:

Weight of feed	= 1 kg
Weight obtained	= 885 g
Weight of uncrushed kernel	= 40 g
Percentage recovery	= 845/1000 × 100
	= 84.5 per cent
Crushing efficiency	= 845/885 × 100
	= 95.4 per cent

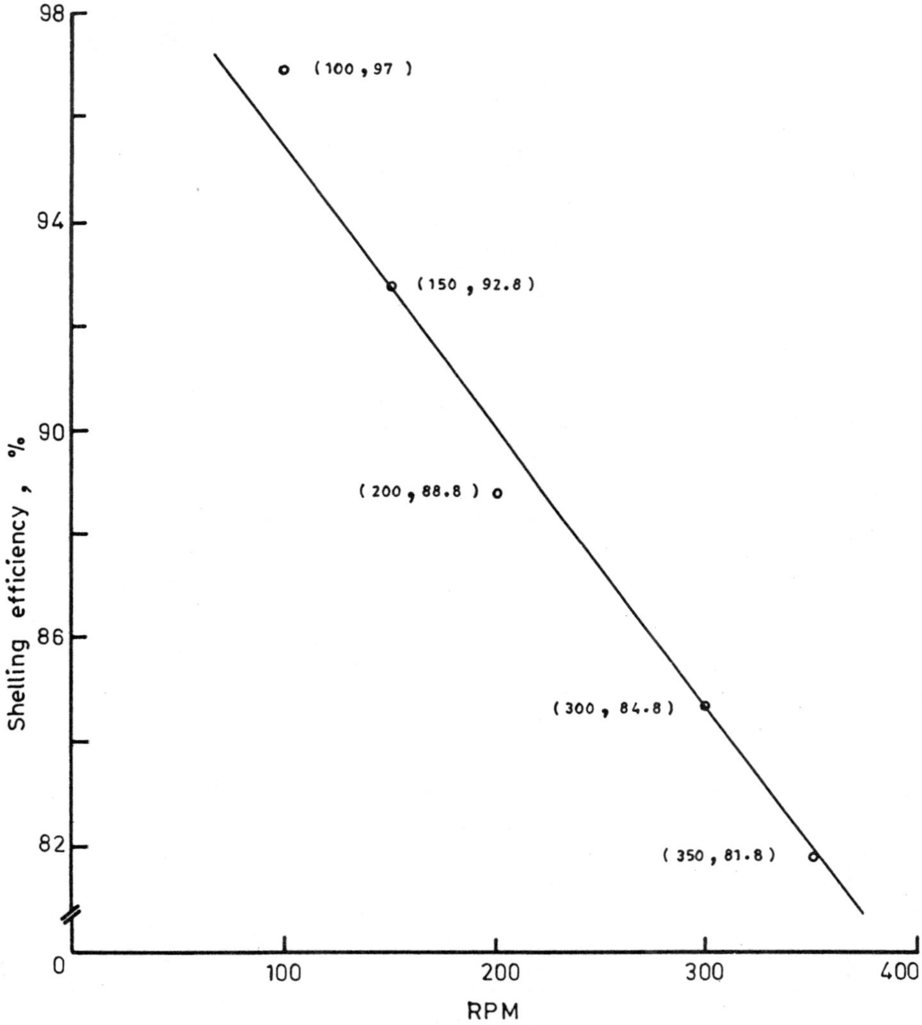

Fig. 4. Trials in decorticator, rpm vs shelling efficiency at moisture content 8.0 % (w.b.)

SUMMARY AND CONCLUSION

The capacity of the sheller is 15 kg/hr and the optimum moisture content of the nut is 8 per cent (w.b.), which results in the maximum shelling efficiency of the machine under the optimum speed of operation of 100 rpm. The power required is a 1 hp motor.

The cost of shelling is Rs 40/hr and the cost of crushing is Rs 15/quintal.

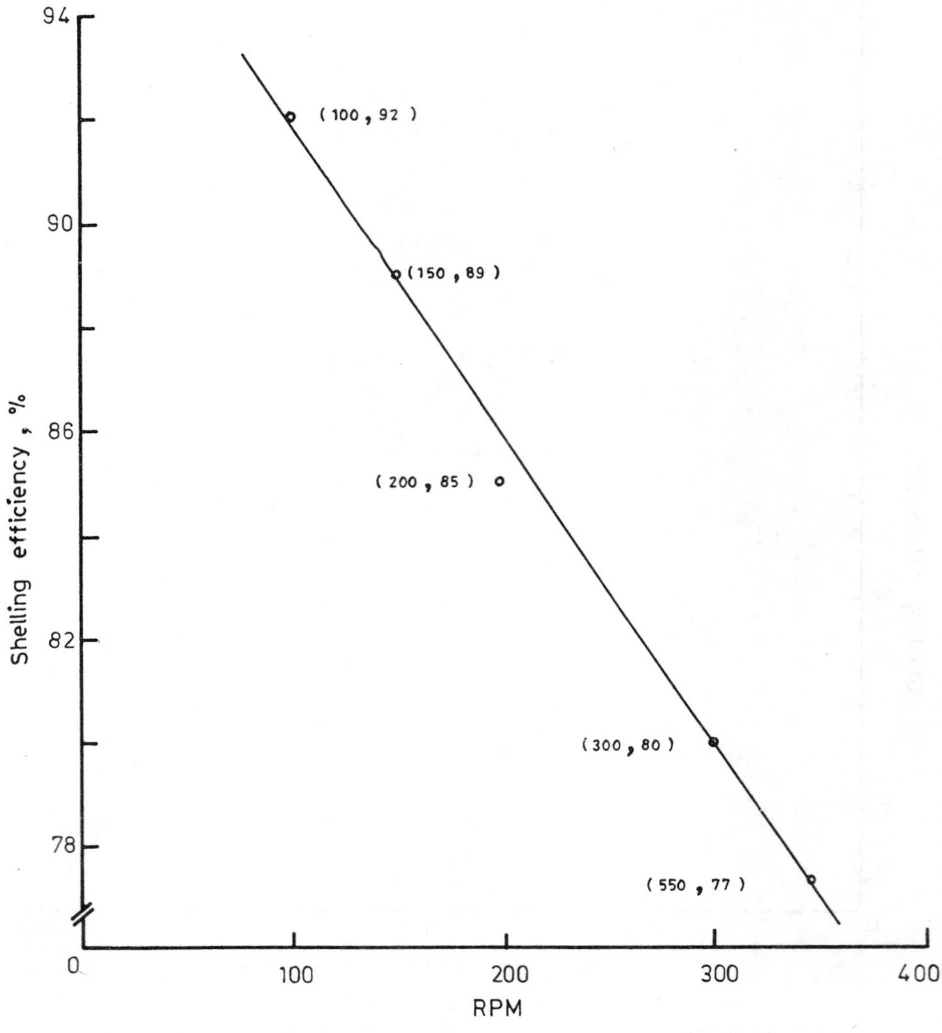

Fig. 5. Trials in decorticator, rpm vs shelling efficiency at moisture content 10% (w.b.)

SUGGESTIONS FOR FURTHER WORK

The neem kernel crusher should be tested and modified so as to suit local use and provisions should be made to collect the oil during crushing.

The neem fruit pulper has to be designed, developed, and tested.

A pulverizer unit should be developed rather of a ball mill, since the ball mill consumes a lot of time and is not economical.

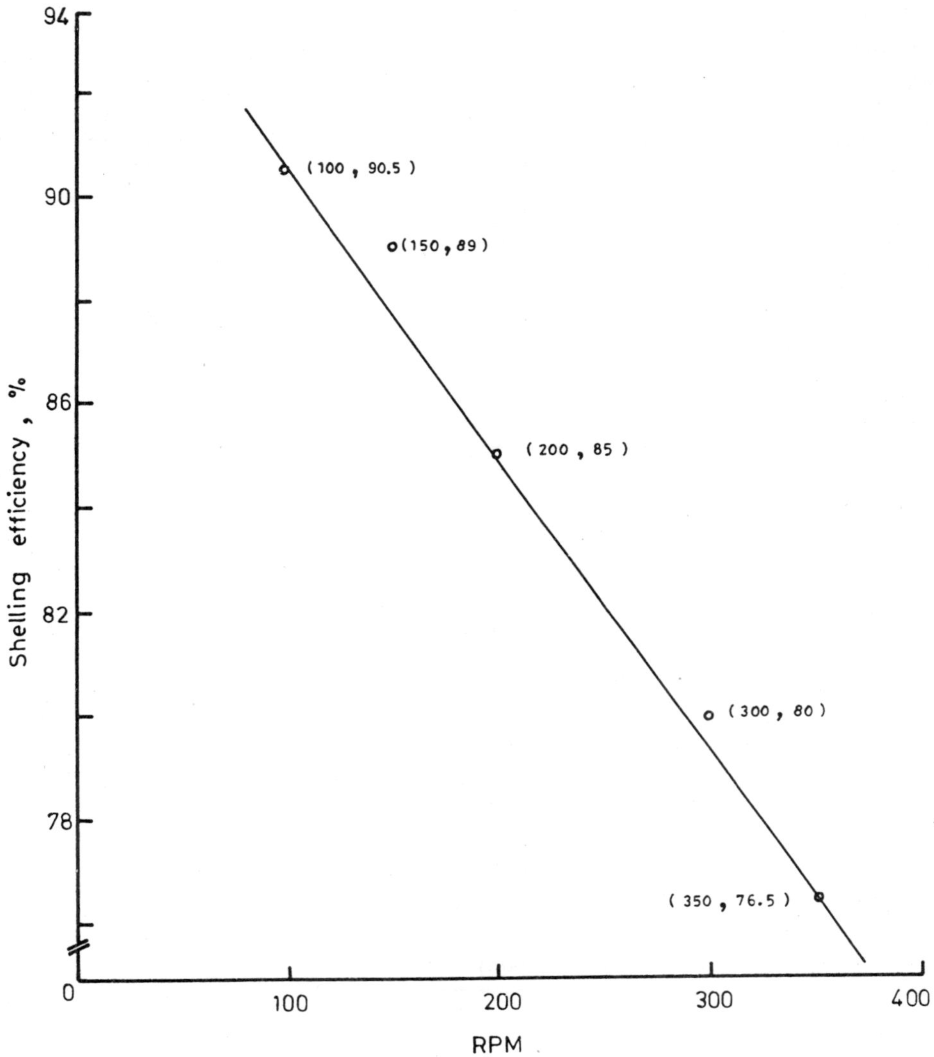

Fig. 6. Trials in decorticator, rpm vs shelling efficiency at moisture content 12 % (w.b.)

ANNEXURE 1

Cost analysis
Data used

Description	Decorticator	Crusher
Cost of machine	Rs 5,500	Rs 5,250
Power required	1 hp	2 hp
Capacity	15 kg/hr	34 kg/hr
Salvage value	10%	10%
Rate of interest	5%	5%
Insurance	1.5%	1.5%
Labour wages	Rs 20/day	Rs 20/day
No. of labourers employed	2	1
Electricity charges	60 paise/unit	60 paise/unit
Working hours	8 hrs/day	8 hrs/day
Working days	300 days/year	300 days/year

Decorticator

I. Fixed cost

A. Depreciation

$$= \frac{\text{Cost} - \text{Salvage}}{\text{Life period in years}}$$

$$= \frac{5500 - 550}{10}$$

$$= \text{Rs } 495$$

B. Rate of interest

$$= \frac{\text{Cost} + \text{Salvage}}{2} \times \text{Rate of interest}$$

$$= \frac{5500 + 550}{2} \times 0.05$$

$$= \text{Rs } 151.25$$

C. Insurance

$$= \frac{15}{100} \times 5{,}500$$

$$= \text{Rs } 82.5$$

Total fixed cost per year $= \text{Rs } 728.75$

Total fixed cost per hour $= \dfrac{728.75}{300 \times 8} = \text{Rs } 0.30/\text{hr}$

II. Variable cost

A. Electricity charges $= \frac{1 \times 736}{1000} \times 0.60$

$= 0.4410$ Rs/hr

B. Labour wages = Rs 40/day

= Rs 5/hr

Total cost = Fixed cost + Variable cost

= Rs. 5.3/hr

Capacity = 15 kg/hr

Total cost $= \frac{5.3}{15}$ = Rs 0.353/kg

= Rs 40/quintal (approx.)

Neem kernel crusher

I. Fixed cost

A. Depreciation $= \frac{5250 - 525}{10}$

= Rs 472.50

B. Rate of interest $= \frac{\text{Cost} + \text{Salvage}}{2} \times \text{Rate of interest}$

$= \frac{5250 + 525}{2} \times 0.05$

$= \text{Rs } 144.375$

C. Insurance $= \dfrac{15}{100} \times 5250$

$= \text{Rs } 78.75$

Total fixed cost $= \text{Rs } 696.63/\text{year}$

$= \dfrac{695.63}{300 \times 8}$

$= \text{Rs } 0.29/\text{hr}$

II. Variable cost

A. Electricity charges $= \dfrac{2 \times 739}{1000} \times 0.60$

$= \text{Rs } 0.8832/\text{hr}$

B. Labour wages $= \text{Rs } 20/\text{day}$

$= \text{Rs } 2.5/\text{hr}$

Total cost = Fixed cost + Variable cost

$= \text{Rs } 3.67/\text{hr}$

Capacity $= 34 \text{ kg/hr}$

Total cost = 15 paise/kg (approx.)

= Rs 15/quintal (approx.)

103

Technology for Production of Insecticides of Plant Origin at Rural Level

A. DAKSHINAMURTHY*

Post Harvest Technology Scheme CIAE, Bhopal, India

ABSTRACT

The leaf bits of *Azadirachta indica*, *Pongamia glabra* and *Vitex negundo* packed in cloth bags were found to protect the wheat from cross infestation by insects. Laboratory testing with 2 per cent extracts of these leaves were observed to cause mortality of *Sitophilus oryzae* to vaying extent. The *P. glabra* kernel extract at 4 per cent concentration was quite effective in reducing insect infestation of wheat up to six months in bag storage. The kernel oil also had a vapour effect at varying concentrations against the storage pests of wheat and green gram. A set of simple processing equipments were found to be suitable for production of insecticides of plant origin at rural level.

INTRODUCTION

Insecticides of plant origin have been used for centuries because they are non-hazardous to non-target organisms. Many workers have reported that mixing of various plant products with grains repels insects (Fry, 1938; Pruthi and Singh, 1950; Ketkar et al., 1976) and enables safe storage of grains (Abraham et al., 1972; Anonymous, 1983; Dakshinamurthy, 1988, 1992). However, insecticides of plant origin have not come into large-scale use because of the lack of technology to produce them in sufficient quantity and the time-consuming and labour-intensive procedure to prepare them. Efforts were therefore made to indentify and test simple equipment for production of dry and liquid formulations of insecticides of plant origin at rural level. These formulations were also evaluated for their effectiveness against insect pests of stored grains.

* Present address: Director, Paddy Processing Research Centre, Thanjavur 613 005

MATERIALS AND METHODS

The leaves of neem (*Azadirachta indica*), pungam (*Pongamia glabra*) and nochi (*Vitex negundo*) were dried and leaf bits of 1–1.5 cm were prepared using a leaf grinder. The leaf bits were packed in cloth bags at the rate of 100 g to be used for 1 quintal of wheat grains. They were distributed to 20 users to be placed in stored wheat. Observations were made at monthly intervals for infestation by insects up to six months from June to December.

Laboratory testing for the effectiveness of water, alcohol, and ether extracts of leaves at 2 per cent concentration (2 g/100 ml) was carried out with the dry film technique in petri dishes against the weevil *Sitophilus oryzae*.

Impregnation of gunny bags with 1–4 per cent concentration of ether extract of pungam kernel was compared with a control. There were five treatments and four replications. Wheat at a moisture content of 10 $\pm$ 1 per cent was stored in gunny bags (20 kg capacity) at ambient conditions for natural infestation. Observations were taken for a period of six months at monthly intervals and the per cent infestation was recorded.

The metal and plastic bins of 1 quintal capacity were surface-treated with 100 ml of 2 per cent pungam kernel extract in ether as a prophylactic measure. Wheat was stored in these treated bins and compared with a control. Obsevations were made for insect infestation at monthly intervals for a period of four months.

The vapour effect of the pungam kernel extract/oil was tried in smaller containers of 1 kg capacity in storage of wheat and green gram. Paper mats of 4 × 2 cm size were treated with 0.5, 1.0, 1.5 and 2.0 ml of the extract and placed *in situ* and compared with a control. There were five treatments and four replications. Observations were made on the per cent infestation of grains after 6–7 weeks.

RESULTS AND DISCUSSION

Use of Leaf Bits

The studies on the use of leaf bits of neem, pungam and nochi in prevention of insect development and cross-infestation in storage of wheat revealed that grains without any initial infestation were safe from cross-infestation up to six months. However, wheat with initial infestation of 0.5–1.0 per cent deteriorated and infestation reached a maximum of 9.4 per cent (Table 1). Thus, it was confirmed that these plant materials are not effective in preventing the development of internal feeders of stored grain such as *S. oryzae* and *Rhizopertha dominica*.

Use of Leaf Extracts

The observations on mortality of *S. oryzae* with different extracts of the leaf materials indicated (Fig. 1) that all the treatments were better than control. The highest mortality of 98.4 per cent was observed in ether extract of nochi, followed by neem (88.8 per cent) and pungam (81.6 per cent). The alcohol and water extracts of the leaves were next in the order of merit ranging from 14 to 89 per cent. Nochi leaves were the most effective, followed by neem and pungam.

Table 1. Effect of leaf bits on insect infestation in wheat

Sample	% initial infestation	% final infestation
1	0.5	6.0
2	—	—
3	0.8	7.8
4	—	—
5	—	—
6	1.0	9.4
7	—	—
8	—	—
9	—	—
10	0.6	4.8
11	0.9	8.7
12	0.6	6.3
13	—	—
14	—	—
15	0.8	8.1
16	—	—
17	—	—
18	—	—
19	—	—
20	—	—

Impregnation of Bags with Kernel Extract

The results indicated that the control recorded the maximum infestation of 6.8 per cent in three months, whereas the treated lots had 0.8–4.6 per cent infestation in wheat by *S. oryzae*, *R. dominica* and *Corcyra cephalonica* (Table 2). In a six month period, the effect of concentration was quite significant and the higher concentration of 4 per cent recorded the lowest infestation of 1.7 per cent (Fig. 2).

Surface Treatment of Bins

Surface treatment of storage bins with pungam kernel extract (2 per cent) revealed that wheat in the untreated bin had an infestation of 1.2–3.8 per cent by *C. cephalonica*, while the wheat in treated bins was free from infestation up to four months.

Vapour Effect of Extract/Oil

The vapour effect of pungam kernel oil was evident (Table 3) from the lower per cent infestation in different concentrations (3.05 to 3.5 in wheat and 3.82 to 5.40 in green gram) as compared to higher per cent infestation in untreated wheat (5.70) and green gram (7.21). The dosage of 2 ml/kg of grain was better than control, but on par with other treatments. Eli and Vladimir (1990) reported the fumigant toxicity of essential oils from plants to be effective against stored grain pests. Credland (1992) also explained the ovicidal effects of oils in relation to structure of bruchid eggs.

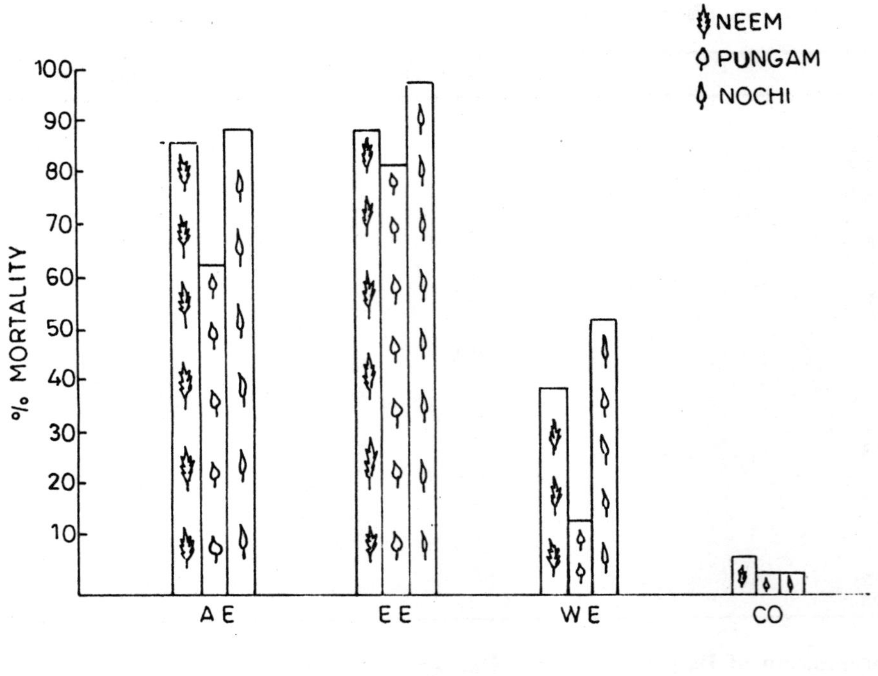

Fig. 1. Effect of leaf extracts on mortality of *S. oryzae*

Table 2. Effect of impregnation of bags with pungam kernel extract on insect infestation in wheat (three months' storage)

Treatments	% infestation
Pungam kernel extract 1%	4.6
Pungam kernel extract 2%	1.0
Pungam kernel extract 3%	0.9
Pungam kernel extract 4%	0.8
Control	6.8

SED = 0.376
CD = 0.81

Technology for Rural Level

It was observed that the lack of simple processing equipment at rural level is one reason for the non-adoption of insecticides of plant origin. So, the equipment required for production of dry and liquid formulations was identified and tested.

The leaf grinder (Fig. 3) is a simple hand-operated machine with a set of mild steel blades and a hopper. The capacity of the unit is 30–35 kg/hr and it can make leaf bits of 1–1.5 cm. The cost of this unit is around Rs 400.

The seeds should be decorticated in a decorticator (Fig. 4) to obtain the kernels for preparation of extracts. A person can decorticate about 2–3 kg/hr by hand shelling. A

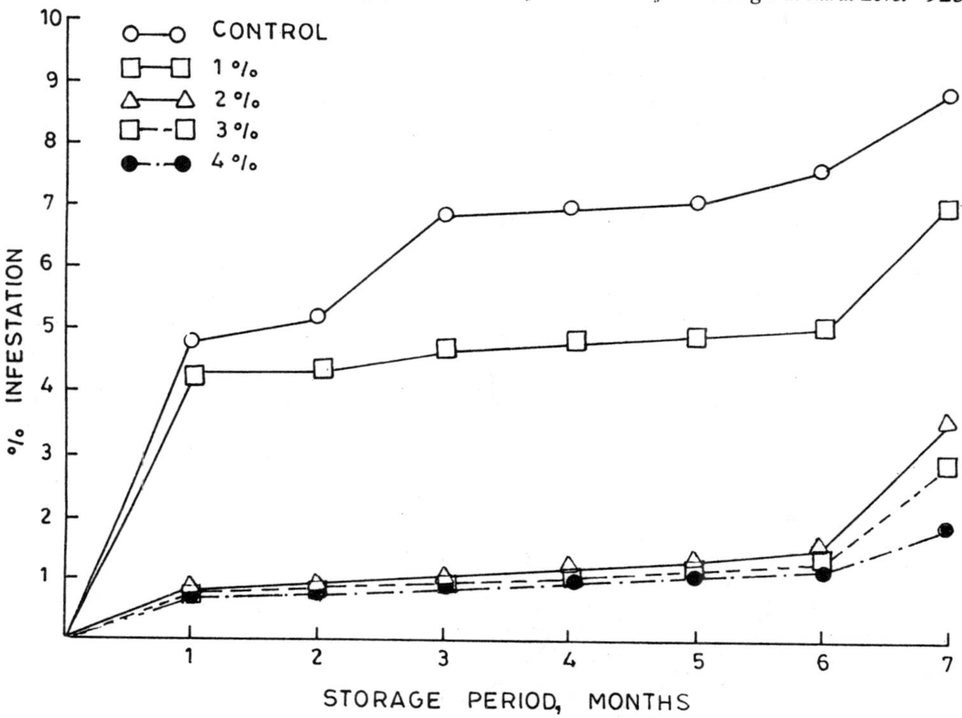

Fig. 2. Effect of impregnation of bags with kernel extract

Table 3. Vapour effect of pungam kernel extract/oil on storage of wheat and green gram

Treatments	% infestation	
	Wheat	Green gram
Control	5.70	7.21
Pungam kernel oil 0.5 ml (0.05%)	3.57	5.40
Pungam kernel oil 1.0 ml (0.10%)	3.50	4.54
Pungam kernel oil 1.5 ml (0.15%)	3.45	3.90
Pungam kernel oil 2.0 ml (0.20%)	3.05	3.82
SED	0.21	0.36
CD	0.45	0.77

hand-operated sheller has a capacity of 30–35 kg/hr and a power-operated unit has a capacity of 90–100 kg/hr. The cost of these units is Rs 1,500 and 2,000 respectively.

The decorticated kernels should be converted to flakes in a flaking machine (Fig. 5) before being used for extraction. The flaking machine consists of a set of cylindrical rollers with grooves on its surface. The capacity of the unit is 40 kg/hr and it costs Rs 750.

The partly deoiled cake can be powdered in the milling unit (Fig. 6), which has a capacity of 20 kg/hr and costs Rs 1,000.

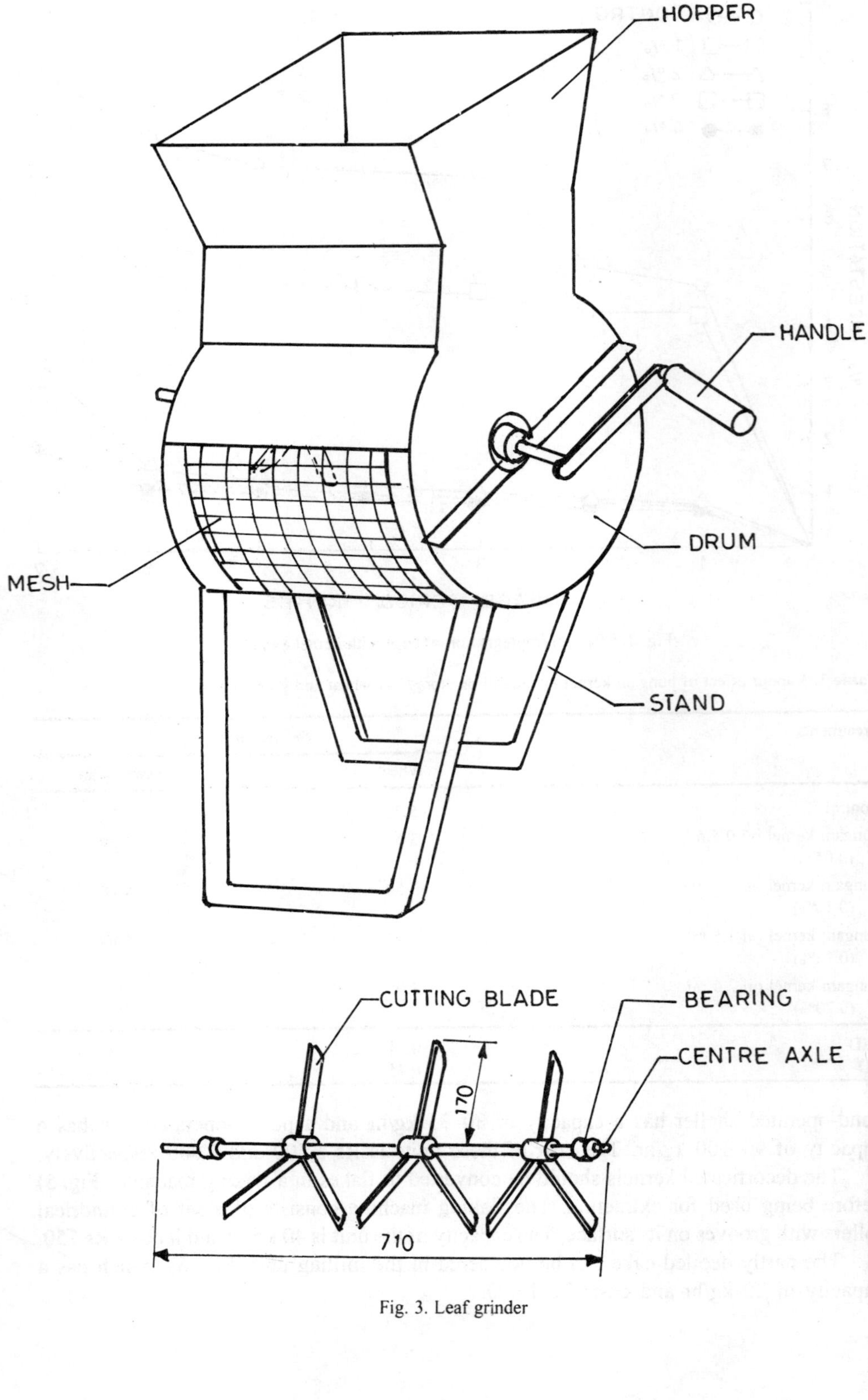

Fig. 3. Leaf grinder

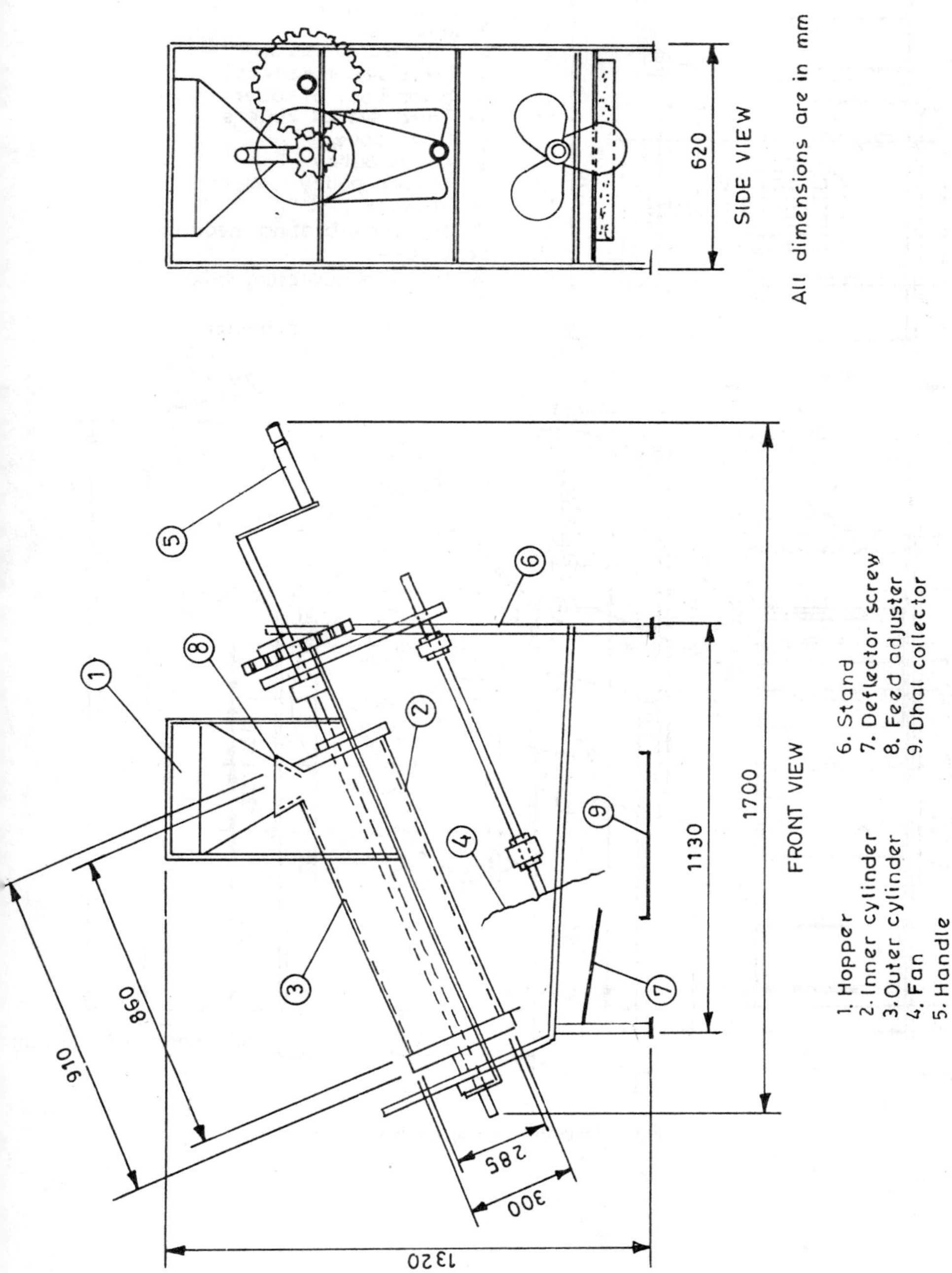

Fig. 4. Manually operated soybean dehuller (neem seed decorticator)

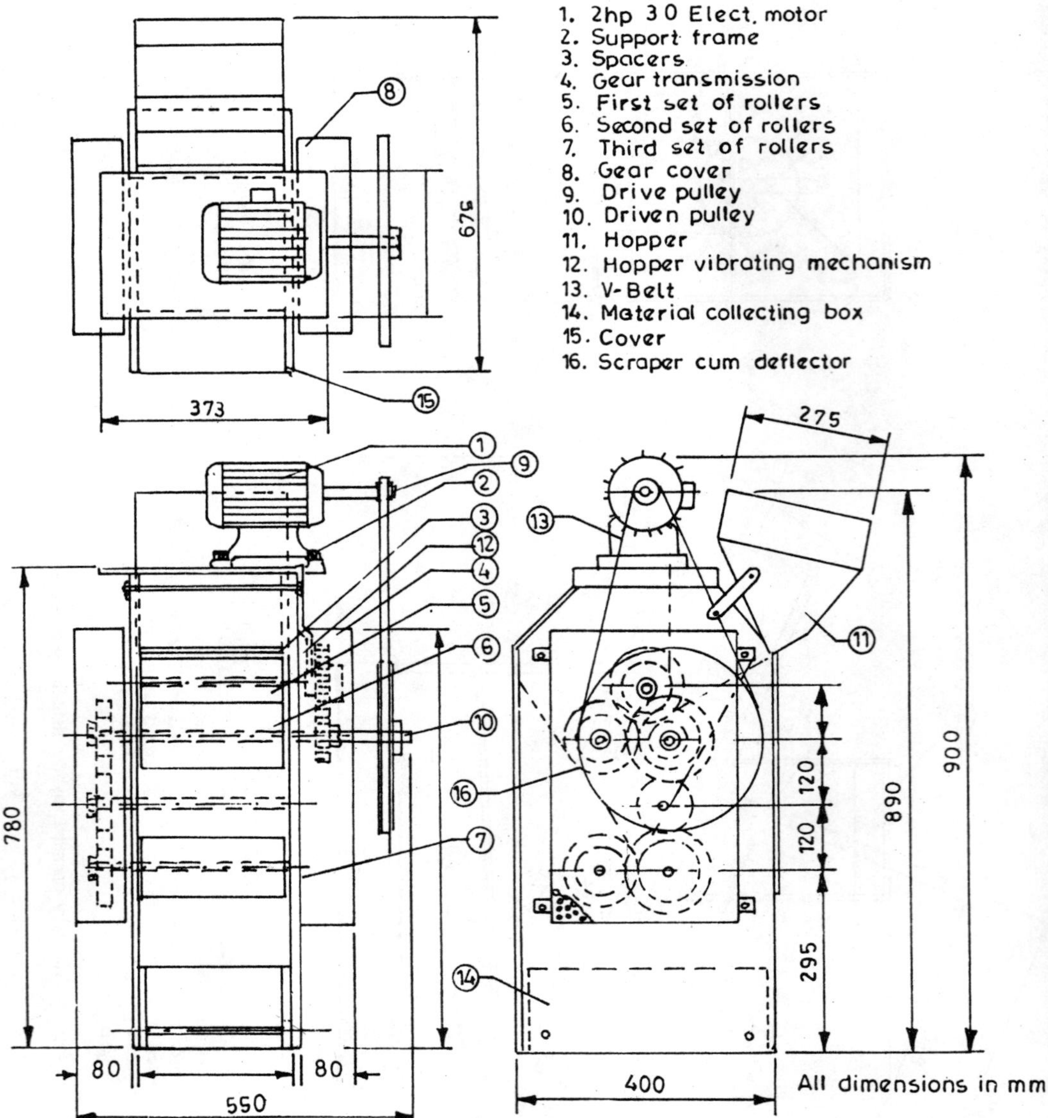

Fig. 5. Five-roller flaking machine

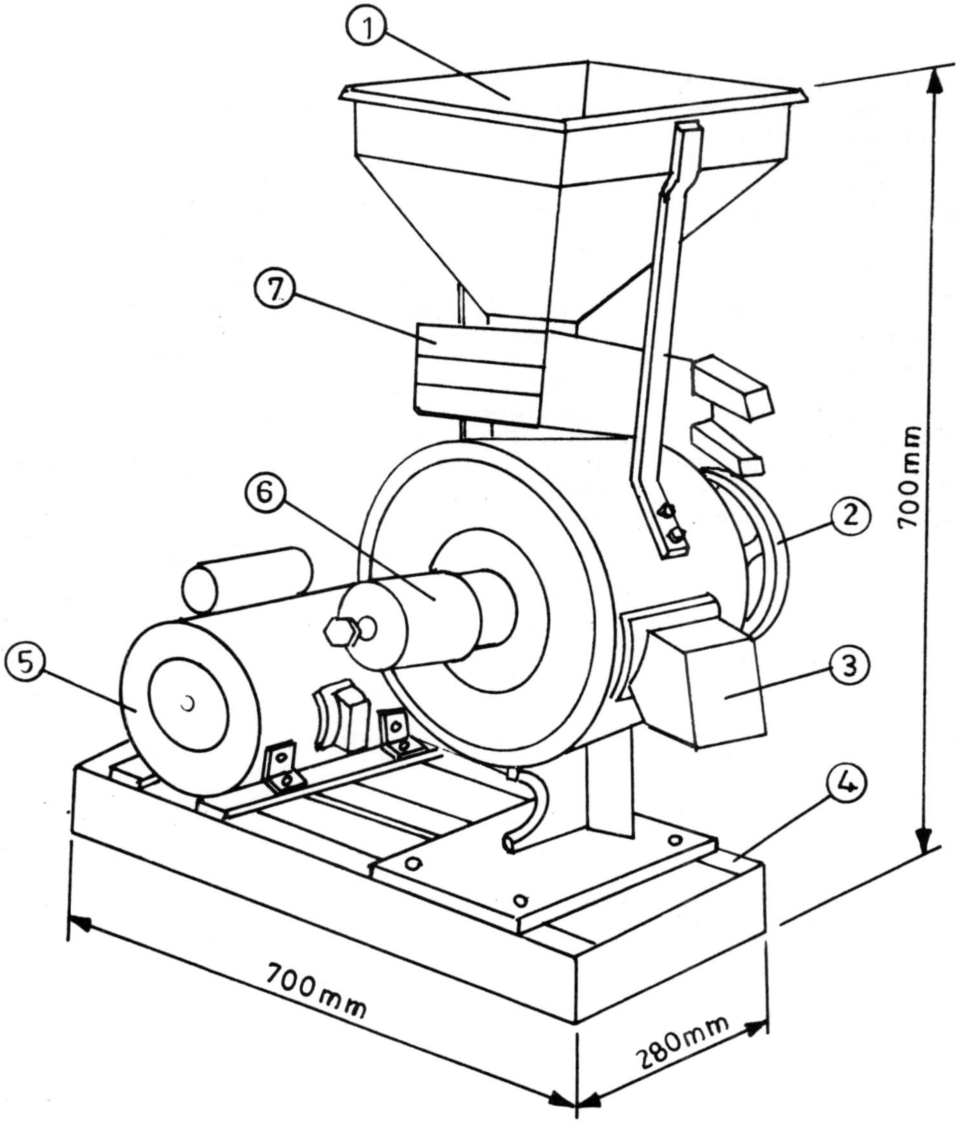

1. FEED HOPPER
2. V-BELT PULLEY
3. PRODUCT OUTLET
4. PLATFORM
5. MOTOR
6. CLEARANCE VARIATOR
7. CLEANING ADJUSTMENT

Fig. 6. Low-cost multipurpose grain mill

REFERENCES

Abraham, C.C., B. Thomas, K. Karunakaran and R. Gopalakrishnan. 1972. Relative efficiency of some plant products in controlling infestations by the Angoumois grain moth *Sitotroga cerealella Olivier* (Gelechidae: Lepidoptera) infesting stored paddy in Kerala. *Bull. Grain Tech.* 10(4): 263–266

Anonymous. 1983. *Decade of Post Harvest Technology (1972–1982)*. PHT Scheme, Tamil Nadu Agricultural University, Coimbatore.

Credland, P.F. 1992. The structure of bruchid eggs may explain the ovicidal effect of oils. *J. Stored Prod. Res.* 28(1): 1–9.

Dakshinamurthy, A. 1988. Effect of certain plant products on storage pests of paddy. *Trop. Sci.* 28: 119–122.

Dakshinamurthy, A. 1992. Insect management in grain and seed storage of wheat using non-hazardous materials. *Proc. Nat. Sem. on Bio-ecology and Control of Insect Pests,* Muzaffarnagar, Oct. 2–4, 1991. Published by Uttar Pradesh Zoological Society, Muzaffarnagar 251 001, India.

Eli, S. and Pisarev Vladimır. 1990. Toxicity of plant oils and their major constituents against stored product insects. *Proc. 5th Int. Working Conf. on Stored Product Protection*, Bordeaux, France, 9–14, 1990. Published by Esculea Agricola Panamericana, Tegucigalpa, Honduras.

Fry, J.S. 1938. Neem leaves as an insecticide. *Gold Coast Farmer* 6 (10): 190.

Ketkar, C.M., G.G. Kale and V.B. Tappire. 1976. Final Technical Report on Uitilization of Neem and Its Products. Khadi and Village Industries Commission, Bombay.

Pruthi, H.S. and M. Singh. 1950. Pests of stored grain and their control. *Indian J. Agric. Sci.* 18(4): 1–52.

104

Cold Processing of Neem Seed

G. RAMAKRISHNA, N.B.L. PRASAD AND G. AZEEMODDIN

Jawaharlal Nehru Technological University
Oil Technological Research Institute
Anantapur 515 001, India

ABSTRACT

An integrated process of cleaning, decortication and crushing of neem seed, in cold, is described. The resulting oil is rich in azadirachtin (1700 ppm).

INTRODUCTION

Azadirachtin-rich neem oils, that is, neem oils with an azadirachtin concentration of 1500–2500 ppm or more, are now in great demand all over the globe, especially in advanced countries because of tremendous scope and unlimited potential in pesticidal formulations. The neem oil-based pesticides are biodegradable, safe, efficient versatile, environment-friendly, and above all low priced. Apart from excellent qualities, these pesticidal formulations are extremely popular among the farming community. Commercial production of azadirachtin-rich neem oil has created a sensational impact on the export scenario of neem oil in India. Azadirachtin-rich neem oil is different from the conventional type of neem oil, although they are identical in colour, taste, and flavour. Azadirachtin is a C-26 triterpene present in neem kernel oil (Bringi and Thakur, 1987). This compound is hypersensitive to heat and if temperature during processing is above 50°C, the azadirachtin in neem oil gets completely denatured or destroyed.

Processing of neem seed in cold for recovery of azadirachtin-rich oil is not only an economic necessity, but is also a technological challenge. Conventional screw pressing and solvent extraction processes involve heat treatment to seed or oil at some stage and the resulting oil does not show the presence of azadirachtin. Hydraulic pressing of pure neem kernel is effective, but highly uneconomical because of low yields of oil. Cold pressing of pure neem kernels in a screw press, keeping the temperature at or below 50°C during pressing, was found to be the only answer to the problem.

As a step towards optimizing the cold pressing of neem kernels at rural level, an efficient decortication of neem seed and a suitable screw press were found to the *sine qua non* for the development of appropriate technology required in the recovery of azadirachtin-rich neem oil.

EXPERIMENT

Seed Cleaning

Commercial neem seed samples are notorious for their high refractions, ranging from 10 to 20 per cent, chiefly containing stones, sticks, sand, dust and mud pellets. As seed cleaning constitutes an important phase in the development of know-how, several approaches, involving machinery and equipment and also process development, were made. Seed cleaning data are given in Table 1.

Mechanical Destoner

The mechanical destoner is a groundut destoner in which sound seeds are separated from stones, sand and other heavier non-seed material. The device consists of a hopper and a rotating feeder. A powerful blower fan is installed at the bottom to create an air blast on an inclined plate. All light particles, including seed material, are lifted and thrown out through the exit hood. Stones, mud pellets and other heavier material are collected at the opposite end.

Mechanical Sieve-shaker Separator Fitted with Air Cyclone System

A shaker-separator fitted with suitable sieves (screens) and air cyclone, cynchronized with the mechine to get continuous separation of fraction, is used.

Sedimentation (Gravitational) Technique

In the sedimentation or gravitational technique, the seed sample is charged into a tank of water. The seed, being light in weight and low in density, floats on the surface. The foreign matter fraction containing stones, sand, sticks, mud pellets and silica sinks to the bottom. The seed fraction is easily and immediately removed from the water tank and allowed to dry in the sun.

Table 1. Cleaning of neem seed

Device/process	Refraction per cent	Efficiency per cent
Mechanical groundnut type destoner	3.5	22.5
Sieve-shaker separator with air-cyclone system	8.3	53.2
Sedimentation technique	15.6	99.0

Decortication of Seed

Neem seed constitutes 53 per cent hull and 47 per cent kernel. The oil content is 21 per cent in whole seed and 50 per cent in kernel.

Dried neem seed is decorticated in a disc-type dehuller fitted with granite and emery discs separately (Figs. 1 and 2).

One disc is static and the other rotates at 600 rpm by a mechanical drive. The clearance between the discs can be varied. A shaker-separator with suitable mesh screens and air cyclone separator are synchronized with the machine to get continuous performance. Data on decortication of neem seed using two different disc hullers are given in Table 2.

Crushing

Kernel fractions were crushed in a table expeller. A good number of approaches were made in crushing neem seed kernel using a table expeller with no cooking and heating provisions. In each run, 10 kg was used. Figures 3 and 4 show a commercial model of table expeller.

Neem seed was directly fed into the hopper of the table expeller in five passes. There was no preconditioning of expeller or pretreatment of raw material. No water was added during crushing.

In the second run, neem kernel were taken and subjected to size reduction in a disintegrator and pressed in the expeller. After the first pass, the entire mass was humidified and crushed. In total six passes were carried out.

Fig. 1. Granite disc huller

Fig. 2. Emery disc huller

Table 2. Decortication of neem seed* in disc hullers

Device	Kernel per cent	Hull per cent	Hull, kernel powder per cent	Uncut seed per cent
Granite disc huller (Sturtevant)	25	40	15	20
Emery disc huller (Bauer)	35	53	11	1.0

*Neem seed has hull content 53 per cent
kernel content 47 per cent

In the third run, neem kernels were flaked by passing through a single-pair roller reduction mill. The flaked kernel were humidified and crushed. In total, seven passes were carried out.

In the fourth run, the expeller was preconditioned by passing oilcake of the third run through the expeller till the shaft mounted worm attained a temperature of 65°C. Neem kernels flaked by passing through a single-pair roller reduction mill, were fed into the preconditioned expeller and then crushed. In total only four passes were carried out.

The oil obtained in each run was filtered in cold using a hand-operated filter press. The filtered oil is assessed for its quality.

Data obtained on crushing of seed and kernels are presented in Table 3.

Neem kernel oil obtained in the fourth run showed the following characteristics.

Free Fatty Acids (as oleic)	11 per cent
Colour in 1/4 inch cell	10.2Y + 1.2R
Refractive index at 40°C	1.4644

Fig. 3. Table expeller

Fig. 4. Table expeller

Table 3. Crushing of neem seed and kernels in table expeller under different processing conditions

Details of the process	Oil		Cake		Processing loss %	Oil Quality	
	Yield %	Exit temperature, °C	Yield %	Exit temperature, °C		Free fatty acids	Lovibond colour in 1/4 inch cell
							Y R
Direct crushing of wholeseed, no addition of water (5 passes)	10	60	89	80	1.0	11	10.2 + 2.2
Size reduction of kernels humidified after Ist pass (6 passes)	34	53	65	60	1.0	11	10.2 + 1.2
Flaked humidified and crushed (7 passes)	35	47	64	60	1.0	11	10.2 + 1.2
Expeller preconditioned, flaked, humidified and crushed (4 passes)	36	45	63	55	1.0	11	10.2 + 1.2

Iodine value (Wijs)	68.5
Saponification value	199
Unsaponifiable matter	1.49 per cent
Azadirachtin content	1700 ppm

DISCUSSION

It is seen from Table 1 that the sedimentation technique is very efficient in the separation of stones and other foreign material from the neem seed.

Data presented in Table 2 show that disc huller fitted with emery discs, sieve-shaker separator, and air cyclone system is most efficient in decortication of neem seed. The granite disc huller is also good, but it needs some more modifications.

Table 3 shows that crushing of neem kernels in cold involves a special treatment. Neem seed crushing gives a poor yield of oil. The exit temperatures of oil and cake are 60°C and 80°C respectively.

Size reduction of neem kernels, humidification, and crushing gave a good yield of oil. The exit temperatures of oil and cake were 53°C and 60°C respectively. Flaking of neem kernels, humidification, and crushing gave a good yield of oil. The exit temperatures of oil and cake were 47°C and 60°C respectively. Size-reduction, flaking, humidification and crushing of neem kernels gave the best yield of oil and exit temperature of oil and cake were 45°C and 55°C respectively. Under these conditions the azadirachtin contents are preserved intact in oil obtained, i.e., 1700 ppm.

In all the runs, a processing loss of 1 per cent material was observed. The quality of oil obtained from crushing neem kernel was uniformly good when compared to whole seed oil.

ACKNOWLEDGEMENT

Our thanks are due to V. Vannappa, Puttaiah Establishments, Kalluru 515 774 (Andhra Pradesh), India, for arranging the free supply of neem seed for carrying out the present work.

REFERENCE

Bringi, N.V. and M.S. Thakur. 1987. *Non-traditional Oilseeds and Oils in India*, N.V. Bringi (ed.), Oxford & IBH Publishing Co. Pvt. Ltd., New Delhi, pp. 118–141.

11

PROPAGATION

105

Site Suitability for Neem in Queensland

AXEL BOSSELMANN

Department of Botany, University of Queensland, Brisbane, Australia 4072

ABSTRACT

While neem is not traditional to Australian society but was brought to the country repeatedly during this century, especially once the insect repellent properties of Azadirachtin became known. In ignorance of neem's siting and growing requirements, neem has often been planted in Australia on not particularly suitable sites where it proved to be sensitive to pests and diseases, slow and not at all prolific in cropping; this was the case even in the state of Queensland, with its variety of arid, tropical and sub-tropical climates in coastal as well as elevated areas, a state which offers many potentially suitable sites for both large- and small-scale cultivation of neem in order to either produce azadirachtin or restore land. By arriving at and describing suitability criteria, growers are being helped to identify proper sites for neem, the primary considerations being temperatures, specifically the average minimum temperatures in winter, the nature of the soil and the spread of rainfall. Good initial ground management and spacing are also likely to ensure a good crop.

INTRODUCTION

Neem (*Azadirachta indica*) is not an accultured or 'social' tree in Queensland as it has always been on the Indian subcontinent or as it has become in tropical Africa within a few generations (Vietmeyer, 1992). Although most neem trees in Australia were deliberately planted and did not result from natural spread, people in the neem industry there did not normally have the benefit of traditional experience. This lack of hands-on expertise is matched by the dearth of scientific knowledge on neem ecotypes. Their respective biomass, and their adaptation to new environments.

Information about neem and insect repellency far outweighs what we know about neem as a tree. The previous three International Conferences on neem dealt almost exclusively with pest control. The World Neem Conference is also to deal with the tree and its place in the environment, an overdue change in emphasis.

The twin lack of general knowledge about neem and of specific know-how did not prevent growers in Queensland from attempting to raise the tree there in order to protect their crops or to sell seeds from which to extract azadirachtin. It did mean that neem, once planted, did not always grow or produce to expectations. The criteria and approaches outlined in this paper should help growers find sites suitable for the purpose for which they intend to grow neem. In practice, we are more likely to find out why neem may not have grown at all or performed to expectations.

The determination of site suitability helps us find out why/how neem adapts to new areas. The majority of neem was planted in Australia in the last twenty years, largely for the purpose of crop protection, directly or through the intended extraction of its byproducts. This was the first time that neem was introduced for a not entirely traditional purpose and by people not previously culturally familiar with the tree.

Brief History of Neem in Australia

Neem had been introduced to Australia in small numbers on previous occasions. The tree first arrived on that continent in the waning years of the Makassar Trade between North Australian Aborigines and East Indies Islanders from what is now Eastern Indonesia. This trade, which legally ended in 1903, influenced Aboriginal culture in its Day more than the Europeans did and it brought many tropical staples to North Australia as well as neem seed the progeny of which constituted the so-called Northern Territories Accession. Two generations later, former members of the Imperial Armed Forces who had become familiar with neem during World War II as a remedy for malaria, often enough as prisoners of war, took neem with them from Southeast Asia.

More recently, neem was brought to Australia for aesthetic reasons, such as roadside trees at the RAAF Base at Darwin after Cyclone Tracy (1975) or for mine rehabilitation, notably by Comalco on their bauxite mine at Weipa near Cape York, in 1968–1970. The majority of the following findings and observations derive from Weipa.

APPROACHES TO SITE SUITABILITY

Three approaches to neem in its environment were used: a geographical and an ecological one and a systems approach derived from engineering. These are the L-approach, the niche approach and the irreducible minimum.

L-Approach

The L-approach is named for L as in latitude and longitude of a given locality in Queensland. It is a straight forward approach; the azimuths are compared to those of a known neem-growing area elsewhere and soil and weather data of each respective site are gathered and made use of, in comparison with established neem-growing areas if possible. Neem's very universality overseas does not mean that site-specific data are always available to compare with those of similar areas of Queensland.

We shall compare neem at Weipa, near Cape York, and at Redlands Bay, east of Brisbane, discuss a planned neem project at Charters Towers, in semiarid North Queensland, and the Atherton Tablelands in Far North Queensland where a large number of neem growers live. Several neem growing areas overseas will also be taken note of (see Table 1).

Niche Approach

This approach may be less relevant to a grower or an extension agent but it fits in more comfortably with an ecologist's view of neem's suitability in a given area.

The tree may prove to suffer from, or else tolerate, competition or it may invade and later monopolize a site. Neem may compete poorly in one area of Queensland but may thrive somewhere else due to geographical differences and our problem is once more dearth of data.

In the following, we shall see how compatible neem is with other trees, weeds, ferns and grass in Weipa and Cleveland—two geographically distinct areas of Queensland, and incorporate some recent findings from Northwest India.

Irreducible Minimum

This approach is designed to ascertain and analyze the minimum demands on a site to be met for neem to survive or to produce which can be two different criteria.

Each area and site—and sections within each site—are subject to a stress load which varies over time and because of human intervention and which the 'irreducible minimum' must match.

This can be done with the help of those findings and observations needed for any of the above approaches. The site has to be well-known, reliably observed and carefully analyzed. Some basic or summarized data and a very good understanding of an area will help approximate this approach which makes use of the skills of tribals, traditional people, and experienced farmers and foresters alike.

Irreducible minima have been approximated in the following for several sites in Queensland, based on site visits as well as interviews with owners. The approach needs to be verified or, more likely, modified over several years, to be more accurate. The soil profile, the ability of the soil to retain heat and moisture, site-specific insolation, air movement, water runoff, land use, plant growth, animal movements, once known, will enable us to arrive at a more exact idea of the prevailing irrecducible minimum. The nature and frequency of disasters also need to be considered.

WEIPA—NEEM IN A CONGENIAL ENVIRONMENT

The Geographical Approach

Queensland is the second-largest state of Australia, of a size double that of present-day Pakistan and in some ways similar, with the exception of that nation's alpine country which is absent in Queensland.

Weipa is in the very far north of Queensland, at 12° 35' S and 141° 51' E, near Cape York, situated in a mangrove belt and on a layer of bauxite, in an area similar to the lateritic belt of the Casamanche in West Africa or that of Bihar and Bengal in

India. Neem was introduced to Weipa for purposes of mine forestry from India from 1968 to 1970 (Nicholson, 1992)

While the original trees are no longer there, due to the expansion of the mine, seedlings from these trees have been successfully raised since 1988 and, in the following two years, material from six other countries was planted out on areas previously mined for bauxite, with trees in their third season showing a minute crop.

The soil is made up of bauxite mine tailings on top of kaolin on some sites and a very sandy loam on others. Nutrient values are extremely low. The long-term average temperatures range from 17°C to 38°C, humidity from 75 to 95 per cent, rainfall from 1,300 mm to 2,600 mm annually (Comalco 1992; Brisbane Weather Bureau, 1991).

This compares with a location west of Zinguinchor in the Western Casamanche, Senegal, (Laurie, 1974), with similar temperatures (Table 1) and rainfall (Danida, 1976). Neem has grown in the Senegal as a common roadside tree for almost forty years (Vietmeyer, 1992). It compares less well with locations in SW Burma at 12° S where neem does grow but does not produce as much as in the drier Central Burma (Oo, 1987), the annual rainfall exceeding 2,000 mm in the one versus about 1,000 mm in the other area.

The oldest and best managed trees in Weipa are at least as high and have more usable timber at age five than neem plantation trees in northern Nigeria at age ten (Laurie, 1974).

Neem and Its Niche in Weipa

Even though neem trees are not very deep-rooted in Weipa, they did withstand cyclones Mark and Nita in 1992. Only one tree was lost and a dozen bent sideways.

The roots form a network of strong and very long lateral roots at a depth of 1 or 2 ft, supported by much finer roots, the thinnest of which can be found 5 ft deep. The most productive trees in Weipa were originally well-spaced, mulched and managed as to ground cover; this was almost as important as irrigation and fertilizer in creating attractive and prolific trees (Benge, 1989).

Neem is stressed on site by vines, grass and undergrowth such as stylo but intergrows naturally with trees such as *Acacia mearnsii* and *L. leucocephala*, and the odd young eucalypt. This matches findings elsewhere, especially in NW India (Benge, 1989; Arya et al., 1992)

Arya et al. (1992) observed an "unusual association" between *Prosopis cineraria*—khejri—and *A. indica*—neem—at Jumpa, about forty kilometres southwest of Hissar at the edge of the Thar Desert, "characterized by low rainfall of about 300 mm, high temperatures of up to 48°C in summer and a sandy clay soil." Both species were growing in very close association, often touching at the base, and the canopies were intermingled. About thirty such associations were distributed throughout the area.

The height and circumference of khejri ranged from 4 to 6 m and 20 to 90 cm, respectively, and that of neem from 2 to 5 m and 15 to 40 cm, respectively. Khejri is slow growing with a root system penetrating to a depth of 233 cm while neem is fast growing and has a superficial root system which spreads horizontally to a 150 cm radius. The canopy of khejri is relatively sparse compared with the dense canopy of neem. It is assumed that there is little competition between the species for moisture and

Table 1. Neem in various tropical environments

1.1. Neem at Several Australian and Overseas Locations

Location	Azimuths	Predominant Soil Type	Annual Averages	
			Temperature (°C)	Rainfall (mm)
Coastal Casamanche	11°N; 15°W	lateritic	26	1,750
Northern Nigeria*	10°N; 10°E	lateritic	28	over 1,000
Tennaserim, Burma	12°N; 99°E	lateritic (3)	31	over 4,000
Cape York, Australia *	12°S; 143°E	lateritic (1)	26	1,724
Northern + Terr., Aust.	15oS: 133°E	lateritic	27	757
Central India *	17°N; 78°E	lateritic (2)	27	754
Central Burma *	20°N; 95°E	lateritic (2)	27	829

Legend: * Towns/Cities: Ziguinchor; Kano; Weipa; Mandalay; Hyderabad; + intended plantation 100 km south of Katherine, North. Terr. (1) bauxitic lateritie; (2) black soil occurs; (3) alluvial soils occur

Sources: M.V. Laurie/FAO 1974; DANIDA 1976; Tla Oo 1987; Weather Bureua Brisbane, 1991; Atlas of Austr. Resources 1980/1988; RCS 1990

1.2. Tropical neem: biomass, growth and production

Area	No and age of trees	Purpose of plantation	Annual increment (3)	Useful wood (kg)	Annual yield: Timber	
					crop: kg/tree	m³/ha
Cape (1) York	100/5 yrs	reclamation	100; 3–5	81	0.5–2	3–5
India	1,000/8yrs	trials	150; 2–3	n.a.	9	3–10
Niger	2,000/8yrs	plantation	75; n.a.	62	n.a.	10.9
Irraw (2) Delta	n.a./5–10	village trees	n.a. n.a.	n.a.	8	n.a.

Legend: (1) Weipa: @ 100 neem/ha; best one hundred trees chosen
(2) Burma: village and roadside trees observed; highest yield reported at over 50 kg from 15 yr-old tree
(3) First column to indicate increment in height; second column to show increment in girth; all in cm

Sources: Comalco Regeneration Internal Data, Weipa 1991-92; Tla Oo, Neem Cultivation in Burma, Terminal Report, 1987; M.V. Laurie/FAO Monograph, p. 50, Rome 1974; NG Hedge, BAIF Poona, in: Neem Newsletter 9/1. p. 3 ff, 1992.

Table 2. Soil in a neem/cashew orchard near Weipa, Australia, 1992

Major nutrients	Depth (cm)	Characteristics at depth indicated	Aggregates (in cm)	pH measured	at depth (cm)
low to very low	50–60	sandy loam, with gravel and ironstone	appr.1	7.5	30
throughout	100–110	bauxitic; with gravel and ironstone	over 5	6.5	60
below	110 to 150	bauxitic laterite	over 10	5.5	120

Source: Author working at Weipa, January 1992.

nutrients. Khejri may help neem obtain nitrogen which would otherwise be a limiting factor in these arid areas. Khejri is also known to add organic matter and increase soil moisture content. Neem may also benefit by being sheltered from intense sunlight and from low temperatures. Neem may, in return, protect khejri from a number of insect pests.

The association between these species may lead foresters to formulate silvicultural systems which rely less heavily on expensive protection by pesticide(s). The authors were "unaware of any attempts to reproduce experimentally the intimate association observed between these naturally established trees" (Arya et al., 1992). The authors made an observsation similar to what one can observe on less-well managed neem sites near Weipa: "In the natural state, the neem tree suffers from insect attack, particularly at the growing tip, producing a bottle brush effect from resprouting" (ibidem).

An Australian native—*Dysoxylon oppositifolium*—Pink Mahagony—itself a *melia*—occurs near Weipa as a canopy species on a so-called Land Unit described as follows: "Soils are lateritic/bauxitic Red Earths. Soil reactions are slightly to medium acid. Infiltration and drainage are rapid so that little runoff occurs . . . . (Table 2). The vegetation is notophyllic vine forest; the occasional deciduous and semi-evergreen trees are conspicuous and wattles (*Acacia* spp.) are present" (Godwin, 1985). These latter interact with neem closely.

Irreducible Minimum for Neem at Weipa

The stress load on a given site is made up of constant as well as occasional stress factors. In Weipa, these are: the delayed and indifferent rains from 1990 to 1992; heavy, at times cyclonic winds, such as in January and December 1992, the occasional fire, such as in September 1991. As far as lack of nutrients and bedrock at low depth are concerned, the frustration of many a neem grower is aptly expressed in the literature (Zech, 1984; Laurie, 1974; Danida, 1976): "Measurements of physical properties of soil under neem in Kano showed no significant differences between sites where neem is growing well and where it failed or was not planted" (Danida, 1976). The irreducible minimum to ensure neem production in Weipa is achieved by these enabling factors: consistently high to very high temperatures, good heat retention, good insolation, high initial rainfalls (over 4,000 mm in 1988/89; Comalco Int., 1992) (Tables 3–4).

THE ATHERTON TABLELANDS — NEEM AT A HIGHER ALTITUDE

The single largest number of neem growers anywhere in Australia live on the so-called Atherton Tablelands in North Queensland. The original motive behind planting neem was environmental awareness and the expectation of on-farm insect control. The numbers of neem trees involved are not normally large, with the possible exception of two sites (Friend, 1993).

On one of these, Dimbulah, at an altitude of 500 m, at 17°S and 143°E, neem planted in 1989 suffered from flood, frost—at minus 10°C, fire and a severe drought within two years of its establishment, yet some 60 per cent of the trees survived. The soil is an inert clay and the site used to be under pasture; tobacco is common in the area.

Table 3. Weather patterns (summarized) at Weipa, 1990–1992

Temperature (°C)			Rainfall (mm)					Wind speeds:	
max.	min.	mean	monthly			annual		1. in km/h	2. Percentage occurrence
			max.	min.		av.	max.		
36.3	16.2	26.2	804	0	(1.2)	1,724 (3)	2,400 (4)	0 to less 1 1 to 5 over 30	36% 14–18% 1–2%

Legend: (1) December 1992; (2) Six months in 1991; (3) average since 1974; (4) exceeded in 1988/89.
Sources: Comalco Internal Regeneration Data, Weipa 1974–1993
Brisbane Weather Bureau/Queensland Met. Services 1991.

Table 4. Ground temperatures (°C) and soil water stress (c/bar); Weipa

Sites (1)	Early morning		Mid-morning		Noon		Afternoon	
Lake Mc Leod	29.1 29.5	26	35.8	36	39.9	40	Not available	
Rock Follies		36	32.1	29	38.3	38	40.8	62 (2)
Growth Trial(2)	31.9	34	34.0	34	37.8	40	48.8 (2)	42
Neem Cashew Orchard	29.0	33	33.3	33	35.2	30	46.8	42

Legend: (1) name of formerly mined sites as given by Comalco, Weipa
(2) growth trial conducted by author in Weipa itself
Source: Author's findings obtained 11.11–1.12. 1992.

Frost or at least low winter temperatures are common to the site as well as to the Tableland in general. The years 1991 and 1992 were dry years on the Tableland, with 13 inches of rainfall in twenty months at Dimbulah. On the positive side, daylength is uniform, temperatures exceed 30°C in summer and there is little weed competition.

As in Weipa, neem appears compatible with native trees; the grower exports native seed from his property. The trees flowered in both 1991 and 1992, under very difficult circumstances, as described above, with a crop of a few hundred seed in 1992/93.

CHARTERS TOWERS — PROPOSED REHABILITATION OF A MINESITE

Charters Towers, at 20°5' S and 143°16' E was once one of the important cities of Australia, with its own Stock Exchange, due to gold which continued to be mined in and around the city till 1930. Nearby reefs continue to be mined. A former minesite at the outskirts of Charters Towers, used by the University of Queensland to house an underground seismograph station, is to be revegetated by neem trees as of 1992. The area is at least semiarid, isnpite of its closeness to the Coral Sea about one hundred kilometres away, with temperatures ranging from 8°C to 40°C; the annual average rainfall of 661 mm is concentrated during the wet season and has failed since 1991. There was frost in 1991 but not since.

The site is on weathered granite, the soil being a yellow duplex of little depth. It consists of a basin between two slopes and is studded with 'mullock'—decomposed mine tailings—it supports some ground vegetation, such as Hummock Grasses, *Trioidea* spp., Rubbervine, *Cryptostegia grandiflora*, which is an arboreal weed, *Albizzia* and Chinapple, *Zyzyphus indica*, a local wattle, *Acacia* sp., and some mango and tamarind deep in the basin and at the fringes, respectively. As a location, it compares with Hyderabad in Central India where neem is very common, but it is much closer to the coast than that area and gets much colder in winter as well (0–8°C versus 16 to 21°C; Brisbane Weather Bureau; 1991; RCS, 1990). The enabling factors likely to maintain the irreducible minimum consist of the high insolation, with a minimum of nine hours of sunshine on site during most of the year, the usually very high temperatures, the relative dearth of competition from groundcover, the loose nature of the rock, and the potential of sufficient occasional rainfall.

The limitations are heat reflection from the ground, extended droughts, frequent low winter temperatures and possible damage by wildlife and by vandals. The hazards of possible contamination with cyanide from the mining days and the effects of competition from rubbervine are unknown and form part of the trial.

REDLANDS BAY—A MARGINAL SITE FOR NEEM IN QUEENSLAND

Geographical Data

Redlands Bay is the name of an area east of Brisbane, the capital of Queensland, at 27° 30' S and 153° 15' E, on red duplex soil, close to mangroves along Moreton Bay. It is also the name of a research station near the township of Cleveland. While rainfall is usually less than in Weipa, it tends to be better distributed; 1991 and 1992 were drier than usual (see Table 5). Temperature averages range from 6 to 36°C, with actual values sometimes dropping to zero at night, e.g., July 1991.

Development of Neem at Cleveland

About 120 neem trees, many of them originally tissue-cultured (Drew, 1993), of six different accessions, were planted in Cleveland in early 1988. Ten of these had to be replanted; 25 died and were removed, with eight of them regrowing as suckers. Cyclone Fran alone felled six trees in March 1992, three of which were measured and weighed (see Table 6). Inflorescence occurs at the same time as in Weipa but a crop will not form till December/January and ripen till mid-year, the Thai variety doing so as early as February. This compares with crop formation in Weipa in October/November and their earliest crop maturing from December onwards. Harvest in Weipa outweigh those in Cleveland by several orders of magnitude—150 kg as against 0.5 kg per 100 trees. After earlier careful ground management in Cleveland, weeds have begun to appear and reach the height of some of the smaller trees at 4 to 5 ft. (A very recent finding suggests that weeds grow less well with neem in containers at high temperatures. This was observed at both Weipa and the University of Queensland.) Unlike at Weipa, neem has not been observed so far to grow close to other trees at Redlands Bay.

Table 5. Cleveland—A marginally subtropical coastal site meteorological data from Redlands Bay Research Station

Rain (mm)	Evap.	cl/ds	Temperatures (°C)		Moisture(%)		Wind (km/h)	Month/Year
210	158	6	33.1	14.2	96	42	24 NE	Dec. 1991
43	190	8	33.2	15.5	97	55	18 NE	Jan. 1992
239	128	6	31.5	16.5	98	61	24 NE	Feb. 1992
10	153	15	29.4	7.8	91	46	24 SE	Oct. 1992
150	177	10	34.0	10.9	94	32	24 NE	Nov. 1992
105	162	5	30.5	19.0	90	50	81 NE	Jan. 1993

Legend: Columns Three and Four indicate the resp. maxima and minima
Column Two: indicates number of near-cloudless days
Column Five: prevailing direction and maximum strength
Column Six: data for Dec. 1992 not avail.; neem 4 yrs. old

Source: Queensland Department of Primary Industries; weather data taken at Redlands Bay Research Station east of Brisbane.

Table 6. Biomass, new growth and neem production at Redlands Bay

Roots	Useful wood	Leaf formation	Inflorescence	Harvest*
4.3 kg(1)	28 kg @ 10 cm	1–3 cm/winter	27% : 1990/91	0.4 kg
3.3 kg(2)	22 kg @ 10 cm	3–7 cm/spring	45% : 1991/92	0.5 kg
1.5 kg(3)	15 kg @ 4 cm	10–12 cm/summer	62% : 1992/93	0.6 kg

Legend: * harvest carried out by Dr R. Drew, Redlands Bay Station (1) at 4 cm diameter; (2) at 2 cm dia.; (3) at 1 cm diameter (4) minimum 4 cm diameter; actual diameters given leaves: more than 3 kg/tree; root exc. 2m; trees 3–4 m new leaf growth measured weekly data from first three columns taken from six fallen trees

Source: Author, working with material from site visits at Redlands Bay Research Station, Cleveland, since 1989/90.

Factors Enabling Neem at Cleveland

The Irreducible minimum is met at Cleveland by the high daytime temperatures during most of the year, the good supply of water by rain or irrigation, and good spacing. The stress factors are cold spells, heavy winds and disease.

Wood Rot which may have originated from previous growth in the ground rotting as well as Pseudomonas solanaceae bacteria harboured by nearby perennial crops caused considerable distress in 1990/91. The latter is known in the literature as Neem Blight—*P. azadirachtae* (Benge, 1989).

Presently, the majority of neem trees there have begun to form into solid trees, at over 4 m high and more than 20 cm across, with plentiful and quite attractive inflorescence.

NEEM IN ALLUVIAL COASTAL QUEENSLAND

Some 400 neem trees originating from Redlands Bay Research Station were planted out on a research station near Ayr, also belonging to the Queensland Department of Primary Industries, to make use of the higher winter temperatures in North Queensland. Current work at that station focusses on abscission and on the formation of azadirachtin during the maturing process of the neem berry.

The trees recovered after a cyclone in 1990 and several floods; they have since started forming crops and promise yields similar to but generally lower than those of Weipa. Once saturated by water, the clayey soil failed to hold the trees firmly when wet and stressed the trees further by compacting when drying out subsequently. Weeds do not invade the well-managed site, nor do trees grow alongside neem there. The enabling factors are evenly warm to hot weather, a relatively uniform daylength—10.5 to 14.5 hours/day, good rainfalls at the beginning (1989–91), as well as the protection which the trees give one another.

The trees appear to have redeveloped a viable root system since having been uprooted by that cyclone in 1990. Private neem growers in or near the coastal city of Ayr have had to contend with that cyclone but not with their neem trees being uprooted during that event; their orchards, raised under identical conditions to the ones described above, vary drastically in output.

COMMENTS AND CONCLUSIONS

Geographical Considerations

Neem has been successfully established in Queensland but there are sites and areas where it will either not grow at all or, more likely, be considerably stressed and not be able to produce a viable amount of insect repellent. One factor which sets at least certain varieties of neem back in terms of bulk and crop production is low winter temperatures. While latitudes in themselves may not cause a loss in production, neem crops form and mature later at higher latitudes. High altitudes or, conversely, the vicinity of a coastline exercise a more indirect influence on the tree through temperatures and the distribution of rain.

Environmental Considerations versus Growers' Expectations

To their own regret, many growers planted neem with the expectation of rapid growth and a sizable crop early in the tree's life where reasons such as those mentioned above rendered that hope unrealistic. Their initial decision to plant neem had been based on their genuine concern for the environment and on our limited understanding of neem's adaptability away from its original tropical habitat.

In the following, we shall briefly repeat the factors enabling or limiting neem in its ability to grow and produce a crop in Queensland.

Charters Towers — Introducing Neem to a New Site

On a new project and in a dry area such as Charters Towers, neem may perhaps be expected to grow as well as in a more established one, such as Weipa but not to yield a comparable crop for years to come. The most severely limiting factor, apart from the dryness, are the low temperatures on some winter mornings. As a result, neem is much less likely here to spread naturally and thus be of concern to ecologists but it is expected to fit on well with the other trees on site, with the possible exception of rubbervine-*Cryptostegia grandiflora*—which is a most pervasive weed. What effect potential site contamination may have on neem seedlings is not yet known. Insolation, even in winter, high temperatures during most of the year and good care should help neem grow on the bare former minesite owned by the University of Queensland.

Neem in a Cold Winter — the Atherton Tableland

The limiting factor throughout the Atherton Tableland, an area in Far North Queensland at high altitude, may well be low night temperatures in winter. The stress factors neem encountered on that most extreme site near Dimbulah were flood, frost and fire. The enabling factors are likely to be heat reflection and the good insolation on the Tableland, improved on by good rains on its eastern half.

Neem in the Tropical Cyclone Belt — Weipa Below Cape York

At Weipa, the warm to very hot climate, coupled with at least occasionally high rainfalls, enable neem to thrive second to nowhere else in Australia and quite as well as in comparable, well-known neem growing areas overseas such as Nigeria and the Casamacche in West Africa. While undergrowth can and does set back younger trees, the proximity of both *Acacia* and *Leucaena* spp.—frequent and unintended—seems to benefit neem on some sites. The inert bauxitic soil does not prove to be a stress factor, nor do fire and cyclones, both of which occurred in 1991–92. The enabling factors may be heat reflection, the near-constant daylight and good wet seasons when the trees were first planted from 1988 to 1990.

Neem on Research Stations in Coastal Queensland

Among these, Redlands Bay east of Brisbane proved to be a marginal site. Winter temperatures are often low, the frequent heavy coastal winds damage the orchard, and the trees themselves proved to be susceptible to insect attacks and diseases which they did not suffer from on more suitable sites. The research station at Ayr, which received most of the neem trees originally raised at Redlands Bay Research Station, is more suited to neem in terms of location and weather; its limiting factors are the effects of cyclones and clayey soils. These are more than matched by warm weather and good rains.

SUMMARY

The three approaches used here to discuss site suitability for neem in Queensland, the L-Approach, the Niche Approach and the Irreducible Minimum, can be used to help a grower decide if neem is likely to grow and produce on a given site and to understand why a tree may have failed to do so in the past. The tree is not yet part of Australia's culture; nor have its growing and siting requirements been well understood. Any of the above methods would be more commonly used to help find out why neem did not grow at all or do as well as expected; for data on neem's biomass, growth, and performance are none too frequent in the literature.

While it is useful to know how and why neem was introduced to Australia during the last one hundred years, the chief contribution this paper seeks to make is by introducing three environmental approaches to neem and to show how neem adapts to a new growing area.

While many more data on neem, its physiology and growth, are yet needed, these approaches help determine the kind of data to be sought and their likely sources. They also allow useful comparisons with other growing areas.

The introduction, or better, reintroduction of neem to Australia during the last decade may have been the first time that a crop was brought to a country for the express purpose of helping to ensure the wellbeing of its environment and its inhabitants, in this case through the provision of Azadirachtin. An admittedly similar situation to this was the spread of pyrethrum in Kenya earlier this century, a plant which can be labelled as neem's "predecessor crop" with some justification (Vietmeyer, 1992).

ACKNOWLEDGEMENTS

Without the help of Comalco Pvt. Ltd, at Weipa in Far North Queensland, I could not have gathered the bulk of the above data and made most of the observations vital to my work with neem; I like to thank especially Paul Warren and his staff. Without a grant of a University of Queensland Research Scholarship, I could not have carried out this work: and without the University of Queensland's owning the site at Charters Towers, I would not have had access to it for my present regeneration trials with neem. I am grateful to my supervisors Martin Rice and David Doley, to Eric Heidecker and Col Lynam, all of the University of Queensland. Without people's help from the Department of Primary Industries, such as, among others, Rod Drew and Ian Baily, at Cleveland and Ayr, respectively, as well as from those within the industry, such as Joe Friend, Ausneem Plantations, Barry and Kevin Ahern, Ayr, and Mr and Mrs Lehmkuhl, of the Neem Growers' Association, Mackay, this work would have been both impossible and meaningless.

I would also like to express my appreciation to Dr. and Mrs. Joarder and their colleagues from Rajshahi University, Bangladesh.

REFERENCES

Atlas of Australian Resources, Vol. 1: Soils and Land Uses, Map 1, 13 pp. Canberra 1980.

Atlas of Australian Resources, Vol. 5: Geology and Minerals, Map 1, 13 pp. Canberra 1988.

Arya, S., O.P. Toky and P. and S. Harris. 1992. A Most Unusual Association, full text of paper, see: Agroforestry Today, Vol. 4, no 3, pp. 9/10; published by ICRAF, Nairobi, Kenya. Also see; P. Harris, Drought Defeaters Report, 21 pp., Henry Doubleday Research Association Newsletter, Coventry, Summer 1992.

Benge, M. Vol 1, ch. 1. p. 12–15. In: *Biological Insecticides*, M. Jacobson (ed.) CRC Press, Baton Rouge/Florida, 1989.

Brisbane Weather Bureau of the Queensland Meteorological Service, 1991 data on Weipa, Charters Towers, Ayr, Tableland.

Comalco. 1992. Regeneration Internal Data, Comalco, Weipa, 1992.

Danida, 1976. *Proc. of the Savannah Afforestation in Africa*—Danida in Kaduna, Nigeria; published by FAO, Rome, 1977.

Drew, R. 1996. Clonal propagation of neem by tissue culture. These volumes.

Friend, J. 1996. Problems and prospects for neem plantations in Australia. These volumes.

Godwin, M.D. 1985. Land Units of the Weipa Region, Unit 3 c; Report compiled for Comalco Ltd., Weipa, by the Queensland National Parks and Wildlife Service, Brisbane.

Hegde, N.G., BAIF, Poona, Research Report. Neem Newsletter, Delhi, January 1992 9(1): 3ff.

Laurie, M.V. 1974. *Treeplanting in African Savannahs*; 50 pp.; Maps 2/3; indicated as: FAO Monograph, Forestry series 14, 1974.

Oo Tla. 1987. Neem Cultivation in Burma, Terminal Report, Burmese-German Project on Plant Protection and Rodent Control, GTZ Rangoon, December 1987, pp. 51/2.

Nicholson, D. 1992. Personal communication, Queensland Forestry Service, retd., at II Ausneem Conference, Mareeba, 1992.

RCS Independent Technical Report. 1990. 43 pp., Darwin; made available courtesy of Greening Australia Queensland, Brisbane, 1991.

Vietmeyer, N. 1992. Neem—A Solution to Global Problems. National Academy Press. Washington D.C., pp. 30, 85

W. Zech. 1984. In: *Potash Review No.* 1, Bern 1984; see also in: *Symposium on Fast Growing Plantation Trees, on Nutrient Deficiencies under Selected Trees*, 399ff pp., Prefotia and Pietermaritsburg, April/May 1984.

106

Somatic Embryogenesis and Organogenesis in Neem (*Azadirachta indica* A. Juss.)

R. ISLAM, N. JOARDER, O.I. JOARDER, M. HOSSAIN, S.M. RAHMAN AND A.T.M. NADERUZZAMAN

Department of Botany, Rajshahi University, Rajshahi, Bangladesh

ABSTRACT

Methods for regeneration of plants *in vitro* from cotyledon and nucellus of *Azadirachta indica* are described. High frequency somatic embryogenesis was achieved from cotyledon explants in MS medium containing different concentrations of NAA. The embryoids developed into shoots when the embryogenic calli were transferred to cytokinin—supplemented media. However, the frequency of embryo conversion was very low. A high concentration of BA (0.5–2.5 mg/l) along with a comparatively low one (0.1–0.2 mg/l) of NAA induced cotyledon and nucellus explants to produced adventitious shoots. Shoot organogenesis best occurred when 2 mg/l BA and 0.2 mg/l NAA were used.

INTRODUCTION

Neem is a member of the mahogany family, Meliaceae, and today it is known by the botanical name *Azadirachta indica*. Neem is in great demand in Bangladesh, India, Australia, and many other countries as a valued timber tree. It is evergreen and often planted on roadsides, and its wood is very hard and durable. The plant yields many strange and varied products which may eventually benefit every person on the planet. Neem extracts are known to affect more than 200 species of insects many of which are resistant to or inherently difficult to control with conventional pesticides (Wulf, 1991). Neem oil is now used by soap, cosmetics, lubricants and fertilizer manufacturers. Neem preparations have been found to show toxicity to cultures of 14 common fungi and several species of pathogenic bacteria.

This report describes a system for *in vitro* plant regenration in *A. indica* through somatic embryogenesis and organogenesis.

MATERIALS AND METHODS

Immature fruits (75–85 days after anthesis) were collected from a 30-year-old neem tree, surface disinfected with 0.1 per cent $HgCI_2$ (w/v) for 12 min., and finally washed four times with autoclaved, distiled water.

Cotyledon Culture for Somatic Embryogenesis

Immature seeds from surface disinfected fruits were removed and cotyledons excluding embryonic axis were excised and cultured on MS (Murashige and Skoog, 1962) medium supplemented with different concentrations of 2,4-D, (5–15 mg/1) and NAA (1.5–15 mg/1). There were 12–15 explants per treatment. The data on per cent explants showing callus formation, degree of callusing, and nature of callus were recorded after eight weeks of culture initiation. The embryogenic callus masses were transferred to MS media with benzyladenine (BA) or kinetin alone or in combination with naphthaleneacetic acid (NAA) or indoleacetic acid (IAA). There were 10 replicated cultures in each treatment and data on number of embryoids per callus mass, number of shoots, and shoot length were recorded after eight weeks of subculture.

Cotyledon Culture for Organogenesis

Cotyledons were excised in the same way and cultured on MS medium with combinations of BA and NAA. There were 12–15 explants per treatment and data on per cent explants showing shoot formation, number of shoots per explant, and shoot length were recorded after 14 weeks of culture initiation.

Nucellus Culture for Organogenesis

Nucellar tissues from surface-disinfected fruits were excised and cultured on MS medium fortified with cytokinin alone or in combination with NAA. There were 10 explants in each treatment and data on shooting behaviour were recorded after 14 weeks of culture initiation.

The explants were cultured singly in 25 × 150 mm tubes with 15 ml medium and subcultured in 250 ml flasks with 50 ml medium during proliferation of shoots. All cultures were grown in a growth room under 16 hour photoperiod with a light intensity of 60–70 $\mu E/m^2/sec$, provided by warm white fluorescent tubular lamps at 27 ± 1°C. All media were solidified with 0.7 per cent Difco bacto-agar and supplemented with 4 per cent sucrose. Media pH values were adjusted to 5.7 ± 0.1°C under 1.1 kg/cm^2 pressure for 20 min. All experiments were repeated twice.

RESULTS

Somatic Embryognensis from Cotyledon Cultures

Morphogenetic changes on the cotyledonary explants were noticed within 10 days of culture. At the initial stage cotyledon attained a maximum size and became dark green.

Callus initiation started from the cut margins and within five to six weeks whole explants transformed into embryogenic or non-embryogenic callus. Embryogenic callus was translucent to opaque in appearance and milky white to yellow green in colour. It appeared as compact nodular regions of non-vacuolated cells which produced small somatic embryos. Embryos further developed into shoots when embryogenic callus masses were subcultured in appropriate media. The surface of the embryogenic callus was often shiny. In contrast, non-embryogenic callus was crystalline and transparent in appearance, was white, tan, brown, or yellow in colour and appeared as loosely packed, friable regions of vacuolated cells which produced more callus and rarely roots, but never embryos. Only NAA-containing media induced embryogenic callus and 100 per cent of induced calli were embryogenic in 3 or 5 mg/l NAA (Table 1).

Stereomicroscopic observations showed that embryogenesis began within embryogenic callus after the explants had been cultured for five to six weeks. Early stages of embryogenesis were no longer visible before five weeks but after that numerous globular, heart-shaped or torpedo-shaped embryoids were observed. These embryoids later on differentiated into characteristic individualized bipolar structures lacking any vascular connections with the parent tissue. Different stages of embryogenic structures, normal and even abnormal were observed. Embryogenic tissues continually transferred to fresh media with the same hormonal combinations retained their embryogenic potential for a long period.

Germination of somatic embryos and formation of plantlets occurred only rarely in the induction media. The transfer of embryogenic tissues with induced embryos to MS medium containing BA or kinetin with or without an auxin led to greening and further development of embryoids into shoots. The frequency of somatic embryo conversion into shoots was very low (Table 2). The majority of cultures produced embryoids which did not germinate.

To increase the frequency of somatic embryo conversion into shoots individual somatic embryos were isolated and cultured on MS medium containing BA or kinetin with or without GA. Most of the somatic embryos developed into green embryogenic

Table 1. Induction of embryogenic callus on cotyledon explants. Data were recorded after eight weeks of culture initiation

Growth regulator (mg/l)		Degree of callusing	% of explants showing callus formation	
			Embryogenic	Non-embryogenic
2,4-D	5.0	+		10.2
	7.0	+		12.5
	10.0	++		
	15.0	+		24.5
NAA	1.5	++	70	30
	2.0	++	58	42
	2.5	+++	30	70
	3.0	+++		100
	5.0	+++		100
	7.0	+++	25	75
	10.0	+	78	22
	15.0	+		100

Table 2. Response of subcultured cotyledon calli only embryogenic to different concentrations and combinations of phytohormones. Data were recorded after 14 weeks of culture initiation (six weeks after subculture)

Growth regulators (mg/l)				No. embryoids per callus	No. shoots per callus	Shoot length (cm)
Cytokinin		Auxin				
BA	0.5			22.5	2.5	5.0
	1.0			41.6	3.6	3.4
	1.5			17.6	2.1	2.6
	2.0			2.2	1.0	1.4
Kinetin	0.5			10.4	1.5	3.2
	1.0			7.6	2.9	2.4
	1.5			21.4	1.6	2.6
	2.0			11.4	1.2	2.5
BA	1.5	NAA	0.1	31.2	3.9	2.9
	1.5		0.2	47.6	5.2	3.2
	2.0		0.1	34.6	2.1	1.6
	2.0		0.2	24.6	2.6	2.2
Kinetin	1.5	NAA	0.1	12.7	1.2	1.8
	1.5		0.2	10.9	2.5	1.5
	2.0		0.1	11.4	1.7	2.0
	2.0		0.2	15.7	1.3	1.2
BA	1.0	IAA	2.0	27.6	2.3	2.5
	2.0		2.0	21.5	1.9	1.8
Kinetin	1.0	IAA	2.0	49.3	6.3	5.2
	2.0		2.0	31.1	3.2	3.2

calli and produced a second crop of somatic embryos (secondary embryogenesis). Only a few well-developed embryos grew into shoots (Table 3).

Organogenesis from Cotyledon Cultures

Irrespective of growth regulator concentrations used, the cotyledon explants when cultured on BA and NAA combinations swelled within one to two weeks of incubation and became dark green. Generally, the adventitious buds became visible within three to four weeks of culture. At the initial stage they appeared in the form of tiny tube like protrusions which resembled embryoids. At the later stage the shoot buds were recognizable as monopolar structures developing procambial strands which established connection with the preexisting vascular tissues dispersed within the callus formed during development of shoot buds. At the end of six to seven weeks most of the adventitious buds developed into shoots. The shoot clumps were transferred to fresh medium for development and growth of shoots. After excision of the elongated shoots, the remaining explant with small shoots could be recultured in the same medium for further elongation. Following this procedure more than 40 shoots could be recovered when 2 mg/l BA and 0.2 mg/l NAA were used in the medium (Table 4).

Organogenesis from Nucellus Cultures

Morphogenetic changes of the nucellar explants were noticed within two weeks of culture. At the initial stage the explants increased in size and became green. In most

Table 3. Effect of subculturing single embryoid on different concentrations of BA or kinetin with or without GA_3. Data recorded after six weeks of embryoid culture

Growth regulators (mg/l)		% of embryoids germinated	% of embryoids showing secondary embryogenesis
BA	0.1	1.5	31.9
	0.2	7.6	21.7
	0.5	1.2	27.6
	1.0	1.9	14.2
	1.5	1.5	19.3
Kinetin	0.1	2.5	24.7
	0.2	3.5	17.6
	0.5	1.7	9.2
	1.0	1.6	7.9
	1.5	1.9	1.2
BA	0.2 + GA_3 0.01	11.7	2.5
	0.5 + GA_3 0.01	7.6	1.7
BA	0.2 + GA_3 0.01	9.2	9.6
	0.5 + GA_3 0.01	1.5	1.4

Table 4. Response of cotyledon explants on BA and NAA in MS medium. Data were recorded after 14 weeks of culture initiation (after six weeks of subculture)

Growth regulators (mg/l)		% of explants showing shoot formation	No. of shoots per explant	Shoot length (cm)
BA	NAA			
0.5	0.1			
	0.2	27.4	3.4	4.1
1.0	0.1	19.3	1.9	3.3
	0.2	19.8	1.8	3.2
1.5	0.1	10.8	2.4	3.2
	0.2	13.9	1.8	2.9
2.0	0.1	21.5	17.3	3.6
	0.2	65.3	41.9	4.8
2.5	0.1	17.6	3.9	1.9
	0.2	19.3	3.7	1.7

of the cases callogenesis and adventitious bud initiation took place simultaneously. However, media fortified with BA or kinetin alone did not induce the explants to produce any callus. No variation was observed between the shoot buds originated from either cotyledon or nucellus cultures with regard to morphological structures and mode of development. Among the different treatment combinations 2 mg/l BA and 0.2 mg/l NAA was found to be the best formulation, where maximum frequency of adventitious shoot proliferating cultures with the highest number of shoots per explant and greatest shoot length were recorded (Table 5).

CONCLUSION

The results of the present investigation demonstrate the potential of neem cotyledon and nucellus for somatic embryogenesis and organogenesis. One problem was low fre-

Table 5. Response of nucellar tissues to different combinations and concentrations of phytohormones. Data were recorded after 14 weeks of culture initiation (after six weeks of subculture)

Growth regulators (mg/l)			% of explants showing shoot formation	Degree of callusing	No. shoots per explants	Shoot length (cm)
Cytokinin		Auxin				
BA	2.0		11.6		1.2	1.2
	3.0		13.7		3.9	2.2
	2.0	NAA 0.2	24.9	+	22.8	4.8
	2.0	0.5	10.3	++	2.5	2.4
	3.0	0.2	17.3	+++	3.9	1.2
	3.0	0.5	11.2	+++	4.2	3.2
Kinetin	2.0		11.5		1.2	1.2
	3.0		12.8		2.4	1.3
	2.0	NAA 0.2	14.0	+	1.6	2.2
	2.0	0.5	11.4	+	4.1	1.5
	3.0	0.2	12.3	+	2.2	1.6
	3.0	0.5	11.4	+	2.3	2.2

quency of somatic embryo germination. Further experiments are needed to achieve a high frequency of somatic embryo conversion to plantlets. Regeneration of plants via somatic embryogenesis is considered to be an efficient approach for clonal plant propagation (Chalupa, 1990). Somatic embryogenesis also provides the basis for the genetic improvement through somaclonal variation or enabling genetic manipulation (Vieitez and Braciela, 1990). Somatic embryogenesis from callus has been reported in a large number of plants belonging to diverse groups (Williams and Maheswaran, 1986; Rangaswamy, 1986; Jasrai et al.,1992). The nucellar culture technique described here may be useful for breeding of horticulturally useful, disease-free neem plants, since nucellar derived plants are free of virus and other disease causing micro-organisms due to absence of vascular connection between the surrounding maternal tissue and the nucellus (Litz, 1984).

REFERENCES

Chalupa, P. 1990. Plant regeneration by somatic embryogenesis from cultured immature embryos of oak (*Quercus robus* L.) and linden (*Tilia cordata* Mill.), *Plant Cell Rep*. 9: 398–401.

Jasrai, Y.T., S.M. Barot and A.R. Mehta. 1992. Plant regeneration through somatic embryogenesis in hypocotyl explants of *Trachyspermum ammi* (L.) Sprague. *Plant Cell Tissue Organ Cult*. 29: 57–60.

Litz, R.E. 1984. *In vitro* somatic embryogenesis from nucellar callus of monoembryonic mango. *Hort. Sci*. 19: 715–717.

Murashige, T. and F. Skoog. 1962. A revised medium for rapid growth and bioassays with tobacco tissue cultures. *Physiol. Plant* 15: 473–497.

Rangaswamy, N.C. 1986. Somatic embryogenesis in angiosperm cell, tissue and organ cultures. *Proc. Ind. Acad. Sci.* (Plant Sci.) 96: 247–271.

Vieitez, A.M. and J. Braciela. 1990. Somatic embryogenesis and plant regeneration from embryonic tissues of *Camellia japonica* L. *Plant Cell Tissue Organ Cult*. 21: 267–274.

Williams, E.G. and G. Maheswaran. 1986. Somatic embryogenesis: Factors influencing coordinated behaviour of an embryonic group. *Ann. Bot*. 57: 44–62.

Wulf, A. 1991. On the effectiveness of neem-extracts against bark beetles. *Plant Res. Dev*. 34: 81–88.

107

Micropropagation of Neem through Axillary Bud Culture

O.I. JOARDER, A.T.M. NADERUZZAMAN, R. ISLAM, M. HOSSAIN, N. JOARDER AND B.K. BISWAS

Department of Botany, Rajshahi University, Rajshahi, Bangladesh

ABSTRACT

A micropropagation method is described using nodal explants from field-grown adult *Azadirachta indica* tree. The explants were established by culturing two weeks in hormone-free MS followed by four weeks in cytokinin supplemented-medium. Established explants upon transfer to medium containing low concentration (0.05–0.3 mg/l) of cytotokinin produced shoots. Multiplication of shoots was achieved by culturing nodal segments from *in vitro* proliferated shoots. Rooting was induced at a high frequency by subculturing individual shoots on medium with 0.5 mg/l IBA. Plantlets were acclimated and successfully established in soil.

INTRODUCTION

Native to India and Burma, neem is a botanical cousin of mahogany. For centuries, millions have cleaned their teeth with neem twigs, smeared skin disorders with neem leaf juice, taken neem tea as a tonic, and placed neem leaves in their beds, books, grain bins, cupboards, and closets to keep away troublesome bugs. The tree has relieved so many different pains, fevers, infections and other complaints that it has been called 'the village pharmacy'. To those millions, neem has always had miraculous powers and now scientists around the world are beginning to think they may be right. Two decades of research have revealed promising results in so many disciplines that this obscure species may be of enormous benefit to countries both poor and rich. Even some of the most cautious researchers are saying that neem deserves to be called a wonder plant.

Micropropagation is being used extensively for rapid clonal propagation of many fruit, nut, timber and ornamental trees (Zimmerman, 1985; Hammerschlag, 1986;

Hansman and Y de Novoa, 1986), since it enables rapid propagation and hastens the availability of new cultivars (Zimmerman, 1981). A micropropagation method could assist in building up the clonal stock of elite genotypes by enabling rapid multiplication after their selection in a breeding programme. This investigation was undertaken to establish an *in vitro* propagation technique for neem, beginning with explants from an adult tree.

MATERIALS AND METHODS

The plant material was collected from a 30 year old field-grown *Azadirachta indica* tree. Shoots 7–10 cm long were collected from new growth on mature branches and from the basal sprouts and brought to the laboratory in water. After thorough washing under running tap water the shoots were cut into pieces and surface disinfected with 0.1 per cent $HgCl_2$ for 12 min., after rapid rinse in 70 per cent ethanol, and were finally washed three times with autoclaved distilled water. Explants with single, two and three nodes were prepared aseptically and either cultured directly on bud sprouting medium for six weeks or in the first, second, or third week on MS (Murashige and Skoog, 1962) basal medium followed by five, four, and three weeks of incubation on bud sprouting medium. MS medium with different concentrations (0.5–2 mg/l) of benzyladenine (BA), kinetin, or isopentenyladenine (2iP) was used for bud sprouting. The explants were cultured in 25 × 150 mm culture tubes each containing 15 ml of medium. After six weeks, explants showing response were subcultured in shoot proliferation medium, MS medium with low concentrations (0.05–0.3 mg/l) of BA, kinetin or 2iP was used for shoot proliferation. Subsequently, the cultures were transferred to fresh shoot proliferation medium every six weeks. There were 15–20 explants per treatment. The data on percentage of explants showing response (sprouting of buds) were recorded after six weeks of culture initiation, and data on number of shoots per explant and shoot length were collected after 12 weeks and 24 weeks of culture initiation.

Microcuttings prepared from shoots formed *in vitro* (5–10 cm long) were cultured on MS medium with 0.2–1.5 mg/l of indolebutyric acid (IBA), naphthaleneacetic acid (NAA), or indoleacetic acid (IAA) for adventitious root induction. There were 10–12 replicated cultures in each treatment and data on root proliferation efficiency of the cultured shoots were recorded after six weeks of culture.

MS medium with full-strength salts and 3 per cent sucrose was used for shoot proliferation and MS with half-strength salts and 1.5 per cent sucrose was used for adventitious root induction. The medium pH was adjusted to 5.6 ± 0.1; all media were gelled with 0.7 per cent Difco Bactoagar and were autoclaved for 15 min. at 121°C and 1.1 kg/cm^2 pressure. The cultures were grown at 26 ± 1°C under a 16 hour photoperiod with a light intensity of 60 to µE/m^2/sec provided by warm white fluorescent tubes.

RESULTS

The cultured nodal explants showed their first response by enlarging and bursting of axillary buds within four to five weeks of culture. Two weeks' culture in MS media followed by four weeks in bud sprouting media was found to be the most beneficial for a satisfactory response of explants in all instances (Table 1). Maximum response was

Table 1. Effect of different treatments on *in vitro* response (showing enlargement and bursting of axillary buds) of nodal segments. Data were recorded after six weeks of culture initiation

Treatment (weeks)		Cytokinin (mg/l)	% cultures showing response		
MS	MS + Cytokinin		Explants with 1 node	Explants with 2 nodes	Explants with 3 nodes
0	6	0.5 BA	11.54	12.26	12.94
1	5		10.25	17.67	14.59
2	4		14.17	12.00	15.67
3	3		19.27	14.13	17.22
0	6	1.0 BA	12.75	11.22	13.76
1	5		12.16	19.22	17.63
2	4		24.75	11.16	17.64
3	3		19.23	12.44	18.13
0	6	1.5 BA	12.24	11.66	14.29
1	5		15.76	11.24	17.29
2	4		60.29	29.26	36.35
3	3		21.14	11.26	15.26
0	6	2.0 BA	11.20	10.95	12.26
1	5		14.16	10.75	16.23
2	4		12.34	15.00	16.95
3	3		17.29	10.99	11.26
0	6	0.5 kinetin	11.55	12.04	11.67
1	5		13.76	14.97	15.29
2	4		18.29	16.33	17.66
3	3		16.24	11.19	12.29
0	6	1.0 kinetin	11.22	12.24	11.12
1	5		14.12	12.14	12.05
2	4		17.69	17.09	14.19
3	3		11.15	13.06	11.03
0	6	1.5 kinetin	11.43	11.67	11.29
1	5		12.29	14.29	15.55
2	4		10.23	16.37	17.29
3	3		16.05	12.43	11.11
0	6	2.0 kinetin	11.67	12.03	11.29
1	5		12.75	12.05	12.15
2	4		14.93	15.55	16.25
3	3		12.23	13.26	11.45
0	6	0.5 2iP	11.05	11.11	12.29
1	5		12.46	13.29	12.23
2	4		17.19	15.57	14.16
3	3		14.70	16.29	14.05
0	6	1.0 2iP	11.55	12.05	11.99
1	5		12.00	11.76	12.76
2	4		19.14	17.29	15.29
3	3		11.16	12.25	11.29
0	6	1.5 2iP	11.39	11.05	11.26
1	5		11.76	11.55	11.92
2	4		15.29	13.39	12.76
3	3		12.75	12.76	11.15
0	6	2.0 2iP	11.76	11.95	11.07
1	5		11.05	12.22	12.26
2	4		14.92	12.05	12.23
3	3		12.70	11.11	11.25

observed in explants containing single node at 1.5 mg/l BA. In BA supplemented media bursting of axillary buds occurred five to seven days earlier than in the media containing kinetin or 2iP. After cleaning basal portion and on transfer to shoot proliferation media new shoot development was found to start from the axillary buds on respective nodes within one to two weeks of culture. From one bud a maximum of two shoots were produced. For maximum production of shoots, the stock cultures were transferred to fresh shoot proliferation medium of the same composition but further increase in shoot number was not observed. However, the length of the shoot increased significantly (Table 2). When the nodal segments taken from shoots proliferated *in vitro* were cultured on the same media, a two-fold increase in number of shoots was observed. It was further observed that the shoot multiplication rate increased with the increasing number of subcultures and then declined after five to six subcultures.

Shoots taken from the cultures proliferated *in vitro* were excised carefully and transferred to the rooting media. Shoots 5, 8 and 10 cm long were tested for their rooting capacity. Shoots 10 cm long showed maximum response (Table 3). Among the three auxins tested IBA was found to be superior to NAA and IAA and 0.5 mg/l IBA was found to be the best concentration for maximum rooting frequency. When the concentration of IBA was increased, a considerable amount of callus developed from basal portion of the cutting. In most of the treatments of NAA a similar response was

Table 2. Effect of different concentrations of cytokinins on shoot proliferation from nodal segments. Data were recorded after 12 weeks (six weeks of first subculture) and 24 weeks (six weeks of third subculture) after culture initiation

Growth regulators (mg/l)	First subculture			Third subculture		
	Explants with 1 node	Explants with 2 nodes	Explants with 3 nodes	Explants with 1 node	Explants with 2 nodes	Explants with 3 nodes
0.05 BA	1.0 4.1	2.0 4.2	2.4 4.7	1.0 5.3	2.0 6.2	2.3 6.7
0.10 BA	1.0 2.3	2.0 7.0	2.9 6.1	1.0 12.2	2.0 10.5	2.9 9.9
0.30 BA	1.0 2.4	1.9 2.3	2.0 1.2	1.0 6.9	1.7 5.1	2.0 3.7
0.05 kinetin	1.0 2.5	1.9 2.6	1.9 2.9	1.0 4.2	1.9 4.9	1.9 5.9
0.10 kinetin	1.0 3.1	2.0 3.6	2.4 2.5	1.0 4.7	2.0 6.2	2.4 4.2
0.30 kinetin	1.0 2.2	1.7 2.4	2.5 2.6	1.0 3.9	1.7 4.1	2.5 4.2
0.05 2iP	1.0 2.4	1.7 3.9	2.1 2.2	1.0 3.7	1.7 2.9	2.1 3.1
0.10 2iP	1.0 3.1	2.0 3.4	2.2 3.9	1.0 4.2	2.0 4.2	2.2 3.7
0.30 2iP	1.0 3.2	1.9 3.6	2.2 2.7	1.0 4.4	1.9 4.7	2.2 3.9

Upper value is no. of shoots per explant
Lower value is shoot length (cm)

Table 3. Adventitious root induction on *in vitro* formed shoots in half-strength MS medium supplemented with different concentrations of IBA, NAA, or IAA. Data were recorded after six weeks of culture

Growth regulator (mg/l)		5 cm shoot		8 cm shoot		10 cm shoot	
		% shoots formed roots	No. roots per shoot	% shoots formed roots	No. roots per shoot	% shoots formed roots	No. roots per shoot
IBA	0.2	15.2	4.2	14.2	4.3	22.2	5.7
	0.5	59.2	7.3	70.3	8.4	82.2	9.0
	1.0	40.2	2.7	52.6	3.0	59.1	8.2
	1.5	10.1	2.0	9.2	3.1	7.7	2.3
NAA	0.2	20.1	2.3	15.2	2.0	17.3	2.1
	0.5	++	—	++	—	++	—
	1.0	+++	—	+++	—	+++	—
	1.5	+++	—	+++	—	+++	—
IAA	0.2	—	—	—	—	—	—
	0.5	7.2	2.1	8.2	2.2	3.7	1.9
	1.0	+	—	+	—	+	—
	1.5	+	—	+	—	+	—

– & + indicates no response and degree of callusing respectively.

noticed. With the increasing numbers of subculture, both the percentages of rooted cuttings and the number of roots per cutting increased considerably (data not shown).

The plantlets regenerated *in vitro* were removed from the culture tubes, washed thoroughly to remove the nutrient medium and transplanted to polybags filled with garden soil and compost (1:1). During the first 7–10 days, the potted plantlets were covered with glass beakers to provide high humidity. Fifty per cent of transferred plantlets adapted in soil within three to four weeks. Plantlets were subsequently transferred to larger pots and gradually acclimated to outdoor conditions.

CONCLUSION

The results presented in the foregoing sections reveal that an adult *A. indica* tree could be propagated through *in vitro* culture of nodal explants. Two weeks culture of nodal explants on growth-regulator free MS medium was essential for successful sprouting of axillary buds. When the explants with sprouted buds were maintained in the same medium initiation and elongation of shoots did not occur. However, when the explants were transferred to low concentration of BA, rapid initiation and elongation of shoot occurred. Hu and Wang (1983) reported that sometimes the elongation of shoots is inhibited by high cytokinin levels; thus, an intermediate shoot elongation stage becomes necessary, Blueberry shoots produced in medium containing 15 mg/l 6-dimethylamino purine showed sufficient elongation when transferred to a medium containing relatively low levels of this cytokinin (Zimmerman and Broome, 1980). It is evident that BA was a suitable cytokinin for proliferation of shoots in neem. The superiority of BA to other cytokinins for multiple shoot formation in trees has also been reported by other investigators (Zaman et al., 1991).

The early response and increased shoot production capacity of nodal segments taken from shoots proliferated *in vitro* in comparison to that of field-grown trees might be due to the absence of lag period between explanting and adaptation of explants to *in vitro* conditions.

By using the method described above, a considerable number of clonal plants could be propagated from nodal segments by continuous subculturing of shoot propagules. The multiplication rate that we achieved was not sufficiently large to be commercially significant, but our results provide a basis for further research in micropropagation of other genotypes of neem. Efforts are being made for refinement of culture methods for consistent high plant yields from other parts of the plant.

REFERENCES

Hammerschlag, F.A. 1986. Temperate fruits and nuts. In: *Tissue Culture as a Plant Production System for Horticultural Crops*. R.H. Zimmerman, F.A. Griesbach, F.A. Hammerschlag and R.H. Lawson (eds.), Martinus Nijhoff, Dordrecht, pp. 221–236.

Hansman, D. and C.O. Y de Novoa. 1986. Micropropagation of temperate nut trees. *Hortic. Abstr*. 56: 403–416.

Hu, C.Y. and P.J. Wang. 1983. Meristem, shoot tip and bud culture. In: *Handbook of Plant Cell Culture*. Vol. I. D.A. Evans, W.R. Sharp, P.V. Amirato and Y. Yamada (eds.), Macmillan, New York, pp. 177–227.

Murashige, T. and F. Skoog. 1962. A revised medium for rapid growth and bioassays with tobacco tissue cultures. *Physiol. Plant* 15: 473–497.

Zaman, M.A., S.M. Rahman, N. Joarder and R. Islam. 1991. Mass propagation of mulberry (*Morus alba* L.) through axillary bud culture. *Plant Tissue Cult.* 1: 75–78.

Zimmerman, R.H. 1981. Micropropagation of fruit plants. *Acta. Hortic.* 120: 217–222.

Zimmerman, R.H. 1985. Application of tissue culture propagation to woody plants. In: *Tissue Culture in Forestry and Agriculture*. R.R. Henke, K.W. Hughes, M.J. Constanin and A. Holaender (eds.), Plenum, New York, pp. 165–177.

Zimmerman, R.H. and O.C. Broome. 1980. Blueberry micropropagation. *Proc. Conf. Nursing Production of Fruit Plants Through Tissue Culture—Applications and Feasibility*. US Dept. Agric. Sci. Educ. Adm., ARR-NE-11, pp. 44–47.

108

Potential Application of *in vitro* Methods for Propagation of Neem (*Azadirachta indica* A. Juss.)

M.S. JOSHI AND S.R. THENGANE

Plant Tissue Culture Division
National Chemical Laboratory, Pune 411 008, India

ABSTRACT

Despite development of synthetic drugs, plants remain a major source of pharmaceuticals and other fine chemicals. Most of the natural products called 'secondary metabolites' are distributed only in specific tissues or organs. During the last few decades, tremendous efforts have been directed to increase the production of these metabolites, either by improved plant type or through *in vitro* techniques. Plant tissue culture offers fascination solutions to either approach.

Neem (*Azadirachta indica*) has multifarious properties and uses in agriculture as an insecticide and nematicide and in the pharmaeceutical and soap industries. Recently, the insecticidal and pesticidal activity of compounds isolated from various parts of the tree has been demonstrated. However, viability of the neem seeds limits propagation.

Studies were initiated for the propagation of neem by organogenesis and somatic embryogenesis. The induction of somatic embryos was achieved directly on explant tissue or from callus. Murashige and Skoog's basal medium supplemented with auxins, indole 3-acetic acid (IAA), or 2,4 dichloro phenoxy acetic acid (2,4D), along with cytokinins, 6-benzyl amino purine, (BAP), and kinetin, were used.

Adventitious multiple shoots (10–15) were induced from excised axillary and terminal buds of neem on medium containing BAP, kinetin, and sucrose. The individual shoots were transferred to sand:soil (1:1) mixture for direct rooting and grown in polyhouse, where 90 per cent survival was observed.

Somatic embryos were induced via callus, as well as directly on the cotyledonary explants. However, type of the auxin was found to be a deciding factor for embryo induction pathway. Direct embryo induction was observed on medium supplemented with 2, 4D, BAP, and casein hydrolysate, while indirect embryogenesis was observed when IAA was used. Seventy per cent embryos showed germination. Survival of the plantlets was 80–90 per cent.

NCL Communication No. 5655

In vitro propagation of neem can be used for the isolation of high-yielding variants, as well as for mass propagation in a short period. The problem of seed viability can also be overcome by this technique.

INTRODUCTION

Mankind is faced today with a number of problems that urgently require solution in areas of agriculture and forestry. The conservation and management of natural resources is one method of maintaining the quality of the environment in the face of population explosion, increased industrial activity, and recurrent use of synthetic insecticides (Higgins et al., 1985). Some of these problems can be circumvented either by use of degraded lands and forests for increasing productivity or by rapid replanting of superior quality planting material. In addition to conventional techniques like genetics and plant breeding, biotechnology is playing an important role in solving some of these problems via improved agriculture, waste treatment procedure, energy conservation, and health care. More than two decades ago, plant cell and tissue culture emerged as a strong branch of plant biology. Besides its use for mass multiplication of superior phenotypes and genotypes, it also provides a means for regenerating novel plants.

Plants produce and accumulate endogenous chemicals that function in many different ways. Most of these natural products, called 'secondary metabolites', are restricted in their distribution to specific tissues or organs of the plants. Tremendous efforts have been directed to increase the production of these metabolites, either by improved plant type or through *in vitro* techniques. Plant tissue culture offers fascinating solutions to either of the approaches.

Azadirachta indica is an indigenous but underexploited tree. The recent spurt of interest in this tree is due to its spectacular biological activity against a wide range of insects and nematodes. Azadirachtin and other terpenoids are natural, environmentally compatible products which could be promising candidates for pest control (Koul et al., 1990).

Mass propagation of neem on a commercial scale has limitations of seed viability. Moreover, vegetative propagation of this tree using conventional methods is not possible (Narayan and Jaiswal, 1985). Tissue culture offers a reliable method for mass production of plants in a shorter time and without seasonal constraints.

Plantlet regeneration is best achieved with adventitious shoots which are then rooted or from somatic embryognesis. The advantages of somatic embryogenesis over the conventional two-stage method has drawn much attention. A large number of plants having an in-built root and shoot system can be produced. Somatic embryos are amenable to automatic handling and their encapsulation with synthetic gels for use as artificial seeds has added advantages.

In the present paper, propagation of neem via adventitious shoot formation as well as from somatic embryogenesis is reported.

MATERIALS AND METHODS

Sterilization of Explant Material

Axillary and terminal buds were used as explant material for adventitious shoot regeneration, whereas cotyledons from unripe mature fruits were used for plantlet production via somatic embryogenesis pathway.

The explants were washed with liquid detergent (10 per cent savlon) and thoroughly rinsed under running tap water. They were then surface-sterilized with 0.15 per cent mercuric chloride for 10 min. and finally washed four or five times with sterile water before inoculation under sterile conditions.

Experimental

The axillary and terminal buds were inoculated on Murashige and Skoog's (MS) medium (Murahige and Skoog, 1962), supplemented with MS vitamins, cytokinins like 6-benzyl amino purine (BAP), kinetin, or 2-isopentenyl adenine (2iP). Sucrose was used as carbohydrate source. The cultures were maintained at 25° ± 1°C and incubated at 16/8 hr photoperiod.

The cotyledons were removed aseptically and embryo axis excised. Sections of cotyledons were cultured on MS medium supplemented with MS vitamins, casein hydrolysate, and sucrose. Auxins like indole-3-acetic acid (IAA) or 2,4 di-chlorophenoxy acetic acid (2,4D), and cytokinins like 6-benzyl amino purine (BAP) or kinetin were added alone or in combination. These cultures were incubated in the dark.

All the media used were solidified with 0.4 per cent agar (bacteriological agar) after the pH was adjusted to 5.7. All growth regulators were incorporated into the medium before autoclaving. The media were autoclaved at 120°C and 15 psi for 20 min.

Healthy plantlets derived through vegetative buds or cotyledons were planted in sand-soil (1:1) mixture and were transferred to a glasshouse for further growth.

RESULTS AND DISCUSSION

Regeneration from Vegetative Buds

After about 10 days of incubation, sprouting was observed from axillary and terminal buds. Table 1 shows the effect of cytokinins alone or in combination on multiple shoot induction. Of the cytokinins tested individually, BAP alone showed maximum number (85 per cent), cultures with multiple shoots followed by kinetin and 2iP. However, combination of BAP with kinetin was found to be best with induction of four to five shoots per explant (Fig. 1).

Application of cytokinins has been commonly found effective in promoting direct or indirect shoot initiation (George and Sherrington, 1984). Kinetin and BAP are the most commonly used cytokinins, while 2iP and zeatin are used less frequently (Thorpe and Patel, 1984). Evans et al. (1981) found that for 75 per cent of the species forming shoots, kinetin/BAP was used in concentration 0.05–46 μM.

Table 1. Effect of cytokinins on multiple shoot induction

Cytokinins (0.5–5.0 mg/l)	Cultures with multiple shoots (%)	Number of shoots per culture
BAP	85.0	2–3
KN	58.6	2–3
2iP	nil	nil
BAP + KN	63.6	4–5
BAP + 2iP	5.2	1–2
KN + 2iP	3.3	1–2

Number of cultures studied = 50/experiment.

Fig. 1. Multiple shoot induction from axillary bud.

Multiple shoots formed from the explants could be isolated and subcultured on fresh medium for further multiplication. Thus, 10–15 adventitious shoots per explant could be induced. The individual shoots were directly transferred to polybags containing sand-soil (1:1) mixture and grown in glasshouse, where 90 per cent survival was observed (Fig. 2). The main advantage of using *ex vitro* conditions for root induction is that rooting and acclimatization can be carried out simultaneously, thereby omitting one of the steps and eventually reducing production cost (Thorpe et al., 1991).

Somatic Embryogenesis

For the induction of embryos, auxin alone and in combination with cytokinins was used. IAA alone (0.5–5.0 mg/1) and in combination with kinetin 0.2–3.0 mg/l showed callusing and rhizogenesis, where callus was smooth and friable. However, IAA in combination with BAP induced compact nodular callus. Upon transfer of the calli to respective media, there was further increase in callusing. The compact, nodular callus showed initiation of globular shiny masses after two subcultures of 20 days each (Fig. 3).

Fig. 2. *In vivo* rooted plants in poly bags.

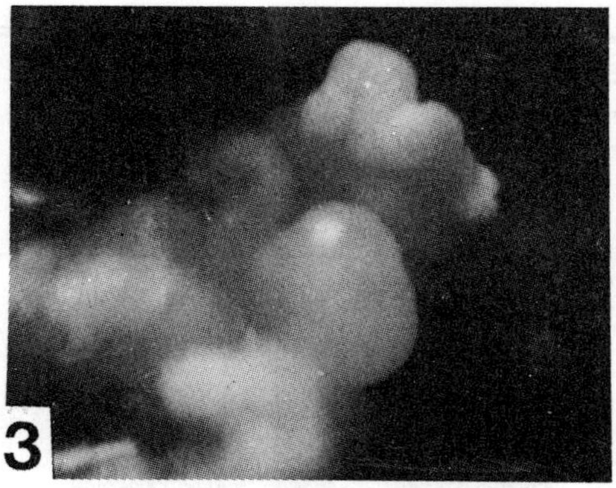

Fig. 3. Callus showing shiny globular masses.

The effect of incubation period on embryogenic response was studied by subculturing the callus on the induction medium up to 16 weeks (Table 2). A gradual increase in embryogenic response with increase in incubation period was observed, with maximum response at the eighth week of incubation. After 16 weeks of incubation on the same medium, cultures did not produce any embryos.

For maturation and germination of the somatic embryos the concentration of casein hydrolysate, IAA, and BAP was tested. On total omission of casein hydrolysate, with increase in BAP, 60–70 per cent of the embryos enlarged and germinated within 20–30 days. Germinated embryos, when transferred to MS half-strength medium, showed elongation and development of complete plantlet with well-developed root and shoot (Fig. 4). The detailed method is published elsewhere (Shrikhande et al., in press).

The type of auxin was found to be a deciding factor for embryo induction pathway. When 2, 4D was used in combination with BAP, somatic embryos were induced directly on the cotyledonary explant (Fig. 5). However, 2, 4D alone as well as in combination with kinetin induced slimy brown callus.

Induction of about 20–25 somatic embryos per explant was observed. Germination of these embryos was found to be bery difficult. Only 2–5 per cent germination was observed. Several experiments are being carried out to increase the induction rate of embryos as well as to promote their germination.

Generally, transfer of culture from high auxin medium to a medium containing low or zero level of auxin promotes increased embryo induction and germination (Kamada and Harada, 1979; Raghavan and Nagmani, 1983). In neem, presence of auxin as well as cytokinin throughout the growth of the somatic embryos seems to be essential. A hormone-free medium is only required to enhance growth of the germinated embryos. Transfer of germinated embryos to half-strength MS medium with 2 per cent sucrose showed elongation and development of complete plantlet with root and shoot. Rooted plantlets were transferred to sand-soil (1:1) mixture and grown in glasshouse (Fig. 6). Survival of the plants was 80–90 per cent.

Table 2. Embryogenic response in relation to incubation period

Period in weeks	Cultures showing	
	Callus (%)	Embryogenic response (%)
2	90	10
4	80	20
8	74	26
12	86	14
16	94	6

Percentage response = response/30 cultures.

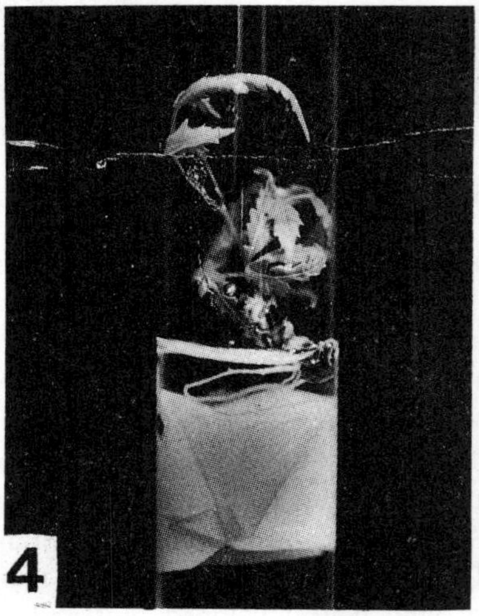

Fig. 4. Complete plantlet on MS half medium.

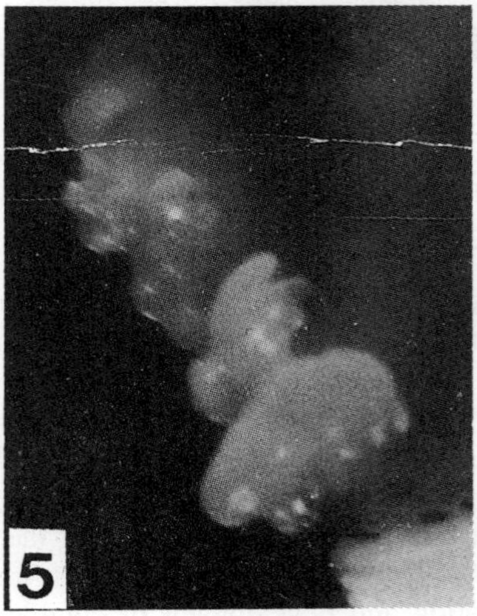

Fig. 5. Direct embryo induction on cotyledon edge.

Fig. 6. One-year-old well-established plant.

CONCLUSION

The application of micropropagation techniques to semitropical and tropical trees is not common and success in obtaining viable plantlets through organogenesis and embryogenesis in these tree species is limited. Tissue culture methods would be useful for propagation of superior neem trees showing faster growth and higher yield of active compounds. In addition, variants with wide adaptability can be isolated.

REFERENCES

Evans, D.A., W.R. Sharp and C.E. Flick. 1981. Growth and behaviour of cell cultures. In: *Plant Tissue Culture—Methods and Application in Agriculture*. T.A. Thorpe (ed.), Academic Press, Orlando, Florida, USA, pp. 45–113.

George, E.F. and P.D. Sherrington, eds. 1984. *Plant Propagation by Tissue Culture. Handbook and Directory of Commercial Laboratories*. Exegetics Ltd., Eversley, England.

Higgins, J., D.J. Best and J. Jones, eds. 1985. *Biotechnology—Principles and Applications*. Blackwell Scientific Publications, California, USA.

Kamada, H. and H. Harada. 1979. Studies on the organogenesis in carrot tissue cultures. I. Effects of growth regulators on somatic embryogenesis and root formation. *Z. Pflanzenphysiol.* 91: 255–266.

Koul, O., Murray B. Isman and C.M. Ketkar. 1990. Properties and uses of neem, *Azadirachta indica*. *Can. J. Bot.* 68: 1–11.

Murashige, T. and F. Skoog. 1962. A revised medium for rapid growth and bio-assays with tobacco tissue culture. *Physiol. Plant.* 15: 473–497.

Narayan, P. and V.S. Jaiswal. 1985. Plantlet regeneration from leaflet callus of *Azadirachta indica* A. Juss. *J. Tree Sci.* 4(2): 65–68.

Raghavan, V. and R. Nagmani. 1983. Morphogenesis of pollen callus cultures of *Hyoscyamus niger*. *Am. J. Bot.* 70: 524–531.

Shrikhande, M., S.R. Thengane and A.F. Mascarenhas (in press). Somatic embryogenesis and plantlet regeneration in neem, *Azadirachta indica* A. Juss. In: *Vitro-Cellular and Dvptl. Biol. Plant.* January 1993, 20 pp.

Thorpe, T.A., I.S. Harry and P.P. Kumar. 1991. Application of micropropagation to forestry. In: *Micropropagation—Technology and Application*. P.C. Debergh and R.H. Zimmerman (eds.), Kluwer Academic Publishers, Dordrecht, Netherlands, pp. 311–336.

Thorpe, T.A. and K.R. Patel. 1984. Clonal propagation: Adventitious buds. In: *Cell Culture and Somatic Cell Genetics*, Vol I. I.K. Vasil (ed.), Academic Press Inc., Orlando, Florida, USA, pp. 49–60.

109

High Frequency Regeneration of Neem *in vitro* (*Azadirachta indica* A. Juss.) from Immature Embryo Explants

M. THIAGARAJAN AND P.M. MURALI

Dalmia Centre for Biotechnology
Siruvani Main Road, Kalampalayam
Coimbatore 641 010, Tamil Nadu, India

ABSTRACT

A study was earlier undertaken to optimize conditions to establish *in vitro* plantlets of neem (*Azadirachta indica*) culturing seed embryos (Thiagarajan and Murali, in press). In this study we report high frequency regeneration of neem plantlets from excised embryos, with intact cotyledon, grouped according to their size from immature fruits (collected from trees 30 days after flowering was observed). The embryos were cultured in basal MS medium supplemented with various combinations of sugars, nitrates, cytokinins, and auxins. The embryos elicited interesting responses with increasing concentrations of indole acetic acid during the regeneration of plantlets. The optimum condition for high frequency regeneration of neem plantlets from intact embryos was in a medium containing 3 per cent sucrose supplemented with 0.1 mg/l naphthalene acetic acid and 0.3 mg/l 6-benzyl amino purine. The regenerated plantlets have been sucessfully established outdoors.

INTRODUCTION

Culture *in vitro* of embryos primarily helps to determine the factors that regulate growth of the primordial organs of the seedling plant and to study metabolic and biochemical aspects of germination. The excision of embryos from seeds of higher plants and their culture in defined media makes it possible to investigate the factors that influence embryonic growth under controlled conditions, which is otherwise difficult to undertake because the study of embryos enclosed within a seed might face interference from accessory tissues. In addition, advances in embryo culture have opened up the possibility of obtaining plantlets from inviable hybrids, which are discarded due to their inability to germinate under normal conditions, and in some types of seeds, dormancy can be overcome by culturing the excised embryos. This type of study also helps to identify endogenous promoters or inhibitors of germination (Raghavan, 1977) A number of studies of excised embryos from ovules and seeds of a variety of species and plants grown under controlled conditions have been reported (Brink and Cooper, 1947; Rappaport, 1954; Sanders and Ziebur, 1963; Narayanaswami and Nortstog, 1964; Maheshwari and Rangaswamy, 1965; Wardlaw, 1965; Degivry, 1966; Raghavan, 1966).

We have earlier reported (Thiagarajan and Murali, in press) the optimum conditions for the regeneration of neem from immature embryos. In this study, we have identified some of the requirements for high frequency germination of neem, culturing neem embryos of different sizes excised from immature fruits collected from trees 30 days after flowering.

The study assumes significance in that *Azadirachta indica* (A. Juss), a member of *Meliaceae* (commonly known as neem), is an important multi-purpose tree in Indian forestry (Troup, 1989) with valuable medicinal, insecticidal, and industrial properties and is an excellent source for a variety of secondary metabolites (Sathyavati et al., 1976; Mitra, 1963; Radwanski and Wickens, 1981).

MATERIALS AND METHODS

Fruits collected from healthy, mature neem trees in and around Madurai, Tamil Nadu, 30 to 50 days after flowering were surface-sterilized with 0.1 per cent mercuric chloride, followed by a wash with 0.05 per cent Tween-20 and several washes with sterile distilled water. The fruits were gently incised with a sterile blade with adequate precaution against accidental damage of embryos. The embryos were scooped out carefully with the help of a scalpel and gently transferred after being grouped by size into basal MS medium supplemented with different combinations of auxins and cytokinins, sugar (sucrose or glucose), and different concentrations of potassium nitrate. The pH of the culture media used was 5.5 with 0.8 per cent agar added prior to autoclaving. The cultures were incubated at 27°C with a 16/8 light/dark cycle with an illumination of 2,000 lux light intensity. Each experiment was repeated at least thrice in batches of 10.

RESULTS AND DISCUSSION

Young embryos with intact cotyledons were excised and grouped according to their size from the 30th day after flowering was observed to the 50th day. The size of these embryos with intact cotyledons varied between 3 mm and 12 mm. They were transferred and cultured in basal MS medium supplemented with various combinations of sugars, nitrogen source, and hormones as discussed below. Preliminary experiments showed that the excised embryos did not germinate in a plain MS basal medium and an auxin, a cytokinin, and a sugar were found to be essential. Based on this observation numerous combinations of auxins and cytokinins either in combination or alone were attempted.

Effect of Hormones

Optimum concentration of naphthalene acetic acid (NAA) and 6-benzyl amino purine (BAP) was arrived at after the embryos were incubated in different concentrations of hormones (Tables 1 and 2) for plantlet production.

Effect of Sugars

Sucrose (3–12 per cent) and glucose (3–12 per cent) were supplemented in basal MS medium and embryos were cultured in different hormone combinations. Sucrose or glucose at concentrations ranging from 3 to 12 per cent did not have any significant effect on the germination percentage of the excised embryos. However, germinated embryos established in the soil when sucrose was used (Tables 3 and 4).

Table 1. Effect of auxin and cytokinin on regeneration of plantlets from embryos. Experiments done in replicates of 10

MS Medium supplements + 12% sucrose	Response	Observation after no. of days
2, 4 D (2.2 mg/l)	No response	45
2, 4 D (0.1 mg/l) and kinetin (0.2 mg/l)	Callus formation	8
2, 4 D (0.5 mg/l) and kinetin (1 mg/l)	Callus formation	8
NAA (0.01 mg/l) and BAP (0.01 mg/l)	Radicle emergence	6

Table 2. Effect of 0.1 mg/l NAA and 0.3 mg/l BAP with MS medium supplemented with 3 per cent sucrose. Embryos were pooled and categorized by size and the experiment was done in replicates of 10

Parameter	Size of the embryos, mm			
	3–5	5–7	7–9	9–12
	Observation in days			
Radicle	5	6	6	5
Lateral roots	34	20	19	17
Shoot	34	19	19	16

Table 3. Germination percentage of embryos *in vitro* in MS medium containing 3 per cent sucrose with 0.1 mg/l NAA, 0.3 mg/l BAP, and 3 per cent glucose. Experiments done in replicates of 10

Size of the embryos,	Percentage of germination		Percentage established in soil	
	Sucrose	Glucose	Sucrose	Glucose
3–5	89	87	85	79
5–7	92	93	90	64
7–9	96	92	96	45
9–12	96	94	95	54

Effect of Different Concentrations of Potassium Nitrate

High concentrations of nitrate have proved to be beneficial with immature embryos of certain plants like sunflower to improve shoot induction. However, elevated concentrations of potassium nitrate (up to 6.5 g/l) did not have any effect on the germination percentage of the excised embryos in sucrose or glucose under the various hormonal supplementations.

High Frequency Regeneration and Effect of Indole Acetic Acid (IAA)

Neem embryos germinated in a medium containing 0.1 mg NAA and 0.3 mg BAP were transferred to an intermediate medium containing different concentrations of IAA to study the effect of this hormone. During the normal course of germination in 0.1 mg NAA and 0.3 mg BAP the appearance of root and shoot was almost simultaneous. But after the emergence of the radicle, when the embryos were subcultured in the same medium with the addition of IAA (0.01–1.0 mg) the shooting and rooting pattern changed. At lower concentrations of EAA (0.01–0.1 mg) the number of shoots per embryo gradually began to increase. Often about five to six shoots were regularly excised from embryos of all sizes.

In this case, the shoots were excised and cultured in a medium without BAP with half sugar to achieve rooting. Higher concentrations of IAA (0.5–1.0 mg) did not yield more shoots and often led to poorer establishment of the plantlets outdoors. It is interesting to note that rooting of neem and its subsequent establishment outdoors is achieved with a better frequency with lower concentrations of IAA.

In the course of this study an attempt has been made to understand the response of neem embryos to different hormones and to optimize the number of plantlets produced from them. Since very little genetic study has been done on this plant techniques of embryo rescue and regeneration similar to that of commercial crops would help researchers to attempt systematic breeding.

ACKNOWLEDGEMENT

The authors express their gratitude to the management of the Dalmia Centre for Biotechnology for financial assistance.

REFERENCES

Brink, R.A. and D.C. Cooper. 1947. *Bot. Rev.* 13: 423–541.

Degivry, M.T. 1966. *Bull. Sci. Bourgogne* 24: 57–87.

Maheshwari, P. and N.S. Rangaswamy. 1965. *Adv. Bot. Res.* 2: 219–321.

Mitra, C.R. 1963. In: *Neem*. Pub. Secretary, Indian Central Oilseeds Committee, Hyderabad. Examiner Press, Bombay.

Naina, N.S., P.K. Gupta and A.F. Mascarenhas. 1989. Genetic transformation and regeneration of transgenic neem (*Azadirachta indica*) plants using *Agrobacterium tumifaciens*. *Curr. Sci.* 58(4): 185–187.

Narayanaswami, S. and K. Nortstog. 1964. *Bot. Rev.* 30: 587–628.

Radwanski, S. and G.E. Wickens. 1981. Vegetative fallow and potential value of the neem tree (*Azadirachta indica*) in the tropics. *Econ. Bot.* 35: 398–414.

Raghavan, V. 1966. *Biol. Rev. Cambridge Philos. Soc.* 41: 1–58.

Raghavan, V. 1977. Applied aspects of embryo culture. In: *Applied and Fundamental Aspects of Plant Cell, Tissue and Organ Culture*. J. Reinert and Y.P.S. Bajaj (eds.), Narosa Publishing House, New Delhi, pp. 375–397.

Rappaport, J. 1954. *Bot. Rev.* 20: 210–225.

Sanders, M.E. and N.K. Ziebur. 1963. In: *Recent Advances in the Embryology of Angiosperms*. P. Maheswari (ed.), Int. Soc. Plant Morphol., Delhi, pp. 297–325.

Sathyavati, G.V., M.K. Raina and M. Sharma. 1976. *Medicinal Plants of India* 1. ICMR.

Thiagarajan, M. and P.M. Murali. (in press). Optimum conditions for regeneration of neem from immature embryos. *Indian Forester.*

Troup, R.S. 1989. *The Silviculture of Indian Trees*, Vol. I. International Book Distributors and Book Sellers, Dehradun, New Delhi.

Wardlaw, C.W. 1965. In: *Encyclopedia of Plant Physiology,* W. Ruhland (ed.), Springer Verlag, Berlin and New York. Vol. XV/1, pp. 424–442.

Tissue Culture Studies in Neem (*Azadirachta indica* A. Juss.)

A. NIRMALAKUMARI, N.M. RAMASWAMY AND S.R. SREE RANGASWAMY

Centre for Plant Molecular Biology, TNAU, Coimbatore 641003, India

ABSTRACT

Experiments to study the *in vitro* response of various explants viz., leaf, stem, bark and seed cotyledon for callus induction and shoot-tip and cotyledons for micropropagation on MS medium with different levels of auxins and cytokinins were conducted. The leaf explant produced nodular and friable calli on MS + 2, 4-D (2.0 mgl^{-1}) + Ki (0.5 mgl^{-1}) + Sucrose (3%). Shoot tip and cotyledons were inoculated on MS + BAP (1.1 mgl^{-1}) + CH (200 mgl^{-1}) + Sucrose (2%) + Sorbitol (1%). Cotyledons produced small shoot buds after 30 days of inoculation, which gave 15–20 leafy shoots per cotyledon. Shoot tips directly produced 6–7 shoots. One month old calli of leaf and stem were subcultured on MS + BAP (2.0 mgl^{-1}) + IAA (0.5 mgl^{-1}) + Sucrose (3%). They differentiated for embryoid formation and produced leafy shoots. Possibility of micropropagation of neem trees for large scale cultivation is indicated.

INTRODUCTION

Neem (*Azadirachta indica* A. Juss) is one of the most valuable but least exploited multipurpose trees among tropical trees. It provides fodder, firewood, medicines, pesticides, mosquito repellents, fertilizers, diabetic food, soaps, lubricants, gums, agricultural implements, toothpaste, toothbrush sticks, and even contraceptives (Tewari, 1992). Over 2,400 plants are known to contain insecticidal and insect-repellent constituents (Jacobson et al., 1983), but only neem holds out the promise of providing a highly effective, non-toxic, and environmentally harmless means of controlling or eliminating insect pests which inflict losses in agricultural production (Govindachari, 1992). That is why neem has attained worldwide prominence in recent years.

Raising plants from seeds of neem is an easy and conventional method of propagation, but the viability of the seed is only 15 days after maturation and wide variability exists among the seedlings (Krikorian et al., 1981). Tissue culture has been investigated for rapid propagation of superior lines as well as uniform neem plantations (Gavinlertvatana et al., 1987). In higher plants regeneration from callus is a necessary step to manipulate the cells and tissues under *in vitro* conditions. Such studies are limited in neem.

As early as 1972, Rangaswamy and Promila studied the morphogenesis of the adult embryo of neem. Sanyal et al. (1981) studied the induced chemical and histological differentiation in stem callus of neem and they did not succeed in regenerating entire plants from callus. Narayan and Jaiswal (1985) regenerated plantlets from callus of neem using mature leaf explants. *In vitro* morphogenesis in *A. indica* using cotyledon explants was studied by Muralidharan and Mascarenhas (1987) and plantlets were obtained through organogenesis. *In vitro* propagation of neem using leaf discs was achieved by Ramesh and Padhya (1990).

In the present investigation experiments were conducted to study the *in vitro* response of cotyledon, leaf, stem, and bark for callus induction and of cotyledon and shoot tip for micropropagation.

MATERIALS AND METHODS

Fresh and healthy leaflets, internodal stem bits, peeled brownishgreen bark from newly developed branches and matured yellow fruits were collected from the matured neem tree at Paddy Breeding Station, Tamil Nadu Agricultural University, Coimbatore. The leaflets were washed under running tap water and then with Tween-20. Leaf bits of 1–2 cm^2 were cut into pieces. The green young internodal stems were cut into pieces. The green young internodal stems were also washed and cut into bits of 1 cm length. The greenish barks were peeled from newly formed current-year branches and cut into pieces of 1 cm length. The cotyledons were obtained from mature, fresh yellow fruits after the fruit coat, pulp, and seed coat were removed. The two cotyledons were split open and the embryo was removed. The shoot tips with meristem and one or two leaves were collected from newly grown shoots. All these explants were surface-sterilized with mercuric chloride (0.1 per cent) and washed thoroughly five to six times with sterile, double-distilled water.

For callus induction, 21 MS (Murashige and Skoog, 1962) media combinations supplemented with various concentrations and combinations of 2, 4-Dichlorophenoxy acetic acid, (2, 4-D), indole-3-acetic acid (IAA), 6-benzyl aminopurine (BAP), and kinetin were used. The pH of the medium was adjusted to 5.8 before addition of 8 g/l Difco bacto agar and then autoclaved for 20 min. at 121°C at 15 lbs pressure. The tubes were sealed with cotton wool. Inoculation was done with one explant per culture tube (25 × 15 mm) and 40 explants were used for each treatment. The tubes inoculated with explants for callus induction were kept in the dark continuously for four weeks. The tubes inoculated with explants for micropropagation in 16 different media combinations were incubated under white fluorescent light (1,000 lux) for 16/8 hr light/dark at 25° ± 1°C.

The calli (22–28 days old) of cotyledon, leaf, stem, and bark explants were subcultured on 16 different regeneration MS media combinations supplemented with various levels of BAP, kinetin, IAA, and casein hydrolysate. For observations, calli from each explant were subcultured in 100 tubes in each media combination. The tubes were kept in 1,000 lux white fluorescent light for 16/8 hr light/dark at 25° $\pm$ 1°C.

The regenerated shoots (7–8 cm in length) with four or five compound leaves from calli in regeneration medium after 25 to 30 days of subculturing and shootlets formed directly from cotyledon and shoot tip explants were transferred to 10 different half strength MS rooting media supplemented with various levels of IAA, IBA, and NAA. Rooted shoots were transferred to pots containing a mixture of soil, sand, and well-decomposed organic manure (1:1:1) and kept in mist for two weeks for acclimatization.

RESULTS AND DISCUSSION

Explants and Callus Induction

For callus induction from cotyledon, leaf, stem, and bark, 21 media combinations were tried. None of the explants produced calli or any morphogenetic response on MS basal medium. But the explants responded for callus induction and proliferation in MS medium supplemented with 2, 4-D and Kinetin alone or in combination and with BAP alone or combination with IAA. Shulz (1983) also achieved callus induction using young leaf explants in MS medium supplemented with IAA.

The callus induction frequency of cotyledon in different media combinations tried ranged from 12.5 to 95.5 per cent. The cotyledon responded for callus induction only in 10 media combinations. The maximum callusing frequency was found in MS + 2, 4-D (2.0 mgl) + Kinetin (0.5 mgl). Similar results were obtained by Rao et al. (1988), who observed maximum callusing in MS medium supplemented with 2, 4-D and kinetin using cotyledon, leaf, and stem explants. The response of leaf and stem bits for callus induction in our study was observed in 18 and 14 media combinations respectively. The callusing frequency ranged from 5.0 to 94.0 per cent in leaf and 7.5 to 93.5 per cent in stem explants. The callusing frequency was maximum in leaf in MS + 2, 4-D (2 mgl^{-1}) + kinetin (0.5 mgl^{-1}) and it was high in MS + BAP (2.0 mgl^{-1}) + IAA (0.5 mgl^{-1}) for stem explant (Table 1). The explant bark responded only in four media combinations for callusing and the frequency was generally low, ranging from 5.0 to 12.5 per cent. The best media combination was MS + 2, 4-D (4.0 mgl^{-1}) + kinetin (0.5 mgl^{-1}). The nature of calli differed with varying explants and media combinations. The media having 2, 4-D and kinetin gave rise to root-like structures after 12 to 14 days of culturing. The media containing BAP and IAA produced greenish friable calli with shoot buds (Fig. 1). The callus induction was early in leaf explant (11 days) and was late in cotyledon (22 days) after inoculation.

Differentiation for Shootlets

Four-weeks-old calli from the four explants were subcultured in 16 MS regeneration media combinations supplemented with varying levels of BAP, kinetin, IAA and CH. The calli of leaf, cotyledon, and stem alone responded for differentiation after 25 to 30

Table 1. Effect of different media combinations on callus induction and shoot production in different explants of neem based on mean values of 40 explants

Explants/MS + (growth regulators, mgl^{-1})		Nature of morphogenetic response	Callusing frequency (%)	Mean. shoot No. per explant
Cotyledon				
Control (MS basal)		—	—	—
MS + 2, 4-D	2	Glossy straw white callus	20.0	—
	4	Glossy straw white callus	12.5	—
IAA	1	—	—	—
	2	—	—	—
BAP	0.5	—	—	—
	1.0	Green shoot buds	—	15–17
	1.5	Green shoot buds	—	15–17
	2.0	Green shoot buds	—	8–9
	2.5	Green shoot buds	—	8–9
Kinetin	0.5	—	—	
	1.0	—	—	—
	2.0	—	—	—
2, 4-D + Kinetin	2 + 0.5	Dull white callus	95.5	—
	2 + 1.0	Dull white callus	92.5	—
	4 + 0.5	Dull white callus	82.5	—
	4 + 1.0	Dull white callus	95.0	—
	4 + 2.0	Dull white callus	80.0	—
IAA + BAP	0.5 + 1.0	Granular light green callus with shoots	92.5	4–8
	0.5 +1.5	Granular light green callus with shoots	90.0	5–6
	0.5 + 2.0	Granular light green callus with shoots	95.0	4–5
Leaf				
Control (MS Basal)		—	—	—
MS + 2, 4-D	2	Compact brown-green callus	7.5	—
	4	—	—	—
IAA	1	—	—	—
	2	—	—	—
BAP	0.5	Compact light green callus	12.5	—
	1.0	Compact light green callus	17.5	—
	1.5	Green callus with shoot buds	40.0	3-4
	2.0	Green callus with shoot buds	47.5	4-5

contd.

Table 1 contd.

Explants/MS + (growth regulators, mgl^{-1})		Nature of morphogenetic response	Callusing frequency (%)	Mean. shoot No. per explant.
	2.5	Green callus with shoot buds	40.5	4–5
Kinetin	0.5	White callus	5.0	—
	1.0	White callus	10.0	—
	2.0	White callus	12.5	—
2, 4-D + Kinetin				
	2 + 0.5	Compact greenish white callus with roots	94.0	—
	2 + 1.0	Compact greenish white callus with roots	80.0	—
	4 + 0.5	Compact greenish white callus with roots	37.5	—
	4 + 1.0	Compact greenish white callus with roots	80.0	—
	4 + 2.0	Compact greenish white callus with roots	62.5	—
BAP + IAA	0.5 + 1.0	Green and white friable callus with shoot buds	37.5	7–8
	0.5 + 1.5	Green and white friable callus with shoot buds	62.5	7–8
	0.5 + 2.0	Green and white friable callus with shoot buds	92.5	10–12
Stem				
Control (MS basal)		—	—	—
MS + 2, 4-D	2	Compact powdery white callus	10.0	—
	4	Compact powdery white callus	7.5	—
BAP	0.5	—	—	—
	1.0	Compact green callus with shoot buds	75.0	2–4
	1.5	Compact green callus with shoot buds	67.5	2–4
	2.0	Compact green callus with shoot buds	80.0	5–6
	2.5	Compact green callus with shoot buds	80.0	7–8

contd.

Table 1 contd.

Explants/MS + (growth regulators, mgl^{-1})		Nature of morphogenetic response	Callusing frequency (%)	Mean. shoot No. per explant
Kinetin	0.5	—	—	—
	1.0	—	—	—
	2.0	—	—	—
2, 4-D Kinetin	2 + 0.5	Friable green callus with roots	90.0	—
	2 + 1.0	Friable green callus with roots	92.5	—
	4 + 0.5	Compact brown callus	67.5	—
	4 + 1.0	Compact brown callus	52.0	—
	4 + 2.0	—	—	—
IAA + BAP	0.5 + 1.0	Light green granular callus with shoots	88.0	8–9
	0.5 + 1.5	Light green granular callus with shoots	86.5	8–9
	0.5 + 2.0	Light green granular callus with shoots	93.5	4–6

Fig. 1. Calli on MS + BAP (2.0 mgl^{-1}) + IAA (0.5 mgl^{-1}) + sucrose (3 per cent) with shoot buds after 30 days in leaf explant

days of subculturing in MS supplemented with BAP, IAA and CH. The bark calli did not respond for differentiation in any of the 16 media combinations use. There was no response in MS basal and MS supplemented with kinetin. In the other media combinations better differentiation was achieved in MS supplemented with BAP and IAA combinations. The shootlet development was enhanced by the addition of CH (200 mg/l) to the medium. The regeneration efficiency of leaf calli ranged from 25.0 to 82.5 per cent. The regeneration efficiency of cotyledon was 7.9 to 69.8 per cent and that of stem calli was 9.0 to 62.1 per cent (Table 2). Using cotyledon explants, *in vitro* organogenesis and plantlet formation in cashew was achieved by Philips (1984). Bhansali and Arya (1978) found differentiation in stem and leaf explant calli of grapefruit in culture. In the three explants of neem in the present study the maximum differentiation was achieved in MS + BAP (2.0 mgl^{-1}) + IAA (0.5 mgl^{-1}) + CH (200 mgl^{-1}) + Sucrose (3 per cent).

Micropropagation

Direct embryogenesis or adventitious bud formation with callogenesis has been considered a successful method for clonal propagation of tree species (Von Arnold and Wallin, 1988). Domecq (1988) used shoot tips of *Melia azadarach* var. *gigantea* for micropropagation. We have observed that shoot tips and de-embryonated fresh fruit seed cotyledons of neem were successful for shoot production in BAP (1–1.5 mgl^{-1}) alone and CH (200–400 mgl^{-1}) on MS medium. After 15 to 20 days of inoculation, initiation of buds (Fig. 2) was seen on the dorsal side of the cotyledon after 25–30 days of

Table 2. Effect of different media combinations on regeneration* of calli from cotyledon, leaf, and stem explants of neem**

Media combinations (mgl^{-1})		Regeneration frequency (%) calli from		
		Cotyledon	Leaf	Stem
MS basal + Sucrose (3%) + BAP	1.0	52.7	61.4	43.5
	1.5	56.4	64.8	32.6
	2.0	46.3	51.1	21.3
	2.5	32.2	47.5	17.6
Kinetin	1.0	—	—	—
	2.0	—	—	—
BAP + IAA	1.0 + 0.5	53.2	76.5	45.5
	1.5 + 0.5	55.5	69.8	42.0
	2.0 + 0.5	59.3	72.9	50.3
	1.0 + 1.0	7.9	25.0	9.0
	1.5 + 1.0	10.2	25.3	12.4
	2.0 + 1.0	20.0	31.6	15.1
BAP + IAA + CH	1.0 + 0.5 + 200	60.5	75.6	55.7
	1.5 + 0.5 + 200	62.7	77.8	58.5
	2.0 + 0.5 + 200	69.8	82.5	62.1

*Shoots with al least one leaf are counted

**Mean of calli subcultured in 100 tubes

Fig. 2. Shootlet formation in regeneration medium MS + BAP (2.0 mgl^{-1}) + IAA (0.5 mgl^{-1}) + CH (200 mgl^{-1}) + sucrose (3 per cent) after 30 days of subculturing of leaf calli

culturing. They gave rise to leafy shoots ranging from 17 to 20 per cotyledon after 45 days (Fig. 3) and their frequency ranged from 48.5 to 92.5 per cent. In loblolly pine more than 3,000 plantlets were obtained from cotyledon explants (Amerson et al., 1988).

Multiple shootlets (12–15) were produced from shoot tips in MS + BAP (1.5 mgl^{-1}) + CH medium. The efficiency of shootlet development was maximum in MS + BAP (1.1 mgl^{-1}) + CH (200 mgl^{-1}) and it ranged from 12.5 to 92.0 per cent for shoot tip. Simultaneous occurrence of callus (Fig. 4) was also seen on the shoot tips at their cut ends. Such calli were separated and cultured on regeneration medium used for micropropagation. The calli differentiated for shoot formation after 18 to 20 days. The differentiated shootlets were allowed to grow further.

Rooting and Acclimatization

Ten media compositions with half strength MS basal and varying levels of different auxins were tried to induce rooting in the regenerated shoots. Regenerated shoots (7 to 8 cm height) with four to five leaves from either direct or indirect embryogenesis were

Fig. 3. Leafy shoots from the cotyledon after 45 days of inoculation

transferred to rooting medium. But rooting was good (39.0 per cent) and quickly achieved (22 days) with IBA (0.5–2.0 mgl⁻¹). The auxins IAA and NAA have no effect on root induction. Root induction occurred in IBA containing half-strength MS medium after 22 to 30 days of transferring and the frequency ranged from 67.5 to 93.0 per cent.

Table 3. Frequency (%) of shootlet formation from cotyledon* and shoot tip ** explants of neem***

	Basal medium MS			
CH (mgl^{-1})	BAP 0.0 (mgl^{-1})	1.1	1.5	2.0
0.0	M	60.0	56.5	52.0
	—	12.5	27.5	32.0
200	—	92.5	90.0	88.0
	—	92.0	88.0	72.5
400	—	88.5	82.5	75.0
	—	82.5	76.0	54.0
600	—	87.5	48.5	52.5
	—	52.5	42.0	32.0

* Values given first place
** Values given second place
*** Observations taken after 45 days of culturing on MS + sorbitol (1.0 per cent) + sucrose (2 per cent); average of 40 explants; shoots with at least one leaf counted.

BAP - benzyl amino purine
CH - Casein hydrolysate

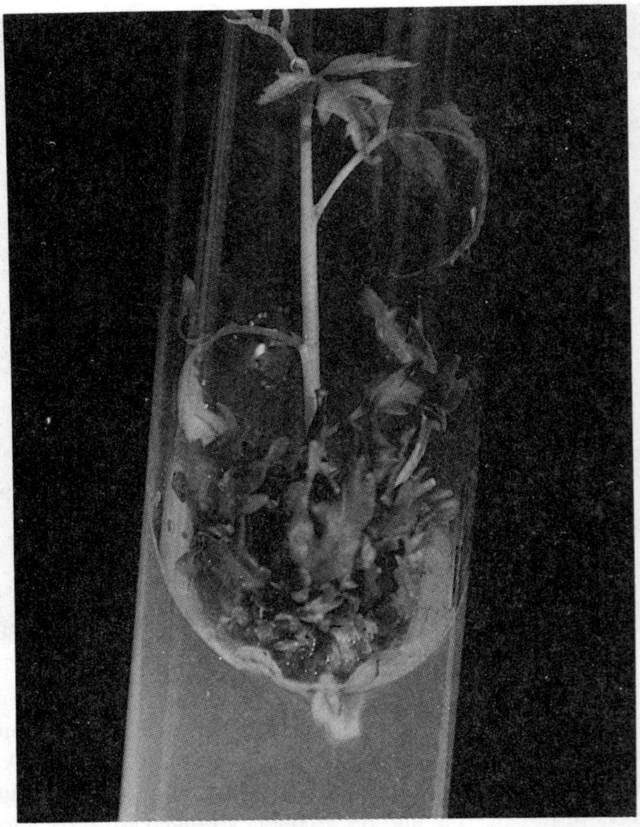

Fig. 4. Multiple shootlets with calli at cut end of the shoot tip explant on micropropagation medium

The number of roots developed ranged from 10 to 19 per shoot. Zaheer Ahmed et al. (1990) also showed that IBA was essential for root induction and IAA was ineffective in neem. The rooted plants were transferred to mist chamber in nursery pots for hardening.

CONCLUSION

This *in vitro* study on neem indicates that callus induction varies with explant apart from media compositions. The media inducing maximum callusing differ from one explant to another. Callus induction requires the presence of an auxin for all the four explants. Cytokinin alone could not induce calli but induced direct embryogenesis. The cotyledons produce only shoot buds when there in BAP alone. But they produce shoot buds as well as calli when the media have both BAP and IAA. The addition of an auxin, such as IAA, even at very low concentrations usually resulted in a significant increase in callus production and a general reduction in the average number of shoots produced.

In the present study we have demonstrated that it is possible to otbain 8 to 10 plantlets from leaf, stem, or cotyledon explants through indirect embryogenesis and 15 to 20 plantlets from cotyledon or 12 to 15 plantlets from shoot tip through direct

Fig. 5. Rooted plant on 1/2 MS + IBA (0.5 mgl^{-1})

embryogenesis in four to five months. This offers great scope for micropropagation of elite neem trees, and the direct embryognesis from cotyledon indicate the possibility of transformations in neem.

REFERENCES

*Amerson, H.V., L.J. Frampton, Jr., R.L. Mott and P.C. Spaine. 1988. Tissue culture of conifers using loblolly pine as a mode. In: *Proc. Con. on Genetic Manipulation of Woody Plants*. J.W. Hanover and Daniel E. Kealthley (eds.), Plenum.

Bansali, R. and H.C. Arya. 1978. Differentiation in explant of *Citrus paradisi* Macf. (grape fruit) grown in culture. *Indian J. Exp. Biol.* 16: 409–411.

Domecq, C.M. 1988. *In vitro* culture of shoot tips of *Melia azadarach* var. *gigantea*. *Phyton*, Argentina, 48(1–2): 33–42.

Govindachari, T.R. 1992. Chemical and biological investigations on *Azadirachta indica* (the neem tree). *Curr. Sci.* 63, 3(10): 117–122.

Gavinlertvatana, P., A.C. Matheson and M.E.P. Sim. 1987. Feasibility study on tissue culture for multipurpose forest tree species. Project No. 1. Plantek International Ltd., Singapore, vi + 59 pp.

Jacobson, M., J.B. Stoke, J.D. Warthern, Jr., R.E. Redfern, D.K. Reed, R.E. Webb and L. Telek. 1983. In: *Proc. 2nd Int. Neem Conf.* Ravischholzhausen, FRG, GTZ, Eschborn, 31 pp.

Krikorian, D., Abraham, Madhav Singh and E. Quinn. 1981. In: *Proc. Int. Sym. on Tissue Culture of Economically Important Plants*, A.N. Rao (ed.), COSTED and ANBS Asian Network for Biological Sciences. 167 pp.

Muralidharan, E.M. and A.F. Mascarenhas. 1987. *In vitro* morphogenesis in *Azadirachta indica* (A. Juss.), from leaf discs. *Indian J. Exp. Biol.* 28: 932–935.

Murashige, T. and F. Skoog. 1962. A revised medium for rapid growth and bioassays with tobacco tissue cultures. *Physiol. Plant* 15: 473–497.

Narayan, P. and V.S. Jaiswal. 1985. Plantlet regeneration from leaflet callus of *Azadirachta indica*. *J. Tree Sci.* 4: 65–68.

Philips, V.J. 1984. *In vitro* organogenesis and plantlet formation in cashew (*Anacardium occidentalis* L.). *Ann. Bot.* 54: 149–152.

Ramesh, K. and M.A. Padhya. 1990. *In vitro* propagation of neem (*Azadirachta indica* A. Juss.), from leaf discs. *Indian J. Exp. Biol.* 28: 932–935.

Rangaswamy, N.S. and Promila. 1972. Morphogenesis of the adult embryo of *Azadirachta indica* A. Juss. *Z. Pfl. Physiol.* 67: 377–379.

Rao, M.V.S., Y.V. Rao, Y.S. Rao and V. Monga. 1988. Induction and growth of callus in *Azadirachta indica* A. Juss. *Crop Improvement* 15(2): 203–205.

Sanyal, M., A. Das., M. Banerjee and P.C. Datta. 1981. *In vitro* hormone induced chemical and histological differentiation in stem callus of neem, *Azadirachta indica* A. Juss. *Indian J. Exp. Biol.* 19: 1067–1068.

Shulz, F.A. 1983. Tissue culture of *Azadirachta indica*. In: *Proc. 2nd Int. Neem Conf.* Rauischholzhausen, FRG GTZ, Eschborn, pp. 539–541.

Tewari, D.N. 1992. *Monograph on neem* (*Azadirachta indica* A. Juss). International Book Distributors, 9/3, Rajpur Road, Dehradun 248001, India.

* Von Arnold, S. and A. Vallin. 1988. *Int. Assoc. Plant Tissue Cult. Newsletter* 56.

Zaheer Ahmed, N. Zaidi and F.H. Shah. 1990. Micropropagation of *Melia azadarach* from mature tissue. *Pak. J. Bot.* 22(2): 172–178.

* Originals not seen

111

Investigations on Neem and *Melia* for Sustainable Production

M.V.R. PRASAD, A.R.G. RANGANATHA, M. SUJATHA, MUKTA DHINGRA, M. RAMACHANDRAM AND G. NAGARAJ

Directorate of Oilseeds Research
Rajendranagar, Hyderabad 500 030, India

ABSTRACT

Research is in progress on neem and *Melia* to estimate their potential for sustainable agricultural development. Tree plantations of these species are raised in different sets of management and observations are recorded. The growth parameters of *Melia* are substantially higher than those of neem, indicating a faster growth and higher biomass production. The results indicated that *Melia* starts bearing from the second year, with continuous bearing facilitating better collection. Research on selection of desirable types resulted in the identification of super early types, which bear fruit in first year (10 months), possessing desirable canopy structure. The plant-to-progeny blocks of five superior types are established and studies are in progress. Studies have also been undertaken to investigate quality parameters, including fatty acid profile. Since trees generally take longer time to produce seeds, the research was undertaken to develop tissue culture protocol for mass propagation of *Melia*. The technique has been standardized for *in vitro* multiplication of *Melia* from nodal explants. The procedure is rapid and promising for large-scale production and for the maintenance of genetic purity of superior trees. The studies on intercropping in tree plantations indicated that horse gram, cowpea, red gram, and niger are better suited for the purpose. Even sesame and castor could be intercropped in the early growth period of trees.

INTRODUCTION

Neem (*Azadirachta indica*) grows wild, thrives best in drier climate in all kinds of soil, and is ideally suited for afforestation in hot areas (Bringi, 1987). Neem trees are generally found in South-East Asia and tropical regions of Australia and Africa. The other species with similar botanical characters is known as persian lilac or backyan (*Melia azedarach*). *Melia* is a native of West Asia but it is now naturalized throughout the warm countries. It withstands cooler climates better than neem. It is a fast-growing species and also is naturalized all over India. It bears early and the fruit-bearing period is longer than that of neem. Seeds yield 30 per cent of drying oil, which is suitable for making soaps and hair oils. The timber is tolerant to attack by white ants. This species is also cultivated for its shade and ornamental value on account of its quick growth (Achaya, 1990).

NEEM AND *MELIA* GROWTH ATTRIBUTES

The plant height, stem girth and canopy attributes of neem and *Melia* are presented in Tables 1–4. The results indicated that *Melia* was fast growing and a high biomass producer. *Melia* also started flowering and fruiting early and the fruiting period was longer than that of neem.

SUPERIOR EARLY-BEARING TYPES OF *MELIA*

Based on the fruiting, canopy, and other attributes, five early superior types were identified. These five types were further raised as separate plant progeny blocks. The

Table 1. Growth attributes of neem

Attribute/Stage	1 year	2 years	3 years	4 years
Plant height (m)	0.91	1.78	2.54	3.12
Stem girth (cm)	11.6	14.4	22.6	35.1
Canopy (m)	0.41	0.85	2.68	3.18
Height of branching (m)	0.83	1.22	1.27	1.33

Table 2. Growth attributes of *Melia*

Attribute/Stage	1 year	2 years	3 years	4 years
Plant height (m)	2.02	3.88	5.44	5.81
Stem girth (cm)	19.8	35.2	46.8	61.2
Canopy (m)	3.56	6.10	9.90	10.88
Height of branching (m)	1.53	1.68	1.73	1.79

Table 3. Neem growth attributes in different managements

Year/Character	Plant height (m)			Stem girth (cm)			Canopy (m)		
	1	2	3	1	2	3	1	2	3
1st year	0.69	0.78	0.91	6.8	8.4	11.6	0.27	0.32	0.41
2nd year	1.32	1.46	1.78	9.7	12.1	14.4	0.53	0.62	0.85
3rd year	1.96	2.12	2.54	14.3	17.9	22.6	1.64	2.09	2.68
4th year	2.23	2.67	3.12	19.2	23.6	35.1	1.97	2.43	3.78

1 = Poor soils with poor management
2 = Normal soils with poor management
3 = Normal soils with normal management

growth attributes are presented in Table 5. These lines started flowering and fruiting in less than a year, leading to the identification of better canopy types.

RAPID MULTIPLICATION OF *MELIA*

For mass propagation and to produce plants all year round, tissue culture technique was resorted to. Nodal segments of approximately 6 cm length were obtained from one- to two-year sexually mature trees. Explants were washed in running tap water for 60 min. and then rinsed with 0.2 per cent Teepol solution. After being washed with distilled water, segments were dipped in 70 per cent ethanol for 30 sec. Surface sterilization was done with 0.1 per cent mercuric chloride solution for 20 min. followed by thorough washing with sterile glass double-distilled water. Segments were separated into 1 cm pieces each bearing a single node. Nodal explants were cultured on solidified MS medium (Murashige and Skoog, 1962) supplemented with various concentrations and cominations of 6-benzyl amino purine (BA), kinetin, and naphthalene acetic acid (NAA). The pH of the medium was adjusted to 5.8 before the medium was solidified with 0.8 per cent agar. All cultures were maintained in a culture room at 26° ± 2°C and a relative humidity of 58–60 per cent with 16/8 hr photoperiod. Each treatment consisted of 24 replicates and each experiment was repeated twice.

Nodal segments cultured on MS medium supplemented with low concentration of kinetin (0.5 mg/l) showed growth of callus from the base of the segments and elongation of a single shoot (Table 6). However, higher concentration of kinetin (2.0 mg/l) was unable to induce any response from the explants. BA (0.5 mg/l) incorporated into MS medium induced the elongation and proliferation of the axillary buds accompanied by slight callus formation at the base. Higher concentration of BA (2.0 mg/l) triggered the

Table 4. *Melia* growth attributes in different managements

Year/Character	Plant height (m)			Stem girth (cm)			Canopy (m)		
	1	2	3	1	2	3	1	2	3
1st year	1.43	1.67	2.02	10.2	13.4	19.8	1.97	2.82	3.56
2nd year	2.39	3.14	3.88	19.7	26.9	35.2	3.26	4.37	6.10
3rd year	3.64	4.38	5.44	26.8	34.3	46.8	4.38	6.38	9.90
4th year	4.13	4.92	5.81	33.4	43.7	61.2	5.12	8.34	10.88

1 = Poor soils with poor management
2 = Normal soils with poor management
3 = Normal soils with normal managemen

Table 5. Growth attributes of selected plant progenies of *Melia*

Attribute/Period	6 months	9 months	15 months
Plant height (cm)	77.8	163.2	231.6
Stem girth (cm)	7.8	17.2	24.6
Canopy (cm)	91.4	228.2	292.8
Height of branching (cm)	27.2	69.4	88.8

Table 6. Response of nodal explants of *Melia azedarach* to various phytohormones

Medium (mg/l)	Response
MS + kinetin (0.5)	Base starts callusing; axillary bud elongates to fifth node.
MS + BA (0.5)	Base forms dark brown callus; axillary bud elongates and proliferates.
MS + kinetin (2.0)	No growth.
MS + BA (2.0)	Slight proliferation of axillary bud, plant fragile and pale green in colour.
MS + BA (2.0) + NAA (0.5)	No growth.

MS = Murashige and Skoog's basal medium.

induction of a few fragile and pale green shoots from the axillary buds. Combinations of BA (1.0 mg/l) and NAA (0.5 mg/l) did not elicit any respone in the explants.

BA at 0.5 mg/l was found to be the best for induction of shoot proliferation. Shoot clusters were separated at the base and the callused region was discarded before subculturing to fresh media. Due to the rapid growth rate, subculturing was done at two-week intervals. Each shoot cluster, at the end of a two-week culture period, consisted of one long shoot of 4.6 cm surrounded by numerous tiny shoots. Axillaries of one to three basal nodes also elongated. Enhanced axillary branching along with rapid proliferation from the base led to a higher rate of multiplication (Dhingra et al., 1991).

For obtaining plantlets, well developed (2 to 6 cm long) shoots were excised and the basal region was trimmed before being put into root induction medium. Of the various auxins tested for root induction, NAA proved to be the most effective. Besides the type and concentration of auxin, root induction was also influenced by the basal media strength and aeration. Half-strength MS medium gelled with 0.7 per cent agar proved to be the best for promoting root growth while shoots failed to root in liquid medium. Solidified half-strength MS medium incorporated with 1.0 mg/l NAA induced good rooting in all cultures with 100 per cent success in one week. The micropropagated shoots were transferred to plastic pots containing sterile vermiculite. These were later transferred to the soil after hardening.

The tissue culture of forest trees has shown promise in obtaining regenerants and rapid clonal multiplication. The basic advantages of *in vitro* propagation lies in the rapidity with which plant multiplication can be achieved in a relatively short time with less space. Another advantage is that micropropagation can be carried out throughout the year, irrrespective of seasonal changes. In the present case, this technique has been useful for maintenance of superior genetic stocks.

The protocol developed for the nodal segment culture can be effectively used for rapid multiplication. Single node explants from matured trees as used in this study are useful for maintaining genetic stability (heterozygosity) which might be otherwise disturbed if regeneration is obtained from callus cultures. Nodal stem segments carrying an axillary bud or excised axillary buds have also been used successfully as explants in some instances. Teak (Gupta et al., 1980) and *Dalbergia sissoo* (Datta and Datta, 1983) have also been propagated similarly using single node culture unlike the conventional shoot tip culture technique. The *Melia* plants raised through tissue culture also started flowering and started yielding fruits.

OIL QUALITY

To study the oil quality, especially the fatty acid profile, of *Melia*, studies were initiated and the preliminary results are shown in Table 7.

The kernel, which itself constituted only 8–10 per cent of the seed, contained 26–28 per cent oil. The deoiled kernel cake is very rich in protein (55.6 per cent). The oil is rich in unsaturated fatty acids, namely, oleic (37.3 per cent) and linoleic acid (32.3 per cent).

INTERCROPPING ANNUAL SPECIES IN NEEM AND *MELIA* PLANTATIONS

To increase land use efficiency and net returns to farmers, intercropping studies in neem and *Melia* plantations were taken up. The annual legumes included were red gram, green gram, horse gram, cowpea, and velvet bean. Niger, castor, sesame, sunflower, and sunhemp were also raised as intercrops. In the early phase, intercrops red gram, horse gram, cowpea, niger, and castor were effective, but after development of canopy coverage of *Melia* (from the fourth year), intercropping was not effective. Both in neem and in *Melia*, among all intercrops, horse gram among pulses and niger in oilseeds were superior. On better soils, sunflower early variety, Morden, gave superior performance. In neem, red gram and castor were superior due to less canopy coverage.

SPECIAL FEATURES OF *MELIA*

At the Directorate of Oilseeds Research, among 40 different species of tree oilseeds tested, *Melia* exhibited the following special features:

1) High biomass production,
2) Early flowering and bearing,
3) Long fruiting period facilitating better collection, and
4) Feasibility of use in afforestation in waste land and forestry and environment programmes.

Based on the experiences, *Melia* seed was supplied for large-scale use in watershed programmes of the Maheswaram Project. The recommendations were also sent to the

Table 7. Oil and protein content and fatty acid composition of *Melia*

	Oil (%)	Protein (%)
Shell	4–8	18
Seed	6–8	20
Kernel	26–28	55

Fatty acid	Per cent
16:0	19.5
18:0	10.9
18:1	37.3
18:2	32.3

Government of India (through NIRD national meet) for popularization of *Melia* in different tree plantation programmes. Among the limitations of neem are its slow growth and short period of fruit collection. In *Melia*, collection of fruits can provide long-term employment.

IMPROVEMENT OF *MELIA* AND NEEM

The following areas should be considered for improvement of *Melia* and neem:

1) Development of different duration types,
2) Higher seed yielding types,
3) Rapid multiplication through tissue culture,
4) Development of high-oil and superior quality types,
5) Development of different canopy architecture types,
6) Development of types suitable to different situations (industrial areas, abandoned mines) and environments,
7) Collection/augmentation, evaluation, and maintenance of different genetic stocks, and
8) Use in afforestation of waste lands, problem soils, and forestry and environment programmes.

REFERENCES

Achaya, K.T. 1990. *Oilseeds and Oil Milling in India*. Oxford & IBH Publishing Co. Pvt. Ltd., New Delhi.

Bringi, N.V. 1987. *Non-traditional Oilseeds and Oils of India*. Oxford & IBH Publishing Co. Pvt. Ltd., New Delhi.

Datta, S.K. and K. Datta. 1983. Auxin induced regeneration of forest tree *Dalbergia sissoo* Roxb. through tissue culture. *Cur. Sci.* 52: 434–436.

Dhingra, M., M. Sujatha and A.R.G. Ranganatha. 1991. Rapid multiplication of *Mellia azedarach* L. through tissue culture. *J. Oilseeds. Res.* 8: 215–219.

Gupta, P.K., A.L. Nadgir, A.F. Mascarenhas and V. Jagannathan. 1980. Tissue culture of forest trees; Clonal multiplication of *Tectona grandis* L. (teak) by tissue culture. *Plant Sci. Lett.* 17: 259–268.

Murashige, T. and F. Skoog. 1962. A revised medium for rapid growth and bioassays with tobacco tissue cultures. *Physiol. Plant.* 15: 473–479.

112

Clonal Propagation of Neem by Tissue Culture

R.A. DREW

Queensland Department of Primary Industries
PO Box 327, Cleveland, Q 4163, Australia

ABSTRACT

A technique for production of micro-cuttings *in vitro* has been developed for clonal propagation of neem lines with high yields of azadirachtin. Neem shoots were established in culture from nodal bud explants of glasshouse-grown plants. Apically dominant shoots were grown on hormone-free MS (Murashige and Skoog, 1962) medium and multiplied by dissection into nodal sections. The use of cytokinins in this medium caused excess callus production on cut sections of the explants. Good root systems were initiated when actively growing axillary shoots were removed from nodal sections and cultured on MS medium plus 10 μM (IBA) for four days, with subsequent root development on hormone-free medium. Longer exposure to IBA reduced rooting percentages, increased time to initiate roots, and produced excess callus at the base of the explants. In lieu of a transfer to hormone-free medium after four days, photo-oxidation of the IBA was achieved by injection of 100 μM riboflavin solution into the medium. This procedure has been used successfully for the culture of neem lines from eight countries (Australia, India, Mauritius, Indonesia, Thailand, Ghana, Gambia, and Kenya). Root initiation of more than 90 per cent has been achieved with all lines. Plants of all lines have been successfully established in the field and have produced seed after two years. As the multiplication system is based on production of micro-cuttings from axillary buds, it should be free of genetic off-types. More than 100 individual clones have been established in culture. Three of these have produced embryogenic callus at the base of nodal stem sections. Embryos have been proliferated on hormone-free MS medium and some have developed shoots and roots.

INTRODUCTION

A method of rapid vegetative multiplication of neem (*Azadirachta indica* A. Juss.) would be useful for clonal propagation of high azadirachtin-yielding genotypes. Micropropagation would also eliminate problems associated with rapid loss of viability of seed and variability in seedlings.

Published techniques for *in vitro* culture of neem involve callus culture followed by organogenesis. Callus production has been obtained from adult embryos (Rangaswamy and Promila, 1972), young seedlings (Naina et al., 1989), leaves, cotyledons, and petioles (Narayan and Jaiswal, 1985; Rao et al., 1990; Sanyal et al., 1983; Schulz, 1983), stem segments (Jaiswal and Narayan, 1984) and stem bark segments (Sanyal et al., 1981). However, occurrence of genetic off-types is well documented in many other species when regenerated from callus cultures (Larkin and Scowcroft, 1981). Therefore, it is not recommended as a method of clonal propagation.

A system for micropropagation which is not prone to production of genetic off-types has been reported for papaya (Drew, 1992), based on rapid growth of micro-cuttings *in vitro*. This paper details experiments which resulted in a similar technique for clonal propagation of neem via production of micro-cuttings dissected from double-node sections of apically dominant shoots.

MATERIALS AND METHODS

Explant Preparation and Establishment *In vitro*

Neem seeds were obtained from eight countries—Australia, Gambia, Ghana, India, Indonesia, Kenya, Mauritius, and Thailand. The seeds were germinated and the seedlings were grown in pots in a glasshouse. When the trees were 6 to 12 months old, the apex plus four to eight nodes were removed. The side shoot which developed on each plant was removed after a further three to four weeks and provided an additional apex plus 10 nodes from each plant.

Nodal sections were washed in running water for 1 hr, then disinfested for 20 min. in 1 or 2 per cent sodium hypochlorite solution containing a few drops of 7X detergent. After three rinses in sterile water, explants were trimmed and placed in culture medium.

Shoot Growth *In vitro*

Nodal sections were placed in 30 ml polycarbonate screw cap tubes. All media had pH adjusted to 5.65 before autoclaving at 121°C for 15 min. Cultures were incubated at 25°C for 16 hr days with cool white fluorescent tubes.

In experiment 1, nodes were cultured on MS (Murashige and Skoog, 1962), 1/2 MS, woody plant medium (Lloyd and McCowan, 1981), nouvelles formule (Quoirin and Lepoivre, 1977) and redwood medium (Ball et al., 1978), each containing 0.3 mg/l 6-benzyl amino purine (BAP) and 0.1 mg/l IBA. In experiment 2, sucrose concentrations of 1, 2, 3, 4, and 5 per cent were added to MS medium.

Shoot Multiplication *In vitro*

Neem shoots grown *in vitro* were dissected into single and double node sections and cultured on MS medium plus BAP, isopentenyl adenine (2iP), and kinetin at 1.0, 3.2, and 5.0 μM. In a second experiment, double node sections were placed on the same media treatments, and shoot numbers, shoot heights, and callus weights were recorded. As a measure of plant quality, shoot colour was rated between 5 (healthy green) and

1 (pale necrotic). A third experiment compared growth of six lines of neem on hormone-free MS medium at incubation temperatures ranging from 20°C to 35°C. In these experiments replicates varied between 16 and 20 per treatment.

Root Initiation and Growth *In vitro*

Actively growing shoots, 5–10 mm high, were excised from nodal sections and used in rooting experiments. All experiments had a minimum of 20 replicates per treatment. Root initiation was recorded daily and root growth was assessed after 21 days. Callus production at the base of each shoot was rated between 0 (no callus) and 5 (maximum callus).

In the first experiment neem shoots were cultured on MS medium containing 0, 0.1, 1, or 10 μM IBA, naphthalene acetic acid (NAA), or indole-3-acetic acid (IAA). In a second experiment shoots were exposed to MS medium containing 10 μM IBA or 1 μM NAA for 0, 1, 2, 3, 5, or 21 days before transfer to hormone-free medium.

RESULTS AND DISCUSSION

Explant Preparation and Establishment *In vitro*

Dissection of nodes lower than six or seven from the apex was difficult when plants were more than six months old as the stem was too woody. The subsequent production of an actively growing side shoot (after three to four weeks) allowed removal of apex plus 10 nodes. One per cent sodium hypochlorite was better for disinfestation. Two per cent produced higher percentages of clean explants; however, it also resulted in damage to the axillary buds. The rate of contamination varied with source of explant, age of trees, and time of year when dissected. Position of node on tree before dissection affected both contamination rates and growth rates of shoots *in vitro* (Table 1). Nodes three to six represented the best compromise between growth and contamination rates.

Shoot Growth *In vitro*

In the first experiment, the best growth and quality of shoots was obtained on MS medium. However, all shoots produced excess callus at the base of the stem. In the second experiment, 3 per cent sucrose produced maximum shoot growth.

Table 1. Effect of node position before dissection on contamination rates (top figure = % not contaminated) and shoot heights (lower figure = mm) *in vitro* after 21 days

Plant source	Node number (from apex)								
	Apex	2	3	4	5	6	7	8	9
Kenya	93	86	64	71	50	36	29	29	21
	3.2	2.3	4.1	6.9	7.6	3.8	3.3	1.5	1.3
Gambia	100	86	100	100	100	57	71	57	50
	2.6	1.8	3.6	7.6	7.6	13.8	7.8	4.5	1.7

Shoot Multiplication *In vitro*

Single node sections had poor survival rates compared to double node sections (Table 2), in terms of growth of shoots from axillary buds. All double node sections produced actively growing axillary shoots on all treatments (Table 2).

In the second experiment, shoot and callus growth were monitored on double node sections. Although 1 μM BAP and 1 μM 2iP gave good shoot growth, callus production was excessive with all cytokinin treatments (Table 3). Of the cytokinin treatments, 1 μM 2iP produced the best quality shoots., however, the best treatment in terms of shoot quality and minimal callus production was the control treatment (nil cytokinin). This treatment was also satisfactory in terms of production of shoots and shoot growth (Table 3).

Table 2. Per cent of nodal sections which produced shoots from axillary buds, as affected by cytokinin component and concentration

Cytokinin treatment (μM)		Source		
		Indian single node	Mauritius single nodes	Indian double nodes
BAP	1.0	54	40	100
	3.2	78	22	100
	5.0	50	88	100
2iP	1.0	100	44	100
	3.2	63	38	100
	5.0	53	31	100
Kinetin	1.0	80	70	100
	3.2	67	73	100
	5.0	50	46	100
Control	(nil)	81	46	100

Table 3. Effect of cytokinin component and concentration on growth and quality of shoots and on callus production from axillary buds of double node neem sections

Cytokinin treatment (μM)		Mean shoot number	Mean shoot height (mm)	Callus wt (g)	Callus production (% of total wt)	Shoot colour rating
BAP	1.0	2.0	10.8	5.66	70	2.9
	3.2	2.0	5.3	1.77	77	1.8
	10.0	1.9	3.0	.44	78	1.4
2iP	1.0	1.7	6.3	1.76	59	4.2
	3.2	1.4	3.9	4.44	77	2.4
	10.0	1.6	3.5	7.06	75	2.8
Kinetin	1.0	1.7	3.9	1.61	57	3.4
	3.2	1.2	3.9	4.18	79	3.2
	10.0	1.6	3.5	4.21	75	2.8
Control	(nil)	1.8	4.2	0.0	0	5.0
LSD	(5%)	36	1.98	1.69		43

Growth of all eight clones was sub-optimal at 20°C. Growth in shoot weight and height increased with increasing temperature between 20°C and 35°C. However, shoot quality was poor at 20°C and 35°C. Thus, 25°C or 30°C are recommended for culture incubation.

Root Initiation and Growth *In vitro*

In the first experiment, the best concentration of each auxin for root initiation was 10 µM IBA, 1 µM NAA, and 10 µM IAA (Table 4). All auxin treatments produced thickened roots and excess callus at the base of the explant. Callus production increased with auxin concentration.

The second experiment, which compared duration of exposure of neem shoots to either 10 µM IBA or 1 µM NAA, showed IBA to be the better treatment and four days to be the optimal duration of exposure to auxin in terms of time to initiate roots (Drew, 1991), per cent rooted, number of roots per shoot, and root quality (Table 5).

In lieu of a transfer to hormone-free medium after four days, good root initiation was also achieved either by four days' dark incubation of media containing 10 µM IBA and 10 µM riboflavin before transfer to the light, or by overlaying the medium containing 10 µM IBA with 100 µM riboflavin after four days' incubation (Drew, 1990, 1991). Riboflavin has been shown to rapidly photo-oxidize IBA in tissue culture media in light (Drew et al., 1991).

CONCLUSIONS

Neem genotypes from eight countries have been multiplied successfully by dissection of apically dominant shoots into double node sections when cultured on hormone-free MS medium. The use of any cytokinin in this medium caused undesirable callus production.

Table 4. Effects of auxin component and concentration on rooting and callus production on Indian and Indonesian neem shoots

Auxin treatment µM		Indian			Indonesian		
		% of shoots with roots	Mean root number per shoot	Callus rating	% of shoots with roots	Mean root number per shoot	Callus rating
NAA	10.0	50	2.9a	3.2a	60	3.3	3.4a
	1.0	65	3.4a	2.3bc	90	3.2	2.9a
	0.1	40	2.4a	0.9bd	75	2.4	1.2bc
IBA	10.0	75	5.7b	3.1a	100	3.1	3.4a
	1.0	65	3.7ac	1.2bd	80	2.6	0.9bc
	0.1	55	1.9ad	0.5bdf	55	1.9	0.5bc
IAA	10.0	65	2.7a	1.7be	65	2.4	1.3bc
	1.0	30	2.0a	0.3bdf	50	1.8	0.6b
	0.1	25	2.8a	0.5bdf	15	1.7	1.7bc
Control (nil)		30	2.7a	0.8bd	35	2.0	0.1bd

a and b, c and d, e and f differ significantly at $P < 0.01$.

Table 5. Effect on rooting and callus production of duration of exposure of two neem lines to 10 μM IBA before transfer to hormone-free medium

Days exposure to 10μM IBA	Indian			Indonesian		
	% rooted after 21 days	Mean no. primary roots	Callus rating	% rooted after 21 days	Mean no. primary roots	Callus rating
0	40	4.4bc	0.0a	40	3.8	0.8ac
1	90	4.1bc	0.9bce	80	4.5	0.9ac
2	70	4.0bc	1.2bc	80	4.5	0.9ac
3	90	3.8bc	1.1bce	80	3.5	1.6ad
4	100	6.1a	1.4bc	85	4.4	1.7ad
5	100	4.7bc	1.8bcf	95	4.1	1.7ad
21	75	7.8bd	3.6bd	95	4.0	3.3b

a and b, c and d, e and f differ significantly at $P < 0.01$.

Shoots of all eight genotypes enhibited more than 90 per cent rooting when cultured for four days on medium containing 10 μM IBA, before transfer to hormone-free medium. This was achieved by day 12 (8 days after transfer to hormone-free medium).

All genotypes have been established in soil; however, acclimatization was difficult because of callus production at the base of the plantlets, which often meant that roots were poorly attached to the stem.

Plants of all lines were established in the field at Ayr Research Station in North Queensland (latitude 19°S). Seed production began two years after planting and has followed a similar pattern to that of seedlings. Yield data is being collected and HPLC analyses are being performed to determine azadirachtin content of these clones.

In studies of organogenesis and embryogenesis, callus has been grown from numerous clones on MS media containing a range of auxins and cytokinins. Best callus growth was obtained on MS medium containing 5 μM BAP. Of 120 individual clones that have been established *in vitro*, only three have produced embryogenic callus. This was produced at the base of nodal stem sections. The resultant embryos were proliferated on hormone-free MS medium and some have developed shoots and roots. Embryogenesis appears to be influenced more by genotype than by media composition.

ACKNOWLEDGEMENTS

This work was funded in part by the Rural Industries Research and Development Corporation. The author wishes to acknowiedge the contribution of J. Neale in experimental work and statistical analysis.

REFERENCES

Ball, E.A., D.M. Morris and J.A. Rydelius. 1978. Cloning of *Sequoia sempervirens* from mature trees through tissue culture. *Conf. In vitro Mult. Woody Species*, Gembloux, Belgium, pp. 181–226.

Drew, R.A. 1990. Micropropagation of neem and regulation of *in vitro* rooting with riboflavin. Abstracts, *VII Int. Cong. Plant Tiss. Cell Cult.*, Amsterdam, pp. A3–59.

Drew, R.A. 1991. Optimising root initiation by controlling exposure to auxin. *Proc. Int. Plant Prop. Soc.* 4: in press.

Drew, R.A. 1992. Improved techniques from *in vitro* propagation and germplasm storage of papaya. *Hort. Science* 27: in press.

Drew, R.A., B.W. Simpson and W.J. Osborne. 1991. Degradation of exogenous indole-3-butyric acid and riboflavin and their influence on rooting response of papaya *in vitro*. *Plant Cell, Tiss. Organ Cult.* 26: 29–34.

Jaiswal, V.S and P. Narayan. 1984. Regeneration of plantlets from stem tissue of *Azadirachta indica* A. Juss. *Proc. Natl. Symp. on Appl. Biotechnology of Medicinal, Aromatic and Timber Yielding Plants*, Calcutta, pp. 132–135.

Lloyd, G. and B. Mccowan. 1981. Commercially feasible micropropagation of mountain laurel, *Kalmia latifolia*, by use of shoot tip culture. *Proc. Int. Plant Prop. Soc.* 30: 421–427.

Murashige, T. and E. Skoog. 1962. A revised medium for rapid growth and bioassays with tobacco tissue cultures. *Physiol. Plant.* 15: 473–497.

Naina, N.S., P.K. Gupta and A.F. Mascarenhas. 1989. Genetic transformation and regeneration of transgenic neem (*Azadirachta indica*) plants using *Agrobacterium tumerfaciens*. *Curr. Sci.* 58(4): 184–187.

Narayan, P. and V.S. Jaiswal. 1985, Plantlet regeneration from leaflet callus of *Azadirachta indica* Juss. *J. Tree Sci.* 4(2): 65–68.

Quoirin, W. and P. Lepoivre. 1977. Etude de milieux adaptes aux cultures *in vitro* de prunus. *Acta Hort.* 78: 437–442.

Rangaswamy, N.S. and Promila. 1972. Morphogenesis of the adult embryo of *Azadirachta indica*. A. Juss. *Z. Pflanzenphysiol.* 67: 377–379.

Rao, M.V.S., Y.V. Rao, Y.S. Rao and V. Manga. 1990. Induction and growth of callus in *Azadirachta indica* Juss. Hort. Abstr. *Hort. Science* 60(9): 872.

Sanyal, M., A. Das, D. Banerjee and P.C. Datta. 1981. *In vitro* hormone induced chemical and histological differentiation in stem callus of neem *Azadirachta indica* A. Juss. *Indian J. Exp. Bot.* 19: 1067–1068.

Sanyal, M., S. Tikadar and P.C. Datta. 1983. Nimbin and sitosterol in cotyledon and cultured tissues of *Azadirachta indica*. *Indian Drugs* 20(12): 479–481.

Schulz, F.A. 1983. Tissue culture of *Azadirachta indica*. *Proc. 2nd Int. Neem Conf.*, Rauischholzhausen. pp. 539–541.

113

Twig Blight — A New Disease of Neem Seedlings

J.C. KAUSHIK, S.S. BISLA AND D.P. SINGH

Department of Agroforestry
CCS Haryana Agricultural University, Hisar 125 004, India

ABSTRACT

Colletotrichum state of *Glomerella cingulata* (Stomen) spaned & Sch. was recorded causing blight of twigs and young branches of neem seedlings. Pre and post inoculation moist period contributed for high infections under screen house conditions. A temperature range of 24° to 32°C combined with high relative humidity (> 70 per cent) increased disease intensity. Fungicidal spray with Derosal or Captaf @ 0.2 per cent controlled the disease effectively.

INTRODUCTION

Most neem trees (*Azadirachta indica* A. Juss.) are reported to be remarkably disease-free, but in India seven fungal and one bacterial pathogen have been recorded to cause leaf spots, stem cankers, and root rot (Bakshi, 1976; Saikia, 1982; Sankaran et al., 1987; Srivastava and Patel, 1969). Recently, a species of *Colletotrichum* was observed to cause twig and shoot blight of neem seedlings resulting in severe mortality of plants in the nursery at the forestry farm, CCS HAU, from July to September 1992. The fungus was isolated, purified, and identified as *C. gloeosporioides* Penz. (*Glomerella cingulata*). This pathogen has been reported for the first time on this host. Other hosts of this pathogen have been reported by Bhat et al. (1988) and Sharma et al. (1985). We report investigations of the pathogenicity to *A. indica*, epidemiology, and management of the disease through fungitoxicants.

MATERIALS AND METHODS

Colletotrichum gloeosporioides (*Glomerella cingulata*) was isolated on potato dextrose agar from twigs of neem showing brown and black lesions and maintained at 5°C for further studies. The neem seedlings were raised from seeds sown in polythene bags (9 × 4 inch) containing loam soil + farmyard manure + sand in equal amounts. Two- to three-month-old seedlings (uninjured and injured by needle pricking) were inoculated with a suspension of 1×10^5 conidia per ml. The conidial suspension was prepared in sterilized distilled water from eight-day-old cultures grown on potato dextrose agar medium. Ten seedlings were kept in each treatment. An equal number of seedlings sprayed with sterilized distilled water alone served as control.

To see the effect of pre- and post-inoculation moist periods, seedlings were exposed to high humidity conditions in moist chamber for different periods before and after inoculation. The temperature and relative humidity inside the moist chamber were recorded continuously. The seedlings were scored 10 days after inoculation when maximum disease developed. The per cent disease index of seedlings was recorded in each treatment.

The disease development on leaves and shoots of *A. indica* was recorded in the field from July to October. Observations on disease severity were recorded at weekly intervals. The data on temperature, relative humidity, and cumulative rainfall were collected from the CCS HAU meteorological laboratory. The infected seedlings were scored on a 0–3 grading scale as given below:

0 = Healthy
1 = Incipient stem discolouration (brown) but not blackening OR 1 to 5 leaf spots per leaflet
2 = Advanced, black, and constricted lesions on stem OR 6 to 10 leaf spots per leaflet
3 = Complete girdling OR death of seedling or blighted leaves

Percentage disease index was worked out by the following formula

$$\text{Per cent disease index} = \frac{\Sigma \text{ of all numerical ratings}}{\text{No. of plants observed} \times 4} \times 100$$

Three-month-old seedlings of *A. indica* were sprayed ensuring coverage of all above-ground surfaces. The chemicals evaluated in the present studies were Derosal, Dithane M-45, Captaf, Blitox, and Difolatan and a control treatment of distilled water. To improve adhesion of the fungicides and water, Tween-20, a wetting agent, was added. Each treatment had three replications and each replication had 10 seedlings. Twenty-four hours later, when dry, seedlings were sprayed with conidial suspension (1×10^{-4}/ ml) and incubated in a moist chamber for 48 hrs and then transferred to the greenhouse. After 10 days, observation on disease severity was recorded on the given scale.

RESULTS AND DISCUSSION

Nursery Survey

Infected seedlings displayed minute, circular to oblong, brown to black lesions on the succulent portion of the twigs or main shoot of two- to three-month-old seedlings. The infection advanced and finally caused girdling and constriction of the enitre region which extended towards the base, resulting in death of seedlings. Acervuli on the host tissue remained submerged, scattered, and black. Isolations from symptomatic tissues consistently yielded *C. gloeosporioides*.

Pathogenicity and Effect of Moist Period

Seedlings were invariably covered with droplets of moisture when removed from the moist chambers. Disease was not observed on inoculated seedlings upon removal from the chambers, but within five to eight days the lesions on leaves and shoots were very pronounced. The injured surfaces of seedlings showed more blighted symptoms than uninjured surfaces especially around the growing tender regions of the main shoot, but the pathogen could cause infection directly on the host surface without any injury. The plants which were not provided with moist period before and after inoculation did not get infection, whereas plants provided with pre- and post-inoculation moist period for 24 hrs and 48 hrs respectively got maximum infection (Table 1). However, the effect of post-inoculation moist period was more pronounced. It is thus clear that the pre-inoculation moist period predisposes the plants to inoculation while the post-inoculation moist period contributes to disease progress.

Table 1. Effect of pre- and post-inoculation moist period on the development of twig blight of neem

Plants kept in moist chamber		Per cent* disease index
Pre-inoculation (hrs)	Post-inoculation (hrs)	
—	—	0
12	—	0
12	12	0
12	24	0
24	24	48.5
24	36	62.6
—	12	0
—	24	32.4
—	36	52.3

*Calculated on the basis of 10 seedlings per treatment.

Environmental Factors

The first symptoms of the disease appeared in the third week of July 1992 when the air temperature and relative humidity ranged from 24.5° to 33.2°C and 60.9 to 86.0 per cent, respectively (Table 2). Up to the first week of August the disease progressed moderately. Then, with the decrease in temperature range (24.3°–32.5°C), increase in relative humidity (71.5–93 per cent), and frequent rains during this period, the disease attained its most severe form in August and early September. From the third week of September onwards, with the decrease in relative humidity and increase in sunshine, the disease progress was reduced. It is thus concluded that a temperature range of 24° to 32°C, relative humidity greater than 70 per cent, cloudy weather, and frequent rains are the factors most conducive to disease progress. Sharma et al. (1985) reported that build-up of infection spores to a critical intensity, under similar environmental conditions, seems to be prerequisite for a sudden and wholly destructive epiphytotic on teak and eucalyptus.

Disease Control

All the five fungicides were found effective to varying degrees. Derosal and Captaf were highly effective in checking the disease and suppressing further spread compared with control (Table 3). Partly infected seedlings sometimes recovered with growth of epicormic shoots free of infection. Carbendazim, Captafol, and Mancozeb have been reported to be effective fungicides against anthracnose of chillies (Chauhan and Duhan, 1977; Motikhaye, 1983).

Table 2. Mean temperature, relative humidity, rainfall (weekly), and incidence of twig blight of neem

Period of observations	Temperature (°C)		Relative humidity (%)		Cummulative rainfall (mm)	Per cent disease index
	Minimum	Maximum	Morning	Evening		
1–7 July	24.5	35.7	73.4	55.2	21.8	—
8–15 July	24.9	36.8	75.5	53.4	6.3	—
16–23 July	24.5	33.2	86.0	60.7	21.2	4.3
24–31 July	24.8	32.9	83.0	66.7	40.3	2.9
1–7 Aug.	24.3	33.8	86.5	68.2	10.1	21.2
8–15 Aug.	24.6	32.7	93.0	71.5	36.0	35.8
16–23 Aug.	24.7	32.5	93.5	75.4	139.5	58.4
24–31 Aug.	24.9	35.3	82.1	55.3	0.0	48.3
1–7 Sept.	24.5	33.5	87.2	73.8	49.5	42.4
8–15 Sept.	22.6	32.1	86.8	60.0	4.2	39.3
16–23 Sept.	19.7	33.5	83.8	38.3	7.2	27.0
24–30 Sept.	16.7	34.8	76.8	33.8	0.0	21.0
1–7 Oct.	18.6	34.6	72.8	32.6	0.0	13.1
8–15 Oct.	18.7	33.9	72.6	34.8	0.0	11.3

Table 3. Effect of fungicidal sprays on the development of twig blight of neem

Treatment	Concentration (%)	Per cent disease index	Per cent disease reduction over cotrol
Derosal (Carbenadazim)	0.2	12.6 (20.74)*	57.71
Difolatan (Captafol)	0.2	23.2 (28.80)	22.14
Captaf (Captan)	0.2	13.3 (21.29)	55.36
Blitox (copper oxychloride)	0.2	24.2 (29.47)	18.79
Dithane M-45 (Mancozed)	0.2	16.8 (24.18)	43.62
Control	—	29.8 (33.07)	—
CD at 5%		(3.40)	

* Figures in parentheses represent angular transform values.

REFERENCES

Bakshi, B.K. 1976. *Forest Pathology. Principles and Practices in Forestry.* FRI (ICAR), Dehradun, 437 pp.

Bhat, M.N., R.K. Hegde and P.C. Hiremath. 1988. A new host record for *Colletotrichum gloeosporioides* (Penz) Penz & Sacc. from India. *Current Science* 57(18): 1012–13.

Chauhan, M.S. and J.C. Duhan. 1977. Fungicidal compounds to control anthracnose and ripe fruit rot of chillies. *Pesticides* 11: 18–19.

Motikhaye, S.G. 1983. Studies on the efficacy of some fungicides against five pathogenic fungi in laboratories. *Hindustan Antib. Bull.* 25(1/2): 46–48.

Saikia, U.N. 1982 . Phytopathogenic fungi of India—I. *Indian Phytopath.* 35(4): 733–734.

Sankaran, K.V., M. Balasundaran and J.K. Sharma. 1987. Seedling diseases of *Azadirachta indica* in Kerala, India. *Eur. J. For. Path.* 16: 324–328.

Sharma, J.K., C. Mohanan and E.J.M. Florence. 1985. Diseases of *Eucalyptus* spp. KFRI, *Res. Rep.* 36: 113–226.

Srivastava, S.K. and P.N. Patel. 1969. Epidemiology and bacterial leaf spot of blight and shot hole diseases of neem in Rajasthan. *Indian Phytopath.* 21(1): 124–125.

114

Vegetative Propagation of *Azadirachta indica* A. Juss.

K.P. SRIVASUKI, N.S. PHILOMINA AND R. VIJAYA KUMAR

Biotechnology Centre for Tree Improvement, Tirupati, India

ABSTRACT

Vegetative propagation of neem (*Azadirachta indica*) assumes significance in view of its growing importance as a multipurpose tree. The Biotechnology Centre for Tree Improvement of Andhra Pradesh is conducting vegetative propagation experiments on this species. Successful propagation of neem through mist propagation and micropropagation has been standardized and the results are discussed in this paper.

INTRODUCTION

Neem (*Azadirachta indica*) is one of the most valuable tree species of the arid and semiarid tropics. It is commonly planted in villages throughout India. This multipurpose tree grows on a wide variety of soils and has great adaptability to adverse site conditions. It is valued as a source of a variety of secondary metabolites with medicinal properties. The wood is used as fuel and also as timber for construction purpose. Neem seeds are viable for short durations and trees produced through seed germination exhibit considerable variations. Production of nursery stock through vegetative propagation is desirable for raising plantations of high productivity. In Andhra Pradesh, the forest department has conducted research on vegetative propagation of neem extensively and standardized techniques of mist propagation and micropropagation.

MATERIALS AND METHODS

Mist Propagation

Terminal cuttings, mid cuttings, and epicormic shoots were treated with indole-3-butyric acid (IBA) at 1000, 2000, and 3000 ppm and subjected to intermittent misting in mist tunnels. Cuttings were treated with 6 per cent Bavistin as a preventive measure against pathogen invasion. The medium used for rooting was vermiculite.

Micropropagation

Shoot tips from mature trees were used as explants. They were washed under running tap water and then with detergent. They were surface sterilized with 0.1 per cent $HgCl_2$ and washed several times with sterile distilled water. They were cultured on modified MS medium with various combinations of growth regulators and vitamins. The cultures were incubated at 23°C for 18 hrs in light.

RESULTS

Mist Propagation

Treatment with 2000 ppm of IBA gave 44 per cent rooting in mid cuttings, whereas maximum rooting (85 per cent) noticed with 1000 ppm treatment in terminal cuttings. Epicormic shoots responded extremely well for rooting at 1000 ppm concentration (95 per cent). Rooting response was found to be good throughout the year except during winter (Novermber, December). Epicormic shoots rooted within 10 to 15 days, whereas others took 20 to 25 days (Table 1).

Micropropagation

Multiple shoot production with good leaf expansion and shoot elongation was observed on MS with 6-benzyl aminopurine (0.2 mg/l), kinetin (0.1 mg/l), and gibberellic acid 3 (0.1 mg/l). Healthy root development was seen in MS 1/2 strength with IBA (0.05 mg/l). The regenerated plants were successfully transferred to earthen pots containing sand and soil mixture (1:1) with 90 per cent survival (Table 2).

Table 1. Results of rooting experiment conducted on neem, July 1991

Type of cuttings	1000 ppm IBA		2000 ppm IBA		3000 ppm IBA	
	No. of cuttings rooted	Percentage	No. of cuttings rooted	Percentage	No. of cuttings rooted	Percentage
Terminal cuttings	85	85	69	69	40	40
Mid cuttings	50	50	44	44	32	32
Epicormic shoots	95	95	90	90	60	60

Table 2. Response of meristems of neem to various growth regulators

Growth regulators mg/l	Nature of response	Period of response (days)	No. shoots/roots per culture
MS2: Kinetin (0.1) + NAA (0.1)	Friable callus	14	
MS3: BAP (0.1) + NAA (0.1)	Compact green callus	12–15	
MS5: BAP (0.2) + Kinetin (0.1)	Axillary shoot formation	20–25	5–6
MSI: BAP (0.1) + IBA (0.3) + NAA (0.1)	Green callus with roots	10–15	
MS4: BAP (0.4) + Kinetin (0.2) + GA (0.1)	Axillary and adventitious shoot formation	30–40	6–8
1/2 MS5 + IBA (0.05)	Profuse rooting	7–10	15–20
1/2 MS6 + IBA (0.1) + NAA (0.1)	Poor rooting	10–15	5–6

NAA: naphthalene acetic acid
BAP: 6-benzyl aminopurine
GA: gibberillic acid

DISCUSSION

The data on rooting experiment show that 1000 ppm concentration on IBA enhanced the rooting percentage in terminal cuttings and epicormic shoots of neem when compared to mid cuttings. The better response of these cuttings can be attributed to their juvenile nature. Auxin absorption would have been better through these tender juvenile tissues than through the secondary tissues of mid cuttings. Mid cuttings withstood higher concentration (2000 ppm) and rooting response improved in percentage of rooting, whereas it slightly declined for the terminal cuttings and epicormic shoots, as they are tender and juvenile. Terminal cuttings responded better than mid cuttings in general, which is also observed in case of *Tamarindus indica* (Srivasuki et al., 1990). This may be due to the interaction of endogenous auxins present in the apical meristems with exogenously applied auxin resulting in favourable balance of physiological factors for rhizogenesis. Martin and Guillot (1982) also reported superiority of terminal cuttings over middle or basal cuttings.

Successful differentiation and multiplication of shoot tip explants and subsequent rooting shows the feasibility of this technique for large-scale production of planting stock. Since the apical meristems from the mature trees are successfully differentiated into plantlets, the technique offers promise in producing true-to-type plants for tree improvement programme. Barring callus regeneration from leaf discs (Ramesh and Padhya, 1982) and callus regeneration from various explant sources (Rangaswamy and Promila, 1972), no work has been done on micropropagation, especially shoot tip cultures of neem. This technique can be effectively employed in tree improvement programme and for production of disease-free stocks.

ACKNOWLEDGEMENT

We express our deep sense of gratitude to Dr. K. Kesava Reddy, IFS, Principal Chief Conservator of Forests, Government of Andhra Pradesh, Hyderabad, for his continuous and constant encouragement in undertaking these studies.

REFERENCES

Martin, R. and J. Guillot. 1982. Quelques essais de bouturage de l' aulne, *Ministere d'Agriculture Revue Forestrire Francaise* 34: 381–391.

Ramesh, K. and M.A. Padhya. 1990. *In vitro* propagation of neem from leaf discs. *Ind. J. Exp. Biol.* 28: 932–935.

Rangaswamy, N.S. and Promila. 1972. *Z. Pflanzen Physiol.* 67: 377.

Srivasuki, K.P., R.D. Reddy and K. Kesava Reddy. 1990. Rooting of terminal cuttings in *Tamarindus indicus*. *Ind. For.* 116: 984–985.

12

OTHER INSECTICIDAL PLANTS AND MISCELLANEOUS

115

Juvenomimetic Activity of Benzene Extracts of Twenty Plants from Kerala on the Red Cotton Bug *Dysdercus cingulatus* (Fb.)

K. SARADAMMA, D. DALE AND N.M. DAS

College of Agriculture, Vellayani, Kerala, India

ABSTRACT

Benzene extracts of 20 plants including neem, from Kerala were screened for their juvenomimetic activity against red cottonbug *D. cingulatus* (Fb.). The extracts were topically applied to the newly moulted fifth-instar nymphs on the abdominal tergites. These insects were fed with fresh cotton-seeds. The duration of the last nymphal stage of the treated bugs significantly varied in different treatments. Extracts of *Azadirachta indica* and *Eupatorium odoratum* could completely inhibit the normal adult emergence. Mean longevity of normal adults varied from 2 to 19 days whereas in control the mean longevity was 14.2 days. Egg-laying was suppressed by the extracts of *Clerodendron infortunatuom*, *Vitex negundo*, *Solanum indicum* and *Ageratum conyzoides*. Hatchability of eggs laid by normal adults varied from 0 to 100 per cent. *Nerium oleander*, *Codiaeum veriegatum* and *Adathoda visica* extracts caused the production of non-viable eggs only. Maximum number of malformed adults was observed in the treatment with *A. indica* followed by *Thevatia neriifolia*. *Narium oleander* and *Lantana camara*. The longevity of malformed adults varied from 0.87 to 7.44 days. Hatchability of eggs laid by the malformed adults was influenced by the treatments. Among the 20 plants, ten plants caused 100% sterility. Many of the plants screened had juvenilizing effects on *D. cingulatus* almost comparable to that of *A. indica*.

INTRODUCTION

Plants are known to yield allelochemicals which modulate the behaviour of herbivores in the ecosystem. Kerala is a repository of plants with allelochemicals and it was with this premise that we attempted to search for such chemicals in plants and to assay them against the red cotton bug *Dysdercus cingulatus*.

Neem, *Azadirachta indica*, has been identified as the best known source for botanical pesticides (Jacobson, 1980). Along with neem, quick-growing herbs or shrubs having a high biomass production which can be grown in wasteland or along fences and borders of cultivated fields were tested for their biological activity against the red cotton bug.

MATERIALS AND METHODS

Selection of Plants

Locally available plants which were known to possess biocidal activity, medicinal property, or poisonous effect on higher animals and/or those which harbour low levels of insect population were screened and 20 plants were selected:

1) Neem—*Azadirachta indica* A. Juss.
2) Clerodendron—*Clerodendron infortunatum* Linn
3) Eupatorium—*Eupatorium odoratum* Linn
(= *Cromolaena odorata*)
4) Tulsi—*ocimum sanctum* L.
5) Oleander—*Thevatia neriifolia* J.
6) Champakam—*Plumeria rubra* Ait
7) Calotropis—*Calotropis gigantea* R.
8) Notchy—*Vitex negundo* Linn
9) Nerium—*Nerium oleander* Linn
10) Panikoorka—*Coleus aromaticus* B.
11) Chunda—*Solanum indicum* Linn
12) Thumba—*Leucas aspera* Linn
13) Tapioca—*Manihot esculenta* C.
14) Lantana—*Lantana camara* Linn
15) Croton—*Codiaeum variegatum* B
16) Adathoda—*Adathoda vesica* N.
17) Papaya—*Carica papaya* Linn
18) Pandanus—*Pandanus odoratissimus* R
19) Ageratum—*Ageratum conyzoides* Linn
20) Mentha—*Mentha viridis* Linn

Application of Plant Extracts

Benzene extracts of these 20 plants were screened for their juvenomimetic activity against penultimate instars of *D. cingulatus*. The plant extracts were applied topically to the newly moulted fifth-instar nymphs on the abdominal tergites using a Hamilton micro-applicator. Groups of 20 insects each treated with 2 μl of the extract confined in petri dishes served as one replication and four such replications formed one treatment. Insects treated with 2 μl of the solvent alone served as control. The treated insects were fed on fresh, untreated, water-soaked cotton seeds.

Assessment of Results

The effect of the plant extracts was assessed on the basis of the extent of morphological and physiological changes subsequent to the treatments. The treated insects were grouped by the morphological changes observed in the treated nymphs (Fig. 1).

The duration of the next moult as adult, adultoid, or supernumerary instars was treated as the last nymphal duration. The moulted insects were maintained till death, allowing the adults or adultoids to mate and lay eggs, and their longevity was assessed. The eggs laid by each isolated insect in different categories were counted daily and the fecundity and hatchability of eggs laid by insects under each category were calculated. From the data relating to the fecundity of insects and hatching percentage of eggs, the per cent sterility was calculated by the formula suggested by Outram (1973). The sterility percentage (S) = $100 - (a \times b) \times 100/A \times B$, where a = number of eggs laid/female in treatment, b = percentage of eggs hatching in treatment, A = number of eggs in control, and B = percentage of eggs hatching in control.

Statistical Analysis

The data were subjected to analysis of variance and the treatments were compared using Duncan's Multiple Range Test.

RESULTS AND DISCUSSION

Data relating to the experiment and the results of the statistical ananlysis are presented in Table 1. The duration of the last nymphal stage of treated bugs varied significantly from 8.22 to 14.33 days in different treatments as against 7.5 days in control. Extracts of *C. papaya, L. aspera, P. rubra, S. indicum, C. gigantea, L. camara*, and *C. variegatum* had prolonging effect on fifth-instar nymphs of *D. cingulatus*. Prolonged larval duration due to the application of juvenoids had also been recorded by Slama (1974) and Koul (1984).

Application of plant extracts caused some mortality in early fifth-instar stage and there was mortality at the next moult mainly due to the failure of moulted insects to extricate themselves from the exuviae, which can be attributed to the juvenomimetic activity of the phytochemicals in the extracts. Mortality up to five days after treatment was observed in eight extracts and ranged from 11.76 to 4.76 per cent. Mortality after the fifth day was significantly higher in 16 treatments than in control. High mortalities were observed in *C. variegatum, C. infortunatum, A. conyzoides*, and *C. gigantes*.

Extracts of *A. indica* and *E. odoratum* completely inhibited the normal adult emergence. Extracts of *T. neriifolia* and *C. infortunatum* ranked next, permitting only 5.55 and 5.88 per cent nymphs to emerge as normal adults. *N. oleander* was ranked fifth; 15.0 per cent of the treated nymphs emerged as normal adults. Except *M. viridis* and *C. aromaticus*, all treatments were significantly superior to control in reducing the emergence of normal adults.

Emergence of malformed adults was maximum with extract of *A. indica* (94.73 per cent) followed by *E. odoratum* (86.66 per cent), *T. neriifolia* (72.22 per cent), *N. oleander*, and *L. camara* (57.64 per cent). Except *A. conyzoides*, *M. viridis*, and *C.*

Fig. 1. Abnormalities in adults of *D. cingulatus* caused by application of plant extracts

aromaticus, all the treatments caused significantly higher percentage of malformed adults than in control. Adultoids emerged in treatments retained nymphal characters such as presence of two torsal segments and abdominal spots. Such morphogenetic effects of juvenoids have been extensively reported earlier (Outram, 1973; Chiu, 1985).

The mean adult longevities of normal and malformed adults in many treatments came below those of corresponding controls, as shown clearly in Fig. 1. The mean longevity of normal adults varied from 19 to 2 days in different treatments while in controls it was 14.2 days. Significant reduction was observed in longevity of bugs emerging from nymphs treated with extract of *A. conyzoides*, *V. negundo*, *C. variegatum*, *C. gigantea*, *S. indicum*, *L. camara*, and *P. odoratissimus*. The longevity of malformed adults varied from 0.87 to 7.44 days.

Fecundity of the bugs was also affected by the treatments. Egg laying was completely suppressed by the extracts of *F. odoratum*, *C. infortunatum*, *V. negundo*, *S. indicum*, and *A. conyzoides*. All plants excluding *O. sanctum*, *C. gigantea*, and *M. esculenta* caused significant suppression of fecundity. Even though malformed adults of nine treatments laid eggs, the fecundity was significantly low with extracts of *A. vesica* (5 per cent), *L. aspera* (6.72 per cent), *A. indica* (11.6 per cent), *T. neriifolia* (12.1 per cent), and *N. oleander* (12.38 per cent).

Hatchability of eggs laid by normal adults varied from 0 to 100 per cent in treatments. *N. oleander*, *C. variegatum*, and *A. vesica* extracts caused the production of non-viable eggs only. The hatchability was significantly lower in *M. esculenta* and *M. viridis* extracts. In the case of malformed adults the same trend was observed. Except in *C. papaya* and *L. aspera*, the eggs laid by malformed adults in all the treatments failed to hatch.

Since the juvenomimetic effect of plant extract takes time to manifest itself and since the effect largely depends on the occurrence of the pest in the receptive stage in the field, the treatment may not emerge as a single effective technique for controlling any pest in the field. In integrated pest control, the strategy is to contain the pest population and not to eliminate it. Hence, the use of allelochemicals regulating the growth and development of insects may emerge as an important component of an integrated control programme for tackling a pest in the field. In this context, the best index for assessing the relative efficacy of the plant extracts will be the possible suppression of the post-treatment population of the pests involved. The number of adults emerging from the treated insects and their sterility percentage will give an index of the possible buildup of the post-treatment population of the test insect. The data relating to these aspects have been summarized in Fig. 3. It can be seen that 10 out of 20 plants tested caused 100 per cent sterility of *D. cingulatus* in benzene extracts. Sterility (used in a broad sense as possible contribution to the ensuing generation) caused by the plant extracts may be due to the suppression of normal adult emergence, lack of egg laying by normal and malformed adults, or lack of egg hatching. Jacobson et al. (1978), Fagoonee (1980), Velusamy et al. (1987), and several others also reported reduced fecundity of various insects as a result of juvenoid application. Slama (1974) observed that the juvenoids do not suppress the egg production, as do chemosterilants, but the females lay eggs of very low or zero hatchability. But in many treatments included in the present investigations, lack of egg production was observed.

Table 1. Hormonal effects of the benzene extracts of different plants on the last-instar nymphs of *D. cingulatus*

Extracts*	Duration of nymphal instar (days)	Nymphal mortality (%)		Adult emerged (%)		Adult longevity (days)	No. eggs laid per female	Percentage of eggs hatched	Sterility percentage of normal + malformed adults
		Up to five days after treatment	After fifth day of treatment						
A. indica	9.50bcde	0.00	5.26fgh	N	0.00a	0.00	0.00	0.00	100.00a
				M	94.73a	3.68	11.60ab	0.00	
C. infortunatum	9.60bcde	11.76	35.30abc	N	5.88ab	12.00def	0.00	0.00	100.00a
				M	47.05cd	4.50	0.00	0.00	
E. odoratum	10.30bcde	6.66	6.66fg	N	0.00a	0.00	0.00	0.00	100.00a
				M	86.66b	2.30	0.00	0.00	
O. sanctum	8.22c	0.00	10.00efg	N	70.00i	14.00def	65.60ef	68.56bcd	26.79f
				M	20.00fg	3.00	0.00	0.00	
P. rubra	12.33abcd	0.00	15.00cdefg	N	30.00def	9.75cd	40.00bcde	50.00abc	83.45bcd
				M	55.00cd	4.55	26.75bc	0.00	
T. neriifolia	10.28bcde	11.11	11.11defg	N	5.55bc	19.00f	25.00abcd	0.00	100.00a
				M	72.22bc	5.30	12.10ab	0.00	
C. gigantea	12.53abcd	4.76	23.81abcde	N	33.33defg	7.57bc	50.00def	70.00bcd	74.12bcde
				M	38.10de	4.87	35.00cd	0.00	
V. negundo	10.18bcde	6.66	20.00bcdef	N	26.66def	3.00ab	0.00	0.00	100.00a
				M	46.66cd	4.80	0.00	0.00	
N. oleander	9.60bcde	0.00	15.00cdefg	N	15.00cd	15.50ef	28.00abcd	0.00	100.00a
				M	70.00bc	3.54	12.38ab	0.00	
C. parviflorus	8.94cde	0.00	5.00fgh	N	90.00j	9.78cd	26.60abcd	88.00cde	61.90de
				M	5.00hi	2.00	0.00	0.00	
S. indicum	12.16abcd	0.00	14.28cdefg	N	38.57fgh	7.50bc	0.00	0.00	100.00a
				M	47.14de	2.75	29.00bc	0.00	
L. aspera	13.22ab	0.00	15.39cdefg	N	38.46efgh	16.00f	36.64bcde	95.00de	70.02cde
				M	46.15d	7.44	6.72a	28.00	
M. esculenta	11.00abcde	7.69	15.38cdefg	N	61.54hi	9.83cd	46.75cdef	21.70a	83.66bc
				M	15.38g	3.16	0.00	0.00	
L. camara	12.33abcd	0.00	14.32cdefg	N	28.63def	8.66c	32.00bcde	93.75de	51.32e
				M	57.04bcd	3.80	0.00	0.00	

C. variegatum	14.33a	0.00	44.44a	N	44.44fgh	6.00abc	25.00abcd	0.00	
				M	11.11gh	5.76	0.00	0.00	100.00a
A. vesica	9.81bcde	5.00	15.00cdefg	N	55.00ghi	9.82ef	11.66ab	0.00	
				M	25.00efg	4.50	5.00a	0.00	100.00a
C. papaya	12.77abc	8.80	35.52ab	N	18.88cde	15.00f	42.00ab	70.00abc	
				M	36.72def	6.88	55.00a	55.00	74.49bcde
P. odoratissimus	9.09cde	0.00	6.25fg	N	50.00fghi	8.50c	4.33a	100.00e	
				M	43.75de	0.87	0.00	0.00	92.95b
A. conyzoides	8.75de	0.00	33.33abcd	N	66.66i	2.00a	0.00	0.00	
				M	0.00i	0.00	0.00	0.00	100.00a
M. viridis	9.25cde	0.00	5.00fgh	N	95.00j	10.74cde	16.66abc	40.00ab	
				M	0.00i	0.00	0.00	0.00	89.15bc
Control	7.50e	10.00	0.00h	N	90.00j	14.70def	71.20f	86.28cde	
				M	0.00i	0.00	0.00	0.00	0.00

*25 g dried plant material extracted with 100 ml solvent, applied topically at 2 μl/specimen.
N = normal, M = malformed
Means followed by a common letter in a column are not significantly different at 5% level (DMRT).
Each treatment consisted of four replications and 20 insects were treated in each replication.

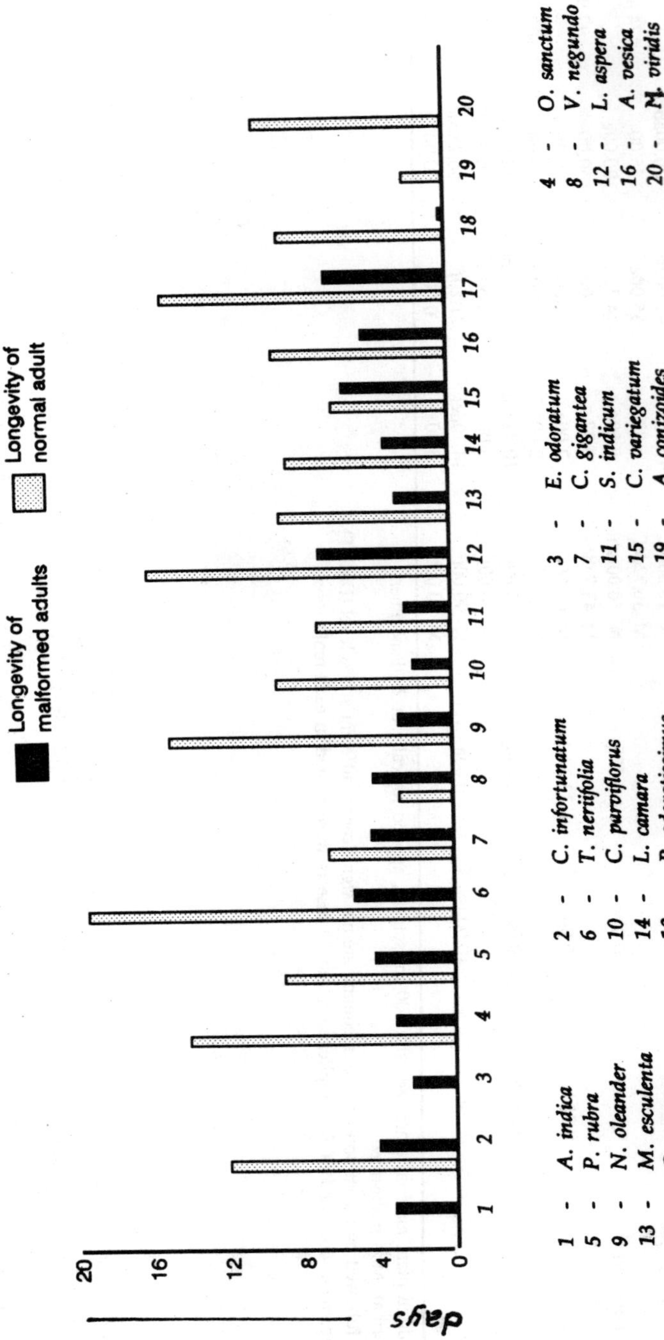

Fig. 2. Longevity of adults of *D. cingulatus* emerged after treatment with plant extracts

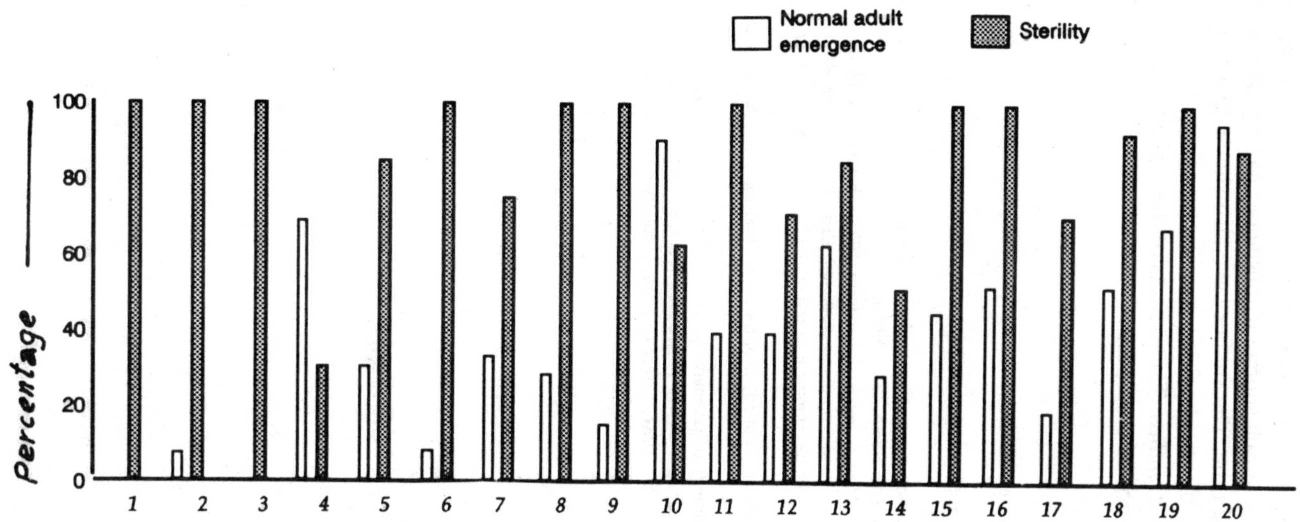

1 -	*A. indica*	2 -	*C. infortunatum*	3 -	*E. odoratum*	4 -	*O. sanctum*
5 -	*P. rubra*	6 -	*T. neriifolia*	7 -	*C. gigantea*	8 -	*V. negundo*
9 -	*N. oleander*	10 -	*C. parviflorus*	11 -	*S. indicum*	12 -	*L. aspera*
13 -	*M. esculenta*	14 -	*L. camara*	15 -	*C. variegatum*	16 -	*A. vesica*
17 -	*C. papaya*	18 -	*P. odoratissimus*	19 -	*A. conizoides*	20 -	*M. viridis*

Fig. 3. Percentage of normal adult emergence and sterility in adults of *D. cingulatus* after treatment with plant extracts

Most of the reports on the growth and development suppressing factors identified in plants earlier were from parts of the plants other than leaves. The identification of the leaves of the potential plants as a rich source of substances having juvenile hormone activity is being done for the first time. The studies have also shown that as growth and development arrestants the local plants identified were as promising as *A. indica* and in some cases they ranked above *A. indica*. The extracts of these plants can be exploited effectively in the integrated control of crop pests.

REFERENCES

Chiu, S.F. 1985. Recent research findings in meliaceae and other promising botanical insecticides in China. 2. *Pflkrankh PHC Shutz* 92: 346.

Fagoonee, I. 1980. Behavioural response of *Gocidolomia binotalis* to neem. *Proc. 1st Int. Neem Conf.*, Rottach-Egern. GTZ, Eschborn, pp. 109–120.

Jacobson, M. 1980. Isolation and identification of insect antifeedants and growth inhibitors from plants. An overview. *Proc. 1st Int. Neem Conf.*, Rottach-Egern. GTZ, Eschborn, pp. 13–20.

Jacobson, M., D.K. Reed, M. Crystal, D.S. Moreno and E. Soderstorm. 1978. Chemistry and biological action of insect feeding deterrents from the weed and crop plants. *Entomol. Exp. Appl.* 24: 448–457.

Koul, O. 1984. Azadirachtin I. Interaction with development of red cotton bugs. *Entomol. Exp. Appl.* 36(1): 85–88.

Outram, I. 1973. Synthetic juvenile hormone effect of pupae of the spruce budworm. *J. Econ. Entomol.* 66: 1033–1035.

Slama, K. 1974. The chemistry and physiology of juvenads. In: *Insect Hormones and Bioanalogues*. K. Slama, M. Romanyak and F. Sorm (eds.), Springer-Verlag-Klien, New York, 477 pp.

Velusamy, R., R. Rajendran and P.C.S. Babu. 1987. Effect of neem products on brown plant hopper. (BPH) oviposition. *Int. Rice Res. Newsletter* 12(2): 36.

116

Biological Effects of Certain Oils and Terpene Fractions of Plant Origin on Insects

B. SUBRAHMANYAM AND P.J. RAO

Division of Entomology
Indian Agricultural Research Institute
New Delhi 110 012, India

ABSTRACT

Selected aromatic oils and terpene fractions of plant origin were tested in a bioassay to evaluate their toxicity to and growth regulatory activity on three insect pests of economic importance, namely the desert locust (*Schistocerca gregaria*), castor semilooper (*Achaea janata*), and Bihar hairy caterpillar (*Spilosoma obliqua*). Terpene fraction of the mint oil (*Mentha arvensis*) has the highest biological activity on the locust, causing deformities at the nymphal adult moult. Such effects were also noticed from *Artemesia* oil and Himalayan cedarwood oil, though to a lesser extent. Terpene fraction of mint oil was also active on *A. janata*, causing deformities at larval-pupal moult, but was toxic to *S. obliqua*. Mint oil was very toxic to *A. janata* larvae, causing desiccation. Oil of sweet basil was moderately toxic to locust hoppers and non-toxic to *A. janata* larvae.

INTRODUCTION

Plants synthesize innumerable chemical substances that are used in industry, medicine, and agriculture. Chemical examination of such natural products for plant protection, over the past fifty years, has been fruitful, yielding insecticides like pyrethrins, rotenone, nicotine, and physostigmine. Extensive toxicological and biological data on natural products are now available (Jacobson, 1958, 1975; Jacobson and Crosby, 1971). Plants have also yielded substances that either mimic insect morphogenetic hormones, namely juvenile hormone and ecdysone (Slama and Williams, 1966; Williams, 1956), or disrupt insect growth by antagonizing juvenile hormone action (Bowers et al., 1976), or work with a mode of action yet to be established (Butterworth and Morgan, 1968; Rembold, 1984).

Though Indian ecosystems are diverse and contain flora that yield a multitude of natural products, only very limited information is available on their possible growth regulatory activity (Banerji et al., 1971; Prabhu and John, 1975; Prakash and Ghoshal, 1979; Saxena and Srivastava, 1972). The present study aims to evaluate some aromatic oils and terpene fractions for their toxic and growth-disrupting activity against three species of insect pests.

MATERIALS AND METHODS

Test Insects

Three species of test insects—the desert locust and two lepidopterous pests—were chosen for the study. The desert locust *Schistocerca gregaria* (Acrididae) has been reared in our laboratory at 28°C (± 2°C) continuously for several years on cabbage, cauliflower, and maize leaves supplemented with wheat bran according to Mehrotra and Rao (1966). Moths of the two lepidopterans—the castor semilooper *Achaea janata* (Noctuidae) and the Bihar hairy caterpillar *Spilosoma obliqua* (Arctiidae)—were collected from light traps during the monsoon and reared on castor leaves under controlled conditions (25 ± 2°C, 16:8 hr photoperiod, and 50–60 per cent relative humidity). One-day-old fifth-instar hoppers, freshly moulted final-instar larvae of the semilooper (day 9), and the 13-day-old larvae of the hairy caterpillar were used for the bioassay. The final instar of the desert locust lasted for 8–12 days and that of the semilooper for 3.5 to 4 days. The hairy caterpillar had a larval stage of 25 days' duration.

Test Materials

Aromatic oils, terpenes and other fractions of plant origin used in this study were obtained from the Division of Chemistry, National Bureau of Plant Genetic Resources (NBPGR), New Delhi.

Bioassay

Bioassay were carried out using test insects of known age. The dose required was determined in a preliminary test and 5, 1.5, and 0.5 as well as 0.6 µl of the test material were used in the case of desert locust, castor semilooper, and Bihar hairy caterpillar respectively. The samples were diluted with equal volume of acetone and applied to the hoppers; for the two lepidopterans, pure samples were applied directly. The test material was applied topically using Hamilton syringe (25, 10, or 5 µl), under mild CO_2 anaesthesia. The sample was applied on the cervical membrane of the hoppers and in case of the larvae, it was evenly spread on the entire dorsal surface and allowed to dry. The immediate reaction, if any, was noted and the test insects were maintained after treatment under the same conditions as detailed for each species. Mortality, deformities at moult, and other abnormalities were recorded and the corrected percentages (Abbott's correction) are tabulated for comparison.

RESULTS

Schistocerca gregaria

Fourteen test materials of plant origin that contained eight samples of oils and three terpene fractions besides two others were tested at a uniform dose of 5 μl per insect on one-day-old fifth-instar desert locust hoppers (Table 1). Linalool fraction from sweet basil and menthone fraction of mint oil caused toxicity in more than 50 per cent of the test samples of hoppers. Two samples of mint oil and the oil of sweet basil were of moderate toxicity (35–40 per cent). *Artemesia* oil, Himalayan cedarwood oil, *Glocimum* oil, oil of marigold, mint oil from Shivalik, and high boiling point fraction of mint oil were non-toxic (mortality less than 15 per cent). While the whole terpene fraction of mint oil was non-toxic, the two purified fractions (1 and 3) were moderately toxic, causing 35 and 30 per cent mortality respectively. More than 50 per cent of the mortality of nymphs in all treatments occurred during the first 72 hrs after topical application. The initial reaction of the hoppers to some of the test materials was remarkable. All the three samples of terpenes, mint oil (EC 41911), menthone fraction, and oil of sweet basil caused knock-down and convulsions within 30–60 min. following the treatment. However, many insects recovered within 12–18 hrs after treatment.

Some of the test materials caused disturbances at the nymphal-adult moult, besides the toxicity detailed above. Oil of the Himalayan cedarwood caused disturbances at moult in 12.5 per cent of the sample of hoppers tested. Such effects due to the whole terpene fraction of mint oil, its fraction 3, and the menthone fraction were to the extent of 17–20 per cent. The characteristic features of the deformity caused by terpenes of mint oil are short forewings due to incomplete expansion, and in some cases these remain almost rudimentary at the base of the thorax (Fig. 1). Such defects are very rare in the laboratory colony and constituted less than 5 per cent in acetone-treated control.

Achaea janata and *Spilosoma obliqua*

Eight samples were tested against *A. janata*, whereas only the terpene fraction of mint oil was tested against *S. obliqua*. The three samples of mint oil and the menthone fraction were toxic to *A. janata*, causing mortality in over 80 per cent of the larvae. The mortality was due to excessive desiccation and occurred within the first 24 hrs after treatment (Table 2). The rest of the test samples, including the oil of sweet basil, were either moderately toxic or non-toxic.

The terpene fraction of mint oil (Fraction 1) was moderately toxic, but caused deformities at the larval-pupal moult (Fig. 2). Larvae exposed to this treatment 6 hrs before or after the last larval moult were more sensitive than the 12- to 24-hr-old last-instar larvae. Ther percentage of normal adults emerging with the whole terpene fraction of mint oil and its fraction 3 as well as the oil of sweet basil was more than 75.

Topical application of terpenes of mint oil (0.6 μl/larva) to *S. obliqua* caused very high toxicity (90 per cent). A lower dose of 0.5 μl caused deformed adults (7.5 per cent) besides toxicity to the extent of 63 per cent. These adults either had crippled wings or could not eclode fully. Further testing on this species was hampered by over-wintering

Table 1. Gross effects of topical application of some plant derivatives on the final instar hoppers of the desert locust (*S. gregaria*)

Treatment (5 μl, topical)	Number of insects treated	Mortality of fifth instar (%)*	Deformity at moult (%)*	Normal adults (%)	Initial symptoms and nature of deformity
Artemesia oil (*A. pallens*)	97	7.17	6.04	80.41	No initial symptoms; forewings slightly deformed.
Cedarwood oil (*Cedrus deodara*)	55	—	12.38	83.64	No initial symptoms; forewings short and remain half open.
Mint oil (*Mentha arvensis*)	56	39.90	1.78	57.14	Initial lethargy in feeding and movement.
Mint oil (Shivalik)	12	14.80	—	83.34	No initial symptoms.
Mint oil (EC 41911)	43	35.93	—	62.79	Initial knock-down; convulsions and recovery.
Oil of sweet basil (*Ocimum basilicum*)	206	36.83	4.05	59.30	Initial knock-down and some recovery; deformed insects had crippled or rudimentary wings.
Oil of marigolds (*Tagetes minuta*)	39	—	—	100	No initial symptoms.
Glocimum oil	80	7.91	—	88.75	No initial symptoms.
Terpenes of mint oil	184	7.68	17.44	69.02	Instant convulsions and recovery, death at moult. Forewings alone or both wings crippled or short; forewings, pronotum short, exposing mesonotum.
Terpenes of mint oil: Fraction 1	11	35.06	—	63.64	Initial convulsions and recovery.
Terpenes of mint oil: Fraction 3	56	30.72	21.21	42.86	Initial knock-down and recovery. Forewings alone or both wings crippled or short; forewings, pronotum short, exposing mesonotum.
High boiling poing fraction of mint oil	112	3.16	3.57	91.07	No initial convulsions. Forewings alone or both wings crippled or short; forewings, pronotum short, exposing mesonotum.'
Menthone-isomenthone fraction of mint oil	23	55.80	17.83	21.75	Initial knock-down; convulsions and slow recovery.
Linalool type of compounds from sweet basil	15	52.52	—	46.67	No initial symptoms.

*Corrected

in pupal stage. No larval mortality or deformed adults were noticed in untreated control of both the species (n = 30).

Fig. 1. Characteristic deformity of wings of *S. gregaria* treated topically with 5 μl of terpene fraction of mint oil (*M. arvensis*)

Table 2. Gross effects of topical application of some plant derivatives on the final instar larvae of the castor semilooper (*A. janata*)

Treatment (1.5 μl topical)	Number of insects treated	Larval mortality (%)	Mortality as pre-pupa, pupa, intermediates*	Deformed adults	Normal adults	Initial symptoms and nature of deformity
Mint oil	60	81.67	5.17	1.66	8.34	Mortality due to excessive desiccation
Mint oil (Shivalik)	30	86.67	—	—	13.33	Mortality due to excessive desiccation
Mint oil (EC 41911)	30	83.33	—	6.67	10.00	Mortality due to excessive desiccation
Oil of sweet basil	30	16.67	3.33	3.33	76.67	Mortality due to excessive desiccation
Terpenes of mint oil	30	6.67	—	3.33	90.00	—
Terpenes of mint oil: Fraction 1	157	24.84	22.91	3.18	46.50	Larval-pupal intermediates predominant
Terpenes of mint oil: Fraction 3	31	3.23	—	3.22	90.32	—
Menthone-isomenthone, Fraction of mint oil	30	100	—	—	—	Mortality due to excessive desiccation

*Corrected

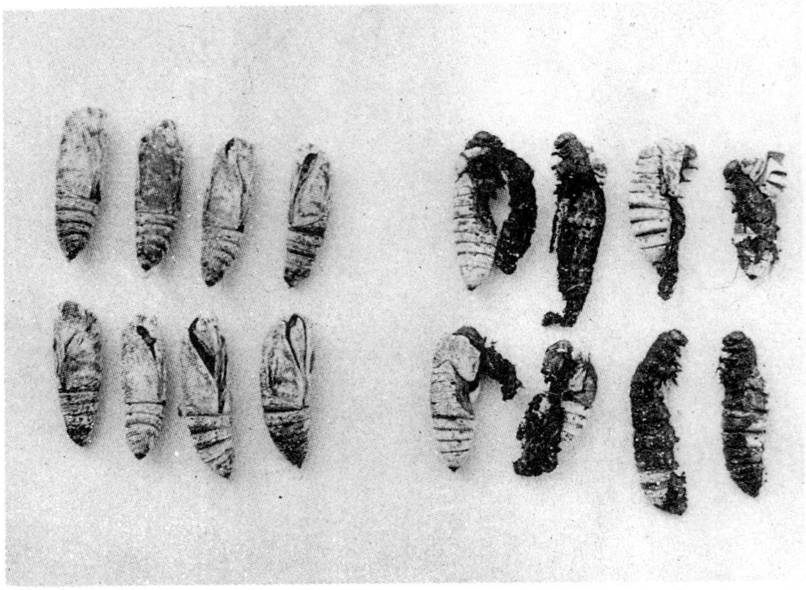

Fig. 2. Deformities at the larval-pupal transformation of *A. janata* treated topically with 1.5 µl of terpene fraction of mint oil (*M. arvensis*) within 6 hrs prior to the last larval moult

DISCUSSION

The present study evaluates some aromatic oils and terpene fractions for their toxicity and growth-disrupting activity on three species of insects. Of the eight samples of oils tested without fractionation, oil of the Himalayan cedarwood showed the highest activity, causing deformity at moult in 12.38 per cent of the hoppers tested. Though mint oil was not as active, its terpene fractions have appreciable growth-disrupting activity on both *S. gregaria* and *A. janata*. These two sources contain sesquiterpene and sesquiterpene alcohol (Anonymous, 1950, 1976) and are the likely compounds for the growth-disrupting activity. Exogenous application of the junvenile hormone of insects, a sesquiterpene, and its analogues caused disruption of moult depending on the dose and time of application (Slama et al., 1974). Bowers and Nishida (1980) isolated two very potent juvenile hormone mimics (Juvocimene I and II) from the oil of sweet basil that were very active against the milkweed bug *Oncopeltus fasciatus*. Such high activity was not noticed on the test insects used in our study involving oil of sweet basil. Similarly, the oil of marigold containing tagetone and having juvenile hormone mimicking activity on the red cotton bug, *Dysdercus koenigii* (Saxena and Srivastava, 1973), was not active on the insect species used in this study.

The menthone-isomenthone fraction of mint oil caused moult disturbances in locust hoppers but not in castor semilooper. Notwithstanding further testing, such a differential reaction may suggest the possibility of selectivity. The terpene as well as the menthone fractions of mint oil are thus the most active on locust hoppers. The terpene fraction of mint oil, though not toxic to semilopper larvae, was extremely toxic to Bihar hairy caterpillar. Toxic effects of the oils on locust hoppers and semilooper larvae are also contrastingly different. While the mint oils are very toxic, leading to desiccaation of *A. janata* larvae, they were not very toxic to the locust hoppers. On the contrary, the oil of sweet basil is moderately toxic to locust hoppers and non-toxic to *A. janata* larvae. These results reflect the differences in permeability barrier and the effect on the nervous system of different insect species. Further studies on these aromatic oils and their fractions are in progress.

ACKNOWLEDGEMENTS

The authors wish to express their gratitude to Dr. M.L. Maheshwari, Principal Scientist (Chemistry), NBPGR, New Delhi, for providing the samples of all the test materials used in this study and fruitful discussions. Thanks are also due to Dr. Swaraj Ghai, Head, Division of Entomology, IARI, New Delhi, for providing necessary facilities and encouragement.

REFERENCES

Anonymous. *The Wealth of India—Raw Materials*. 1: 122, 1950; 2: 106–110, 1966; 6: 337–346, 1966; 7: 79–89, 1966; 10: 109, 1976. Council of Scientific and Industrial Research, New Delhi.

Banerji, A., G.J. Chintalwar, N.K. Joshi and M.S. Chadha. 1971. Isolation of ecdysones from Indian plants. *Phytochemistry* 10: 2225–2226.

Bowers, B.W., T. Ohta, J.S. Cleere and P.A. Marsella. 1976. Discovery of insect anti-juvenile hormones in plants. *Science* 193: 542–547.

Bowers, W.S. and R. Nishida. 1980. Juvocimenes: Potential juvenile hormone mimics from sweet basil. *Science* 209: 1030–1032.

Butterworth, J.H. and E.D. Morgan. 1968. Isolation of a substance that suppresses feeding in locusts. *J. Chem. Soc. Chem. Commun.* 23–24.

Jacobson, M. 1958. *Insecticides from Plants—A Review of Literature 1941–53*. USDA Agric. Handbook No. 154, 300 pp.

Jacobson, M. 1975. *Insecticides from Plants—A Review of Literature 1954–71*. USDA Agric. Handbook No. 461. 138 pp.

Jacobson, M. and D.G. Crosby. 1971. *Naturally Occurring Insecticides*. M. Dekker, New York, pp. 307–457.

Mehrotra, K.N. and P.J. Rao. 1966. Phagostimulants for locusts. *Indian J. Exp. Biol.* 4: 56–57.

Prabhu, V.K.K. and M. John. 1975. Juvenomimetic activity of some plants. *Experientia* 31: 913.

Prakash, A. and S. Ghoshal. 1979. Phytoecdysones. *J. Sci. Indus. Res.* (India) 38: 632–647.

Rembold, H. (1984). Secondary plant products in insect control, with special reference to the azadirachtins. In: *Advances in Invertebrate Reproduction*, Vol. III. W.E. Engels. (ed.), Elsevier, Amsterdam, pp. 481–491.

Saxena, B.P. and J.B. Srivastava. 1972. Studies on plants with juvenile hormone activity: Effect of *Iris ensata* Thumb. (Iridaceae) on *Dysdercus koenigii* (Pyrrhocoridae). *Experientia* 28: 112–113.

Saxena, B.P. and J.B. Srivastava. 1973. *Tagetes minuta* L. oil—A new source of juvenile hormone mimicking substance. *Indian J. Exp. Biol.* 11: 56–58.

Slama, K., M. Romanuk and F. Sorm. 1974. *Insect Hormones and Bioanalogues*. Springer-Verlag, New York, 91 pp.

Slama, K. and C.M. Williams. 1966. 'Paper factor' as an inhibitor of the embryonic development of the European bug, *Pyrrhocoris apterous*. *Nature* 210: 329–330.

Williams, C.M. 1956. The juvenile hormone of insects. *Nature* 178: 212–213.

Antifeedant Effect of *Jatropha curcas* L. Seed Oil and Extracts on the Variegated Grasshopper *Zonocerus variegatus* and the Eggplant Defoliator, *Selepa docilis*

J.R. COBBINAH

Forestry Research Institute of Ghana
Univ. Science and Technology Campus, Kumasi, Ghana

ABSTRACT

Studies were conducted to determine the effect of *Jatropha curcas* L. seed-oil and various solvent-extracts on survival and feeding rates of the variegated grasshopper, *Zonocerus variegatus* L. Although *J. curcas* seed-oil emulsion was less effective, when compared with 0.05 per cent Cymbush (a synthetic pyrethroid), it nevertheless reduced the damage caused by *Z. variegatus* to *Cassava foliage* (Manihot esculenta Crantz.). The level of feeding suppression achieved with *J. curcas* extract (42.91 per cent) was, however, similar to neem extrat (40.89 per cent) at 5 per cent significance level. The hexane extract, which gave the highest activity among the four solvent systems used in the extraction of *J. curcas* seeds, was separated into four fractions by column chromatography. Physico-chemical characteristics of the most active fraction were determined. A complex of functional groups were indicated. Further purification and spectroscopic analyses are required to establish the structure(s) of the bioactive components in the seed.

INTRODUCTION

Jatropha curcas L. is a large shrub commonly found in the tropical and subtropical regions of the world. It is found in very diverse vegetational areas from wet evergreen forest to arid region and is well adapted for cultivation within marginal and degraded lands. It probably owes its successful cultivation to having no known major pests.

In Ghana it is known extensively for its many medicinal uses and commonly grown as a living fence because sheep and goats do not browse on it. It is also used to a limited extent in the manufacture of a local soap. Cobbinah and Appiah-Kwarteng (1989) demonstrated that the oil is an effective protectant of maize and cowpea against stored pests. The effect of the emulsified seed extracts on the ployphaguous grasshopper, *Zonocerus variegatus* L, and the eggplant defoliator, *Selepa docilis*, is reported in the paper.

MATERIALS AND METHODS

The study was conducted in the laboratories of FORIG at the University of Science and Technology campus, Kumasi, Ghana. Nymphs of *Z. variegatus* and larvae, prepupae, and pupae of *S. docilis* were used. The test insects were usually conditioned in the laboratory (temperature 25°–30°C and humidity 65–85 per cent) before use. For eighteen hours before use they were given no food to enable them to evacuate their intestines.

Plant extracts were prepared according to methods described earlier (Cobbinah and Osei-Owusu, 1988). Emulsion solutions were prepared to the desired concentrations using Triton-X or Sparkle solution.

Zonocerus variegatus Trials

Topical application bioassay procedure was used. Five microlitres of test solutions were applied to abdominal segments of test animals with Gallenkamp microlitre transferpipette. Water and water-emulsifier mixture served as control and standard treatments. Twenty-five treated nymphs were introduced into cages holding food plants (cassava leaves). Mortality and food consumption were usually determined after 24 hrs. The effect of treatments on food consumption was evaluated by determining feeding suppression. Feeding suppression = (per cent eaten in control unit – per cent eaten in treated unit)/(per cent eaten in control unit) × 100.

Selepa docilis Trials

The effect of Jatropha extracts on establishment of first-instar larvae and growth of fourth-instar larvae, prepupae, and pupae were determined. First-instar larvae from the same egg batch were divided into three groups and placed on foliage that had been dipped for 30 sec in water, water + 0.2 per cent sparkle, or extract of Jatropha. Egg batches containing less than 30 eggs were discarded. Larvae from the same egg batch were used within replicates (block) to reduce innate differences that could affect their response to extract. Larval establishment measured as per cent survival or conversely per cent mortality to the end of second-instar stage was recorded.

The fourth instar is the most voracious larval stage and causes the greatest damage to crop (Frempong, 1981). The effect of Jatropha extract on this stage was determined by placing 10 newly moulted fourth-instar larvae on foliage dipped in treatment solutions. Treatment effect on development was measured by recording number of prepupae and pupae four days post-treatment.

The effect of Jatropha extract on development of prepupae and pupae was determined by topically applying 5 μl extracts to the prepupa or pupa body after exposing the body by cutting the cocoon on the tail end. Treatment effect was assessed by recording mean developmental period and per cent adult emergence. There were 10 prepupae or pupae in each treatment. All studies described above had four replicates each and were conducted in the laboratory.

STATISTICAL ANALYSES

Data were analysed using analysis of variance (ANOVA). Percentage data were transformed (angular transformation) prior to analyses. Significant treatment means were separated using Fisher's least significance (Snedecor and Cochran, 1967) and significance was determined at $p = 0.05$ level.

RESULTS AND DISCUSSION

Table 1 compares the effect of Jatropha extract to neem extract and the synthetic pyrethroid, Cymbush (100 g cypermethrin, Imperial Chemical Industries, Plant Protection Industries, Survey, England). The plant extracts had similar levels of activity on food consumption of *Z. variegatus*. Neem and Jatropha oils had similar levels of activity on *Sitophulus zeamais* and *Callosobruchus maculatus*, major stored grain pests of maize and cowpea respectively (Cobbinah, unpublished). The effect of the two plant extracts were, however, minimal compared to that of Cymbush, which caused high mortalities and reduced food consumption by 86 per cent within 24 hrs.

The potency of an extract is reflected by the observed distribution pattern in cages (Fig. 1). The more potent an extract or treatment solution, the lesser the number of nymphs on plant. Usually nymphs on cage floor were either dead or so weak that they could hardly cause any damage to the test plant. In these studies we found that the plant extracts, unlike the synthetic insecticide, largely incapacitate the hoppers; reversible paralysis was observed, whereas the effect of Cymbush was largely irreversible. This observation was corroborated when the level of feeding suppression of Jatropha extracts was determined over a four-day period (Table 2).

Figure 2 shows the effect of the various treatments on the establishment and early survivals of eclosing *S. docilis* larvae. Five days after eclosing, over 95 per cent of larvae eclosing on 7.5 per cent Jatropha-treated foliage had died. On the control (water) and the standard (water + 0.2 per cent Sparkle) treated foliage the mortalities were 55 and

Table 1. Mortality and food consumption of fourth-instar *Z. variegatus* topically treated with emulsion of neem oil, Jatropha oil, and Cymbush

Treatment	% corrected mortality	Mean faecal weights	% feeding suppression
Water + 0.2 Sparkle	0.0	1.1051[a]	
5% Neem emulsion	16.0 ± 3.2^a	0.6885[b]	40.89[a]
5% Jatropha emulsion	24.1 ± 4.4^b	0.6096[b]	42.91[a]
0.05% Cymbush	75.3 ± 2.5^a	0.1534[c]	86.03[b]

Values followed by the same letters are not significantly different at 5 per cent level.

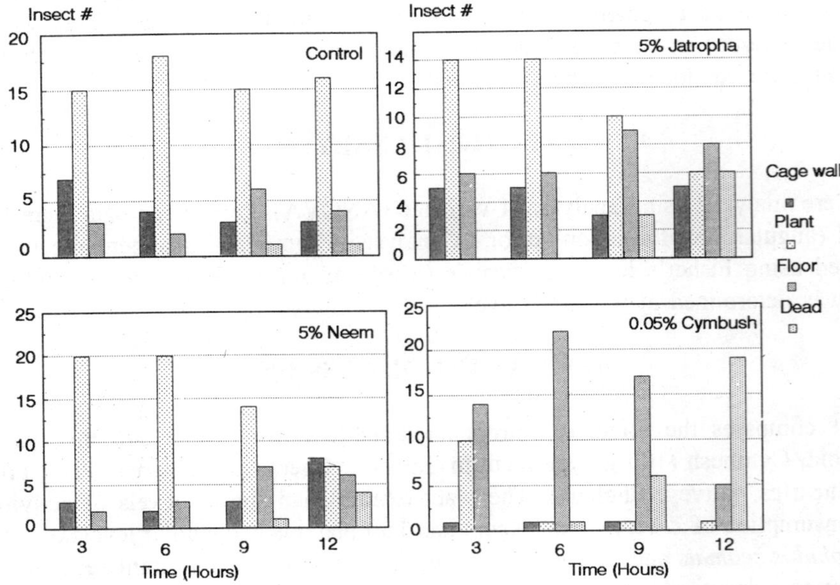

Fig. 1. Distribution of *Z. variegatus* nymphs in cage

Table 2. Influence of different concentrations of Jatropha extracts on per cent mortality and per cent feeding suppression of fourth-instar *Z. variegatus*

Treatment	Mortality and feeding suppression after							
	24 hrs		48 hrs		72 hrs		96 hrs	
	% corr. mortality	% feeding suppression	% corr. mortality	% feeding suppression	% corr. mortality	% feeding suppression	% corr. mortality	% feeding suppression
Water 0.2% sparkle	0[a]	34.53[a]	0[a]	13.06[a]	3.34[a]	14.76[a]	3.5[a]	1.61[a]
2.5% Jatropha emulsion	0[a]	51.16[a]	0[a]	30.72[b]	3.34[a]	16.34[a]	3.5[a]	3.46[b]
5% Jatropha emulsion	0[a]	61.25[b]	0[a]	45.02[b]	3.34[a]	46.13[b]	3.5[a]	8.73[c]
7.5% Jatropha emulsion	0[a]	68.96[b]	0[a]	60.93[c]	3.34[a]	53.00[b]	3.5[a]	40.00[d]
10% Jatropha emulsion	6.67[b]	75.01[b]	6.67[b]	62.61[c]	6.67[b]	63.50bc	7.14[b]	43.42[d]

Values followed by the same letters are not significantly different at 5 per cent level.

60 per cent respectively. Cobbinah (1985) has recorded low to moderate mortalities in the establishment of newly eclosed lepidopterous larvae on host plants. However, the high mortality recorded on the Jatropha-treated foliage indicates the presence of toxic or feeding deterrent factors. In our studies the cumulative effect of feeding deterrents affecting adequacy of food intake may be responsible for the sharp increase in mortality between days three and five.

Figure 3 shows the effect of various treatments on the development of fourth-instar *S. docilis*. Four days after treatment 80 and 70 per cent larvae developed to the pupal

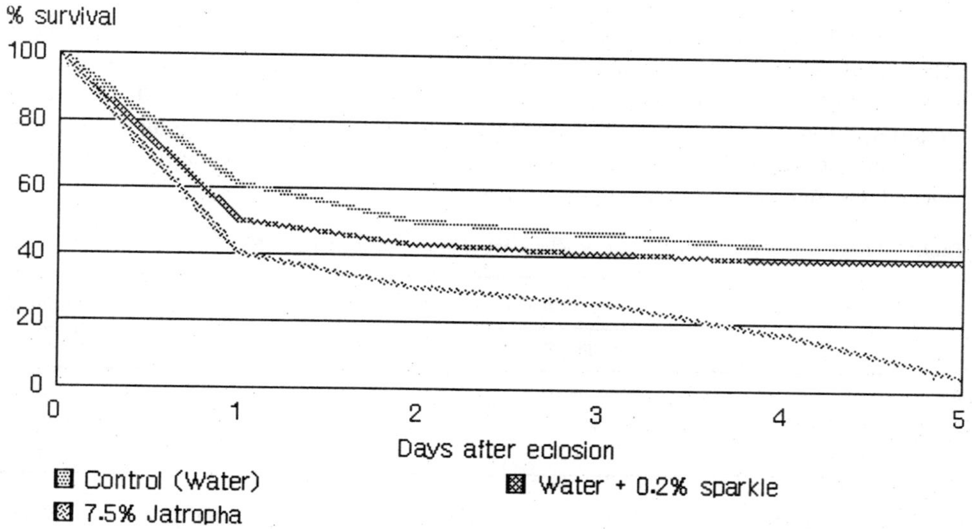

Fig. 2. Establishment and early survival of *S. docilis* on eggplant foliage treated with three solutions

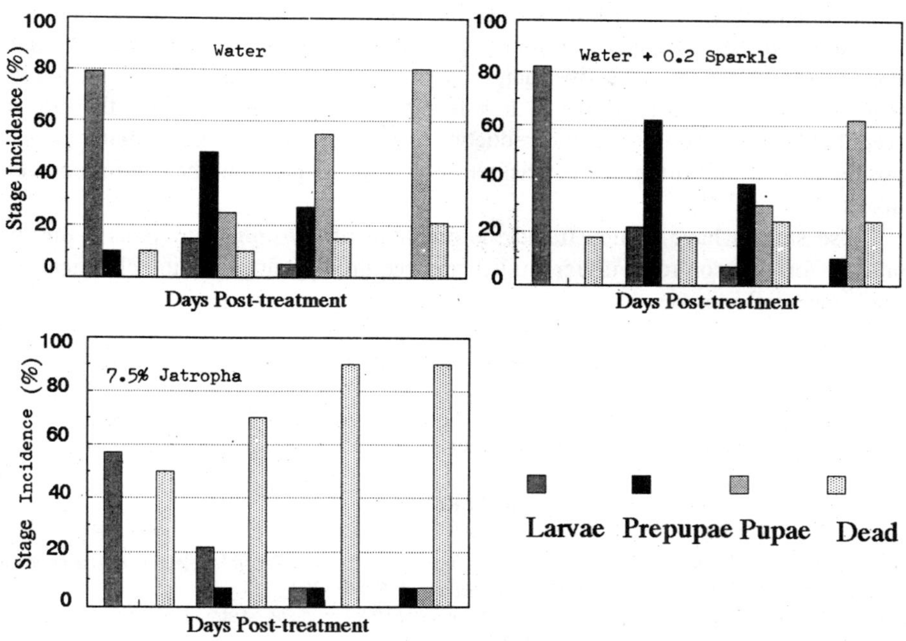

Fig. 3. Rates of development of fourth-instar larvae of *S. docilis* on different treatments

stage on the control and the standard treatments. Only 5 per cent developed into prepupae and pupae on the foliage treated with 7.5 per cent Jatropha extract. When prepupae and pupae were treated none of the prepupae treated with the Jatropha extract emerged as adults. However, 5.5 per cent of the Jatropha-treated pupae emerged as adults. Their development was, however, significantly extended (Table 3).

Treated prepupae usually crawled out of their cocoon unable to develop pupal cuticle or make a new cocoon and they all died. The inability of the prepupae to develop to pupae and put on pupal cage suggests that Jatropha extract has disruptive effect on metamorphosis. No unusual characteristics were, however, observed among those pupae which did survive to adulthood. Sparkle, used as emulsifier (standard) significantly reduced the per cent adult emergence (Table 3).

Table 3. Effect of Jatropha extract on the development of prepupae and pupae of *S. docilis*

Treatment	Mean development period		Mean	% emergence
	Prepupae	Pupae	Prepupae	Pupae
Water	11.83 + 0.98	9.0 + 3.91	77.7	70.58
Water + 0.2% Sparkle	12.51 + 1.51	9.5 + 2.15	40.5	54.5
7.5% Jatropha emulsion	0	11.6 + 0.54	0	5.55

This is the first report of the effect of Jatropha seed extract on *S. docilis*. Neem seed extracts have been found to affect food intake of *Z. variegatus* and *S. docilis* (Olaifa and Akingbohungbe, 1986; Dreyev, 1986; Cobbinah and Osei-Owusu, 1988).

The potential activity of Jatropha seed extract can be seen from its effects on food intake of *Z. variegatus* and larval and pupal developments of *S. docilis*. Although the Jatropha extract was less effective than 0.05 per cent Cymbush, nevertheless, the deterrent action of the extracts results in low feeding and less active insects that are more susceptible to predation. Also, the reduction in activity of the extract with increasing exposure time would minimize the risks to non-target organisms that might contact the residues.

These studies have demonstrated the potential of *Jatropha curcas* as a promising candidate for pest control. Further studies are, required to elucidate the chemical nature of the bioactive compounds.

ACKNOWLEDGEMENT

This work was supported by research grant No.F/982-2 from the International Foundation for Science (IFS). IFS also provided supplementary funds for my participation in the World Neem Conference.

REFERENCES

Cobbinah, J.R. 1985. The gum leaf skeletonizer *Uraba lugens*, and its hosts. Possible selection of strains of insects that are able to feed on resistant trees. *Insect Sci. Appl.* 6: 291–299.

Cobbinah, J.R. and J. Appiah-Kwarteng. 1989. Effects of some neem products on stored maize weevil, *Sitrophilus zeamais* Motsch. *Insect Sci. Appl.* 10: 89–92.

Cobbinah, J.R. and K. Osei-Owusu. 1988. Effects of neem seed extracts on insect pests of eggplant, okra and cowpea. *Insect Sci. Appl.* 9: 901–907.

Dreyer, M. 1986. Field and laboratory trials with simple neem products as protectants against pests of vegetable and field crops in Togo. *Proc. 3rd Int. Neem Conf.* Nairobi, 1986, GTZ, Eschborn, pp. 431–444.

Frempong, E. 1981. The biology of *Selepa docilis* Butler (Lepidoptera: Noctuidae) as egg plant defoliator in Ghana. *Bulletin de l' IFAN* T43 (1981) Ser. App. 174–186.

Olaifa, J.I. and A.E. Akingbohungbe. 1986. Antifeedants and insecticidal effects of extracts of *Azadirachta indica*, *Petiveria alliacea* and *Piper guineense* on the variegated grasshopper, *Zonocerus variegatus*. *Proc. 3rd Int. Neem Conf.* Nairobi, 1986, GTZ, Eschborn, pp. 405–418.

Snedecor, G.W. and W.G. Cochran. 1967. *Statistical Methods*. Iowa State University, Ames.

118

Phytophagous Insects Living on *Ajuga* Species *A. bracteosa*, *A. chamaepitys*, *A. genevensis*, *A. pyramidalis*, *A. reptans* var. *reptans* and *A. reptans* var. *atropurpurea*

BÉLA DARVAS[1], LÁSZLÓ ALFONZ POLGÁR[1], ABDEL-MONEM MOKHTAR[1,2], PIROSKA SZABÓ[3], MÁRIA TORMA-GAZDAG[3], ZOLTÁN ILOVAI[3], EDE PETRÓ[4], CHIH-HUA TSOU[5], YAW-HUEI LIN[5] AND ARILD ANDERSEN[6]

[1]Plant Protection Institute, Hungarian Academy of Sciences
Budapest, PO Box 102, H-1525, Hungary
[2]Plant Protection Institute, Dokki-Cairo, Egypt
[3]Plant Health and Soil Conservation Station of Csongrárd-county
Hódmezővásárhely, Rárósi u. 102, H-6801, Hungary
[4]Plant Health and Soil Conservation Station of the Capital
Budapest, Higany u. 2. H-1118, Hungary
[5]Institute of Botany, Academia Sinica,
Nankang, Tapei, Taiwan 11529, Republic of China
[6]Norwegian Plant Protection Institute
Ås, Fellesbygget, N-1432, Norway

ABSTRACT

Ajuga species are known to contain allelochemicals which provide a degree of resistance to non-adapted insects.

Poecilimon schmidti Fiber (Orthoptera, Phaneropteridae) second instar is able to develop fully on *A. reptans* var. *reptans*, although egg production is four times less than that of the control living on *Taraxacum officinale*.

In early spring, under greenhouse condition, *Aphis nasturtii* Kaltenbach (Homoptera, Aphididae) can live on *A. reptans* var. *reptans* and *Aulacorthum solani* Kaltenbach (Homoptera, Aphididae) on *A. chamaepitys*. *Myzus persicae* Sulzer (Homoptera, Aphididae) and *Aphis gossypii* Glover (Homoptera, Aphididae) also attack *Ajuga* species, but *A. gossypii* produces fewer progeny on *Ajuga* species than on cucumber. *Myzus persicae* can easily adapt to *A. genevensis*, *A. reptans* var. *reptans*, and *A. reptans* var. *atropurpurea*. In autumn *Ajuga* species are free from these ploYphagous aphids, possibly because of the increased level of allelochemicals. A predator of aphids, *Aphidoletes aphidimyza* Rond. (Diptera, Cecidomyiidae), and hymenopterous parasitoids (*Aphidius matricariae* Hal., Hymenoptera, Aphydiidae; *Aphelinus chaonia* Walk., *A. asychis* Walk., Hymenoptera, Aphelinidae) accepted aphids living on *Ajuga* species. *Ajuga reptans* var. *reptans*, *A. reptans* var. *atropurpurea*, and *A. genevensis* are suitable hosts for *Trialeurodes vaporariorum* Westwood (Homoptera, Aleurodidae) adults, but they are not appropriate for larval development. *Emelyanoviana mollicula* (Boheman) and *Eupteryx vittata* L. (Homoptera, Cicadellidae) attack *A. bracteosa*, *A. chamaepitys*, and *A. genevensis*.

Cnephasia communana H.-S. (Lepidoptera, Tortricidae) is able to develop on *A. reptans* var. *reptans* in early spring. *Mamestra* sp. (Lepidoptera, Noctuidae) larvae may feed on *A. reptans* var. *reptans* and *A. reptans* var. *atropurpurea*, but the pupal stage is not reached.

Chromatomyia horticola Goreau (Diptera, Agromyzidae) attacks early leaves of *A. chamaepitys* but does not accept this species following flowering when leaf-shape changes to the narrow type. *Liriomyza bryoniae* Kaltenbach (Diptera, Agromyzidae) attacks *A. bracteosa* (syn. *A. remota*), *A. chamaepitys*, *A. genevensis*, *A. reptans* var. *reptans*, *A. reptans* var. *atropurpurea* leaves. Parasitoids of *C. horticola* (*Hemiptarsenus dropion* Walk., *Diglyphus isaea* Walk., Hymnoptera, Eulophidae), and *L. bryoniaea* (*Hemiptarsenus unguicellus* Zett., Hymnoptera, Eulophidae; *Opius pallipes* Wesm., Hymnoptera, Braconidae) are able to develop in hosts living on *Ajuga* plants.

INTRODUCTION

During approximately 200 million years of coexistence, the plant and animal kingdoms have affected each other in different ways (Jermy, 1984). Many plants produce allelochemicals (secondary plant substances) which affect the life of herbivores. Some of the allelochemicals can interfere with the hormonal balance of the insect, inducing symptoms of insect developmental inhibitor (IDI, = insect growth regulators, IGR).

Juvabiones isolated from *Abies balsamea* (Pinaceae) (Slama and Williams, 1966; Bowers et al., 1966), juvocimenes (which are clerodanes) separated from *Oscimum basilicum* (Labiatae) (Bowers and Nishida, 1980), and echinolon originated from *Echinaceae angustifolia* (Asteraceae) (Jacobson et al., 1975) belong to the natural compounds with juvenile hormone agonist activity. Asarones are produced by *Acorus calamus* (Araceae), and ageratochromenes, also termed 'precocenes', isolated first from *Ageratum houstonianum* (Asteraceae), have junvenile hormone antagonist activity disrupting *Corpora allata* of the precocene-sensitive insects (Pratt and Bowers, 1977).

Zooecdysteriods play significant roles on insect spermiogenesis, ovogenesis,and embryogenesis, organization of moults, and diapause (Koolman, 1989). Plants also synthesize ecdysteroid-type compounds called 'phytoecdysteroids'. Ponasterone A was the first phytoecdysteroid isolated by Nakanishi et al. (1966) from a plant species, *Podocarpus nakaii* (Podocarpaceae). At present, plant species belonging to 100 families are known to produce approximately 100 phytoecdysteroids (Slama, 1979; Bergamasco and Horn, 1983; Lafont and Horn, 1989; Camps, 1991). Surprisingly, the level of ecdysteroids is lower in insects than in plants. The richest known source of zooecdysteroids is the locust embryo (250 ppm). Generally, 1 ppm ecdysteroid is frequently found in insect samples. In plants, higher phytoecdysteroid contents are often detected, for example:

Diploclisia glaucescens (Menispermaceae) shoots include 32,000 ppm 20-OH ecdysone (Bandara et al., 1989)

Cyanothis arachnoidae (Commelinaceae) roots produce 29,000 ppm 20-OH ecdysone (Chou and Lu, 1980)

Dacrydium intermedium (Podocarpaceae) contains 22,000 ppm phytoecdysteroids (Russell et al., 1972).

Serratula (*S. inermis, S. xeranthemoides, S. tinctoria*) (Asteraceae) flowers and seeds include 20,000 ppm ecdysteroids (Yatsyuk and Segel, 1970; Kholodova et al., 1979; Bathory et al., 1990).

Vitex (*V. madiensis, V. thyrsiflora*) (Verbenaceae) roots and fruits contain 5000–7000 ppm 20-OH ecdysone and ajugasterone C (Kubo et al., 1984, 1985).

Polypodium vulgare (Polypodiaceae) rhizomes produce 2000–4000 ppm ecdysteroids (Camps et al., 1990a 1990b).

Silene species (*S. otites* var. *hungarica, S. nutans, S. scabrifolia*) (Caryophyllaceae) contain 1000–4000 ppm ecdysteroids (Saatov et al., 1986; Bathory et al., 1987).

Ajuga species (*A. incisa, A. decumbens, A. nipponensis, A. japonica*, etc.) include 1800–2200 ppm ecdysteroids (Matsuoka et al., 1969).

In plant samples, the sequence of the phytoecdysteroids in order of abundance seems to be 20-OH ecdysone, ponasterone A, polypodine B (= ajugasterone A), ecdysone, and pterosterone (Bergamasco and Horn, 1983).

More than 100 *Ajuga* species (Table 1) are known in the world (Darvas, 1991). *Ajuga* species synthesize the aglycones luteolin and apigenin and their glycosides; iridoides (harpagid, harpagid-acetate); catechin-like compounds; caffeic, chlorogenic, and 4-caffeoylquinic acids (Litvinenko et al., 1970); neo-clerodanes (see below), and phytoecdysteroids.

20-OH ecdysone, cyasterone, ploypodine B, ajugasterone B-C, and ajugalactone were isolated by Koreeda et al. (1972) from *A. decumbens, A. incisa, A. nipponensis,* and *A. japonica*. Chou and Lu (1980) discovered ajugasterone D from *A. nipponensis*. Usmanov et al. (1975) isolated turkesterone and 22-acetyl-cyasterone from *A. Trukestanica.* Camps et al. (1981a, 1982, 1983, 1985a, b, 1990b) and Tomas et al. (1992) working with *A. reptans* var. *reptans* isolated 29-norcyasterone, sengosterone, 29-norsengosterone, makisterone A, polypodine B, 2-, while 3-acetyl-norcyasterone, cyasterone, makisterone A, 20-OH ecdysone, and ajugalactone were isolated from *A. chamaepitys* (Camps et al., 1985c).

Ajuga species produce different neoclerodanes with deterrent activity to insects. *Ajuga bracteosa* (syn. *A. remota*) produces ajugarin I-IV (Kubo et al., 1976, 1982, 1983). From *A. reptans*, Camps et al. (1981a,b) and Camps (1990) isolated ajugareptansin and ajugareptanson A-B. In *A. iva*, Camps et al. (1981a,b) have found ivain 1-4. In *A. nipponensis* ajugamarin and related compounds were found by Shimomura et al. (1981). From *A. chamaepitys*, ajugapitins and chamaepitin were isolated by Hernandez et al. (1982) and Camps et al. (1984, 1987).

Exogenous phytoecdysteroids, which are structurally related to ecdysone, are thought to be rapidly metabolized in insects (Hikino et al., 1975). Plant species having high steroid contents (> 1000 ppm) are not frequently visited by insects. In Norway, we (Darvas and Andersen) found that *Subcoccinella vigintiquatuorpunctata* L. (Coleoptera, Chrysomelidae) could defoliate *Silene vulgaris*. Insects living on *Ajuga* species have

Table 1. *Ajuga* species of the world based on the Index Kewensis (1885–1991)

Ajuga species	AF	AM	AU	AS	JA	EU	f./var/ssp.; (hybrid)
abnormis Rouy	—	—	—	—	—	+	
× *adulterina* Wallr.	—	—	—	—	—	+	*(genevensis × pyramidalis)*
arabica Davis	—	—	—	+SA	—	—	
argyrea Stapf	—	—	—	—	—	+A	
astalanosa Schur.	—	—	—	—	—	+	
austro-iranica Rech.	—	—	+S	—	—	—	
australis R.	—	—	+	—	—	—	
bastarda Makino	—	—	—	—	+	—	
bombycina Boiss.	—	—	—	+	—	—	
brachystemon Maxim.	—	—	—	+I	—	—	alba
bracteosa Benth.	+m	—	—	+I	+	—	
calantha Diels	—	—	—	+C	—	—	
campylantha Diels	—	—	—	+C	—	—	
campylanthoides Wu & Chen	—	—	+C	—	—	—	*subacaulis*
candolleana Rouy	—	—	—	—	—	+F	
chamaecystus Ging.	—	—	—	+S	+	—	*bamianica, euprasioides, multisecta, scoparia, tomentella*
chamaepitys Schreb.	+	e	—	+S	—	+	*cuneatifolia, cyria, euphratica, glareosa, laevigata, latiloba, libanotica, mardinensis, pseudochia, rechingeri, suffrutescens, trichodactylites*
chia Schreb.	—	—	—	+S	—	—	
chosmophila Davis	—	—	—	+Sy	—	—	
ciliata Bunge.	—	—	—	+C	—	—	*chanetii, hirta, ovatisepala, pauciflora*
comata Stapf	—	—	—	—	—	+A	
contraversa Bor.	—	—	—	—	—	+F	
crenata Hochst.	+Et	—	—	—	—	—	
cuneatifolia Stapf.	—	—	—	—	—	+A	
decumbens Thunb.	—	—	—	+Ko	+	—	*glabrata, rosa*
davisiana Kit T. & Y.	—	—	—	—	—	+Tu	
devestita Lév. & Van.	—	—	—	+Ko	—	—	
dityocarpa Hayata	—	—	—	+Fo	—	—	
elatior Ohwi	—	—	—	—	+	—	
faureii Lév. &Van.	—	—	—	+Ko	—	—	
flaccida Baker	+M	—	—	—	—	—	
formosana Hayata	—	—	—	+Fo	—	—	
forrestii Diels	—	—	—	+C	—	—	
genevensis L.	—	e	—	+S	—	+	*glabrifolia, foliosa*
glabrescens Makino	—	—	—	—	+	—	
grandiflora Stapf	—	—	+	—	—	—	
grosse-serrata F&S	—	—	—	—	+	—	
halacsyana Zahlbr.	—	—	—	—	—	+	
× *hampeana* A. B&V	—	—	—	—	—	+	*(pyramidalis × reptans)*
humlis Miq.	—	—	—	—	+	—	
x *hybrida* Kerner	—	—	—	—	—	+*A*	*(genevensis × reptans)*
incisa Maxim	—	—	—	—	+	—	
integrigoli B&H	—	—	—	+Ne	—	—	
iva Schreb.	+N	—	—	+S	—	+S	*punctata, pseudoiva*
japonica Miq.	—	—	—	—	+	—	
knafii Rouy	—	—	—	—	—	+	

contd.

Table 1. contd.

Ajuga species	AF	AM	AU	AS	JA	EU	f./var/ssp.; (hybrid)
labordei Vaniot	—	—	—	+C	—	—	
laevigata Boiss.	—	—	—	+S	—	—	
lanosa Sun	—	—	—	+C	—	—	
laxmannii Benth.	—	—	—	+W	—	+E	*longifolia, isophylla*
linearifolia Pampan	—	—	—	+C	—	—	
lobata D.D.	—	—	—	+I	—	—	
lucida Bor.	—	—	—	—	—	+It	
lupulina Maxim	—	—	—	+C	—	—	
lycia Stapf	—	—	—	—	—	+A	
lycopsiformis Lajac	—	—	—	—	—	+It	
macrosperma Wall	—	—	—	+I	—	—	
mairei Lévilleé	—	—	—	+C	—	—	
makinoi Nakai	—	—	—	—	+	—	
matsumurana Kudo	—	—	—	+Fo	—	—	
microphylla Heget.	—	—	—	—	—	+Sw	
× *mixta* Makino	—	—	—	—	+	—	
mollis Gladkova	—	—	—	+Kr	—	—	
morii Nakai	—	—	—	+Ko	—	—	
multiflora Bunge	—	—	—	+S	—	—	*brevispicata, rosea*
nana Gilib.	—	—	—	—	—	+	
nanti Bor.	—	—	—	—	—	+F	
nipponensis Makino	—	—	—	—	+	—	*pallescens*
nubigena Diels	—	—	—	+C	—	—	
oocephala Baker	+M	—	—	—	—	—	
ophrydis Burch.	+	—	—	—	—	—	
orientalis L.	—	—	—	+S	-	+S	*aenesia*
ovalifolia Bur. & Fr.	—	—	—	+C	—	—	*albiflora, angustifolia*
ovata Wall.	—	—	—	+B	—	—	
pachyrrhiza Kitag.	—	—	—	+C	—	—	
parviflora Benth.	—	—	—	+I	—	—	
piskoi Deg. & Bal.	—	—	—	—	—	+Al	
postii Briq.	—	—	—	+Sy	—	—	
× *pseudo-pyramidalis* Sch.	—	—	—	—	—	+	(*pyramidalis* × *reptans*)
pantantha Hand.-M.	—	—	—	+C	—	—	
pumila Schur.	—	—	—	—	—	+	
pygmaea A. Gray	—	—	—	—	+	—	
pyramidalis L.	—	—	—	+W	—	+	*meonantha, occidentalis, rotundifolia*
relicta Davis	—	—	—	+	—	+Tu	
reptans L.	+N	e	—	+W	—	+	*alpina, alpestris, atropurpurea, lasiotemon*
robusta Baker	—	—	—	+M	—	—	
rhodopea Velen.	—	—	—	—	—	+Bl	
salicifolia Schreb.	—	—	—	+W	—	+E	bassarabica
saxiola Assadi	—	—	—	+Ir	—	—	
sciaphila Smith	—	—	—	+C	—	—	
shikotanensis M. & T.	—	—	—	+Ku	—	—	
spectabilis Nakai	—	—	—	+Ko	—	—	
stalonifera J&T	—	—	—	—	—	+	
taiwanensis Nakai	—	—	—	+S	+	—	
tenorii Presl.	—	—	—	—	—	+S	
turkestanica Briq.	—	—	—	+T	—	—	
typica Mat. & Kudo	—	—	—	—	+	—	
vesiculifera Herd	—	—	—	+m	—	—	

contd.

Table 1. contd.

Ajuga species	AF	AM	AU	AS	JA	EU	f./var/ssp.; (hybrid)
vestita Boiss	—	—	—	—	—	+G	
vulgaris Rouy	—	—	—	—	—	+F	
xylorrhiza Lit Tan	—	—	—	—	—	+Tu	
yesoensis Maxim.	—	—	—	—	+	—	
zakhoensis Rech.	—	—	—	+Ir	—	—	
zambeziaca Kupic.	+S	—	—	—	—	—	

AF = Africa, AM = America, AU = Australia, AS = Asia, JA = Japan, EU = Europe; e = exotic, N = North, S = South, E = East, W = West, m = middle x = hybrid; A = Austria, Al = Albania, B = Burma, Bl = Bulgaria, C = China, Et = Ethiopia, F = France, Fo = Formosa, G = Greece, I = India, Ir = Iran, It = Italy, Ko = Korea, Kr = Krimea p., Ku = Kuril I., M = Madagascar, Ne = Nepal, SA = SaudiArabia, Sv = Switzerland, Sy = Syria, T = Turkmenia, Tu = Turkey.

been scarcely studied. Spencer (1990) published that some agromyzid species, *Liriomyza trifolii* Burgess, *L. bryoniae* Kaltenbach, *L. strigata* (Meigen), *Chromatomyia horticola* Goreau, and *Amauromyza labiatarum* Hendel (Diptera, Agromyzidae), can develop on *Ajuga* species. Our aim was to study insects living on *Ajuga* species and known to tolerate high neoclerodane content with deterrent activity and phytoecdysteroid content with IDI-type activity.

MATERIALS AND METHODS

Insects living on *Ajuga* species in field and greenhouse conditions were collected and examined during 1991 and 1992 in Hungary. *Poecilimon schmidti* (our stock originated from Caucasus) second instars which visited *Ajuga* species were collected and were reared on *A. reptans* var. *reptans* leaves. *Aphis nasturtii* and *Aulacorthum solani* visiting *Ajuga* species in greenhouses were counted on 10 leaves. Data were analysed with one-way analysis of variance (ANOVA). Five to nine *Aphis gossypii* alate fourth-instar nymphs were placed on *Ajuga* species in May, June, and August. Progenies of the isolated females (developed from fourth-instar alate) were counted. *Trialeurodes vaporariorum* freshly emerged female and male (from *Salvia*), a couple, was placed on *Ajuga* species, repeated four times. Lifetimes, egg production of adults, and larval development of the new generation were examined. Data were analysed with ANOVA.

RESULTS AND DISCUSSION

Ajuga species have repellent activity against domesticated and wild mammalian herbivores (cow, sheep, etc.), but some snail species (Gastropoda) can attack them. Under greenhouse conditions (Budapest), *A. chamaepitys* were attacked by *Vallonia costata* (O.F. Müller), *Zonitoides arborescens* (Say), *Deroceras laeve* (O.F. Müller), and *D. sturanyi* (Simroth) (Darvas and Petró, unpublished results). These snail species reject *A. reptans* var. *reptans* and *A. reptans* var. *atropurpurea*, but *Deroceras* species attack *A. bracteosa* as well. Insects living on our *Ajuga* stocks were as follows:

Poecilimon schmidti **Fieber (Orthoptera: Phaneropteridae)**

First-instar *P. schmidti* larvae, emerged in May, visited *A. reptans* var. *reptans* and feed on it. This polyphagous species usually develops on dicotyledonous plants. In June, *P. schmidti* females living on *A. reptans* var. *reptans* laid four times less eggs (9–18/female), than those which fed on *Taraxacum officinale* (Compositae) (60–70/female).

Aphis nasturtii **Kaltenbach (Homoptera: Aphididae)**

In January, under greenhouse conditions, *A. nasturtii* could attack *A. reptans* var. *reptans* (Fig. 1), although it might choose *Solanum tuberosum* or *Capsicum annuum* as well. After April, *A. nasturtii* rejected *A. reptans* var. *reptans*. This aphid could attack a closely related plant species, *Prunella vulgaris* (Labiatae), in July, which produces steroid glycosides (Kojima et al., 1990). It did not attack *A. chamaepitys* and *A. reptans* var. *atropurpurea*.

During spring, the phytoecdysteroid content of mature *A. reptans* var. *reptans* is lower (339 ppm) than in summer (724 ppm) (Tomas et al., 1992). Ajugalactone content rapidly increases from spring (181 ppm) to summer (416 ppm). Under greenhouse conditions, phytoecdysteroid content is the lowest (195 ppm). It is likely that low ecdysteroid contents induced by greenhouse conditions cannot defend *A. reptans* var. *reptans* against *A. nasturtii*. On the other hand, *A. nasturtii* can tolerate only a limited amount of phytoecdysteroids produced by *A. reptans* var. *reptans*.

Aphidolethes aphidimyza Rond. (Diptera: Cecidomyiidae) larvae can develop on *A. nasturtii* living on *A. reptans* var *reptans*. This indicates that *A. nasturtii* catabolizes *A.*

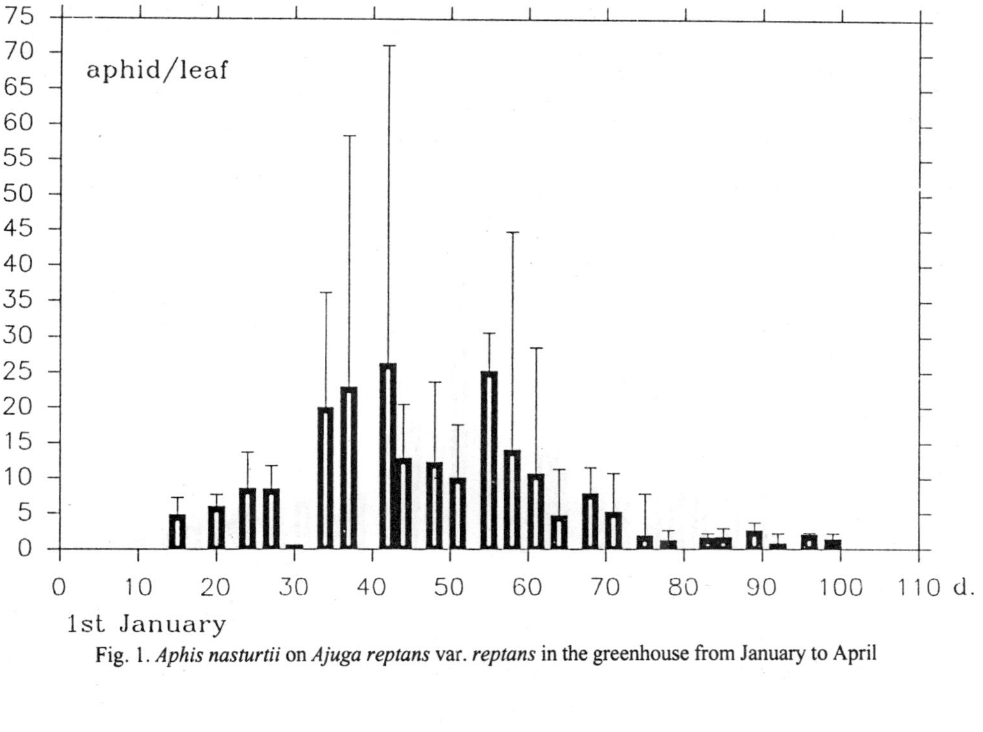

Fig. 1. *Aphis nasturtii* on *Ajuga reptans* var. *reptans* in the greenhouse from January to April

reptans var. *reptans* type of allelochemicals or the sensitivity of *A. aphidimyza* against them is low.

Aulacorthum solani Kaltenbach (Homoptera: Aphididae)

In January, under geenhouse conditions, *A. solani* could attack *A. chamaepitys* (Fig. 2), although it might choose *Solanum tuberosum* or *Lycopersicom esculentum* as well. After April, *A. solani* rejected *A. chamaepitys*. It did not attack *A. reptans* var. *reptans* or *A. reptans* var. *atropurpurea*.

Myzua persicae Sulzer (Homoptera: Aphididae)

Myzus persicae, an extermely polyphagous aphid species, did not colonize *A. reptans* var. *reptans*, *A. reptans* var. *atropurpurea*, or *A. chamaepitys* during 1991. In May 1992, natural infestations occurred under field conditions and heavily damaged *A. reptans* var. *atropurpurea* and *A. genevensis*. Anholocyclic *M. persicae* could accept *A. reptans* var. *reptans* during winter, but not during spring or summer. Stocks of *M. persicae* living on *A. reptans* var. *atropurpurea* and *A. genevensis* were parasitized by *Aphidius matricariae* Hal. (Hymenoptera: Aphidiidae). Some of the *A. matricariae* were hyperparasitized by *Alloxysta victrix* Westwood (Hymenoptera: Charipidae). From July, there was another heavy *M. persicae* infestation. Colonies of *M. periscae* living on *Ajuga* species were parasitized this time by *Aphelinus asychis* Walk. and *A. chaonia* Walk. (Hymenoptera: Aphelinidae). From September, *M. persicae* rejected *Ajuga* species.

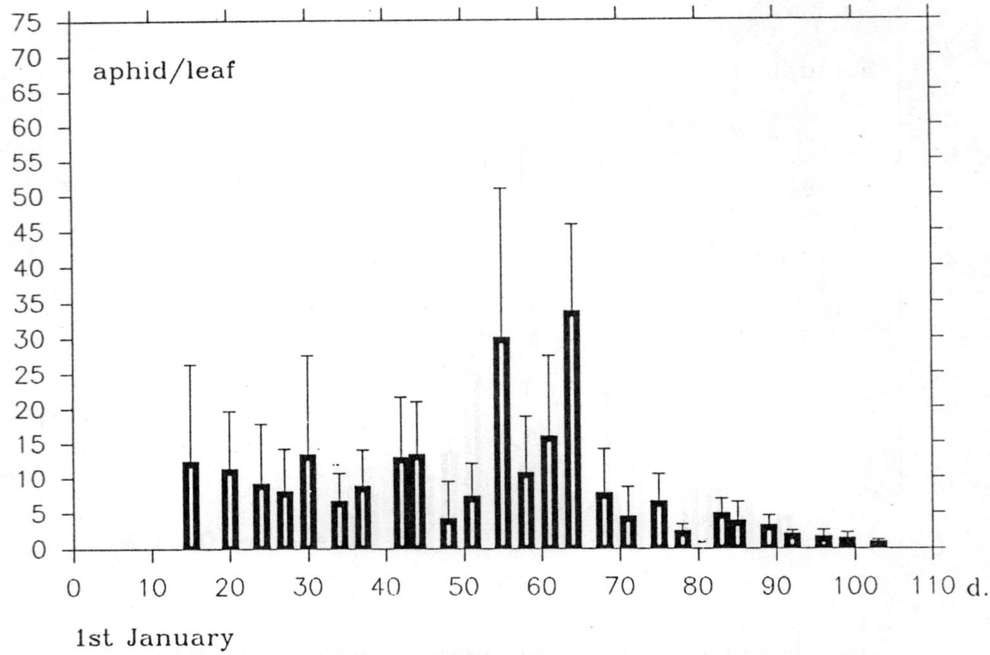

Fig. 2. *Aulacorthum solani* on *Ajuga chamaepitys* in the greenhouse from January to April

Aphis gossypii Glover (Homoptera: Aphididae)

Anholocyclic *A. gossypii* stock, overwintered on cucumber under greenhouse conditions, could accept *A. reptans* var. *atropurpurea*, *A. reptans* var. *reptans*, and *A. chamaepitys*. This *A. gossypii* biotype, living on *Cucumis sativus*, *C. melo*, or *Citrullus vulgaris*, did not accept *Capsicum annuum*, *Solanum tuberosum*, *S. melongena*, *Cucurbita ficifolia*, and *Gossypium hirsutum*. Progeny produced by alate female living on *A. reptans* var. *reptans* and *A. reptans* var. *atropurpurea* decreased from spring (Table 2) to summer (Table 3).

Trialeurodes vaporariorum Westwood (Homoptera: Aleurodidae)

Average lifetime of *T. vaporariorum* adults was 12 days, independent of the host plant. Fermales produced similar numbers of fertile eggs on *A. reptans* var. *reptans*, *A. reptans* var. *atropurpurea*, *A. genevensis* (Table 4), and cucumber. Emerged first-instar of greenhouse white-fly settled on *Ajuga* plants, but died mostly before the first moult (Mele et al., 1992). A limited number of adults emerged on *A. genevensis*. A relative

Table 2. Progeny of *Aphis gossypii*, living on *Ajuga* species during May (in the greenhouse) and June (in the field)

Ajuga stock	Progeny of *A. gossypii*			
	May		June	
	mean ± SD/ female	%	mean ± SD/ female	%
ARR-HH4	10.88 ± 1.54bc	45	9.61 ± 4.91bc	47
ARR-BT32	6.94 ± 1.85bc	29	5.88 ± 4.10c	29
ARA-BB1	13.06 ± 4.54b	54	5.14 ± 1.49c	25
ARA-HH3	9.63 ± 4.70b	40	5.81 ± 2.67c	28
AC-BS3	3.37 ± 1.45c	14	4.89 ± 1.41c	24
Cucumis sativus	24.11 ± 6.31a	100	20.62 ± 0.27a	100
LSD 1%	6.59		6.59	

ARR = *A. reptans* var. *reptans*, *ARA* = *A. reptans* var. *atropurpurea*, *AC* = *A. chamaepitys*. Values followed by the same letter within a column do not differ significantly ($p < 1\%$) based on ANOVA.

Table 3. Progeny of *Aphis gossypii* living on *Ajuga* species in August

Ajuga stock	Progeny of *A. gossypii*	
	mean ± SD/ female	%
ARR-BV36	4.85 ± 2.86b	20
ARR-BT33	5.77 ± 0.58b	24
AG-BS1	7.88 ± 3.38b	33
AC-BS3	5.39 ± 1.87b	23
Cucumis sativus	23.77 ± 4.49a	100
LSD 1%	5.62	

AG = *A. genevensis*. Values followed by the same letter within a column do not differ significantly ($p < 1\%$) based on ANOVA.

Table 4. Egg production and development of *Trialeurodes vaporariorum* on *A. reptans* var. *reptans*, *A. reptans* var. *atropurpurea*, and *A. genevensis*

Ajuga stock	Egg production during 12 days/female (mean ± SD)	Died L_1 (mean ± SD)	Emerged adult (mean ± SD)
ARR-HS9	35.50 ± 11.90	34.24 ± 10.21b	0.00 ± 0.00b
ARR-HP15	26.25 ± 5.91	26.00 ± 5.60b	0.00 ± 0.00b
ARA-HK6	21.75 ± 8.66	20.75 ± 7.18b	0.00 ± 0.00b
AG-HM8	18.75 ± 7.93	13.75 ± 6.90ab	0.25 ± 0.50b
AG-HK3	24.00 ± 5.10	23.25 ± 5.32b	0.00 ± 0.00b
Cucumber	34.50 ± 21.67	6.25 ± 3.77a	16.50 ± 9.68a
LSD 1%		13.84	8.06

Values followed by the same letter within a column do not differ significantly ($p < 1\%$) based on ANOVA.

of *T. vaporariorum*, *Bemisia tabaci* (Gennadius) (Homoptera, Aleurodidae) can develop on *A. reptans* var. *atropurpurea* (Melé, personal communication).

Our results indicate that *A. reptans* var. *reptans*, *A. reptans* var. *atropurpurea*, and *A. genevensis* are suitable hosts for *T. vaporariorum* adults, but they are not appropriate for the larval development of this species.

Emelyanoviana mollicula (Boheman) (Homoptera: Ciccadellidae)

Ajuga bracteosa (AB-BTI), *A. chamaepitys* (AC-BS3), and *A. genevensis* (*AG*-BS1, *AG*-BT6) were infected by *E. mollicula*. Larvae of this species can develop on the *Ajuga* species mentioned. *Emelyanoviana mollicula* did not choose *A. reptans* var. *reptans* and *A. reptans* var. *atropurpurea* stocks.

Eupterix vittata L. (Homoptera: Ciccadellidae)

Ajuga bracteosa (*AB*-BT1), *A. chamaepitys* (*AC*-BS3), and *A. genevensis* (*AG*-BS1, *AG*-BT6) were infected by *E. vittata*. Larvae of this viurs vector species can develop on the *Ajuga* species mentioned. *Eupteria vittata* did not choose *A. reptans* var. *reptans* and *A. reptans* var. *atropurpurea* stocks.

Cnephasia communana H.S. (Lepidoptera: Tortricidae)

Cnephasia communana females laid eggs and could develop on *A. reptans* var. *reptans* (*ARR*-BT32) in May. Following pupation, normal adults emerged.

Mamestra sp. (Lepidoptera: Noctuidae)

Mamestra sp. laid eggs on *A. reptans* var. *reptans* (*ARR*-BT32, *ARR*-BT33) and *A. reptans* var. *atropurpurea* (*ARA*-BB1, *ARA*-BT2) in May. Two of five second-instar caterpillars died at the second moult. Three larvae moulted to fifth instar (May 18), but only one pupated after sixth instar (June 8) and that died as a larval-pupal intermediate.

Chromatomyia horticola Goreau (Diptera: Agromyzidae)

The second generation of the polyphagous agromyzid fly *C. horticola* attacked lower, broader leaves of *A. chamaepitys* (*AC*-BC1, *AC*-BS3, *AC*-BD4, *AC*-BO5, *AC*-BC15) in June. Five per cent of the *A. chamaepitys* leaves were infected. The first generation developed on *Lepidium draba* (Cruciferae). The second generation of the females accepted *Prunella vulgaris* (Labiatae) leaves for feeding, but they did not lay eggs on them. Approximately 25 per cent of the larvae died in the first moult in *A. chamaepitys* leaves. In a few cases, larval-puparial intermediates were found at the time of pupariation. Some larvae pupariated and later emerged as normal adults. One-third of the survivors were parasitized by *Hemiptarsenus dropion* Walk. and *Diglyphus isaea* Walk. (Hymenoptera: Eulophidae). The third generation of *C. horticola* females did not prefer to feed on the new, narrow leaves (appearing during flowering) of *A. chamaepitys* (leaf polymorphism exists) and did not lay eggs on them. This species rejected *A. reptans* var. *reptans* and *A. reptans* var. *atropurpurea* stocks. Our results complete Spencer's (1990) data.

Liriomyza bryoniae Kaltenbach (Diptera: Agromyzidae)

The late generations of this polyphagous species can attack *A. bracteosa*, *A. chamaepitys* (this species accepted the narrow leaves as well), *A. gnevensis*, *A. reptans* var. *reptans*, and *A. reptans* var. *atropurpurea* in August and September.

Some of the *L. bryoniae* larvae living in *A. reptans* var. *reptans* were parasitized by *Hemiptarsenus unguicellus* Zett., (Hymenoptera: Eulophidae) and *Opius pallipes* Wesm. (Hymenoptera: Braconidae).

Larvae of *L. trifolii*, an American relative of *L. bryoniae*, developed in older *A. reptans* var. *reptans* and *A. reptans* var. *atropurpurea* leaves, but many of them died in younger leaves (Mele, personal communication).

CONCLUSION

It seems that some widely polyphagous insect species can adapt to *Ajuga* species. Polyphagous insects, which can develop on Cucurbitaceae species, are usually involved in this group. Cucurbitaceae species produce a significant amount of phytosteroids (Akihisa et al., 1987), and probably the insect species living on them have evolved an efficient phytosteroid metabolism. There are significant differences among insect species in the ability to develop on a single *Ajuga* species but the time when a special plant-insect relationship exists seems to be very narrow.

ACKNOWLEDGEMENTS

This work was supported by a grant from OMFB (No. 3095/91), Hungary. The authors thank Dr. A. Orosz, Dr. J. Papp (Hungarian Natural History Museum, Budapest), and Dr. Cs. Thuróczy (Savaria Museum, Szombathely) for identification, Prof. A. Terpó (Hungarian Natural History Museum, Budapest) for botanical help, Dr. E. Melé (Dept. Plant Genetics, IRTA, Cabrils) for the invaluable discussion, Dr. J. Hardie (Imperial college, Ascot) and Dr. A. Szentesi (Plant Prot. Inst., Budapest) for critical reading of the manuscript, and Mr. A.S. Bream (Al-Azhar Univ., Egypt) for technical assistance.

REFERENCES

Akihisa, T., N. Shimizu, P. Ghosh, S. Thakur, F.U. Rosenstein, T. Tamura and T. Matsumoto. 1987. Sterols of the Cucurbitaceae. *Phytochemistry* 26: 1693–1700.

Bandara, B.M.R., L. Jayasimghe, V. Karunaratna, G.P. Wannigama, M. Bokel, W. Kraus and S. Sotheswaran. 1989. Ecdysterone from stem of *Diploclisia glaucescens. Phytochemistry* 28: 1073–1075.

Bathory, M., I. Máthé, P. Solymosi and K. Szendrei. 1987. Phytoecdysteroids in some species of Caryophyllaceae and Chenopodiaceae. *Acta Botanica Hung.* 33: 377–385.

Báthory, M., K. Szendrei, D. Rudel, R. Lafont, J. Gharib and J.P. Girault. 1990. *Serratula tinctoria* L., an adequate plant source of ecdysteroids. *Invert. Reprod. Devpt.* 18: 104.

Bergamasco, R. and D.H.S. Horn. 1983. Distribution and role of insect hormones in plants. In: *Endocrinology of Insects.* R.H.G. Downer and H. Laufer (eds.), Alan R. Liss, Inc., New York, pp. 627–654.

Bowers, W.S., H.M. Fales, M.J. Thompson and E.C. Uebel. 1966. Juvenile hormone: Identification of an active compound from balsam fir. *Science* 154: 1020–1021.

Bowers, W.S. and R. Nishida. 1980. Juvocimenes: Potent juvenile hormone mimics from sweet basil. *Science* 209: 1030–1032.

Camps, F. 1990. Allelochemicals from *Ajuga* plants. ESF Workshop, Chemistry of Insect-Plant Interaction, Boldern, 4–7 Nov.

Camps, F. 1991. Plant ecdysteroids and their interaction with insects. In: *Ecological Chemistry and Biochemistry of Plant Terpenoids.* J.B. Harborne and F.A. Tomas-Barberan (eds.), Clarendon Press, Oxford, pp. 331–376.

Camps, F., J. Coll and A. Cortel. 1981a. Allelochemicals on insect isolated from *Ajuga* plants (Labiatae). *Rev. Latinoam. Quim.* 12: 81–88.

Camps, F., J. Coll and A. Cortel. 1981b. Two new clerodane diterpenoids from *Ajuga reptans* (Labiatae). *Chem. Lett.* 1093–1096.

Camps, F., J. Coll and A. Cortel. 1982. 29-Norsengosterone and 29-norcyasterone, new C-28 phytoecdysteroids from *Ajuga reptans* (Labiatae). *Chem. Lett.* 1313–1316.

Camps, F., J. Coll and A. Cortel. 1983. Sterols from *Ajuga reptans. Ann. Quimica* 79: 228–229.

Camps, F., J. Coll and O. Dargallo. 1984. Neo-clerodane diterpenoids from *Ajuga chamaepitys. Phytochemistry* 23: 2577–2579.

Camps, F., J. Coll, E. Molins and C. Miravitlles. 1985a. 2-Acetyl and 3-acetyl-29-norcyasterone, new minor phytoecdysteroids from *Ajuga reptans* (Labiatae). *J. Chem. Res. S.* 1: 14–15.

Camps, F., J. Coll, A. Cortel, E. Molins and C. Miravitlles. 1985b. 2-acetyl- and 3-acetyl-29-norcyasterone, new minor phytoecdysteroids from *Ajuga reptans* (Labiatae). *J. Chem. Res. M.* 1: 241–266.

Camps, F., J. Coll and O. Dargallo. 1985c. Phytoecdysteroids from *Ajuga chamaepitys. Ann. Quimica* 81: 74–75.

Camps, F., J. Coll, O. Dargallo, J. Rius and C. Miravitlles. 1987. Clerodane diterpenoids from *Teucrium* and *Ajuga* plants. *Phytochemistry* 26: 1475–1479.

Camps, F., E. Claveria, J. Coll, M.P. Marco, J. Messequer and E. Mele. 1990a. Ecdysteroid producttion in tissue culture of *Polypodium vulgare. Phytochemistry* 29: 3819–3821.

Camps, F., J. Coll, M.P. Marco and J. Tomas. 1990b. Efficient determination of phytoecdysteroids from *Ajuga* species and *Polypodium vulgare* by high-performance liquid chromatography. *J. Chromatog.* 541: 199–207.

Chou, W.-S. and H.-S. Lu. 1980. Growth regulators and silk production of *Bombyx mori* L. from phytogenous ecdysteroids. In: *Progress in Ecdysone Research.* J.A. Hoffmann (ed.), Elsevier, Amsterdam, pp. 281–297.

Darvas, B. 1991. Phytoecdysteroids of *Ajuga* spp., as insect growth regulator-type botanical insecticides. *Növényvédelem.* 27:481–498 (in Hungarian).

Hernandez, A., C. Pascual, J. Sanz and B. Rodriguez. 1982. Diterpenoids from *Ajuga chamaepitys*: Two neo-clerodane derivatives. *Phytochemistry* 21: 2909–2911.

Hikino, H., Y. Ohizumi and T. Takemoto. 1975. Detoxification mechanism of *Bombyx mori* against exogenous phytoecdysone ecdysterone. *J. Insect Physiol.* 21: 1953–1963.

Index Kewensis, Vols. 1885–1991, Suppl. I–XVIII. Univ./Clarendon Press, Oxford.

Jacobson, M., R.E. Redfern and G.D. Mills, Jr. 1975. Naturally occurring insect growth regulators III. Echinolone, a highly active juvenile hormone mimic from *Echinosea anglutifolia* roots. *Lloydia* 38: 473–476.

Jermy, T. 1984. Evolution of insect/host plant relationships. *Amer. Natur.* 124: 609–629.

Kholodova, Yu.D., U. Baltaev, V.O. Volovenko, M.B. Gorovits and N.K. Abubakirov. 1979. Phytoecdysones of *Serratula xeranthemoides. Khim. Prir. Soedin.* 2: 171–174.

Kojima, H., N. Sato, A. Hatano and H. Ogura. 1990. Sterol glucosides from *Prunella vulgaris. Phytochemistry* 29: 2351–2355.

Koolman, J., ed. 1989. *Ecdysone. From Chemistry to Mode of Action.* Georg Thieme Verlag, Stuttgart, 482 pp.

Koreeda, M., K. Nakanishi and M. Goto. 1972. Ajugalactone, an insect moulting inhibitor as tested in the *Chilo* dipping method. *J. Amer. Chem. Soc.* 92: 7512–7513.

Kubo, I., Y.-W. Lee, V. Balogh-Nair, K. Nakanishi and A. Chapya. 1976. Structure of ajugarins. *J. Chem Soc., Chem. Comm.* 22: 949–950.

Kubo, I., J.A. Klocke, I. Miura and Y. Fukuyama. 1982. Structure of ajugarin-IV. *J. Chem. Soc., Chem. Comm.* 90: 618–619.

Kubo, I., Y. Fukuyama and A. Chapya. 1983. Structure of ajugarin-V. *Chem. Lett.* 91: 223–224.

Kubo, I., A. Matsumoto and J.F. Ayafor. 1984. Efficient isolation of a large amount of 20-hydroxyecdysone from *Vitex madiensis* (Verbenacea) by droplet counter-current chromatography. *Agric. Biol. Chem.* 48: 1683–1684.

Kubo, I., A. Matsumoto, F.J. Hanke and J.F. Ayafor. 1985. Analytical droplet counter-current chromatography isolation of 20-hydroxyecdysone from *Vitex thyrsiflora* (Verbenaceae). *J. Chromatog.* 321: 246–248.

Lafont, R. and D.H.S. Horn. 1989. Phytoecdysteroids: Structures and occurrence. In: *Ecdysone. From Chemistry to Mode of Action* J. Koolman (ed.), Georg Thieme Verlag, Stuttgart, pp. 39–64.

Litvinenko, V.I., I.G. Zoz and V.S. Sokolov. 1970. Chemotaxonomische Untersuchungen zur Unterfamilie Ajugoideae Benth. der Lamiaceae Lindley. *Planta Medica* 3: 242–254.

Matsuoka, T., S. Iami, M. Sakai and M. Kamada. 1969. Studies on phytoecdysones. *Ann. Rep. Takeda Res. Labs* 28: 221–271.

Melé, E., J. Messeguer, R. Gabarra, J. Tomás, J. Coll and F. Camps. 1992. *In vitro* bioassay for the effect of *Ajuga reptans* phytoecdysteroids on *Trialeurodes vaporariorum* larval development. *Entomol. Exp. Appl.* 62: 163–168.

Nakanishi, K., M. Koreeda, S. Sasaki, M.L. Chang and H.Y. Hhu. 1966.The structure of ponasterone A, an insect moulting hormone from the leaves of *Podocarpus Nakaii* Hay. *Chem. Comm.* 915–917.

Pratt, G.E. and W.S. Bowers. 1977. Precocene II inhibits juvenile hormone biosynthesis by cockroach *Corpora allata in vitro. Nature* 265: 548–550.

Russell, G.B., P.G. Fenemore, D.H.S. Horn and E.J. Middleton. 1972. Insect moulting hormones: The phytoecdysones of *Dacrydium intermedium. Aust. J. Chem.* 25: 1935–1941.

Saatov, Z., M.B. Gorovits, S. Melibaev and N.K. Abubakirov. 1986. Phytoecdysteroids from plants of the genus *Silene.* IX. *Khim. Prir. Soedin.* 77–80.

Shimomura, H., Y. Sashida, K. Ogawa and Y. Iitaka. 1981. Ajugamarin, a new bitter diterpene from *Ajuga nipponensis* Makino. *Tetrahedron Lett.* 22: 1367–1368.

Sláma, K. 1979. Insect hormones and antihormones in plants. In: *Herbivores: Their Interaction with Secondary Plant Metabolites.* G.A. Rosental and D.H. Janzen (eds.), Academic Press, New York, pp. 683–700.

Sláma, K. and C.M. Williams. 1966. 'Paper factor' as an inhibitor of the embryonic development of the European bug, *Pyrrhocoris apterus. Nature* 210: 329–330.

Spencer, K.A. 1990. *Host Specialization in the World Agromyzidae* (Diptera). Kluwer Acad. Publ., Dordrecht.

Tomás, J., F. Camps, E. Claveria, J. Coll, E. Melé and J. Messeguer. 1992. Composition and localization of phytoecdysteroid in *Ajuga reptans* L. *in vivo* and *in vitro* cultures. *Phytochemistry* 31: 1585–1591.

Usmanov, B.Z., M.B. Gorovits and N.K. Abubakirov. 1975. Phytoecdysones of *Ajuga turkestanica.* III. Structure of turkesterone. *Khim. Prir. Soedin.* 466–470.

Yatsyuk, J.K. and G.M. Segel. 1970. On the isolation of ecdysterone. *Khim. Prir. Soedin.* 6: 281.

119

Biological Activity of the Constituents of *Dalbergia sissoo* and Their Derivatives

P. VENKAT KUMAR, D.B. SAXENA AND G.T. GUJAR*

Division of Agricultural Chemicals
*Division of Entomology
Indian Agricultural Research Institute
New Delhi 110 012, India

ABSTRACT

Seven compounds—dalbergiphenol, dalbergiquinone, dalbergichromene, dalbergin, isodalbergin, methyl dalbergin, and melannein—along with an unidentified ester of dalbergiphenol with a higher fatty acid were isolated. Melannein has been isolated for the first time from *Dalbergia sissoo*. Dalbergiphenol was derivatized to methyl and ethyl ether and acetate and also isomerized to isodalbergiphenol with a view to study structure–activity relationship. These compounds were identified and characterised by infrared, NMR, and mass spectroscopy. Bioassay of these major compounds was conducted on the neonate nymphs of *Dysdercus koenigii* and *D. similis*. Melannein and dalbergiphenol showed highest toxicity towards both insect species. When tested for anti-fungal activity, dalbergiphenol showed maximum inhibition of the fungi *Alternaria alternata, Rhizoctonia bataticola*, and *Sclerotium rolfsii*.

INTRODUCTION

In view of growing awareness of the hazards of synthetic pesticides, the development of newer pesticides from natural sources is being emphasized.

Botanical pesticides are expected to be free from harmful effects such as environmental pollution, human health hazards, pest resistance, and development of minor pests into major ones encountered with the synthetic pesticides. Moreover, they serve as lead chemicals to give a new hope for future pest control agents. The wood of *Dalbergia*

Contribution No. 506 of Division of Agricultural Chemicals.

sissoo, commonly known as shisham, is most valued for making furniture on account of its durability and resistance to insect and microbial attack. Although the heartwood of *D. sissoo* has been subjected to exhaustive phytochemical investigations (Ahluwalia and Seshadri, 1963; Mukerjee et al., 1971; Kulshrestha and Mukerjee, 1974), no systematic work is available on biological activity of the chemical constituents of this plant. Studies were, therfore, carried out on the insecticidal activity of the constituents against cotton stainer, *Dysdercus koenigii* and *D. similis*, and also on antifungal activity of major constituents and extracts against *Alternaria alternata, Rhizoctonia bataticola*, and *Sclerotium rolfsii*.

MATERIALS AND METHODS

The trunk of *D. sissoo* was collected from the Indian Agricultural Research Institute's campus in July 1990. The bark was removed and heartwood was chopped into small pieces. The chips (5 kg) were extracted with 95 per cent ethanol under cold and dark conditions. The solvent was removed under vacuum. The concentrate (25 g) was subjected to column chromatography over silica gel (600 g) and was eluted successively with solvents of increasing polarity starting with hexane, hexane-benzene, benzene, benzene-chloroform and chloroform. Fig. 1 shows the structures of the compounds studied. All melting points are uncorrected. Infrared spectra were recorded as solution in chloroform or carbon tetrachloride or as KBr pellet on Perkin-Elmer Model 405, Grating IR spectrophotometer. Nuclear magnetic resonance (^1H NMR) spectra were recorded on Varian 60-MHz (Model EM 3601) instrument using $CDCl_3/CCl_4$ as solvent and TMS as an internal reference. The chemical shifts are denoted in δ values. Mass spectra were recorded on Jeol-JMS D-300 instrument at 70 ev in the electron impact mode.

Major Constituents

Dalbergiphenol (Ia)

Fractions eluted with hexane-benzene mixture were dissolved in chloroform and treated with 5 per cent aqueous sodium hydroxide to isolate phenols. The aqueous portion was treated with 10 per cent hydrochloric acid to regenerate phenols and extracted with chloroform. Most of the crude product distilled as a colourless liquid, b.p. 148°–151°/0.3 mm Hg; Rf 0.7 in 10 per cent acetone in benzene; ν max (CCl_4): 3560, 2900, 1640, 1590, 1300, 1250, 1190, 1020, 900, 860, 690cm^{-1}. Elemental analysis, mass fragmentation, and NMR data of this compound and the following compounds are given in Tables 1 and 2.

Dalbergiquinone (II)

Fractions eluted with benzene were combined and refluxed with Girrard's reagent-T (0.94 g), acetic acid (1.7 ml), and ethanol (95 per cent, 7.5 ml) for 24 hrs on an oil bath. After cooling ethanol was removed under vacuum and extracted with ether. After washing the ethereal layer with water and drying over anhydrous sodium sulphate, the solvent was distilled off to give dalbergiphenol (1.6 g).

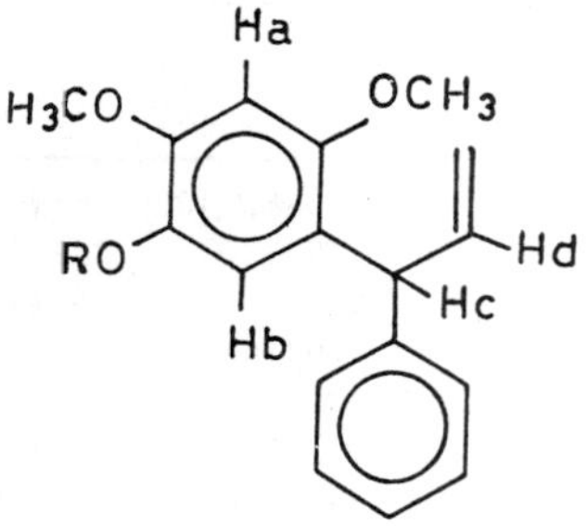

Ia R = H
Ib R = CH_3
Ic R = CH_2CH_3
Id R = $COCH_3$

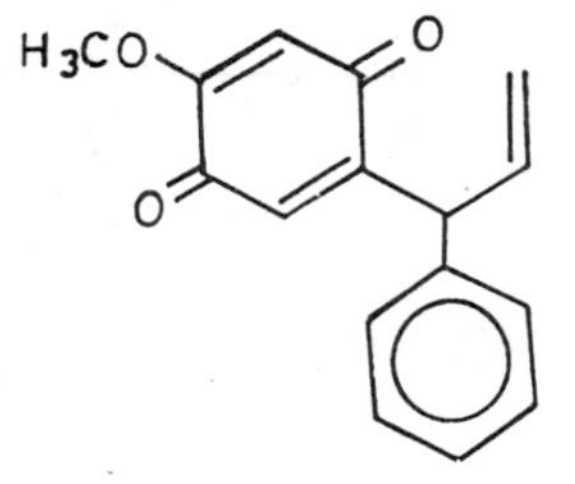

II

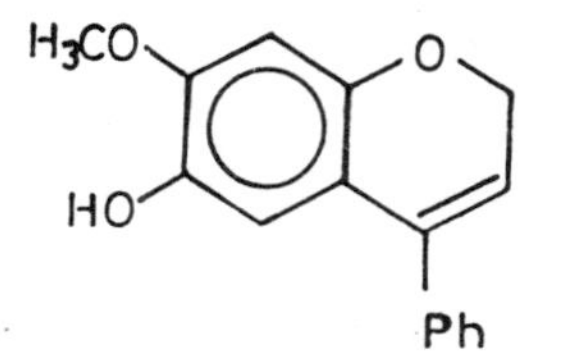

III

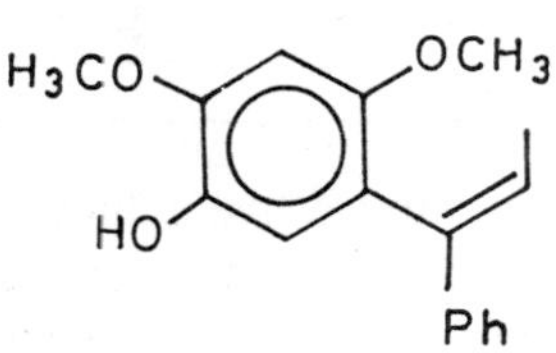

Ie

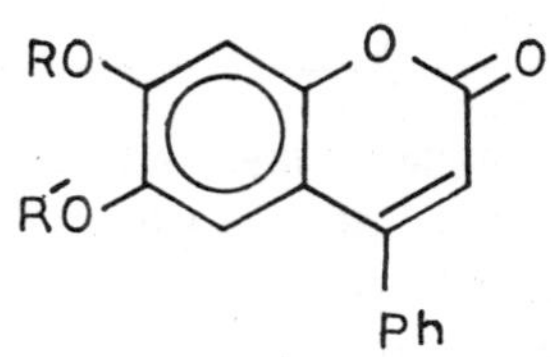

IV a R = CH_3 , R' = H
IV b R = H , R' = CH_3
IV c R = CH_3 , R' = CH_3

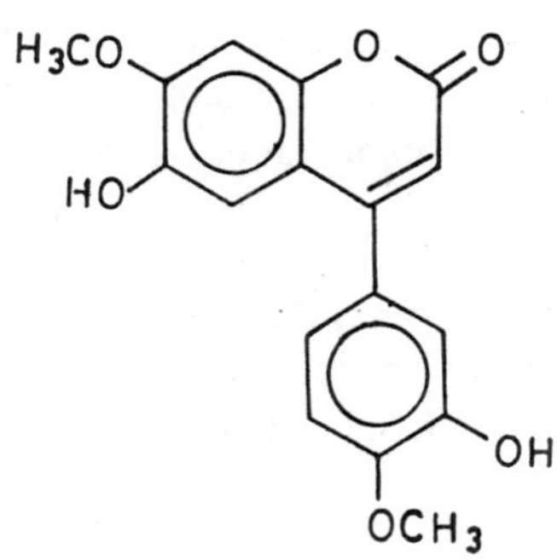

IV d

Fig. 1. Structures of the compounds studied

Table 1. Elemental anlysis (%) and mass fragmentation of chemical constituents of *D. sissoo*

Compound	Molecular formula	Calculated		Found		m/z (%)
		C	H	C	H	
Ia	$C_{17}H_{18}O_3$	75.55	6.67	75.38	6.77	270 (M^+,100), 254 (65.4), 240 (33.3), 234 (33.3), 196 (46.15), 150 (30.7) 126 (56.4)
II	$C_{16}H_{14}O_3$	75.5	5.51	75.1	5.7	255 (M^+,100), 240 (30.1), 212 (20.5), 166 (22.1), 144 (30.2), 115 (45.5), 92 (31.3), 69 (77.8)
III	$C_{16}H_{14}O_3$	75.5	5.51	75.0	5.4	
IVa	$C_{16}H_{12}O_4$	71.6	4.4	72.0	4.01	268 (M^+,100), 240 (77.7), 225 (100), 197 (7.6), 170 (4.8), 156 (11.9) 142 (11.9), 128 (19.5) 77 (18.0)
IVb	$C_{16}H_{12}O_4$	71.6	4.4	71.9	4.3	
IVc	$C_{17}H_{14}O_4$	72.34	4.96	72.29	5.1	282 (M^+,100), 254 (100), 239 (70.6), 238 (85.3), 213 (22.8), 184 (27.2)
IVd	$C_{17}H_{14}O_6$	65.0	4.5	65.2	4.3	

The aqueous layer was heated on a water bath with concentrated hydorchloric acid (1.2 ml) for 4 hrs and was extracted with solvent ether after cooling. The ethereal layer was washed with water and dried over anhydrous sodium sulphate, and the solvent was removed to give ketones (0.7 g). On cystallization from ethyl alcohol, compound II was obtained as light yellow needles, m.p. 118°–120°C; Rf 0.62 in 10 per cent acetone in benzene; νmax (KBr): 3060, 1720, 1640, 1440, 1350, 1280, 1220, 1190, 1170, 1060, 970, 910, 860cm^{-1}.

Dalbergichromene (III)

Removal of solvent from the fractions eluted with 50 per cent benzene in chloroform left yellow, needle-shaped crystalline solid, m.p. 99°–100°C; Rf 0.36 in 10 per cent acetone in benzene. νmax (nujol): 3600, 1620, 815cm^{-1}.

Dalbergin (IVa)

From the last fractions eluted from the column with chloroform, a polar compound IVa was isolated as colourless crystalline solid, m.p. 210°–216°C; Rf, 0.22 in 10 per cent acetone in benzene; νmax (KBr): 3250, 1680, 1605, 1545, 1445cm^{-1}.

Minor Constituents

Isodalbergin (IVb)

Isodalbergin (IVb) was obtained as a colourless solid during the purification of dalbergichromene (III). It was cystallized from ethanol as colourless needles, m.p. 190°–193°C; Rf, 0.45 in 10 per cent acetone in benzene; νmax (nujol): 1665cm^{-1}.

Table 2. NMR data of compounds I–IV of *D. sissoo*

Compound	Ha	Hb	Hc	Hd	= CH_2	$-OCH_3$	Ring B	Others
Ia	6.37(s)	6.67(s)	4.66 (d, J = 2 Hz)	6.08 (m)	5.07 (m)	3.5, 3.6(2s)	7.2(s)	
Ib	6.5(s)	6.6(s)	4.73 (d, J = 2 Hz)	6.05 (m)	5.33 (m)	3.6, 3.7, 3.8(3s)	7.22(s)	
Ic	6.6(s)	6.95(s)	4.76 (d, J = 2 Hz)	6.05 (m)	5.32 (m)	3.92(m, 5H, OCH_3; OCH_2), 3.8(S, OCH_3)	7.3(s)	1.4(t, J = 7 Hz) CH_3CH_2
Id	6.5(s)	6.81(s)	4.74 (d, J = 2 Hz)	6.3 (m)	5.18 (m)	3.7, 3.81(2s)	7.23(s)	2.28(s) ($COCH_3$)
Ie	6.53(s)	6.88(s)		6.4 (q, J = 7 Hz)		3.6, 3.88(2s)	7.2(s)	1.66(d, 3H, J = 7 Hz, CH_3-CH)
II	5.80(s)	6.40(s)	4.9 (d, J = 2 Hz)	6.13 (m)	5.13, 5.33 (d,d,J = 2 Hz)	3.75(s)	7.21(s)	
III	6.60(s)	6.65(s)		5.64 (t, J = 4 Hz)		3.8(s)	7.28(s)	4.78, 6.0(d,d, J = 4 Hz, -CH_2)
IVa	6.93(s)	7.03(s)		6.27(s)		4.0(s)	7.5(s)	9.4(bs, OH) collapsed on D_2 exchange
IVb	6.93(s)	7.03(s)		6.27 (s)		3.97(s)	7.5(s)	9.56(bs, OH)
IVc	6.73(s)	7.23(s)		6.20 (s)		3.73, 3.93(2s)	7.4(s)	
IVd	6.90(s)	7.03(s)		6.20 (s)		3.85, 3.90(2s)	7.0–7.28 (m, ArH)	9.3(bs, OH)

Methyldalbergin (IVc)

While dalbergichromene (III) was purified, methyl dalbergin (IVc) was obtained as a dark solid. It crystallized as brown rods from ethanol, m.p. 142°–143°C; Rf, 0.48 in 10% acetone in benzene; νmax (nujol): 1720, 1610, 1540, 1450, 1370, 1340, 1265, 1220, 1200, 1140, 1105, 1020, 995, 950, 860, 810, 765, 700cm^{-1}.

Melannein (IVd)

After isolation of dalbergin (IVa) from last fractions of column chromatography over silica gel, melannein (IVd) was obtained as light brown solid on rigorous column chromatography over silica gel. It crystallized from ethanol as yellow rhombs, m.p. 219°–220°C; Rf, 0.28 in 10 per cent acetone in benzene. It gave emerald green colour changing to permanent red with Mg/HCl. νmax (KBr): 3290, 2950, 1660, 1280, 1240, 1175, 1135, 815, 780cm^{-1}.

Derivatives

Methyl ether of Dalbergiphenol (Ib)

Dalbergiphenol (2.5 g) was dissolved in dry acetone (25 ml) and refluxed for 5 hrs with dimethyl sulphate (2 ml) and anhydrous potassium carbonate (2 g). After the solvent was distilled off, the residue which was left overnight in water was extracted with ether. The ether layer was washed with water and dried over anhydrous sodium sulphate, and the solvent was removed to give yellowish plates (1.92 g) from ethanol, m.p. 58°–59°C.

Ethyl ether of Dalbergiphenol (Ic)

Dalbergiphenol (2.5 g) solution in dry acetone (25 ml) was refluxed with diethyl sulphate (2 ml) for 14 hrs and worked up as above to give a mixture which was subjected to column chromatography over silica gel. Fractions eluted with 50 per cent hexane in benzene on removal of solvent gave yellow plates (2.8 g) from ethanol, m.p. 54°–55°C.

Acetate of Dalbergiphenol (Id)

Acetate of dalbergiphenol was prepared by the reaction of dalbergiphenol (1.25 g) with acetic anhydride (7.5 ml) and pyridine (5 ml) under anhydrous condition at room temperature for 40 hrs. Ice cold water was added to the reaction mixture. It was extracted with ether, washed with water, and dried over anhydrous sodium sulphate, and the solvent was removed to give a solid. It crystallized from ethanol as colourless plates (1.3 g), m.p. 96°C.

Isodalbergiphenol (Ie)

Sodium tertiary butoxide (5 g) was mixed with dalbergiphenol (1.5 g) and stirred in dimethyl sulphoxide at 70°C for 3 hrs. The mixture was poured over ice and hydrochloric acid, contained in a beaker, and then extracted with ether. The ethereal layer was washed with aqueous sodium bicarbonate and ether was evaporated. The residue (0.9 g) was chromatographed over silica gel (40 g). The fractions eluted with benzene gave isodalbergiphenol which was crystallized from ethanol (1.02 g), m.p. 71°–72°C.

Insecticidal Activity

The chemical constituents from heartwood of *D. sissoo* were assayed against the neonate first-instar nymphs of red cotton stainer, *Dysdercus koenigii* and *D. similis* (Heteroptera, Pyrrhocoridae). Red cotton stainer, used in the present studies were reared at 26°C and 50–60 per cent relative humidity in the Division of Entomology. Adults and nymphs were fed on soaked cotton seeds which were changed daily to maintain cleanliness. Adult females laid eggs after two days of mating. Eggs hatched within a week. The first-instar nymphs were initially cream-coloured and turned red after 24 hrs. First-instar nymphs 24–48 hrs old were used to test toxicity of chemicals by topical application with the help of electrically operated microapplicator (M 1025, Instrumentation Specialities Co. Inc., Nebraska, USA) fitted with 1 ml tuberculin syringe 27 gauge needle. The five to six different concentrations of test chemical were prepared in acetone. About 20–30 insects were individually treated with 1 μl of solution applied on dorsum. Thus insects were treated at five to six different doses along with acetone as control. The solution was allowed to dry before the insects were transferred to jars. They were allowed to feed on soaked cotton during the experimental period. Observations on nymphal mortality were taken after 24 hrs. The data were then subjected to probit analysis using computer programme for calculation of LD_{50} (μg/insect) values.

Bioassay for Fungicidal Activity

Three compounds—dalbergiphenol (Ia), dalbergiquinone (II), and dalbergin (IVa)—along with hexane, acetone extracts, and hexane portion of acetone extract were assessed for fungicidal activity against *A. alternata*, *R. bataticola*, and *S. rolfsii* by poisoned food technique (Nene and Thapliyal, 1979).

The required quantity (260 mg) of the test chemical was weighed and dissolved in acetone. Requisite concentration of this solution was added to Potato-Dextrose-Agar medium in a conical flask and mixed well so as to get different concentrations. The contents of each conical flask were poured equally into two petri dishes. The medium was allowed to set under UV light in a laminar flow chamber and 55 mm discs of the fungi were transferred aseptically to the petri dishes. The petri dishes were kept in a B.O.D. incubator at 25°C and incubated for seven days. The diameter of the fungal colony in each petri dish was measured after seven days. The fungicidal activity was expressed as per cent inhibition over control after correction.

RESULTS AND DISCUSSION

Characterization of Chemical Constituents

Dalbergiphenol was analysed for $C_{17}H_{18}O_3$ (M^+, m/z 270). The IR spectrum showed the presence of vinylic double bond (1640, 900cm^{-1}) besides phenolic hydroxyl group (3560cm^{-1}) which was supported by its NMR spectrum displaying multiplets at 5.07 and

6.08 for two protons and one proton respectively besides a doublet at 4.66 (J = 2Hz) for a single proton. The above peaks in conjugation with three singlets at 6.37, 6.67, and 7.20 integrating for one, one, and five protons respectively suggested it to be a diphenyl compound with a vinyl group. Two singlets at 3.5 and 3.6 integrating for three protons each suggested the presence of two methoxyl groups. On the basis of the spectral data and its molecular formula the compound was identified as dalbergiphenol reported earlier by Kulshrestha and Mukerjee (1974).

During the purification of dalbergiphenol, another compound, a viscous liquid, was obtained from neutral portion. It had high molecular weight (M^+, m/z 528) but the fragmentation pattern from m/z 270 onwards was similar to that of dalbergiphenol except for difference in the intensity of various peaks. Its IR spectrum did not exhibit a peak around 3560cm^{-1} for a phenolic hydroxyl group. Rather strong bands observed at 1740 and 1235cm^{-1} in addition to 3060, 1630, 910cm^{-1} (for vinylic group also observed in case of dalbergiphenol) and 745cm^{-1} for unsubstituted phenyl group as also observed in dalbergiphenol indicated it to be an ester. From the above spectral data in conjugation with absorption peaks displayed at 1.2 and 2.2 in addition to all other peaks resembling dalbergiphenol, the least polar compound has been presumed to be an ester of higher fatty acid with dalbergiphenol. Complete structure could not be assigned because the material was too small to allow microsaponification.

The yellow colour coupled with a peak at 1720cm^{-1} in IR spectrum indicated II, $C_{16}H_{14}O_3$, to be a quinone. Therefore, m.p. and spectral data of this quinone were compared with those reported earlier for dlabergiquinone (II) (Ahluwalia and Seshadri, 1963). The two were identical. Finally, the structure was confirmed with mass fragmentation pattern.

Two doublets at 4.78 and 6.0 (J = 4 Hz) for one proton each and a triplet at 5.64 (J = 4 Hz) for one proton in its NMR spectrum suggested the presence of a chromene structure in compound III. Therefore, the spectral data of this compound was compared with those reported for dalbergichromene. A singlet integrating for three protons at 3.8 for one methoxyl group, a singlet each at 6.60 and 6.65 integrating for single aromatic proton, a singlet recorded at 7.28 for five protons in its NMR spectrum in conjugation with a peak at 3600cm^{-1} in its IR spectrum for phenolic hydroxyl group and molecular formula $C_{16}H_{14}O_3$ proved it to be dalbergichromene reported earlier by Mukerjee et al. (1971) on the basis of degradation reaction.

A peak appearing at 3250cm^{-1} in the IR spectrum of IVa and a broad band appearing at 9.4 in NMR spectrum which collapsed on exchange with D_2O suggested the presence of a phenolic hydroxyl group. A peak at 1680cm^{-1} in its IR spectrum, a singlet displayed at 6.27 for single proton, and the positive colour test with Mg/HCl led to its identification as 4-phenylcoumarin. On the basis of NMR spectrum which showed presence of a methoxyl group (singlet at 4.0), two aromatic protons Ha and Hb which appeared as singlets at 6.93, 7.03 for one proton each and five aromatic protons appearing as singlets at 7.5, IVa, $C_{16}H_{12}O_4$, was thought to be dalbergin, a 4-phenylcoumarin, reported earlier (Ahluwalia and Seshadri, 1957). The m.p. and spectral data of this compound and those reported for dalbergin were found to be identical. The structure was confirmed by mass fragmentation pattern. Molecular ion peak (M^+, m/z 268) was most abundant, normally observed in case of coumarins, and loss of neutral molecule of carbon monoxide (M^+–CO; m/z 240) followed by loss of m/z 15 for a methyl group m/z 225) favoured methoxyl group at C_7 position, presumably due to relative stability of quinonoid oxonium ion. This species further fragments by two consecutive losses of carbon monoxide, ultimately forming a phenyl ion (m/z 77). The above fragmentation is in line with those reported for 7-methyoxy coumarins (Porter and Baldas, 1971).

A peak at 1665cm^{-1} in IR spectrum of IVb for a carbonyl group and a singlet at 6.27 suggested IVb to be a 4-phenylcoumarin. The other peaks appearing as singlet at 3.97 for three protons of one methoxyl group, two singlets at 6.93 and 7.03 for two aromatic protons (Ha and Hb) and a singlet at 7.5 for five aromatic protons, besides a phenolic group at 9.56 and the molecular formula $C_{16}H_{12}O_4$, suggested IVb to be dalbergin. Therefore, the m.p. and spectral data of IVb were compared with those of dalbergin (IVa). The two were not superimposable. It was found to be an isomer of dalbergin, isodalbergin with methoxyl at C_6 and hydroxyl at C_7 position (Mukerjee et al., 1971).

The IR spectrum showing absorption at 1720cm^{-1} and positive colour test with Mg/HCl suggested IVc to be a 4-phenylcoumarin. Presence of a singlet at 6.2 for single proton also supported it to be a 4-phenylcoumarin. Considering the presence of two methoxyl groups, which appeared as two singlets at 3.73 and 3.93 integrating for three protons each, two singlets at 6.73 and 7.23 for two aromatic protons Ha and Hb and a singlet at 7.40 for five aromatic protons in conjugation with its molecular formula ($C_{17}H_{14}O_4$, M^+, m/z 282) , this compound was suspected to be methyldalbergin prepared earlier by Ahluwalia and Seshadri (1957) from dalbergin as its methyl ether. Finally, the structure was confirmed by its mass fragmentation pattern. Molecular ion peak was most abundant, normally observed in case of coumarins.

Positive colour test with Mg/HCl, peaks at 1660 and 815cm^{-1} in IR spectrum, and a singlet at 6.2 in NMR spectrum indicated IVd to be a 4-phenylcoumarin. Other features recorded from its NMR spectrum indicate the presence of two methoxyl groups (singlets at 3.85 and 3.9 integrating for three protons each) besides five aromatic protons (a multiplet from 7.01 to 7.28) and a phenolic hydroxyl, 3290cm^{-1}, 9.3 (bs, lH, –OH, collapsed on exchange with D_2O). The above spectral data and molecular formula, $C_{17}H_{14}O_6$, did not compare with any of the 4-phenylcoumarins reported from *D. sissoo* in literature. Further perusal of literature in respect of other *Dalbergia* sp. showed the presence of 4-phenylcoumarin, melannein with above characteristics in *D. melanoxylon* (Donelly et al., 1968). Hence, structure IVd was assigned to this minor component.

NMR spectrum displaying three singlets at 3.6, 3.7, and 3.8 confirmed the formation of methyl ether (Ib). A triplet at 1.40 (J = 7 Hz) integrating for three methyl protons and a multiplet at 3.92 for five protons including two protons of OCH_2 indicated the formation of ethyl ether (Ic). A singlet at 2.28 for three protons in NMR spectrum confirmed the formation of an acetate (Id).

A doublet at 1.66 (J = 7 Hz) integrating for three protons of methyl group and a quartet at 6.4 (J = 7 Hz) integrating for single proton in place of two multiplets appearing at 5.07 for two protons of = CH_2 group and at 6.08 for Hd proton in the NMR spectrum of Ie, confirmed the isomerization of vinyl side chain of dalbergiphenol into propenyl side chain. Value of absorption of other peaks remained unchanged in isodalbergiphenol (Ie).

Insecticidal Activity

The results of toxicity of constituents of *D. sissoo* to the cotton stainer showed a wide variation in toxicity of compounds ranging from LD_{50} 3.46 μg/insect to 38.61 μg/insect (Table 3). The decreasing order of toxicity of products of *D. sissoo* was melannein, isodalbergiphenol, dalbergiphenol, methyl ether of dalbergiphenol, ehtyl ether of dalbergiphenol, dalbergin and dalbergiquinone.

Melannein and isodalbergiphenol did not show any significant differences in their activity against *D. koenigii*. However, both of them showed significantly higher toxicity in comparison to the remaining five compounds isolated from *D. sissoo*. Among the derivatives of dalbergiphenol, isodalbergiphenol showed significantly higher toxicity than methyl and ethyl ethers of dalbergiphenol. However, methylether of dalbergiphenol, ethylether of dalbergiphenol, and dalbergiphenol were at par in their toxicity.

The decreasing order of toxicity of compounds of *D. sissoo* against *D. similis* was melannein, isodalbergiphenol, methyl ether of dalbergiphenol, dalbergiphenol, ethyl ether of dalbergiphenol, dalbergin, and dalbergiquinone. Melannein, isodalbergiphenol, methylether of dalbergiphenol, and dalbergiphenol were at par in their toxicity. However, their toxicities were significantly higher than those of ethylether of dalbergiphenol, lalbergin, and dalbergiquinone. Dalbergiphenol, ethyl ether of dalbergiphenol, dalbergin, and dalbergiquinone were at par in their toxicity to *D. koenigii*.

Both insect species did not show significant difference in their susceptibilities towards most compounds of *D. sissoo* except for dalbergiquinone, which showed higher toxicity to *D. similis* than to *D. koenigii*.

The coumarin constituents of *D. sissoo*—dalbergin and melannein—were insecticidally active, melannein being more active. Simple coumarin was earlier shown to be a phagostimulant and a deterrent and furano-coumarins such as xanthotoxin and angelicin were well known for their insecticidal activity (Towers, 1980). The role of *Dalbergia* neoflavanoids in imparting defence against insect pest attack to the heartwood was earlier reported by Marini Bettolo (1983).

Table 3. Toxicity of constituents of *D. sissoo* against *Dysdercus* spp.

Constituents	*Dysdercus koenigii*			*Dysdercus similis*		
	LD_{50} (μg/insect)	X^2 (d.f.)	Fiducial limits	LD_{50} (μg/insect)	X^2 (d.f.)	Fiducial limits
Melannein	3.46	3.345(3)	2.18–4.93	6.16	1.07(5)	4.15– 9.61
Isodalbergiphenol	4.54	1.34 (3)	3.20–4.23	8.26	0.48(3)	5.86–12.31
Dalbergiphenol	10.67	0.32 (3)	7.78–16.49	12.69	2.67(3)	9.10–20.98
Mehylether of dalbergiphenol	11.91	0.79 (3)	6.12–85.10	10.53	0.90(4)	6.18–28.41
Ethylether of dalbergiphenol	15.27	3.06 (3)	10.23–32.73	13.96	3.50(4)	8.94–32.10
Dalbergin	16.90	0.32 (3)	11.89–35.29	14.81	3.22(5)	9.36–37.10
Dalbergiquinone	38.61	3.57 (3)	20.79–50.86	17.37	0.45(3)	10.82–55.99

Phenolics have been reported to have a wide range of biological activity against insect species. The simple phenolics like β-asarone or eugenol possess behaviour-modifying activity in addition to growth-inhibiting and toxicity effects (Rameshbabu et al., 1991). The presence of phenolics like dalbergiphenol along with other similar compounds could also be a contributing factor for the resistance of heartwood to the attack of insect pests. Isodalbergiphenol, an isomer of dalbergiphenol, was found to be twice as effective as dalbergiphenol. This compound showed close resemblance with the basic structure of β-asarone, wherein toxicity was attributed to *cis* configuration of propenyl side chain (Koul et al., 1991).

The low toxicity of dalbergiquinone could be due to the presence of 1, 4-benzoquinone system which has also been found to be less toxic than 1, 4-naphthoquinone system found in plumbagin (Hassanali and Lwande, 1989).

The wood of *Dalbergia retusa* has been found to inhibit the growth of shipworms (Santhakumaran, 1991) and teredinids, which was attributed to presence of obtusaquinone (Turner, 1974; Bultman et al., 1978), thereby supporting our results on insecticidal activity of constituents of *D. sissoo*.

Fungicidal activity

Perusal of the data on fungicidal activity (Table 4) revealed that hexane portion of the acetone extract and dalbergiphenol (Ia) exhibited high fungicidal activity against *R.*

Table 4. Fungicidal activity of extracts and other constituents of *D. sissoo*

Extract/chemical constituent	Concentration (ppm)	Per cent inhibition of		
		R. bataticola	*S. rolfsii*	*A. alternata*
Hexane extract	1000	90	100	100
	500	85	87	80
	250	62	75	73
	125	50	44	60
	62.5	37	43	50
Hexane portion of acetone extract	1000	100	87	75
	500	87	69	65
	250	75	45	50
	125	62	22	34
	62.5	50	16	21
Acetone extract	1000	87	54	60
	500	77	40	55
	250	50	12	43
Dalbergiphenol	1000	87	55	70
	500	83	46	57
	250	76	25	36
	125	62		
	62.5	50		
Dalbergiquinone	1000	71	50	61
	500	66	41	54
	250	52	25	42
Dalbergin	1000		52	
	500		36	
	250		17	

bataticola with LC_{50} of about 39.8 and 50.1 ppm respectively. This order of toxicity was followed by hexane extract showing LC_{50} of about 63.0 ppm. However, against *A. alternata* and *S. rolfsii*, hexane extract was the most effective of all constituents and extracts with LC_{50} of about 44.7 and 56.2 ppm respectively. The high activity of hexane portion of actone extract could be largely due to the presence of dalbergiphenol, a chief constituent. Phenolics have earlier been implicated in imparting resistance to the attack of insect pests and pathogens by Marini Bettolo (1983).

ACKNOWLEDGEMENT

One of us (P.V.) is grateful to the Indian Agricultural Research Institute, New Delhi, for the award of Junior Fellowship.

REFERENCES

Ahluwalia, V.K. and T.R. Seshadri. 1957. Constitution of dalbergin. Part II. *J. Chem. Soc.* 41: 970–972.

Ahluwalia, V.K. and T.R. Seshadri. 1963. Dalbergiquinone from the heartwood of *Dalbergia sissoo*. *Curr. Sci.* 34: 955.

Bultman, J.D., E.C. Haderlie and J.R. DePalma. 1978. *Proc. 4th Int. Biodeterioration Symp.*, Berlin. T.A. Oxley, G. Becker and D. Allsopp (eds.), Pitman Publishing Ltd., London, 199 pp.

Donelly, B.J., D.M.X. Donelly and A.M. Sullivan. 1968. The occurrence of melannein in the genus *Dalbergia*. *Tetrahedron* 24: 2617–2622.

Hassanali, A. and W. Lwande. 1989. Antipest secondary metabolites from African plants. Insecticides of plant origin. *ACS Symp. Ser.* 387: 121–140.

Koul, O., M.J. Smirle and M.B. Isman. 1991. Asarone from *Acorus calamus* L. oil: Their effect on feeding behaviour and dietary utilization in *Peridroma saucia*. *J. Chem. Ecol.* 16: 1911–1920.

Kulshrestha, S.K. and S.K. Mukerjee. 1974. Dalbergiphenol, a new constituent of the heartwood of *Dalbergia* species. *Indian J. Chem.* 2: 10–14.

Marini Bettolo, G.B. 1983. The role of natural products in plant, insects and plant, fungi interaction. In: *Natural Products for Innovative Pest Management*. D.L. Whitehead and W.S. Bowers (eds.), Pergamon Press, Oxford, pp. 187–222.

Mukerjee, S.K., T. Saroja and T.R. Seshadri. 1971. Dalbergichromene. A new neoflavonoid from stem bark and heartwood of *Dalbergia sissoo*. *Tetrahedron* 27: 803.

Nene, Y.L. and P.N. Thapliyal 1979. *Fungicides in Plant Diseases Control*. Oxford and IBH Publ. Co. Pvt. Ltd., New Delhi, 413 pp.

Porter, Q.N. and J. Baldas. 1971. Mass spectroscopy of heterocyclic compounds. Wiley Eastern, 147 pp.

Rameshbabu, B., D.B. Saxena and G.T. Gujar. 1991. Asarone derivatives as contact insecticides against red cotton bug (*Dysdercus koenigii* F.). *Pesticide Res. J.* 3: 145–150.

Santhakumaran, L.N. 1991. Natural resistance of twelve timber species to marine borers in Goa waters. *Proc. Nat. Acad. Sci. India* LXI: 49–58.

Towers, G.H.N. 1980. Photosensitizers from plants and their photodynamic action. *Phytochemistry* 6: 183–202.

Turner, R.D. 1974. *Proc. Workshop on the Biodeterioration of Tropical Woods: Chemical Basis for Natural Resistance*. J.D. Bultman (ed.), Naval Research Laboratory, Washington, DC.

120

Effect of *Acorus calamus* Oil and Its Main Compound β-Asarone on the Larger Corn Borer, *Prostephanus truncatus* (Horn), the Main Pest of Corn in Central Africa

GERHARD H. SCHMIDT AND MARTIN STRELOKE

Department of Zoology–Entomology, Faculty of Biology, University of Hannover
Herrenhäuser Strasse 2, D-3000 Hannover 21, Germany

ABSTRACT

The essential oil of Indian *Acorus calamus* (L.) is very effective in stored-product pest control. Its ingredient, β-asarone, has toxic and sterilizing effects. *Prostephanus truncatus* (Horn) was imported to Africa in the 1970s from Central America. Now, the beetle is widespread in Central Africa, where about 30 per cent of the maize harvested is destroyed. A regulation of its populations to protect the environment is necessary. For that purpose *A. calamus* oil and β-asarone were tested. For direct treatment, corn grains were covered with acetone solutions. After evaporation of the solvent, the grains were divided into portions and infested with beetles. Additionally, grains were treated with rhizome powder of *A. calamus*. The damage was calculated from the weight of corn dust received from the infested grains. A treatment with only 0.01 per cent oil reduced the grain loss within 21 days significantly; the amount of feeding was lowered to 50 per cent compared to the control. After treatment with β-asarone vapours a strong effect was observed first, beyond 21st day. Very important was the temperature; an effect was observed only above 30°C. An admixture of *A. calamus* rhizome powder to the corn reduced feeding activity by up to 83 per cent.

INTRODUCTION

Acorus calamus (L.) (Araceae) is wide spread in Asia, North America, and Europe. The plant prefers surroundings of lakes with swampy or marshy habitats. There are three caryotypes of *A. calamus* (Roest, 1978, 1979; Roest and Bos, 1979; Cavazza, 1976; Streloke et al., 1989).

The diploid caryotype (2n = 24) grows in North America and in special regions of Asia (Siberia) and has rhizomes very poor in the most effective ingredient, β-asarone.

The triploid caryotype (3n = 36) is present in Central Europe and Kashmir, and the β-asarone content of its rhizomes varies from 9 to 13 per cent.

The tetraploid caryotype (4n = 48) was found in India, East Asia, and Japan. The essential oil of the rhizomes consists 70–96 per cent of β-asarone.

The essential oil obtained by steam distillation from Indian *A. calamus* rhizomes shows pronounced insecticidal properties. The first systematic research seems to have been conducted at the Agricultural College and Research Institute at Coimbatore, India, from 1938 onwards (Subramaniam, 1942). There is an extensive literature covering the whole spectrum of the insecticidal activity of *A. calamus* rhizomes. The last review was published by Saxena and Koul (1978).

Toxic and sterilizing effects of the vapours of the essential oil of rhizome of *A. calamus* from India on various insects have been reported by Saxena and Mathur (1976), Koul et al. (1977a, b), Saxena et al. (1976, 1977a, b), Tikku et al. (1978), and Schmidt and Borchers (1981). It can be used as chemosterilant on both sexes of insects.

In more recent studies, El-Nahal et al. (1989) reported on the toxic effects of the oil vapours on the adults and Risha et al. (1990) on the eggs of *Sitophilus granarius* (L.), *S. oryzae* (L.) and *Callosobruchus chinensis* (L.); larvae and pupae did not show any appreciable susceptibility to the oil vapours. Schmidt et al. (1991) summarized that the number of offspring emerging from adult diets used during and after treatment with *A. calamus* vaopours against these stored-product beetles was considerably lower than that in the respective controls. This response was correlated to the increase of exposure time rather than to the increase in dose.

The insecticidal effect of the essential oil was always due to the ingredient β-asarone (cis-isomer) (Fig. 1). In many insects a repellent effect was also reported (Banerji et al., 1982; Muckensturm et al., 1982; Tewari and Krishna Moorthy, 1985; Koul, 1987; Koul et al., 1990). In contrast to these observations, Jacobson et al. (1976) found attractant effects of Indian *A. calamus* oil and β-asarone to *Ceratitis capitata* (Wiedemann), *Dacus cucurbita* Coquillett, and *D. dorsalis* Hendel. α-Asarone (trans isomer, Fig. 1) was attractive to *Psilla rosae* F. (Guerin et al., 1983) and repellent to *Peridroma saucia* (Huebner) (Koul et al., 1990).

The aim of this research work was to find out whether Indian *A. calamus* oil or β-asarone can be used for the control of the larger corn Borer, *Prostephanus truncatus* (Horn), the main storage pest in the tropical region of Central Africa.

MATERIALS AND METHODS

Prostephanus truncatus (Horn) was introduced to Tanzania and later on to West Africa in the 1970s from Central America (Pierce and Schmidt, 1992). Recently the beetle was found in India (Verma and Lal, 1987). In Africa, the corn reduction due to the beetle was calculated at 30–35 per cent (Laborius et al., 1985). *Prostephanus truncatus* may be controlled by methods normally used in stored product protection, like conventional insecticides (permethrin and pirimiphos-methyl) (Golob et al., 1985). But such pesticides are very expensive for developing countries like Tanzania and Togo. Therefore, a search was started for available plant compounds with strong insecticidal effects.

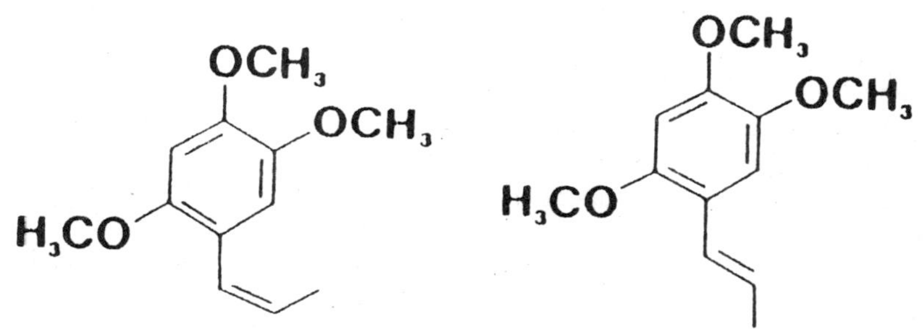

Fig. 1. Structure of β-arasone and α-arasone

Rearing of *P. truncatus*

The test animals were received from a breeding stock in the Institute of Stored-Product Protection of Biologische Bundesanstalt, Berlin-Dahlem. At Hannover, the rearing occurred in 1.5 l glasses filled half with corn grains and infested with about 200 adult beetles each. When the corn was destroyed, 200 adults were removed and given the same amount of fresh corn grains. This procedure was repeated as often as needed. The rearing was carried out in a dark incubator at 30°C and 75 per cent relative humidity.

Formulation of the Essential Oil

The *A. calamus* oil used for the tests was obtained from dried and pulverized rhizomes of Indian plants by steam distillation. The rhizomes were commercially available from Fa. Caelo, Hilden, Germany. The oil contained only β-asarone as effective substance, an oily liquid. A two-year-old product from India, received from B.P. Saxena, RRL Jammu-Tawi, showed a β-asarone content of only about 80 per cent, a little higher than commercially available β-asarone; the α-asarone content was relatively high. The trans-isomer is solved in the commercial β-asarone. In the *A. calamus* oil, the latter was enriched in time by isomerization. Acetone was used as solvent for β-asarone and *A. calamus* oil. The control animals were treated with the solvent only.

Toxicological Experiments

The tests were carried out using adult beetles of various age. Twenty beetles (50 beetles in experiments with oil vapours) were exposed to 50 g corn for 21 to 42 days. The experiments with corn grains treated with the essential oil of *A. calamus*, commercial β-asarone, and rhizome powder were carried out in closed, commercially available honey glasses of 400 ml volume. In contrast, 1.5 l glasses were used for the experiments with oil vapours. A filter paper (55 mm diameter) with 45 mg β-asarone which was pipetted dropwise with an Eppendorf pipette around the middle of the filter paper disc was hung in the screw-cap glass jar and closed. On the bottom of the bigger glass an open honey glass of 400 ml was exposed in an over-saturated NaCl solution containing corn and beetles as mentioned before, so that the relative humidity was 75 per cent inside the jar.

For the direct treatment 200 g corn grains were filled into a 400 ml honey jar and treated with the active substance dissolved in 150 ml acetone. When the upper layer of the corn was dried up, the whole content was put in a petri dish (13.7 cm diameter) from which the acetone evaporated overnight. Then, the corn was divided into 50 g portions each put into a clean honey jar and exposed to 20 beetles.

The rhizome powder was added to the corn in various amounts and mixed by shaking.

The experiments were carried out at 30°C and 75 per cent relative humidity in the dark. The damage of the corn grains by the beetles was measured by weighing the received corn dust and sieving of the corn-beetle mixture for 0.5 min. (pore diameter 0.71 mm). In the rhizome powder tests, the weight of the corn dust was corrected by the rhizome powder added before.

Statistics

For each treatment, control experiments were carried out with untreated insects and an analysis of variance was applied.

RESULTS

Direct Treatment of Corn Grains

Figure 2 shows the results of experiments with two samples of *A. calamus* oil and commercial β-asarone. The differences between the various test groups are significant ($p < 0.01$). The differences between the amounts of the active compound of the formulations tested were not significant concerning the doses and amount of corn dust. There was no direct influence of the content of β-asarone. The content of β-asarone of the *A. calamus* oil distilled at Hannover was highest (around 98 per cent) and little more active than older Indian oil (around 80 per cent β-asarone) and the commercial β-asarone (73 per cent). Even the treatment of the corn with the lowest amounts of 0.01 and 0.025 per cent (w/w) of the oils showed a distinct effect within 21 days. In the lowest treatments of 0.01 per cent the reduction of corn dust was still significantly lower than in the control.

In the period of 21 to 42 days the amount of corn dust increased to only 11 per cent in the treatment with 0.1 per cent oil and in that with 0.01 per cent oil to 87 per cent, which was significantly lower than in the control (158 per cent).

Effect of Vapours of Commercial β-Asarone

Figure 3 demonstrates the toxic effect of vapours of the active compound during different periods of treatment. After 21 days of treatment, there was no difference between treatment and control, but after 42 days the differences were highly significant (*t* test). Between 21 and 42 days an increase of more than 100 per cent corn dust (related to that after 21 days) was obtained in the control, but only 22 per cent in the experiments

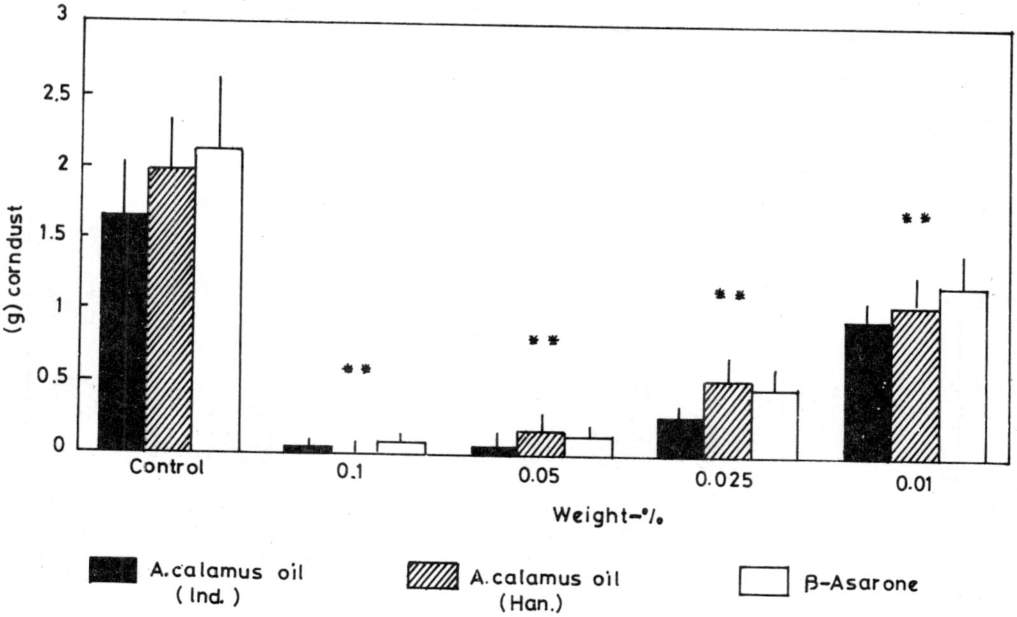

Fig. 2. Reproduction of corn dust due to the feeding activity of *P. truncatus* on corn grains directly treated with *A. calamus* oil and commercial β-asarone during a period of 21 days; $^{**}p < 0.01$ for significant differences of all three bars in comparison to the control.

with β-asarone treatment. Also, commercial β-asarone vapours have a time-related reducing effect on the feeding activity of *P. truncatus*.

To exclude the possibility that small drops of β-asarone dripping down from the filter paper affixed to the inside of the screw-cap glass jar would simulate a vapour effect, the test vessels were laid on their sides (rectangle turned). In this position the small drops drip on the glass wall without touching the corn or the beetles. The results were the same as received from normal position of the test glasses (Fig. 4). There was no drop effect.

Because the rim of the petri dish containing the corn and the beetles was on the same level as the highest border of the filter paper treated with the active substance, the vapours of β-asarone should distribute evenly to the glass volume. No heavy vapour was found which may not pass the rim of the petri dish.

These results were recorded at a temperature of 30°C. In further experiments the beetles were tested at 25°C (Fig. 5). After 42 days of treatment the amount of corn dust was lower than at 30°C in all cases, and there was no difference from the control at this lower temperature.

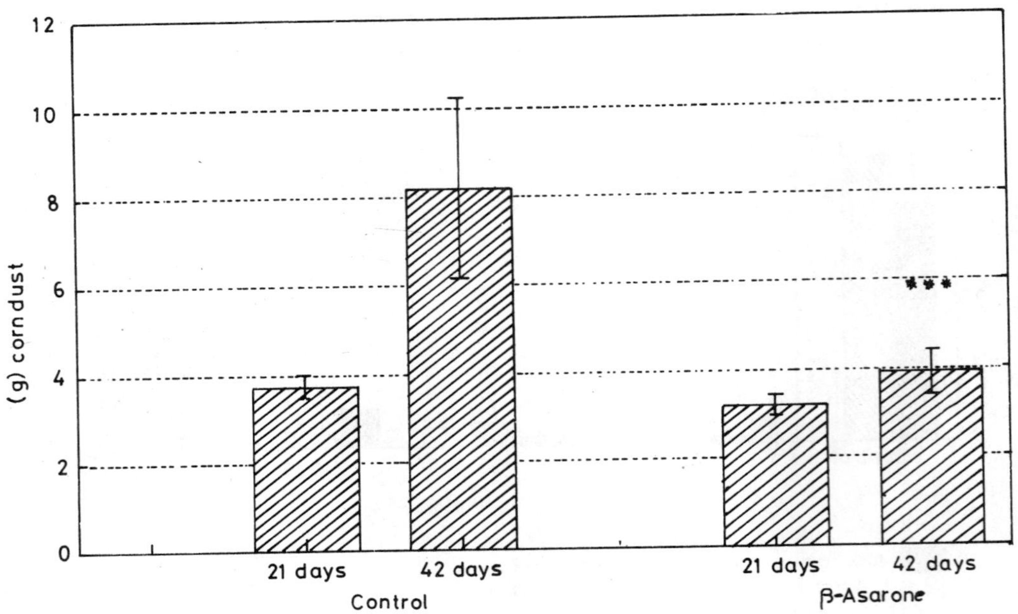

Fig. 3. Vapour effect of commercial β-asarone on the amount of corn dust produced by *P. truncatus* at 30°C between 21 and 42 days; ***$p < 0.01$ for significant difference between the control and treatment period of 42 days.

Effect of Rhizome Powder

Figure 6 shows that an addition of 0.5 per cent rhizome powder to the test corn already had an effect on the feeding activity of the beetles. A mixture of 2 per cent rhizome powder showed a feeding reduction of 83.8 per cent compared to the control ($p < 0.01$).

Comparison of the Amount of Corn Dust Produced by Feeding and the Number of Live Beetles

The parameter corn dust produced by feeding was chosen because it can be easily determined and gave reproducible results. The determination of the number of dead beetles, especially after 42 days, was more difficult because of the cannibalism taking place; often body parts of the dead insects were missing or destroyed. The live beetles were found mostly inside the corn grains if large amounts of corn were used as test materials. This makes the determination of the number of live beetles time consuming. To calculate at least approximately the relation of corn dust produced to population development of *P. truncatus*, the number of live beetles was determined after 42 days in one experiment treated by oil vapours. Whereas the corn dust produced by feeding activity of the beetles was slowed to 50.2 per cent compared to the control, the number of live beetles was reduced to 33 per cent. The influence of β-asarone on population development was rather higher than calculated from the corn dust. On the other hand,

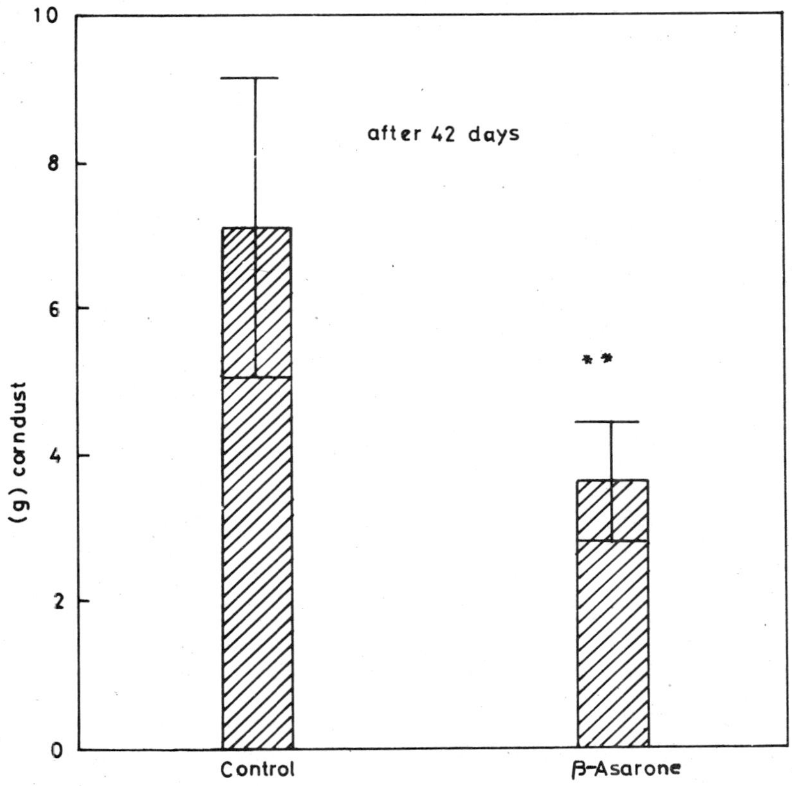

Fig. 4. Vapour effect of commercial β-asarone on the amount of corn dust produced by *P. truncatus* after 42 days at 30°C when the test jar was rectangular, turned and laid on side; the treated filter paper was deposited next to the test glass with the beetles; $^{**}p < 0.01$ for significant difference from the control.

the amount of corn dust resulting from feeding activity can be used to determine the degree of infestation.

DISCUSSION

An application of active compounds of *A. calamus* oil or rhizome powder to corn grains infested with *P. truncatus* showed that the essential oil and powder can be used in rural tropic and subtropic stored-product protection. A direct treatment of the corn was most effective; the oil has either a contact and feeding or vapour effect. An application of 1000 ppm commercial β-asarone to the corn reduced the feeding activity significantly. In the long term, a concentration of 100 ppm seems to have a similar effect.

The results agree well with those of Chander and Ahmed (1986) obtained for *C. chinensis* under similar conditions; concentrations of about 230 and 450 ppm *A. calamus* oil caused 100 per cent mortality within two days. Also, and ovicidal effect and a

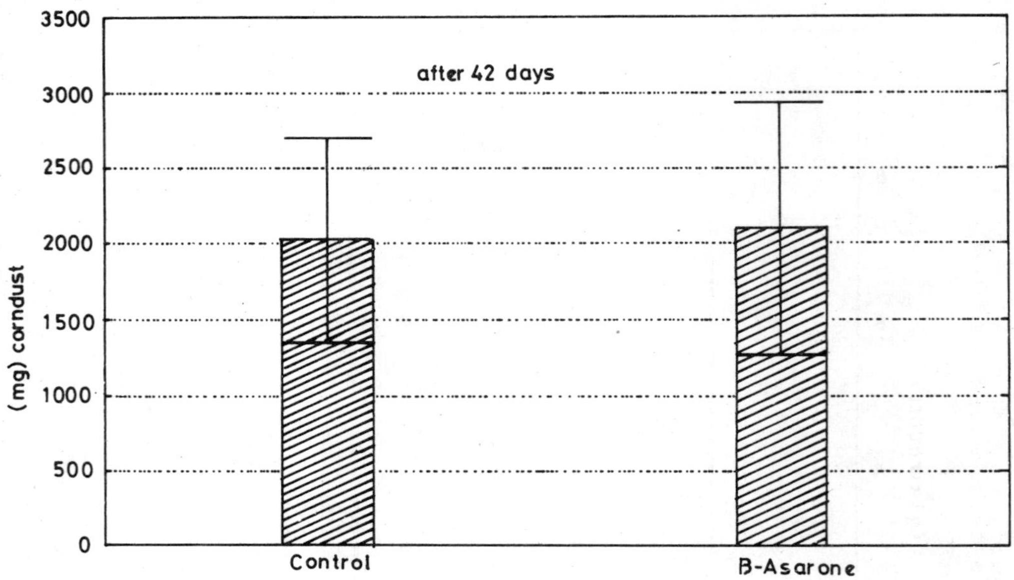

Fig. 5. Vapour effect of commercial β-Asarone on the amount of corn dust produced by *P. truncatus* after 42 days at 25°C; no significant difference from the control.

long term activity (135 days) were observed, but only when the essential oil was not exposed to the sunlight (Yadava, 1971; Saxena and Mukerjee, 1985). Using organochlorine solvents the effect of *A. calamus* oil was reduced rapidly (Mazza, 1985).

The knockdown effect of Indian *A. calamus* oil was independent of its β-asarone content. This result can be compared with the findings of E1-Nahal et al. (1989), who demonstrated that the period of exposure is a more relevant factor than the dosage in efficacy of the treatments. The β-Asarone content of the oil may be affected by an isomerization to α-asarone which increases in older *A. calamus* oil (Stahl and Keller, 1981); this isomerization occurs faster at 30°C than at room temperature of 20°C or lower.

Direct treatment of corn with *A. calamus* oil was significantly more effective than treatment with rhizome powder mixed with the corn grains. Comparative amounts of corn dust were produced by feeding activity (trreatment with 250 ppm *A. calamus* oil or 2 per cent rhizome powder). In both cases 50 mg of the active compound was in the test glass if a β-asarone content of 5 per cent dried rhizomes was calculated (Roest and Bos, 1979); with direct treatment it was only 12.5 mg, but more effective.

The effect of an admixture of rhizome powder to the corn grains is very important in view of the slow starting effect of *A. calamus* oil. In longer treatment periods the importance of this factor is reduced as it usually takes place in practise.

The effective amounts of rhizome powder calculated for *P. truncatus* were similar to those for other storage-product pests, such as *C. chinensis* (Pandey et al., 1976). In *Sitotroga cerealella* (Oliv.), *Rhizopertha dominica* (F.), *S. oryzae* (L.), and *Tribolium*

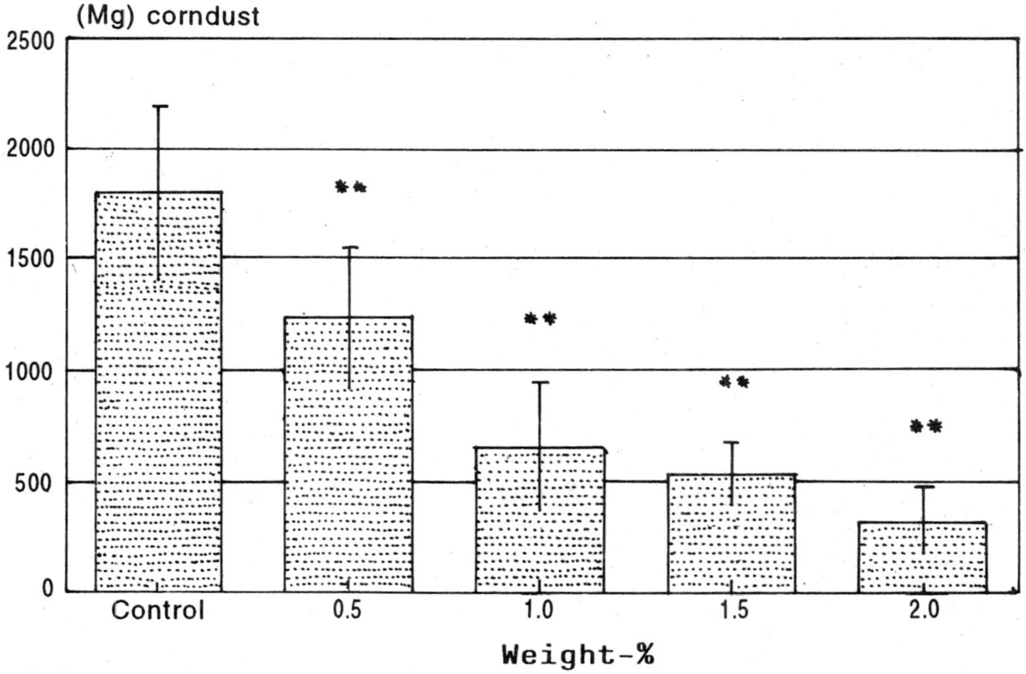

Fig. 6. Effect of admixture of various amounts of *A. calamus* rhizome powder to corn grains on the amount of corn dust produced by *P. truncatus* after 42 days at 30°C; $^{**}p < 0.01$ for significant differences between the control and treatments.

castaneum (Herbst), an application of rhizome powder can result in different effects (Mammen et al., 1968). Good results were reported for unpeeled rice (paddy) (Anon., 1954). The efficiency of the rhizome powder was reduced for *S. cerealella* on the same stored product (Abraham et al., 1972); the powder had a low effect (Mammen et al., 1968). The highest effect of the rhizome powder was found in *C. chinensis* (Pandey et al., 1976; Khan, 1986) and various other pests of stored products (Golob and Webley, 1980).

As an active space treatment, the vapour of *A. calamus* oil is of high interest for control of *P. truncatus*, because fumigation is a common method of treatment in stored-product protection. Using β-asarone, a significant vapour effect was found in 21 to 42 days after treatment in *P. truncatus*. The feeding activity of the beetles was almost stopped. This toxic effect of commercial β-Asarone was also found with *A. calamus* oil in adults of *C. chinensis*, *S. granarius*, and *S. oryzae* (El-Nahal et al., 1989; Schmidt et al., 1991) and *E. kuehniella* and *T. confusum* (Smet et al., 1986). All the authors reported on a slow starting effect of the oil vapours; most important was the length of the period of treatment (Koul et al., 1977b). *Rhizopertha dominica*, a species closely related to *P. truncatus*, was tolerant to vapours of *A. calamus* oil (El-Nahal et al., 1989), but the treatment period lasted only 11 days; this may be an explanation for the insensibility of *R. dominica*. Influence of the temperature seems to be very important in a range of 25°–30°C in stored-product protection. Some authors reported on-contact applications of *A. calamus* oil to stored-product pests at a temperature of 27°C (Teotia and Tewari, 1977; Teotia and Pandey, 1979; Chauhan et al., 1987). In these cases, there is not only a contact but also a vapour effect (Streloke et al., 1989). Smet et al. (1986) found that *E. kuehniella* is sensitive also at a temperature of 25°C, unlike *P. truncatus*.

Streloke et al. (1989) showed that the temperature effect is not related to the physicochemical properties of β-asarone; according to Smet et al. (1986) it is possible that differences in insect activity are due to this phenomenon. The feeding activity of *P. truncatus* is significantly lower at 25°C. *Ephostia kuehniella* is most active outside of the substrate and β-asarone vapours may not penetrate easily the feeding substrate of the larvae, as was assumed by Risha et al. (1990).

The vapours of β-Asarone have probably a density similar to that of the atmosphere. This may be concluded from the experiments with the side-laid test glasses, in which no difference was found from the effect of upright glasses. Also, Saxena and Srivastava (1972) found that the position of the active compound inside the test glass has no influence on the vapour effect; this was observed for *A. calamus* oil and rhizome pieces.

The high differences in the feeding activity of *P. truncatus* in the treatment period from 1 to 21 days and that from 21 to 42 days may be related to a sterilizing effect of β-asarone in the adult beetles and to the toxic effect of the offspring (Smet et al., 1986; Koul et al., 1977a; Tikku et al., 1978). To answer this question more knowledge is needed on the population development of *P. turncatus*.

REFERENCES

Abraham, C., B. Thomas, K. Karunakaran and R. Gopalakrishnan. 1972. Relative efficiency of some plant products in controlling infestation by the Angoumois grains moth *Sitotroga cerealella* Olivier (Gelenchiidae: Lepidoptera) infesting stored paddy in Kerala. *Argic. Res. J. Kerala* 10: 59–60.

Anonynous. 1954. What's new in farming. *Indian Farming* 23.

Banerji, R., G. Misra, S. Nigam, N. Prasad, R.S. Pandey and Y.K. Mathur. 1982. Indigenous plants as antifeedants. *Indian J. Ent.* 44: 71–76.

Cavazza, G. 1976. Les chimiotypes parmi les plants aromatiques cas du calamus polyploied. *Ann. Fals. Exp. Chim.* 69: 833–844.

Chander, H. and S.M. Ahmed. 1986. Efficacy of oils from medicinal plants as protectants of green gram against the pulse beetle *Callosobruchus chinensis. Entomon.* 1i: 21–28.

Chauhan, S.P.S., A. Kumar, C.L. Singh and U.K. Pandey. 1987. Toxicity of some plant extracts against rice moth *Corcyra cephalonica* (Stainton) (Lepidoptera). *Indian J. Entomol.* 49: 532–534.

El-Nahal, A.K.M., G.H. Schmidt and E.M. Risha. 1989. Vapours of *Acorus calamus* oil—A space treatment for stored-product insects. *J. Stored Prod. Res.* 25: 211–216.

Golob, P., P. Changjaroen, A. Ahmed and J. Cox. 1985. Susceptibility of *Prostephanus truncatus* (Horn) (Coleoptera: Bostrichidae) to insecticides. *J. Stored Prod. Res.* 21: 141–150.

Golob, P. and D.L. Webley. 1980. The use of plants and minerals as traditional protectants of stored products. Report of the Tropical Products Institute, G138, London, vi + 32 pp.

Guerin, P.M., E. Stadler and H.R. Buser. 1983. Identification of host plant attractants for the carrot fly, *Psila rosae*. *J. Chem. Ecol.* 9: 843–861.

Jacobson, M., I. Keiser, D.H. Miyashita and E.J. Harris. 1976. Indian calamus root oil: Attractiveness of the constitutents to oriental fruitflies, melon flies, and Mediterranean fruit flies. *Lloydia* 39: 412–415.

Khan, M.I. 1986. Efficacy of *Acorus calamus* L. rhizome powder against pulse beetle (*Callosobruchus chinensis*). *PKV Res. J.* (India) 10: 72–74.

Koul, O., M.J. Smirle, M.B. Isman and Y.S. Szeto. 1990. Synergism of a natural insect growth inhibitor is mediated by bioactivation. *Experientia* 46: 1082–1084.

Koul, O. 1987. Antifeedant and growth inhibitory effects of calamus oil and neem oil on *Spodoptera litura* under laboratory conditions. *Phytoparasitica* 15: 169–180.

Koul, O., K. Tikku and B.P. Saxena. 1977a. Follicular regression in *Trogoderma granarium* due to sterilizing vapours of *Acorus calamus* L. oil. *Curr. Sci.* 46: 724–725.

Koul, O., K. Tikku and B.P. Saxena. 1977b. Mode of action of *Acorus calamus* L. oil vapours on adult male sterility in red cotton bug. *Experientia* 33: 29–31.

Laborius, G.A., B. Leliveldt and H. Keil. 1985. Der GroBe Kornbohrer, *Prostephanus truncatus* (Horn). Ein neuer Vorratsschädling in Afrika. *Prakt. Schädlingsbekämpf.* 37: 179–186.

Mammen, K.V., A. Visalakshi and M.R.G.K. Nair. 1968. Insecticidal dusts to control pests of stored paddy. *Agric. Res. J. Kerala* 6: 59–60.

Mazza, G. 1985. Gas chromatographic and mass spectrometric studies of the constitutents of the rhizome of calamus. *J. Chromatog.* 328: 179–194.

Muckensturm, B., D. Dupley, F. Mahammadi, A. Moradi, P.C. Robert, M.T. Simonis and J.C. Kienlen. 1982. Le role des phenylpropanoides naturels en tant qu' antiappetans pour insects. In: *Les Mediateurs Chimiques* (Agissant sur le Compotement des Insect, Symp. Int., Versailles, 1981). INRA, Paris, pp. 131–135.

Pandey, N.D., S.R. Singh and G.C. Tewari. 1976. Use of some plant powders, oils and extracts as protectants against pulse beetle, *Callosobruchus chinensis* Linn. *India J. Ent.* 38: 110–113.

Pierce, S. and G.H. Schmidt. 1992. Zum Stand der Kenntnis über *Prostophanus truncatus* (Horn) (Col., Bostrichidae), einen gefährlichen Maisschädling in Africa. *Anz. Schädlingskde., Pflanzenschutz, Umweltschuts* 65: 25–32.

Risha, E.M., A.K.M. El-Nahal and G.H. Schmidt. 1990. Toxicity of vapours of *Acorus calamus* L. oil to the immature stages of some stored-product coleoptera. *J. Stored Prod. Res.* 26: 133–137.

Roest, L.C.M. 1978. Biosystematic investigations with Acorus L. (Araceae). *Proc. K. Ned. Akad. West., Ser. C.: Biol. Med. Sci.* 81: 428–441.

Roest, L.C.M. 1978. Biosystematic investigations with Acorus L. 4. Communication. A synthetic approach to the classification of the genus. *Planta Medica* 37: 289–307.

Roest, L.C.M. and R. Bos. 1978. Biosystematic investigations with Acorus L. 3. Communication. Constituents of essential oils. *Planta Medica* 36: 350–361.

Saxena, B.P. and O. Koul. 1978. Utilization of essential oils for insect control. *Ind. Perfumer* 22: 138–149.

Saxena, B.P., O. Koul and K. Tikku. 1976. Non-toxic protectant against the stored insect pests. *Bull. Grain Technol.* 14: 190–193.

Saxena, B.P., O. Koul and K. Tikku. 1977a. Inhibition of embryonic development by oil of *Acorus calamus* L. treated males in *Dysdercus koenigii* F. *Indian Perfumer* 21: 73–78.

Saxena, B.P. and A.C. Mathur. 1976. Loss of fecundity in *Dysdercus koenigii* F. due to vapours of *Acorus calamus* L. oil. *Experientia* 32: 314–316.

Saxena, B.P. and S.K. Mukerjee. 1985. Photo-oxidation of beta-Asarone. *Indian J. Chem.* 24B: 683–684.

Saxena, B.P. and J.B. Srivastava. 1972. Effect of *Acorus calamus* L. oil vapour on *Dysdercus koenigii* F. *Indian J. Exp. Biol.* 10: 391–393.

Saxena, B.P., K. Tikku and O. Koul. 1977b. Spermatogenesis in *Dysdercus koenigii* and induced sterility by *Acorus calamus* oil vapours. *Acta. Ent. Bohemoslov.* 74: 381–387.

Schmidt, G.H. and D. Borchers. 1981. Untersuchungen zur sterilisierenden Wirkung von indischem Kalmusöl bei Ameisen. *Mitt. Dtsch. Ges. Allg. Angew. Ent.* 3: 201–213.

Schmidt, G.H., E.M. Risha and A.K.M. El-Nahal. 1991. Reproduction of progeny of some stored-product Coleoptera by vapours of *Acorus calamus* oil. *J. Stored Prod. Res.* 27: 121–127.

Smet, H., H. van Mellaert, M. Rans and A. De Loof. 1986. The effect on mortality and reproduction of beta-asarone vapours on two insect species of stored grain: *Ephestia kuehniella* (Lepidoptera) and *Tribolium confusum* du Val (Coleoptera). *Med. Fac. Landbouw. Rijksuniv, Gent* 51/3b: 1197–1203.

Stahl, E. and K. Keller. 1981. Zur Klassifizierung handelsublicher Kalmusdrogen. *Planta Medica* 43: 128–140.

Stroloke, M., K.R.S. Ascher, G.H. Schmidt and W.P. Neumann. 1989. Vapor pressure and volatility of beta-asarone, the main ingredient of an indigenous stored-product insecticide, *Acorus calamus* oil. *Phytoparasitica* 17: 299–313.

Subramaniam, T.V. 1942. *Acorus calamus*—The sweet-flag—A new indigenous insecticide for the household. *Indian J. Ent.* 4: 238.

Teotia, T. and G.P. Pandey. 1979. Insecticidal properties of rhizomes of sweet flag, *Acorus calamus* against rice weevil, *Sitophilus oryzae* Linn. *Indian J. Ent.* 41: 91–94.

Teotia, T.P.S. and G.C. Tewari. 1977. Insecticidal properties of drupes of dharek (*Melia azederach*) and rhizomes of sweetflag (*Acorus calamus*) against adults of *Sitotroga cerealella* Oliv. *Indian J. Ent.* 39: 222–227.

Tewari, G.C. and P.N. Krishna-Moorthy. 1985. Plant extracts as antifeedants against *Henosepilachna vigintiochopunctata* (Fabricius) and their effect on its parasite. *Indian J. Agric. Sci.* 55: 120–124.

Tikku, K., O. Koul and B.P. Saxena. 1978. The influence of *Acorus calamus* L. oil vapours on the histocytological pattern of the ovaries of *Trogoderma granarium* Evert. *Bull. Grain Technol.* 16: 3–9.

Verma, B.R. and B. Lal. 1987. Record of *Prostephanus truncatus* (Horn) from India with report of a new host. *FAO Plant Prot. Bull.* 35: 100.

Yadava, R.L. 1971. Use of essential oil of *Acorus calamus* L. as an insecticide against the pulse beetle, *Bruchus chinensis* L. *Z. Angew Ent.* 68: 289–294.

121

Efficacy of Botanical Pesticides in the Control of Citrus Butterfly *Papilio demoleus* Linn. (Papilionidae: Lepidoptera)

P. KAMESWARA RAO AND CH. BALAGANGADHAR TILAK

Department of Entomology (APAU)
S.V. Agricultural College, Tirupati 517 502, A.P., India

ABSTRACT

A laboratory experiment was conducted for two generations with nine botanical pesticides — Neemguard, Vepanic, Margosol and Necknool (neem products), Allitin (garlic synthetic extract), Annmet and Sitpal (*Annona* products), Karrich (*Pongamia* product), and RD 9 Repelin (mixture of neem, karanjin, castor, mahua, and sesame oils) for the control of first, second, and fourth instar larvae of citrus butterfly *Papilio demoleus* Linn. The experiment was carried out by non-choice method. Based on per cent protection of leaf area over control and per cent larval mortality, Allitin was found to be superior, followed by RD 9 Repelin and Margosol, neem, *Pongamia*, and *Annona* products at 1.0 per cent concentration.

INTRODUCTION

The citrus butterfly, *Papilio demoleus* Linn., is a destructive pest causing severe defoliation in citrus nurseries as well as orchards. Due to the adverse effects of application of synthetic insecticides, new environmentally safe and non-hazardous botanical pesticides were suggested in recent pest management programmes. Among the botanical pesticides neem extracts were found to be superior against locusts (Pradhan et al., 1963). *Lymantria dispar* L. (Skatulla and Meisner, 1975), and *Heliothis zea* Boddie (Reed and Reed, 1985). Garlic extracts were also found to be effective against *Henosepilachna vigintioctopunctata* (Chitra et al., 1990). Solunke and Deshpande (1991) stated that the plant extracts of *Ficus* sp., beshram, *Parthenium* sp., neem, and *Datura* significantly reduced the pest population of *P. demoleus*. The information on the control of *P. demoleus* with botanicals was meagre and hence the present studies were undertaken.

MATERIALS AND METHODS

The larvae of citrus butterfly were reared in the laboratory on fresh leaves of citrus. Four neem products—Neemguard, Vepanic, Margosol, and Necknool—garlic synthetic extract Allitin, two *Annona* products—Annmet and Sitpal—Pongamia product Karrich, and RD 9 Repelin (mixture of neem, karanjin, castor, mahua, and sesamum) were tested at 1.0 per cent concentration against first, second, and fourth instar larvae of *P. demoleus* by leaf disc method for two generations.

Antifeedant activity of the botanicals was tested by non-choice method. For this purpose citrus leaf discs of 3.6 cm were dipped in different botanical pesticides and air dried. They were kept in petri plates containing moist filter paper to avoid drying. Each treatment was replicated thrice. One larva prestarved for 6 hrs was released in each petri plate. One leaf disc was used for first and second instar larvae and three leaf discs were used for fourth instar larvae depending on their feeding capacity. The leaf area fed by larvae was assessed after 48 hrs by graph paper method. The per cent leaf area protected over control was calculated by the following formula:

$$\text{Per cent leaf area protected over control} = \frac{\text{Per cent leaf area fed in control} - \text{Per cent leaf area fed in treatment}}{\text{Per cent leaf area fed in control}} \times 100$$

To study the larval mortality, 10 larvae prestarved for 6 hrs were released into each petri plate containing treated citrus leaves. Per cent larval mortality was recorded after 72 hrs. Each treatment was replicated thrice and observations were made for two generations.

RESULTS AND DISCUSSION

Allitin 1.0 per cent was found to be superior among all the botanicals, recording 94.55, 95.82, and 93.22 per cent protection over control against first, second, and fourth instar larvae respectively, followed by RD 9 Repelin (75.07, 89.47, and 78.6 per cent) (Table 1). Among the neem products, Margosol 1.0 per cent was found to be effective, recording 77.46, 87.70, and 75.24 per cent protection over control against first, second, and fourth instar larvae respectively, which was on par with other neem products. Based on overall efficacy (mean of three instars) Allitin recorded highest per cent protection of leaf area (94.56) followed by RD 9 Repelin (81.05), which was on par with Margosol (80.13), Neemguard (78.25), and Necknool (77.05). Lowest per cent protection was recorded in Sitpal (60.96), which was on par with Annmet (63.77) and Karrich (65.07).

In the larval mortality studies, Allitin recorded 93.33, 75.00, and 70.00 per cent larval mortalities after 72 hrs against first, second, and fourth instar larvae respectively (Table 1). Based on the overall efficacy (mean of three instars) Allitin recorded highest per cent larval mortality (79.44) among all the botanicals. RD 9 Repelin recorded 66.67 per cent larval mortality, which was at par with Margosol (57.22). Lowest per cent

Table 1. Efficacy of botanical pesticides in the control of *P. demoleus* (mean of two generations)

Treatment (% concentration)	Per cent leaf area protected over control after 48 hrs (mean of two generations)				Per cent larval mortality after 72 hrs (mean of two generations)			
	First instar larvae	Second instar larvae	Fourth instar larvae	Overall efficacy (mean of three instars)	First instar larvae	Second instar larvae	Fourth instar larvae	Overall efficacy (mean of three instar)
RD 9 Repelin (1.0)	75.07 (60.14)^{bc}	89.47 (70.95)^{b}	79.61 (62.41)^{b}	81.05 (64.14)^{b}	83.33 (68.33)^{b}	63.33 (52.76)^{ab}	53.33 (46.92)^{b}	66.67 (56.00)^{b}
Neemguard (1.0)	72.70 (58.77)^{bc}	87.37 (69.17)^{b}	75.19 (60.13)^{c}	78.25 (62.69)^{b}	58.33 (49.84)^{c}	56.00 48.84)^{bc}	45.00 (42.07)^{bc}	53.33 (46.11)^{c}
Allitin (1.0)	94.55 (76.72)^{a}	95.82 (78.22)^{a}	93.22 (74.80)^{a}	94.56 (76.58)^{a}	93.33 (79.40)^{a}	75.00 (60.10)^{a}	70.00 (56.88)^{a}	79.44 (65.46)^{a}
Annmet (1.0)	56.80 (49.30)^{e}	76.12 (60.79)^{c}	58.41 (49.82)^{de}	63.77 (53.31)^{c}	38.33 (38.22)^{d}	23.33 (28.53)^{ef}	18.33 (25.20)^{d}	26.66 (30.65)^{d}
Vepanic (1.0)	72.40 (58.64)^{bc}	85.99 (68.01)^{b}	71.68 (57.84)^{c}	76.71 (61.69)^{b}	58.33 (49.80)^{c}	50.00 (45.00)^{e}	38.33 (38.19)^{c}	48.88 (44.66)^{c}
Margosol (1.0)	77.46 (60.96)^{b}	87.70 (69.44)^{b}	75.24 (60.19)^{bc}	80.13 (63.53)^{b}	61.67 (51.80)^{c}	60.00 (50.80)^{bc}	50.00 (45.00)^{bc}	57.22 (49.20)^{bc}
Karrich (1.0)	54.48 (48.23)^{e}	79.07 (61.80)^{c}	61.67 (51.50)^{d}	65.07 (53.86)^{c}	40.00 (39.14)^{d}	35.00 (36.21)^{d}	21.67 (27.17)^{d}	32.22 (34.17)^{d}
Sitpal (1.0)	52.74 (47.15)^{e}	76.36 (60.99)^{c}	53.78 (47.17)^{e}	60.96 (51.77)^{c}	35.00 (35.96)^{d}	25.00 (29.88)^{de}	21.67 (27.17)^{d}	27.22 (31.00)^{d}
Necknool (1.0)	72.38 (57.68)^{b}	85.94 (68.00)^{b}	72.84 (58.73)^{c}	77.05 (61.47)^{b}	56.67 (48.84)^{d}	50.00 (45.00)^{e}	41.67 (40.19)^{c}	49.44 (44.67)^{c}
Control	0.00 (0.00)^{f}	0.00 (0.00)^{d}	0.00 (0.00)^{f}	0.00 (0.00)^{d}	0.00 (2.50)^{e}	0.00 (2.50)^{g}	0.00 (2.50)^{e}	0.00 (2.50)^{e}
F-Test	*	*	*	*	*	*	*	*

*Significant.
Figures in parentheses are mean of angular transformed values.
Means denoted by same letter in each column are not statistically different according to DMRT at P = 0.05.

larval mortality was observed in Annmet (26.66), which was at par with Sitpal (27.22) and Karrich (32.22). The superiority of Allitin was also observed by Chitra et al. (1990) in the control of *H. vigintioctopunctata*. According to Chari et al. (1990) RD 9 Repelin was found to be superior over Margoson 'O', Neemguard, and Wellgro at 2.0 per cent in the control of *Spodoptera litura*.

On the basis of the present investigations it was concluded that Allitin 1.0 per cent was superior to the other botanicals followed by RD 9 Repelin and Margosol. Other neem products, *Pongamia* product, and *Annona* products were next effective in the control of citrus butterfly *P. demoleus* Linn.

ACKNOWLEDGEMENTS

The authors are highly thankful to Dr. G. S. Srimannarayana, Osmania University of Hyderabad, ITC of Rajahmundry, BARC of Trombay, Bombay, Sovereign Agro Industries of Mutlur, and Gharda Chemicals of Bombay for supplying their products for the present investigations. Thanks are due to Dr. K.C. Chitra, Associate Professor in Entomology, S.V. Agricultural College, Tirupati for her keen interest in the present studies.

REFERENCES

Chari, M.S., G. Ramprasad, S. Sitarmaiah and P.S.N. Murthy. 1990. Bioefficacy of neem formulations against *Spodoptera litura* in tobacco nurseries. In: *Botanical Pesticides in Integrated Pest Management*. M.S. Chari and G. Ramprasad (eds.), Indian Society of Tobacco Science, Rajahmundry, pp. 145–153.

Chitra, K.C., P. Venkatarami Reddy and P. Kameswara Rao. 1990. Effect of certain plant extracts in the control of brinjal spotted leaf beetle *Henosepilachna vigintioctopunctata* Fabr. *J. Appl. Zool. Res.* 1(1): 39–41.

Pradhan, S., M.G. Jotwani and B.K. Rai. 1963. The repellent properties of some neem products. *Bull. Reg. Res. Lab.* 1: 140–151.

Reed, D.K. and G.I. Reed. 1985. Control of vegetable insects with neem seed extracts. *Proc. Indian Acad. Sci.* 94: 335–339.

Skatulla, U. and J. Meisner. 1975. Laboratory studies with neem seed extracts for the control of gypsy moth *Lymantria dispar* L. *Angeiger fur Schadlingskunde Pflanzenschutz Umwetschutz* 48: 38–40.

Solunke, B.R. and S.V. Deshpande. 1991. Studies on the use of plant products for the control of lemon butterfly larvae. *J. Marathwada Agric. Univ.* 16: 302–303.

Efficacy of *Ajuga* (*A. chamaepitys*, *A. reptans*, var. *reptans*, and var. *atropurpurea*) Extracts on a Wide Variety of Non-adapted Insect Species

BÉLA DARVAS[1], LÁSZLÓ ALFONZ POLGÁR[1], AHMAD SABER BREAM[1], IMRE CSATLÓS[2], ALI IBRAHIM FARAG[3], MÁRIA TORMA-GAZDAG[2], ZOLTÁN ILOVAI[2], MARIA PIA CALCAGNO[4] AND JOSEP COLL TOLEDANO[4]

[1]Plant Protection Institute, Hungarian Academy of Sciences, Budapest, PO Box 102 H-1525, Hungary
[2]Plant Health and Soil Conservation Station of Csongrád-county, Hódmezővásárhely Rárósi u. 102, H-6801, Hungary
[3]Plant Protection Department, Faculty of Agriculture, Menoufia University, Egypt
[4]CSIC Department of Organic and Biological Chemistry, Jordi Girona 18–26, 08034 Barcelona, Spain

ABSTRACT

Ajuga species produce significant quantities of diterpenoid neoclerodanes (e.g., ajugarins, ajugareptansin, ajugareptansons, ajugapitins) with antifeedant activity and phytoecdysteroids (e.g., 20-OH ecdysone, cyasterone, sengosterone, ajugalactone) with insect developmental inhibitor-type (IDI) activities.

Acetonitrile extracts of *Ajuga* species, collected in Hungary, were made from whole plants after flowering. Residues were dissolved in ethanol (1 ml representing 1 g dry *Ajuga* plants). These extracts were tested in bioassays. The average phytoecdysteroid contents of the extracts were as follows: (1) *A. reptans* var. *reptans* (ARR): norcyasterone (NC) 8 ppm, 20-OH exdysone (20 E) 42 ppm, cyasterone (CY) 30 ppm, ajugalactone (AJL) 25 ppm; (2) *A. reptans* var. *atropurpurea* (ARA): norsengosterone (NS) 9–19 ppm, NC 14–18 ppm, 20 E 57–103 ppm, CY 17–27 ppm, AJL 28–40 ppm; (3) *A. chamaepitys* (AC): 20 E 2–9 ppm, AJL 14–21 ppm.

Extracts of ARR (1 per cent, applied in water, cause developmental disturbances during larval-pupal moult in *Aedes aegypti* (L.) (Diptera, Culicidae). 20 E had a similar activity at 1000 ppm. Food treated with ARR extract (1 per cent), applied from birth, caused moulting disturbances during the larval development of *Neobelieria bullata* Parker (Diptera, Sarcophagidae). Extracts of ARR and ARA (2.5–25 per cent) caused early head sclerotization during wandering larval stage when used with ohtaki-type wet synchronization. 20 E induced this effect at 0.34 ppm (70 μM).

Extracts of AC (10 per cent, which were incorporated in the food, had strong deterrent effects on *Periplaneta americana* L. (Dictyoptera, Blattidae) resulting in a decrease of production and weight of oothecae. AT 5000 ppm, 20 E alone did not show any activity. Leaves and extracts of AC (0.2–2 per cent) have strong antifeedant effect (ajugapitins may be involved) on *Spodoptera littoralis* (Boisd.) (Lepidoptera Noctuidae).

INTRODUCTION

Ajuga species produce numerous secondary plant substances: aglycones, luteolin, apigenin, and their glycosides; iridoides (harpagid, harpagid-acetate); catechin-like compounds; carffeic, 4-caffeoylquinic, and chlorogenic acids (Litvinenko et al., 1970); a significant quantity of diterpenoid neo-clerodanes (e.g. ajugarins, ajugareptansin, ajugareptansons, ivains, ajugapitins, ajugamarins) (Kubo et al., 1976, 1982, 1983a; Camps et al., 1981a, b, 1984, 1987; Shimomura et al., 1981; Hernandez et al., 1982) with phagodeterrent activity and phytoecdysteroids (e.g. 20-OH ecdysone, makisterone A, ajugasterone A-D, turkesterone, cyasterone, 29-norcyasterone, sengosterone, 29-norsengosterone, ajugalactone) (see Table 1, edited after Matsuoka et al., 1969; Ikan and Ravid, 1971a, b, Koreeda et al., 1972; Usmanov et al., 1975, 1978; Khafagy et al., 1979; Chou and Lu, 1980; Camps et al., 1981a, 1982a, b, 1983, 1985a,b,c; Nazmi Sabri et al., 1981; Bergamasco and Horn, 1983; Kubo et al., 1983a; Dargallo, 1985; Lafont and Horn, 1989; Matsumoto and Tanaka, 1991; Darvas, 1991; Tomas et al., 1992; Wessner et al., 1992) with insect developmental inhibitor (IDI, = insect growth regulators, IGR) type activity. There are differences between the neo-clerodane and phytoecdysteroid profiles of *Ajuga* species.

The effects of phytoecdysteroids in insects are as follows.

Developmental/moulting disturbances on larvae and sterilizing effects on females were found (Ohtaki et al., 1967; Robbins et al., 1968; Matsuoka et al., 1969; Shigematsu et al., 1974; Hikino et al., 1975; Mango et al., 1976; Singh et al., 1982). Applying phytoecdysteroids *per os*, the relative sensitivity (LC_{50}) on different insects orders generally is: Nematocera (3–50 ppm), ticks (5–50 ppm), Lepidoptera (25–200 ppm), Aleurodidae (50–200 ppm), Brachicera (100–300 ppm), Coleoptera (500–5000 ppm).

There are significant differences in sensitivity, depending on insect age. AT 100–150 ppm, polypodine B (= ajugasterone A), 20-OH ecdysone, poststerone, cyasterone, and ponasterone A caused developmental disturbances at pupariation and emergence (Singh et al., 1968), while 500–1000 ppm (Robbins et al., 1968) caused sterility in *Musca domestica* L. females. Last-instar insects (*Chilo suppressalis* (Walker), *Culex pipiens molestus* L., *Trialeurodes vaporariorum* Westwood) are more sensitive than younger ones for 20-OH ecdysone (Matsuoka et al., 1969; Melé et al., 1991).

Depending on the sublethal concentrations, phytoecdysteroids have agonist or antagonist effects. Five ppm 20-OH ecdysone or 10 ppm inokosterone reduced the time of larval development, while double concentrations increased it twice (Shigematsu et al., 1974).

Table 1. Phytoecdysteroids in *Ajuga* species (ppm)

Ajuga species	Tissue	20 E	CY	MKA	Ajugasterone A	Ajugasterone B	Ajugasterone C	Ajugasterone D	AJL	Other	Reference
australis		+	+	+							Bergamasco and Horn, 1983
bracteosa	l	+	+								Kubo et al., 1981
	l	+	+				+				Kubo et al., 1983a
chamaecystus	l	+	+				+ ?				Kubo et al., 1983a
chamaepitys	w, d	38	160	40					43		Camps et al., 1985c
chia	l, s, f		19								Ikan and Ravid, 1971b
	w	100									Abubakirov, 1982
decumbens	w, f	1200	800			10	15				Matsuoka et al., 1969
	w	1250	600			15	15		100		Koreeda et al., 1972
incisa	w, f	1200	800		140	80					Matsuoka et al., 1969
iva	w, f	121					+				Ikan and Ravid, 1971a
	w, f	50	1								Khafagy et al., 1979
	w, d			1070							Khafagy et al., 1979
	r, d	2200	1330	280						23-OH-CY	Nazmi Sabri et al., 1981
	s, d	800	530	640							Nazmi Sabri et al., 1981
	l, d	520	440	440							Nazmi Sabri et al., 1981
	w, d	1600	1300	1600						22-OXO-CY	Wessner et al., 1992
										24(28) MKA	Wessner et al., 1992
										24PRECY	Wessner et al., 1992
var. *pseudoiva*		44	70	35							Dargallo, 1985
japonica	l, d	1800	580				130				Matsuoka et al., 1969
multiflora		+									Bergamasco and Horn, 1983
nipponensis	w, f	1200	800								Matsuoka et al., 1969
	w, d	1000			1000			1000		STD:1000	Chou and Lu, 1980
pyramidalis											
ssp. *occidentalis*		170	70	50				90		NC:180	Dargallo, 1985
reptans	w, d	66	26						94	NC:120	Camps et al., 1982b
										NS:110	Camps et al., 1982b
	l	+							+		Kubo et al., 1983a
	w, d									2Ac-Cy	Camps et al., 1985a
										3Ac-Cy	Camps et al., 1985a
	w, d									SG	*Camps et al., 1992

Contd.

Table 1. contd.

Ajuga species	Tissue	20 E	CY	MKA	Ajugasterone				AJL	Other	Reference
					A	B	C	D			
var. *atropurpurea*	r, f	47	15							NC:21	Matsumoto and Tanaka, 1991
										iCY:5	Matsumoto and Tanaka, 1991
turkestanica		480	100			+				TU:520	Usmanov et al., 1975
										22Ac-CY:500	Usmanov et al., 1978

20 E = 20-OH ecdysone, MKA = makisterone A, TU = turkesterone, CY = cyasterone, NC = 29-norcyasterone, iCY = isocyasterone, SG = sengosterone, NS = 29-norsengosterone, AJL = ajugalactone, 24(28)MKA = 24(28)-dehydro-MKA, 24PRECY = 24-dehydro-PRECY, STD = stachisterone D, 2Ac-CY = 2-acetyl-CY, 3Ac-CY = 3-acetyl-CY, 22Ac-CY = 22-acetyl-CY, l = leaves, r = roots, w = whole plants, d = dried plants, f = fresh plants, s = stems. * Camps, Claveria, Coll, Melé, Meeseguer & Tomás (1992) Xth Ecd. Workshop, Liverpool (lecture), *A. bracteosa* Benth. (Wall. Pl. As. Rar., i. 59) = *A. remota*, *A. chamaecystus* Ging. (Bent. Lab. Gen. Sp., 698.) = *A. camaecistus* (K.H. Rechinger, Fl. Iran, 150:11) ≈ *A. chameacistus* (Kubo et al., 1983a).

Ponasterone A, applied on fourth-instar *Bombyx mori* L. larvae at 100 ppm, induced larval-superlarval moult (Nakanishi, 1971). Ponasterone A and 20-OH ecdysone, applied on *Ornithodoros moubata* complex (Acarina, Argasidae) at 5 ppm, had similar effects (Mango et al., 1976). Superlarval moult is known as a syndrome of hyper-juvenilism. As a special symptom, premature moult was found in *Drosophila melanogaster* L. (Fourche, 1967) and in a termite species (Lüscher and Karlson, 1958). RH 5849, as an ecdysteroid agonist, also induced premature moult in lepidopterous larvae (Darvas et al., 1992) and diapause-like longer development in an aphid parasitoid, *Aphidius matricariae* Hal. (Polgar and Darvas, 1991).

Ajuga extracts as botanical insecticides were tested in some cases. Schmutterer and Tervooren (1980) using methanolic extracts of *A. bracteosa* and *A. reptans* found deterrent and moulting inhibitor activities on *Epilachna varivestis* Muls. (Coleoptera Coccinellidae). Schauer and Schmutterer (1981) found that *A. bracteosa* extracts had repellent and IDI-type activity (decreased fecundity, abnormal first-instar nymph) on *Tetranychus urticae* Koch. (Acarina, Tetranychidae). Marcard et al. (1986) established that *A. reptans* was more effective on Culicidae species (*Aedes aegypti* (L.), *A. togoi* (Theobald), *Culex quinquefasciatus* Say) than *A. bracteosa* extracts. Ethanolic or petroletheric extracts of *A. reptans* inhibited the last larval moult of *Periplaneta americana* L. (Orthoptera, Blatellidae), while water extracts do not show significant activities. Ethanolic extracts lowered the ecdysteroid level in larvae (Richter and Birkenbeil, 1987) and Richter and Birkenbeil (1989) proposed that PTTH production is suppressed after *Ajuga* extract treatment. Premature moult was found in lepidopteran species *B. mori*, *Platyedra gossypiella* (Lepidoptera, Gelechiidae), and *Spodoptera frugiperda* (J.E. Smith) following their feeding on *A. bracteosa* extracts. Head capsules were doubled or tripled (Kubo et al., 1981, 1983b) on the treated caterpillars. *Heliothis zea* (Boddie) and *H. virescens* F. did not show significant sensitivity.

Ajugasterone B and C and makisterone A have similar activities as 20-OH ecdysone on *C. suppressalis*, while polypodin B is four times more active in *Calliphora* assays than the compounds mentioned. Cyasterone acts at the same level as 20-OH ecdysone on Diptera (*M. domestica, Sarcophaga peregrina*) and Lepidoptera (*B. mori*, *Samia cynthia* Donovan) species (Wilson, 1987).

Neo-clerodanes, produced by *Ajuga* species, have phagodeterrent activity on *Spodoptera littoralis* (Boisd.) (Bellés et al., 1985). Ajugapitin series, found in *A. chamaepitys*, have the strongest activity. Jacobson (1989) supposed that clerodanes have juvenilizing effects as well.

Our aim was to test the effects of acetonitrile extracts of different stocks of *A. reptans* var. *reptans*, *A. reptans* var. *atropurpurea*, and *A. chamaepitys* on various insect orders.

MATERIALS AND METHODS

Extraction

Whole *Ajuga* plants at flowering, in summer, were cleaned with water and dried at 70°C. Dry plant material was ground with Labor Min grinder. Ten grams ground material was extracted in 150 ml acetonitrile with Ultraturx TP 45. Extraction was repeated twice in 50 ml acetonitrile. Lipids and apolar materials were cleaned by 100 ml petrolether extraction (three times). Solvent was evaporated under vacuum at 60°C. A single sample was dissolved in 10 ml ethanol (1 g dry plant material/ml conc.) and used for analysis or in biological assays.

Determination of Phytoecdysteroids

Sample Cleaning with Sep-Pak C_{18} Column

One millilitre ethanolic sample was evaporated under N_2 at room temperature. Sep-Pak C_{18} cloumn was cleaned and activated with 10 ml mthanol and 10 ml double-distilled, deionized water. The sample was dissolved in 5 ml distilled water twice and put into the column. The sample in the column was cleaned with 10 ml 15 per cent methanol. Phytoecdysteroids were collected with 5 ml 85 per cent methanol and the sample was dried under N_2.

HPLC Method for Phytoecdysteroids Determination

Sample was dissolved in 5 ml 85 per cent methanol. Applied Biosyst. 491 Dynamic Mixer Injector, Applied Biosyst. 400 Solvent Delivery System (1.2 ml/min., isopropyl alcohol: water = 7:93), 10 × 0.4 cm internal diameter (5 μm) Spherisorb ODS-2 C_{18} (Tracer analitica) column, Spark Holland Column Thermostat (55 °C), Applied Biosyst. 1000S Diode Array Detector (standard: methylanthranilate) were used for determination. Standard retention times were generally: norsengosterone 9.45 min., norcyasterone (249 nm) 11.08 min., polypodine B (245 nm) 11.80 min., sengosterone (248 um) 13.04 min., 20-OH ecdysone (249 nm) 13.92 min., cyasterone 15.49 min., and ajugalactone (244 nm) 26.50 min. A sample of 5–20 μl was injected for qualitative and 30 μl (6 μl sample + 24 μl standard mix) for quantitative determination.

Biological Assays

Phagodeterrent Activity

Periplaneta americana L. (Dictyoptera: Blattidae adults): Five g cockroach food (wheat germ: peanut: saccharose: milk powder: corn flour: yeast: wheat bran = 2:1:1:1:1:1:0.5) was treated with an appropriate ethanolic sample (ethanol was evaporated on the food). Test dish (one litre vol. glass) contained 1.5 cm sand, a piece of thick paper, 5 g treated food, and untreated water. Two freshly moulted females and two males were placed into the dish. Test dishes were placed in a dark room at 30°C. Number of ootheca were counted 35 days later. Weights of oothecae were measured individually. Data were analysed using one-way analysis of variance (Freund, 1962).

Spodoptera littoralis (Boisd.) (Lepidoptera : Noctuidae) L_6: Freshly moulted sixth-instar larva of *S. littoralis* was placed on *Ajuga* leaves at room temperature and long-day conditions during September. Six replicates were used and the weight of larva was measured daily.

Freshly moulted sixth-instar larva of *S. littoralis* was placed on a piece of *Ajuga* extract-treated *Mamestra brassicae* (MABR) diet (400 g saccharose, 200 g agar, 400 g yeast, 1.5 kg wheat germ, 1 kg alfalfa powder, 50 ml acetic acid, 40 g ascorbic acid, 40 g nipagine, in 7 l water) at room temperature and long-day conditions during September (as above).

In a separate experiment freshly moulted sixth-instar larva of *S. littoralis* was placed on 0.5 g *Ajuga* extract-treated MABR diet offered in a small aluminium dish. Ten replicates were used at 24°C and long-day condition (L/D = 18/6). After 42 hrs (two light periods and one dark) remaining diet was dried and consumption was calculated from the dried weight of food remaining.

Insect Developmental Inhibitor Activity

Aedes aegypti (L.) (Diptera: Culicidae) third-instar larvae: Ethanolic *Ajuga* sample was placed in a 25 ml glass and evaporated under N_2, at room temperature. The residue was dissolved in 10 ml tap water, sonicated for 5 sec. After the suitable dilution, 10–20 *A. aegypti* third-instar larvae were placed into the dish with some laboratory mouse food. The experiment was checked daily. Dead larvae was treated with lactic acid and the state of moult was examined. Experiment was repeated twice.

Neobellieria bullata Parker (Diptera: Sarcophagidae) first-instar larvae: Stock culture of *N. bullata* was reared in a constant temperature room at 25°C and 50 per cent relative humitidy, under 16:8 hr light:dark condition. Groups of about 200 adults were fed on a mixture (1:1, w/w) of saccharose and milk powder. Larvae were reared under the same conditions on homogenized beef liver and 1 per cent agar (1:1, v/v). Bran placed under the diet prevented the larvae from being drowned. First instars were ontained from 14-day-old females by squeezing their abdomens. Larval diets were treated with *Ajuga* extracts. Twenty-five *N. bullata* were placed on the treated diet. Four replicates were used. Weight and state of larvae were recorded daily.

Neobellieria bullata third instars at wandering stage: Progenies of 11 females were used for this type of experiment. At 96 hrs after birth, at wandering stage, the progenies of a female were separated into four groups (15 larvae in each). *Ajuga* extract (ethanol was evaporated under N_2) was added to 2 ml water in 100 ml Erlenmeyer flask for wet synchronization (Ohtaki, 1966). Progeny of a single female shared a similar repetition. Duration of the pupariation was measured until the white prepuparial stage. Pupariation time, when half of the tested insects pupariated, was calculated by linear regression (Darvas et al., 1993). Data were analysed with ANOVA after four replicates.

RESULTS AND DISCUSSION

Phytoecdysteroid Content of Acetonitrile Extracts of *Ajuga* Species

Results are summarized in Table 2.

Acetonitrile resulted in a relatively clean extraction, but phytoecdysteroid contents seem to be approximately 5–10 times lower than with methanol extraction (Camps et al., 1981a,b, 1985a,b,c, 1990; Camps, 1991; Tomas et al., 1992). Ajugalactone extraction by acetonitrile is relatively efficient.

Table 2. Phytoecdysteroid content of acetonitrile extracts of *Ajuga* species

Steroid	Amount in ppm in stock of						
	ARR-HH4	ARA-HH3	ARA-HH5	AC-HH6	AC-HF8	AC-HH10	AC-HT14
Norsengosterone	0	19	9	0	0	0	0
Norcyasterone	8	18	14	0	0	0	0
Polypodine B	0	0	0	0	0	0	0
Sengosterone	0	0	0	0	0	0	0
20-OH ecdysone	42	103	57	2	7	2	9
Cyasterone	30	27	17	0	0	0	0
Ajugalactone	25	28	40	21	21	14	19
Others		a	b	c	d	e	

ARR = *A. reptans* var. *reptans*, ARA = *A. reptans* var. *atropurpurea*, AC = *A. chamaepitys*; unidentified steroids at : a: B31 13 ppm, B135 7 ppm, A147 14 ppm; b: B31 11 ppm, B135 7 ppm, A147 10 ppm, A188 15 ppm; c: B121 (makisterone A?) 32 ppm; d: B121 32 ppm: e: B121 21 ppm.

Phagodeterrent Activity of *Ajuga* Extract

Periplaneta americana adults

Results are summarized in Tables 3–6.

Ajuga chamaepitys extracts showed strong phagodeterrent activity on *P. americana*. Ootheca production was reduced by the chronic application of *A. chamaepitys* extracts.

Ajuga chamaepitys extracts significantly reduced the weight of ootheca. There is a strong correlation between food consumption and ootheca production: Y = 49.28 + 0.38X, R = 0.9121 (Y = food consumption mg/35 days by two females and two males; X = ootheca production mg/35 days by two females). 20-hydroxy ecdysone had no activity at 5000 ppm. Phagodeterrent activity, likely due to neo-clerodanes, does play a stronger role in reduction of ootheca production than phytoecdysteroids, as suggested by Richter and Birkenbeli (1987–1989).

Spodoptera littoralis sixth instar

Caterpillars consumed only a small amount of *Ajuga* leaves. *Ajuga* leaves blocked larval growth soon after consumption (Table 5). Larvae consuming cabbage started pupation between fifth to sixth day, while caterpillars feeding on *Ajuga* leaves started to die on days 9–10.

MABR diet containing *A. chamaepitys* extract (at 2 per cent) inhibited larval growth. Sixth-instar *S. littoralis* died before pupation (Tables 7 and 8).

Ajuga chamaepitys extract (AC-HG11) at 0.5 per cent showed significant phagodeterrent activity, and decreased the food consumption to 75 per cent of controls. Bellés et al. (1985) found that ajugapitin series, produced by *A. chamaepitys*, had the strongest phagodeterrent activity in the case of *S. littoralis*.

Table 3. Effect of *Ajuga* on *P. americana* food consumption and ootheca production

Ajuga stock	Conc.(%)	Ootheca no./35 days /2 females	Food remaining g/35 days/2 females + 2 males
ARR-HP15	10	9	1.35
ARA-HH5	10	14	1.79
AC-HH6	10	8	0.66
AC-HH7	10	5	0.81
AC-HH8	10	9	0.61
AC-HS9	10	6	1.16
AC-HH19	10	7	0.89
AC-HT13	10	6	1.35
AC-HT14	10	5	0.87
AC mean		6.57 ± 1.51b	0.91 ± 0.26b
Ethanol 1		19	3.59
Ethanol 2		17	3.11
Ethanol 3		18	3.06
Ethanol 4		17	3.43
Ethanol mean		17.75 ± 0.96a	3.30 ± 0.25a
Untreated 1		16	3.67
Untreated 2		18	3.17
Untreated 3		16	3.85
Untreated 4		18	3.19
Untreated mean		17.00 ± 1.15a	3.47 ± 0.34a
LSD 1%		2.83	0.61

Values followed by the same letter within a column do not differ significantly ($p < 1\%$) based on ANOVA.

Insect Developmental Inhibitor Effects of *Ajuga* Extracts

Third-instar Aedes aegypti

There are differences on the efficacy of *Ajuga* extracts related to plant species and stocks (Table 9). Between *A. reptans* var. *reptans* stocks, ARR-HP15 and ARR-HP17 have the strongest activity and at 0.5 per cent (0.5 ppm phytoecdysteroids) inhibited the development of *A. aegypti* (Table 9). These two stocks have unusually big leaves. Larvae died during moulting or some of them as larval-pupal intermediates (Matsuoka et al., 1969). Such effects reflect on IDI activity and characteristic of phytoecdysteroid content. *Ajuga reptans* var. *reptans* extracts were approximately 1000 times more active (calculated by its phytosteroid content) than 20-OH ecdysone. *Ajuga reptans* var. *atropurpurea* and *A. chamaepitys* extracts have weaker activity.

Neobellieria bullata first-instar

Ajuga reptans var. *reptans* with big leaves, especially the ARR-HT17 stock has strong activity on *N. bullata* 1st instas (Table 10). This treatment reduced the larval growth and resulting in miniature puparia, which died before emergence.

Table 4. Effects of *Ajuga* extracts on *P. americana* ootheca weight

Ajuga stock	Conc.(%)	Ootheca weight	
		Mg ± SD	% of control
ARR-HP15	10	67.25 ± 16.88ab	85
ARA-HH5	10	74.60 ± 11.02ab	94
AC-HH6	20	49.40 ± 19.01b	62
AC-HH6	10	50.38 ± 13.86b	64
AC-HH7	10	48.40 ± 18.69b	61
AC-HF8	10	53.89 ± 10.73b	68
AC-HS9	10	57.00 ± 11.88b	72
AC-HH10	10	46.33 ± 9.89b	58
AC-HT13	10	63.83 ± 10.69ab	81
AC-HT14	10	55.80 ± 8.35b	70
Untreated 1		77.00 ± 11.59a	97
Untreated 2		79.20 ± 6.61a	100
LSD 1%		18.64	

Values followed by the same letter within a column do not differ significantly ($p < 1\%$) based on ANOVA.

Table 5. Effects of *Ajuga* extracts on *P. americana* ootheca production

Ajuga stock	Conc.(%)	Ootheca (35 day)			Food cons. g/35 days/ 2 f + 2 m	Factor prod./ food. cons.
		Weight (mg)	No./2f	Prod. (mg/2 f)		
ARR-HP15	10	67.25	9	605.25	1.35	2.23
ARA-HH5	10	74.60	14	1044.40	1.79	1.71
AC-HH6	20	49.40	5	247.00	0.73	2.96
AC-HH6	10	50.38	8	403.04	0.66	1.63
AC-HH7	10	48.40	5	242.00	0.81	3.35
AC-HF8	10	53.89	9	485.01	0.61	1.26
AC-HS9	10	57.00	6	342.00	1.16	3.39
AC-HH10	10	46.33	7	324.31	0.89	2.74
AC-HT13	10	63.83	6	382.98	1.35	3.52
AC-HT14	10	55.80	5	279.00	0.87	3.12
Untreated 1		77.00	16	1232.00	3.67	2.89
Untreated 2		79.20	18	1425.60	3.17	2.22

Neobellieria bullata third-instar wandering stage

Ajuga reptans var. *reptans* stocks with big leaves (ARR-HP15) have the strongest activity on *N. bullata* wandering larvae (Table 11). *Ajuga reptans* var. *reptans* and *A. reptans* var. *atropurpurea* extracts also reduced the period required for larval/puparial intermediate development (Table 12). After premature head sclerotization, these extracts caused larval-type puparia, which died before pupation. 20-OH ecdysone induces this effect at 0.34 ppm (Darva et al., 1993).

Table 6. Growth of sixth-instar *S. littoralis* after *Ajuga* leaves consumption

Ajuga stock	Larval weight (mg ± SD)		
	Before treatment	After treatment	
		2 days	4 days
ARR-BT32	134.8	121.40 ± 34.67a	129.50 ± 26.54a
ARA-BB1	146.2	137.00 ± 17.87a	118.60 ± 21.56a
AC-BS3	139.8	135.00 ± 15.57a	137.80 ± 36.21a
AG-BS1	136.2	125.80 ± 16.90b	125.75 ± 19.17a
Cabbage	140.1	301.40 ± 42.49b	423.50 ± 26.89b
LSD 1%		49.87	48.68

AG = *A. genevensis*; values followed by the same letter within a column do not differ significantly ($p < 1\%$) based on ANOVA.

Table 7. Growth of sixth-instar *S. littoralis* after *Ajuga* extract consumption

Ajuga stock	Larval weigth (mg ± SD)			Mortality		Emerged	
	Before treatment	After treatment		Larv.	Pupa	Pupa Inter-med.	Inter-med.
		2 days	4 days				
ARR-HP15 2%	147.8	208.80 ± 55.66a	256.00 ± 72.38b	1	3		1[a]
ARR-HP15 0.2%	137.2	188.60 ± 38.77b	300.80 ± 62.12b	1			4[a]
ARA-HH3 2%	133.6	190.20 ± 37.90b	247.80 ± 84.55b			2	3
ARA-HH3 0.2%	153.6	291.00 ± 14.54a	486.80 ± 39.27a				5
AC-HT14 2%	144.4	149.20 ± 16.30b	195.20 ± 56.92b	2	2	1	0
AC-HT14 0.2%	136.8	269.80 ± 48.77a	514.20 ± 123.19a			1	4
Untreated	141.4	276.60 ± 50.66a	526.00 ± 122.68a				5
LSD 1%		70.63	149.38				

Values followed by the same letter within a column do not differ significantly ($p < 1\%$) based on ANOVA; [a]adults with curled wings.

Table 8. Effect of *Ajuga* extracts on sixth-instar *S. littoralis* feeding

Ajuga stock	Conc. (%)	Food consumption (mg dry weight ± SD/42 hrs)
ARR-HH4	1.0	47.51 ± 9.86ab
ARR-HH4	0.5	56.13 ± 4.86a
ARR-HH4	0.1	56.64 ± 7.46a
ARA-HH5	1.0	52.69 ± 6.60ab
ARA-HH5	0.5	50.39 ± 6.77ab
ARA-HH5	0.1	62.90 ± 9.58a
AC-HG11	1.0	45.70 ± 6.43b
AC-HG11	0.5	44.04 ± 3.60b
AC-HG11	0.1	66.00 ± 8.30a
Untreated		60.43 ± 6.07a
LSD %		10.22

Values followed by the same letter within a column do not differ significantly ($p < 1\%$) based on ANOVA.

Table 9. Effect of *Ajuga* extracts on third-instar *A. aegypti*

Ajuga stock	Conc. (%)	Mort. (%)	Intermed. (%)
ARR-HP15	5.00	100	
ARR-HP15	0.50	100	
ARR-HP15	0.05	3	3
ARR-HP16	5.00	100	
ARR-HP16	0.50	35	35
ARR-HP16	0.05	0	0
ARR-HP17	5.00	100	
ARR-HP17	0.50	100	
ARR-HP17	0.05	3	3
ARA-HH5	5.00	29	29
ARA-HH5	0.50	0	0
ARA-HH5	0.05	0	0
AC-HH2	5.00	83	43
AC-HH2	0.50	11	11
AC-HH2	0.05	7	7
AC-HH6	5.00	65	5
AC-HH6	0.50	5	5
AC-HH6	0.05	0	0
AC-HH7	5.00	59	18
AC-HH7	0.50	17	10
AC-HH7	0.05	10	10
AC-HF8	5.00	76	40
AC-HF8	0.50	5	0
AC-HF8	0.05	0	0
AC-HS9	5.00	73	14
AC-HS9	0.50	0	0
AC-HS9	0.05	0	0
AC-HH10	5.00	97	10
AC-HH10	0.50	6	6
AC-HH10	0.05	2	2
20-OH ecdysone	1000 ppm	100	
20-OH ecdysone	100 ppm	50–55	20–45
20-OH ecdysone	10 ppm	0	0
20-OH ecdysone	1 ppm	0	0
Solvent 1	0.500 ml	0	0
Solvent 2	0.050 ml	0	0
Solvent 3	0.005 ml	0	0
Untreated 1		0	0
Untreated 2		0	0

Table 10. Effects of *Ajuga* extracts on first-instar *N. bullata*

Ajuga stock	Conc. (%)	Mort. (%) at emergence	Weight of Puparium	
			mg	%
ARA-HH5	1	5	119.4bc	87
AC-HH2	1	63	104.6c	77
AC-HH6	1	5	120.4b	88
AC-HH7	1	11	117.5bc	86
AC-HF8	1	0	127.2ab	0
AC-HS9	1	26	122.4ab	90
AC-HH10	1	11	132.4ab	97
Ethanol		0	125.2ab	92
Untreated 1			136.6a	100
LSD 1%			15.34	
ARR-HH4	1	75	85.0b	79
ARR-HP15	1	96		
ARR-HP16 8	1	100		
ARR-HP17	1	100*		
ARA-HH3	1	18	103.4a	96
AC-HG11	1	0	114.3a	106
AC-HT13	1	32	117.3a	108
AC-HT14	1	0	105.7a	98
Untreated 2			108.3a	100
LSD 1%			16.36	
ARR-HH4	0.5	0	121.0a	94
ARR-HP15	0.5	93*	72.1b	56
ARR-HP16	0.5	93	77.8b	61
ARR-HP17	0.5	100**	58.1b	45
Untreated 3			128.3a	100
LSD 1%			21.20	

Values followed by the same letter within a column do not differ significantly ($p < 1\%$) based on ANOVA.
*Died as larva; **died after third instar as a miniature puparium.

Ajuga species produce promising compounds against some insects. The effectivities of acetonitrile extracts of *Ajuga* stocks are changeable on species. Experiments for determination and isolation of the active components are going on.

ACKNOWLEDGEMENTS

This work was supported by a grant from OMFB (No. 3095/91), Hungary. The authors thank Prof. Lafont (Ecole Normale Superior, CNRS, Paris), Prof. F. Camps (Dept. Biol. Org. Chem., CID-CSIC, Barcelona), and Dr. M. Bathory (SOTE, Szeged) for invaluable discussions, and Dr. J. Hardie (Imperial College, Ascot) and Dr. A. Szentesi (Plant Prot. Inst., Budapest) for critical reading of the manuscript.

Table 11. Effects of *Ajuga* extracts on third-instar *N. bullata* wandering stage with Ohtaki-type wet synchronization

Ajuga stock	Conc. (%)	Puparium (% ± SD)	Imago (% ± SD)
ARR-HP15	25.00	2.50 ± 5.00c	0.00 ± 0.00d
ARR-HP15	2.50	37.50 ± 5.00b	2.50 ± 5.00d
ARR-HP15	0.25	52.50 ± 15.00b	22.50 ± 18.93c
ARA-HH3	25.00	25.00 ± 5.77c	0.00 ± 0.00d
ARA-HH3	2.50	60.00 ± 18.26b	32.50 ± 9.57c
ARA-HH3	0.25	45.00 ± 19.15bc	32.50 ± 5.00c
AC-HT14	25.00	25.00 ± 12.91c	0.00 ± 0.00d
AC-HT14	2.50	50.00 ± 23.09bc	17.50 ± 17.08cd
AC-HT14	0.25	50.00 ± 11.55bc	20.00 ± 8.16c
Water		82.50 ± 17.08ab	55.00 ± 10.00b
Dry		100.00 ± 0.00a	92.50 ± 5.00a
LSD 1%		26.98	18.32

Notes: ARR = *A. reptans* var. *reptans*, ARA = *A. reptans* var. *atropurpurea*, AC = *A. chamaepitys*; values followed by the same letter within a column do not differ significantly (p < 1%) based on ANOVA.

Table 12. Effects of *Ajuga* extracts on third-instar *N. bullata* wandering stage with Ohtaki-type wet synchronization

Ajuga stock	Conc. (%)	Larval-puparial intermediate (% ± SD)	
		at 80th hr	developmental time (hrs)
ARR-HP15	25.00	92.50 ± 9.57c	56.25 ± 10.54ab
ARR-HP15	2.50	30.00 ± 8.16b	68.17 ± 13.22b
ARR-HP15	0.25	15.00 ± 10.00ab	69.33 ± 12.44b
ARA-HH3	25.00	40.00 ± 8.16b	56.00 ± 9.60a
ARA-HH3	2.50	15.00 ± 12.91ab	59.67 ± 18.29ab
ARA-HH3	0.25	2.50 ± 5.00a	
AC-HT14	25.00	5.00 ± 5.77a	
AC-HT14	2.50	5.00 ± 10.00a	
AC-HT14	0.25	7.50 ± 5.00a	
Water		2.50 ± 5.00a	179.18 ± 11.52c*
Dry		0.00 ± 0.00a	44.65 ± 4.80a*
LSD 1%		16.27	11.48

Values followed by the same letter within a column do not differ significantly (p < 1%) based on ANOVA.
*Time needed for pupariation.

REFERENCES

Abubakirov, N.K. 1982. Ecdysteroids of flowering plants (Angiospermae). *Proc. Indian Sci. Acad.*, 48A (1) : 122–138.

Bergamasco, R. and D.H.S. Horn. 1983. Distribution and role of insect hormones in plants. In: *Endrocrinology of Insects*. R.G.H. Downer and H. Laufer (eds.), Alan R. Liss, Inc., New York, pp. 627–654.

Bellés, X., F. Camps, J. Coll and D.M. Piulachs. 1985. Insect antifeedant activity of clerodane diterpenoids against larvae of *Spodoptera littoralis* (Boisd.) (Lepidoptera). *J. Chem. Ecol.* 11: 1439–1445.

Camps, F. 1991. Plant ecdysteroids and their interaction with insects. In: *Ecological Chemistry and Biochemistry of Plant Terpenoids*. J.B. Harborne and F.A. Tomas-Barberan (eds.), Clarendon Press, Oxford, pp. 331–376.

Camps, F., J. Coll and A. Cortel. 1981a. Allelochemicals on insect isolated from *Ajuga* plants (Labiatae). *Rev. Latinoam. Quim.* 12: 81–88.

Camps, F., J. Coll and A. Cortel. 1981b. Two new clerodane diterpenoids from *Ajuga reptans* (Labiatae). *Chem. Lett.* 1093–1096.

Camps, F., J. Coll and A. Cortel. 1982a. New clerodane diterpenoids from *Ajuga iva* (Labiatae). *Chem. Lett.* 1053–1056.

Camps, F., J. Coll and A. Cortel, 1982b. 29-Norsengosterone and 29-norcyasterone, new c-28 phytoecdysteroids from Ajuga reptans (Labiatae). *Chem. Lett.* 1313–1316.

Camps, F., J. Coll and A. Cortel. 1983. Sterols from *Ajuga reptans*. *An. Quim.* 79: 228–229.

Camps, F., J. Coll and O. Dargallo. 1984. Neo-clerodane diterpenoids from *Ajuga chamaepitys*. *Phytochemistry* 23: 2577–2579.

Camps, F., J. Coll, E. Molins and C. Miravitlles. 1985a. 2-Acetyl and 3-acetyl-29-norcyasterone, new minor phytoecdysteroids from *Ajuga reptans* (Labiatae). *J. Chem. Res. S.* 1: 14–15.

Camps, F., J. Coll, A. Cortel, E. Molins and C. Miravitlles. 1985b. 2-acetyl- and 3-acetyl-29-norcyasterone, new minor phytoecdysteroids from *Ajuga reptans* (Labiatae). *J. Chem. Res. M.* 1: 241–266.

Camps, F., J. Coll and O. Dargallo. 1985c. Phytoecdysteroids from *Ajuga chamaepitys*. An. Quim. 81: 74–75.

Camps, F., J. Coll, O. Dargallo, J. Rius and C. Miravitlles. 1987. Clerodane diterpenoids from *Teucrium* and *Ajuga* plants. *Phytochemistry* 26: 1475–1479.

Camps, F., J. Coll, M.P. Marco and J. Tomás. 1990. Efficient determination of phytoecdysteroids from *Ajuga* species and *Polypodium vulgare* by high-performance liquid chromatography. *J. Chromatog.* 541: 199–207.

Chou W.-S. and H.-S. Lu. 1980. Growth regulators and silk production of *Bombyx mori* L. from phytogenous ecdysteroids. In: *Progress in Ecdysone Research*. J.A. Hoffman (ed.), Elsevier, Amsterdam, pp. 281–297.

Dargallo, O. 1985. Ph. D. Thesis, Fac. Pharmacy, Univ. Barcelona.

Darvas, B. 1991. Phytoecdysteroids of *Ajuga* spp., as insect growth regulator-type botanical insecticides. *Növényvédelem* 27: 481–498. (in Hung.)

Darvas, B., L. Polgár, M.H. Tag El-Din, K. Eröss and K.D. Wing. 1992. Developmental disturbances in different insect orders caused by an ecdysteroid agonist, RH 5849. *J. Econ. Entomol.* 85: 2107–2112.

Darvas, B., H.H. Rees, N. Hoggard, A.I. Farag, G. O'Hanlon and J. Mercer. 1993. Effects of wet environment on ecdysone 20-monooxygenase and ecdysteroid level during wandering behaviour of *Neobellieria bullata* and *Parasarcophaga argyrostoma* larvae. *Physiol. Entomol.* (submitted).

Fourche, J. 1967. Action de l'ecdysone sur les larves de *Drosophila melanogaster* soumises au jeune. Existence d'un double conditionnement pour la formation du puparium, *C.R. Acad. Sci. Paris*, Ser. D., 264: 2398–2399.

Freund, J.E. 1962. *Mathematical Statistics*. Prentice Hall, Englewood Cliffs, New Jersey.

Hernandez, A., C. Pascual, J. Sanz and B. Rodriguez. 1982. Diterpenoids from *Ajuga chamaepitys:* Two neo-clerodane derivatives. *Phytochemistry* 21: 2909–2911.

Hikino, H., Y. Ohizumi and T. Takemoto. 1975. Detoxification mechanism of *Bombyx mori* against exogenous phytoecdysone ecdysterone. *J. Insect Physiol.* 21: 1953–1963.

Ikan, R. and U. Ravid. 1971a. The isolation and identification of ecdysterone from *Ajuga iva*. Planta Medica 20: 33–34.

Ikan, R. and U. Ravid. 1971 b. The isolation and identification of cyasterone from *Ajuga chia* (Labiatae). *Phytochemistry* 10:1659–1661.

Jacobson, M. 1989. Botanical pesticides, past, present, and future. In: *Insecticides of Plant Origin*. J.T. Arnason, B.J.R. Philogeny and P. Morand (eds.), ACS Symp. Ser. 387: 1–5.

Khafagy, S.M., N. Nazmi Sabri, N. El-Sebakhy, B. Blessington and A. Asaad. 1979. A C-28 ecdysone-like substances from *Ajuga iva*. *Planta Medica* 35: 184–185.

Koreeda, M., K. Nakanishi and M. Goto. 1972. Ajugalactone, an insect moulting inhibitor as tested in the *Chilo* dipping method. *J. Amer. Chem. Soc.* 92: 7512–7513.

Kubo, I. Y.-W. Lee, V. Balogh-Nair, K. Nakanishi and A. Chapya. 1976. Structure of ajugarins. *J. Chem. Soc. Chem. Commun.* 22: 949–950.

Kubo, I., J.A. Klocke and S. Asano. 1981. Insect ecdysis inhibitors from the East African medicinal plant *Ajuga remota* (Labiatae). *Agric. Biol. Chem.* 45:1925–1927.

Kubo, I., J.A. Klocke, I. Miura and Y. Fukuyama. 1982. Structure of ajugarin-IV. *J. Chem. Soc. Chem. Comm.* 90: 618–619.

Kubo, I., J.A. Klocke, I. Ganjian., N. Ichikawa and T. Matsumoto. 1983a. Efficient isolation of phytoecdysones from *Ajuga* plants by high-performance liquid chromatography and droplet counter-current chromatography. *J. Chromatog.* 257: 157–161.

Kubo, I., J.A. Klocke and S. Asano. 1983b. Effects of ingested phytoecdysteroids on the growth and development of two lepidoterous larvae. *J. Insect Physiol.* 29: 307–316.

Lafont, R. and D.H.S. Horn. 1989. Phytoecdysteroids: Structures and occurrence. In: *Ecdysone. From Chemistry to Mode of Action.* J. Koolman (ed.), Georg Thieme Verlag, Stuttgart, pp. 39–64.

Litvinenko, V.I., I.G. Zoz and V.S. Sokolov. 1970. Chemotaxonomische Untersuchungen zur Unterfamilie Ajugoideae Benth. der Lamiaceae Lindley. *Planta Medica* 3: 243–253.

Luscher, M. and P. Karlson. 1958. Experimentelle Auslösung von Hautungen bei der Termite *Kalotermes flavicollis* (Fabr.). *J. Insect Physiol.* 1: 341–345.

Mango, C., T.R. Odhiambo and R. Galun. 1976. Ecdysone and the supertick. *Nature* 260: 318–319.

Marcard, M., C.P.W. Zebitz and H. Schmutterer. 1986. Wirkung von methanolischen Rohextrakten aus *Ajuga* spp. auf Entwicklungsstadien verschiedener Stechmückenarten. *J. Appl. Entamol.* 101: 146–154.

Matsumoto, T. and N. Tanaka. 1991. Production of phytoecdysteroids by hairy root cultures of *Ajuga reptans* var. *atropurpurea. Agric. Biol. Chem.* 55: 1019–1025.

Matsuoka, T., S. Imai, M. Sakai and M. Kamada. 1969. Studies on phytoecdysones. *Ann. Rep. Takeda Res. Labs* 28: 221–271.

Mele, E., J. Messeguer, R. Gabarra, J. Tomas, J. Coll and F. Camps. 1991. *In vitro* bioassay for the effect of *Ajuga reptans* phytoecdysteroids on *Trialeurodes vaporariorum* larval development. *Entmol. Exp. Appl.* 62: 163–168.

Nakanishi, K. 1971. The ecdysones. *Pure Appl. Chem.* 25: 167–195.

Nazmi Sabri, N., A. Assad and S.M. Khafagy. 1981. Isolation of four ecdysones from *Ajuga iva* roots and a rapid semiquantitative method for ecdysones determination. *Planta Medica* 42: 283–290.

Ohtaki, T. 1966. On the delayed pupation of the fleshfly, *Sarcophaga peregrina* Robineau-Desvoidy. *Jap. J. Med. Sci. Biol.* 16: 97–104.

Ohtaki, T., R.D. Milkman and C.M. Williams. 1967. Ecdysone and ecdysone analogues: Their assay on the fleshfly *Sarcophaga peregrina. Proc. Nat. Acad. Sci.* 58: 981–984.

Polgar, L. and B. Darvas. 1991. Effects of nonsteroidal agonist, RH 5849 on a host/parasite system, *Myzus persicae/Aphidius matricariea.* In: *Behaviour and Impact of Aphidophaga.* L. Polgar, R.J. Chambers, A.F.G. Dixon and I. Hodek (eds.), Academic Publ. Bv, The Hague, pp. 323–327.

Richter, K. and H. Birkenbeil. 1987. The effect of extract of *Ajuga reptans* on moult regulation in *Periplaneta americana. J. Insect Physiol.* 33: 933–939.

Richter, K. and H. Birkenbeil. 1989. The effect of extract from *Ajuga reptans* on moult regulation in the cockroach, *Periplaneta americana.* Tag. -Ber., Akad. Landwirtsch. -Wiss. DDR, 274: 145–150.

Robbins, W.E., J. Kaplanis, M.J. Thompson, T.J. Sortino, C.F. Cohen and S.C. Joyner. 1968. Ecdysones and analogs: Effects on development and reproduction of insects. *Science* 161: 1158–1160.

Schauer, M. and H. Schmutterer. 1981. Wirkung von Frischpressaften and Rohextrakten aus der Labiatae *Ajuga remota* auf die Gemeine Spimmilbe *Tetranychus urticae* Koch. *Z. angew. Ent.* 91: 425–433.

Schmutterer, H. and G. Tervooren. 1980. Die Wirkung von Rohpressaften and Rohextrakten aus *Ajuga*-arten auf Frassaktivität and Metamorphose von *Epilachna varivestis. Z. angew. Ent.* 89: 470–478.

Shigematsu, H., H. Moriyama and N. Agui. 1974. Growth and silk formation of silkworm larvae influenced by phytoecdysones. *J. Insect Physiol.* 20: 867–875.

Shimomura, H., Y. Sashida, K. Ogawa and Y. Iitaya. 1981. Ajugamarin, a new bitter diterpene from *Ajuga nipponensis* Makino. *Tetrahedron Lett.* 22: 1367–1368.

Singh, P., G.B. Russel and S. Fredericksen. 1982. The dietary effects of some ecdysteroids on the development of housefly. *Entomol. Exp. Appl.* 32: 7–12.

Tomás, J., F. Camps, E. Claveria, J. Coll, E. Melé and J. Messeguer. 1992. Composition and localization of phytoecdysteroid in *Ajuga reptans* L. *in vivo* and *in vitro* cultures. *Phytochemistry* 31: 1585–1591.

Usmanov, B.Z., M.B. Gorovits and N.K. Abubakhirov. 1975. Phytoecdysones of *Ajuga turkestanica.* III. Structure of turkesterone. *Khim. Prir. Soedin.* 466–470.

Usmanov, B.Z., Y.V. Rakhes and N.K. Abubakhirov. 1978. Phytoecdysones of *Ajuga turtkestanica.* VI 22-acetylcyasterone. *Khim. Prir. Soedin.* 215–219.

Wessner, M., B. Champion, J.-P. Girault, N. Kaouadji, B. Saidi and R. Lafont. 1992. Ecdysteroids from *Ajuga iva. Phytochemistry* 31: 3785–3788.

Wilson, I.D. 1987. The ecdysteroids. In: *Handbook of Natural Pesticides.* E.D. Morgan and N.B. Mandava (eds.), CRC Press, Boca Raton, pp. 15–86.

123

Utility of Botanicals and Biocides as Additives to Insecticides in *Helicoverpa armigera* Hübner Management on Cotton

B. ROSAIAH[1] AND A.S. REDDY[2]

[1]Regional Agricultural Research Station, Lam, Guntur
[2]Andhra Pradesh Agricultural University, R. Nagar, Hyderabad, India

ABSTRACT

The botanicals Repelin, Neemguard, Biosol and Indiara of neem origin at 0.5 per cent concentration and biocides *Bacillus thuringiensis* at 0.1 per cent and nuclear polyhedrosis virus 250 LE/ha were evaluated alone and in combination with quinalphos at 0.05 per cent against *Helicoverpa armigera* (Hub.) on cotton. The botanicals and biocides performed better in combination with insecticide than they did alone. The performance of Repelin with quinalphos was most effective and consistent during the two years of study in reducing the damage and contributed to maximum yield. Repelin either alone or with quinalphos has shown highest ovipositional repellance to the adult moth. The superiority of Repelin with quinalphos as larvicide was also confirmed under laboratory testing. Neemguard, Biosol, and nuclear polyhedrosis virus with quinalphos were next in efficacy. The utility of botanicals and biocides as additive tools in the chemical management of *H. armigera* on cotton is discussed.

INTRODUCTION

Helicoverpa armigera Hubner became a serious pest on cotton and several other crops of economic importance in recent years in the Nagarjuna Sagar Project (NSP) area of Andhra Pradesh, India. Several attempts have been made to check the incidence of *H. armigera* but all the known control interventions have failed. The failure of control was attributed to the development of insecticidal resistance by *H. armigera* in the NSP area (Dhingra et al., 1988; Reddy, 1990; Armes et al., 1992). The natural plant products are considered to be safer to the ecosystem and compatible with insecticides and may also help in dilution of resistance. The utility of neem products in the suppression of pod borers (Singh et al., 1985), *Spodoptera litura* (Shashi Gupta and Rao, 1990; Chari et al., 1990) and *Heliothis virescens* (Barnby and Klocke, 1990) were reported by various workers. However, information on plant products and biocides against *H. armigera* in the NSP area is lacking. Therefore, some botanicals and biocides were evaluated alone and in combination with quinalphos, a popular insecticide, for their effectiveness against *H. armigera* on cotton.

MATERIALS AND METHODS

Field Studies

The field efficacy of the neem based insecticides Repelin, Neemguard, Indiara and Biosol each at 0.5 per cent concentration, *Bacillus thuringiensis* (Bt) at 1 kg/ha, nuclear polyhedrosis virus (NPV) 250 LE/ha alone and in combination with quinalphos 0.05 per cent and quinalphos 0.05 per cent alone was studied during 1988–89 and 1990–91 seasons at the Regional Agricultural Research Station, Lam. The treatments were replicated thrice. The test variety was MCU-5. A plot size of 37.8 m^2 was maintained for each treatment. The treatments were repeated at 10 day intervals. The bollworm infestation data on egg and larval population and damage in squares, bolls, and flowers were recorded before each spraying. The locule damage and yields were recorded at each picking.

Influence of Botanicals and Biocides on the Oviposition by Adult Moth

Three 60-day-old potted plants sprayed with botanicals and biocides and later covered with fine nylon mesh were exposed to five pairs of newly emerged adult moths for egg laying. There were three replications. On the fifth day of release the eggs were counted in each treatment and the effect of oviposition was assessed.

Ovicidal and Larvicidal Action

Fifteen freshly laid eggs were kept in petri dish and exposed of various treatments using atomizer sprayer. Observations on the hatchability of eggs were recorded four days after treatment to find out the ovicidal action.

A batch of ten third and fifth-instar larvae were fed separately in the laboratory with the leaves sprayed in the field with the treatments. The feeding continued for four days and mortality counts were recorded to find out the larvicidal action.

The data collected in various studies were subjected to statistical significance test.

RESULTS AND DISCUSSION

The pooled data on the performance of the treatments in field studies of two years are presented in Table 1. There were significant differences among the various treatments with respect to egg and larval population and damage in squares, flowers, bolls, locules, and yield.

Table 1. Effect of different botanicals and biocides against *H. armigera* and seed cotton yield (pooled data of 1988-89 and 1990-91)

Treatment	Dose (%)	Population/5 plats		Percent damage in				Yield
		Eggs	Larvae	Squares	Flowers	Bolls	Locules	Kg/ha
Biosol	0.5	6.84	4.15	28.51	10.15	24.83	26.00	601
		(2.68)	(2.17)*	(32.10)	(18.53)	(29.10)	(30.70)**	
Neemguard	0.5	5.50	4.76	23.81	7.36	26.09	29.86	694
		(2.43)	(2.25)	(29.23)	(15.53)	(30.20)	(31.80)	
Repelin	0.5	4.42	3.35	23.53	9.13	20.24	26.88	799
		(2.10)	(1.94)	(29.10)	(16.76)	(26.40)	(30.83)	
Indiara	0.5	6.14	3.36	28.26	7.66	17.99	28.74	594
		(2.62)	(1.95)	(31.83)	(15.60)	(24..93)	(32.23)	
Bacillus thurin-	0.1	8.58	5.99	36.20	14.13	30.60	32.81	515
gienses (Bt)		(2.99)	(2.56)	(36.30)	(21.76)	(33.03)	(35.10)	
NPV 250 LE/ha		8.33	2.83	24.26	7.35	22.19	28.20	740
		(2.96)	(1.78)	(30.06)	(17.30)	(28.00)	(30.93)	
Biosol	0.5	5.16	2.83	20.58	7.85	14.95	23.26	1236
+quinalphos	0.05	(2.36)	(1.78)	(26.73)	(16.30)	(21.83)	(28.30)	
Neemguard	0.5	4.45	2.14	18.36	5.45	14.21	17.63	1117
+ quinalphos	0.05	(2.11)	(1.60)	(25.10)	(13.06)	(21.63)	(24.50)	
Repelin	0.5	4.53	1.39	15.06	5.20	10.60	14.00	1262
+quinalphos	0.05	(2.13)	(1.34)	(23.16)	(12.80)	(18.40)	(21.86)	
Indiara	0.5	5.66	2.60	20.19	5.95	14.41	20.03	1004
+ quinalphos	0.05	(2.47)	(1.66)	(26.96)	(14.36)	(21.72)	(25.20)	
B t	0.1	5.73	2.83	28.19	8.55	15.65	12.63	971
+ quinalphos	0.05	(2.50)	(1.82)	(32.70)	(16.13)	(22.33)	(20.90)	
NPV 250 LE/ha		7.79	2.08	20.83	8.91	11.24	13.38	1176
+ quinalphos	0.05	(2.86)	(1.53)	(27.50)	(16.26)	(19.46)	(21.73)	
Quinalphos	0.05	6.34	3.29	25.44	8.98	22.11	23.85	819
		(2.60)	(1.92)	(30.13)	(16.40)	(27.50)	(28.96)	
Control		7.33	6.31	35.41	10.17	35.81	39.84	403
		(2.76)	(2.88)	(36.40)	(19.96)	(38.06)	(39.03)	
CD (0.05)		(0.21)	(0.19)	(3.86)	(3.38)	(5.94)	(3.39)	(231)
CV (%)		10.90	11.80	10.71	12.25	13.70	13.70	16.26

*figures in parentheses are the means of $\sqrt{X+0.5}$ values.
**figures in parentheses are the arcsin values.

The egg population per 5 plants was significantly lowest with Repelin (4.42) which was on par with Neemguard + quinalphos (4.45) and Repelin + quinalphos (4.53) as compared to quinalphos (6.34) and control (7.33).

The larval population per 5 plants was lowest with Repeline + quinalphos treatment (1.39) which was on par with NPV + quinalphos (1.53) and superior to quinalphos (3.29) and control (6.31) (Fig. 1).

The per cent damage in squares ranged from 15.06 to 35.41 among the treatments. Repelin + quinalphos recorded lowest square damage (15.06) and it did not significantly differ with Neemguard + quinalphos (18.36), Biosol + quinalphos (20.58) and Indiara + quinalphos (20.19) but it was superior to quinalphos (25.44) and control (35.41).

The flower damage varied from 5.20 per cent in Repelin + quinalphos to 14.13 per cent in the treatment where B t involved.

The per cent damage in bolls was lowest with Repelin + quinalphos (10.60), which was on par with NPV + quinalphos (11.24), Neemguard + quinalphos (14.21), Indiara

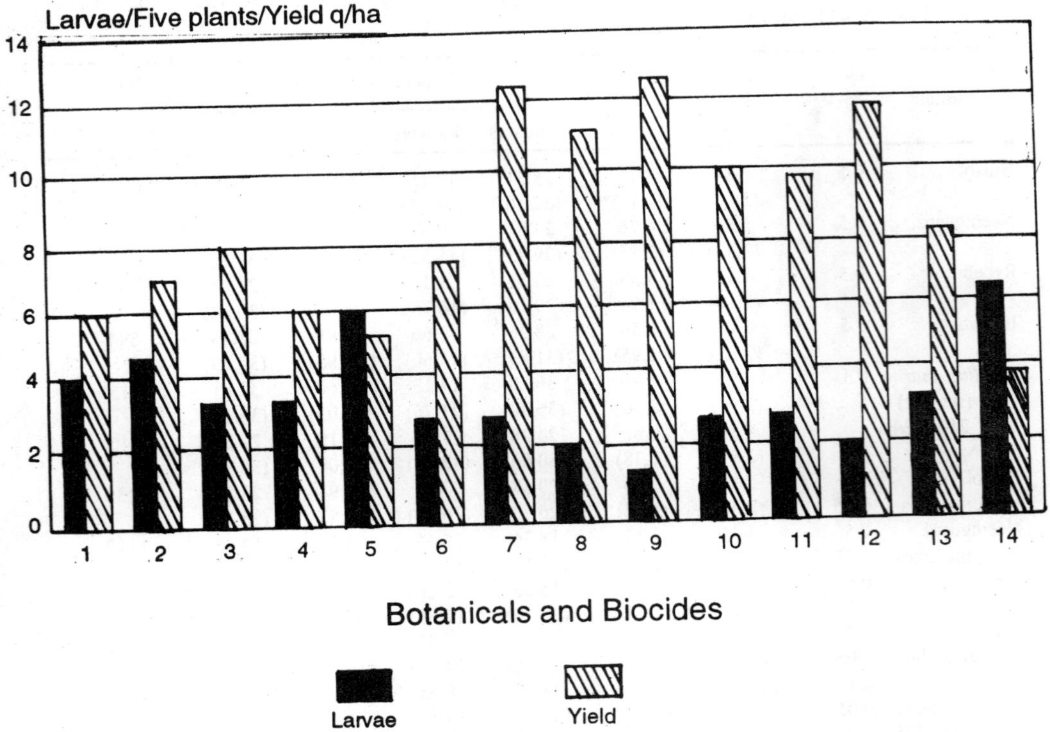

Fig. 1. Effect of botanicals and biocides against *H. armigera* and yield

1. Biosol, 2. Neemguard, 3. Repelin, 4. Indiara, 5. Bt, 6. NPV, 7. Biosol + quinalphos, 8. Neemguard + quinalphos, 9. Repelin + quinalphos, 10. Indiara + quinalphos, 11. Bt + quinalphos, 12. NPV + quinalphos, 13. Quinalphos, 14. Control.

+ quinalphos (14.41), Biosol + quinalphos (14.95), and Bt + quinalphos (15.65) but it was superior to quinalphos (22.11) and control (35.81).

The per cent damage in locules varied between 12.63 in B t + quinalphos and 39.84 in untreated control.

The yield varied from 403 to 1262 kg/ha among the treatments. Maximum yield of 1262 kg/ha was recorded in the plot where Repelin + quinalphos was used and it was on par with Biosol + quinalphos (1236 kg/ha), NPV + quinalphos (1176 kg/ha) and Neemguard + quinalphos (1117 kg/ha) and all these treatments were superior to quinalphos (819 kg/ha) and control (403 kg/ha) (Fig. 1).

From the foregoing it is evident that the plant products and biocides performed better in combination with quinalphos. Repelin in combination with quinalphos effectively reduced the *H. armigera* damage and contributed to maximum yields. Studies conducted at the Regional Agricultural Research Station, Lam, also indicated that Repelin when used alongwith insecticide was effective in reducing the *H. armigera* damage and recorded maximum yields (Anon., 1992). Neemguard + quinalphos and Biosol + quinalphos were also promising next to Repelin + quinalphos.

The treatment with NPV 250 LE/ ha + quinalphos was also found effective in reducing the damage in bolls and locules, and contributed to better yields similar to those obtained with Repelin + quinalphos. It is well known that NPV infects fat body, blood cells, and hypodermis apart from other organs. The infectivity to these organs by the virus may have played a major role in increasing the pest's susceptibility to insecticides (Rabindra and Jayaraj, 1990; Sachan, 1990). Subacute infection of polyhedrosis virus in the larvae of cabbage looper was found to enhance its susceptibility to insecticides (Girardeau and Mitchell, 1968).

Repelin exhibited maximum ovipositional repellance (32.33 eggs/plant) and was on par with Neemguard + quinalphos (33.66/plant), Biosol + quinalphos (42.33/plant), Repelin + quinalphos (43.66/plant), Neemguard (49.00/plant), and Biosol (53.33/plant), as compared to quinalphos alone (85.66/plant), and control (137.66/plant). Ayyangar and Rao (1989) found total inhibition of oviposition by *Spodoptera litura* moths on the area treated with methanol extract of neem. The observations made by Ramachandra Rao et al. (1990) on the ovipositional repellent effects of neem products also strengthen the view that botanicals could be successfully exploited as ovipositional repellents at the time of brood emergence (Table 2).

The data on the ovicidal action of different botanicals and biocides revealed that the suppression of egg hatching ranged from 4.03 to 37.00 per cent among the sole treatments and 70.70 to 82.16 per cent in combination treatments as compared to 64.76 per cent in quinalphos alone. Neemguard + quinalphos was most toxic (82.16 per cent) to the eggs and was on par with other combination treatments and quinalphos alone (Table 3).

The larvicidal action of botanicals and biocides indicated that Repelin + quinalphos recorded maximum mortality of 78.23 per cent in third instar and 60.83 per cent in fifth

Table 2. Influence of botanicals and biocides on the oviposition of *H. armigera* (field cage)

Treatment	Dose (%)	Number of eggs/plant	
Repelin	0.5	32.33	(5.65)*
Neemguard	0.5	49.00	(6.77)
Indiara	0.5	65.33	(8.08)
Biosol	0.5	53.33	(7.23)
Bt	0.1	115.66	(10.71)
NPV 250 LE/ha		129.33	(11.37)
Repelin +quinalphos	0.5 0.05	43.66	(6.50)
Neemguard +quinalphos	0.5 0.05	33.66	(5.81)
Indiara + quinalphos	0.5 0.05	62.00	(7.87)
Biosol +quinalphos	0.5 0.05	42.33	(6.48)
B.t +quinalphos	0.1 0.05	72.66	(8.44)
NPV 250 LE/ha +quinalphos	0.05	99.33	(9.54)
Quinalphos	0.05	85.66	(9.13)
Control		137.60	(10.82)
	CD (0.05)	(1.98)	
	CV (%)	(14.52)	

* Figures in parentheses are means of $\sqrt{x + 0.5}$ values

Table 3. Toxicological effects of botanicals and biocides on the eggs and larvae of *H. armigera* (Lab study)

Treatment	Dose (%)	Corrected percent mortality of		
		Eggs	Third instar larvae	Fifth instar larvae
Repelin	0.5	37.00 (37.40)	28.90 (32.49)	12.50 (20.70)*
Neemguard	0.5	18.66 (22.46)	16.56 (20.96)	12.50 (20.70)
Indiara	0.5	11.26 (17.23)	15.16 (20.23)	8.50 (15.15)
Biosol	0.5	11.26 (17.23)	7.16 (11.70)	8.50 (9.60)
Bt	0.1	4.03 (9.23)	7.16 (11.70)	4.50 (15.15)
NPV 250 LE/ha		7.20 (14.00)	40.83 (39.50)	25.00 (29.48)
Repelin + quinalphos	0.5 0.05	78.10 (63.03)	78.23 (66.93)	60.83 (51.43)
Neemguard + quinalphos	0.5 0.05	82.16 (65.23)	70.00 (61.93)	58.53 (50.00)
Indiara + quinalphos	0.5 0.05	70.70 (57.76)	51.23 (45.60)	47.50 (43.48)
Biosol + quinalphos	0.5 0.05	74.43 (59.46)	53.33 (46.93)	50.23 (45.86)
Bt +quinalphos	0.1 0.05	78.33 (63.43)	50.20 (45.09)	41.67 (40.17)
NPV 250 LE/ha + quinalphos	0.05	71.43 (59.43)	65.30 (60.15)	54.16 (47.41)
Quinalphos	0.05	64.76 (53.83)	46.60 (43.06)	41.66 (40.00)
CD		(16.74)	(23.58)	(13.24)
CV (%)		(23.95)	(17.13)	(15.31)

* Figures in parentheses are arcs in values

instar and was on par with other combination treatments for both the instars while it was superior to the sole application of quinalphos only in respect of third instar larvae. The neem products as sole treatments have exercised poor control with larval mortality of 7.16 to 28.90 per cent and 4.50 to 12.50 per cent in respect of third and fifth instar larvae respectively. It is likely that the combined antifeedant action of Repelin and neurotoxic action of quinalphos resulted in higher toxicity (Table 3).

The present findings clearly reveal that botanicals and biocides in combination with quinalphos could be exploited as additive control tools in the management of *H. armigera*. The observations of Singh et al. (1985), Chari et al. (1990), Barnby and Klocke (1990), Shashi Gupta and Rao (1990) and Rabindra and Jayaraj (1990) on the effectiveness of botanicals and biocides also support the present understanding on the need for exploiting them as additive tools in the management of *H. armigera* in view of their ecological safety and compatibility with insecticides. Furthermore, the use of botanicals and biocides may help in dilution of resistance by *H. armigera*.

REFERENCES

Annoymous. 1992. *Annual Report.* Heliothis Management Scheme, Regional Agricultural Research Station, Lam, Guntur.

Armes, N.J., DR. Jadhav, G.S. Bond and A.B.S. King. 1992. Insecticide resistance in *Helicoverpa armigera* in South India. *Pesticide Sci.*, 34: 355–364.

Ayyangar, G.S.G., and P.J. Rao. 1989. Neem (*Azadirachta indica*) extract as larval repellants and ovipositional deterrants to *Spodoptera litura* F. *Indian J. Entomol.*, 51: 121–124.

Barnby, M.A., and J.A. Klocke. 1990. Effects of azadirachtin on the levels of ecdysteroids and prothoracic hormone like activity in *Heliothis virescens* F. larvae. *J. Insect Physiol.*, 36: 125–131.

Chari, M.S., G. Ramaprasad, S. Sitaramaiah and P.S.N. Murthy. 1990. Bioefficacy of neem formulations against *Spodoptera litura* F. in tobacco nurseries. *Proc. Nat. Symp. on problems and prospects of Botanical Pesticides in Integrated Pest Management*, (Eds. Chari and Ramaprasad) published by ISTS, 1993, CTRI, Rajahmundry pp. 145–153.

Dhingra, S., A. Phokela and K.N. Mehrothra. 1988. Cypermethrin resistance in the population of *Heliothis armigera* H. *Nat. Acad. Sci. Letters*, II (4): 123–125.

Girardeau, J.H., Jr., and E.R. Mitchell. 1968. The influence of subacute infection of polyhedrosis virus in cabbage looper on susceptribility to chemical insecticides. *J. Econ. Entomol.*, 61: 212–213.

Rabindra, R.J., and S. Jayaraj. 1990. Effect of nuclear polyhedrosis virus infection on the insecticide susceptibility of *Heliothis armigera* H. and *Spodoptera litura* F. larvae. *J. Biol. Control.*, 4: 31–34.

Ramachandra Rao, G.G., G. Raghavaiah and B. Nagalingam. 1990. Effect of botanicals on certain behavioural responses and on growth inhibition of tobacco caterpillar *Spodoptera litura* F. *Proc. Nat. Symp. on Problems and Prospects of Botanical Pesticides in Integrated Pest Management*, (Eds. Chari and Ramaprasad) published by ISTS, 1993, CTRI, Rajahmundry, pp. 175-182.

Reddy, A.S. 1990. *Heliothis armigera* (Hub.) a serious threat to cotton cultivation in Andhra Pradesh. *Proc. National Workshop on Heliothis Management* (Eds. S. Jayaraj, S. Uthamasami, M. Gopalan and R.J. Rabindra) published by Tamil Nadu Agricultural University, Coimbatore, pp. 249-263.

Sachan, J.N. 1990. Present status of *Heliothis armigera* on pulses and strategies for management. *Proc. First Nat. Workshop on Heliotis Management, Current Status and Future Strategies.* (Ed. J.N. Sachan, Directorate of Pulses Research, Kanpur) pp. 8–33.

Shashi Gupta and P.J. Rao. 1990. Effect of Azadirachtin on neuroendocrine system of *Spodoptera litura* F. *Indian J. Entomol.*, 52: 589–594.

Singh, R.P., Yasbhir Singh and S.P. Singh. 1985. Field evaluation of neem (*Azadirachta indica*) seed kernel extracts against pod borers of pigeonpea *Cajanas cajan. Indian J. Entomol.*, 47: 111–112.

124

Performance of Some Plant Products as Additives to Insecticides against *Helicoverpa armigera* Hübner (Noctuidae: Lepidoptera)

N. VENUGOPAL RAO[1], P. RAJA SEKHAR[1], M. VENKATAIAH[1] AND A.S. REDDY[2]

[1]Heliothis Management Scheme, Andhra Pradesh Agricultural University
Regional Agricultural Research Station, Lam, Guntur 522 034, India
[2]Associate Director of Research and Principal Investigator, HMS (Retd.)
A.P. Agricultural University, Hyderabad, India

ABSTRACT

The field studies conducted at Regional Agricultral Research Station, Lam of Andhra Paradesh Agricultural University, indicated that some of the plant products, sesame oil, neem oil based formulation such as 'Repelin', 'Neem-guard', 'Bio-bitters', nicotine sulphate, parthnium extract, diflubenzuron, an insert growth regulator and the insect pathogen, Helio-NPV are useful as additives to conventional insecticides to manage the insecticide resistant bollworm, *Helicoverpa armigera*.

INTRODUCTION

Heliothis bollworm has become a pest of world-wide importance, because of its association with many food, fibre and vegetable crops of both tropical and temperate climate. Of late its outbreak was felt in Sundan, Egypy, Thailand, Zimbabwe, Australia, the United States and India with reference to Cotton (*Gossypium* spp.), an important commercial crop, grown in more than 35 million ha. In recent seasons the continuous failure of Cotton crop in Andhra Pradesh, Karnataka, Tamil Nadu, Maharastra, Gujarat and the Punjab could be directly attributed to the explosion of Heliothis bollworm. Observations and documentary evidence indicated that Heliothis has acquired "insecticide resistance" in Andhra Pradesh (Reddy, 1988; Mehrotra, 1988; Jayaraj, 1988; Sawicki, 1989 and Anonymous, 1990). In the process of indentifying methods of insecticide resistance management of the bollworm, a few natural products were evaluated as additives to conventional insecticides during the 1990-91 and 1991-92 seasons at the Regional Agricultural Research Station, Lam, Guntur.

MATERIALS AND METHODS

Fields experiment with neem oil and sesame oil, insect growth regulator-diflubenzuron and insect pathogen-Hellio NPV as additives to commonly used insecticides like endosulfan and cypermethrin was conducted during 1990-91 season. Cotton crop (CV. L 389) was raised with recommended agronomic practices to evaluate these selective treatments (Table 1) in randomised block design. An individual plot size of 31.5 m^2 was maintained with a spacing of 1.05 m and 0.6m between rows and plants, respectively. The treatments were replicated thrice. This experiment was repeated during 1991-92 including one more plant product-Parthenium extract since preliminary studies in the laboratory revealed the insecticidal property of this plant product (Anon., 1990). During 1991-92, a separate field experiment was also conducted to evaluate the efficacies of Biobitters (Neem seed kernal Extract nutrients) Repelin (Neem, Mahua and Pongamia oils), Nicotine sulphate and neemguard (Neem oil) in combination with the commonly used insecticide schedule (Is) comprising of monocrotophos, endosulfan, chlorpyriphos and quinalphos in september and acephate, triazophos and fenvalerate in October on cotton under Andhra Pradesh conditions. Data pertaining to the incidence in terms of eggs and larvae on five randomly selected plants in each treatment and damage to squares and bolls were collected in all the three experiments synchronizing with each round of insecticide application (seven to nine times). The efficacy of the treatments was taken into consideration with the cumulative impact (average of all the post spraying counts at weekly intervals) of the selective treatments imposed. Data on the seed Cotton yields collected through three to four pickings were also compared to make clear the ultimate influence on the crop. The data are presented in Tables 1, 2 and 3.

RESULTS AND DISCUSSION

During 1990–91, when the treatment impact, in terms of mortality of third instar larvae, was as low as 31.2. and 24.4 per cent with endosulfan and cypermethrin respectively, the action of these two was observed to be sufficiently high showing 54.4 and 50.0 per cent larval mortality with sesams oil. The efficacy of cypermethrin was also found significantly high when it was mixed with neem oil (40.0 per cent) or diflubenzuron (54.0 per cent) or NPV (60.0 per cent). This trend was also reflected in square and boll damage due to Heliothis excepting when it was mixed with NPV with regard to square damage. The square damage was as low as 8.6 per cent in the plots where sesame oil was used along with cypermethrin. The utility of all the four additives was felt in terms of boll damage also. The reduction in the larval population and comparatively low egg laying in the plots where sesame oil was mixed along with these selective insecticides strengthen the view that this plant product is advantageous against the bollworm. The

Table 1. Relative impact of insecticidal additives on the incidence of bollworm, damage levels and yields of cotton (RARS, Lam, 1990–91)

Treatment	Mortality of third Instar larvae in laboratory (% Leaf dip method)	incidence* (No.15 plants)		Damage (%)*			Seed cotton yield (kg/ha)	Yield increase due to additive treatment (kg/ha)
		Eggs	Larvae	Square	Flower	Boll		
Endosulfan (0.07%)	31.2^{de}	4.3^{b}	5.7^{b}	31.2^{b}	8.3^{b}	24.0^{c}	58^{bc}	—
Cypermethrin (0.006%)	24.4^{c}	2.3^{a}	3.7^{b}	17.4^{a}	3.7^{b}	15.0^{ab}	677^{bc}	—
Sesame oil (0.5%)	29.9^{cde}	3.0^{a}	4.7^{b}	39.3^{bc}	33.3^{a}	18.5^{b}	530^{cd}	—
Diflubenzuron (0.01%)	9.3^{b}	5.3^{b}	6.0^{b}	28.1^{a}	16.6^{b}	9.7^{a}	767^{c}	—
NPV (500 LE/ha)	12.6^{b}	6.0^{b}	5.7^{b}	30.8^{a}	6.7^{b}	9.3^{a}	581^{c}	—
Neem oil (0.5%)	11.8^{b}	2.0^{a}	7.3^{b}	24.3^{a}	25.0^{a}	10.9^{a}	645^{bc}	—
Endosulfan + Sesame oil	54.4^{fg}	2.0^{a}	3.0^{a}	20.9^{a}	0.1^{b}	9.7^{a}	1043^{a}	454
Cypermethrin + Sesame oil	50.0^{f}	1.0^{a}	2.7^{a}	8.6^{a}	1.8^{b}	7.4^{a}	1285^{a}	608
Endosulfan + Diflubenzuron	30.6^{cde}	3.3^{a}	4.7^{b}	16.4^{a}	3.7^{b}	8.2^{a}	988^{ab}	399
Cypermethrin + Diflubenzuron	54.0^{fg}	3.7^{b}	4.6^{b}	11.6^{a}	3.7^{b}	6.9^{a}	1062^{a}	385
Endosulfan + NPV	35.5^{de}	8.3^{b}	5.0^{b}	24.0^{b}	6.7^{b}	8.1^{a}	1013^{a}	424
Cypermethrin + NPV	60.0^{g}	8.1^{b}	3.3^{b}	24.8^{b}	1.1^{b}	10.0^{a}	1183^{a}	594
Endosulfan + Neem oil	27.2^{cd}	2.7^{a}	4.0^{b}	16.8^{a}	11.1^{ab}	5.2^{a}	924^{ab}	335
Cypermethrin + Neem oil	40.0^{e}	2.3^{a}	2.7^{a}	21.0^{a}	1.8^{b}	10.1^{a}	1252^{a}	575
Untreated check	0.0^{a}	3.7^{b}	6.3^{b}	30.4^{b}	33.3^{ab}	4.2^{a}	534^{cd}	—

* Mean of nine post spraying counts
Means followed by the same latter do not differ significantly
(CP = 0.05) by DMRT

Table 2. Cumulative performance of different treatment on bollworm of cotton (1991–92)

Treatment	Incidence/10 plants*		Damage (%)*		Seed cotton yield (kg/ha)	Yield increase due to additive treatment (kg/ha)
	Eggs	Larvae	Square	Boll		
Endosulfan (9.70%)	7.3^{a}	8.4^{a}	28.5^{a}	15.6^{b}	967^{b}	
Cypermethrin (0.006%)	7.5^{a}	8.0^{a}	26.3^{a}	13.5^{a}	983^{b}	
Sesame Oil (1.0%)	8.0^{b}	12.1^{a}	30.9^{a}	16.9^{b}	833^{c}	
Diflubenzuron (0.01%)	6.1^{a}	8.7^{a}	30.2^{a}	16.9^{b}	933^{c}	
NPV @ 500 LE/ha	4.0^{a}	7.9^{a}	24.7^{a}	15.4^{b}	927^{c}	
Neem Oil (1.0%)	4.5^{a}	8.2^{a}	28.4^{a}	14.7^{b}	920^{c}	
Parthenium (1.0%)	8.2^{b}	10.3^{a}	29.1^{a}	15.2^{b}	937^{c}	
Endosulfan + Sesame Oil (0.5%)	6.2^{a}	6.0^{a}	23.5^{a}	14.0^{a}	1170^{a}	203
Cypermethrin + Sesame Oil	4.8^{a}	5.2^{a}	22.0^{a}	8.5^{a}	1233^{a}	250
Endosulfan + Diflubenzuron (0.01%)	7.4^{a}	11.0^{a}	32.5^{a}	15.8^{b}	1160^{a}	193
Cypermethrin + Diflubenzuron	6.1^{a}	9.0^{a}	30.5^{a}	8.2^{a}	1150^{a}	167
Endosulfan + NPV (500 LE/ha)	10.3^{b}	6.9^{a}	24.4^{a}	13.7^{a}	1153^{a}	186
Cypermethrin + NPV	8.5^{b}	7.9^{a}	25.2^{a}	10.5^{a}	1133^{a}	150
Endosulfan + Neem Oil (0.5%)	6.1^{a}	7.9^{a}	28.5^{a}	18.3^{b}	1120^{a}	153
Cypermethrin + Neem Oil	5.2^{a}	9.2^{a}	26.7^{a}	10.8^{a}	1180^{a}	197
Endosulfan + Parthenium (1.0%)	7.1^{a}	8.9^{a}	26.4^{a}	11.7^{a}	1083^{b}	116
Cypermethrin + Parthenium	6.3^{a}	8.7^{a}	28.0^{a}	10.6^{a}	1143^{a}	160
Untreated check	3.8^{a}	6.4^{a}	34.4^{a}	21.3^{c}	677^{d}	

* Mean of seven post spraying counts
Means followed by the same letters are not significantly different ($p = 0.05$) by DMRT

Table 3. Evaluation of botanical insecticides against bollworm of Cotton (1991–92)

Treatments	Incidence (No./10 Plants)*		Damage (%)*		Seed Cotton Yield (kg/ha)	Yield increase due to additive (kg/ha)
	Eggs	Larvae	Square	Boll		
Parthenium (2.%)	4.7^{a}	19.9^{c}	34.8^{a}	11.3^{a}	1133^{d}	—
Parthenium + IS	13.6^{defg}	19.1^{c}	34.1^{a}	12.7^{a}	1267^{b}	134
Biobitters (1%)	10.3^{bcd}	18.2^{c}	37.8^{a}	8.1^{a}	867^{f}	—
Biobitters + IS	13.3^{cdef}	13.3^{a}	29.1^{a}	6.7^{a}	1067^{de}	200
Neemguard (1%)	6.1^{a}	20.0^{c}	36.4^{a}	7.4^{a}	1200^{cd}	—
Neemguard + IS	13.7^{defg}	13.0^{a}	30.0^{a}	5.2^{a}	1267^{b}	67
Nicotine sulphate (0.5%)	8.0^{abc}	14.4^{a}	31.9^{a}	7.8^{a}	1200^{cd}	—
Nicotine sulphate + IS	14.4^{bcde}	9.9^{a}	35.5^{a}	5.4^{a}	1333^{b}	133
Repelin RD-9 (1.5%)	6.6^{a}	12.7^{ab}	26.3^{a}	8.3^{a}	1267^{b}	—
Repelin + IS	13.6^{cdefg}	9.1^{a}	27.8^{a}	8.1^{a}	1533^{a}	266
Untreated check	8.3^{abc}	10.5^{ab}	29.4^{a}	7.4^{a}	933^{e}	—
IS alone	9.7^{bcd}	10.8^{ab}	29.6^{a}	8.2^{a}	1200^{cd}	—

* Mean of seven counts recorded after each spraying at weekly intervals
Means followed by the same letter do not differ significantly (P=0.05) by DMRT IS = Insecticide schedule

highest yield recorded with cypermethrin + sesame oil (12.85 o/ha) and cypermethrin + neem oil (12.52 o/ha) which was 608 and 575 kg/ha more than the lone treatments (Table 1) further highlighted the utility of these plant products under the existing situations.

The results of 1991–92 also revealed the increased action of cypermethrin and endosulfan when mixed with sesame oil, neem oil or other additives. Though there were no significant differences in the larval levels and square damages among treatments the pronounced action of seasame oil and neem oil either with cypermethrin or endosulfan resulted in an incerease of 153 to 250 kg of seed cotton/ha. Interestingly the action of parthenium extract was also observed when applied either with endosulfan or cypermethrin (Table 2).

The data collected through another field experiment also indicated the advantageous role of plant products. The pest incidence in terms of larvae per 10 plants was low in repelin + IS (9.1), followed by nicotine sulphate + IS (9.9) and neemguard + IS (13.0). The differences in terms of square and boll damage could not be felt due to interference of other factors. However, neemguard, necotine sulphate and biobitters in combination with insecticidal schedule recorded low boll damage comparatively. Repelin when mixed with insecticide schedule gave maximum seed cotton yield of 1533 kg/ha and it was superior to all other treatments. Among others, Repelin alone, the combination treatments of neemguard, parthenium and nicotine sulphate gave superior yields over others. The seed cotton yields were also found to be high in all the botanicals when mixed with insecticides. The increase was 266, 200, 134, 133 and 67 kg/ha due to the use of repelin, biobitters, parthenium extract, nicotine sulphate and neemguard respectively as compared to their lone treatments (Table 3).

These results are in agreement with the observations of Tirumala Rao et al. (1990), Butani and Mittal (1990), Dandapani (1985), Phadke (1988) and suggestions given by Abdual Kareem (1984), Jayaraj (1986), Saxena (1990), Chari et al. (1990), Ketkar and Ketkar (1990), Parmar (1992) and Subramaniam (1992) with regard to neem-based products. However, the insecticide synergism identified with sesame oil and parthenium extract is an observation to be further probed as a solution for insecticide failure, on pests like Heliothis bollworm.

REFERENCES

Abdual Kareem, A. 1984. Neem products: Their use in insect pest control. *Neem Newsletter* 1: 6

Anonymous. 1990. Heliothis Management Scheme (ICAR). RARS, Lam, Annual Progress Report, 1989–90.

Butani, P.G. and V.P. Mittal. 1990. Comparative efficacy of botanical insecticide (NSKS) and other insecticides against gram pod borer, *H. Armigera. Absts. Nat. Symp. on Problems and Prospects of Botanical Pesticides in IPM.* CTRI, Rajahmundry (A.P), 21–22 Jan. 1990. 12 pp.

Chari, M.S., S.R. Prabhu and G. Rama Prasad. 1990. Role of botanicals in pest management of tobacco in India. *Nat. Symp. on Problems and Prospects of Botanical Pesticides in IPM*, CTRI, Rajahmundry (AP), 21–22 Jan. 1990. pp. 1–4.

Jayaraj, S. 1986. Neem products for the management of pests and diseases in rice based cropping systems in Tamilnadu, India. *Abstr. 3rd Int. Neem Conf.* 10–15 July, Nairobi, Kenya. Published by GTZ, Eschborn.

Jayaraj, S. 1988. Keynote address delivered at the National Workshop on *Heliothis* Management, Tamil Nadu Agricultural University, Coimbatore, 18–19, Feb. 1988.

Ketkar, C.M. and M.S. Ketkar. 1990. Versatile neem—A source for plant protection. *Nat. Symp. on Problems and Prospects of Botanical Pesticides in IPM.* CTRI, Rajahmundry, 21–22 Jan., 1990. pp. 11–14.

Mehrotra, K.N. 1988. Another case of insecticide resistance *Heliothis* bollworm of cotton—A report submitted to the Vice-Chancellor of Andhra Pradesh Agricultural University on his visit to Guntur region of Andhra Pradesh.

Parmar, B.S. 1992. Use and limitations of botanical pesticides. *Nat. Sem. on Changing Scenario in Pests and Pest Management* in India. Plant Protection Association of India, CPPTI, Rajendranagar, Hyderabad, Andhra Pradesh. Jan. 31 to Feb. 1. 1992.

Phadke, A.D. 1988. Use of neem product in IRM in cotton. *Pesticides* 22 (4) : 36–37

Reddy, A.S. 1988. *Heliothis armigera* Hub. a serious threat to cotton cultivation in A.P. *Nat. Workshop on Heliothis Management*, Tamil Nadu Agricultural University, Coimbatore, Feb. 18–19, 1988.

Saxena, R.C. 1990. Neem as a source of natural insecticides. An update. *Abstr. Nat. Symp. on Problems and Prospects of Botanical Pesticides in Integrated Pest Management*. CTRI, Rajahmundry, India, Jan. 21–22, 1990, p. 1.

Sawicki, R.M. 1989. Report on the control of cotton pests in central and southern India. A tour report of an expert sponsored by Royal Society, London, U.K. and Indian National Science Academy.

Subramaniam, T.S. 1992. Neem and RD-9 repelin in pest control. *Nat. Sem. on Changing Scenario in Pests and Pest Management in India*. Plant Protection Association of India, CPPTI, Hyderabad, Jan. 31 to Feb. 1, 1992.

Tirumala Rao, K.,N. Venugopal Rao and A. Satyanarayana Reddy. 1990. RD-9 repelin, and effective plant origin insecticide against pigeonpea pod borer, *H. armiger*. *Abstr. of Nat. Symp. on Problems and Prospects of Botanical Pesticides in IPM*. CTRI, Rajahmundry, Jan. 21–22, 1990. 22 pp.

125

Effect of Botanicals on Certain Parasitoids

K. SRINIVASA BABU, M.S.N. MURTHY AND T. RAMESH BABU

Department of Entomology, College of Agriculture, Andhra Pradesh Agricultural University, Hyderabad 500 030, India

ABSTRACT

The effect of botanical insecticides Repelin and Neemguard were evaluated in laboratory as well as field condition along with conventional insecticides endosulfan and phosalone on parasitisation and subsequent development of *Trichogramma australicum* Girault, *Bracon hebetor* Say and *Tetrastichus israeli* (Mani and Kurian) the egg, larval and pupal parasitoids of *Opisina arenosella* Walker, respectively and their laboratory host *Corcyra cephalonica* Stainton during 1991–92.

Botanical insecticides were relatively safe at lower concentrations to all the three parasitoids and subsequent development of parasitoids in the laboratory and under field conditions. However, higher concentrations of botanical insecticides affected the parasitoids adversely. Endosulfan and phosalone were highly toxic to all the parasitoids. The botanical and conventional insecticides at recommended (field) concentrations had severe adverse effects on the egg and pupal stages of the host, *C. cephalonica* and were much less on the larval stage.

INTRODUCTION

Indiscriminate use of synthetic organic insecticides is associated with toxicity to non-target organisms, pest resurgence, development of resistance and eco-pollution. The problems cropped up owing to their imprudent usage have warranted botanical insecticides which were environmentally sound and eco-friendly. The effect of these botanical insecticides on the crop pests was well researched and documented. Schumutterer (1990) revived the antifeedent, repellent actions, effect on fecundity, growth and development of various neem based products against various crop pests. However, the literature pertaining to influnce of botanical insecticides against natural enemies of crop pests in scanty. In this context, it was contemplated to initiate laboratory cum field experiments to establish the effects of Repelin and Neemguard on parasitisation and development of *Trichigramma australicum* Girault, *Bracon hebetor* Say and *Tetrastichus israeli* Mani and Kurian, the egg, larval and pupal parasitoids, respectively of *Opisina arenosella* Walker, the black headed caterpillar of coconut.

MATERIALS AND METHODS

Two botanical insecticides each at three concentrations were used along with conventional insecticides. The common, trade and chemical name of these insecticides, their source of supply and concentrations used are listed in Table 1.

The cultures of test parasitoids were maintained on their respective stages of alternate host, *Corcyra cephalonica* Stainton under laboratory conditions at 25–27 °C and 70–80 per cent R.H.

Three different sets of experiments were carried out viz., (i) treating the parasitoids alone, (ii) treating the parasitoids and host and (iii) treating the host alone with a view to establish the effects of botanical insecticides against parasitoids.

The adult parasitoids of *T. australicum* (25 adults/100 eggs), *B. hebetor* (5 pairs/10 larvae) and *T. israeli* (30 paris/10 pupae) were released on their respective stages of host. Each treatment was replicated thrice. Studies were also undertaken to assess the effect of botanical insecticides on the development of eggs (100 / treatment), larvae (10/treatment) and pupae (10/treatment) of the host, *C. cephalonica*, with three replications.

The field experiments were conducted in the coconut garden to evaluate the effect of botanical insecticides on the same test parasitoids in three replications. Hundred freshly laid eggs of *C. cephalonica* were glued to the egg card was stapled to a leaflet of coconut frond and caged. Later, twenty five adult parasitoids of *T. australicum* were released into the cage. Ten third instar larvae of *Opisina arenosella* Walker were caged on a leaflet of coconut frond. After the larvae have constructed the galleries five pairs of *B. hebetor* were released into the cage. The pre-pupal stage of *O. arenosella* were allowed to pupate on a coconut leaflet by caging. After pupation adult parasitoids of

Table 1. Details of the insecticides used

Common name (Trade name)	Concentration (%)	Chemical name	Formulation	Source of supply
(R.D.9 Repelin)	0.50 1.00 2.00	Oils of Neem, Karanj, Mahua, Castor and Gingelly	100% EC	M/S ITC Ltd. ILTD Division
(Neemguard)	0.25 0.50 1.00	Neem based product	100% EC	M/S Gharda Chemicals
Phosalone (Zolone)	0.07	S (6-Chloro-2-Oxo-3 (2H) benzoxazol) methyl 0,0 diethyl dithionate	35% EC	M/s Voltas India Ltd.
Endosulfan (Thiodan)	0.07	1,4,5,6,7,7 Hexachloro, 2,3, bis(Hydroxy methyl) bicyclo (2,2,1) heptene-5	35% Ec	M/s Hoechst India Ltd.

T. israeli @ three pairs per pupa were released into the cage. In each cage only ten pupae were allowed to remain.

In the field the parasitoids were treated with the help of an atomizer. First the parasitoids were released into the cage and then the chemical is sprayed uniformly all over the cage. The larvae and pupae were treated *in situ* i.e. when they were in the galleries on the leaflets of coconut frond.

The data were collected on per cent host eggs parasitised, per cent adults emerged from parasitised eggs, per cent fresh eggs parasitized by resultant progeny and per cent adult parasitoids emerged from the parasitised eggs of *T. australicum*, an egg parasitoid. Similarly the data related to per cent host larvae parasitized, number of eggs laid per female, percentage eggs developed into adults, per cent fresh host larvae parasitized by resultant progeny and subsequent development of larval parasitoid, *B. hebetor* and per cent pupae parasitized, number of adults emerged from a parasitized pupa, per cent fresh host pupae parasitized by resultant progeny and number of adults emerged from a pupa parasitized by *T. israeli* were collected. The hatchability, larval and pupal development of the host *C. cephalonica* were also recorded. The data were subjected to randomized block design analysis.

RESULTS AND DISCUSSION

Trichogramma australicum Girault

Severe adverse effects were manifested wherever the treatment of adults parasitoids was involved sequelling in death of the parasitoids (Table 2). However, when treated host eggs were offered to untreated adult parasitoids, the parasitoids survived and the per cent parasitisation ranged from 41 to 72 with botanical insecticides whereas, the conventional insecticides inflicted adverse effects. The per cent adult emergence of *T. australicum* from parasitized eggs in botanical insecticides (around 68-80) though significantly less compared to untreated control (91 per cent) the differences are marginal. Similar trend was observed in the performance of the resultant progeny (Table 2) indicating that the adverse effects were carried over to F1 progency.

Rao (1990) reported low adult emergence of *T. chilonis* from Repelin treated eggs of *C. cephalonica*. *Trichogramma minutum* Riley and *Telonomus sphingis* (Ashmead) did not parasitize the eggs of hornworm, *Manduca sexta* (Johannson) on tobacco (Rabb and Bardley, 1968). However, Joshi et al. (1982) reported no adverse effects with the neem kernel suspension on the development of *Telonomus remus*. All the test chemicals at recommended concentrations were found to be highly toxic to the parasitoid in field and the per cent parasitization was as low as 39.6 per cent even with untreated control as against 90 per cent under laboratory conditions (Table 2).

Bracon hebetor Say

The recommended concentrations of all the test chemicals were found to be fatal to *B. hebetor* adult parasitoids under laboratory conditions (Table 3). The mean number of eggs laid by a female parasitoid on the treated host larva was recorded to be 12.2 and 10.3 with Neemguard 0.5 per cent and 1.0 per cent, respectively with a maximum of 15.3 in control. However, further development of the parasitoid was drastically affected as none of the eggs laid on the treated host larva developed into adults in the

Table 2. Effect of insecticides on *T. australicum* in laboratory and field

Treatment	Exp.3 Host alone treated					
	Parasitized host eggs (%)		Adults emerged from parasitized eggs (%)		Fresh host eggs parasitized by resultant progeny (%)	Adults emerged from eggs parasitized by resultant progeny (%)
	Lab	Field	Lab	Field		
Repelin 1.0%	65.0 (53.7)	15.7 (23.2)	75.4 (66.0)	70.2 (56.9)	68.7 (56.8)	80.5 (63.8)
Repelin 2.0%	41.3 (40.2)	10.7 (19.0)	68.6 (57.2)	65.5 (54.1)	43.3 (41.2)	71.5 (57.7)
Neemguard 0.5%	72.3 (58.2)	26.3 (30.8)	80.2 (63.6)	76.0 (60.7)	78.0 (62.1)	83.1 (65.8)
Neemguard 1.0%	44.6 (41.9)	13.3 (21.3)	71.7 (57.8)	68.5 (55.9)	49.7 (44.8)	72.5 (58.4)
Phosalone 0.07%	31.3 (34.0)	9.3 (17.8)	66.9 (54.9)	64.2 (53.2)	38.0 (38.1)	69.3 (56.3)
Endosulfan 0.07%	0.0	0.0	0.0	0.0	0.0	0.0
Control	88.4 (70.1)	33.7 (35.5)	91.34 (72.9)	86.1 (68.1)	88.0 (69.8)	92.4 (74.2)
C.D. (P = 0.05)	3.3	3.1	2.5	3.1	3.3	1.2

Values in parentheses are angular transformed

Table 3. Effect of insecticides on *B. hebetor* in laboratory

Treatment	Mean number of eggs laid by a female			Eggs developed into adults (%)		Fresh host larvae parasitized by resultant progeny (%)		Mean no. of eggs laid by a female		Eggs developed into adults (%)	
	Exp 1	Exp 2	Exp 3	Exp 1	Exp 2	Exp 1	Exp 2	Exp 1	Exp 2	Exp 1	Exp 2
Repelin 0.5%	11.27	10.00	—	76.97	76.06	86.67	80.0	12.53	11.80	79.7	79.1
				(62.17)	(60.72)	(68.86)	(63.4)			(63.3)	(62.9)
Repelin 1.0%	0.0	0.0	10.3	0.0	0.0	0.0	0.0	0.0	0.0	0.0	0.0
Repelin 2.0%	—	—	7.0								
Neemguard 0.25%	16.53	13.87	—	81.40	79.40	93.3	86.7	17.47	14.13	83.6	81.1
				(64.45)	(63.01)	(77.7)	(68.5)			(66.1)	(64.2)
Neemguard 0.5%	8.86	8.47	12.2	74.30	0.00	76.7	0.0	9.40	0.00	75.6	0.0
				(59.56)	(0.00)	(61.2)				(60.0)	
Neemguard 1.0%	—	—	8.4								
Phosalone 0.035%	0.00	0.00	—	0.0	0.0	0.0	0.0	0.00	0.00	0.0	0.0
Phosalone 0.07%	—	—	0.00								
Endosulfan 0.035%	0.00	0.00	—	0.0	0.0	0.0	0.0	0.00	0.00	0.0	0.0
Endosulfan 0.07%	—	—	0.0								
Control	18.80	17.67	15.30	89.37	84.93	100.0	100.0	18.47	17.80	85.4	83.3
				(70.78)	(67.12)	(90.0)	(90.0)			(67.5)	(66.3)
C.D (P = 0.05)	0.85	0.99	1.71	3.0	2.1	7.9		0.81	3.05	1.39	2.53

Values in parentheses are angular transformed
Exp 1 : Parasitoids alone treated. Exp 2: Parasitoids and host treated. Exp 3: Host alore treated.

botanical insecticides at field concentrations. Similar results were reported by Campbell and Duffey (1979) with another larval parasitoid, *Hyposoter exigua* when treated with alkaloid, tomatine. Conventional insecticides were proved to be detrimental as the female parasitoids did not lay the eggs even at reduced concentration (Table 3).

The botanical insecticides were found to be relatively safe to *B. hebetor* under field conditions, irrespective of the parasitoid or host treated (Table 4). The relatively severe adverse in the laboratory could be attributed to the restricted movement of adult parasitoids on the treated surface due to stickyness of Repelin and Neemguard and direct contact action of phosalone and endosulfan.

As the recommended concentrations were found to be fatal to the adults of *B. hebetor*, half the recommended concentrations were tried in the laboratory to establish the toxic effects. Botanical insecticides at reduced concentration registered 90 and 80 per cent parasitisation with Neemguard 0.25 per cent and Repelin 0.5 per cent, respectively, whereas, conventional insecticides at low concentration also exhibited detrimental effects. Subsequent development and performance of F1 progeny was also unaffected in botanical insecticidal treatments. Schmutterer (1990) stated that azadirachtin influence harmonal system and exert a dose dependant influence on fecundity. Half the recommended concentrations of the insecticides are suggested for combination treatments in the pest management programmes.

Tetrastichus israeli Mani and Kurian

The adult parasitoids of *T. israeli* also succumbed to recommended concentration of test chemicals under direct treatment in the laboratory (Table 5). The reduction in the parasitisation when treated host offered to untreated adult parasitoid was also found to be substantial in comparision to control in laboratory (Table 5) whereas, the differences were marginal under field conditions, irrespective of the parasitoid/host treated. Half the recommended concentration of botanical insecticide was however found to be relatively safe against parasitisation, subsequent development and performance of F1 progeny of *T. israeli* under laboratory conditions.

Corcyra cephalonica Stainton

Pupal stage was most vulnerable followed by egg and larval stages of *C. cephalonica* to botanical as well as conventional insecticides (Fig. 1). The adverse effects were however more severe at higher concentration. Similar results were also reported by Saxena et al. (1981) with the eggs of *Cnaphalocrocis medinalis* and Koul (1984) with *Dysdercus koengii*. The inhibition of egg hatch might be due to chocking effect of the oils present in the botanical formuiations (Schmutterer, 1990). Sharma (1992) reported that azadirachtin 0.5 to 2.5 1 topical application on the larvae of *C. cephalonica* has no effect on the development. The severe adverse effects on pupae observed were in confirmity with the results of Gujar and Mehrotra (1983).

It is quite evident from the present studies that the adverse effects of botanical formulations against natural enemies are less severe compared to conventional insecticides. However, further detailed investigations are necessary to establish the mode of action of botanical insecticides on natural enemies and their respective hosts, so that environmentally sound integrated pest management programmes can be developed, by including eco-friendly botanical insecticides as one of the components.

Table 4. Effect of insecticides on *B. hebetor* in the field

Treatment	Host larvae parasitized			Mean number of eggs laid by a female			Eggs developed into adults (%)		
	Exp 1	Exp 2	Exp 3	Exp 1	Exp 2	Exp 3	Exp 1	Exp 2	Exp 3
Repelin 1.0%	40.0 (39.2)	36.7 (37.2)	40.0 (39.2)	8.2	6.1	6.5	76.3 (60.8)	76.8 (61.3)	74.3 (59.5)
Repelin 2.0%	30.0 (33.2)	26.7 (30.9)	26.7 (31.1)	5.0	3.4	4.4	70.4 (57.0)	72.1 (58.2)	59.7 (56.6)
Neemguard 0.5%	50.0 (45.0)	46.7 (43.1)	46.7 (43.1)	9.6	8.0	9.0	80.6 (63.9)	80.0 (63.4)	80.8 (63.9)
Neemguard 1.0%	36.7 (34.2)	30.0 (33.0)	33.3 (35.2)	6.8	4.8	5.7	71.0 (57.5)	73.5 (59.1)	7.3 (57.0)
Phosalone 0.07%	0.0	0.0	0.0	0.0	0.0	0.0	0.0	0.0	0.0
Endosulfan 0.07%	0.0	0.0	0.0	0.0	0.0	0.0	0.0	0.0	0.0
Control	53.3 (46.9)	50.0 (46.9)	56.7 (48.9)	10.8	10.8	10.7	82.2 (65.0)	84.5 (66.8)	84.7 (67.0)
C.D (P = 0.05)	5.5	5.6	4.9	1.2	1.5	0.89	2.7	2.1	1.9

Values in parentheses are angular transformed
Exp 1: Parasitoids alone treated. Exp 2: Parasitoids and host treated. Exp 3: Host alone treated.

Table 5. Effect of insecticides on *T. israeli* in the laboratory

Treatment	Host pupae parasitized (%)			Mean number of adults emerged			Fresh pupae parasitized by resultant progeny (%)			Mean number of adults emerged		
	Exp 1	Exp 2	Exp 3	Exp 1	Exp 2	Exp 3	Exp 1	Exp 2	Exp 3	Exp 1	Exp 2	Exp 3
Repelin 0.5%	66.7 (55.0)	46.7 (43.1)	—	27.0	23.9	—	73.3 (59.0)	50.0 (45.0)	—	28.3	24.6	—
Repelin 1.0%	0.0	0.0	53.3 (46.9)	0.0	0.0	26.0	0.0	0.0	63.3 (52.7)	0.0	0.0	27.2
Repelin 2.0%	0.0	0.0	26.7 (31.6)	0.0	0.0	23.3	0.0	0.0	30.0 (33.0)	0.0	0.0	24.1
Neemguard 0.25%	76.7 (61.2)	63.3 (52.8)	—	27.7	25.4	—	80.0 (63.5)	66.7 (55.0)	—	29.5	26.6	—
Neemguard 0.5%	40.0 (39.1)	30.0 (33.3)	67.7 (55.0)	25.1	22.4	28.5	43.3 (41.1)	33.3 (35.2)	73.3 (58.9)	27.6	23.1	28.2
Neemguard 1.0%	0.0	0.0	30.0 (33.0)	0.0	0.0	23.4	0.0	0.0	36.7 (37.2)	0.0	0.0	25.0
Phosalone 0.035%	0.0	0.0	—	0.0	0.0	—	0.0	0.0	—	0.0	0.0	—
Phosalone 0.07%	0.0	0.0	16.7 (19.9)	0.0	0.0	14.6	0.0	0.0	23.3 (28.9)	0.0	0.0	16.7
Endosulfan 0.035%	0.0	0.0	—	0.0	0.0	—	0.0	0.0	0.0	0.0	0.0	—
Endosulfan 0.07%	0.0	0.0	0.0	0.0	0.0	0.0	0.0	0.0	0.0	0.0	0.0	0.0
Control	93.3 (77.7)	86.7 (72.2)	86.7 (68.9)	30.1	29.5	29.7	93.3 (77.7)	90.0 (75.0)	86.7 (68.6)	30.9	29.3	30.6
C.D (P=0.05)	8.01	9.66	15.8	1.8	1.7	8.4	8.3	6.0	17.5	1.4	2.1	9.7

Values in parentheses are angular transformed

Exp 1: Parasitoids alone treated. Exp 2: Parasitoids and host treated. Exp 3: Host alone treated.

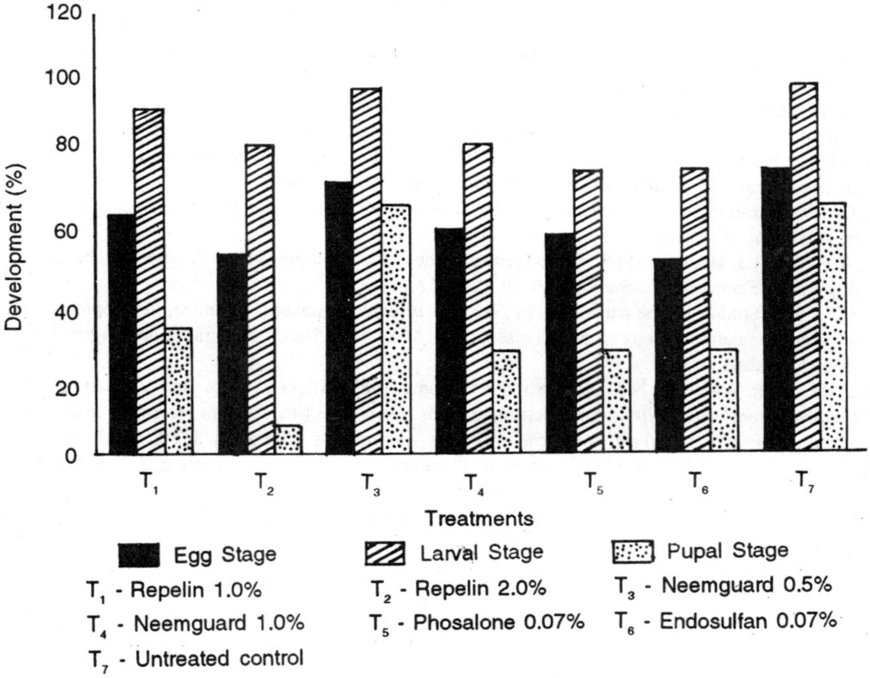

Fig. 1. Effect of insecticides on development stages of *C. cepholonica*

Table 6. Effect of insecticides on *T. israeli* in the field

Treatment	Host pupae parasitized (%)			Mean number of adults emerged		
	Exp 1	Exp 2	Exp 3	Exp 1	Exp 2	Exp 3
Repelin 1.0%	30.0 (33.0)	30.0 (33.0)	33.3 (35.2)	41.0	39.2	41.4
Repelin 2.0%	16.7 (23.8)	16.7 (18.4)	23.3 (28.7)	38.7	35.8	37.9
Neemguard 0.5%	46.7 (43.0)	33.3 (28.8)	46.7 (43.0)	45.1	41.8	46.0
Neemguard 1.0%	23.3 (28.8)	23.3 (28.8)	30.0 (33.0)	40.8	38.8	39.6
Phosalone 0.70%	0.0	0.0	16.7 (19.9)	0.0	0.0	25.2
Endosulfan 0.07%	0.0	0.0	0.0	0.0	0.0	0.0
Untreated Control	53.3 (46.9)	46.7 (43.0)	56.7 (48.9)	47.1	46.5	48.1
C.D	8.2	7.7	14.3	2.7	2.6	15.2

Values in the parentheses are angular transformed
Exp 1: Parasitoids alone treated. Exp 2: Parasitoids and host treated. Exp 3: Host alone treated

REFERENCES

Campbell, B.C. and S.S. Duffey. 1979. Tomatine and parasitic wasps: potential incompatibility of plant antibiosis with biological control. *Science* 205: 700–702.

Gujar, G.T. and K.N. Mehrotra. 1983. Inhibition of growth and development of the tobacco caterpillar *Spodoptera litura* Fabr. due to azadirachtin and other neem products. *Indian Journal of Entomology* 45(4): 431–34.

Joshi, B.G., G. Ramaprasad and S. Sitaramaiah. 1982. Effect of neem seed kernel suspension on *Telenomus remus* an egg parasite of *Spodeptera litura. Phytoparasitica* 10(1): 61–3.

Koul, O. 1984. Azadirachtin: I-interaction with the development of red cotton bugs. *Entomologia Experimentalis et. Applicata* 36: 85–8.

Rabb, R.L. and R.J. Bradley. 1968. The influence of host plants on parasitism of eggs of tobacco hornworm. *Journal of Economic Entomology* 60(5): 1249–52.

Rao, R.G. 1990. Studies on the parasitoids of *Heliothis armigera* Hubner (Noctuidae, Lepidoptera) with special reference to relative toxicity of certain insecticides. M.Sc. (Ag) Thesis, Andhra Pradesh Agricultural University, Hyderabad.

Saxena, R.C., G.P. Waldbaur, N.J. Liquido and B.C. Puma. 1981. Effect of neem seed oil on the rice leaf folder *Cnaphalocrocis medinalis* Natural Pesticides from Neem Tree (*Azadirachta indica* A. Juss.) *Proceedings of the I International Neem Conference,* Rottach - Egern, West Germany 1980, Eschborn, GTZ.

Schmutterer, H. 1990. Properties and potential of natural pesticides from neem tree (*Azadirachta indica*). *Annual Review of Entomology* 35: 271–94.

Sharma, G.K. 1992. Growth Inhibiting activity of azadirachtin on *Corcyra cephalonica. Phytoparasitica* 20(1): 47–50.

126

Effectiveness of Neem-based Products (Azadirachtin) against *Tetranychus macfarlanei* (Acarina: Tetranychidae) on Okra and Their Safety to Predatory Mites

A.B. RAI, G.R. BHANDERI AND C.B. PATEL

Department of Entomology
N. M. College of Agriculture
Gujarat Agricultural University
Navsari Campus, Navsari 396 450, India

ABSTRACT

Four azadirachtin-rich products in comparison to six conventional chemicals were evaluated under laboratory and field conditions for their toxicity to spider mite, *Tetranychus macfarlanei* Baker and Pritchard, and predatory mite complex, during 1990–91. Among phytoseiid, stigmaeid, and tydeid mite predators associated with prey mite, the first was the effective and dominant group. The results revealed significant superiority of the equally effective chemicals dicofol 0.03 per cent, endosulfan 0.07 per cent, and monocrotophos 0.04 per cent in reducing mite population (60.48 to 68.39 per cent). However, predatory mites markedely diminished in plots treated with dicofol, monocrotophos, and methyl-o-demeton (39.81 to 43.27 per cent reduction). Mite predators also diminished in the plot treated with plant products, but to a lower degree (22.89 to 25.87 per cent reduction) than with all other chemical treatments. Among the plant products, neemark 0.5 per cent was comparable with tetradifon, ethion, and methyl-o-demeton in controlling prey mite. The same products were tested in laboratory for their ovicidal, contact, and residual actions employing leaf dip technique. Repelin 0.5 per cent, margocide OK 0.8 per cent, and neemark 0.5 per cent showed comparable ovicidal action to promising chemical acaricides like dicofol, tetradifon, and ethion. Plant products had comparatively less acaricidal effect. Dicofol performed best, followed by monocrotophos. When residual action was studied 72 hrs after treatment, monocrotophos ranked first followed by dicofol, neemark, and endosulfan. As ovicides, plant products were comparable to promising chemicals. In addition, neemark in view of its good acaricidal action, residual action, and greater safety to predators can be manipulated and offers great scope in integrated pest management strategies against okra mite.

INTRODUCTION

Among the vegetable crops, perhaps okra and brinjal are the most affected by mites causing colossal economic loss throughout the country. Of 10 different species of mites reported on okra in India (Sadana, 1985), *Tetranychus macfarlanei* Baker and Pritchard, hitherto recorded as a minor pest, has now assumed a greater importance, causing appreciable damage to okra, cotton, and brinjal in southern and central Gujarat (Jose and Shah, 1989; Rai et al., 1991). There are a few reports on host range of this mite in India and abroad (Gupta, 1985). Though the pest can be effectively controlled by foliage application of chemicals (Pande and Yadav, 1983; Rai et al., 1991), the deleterious side effects of chemicals cannot be overlooked. In view of the danger involved in the use of synthetic chemicals, emphasis is now shifting in favour of biological control and use of botanical chemicals. Of the botanicals, neem has received greater attention from scientists all over the world. The technology for scientific extraction and enrichment of neem extracts to make it rich in azadirachtin have been developed. Many studies have focussed on effectiveness of neem products against insect pests in India (Pradhan et al., 1962; Jotwani and Srivastava, 1984). Practically no information is available on effectiveness of neem against mite pests, expect for a few scattered data (Sharma and Pande 1986; Devraj Urs, 1990). The present studies were carried out to observe the effectiveness of azadirachtin-rich products compared to conventional chemicals against *T. macfarlanei* on okra.

MATERIALS AND METHODS

A field trial with three replications was conducted at Agriculture Research Farm, Gujarat Agricultural University, Navsari, in randomized block design with 52 plants per plot (4 × 2 m) to study the relative efficacy of three azadirachtin-rich products in comparison to six conventional chemicals and control (with and without water spray) against *T. macfarlanei* and the predatory mites during 1990. The treatments were applied at recommended concentrations (Tables 1 and 2) synchronized with peak incidence of the mite (May 1990). For sampling, three leaves representing top, middle and bottom canopy were chosen randomly from each of five plants selected at random and tagged with luggage labels to count the number of spider mites (all post-embryonic stages combined) from whole leaf including upper and lower surfaces. These samples were also used to count mite predators (only adults combined of all predators). A hand lens (10x) was used to count the mites. The pretreatment counts were made a day before application of treatments and post-treatment counts were made 24 and 72 hrs after treatment. Further observations could not be made following considerable rains (120 mm). The percentage reduction of spider mite population as well as mite predators was calculated based on initial counts. The data were statistically analysed after angular transformation.

The same treatments incorporating one more treatment of plant products (RD-9 Repelin), with three replications, were evaluated in laboratory against eggs and adult stages of *T. macfarlanei* for their ovicidal, acaricidal, and residual action employing leaf dip technique (Tables 3 to 5). The excised okra leaf (4 × 3 cm) was dipped in treatment

Table 1. Field efficacy of different chemicals including biopesticides against *T. macfarlanei*

Treatments (T)	Per cent reduction after indicated period (P) of treatment		Pooled
	24 hrs	72 hrs	
Margocide OK 0.8%	42.87[d] (46.29)	45.55[ef] (50.96)	44.21[ef] (48.62)
Margocide CK 0.1%	42.37[d] (45.42)	44.10[f] (48.43)	43.23[f] (46.91)
Neemark 0.5%	44.93[cd] (49.88)	48.11[de] (55.42)	46.52[def] (52.65)
Dicofol 0.03%	52.93[a] (63.66)	58.64[a] (72.92)	55.79[a] (68.39)
Methyl-o-demeton 0.025%	47.74[bc] (54.77)	51.16[cd] (60.67)	49.45[bcd] (57.74)
Tetradifon 0.07%	43.75[d] (47.82)	50.89[cd] (60.21)	47.32[cde] (54.04)
Endosulfan 0.07%	48.56[bc] (56.20)	53.54[bc] (64.68)	51.05[abc] (60.48)
Monocrotophos 0.04%	49.03[b] (57.01)	56.50[ab] (69.54)	52.76[ab] (63.38)
Ethion 0.1%	48.08[bc] (55.37)	51.37[cd] (61.03)	49.73[bcd] (58.23)
Water spray	32.40[e] (28.71)	34.93[g] (32.78)	33.66[g] (30.72)
Control	24.16[f] (16.75)	25.14[h] (18.05)	24.65[h] (17.40)
SEM ± T	1.10	0.96	1.065
P			0.312
P × T Interaction			1.034

Figures in parentheses are retransformed values of arcsin transformation.
Mean values followed by same letters are not different at 5 per cent level of significance in respective column according to Duncan's New Multiple Range Test.

emulsions separately for a few seconds, hung vertically to remove excess of treatment emulsions, and then fan dried. The leaf bits were kept upside down on wet filter paper overlying a cotton wad in a petri dish. Twenty-five adult spider mites obtained from laboratory culture were released per leaf disc. After 24 and 48 hrs, mortality of the mite was counted and per cent mortality was calculated. The residual effect of treatments was determined after three days of treatment adopting the same technique. Twenty five adult mites were released after 72 hrs of treatment. The mortality of mites was estimated 24 hrs after release. In order to test ovicidal action, 10 adult females were released on the leaf bit. The next day, eggs were counted and adults removed. A similar procedure was followed by immersing leaf bits containing eggs in different treatment emulsions separately. Eggs were observed daily up to seven days. The unhatched eggs were considered unviable. The data were analysed statistically and interpreted.

Table 2. Toxicity of chemicals including biopesticides to predatory mites on okra

Treatment (T)	Per cent reduction after indicated period (P) of treatment		Pooled
	24 hrs	72 hrs	
Margocide OK 0.8%	29.84[d] (24.76)	31.30[c] (26.99)	30.57[e] (25.87)
Margocide CK 0.1%	27.59[d] (21.45)	29.56[c] (24.34)	22.89[e] (22.89)
Neemark 0.5%	29.32[d] (23.98)	31.58[c] (27.42)	30.45[e] (25.68)
Dicofol 0.03%	37.79[a] (37.89)	43.49[a] (47.37)	40.74[ab] (42.59)
Methyl-o-demeton 0.025%	39.98[a] (41.28)	42.28[a] (45.26)	41.13[a] (43.27)
Tetradifon 0.07%	34.49[bc] (32.07)	36.56[b] (35.48)	35.53[d] (33.77)
Endosulfan 0.07%	34.84[bc] (32.64)	36.55[b] (35.46)	35.70[d] (34.05)
Monocrotophos 0.04%	36.66[ab] (35.65)	41.58[a] (44.05)	39.12[abc] (39.81)
Ethion 0.1%	33.57[c] (30.58)	36.09[b] (34.70)	34.83[d] (32.62)
Water spray	22.20[e] (14.28)	23.59[d] (16.02)	22.89[f] (15.13)
Control (Without water)	14.24[f] (6.05)	17.21[c] (8.75)	15.72[g] (7.34)
SEM ± T	0.82	0.85	0.609
P			0.252
P × T Interaction			0.837

Figures in parentheses are retransformed values of arcsin transformation.
Mean values followed by same letters are not different at 5 per cent level of significance in respective column according to Duncan's New Multiple Range Test.

RESULTS AND DISCUSSION

Field Study: Effect of Application of Pesticides/Biopesticides on *Tetranychus macfarlanei* and Its Mite Predators

All the treatments provided significant reduction of spider mite population over two control treatments. However, among the controls, the water spray alone significantly reduced the mite population as compared to the plot which received no water spray. The reciprocal interaction between period of observation and treatment was significant, indicating treatments to be non-consistent in their effectiveness in different periods of observation after treatment. Dicofol provided maximum reduction of spider mite population (68.39 per cent) and was comparable with the treatments of monocrotophos and endosulfan, but it remained significantly superior to the rest of the treatments. Among the plant products, neemark (52.65 per cent reduction) was comparable with chemical

Table 3. Acaricidal test of chemicals including plant products against *T. macfarlanei* on okra

Treatments (T)	Per cent reduction of mite population at indicated period (P) of treatment		Pooled
	24 hrs	48 hrs	
Margocide OK 0.8%	29.95^{ef} (21.97)	40.61^{fg} (42.37)	34.28^{gh} (31.72)
Margocide CK 0.1%	38.41^{def} (38.60)	47.81^{efg} (54.90)	43.11^{efg} (46.70)
Neemark 0.5%	46.17^{cd} (52.04)	63.22^{cde} (79.70)	54.70^{de} (66.61)
Dicofol 0.03%	90.00^{a} (100.00)	90.00^{a} (100.00)	90.00^{a} (100.00)
Methyl-o-demeton 0.025%	41.22^{de} (43.42)	65.73^{bcd} (83.10)	53.48 (64.59)
Tetradifon 0.07%	28.29^{ef} (22.46)	44.61^{fg} (49.32)	36.45^{gh} (35.30)
Endosulfan 0.07%	57.26^{bc} (70.75)	67.81^{bcd} (85.74)	62.53^{cd} (78.72)
Monocrotophos 0.04%	69.73^{b} (88.00)	80.68^{ab} (97.38)	75.20^{ab} (97.47)
Ethion 0.1%	67.52^{b} (85.38)	72.29^{abc} (90.75)	69.61^{bc} (87.86)
Repelin	34.02^{def} (31.30)	51.14^{def} (60.63)	42.58^{fg} (45.78)
Control	23.47^{f} (15.86)	30.20^{g} (25.30)	26.84^{h} (20.39)
SEM ± T	4.46	4.95	3.343
P			1.420
P × T Interaction			4.709

Figures in the parentheses are retransformed values of arcsin transformation.
Mean values followed by same letters are not different at 5 per cent level of significance in respective column according to Duncan's New Multiple Range Test.

treatments like tetradifon, methyl-o-demeton, and ethion (54.04 to 58.23 per cent reduction: Table 1).

A fairly large complex of mite predators consisting of *Amblyseius* (*Euseius*) *alstoniae* Gupta, *A. (Proprioseiopsis) peltatus* Van der Merwe, and *A. (Typhlodromips) tetranychivorus* (Gupta) of phytoseiide, *Pronematus fleschneri* Baker and *P. ubicuitus* (McGregor) of tydeidae, and *Agistemus* sp. of stigmaeidae were encountered in association with *T. macfarlanei*. The phytoseiids were effective and the dominant group studies on the toxicity of different treatments against predatory mites revealed that predatory mites markedly diminished in plots treated with methyl-o-demeton, which was most toxic (43.27 per cent mortality) and comparable in its toxicity to predators with dicofol and monocrotophos (39.81 to 42.59 per cent mortality: Table 2). Endosulfan was found to cause comparatively low mortality of predators (34.05 per cent) and at the same time it was as highly effective as dicofol and monocrotophos against spider mite. All the tested plant products showed less adverse effect on predators (22.89 to 25.87 per cent mortality) than all other chemical treatments and also registered low mortality of the spider mites except neemark, which was as effective as moderately toxic chemicals such as tetradifon, ethion, and methyl-o-demeton. In the past many workers have found

Table 4. Residual toxicity of chemicals including plant products against *T. macfarlanei*

Treatments	Per cent reduction	
Margocide OK 0.8%	52.77^{cde}	(63.40)
Margocide CK 0.1%	55.33^{cd}	(67.64)
Neemark 0.5%	65.49^{abc}	(82.79)
Dicofol 0.03%	70.81^{ab}	(89.20)
Methyl-o-demeton 0.025%	56.17^{bcd}	(69.01)
Tetradifon 0.07%	45.97^{de}	(51.69)
Endosulfan 0.07%	63.27^{abc}	(79.77)
Monocrotophos 0.04%	73.29^{a}	(91.73)
Ethion 0.1%	54.67^{cd}	(66.56)
Repelin 0.5%	46.77^{de}	(53.09)
Control	38.93^{e}	(39.48)
SEM ±	4.32	

Figures in the parentheses are retransformed values of arcsin transformation.
Mean values followed by same letters are not different at 5 per cent level of significance in respective column according to Duncan's New Multiple Range Test.

dicofol (Bindra and Goyal, 1966; Nangia and Channabasavanna, 1983; Sharma and Pande, 1986; Shah et al., 1989; Rai et al., 1991), monocrotophos (Singh et al., 1990), and endosulfan (Jagdish and Channabasavanna, 1983) to be effective against spider mite infesting okra and other vegetable crops, which closely agrees with the present findings. Similarly, high toxicity of dicofol 0.002–0.01 per cent to predatory mites *Amblyseius tetranychivorus* (Nangia and Channabasavanna, 1983) and *A. alstoniae* (Jose, 1983) methyl-o-demeton 0.03 and monocrotophos 0.04 per cent (Jose, 1983) are also advocated which is in agreement with the present findings. Contrary to the present findings Jagdish and Channabasavanna (1989) found *A. tetranychivorus* to be most susceptible to endosulfan 0.075 per cent. However, in the present study, endosulfan was found moderately toxic.

Laboratory Study

Acaricidal Effect against Tetranychus macfarlanei

The results (Table 3) revealed that all the treatments were significantly superior over control except margocide OK (31.72 per cent mortality) and tetradifon (35.30 per cent mortality), which were at par with control (20.39 per cent mortality). Dicofol maintained its superiority, recording maximum kill of mite (100 per cent). It was comparable to the next best treatment, monocrotophos (97.47 per cent mortality), which was at par with ethion (87.86 per cent mortality). Amongst plant products, neemark ranked first (66.61 per cent mortality) and was comparable with endosulfan (78.72 per cent) and methyl-o-demeton (61.40 per cent mortality). Previously dicofol, monocrotophos, and endosulfan were reported as good acaricides to control the spider mite (Jose and Shah, 1989b; Rai et al., 1991). The effectiveness of neemark in the present study also conformed more or less with the findings of Pande et al. (1987) and Devraj Urs (1990), who found neem leaf extract 1 per cent and neem seed kernel 5 per cent to be effective

Table 5. Ovicidal effect of chemicals including plant products against *T. macfarlanei*

Treatments	Per cent hatching	
Margocide OK 0.8%	17.47[ab]	(9.01)
Margocide CK 0.1%	47.57[cd]	(54.48)
Neemark 0.5%	14.77[ab]	(6.50)
Dicofol 0.03%	15.38[ab]	(7.03)
Methyl-o-demeton 0.025%	57.90[d]	(71.76)
Tetradifon 0.07%	21.35[ab]	(13.25)
Endosulfan 0.07%	42.28[c]	(45.26)
Monocrotophos 0.04%	49.74[cd]	(58.24)
Ethion 0.1%	25.50[b]	(18.53)
Repelin 0.5%	8.89[a]	(2.39)
Control	71.25[e]	(89.67)
SEM ±	3.72	

Figures in the parentheses are retransformed values of arcsin transformation.
Mean values followed by same letters are not different at 5 per cent level of significance in respective column according to Duncan's New Multiple Range Test.

against *T. neocaledonicus* and *T. urticae*, respectively, showing the scope of azadirachtin in reducing pray mite population.

Residual Toxicity against T. macfarlanei

Monocrotophos 0.04 per cent was found to be the best treatment in its residual toxicity as it recorded the maximum kill of spider mite (91.73 per cent) 72 hrs after treatment. It was at par with dicofol, endosulfan, and neemark (79.77 to 89.20 per cent). The treatments of tetradifon and Repelin were least effective and comparable to the control in their residual toxicity (Table 4). Practically no information is available with regard to residual toxicity of plant products. The best residual action of monocrotophos 0.04 per cent and dicofol 0.03 per cent have also been reported by Jose and Shah (1989b).

Ovicidal Action against T. macfarlanei

The best ovicidal action of dicofol, tetradifon, and ethion as recorded in the present study closely agrees with the findings of Rai et al. (1991), who found tetradifon 0.1 and 0.5 per cent, dicofol 0.02 and 0.0.4 per cent and ethion 0.1 and 0.5 per cent to be best in ovicidal action against *T. macfarlanei*. Repelin, neemark and margocide OK showed best ovicidal action among plant products, recording minimum eggs hatching (2.39 to 13.25 per cent). Margocide CK methyl-o-demeton, endosulfan, and monocrotophos showed least ovicidal action as they recorded considerable hatching (45.26 to 71.76 per cent Table 5). There are many reports on effectiveness of dicofol, tetradifon, and ethion as ovicides (Gupta et al., 1972; Jose and Shah, 1989b; Rai et al., 1991), but no reports are available on ovicidal action of plant products. The ovicidal action of plant products in the present study may be due to persistence of oil film enveloping the eggs which may stifle the embryo, leading to adverse effects on egg hatching.

From the foregoing results, it can be stated that although plant products were relatively less effective as acaricides, their ovicidal action, particularly of neemark,

Repelin, and margocide OK, was as good as that of all the promising chemicals. They also showed least adverse effect on mite predators. Neemark in view of its good acaricidal action, residual action, and ovicidal action offers great scope in integrated past management strategies against okra mite. In chemical treatments endosulfan was comparatively safer to predators and as effective as the most promising acaricides (dicofol and monocrotophos), which were highly toxic to predatory mites, in reducing spider mite population.

REFERENCES

Bindra, D.S. and N.P. Goyal. 1966. Preliminary studies on the control of vegetable mite on bhindi. *J. Res. Punjab Agric. Univ*. 3: 174–177.

Devraj Urs, K.C. 1990. Efficacy of certain plant products in the control of brinjal red spider mite, *Tetranychus urticae. Fourth Nat. Sym. in Acarology*, Kerala, 12–14 Oct. 1990, Abstract: 68. Division of Acarology, Department of Zoology, University of Calicut, KTC Offset Press, Calicut.

Gupta, S.K. 1985. *Handbook of Plant Mites of India*. Zoological Survey of India, Calcutta, 520 pp.

Gupta, S.K., A.S. Sidhu, Gurdipsingh and M.S. Dhooria. 1972. Evaluation of new pesticides for the control of *T. cucurbitae* Rahman and Sapra infesting brinjal in Punjab. *Indian J. Agric. Sci*. 42: 921–924.

Jagdish, P.S. and G.P. Channabasavanna. 1983. Toxicity of different pesticides to *Tetranychus ludeni* in laboratory. *Acar. Newsl*. 12: 8–9.

Jagdish, P.S. and G.P. Channabasavanna. 1989. Toxicity of pesticides on *Amblyseius tetranychivorus* (Acari: Phytoseiidae), an effective predator of *T. ludeni*. In: *Progress in Acarology*, Vol. II. G.P. Channabasavanna and C.A. Viraktamath (eds.), Oxford & IBH Publishing Co. Pvt. Ltd., New Delhi, pp. 449–452.

Jose, V.T. 1983. Studies on biology of the spider mite *Tetranychus macfarlanei* B & P., its important predators and their population dynamics as influenced by agricultural chemicals. Ph.D. thesis, Gujarat Agricultural University, Ahmedabad.

Jose, V.T. and A.H. Shah. 1989a. Carryover of spider mite *Tetranychus macfarlanei* through alternate host plants in cotton growing areas of South and Central Gujarat, India. In: *Progress in Acarology*, Vol. II. G.P. Channabasavanna and C.A. Viraktamath (eds.), Oxford & IBH Publishing Co. Pvt. Ltd., New Delhi, pp. 29–31.

Jose, V.T. and A.H. Shah. 1989b. Chemical control of spider mite, *Tetranychus macfarlanei* (Acari: Tetranychidae) on cotton. In: *Progress in Acarology*, Vol. II. G.P. Channabasavanna and C.A. Viraktamath (eds.), Oxford & IBH Publishing Co. Pvt. Ltd., New Delhi, pp. 449–452.

Jotwani, M.G. and K.P. Srivastava. 1984. Neem in insect control. *Neem Newsletter* 1: 4–5.

Nangia, N. and G.P. Channabasavanna. 1983. Effect of sulphur and dicofol on *Tetranychus ludeni* and its predator *Amblyseius tetranychivorus*. In: *Plant Protection for Human Welfare*. British Crop Protuction Council, Cryodon, UK, 1028 pp.

Pande, Y.D., M. Majumdar and H.L. Roy. 1987. Laboratory evaluation of various concentrations of neem (*Azadirachta indica*) leaf extract against *Tetranychus neocaledonicus* Andre infesting okra in Tripura. *First Nat. Sem. on Acarology*, Kalyani, West Bengal, Oct. 29–31, Abstract: 43.

Pande, Y.D. and S.R.S. Yadav. 1983. Acaricidal test of seven pesticides against *Tetranychus macfarlanei* B & P. (Acari: Tetranychidae). *Pestology* 7(5): 17–20.

Pradhan, S., M.G. Jotwani and B.K. Rai. 1962. The neem seed deterrent to locusts. *Indian Fmg*. 12(8): 7–11.

Rai, A.B., A.S. Sejalia, C.B. Patel and A.H. Shah. 1991. Studies on okra mite *Tetranychus macfarlanei* (Acari: Tetranychidae) and its chemical control. In: *Modern Acarology*, Vol. I. F. Dusbabek and V. Bukva (eds.), Academia, Prague and SPB Academic Publishing, The Hague, pp. 571–580.

Sadana, G.L. 1985. *Plant Feeding Mites of India* (*Tetranychidae*). Kalyani Publishers, New Delhi, 29 pp.

Singh, O.P., J.J. Singh and N.N. Kapoor. 1990. Seasonal incidence and chemical control of red spider mite, *Tetranychus telarius* Linn. on soybean in Madhya Pradesh, India. *Indian J. Entomol*. 52: 57–62.

Shah, A.H., R.C. Jhala, C.B. Patel and J.R. Patel. 1989. Field efficacy of sulphur 70 w.p. in comparison to other acaricides against mites on brinjal, Indian bean, red gram, cotton, sorghum, mango and coconut. *Proc. Int. Symp. on Elemental Sulphur in Agriculture*, Vol. I. France, 25–27 Mar. pp. 283–289. Acropolis, Nice, France, Syndicate Francais du Soufre, La, Canebiere, Marseille, Cedex I.

Sharma, B.L. and Y.D. Pandey. 1986. Persistence and relative residual toxicity of some pesticides against *Tetranychus neocaledonicus* Andre on bottle gourd. *Pesticides* 20: 39–40.

127

Use of Botanical Formulations as Potential Repellents in Bird Pest Management

P. SYAMSUNDER RAO, K. SESHACHALAM BABU, M. ANAND RAO AND B.P. VARDHANI

All India Coordinated Research Project on Agricultural Ornithology, Andhra Pradesh Agricultural University, Rajendranagar, Hyderabad, India

ABSTRACT

Development of non-lethal, environmentally safe botanical formulations as bird repellents is needed to prevent heavy losses. Repellence tests conducted in captivity and field for the efficacy of botanical formulations Repelin, Wellgro and Neemostar with different concentrations at Hyderabad gave encouraging results. House sparrow (*Passer domesticus*), baya weaver bird (*Ploceus philippinus*) and roseringed parakeet (*Psittacula krameri*) showed feeding aversions and avoided consumption of the treated grains in the captivity and in the field difference in yield was recorded. In view of antifeedent properties and environmental safety, the botancial formulations may prove an alternative to chemical sprays in controlling the birds and warrant further investigations.

INTRODUCTION

Vertebrate pests play an important role in limiting agricultural production in many parts of the world. Among the vertebrate pests granivorous birds, next to rodents, attract the concern of farmers and scientists. Any factor that brings about reduction in food production has to be viewed seriously and warrants appropriate action. The loss due to bird depredation can be quite considerable in crops raised under isolation or around roosting and perching sites and loss is also known to vary in different seasons and in different regions. Many insectivorous birds seen in and around farmland and few crop depredators are beneficial to agriculture as they also feed on insect pests of crops. The birds thus play a dual role in the agricultural ecosystem.

Farmers mostly depend on the synthetic pesticides for the control of insects pests and increase in the use of these pesticides has caused considerable concern about health and environment besides adversely affecting birds and upsetting the ecological balance. Successful management of birds will not only reduce heavy crop losses but also help maintain the ecological balance. The development of non-lethal, not-toxic, economically viable and ecologically safe botanical formulations for pest management is needed.

BIRD PEST MANAGEMENT

Bird pest management has been a highly complex problem for both farmers and scientists mainly because of the dual role of bird pests in agriculture and because of the need to develop safe and economical measures that would not have an adverse effect on our ecosystem. The current practice in bird pest management include population regulation through lethal control habitat manipulation, chemical control (repellent and toxic baits) and various scaring techniques. It may be necessary to adopt a variety of methods to solve the bird damage problem; no single method provides full protection. The use of chemical as repellents is receiving much attention.

In India, studies were conducted on chemical repellents and the results obtained are not encouraging. Screening of chemicals for bird repellence on rice, sorghum, maize and sunflower was done with dimethoate (Rogor), carbaryl (Quinalphos) monocrotophos (Nuvacron), malathion, fenvalerate and fish oil rosin soap. These insecticides did not repels the birds. However, bird repellents like methiocarb (mesural), Curb and Thiram which were received as samples from foreign countries, when tested in experimental fields, showed antifeedent and repellent properties (Anon., 1980). Due to the increasing cost of these imported repellents further studies could not be carried out.

USE OF BOTANICAL FORMULATIONS AS BIRD PEST REPELLENTS

Deoiled neem cake solution (expeller cake) at 250 g/ 300 g/ was able to reduce the level of damage in maize (Shivanarayan and Anand Rao, 1988). The preliminary studies suggested concentration on the use of botanical formulations on cereal crops eaten by roseringed parakeet (*Psittacula krameri*), house sparrows (*Passer domesticus*), munias (*Lonchura* sp) and baya weaver bird (*Ploceus philippinus*). RD9 Repelin (a product from ITC) a plant extract concentrate of neem (*Azadirachta indica*) Karamji (*Pongamia glabra*), castor (*Ricimus communis*), Mahua (*Madhuca indica)* and sesame (*Sesamum indicus*) was tried in the laboratory and also in the field. Efficacy of Repelin 3-5 per cent concentration tried in cages provoked behavioural response from rose ringed parakeet, house sparrows, baya weaver birds and munias. Antifeedant properties with the treated grain were observed during the experimental period (Anand Rao et al., 1990). In the field the results were not satisfactory. Five per cent of concentration of Repelin RD9 tried in field on common millet at Punjab Agricultural University, Ludhiana, gave satisfactory results on small birds like house sparrow, and weaver birds but not on roseringed parakeet on sunflower crop (Sandhu and Toor, 1991). Similarly, other botaincal formulations like Necrep, an extract of neem kernel and Vitpomol ES from the plant *Vitex negundo* on house sparrow and baya weaver bird at 5 ml/kg paddy seed under caged conditions with choice and no choice showed repellence of the treated seed (Vardhani et al., 1992).

MATERIALS AND METHODS

The field experiment was conducted during Rabi and Kharif 1991–92 at the college farm, A.P. Agricultural University, Rajendranagar, Hyderabad and another at the farmers field in Nalgonda Distric (A.P.). There were six treatments and each one was replicated three times in randomised block division (RBD) including the control. The botanical formulations like Repelin, Wellgro (ITC) Nemostar (Shaw wallace) Neem cake solution and Necrep (Plant extract) were used as repellents and an equal area was kept with the Nylon net. The spraying was carried out at the time of milky stage of the grain and second spray was carried out after a gap of one week. Bird observation were made one day before the each spray and five days after the spray. Number of bird visitants both beneficial and harmful were observed in the morning and evening hours. Field labels were tied to healthy ear-heads in each treatment at random before the spray to test the efficacy of the chemical sprayed. At the time of harvest healthy and damaged earhead count was taken the yield was analysed treatment-wise. In the farmers field one acre of crop was taken with two treatments i.e. one botanical and physical method (Reflective ribbon) and results were analysed and compared with the control plot (usual farmers practice plots).

RESULTS AND DISCUSSION

The results revealed that among the treatment tested during Rabi 1991 neem cake solution with 250g/l gave maximum yield followed by 275g/l, 300g/l, Repelin 2, 4 and 5 per cent solutions. In Kharif 1991 Wellgro (an ITC product) showed maximum yield compared to Necrep 3 per cent, Nemostar 5 per cent, Repelin 5 per cent. These yields were compared with the control plots (Table 1). During Kharif 1992 enriched Repelin (3000 ppm) 10 ml/l recorded highest yields followed by 6 ml, 3.5 ml Nemostar 5 ml/l. A verification trial was conducted in farmers field to test the efficacy in natural situations also revealed high yields to those in the control plots (Table 1). The bird observations made before spraying and after spraying revealed that the number of birds recorded was less after each spraying (Table 2) However they started feeding after 5th day onwards.

The effect of chemical decreases after 4 days and gradually reduces its repellence. The effect of botanicals pesticide on birds indicate certain physiological changes or post ingestional disturbances causing aversions to feeding, thus given the effect of repellence. It has been claimed that 75 per cent degradation is still observed with the use of plant extracts. Hence the efficacy remains for 4–5 days (depending on day temperature). To achieve everlasting acceptance by farmers these plant extracts should be supplemented with one physical method (reflective ribbon). Experiment conducted in this direction in the farmers field showed satisfactory result in terms of yield compared to the control plot. Sandhu and Toor (1991) reported similar results with sunflower and pearl millets tried with one botanical and one physical method.

Table 1. Evaluation of botanical pesticides for bird pest repellency on sorghum crop

Season	Treatment	Yield kg/ha
Rabi 1991	Neem cake 250 g/l	825.00
	275 g/l	687.00
	300 g/l	587.00
	Repelin 2%	512.50
	4%	550.00
	5%	487.50
	Control	350.00
Kharif 1991	Wellgro 5%	1325.25
	Necrep 3 ml/l	1188.75
	Neem cake 300 g/kg	1162.50
	Repelin 5%	963.75
	Neemostar 5 ml/l	862.50
	Control	287.50
Kharif 1992	Welgrow 5%	2400.00
	RD 9 Repelin 10 ml/l (enriched 3000 ppm)	3262.50
	-do- 6 ml/l	3012.50
	-do- 3.5 ml/l	2462.50
	Neemostar 5 ml/l	2362.00
	Control	2425.00
In the farmer's field		
Kharif 1992	RD 9 Repelin 10 ml/l	1250.00
	6 ml/l	1150.00
	3.3 ml/l	1050.00
	Reflective ribbon	1250.00
	Control	925.00

Table 2. Showing the bird visitation data on sorghum crop before and after spraying of botanical pesticides

	WTM	BHM	SM	IB	RRP	IM	HC	Total
Kharif 1991	59 (32)	44 (46)	39 (25)	27 (28)	26 (20)	5 (0)	1 (0)	201 (148)
Rabi 1992	—	—	—	44 (20)	22 (39)	38 (64)	9 (0)	113 (103)
Kharif 1992	—	24 (23)	13 (6)	46 (26)	3 (–)	—	3 (0)	86 (55)

WTM = White Throated Munia; BHM = Black Head Munia; SM = Spotted Munia, RRP = Roseringed Parakeet IM = Indian Myna; HC = House Crow.
Figures in parentheses show the number of birds visited after the spraying.

INTEGRATED PEST MANAGEMENT

In view of their antifeedant properties, safe application and propagation of botanical pesticides in integrated pest management programmes is essential. With the increased awareness among farmers and scientists of the dual role of birds in agriculture, the study of their role with particular reference to their population in different agroecological regions, farming systems, and bird pest management programmes is of national importance.

CONCLUSION

For complete protection of crops from bird damage one method may not be foolproof. It would be more appropriate to supplement chemical repellents with scaring devices such as reflective ribbon or pyrotechnic method in an integrated pest management programme (Syamsunder Rao et al., 1991). Botanical formulations can be used as an alternative to chemical sprays which is ecologically safe and non-hazardous to birds and the environment.

REFERENCES

Anand Rao, M., B.P. Vardhani and P. Syamsunder Rao. 1990. Effect of botanical pesticides as repellents on bird predators. Proc. *National Symp. on Problems and Prospects of Botanical Pesticides in Integrated Pest Management*. Eds. Chare and Ramaprasad. Published by ISTS, 1993, CTRI, Rajahmundry, pp. 473–477.

Anonymous. 1980. Biological Control of Bird Pests. Andhra Pradesh Agricultural University, Final Report.

Sandhu, P.S. and H.S. Toor. 1991. Evaluation of one physical and two plant based chemical repellents against bird damage on ripening sunflower and common millet. *Int. Symp. on Environmental and Hormonal Approaches to Ornithology*. Nov. 27–Dec. 1, 1991. Srinagar Garhwal, V.P. (Abst.) p. 8.

Shivanarayan, N., and M. Anand Rao. 1988. Efficacy of neem cake as bird repellent on Maize. *Pavo* 26(1&2): 49–52.

Syamsunder Rao, P., K. Seshachalam Babu, C. Nagender Rao and M. Anand Rao. 1991. An integrated approach to the bird pest management. *International Symposium on Environmental and Hormonal Approaches to Ornithology*. Nov. 27–Dec. 1, 1991. Srinagar, Garhwal, V.P. (Abst.) p. 7.

Vardhani, B.P., P. Syamsunder Rao and G. Srimanarayana. 1992. The efficacy of certain plant extracts as repellents against house sparrow (*Passer domesticus*) and Baya weaver bird (*Ploceus philippinus*). *J. Appl. Zool. Res.* Vol. 3(2). Published by AZRA, 1992. Cuttack pp. 193–194.

13

THERAPEUTIC POTENTIAL AND OTHER MEDICAL APPLICATION

128

Azadirachta indica A. Juss. Stem Bark as an Anti-Leprosy Source

M.S. SUBRAMANIAN AND K.K. LAKSHMANAN

Department of Botany
Kongunadu Arts & Science College
Coimbatore 641 029, Tamil Nadu, India

ABSTRACT

Following the *siddha* system of Indian medicine the stem bark of *Azadirachta indica* A. Juss. was added as a constituent of an anti-leprosy drug SULAK along with parts of 21 plants. Results of pharmaco-phytochemical studies on the stem bark are presented.

Pharmacognosical characters of the bark are studied with the following parameters: taxonomy, anatomy, and powder study based on conventional methods. Phytochemical characters are studied with the following parameters: estimation of sugar, protein, free amino acids, and oil based on standard methods; the geochemicals copper, ferrous, manganese, potassium, sodium, zinc, phosphorus and sulphur (atomic absorption spectrophotometer-AAs, Varian Techtron Model Spectra 20 BA); the active principles alkaloid, anthraquinone heteroside, cardiac glycoside, glycone, phenol, sterol, tannin, and triterpene physical constants and chemical identity of the bark with that of standard anti-leprosy agents like clofazimine, dapsone, and rifampicin (Indian pharmacopoel methods).

The bark has high oil content (17.5 per cent); alanine, methionine, and phenylalnine are the free amino acids present in lesser amount. It is rich in ferrous (325 ppm) and sulphur (425.93 ppm), but copper is absent. Active principles like unsaturated sterols, triterpene, saponin, cardiac glycoside, 2-deoxy sugar, and phenol are determined. Among these phenols, unsaturated sterols, triterpenes, and saponin are rich. Maximum alkaloid precipitation is observed from water extractive. Ethanolic extractive gives rich tertiary alkaloid (0.070 mEq/100 g dry wt). Total solid and ash are 83 and 6.01 per cent respectively.

Chemical identity comparison of bark with standard antileprosy agents reveals that the activity is equivalent to that of clofazimine (4.3 per cent), dapsone (8.1 per cent), and rifampicin (2.2 per cent).

INTRODUCTION

Neem (*Azadirachta indica* A. Juss., Meliaceae) has been recognized for the unique insecticidal properties, bioactive limonoids, systemic action, and non-toxic, nonmutagenic, and boidegradable compounds. The wide range of insects affected has made the neem tree the focus of worldwide attention, including three international neem conferences (Daniel and Nakanishi, 1987). In India neem preparation has been in use for several millennia and Ayurveda regards it as 'sarva roga nivarini' (Jacobson, 1987).

Neem products are used for treatment of a whole gamut of diseases, including skin infection, cardiovascular disorder, diabetes, and cancer (Govindachari, 1992). It has important fungicidal and anti-malarial properties. Nimbidin from neem oil is effective in various skin diseases (Satyavathi et al., 1976). Neem oil inhibits the growth of all the three strains of *Mycobacterium tuberculosis* and *M. pyrogens* var. *aureus* (Chopra et al., 1956). Neem is used in the treatment of leprosy (Ojha, 1960). The water extract (10 per cent) of leaves shows antiviral activity (Rao et al., 1969). The gum from bark is a stimulant and demulcent tonic. It possesses anti-leprosy, antispirochaetal, and immenagogue properties (Nadkarni, 1954; Dastar, 1970; Satyavathi et al., 1976; Subramanian, 1986).

The earlier reports reveal that neem preparations work well against a variety of skin diseases and such reports are supported by present investigations. The stem bark is used as one of the ingredients in the preparation of the drug SULAK and its ointment to treat leprosy (Subramanian and Lakshmanan, 1988; Subramanian, 1990). The pharmaco-phytochemical characters and chemical identity comparison of bark with that of standard anti-leprosy agents like clofazimine, dapsone, and rifampicin (administered together in multi-drug therapy) are studied.

MATERIALS AND METHODS

The plant material was collected from Coimbatore at flowering stage and its voucher specimen was deposited in the herbarium of the Botany Department, Bharathiar University, Coimbatore, with Voucher No. MSS.1. The shade-dried bark was powdered with Wiley mill (5.0 mm) and processed for pharmaco-phytochemical studies by fluorescence evaluation (Chase and Pratt, 1949; Kokoshi et al., 1958) and by estimation of sugar (Somagyi, 1952); protein (Lowry et al., 1951), free amino acids (Jayaraman, 1985), and oil (Anon., 1975). It was processed for geochemical studies by atomic absorbtion spectrophotometer. Spectra 20 BA and flamephotometer, EEL Evans electroselinium (Paech and Tracey, 1979; Jackson, 1967), and determination of active principles (Fong, 1973; Viladomat et al., 1984). Physical constants were evaluated and chemical identity of bark was compared with that of standard anti-leprosy agents like clofazimine, dapsone, and rifampicin (Anon., 1070).

Neem bark is one of the ingredients in the preparation of the drug SULAK and its ointment along with parts of 21 plants (Table 1). The mode of administration is described elsewhere (Subramanian and Lakshmanan, 1988; Subramanian, 1990).

Table 1. Plant parts used for preparation of SULAK

Plant	Family	Part used
Ailanthus excelsa Roxb	Simaroubaceae	Stem bark
Andrographis alata (Vahl) Nees	Acanthaceae	Whole plant
A. paniculata (Burm.f) Wall. ex.Nees	Acanthaceae	Whole plant
Aristolochia bracteolata Lam.	Aristolochiaceae	Whole plant
Aristolochia indica L.	Aristolochiaceae	Whole plant
Azadirachta indica A. Juss.	Meliaceae	Stem bark
Citrullus colocynthis (L.) Schrades	Cucurbitaceae	Root
Clitoria ternatea L.	Leguminosae	Root
Corollacarpus epigaeus (Rott. & Willd.) Clarke	Cucurbitaceae	Tuber (root)
Evolvulus alsinoides (L.)	Convolvulaceae	Whole plant
Gymnema sylvestre (Retz.) R.Br. ex. Shult	Asclepiadaceae	Stem and leaf
Indigofera aspalathoides Vahl ex DC	Leguminosae	Whole plant
Leptadenia reticulata (Retz.) Wight and Arn.	Asclepiadaceae	Whole plant
Leucas aspera (Willd) Link	Labiatae	Whole plant
Ocimum tenuifolium L.	Labiatae	Whole plant
Phyllanthus amarus Schum & Thonn	Euphorbiaceae	Whole plant
Piper cubeba L.	Piperaceae	Fruit
Rhinacanthus nasutus L. Kurz	Acanthaceae	Whole plant
Strychnos nux-vomica L.	Loganiaceae	Stem bark
Toddalia asiatica (L.) Lam.	Rutaceae	Root
Wattakaka volubilis (L.F.) Stapf.	Asclepiadaceae	Stem and leaf
Zingiber officinale Rosc	Zingiberaceae	Rhizome

OBSERVATION AND RESULTS

Pharmacognosy

Macromorphology of Bark

The bark originates from the trunk is curved slightly, and has lichens on the outer surface. It is dark brown on the outer and light yellow on inner surface, has clean fissures, flat edges, longitudinal furrows, and long fractures. It is hard and stony.

Micormorphology of Bark

The bark is covered externally by cork, composed of radially arranged cells of variable size and form. The secondary cortex and phloem are predominant. The rays extend to inner boundary of cork. Tannin cells are large and conspicuous.

Powder Study

Organoleptic evaluation showed a copper leaf colour, woody odour, bitter taste, fine and fibrous nature, and heavy weight.

Microscopic evaluation showed fragments of epidermal cells, cork cells, macro and branchy scleroids, septate and aseptate fibres, tannin cells, and brown mass.

The fluorescence property of the powder as such, with chemical reagents and extracts from different solvents observed, is characteristic under 200 ~ 280, 280 ~ 320, and 320 ~ 400 nm UV light and is presented in Tables 2, 3, 4.

Physical constants evaluation (percentage on dry weight basis) showed that the bark powder has total solid 83 per cent, ash 6.01 per cent, and sulphated ash 8.32 per cent. The ash is insoluble in ethanol 3.58 per cent, water 2.01 per cent, and 1 N NaOH 0.77 per cent and is soluble in 50 per cent HCl.

The extractive values from different solvents are as follows: petroleum ether 10.92 per cent, solvent ether 16.05 per cent, benzene 14.07 per cent, chloroform 8.33 per cent, acetone 30.45 per cent, ethylacetate 5.25 per cent, ethanol 60 per cent, and water 38 per cent.

Phytochemistry

Biochemical Estimation (Percentage on Dry Weight Basis)

The bark powder has sugar 14 per cent, protein 14.3 per cent, free amino acid 2 per cent, and oil 17.5 per cent. Among free amino acids, alanine, methionine, and phenylalanine are distributed poorly.

Geochemical Estimation

Amount of mineral from the bark powder is as follows: ferrous 325 ppm, manganese 8.33 ppm, potassium 14 ppm, sodium 2.5 ppm, zinc 5.95 ppm, phosphorus 1.8 ppm, and sulphur 425.93 ppm.

Active Principle Determination (from Ethanolic Extract)

The results of the screening by colouration/precipitation method are presented in Tables 5, 6, 7.

Table 2. Behaviour of powder of *A. indica* stem bark with chemical reagents

Treatment of powder with	Colour under daylight
Powder as such	Copper leaf
1N NaOH (aqueous)	Truck brown
1N NaOH (in methanol)	Truck brown
Picric acid (saturated solution)	Mid-buff
Acetic acid (concentrated)	Brown
Hydrochloric acid (concentrated)	Truck brown
Nitric acid (concentrated)	Post office red
Sulphuric acid (concentrated)	Truck brown
Saliwanoff's reagent	Truck brown
Ferric chloride (5% solution)	Sandstone
40% NaOH (aqueous) + 10% lead acetate	Golden brown
Iodine solution (5%)	Brown
Sudan III (in ethanol)	Crimson red
Nitric acid (concentrated) + Ammonia solution	Golden yellow

NaOH : Sodium hydroxide

Table 3. Behaviour of powder of *A. indica* stem bark with chemical reagents under UV light

Treatment of powder with	Colour under UV light		
	200–280 nm	280–320 nm	320–400 nm
Powder as such	Sandstone	Copper leaf	Mid buff
Nitrocellulose (in amyl acetate)	Truck brown	Copper leaf	Copper leaf
1N NaOH (aqueous)	Leaf brown	Truck brown	Leaf brown
1N NaOH (aqueous) + Nitrocellulose in amyl acetate	Leaf brown	Leaf brown	Leaf brown
IN NaOH (in methanol)	Leaf brown	Leaf brown	Leaf brown
NaOH (in methanol) + Nitrocellulose in amyl acetate	Truck brown	Leaf brown	Leaf brown
IN hydrochloric acid	Leaf brown	Golden brown	Leaf brown
IN hydrochloric acid + Nitrocellulose (in amyl acetate)	Truck brown	Leaf brown	Leaf brown
50% nitric acid	Sandstone	Sandstone	Light grey
50% sulphuric acid	Olive green	Dark grey	Dark grey
Methanol	Golden brown	Golden brown	Leaf brown
Saliwanoff's reagent	Truck brown	Truck brown	Truck brown

Table 4. Behaviour of powder extractives of *A. indica* stem bark under day and UV light

Name of extractive	Colour under day light	Colour under UV light		
		200–280 nm	280–320 nm	320–400 nm
Petroleum ether	Colourless	Off-white	Colourless	Colourless
Solvent ether	Colourless	Olive green	Pistachio green	Opaline green
Benzene	Pale green	Opaline green	Pistachio green	Golden yellow
Chloroform	Lemon yellow	Sandstone	Opaline green	Eucalyptus green
Acetone	Copper leaf	Mid-buff	Golden brown	Mid-buff
Ethyl acetate	Copper leaf	Crimson red	Brown	Crimson red
Ethanol	Cherry red	Crimson red	Brown	Crimson red
Water pH 5	Cherry red	Post office red	Copper brown	Post office red
Water pH 7	Cherry red	Crison red	Truck brown	Crimson red
Water pH 9	Cherry red	Gulf red	New post office red	Gulf red

Chemical Identity Comparison

The result of the chemical identity comparison of the bark with the standard anti-leprosy agents clofazimine, dapsone, and rifampicin is presented in Table 8.

DISCUSSION

The bark has high oil content. The content and distribution of free amino acids are poor. Among minerals ferrous and sulphur are rich while copper is absent.

Bhandari and Mukerji (1959) described the morphology and chemistry of trunk bark. Sanyal et al. (1986) evaluated the root bark. Nimbin, nimbidin, and nimbosterol (Mitra et al., 1953), nimbin nimbidin, essential oil, and bitter principle (Chopra et al.,

Table 5. Phytochemical evaluation of *A. indica* stem bark powder

Active principle	Name of Test	Degree of colouration/precipitation
Leucoanthocyanins	–	–
Flavones	–	–
Glycones	–	–
Aglycones	–	–
Unsaturated Sterols and Triterpenes	Liebermann-Burchard Test	3+
	Salkowski Test	5+
Anthraquinone	Borntrager Test BZ	–
Heterosides	AZ	–
	Modified Borntrager Test BZ	2+
	AZ	–
Saponin	Froth Test	3+
	Liebermann-Burchard Test	3+
2-deoxy-sugar	Killer-Killiani Test	3+
Tannin		+
Phenol		7+
Cardiac	Keddi Reagent Test	3+
Glycosides	Modified Keddi Reagent Test	3+

Table 6. Alkaloids evaluation of *A. indica* stem bark powder

Extract	Degree of precipitation				
	DROG	HAG	SCH	VAL	WAG
Petroleum ether	+	+	–	+	–
Solvent ether	+	+	–	+	–
Benzene	+	+	–	–	–
Chloroform	+	+	–	–	–
Acetone	+	+	–	–	–
Ethyl acetate	+	+	–	–	–
Ethanol	+	+	–	+	+
Water	++	+	+	+	+

+ positive; – negative;
No. + indicates intensity of precipitation.

Table 7. Alkaloid estimation by titration from ethanolic extract of A. *indica* stem bark powder

Alkaloid group	Amount in mEq/100 g dry wt
Tertiary	0.070
Quaternary	0.030

The values are average of three replicates.

1956), and diterpenoids, nimbonone, and nimbonolone (Siddiqui et al., 1988; Ara et al., 1989) were isolated from stem bark. Essential oil and tannin (Bhattacharji et al., 1953), nimbinene (Wolfgang and Rudolf, 1978, 1981), and nimbidiol (Majumdar et al., 1987)

Table 8. Chemical comparison of *A. indica* stem bark powder with anti-leprosy agents

Anti-leprosy agents	Preliminary tests				Confirmatory test (percentage by titration)*
	A	B	C	D	
Clofazimine	4+	4+			4.3
Dapsone	2+	2+	5+	3+	8.1
Rifampicin	3+				2.2

* The values are average of three replicates.

ABCD. Colouration/precipitation reactions as per IP method. Dapsone has ABCD reactions in the tests, clofazimine has A and B, and rifampicin has A only.

were isolated from root bark. The usefulness of neem products and neem preparation was studied exhaustively by Jacobson (1987) and Govindachari (1992).

In the present study unsaturated sterol, triterpene, saponin, cardiac glycoside, 2-deoxy sugar, and phenol are determined. Among them phenol, unsaturated sterol, triterpene, and saponin are rich (Table 5). Maximum alkaloid precipitation was observed from water extractive (Table 6). The ethanolic extractive gave more tertiary alkaloid content from quaternary alkaloid (Table 7).

The present investigation also reveals that the bark powder has chemical identity with clofazimine (4.3 per cent), dapsone (8.1 per cent), and rifampicin (2.2 per cent) which are given in a multi-drug therapy at present to cure leprosy and hence the bark is an anti-leprosy source. Other reports have confirmed the anti-leprosy property of the bark (Nadkarni, 1954; Dastar, 1970; Satyavathi et al., 1976; Subramanian, 1986, 1990). Neem oil was effective in various skin diseases (Satyavathi et al., 1976; Govindachari, 1992) and inhibited the growth of all the three strains of *Mycobacterium tuberculosis* and *M. pyrogens* var. *aureus* (see Chopra et al., 1956). Neem products inhibit the growth of *M. leprae* in the mouse foot pad (Ojha, 1960).

ACKNOWLEDGEMENT

The first author is thankful to Dr. M. Aruchami, Principal and Mr. A. Thomas, Head, Dept. of Botany, Kongunadu Arts and Science College, Coimbatore for the facilities provided.

REFERENCES

Anon. 1970. *The Pharmacopoea of India.* Govt. of India, New Delhi.

Anon. 1975. *Official and Tentative Methods of the American Oil Chemists Society.* Vol. 1. AOCS, Illinois, Ohio. 375 pp.

Ara, I., B.S. Siddiqui, S. Faizi and S. Siddiqui. 1989. Diterpenoids from the stem bark of *Azadirchta indica. Phytochemistry* 28: 1177–1180.

Bhandari, P.R. and B. Mukerji. 1959. *The Neem Indian Lilac.* Eastpharmacist-1, Delhi.

Bhattacharji, S., C. Mitra and S. Siddiqui. 1953. Chemical examination of the trunk of neem. *J. Sci. Ind. Res.* (India). 12B: 154–156.

Chase, P.R. and R. Pratt. 1949. Fluorescence of powdered vegetable drugs with particular reference to development of a system of identification. *J. Am. Pharm. Assoc.* 38: 324–331.

Chopra, R.N., S.L. Nayar and I.C. Chopra. 1956. *Glossary of Indian Medicinal Plants.* CSIR, New Delhi.

Chopra, R.N., I.C. Chopra and B.S. Varma. 1969. *Supplementary to Glossary of Indian Medicinal Plants,* CSIR, New Delhi.

Daniel, R.S. and K. Nakanishi. 1987. A simplified isolation procedure for azadirachtin. *J. Nat. Prod.* 50: 241–244.

Dastar, J.F. 1970. *Medicinal Plants of India and Pakistan.* Taraporevala Sons and Co. Pvt. Ltd., Bombay.

Fong, H.H.S. 1973. I~VII Phytochemical screening. In: *Experiments in Pharmaceutical Sciences.* M.H. Malone and S.L. McLaughlin (eds.), Uni Pacific Pub. Services, California, Stockton, pp. 43–51.

Govindachari, T.R. 1992. Chemical and biological investigations on *Azadirachta indica* (the neem tree). *Curr. Sci.* 63: 117–122.

Jackson, M.L. 1967. *Soil Chemical Analysis.* Prentice Hall of India (P) Ltd., New Delhi.

Jacobson, M. 1987. In: *Handbook of Natural Pesticides*, Vol. 3. *Insect Growth Regulators*, Part B. E.D. Morgan and N.B. Mandava (eds.), CRC Press Inc., Boca Raton, Florida.

Jayaraman, J. 1985. *Laboratory Manual in Biochemistry.* Wiley Eastern Ltd., Madras, pp. 63–66.

Kokoshi, J., R. Kokoshi and F.J. Slama. 1958. Fluorescence of powdered vegetable drugs under ultraviolet radiation. *J. Am. Pharm. Assoc.* 47: 715.

Lowry, D.H., N.J. Rosebrough, A.L. Jan and R.J. Randed. 1951. Protein measurement with folin phenol reagent. *J. Biol. Chem.* 193: 265–275.

Majumdar, P.L., D.C. Matti, W. Kraus and M. Bobel. 1987. Nimbidiol modified diterpenoid of the root bark of *Azadirachta indica. Phytochemistry* 26: 3021–3023.

Mitra, C., P. Narasimha and S. Siddiqui. 1953. Chemical examination of root bark of neem *Azadirachta indica*, *Melia azadirachta. J. Sci. Ind. Res.* (India) 12: 152–153.

Nadkarni, K.M. 1954. *Indian Materia Medica.* Karnataka Printing Press, Bombay.

Ojha, D. 1960. Nimbadi vepa in the treatment of leprosy. A preliminary report. *Indian J. Med. Sci.* 20: 217–221.

Paech, K. and M.V. Tracey. 1979. *Modern Methods of Plant Analysis*, Vol. 1. Narosa Publishing House, New Delhi.

Rao, A.R., S. Sukumar, T.B. Paramasivam, S. Kamalashi, A.R. Parasuraman and M. Shantha. 1969. Study of anti-viral activity of tender leaves of margosa tree (*Melia azadirachta*) on vaccina and variola virus. A preliminary report. *Indian J. Med. Res.* 57: 495.

Sanyal, M., A. Mukerjee and P.C. Datta. 1986. Evaluation of root bark of *Azadirachta indica* (Meliaceae). Medicinal identification morphology and chemical constituents. *J. Plant Anat. Morphol.* (Jodhpur) 3: 13–20.

Satyavathi, G.V., M.K. Raina and M. Sharma, Ed. 1976. *Medicinal Plants of Indica*, Vol. 1. ICMR, New Delhi. pp. 112–117.

Siddiqui, S., B.S. Siddiqui, S. Faizi and T. Mahmood. 1988. Tetracyclic triterpenoids and their derivatives from *Azadirachta indica. J. Nat. Prod.* (Lloydia) 51: 30–43.

Somagyi, M. 1952. Notes on sugar estimation. *J. Biol. Chem.* 195: 19–23.

Subramanian, M.S. 1986. An overview on anti-leprotic plants. J. *Swamy Bot.* Cl. 3: 1–7.

Subramanian, M.S. 1990. Pharmacognosy and phytochemistry of some anti-leprosy plants. Ph.D. thesis, Bharathiar University, Coimbatore.

Subramanian, M.S. and K.K. Lakshmanan. 1988. Thozhu noi neekkum davara janjeevi (in Tamil). In: *Biology and Biotechnology*, Part I. Anandavalli Mahadevan, ed., M.K. University, Madurai, pp. 48–54.

Viladomat, F., C. Codina, J. Bastida, M. Calobardes and M. Serrano. 1984. Alkaloid screening of Catalonia (Spain) Plants, I. *J. Nat. Prod.* 47: 64–69.

Wolfgang, K., and C. Rudolf. 1978. 17-Epi azadiradion and 17-hydroxy azadiradion, two new constituents from *Azadirachta indica* A. Juss. *Tetrahedron Lett.* 27: 2395–2398.

Wolfgang, K., and C. Rudolf. 1981. Pentanortriterpenoids from *Azadirachta indica* A. Juss. (Meliaceae). *Biol. Act. Nat. Prod.* (Proc.) 3: 27–29.

129

Identification and Characterization of the Immunomodulatory Fraction from Neem Seed Extract Responsible for Long-term Antifertility Activity

SANJAY GARG, G.P. TALWAR AND S.N. UPADHYAY

National Institute of Immunology, New Delhi 110 067, India

ABSTRACT

The extracts from neem seeds are known to have potent immunomodulatory activities. A single intrauterine administration of a purified extract of *Azadirachta indica* (neem) seeds was found to cause a long-term reversible block of fertility in rodents and monkeys. To identify and characterize the active fraction responsible for this particular activity, the seeds were extracted with a series of nonpolar to polar solvents and three fractions so obtained were studied for antifertility activity after intrauterine administration. Peanut oil was used as control. The nonpolar fraction showing the biological activity was further fractionated using solvent partitioning method. The fractionation and bioactivity study were continued till the last active fraction, which on preliminary characterization, appeared to be lipid-terpenoid associate. Further characterization of this fraction is in progress.

INTRODUCTION

Azadirachta indica (syn. *Melia azadirachta*, margosa) known as neem or nimba, which in Sanskrit means 'the health giving' belongs to the family Meliaceae, is indigenous to India, and is widely distributed in Asia, Africa and other tropical parts of the world. Neem oil, an extract from neem seeds, has been reported to possess a number of biological activities including anti-inflammatory, antipyretic (Okapanyi and Ezenknw, 1981), hypoglycemic (Pillai and Santhamukari, 1981) antifungal and antibacterial (Rao et al., 1986) activity. In addition, as an antifertility agent neem oil is reported to possess spermicidal (Sinha et al., 1984a), anti-implantation, and abortifacient (Sinha et al., 1984b) activity indicating its potential as an intravaginal contraceptive.

In recent years, much attention has been paid by many international authorities to one of the most serious problems faced by the human race, population explosion. Along with all possible approaches to contraception, there has been a tremendous interest in exploring the exploitation of the basic traits of immunological approach such as specificity, recognition, and memory to intercept events critical to reproduction. Various feasible immunological approaches such as use of sperm or ova antigens and antibodies to hormones of the reproductive system are under extensive investigation (Talwar, 1986).

While humoral immunity against antigens of reproductive system has received due consideration for immunocontraception, cell-mediated immunity has been totally ignored, probably due to lack of our understanding of the cellular mechanisms involved in reproduction. From our current understanding of the immune system it appears that it is unlikely to induce selective and specific response in the genital tract using conventional immunization procedures. However, many traditional plant products, which are known to have immunomodulatory activities generally act by non-specific stimulation of the immune system, implying non-antigen dependent stimulation of the function and efficiency of macrophages, granulocytes, and complement and natural killer cells.

Upadhyay et al. (1990) have reported a novel approach for immunocontraception by intervention of local cell mediated immunity in the reproductive system by using *Azadirachta indica* (neem) seed oil for long-term and reversible blocking of fertility after a single intrauterine application. The mechanism suggested is that the neem oil activates the local population of uterine leucocytes and that subsequent exposures to allogenic sperm or blastocytes, after normal mating, initiate a cellular immune response within the uterine compartment leading to blocking of implantation.

The combined ethnopharmacognostic research, natural products chemistry, and modern immunological screening techniques facilitate an efficient analysis of immunomodulatory activities of plants used in traditional medicine (Hadden et al., 1988; Hamburger and Hostetmann, 1991). Moreover, this approach may provide a useful basis for extended research aimed at the rationalization of the therapeutic use of such plants and the isolation of immunoactive constituents.

The chemical and therapeutic studies on neem undertaken since about the beginning of this century were initially concerned with the fatty acid components and amorphous bitter substances of the oil. More than 100 constituents (Siddiqui et al., 1988) have so far been isolated from different parts of the tree and their structure elucidated. These studies were directed towards antifeedant and insect repellent activities. Further investigations of the development of neem oil or the extract from the seeds as a long term contraceptive are limited by the question of finding the active principle from the total constituents of the seeds.

MATERIALS AND METHODS

In order to identify the immunocontraceptive fraction from neem seeds, detailed chemical investigation along with biological testing at each step was carried out. For testing the *in vivo* antifertility activity, the fraction after solubilization or emulsification (as the case may be), using gum acacia or polyvinyl alcohol as an emulsifying agent, was injected into the right uterine horn of adult female Wistar rats of proven fertility. In the contralateral horn peanut oil was injected to serve as control. The animals were allowed to recover for 15 days and then left to mate with males of proven fertility. They are observed daily for cyclicity, sperm positivity and pregnancy. The animals were sacrificed on days 15–20 of sperm positive mating and observed for number of implantations on the treated side versus control side. In cases, where no implantation was observed on the treated side and normal implantations were observed on the control side, the fraction was considered to be biologically active as an antifertility agent. The experiments were carried out on a statistically significant number of animals.

In the initial stages, fractionation of mechanically expressed oil was carried out using different techniques of separation such as solvent partitioning, saponification, and column chromatographic fractionation. The fractions obtained by these methods were not biologically active. To avoid the inconsistencies of the mechanically expressed oil, solvent extraction was used in further studies.

RESULTS

Extraction of neem seeds using hexane, ethanol, and water in a sequence yielded three fractions. Repeated experiments on extraction and biological activity have revealed that the hexane fraction retains all the activity and is devoid of any toxicity. Alcohol and water fractions were inactive and the former caused some pathological changes in the uterine epithelium.

Hexane fraction on fractionation using column chromatography yielded a fraction with significant antifertility activity but it was not completely active. Saponification resulted in inactive fractions. Solvent partitioning using aqueous methanol yielded aqueous methanolic extract (10.5 per cent) of the total solvent extractable matter was found to be biologically active fraction and the residue was inactive. The detailed investigations on physicochemical properties of active fraction have shown it to be highly rich in fatty acids, both in the free form and as glycerides. Terpenoids are also present as lipidterpenoid associate in this fraction. Studies on the antifertility activities of subfractions of this fraction are in progress.

CONCLUSION

It can be concluded from the present study that neem oil can be fractionated to yield immunocontraceptive fraction or pure principle, which will definitely have better stability and lesser toxicity, and probably have better activity than the total crude oil.

REFERENCES

Hadden, J.W., F. Spreafico, Y. Yamamura, K.F. Austen, P. Dukor and K. Masek, Eds. 1988., *Advances in Immunopharmacology*, 4, *Proc. 4th Int. Conf. Immunopharmacology*, Osaka, Japan, 16 to 19 May, Pragmon city.

Okpanyi, S.N. and G.C. Ezenknw. 1981. Anti-inflammatory and antipyretic activities of *Azadirachta indica*, *Planta Med.* 41: 34.

Pillai, N.R. and G. Santhakumari. 1981. Hypoglycemic activity of Neem oil, *Indian J. Pharmacol.* 13: 91.

Siddiqui, S., B.S. Siddiqui, S. Faizi and T. Mahmood. 1988. Tetracyclic triterpenoids and their derivatives from *Azadirachta indica*, *J. Nat. Products* 51: 1, 30.

Sinha, K.C., S.S. Riar, R.S. Tiwary, A.R. Dhawan, J. Bardhan, P. Thomas, A.R. Kain and R.K. Jain. 1984a. Neem oil as a vaginal contraceptive. *Indian J. Med. Res.* 79: 131.

Sinha, K.C., S.S. Riar, J. Bardhan, P. Thomas, A.R. Kain and R.K. Jain. 1984b. Antiimplantation effects of neem oil, *Indian J. Med. Res.* 80: 708.

Rao, D.K., K. Singh, P. Chopra, P.C. Chabra and G. Ramanajalu. 1986. *In vitro* antibacterial activities of neem oil, *Indian. J. Med. Res.* 84: 314.

Talwar, G.P., Ed. 1986. Contraception Research for Today and the Nineties. Springer Verlag.

Upadhyay, S.N., C. Kaushic and G.P. Talwar. 1990. Antifertility effect of neem by single intrauterine administration: a novel method of contraception. *Proc. Roy. Soc. Lond. B.* 242: 175.

Hamburger, M. and K. Hostettmann. 1991. Bioactivity in plants: the link between phytochemistry and medicine, *Phytochemistry* 30, 12: 3864.

130

Clinical Studies with Praneem Polyherbal Cream in Chlamydial Cervicitis

SANJAY GARG[1], A. MITTAL[2], S.K. KAPUR[2], S. GUPTA[3], S.N. UPADHYAY[1] AND G.P. TALWAR[1]

[1]National Institute of Immunology, New Delhi, India
[2]Institute of Pathology, ICMR, New Delhi, India
[3]Safdarjung Hospital, New Delhi, India

ABSTRACT

The extracts from the seeds of *Azadirachta indica* are reported to be potent modulator of immune reactions. To investigate the clinical implications of this property, the purified extract of neem seeds (Praneem) has been formulated as Praneem Polyherbal cream, along with two other components of plant origin. This cream is studied to assess the clinical efficacy in the treatment of proven cases of leucorrhea caused by *Chlamydia trachomatis*. Sixteen cases with proven Chlamydial cervicitis having a score of 3 to 4 plus on severity scale of 0–4 were asked to use 5 ml of the cresam daily, intravaginally, one hour before going to bed. Smears and cultures were studied on day 8, 15 and 22 of application of the cream. Out of the 16 patients enrolled, 14 became negative to C. *trachomatis* after 7–21 days of treatment. One patient discontinued the treatment because of mild irritation and another was lost to follow up. The formulation has proved to be highly effective in alleviating the *C. trachomatis* from the cervicovaginal region.

INTRODUCTION

Neem (*Azadirachta indica* A. Juss. family Meliaceae) is unique in its amazing array of properties and uses. Neem oil has been demonstrated to have potent immunomodulatory activities (Upadhyay et al., 1992). It has been shown that its action is mediated through activation of macrophages, which pick up the oil on exposure and induce the production of gamma interferon by T lymphocytes.

Extract from neem seeds, along with Reetha saponins (extract from the pericarp of *Sapindus mukrossi*) and quinine hydrochloride, have been formulated as water washable emulsion based polyherbal cream. The cream has been developed using ingredients approved in the Indian Pharmacopoeia and tested extensively for acute and subacute toxicity in monkeys, Draize test in rabbits, and cumulative skin sensitivity in human volunteers. The ingredients of the polyherbal cream have a direct action on a number of microorganisms, supported by an indirect action through activation of local cell-mediated immune reactions. Keeping in mind these activities the polyherbal cream was studied in 16 patients of Chlamydial cervicitis in Safdarjung Hospital in New Delhi.

The female genital tract can be infected by a number of sexually transmitted organisms (McGregor, 1989). Those commonly isolated from symptomatic patients include *Neisseria gonorrhea, Trichomonas vaginalis, Candida albicans, Gardenella vaginalis* and *Chlamydia trachomatis*.

Chlamydia trachomatis is a genital pathogen responsible for a variety of clinical syndormes and is one of the common organisms isolated in women with leucorrhea (Mittal, 1992). Studies have reported that the prevalence rate of chlamydia from cervix of women attending venereal disease clinic was 15–30 per cent (Stamm and Holmes, 1984) and *C. trachomatis* antigen has been demonstrated in 31 per cent of the patients (Bhujwala et al., 1992) suffering from pelvic inflammatory diseases and infertility. The symptoms include a friable cervix, mucopurulent cervical discharge, and oedema of an area of ectopy (Mittal et al., in press).

The purpose of the present study was to assess the clinical efficiency of the polyherbal cream in confirmed cases of Chlamydial cervicitis in patients attending a gynaecology clinic. The subjects with confirmed *C. trachomatis* infection were asked to use the cream and observed at weekly intervals for microbiological and clinical status after daily application of the cream.

MATERIALS AND METHODS

Polyherbal Cream

The polyherbal cream was developed from pharmaceutically acceptable ingredients including the vehicle, stabilizers, antioxidants, and perfume to mask the unpleasant odour of neem extract. The formulation is packed in 50 ml collapsible tubes and an applicator is supplied for intravaginal administration of a measured dose.

Patients

The study was conducted in women attending the gynaecology outpatient department of Safdarjung Hospital in New Delhi with clinical signs and symptoms of Chlamydial cervicitis. Most of the patients had a history of vaginal discharge, dyspareumia and pain in the lower abdominal region. Endocervical smear on a cotton swab was used to perform the microbiological investigations.

Diagnostic Methods for *C. trachomatis*

For diagnosis of *C. trachomatis* infection (Mittal et al., in press), three different methods were used. 1) Direct examination of smear (Giemsa cytology), 2) antigen detection by

ELISA and 3) isolation in a cell culture system (Ripa and Mardh, 1977). The patients were scored on a severity scale of 0 to 4 and only the confirmed cases were selected for trials.

Dose Schedule

The subjects were prescribed administration of 5 ml of the polyherbal cream intravaginally, daily before going to bed.

Follow-up

The patients were asked to report on days 8, 15 and 22 of application of the cream. They were observed for clinical and microbiological status, and also for the side effects, if any. On each visit they were scored on the same severity scale.

RESULTS

The trials were completed in 14 patients out of the 16 enrolled. All the patients recovered completely with the treatment, both clinically and microbiologically. One patient discontinued treatment because of mild irritation and another was lost to follow up. No side effects were observed in the patients treated.

DISCUSSION

Direct and indirect (mediated through immune systems) effects of traditionally used herbal remedies can be used in prophylaxis and therapeutics of a number of diseases in humans. Neem seed extract, reported to have an immunomodulatory activity, also has an antimicrobial activity. The present study reports the development and clinical studies of a neem-based formulation. The polyherbal cream was found to be highly effective in the treatment of Chlamydial cervicitis. Out of the 14 patients, seven recovered with seven applications, three with 14 applications and one with 21 daily applications. Three patients came for follow up only after 28 days.

Although the mechanism of action of polyherbal cream is still unknown, the present study signifies the efficacy of the cream in Chlamydial cervicitis, wherein it was observed that the formulation is highly effective in alleviating *C. trachomatis* in the cervicovaginal region. The modified protocol, which involves studying the effect of this cream on other microorganisms, also is in clinical stages.

REFERENCES

Bhujwala, R.A., P. Seth, A. Gupta and N.C. Bhargawa. 1992. Nongonococcal uerthritis in males, a preliminary study, *Indian J. Med. Res.* 75, 485.

McGregor, J.A. 1989. Chlamydia infections in women and other sexually transmitted diseases. *Obstet. Gynecol. Clin. North Am.* 16, 289.

Mittal, A. 1992. Genital chlamydias, *ICMR Bull.* 22, (10): 103–105.

Mittal, A., S. Kapoor and S. Gupta. (in press). Chlamydial cervicitis: Role of culture, enzyme immunoassay and Giemsa cytology in diagonsis APMIS (in press).

Ripa, K.T. and P.A. Mardh. 1977. New simplified culture technique for *Chlamydia trachomatis*, non-gonococcal urethritis and related infections. *Am. Soc. Microbiol.* 6, 323–327.

Stamm, W.E. and K.K. Holmes. 1984. *Chlamydia trachomatis* infections in the adult. In: *Sexually Transmitted Diseases* . K.K. Holmes and P.A. March (eds.), Mcgraw-Hill Inc., New York, 259 pp.

Upadhyay, S.N., S. Dhawan, S. Garg and G.P. Talwar. 1992. Immunomodulatory effects of neem oil, *Int. J. Immunopharmacol.* 14(7), 1187–1193.

131

Rat Toxicity Studies with Neem Oil

P. BADRI SRIMAN NARAYANAN, R. SAMPATHRAJ AND G. VANITHA KUMARI

Division of Endocrinology, Department of Zoology, Bharathiar University
Coimbatore 641 046, India

ABSTRACT

Chronic administration of neem oil to adult albino rats for eight days was not well tolerated and there was clear cut evidence of a very mild microscopic lesions in both the liver and the kidney as observed histologically. Liver, kidney function tests were performed by determining the glutamate oxaloacetic and glutamic pyruvic transaminase enzymes and cholesterol in these tissues. Blood/serum parameters like haemoglobin, blood glucose, serum proteins and transaminases, and serum cholesterol were also determined. All the parameters showed marked changes indicative of toxic effect of the oil.

INTRODUCTION

Neem oil is traditionally known to possess a wide spectrum of medicinal properties (Kirtikar and Basu, 1965) and it is reported to exhibit spermicidal activity (Sinha et al., 1984). The intravaginal application of this oil before and after coitus prevents pregnancy (Sinha et al., 1984). Neem oil after absorption enters the circulation through vaginal mucosa and manifests its antifertility effects (Riar et al., 1988). Further, neem oil administration to rats caused a significant alteration in testicular cell population resulting in the inhibition of androgen production (Ashok Purohit and Dixit, 1991). In the view of the potential spermicidal and therapeutic value of neem oil, it is imperative to ascertain the toxic effect, if any, of the compound; otherwise, untoward side effects may occur. Since the liver and kidney are important organs involved with drug metabolism and excretion, it is essential to study the effect of any pharmaceutical agents on these organs. Hence, in the present investigation, an attempt has been made to study the probable effects of neem oil on the liver and kidney functions as well as on their histological architecture.

MATERIALS AND METHODS

Neem Oil

The compound neem oil was obtained from Bio-Organic Ltd., Madras, with a label purity of 99 per cent.

Animals

Albino rats of Wistar strain weighing 150–200 g were obtained from our department animal house and kept in individual cages under controlled light and temperature conditions (20°–22°C). They were provided with standard diet (Hindustan Lever Ltd., India) and water *ad libitum*. All animals were kept in quarantine for 20 days before drug treatment and grouped as follows:

Group I served as control and received olive oil as vehicle for eight consecutive days intramuscularly.

Group II received 25 mg of neem oil/100 g body weight/day for eight consecutive days intramuscularly.

Group III received 50 mg of neem oil/100 g body weight/day for eight consecutive days intramuscularly.

Twenty-four hours after completion of treatment, the animals were killed by decapitation. About 4 ml of blood was collected in heparinized tubes. After clotting for 4 hrs at room temperature, blood was centrifuged at 3000 rpm for 10 min., serum was separated and stored at –20°C until assayed for biochemical parameters.

Biochemical Studies

Required amounts of tissue from each organ were homogenized in 0.01 M tris buffer (pH 7.4) and centrifuged at 3000 rpm, and the supernatant was used for biochemical analysis. Total protein in serum and tissues were estimated by the method of Lowry et al. (1951). Serum and tissue glutamate oxaloacetic transaminase (GOT) and glutamic pyruvic transaminase (GPT) were measured following the method of Reitman and Frankel (1957). Blood haemoglobin was determined by the method of Gambino and Waraksa (1969) and blood glucose by the method of Sasaki et al. (1972). Serum cholesterol was estimated by the method adopted by Hanel and Dam (1955).

Histological Studies

After decapitation, liver and kidneys were dissected out, free from adjoining fatty tissue and mesenteries, washed in chilled saline, and blotted on a filter paper. A sample of tissue from each organ was fixed in Bouin's fluid. After routine processing of each tissue, paraffin sections (7 µm) were stained using Harris' haemotoxylin-eosin.

Statistical Analysis

The data were analysed statistically using Student's *t* test.

RESULTS

Tissue Biochemical Parameters

Table 1 depicts the effect of neem oil on the activities of liver and kidney transaminases and total protein concentrations. There was a marked decrease in the protein concentration in both the tissues studied. While GOT and GPT were the decreased in the liver, a reverse trend was observed in the kidney.

Blood and Serum Parameters

Serum protein concentration as well as the activity of GOT were markedly reduced after neem oil treatment at all the doses administered (Table 2). However, an elevated serum GPT activity was noticed in the drug-treated groups. The blood haemoglobin concentrations were increased only in the high dose drug-treated group with no significant change in the blood glucose levels.

Histological Studies

Very mild lesions only were noticed in both the liver and kidney tissues after neem oil administration.

Table 1. Effect of neem oil on the total protein concentrations and activities of transaminases in the liver and kidney of rats

Tissue parameter	Group I	Group II	Group III
Liver			
Total protein	3.56 ± 0.25	2.96 ± 0.13 *	2.61 ± 0.13*
GPT (units/min./mg protein)	24.54 ± 0.2	19.16± 0.13**	15.58 ± 0.003***
GOT (units/min./mg protein)	32.14 ± 0.3	23.39 ± 0.38***	17.54 ± 0.13***
Kidney			
Total protein (mg/ml)	3.98 ± 0.008	3.84 ± 0.005	3.02 ± 0.13*
GPT (unit/min./mg protein)	2.04 ± 0.13	7.31 ± 0.13***	15.36 ± 0.13***
GOT (unit/min./mg protein)	15.54 ± 0.36	19.84 ± 0.006***	26.38 ± 0.25***

Each value is mean ± SEM of five estimations.
*P < 0.05, **P < 0.001, and ***P < 0.001 : Control vs experimentals.
Group I : Vehicle treated.
Group II : Received 25 mg of neem oil/100 g body wt./day for eight consecutive days intramuscularly.
Group III : Received 50 mg of neem oil/100 g body wt./day for eight consecutive days intramuscularly.

Table 2. Effect of neem oil on a few serum and blood biochemical parameters

Tissue parameter	Group I	Group II	Group III
Serum			
Cholesterol (mg/ml)	202.6 ± 0.17	192.4 ± 0.32	181.9 ± 0.1*
Total protein	29.02 ± 0.125	22.06 ± 0.18***	21.61 ± 0.17***
GPT (units/min./mg protein)	59.50 ± 0.13	67.10 ± 0.13***	74.50 ± 0.10***
GOT (units/min./mg protein)	148.40 ± 0.10	116.21 ± 1.27***	72.14 ± 1.54***
Blood			
Haemoglobin (mg/ml)	10.4 ± 0.05	11.2 ± 0.14	11.4 ± 0.15*
Glucose (mg/ml)	63.5 ± 0.16	64.8 ± 0.18	66.2 ± 0.41

Each value is mean ± SEM of five estimations.
*p < 0.05 and ***p < 0.001: Control vs experimentals.
Group I : Vehicle treated.
Group II : Received 25 mg of neem oil/100 g body wt./day for eight consecutive days intramuscularly.
Group III : Received 50 mg of neem oil/100 g body wt./day for eight consecutive days intramuscularly.

DISCUSSION

In the present investigation, no mortality was observed in any of the groups tested with neem oil. As liver and kidney are physiologically related in regard to excretory, metabolic, and vasoregulatry functions, they have common susceptibilities to injurious factors. Histological studies did reveal mild changes. Gross inspection of these organs revealed mild lesions attributable to the neem oil treatment.

Proteins are the chief components of cellular structure and organizations. The decrease in protein concentration observed in the present investigation may be due to the enhanced proteolysis or decreased utilization of amino acids for protein synthesis in the liver. As most of the proteins are formed in the liver, even a mild lesion in this organ may decrease the synthesis of protein (Oser, 1965). The altered protein concentration may also be responsible for biochemical, hematological, and even histopathological lesions in the animal body.

GPT and GOT are present in high concentration in the hepatocyte. GPT catalyses the interconversion of alanine to oxoglutarate to glutamate and pyruvate. GOT plays a major role in malate-aspartate shuttle which appears to be the primary mechanism for the transfer of reducing equivalents from cytosol to mitochondria in rat liver (Williamson et al., 1971) and kidney (Rose et al., 1977).

These enzymes have also been related to the regulation of tissue glycolysis and further utilization of glutamic acid. The observed decrease in the activity of GPT and GOT enzymes in liver could be due to the decreased glycolytic flux. On the other hand, the occurrence of gluconeogenesis in kidney is still enigmatic; the GPT and GOT enzymes may have some other metabolic functions. Therefore, in the present study, the observed increase in the GPT and GOT enzymes in the kidney might be due to active operation of alanine-glucose cylce or aspartate-malate shuttle, probably not conducive to tissue functioning.

Blood serum contains a number of organic constituents such as haemoglobin, protein, bilirubin, cholesterol, and enzymes whose concentrations vary with changes in the physiological states. Serum enzymes, especially GOT and GPT, are often considered sensitive indicators to adverse drug effects or to the presence of disease (Tietz, 1986). In the present investigation, all the organic constituents of the blood and serum showed marked changes with the exception of blood glucose and haemoglobin levels.

The multiple changes in the serum protein levels may have resulted from altered rates of anabolism and catabolism. Similar changes in the serum biochemical constituents were observed after administration of sitosterol, a phytosterol, in rats (Malini and Vanithakumari, 1990).

Thus, from the foregoing data it has been observed that neem oil exerts an adverse effect on the liver and kidney especially at the high doses administered.

ACKNOWLEDGEMENTS

We thank Prof. G. Sundarajulu, Head Department of Zoology, Bharathiar University, Coimbatore, and one author (`.B.) gratefully acknowledges Prof. N. Namasivayam, Principal, NGM College, Pollachi, for his continuous encouragement.

REFERENCES

Ashok Purohit and V.P. Dixit. 1991. Antispermatogenic efficacy of neem (*Azadirachta indica* A. Juss) materials in rats. *Neem Newsletter* 8(2): 13–14.

Gambino S.R. and A.J. Waraksa. 1969. An automatic instrument for the measurement of haemoglobin concentratio. *Am. J. Pathol.* 52: 557–564.

Hanel, H.K. and H. Dam. 1955. The estimation of cholesterol. *Acta Chem. Scan.* 9: 677–682.

Kirtikar, K.R. and B.D. Basu. 1965. *Indian Medicinal Plants*, Vol. 1. Periodical Experts, New Delhi. pp. 436–441.

Lowry, O.H., N.J. Rosebrough, A.L. Farr and R.J. Randall. 1951. Protein measurements with the folin phenol reagent. *J. Biol. Chem.* 193: 265–275.

Malini, T. and G. Vanithakumari. 1990. Rat toxicity studies with B-sitosterol. *J. Ethnopharmacol.* 28: 221–234.

Oser, B.L. 1965. *Hawks Physiological Chemistry*, 14th ed. Tata McGraw Hill Pub. Co., New Delhi, pp. 499–509.

Reitman, S. and S. Frankel. 1957. A colorimetric method for the determination of serum glutamate oxaloacetic and glutamic pyruvic transaminases. *Am. J. Clin. Pathol.* 28: 56–63.

Riar, S.S., J. Bardhan, P. Thomas, A.K. Kain and R. Parshad. 1988. Mechanism of antifertility action of neem oil. *Indian J. Med. Res.* 88: 339–342.

Ross, R., P. Silva and S. Bullock. 1977. role of malate aspartate shuttle in renal sodium transport in the rat. *Clin. Sci.* 60: 419.

Sasaki, J., S. Matsuy and A. Sanal. 1972. Effect of acetic acid concentration on the colour reaction in the O-toluidine boric acid method for blood glucose determination. *Rinsho Kagaku* 1: 346–353.

Sinha, K.C., S.S. Riar, R.S. Tiwary, A.K. Dhawan, J. Bardhan, P. Thomas, A.K. Kain and R.K. Jain. 1984. Neem oil as a vaginal contraceptive. *Indian J. Med. Res.* 79: 131–136.

Tietz, N.W. 1986. *Fundamental of Clinical Chemistry.* W.B. Saunders, Philadelphia and London, 447 pp.

Williamson, J.R., R. Jokob and C. Refino. 1971. Control of the removal of reducing equivalents into mitochondria. *Arch. Biochem. Biophys.* 158: 263–268.

132

Alteration in Testicular Histoarchitecture Following Neem Oil Administration in Albino Rats

M.P. MANORANJITHAM, A. PREM ANANDHI, R. SAMPATHRAJ AND C. VANITHAKUMARI

Division of Endocrinology, Department of Zoology, Bharathiar University
Coimbatore 641 046, India

ABSTRACT

Two different doses of neem oil (*Azadirachta indica* A. Juss.) administered to adult albino rats caused marked structural alteration in the testes. Spermatogenesis was impaired drastically. Biochemical parameters like protein and glycogen concentrations as well as the activities of enzymes like lactate dehydrogenase and isozymes and phosphomonoesterases were also altered, indicating adverse effect of the oil on testicular function.

INTRODUCTION

The male sexual function is very sensitive to pharmacological agents of all sorts (Mann and Lutwak Mann, 1981). The primary male sex organ is the testes. The main functions of testes are spermatogenesis and steroidogenesis (Steinberger and Steinberger, 1974; Sharp, 1987). Androgen is essential for most stages, particularly meiosis, of spermatogenesis. The increased concentration of testosterone or the interruption in the testosterone production leads to the atrophy of the organ and impairment of spermatogenesis (Mann and Lutwak Mann, 1981).

Azadirachta indica is an indigenous plant that is grown over the greater part of India and Burma (Mendulkar et al., 1981). Almost all parts of neem tree are widely used for medicinal purposes (Kirtikar and Basu, 1965). Neem seed oil showed antispermatogenic activity by affecting leydig cell number and function (Ashok Purohit and Dixit, 1991). The present work assesses the possible antispermatogenic activity, alteration in testicular histoarchitecture, and some androgen-sensitive biochemical markers following neem oil administration to albino rats.

MATERIALS AND METHODS

Neem Oil

The compound neem oil was obtained from Bio-Organic Ltd., Madras, India with label purity of 99 per cent.

Animals

Adult male rats of Wistar strain (weighing 150–200 g) were obtained from our department animal house and kept in individual cages under controlled light and temperature conditions (20°–22°C). They were provided with standard diet (Hindustan Lever Ltd., India) and water *ad libitum*. All animals were kept in quarantine for 20 days before the drug treatment and were grouped as follows. The doses were fixed based upon the work of Ashok Purohit and Dixit (1991).

Group I served as control and received olive oil as vehicle.
Group II received 25 mg of neem oil/100 g body weight/day for eight consecutive days.
Group III received 50 mg of neem oil/100 g body weight/day for eight consecutive days.

Twenty-four hours after completion of the treatment schedule, the animals were killed by decapitation. The testes were dissected out, cleaned, blotted, and then weighed. They were subjected to biochemical analysis and histological studies.

Histological Investigation

A sample of testicular tissue from each organ was fixed in Bouin's fluid. After routine processing of each tissue, paraffin sections (7 μm) were stained using Harris' haematoxylin eosin. Histometric observations were made with an occulometer.

Biochemical Estimations

Required amounts of tissue were homogenized in 0.01 M tris buffer (pH 7.4) and centrifuged at 3000 rpm, and the supernatant was used for the estimation of protein (Lowry et al., 1951), glycogen (Morales et al., 1973), alkaline (ALP) and acid (ACP) phosphatases (Andersch and Szczypinsky, 1947), and for total lactate dehydrogenase (LDH) and its isozymes (King, 1965; Dietz et al., 1970).

Statistical Analysis

Student's *t* test was applied for statistical comparison of the differences in data between the test groups.

RESULTS

After neem oil treatment the histoarchitecture of the testis showed variations in the low dose group. The number of leydig cells was reduced. Seminiferous tubules showed variable spermatogenic activity and underwent pronounced shrinkage in diameter. Normal

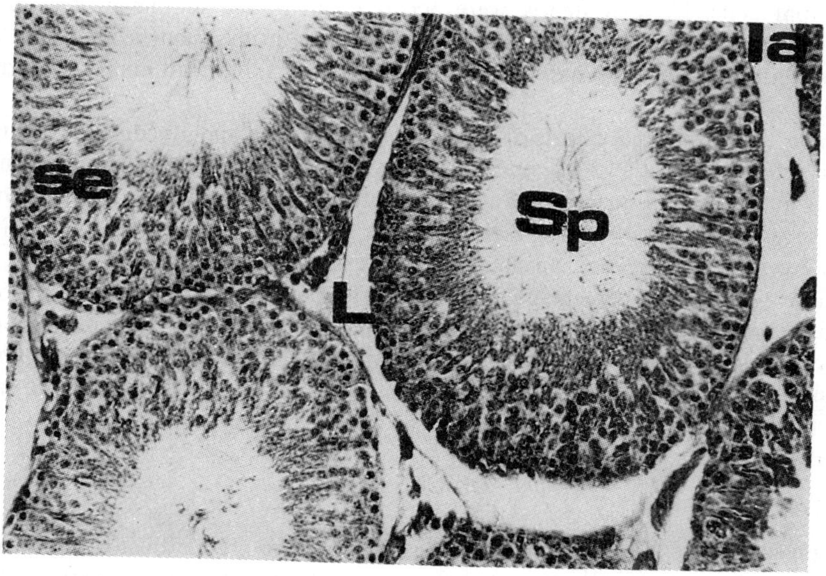

Fig. 1a. Testis Control : Normal spermatogenesis is seen within the seminiferous tubules; Haematoxylin Cosin (HE) 200 ×

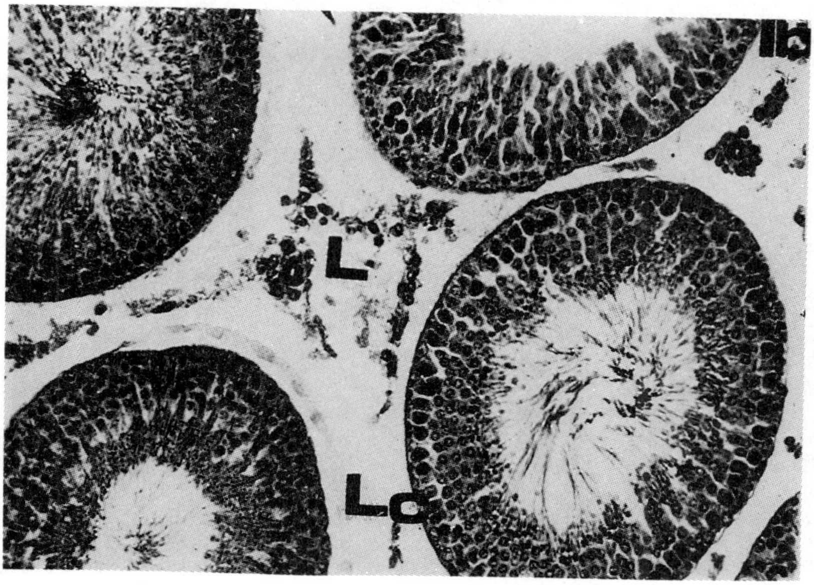

Fig. 1b. Testis treated with neem oil (25 mg/100 g body wt./day/rat) intramuscularly. Note asynchronization of germinal layers and the affected primary and secondary spermatocyte stages. Sertoli cells are darkly stained and leydig cells are sparsely distributed; HE 200 ×

spermatogenesis was seen but only in a limited number of the tubules (Fig. 1a) and in some tubules the lumen was filled with immature germ cells (Fig. 1b).

In the group treated with high dose neem oil the intercellular spacing became wider. Leydig cells were reduced in number. Basement membrane was thickened and disrupted. Spermiogenesis was not seen. Spermatogenesis was adversely affected except for spermatogonial cells. Cellular debris were accumulated in the lumina of the collapsed seminiferous tubules. Numerous giant cells were seen in the tubules. In many tubules only the sertoli cells were prominent and the lumen was filled with immature germ cells (Fig. 1c).

Table 1 illustrates the effects of neem oil on the protein and glycogen concentrations and the activities of ACP, ALP, and LDH enzymes. Neem oil administration caused an increase in the protein concentration and activities of ALP and total LDH enzymes in low dose group. Contrary to these changes, the glycogen concentration and activity of ACP were markedly lowered in all the drug-treated groups.

Table 2 depicts the percentage distribution of LDH isoenzymes in the testes of adult rats after neem oil treatment. In the control testis the M-subunit was predominant and this was mainly due to the contribution from LDH_5 and LDH_1 subunits. On drug treatment, the H/M ratio was raised and there was a shift towards the H-type subunits of the total LDH.

DISCUSSION

In the present study, no mortality was seen in any of the groups treated with neem oil. The histoarchitecture of the testes, especially in the groups treated with high dose,

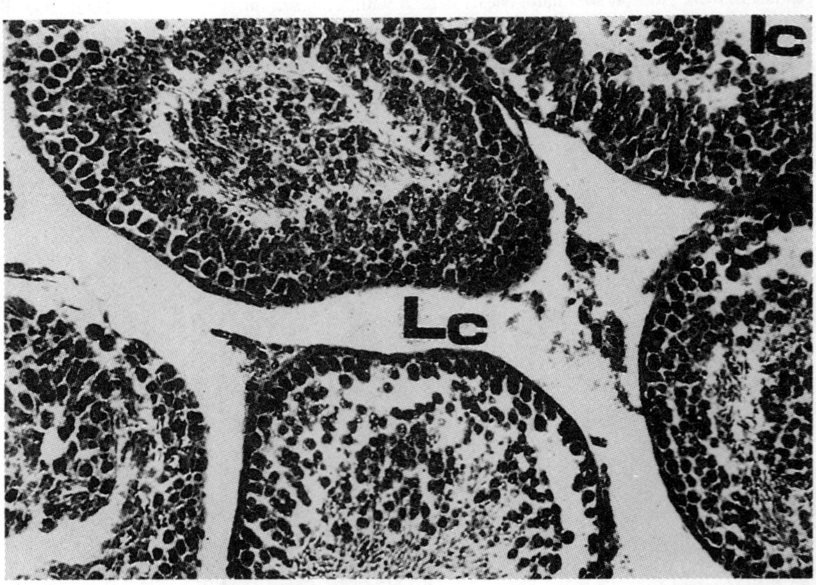

Fig. 1c. Testis treated with neem oil (50 mg/100 g body wt./day/rat) intramuscularly. Note the desquamation of seminiferous tubular epithelia and asynchronization of the different cellular layers. The seminiferous tubules were reduced in size and diameter. The lumen of all the seminiferous tubules are filled with cellular debris. Interstitial space was increased; HE 200 × (Se-Secretory epithelium, Sp—sperms, Lc—light cells, L—leydig cells

Table 1. Effect of neem oil on protein and glycogen concentration and on a few enzymes of the testes of rats

Tissue parameters	Group I	Group II	Group III
Protein/mg/100 g tissue	0.52 ± 0.003	0.72 ± 0.050	0.54 ± 0.056
Glycogen, mg/100 g tissue	0.037 ± 0.001	0.030 ± 0.0003*	0.028 ± 0.003*
Acid phosphatase[a]	0.061 ± 0.005	0.040 ± 0.004*	0.034 ± 0.006**
Alkaline phosphatase[a]	0.136 ± 0.010	0.298 ± 0.057*	0.345 ± 0.036***
LDH (mIU/min./mg protein)	0.36 ± 0.006	0.41 ± 0.004*	0.28 ± 0.006 ***

Each value is mean ± SEM of five estimations.
*$p < 0.05$, **$p < 0.01$, and ***$p < 0.001$: Control vs experimentals.
Group I : Vehicle treated.
Group II : Received 25 mg of neem oil/100 g body wt./day for eight consecutive days intramuscularly.
Group III : Received 50 mg of neem oil/100 g body wt./day for eight consecutive days intramuscularly.
[a] The values are expressed as μ moles of p-nitrophenol formed/hr/mg protein.

revealed degenerating seminiferous tubules and reduced leydig cell number, suggesting lowered gonadotrophic levels and an antiandrogenic effect (Dixit et al., 1975). Spermiogenesis and spermatid formation are known to be under the control of both androgen and gonadotrophins (Dym and Clermont, 1967; Chemes et al., 1976; Ritzen et al., 1981). Thus, neem oil appears to exert a direct inhibitory effect on the spermatogenic process of the testes.

In the present study the increase in the protein levels in the testes may be due to alteration in the testicular metabolic pathways.

Glycogen is found in the sertoli cells and spermatogonia of the testis and serves as a source of glucose for the tubular cells (Means and Hall, 1968). Reduced glycogen levels are known to affect protein synthesis and indicate inhibited spermatogenesis. In the present study the reduction in the glycogen levels in the testes may be due to impaired glycolysis (Mohri et al., 1975) and in addition suggests impaired spermatogenesis. Similar observations have been made by Dixit and Gupta (1982) in the testis of dogs treated with solasodine compound from *Sapindus trifoliatus*.

Acid and alkaline phosphatases regulate the secretory activity of the testis and are present in the seminiferous tubules and leydig cells (Elkington and Blackshaw, 1970). In the present study, a decrease in the activity was seen. Such a decrease in ACP activity has been correlated with increased spermatogenic destruction as a consequence of inadequate androgen supply within the testes (Blackshaw, 1973; Elkington and Blackshaw, 1974). Alkaline phosphatase enzyme is a histochemical marker for primordial germ cells (Mintz, 1957). An increased activity of this enzyme as observed in the present study after neem oil treatment may indicate the predominance of primordial germ cells and not the other more advanced cells of spermatogenesis. An elevated ALP activity associated with testicular damage has been reported by Bhardwaja and Garg (1984) after treating the rats with crude extracts of *Marsilea drummonodii*.

The testis-specific isozymes of LDH markers for active spermatogenesis (Blackshaw and Elkington, 1967). Lactate dehydrogenase was found in the seminiferous tubules, in spermatocytes, and in the midpiece of spermatozoa (Ambadkar and George, 1984). In the present study, after neem oil treatment the levels of LDH show variations. In the low dose group, it is increased and in the high dose group it is decreased; this may indicate altered metabolism and tissue respiration. The decline in the M-subunits in the

Table 2. Effect of neem oil on the percentage distribution of LDH isozymes in the testes of rats

Status animals	LDH_1	LDH_2	LDH_3	LDH_4	LDH_5	%M	%H	H/M ratio
Group I	28.36 ± 0.27	13.74 ± 0.10	7.36 ± 0.10	16.06 ± 0.005	33.01 ± 0.28	52.18	46.37	0.89
Group II	25.81 ± 0.28**	22.75 ± 0.13***	7.75 ± 0.13	18.69 ± 0.18**	22.56 ± 0.31 ***	46.13	51.42	1.11
Group III	45.43 ± 0.18***	32.61 ± 0.17***	5.92 ± 0.14 ***	5.32 + 0.04***	12.52 ± 0.15***	27.62	74.18	2.68

Each value is mean ± SEM of five estimations.
**$p < 0.01$ and ***$p < 0.001$: Control vs experimentals.

Group I : Vehicle treated.
Group II : Received 25 mg of neem oil/100 g body wt./day for eight consecutive days intramuscularly.
Group III: Received 50 mg of neem oil/100 g body wt./day for eight consecutive days intramuscularly.

drug-treated groups probably indicates the altered cellular respiration, that is, a shift from anaerobic to aerobic type of respiration and consequent impairment of spermatogenesis. Similar results were obtained in the testis of rats treated with the antiandrogen estradiol by Blackshaw and Elkington (1967). Vanithakumari (1986) observed a decrease in the activity of LDH in the testis of P-chlorophenylalanine-treated rats, which was suggested to be due to the accumulation of lactate and a decrease in cellular respiration (Free et al., 1969).

Thus, in the present study, the altered biochemical parameters and the histoarchitecture reveal the adverse effect of neem oil on the testis. This may be due to curtailing of androgen supply within the testes or it may be a direct effect of the neem oil on the tissue.

ACKNOWLEDGEMENT

We thank Prof. G. Sundarajulu, Head, Department of Zoology, Bharathiar University, Coimbatore, and one author (R.S.) gratefully acknowledges the financial assistance received from the Director, DGHS, New Delhi.

REFERENCES

Ambadkar, P.M. and J.C. George. 1984. Histochemical localisation of certain oxidative enzymes in the rat testis. *J. Histochem. Cytochem.* 12: 587.

Andersch, M.A. and A.J. Szczypinsky. 1947. Use of p-nitrophenyl phosphate as substrate in the determination of serum acid phosphatases. *Am. J. Clin. Pathol.* 17: 571–574.

Ashok Purohit and V.P. Dixit. 1991. Antispermatogenic efficacy of neem (*Azadirachta indica* A. Juss) materials in rats. *Neem Newsletter* 8(2): 13–14.

Bhardwaja, T.N. and A. Garg. 1984. The antifertility effects of an Australian species of the aquatic fern Marsilea. *Indian Fern J.* 1: 75.

Blackshaw, A.W. 1973. Testicular enzymes and spermatogenesis. *J. Reprod. Fertil.* 18: 55.

Blackshaw, A.W. and J.S.H. Elkington. 1967. Hormones and the isozymes of lactate dehydrogenase. *Med. J. Aust.* II: 520.

Chemes, H.E., M. Dym, D.W. Fawcett, J. Hasser and R.J. Sherins. 1976. Patho-physiological observations of sertolicells in patients with seminal cell aplasia and severe germ cell depletion. Ultrastructural findings and endocrine levels. *Biol. Reprod.* 17: 108.

Dietz, A.A., T. Lubrano and H.M. Rubinstein. 1970. Disc electrophoresis of lactate dehydrogenase. *Clin. Chim. Acta.* 27: 225–232.

Dixit, V.P., N.K. Lohiya and M. Arya. 1975. Inhibition of spermatogenesis in the pigeon *Columba livia* Gmelin, following the administration of Methallibure (lck-33827) confirmation. *Indian J. Exp. Biol.* 13: 490.

Dixit, V.P. and R.S. Gupta. 1982. Antispermatogenic and antiandrogenic activity of *Sapindus trifoliatus* fruit extract in intact and castrated male gerbils. *Planta Med.* 46: 242.

Dym, M. and Y. Clermont. 1967. Effects of X-ray on type A spermatogonia in the rat. *Anat. Rec.* 157: 238.

Elkington, J.S.H. and A.W. Blackshaw. 1970. The effects of testosterone, oestradiol andpregnant mare serum gonadotrophin on growth and enzyme in the rat testis. *J. Reprod. Fertil.* 23: 1.

Elkington, J.S.H. and A.W. Blackshaw. 1974. Studies in testicular function. I. Quantitative effects of FSH, LH, testosterone and dihydro testosterone on restoration and maintenance of spermatogenesis in the hypophysectomised rats. *Aust. J. Biol. Sci.* 27: 47.

Free, M.J., E.D. Massie and N.L. Vandemark. 1969. Glucose metabolism by the cryptorchid rat testis. *Biol. Reprod.* 1: 354.

King, J. 1965. *Practical Clinical Enzymology*. D. Van Nostrand Co., London, pp. 83–93.

Kirtikar, K.R. and B.D. Basu. 1965. *Indian Medicinal Plants*, Vol. 1. Periodical Experts, New Delhi, pp. 436–441.

Lowry, O.H., N.J. Rosebrough, A.L. Farr and R.J. Randall. 1951. Protein measurements with the folin phenol reagent. *J. Biol. Chem.* 193: 265–275.

Mann, T. and C. Luwak-Mann. 1981. In: *Testis and testicular semen. Reproductive Function and Semen. Themes and Trends in Physiology, Structure and Investigative Andrology*. T. Mann (ed.), Bealine, Heidelberg, Springer, New York.

Means, A.R. and P.F. Hall. 1968. Protein synthesis in the testis: I. Comparison between stimulation by FSH and glucose. *Endrocrinology* 82: 597.

Mendulkar, K.N., V.Y. Deshpande, D.H. Nandal and N.L. Sadre. 1981. *In vitro* action of *Azadiracta indica* in sperm motility and viability. *Indian J. Physiol. Pharmcol.* 25(7): 12–13.

Mintz, B. 1957. Embryological development of primodial germ cells in the mouse: Influence of a new mutation. *WJ. J. Embryol. Exp. Morphol.* 5: 396.

Mohri, H., D.A.I. Suter, P D.C. Brown Woodman, I.G. White and D.D. Ridley. 1975. Identification of the biochemical lesion produced by chlorohydrin in spermatozoa. *Nature* 255: 75.

Morales, M.A., A.J. Jabbagy and H.D. Terenzi. 1973. Mutations affecting accumulation of glycogen. *Neurospora Newsletter* 20: 24.

Ritz, E.M., V. Hanson and F.S. French. 1981. The sertoli cell. In: *The Testes.* H. Burger and D.M. Dekrester (eds.), Raven Press, New York, 171 pp.

Sharp, R.M. 1987. Testosterone and sprmatogenesis. *J. Endocrinol.* 113: 1.

Steinberger, E. and A. Steinberger. 1974. Hormonal control of testicular function in mammals. In: *Handbook of Physiology*, Section 7, *Endocrinology*, Vol. 4. *The Pituitary Gland,* part 2. E. Krobiol and W.H. Sawyer (eds.), American Physiological Society, Washington, 225 pp.

Vanithakumari, G. 1986. Effect of P-chlorophenylalanine on male rats: Histopathological and biochemical changes in the testes. *Indian J. Physiol. Pharmacol.* 30: 223.

133

Effect of Neem Oil: Structural and Functional Changes in the Epididymis of Rats

R. SAMPATHRAJ, P. BADRI SRIMAN NARYANAN AND G. VANITHAKUMARI

Division of Endocrinology, Department of Zoology, Bharathiar University
Coimbatore 641 046, India

ABSTRACT

Neem oil was administered at two different concentrations (25 mg and 50 mg, respectively/100 g body wt/day for eight consecutive days intramuscularly to adult male albino rats to assess epididymal structural and functional activity. Sperm motility decreased markedly in the cauda epididymal fluid. Sperm counts were also decreased significantly in both caput and caudal segments of the epididymis. Histologically, a reduction in epididymal and stereocilia height and luminal diameter was noted. Epididymal weights and activities of acid phosphatase and total lactate dehydrogenase were reduced in both the epididymal segments. Protein concentration was appreciably increased only in the caudal segments. Administration of neem oil impaired the structural and functional integrity of the epididymis.

INTRODUCTION

The epididymis comprises the convoluted tubule which connects the testis to the vas deferens and provides a favourable milieu for the acquisition of the fertilizing activity, motility, storage, and survival of spermatozoa (Jehan et al., 1973; Brooks, 1979a). The necessary substance for sperm maturation and survival has been provided by the epididymal secretions (Orgebin-Crist et al., 1975), which are mainly controlled by androgen (Brook, 1979b; Setty et al., 1977). Agents with antiandrogenic and antifertility effect have been widely used to alter the specific chemical environment, which is peculiar for the sperm to mature in the epididymis (Kalla and Singh, 1979) and epdidymis is a specific site for male contraception. Neem seed oil is reputed to have many medicinal properties in the traditional system of medicine. The oils is also reported to have an antiflagellal action. Further, neem oil in its natural form is a potent spermicidal agent (Sinha et al., 1984). More recently neem seed oil has been used as a vaginal contraceptive (Sinha et al., 1984; Baradan et al., 1991). Very few studies on the spermicidal and antifertility effects of neem oil on rats and monkeys are available (Sinha et al., 1984; Ashok Purohit and Dixit, 1991). Hence, the present investigation is an attempt to study the probable effects of neem oil on the various biochemical parameters as well as histological changes in the epdidymis.

MATERIALS AND METHODS

Neem Oil

The compound neem oil was obtained from Bio-Organic Ltd., Madras, India, with label purity of 99 per cent.

Animals

Adult male rats of Wistar strain (weighing 150–200 g) obtained from our department animal house and kept in individual cages, under controlled light and temperature conditions (20°–22°C), were used for the present studies. They were provided with standard diet (Hindustan Lever Ltd., India) and water *al libitum*. All animals were kept in quarantine for 20 days before drug treatment and grouped as follows and the doses were based upon the work of Ashok Purohit and Dixit (1991).

Group I served as control and received olive oils as vehicle intramuscularly.
Group II received 25 mg of neem oil /100 g body weight/day for eight consecutive days (intramuscularly).
Group III received 50 mg of neem oil /100 g body weight/day for eight consecutive days intramuscularly.

Twenty-four hours after completion of the treatment, the animals were sacrificed by decapitation. The epididymis were dissected out and weighed. The number of spermatozoa and their motility were determined in one epididymis while the other epididymis was processed for both histological and biochemical investigation.

Sperm Motility and Sperm Count

The epididymal spermatozoa were quickly obtained by puncture of both caput and cauda epididymis with a disposable hypodermic needle (gauge 21). Motility of cauda epididymal spermatozoa was determined in a haemocytometer chamber after 5 minutes' incubation at room temperature described by Pholpramool and Chaturapanich (1979). The epididymal fluid from both caput and cauda segments was then subjected to a 20-fold dilution with a spermicidal solution containing 5 g of $NaHCo_3$ and 1 ml of 35 per cent formaldehyde in 100 ml physiological saline. The sperms were counted using a Neubaur haemocytometer.

Biochemical Estimations

After collecting the epididymal fluid, the caput and caudal tissues were processed for the estimation of protein (Lowry et al., 1951) and alkaline (ALP) and acid (ACP) phosphatases (Andersch and Szczypinsky, 1947)) and for total lactate dehydrogenase (LDH) by the method of King (1965).

Histological Investigation

The epididymis was fixed in Bouin's fluid and sections (7 μm) were stained with Harris' haematoxylin-eosin. Histometric observations were made with an occulometer.

Statistical Analysis

Student's *t* test was applied for statistical comparison of the differences in data between the test groups.

RESULTS

Intramuscular administration of neem oil resulted in a marked reduction of wet weights of both caput and cauda epididymis (Table 1). Epididymal spermatozoal motility and count also declined markedly during both low and high dose drug treatments (Table 1).

Table 2 summarizes the effect of neem oil on protein concentration and ALP, ACP, and LDH enzyme activities. Neem oil treatment caused a marked increase in protein concentration of cauda epididymis only. In both the tissues, ACP recorded a significant decrease in activity whereas ALP showed a marked increase in all the drug-treated groups. Similarly, a significant fall in the activities of total LDH in both the caput and cauda epididymis was observed after the low and high dose drug treatment.

The histoarchitecture of the caput after low dose neem oil treatment showed a decrease in the sperm concentration in the epididymal lumen and thinning down of the pseudostratified epithelial layers when compared to control. An increase in the basal cell number in many tubules was also noted (Fig. 1a, b, c).

In the high dose group the pseudostratified epithelial cells were reduced considerably in height and length with a concomitant decrease in number and height of the stereocilia. The tubules were reduced in diameter and were with a narrower lumen (Fig. 1c).

Neem oil caused a marked reduction in the heights of the epithelial cells and stereocilia and in the luminal diameter compared to control in the cauda of low dose groups (Fig. 2a, b). In the high dose group a disruption in the basement membrane and decrease in cell number and volume were also noted in the caudal tissue. In addition, appearance of numerous light cells and basal cells and reduction in cytoplasm and epithelial cells were noted (Fig. 2c).

DISCUSSION

The structure and function of the epididymis, the histochemical differences between the segments, and the influence of the different zones upon the passage and function of the

Table 1. Effect of neem oil on sperm concentration, motility, and epididymal weights of adult albino rats

Parameters	Group I		Group II		Group III	
	Caput	Cauda	Caput	Cauda	Caput	Cauda
Epididymal fluid Sperm concentration (Count 10^4 /ml)	494 ± 0.8	556 ± 0.8	238 ± 1.02***	314 ± 0.57***	193 ± 0.4***	284 ± 0.57***
Sperm motility[a] (FMI)		106 ± 0.85		45 ± 0.74***		21 ± 40.40***
Organ weight (mg/100 g body wt)	234.6 ± 0.56	151.3 ± 0.14	204.4 ± 0.56***	131.8 ± 0.18***	191.5 ± 0.10**	112.4 ± 0.32***

[a]Motility is the movement recorded after 5 min. suspension of the caudal epidymal spermatozoa in phosphate buffered saline, N = 5. Each value is the mean ± SEM of five estimations. Significance relative to Group I : ***$p < 0.001$.

Group I: Vehicle treated.

Group II: Received 25 mg/100 g body wt/day for eight consecutive days intramuscularly.

Group III: Received 50 mg/100 g body wt/day for eight consecutive days intramuscularly.

Table 2. Effect of neem oil on protein concentration and a few enzymes of the caput and caudal epididymis in albino rats[a]

Tissues/groups	Protein (mg/100 mg tissue)	Acid[+] phosphatase	Alkaline[+] phosphatase	Total LDH (mIU/hr/mg protein
Caput epididymis				
Group I	0.069 ± 0.0010	0.074 ± 0.0060	0.131 ± 0.0010	2214 ± 1.0600
Group II	0.073 ± 0.0010	0.034 ± 0.0010***	0.181 ± 0.0010***	1831 ± 2.9200***
Group III	0.076 ± 0.0010	0.026 ± 0.0020***	0.192 ± 0.0010***	1544 ± 2.9700***
Cauda epididymis				
Group I	0.031 ± 0.0010	0.183 ± 0.0010	0.274 ± 0.0010	5892 ± 3.0100
Group II	0.074 ± 0.0000***	0.074 ± 0.0000***	0.284 ± 0.0000*	2948 ± 0.9900***
Group III	0.081 ± 0.0010***	0.063 ± 0.0010***	0.299 ± 0.0020***	2192 ± 1.1700***

[a]Each value is the mean ± SEM of five estimations. Significance relative to Group I: *$p < 0.05$; **$p < 0.05$; ***$p < 0.001$.

[+]The values are expressed as μmol of P-nitrophenol formed 1 hr/mg protein.

Group I : Vehicle treated

Group II: Received 25 mg/100 g body wt/day for eight consecutive days intramuscularly.

Group III: Received 50 mg/100 g body wt/day for eight consecutive days intramuscularly.

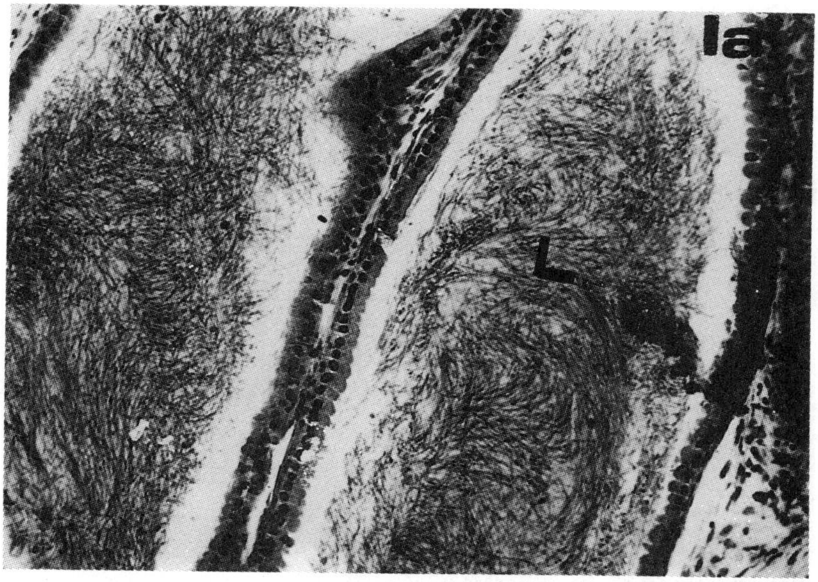

Fig. 1a. Caput epididymis control: Note the normal histoarchitecture with compact pseudostratified epithelium and the lumen filled with sperms; Haematoxylin Eosin (HE) 200 ×

epididymal spermatozoa have all been extensively studied (Bedford, 1966; Orgebin-Crist, 1969; Hamilton, 1972; Brooks, 1979b). A marked reduction in the epididymal weights on administration of neem oil suggests reduced secretory activity and impaired growth of the tissues. This may be due to a deprived androgen supply, since this tissue is dependent on androgen for its growth and activity. A similar reduction in epididymal weight was also observed with many plant products (Tso and Lee, 1982; Manonayagi

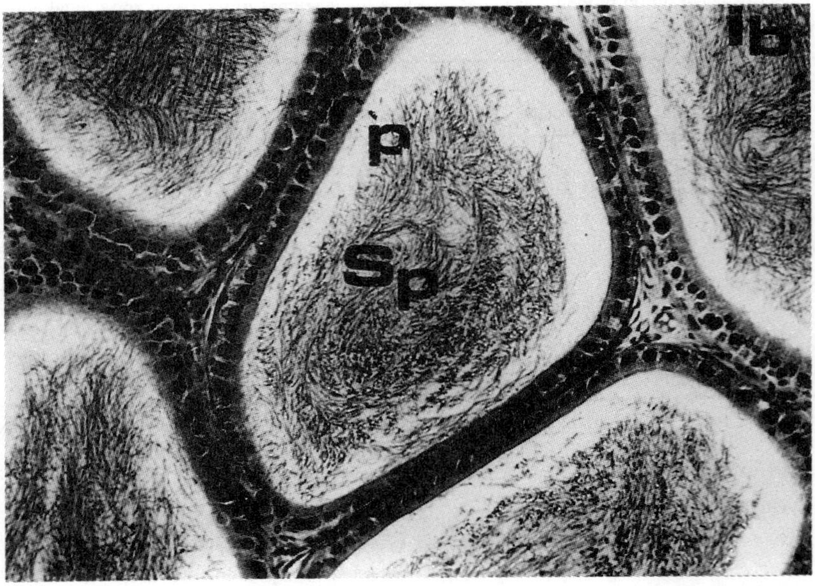

Fig. 1b. Caput epididymis treated with neem oil 25 mg/100g body wt/day/rat intramuscularly. Note the reduction in the height of pseudostratified epithelial cells and number of sperms and the increase in fibromuscular elements; HE 200 ×

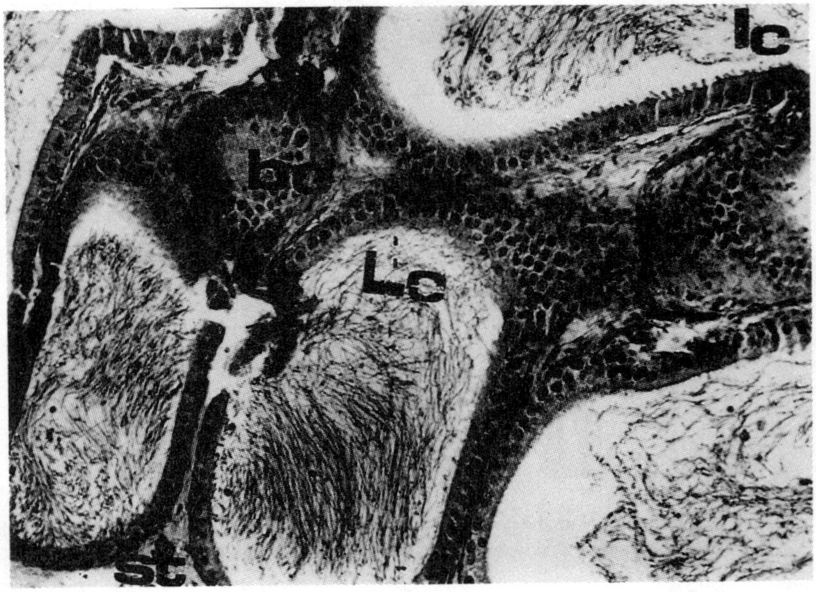

Fig. 1c. Caput epididymis treated with neem oil 50 mg/100g body wt/day/rat intramuscularly. Note the reduction in tubular diameter and sperm concentration. The fibromuscular stroma seemed to be odematous. There is increase in the number of light cells; HE 200 ×. (St— Fibromuscular stroma, P — Pseudostratified epithelium, Sp — Sperm, L — Lumen, LC — Light cells, BC — Basal cells)

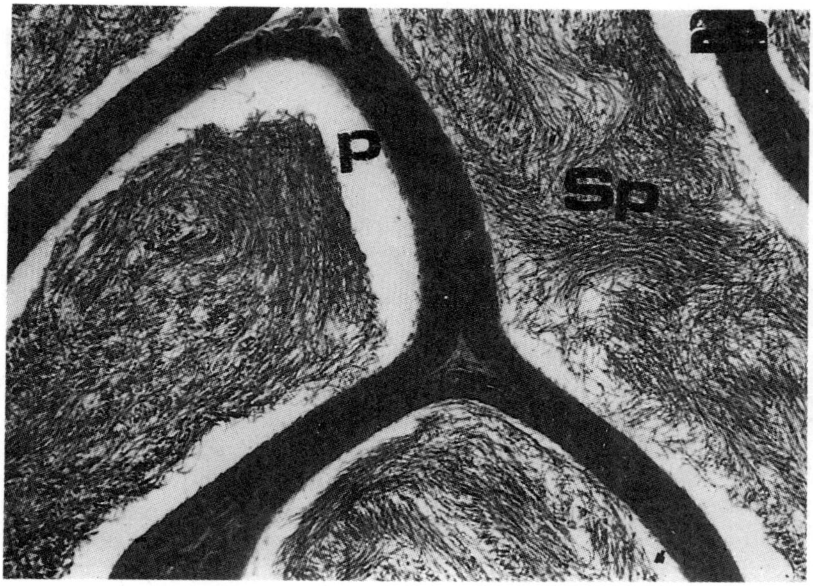

Fig. 2a. Cauda epididymis control showing the normal histoarchitecture; Haematoxylin-Eosin (HE) 200 ×

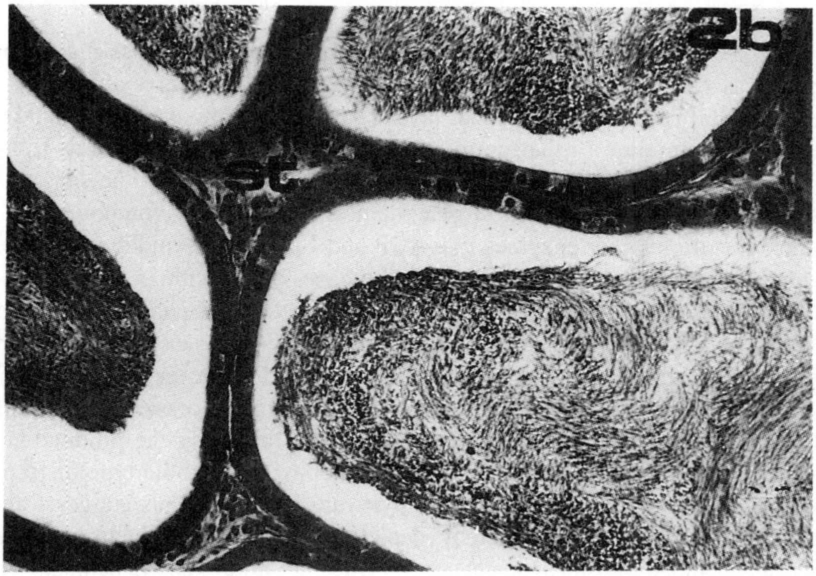

Fig 2b. Cauda epididymis treated with neem oil 25 mg/100 g body wt/day/rat intramuscularly. Note the reduction in the height of pseudostratified epithelial cells and the decrease in sperm concentrations; HE 200 ×

et al., 1989) and a decrease in androgen production has been suggested to be the cause for the rapid decline in the number of spermatozoa within the cauda epididymis. This severely impaired histoarchitecture of both the caput and cauda epididymis may well indicate the interference of the neem oil with the probable androgen supply to this

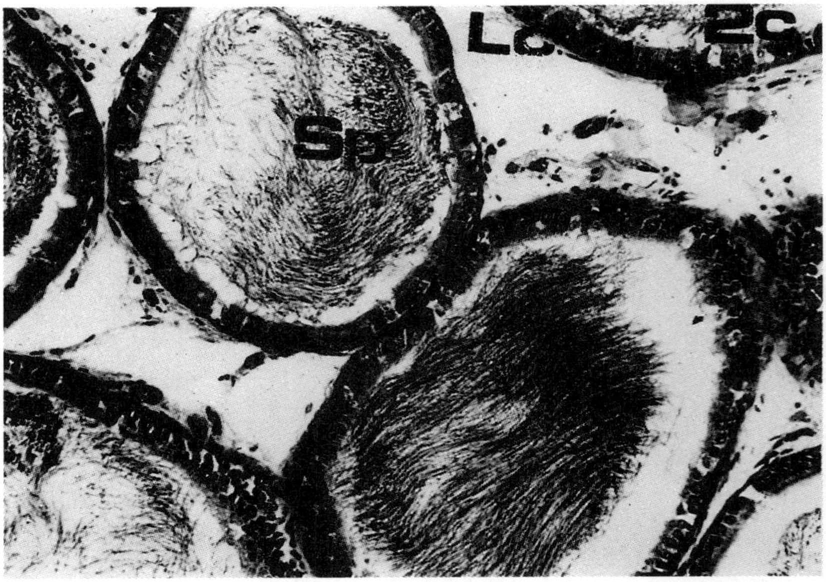

Fig. 2c. Cauda epididymis treated with neem oil 50 mg/100g body wt/rat intramuscularly. Note the shrinkage in the tubular diameter and appearance of numerous light cells and basal cells HE 200 ×.
(St — Fibromuscular stroma, P — Pseudostratified epithelium, SP — Sperm, LC — Light cells)

tissue. A marked reduction in sperm counts and weights of the various male reproductive organs was found with *Ocimum sanctum* treatments by Seth et al. (1976).

The peripheral cells of the epithelial layer of the epididymis synthesize and secrete proteins for the maturation of the sperm in the epididymis. Neem oil seems to have no adverse effects on protein levels in caput but induced an increase of the same in cauda and this result agrees with the action of certain anti-androgenic compounds.

Further, specific marker enzymes like ACP and LDH in the epididymis were found to be depleted on neem oil treatment of both the doses along with an increased activity of ALP enzyme. Emblin treatment was shown to induce similar changes in the epididymis by Gupta et al. (1990). Lactate dehydrogenase has been shown to play an important role in the mitochondrial lactate pyruvate shuttle system to transfer reducing equivalent from cytosol to mitochondria for the generation of energy necessary for motility and survival of germ cells. The decreased activity of LDH favouring the production of the substrate lactate clearly indicates a switch from aerobic to anaerobic type of respiration in this tissue. The consequent inhibition of pyruvate formation thus suggests abolition of citric acid cycle activity and thereby depletion of energy production. Since both glycolytic and pentose phosphate pathway enzymes are androgen dependent (Brooks, 1976), a decrease in the glycolytic enzyme LDH in the present study further confirms a probable deficiency of androgen supply to the epididymis.

Thus, in the present study, a decrease in sperm count and sperm motility and alteration in histoarchitecture and biochemistry of epididymal segments indicate that neem oil treatment either has a direct toxic effect on epididymal tissues or affects the androgen supply to this tissue.

ACKNOWLEDGEMENTS

We thank Prof. G. Sundarajulu, Head, Department of Zoology, Bharathiar University, Coimbatore, and one author (R.S.) gratefully acknowledges the financial assistance received from the Director, DGHS, New Delhi.

REFERENCES

Andersch, M.A. and A.J. Szczypinsky. 1947. Use of P-nitrophenyl phosphate as substrate in the determination of serum acid phosphatases. *Am. J. Clin. Pathol.* 17: 571–574.

Ashok Purohit and V.P. Dixit. 1991. Antispermatogenic efficiency of neem (*Azadirachta indica* A. Juss) materials in rats. *Neem Newsletter.* 8(2): 13–14.

Baradan, J., S.S. Riar, R.C. Sawhney, A.K. Kain, P. Thomas and G. Ilavazhagan. 1991. Neem oil—A fertility controlling agent in rhesus monkey. *Indian J. Physiol. Pharmacol.* 35: 278–280.

Bedford, J.M. 1966. Development of the fertilizing ability of spermatozoa in the epididymis of the rabbit. *Exp. Zool.* 163: 319.

Brooks, D.E. 1976. Activity and androgenic control of glycolytic enzymes in the epididymis and epididymal spermatozoa of the rat. *Biochem. J.* 156: 527.

Brooks, D.E. 1979a. Influence of androgens on the weights of the male accessory reproductive organs and on the activities of mitochondrial enzymes in the epididymis. *J. Endocrinol.* 82: 293–303.

Brooks, D.E. 1979b. Biochemical environment of sperm maturation. In: *The Spematozoon. Maturation, Mortility, Surface Properties and Comparative Aspects.* D.W. Fawcett and J.M. Bedford (eds.), Urban Schwarzenberg, Baltimore, 23 pp.

Gupta, S., U. Kanwar and S.N. Sanyal. 1990. Inhibition of reproductive tissue carbohydrate metabolism and reversibility of the effects of embelin, a plant benezoquinone of antiferfility potential. *Fitoterapia.* Vol. LXI.: 133–143.

Hamilton, D.W. 1972. The mammalian epididymis. In: *Reproductive Biology.* H. Balin and S. Glager (eds.), *Excerpta Medica*, Amsterdam, 263 pp.

Jehan, O., B.S. Setty and A.B. Kax. 1973. Studies on physiology and biochemistry of mammalian epididymis: Effect of castration and steroid hormone replacement on sperm survival in rat epididymis. *Indian J. Exp. Biol.* 11: 270–273.

Kalla, N.R. and B. Singh. 1979. Effect of sodium nitrite on the chlorohydrin-induced lesion of the testis-epidydymis complex in the rat. *J. Reprod. Fertil.* 56: 149.

King, J. 1965. *Practical Clinical Enzymology.* D. Van Nostrand Co., London, pp. 83–93.

Lowry, O.H., N.J. Rosebrough, A.L. Farr and R.J. Randall. 1951. Protein measurements with the folin phenol reagent. *J. Biol. Chem.* 193: 265–275.

Manonayagi, S., G. Vanithakumari, S. Padma and T. Malini. 1989. Effects of bamboo buds: Structural and functional changes in the epididymis of rats. *J. Ethnopharmacol.* 25: 201–212.

Orgebin-Crist, M.C. 1969. Studies on the function of the epididymis. *Biol. Reprod.* (Suppl.) 1: 155.

Orgebin-Crist, M.C., B.J. Danzo and J. Davies. 1975. Endocrine control of the development of maintenance of sperm fertilizing ability in the epididymes. In: *Handbook of Physiological Sect. 7, Endocrinology* Vol. 5. R.O. Greep and E.B. Astwood (eds.), American Physiology Society, Washington, D.C., 319 pp.

Pholpramool, C. and G. Chaturapanich. 1979. Effect of sodium and potassium concentrations and pH on the maintenance of motility of rabbit and rat epididymal spermatozoa. *J. Reprod. Fertil.* 57: 245–251.

Seth, S.D., N. Johri and K.R. Sundram. 1976. Antispermatogenic effect of *Ocimum sanctum. Indian J. Exp. Biol.* 19: 975, 1976.

Setty, B.S., S.S. Riar and A.B. Kar. 1977. Androgenic control of epididymal function in rhesus monkey and rabbit. *Fertil. Steril.* 28: 674–681.

Sinha, K.C., S.S. Riar, R.S. Tiwary, A.K. Dhawan, J. Bardhan, P. Thomas, A.K. Kain and R.K. Jain. 1984. Neem oil as a vaginal contraceptive. *Indian J. Med. Res.* 79: 131–136.

Tso, W.W. and C.S. Lee. 1982. Lactate dehydrogenose-x, an isozyme particularly sensitive to gossypol inhibition. *Int. J. Androl.* 5: 205–209.

An Experimental Evaluation of *Azadirachta indica* (A. Juss.) for Adaptogenic Potential

P. SEN, P. MEDIRATTA AND A. RAY

Department of Pharmacology, University College of Medical Sciences and G.T.B. Hospital, Delhi 110 095, India

ABSTRACT

The effect of *Azadirachta indica* (neem) on some biochemical, enzymatic, immunological, and visceral parameters were investigated in normal and stressed rats and mice in order to elucidate the probable biological mechanism involved in the versatile beneficial effects produced by this plant.

In normal rats *A. indica* (100 mg/kg) lowered the blood glucose, triglyceride, and serum glutamate oxaloacetate transaminase (SGOT) levels significantly and enhanced humoral as well as cellular immune responses.

In stressed rats, *A. indica* (1) lowered the raised cholesterol and urea levels and brought the blood glucose level to normal, (2) markedly reduced the frequency and severity of gastric mucosal lesions, and (3) attenuated the stress-induced immunosuppression of the secondary antibody responses to sheep red blood cells.

It is suggested that the use of *A. indica* in cardiovascular and hepatic disorders and natural and acquired immunodeficiency disorders should be explored.

INTRODUCTION

The medicinal and other properties of *Azadirachta indica* (neem) have been known for ages. It is reported to have versatile beneficial properties in humans. It has a wide range of antibacterial, antiviral, antifungal and antiprotozoal activities, and it is reported to be effective in various skin diseases, ulcers due to burns, and dental disorders. Isolated studies have also been reported on its hypoglycaemic, antiarthritic, antiulcerogenic, antiinflammatory, and antitumour effects (Chopra et al., 1956; Satyavati et al., 1976).

Herbal plants used in traditional medicine have been reported to possess antistress or adaptogenic properties in animals and humans, and induce a non-specific increase in resistance against infections and promote positive health (Bhargava and Singh, 1981; Brekhman and Dardymov, 1969). The restoration of normal milieu in stress and rejuvenating power of these drugs might be due to their action on the biochemical, neurological, and immune system (Sen et al., 1988, 1992). The versatile beneficial effects of *A. indica* strongly indicate its possible antistress effects, but no data in this aspect are available. The present study was, therefore, undertaken to investigate the effects of *A. indica* on biochemical, visceral, and immunological parameters in both normal and stressed rats and mice to explore its possible adaptogenic or antistress effects.

MATERIALS AND METHODS

Animals

Male Wistar rats (200–250 g) and male albino mice (20–30 g) obtained from local sources were used for the study. They were housed in light- and temperature-controlled conditions with free access to food (Hindustan Lever Laboratory Animal Food) and water.

Drugs

Extract of leaves of *A. indica* (Dabur India Ltd.) were used as an aqueous suspension with Tween-80, and were injected intraperitoneally in a dose of 100 mg/kg and in a volume of 1 ml/kg. Control groups were treated with Tween-80 in all experiments. The dose was selected after initial pilot studies with graded doses (10, 100, 200, and 500 mg/kg) of the drug given intraperitoneally daily for one week, and observated that 100 mg/kg produced no mortality and was safe.

BIOCHEMICAL ASSAY

Blood was collected 24 hrs after drug injection from retroorbital plexus by microcapillary technique, and glucose, cholesterol, triglycerides, total protein, albumin, urea, serum glutamate oxaloacetate transaminase (SGOT), and serum glutamate pyruvate transaminase (SGPT) were estimated by Enzokits and Chemkits (Ranbaxy), spectroflourimeter, and UV spectrophotometer (Shimadzu, Japan).

Humoral Immune Response

Humoral immune response was studied in rats immunized with sheep red blood cells (SRBC, 0.5×10^9 cells/ml/100 g) on day 0. On days 1–5, they received vehicle or *A. indica* (100 mg/kg, intraperitoneally). On day 6, the rats were lightly anaesthetized with ether and blood was collected as above; the serum was assayed for haemagglutination titre using microtitre plates, and the mean ranks of different groups were compared for statistical significance (Ray et al., 1991).

Cell-mediated Immune Responses

Cell-mediated immune responses were studied in mice by footpad thickness test (*in vivo*) and in rats by leucocyte migration inhibition test (*in vitro*) (Sen et al., 1991).

Stress Procedure

The rats were immunized with SRBC (0.5×10^9 cells/ ml/100 g) on day 0, and received either vehicle or *A. indica* (100 mg/kg) on days 1–5. On day 7 the booster dose of the antigen was given, and immediately the rats were deprived of food (but not water) for 18 hrs and then subjected to restraint stress in Plexiglass restrainers for 24 hrs at room temperature (22° ± 2°C). After the termination of the stress, blood was collected under light ether anaesthesia for biochemical and immunological studies, the rats were sacrificed by an overdose of ether, and the gastric pathology was studied under a dissection microscope for ulcer number and ulcer severity (mm) (Ray et al., 1991).

The data were analysed using Mann-Whitney 'U' test (two tailed) and Chi square test, wherever appropriate.

RESULTS AND DISCUSSION

There was no significant effect of one week of drug treatment on the general behaviour, food and water intake, and body weight of albino rats. The control group of rats subjected to restraint stress showed delayed entry into the restrainers, struggle, and prompt and profuse defecation, which were remarkably reduced in rats pretreated with *A. indica.*

In normal rats, *A. indica* lowered the glucose, total protein, and triglyceride levels and raised serum cholesterol level marginally as well as the urea level. In rats exposed to restraint stress, *A. indica* pretreatment lowered the stress-induced elevation of cholesterol and urea levels. The SGOT and SGPT levels were also significantly lower in rats pretreated with *A. indica*, when compared to the vehicle-treated group (Table 1).

In the prolonged stress procedure followed in these experiments, the rats were starved for (18 + 24 hrs) 42 hrs, which caused lowering of the basal glucose levels possibly due to depletion of hepatic glycogen store. *Azadirachta indica* appears to have marginally elevated blood glucose level in stressed rats in order to bring back the normal level (Table 1).

Lowering of triglyceride, SGOT, and SGPT levels in normal rats is of particular significance, and it indicates possible beneficial effects of *A. indica* in cardiovascular and hepatic disorders.

In rats immunized with SRBC on day 0 and treated with *A. indica* from days 1 to 5, the primary antibody response to SRBC was significantly higher than in the vehicle-treated rats. In rats given a booster injection of the antigen, a markedly higher anti-SRBC antibody response was observed, which could not be further enhanced by pretreatment with *A. indica*, probably because a near maximal response was already obtained. The anti-SRBC antibody titres were significantly lower in rats exposed to restraint stress as compared to the non-stressed vehicle-treated group; pretreatment with *A. indica* clearly prevented this immunosuppressant effect of restraint stress on the antibody response to the booster SRBC injection (Table 2).

Table 1. Effect of *A. indica* (100 mg/kg) on some biochemical parameters in rats

Treatment	Biochemical parameters (mg %)							
	Glucose	Cholesterol	Triglyceride	Total protein (g %)	Albumin (g %)	Urea	SGOT (IU/l)	SGPT (IU/l)
Normal rats								
Vehicle	130.3 ± 5.0	52.8 ± 5.0	108.5 ± 11	9.3 ± 1.2	3.0 ± 0.3	30.0 ± 1.7	158.1 ± 22.3	28.1 ± 6.0
A. indica	106.0 ± 4.8*	83 ± 18.0	61.1 ± 3.6*	7.0 ± 0.2	2.5 ± 0.1	45.3 ± 6.2*	64.6 ± 3*	21.2 ± 3.5
Stressed rats								
Vehicle	111.5 ± 7.0	104.4 ± 8.7	73.2 ± 8.1	9.7 ± 1.0	2.7 ± 0.2	72.2 ± 13.5		
A. indica	135 ± 3.4	85 ± 10.0	112.6 ± 18.9	7.6 ± 0.5	2.7 ± 0.7	47 ± 4.00		

*$p < 0.02$ (compared to the control groups).
n = 6–10 animals per group.

Table 2. Effects of *A. indica* (100 mg/kg) on humoral antibody responses in rats

Treatment	Anti-SRBC antibody titre (rank) (mean ± SE)	
	Primary antibody titre	Secondary antibody titre
Normal rats		
Vehicle	4.1 ± 0.5	7.0 ± 0.5
A. indica	6.3 ± 0.3 *	6.7 ± 0.6
Stressed rats		
Vehicle		5.3 ± 0.3*
A. indica		6.1 ± 0.7

*$p < 0.05$ (compared to vehicle group).
n = 6–10 animals per group.

When these animals were sacrificed and the gastric pathology was examined, it was observed that the rats exposed to restraint stress developed gastric ulcers which were more numerous as well as more severe than in the control non-stressed, food-deprived group. This adverse effect of restraint stress on gastric mucosal integrity could be significantly attenuated when these animals are treated with *A. indica* prior to the exposure to stress (Table 3).

The effects of *A. indica* on cell-mediated immune responses were assessed by study of footpad thickness test (*in vivo*) in immunized mice and leucocyte migration inhibition test (*in vitro*) in immunized rats. There was a significant enhancement in the paw volume in immunized mice pre-treated with *A. indica* when compared to the vehicle-treated group, and in the *in vitro* test the per cent inhibition of migration of lymphocytes was significantly lower in the rats treated with *A. indica* than in the vehicle-treated group (Table 4). The opposite effects observed in these two tests could be due to the fact that different lymphokines could be released from different cell types, and such discrepancy in the results of these two types of tests was also reported earlier (Tewary et al., 1982).

The T lymphocytes are known to be primarily involved in the cell-mediated immune responses like contact dermatitis, organ transplant, and rejection of grafts; and T lymphocytes through the phenomenon of cellular cooperation stimulate the activity of B lymphocytes.

Stress is known to be a stimulus which disturbs homeostasis, and breakdown in adaptational processes results in various visceral, endocrinal, and behavioural changes (Ader, 1984) causing psychosomatic disorders like peptic ulcer, ulcerative colitis, cardiovascular disorders, hyper-lipidemia, bronchial asthma, contact dermatitis, rheumatoid arthritis, and psychosis. These conditions are also known to be associated with immunosuppression.

In cardiac disorders of ischemic origin, SGOT enzyme level is elevated; in hepatic disease, the level of SGPT is elevated. Levels of both are increased in stress-induced diseases like rheumatoid arthritis. It is significant that *A. indica* attenuates stress-induced gastric ulcer, immunosuppression, and rise in cholesterol and urea levels, and also lowers the SGOT and SGPT levels. All these facts taken together strongly suggest that *A. indica* has an antistress or adaptogenic property, and many of beneficial effects

Table 3. Effects of *A. indica* (100 mg/kg) on restraint stress (RS) induced gastric ulcer in rats

Treatment	Gastric pathology (mean ± SE)	
	No. of Ulcers	Severity (mm)
Vehicle	0.4 ± 0.2	0.04 ± 0.02
A. indica	0.9 ± 0.5	0.1 ± 0.05
Vehicle + RS	9.2 ± 2.6 *	1.8 ± 0.8*
A. indica + RS	1.4 ± 0.8 *	0.2 ± 0.1*

* $p < 0.05$ (compared to respective vehicle groups)
n = 6–8 animals per group.

Table 4. Effect of *A. indica* (100 mg/kg) on cellular immune responses in mice and rats

Treatment	Vehicle	*A. indica* (100 mg/kg)
Footpad thickness in mice (increase in paw volume)	0.07 ± 0.01	0.09 ± 0.01*
Leucocyte migration inhibition in rats (% inhibition)	54%	4.3%

n = 10–13 animals per group.
* $p < 0.02$.

produced by *A. indica* may be explained on this basis. On the basis of the above findings, it will also be of interest to explore the possibility of its effectiveness in stress-induced cardiac and hepatic disorders, rheumatoid arthritis, or some natural or acquired immundeficiency diseases.

ACKNOWLEDGEMENTS

The generous gift samples of *Azadirachta indica* from Dabur Research Foundation, Delhi, and technical assistances by S. Bhatia and V.S. Gupta are gratefully acknowledged.

REFERENCES

Ader, R. 1984. In: *Breakdown in Human Adaptation to Stress—Towards a Multi-disciplinary Approach*. Ballieux, R.E., J.F. Felding, and A.L. L'Affate, eds., Martinus-Ninjhoff, Boston, 653 pp.

Bhargava, K.P. and N. Singh. 1981. Anti-stress activity of Ocimum Lin., *Indian J. Med. Res.* 73: 443.

Brekhman, I.I. and I.V. Dardymov. 1969. New substances of plant origin which increase non specific resistance. *Ann. Rev. Pharmacol.* 9: 419.

Chopra, R.N., S.L. Nayar and I.C. Chopra. 1956. In: *Glossary of Indian Medicinal Plants*. CSIR, New Delhi.

Ray, A., P.K. Mediratta, S. Puri and P. Sen. 1991. Effects of stress on immune responsiveness, gastric ulcerogenesis and plasma corticosterone in rats: Modulation by diazepam and naltrexone. *Indian J. Exp. Biol.* 29: 233.

Satyavati, G.V., M.K. Raina and M. Sharma. 1976. Azadirachta (Meliaceae). In: *Medicinal Plants of India*, Vol. 2. ICMR, New Delhi, 363 p.

Sen, P., A.K. Chakravarti and J. Kohli. 1991. Effects of some imidazoles on cellular immune response—an experimental study. *Indian J. Exp. Biol.* 29: 867.

Sen, P., V. Dewan, S.K. Bhatacharya, V.S. Gupta, P.C. Maiti and P.K. Mediratta. 1988. Modification of humoral immune reponse by Ocimum sanctum lin—an anti-stress agent. In: *Brain and Psychophysiology of Stress*. ICMR, New Delhi, 245 p.

Sen, P., P.C. Maiti, S. Puri, A. Ray, N.A. Avdulov and A.V. Valdman. 1992. A study on the mechanisms ofthe anti-stress activity of Ocimum Sanctum lin, engenol and Tinospora Malabarica in experimental animals. *Indian J. Exp. Biol.* 30: 592.

Tewari, S., M. Seshadri and T.B. Poduval. 1982. Migration inhibition of normal rat thymocytis as an in vitro method for detecting cell mediated immunityin rat and mouse. *J. Immunol. Methods* 51: 231.

135

Thyroid Hormones and Intermediary Metabolism in Fish: Influence of Neem Kernel Extract

M.C. SUBASH PETER

Department of Zoology, Fatima Mata National College
Kollam 691 001, Kerala, India

ABSTRACT

Recent use of neem products as an effective natural pesticide in agriculture has evoked special attention due to its probable influence on fish metabolism. In the present study, ethanolic extract of neem kernel (NKE) with or without thyroid hormones was administered in a freshwater teleost, *Anabas testudineus*, and the activity of certain enzymes and the concentration of a few biochemical indices were investigated in detail. Intraperitoneal injection and immersion treatment of NKE significantly altered the glucose, total protein, triglycerides, urea, RNA and DNA concentrations and the activity of lactate dehydrogenase, aspartate transferase, alanine transferase, and alkaline phosphatase in serum and liver. Injection of thyroxine and triiodothyronine to fish treated with NKE (by both injection and exposure) reversed the NKE-induced decrease or increase in the activity or concentration of various parameters studied. This suggests the regulatory role of thyroid hormones in neem-treated fishes.

INTRODUCTION

The Indian neem tree, *Azadirachta indica* A. Juss. (Meliaceae), has long been recognized for its unique medicinal properties. Although considerable research has been done on the neem-insect interaction, little is known about the influence of neem products on vertebrate physiology. However, there is some information regarding the pharmacological and toxicological effects of neem on warm-blooded animals (Jacobson, 1986). Recent use of neem products as an effective alternative to chemical pesticide in agriculture has evoked special interest among fish physiologists in view of the pollutional hazards, if any, posed by these products. Neem-induced alterations in the turnover of biochemical constituents and enzyme activities may therefore have diagnostic significance in evaluation of adverse health effects of these products.

It is well known that thyroxine (T_4) and triidothyronine (T_3), the important thyroid hormones, regulate a wide variety of metabolic functions in fishes. Extensive investigations have shown that thyroid hormones exert a stimulatory influence on oxidative metablism in fishes (Peter and Oomen, 1989a, b). Recently it has been shown that thyroid hormones exhibit a regulatory role in oxidative metabolism of fish treated with the fish poison, rotenone (Peter and Oomen, 1991). The present study, therefore, examines the influence of neem kernel extract (NKE) on intermediary metabolism of fish and how thyroid hormones regulate the metabolic processes of stress-induced fish. To achieve these goals, different doses of NKE were administered (both intraperitoneally and by exposure) with or without T_4 and T_3 and the levels and activity of certain biochemical constit ents and enzymes were investigated in blood and liver tissues in a teleost, *Anabas testudineus* Bloch.

MATERIALS AND METHODS

Healthy adult Indian climbing perch (*A. testudineus*) weighing 45 to 50 g were collected and kept in aquaria at water temperature 28° ± 1°C (12: 12 hr L:D). The experiments were performed in May-June, which is the prespawning period. Fish were fed every day with commercial fish feed and feeding was stopped for 24 hrs prior to sacrifice. Two sets of experiments were done independently. The first set of experiments examined the effect of intraperitoneal injection of ethanolic extract of neem kernel with or without T_3 or T_4. The laboratory-acclimated fish were divided into seven groups of six fish each. Group I received equal volume (0.1 ml) of hormons vehicle (saline, pH 9.0) and drug vehicle (ethanol:water, 1:200) and served as control. Group 2 received 1 μg of NKE. Group 3 received 5 μg of NKE. Groups 4 and 5 received 2 μg T_3 and 5 μg T_4 respectively. Groups 6 and 7 received 5 μg of NKE; group 6 also received 2 μg of T_3 and group 7 also received 5 μg of T_4.

The second set of experiments was planned to study the effect of T_3 and T_4 in fish immersed in NKE treated water. Acclimated fish were divided into seven groups of six fish each, and were housed in a glass tank of 60 cm length, 20 cm height, 30 cm breadth, and 36 l capacity. Fish of group 1 served as controls which received hormone vehicle only on the 11th day. Group 2 was given 2 mg NKE/1 on alternate days for a period of 10 days (10 mg/l). Group 3 was immersed in water treated with 2 mg of NKE/1 every day for a period of 10 days (20 mg/l). Group 4 received 2 μg T_3. Group 5 received 5 μg T_4 on the 11th day. Groups 6 and 7 were immersed in NKE-treated water (10 mg/1) and they were given 2 μg T_3 and 5 μg T_4 respectively on the 11th day.

All the injections were given intraperitoneally between 8.30 and 9.00 a.m. and the daily dose of the hormone(s)/drug(s) administered was in 0.1 ml of the respective vehicle. Twenty-four hours after the last injection, blood was drawn directly from the heart and fish were then sacrificed by decapitation. The liver tissue was quickly removed, chilled, and weighed and a 10 per cent homogenate was prepared in cold 0.25 M sucrose in a Potter-Elve jam homogenizer. The supernatant was collected by centrifuging the homogenate at 5000 g for 10 min. in a refrigerated centrifuge at 4°C. Blood of three fish was pooled and centrifuged at 5000 g for 10 min. to separate serum.

The enzyme activities and concentration of biochemical indices were determined photometrically at 30°C on a vital lab autoanalyser adopting standard procedures (Bergmeyer et al., 1983). Suppliers' instructions with regard to the pH, incubating time and temperature specified for individual enzymes were strictly followed during enzyme assays. The biochemical constituents studied were glucose, triglycerides, and urea. The enzymes were alkaline phosphatase (ALP) (orthophosphoric-monoester phosphohydroxylase, alkaline optimum, EC 3.1.3.1), lactate dehydrogenase (LDH) (L-lactate: NAD + oxidoreductase, EC 1.1.1.27), asparartate transferase (AST) (L-aspartate: 2-oxyglutarate aminotransferase EC 2.6.1.1), and alanine transferase (ALT) (L-alanine: 2-oxoglutarate aminotransferase, EC 2.6.1.2). A part of liver tissue was homogenized in 5 per cent $HClO_4$ and the total protein (Lowry et al., 1954), RNA (Mejbaum, 1959), and DNA (Burton, 1956) contents were estimated.

Six tests were done for each tissue parameter and four tests were carried out for each serum parameter. Each blood test, however, contained blood of three fish. Data were collected and statistical analyses were done by one-way classification of analysis of variance (Snedecor and Cochran, 1967) supplemented by multiple range test (Kramer, 1956).

RESULTS

Effect of T_3 or T_4 in Fish Injected with NKE

Intraperitoneal injection of 1 and 5 µg NKE, while significantly decreasing the glucose concentration in blood, elevated its level in liver tissue. Injection of T_3 or T_4 in NKE-injected fish significantly increased the glucose concentration in blood and liver, when compared to NKE-treated fish. Administration of various doses of NKE significantly increased total protein and RNA contents in liver. T_3 and T_4 injections in NKE-injected fish further augmented the concentration of these parameters. Although NKE injections did not alter liver DNA concentration, T_3 and T_4 injections increased its level in NKE-treated fish. Injections of NKE decreased the concentration of triglycerides in blood and liver tissue. T_3 or T_4 injection intensified the decrease in triglyceride content in drug-injected fish. The urea concentrations in serum and liver were substantially increased by NKE treatment and T_3 and T_4 injections further elevated its level in serum when given to NKE-injected fish (Table 1).

The activity of LDH in serum and liver increased after NKE injection. Administration of T_3 and T_4 in NKE-injected fish, while intensifying the increase in LDH activity in blood, prevented the activity in liver tissue when compared to NKE-treated fish. The activity of AST, ALT, and ALF decreased in serum and liver after NKE injections. However, these enzyme activities were increased substantially by T_3 or T_4 injection in NKE-injected fish (Table 2).

Table 1. Effect of T_3 or T_4 on the concentration of certain biochemical indices of intermediary metabolism in NKE-injected *A. testudineus*. Fish were killed 24 hrs after the last injection. Each value is mean ± SD of six tests

Treatment	Glucose		Total protein	Triglycerides		Urea		RNA	DNA
	Blood (mg/dl)	Liver (mg/g wet tissue)	Liver (µg/100 mg wet tissue)	Blood (mg/dl)	Liver (µg/100 mg wet tissue)	Blood (mg/dl)	Liver (µg/100 mg wet tissue)	Liver (mg/g wet tissue)	Liver (mg/g wet tissue)
Control	233.15 ± 4.5	16.38 ± 0.2	14.01 ± 0.9	423 ± 6.5	250 ± 4.1	6.2 ± 0.03	47.5 ± 0.2	5.5 ± 0.03	1.87 ± 0.01
1 µg NKE	93.34** ± 2.7	24.96** ± 0.5	18.43** ± 0.8	341** ± 7.8	211** ± 5.1	8.6** ± 0.03	56.5** ± 0.3	11.69** ± 0.04	1.89** ± 0.01
5 µg NKE	89.43** ± 2.8	26.59** ± 0.5	21.34** ± 0.9	267** ± 5.1	189** ± 6.1	8.5** ± 0.01	59.5** ± 0.2	12.73** ± 0.03	1.93** ± 0.01
2 µg T_3	275.48** ± 4.5	39.46** ± 0.3	19.48** ± 0.7	115** ± 3.5	155** ± 5.1	8.4** ± 0.03	31.6** ± 0.2	10.48** ± 0.04	2.58** ± 0.02
5 µg T_4	289.32** ± 4.1	48.63** ± 0.6	20.56** ± 0.8	104** ± 3.4	115** ± 3.3	8.7** ± 0.04	29.3** ± 0.1	9.65** ± 0.03	2.64** ± 0.02
5 µg NKE + 2 µg T_3	189.68** ± 4.8	30.46** ± 0.7	20.43** ± 0.7	123** ± 4.1	165** ± 3.9	9.8* ± 0.04	38.8** ± 0.3	13.46** ± 0.04	2.53** ± 0.02
5 µg NKE + 5 µg T_4	176.56** ± 5.1	29.52** ± 0.5	21.16** ± 0.9	110** ± 4.5	125** ± 4.1	10.5** ± 0.04	40.5** ± 0.3	11.49** ± 0.4	2.85** ± 0.04

* $p < 0.05$, ** $P < 0.01$

Table 2. Effect of T_3 or T_4 on the activity of certain enzymes involved in intermediary metabolism of *A. testudineus*. Fish were killed 24 hrs after the injection. Each value is mean ± SD of six tests

Treatment	LDH (IU)		AST (IU)		ALT (IU)		ALP (IU)	
	Blood	Liver	Blood	Liver	Blood	Liver	Blood	Liver
Control	1064 ± 26	1299 ± 31	360 ± 10.1	467 ± 9	68 ± 4.3	196 ± 8.1	8.4 ± 0.04	8.3 ± 0.03
1 μg NKE	1250** ± 21	2393** ± 29	272** ± 8.5	473 ± 9	58** ± 2.5	183 ± 7.5	7.1** ± 0.03	3.2** ± 0.01
5 μg NKE	2309** ± 48	3205** ± 48	250** ± 9.3	312** ± 8	51** ± 3.3	146** ± 5.1	4.2** ± 0.02	2.3** ± 0.01
2 μg T_3	1450** ± 22	1350** ± 24	512** ± 10.2	658** ± 14	98** ± 2.2	398** ± 8.3	10.5** ± 0.03	9.4** ± 0.05
5 μg T_4	1550** ± 18	1241** ± 49	545** ± 11.3	761** ± 13	120** ± 3.3	415** ± 11.2	14.6** ± 0.04	11.5** ± 0.03
5 μg NKE + 2 μg T_3	2890** ± 38	1510** ± 28	485** ± 9.2	549** ± 10	108** ± 3.1	289** ± 8.4	8.3** ± 0.03	6.6** ± 0.02
5 μg NKE + 5 μg T_4	2495** ± 33	1683** ± 29	547** ± 12.1	683** ± 12	127** ± 4.3	393** ± 9.3	9.4** ± 0.04	7.4** ± 0.03

*$P < 0.05$, **$P < 0.05$

Effect of T_3 and T_4 in Fish Immersed in NKE-treated Water

Fish immersed in different concentrations (10 mg or 20 mg/l) of NKE-treated water exhibited a response similar to the fish injected with NKE. It is also noticed that injection of T_3 or T_4 restored or intensified the NKE-induced decrease or increase in the activity and concentrations of various parameters studied. This response again coincides with the results obtained from intraperitoneal injection treatments (Tables 3, 4).

DISCUSSION

The present study clearly reveals that although NKE treatment exerts a profound change in the intermediary metabolism towards anabolic side, exogenous thyroid hormones restore or regulate the NKE-induced changes in intermediary metabolism of fish.

The NKE treatments (both injection and exposure) decreased the glucose concentration in serum, suggesting hypoglycemic effect of neem in fishes. Intraperitoneal injection of neem seed oil reduced blood glucose level in intact and diabetics-induced mice (Purohit et al., 1990). Bhargava (1987) has observed in human patients that neem oil acted as a synergist with most of the oral antidiabetics, and helped to gradually reduce their intake. It seems that hypoglycemic action of NKE may be due to its direct metabolic effect on tissue and or due to increase in insuline secretion. On the contrary, besides hypoglycemia, NKE treatment produces increased availability of glucose in tissues which may account for the increased glucose oxidation. Increased level of blood glucose after thyroid hormone injection reveals its hyperglycemic action in fishes as has been reported in other lower vertebrates (Jacob and Oommen, 1990). The observed increase in liver glucose concentration also supports the glucose-forming effect of thyroid hormones. Administration of T_3 or T_4 in NKE-treated fish prevented the neem-induced hypoglycemia, suggesting the restorative role of T_3 and T_4 in the turnover of glucose in fish.

Nothing is known about the proteogenic role of neem in vertebrates, especially in fishes. Neem kernel extract treatments increase total protein content in the liver, which reflects a probable increased protein synthesis. It has been reported that neem products have little influence on biochemical parameters of serum in rats (Qadri et al., 1984). The role of thyroid hormones in protein metabolism of fishes is well established. Recently it has been shown that physiological doses of T_3 or T_4 stimulate oxidative metabolism in teleosts (Peter and Oommen, 1988, 1989a) and the increase was later attributed to an increase in protein synthesis (Peter and Oommen, 1989b). Simultaneous injection of NKE and T_3 or T_4 also increases the protein content, thereby confirming their proteogenic role in fishes.

The triglyceride concentrations in serum and liver tissue were decreased by different doses of NKE treatments, which probably reflects decreased triglycerides turnover. Since triglycerides hydrolyse into fatty acids and glycerol, their observed decrease after T_3 or T_4 injection in NKE-treated fish suggests that thyroid hormones also favour catabolism of triacylglycerols.

The level of urea indicates nitrogen balance. It appears that there is an increase in urea content in blood after NKE treatment and thyroid hormone administration, suggesting an increased turnover of protein catabolism. Stimulation of ammonia excretion by

Table 3. Effect of T_3 or T_4 on the concentration of certain biochemical indices of intermediary metabolism in fish exposed to NKE-treated water. Group 2 was exposed to 2 mg/l NKE on alternate days over a period of 10 days, thus receiving total 10 mg/l. Group 3 was exposed to 2 mg/l NKE daily over a period of 10 days, thus receiving total 20 mg/l. Fish were killed 24 hrs after T_3 or T_4 injection. Each value is mean ± SD of six tests

Group	Treatment	Glucose		Total protein	Triglycerides		Urea		RNA	DNA
		Blood (mg/dl)	Liver (mg/g wet tissue)	Liver (µg/100 mg wet tissue)	Blood (mg/dl)	Liver (*µg/100 mg wet tissue)	Blood (mg/dl)	Liver (*µg/100 mg wet tissue)	Liver (mg/g wet tissue)	Liver (mg/g wet tissue)
1.	Control	212.35 ± 3.8	18.43 ± 0.3	15.31 ± 0.8	441 ± 7.3	295 ± 3.5	6.8 ± 0.03	42.3 ± 0.2	5.9 ± 0.03	1.95 ± 0.01
2.	10 mg NKE/l over 10 days	165. 5** ± 2.5	23.56** ± 0.4	21.59** ± 0.9	316** ± 4.5	261** ± 4.5	8.1** ± 0.03	57.3** ± 0.3	8.6** ± 0.04	1.98** ± 0.01
3.	20 mg/l NKE over 10 days	151.35** ± 3.1	20.52** ± 0.3	24.65** ± 0.9	251** ± 4.4	221** ± 4.0	10.4** ± 0.04	61.5** ± 0.3	9.5** ± 0.03	2.05** ± 0.02
4.	2 µg T_3/ 24 hrs	268.59** ± 3.3	33.41** ± 0.3	21.48** ± 0.8	125** ± 3.5	165** ± 3.1	8.2** ± 0.03	33.4** ± 0.2	10.2** ± 0.03	2.59** ± 0.02
5.	5 µg T_4/ 24 hrs	278.35** ± 3.8	39.62** ± 0.4	20.51** ± 0.7	113** ± 4.1	125** ± 3.5	8.6** ± 0.04	28.5** ± 0.2	9.8** ± 0.03	2.61** ± 0.02
6.	10 mg NKE + 2 µg	221.5** ± 4.5	24.39** ± 0.3	22.69 ± 0.9	151** ± 3.8	143** ± 3.3	9.3** ± 0.03	43.3** ± 0.3	9.86** ± 0.04	2.35** ± 0.02
7.	10 mg NKE + 5 µg	215.13** ± 4.3	28.41** ± 0.4	21.65 ± 0.8	141** ± 3.5	151** ± 4.1	10.2** ± 0.04	49.5** ± 0.9	9.9** ± 0.04	2.41** ± 0.03

*p < 0.05, **P < 0.01

Table 4. Effect of T_3 or T_4 on the activity of certain enzymes involved in intermediary metabolism of fish exposed to NKE-treated water. Group 2 was exposed to 2 mg/l NKE on alternate days over a period of 10 days, thus receiving total 10 mg/l. Group 3 was exposed to 2 mg/l NKE daily over a period of 10 days, thus receiving total 20 mg/l. Fish were killed 24 hrs after T_3 or T_4 injection. Each value is mean ± SD of six tests

Group	Treatment	LDH (IU)		AST (IU)		ALT (IU)		ALP (IU)	
		Blood	Liver	Blood	Liver	Blood	Liver	Blood	Liver
1	Control	1112 ± 22	1350 ± 28	372 ± 8.4	456 ± 8.9	59 ± 3.4	202 ± 7.3	8.1 ± 0.04	6.3 ± 0.04
2	10 mg/l NKE over 10 days	1315** ± 25	1740** ± 30	260** ± 9.5	318** ± 9.5	48** ± 2.8	185** ± 6.3	6.3** ± 0.03	4.2** ± 0.03
3	20 mg/l NKE over 10 days	1420** ± 30	2132** ± 38	242** ± 8.3	310** ± 8.6	41** ± 2.2	159** ± 5.1	4.1** ± 0.04	2.5** ± 0.03
4	2 μg T_3/24 hrs	1510** ± 42	1380 ± 40	498** ± 12.3	672** ± 12.5	89** ± 3.5	403** ± 8.1	11.15** ± 0.05	9.1** ± 0.04
5	5 μg T_4/24 hrs	1612** ± 43	1390 ± 35	512** ± 12.5	708** ± 11.5	101** ± 4.5	428** ± 8.5	13.16** ± 0.05	10.5** ± 0.04
6	10 mg NKE + 2 μg T_3	1590** ± 30	1420** ± 32	385** ± 8.5	518** ± 10.5	91** ± 4.8	221** ± 7.5	8.8** ± 0.04	6.5** ± 0.02
7	10 mg NKE + 5 μg T_4	1415** ± 31	1442** ± 33	412** ± 10.5	623** ± 10.9	98** ± 4.7	308** ± 8.1	9.1** ± 0.05	7.1** ± 0.02

**$p < 0.01$

thyroid hormone has been confirmed in fishes (Ray and Medda, 1976). Teleosts excrete about 6–26 per cent of nitrogen as urea through purine degradation and substantial production of urea requires amino acids as its nitrogen donors (Thornburn and Matty, 1963; Velas and Serfaty, 1974). The substantial increase in urea level in blood observed after thyroid hormone injection in NKE-treated fish was probably due to increased activity of transferases which contributes towards urea formation by channelling the nitrogen of amino acids into the irreversible pathway of urine formation.

Nucleic acids are the important determinants of protein synthesis. The increase in RNA content after NKE treatment can be correlated with increased protein synthesis in the liver. Although NKE treatments did not elicit significant increase in DNA contents, the proteogenic role of thyroid hormone is confirmed by its stimulatory influence on DNA concentration. An increase in DNA content and a decrease in RNA content were observed in liver after T_3 and T_4 injection (44 ng/g) in rainbow trout (Oommen and Matty, 1991). The increased contents of RNA and DNA after thyroid hormone administration in NKE-treated fish revealed the regulatory influence of thyroid hormones on nulceic acid metabolism of fish.

The enhanced activity of LDH in serum and liver after NKE treatment suggests increased lactate oxidation. Administration of thyroid hormone, while increasing LDH activity in the serum, prevents the NKE-induced enzyme activity in liver. This may indicate restorative action of thyroid hormones in liver tissue, which functions as a target organ of thyroid hormone action.

Aspartate transferase and ALT are transferases concerned with non-essential amino acid metabolism and gluconeogenesis. The observed decrease in the activities of AST and ALT after NKE treatments may be due to the probable decrease in the level of aspartic acid and alanine respectively in the tissues, suggesting reduced protein breakdown in NKE-treated fish. The significant increase in the activities of these two enzymes after thyroid hormone administration in NKE-treated fish reveals the restorative role of these hormones in amino acid metabolism. It is also evident that an increase in the ratio of AST activity to that of ALT causes increased urea production. Janicki and Lingis (1970) have presented evidence that AST is involved in ammonia production in the teleost liver.

Alkaline phosphatase is a lysosomal enzyme hydrolysing organic monophosphate esters with optimal rate at an alkaline pH. Neem Kernel extract treatments decrease the ALP activity indicating decreased lysosomal activities in blood and liver. Thyroid hormones significantly increase the ALP activity in NKE-treated fish, emphasizing the mobilization of metabolites of trophic importance for maintaining cellular integrity. It seems likely that thyroid hormones exert a restorative action in transamination and lysosomal activities to a certain extent which appears to be most advantageous to the fish in facing neem product exposure.

ACKNOWLEDGEMENTS

The author is grateful to Dr. O.V. Oommen, Department of Zoology, University of Kerala, for his generous gift of thyroid hormones. Thanks are due to Professor G.K. Karanavar, Head of the Department of Zoology, University of Kerala, for providing necessary facilities. I would also like to thank Fr. T.M. Joseph and Mr. Trilok for their excellent assistance.

REFERENCES

M. 1956. A study for the conditions and mechanism for the DPA reaction for the colorimetric estimation of DNA. *J. Biol. Chem.* 66: 315–323.

Bergmeyer, H.U., J. Bergmeyer and M. Grabel. 1983. *Methods of Enzymatic Analysis*. Vol. II. Verlag Chemie, Weinheim.

Bhargava, A.K. 1987. Neem oil as a synergist to antidiabetic drugs for management of secondary hyperglycemia. *Neem Newsletter* 4(3): 31–32.

Jacob, V. and O.V. Oomen. 1990. Intermediary metabolism in a lizard, *Calotes versicolor*: Role of thyroid hormones. *Gen. Comp. Endocrinol.* 77: 324–336.

Jacobson, M. 1986. Pharmacological and toxicological effects of neem and chinaberry on warm blooded animals. *Neem Newsletter* 3(4): 39–43.

Janicki, R. and J. Lingis. 1970. Mechanism of ammonia production from aspartate in teleost liver. *Comp. Biochem. Physiol.* 37: 101–105.

Kramer, C.Y. 1956. Extension of multiple range tests to group means with unequal numbers of replications. *Biometrics* 12: 307–310.

Lowry, O.H., N.R. Roberts, W.U. Mei-ring, W.S. Hixon and E.J. Crawford. 1954. Phosphate studies: Acid and alkaline phosphatase. *J. Biol. Sci.* 207: 19–37.

Mejbaum, W. 1959. Estimation of small amounts of pentose especially in derivatives of adenylic acids *Z. Physiol. Chem.* 258: 117–120.

Oomen, O.V. and A.J. Matty. 1991. The effects of thyroid hormones and starvation on hepatic mitochondrial nucleic acids of rainbow trout (*Oncorhynchus mykiss*) *Gen. Comp. Endocrinol.* 83: 468–472.

Peter, M.C.S. and O.V. Oomen. 1988. Influence of thyroid hormones on hepatic mitochondrial enzymes of thio-uracil-treated telecost, *Anabas testudineus* (Bloch). *Indian J. Exp. Biol.* 26: 651–653.

Peter, M.C.S. and O.V. Oomen. 1989a. Effect of thyroid and gonadal hormones *in vitro* on hepatic succinate dehydrogenase activity of the teleost, *Anabas testudineus* (Bloch). *Zool. Sci.* 6: 185–189.

Peter, M.C.S. and O.V. Oomen 1989b. Oxidative metabolism in a teleost, *Anabas testudineus* Bloch: Effect of thyroid hormones on hepatic enzyme activities. *Gen. Comp. Endocrinol.* 73: 96–107.

Peter, M.C.S. and O.V. Oomen. 1991. Role of thyroid hormones in rotenone-treated teleost, *Anabas testudineus* Bloch: Effects on oxidative enzyme activities. *Indian J. Exp. Biol.* 29: 481–483.

Purohit, A.K., V.B. Joshi and S.A. Vyas. 1990. Hypoglycemic activity of neem seed oil (*A. indica*) on streptozotocin-induced diabetic mice. *J. Diab. Associ. India* 30(2): 36–37.

Qadri, S.S.H., G. Usha and K. Jabeen. 1984. Subacute dermal toxicity of neem-rich 100 (tech) to rats. *Int. Pest Control* 26: 18–20.

Ray, A.K. and A.K. Medda. 1976. Effect of thyroid hormones and analogues on ammonia and urea excretions in latafish (*Ophicephalus punctatus*) *Gen. Comp. Endrocrinol.* 29: 190–197.

Snedecor, G.W. and W.G. Cochran. 1967. *Statistical Methods*, 6th ed. Oxford & IBH Publ. Pvt. Ltd. New Delhi.

Thornburn, C.C. and A.J. Matty. 1963. The effect of thyroxine on some aspects of nitrogen metabolism in the gold fish (*Carassius auratus*) and the trout (*Salmo trutta*). *Comp. Biochem. Physiol.* 1: 1–12.

Velas, F. and A. Serfaty. 1974. La jeune et la realimentation chez la carpe. *Cyprinus carpio*. *J. Physiol.* 68: 245–260.

The Contributors and Keynote Speakers

Abbas, S.R.
Central Institute of Horticulture for Northern Plains,
B-217 Indira Nagar, Lucknow 226 016, India

Abdul Wajid, S.M.
CTRI Research Station, Hunsur 571 105, Karnataka, India

Acharya, A.
All India Coordinated Research Project on Betelvine,
Indian Institute of Horticultural Research, Hessaraghatta Lake,
Bangalore 560 089, India

Ahmad, Anjum
Department of Botany, Aligarh Muslim University,
Aligarh 202 002, India

Akhauri, R.K.
Department of Entomology, Rajendra Agricultural University, Pusa,
Samastipur 848 125, Bihar, India

Alagianagalingam, M.N.
Agricultural College and Research Institute, Tamil Nadu Agricultural University,
Killikulam 627 252, India

Anand Rao, M.
All India Coordinated Research Project on Agricultural Ornithology,
Andhra Pradesh Agricultural University, Rajendranagar, Hyderabad, India

Anderson, Arild
Norwegian Plant Protection Institute, Fellesbygget, N-1432,
Norway

Arumugasamy, K.
Department of Botany, Bharathiar University,
Coimbatore 641 046, India

Azeemoddin, G.
Jawaharlal Nehru Technological University,
Oil Technological Research Institute, Anantapur, 515 001, India

Babu, R.
Division of Entomology, Department of Zoology, Bharathiar University,
Coimbatore 641 046, India

Balasubramanian, C.
Department of Chemistry, Bharathiar University,
Coimbatore 641 046, India

Bandara, K.A.N.P.
Division of Entomology, Central Agricultural Research Institute,
Peradeniya, Sri Lanka

Baumgart, Martin
Institut fur Phytopathologie und Angewandte Zoologie,
Justus-Liebig-Universitat, 6300 Giessen, Germany

Bhandari, A.L.
Department of Agronomy, Punjab Agricultural University,
Ludhiana 141 004, India

Bhanderi, G.R.
Department of Entomology, N.M. College of Agriculture, Gujarat
Agricultural University, Navsari Campus, Navsari 396 450, India

Bhathal, Surinder S.
Department of Entomology, Punjab Agricultural University,
Ludhiana 141 004, India

Bhatnagar, Deepak
USDA/Agricultural Research Service, Southern Regional Research Center,
New Orleans, Louisiana 70124, USA

Bhatnagar, G.S.
Agricultural Research Station, Rajau,
Banswara 327 001, India

Bisla, S.S.
Department of Agroforestry, CCS Haryana Agricultural University,
Hisar 125 004, India

Biswas, B.K.
Department of Botany, Rajshahi University,
Rajshahi, Bangladesh

Bosselmann, Axel
Department of Botany, University of Queensland, Brisbane,
Australia 4072

Brar, R.S.
Department of Agronomy, Punjab Agricultural University,
Ludhiana 141 004, India

Bream, Ahmad Saber
Plant Protection Institute, Hungarian Academy of Sciences, Budapest,
PO Box 102, H-1525, Hungary

Calcagna, Maria Pia
CSIC Department of Organic and Biological Chemistry,
Jordi Girona 18-26, 08034 Barcelona, Spain

Chandrasekar, S.S.
Indian Cardamom Research Institute, Myladumpara,
Kailasanadu P.O. 685 553, Kerala, India

Chaplot, P.S.
Agricultural Research Station, Rajau,
Banswara 327 001, India

Chari, M.S.
Central Tobacco Research Institute,
Rajahmundry 533 105, India

Chendramohan, B.
Department of Plant Pathology, S.V. Agricultural College,
Andhra Pradesh Agricultural University, Tirupati 517 502, India

Chih-Hua Tsou
Institute of Botany, Academia Sinica, Nankang, Tapei, Taiwan 11529,
Republic of China

Cobbinah, J.R.
Forestry Research Institute of Ghana, University Science and
Technology Campus, Kumasi, Ghana

Csatlos, Imre
Plant Health and Soil Conservation Station of Csongrad-county,
Hodmezovasarhely, Rarosi u. 102, H-6801, Hungary

Dakshinamurthy, A.
Post Harvest Technology Scheme CIAE,
Bhopal, India

Dale, D.
College of Agriculture, Vellayani,
Kerala, India

Darvas, Bela
Plant Protection Institute, Hungarian Academy of Sciences,
Budapest, PO Box 102, H-1525, Hungary

Das, N.M.
College of Agriculture, Vellayani, Kerala, India

Dash, D.D.
Department of Entomology, Orissa University of Agriculture and Technology,
Bhubaneswar 751 003, India

Dayanandan, P.
Department of Botany, Madras Christian College, Tambaram,
Madras 600 059, India

Dhaliwal, G.S.
Section of Ecology, Punjab Agricultural University,
Ludhiana 141 004, India

Dhanapal, K.
ICRI Regional Research Station,
Thadiyankudisai 624 212, India

Dhawan, Ashok K.
Department of Entomology, Punjab Agricultural University,
Ludhiana 141 004, India

Dhingra, Mukta
Directorate of Oilseed Research, Rajendranagar,
Hyderabad 500 030, India

Dhruj, I.U.
Department of Plant Pathology, College of Agriculture,
Junagadh 362 001, India

Dilawari, V.K.
Section of Ecology, Punjab Agricultural University,
Ludhiana 141 004, India

Dimetry, Nadia Z.
Pests and Plant Protection Department, National Research Centre,
Dokki, Cairo, Egypt

Dorn, A.
Institute of Zoology, University of Mainz, Saarsts. 21,
D-6500 Mainz, Germany

Drew, R.A.
Queensland Department of Primary Industries,
PO Box 327, Cleveland Q 4163, Australia

Eswaramurthy, S.
Agricultural College and Research Institute, Tamil Nadu Agricultural University,
Killikulam 627 252, India

Facknath, S.
Faculty of Agriculture, University of Mauritius,
Reduit, Mauritius

Farag, Ali Ibrahim
Plant Protection Department, Faculty of Agriculture,
Menoufia University, Egypt

Fazi, Shaheen
H.E.J. Research Institute of Chemistry, University of Karachi,
Karachi 75270, Pakistan

Ganesan, S.
Godrej Agrovet Ltd.,
Madras 600 098, India

Geethanjali, Y.
Osmania University,
Hyderabad 500 001, India

Ghiasuddin
H.E.J. Research Institute of Chemistry, University of Karachi,
Karachi 75270, Pakistan

Godrej, N.B.
Godrej Soaps Limited, Vikhroli,
Bombay 400 079, India

Gogate, M.G.
Conservator of Forests (Research), Maharashtra State PMT Building,
Pune, India

Gomez, D.
Proyecto Catie-Mag/Mip,
Nicaragua

Gopakumar, B.
Indian Cardamom Research Institute, Myladumpara,
Kailasanadu P.O. 685 553, Kerala, India

Gothandapani, L.
Department of Agricultural Processing, College of Agricultural Engineering,
Tamil Nadu Agricultural University, Coimbatore 641 003, India

Gour, T.B.
Department of Entomology, College of Agriculture, Rajendranagar,
Hyderabad 500 030, India

Guerrero, B.
Estacion Experimental del Valle de Sebaco, Nicaragua
Guharay, F.
Proyecto Catie-Mag/Mip, Nicaragua

Gujar, D.R.
Asst. Silviculturist, Ravinagar Nursery,
Nagpur 440 001, India

Gujar, G.T.
Division of Entomology, Indian Agricultural Research Institute,
New Delhi 110 012, India

Gupta, G.P.
Division of Entomology, Indian Agricultural Research Institute,
New Delhi 110 012, India

Gupta, P.
Allahabad Agricultural Institute,
Allahabad 211 007, India

Gupta, P.K.
Rajasthan Agricultural University, Agricultural Research Station,
Borkhera, Kota 324 001, India

Gupta, R.
I.G.A.U., Raipur 492 012,
Madhya Pradesh, India

Gupta, S.
Safdarjang Hospital,
New Delhi, India

Gupta, Shashi
Division of Entomology, Indian Agricultural Research Institute,
New Delhi 110 012, India

Haseeb, M.
Central Institute of Horticulture for Northern Plains,
B-217 Indira Nagar, Lucknow 226 016, India

Hegde, Narayan G.
BAIF Development Research Foundation,
Pune, India

Hellpap, Carsten
GTZ/Institut fur Phytopathologie und Angewandte Zoologie,
Ludwigstr. 23, 6300 Giessen, Germany

Hossain, M.
Department of Botany, Rajshahi University,
Rajshahi, Bangladesh

Hulagur, B.F.
Station Agronomique Laon, (Inra), France

Ilovai, Z.
Plant Health and Soil Conservation Station of Csongrad-county,
Hodmezovasarhely, Rarosi u. 102, H-6801, Hungary

Immaraju, John
AgriDyne Technologies Inc., 2401 South Foothill Dr., Salt Lake City,
Utah 84109, USA

Islam, R.
Department of Botany, Rajshahi University,
Rajshahi, Bangladesh

Isman, Murray B.
Department of Plant Science, University of British Columbia,
Vancouver, B.C., Canada V6t 1z4

Jahanmohini, P.
Division of Entomology, Department of Zoology, Bharathiar University,
Coimbatore 641 046, India

Jain, R.K.
Department of Botany, University of Lucknow,
Lucknow 226 007, India

Jayaraj, S.
Agricultural Research Station,
Paramakudi 623 707, Tamil Nadu, India

Jhansi, B.R.
Division of Entomology, I.I.H.R., Hessarghatta Lake,
Bangalore 560 089, India

Joarder, N.
Department of Botany, Rajshahi University,
Rajshahi, Bangladesh

Joarder, O.I.
Department of Botany, Rajshahi University,
Rajshahi, Bangladesh

Jonathan, E.I.
All India Coordinated Research Project on Betelvine,
Indian Institute of Horticultural Research, Hessaraghatta Lake,
Bangalore 56 089, India

Joseph, Thomas
Indian Cardamom Research Institute, Myladumpara, Kailasanadu P.O.,
Idukki District 685 553, Kerala, India

Joshi, M.S.
Plant Tissue Culture Division, National Chemical Laboratory,
Pune 411 008, India

Joshi, S.P.
National Chemical Laboratory,
Pune 411 008, India

Kalode, M.B.
Directorate of Rice Research, Rajendranagar,
Hyderabad 500 030, India

Kameshwara Rao, P.
Department of Entomology (APAU), S.V. Agricultural College,
Tirupati 517 502, A.P., India

Kameswara Rao, B.V.
Central Tobacco Research Institute,
Rajahmundry 533 105, India

Kapur, S.K.
Institute of Pathology, ICMR,
New Delhi, India

Karunakaran, K.R.
Tamil Nadu Agricultural University,
Coimbatore 641 003, India

Katti, Gururaj
Indian Institute of Pulses Research (ICAR),
Kanpur, India

Kaushik, J.C.
Department of Agroforestry, CCS Haryana Agricultural University,
Hisar 125 004, India

Kaushik, Nutan
Division of Agricultural Chemicals, Indian Agricultural Research
Institute, New Delhi 110 012, India

Kaushik, U.K.
I.G.A.U., Raipur 492 012,
Madhya Pradesh, India

Khalafalla, A.
Shambat Research Station, PO Box 30, Khartoum, North Sudan

Khan, R.M.
Division of Entomology and Nematology, Indian Institute of
Horticultural Research, Hessaraghatta, Bnagalore 560 089, India

Kiarie, M.
Regional Research Centre — Mtwapa, PO Box 16,
Kikambala, Kenya

Kleeberg, Hubertus
Trifolio-M-GmbH, D-35633 Lahnau, Sonnenstr. 22,
Germany

Knauss, J.F.
Washington Research Center, 7379 Route 32, Columbia, MD 21044,
USA

Koul, Opender
Regional Research Laboratory,
Jammu 180 001, India

Krall, S.
Deutsche Gesellschaft fur Technische Zusammenarbeit (GTZ)
GmbH, 6236 Eschborn, Germany

Kranthi, K.R.
Division of Entomology, Indian Agricultural Research Institute,
New Delhi 110 012, India

Kraus, W.
Department of Chemistry, University of Hohenheim, 7000,
Stuttgart 70, Germany

Krishnaiah, N.V.
Directorate of Rice Research, Rajendranagar, Hyderabad 500 030,
India

Krishnakumari, R.
Department of Zoology, Annamalai University, Annamalai Nagar 608 002,
Tamil Nadu, India

Krishna Moorthy, P.N.
Indian Institute of Horticultural Research, Hessarghatta Lake P.O.,
Bangalore 560 089, India

Kudagamage, C.
Division of Entomology, Central Agricultural Research Institute,
Peradeniya, Sri Lanka

Kumar, Jitendra
Division of Agricultural Chemicals, Indian Agricultural Research
Institute, New Delhi 110 012, India

Kumaran, K.
Forest College and Research Institute,
Tamil Nadu, India

Kumaresen, D.
Indian Cardamom Research Institute, Myladumpara,
Kailasanadu P.O. 685 553, Kerala, India

Lakhataria, R.P.
Department of Plant Pathology, College of Agriculture,
Junagadh 362 001, India

Lakshmanan, K.K.
Department of Botany, Kongunadu Arts & Science College,
Coimbatore 641 029, Tamil Nadu, India

Lowery, D. Thomas
Agriculture canada, Research Station, Summerland B.C.,
Canada V0H 1Z0

Maharishi, R.P.
Agricultural Research Station, Rajasthan Agricultural University,
Mandor-Jodhpur, Rajasthan, India

Maiti, Satyabrata
All India Coordinated Research Project on Betelvine,
Indian Institute of Horticultural Research, Hessaraghatta Lake,
Bangalore 56 089, India

Manian, S.
Department of Botany, Bharathiar University,
Coimbatore 641 046, India

Manoranjitham, M.P.
Division of Endocrinology, Department of Zoology, Bharathiar
University, Coimbatore 641 046, India

Mariappan, V.
Tamil Nadu Agricultural University,
Coimbatore 641 003, India

Mashkoor Alam, M.
Department of Botany, Aligarh Muslim University,
Aligarh 202 002, India

Mathur, Y.K.
Department of Entomology, C.S.A. University of Agriculture
& Technology, Kanpur 208 002, India

Mediratta, P.
Department of Pharmacology,
University College of Medical Sciences and G.T.B. Hospital, Delhi 110 095, India

Mercado, J.
Centro Nacional de Proteccion Vegetal, Mag
Apartado Postal P-116, Managua, Nicaragua

Miranda, F.
Universidad Nacional Agraria, Nicaragua

Mishra, S.D.
Division of Nematology, Indian Agricultural Research Institute,
New Delhi 110 012, India

Mishra, S.M.
Allahabad Agricultural Institute,
Allahabad 211 007, India

Mistry, K.B.
Godrej Soaps Limited, Vikhroli,
Bombay 400 079, India

Mittal, A.
Institute of Pathology, ICMR,
New Delhi, India

Mohan, P.S.
Department of Chemistry, Bharathiar University,
Coimbatore 641 046, India

Mohan Raj, D.
Division of Entomology, Indian Agricultural Research Institute,
New Delhi 110 012, India

Mojumdar, Vijayalakshmi
Division of Nematology, Indian Agricultural Research Institute,
New Delhi 110 012, India

Mokhtar, Abdel-Monem
Plant Protection Institute, Hungarian Academy of Sciences, Budapest,
PO Box 102. H-1525, Hungary

Moses, G.J.
Department of Plant Pathology, S.V. Agricultural College,
Andhra Pradesh Agricultural University, Tirupati 517 502, India

Mukherjee , S.N.
Entomology Laboratory, National Chemical Laboratory,
Pune 411 008, India

Murali, P.M.
Dalmia Centre for Biotechnology, Siruvani Main Road, Kalampalayam,
Coimbatore 641 010, Tamil Nadu, India

Murthy, M.S.N.
Department of Entomology, College of Agriculture,
Andhra Pradesh Agricultural University, Hyderabad 500 030, India

Murthy, P.S.N.
Central Tobacco Research Institute,
Rajahmundry 533 105, India

Murugan, K.
Division of Entomology, Department of Zoology, Bharathiar University,
Coimbatore 641 046, India

Muruganandam, B.
Department of Botany, Madras Christian College, Tambaram,
Madras 600 059, India

Muthusamy, M.
Agricultural College and Research Institute, Tamil Nadu Agricultural University,
Killikulam 627 252, India

Naderuzzaman, A.T.M.
Department of Botany, Rajshahi University,
Rajshahi, Bangladesh

Nagaiah, K.
Department of Chemistry, Osmania University,
Hyderabad 500 007, India

Nagaraj, G.
Directorate of Oilseed Research, Rajendranagar,
Hyderabad 500 030, India

Nagasampagi, B.A.
National Chemical Laboratory, Pune 411 008, India

Nageswara Rao, S.
Division of Crop Protection (Entomology), Central Tobacco Research
Institute, Rajahmundry 533 105, India

Naidu, R.
Indian Cardamom Research Institute, Myladumpara, Kailasanadu P.O.,
Idukki District 685 553, Kerala, India

Nakat, R.V.
All India Coordinated Research Project on Betelvine,
Indian Institute of Horticultural Research, Hessaraghatta Lake,
Bangalore 56 089, India

Nanda, B.
Department of Entomology, Orissa University of Agriculture and
Technology, Bhubaneswar 751 003, India

Nanda, U.K.
Department of Entomology, Orissa University of Agriculture and Technology, Bhubaneswar 751 003, India

Narasimha Rao, B.
Department of Entomology, College of Agriculture, Andhra Pradesh Agricultural University, Hyderabad 500 030, India

Narasimha Rao, C.V.
Division of Crop Chemistry and Soil Science, Central Tobacco Research Institute, Rajahmundry 533 105, India

Narayanan, P.Badri Sriman
Division of Endocrinology, Department of Zoology, Bharathiar University, Coimbatore 641 046, India

Narayana Rao, M.
Department of Chemistry, Osmania University, Hyderabad 500 007, India

Nasseh, O.
Deutsche Gesellschaft fur Technische Zusammenarbeit (GTZ) GmbH, 6236 Eschborn, Germany

Nelson, S.J.
Tamil Nadu Agricultural University, Agricultural College and Research Institute, Madura 625 104, India

Nelson, Reese
AgriDyne Technologies Inc., 2401 South Foothill Dr., Salt Lake City, Utah 84109, USA

Nigam, Suman
Department of Entomology, C.S.A. University of Agriculture & Technology, Kanpur 208 002, India

Nirmalakumari, A.
Centre for Plant Molecular Biology, TNAU, Coimbatore 641 003, India

Padhya, M.A.
Tissue Culture Laboratory, Department of Botany, M.S. University, Baroda 390 002, India

Palanisamy, P.T.
Department of Agricultural Processing, College of Agricultural Engineering, Tamil Nadu Agricultural University, Coimbatore 641 003, India

Parija, B.
Department of Entomology, Orissa University of Agriculture and Technology, Bhubaneswar 751 003, India

Parmar, Balraj Singh
Division of Agricultural Chemicals, Indian Agricultural Research Institute, New Delhi 110 012, India

Parvatha Reddy, P.
Division of Entomology and Nematology, Indian Institute of Horticultural Research, Hessaraghatta, Bnagalore 560 089, India

Patel, C.B.
Department of Entomology, N.M. College of Agriculture, Gujarat Agricultural University, Navsari Campus, Navsari 396 450, India

Patel, D.J.
Department of Nematology, Gujarat Agricultural University, Anand Campus, Anand 388 110, Gujarat, India

Patel, H.V.
Department of Nematology, Gujarat Agricultural University, Anand Campus, Anand 388 110, Gujarat, India

Patel, S.K.
Department of Nematology, Gujarat Agricultural University, Anand Campus, Anand 388 110, Gujarat, India

Petro, E.
Plant Health and Soil Conservation Station of the Capital Budapest, Higany v. 2. H-1118, Hungary

Philomina, N.S.
Biotechnology Centre for Tree Improvement, Tirupati, India

Polgar, L.A.
Plant Protection Institute, Dokki-Cairo, Egypt

Porwal, M.K.
Agricultural Research Station, Rajau, Banswara 327 001, India

Prabhu, S.R.
Division of Crop Chemistry and Soil Science, Central Tobacco Research Institute, Rajahmundry 533 105, India

Pradhan, N.C.
Department of Entomology, Orissa University of Agriculture and Technology, Bhubaneswar 751 003, India

Prakasa Rao, E.V.S.
Central Institute of Medicinal and Aromatic Plants, regional Centre, GKVK Post Office, Bangalore 560 065, India

Prasad, M.V.R.
Directorate of Oilseed Research, Rajendranagar, Hyderabad 500 030, India

Prasad, N.B.L.
Jawaharlal Nehru Technological University, Oil Technological Research Institute, Anantapur, 515 001, India

Prasad, V.V.L.N.
Agricultural Research Centre, I.T.C. Ltd., International Business Division, Amrutha Topaz, Punjagutta, Hyderabad 500 482, India

Prem Anandhi, A.
Division of Endocrinology, Department of Zoology, Bharathiar University, Coimbatore 641 046, India

Raguraman, S.
Agricultural Research Station, Paramakudi 623 707, Tamil Nadu, India

Raheja, A.K.
Indian Council of Agricultural Research, Krishi Bhavan New Delhi 110 001, India

Rahman, S.M.
Department of Botany, Rajshahi University, Rajshahi, Bangladesh

Raghunatha Rao, D.
Centre for Research in Medical Entomology (ICMR), Madurai 625 002, India

Rai, A.B.
Department of Entomology, N.M. College of Agriculture, Gujarat Agricultural University, Navsari Campus, Navsari 396 450, India

Rai, R.S.V.
Agricultural University, Mettupalayam 641 301, Tamil Nadu, India

Raja Sekhar, P.
Heliothis Management Scheme, Andhra Pradesh Agricultural University,
Regional Agricultural Research Station, Lam, Guntur 522 034, India

Rajeswari, E.
Centre for Plant Protection Studies, Tamil Nadu University,
Coimbatore 641 003, India

Rajvel, D.S.
Tamil Nadu Agricultural University, Agricultural College and Research Institute,
Madura 625 104, India

Ramakrishna, G.
Jawaharlal Nehru Technological University,
Oil Technological Research Institute, Anantapur, 515 001, India

Ramalingam, K.
Department of Zoology, Annamalai University, Annamalai Nagar 608 002,
Tamil Nadu, India

Ramachandram, M.
Directorate of Oilseed Research, Rajendranagar,
Hyderabad 500 030, India

Raman, K.
Godrej Agrovet Ltd.,
Madras 600 098, India

Ramaprasad, G.
Central Tobacco Research Institute,
Rajahmundry 533 105, India

Ramaswamy, N.M.
Centre for Plant Molecular Biology, TNAU,
Coimbatore 641 003, India

Ramesh, K.
Tissue Culture Laboratory, Department of Botany,
M.S. University, Baroda 390 002, India

Ramesh Babu, T.
Department of Entomology, College of Agriculture, Andhra Pradesh
Agricultural University, Hyderabad 500 030, India

Rana, D.S.
Department of Agronomy, Punjab Agricultural University,
Ludhiana 141 004, India

Ranganatha, A.R.G.
Directorate of Oilseed Research, Rajendranagar,
Hyderabad 500 030, India

Rao, M.S.
Division of Entomology and Nematology, Indian Institute of
Horticultural Research, Hessaraghatta, Bnagalore 560 089, India

Rao, P.J.
Division of Entomology, Indian Agricultural Research Institute,
New Delhi 110 012, India

Ravichandran, P.
Department of Botany, Madras Christian College, Tambaram,
Madras 600 059, India

Ray, A.
Department of Pharmacology,
University College of Medical Sciences and G.T.B. Hospital, Delhi 110 095, India

Reddy, A.S.
Andhra Pradesh Agricultural University,
Rajendranagar, Hyderabad, India

Reddy, V.V.S.
Department of Entomology, College of Agriculture,
Andhra Pradesh Agricultural University, Hyderabad 500 030, India

Rembold, Heinz
Max-Plank-Institute for Biochemistry, Insect Biochemistry Unit,
D-8033, Martinsried, Germany

Rengasamy, Seeni
Division of Agricultural Chemicals, Indian Agricultural Research Institute,
New Delhi 110 012, India

Reuben, R.
Centre for Research in Medical Entomology (ICMR),
Madurai 625 002, India

Rice, Martin
Department of Entomology, University of Queensland,
Queensland 4072, Australia

Rivas, I.
Centro Nacional de Proteccion Vegetal,
Mag Apartado Postal P-116, Managua, Nicaragua

Rojatkar, S.R.
National Chemical Laboratory,
Pune 411 008, India

Rosaiah, B.
Andhra Pradesh Agricultural University, Regional Agricultural Research Station,
Lam, Guntur 522 034, India

Roychoudhury, R.
Department of Botany, University of Lucknow,
Lucknow 226 007, India

Ruggero, William
AgriDyne Technologies Inc., 2401 South Foothill Dr., Salt Lake City,
Utah 84109, USA

Ruiter, Arnold A.
Rasun Natuurprodukten, Postbus 740, 9400 As Assen, Holland

Sachan, J.N.
Indian Institute of Pulses Research (ICAR),
Kanpur, India

Sampathraj, R.
Division of Endocrinology, Department of Zoology, Bharathiar
University, Coimbatore 641 046, India

Sandhu, S.S.
Section of Ecology, Punjab Agricultural University,
Ludhiana 141 004, India

Saradamma, K.
College of Agriculture,
Vellayani, Kerala, India

Sawaikar, D.D.
National Chemical Laboratory,
Pune 411 008, India

Saxena, D.B.
Division of Entomology, Indian Agricultural Research Institute,
New Delhi 110 012, India

Saxena, R.C.
International Centre of Insect Physiology and Ecology,
P.O.Box 30, Mbita, Kenya

Schmidt, Gerhard H.
Department of Zoology-Entomology, Faculty of Biology, University of Hannover, Herrenhauser Strasse 2, D-3000 Hannover 21, Germany

Schmutterer, Heinrich
Institut fur Phytopathologie und Angewandte Zoologie, Justus-Liebig-Universitat, 6300 Geissen, Germany

Sebastian, S.
Allahabad Agricultural Institute,
Allahabad 211 007, India

Selvi Sabhanayakam, N.
Department of Zoology, Annamalai University,
Annamalai Nagar 608 002, Tamil Nadu, India

Selvi, Senthamil
Department of Zoology, Annamalai University,
Annamalai Nagar 608 002, Tamil Nadu, India

Sen, P.
Department of Pharmacology,
University College of Medical Sciences and G.T.B. Hospital, Delhi 110 095, India

Seshachalam Babu, K.
All India Coordinated Research Project on Agricultural Ornithology,
Andhra Pradesh Agricultural University, Rajendranagar, Hyderabad, India

Sethuraman, K.
Tamil Nadu Agricultural University,
Coimbatore 641 003, India

Sharma, K.N.
Department of Agronomy, Punjab Agricultural University,
Ludhiana 141 004, India

Sharma, R.N.
Entomology Laboratory, National Chemical Laboratory,
Pune 411 008, India

Sharma, V.P.
Malaria Research Centre,
22 Sham Nath Marg, Delhi 110 054, India

Shenoi, M.M.
CTRI Research Station, Hunsur 571 105,
Karnataka, India

Shukla, B.C.
I.G.A.U., Raipur 492 012,
Madhya Pradesh, India

Siddiq, S.A.
Shambat Research Station, PO Box 30,
Khartoum, North Sudan

Siddiqui, Bina Shaheen
H.E.J. Research Institute of Chemistry, University of Karachi,
Karachi 75270, Pakistan

Siddiqui, Mansoor A.
Department of Botany, Aligarh Muslim University.
Aligarh 202 002, India

Siddiqui, Salimuzzaman
H.E.J. Research Institute of Chemistry, University of Karachi,
Karachi 75270, Pakistan

Simwat, G.S.
Department of Entomology, Punjab Agricultural University,
Ludhiana 141 004, India

Singh, D.P.
Department of Agroforestry, CCS Haryana Agricultural University,
Hisar 125 004, India

Singh, Darshan
District Extension Specialist (Plant Protection), PAU FASS Office,
Club Road, Sangrur (Punjab), India

Singh, Jaswant
Punjab Agricultural University,
Ludhiana 141 004, India

Singh, Muneshwar
Indian Institute of Soil Science, Bhopal 462 011, India

Singh, R.P.
Division of Entomology, Indian Agricultural Research Institute
New Delhi 110 012, India

Singh, T.G.M.
Department of Entomology, College of Agriculture, Rajendranagar,
Hyderabad 500 030, India

Sinha, M.M.
Department of Entomology, Rajendra Agricultural University, Pusa,
Samastipur 848 125, Bihar, India

Sinha, S.N.
Indian Agricultural Research Institute, Regional Station,
Karnal 132 001, India

Sitaramaiah, K.
All India Coordinated Research Project on Betelvine,
Indian Institute of Horticultural Research, Hessaraghatta Lake,
Bangalore 56 089, India

Sitaramaiah, S.
Central Tobacco Research Institute,
Rajahmundry 533 105, India

Sivakumar, S.S.
Department of Agricultural Processing, College of Agricultural Engineering,
Tamil Nadu Agricultural University, Coimbatore 641 003, India

Sivaprasad, B.
Agricultural Research Centre, I.T.C. Ltd., International Business Division,
Amrutha Topaz, Punjagutta, Hyderabad 500 482, India

Sobita, L.R.
Allahabad Agricultural Institute,
Allahabad 211 007, India

Sombatsiri, Kwanchai
Department of Entomology, Kasetsart University,
Bangkok 109 000, Thailand

Sree Rangaswamy, S.R.
Centre for Plant Molecular Biology, TNAU,
Coimbatore 641 003, India

Sreemannarayana, G.
Osmania University, Hyderabad 500 001, India

Sreenivas, S.S.
CTRI Research Station, Hunsur 571 105, Karnataka, India

Srimannarayana, G.
Department of Chemistry, Osmania University, Hyderabad 500 007, India

Srinivasa Babu, K.
Department of Entomology, College of Agriculture,
Andhra Pradesh Agricultural University, Hyderabad 500 030, India

Srinivasan, K.
Indian Institute of Horticultural Research, Hessarghatta Lake P.O.,
Bangalore 560 089, India

Srivastava, R.P.
Central Institute of Horticulture for Northern Plains,
B-217 Indira Nagar, Lucknow 226 016, India

Srivasuki, K.P.
Biotechnology Centre for Tree Improvement, Tirupati, India

Steinhauer, B.
Institut fur Phytopathologie und Angewandte Zoologie,
Bismark str., 16, D-6300 Giessen, Germany

Stephen, A.
Department of Botany, Madras Christian College, Tambaram,
Madras 600 059, India

Streloke, Martin
Department of Zoology-Entomology, Faculty of Biology,
University of Hannover, Herrenhauser Strasse 2, D-3000 Hannover 21, Germany

Subba Rao, D.
All India Coordinated Research Project on Betelvine,
Indian Institute of Horticultural Research, Hessaraghatta Lake,
Bangalore 56 089, India

Subhash Peter, M.C.
Department of Zoology, Fatima Mata National College, Kollam 691 001, Kerala, India

Subramaniam, K.S.
Agricultural College and Research Institute,
Tamil Nadu Agricultural University, Killikulam 627 252, India

Subramaniam, T.S.
Agricultural Research Centre, I.T.C. Ltd., International Business Division,
Amrutha Topaz, Punjagutta, Hyderabad 500 482, India

Subramanian, M.S.
Department of Botany, Kongunadu Arts & Science College,
Coimbatore 641 029, Tamil Nadu, India

Subrahmanyam, B.
Division of Entomology, Indian Agricultural Research Institute,
New Delhi 110 012, India

Sujatha, M.
Directorate of Oilseed Research, Rajendranagar,
Hyderabad 500 030, India

Sundarababu, P.C.
Tamil Nadu Agricultural University, Agricultural College and Research Institute,
Madura 625 104, India

Surendran, C.
Agricultural University, Mettupalayam 641 301, Tamil Nadu, India

Swaminathan, K.R.
Department of Agricultural Processing, College of Agricultural Engineering,
Tamil Nadu Agricultural University, Coimbatore 641 003, India

Syamsunder Rao, P.
All India Coordinated Research Project on Agricultural Ornithology,
Andhra Pradesh Agricultural University, Rajendranagar, Hyderabad, India

Szabo, P.
Plant Health and Soil Conservation Station of Csongrad-county,
Hodmezovasarhely, Rarosi u. 102, H-6801, Hungary

Takkar, P.N.
Indian Institute of Soil Science, Bhopal 462 011, India

Talwar, G.P.
National Institute of Immunology, New Delhi 110 067, India

Tanzubil, Paul B.
Nyankpala Agricultural Experiment Station, P.O. Box 483
Tamale, Ghana

Thengane, S.R.
Plant Tissue Culture Division, National Chemical Laboratory,
Pune 411 008, India

Thiagarajan, M.
Dalmia Centre for Biotechnology, Siruvani Main Road, Kalampalayam,
Coimbatore 641 010, Tamil Nadu, India

Tilak, Ch. Balagangadhar
Department of Entomology (APAU), S.V. Agricultural College,
Tirupati 517 502, A.P., India

Tiyagi, Sartaj A.
Department of Botany, Aligarh Muslim University,
Aligarh 202 002, India

Toledano, Josep Coll
CSIC Department of Organic and Biological Chemistry,
Jordi Girona 18-26, 08034 Barcelona, Spain

Tomar, S.S.
Rajasthan Agricultural University, Agricultural Research Station,
Borkhera, Kota 324 001, India

Torma-Gazdag, Maria
Plant Health and Soil Conservation Station of Csongrad-county,
Hodmezovasarhely, Rarosi u. 102, H-6801, Hungary

Trumm, P.
Institute of Zoology, University of Mainz, Saarsts. 21,
D-6500 Mainz, Germany

Udaiyan, K.
Department of Botany, Bharathiar University, Coimbatore 641 046, India

Upadhyay, S.N.
National Institute of Immunology, New Delhi 110 067, India

Vaishnav, M.U.
Department of Plant Pathology, College of Agriculture,
Junagadh 362 001, India

Vanitha Kumari, G.
Division of Endocrinology, Department of Zoology,
Bharathiar University, Coimbatore 641 046, India

Varadarasan, S.
Indian Cardamom Research Institute, Myladumpara,
Kailasanadu P.O. 685 553, Kerala, India

Vardhani, B.P.
All India Coordinated Research Project on Agricultural Ornithology,
Andhra Pradesh Agricultural University, Rajendranagar, Hyderabad, India

Varadharaju, N.
Postharvest Technology Scheme, College of Agricultural Engineering,
Tamil Nadu Agricultural University, Coimbatore 641 003, India

Venkat Kumar, P.
Division of Agricultural Chemicals, Indian Agricultural Research Institute,
New Delhi 110 012, India

Venkataiah, M.
A.P. Agricultural University, Hyderabad, India

Venkateswara Rao, S.
Andhra Pradesh Agricultural University,
Regional Agricultural Research Station, Lam, Guntur 522 034, India

Venugopal Rao, N.
Heliothis Management Scheme, Andhra Pradesh Agricultural University,
Regional Agricultural Research Station, Lam, Guntur 522 034, India

Vidya, F.
Tamil Nadu Agricultural University, Coimbatore 641 003, India

Vidyasekaran, P.
Tamil Nadu Agricultural University, Coimbatore 641 003, India

Vijaya Kumar, R.
Biotechnology Centre for Tree Improvement, Tirupati, India

Vishwanathan, R.
Department of Agricultural Processing, College of AgriculturalEngineering,
Tamil Nadu Agricultural University, Coimbatore 641 003, India

Vyas, B.N.
Godrej Soaps Limited, Vikhroli, Bombay 400 079, India

Vyas, S.C.
Department of Plant Pathology, J.N. Agricultural University,
Indore, Madhya Pradesh 452 001, India

Walter, J.F.
W.R. Grace & Co.- Conn., 7379 Route 32, Columbia, MD 21044, USA

Wasnikar, A.R.
All India Coordinated Research Project on Betelvine,
Indian Institute of Horticultural Research, Hessaraghatta Lake,
Bangalore 56 089, India

Wells, Samuel
AgriDyne Technologies Inc., 2401 South Foothill Dr., Salt Lake City, Utah 84109, USA

Wilps, H.
Deutsche Gesellschaft fur Technische Zusammenarbeit (GTZ) GmbH, 6236 Eschborn, Germany

Wood, Timothy
AgriDyne Technologies Inc., 2401 South Foothill Dr., Salt Lake City, Utah 84109, USA

Yadav, R.P.
Department of Entomology, Rajendra Agricultural University, Pusa, Samastipur 848 125, Bihar, India

Yadav, T.D.
Division of Entomology, Indian Agricultural Research Institute, New Delhi 110 012, India

Yaw-Huei Lin
Institute of Botany, Academia Sinica, Nankang, Tapei, Taiwan 11529, Republic of China

Zeringue, Hampden J. Jr.
USDA/Agricultural Research Service, Southern Regional Research Center, New Orleans, Louisiana 70124, USA

Zongo, Joanny O.
Laboratoire de Recherches Protection des Vegetaux, BP 403 Bobo-Dioulasso, Burkina Faso, West Africa

